*Semimicro Qualitative
Organic Analysis*

**The Systematic Identification
of Organic Compounds**

Semimicro Qualitative Organic Analysis

The Systematic Identification of Organic Compounds

THIRD EDITION

NICHOLAS D. CHERONIS
Late Professor of Chemistry
Brooklyn College

JOHN B. ENTRIKIN
Professor of Chemistry
Centenary College of Louisiana

ERNEST M. HODNETT
Professor of Chemistry
Oklahoma State University

QD
98
C45
1965

INTERSCIENCE PUBLISHERS
a division of John Wiley & Sons
New York · London · Sydney

Copyright 1947, by Thomas Y. Crowell Company
Copyright © 1957, by Nicholas D. Cheronis and John B. Entrikin

Copyright © 1965 by John Wiley & Sons, Inc.
FIRST CORRECTED PRINTING, AUGUST, 1968
All Rights Reserved

This book or any part thereof
must not be reproduced in any form
without the written permission of the publisher.

Library of Congress Catalog Card Number: 64-25892
Printed in the United States of America

Preface

The third edition of this work, although based on the second, has been brought up to date by being reorganized, rewritten, and decidedly expanded. Plans for this revision were made before the accidental death of Dr. Nicholas D. Cheronis. These plans, with some modifications, have been carried out by us. Because an abridged version of the second edition has been published as a textbook for students (Cheronis and Entrikin, *Identification of Organic Compounds*, Wiley-Interscience, New York, 1963), this third edition has been expanded to make it even more useful as a reference work in the field of qualitative organic analysis. Although instrumental methods are not ignored, this text is devoted primarily to chemical methods. Detailed procedures are given for the purification of organic substances and the determination of their physical properties together with methods for the detection of the functional groups that are present. Procedures are given in some detail for the characterization of individual compounds.

Important changes from the previous edition are as follows.

(1) The discussions of laboratory techniques have been considerably expanded. New and more specialized equipment is described and illustrated. Equipment that will allow the use of even smaller amounts of material for chemical reactions has received special attention. Methods that require some instrumentation have been introduced when their use is indicated.

(2) The chapters dealing with separation techniques have been enlarged. Thin-layer chromatography, a new and highly useful procedure, is described in some detail; illustrative examples are given.

(3) The chapter dealing with the determination of physical constants has been enlarged by the addition of some theoretical discussions and also the introduction of some methods involving instrumentation.

(4) The chapter dealing with the separation of mixtures has been relocated in the text so that it follows the consideration of separation techniques and precedes methods for detecting the chemical identity of the individual compounds.

(5) More theory and some new examples of the theory are given in the chapters dealing with the classification by solubility and acid-base character. A new flow chart has been prepared for the use of Davidson's indicators.

(6) The chemical tests used to detect chemical structures are now

covered in a single chapter rather than in two chapters. A number of new tests have been added and many more have been referred to in the footnotes. In addition, 245 references to pertinent literature have been provided at the end of this chapter.

(7) The chapters dealing with the preparation of derivatives (Chapters 11 to 21) have been reorganized, revised, and somewhat enlarged.

(8) The tables of compounds and their derivatives have been checked against recent literature. Approximately 7100 compounds are included in these tables as compared with about 4100 compounds in the second edition. The tables are supplemented by numerous references to literature from which information may be found on additional compounds, including classes not covered by these tables.

(9) The references given at the ends of the chapters have been revised to include recent original publications and reviews.

Although all three of us have contributed to this third edition, I (JBE) have been primarily responsible for the revision of Chapters 6 to 10 and for all of the tables, and I (EMH) have revised Chapters 1 to 5 and 11 to 21.

We express sincere appreciation to all those who have assisted in any way in the preparation of this book. We are especially indebted to the hundreds of chemists who have published the results of their experiments, since their work serves as the basis for most of the procedures that we use and the data provided in the tables. We express our sincere appreciation to those of our students and assistants who have made library searches, tested procedures and, in several cases, developed new tests and improved the techniques previously used.

JOHN B. ENTRIKIN
ERNEST M. HODNETT

October 1964

Contents

PART ONE

Techniques of Organic Analysis

1. *Introduction*   3

   Identification, 3   The Systematic Approach, 3   Physical Methods, 4   Microscopic Methods, 5   Sample Size, 6   References, 6

2. *Equipment and Procedures for Small-Scale Work*   8

   Weighing, 9   Measuring Volumes, 12   Heating, 19   Handling, Stirring and Grinding, 24   Separating of Immiscible Solvents, 29   Filtering, 30   Drying, 31   Evaporating, 31   Heating under Reflux, 34   Reaction Vessels, 35   High Vacuum Line, 39   Apparatus and Procedure for Microhydrogenation, 40   Apparatus and Procedure for Preparation of Grignard Reagents, 45   References, 48

3. *Fractionation Procedures*   54

   Crystallization: Selection of Solvents, 55   Determination of Solubility, 60   Preparation of Solutions for Crystallization, 61   Filtration of the Hot Solution and Formation of Crystals, 63   Other Apparatus for Crystallization, 74   Drying of Crystals, 75   *Distillation:* Simple Distillation at Atmospheric Pressure, 83   Fractional Distillation at Atmospheric Pressure, 87   Distillation under Reduced Pressure, 90   Distillation of Small Quantities at Atmospheric or Reduced Pressures, 96   Steam Distillation, 99   *Sublimation:* Fractional Sublimation under Reduced Pressure, 105   *Extraction*, 109   References, 111

4. *Fractionation Procedures*   117

   Chromatographic Procedures: Principles of the Chromatographic Separation, 117   Paper Chromatography, 120   Column Chromatography, 131   Thin-Layer Chromatography, 145   Ion-Exchange Chromatography, 152   Gas Chromatography, 157   Other Methods of Fractionation, 159   References, 162

5. *Physical Properties of Organic Compounds*   171

   Melting Point, 172   Boiling Point, 202   Refractive Index, 213   Density, 219   Optical Rotation, 225   Molecular Weight, 228   Molar Refraction and Dispersion, 234   References, 237

Contents

6. The Separation of Mixtures   244

     Mixtures: General Principles, 246   Preliminary Tests for a General Mixture, 246   A General Procedure for the Separation of Mixtures, 251   Suggestions for Separating Intraclass Mixtures, 257   Alternate Methods for the Separation of Mixtures, 260   *The Use of Other Methods of Separation*, 261   References, 264

PART TWO
Procedures for Tentative Identification of an Unknown

7. *Preliminary Examination of the Pure Compound*   275

     Ignition and Preliminary Tests with Reagents, 276   Observations During Fusion and Cooling, 277   *Analysis for Elements:* Detection of Carbon and Hydrogen, 282   Decomposition of Organic Compounds, 286   Detection of the Elements in the Sodium Fusion Filtrate, 291   References, 299

8. *Classification by Solubility*   303

     The Classification of Solvents, 306   Comparison of Water and Ether as Solvents, 307   Solubility in Water, 308   Solubility in Ethyl Ether, 312   Solubility in Dilute Hydrochloric Acid, 313   Solubility in Dilute Sodium Hydroxide, 315   Solubility in Dilute Sodium Bicarbonate, 317   Solubility in Concentrated Sulfuric Acid, 318   Solubility Determinations, 320   Designation for the Solubility Divisions, 322   Table of Solubility Classifications, 324   References, 326

9. *Classification by the Indicator Method*   328

     Indicator Method of Classifying Acids and Bases, 335   Procedures for Testing, 337   Table of Indicator Classifications, 338   References, 341

10. *Tests for the Classification of an Unknown*   343

     Part I: An Inventory and a Forward Look: P-1 Gross Observations, 347   P-2 The Ignition Test, 348   P-3 Tests for Salts, 349   P-4 Tests for Aromatic Structure, 350   P-5 Tests for Active Unsaturation, 352   P-6 Tests by Selected Oxidizing Agents, 355   P-7 Detection of Acidic Substances, 359   P-8 Test for Compounds that have Nitrogen and Oxygen in the Same Group, 360   P-9 Iodoform Formation Test, 361   *Part II: Tests for Special Classes:* 10.01 Acids, 363   10.02 Acid Anhydrides, 366   10.03 Acid Halides, 366   10.04 Alcohols, 367   10.05 Alkyl and Aryl Halides, 372   10.06 Unsubstituted Amides, 373

10.07 Substituted Amides, 376   10.08 Amines, 378   10.09 Carbohydrates, 388   10.10 Carbonyl Compounds, 392   10.11 Esters, 397   10.12 Ethers, 398   10.13 Hydrazines, 399   10.14 Hydrocarbons, 400   10.15 Nitrates and Nitrites, 401   10.16 Nitriles, 402   10.17 Nitro Compounds, 402   10.18 Nitroso Compounds, 405   10.19 Oximes, Hydrazones, and Semicarbazones, 406   10.20 Phenols, 406   10.21 Sulfides, Disulfides, and Sulfones, 410   10.22 Thiols (Mercaptans and Thiophenols), 411   Coordination of Data, 413   References, 414

PART THREE
Procedures for Final Characterization of an Unknown

Index to Preparation of Derivatives, 427

11. *Problems in the Derivatization of Organic Compounds*   429

    Selection of Derivatives, 430   General Procedure for Preparation of Small Quantities of Derivatives, 434   Evaluation of the Melting Points of Derivatives, 436

12. *Derivatives of Carboxylic Acids and Acid Derivatives*   437

    Carboxylic Acids: Amides, *p*-Toluidides, Anilides, *p*-Bromoanilides, and Other Substituted Amides of Carboxylic Acids, 439   Solid Esters of Carboxylic Acids, 447   Salts and Other Derivatives of Carboxylic Acids, 449   *Amino Acids:* N-Acyl and N-Aroyl Derivatives of Amino Acids, 452   α-Naphthylureido Derivatives and Hydantoins of Amino Acids, 455   Salts of Amino Acids, 456   Chromatographic Detection of Amino Acids, 457   *Acid Halides*, 458   *Acid Anhydrides*, 459   References, 459

13. *Derivatives of Alcohols and Phenols (Monohydric and Polyhydric)*   465

    Alcohols: 3,5-Dinitrobenzoates and *p*-Nitrobenzoates, 467   3-Nitrophthalates, 473   α-Naphthylurethans, 474   Other Substituted Urethans, 479   Other Derivatives, 480   Chromatographic Detection of Small Amounts of Alcohols, 482   *Phenols:* 3,5-Dinitrobenzoates, *p*-Nitrobenzoates, Benzoates, and Acetates, 485   Urethans, 487   Aryloxyacetic Acids, 489   Other Derivatives, 490   Chromatography of Phenols, 491   References, 492

14. *Derivatives of Aldehydes, Ketones, and Acetals*   496

    Aldehydes: Phenylhydrazones, 497   2,4-Dinitrophenylhydrazones and *p*-Nitrophenylhydrazones, 499   Semicarbazones and Thiosemicarbazones, 502   Dimethone Derivatives, 504   Other Derivatives, 506   Chromatographic Identification and Separation of

Aldehydes, 509   *Ketones:* Substituted Hydrazones, 510   Semicarbazones and Thiosemicarbazones, 511   Oximes, 512   Other Derivatives, 513   *Acetals:* Hydrolysis of Acetals, 514   References, 515

15. Derivatives of Carbohydrates   519

Substituted Phenylhydrazones, Osazones, and Osotriazoles, 520   Other Derivatives, 525   Specific Rotation of Carbohydrates and Their Derivatives, 528   Chromatography of Sugars, 528   References, 530

16. Derivatives of Esters and Ethers   532

Esters: Derivatization of the Acidic Part of an Ester, 534   Derivatization of the Acidic and Alcoholic Components after Hydrolysis, 537   Derivatization of the Alcoholic Component, 538   Examples of Complete Characterization of Esters, 538   Derivatization of Lactones and Esters of Inorganic Acids, 540   *Ethers:* Derivatives from Aliphatic Ethers, 542   Derivatives from Aromatic Ethers, 544   References, 547

17. Derivatives of Halogen Compounds   548

Alkyl and Cycloalkyl Halides: S-Alkylthiuronium Picrates, 549   Picrates of β-Naphthyl Ethers, 551   Anilines, *p*-Toluidides, α-Naphthalides, and Alkylmercuric Salts, 551   3,5-Dinitrobenzoates, 555   Other Derivatives of Alkyl and Cycloalkyl Halides, 555   *Aryl Halides:* Nitro Derivatives of Aryl Halides, 559   Sulfonamides of Aryl Halides, 564   *Fluorine Compounds:* Identification and Characterization of Fluorocarbons, 565   Specific Class Tests for Fluorine Compounds, 567   Derivatization of Fluorine Compounds, 567   References, 569

18. Derivatives of Hydrocarbons   571

Alkanes and Cycloalkanes: Characterization by Means of Physical Constants, 572   Oxidation of a Cycloalkane to a Dicarboxylic Acid, 573   *Alkenes and Cycloalkenes, Alkynes, Dienes,* 574   *Aromatic Hydrocarbons:* Nitration of Aromatic Hydrocarbons, 578   Acetamido Derivatives of Aromatic Hydrocarbons, 581   Picrates and 2,4,7-Trinitrofluorenone Adducts of Aromatic Hydrocarbons, 582   Aroylbenzoic Acids from Aromatic Hydrocarbons, 583   Preparation of Derivatives of Aromatic Hydrocarbons with 2,4-Dinitrobenzenesulfenyl Chloride, 585   Oxidation of Side Chains, 585   References, 586

19. Derivatives of Amino-Nitrogen Functions   589

Amines: Acetamides and Benzamides of Amines, 589   Sulfonamides and Sulfenamides of Amines, 595   Substituted Ureas and Thioureas of Amines, 598   General Method for Preparation, 599   Quaternary Ammonium Salts of Amines, 601   Other Derivatives of Amines, 604   Chromatographic Detection of Amines, 606   *Hydrazines*, 607   *Amides, Imides, and Ureas:* Characterization by Hydrolysis, 608   Xanthyl Derivatives of Amides, 609   Mercuric Salts and Other Derivatives of Amides, 610   References, 611

20. *Derivatives of Other Nitrogen Functions*   615

    Azo Compounds, 615   *Azoxy and Hydrazo Compounds*, 617   *Isocyanates and Isocyanides*, 617   *Nitriles:* Hydrolysis of Nitriles to Carboxylic Acids and Amides, 618   Reduction of Nitriles to Amines and Their Characterization by Preparation of Substituted Thioureas, 621   Other Derivatives of Nitriles, 622   *Nitro Compounds:* Derivatization of Nitro Compounds by Reduction to Amines, 625   Other Derivatives of Nitro Compounds, 626   *Nitroso Compounds*, 628   References, 629

21. *Derivatives of Sulfur Functions*   631

    Sulfonamides, 631   *Sulfonyl Chlorides*, 634   *Sulfonic Acids:* S-Benzylthiuronium Derivatives of Sulfonic Acids, 635   Arylamine Salts of Sulfonic Acids, 636   Preparation of Sulfonyl Chlorides from Sulfonic Acids and Conversion to Sulfonamides and N-α-Naphthylsulfonamides, 637   *Thiocyanates and Isothiocyanates*, 640   *Thioethers*, 641   *Thiols (Mercaptans and Thiophenols):* Thioethers, 642   Thioesters, 643   References, 644

22. *Instrumental Methods*   646

    Advantages of the Instrumental Methods, 646   Types of Information Obtained, 647   Sample Handling, 648   Special Procedures for Reducing the Size of the Sample, 649   References, 651

PART FOUR
Tables of Organic Compounds with Their Constants and Derivatives   657

    Preface to Tables: Nomenclature, 659   General References, 661   General Comments, 663   Abbreviations, 664   *List of Tables*, 665

Appendix, 969

Subject Index of Text, 981

Compound Index of Tables, 995

PART ONE

Techniques of Organic Analysis

1

Introduction

Identification

The purpose of qualitative organic analysis is to prove that an unknown substance under investigation bears complete similarity with a pure substance for which the chemical and physical properties are known and to which a specific molecular structure has been assigned. In practice, efforts to prove *complete* similarity are not necessary. Depending on how rigorous the proof of identity needs to be in a given case, a variable but limited number of the physical and chemical properties of the unknown may be determined. The information thus obtained can then be compared with the physical and chemical properties of organic compounds reported in the literature. If a reasonable number of the physical and chemical properties of the unknown substance are essentially identical with the same properties of a recorded pure compound, identity of the unknown has been tentatively established. Rigorous proof of identity normally requires that derivatives be prepared from the unknown and that the data for these derivatives duplicate the data for similar derivatives for the known compound. There are three general approaches for identification: (1) the systematic approach, (2) the use of physical methods, and (3) microscopic methods. Such a division is, of course, arbitrary and there is an advantage in using more than one type of approach in many cases.

The Systematic Approach

The systematic scheme involves a series of steps which may be briefly summarized.

1. The sample is first fractionated to isolate relatively pure components. A component is regarded as pure when further fractionation results in no variation of its physical constants. Melting points are usually employed as criteria of purity of solids, and boiling points and refractive indices as criteria of purity for liquids. Therefore, in this step one or more physical constants of the unknown are determined.

2. The pure unknown is then subjected to gross examination, analysis of the elements present, tests of solubility in a few selected solvents, and tests of reaction to a few pH indicators. On the basis of these tests the unknown is assigned to a large division, which contains several classes of organic compounds.

3. Small samples of the unknown substance are subjected to tests with a number of reagents in order to detect the presence or absence of functional groups, such as the carboxyl (COOH), carbonyl (CO), hydroxyl (OH), nitro (NO_2), amino (NH_2), and the like. By means of these tests the classification of the unknown is restricted to a *single class* of organic compounds.

4. All the data obtained in the above tests are carefully coordinated; the literature is then consulted and a list of possible compounds is prepared. The compound that best fits the experimental data is selected as the most probable one to be identical with the unknown.

5. By means of chemical reactions one or more suitable solid derivatives of the unknown are prepared. If these derivatives melt within 1-2° of the melting points recorded in the literature for the same derivatives of the compound tentatively identified as the unknown, the identification is regarded as complete.

Literature from the nineteenth century, and even earlier, presents very numerous procedures for the identification of specific compounds and for the detection of certain types of compounds. However, it was not until the twentieth century that chemists, particularly teachers of chemistry, began to coordinate the known information and write books that presented some systematic procedure for the identification of an unknown organic compound. The texts cited at the end of this chapter are representative.

Physical Methods

Physical methods that are widely used in qualitative organic analysis may be classified as (a) based on differential migration rates, or (b) using some adaptation of spectroscopy. Differential migration methods include all of the very useful chromatographic techniques, electrophoresis, and the use of ion-exchange resins. In reality, these methods relate to technique for the separation of mixtures. They are discussed in Chapter 4 as fractionation procedures. However, these methods do have application to identification work, provided that the records from pure, known compounds, which have been treated in exactly the same way as the unknown ones, are available for comparison. Preferably, chemical methods for detection and identification should be used on substances that have been separated and purified by an appropriate differential migration technique.

Spectroscopy has been a tremendous aid to the organic chemist by providing him with information about the internal structure of molecules and the bond energies available in different parts of the molecule. Brief discussions are included in Chapter 22, concerning several spectrographic methods as they related to qualitative organic analysis. For details of procedures, more specialized texts must be consulted.

Microscopic Methods

Table 1.01 gives a summary of the microscopic methods. Procedure I is based on the reaction of the unknown compound with a definite number

TABLE 1.01
MICROSCOPIC METHODS FOR IDENTIFICATION OF ORGANIC COMPOUNDS

Procedure	Basis of Procedure
I. Reactions under the microscope	Comparison of reactions of known and unknown with the same reagent
II. Crystallographic properties	Comparison of known and unknown and their derivatives
III. Fusion techniques	Comparison of melting points, eutectics, and refractive indices of melts of known and unknown

of reagents under the microscope and the identification of the resulting products by their crystal structure. Procedure II employs the accurate determination of the crystallographic properties of the unknown and the assumed known and their derivatives. Procedure III depends on the determination of the melting point of the unknown compared to a known and the behavior of each when they are "fused" with known standards.

While the specific applications of the microscopic methods are not covered in this book, literature on the subject is available. A general idea of the applicability of these techniques to qualitative organic analysis may be gained by consulting the publications of Dunbar,[1] Kofler,[2] and McCrone,[3] and by reading the ten papers on microscopy published in *Microchemical Techniques*.[4]

Sample Size

In the earlier classical methods for organic analysis, it was assumed that several grams of the material to be investigated and identified were available. The procedures for purification, chemical testing, and derivatization, allowed by these large quantities, failed to be useful when only very small samples were available. Early workers, in developing methods for the characterization of very small quantities, used the microscope. From that beginning, the prefix *micro* has been used to indicate *very small* quantities. However, the term "very small" is not specific, and may imply milligram quantities to one person and only nanogram or picogram amounts to another chemist. The terms *macro*, *semimicro*, *micro*, and *submicro*, as commonly used in analytical chemistry, attempt to indicate the size of sample required. In current usage the terms generally imply the following quantities: *macro*, grams; *semimicro*, milligrams; *micro*, micrograms; and *submicro*, nanograms or less. Cheronis[5] has discussed the present, and forecast the future in an article entitled "Characterization of Organic Compounds at the Microgram and Submicrogram Ranges."

The quantity and purity of the sample available to the analyst often dictates which approach may be undertaken for its purification and final characterization. This text provides proven methods for the purification, detection, and characterization of substances by the systematic chemical approach when the minimum quantity of sample available is in the range from 50 mg to 1 g (depending on purity and identity).

Texts on Qualitative Organic Analysis

N. Campbell, *Qualitative Organic Chemistry*, Van Nostrand, New York, 1939.
N. D. Cheronis and J. B. Entrikin, *Semimicro Qualitative Organic Analysis*, 2nd ed., Interscience, New York, 1957.

[1] R. E. Dunbar *et al.*, *Microchem. J.*, **1**, 17 (1957); **2**, 113 (1958); **3**, 65, 143 (1959); and **4**, 59, 167 (1960).
[2] L. Kofler and A. Kofler, *Thermomikro-methoden zur Kennziechung Organischer Stoffe and Stoffgemisch*, Wagner, Innsbruck, 1954.
[3] W. C. McCrone, Jr., *Fusion Methods in Chemical Microscopy*, Interscience, New York, 1957.
[4] *Microchemical Techniques*, N. D. Cheronis, ed., Interscience, New York 1962, pp. 153–281.
[5] N. D. Cheronis, in *Microchemical Techniques*, pp. 117–149 (cited in Reference 4).

N. D. Cheronis and J. B. Entrikin, *Identification of Organic Compounds*, Interscience-Wiley, New York, 1963.

H. T. Clarke, *Handbook of Organic Analysis*, Arnold, London, 1928, reprinted 1937.

D. Davidson and D. Perlman, *A Guide to Qualitative Organic Analysis*, 2nd ed. Brooklyn College Bookstore, Brooklyn, 1958.

F. Feigl, *Spot Tests in Organic Analysis*, 6th ed., Elsevier, Amsterdam, 1960.

C. H. Huntress and S. P. Mulliken, *Identification of Pure Organic Compounds: Tables of Data on Selected Compounds of Order I*, Wiley, New York, 1941.

C. H. Huntress, *Identification of Organic Compounds: Tables of Data on Selected Compounds of Order III*, Wiley, New York, 1948.

O. Kamm, *Qualitative Organic Analysis*, 2nd ed. Wiley, New York, 1932.

R. P. Linstead and B. C. L. Weedon, *A Guide to Qualitative Organic Analysis*, Butterworths, London, 1956.

S. M. McElvain, *Characterization of Organic Compounds*, 2nd. ed. Macmillan, New York, 1953.

A. McGookin, *Qualitative Organic Analysis and Scientific Method*, Reinhold, London, 1955.

H. Meyer, *Nachweis und Bestimmung organischer Verbindungen*, Springer, Berlin, 1933.

H. Middleton, *Systematic Qualitative Organic Analysis*, 2nd. ed. Arnold, London, 1943.

S. P. Mulliken, *The Identification of Pure Organic Compounds*, Wiley, New York, 4 Vols., 1904–1922.

H. T. Openshaw, *Laboratory Manual of Qualitative Organic Analysis*, 3rd ed., University Press, Cambridge, 1955.

M. Pesez and P. Poirier, *Methodes et Reactions de l'Analyse Organique*, Vol. III, Masson, Paris, 1954.

L. Rosenthaler, *Der Nachweis organischer Verbindungen*, Enke, Stuttgart, 1923.

F. Schneider, *Organic Qualitative Microanalysis*, Wiley, New York, 1946.

R. L. Shriner, R. C. Fuson, and D. Y. Curtin, *The Systematic Identification of Organic Compounds*, 4th ed., Wiley, New York, 1956.

S. Siggia and H. T. Stolten, *Introduction to Modern Organic Analysis*, Interscience, New York, 1956.

F. J. Smith and E. Jones, *A Scheme of Qualitative Organic Analysis*, Blackie and Son, London, 1953.

H. Staudinger, *Anleitung zur organischen qualitativen Analyse*, 6th ed., Springer, Berlin, 1955.

S. Veibel, *The Identification of Organic Compounds*, 2nd English ed. Gad, Copenhagen, 1959.

A. I. Vogel, *Elementary Practical Organic Chemistry: Part 2, Qualitative Organic Analysis*, Longmans-Green, London, 1957.

F. Wild, *Characterization of Organic Compounds*, 2nd ed., Cambridge Univ. Press, 1958.

2

Equipment and Procedures for Small-Scale Work

Semimicro techniques involve the use of apparatus and tools that are smaller than those used in the ordinary chemical laboratory. This smaller equipment is used in order to minimize loss of material by adsorption on the surfaces of the apparatus. In the case of test tubes, for example, less material will be retained on the walls of a 3-in. tube than on the walls of a 6-in. tube. A drastic reduction in the quantity of starting material for an experiment usually requires also that the procedure be modified, especially for separation and purification.

The procedures described in this book can be followed with the apparatus normally found in college and industrial laboratories if the quantities of material are increased by some factor between two and five. However, standard laboratory equipment can be used for semimicro work with the addition of a few pieces, which are available from most supply houses. The more specialized apparatus, described here, has many advantages in the saving of time and labor; when the quantity of material is very small, it must be used. Semimicro techniques are highly recommended, not only to save time and labor, but also to obtain better results.

A book of this type cannot discuss (or even list) all of the apparatus that has been described in the literature for small-scale experimentation. Frequent references will be made to a more extensive description and discussion published by one of us,[1] and to original references in the journals where more details will be found.

[1] N. D. Cheronis, "Micro and Semimicro Methods," in *Technique of Organic Chemistry*, Vol. 6, A. Weissberger, ed., Interscience, New York, 1954.

Weighing

The use of correct weights of materials is very important in all laboratory work; semimicro qualitative organic analysis is no exception. Too often a worker estimates the quantity of material needed for an experiment and does not achieve the results that he would like. His estimate may easily be incorrect by a factor of two or more, unless he has had much experience with that particular compound. If possible, all solids and many liquids used as reagents in an experiment should be weighed.

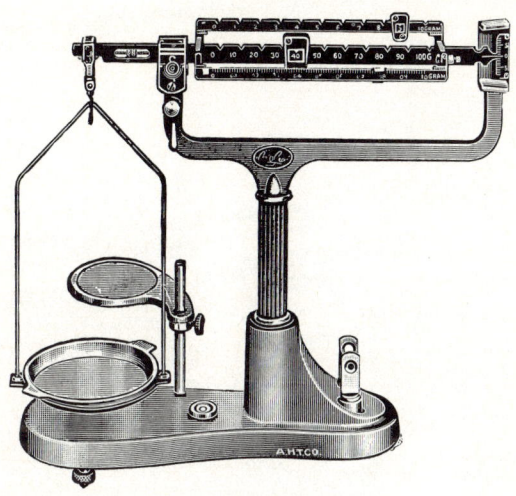

FIGURE 2.01

Triple-beam balance for experimentation with small amounts of material. Sensitivity, 0.01 g. Weighings as high as 111 g use only sleeve riders. Two removable weights. Self-aligning stirrup. (Courtesy A. H. Thomas Co.)

For most preparative experiments on the semimicro scale, sufficient accuracy can be achieved by the use of common laboratory balances. A form of the triple-beam balance (shown in Figure 2.01) is very useful for weighing small quantities of material. The capacity of this balance is normally 111 g; it has a sensitivity of 10 mg, so it can weigh 40 to 50 mg with reasonable accuracy.

The torsion balance[1a] (of which an example is shown in Figure 2.02) is higher in price than some other laboratory balances, but its high sensitivity and rugged construction provide additional advantages. There is no

---

[1a] The Torsion Balance Company, Clifton, N.J.

knife-edge fulcrum to become dull and no bearings, so the sensitivity remains constant. Balances are available with various capacities; some do not require additional weights.

For weighing samples of 5 to 40 mg, a more accurate balance is needed. The usual analytical balance with a sensitivity of 0.1 mg is quite satisfactory even for quantitative experiments. For rougher work, a horn-pan balance and a set of weights are fine, if they are available.

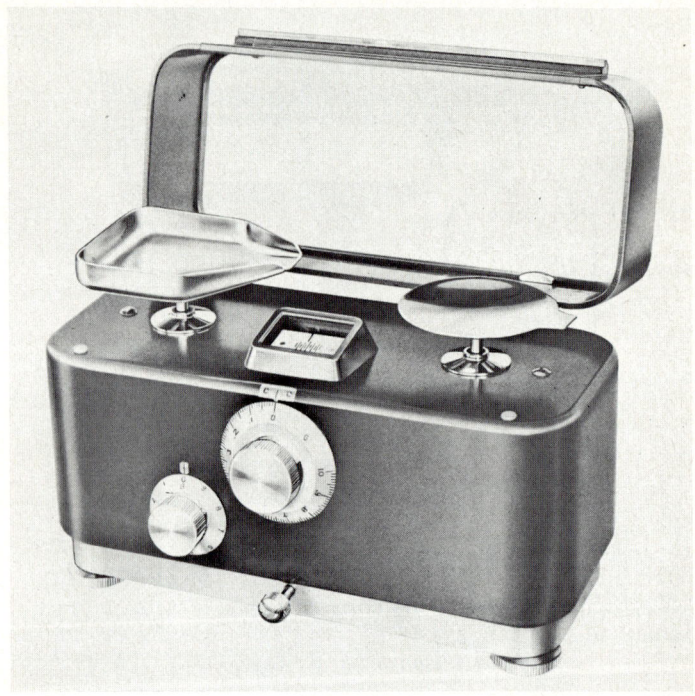

FIGURE 2.02

Torsion balance. Capacity, 120 g; dial, 1 g in 10 mg graduations; readability of dial, 2 mg; weight loader, 9 g in 1 g increments. (Courtesy The Torsion Balance Company.)

The Roller-Smith[2] dial-reading balance shown in Figure 2.03 satisfies most of the requirements for rapid routine work done when small quantities are employed for experimentation. This balance costs no more than a good analytical balance. The dial is graduated up to 500 mg in 2-mg divisions readable to 0.2 mg by means of the vernier scale. The capacity may be increased to 1500 mg without impairing sensitivity by attaching

[2] A. H. Thomas Co., Philadelphia, Pa.

Equipment and Procedures for Small-Scale Work   11

extra weights to the left-hand hook. Tare weights and weighing pan can also be attached to the hooks. Stock and Fill have described[3] a simple direct-reading micro balance for preparative work.

A balance useful for weighings in the range of 5 to 130 mg can be easily and cheaply made from equipment available in many laboratories.[4] The balance is direct-reading, damps in about 5 sec, and has an accuracy of ±0.25 mg. It consists essentially of a "light" spring from a Jolly specific-gravity balance, a pan, an oil-damper, and a microscope light.

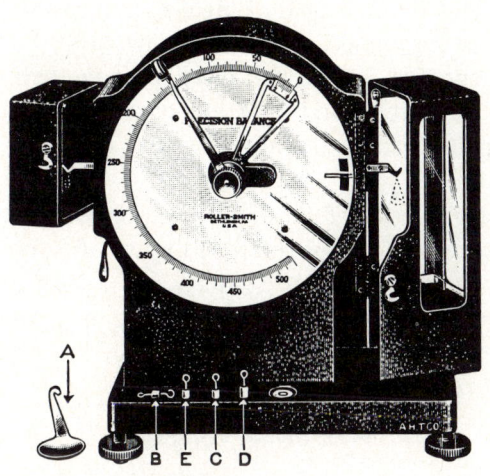

FIGURE 2.03

Dial reading (spiral spring) balance. The dial is graduated to 500 mg in 2-mg divisions, readable to 0.2 mg by means of vernier. Capacity may be increased to 1500 mg without impairing sensitivity by attaching extra weights to the left-hand hook. The weights and weighing pan can also be attached to the hooks. (Courtesy A. H. Thomas Co.)

Solids are weighed on pieces of glazed paper or, if glazed paper is not available, on pieces of ordinary paper. For weighing 50 to 500 mg of solids on a horn-pan balance an $8\frac{1}{2} \times 11$ sheet of paper is cut into eight equal pieces; for quantities of 1 to 5 g the $8\frac{1}{2} \times 11$ sheet is cut into four equal pieces. A piece of paper is placed on each pan of the horn-pan balance and counterbalanced if necessary by the addition of small bits of paper to one pan or the other. When the triple-beam balance is used, a single piece of paper is first weighed and then the proper weights are added. The solid is placed on the paper with a spatula, care being taken not to spill any solid on the pan of the balance. After the weighing is

[3] J. T. Stock and M. A. Fill, *Metallurgia* **37,** 108 (1947).
[4] D. L. Harris, *J. Chem. Educ.*, **38,** 469 (1961).

complete, the paper on which the solid was placed is washed with water and thrown in the waste jar.

Weighing of small quantities of solids or liquids directly in 6- or 8-in. tubes is easily accomplished by means of *weighing rings*. To make a weighing ring, take a piece of No. 20 or No. 22 gauge iron or copper wire 200 mm in length and wrap two loops around the test tube under the lip. Twist the two ends of the wire together and cut one end near the loop. Bend the other end at right angles to the loop so that it projects upward about 30 mm; bend the end of the single wire into the form of a hook. The tube becomes suspended at the lip. The hook of the ring is then inserted into the support from which the pan of the balance is suspended from the beam. The tube and ring are counterbalanced by means of weights and then the proper quantity of solid or liquid is added to the tube.

A weighing tube designed for microdetermination of carbon-hydrogen and nitrogen[5] may prove useful for weighing small samples of liquid for other purposes. It consists of a capillary tube, open at both ends, but containing a piece of bent wire. The sample is easily introduced by capillary action, and does not evaporate significantly during weighing.

Measuring Volumes

For most semimicro procedures, the measurement of volumes down to 0.1 ml is adequate, although measurements of volume as small as 1 λ (1 microliter or 0.001 ml) is sometimes necessary. Cylinders of 10-ml capacity graduated in 0.1 ml are recommended for the measurement of volumes of 0.5 ml or more.

Droppers

For most preparative work the measurement of liquids is accomplished by means of droppers. Practically all solvents and reagents used in semimicro work are dispensed in our laboratories from reagent bottles having a capacity of 15 or 30 ml and provided with plastic caps holding a glass dropper and rubber bulb. The glass droppers have a tip about 2.5 to 3 mm in outer diameter and deliver approximately 0.5 ml of liquid if the rubber bulb is pressed on the upper part. A dropper calibrated to 0.25 ml, 0.5 ml, 0.75 ml, and 1.0 ml is shown in Figure 2.04. This has been found extremely useful in our laboratories. A more accurate pipet dropper,

[5] J. Mitsui and C. Furuki, *Mikrochim. Acta*, **1960**, 169–174.

calibrated in 0.1 ml divisions to a total capacity of 1.0 ml, is shown in Figure 2.05. This micropipet dropper may be employed for rapid titrations of small amounts of liquids. Micropipet droppers should be cleaned immediately after use.

When graduated droppers are not available, ordinary medicine droppers may be employed and the volume estimated by counting the number of drops. For such measurements the weight of each drop is of importance. The chief factors determining the weight of a drop of liquid are: (a) the outside diameter of the dropper at the tip; (b) the surface tension of the liquid; and (c) the density of the liquid. Most commercial droppers have a tip with an outside diameter close to 3 mm. Measurement of forty-eight droppers of about 100 mm length from four different manufacturers disclosed that forty-six droppers had tips with diameters between 2.9 mm and 3.2 mm. The graduated pipet droppers developed by us (Figure 2.04) have tips with a diameter of 2.4 to 3 mm, with an average of 2.6 to 2.8 mm.

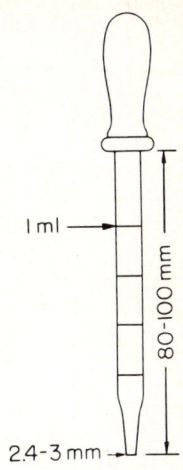

FIGURE 2.04

Calibrated dropper for measuring volumes (1 ml in 0.25-ml divisions).

Table 2.01 shows the influence of the diameter of the orifice of the dropper on the weight of drops. Of three droppers selected for this series of experiments, the first dropper, with a tip 2.4 mm in diameter, represents the average of the droppers of the semimicro reagent bottles. The second

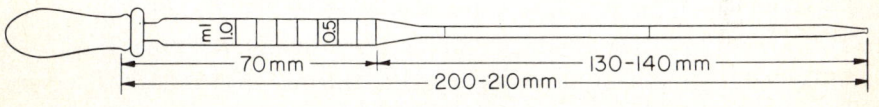

FIGURE 2.05

Calibrated micropipet dropper (1 ml in 0.1-ml divisions).

dropper, with a tip of 3.0 mm represents the average from the commercial droppers. The third dropper, which has a tip of 4.0 mm, represents an extreme case of the commercial dropper. The results indicate that large errors are introduced by measuring reagents in drops unless such information as shown in Table 2.01 is available.

In general, for most preparative work, the measurement need only be approximate. For example, if it is desired to wash a derivative with 1 ml of 50 per cent methanol, no serious error is introduced by the use of one dropperful (from the semimicro reagent bottle) of methanol and one dropperful of water. When the amounts are calculated from the data given in Table 2.01, the volume of the alcohol-water mixture produced is 1.3 ml, containing 50 per cent alcohol by weight.

TABLE 2.01

INFLUENCE OF DIAMETER OF ORIFICE ON MEASUREMENT OF LIQUIDS BY DROPPERS*

Liquid	Density, g/ml	Dropper A Diameter, 2.4 mm		Dropper B Diameter, 3.0 mm		Dropper C Diameter, 4.0 mm	
		I	II	I	II	I	II
Acetone	0.792	70	36	60	45	48	42
Benzene	0.879	60	40	55	56	40	44
Bromobenzene	1.495	50	45	45	73	34	61
Ethanol (95%)	0.816	75	44	65	50	48	40
Ethyl ether	0.708	100	41	88	70	64	49
Methanol	0.792	75	44	65	50	50	40
Water	1.000	25	16	22	25	17	20

* I, Number of drops per gram of substance; II, Number of drops delivered by dropper by filling it with a single pinch on top of the rubber bulb.

The measurement of liquid reagents, however, should be as accurate as possible. If calibrated droppers are not available, two or three may be calibrated in a short time. A convenient method is to take off the rubber bulb of the dropper, close the outlet by means of a little paraffin wax, then add, by means of a 1-ml pipet (graduated in 0.1 ml), 0.5 ml to the dropper. The level is marked by means of white ceramic ink; then another 0.5-ml portion is added and marked; finally, the pipet is drained and heated slightly over a free flame to dry the ink. If it is desired, a mark can be made to indicate delivery of 0.25 ml.

Droppers and Pipets with Long Capillary Tips

Pipets with long capillary tips are indispensable for separating small volumes of immiscible liquids and for transferring small volumes of liquids. Figure 2.06 shows an ungraduated pipet with a long capillary tip; the tip can be sturdy so that it can be cleaned and used again, or thin so that it must be discarded after use. We recommend that these pipets be kept on hand at all times, ready for use.

To prepare a pipet with a long capillary tip, heat a piece of glass tubing 200 mm in length and 6 to 8 mm in diameter with a flame until the glass softens enough to bend easily. Draw gently and steadily, lengthwise, until the tubing reaches the desired diameter. Cut the capillary to give a tip of the desired length and bore. Heat the wide end of the pipet until it is fire polished. If the glass tubing used is of 6-mm bore, the wide end should

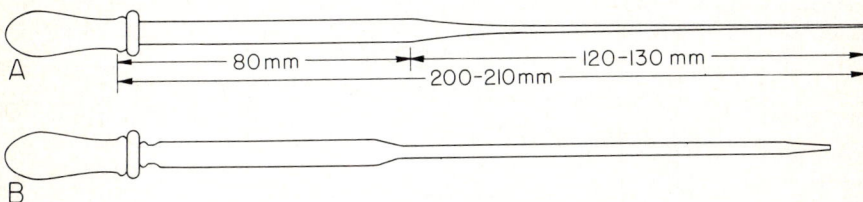

FIGURE 2.06

Ungraduated pipet dropper with long capillary tip. (*A*) With thin tip (0.1 mm). (*B*) With sturdy tip.

be flanged in order to form a tight fit with the rubber bulb. To flange the end, heat it in the flame until the tube has softened; then press firmly against an asbestos pad. The operation is repeated until a flange 7 to 8 mm in diameter is formed. Pipet droppers of various sizes are made by a method similar to that described for the capillary pipet.

Micropipets

Occasionally it becomes necessary to transfer samples of liquids in the range of 5 to 500 λ (0.005 to 0.500 ml) with good precision. The best pipets for this purpose are the micropipets shown in Figure 2.07, which may be obtained with a wide range of volumes. Because of their small diameters that result in variable drainage, these pipets are best used to "contain" rather than to "deliver" a known volume of liquid. Their proper use requires them to be rinsed with a diluting liquid after they are drained, and the rinse added to the other material being diluted. Some assistance in filling the pipet is usually required, such as the glass syringe shown in Figure 2.08. Among the many other types of pipettors that are available, the screw type is one of the best. One of this type (shown in Figure 2.09) uses a threaded piston which is

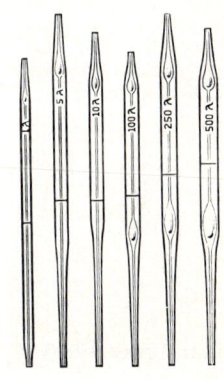

FIGURE 2.07

Microliter pipets, calibrated to contain at 20° volumes of 1 to 500 microliters (0.001 to 0.500 ml) with a volumetric tolerance of 1%, or less, of total capacity. Also available with smaller tolerances and with certified volumes. (Courtesy A. H. Thomas Co.)

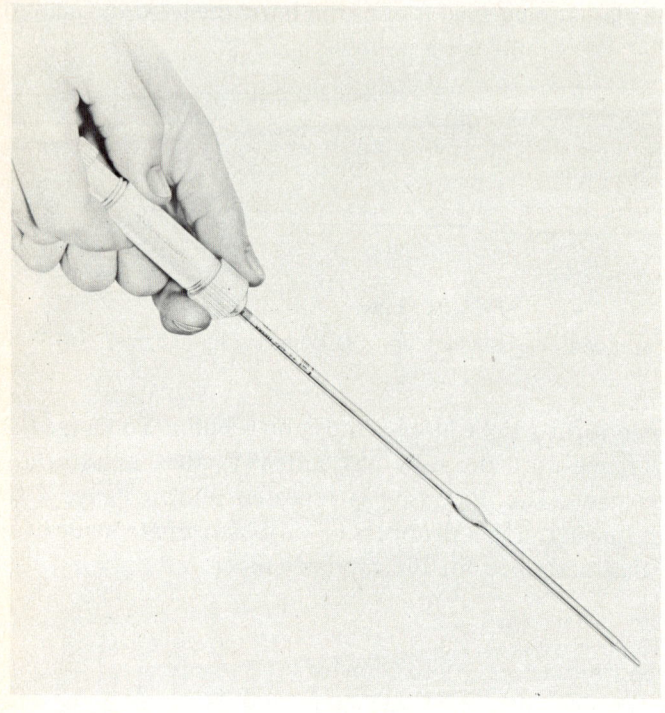

FIGURE 2.09

Screw-type pipettor for microliter pipets. It can be obtained in larger sizes for larger pipets. It allows one-hand operation, and is constructed of plastic. (Courtesy Emil Greiner Co.)

FIGURE 2.08

Syringe-type pipettor for microliter pipets. It can be obtained in different sizes, suitable for control of pipets from 1 μl to about 10 ml. (Courtesy A. H. Thomas Co.)

Equipment and Procedures for Small-Scale Work 17

withdrawn by being rotated. Careful control is possible, even with one-hand operation. For use with pipets of larger volumes the devices shown in Figure 2.10 are desirable and quite satisfactory.

A disposable, self-filling, self-measuring dilution micropipet has been described in the literature,[6] and is available commercially.[7] Although it

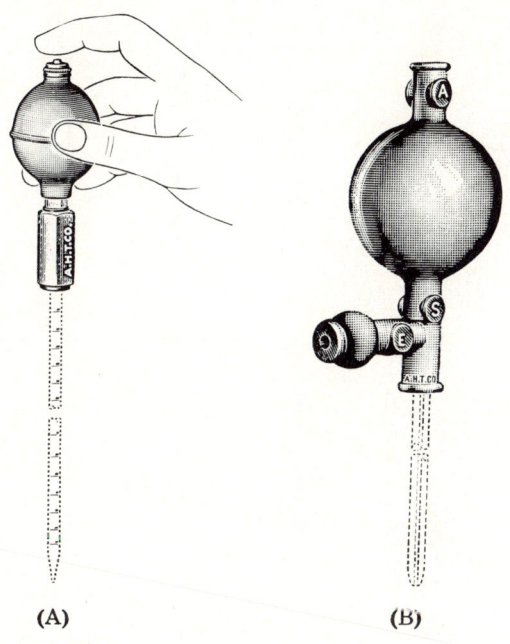

FIGURE 2.10

(A) Caulfield pipettor. Rubber bulb and enclosed valve (same type as used in automobile tires) allow careful control of delivery and uptake. One-hand control. (B) Propipette. One-piece construction, except for three agate ball valves, which are opened by pinching. Sensitive control. (Courtesy A. H. Thomas Co.)

was designed to collect and to dilute measured small quantities of blood, it can be used for other purposes in the chemical laboratory. The pipet is a straight, thin-wall uniform-bore glass capillary tube, which fills completely by capillary action.

Disposable micropipets may be purchased[8] in vials of 100 in sizes of 1 to 100 microliters. They are so inexpensive that they can be discarded

---

[6] H. W. Gerade, "A Disposable, Self-Filling, Self-Measuring Dilution Micropipet," in *Microchemical Techniques*, N. D. Cheronis, ed., Interscience, New York, 1962, pp. 1009–1026.
[7] Unopette, Becton-Dickinson, Rutherford, N.J.
[8] Scientific Glass Apparatus Co., Inc., Bloomfield, N.J.

18 Techniques of Organic Analysis

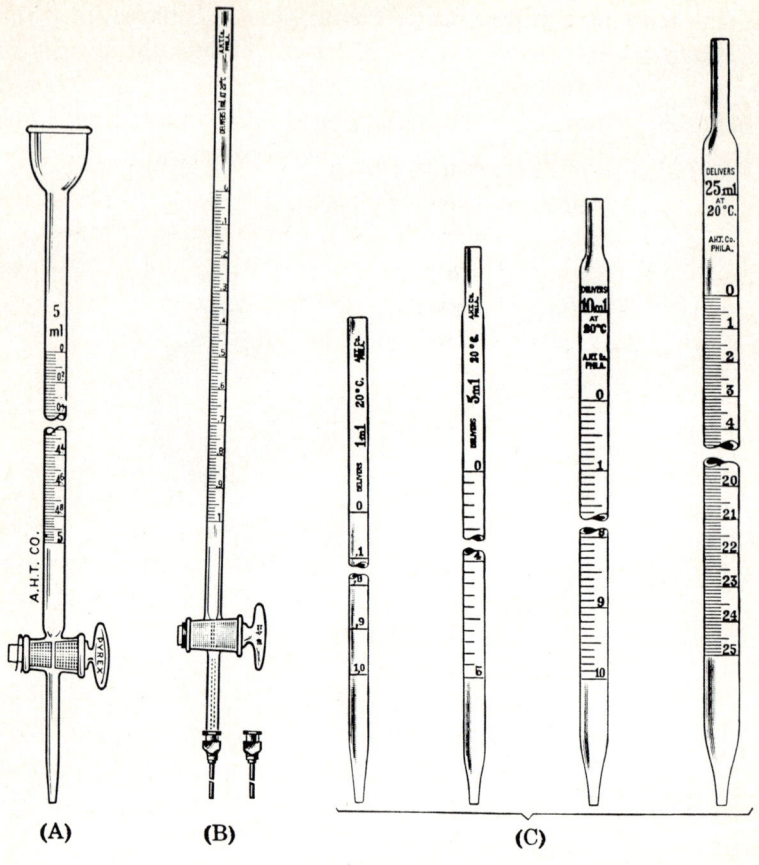

FIGURE 2.11

(A) Microburet. Funnel top and precision bore. Available in 5- and 10-ml sizes. (B) Shohl microburet. It has a metal tip from the syringe, which may be interchanged to obtain different size drops. It has 2-, 3-, and 5-ml capacities. (C) Mohr measuring pipet. Calibrated between points along the main body, in the same manner as a buret. It may be obtained with close tolerances. (Courtesy A. H. Thomas Co.)

after use, and are accurate to within 1% of the stated volume. The pipets are made from constant-bore precision tubing.

Cleaning and drying of micropipets is somewhat difficult because of their small bores. If many micropipets are used in a laboratory, it is desirable to have an apparatus to facilitate cleaning and drying. One of these, which is available commercially,[9] holds 12 pipets. Rinse liquid is drawn through the pipets by suction and, then, the liquid is evaporated

[9] Micro-Metric Instrument Co., Cleveland 22, Ohio.

by drawing air through the pipets. Heat is applied to offset the heat loss due to evaporation.

Microburets

Transfer of a variable known volume of liquid may be accomplished by means of a microburet, as shown in Figure 2.11(*A*) and (*B*), or a small Mohr pipet as shown in Figure 2.11(*C*). Microburets of this type have volumes of 1 to 10 ml and are graduated in intervals of 0.01 or 0.02 ml. Various Mohr pipets may be obtained with capacities of 0.1 ml (graduated at intervals of 0.001 ml) to 50 ml (graduated at intervals of 0.2 ml).

Heating

Burners

Burners for semimicro use can be obtained from most laboratory equipment companies. The maximum flame with well-defined inner and outer cones that can be obtained from such a burner is about 75 mm in height and 15 mm in diameter at the middle; the minimum flame is about 15 mm in height and 10 mm in diameter at the middle. The disadvantage of burners of this type is that a minute flame, which is sometimes required for heating very small quantities, is difficult to obtain.

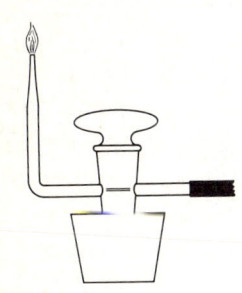

FIGURE 2.12
Microburner constructed from glass stopcock.

Whenever a minute flame is needed, a true micro burner must be used. The flame obtained from these microburners has a maximum height of 40 mm and a minimum height of 5 to 7 mm, with a diameter of about 5 mm. These burners may also be purchased from most laboratory supply houses, or one similar to that shown in Figure 2.12 may be constructed in the laboratory.

For this purpose, a glass stopcock is heated at one end and drawn out to give a capillary about 3 to 4 mm in diameter. The capillary is cut and fire-polished so that the orifice is 1 to 2 mm in diameter. The end of the stopcock having the capillary is then bent upward as shown in the figure. The lower end of the stopcock is inserted into the enlarged opening of a No. 8 to No. 10 rubber stopper. This "home-made" microburner will do for most purposes, although a nonluminous flame is difficult to obtain.

20 Techniques of Organic Analysis

To achieve greater sensitivity of control in heating with a gas burner, more elaborate devices have been developed.[10] One of these is essentially a chimney in which the burner rests, and to which the flask to be heated is clamped. A butterfly valve between the two gives good control over the heat applied to the flask. Another device uses a miniature hot plate heated by a steady flame. The hot plate is raised or lowered, gradually, to give exact control of the heat input.

Hot Plates

The micro hot plate[11] shown in Figure 2.13 is provided with a glass chimney and a special clamp, fitting on the glass, by which a tube or a small flask can be adjusted to a desired height and heated without contact with the hot plate. This type of arrangement may be employed for heating reaction vessels, for evaporations, and for distillations.

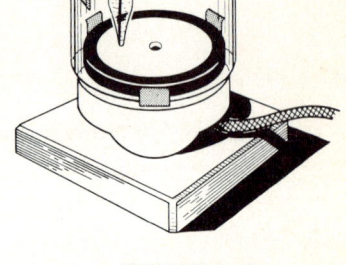

FIGURE 2.13

Micro hot plate with chimney.

Heating Blocks and Stages

Of the commercially available heating blocks and stages the Benedetti-Pichler metal block[12] and the Ma-Schenck heating stage[13] can be employed in small-scale work for various procedures that involve heating. The Benedetti-Pichler heating block has been designed primarily for inorganic microanalysis.

The Ma-Schenck heating stage, shown in Figure 2.14, is more suitable for micro and semimicro preparative and analytical work in the organic laboratory. The stage is electrically heated and consists of a solid aluminum block about 4 in. in diameter and 2.5 in. high. The upper working surface is pierced by 15 holes having diameters from 1.5 mm to 27 mm and depths up to 50 mm for the accommodation of all standard equipment employed in small-scale work: capillaries, test tubes, centrifuge cones, thermometers, beakers, flasks, and other glass and porcelain ware. The periphery is wound with Chromel wire, insulated with asbestos. On the underside is inserted a temperature-sensitive device to provide temperature

---

[10] J. T. Stock and M. A. Fill, *J. Chem. Educ.*, **33**, 619–620 (1956).

[11] Micro-Ware, Inc., Vineland, N.J.

[12] A. A. Benedetti-Pichler, *Microtechnique of Inorganic Analysis*, Wiley, New York, 1942, p. 17. Apparatus available through A. H. Thomas Co., Philadelphia, Pa.

[13] T. S. Ma and R. T. Schenck, *Mikrochim. Acta*, **1953**, 245. Apparatus available through Micro-Ware, Inc, Vineland, N.J.

control through an automatic electronic circuit. The stage is mounted on an insulated metal base and can be employed for temperatures up to 300°. It can be used with direct current if controlled by a rheostat or with alternating current if controlled through a variable transformer or by means of an automatic controller.

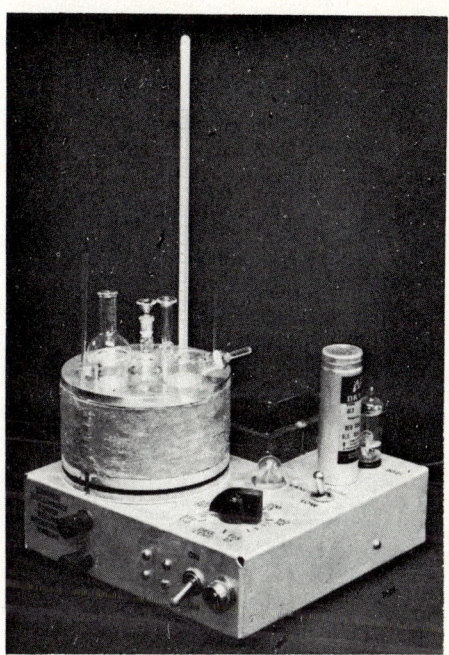

FIGURE 2.14
Universal micro heating stage (Ma and Schenck).

Another electrically heated reaction block has been described,[14] which is useful for many operations on the semimicro scale, including heating under reflux, distilling, recrystallizing, subliming, drying, and determining melting points. It is essentially a block of aluminum alloy, which has been drilled and cut in various ways to accommodate sealed reaction tubes, a thermometer, test tubes, and capillary melting point tubes. The block is heated by means of a heating tape, and is provided with a light for viewing the melting point tubes. Typical uses of the reaction block are illustrated in detail in the literature.[15]

[14] H. P. Schultz, *J. Chem. Educ.*, **35**, 564–565 (1958).
[15] H. P. Schultz, *Microchem. J.*, **5**, 233–241 (1961).

Water Baths

When available, standard steam cones or baths can be employed for heating at temperatures of 40 to 100°. Beakers of 125-ml and 250-ml capacity may be used as water baths in semimicro work, although small tin cans 2 to 3 in. in diameter will serve the same purpose without danger of breakage. These baths may be used for many purposes. However, when the heating is done with a flame, they should not be used for the evaporation of even small amounts of *flammable solvents*. In such cases the beaker or can is heated by means of a hot plate.

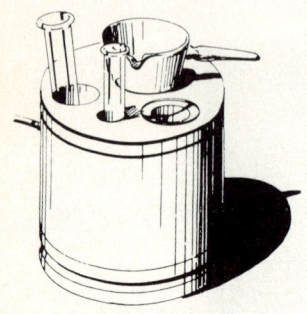

FIGURE 2.15
Heating bath for small-scale experimentation.

The bath shown in Figure 2.15 is made of metal, and is about 70 to 75 mm in depth and 85 to 90 mm in diameter. The removable top, which fits snugly, contains four holes: one is 40 mm in diameter for evaporating dishes or large tubes; two openings are 23 mm in diameter for 6- and 5-in. tubes or very small watch glasses; and the fourth is 13 mm in diameter for 3-in. tubes and microcones.

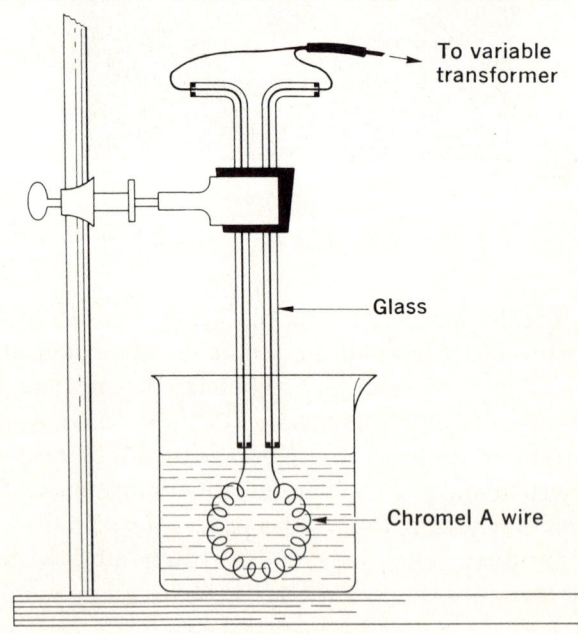

FIGURE 2.16
Immersion heater for oil baths.

Oil Baths and Immersion Heaters

The bath just described (Figure 2.15) can be employed as a convenient oil bath for semimicro work suitable for heating 3-, 6-, and 8-in. tubes and 5-, 10-, and 25-ml flasks. Heating is accomplished either by

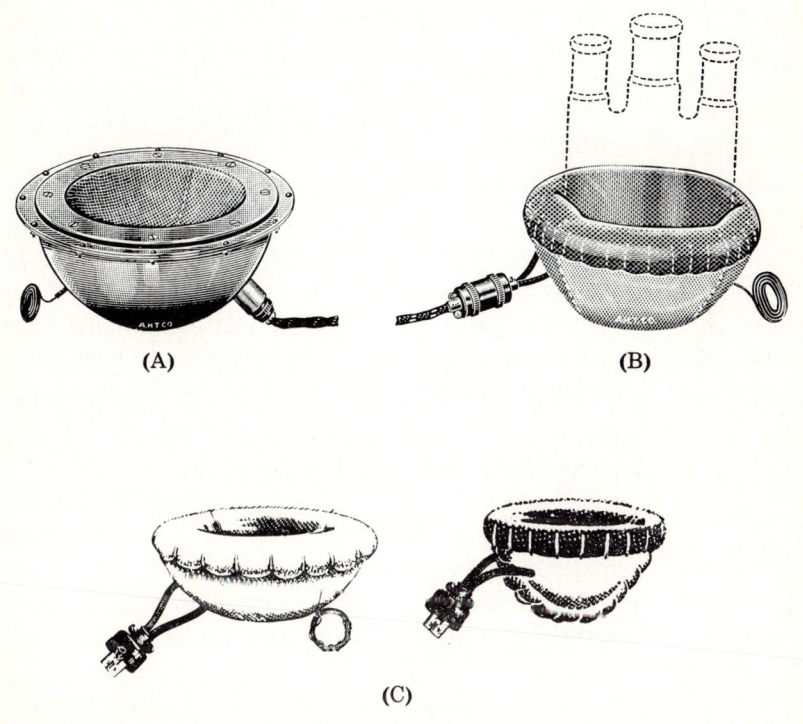

FIGURE 2.17

(*A*) Glas-Col hemispherical heating mantle. Spun aluminum shell, with thermocouple. Sizes to fit 300-ml to 12-l round-bottom flasks. (*B*) Also a Glas-Col hemispherical heating mantle. All glass-fiber cloth exterior. Sizes to fit 50-ml to 5-l round-bottom flasks. (Courtesy A. H. Thomas Co.) (*C*) Glass-fiber heating mantle. Sizes to fit 2- to 200-ml round-bottom or pear-shape flasks. (Courtesy Metro Industries.)

a small burner or, better by an immersion heater (Figure 2.16), operated through a variable transformer to give controlled temperatures up to 250°.

The best liquids for heating baths are the silicone oils, which withstand high temperatures for long periods of time without decomposing or even darkening. Dow Corning silicone fluid 702, for example, can be heated to 250° indefinitely in the presence of air, moisture, and metals; it boils at

24 Techniques of Organic Analysis

400° and freezes at −40°. Although these oils are expensive for large-scale use in the ordinary laboratory, they will usually endure conditions that cause most other fluids to fail. The oils can be purchased through most laboratory supply houses.

Heavy paraffin oil, solid paraffin wax, and highly hydrogenated vegetable fats are satisfactory liquids for heating baths if the operating temperature is below 200°. Other liquid heat-transfer media are glycerol and tetracresyl silicate. For an extensive discussion of liquid heat-transfer media, reference is made to the literature.[16]

Heating Mantles

Electrically operated heating mantles, such as those shown in Figure 2.17, are very convenient for the heating of reaction vessels and distillation flasks where careful control is needed. These heaters are usually made of glass fiber, and fit standard sizes of flasks and beakers. The electrical input must be controlled, usually with a variable transformer. If precaution is taken always to use the correct size of heating mantle and never to overheat an empty container, the heaters have excellent durability.

Handling, Stirring, and Grinding

Spatulas

Two types of spatulas often used in semimicro work are shown in Figure 2.18. The rounded end of the upper one is used for the transfer and handling of crystals in preparative work, and the pointed end for the

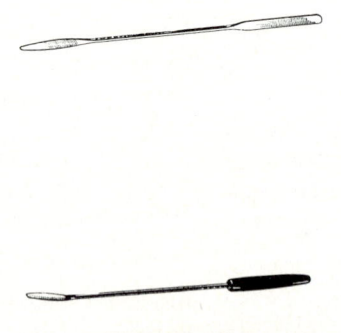

FIGURE 2.18

Microspatulas. *Upper one* is stainless steel, and has one end tapered and the other rounded. *Lower one* has a spoonlike end, and a plastic handle. (Courtesy Fisher Scientific Co.)

[16] R. S. Egly, "Heating and Cooling," in *Technique of Organic Chemistry*, Vol. 3, 2nd ed., A. Weissberger, ed., Interscience, 1957, p. 109.

manipulation of minute quantities of material. The lower spatula in Figure 2.18 has a plastic handle and a blade in the form of a scoop, which is particularly useful for transferring solid samples.

Other micro spatulas may be prepared from wire of suitable diameter, stiffness, and corrosion-resistance. A piece of wire about 6 in. in length is flattened with a hammer at one end and filed to a point at the other end.

Any of these spatulas may be bent at an angle to increase the ease of manipulation of solids. For example, the rounded end of the upper spatula in Figure 2.18 may be bent at a right angle in order that the spatula may be used to press crystals in a Buchner funnel while they are being dried by suction.

Handling in an Inert Atmosphere

Sometimes it is necessary to handle toxic, volatile substances, or those that are easily attacked by the moisture or oxygen of the air. If we wish to weigh accurately samples of these substances, or even to transfer them from one container to another, it is desirable to have an inert atmosphere box for the purpose. These may be obtained from commercial firms in a variety of sizes and designs, but they may also be constructed to fit the particular needs of the worker.

An elaborate dry-box, which can be fabricated in most college or industrial machine shops,[17] compares favorably with those available commercially. It is constructed of 1/8-in. sheet steel, with 1/4-in. Plexiglass windows. An airlock is provided for moving materials into the box without letting moist air inside. To reduce the moisture content in the airlock, it is rapidly evacuated and flushed with a dry gas, several times. The air inside the box is continuously circulated through activated alumina to take out moisture. Materials are handled by means of rubber gloves, which are attached to the side of the dry box.

A somewhat less elaborate dry box has been described.[18] A dry, inert gas, such as nitrogen, is passed through the box at all times. Once the box is filled with room air, however, a week of flushing is needed to attain the desired dryness. Equipment and materials can be inserted into and removed from the box by means of an airlock, which can be flushed separately with dry gas.

A simple glove box for handling chemicals in an inert atmosphere can be constructed from a polyethylene bag, a rigid frame, two gloves, and two

---

[17] R. E. Johnson, *J. Chem. Educ.*, **34**, 80–81 (1957).
[18] S. Y. Tyree, Jr., *J. Chem. Educ.*, **34**, 603–605 (1954).

embroidery hoops.[19] A plastic cylinder (the bag) is placed over the frame, and each end is drawn together about a tube, through one of which nitrogen is admitted and through the other nitrogen escapes. Two holes are made in the side of the polyethylene bag, and the gloves are joined to the edges of the holes by the embroidery hoops.

A glove bag somewhat similar to the one just described can be purchased;[20] they are so inexpensive that they are sold by the half dozen. Gloves are provided with each bag, and are, in fact, an integral part of the bag. The glove bag may be easily discarded when it becomes contaminated or worn.

A controlled-atmosphere jar can be made from a 2-liter resin reaction kettle and other items, which are fairly readily available.[21] With this apparatus, it is possible to remove the cap from a bottle, take out a liquid sample, transfer it to a receiver, and recap both the original bottle and the receiver by remote handling and, entirely, in an inert atmosphere. The cited work should be consulted for details.

Stirring

In small-scale experimentation, mechanical stirring devices are not always essential. For example, in the preparation of 5 to 50 mM of a Grignard reagent, mechanical stirring can be eliminated without appreciably lengthening the reaction time or reducing the yield. Most of the procedures in this book do not require stirring devices. However, in some cases, stirring is desirable, even for small quantities of reactants.

One of the most satisfactory methods of stirring solutions, whether in a closed system or an open one, is the magnetic stirring apparatus and bar shown in Figure 2.19. The magnetic bars are coated with an inert material such as Teflon, and are available in various sizes and shapes (cylindrical, spherical, egg-shape). A magnetic bar is placed in the reaction vessel and the apparatus is assembled in the usual manner, except that the stirring apparatus is placed beneath the vessel. The stirring apparatus, consisting of a motor and rotating magnet, causes the magnet inside the vessel to turn at varying speeds. The reaction mixture may be heated by means of a glass-fiber heating mantle and stirred with the magnetic stirring apparatus at the same time.

Another type of apparatus for simultaneously heating and stirring a solution combines an electric hot plate with the magnetic stirrer. Usually,

---

[19] S. G. Shore, *J. Chem. Educ.*, **39**, 465 (1962).
[20] Instruments for Research and Industry, 108 Franklin Ave., Cheltenham, Penn.
[21] M. B. Naff, *Chemist Analyst*, **50**, 54 (1961).

the temperature of the hot plate and the speed of stirring can be controlled independently. This type of apparatus is ideal for stirring and heating the contents of a beaker or other flat-bottom container. Special stirrers are available for small equipment.[22]

If the solution to be stirred is too large or too viscous, other apparatus must be used to effect stirring. Stirrers of many shapes and kinds have been described in the literature. Among the most useful and effective in

FIGURE 2.19

Magnetic stirrer. An enclosed, variable-speed motor causes the magnetic bar in the flask to rotate. It may be combined with a hot plate for heating. Bars are coated with an inert plastic. (Courtesy A. H. Thomas Co.)

stirring contents of flasks is that described by Hershberg[23] and illustrated in Figure 2.20A. This stirrer shaft is made from a glass rod and Nichrome wire of the proper sizes; it can be inserted into a flask with a small neck, and sweeps the bottom of the flask free of solids that might settle there.

Teflon makes an excellent bearing for a stirrer, since it may be used without added lubrication. A sleeve of Teflon, which fits the stirrer shaft snugly, is placed inside a glass tube of appropriate size.[24] The bearing will operate at high speeds for a long time without trouble.

For many reactions carried out in the laboratory, the stirrer shown in Figure 2.20B is quite satisfactory.[25] The assembly consists of a standard

[22] Micro-Metric Instrument Co., Cleveland 22, Ohio; Tri-R Instruments, Jamaica 35, N.Y.
[23] E. B. Hershberg, *Ind. Eng. Chem., Anal. Ed.*, **8**, 313 (1936).
[24] M. L. Kaplan, R. E. Crocker, and H. Sargent, *J. Chem. Educ.*, **38**, 577 (1961).

ground-glass ball and socket joint; the upper (ball) joint is attached to the stirrer shaft by means of a sleeve of rubber tubing, and the lower (socket) joint is attached by a stopper or standard taper joint to the neck of the flask, and serves as a bearing for the stirrer shaft. Consequently the upper

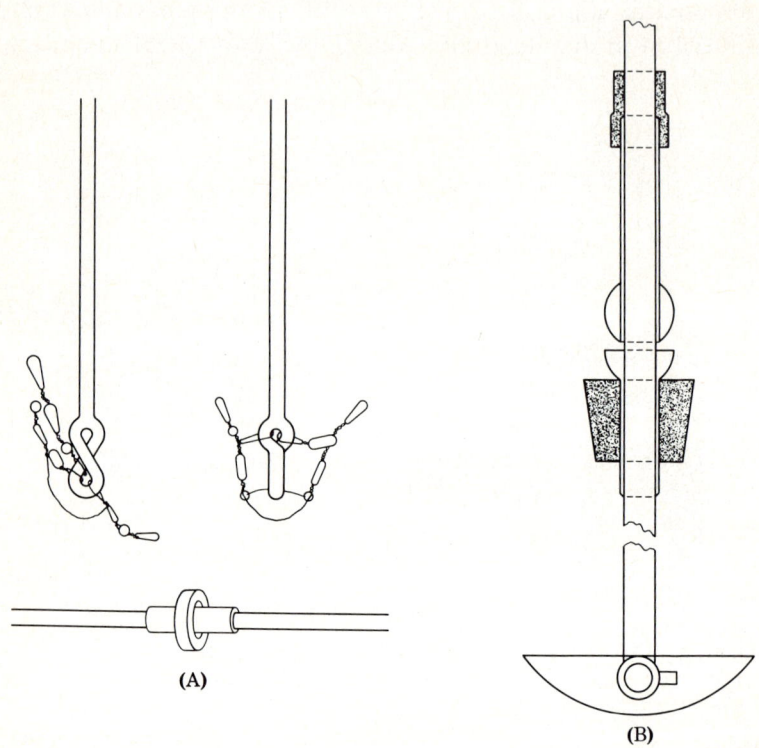

FIGURE 2.20

(A) Stirrer shaft. Nichrome wire twisted on a glass loop.[23] (B) Bearing for stirrer. Ball and socket joint.[25]

part of the ball and socket joint turns on the lower part, so the surface must be kept well lubricated at all times. The assembly is not only relatively inexpensive, but is fairly leak-proof.

For mixing one or more tubes at a time for a few seconds or a few hours, the vortex mixer[26] is very useful. The tube is gently swirled to create a vortex in the liquid. If a water-jacketed tube is used, the tube may be kept at a constant temperature.

[25] J. T. Patton, *J. Chem. Educ.*, **28**, 207 (1951).
[26] Scientific Industries, Inc., 220-05 97th Avenue, Queens Village, N.Y.

Stirring rods are used in many small-scale operations, and should be kept at hand. Stirring rods may be of various lengths and diameters of glass rod, and are easily made by heating each end to fire-polish it. If one end is flattened somewhat by contact with an asbestos board while the glass is hot, it will be useful in crushing solids against flat surfaces.

Mortars

Grinding of small quantities of crystalline solids is usually accomplished by means of the flat end of the microspatula on the glazed surface of porcelain or special paper. Common mortars, such as agate or small porcelain ones, are not recommended for pulverizing small quantities of material since their use entails a considerable loss. For example, if a small agate mortar is employed to pulverize 50 mg of crystals, the amount lost in operation varies from 5 to 20 mg, depending on the nature of the crystals and the care with which the operation and transfer are performed. If a mortar is employed, it is advisable to wash the vessel with a solvent in which the crystals are soluble and evaporate the washings on a small watch glass in order to recover the material adhering to the walls of the pulverizing vessel.

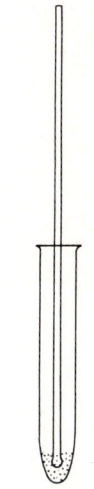

FIGURE 2.21
Pulverizing by means of a rod in a cone or tube.

In the preparation of derivatives for the final step of the characterization of organic substances, the grinding of the crystalline derivative during the purification step is of fundamental importance (an example of this procedure is described on page 470). The grinding of crystalline derivatives during the purification step is best accomplished in the same cone or tube in which the derivative was prepared, as shown in Figure 2.21. Either a microspatula or a rounded glass rod that fits well in the bottom of the tapered tube or cone is slowly rotated until the crystals have been pulverized to the desired fineness.

Separating of Immiscible Solvents

Separatory Funnels

Two types of separatory funnels that may be employed for the separation of small quantities of immiscible liquids in the extraction of solutions are shown in Figure 2.22. At the left is a 10-ml microseparatory funnel graduated in 0.1-ml divisions. The same type of funnel is available either ungraduated or graduated in 0.5-ml divisions. Figure 2.22 shows a regular pear-shape separatory funnel of 30-ml capacity.

Capillary Pipets

Separation of very small (2-ml) quantities of immiscible liquids may be readily accomplished by means of the pipet droppers with long capillary tips described in the section on measuring volumes (p. 14). This method is recommended for separation of solutions contained in a test tube. Ether, benzene, or other solvent is added to the solution, and the tube is closed with a solid rubber stopper. The thumb of the hand holding the tube is placed on top of the rubber stopper to hold it down firmly, and the contents of the tube are shaken. The pressure of the thumb is gradually released with the tube held at an angle and toward the opening of a hood, the stopper is removed, and the tube placed on a rack or stand. After the two layers have separated, the tube is inclined slightly and the desired layer is removed with the capillary pipet dropper and transferred into another tube (Figure 2.23). The bulb of the dropper is pressed and the capillary tip is inserted into the tube until it reaches 1 to 2 mm above the junction of the two liquids. The pressure on the bulb is then released gradually until the desired layer has been withdrawn.

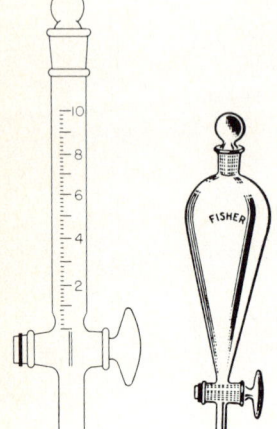

FIGURE 2.22

Left. A 10-ml microseparatory funnel graduated in 0.1-ml divisions.[25] *Right.* Pear-shape regular separatory funnel. (Courtesy Fisher Scientific Co.)

Even in small-scale experimentation, two immiscible liquids may not separate very easily. Consequently there is some advantage in carrying out the extraction process in a small centrifuge tube, since separation of the layers can be speeded by centrifuging the tube for a short time.

Filtering

Apparatuses desirable for the separation of suspended solids and liquids are more fully described in Chapter 3 in the section on crystallization (p. 66). The appropriate part of that section, depending upon the quantity of material to be filtered, should be consulted.

Equipment and Procedures for Small-Scale Work 31

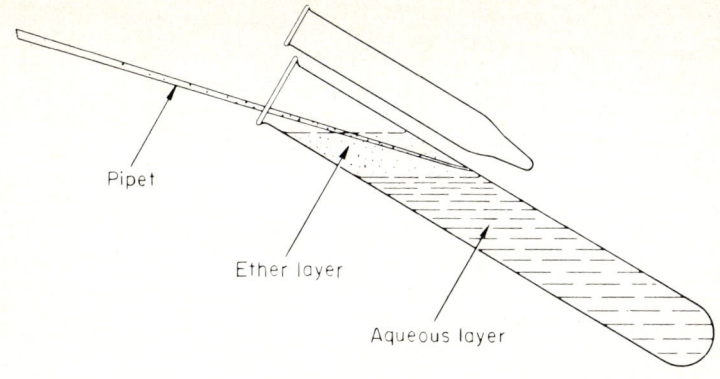

FIGURE 2.23
Separation by means of a capillary pipet.

Drying

Since the drying of solids is most often done in connection with recrystallization, this process is described in the section on crystallization (p. 77). The pertinent part of that section should be consulted for equipment used for drying of solids.

Evaporating

The most common arrangement for evaporation is the use of a steam bath or water bath (Figure 2.15) and a small evaporating dish or watch glass to contain the solution or liquid whose volume is to be reduced. Improvised baths can be made by using small beakers (150- to 400-ml capacity) half filled with water and heated in a hood. With proper care, such an arrangement can be employed for the evaporation of small amounts of flammable solvents. However, a more useful arrangement is a hot plate with a chimney, such as was shown in Figure 2.13.

The traditional steam-bath evaporating procedures can readily be adapted to small evaporating dishes, watch glasses, or vessels of any size. In addition, the setup shown in Figure 2.24 is useful for the evaporation of small amounts of liquid.[27] It consists of a 250-watt infrared bulb with a built-in reflector and an aluminum-sheathed socket attached to a clamping device so that it can be rotated through 360° vertically and about 300° horizontally. The distance of the lamp above the vessel containing the liquid can be varied by raising or lowering the heater so as to obtain

[27] Fisher Scientific Co., Pittsburgh, Pa., and New York, N.Y.

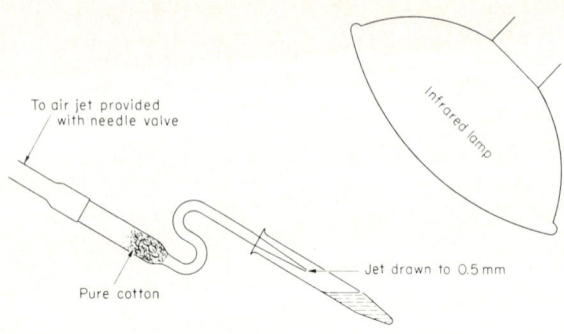

FIGURE 2.24

Evaporation by means of infrared heat.

the desired temperature as recorded by the thermometer placed near the vessel. If it is desired to increase the rate of evaporation, a stream of air is gently blown over the surface of the liquid.

All of the above arrangements involve some loss of material in transfer. Therefore, when the quantity of material involved is small—as, for example, a few milligrams of a substance in a mother liquor or a very dilute solution—concentration by evaporation is best accomplished by the evaporation arrangement shown in Figure 2.25. The vessel consists of a pear-shape flask of 5-, 10-, or 25-ml capacity or the tube in which the solution or filtration was originally made. The flask or tube is fitted with a rubber stopper carrying a microcondenser with the lower end open and the inner tube drawn to a pointed capillary. The upper end of the inner tube is connected by means of rubber tubing to a calcium chloride tube. The stream of dry air admitted through the tube is regulated by means of a screw clamp; the distance of the capillary from the surface of the liquid is adjusted so that, when air is drawn, a ripple is produced.

A small volume of solution can be evaporated in a test tube very easily and conveniently by means of a side arm adapter[28] as illustrated in Figure 2.26(*C*). A current of air may be impinged on the surface of the liquid by connecting the side arm to an aspirator and using the inlet shown at (*A*). A tank of inert compressed gas may be connected to the inlet.

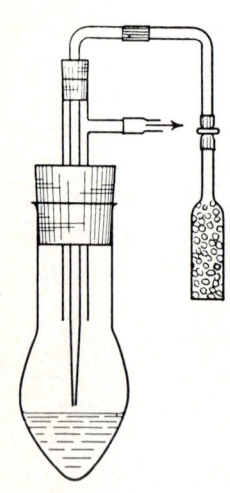

FIGURE 2.25

Evaporation by means of dry air.

[28] M. T. Bush, *Microchem. J.*, **1**, 109–110 (1957).

Evaporation occurs very rapidly, especially if the test tube holding the solution is placed in a beaker of warm water. To complete the process of drying, a vacuum may be applied to the outlet, in which case the capillary tube should be turned up as shown at (B) in Figure 2.26.

A simple, effective apparatus for the evaporation of small volumes (3 to 10 ml) of liquid can be easily constructed. The solution is placed in a vial or test tube, which is closed with a one-hole stopper fitted with a

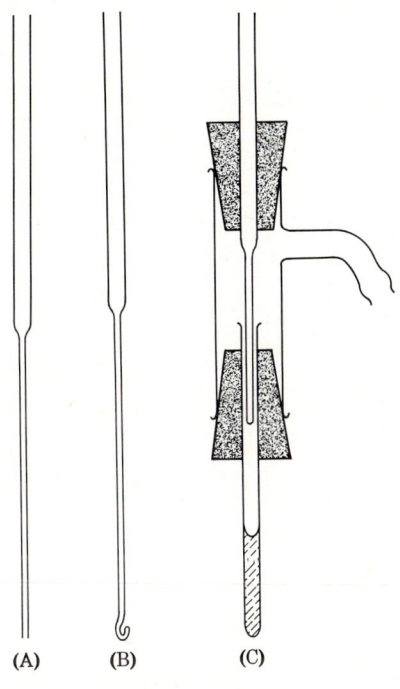

FIGURE 2.26

Small-scale impingement evaporator.[28]

short piece of glass tubing. The apparatus is suspended by a piece of flexible tubing through which a mild vacuum is applied. The container is placed in a beaker of water which is stirred magnetically. As the water swirls, the tube also moves and, thus, bumping of the liquid being evaporated is avoided.[29]

The rotary evaporator is useful, particularly for large amounts of solution, but may be used to advantage for small quantities. The container is a round-bottom flask of appropriate size, which is caused to rotate as a

[29] R. L. Clements, *Chemist Analyst*, **48**, 20 (1959).

34 Techniques of Organic Analysis

gentle vacuum is applied to the liquid. The tendency of the liquid to bump is reduced, and a larger surface is exposed for evaporation as the flask rotates. These may be purchased from many equipment companies or made in the laboratory.[30]

Heating Under Reflux

Figure 2.27 shows two apparatuses for heating under reflux. The flasks may have a capacity of 10, 25, or 50 ml. The most useful tubes are the

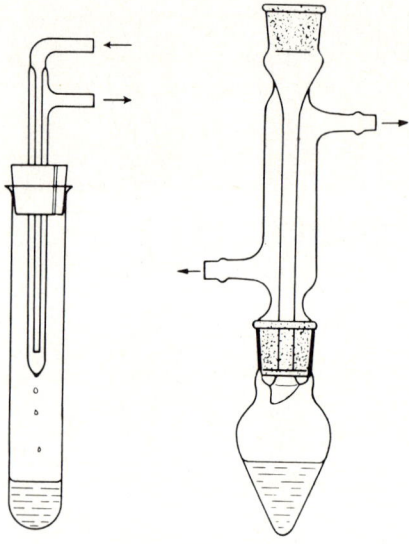

FIGURE 2.27

Left. Apparatus for heating under reflux, using tube and finger microcondenser. *Right.* Apparatus for heating under reflux, using 25-ml pear-shape flask and Liebig microcondenser.

25 × 200 mm and the 25 × 150 mm ones. Six-inch tubes (20 × 150 mm and 18 × 150 mm) cannot be stoppered readily. Whenever a 6-in. or smaller tube must be employed, as is the case when the volume of the liquid is 1 to 2 ml, then a microcondenser of the finger type (Figure 2.27, *left*) should be employed. A cork larger than 15 mm is employed, and it is allowed to rest on the flared top of the tube.

The heating of the vessels is accomplished by means of a bath, a hot plate, or a burner. The vessel can be clamped at such a distance from the

---

[30] R. O. Carleton, *J. Chem. Educ.*, **39**, 256–257 (1962); V. C. Runeckles, *Chemist Analyst*, **50**, 23 (1961); E. M. Arnett, *J. Chem. Educ.*, **37**, 247 (1960); and F. Greef and C. L. Larsen, *J. Chem. Educ.*, **33**, 556 (1956).

flame that gentle refluxing takes place. Boiling stones of the size of grape seeds can be made by breaking tiles or dishes of unglazed porcelain. Commercially available small boiling stones have been found useful.[31]

Reaction Vessels

The reaction vessels generally employed in connection with work for the characterization of organic substances are (a) common 3-, 6-, and 8-in. tubes; (b) 5-, 10-, and 25-ml pear-shape flasks, which are usually fitted with condenser assemblies; (c) Universal assemblies; and (d) special reaction vessels. In categories (a) and (b), which require no description, are the vessels that by far, are the most used ones. A brief description will be given of a type of Universal assembly and two reaction vessels and, for more information, reference is made to the literature.[32]

Figure 2.28 shows a Universal assembly, with interchangeable joints, which can be arranged for several operations without vessel-to-vessel transfer and the resulting loss of material. The various arrangements are convenient for quantities of 45 to 1000 mg of material. Arrangements of the apparatus are shown for reflux (A), separation of immiscible liquids (B), distillation (C), filtration (D), and filtration without transfer except inversion of the assembly (E). This type of apparatus is recommended for research.

Figure 2.29 shows a reaction-vessel assembly, suitable for semimicro work. It consists of a three-neck flask, provided with stirrer, calcium chloride tube, condenser, and dropping funnel. By replacing a stopper on the funnel with a stopcock, the apparatus can be used for reactions that must be carried out in an inert atmosphere. This type of reaction unit is convenient for quantities of 50 to 1000 mg or more. For quantities of 5 to 25 mg, the reaction vessel shown in Figure 2.30 is being used by us. All of the assemblies shown in Figures 2.28 to 2.30 are commercially available.[33] Assemblies with standard tapered and spherical joints, for use with materials in the range between semimicro and macro scales, have been made available in recent years because of the increase of small-scale experimentation.[34]

---

[31] "Technibubbles," Techniservice Co., New York, N.Y.

[32] N. D. Cheronis, "Micro and Semimicro Methods," in *Technique of Organic Chemistry*, Vol. 6, A. Weissberger, ed., Interscience New York, 1954, pp. 132–142.

[33] Micro-Ware, Vineland. N.J.

[34] Ace Glass, Inc., Vineland, N.J. ("Mini-Lab. Assembly"); Quickfit and Quartz Ltd., Heart of Stone, Staffs., England, "Quikfit Semimicro Assemblies"; Scientific Glass Apparatus, Bloomfield, N.J.; Metro Industries, Long Island City 6, N.Y.; Schuco Scientific, 250 West 18th Street, New York 11, N.Y.; U.M.F. Science Corporation Arcadia, Calif.; Kontes Glass Company, Vineland, N.J.

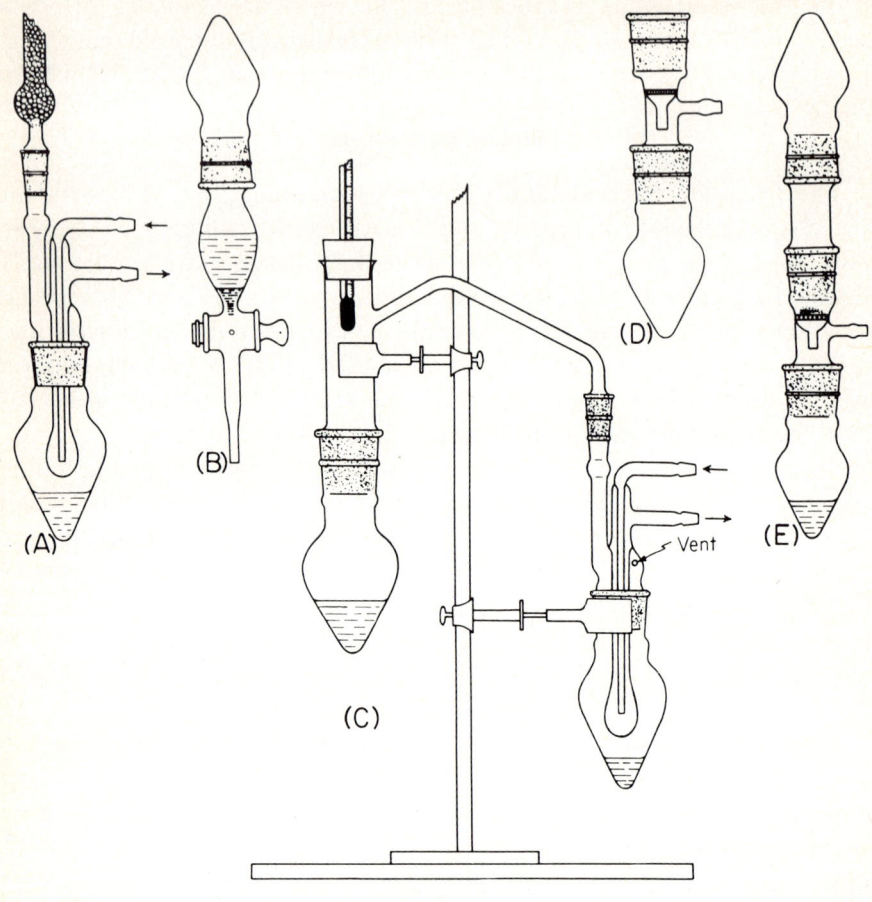

FIGURE 2.28

The Cheronis Universal apparatus consists of 2- to 25-ml flasks, condenser, fritted glass funnel, and sleeve, all having interchangeable joints. Shown above are arrangements for five procedures. (*A*) Refluxing. (*B*) Separation of immiscible liquids. (*C*) Distillation. (*D*) Filtration. (*E*) Crystallization and filtration without transfer.

Small-scale reactions may be conducted in an H-shape flask,[35] shown in Figure 2.31, which is also useful for distillations and filtrations.

Reactions conducted with 10 to 30 millimoles of reactants not in dilute solution, but needing careful temperature control and stirring, require special equipment, such as the reaction vessel shown in Figure 2.32. The thermometer is inserted into a side arm to keep it out of range of the stirrer,

[35] J. B. Polya, *J. Chem. Educ.*, **39**, 294–295 (1962).

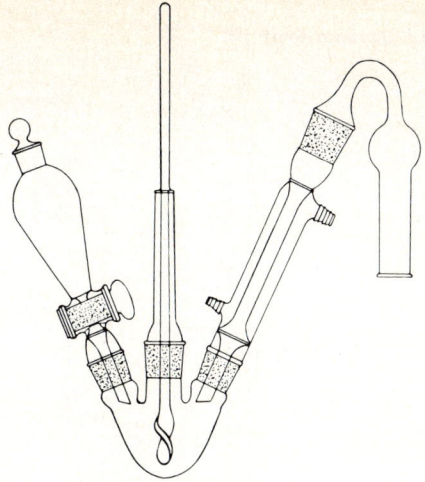

FIGURE 2.29

Reaction vessel for semimicro quantities.

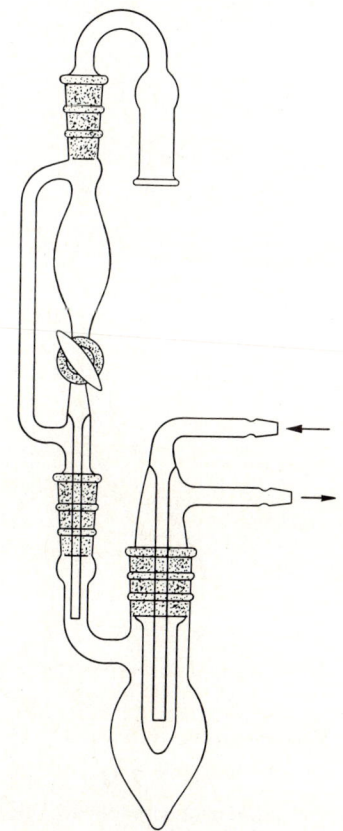

FIGURE 2.30

Reaction vessel for quantities of 5 to 25 mg.

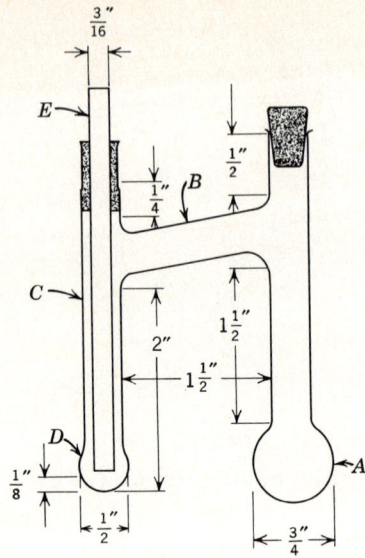

A,B,C $\frac{3}{8}''$ OD

FIGURE 2.31

H-shape flask for reactions.[35]

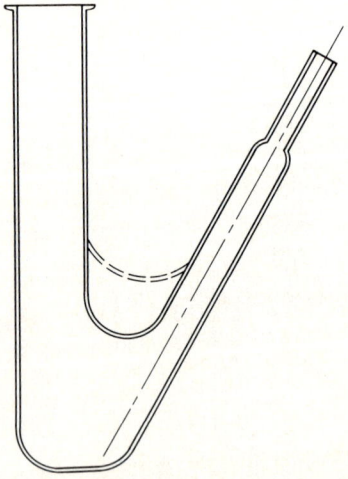

FIGURE 2.32

A small stirred reactor with thermometer well.[36]

and is held in place by a sleeve of rubber tubing. The vessel can be made in different capacities and with more side arms as needed.[36]

Perforated composition board, which can be purchased from hardware or lumber dealers, makes a useful mounting board for semimicro reactions or other operations.[37] Apparatus is easily arranged and clipped to the board, on both sides if necessary.

High Vacuum Line

If a great deal of small-scale preparative work needs to be done, the desirability of using a high vacuum line should be considered. This apparatus is generally all glass, or metal and glass, and is mounted

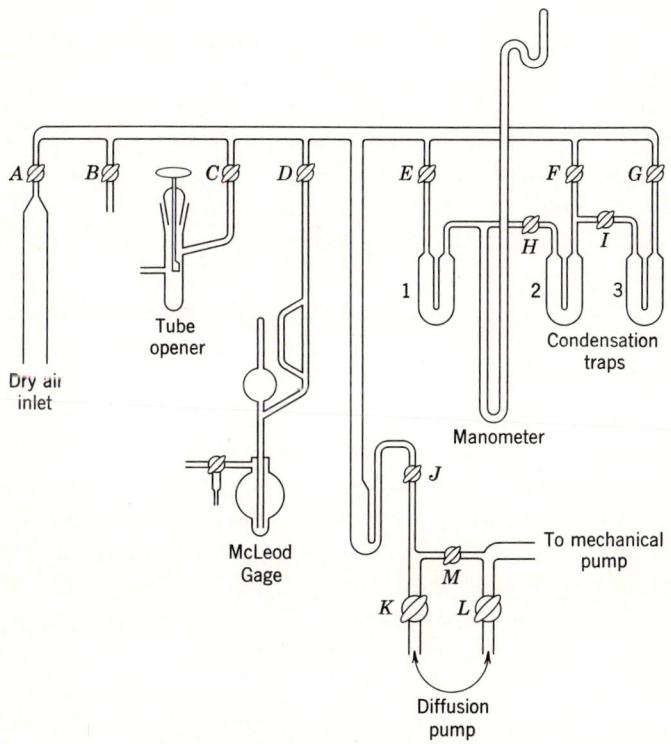

FIGURE 2.33

Typical high-vacuum apparatus.[37a]

---

[36] S. Kasman, *J. Chem. Educ.*, **37**, 150 (1960).
[37] J. A. Beech, *J. Chem. Educ.*, **39**, 293 (1962).
[37a] R. T. Sanderson, *Vacuum Manipulation of Volatile Compounds*, Wiley, New York, 1948.

permanently on a stand. To consistently obtain a pressure of 1 micron of mercury, a diffusion pump is needed, which must be backed up with a rotary oil pump. A pressure gauge, such as a McLeod gauge, capable of indicating down to 1 micron, is needed together with cold traps and precision-ground stopcocks. Transfer of material from one part of the apparatus to another is achieved by cooling the receiver in liquid nitrogen or dry ice. The transfer process is essentially quantitative, even for a few drops of liquid. Unfortunately, materials that boil above 200° at atmospheric pressure have little vapor pressure at room temperature, so that their transfer is very slow, even at 1-micron pressure. The equipment is ideal for compounds that are toxic, or those that react with the constituents of air. A typical high-vacuum setup is shown in Figure 2.33.

Apparatus and Procedure for Microhydrogenation

Hydrogen at atmospheric pressure and in the presence of a catalyst is very useful in the reduction of nitro compounds to amines, the reduction of unsaturated linkages and, to a smaller degree, in the reduction of carbonyl compounds to alcohols. Aromatic compounds are not reduced. Applications of the method are given in Examples 14.08 (page 509) and 20.05 (page 626).

In this method hydrogen gas[38] is dispersed in the form of minute bubbles through a solution of the compound to be reduced. The catalyst suspended in the solution is in constant agitation, and the shaking devices commonly used in catalytic hydrogenations are avoided.

The hydrogenator shown in Figure 2.34 consists of the reaction vessel, connected by means of interchangeable glass joints to the microporous disperser, and a finger condenser in the upper arm, which has an outlet for the excess of hydrogen. The disperser is connected through a piece of rubber tubing to bottle C, shown in Figure 2.35. The outlet is connected by means of rubber tubing to the exhaust system. In the initial stages of hydrogenation the injection of hydrogen gas through the microdisperser is at a rapid rate; the frothing that results can be allowed to rise to the junction of the sidearm and the reaction chamber without impairing the efficiency of the reaction. If a hydrogen tank with a reducing valve is available, it is connected directly to bottle B (Figure 2.35).

---

[38] N. D. Cheronis and M. Koeck, *J. Chem. Educ.*, **20**, 488 (1943); N. D. Cheronis and N. Levin, *J. Chem. Educ.*, **21**, 603 (1944); N. D. Cheronis, "Micro and Semimicro Methods," in *Techniques of Organic Chemistry*, Vol. 6, A. Weissberger, ed., Interscience, New York, 1954, pp. 236–249.

If no hydrogen tank is available, the apparatus shown in Figure 2.35 is employed. Flask A, with a capacity of 500 or 1000 ml, is charged with 20 to 25 g of technical zinc pellets or mossy zinc and 50 ml of water. Dilute (25 per cent) sulfuric acid is added through the dropping funnel. About 10 to 15 ml of acid are added to start the reaction. When the flow of hydrogen in the wash bottle B falls off to 1 to 2 bubbles per second, 10 ml more of acid are added. About 10 to 15 g of zinc and 60 to 80 ml of dilute sulfuric acid are required for the reduction of 1 g of most nitro compounds, provided that the reaction is complete within 30 minutes.

The hydrogen source is connected through rubber tubing (preferably red) to a wide-mouth bottle B, with a capacity of 0.5 to 1 liter, containing 250 to 300 ml of concentrated sulfuric acid. The inlet and outlet and the safety valve are made of 6-mm tubing; the safety tube is about 1 m in length. The height of the sulfuric acid above its normal level in the bottle is a measure of the gas pressure. This height should not exceed 30 to 40 cm. During some hydrogenations, particularly when the compound formed is not very soluble in the solvent, the disperser becomes slightly clogged. In such cases the sulfuric acid in the safety tube begins to rise, and the disperser must be either changed, or cleaned by immersion in a tube containing about 5 ml of warm solvent. The solvent is drawn—by mild suction—a few times into the porous tube until air may be blown easily through the solution. The wash bottle B is connected to a 250-ml bottle C, containing some glass wool and a small layer of calcium chloride. Traces of moisture in the hydrogen do not interfere with the reaction. Insertion of a stopcock in the stopper of bottle C connecting to the exhaust is recommended.

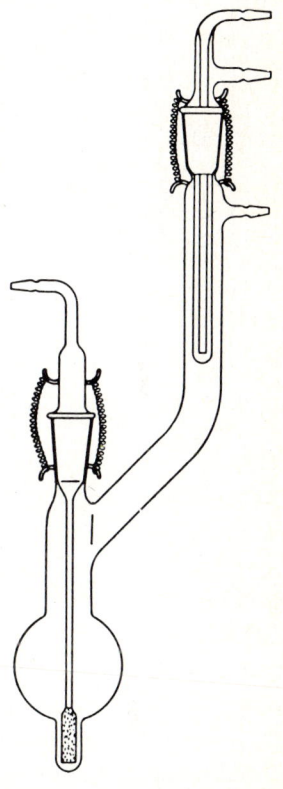

FIGURE 2.34
Microhydrogenator with ground-glass joints.

When a hydrogenator with glass joints is not available, the following arrangement will be found useful. A regular 8-in. tube with side arm is closed with a two-hole rubber stopper through which are attached a regular microcondenser and a simple microporous disperser without a glass joint. The microporous disperser is connected through a piece of rubber

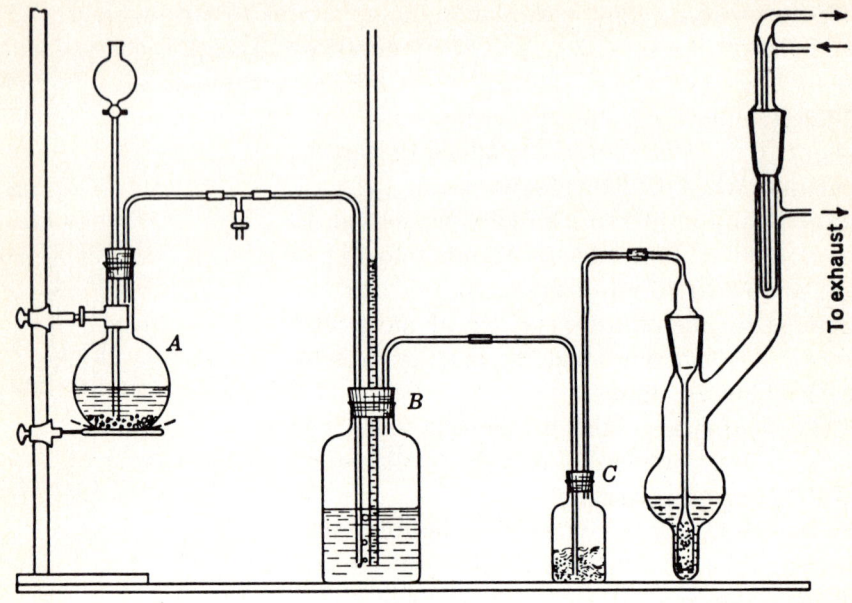

FIGURE 2.35

Assembly for microhydrogenation at atmospheric pressure. When hydrogen gas in a tank is available, bottle B is connected directly to the reducing valve of the tank.

tubing 3 to 4 mm in diameter and 30 to 40 cm in length with the bottle C. The microcondenser is inserted about 80 to 90 mm in the tube. Special precautions should be used to make sure that the opening in the rubber stopper for the microcondenser is of sufficient diameter to permit the insertion without undue pressure, and the insertion should be made using toweling on both hands to avoid injury in case of breakage. The side arm is connected with a piece of tubing that leads into a hood or, through a small opening in a window, to the outside. Alternately, the outlet may be connected to a 2-liter bottle, filled with water to enable the excess of hydrogen to be collected by water displacement. Since the volume of excess of hydrogen is not very great, it may be washed down the drain, provided that only one hydrogenation is being performed in the laboratory.

The rubber stoppers and rubber tubing should be washed before use, first with soap and water and, then, with alcohol. The microporous disperser is cleaned after each run by placing it in a 6-in. test tube with a few milliliters of ethanol or methanol. The alcohol is sucked several times into the tube, just past the glass junction and, then, blown out again. If the disperser is to be used again for reduction of the same compound, no further cleaning is necessary; for complete cleaning, it is removed from the alcohol rinse and air is blown through to drain it. It is then placed in a 6-in. tube containing a small amount of aqua regia and heated in that position until vigorous evolution of bubbles begins, and the tube is allowed to stand for a few minutes. The disperser is removed from the acid and washed by aspirating water through it. If the apirator is made of metal, the disperser is immersed in a 5 per cent solution of sodium hydroxide before aspiration. After the disperser is washed with water, the process is repeated with two successive 10-ml portions of alcohol, and the disperser is allowed to dry. If the cement joining the glass to the porous tube cracks or chips, it may be patched with fresh cement.

The factors that affect the rate of hydrogenation and the purity of the product are (a) the nature and activity of the catalyst and the presence of other substances which may act either as promoters or as retarders; (b) the nature of the solvent; (c) the temperature of the reaction; and (d) the nature of the compound to be reduced.

Several catalysts for microhydrogenation are commercially available. The most suitable for routine work is 5 per cent palladium carbon. Directions for the preparation of platinum, palladium, nickel, and mixed catalysts have been given.[39] The 5 per cent palladium carbon has more consistent activity than the "reduced black" obtained by the reduction of platinic oxide. The amount required for most hydrogenation is 100 to 250 mg of 5 per cent palladium carbon.

The efficiency of various solvents for most hydrogenations is in the following order: ethanol > 2-propanol > esters > ethers > hydrocarbons. In the reduction of olefinic linkages it appears that 80 to 90 per cent ethanol is more efficient than absolute or 95 per cent ethanol. Whenever possible, 80 to 90 per cent ethanol should be used for hydrogenation. However, there are cases in which the use of alcohols is undesirable; for example, the microreduction of carbonyl compounds for derivatization. In such cases, isopropyl or butyl ether or, possibly, glacial acetic acid may be employed. The pH of the medium is important in the microhydrogenation of carbonyl compounds; an initial pH above 7.0 increases the rate of reduction, while an initial pH below 7.0 inhibits the reaction. The temperature of the reaction mixture in the microdisperser apparatus should be 10 to 15° below the boiling point of the solvent; when ethanol is used, 60 to 65° is the best temperature. Temperatures up to 200° have been employed by using solvents with high boiling points; such a procedure, however, involves difficulties in the removal of the solvent.

[39] N. D. Cheronis, op. cit., pp. 239–241.

For the reduction of 100 to 500 mg of substance, the sample is dissolved in 20 to 25 ml of solvent and 100 to 250 mg of 5 per cent palladium carbon (or 25 to 50 mg of platinic oxide) are added. For 1 to 50 mg of substance, 10 ml of solvent and 50 mg of palladized carbon are sufficient. It is advisable to place the solvent and catalyst in the hydrogenating tube and to pass hydrogen through for 2 to 3 minutes before adding the substance to be reduced. To illustrate the detailed procedure, the reduction of piperonal to piperonyl alcohol is described.

The hydrogenating tube is charged with 200 mg of 5 per cent palladium carbon and 25 ml of ethanol. The disperser is inserted, and the outlet tube is connected to the exhaust. The gas is allowed to pass through the suspension of palladium carbon to activate it. During the activation of the catalyst, the hydrogenating tube is immersed in a 600 to 800 ml beaker, containing water at about 60°, and the temperature is maintained at 50 to 60° by the addition of hot water from time to time. When a series of hydrogenations is to be performed, a small immersion heater may be more convenient.

When the catalyst has been activated, 500 mg of piperonal is added by raising the disperser momentarily. The stopper is replaced, and the tube is shaken to wash down the compound adhering to the sides of the tube. The flow of hydrogen at the beginning of the reaction should be at the rate of 4 to 6 bubbles per second, emerging in the wash bottle. The flow is reduced after the first 10 minutes to about 3 to 4 bubbles per second, and is continued for 15 minutes to complete the reduction.

The stopcock connecting the bottle C to the exhaust is opened, and the hydrogenator is allowed to stand for 5 minutes to permit settling of the catalyst. Water is added to the generator, and the liquid is poured off; the zinc is washed twice with water and the flask is again placed in position. The reaction mixture in the hydrogenator is filtered and the residue is washed with two 5-ml portions of alcohol. The filtrates are placed in a dish and evaporated over a water bath until a small amount of solvent remains; from then on, the evaporation is conducted slowly. The filter paper is washed and stored; when several papers have been collected they can be burned, and the ash returned for credit to the manufacturer for recovery of the palladium.

The residual oil from the filtrate is cooled by placing the dish in a freezing mixture. The oil solidifies when it is cooled and scratched with the microspatula. The crude mass is scraped and transferred to a tube; 12 to 15 ml of hexane, heptane, or petroleum ether is added and solution is effected by heating. The hot solution is filtered through the semimicro suction funnel into an 8-in. tube with a side arm. The residue remaining in the solution tube is extracted again, with the filtrates of the first crystallization serving as the solvent. The filtered solution is cooled, and the walls of the tube are scratched with a glass rod. The mixture is allowed to stand for 15 minutes and is filtered. The filtrates are used to make successive extractions of the crude piperonyl alcohol until all is exhausted. The crystals on the suction funnel, when all filtrations have been completed, are washed twice with 2 ml of pure solvent and then, transferred to a drying disc. The yield is about 400 mg of crystals that melt at 52 to 53°.

Apparatus and Procedure for Preparation of Grignard Reagents

Grignard reagents are employed in the preparation of derivatives for the characterization of halides (page 553) and esters (page 537). For semimicro work with only 100 to 200 mg of reagents, it is possible to eliminate all special apparatus and a number of the classical precautions that are usually taken in the preparation of the Grignard reagent. Stirring is not necessary in the reaction of 50 to 100 mg of magnesium metal. The reaction of this amount of metal usually takes about 5 to 10 minutes. Absolute ether, preferably freshly prepared, is employed. However, absolute ether prepared according to standard procedures may be stored in a 250-ml Erlenmeyer flask containing a few pieces of fused calcium chloride and sodium cuttings or a few grams of calcium hydride. The flask is stoppered with a cork covered with tin foil.

When needed, 5 to 7 ml are poured into the previously dried reaction tube through a clean dry funnel fitted with a small plug of cotton. The magnesium turnings should be clean and without a coating of oxide. It is advisable to cut the metal into small pieces so as to expose some fresh and clean surface. The purity of the halide is, at times, of vital importance. Traces of impurities may exert a retarding effect, in which case the reaction either does not start or is sluggish. On the other hand, it is our experience that technical grades of some halides have given good results.

The use of other solvents (such as isopropyl and n-butyl ethers, pyridine, and others) in place of ethyl ether is not recommended for semimicro work. For instance, 5 millimoles each of magnesium and methyl iodide (0.32 ml) show no reaction after 15 minutes if isopropyl ether is used, and only a meager reaction when pyridine is used. Two types of assemblies may be employed by beginners for the preparation of the small quantities of Grignard reagent used in characterization work.

One apparatus is shown in Figure 2.36. The assembly consists of an 8-in. tube, a microcondenser, and a drying tube. The tube is heated over a flame until it is thoroughly dry, then stoppered and allowed to cool. The microcondenser is wiped with a perfectly dry cloth; then the cork, which holds the condenser and the calcium chloride tube, is inserted in the mouth

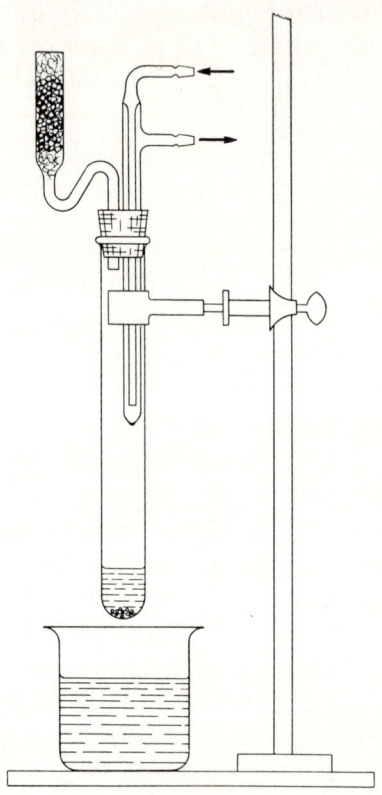

FIGURE 2.36
Simple semimicro assembly for preparation of Grignard reagents.

of the tube. The cork has two holes and fits tightly into the mouth of the tube. The charge is made by momentarily lifting the cork.

Another type of assembly is shown in Figure 2.37. It consists of a 25-ml pear-shape funnel with a short stem, on which is fitted a piece of rubber tubing closed at one end. A piece of glass rod fits into the end of the rubber tubing. This protects the inside of the short stem from getting wet when a beaker of water is raised under the apparatus to cool the reaction mixture.

After the reaction is completed, the rubber guard is removed and a short microporous disperser is attached to filter the Grignard reagent into the mixture containing another reactant. It is very essential to have the reaction vessel and condenser perfectly dry; this is accomplished by heating in an oven at 105° and then assembling it rapidly, using a dry

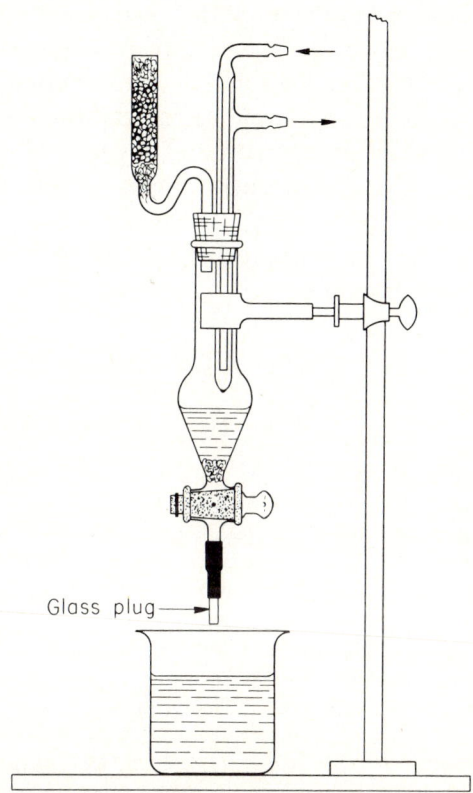

FIGURE 2.37

Semimicro assembly for preparation of Grignard reagents. Apparatus provides delivery of filtered reagents into another vessel with minimum exposure to air.

towel to handle the apparatus. After being cooled, the condenser is connected, and the charge is made. This usually consists of about 0.5 millimoles (120 mg) of magnesium turnings, 0.55 millimoles of the halide, 5 ml of absolute ether, and a crystal of iodine. The reaction starts within a few minutes, and requires 5 to 10 minutes for completion. If the reaction is slow in starting, a small beaker containing warm water (50 to 60°) is raised over the outside of the tube for about 1 minute. If, after the color of iodine disappears, the reaction proceeds very slowly and requires constant heating, moisture is present in the apparatus or in the reagents.

If the tube shown in Figure 2.36 has been employed as a reaction vessel, the Grignard reagent is filtered into the next reaction vessel through a dry funnel having a cotton plug. This entails reaction of the Grignard reagent with moisture and oxygen in the air and, hence, lowering of the yield. If the funnel apparatus shown in Figure 2.37 is employed, the Grignard reagent is allowed to filter into the solution or suspension of the substance with which it is to react; for example, in carbonation, the Grignard reagent is allowed to filter into a mixture of dry ice and ether. The vessel should be selected with the subsequent operations in mind, so that the numbers of transfers will be minimal.

Reaction vessels for the preparation of a Grignard reagent in an inert atmosphere have been described.[40]

REFERENCES

Weighing

L. Cahn, H. Schultz, and P. Gaskins, "A New Electromagnetic Balance with Elastic Ribbon Suspension," in *Microchemical Techniques*, N. D. Cheronis, ed., Interscience, New York, 1962, pp. 1027–1032.

A. H. Corwin, "Weighing," in *Technique of Organic Chemistry*, Vol. 1, Part 1, 3rd ed., A. Weissberger, ed., Interscience, New York, 1959, pp. 72–130. A review with 117 references.

A Friedrich, *Mikrochemie*, **15**, 36 (1934). Salvioni balance.

D. L. Harris, *J. Chem. Educ.*, **38**, 469 (1961). A rapid, direct-reading microbalance.

G. Ingram, *Ind. Chemist*, **37**, 343 (1961). The microgram balance and its application in chemical analysis.

G. Ingram, *Metallurgia*, **39**, 224 (1949); *C.A.* **43**, 3245 (1949). Submicrobalance.

J. Mitsui and C. Furuki, *Mikrochim. Acta*, **1960**, 169. Convenient weighing tube.

J. T. Stock and M. A. Fill, *Mikrochim. Acta*, **1953**, 101. Microbalance for preparative work.

J. T. Stock and M. A. Fill, *Metallurgia*, **37**, 108 (1947). Microbalance for preparative work.

A. F. Williams and J. O. Park, *Analyst*, **85**, 126 (1960). Method of obtaining weighed microsamples of moisture- or oxygen-sensitive compounds.

Measuring Volumes

T. Brindle and C. L.Wilson, *Mikrochemie*, **40**, 141, 310 (1952). Syringe for measuring small volumes.

[40] N. D. Cheronis, "Micro and Semimicro Methods," in *Technique of Organic Chemistry*, Vol. 6, A. Weissberger, ed., Interscience, New York, 1954, p. 347.

N. D. Cheronis, "Micro and Semimicro Methods," in *Techniques of Organic Chemistry*, Vol. 6, A. Weissberger, ed., Interscience, New York, 1954 pp. 118–119. Bibliography of 74 references on microburets.

H. W. Gerade, "A Disposable, Self-Filling, Self-Measuring Dilution Micropipet," in *Microchemical Techniques*, N. D. Cheronis, ed., Interscience, New York, 1962, p. 1009.

G. Gorbach, *Mikrochim. Acta*, **31**, 109 (1943). Bulb-type micropipet.

P. L. Kirk, *Quantitative Ultramicroanalysis*, Wiley, New York, 1950, p. 18. Ultramicro transfer pipet.

A. Krogh, *Ind. Eng. Chem., Anal. Ed.* **7**, 130–131 (1935). Automatic pipet.

A. Krogh and A. Keys, *J. Chem. Soc.* (*London*), **1931**, 2436. Automatic pipet.

K. Linderstrom-Lang and H. Holter, *Compt. rend. trav. lab. Carlsberg, Sér. chim.*, **19**, No. 4, 1 (1931): *Z. Physiol. Chem.*, **201**, 9 (1931). Automatic pipet.

W. Schöniger, *Mikrochemie*, **39**, 401 (1952). Micropipet.

R. C. Sisco, B. Cunningham, and P. L. Kirk, *J. Biol. Chem.*, **139**, 1 (1941). Ultramicro transfer pipet.

R. Steinmann, *Mikrochim. Acta*, **1953**, 490. Bibliography on automatic pipets.

J. T. Stock and M. A. Fill, *Metallurgia*, **33**, 272 (1946); **34**, 225 (1947). Automatic micropipet for spot and drop tests.

Heating

A. A. Benedetti-Pichler, *Microtechnique of Inorganic Analysis*, Wiley, New York, 1942, p. 17. Benedetti-Pichler metal heating block.

H. G. Cassidy, *Ind. Eng. Chem., Anal. Ed.*, **10**, 456 (1938). Small constant-temperature bath.

T. Cifonelli, *Ind. Eng. Chem., Anal. Ed.*, **16**, 134 (1944). Electric microheater.

E. P. Clark, *J. Assoc. Offic. Agr. Chemists*, **16**, 418 (1933). Electric sand bath.

E. N. Dacus, *Ind. Eng. Chem., Anal. Ed.*, **16**, 142 (1944); *C.A.*, **38**, 1398 (1944). Microtorch.

S. S. T. Djang, *Anal. Chem.*, **21**, 873 (1949). Aluminum bath.

R. S. Edly, "Heating and Cooling" in *Techniques of Organic Chemistry*, Vol. 3, 2nd. ed., A. Weissberger, ed., Interscience, New York, 1957, p. 109. Liquid media for heating baths.

J. Erdös, *Mikrochemie*, **35**, 353 (1950). Steel microbomb.

B. W. Gamson, G. Thodos, and O. A. Hougen, *Trans. Am. Inst. Chem. Eng.*, **39**, 1 (1943).; *C.A.*, **37**, 2226 (1943). Micro heating device.

P. F. Holt, *Metallurgia*, **37**, 48 (1947); *C.A.*, **42**, 800 (1948). Micro steam oven.

G. Ingram, *Metallurgia*, **38**, 239 (1948); *C.A.*, **42**, 8031 (1948). All-purpose micro drying oven.

P. L. Kirk, *Quantitative Ultramicroanalysis*, Wiley, New York, 1950, p. 103. Metal and glass microtorch.

T. S. Ma and R. T. Schenck, *Mikrochim. Acta*, **1953**, 236, 245. Micro heating stage; micro drying oven.

A. P. Marion, *Ind. Eng. Chem., Anal. Ed.*, **18**, 82 (1946); *C.A.*, **40**, 2743 (1946). Electrolyte-solution heating element for steam microbath.

O. A. Nelson and H. L. Haller, *Ind. Eng. Chem., Anal. Ed.*, **9**, 402 (1937). Small constant-temperature bath.

C. L. Rulfs, *Anal. Chem.*, **19**, 1046 (1947). Steam microbath.

W. Schöniger, *Mikrochemie*, **34**, 316 (1949). Semimicro autoclave.

H. P. Schultz, *Microchem. J.*, **5**, 233 (1961). Microreaction heating block.

H. P. Schultz, *J. Chem. Educ.*, **35**, 564 (1958). Microreaction heating block.

P. I. Start and M. W. Thring, *J. Sci. Instr.*, **37**, 17 (1960). Laboratory furnaces.

R. Steinman, *Mikrochim. Acta*, **1953**, 496. Bibliography on micro heating devices to 1949.

J. E. Still, *Chem. & Ind. (London)*, **1944**, 294. Immersion heater.

J. T. Stock and M. A. Fill, *J. Chem. Educ.*, **33**, 619 (1956). Heating devices using gas.

J. T. Stock and M. A. Fill, *Anal. Chim. Acta*, **2**, 281 (1948); *C.A.*, **42**, 8534 (1948); *Analyst*, **74**, 123 (1949); *C.A.*, **43**, 5233 (1949). Microblowpipe.

Handling, Stirring, and Grinding

H. K. Alber, *Ind. Eng. Chem., Anal. Ed.*, **13**, 656 (1941). Micro mortar and pestle.

M. A. Fill and J. T. Stock, *Analyst*, **60**, 212 (1944); *Mikrochim. Acta*, **1**, 89 (1953). Microstirrer in vacuum system.

E. B. Hershberg, *Ind. Eng. Chem., Anal. Ed.*, **8**, 313 (1936); *Org. Syntheses*, **17**, 31 (1937). Stirring shaft.

R. E. Johnson, *J. Chem. Educ.*, **34**, 80 (1957). Dry box.

M. L. Kaplan, R. E. Crocker, and H. Sargent, *J. Chem. Educ.*, **38**, 577 (1961). Teflon bearing for stirrers.

P. L. Kirk, *Quantitative Ultramicroanalysis*, Wiley, New York, 1950, p. 38. Stirring device for small volumes.

J. L. Margrave, *J. Chem. Educ.*, **30**, 623 (1953). Inert dropper bulbs.

N. B. Mehta and J. Zupicich, *Chemist Analyst*, **50**, 55 (1961). Solid addition funnel.

M. B. Naff, *Chemist Analyst*, **50**, 54 (1961). Controlled atmosphere transfer jar.

J. T. Patton, *J. Chem. Educ.*, **28**, 207 (1951). Ball and socket stirrer assembly.

J. L. Rabinowitz, *Anal. Chem.*, **24**, 1234 (1952). Magnetic stirrer.

V. R. Shellman and B. J. Magerlein, *Anal. Chem.*, **25**, 1285 (1953). Small motors with stirrers.

S. G. Shore, *J. Chem. Educ.*, **39**, 465 (1962). Glove box.

R. Steinman, *Mikrochim. Acta*, **1953**, 497. Bibliography to 1949 on micro grinders and mortars.

J. T. Stock, M. A. Fill, and R. G. Bjork, *J. Chem. Educ.*, **38**, 524 (1961). Device for charging microtubes with toxic reagents.

J. T. Stock and M. A. Fill, *Metallurgia*, **38**, 118 (1948); *Mikrochim. Acta*, **1953**, 89. A variety of stirring and grinding microdevices.

R. E. Taylor and F. O. Hatfield, *Chemist Analyst*, **50**, 53 (1961). Inert dropper bulbs.

S. Y. Tyree, Jr., *J. Chem. Educ.*, **34**, 603 (1954). Dry box.

B. M. Williams, *Chemist Analyst*, **50**, 21 (1961). Simple laboratory grinder.

Filtering

E. L. Anderson, *Analyst*, **85**, 228 (1960). Centrifugal filtration apparatus.

M. T. Bush, *Microchem. J.*, **1**, 105 (1957). Small-scale all-glass centrifugal filters.

M. T. Bush, *Ind. Eng. Chem., Anal. Ed.*, **18**, 584 (1946). Centrifugal filter.

N. D. Cheronis, "Micro and Semimicro Methods," in *Technique of Organic Chemistry*, Vol. 6, A. Weissberger, ed., Interscience, New York, 1954, pp. 41–44, 384–386.

J. M. Corliss, *Chemist Analyst*, **48**, 16 (1959). Expedient for microfiltration.

L. C. Craig, *Ind. Eng. Chem., Anal. Ed.*, **12**, 773 (1940). Microfiltration.

L. C. Craig and O. W. Post, *Ind. Eng. Chem., Anal. Ed.*, **16**, 413 (1944). Centrifugal filtration.

A. B. Cummins and F. B. Hutto, Jr., "Filtration," in *Technique of Organic Chemistry* Vol. 3, Part 1, 2nd ed., A. Weissberger, ed., Interscience, New York, 1956, pp. 607–786. Discussion of filtration with over 850 references.

P. Dickens, *Chem. Fabrik*, **1**, 323 (1928). Crystallization and filtration in an inert atmosphere.

S. D. Elek, *Mikrochemie, Pregl Festschr.*, **19**, 129 (1936). Use of capillary and special filtering devices.

J. English, Jr., *Ind. Eng. Chem., Anal. Ed.*, **16**, 478 (1944). Centrifugal filtration.

S. W. Gaddis, *J. Chem. Educ.*, **38**, 5 (1961). A simple filtration technique for gravimetric determinations.

H. R. Ing and M. Bergmann, *J. Biol. Chem.*, **129**, 603 (1939). Use of capillary and special filtering devices.

J. J. Kolb, *Chemist Analyst*, **50**, 23 (1961). Microfiltration under pressure.

T. S. Ma, I. Kaimowitz, and A. A. Benedetti-Pichler, *Mikrochim. Acta*, **1954**, 648. Inorganic fiber filter media.

M. J. Mitchell, *Chemist Analyst*, **49**, 30 (1960). Simple semimicro suction filter.

I. Nussbaum, *Chemist Analyst*, **47**, 49 (1958). Supports for rapid vacuum filtration.

O. Romanus, *Chemist Analyst*, **48**, 112 (1959). Expedient for suction filtration.

C. A. Roswell, *Ind. Eng. Chem., Anal. Ed.*, **12**, 350 (1940). Semimicro filtration.

E. L. Skau, *J. Phys. Chem.*, **33**, 951 (1929). Centrifugal filtration.

R. Steinmann, *Mikrochim. Acta*, **1953**, 492–496. Bibliography to 1949 on semimicro and micro methods of filtration and drying.

A. R. Ronzio, *Microchem. J.*, **5**, 19 (1961). Various devices for filtering small amounts of material.

G. F. Wright, *Can. J. Res.*, **B17**, 302 (1939). Semimicro filtration.

G. H. Wyatt, *Analyst*, **71**, 122 (1946). Microfiltration.

H. Yagoda, *Mikrochemie*, **18**, 299 (1935). Microfiltration.

Drying

H. K. Alber, *Mikrochemie*, **25**, 47 (1938). Vacuum drying apparatus.

K. C. Barraclough, *Metallurgia*, **31**, 269 (1945); *C.A.*, 39, 4018 (1945). Desiccants.

G. Broughton, "Solvent Removal, Evaporation, and Drying," in *Technique of Organic Chemistry*, Vol. 3, Part 1, 2nd. ed, A. Weissberger, ed., Interscience, New York, 1956, pp. 787–839. A review with 128 references.

N. D. Cheronis, "Micro and Semimicro Methods," in *Technique of Organic Chemistry*, Vol. 6, A. Weissberger, ed., Interscience, New York, 1954, pp. 50–56. A review with 22 references.

S. Colliander, *Microscope*, **7**, 138 (1949); *C.A.*, 43, 3873 (1949). Silica gel as a dehydrator.

G. Gorbach, *Mikrochemie*, **31**, 116 (1943). Drying apparatus.

F. Hecht, *Mikrochim. Acta*, **3**, 129 (1938). Drying apparatus.

P. F. Holt, *Metallurgia*, **37**, 48 (1947); *C.A.*, **42**, 800 (1948). Drying apparatus.

G. Ingram, *Metallurgia*, **38**, 239 (1948). Drying apparatus.

T. S. Ma and R. T. Schenck, *Mikrochemie*, **40**, 245 (1952). Micro drying oven.

C. R. Noller, *Ind. Eng. Chem., Anal. Ed.*, **14**, 834 (1942). Drying apparatus (heated block).

F. Pavelka, *Mikrochemie*, **32**, 141 (1944); *C.A.*, **41**, 4399 (1947). Micro extraction.

J. D. Reinheimer, *J. Chem. Educ.*, **30**, 139 (1953). A vacuum drying apparatus for small samples.

W. G. Whittleston, *New Zealand J. Sci. Technol.*, **21B**, 162 (1939); *C.A.*, **34**, 2212 (1940). Drying small amounts of heat-sensitive solids.

Evaporating

E. M. Arnett, *J. Chem. Educ.*, **37**, 247 (1960). Rotary evaporator.

G. Broughton, "Solvent Removal, Evaporation, and Drying," in *Technique of Organic Chemistry*, 2nd ed., Vol. 3, Part 1, A. Weissberger, ed., Interscience, New York, 1956, p. 787. A review with 128 references.

M. T. Bush, *Microchem. J.*, **1**, 109 (1957). Small-scale impingement evaporator.

R. O. Carleton, *J. Chem. Educ.*, **39**, 256 (1962). Rotary evaporator.

P. S. Chen, Jr., *Chemist Analyst*, **50**, 84 (1961). Manifold for solvent evaporation.

R. L. Clements, *Chemist Analyst*, **48**, 20 (1959). Small-scale evaporator.

A. F. Colson, *Analyst*, **71**, 322 (1943); *C.A.*, **40**, 5302 (1946). Microchemical apparatus.

J. Donaldson, *Chemist Analyst*, **50**, 54 (1961). Simple air pressure arrangement for solvent evaporation.

J. Donaldson, *Chemist Analyst*, **50**, 53 (1961). An easily constructed evaporating block.

J. Erdös, *Mikrochemie*, **33**, 385 (1948); *C.A.*, **37**, 6169 (1943). Drying and heating a gas.

S. Gaddis, *J. Chem. Educ.*, **20**, 28 (1943); *C.A.*, **37**, 1302 (1943). Semimicro evaporation apparatus.

F. Greef and C. L. Larsen, *J. Chem. Educ.*, **33**, 556 (1956). Rotary evaporator.

L. T. Kurtz, *Ind. Eng. Chem., Anal. Ed.*, **14**, 191 (1942); *C.A.*, **36**, 1809 (1942). Micro evaporation apparatus.

G. H. Perold, *Mikrochim. Acta*, **1959**, 251. Dust-free evaporation of solutions for crystallizations on a small scale.

V. C. Runeckles, *Chemist Analyst*, **50**, 23 (1961). Multiple-unit, all-glass rotary evaporator.

Reaction Vessels

J. A. Beech, *J. Chem. Educ.*, **39**, 293 (1962). Mounting board for small-scale equipment.

N. D. Cheronis, "Micro and Semimicro Methods," in *Technique of Organic Chemistry*, Vol. 6, A. Weissberger, ed., Interscience, New York, 1954, p. 133. A review.

J. M. Connolly and G. Oldham, *J. Chem. Educ.*, **29**, 310 (1952). Equipment for semimicro preparations.

S. Kasman, *J. Chem. Educ.*, **37**, 150 (1960). Vessel for small-scale reactions.

J. B. Polya, *J. Chem. Educ.*, **39,** 294 (1962). H-shaped flask for reactions on small scale.

J. T. Stock and M. A. Fill, *J. Chem. Educ.*, **36,** 194 (1959). Portable reaction assembly.

J. T. Stock and M. A. Fill, *J. Chem. Educ.*, **31,** 144 (1954). Small-scale units for organic identifications.

J. T. Stock and M. A. Fill, *J. Chem. Educ.*, **30,** 296 (1953). Small-scale techniques in the teaching of organic chemistry.

Vacuum Lines

B. D. Dayton, "Vacuum Techniques and Analysis," in *Physical Methods of Analysis*, Vol. 2, W. G. Berl, ed., Academic Press, New York, 1951, p. 334. Theory and practice of vacuum technique.

R. T. Sanderson, *Vacuum Manipulation of Volatile Compounds*, Wiley, New York, 1948.

3

Fractionation Procedures

CRYSTALLIZATION—DISTILLATION—
SUBLIMATION—EXTRACTION

CRYSTALLIZATION

Crystallization procedures are very useful in the characterization of organic compounds, regardless of the purity of the original sample. In the last step of any systematic procedure it is necessary to prepare a *solid derivative*. In practically all cases, even if the reactants are pure, the resulting solid derivative is impure and must be purified because in most cases side reactions give rise to impurities. Consider, for example, the reaction between methanol and 3,5-dinitrobenzoyl chloride. Assume that 1 millimole of the acid halide is heated with an excess (3 millimoles) of the alcohol. The usual procedure is to heat the mixture for 5 to 10 minutes and, then, add water to separate the solid ester. The ester, thus obtained, contains enough 3,5-dinitrobenzoic acid and traces of its anhydride to cause a depression of several degrees in the melting point of the solid ester. The formation of the acid results from (a) reaction with the small amount of water present in the methanol; (b) absorption of moisture by the halide from the walls of the vessel and from the air during weighing; or (c) action of water on the unreacted halide since, in most cases, the conversion of the alcohol to the ester is not complete. It is a common practice to remove the small amount of dinitrobenzoic acid by washing with a dilute solution of sodium hydroxide or sodium carbonate. The rate of solution is not great at ordinary temperatures and, if the mixture is heated to accelerate solution, hydrolysis of the ester may occur, thus forming more dinitrobenzoic acid. Therefore, the crystalline ester is washed and then purified by crystallization from an appropriate solvent.

The term *crystallization* means the deposition of crystals from a solution or a melt of a substance or a mixture of substances. As used in organic chemistry, the process includes dissolving a solid by heating it in an appropriate solvent (such as methanol, water, or benzene), filtering off any undissolved impurity, and causing the separation of crystals by cooling the solution. The steps in any crystallization procedure are (a) selecting the solvent, (b) preparing a nearly saturated hot solution, (c) cooling the solution to induce the formation of crystals, (d) separating the liquid (mother liquor) from the crystals, and (e) drying the crystals.

Selection of Solvents

The choice of the solvent in the crystallization of solids is very important.[1] In some cases the best solvent cannot be selected on the basis of rules or theoretical considerations, but must be experimentally determined; therefore, in this book, the solvent is usually specified for recommended derivatives.

The literature on preparation of derivatives usually indicates the solvent used for recrystallization, but seldom gives solubility data. In such cases, we should proceed cautiously to determine roughly the solubility, as described on page 60. The following general rules with reference to solvents will be useful.

(*1*) A solid usually dissolves best in a liquid that it resembles in structure. For solid esters, such solvents as methanol, ethanol, and ethyl acetate should be among the first tried.

(*2*) It is desirable to select a solvent that will dissolve the crude solid when hot, but only sparingly when cold.

(*3*) If a solvent dissolves a solid very readily in the cold, it may be useful as a crystallizing medium if it is mixed with another solvent in which the compound is sparingly soluble. Thus, if a substance is very soluble in alcohol in the cold and sparingly soluble in water, it is dissolved in a *small amount of hot alcohol*, the solution is filtered, and water is added until cloudiness develops. The solution is then heated until clear, filtered if

---

[1] A. Weissberger and E. S. Proskauer, *Organic Solvents*, 2nd ed., revised by J. A. Riddick and E. E. Toops, Jr., in *Technique of Organic Chemistry*, Vol. 7, A. Weissberger, ed., Interscience, New York, 1955; O. Jordan, *The Technology of Solvents*, Hill, London, 1940; I. Mellan, *Industrial Solvents*, Reinhold, New York, 1957; and J. H. Hildebrand, *Solubility of Nonelectrolytes*, Reinhold, New York, 1950.

necessary, and cooled slowly. The use of *solvent pairs* is very helpful in many crystallizations.

(4) The impurities present should be either very soluble or as sparingly soluble as possible in the solvent. For example, in the case of 3,5-dinitrobenzoates ($C_6H_3(NO_2)_2COOR$) used for the identification of hydroxy compounds, the impurity (dinitrobenzoic acid) is completely soluble in the methanol or ethanol used as a solvent for the crystallization. Another derivative for hydroxy compounds may be prepared by allowing the alcohol to react with an isocyanate, for example, phenyl isocyanate (C_6H_5NCO). The desired derivative is known as a *urethan*, $C_6H_5NHCOOR$, and the impurity is diphenylurea, $C_6H_5NHCONHC_6H_5$, produced by the reaction of the isocyanate with the moisture present in the hydroxy compound or on the walls of the vessel. Hot petroleum ether dissolves the urethan but not the diphenylurea.

(5) The solvent should be chemically inert to the compound to be crystallized. In some cases, however, a solvent may be chosen because it reacts chemically with the compound. For example, some aromatic acids may be purified by being dissolved in dilute sodium hydroxide and then, after filtration of the solution, precipitated by neutralization with dilute hydrochloric or sulfuric acid.

(6) In the event that a solvent is not reported in the literature, it is recommended that the procedure outlined on page 60 be tried with solvents in the following order: methanol or ethanol, mixture of lower alcohols and water, acetone or mixture of acetone and alcohol, benzene or mixtures of benzene and toluene, petroleum ether or benzene and petroleum ether, glacial acetic acid or aqueous acetic acid. Table 3.01 shows a number of solvents and solvent pairs used for crystallization of derivatives. The general procedure for the use of solvent pairs is to dissolve the compound in a solvent in which the derivative is very soluble and, then, to add cautiously a solvent in which it is less soluble. The two solvents must be miscible, and should be used hot.

(7) If several solvents are found to be applicable, additional factors, such as volatility, cost, and availability, should be considered. A higher boiling point is usually an advantage because evaporation is minimal and because there is a larger interval between the solution and crystallization temperatures. This generalization is valid only for solvents boiling between 60° and 150°, since beyond this range the difficulty of removing the solvent and the danger of decomposition of the substance offset the advantages.

Some factual knowledge is useful in lessening the number of solvents to be tried in a given case. Substances with hydroxyl groups are likely to be soluble in methanol or water. Substances soluble in methanol are slightly less soluble in ethanol and even less soluble in higher alcohols.

TABLE 3.01
COMMON SOLVENTS FOR CRYSTALLIZATION OF DERIVATIVES

Derivatives	Solvent or Solvent Pair
Some carboxylic acids, amides, and substituted amides	Water
Most derivatives: benzoates, 3,5-dinitrobenzoates, amides, *p*-toluidides, nitro and bromo compounds, etc.	Methanol
p-Nitrobenzyl esters, sulfonamides, anilides, picrates, semicarbazones, hydrazones, substituted hydrazones, etc.	Methanol-water
Same as for methanol and methanol-water mixtures, molecular complexes	Ethanol
Xanthylamides	Dioxane-water
Phenylurethans, α-naphthylurethans	Petroleum ether[a]
p-Nitrophenylurethans, 3,5-dinitrophenylurethans	Petroleum ether-benzene
Osazones, bromo compounds, nitro compounds	Acetone-alcohol
Quaternary ammonium salts	Isopropyl ether
Quaternary ammonium salts, esters	Ethyl acetate
Picrates, molecular complexes	Benzene
Sulfonyl chlorides, acid chlorides, anhydrides	Chloroform and carbon tetrachloride

[a] The original fractions of petroleum distillates such as ligroin or petroleum ether (bp 30 to 60°, 60 to 80°, or 90 to 110°, etc.) have been displaced in recent years by the commercial availability of hydrocarbons, such as pentane, hexane, and heptane, with boiling ranges within 5 to 10° from the boiling points of the pure hydrocarbons.

Polyhydroxy compounds may be more soluble in water than in methanol. Alcohol-water is usually a good solvent pair for compounds containing hydroxyl groups. Substances with aromatic structures are likely to be soluble in benzene, toluene, and ether. Substances soluble in aliphatic hydrocarbons (ligroin, petroleum ether, and the like) have low solubilities in water. Heterocyclic compounds, soluble in alcohols, may often be precipitated by the addition of ether. When a substance is found to be sparingly soluble in the common organic solvents, glacial acetic acid or pyridine should be tried.

These facts depend upon the principles of hydrogen bond formation and become reasonable when these principles are understood. Hydrogen bonds are formed between small electronegative atoms through hydrogen atoms. Although the hydrogen atom is formally bound to one of the electronegative atoms by an ordinary chemical bond, it is attracted by the other electronegative atom. Thus a bond with an energy of about

$$—O—H \text{ - - - } O—$$

5 kcal/mole holds the electronegative atoms together. Since covalent chemical bonds usually have energies of 50 to 100 kcal/mole, it can be seen that the hydrogen bond is a much weaker bond than these.

Although the hydrogen bond is relatively weak, it has a great influence on the physical properties of compounds. Compounds such as alcohols can form hydrogen bonds between molecules as follows.

$$R—CH_2—O—H \text{ - - - } O—CH_2—R$$
$$|$$
$$H$$

Carboxylic acids can form hydrogen bonds between two molecules:

$$\begin{array}{c} O \text{ - - - } H—O \\ R—C \diagup \qquad \diagdown C—R \\ O—H \text{ - - - } O \end{array}$$

The result in each case is that the molecules are somewhat associated, and are not independent of one another. In the case of carboxylic acids, the molecular weight, determined in an inert solvent such as benzene, is almost exactly twice the formula weight.

This association between molecules results in higher boiling points than would be expected on the basis of molecular size and shape alone. This will be discussed in more detail in the section on physical properties in Chapter 5. Association of molecules also results in higher melting points, density, and viscosity.

When two substances are mixed, the hydrogen bonds of the solute, the solvent, and the solute-solvent pair should be considered. If on mixing two substances, more hydrogen bonds are formed than are broken, the substances will likely be very soluble. On the other hand, if the net result during mixing of the two is the breaking of hydrogen bonds, the solubility will be low. If there is no change in the number of hydrogen bonds as the two compounds come together, the compounds will be mutually soluble.

For example, compounds with hydroxyl groups will be fairly soluble in methanol and in ethanol, since these compounds can form hydrogen bonds with the alcohols. Water itself is so thoroughly hydrogen-bonded that only small molecules capable of forming hydrogen bonds will be highly soluble in it. However, polyfunctional molecules, such as glycols, glycerol, dibasic acids, hydroxy acids, and amino acids, will dissolve in water, since these molecules have great ability to form hydrogen bonds. The mixtures have about the same number of hydrogen bonds as the original compounds themselves. Monocarboxylic acids with more than six carbon atoms are not so soluble in water, but are fairly soluble in ethanol.

Ketones, aldehydes, esters, and tertiary amines are not associated by hydrogen bonding because none of these compounds has a hydrogen atom attached to an electronegative atom. However, they are soluble in alcohols because they have the possibility of forming hydrogen bonds with compounds having active hydrogen atoms. These compounds are also soluble in most other compounds, which are not associated by hydrogen bonding; there is no change in the amount of hydrogen bonding on mixing. For example, 2-butanone is highly soluble in benzene, and N,N-dimethylaniline is very soluble in carbon tetrachloride.

A good solvent for recrystallization is one in which the compound has little solubility at room temperature or below. Consequently, many monocarboxylic acids may be crystallized from water in which they have low solubility. On the other hand, most polyfunctional compounds, such as dibasic acids, hydroxy acids, and amino acids, are too soluble in water, but may be crystallized from ethanol or an ethanol mixture. Many ketones and aldehydes are difficult to crystallize because they are too soluble in most solvents. However, hydrocarbons are useful solvents here: aliphatic hydrocarbons for the aromatic compounds and aromatic hydrocarbons for the aliphatic compounds.

These principles are discussed in more detail in an article by Ewell et al.,[2] which deals primarily with the formation of azeotropic mixtures. At best, these principles are guides to experimental procedures. The predictions should be checked by using small amounts of material before committing a sizable portion of a solid to be recrystallized. If the compound turns out to be more soluble than expected, it will require extra time and trouble to obtain a pure product.

[2] R. H. Ewell, J. M. Harrison, and L. Berg, *Ind. Eng. Chem.*, **36**, 871–875 (1944).

Determination of Solubility

When it is impossible to determine from the literature what solvent to use, or if the solvent is known but not the amount to be used, the solubility of the unknown substance or derivative should be determined roughly, both at room temperature and at the boiling point of the solvent. About 10 to 50 mg of the solid is placed in a 3-in. test tube (preferably with a tapered end) and, by means of a graduated pipet dropper, 0.1 to 0.5 ml of the solvent is added and the tube is shaken to bring the solvent in contact with all the solid in the tube and then allowed to stand 1 to 2 minutes. The ratio of solid to solvent in the initial test should be about 1 : 10. If, at this ratio, the solid dissolves completely at room temperature, the solvent is not useful for crystallization unless it is miscible with another solvent in which the solute is sparingly soluble. This can be determined by a separate test in which the same quantities of solute and first solvent are used and between one fifth and one tenth as much of the second solvent is added.

If the solid does not dissolve, successive portions of 0.1 to 0.5 ml of solvent may be added, the tube being shaken after the addition of each portion, until the solid dissolves. The number of milliliters of solvent used, divided into the number of milligrams of solid taken for the solubility test, gives the number of milligrams of solid dissolved by each milliliter of solvent at room temperature.

The solubility of the solid at or near the boiling point of the solvent is determined by a similar procedure. If, for example, the solubility tests are performed on 50 mg of the sample, after the addition of the first 0.5 ml of solvent, the tube is cautiously heated until the solvent just begins to boil. If the solid dissolves completely, the test is repeated, 0.3 ml of solvent being used. If some of the solid remains undissolved, more solvent is added in portions of 0.1 to 0.2 ml, the tube being heated after the addition of each portion.

In this manner the solubility of the solid at or near the boiling point of the solvent and also at room temperature is determined. This information is used to choose the solvent and the amount to be used. Let it be assumed, for example, that 8 ml of methanol was used to dissolve 50 mg of a derivative at 20°, and 0.9 ml at the boiling point of alcohol. The solubility, therefore, at 20° is about 6 mg per milliliter and, at or near the boiling point of methanol, 55 mg per milliliter. If the amount of derivative to be crystallized is 200 mg, the amount of methanol to be used is 4 to 4.5 ml. After crystallization, about 25 mg of the derivative will remain in the mother liquor. Generally, the solvent is not particularly useful unless the solubility near the boiling point of the solvent is *at least 5 times the solubility at room temperature*. If, in the above example, the solubility at room temperature had been 12 mg per milliliter, the loss would have been at least 50 mg in the first crystallization.

The solvent used most extensively in the experimental part of this work is methanol. Wherever possible, the crude derivative is dissolved in methanol and precipitated after filtration by cautious addition of water. Although methanol is toxic when absorbed in the tissues in appreciable quantities, the handling of small amounts has been found entirely safe—provided that caution is used. A number of other factors make the use of methanol more desirable than ethanol. In general, the solubilities of organic compounds are not greatly different in the two homologs. The commercial grade of methanol is almost anhydrous and more pure than the commercial grade of ethanol. In addition, the ease with which methanol is obtained in the market, plus its low price, makes its use desirable wherever possible.

Preparation of Solutions for Crystallization

Figure 3.01 shows four types of test tubes that have been found most useful in the preparation of solutions for the crystallization of solid organic substances. In each case both the size of the tube and the range of quantities for which it is suitable are given. For example, in the preparation and purification of 100–500 mg of derivatives, 8-inch tubes are used throughout. If the amount of solvent to be used in the preparation of the nearly saturated hot solution is less than 5 ml, a 6-in. tube (20 × 150 mm) is preferable. For amounts of 10 to 30 ml, 8-in. tubes (25 × 200 mm) are used. Small flasks of 25 to 50 ml capacity may be used, but they cannot be cleaned so easily as test tubes. If flasks are used, preference is given to the pear-shape type, particularly when the amount of solution is small.

When the amount of solid to be crystallized is about 50 mg or less, the 3-in. tapered tube shown in Figure 3.01E is used with a special procedure for filtration and transfer of crystals described on page 70. The compound to be crystallized is placed on a piece of shiny paper (50 to 60 mm in width and 75 to 100 mm in length), creased at the middle along about one half its length. The paper containing the crystals is held at the edges by the thumb and index finger, and is tilted so that the crystals roll through the creased end into the mouth of the tube. By means of a pipet dropper (see Figure 2.04) the appropriate amount of hot solvent is added to the upper part of the tube in order to wash down the crystals adhering to the sides of the vessel. The solid is crushed with a rod against the walls of the vessel to help dissolve it. The solvent may be heated in another vessel and added by means of an ordinary dropper as outlined on page 12. In this manner, the amount of solvent added is controlled. The tube is inclined at an angle

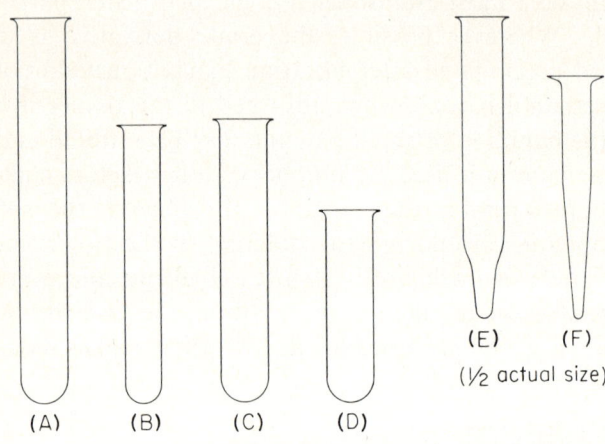

FIGURE 3.01

(*A*) Tube 25 × 200 mm (8 in.) employed in crystallization of 100 to 500 mg or more of organic compounds. The 8-in. tube with a side arm is employed as a filter vessel in most semimicro crystallizations (see Figure 3.03). (*B*) and (*C*) Tubes 18 × 150 mm and 20 × 150 mm (common 6-in. tubes) used for crystallization of 50 to 100 mg of material. (*D*) Tubes 25 × 100 mm, often employed as reaction vessel. (*E*) and (*F*) Tubes 15 × 75 mm (3-in.) with tapered end employed for crystallization of milligram quantities of material and recommended for all types of tests. The cone shown at (*F*) is available in 1- and 2-ml capacities.

of 65 to 75°, with the mouth directed away, and is moved gently over a small flame of a microburner or over a hot plate. Since most organic solvents are flammable, great care should be exercised if a flame is used for heating. We have found that in dealing with small amounts of solvent it is possible, with care, to avoid accidents. The vessel should never be more than one third full; the flame should be directed, first, at the top of the liquid layer and, then, slowly moved downward. If complete solution of the solid is not effected when the liquid just begins to boil, more solvent is added. Finally, when the solid dissolves slowly or the amount of solvent exceeds 10 ml, the apparatus shown in Figure 2.28 is used. This arrangement provides for effecting solution by heating under reflux.

A compact apparatus[3] for dissolving a small amount of solid in a heated liquid and filtering the hot solution is shown in Figure 3.02. The bulb may be of any capacity, but the most convenient size is 1 to 5 ml.

The solid to be dissolved is placed in the bulb through the long neck, the solvent is added, and solution effected by warming the bulb. To filter the solution from undissolved impurities or from charcoal added to decolorize the solution, the apparatus is turned so that the long stem points downward. A rubber bulb is joined to the standard-taper joint, air is pumped into the bulb, and the solution is forced through the fritted

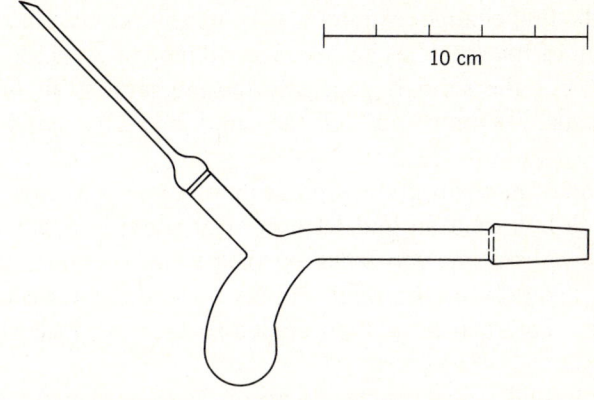

FIGURE 3.02

Apparatus for preparing and filtering a hot solution for crystallization.[3]

glass disc into a convenient vessel. Since handling is at a minimum, little of the solution is lost and, consequently, more concentrated solutions can be used. More volatile solvents can be used, since pressure is used for filtration rather than suction.

Filtration of the Hot Solution and Formation of Crystals

The hot solution for crystallization should be saturated (or nearly so) at a temperature 10° below the boiling point of the solvent. If the solution is saturated at the boiling point of the solvent, crystallization will occur during filtration; the filter will become clogged and the troublesome transfer and preparation of the hot solution must be done over again. Crystallization during filtration may sometimes be prevented by preheating the filtering apparatus. If the hot solution is free from suspended or insoluble particles or colored impurities, it may be cooled directly without filtration. For the removal of colored impurities, charcoal, diatomaceous earth, and silica gel are recommended for microcrystallization. The amount of charcoal to be added for 100 to 200 mg of solid should not exceed 20 mg.

[3] M. Martin-Smith, *Lab. Practice*, **7**, 572–574 (1958).

The separation of crystals from hot solutions is commonly induced by cooling and stirring. If crystallization does not occur under these conditions, the sides of the vessel are sometimes scratched with a glass rod to start crystallization. An alternate method is to stopper the vessel containing the hot solution and to shake it vigorously as the solution cools. The first evidence of crystallization is usually the formation of minute crystalline particles (nucleates), which then grow into crystals.[4] However, sometimes, the first change seen in the cooling solution is a haze caused by the separation of the solute as an oil. The oil may later solidify at a lower temperature, but the solid is generally impure, and is in lumps rather than in crystals. "Oiling out" of the solid must be avoided for best results.

One method of preventing the separation of an oil is to cover the beaker containing the hot solution and let it cool to room temperature without stirring. The walls of the container are then rubbed with a stirring rod to induce crystallization, or the solution may be seeded with crystals if they are available. The mixture is then cooled further to cause more crystals to form.

If the compound still separates as an oil in spite of these precautions, the solution should be warmed until the cloudiness disappears, and more solvent added. If the addition of more solvent is not effective in preventing oiling, it may be necessary to evaporate the liquid and try another solvent.

Another successful method of treating solutions that have "oiled out" is to separate the supernatant liquid from the oil, add fresh solvent, and reheat to dissolve the oil. The process may have to be repeated two or three times to obtain crystals. "Oiling out" of alcohol-water solutions is usually due to the addition of too much water to the solution.

The most important problem encountered in the course of crystallization procedure is the elimination of impurities, the type and amount of which influence the shape and size of the crystals and, usually, the rate of crystallization as well. Some impurities, particularly when present in minute amounts, may be removed by treatment of the hot solution with such adsorbing media as activated carbon or diatomaceous earths (Filter-cel and the like). The bulk of the impurities, however, remains in the solution. Therefore, conditions must be chosen that allow the formation of pure

---

[4] R. S. Tipson, "Crystallization and Recrystallization" in *Technique of Organic Chemistry*, Vol. 3, Part 1, 2nd ed., A. Weissberger, ed., Interscience New York, 1956, pp. 395–562; R. S. Tipson, *Anal. Chem.*, **22,** 628–636 (1950); N. D. Cheronis, "Micro and Semimicro Methods" in *Technique of Organic Chemistry*, Vol. 6, A. Weissberger, ed., Interscience, New York, 1954, pp. 14–17.

crystals and their separation from the mother liquor without adherent impurities and with minimum loss of the substance being crystallized.

In the following sections, three procedures and apparatuses are described in detail. The first is to be used when the quantity being crystallized is 50 mg or more; the second when the quantity is 10 to 50 mg; and the third when the quantity is less than 10 mg. The principles underlying all three procedures are the same, for the essential objective in each case is the separation of the crystals from the mother liquor.

The primary difference between the first two is in the size of the apparatus; both employ filtration with suction. As is usual, the removal of the mother liquor by filtration with suction is followed by washing of the crystals with solvent. Although the practice of removing the crystals from the filter and suspending them in the solvent is generally regarded as more desirable than washing the crystals on the filter, the latter procedure is preferable in dealing with microcrystallizations in order to avoid transfers. Therefore, conditions must be chosen that will insure removal of the adhering mother liquor. Since the quantity of crystals is usually small, channeling of the solvent does not occur readily. Moreover, if several small portions of solvent are added to the crystals without any suction being applied to the filter and the crystals are stirred, removal of the mother liquor is complete.

When the quantity being crystallized is less than 10 mg, a radical change in procedure as well as apparatus is required. Centrifugation is employed, rather than filtration, both for the separation of the hot clear solution from the solid impurities and added adsorbents and for the separation and washing of the crystals. In addition, to minimize losses, the crystals are dried without transfer to a drying plate, paper disc, or watch glass.

Regardless of the procedure used, the steps in the sequence of purification are the same: solution, crystallization, removal of the mother liquor, removal of a small quantity for determination of the melting point, and repetition of the cycle until samples from two successive crystallizations exhibit essentially the same melting temperature (within 0.5°).

Quantities of 50 mg or More

The hot solution, prepared according to the directions given in the preceding section, is filtered by means of an apparatus such as that shown in Figure 3.03. The funnel[5] has a diameter of 50 mm at the top, and its stem is 55 mm long. The inside of the funnel is slightly etched or ground so as to provide a firm seat for a porcelain perforated disc,[5] 20 mm in diameter

---

[5] Wilkens-Anderson Co., Chicago, Ill.

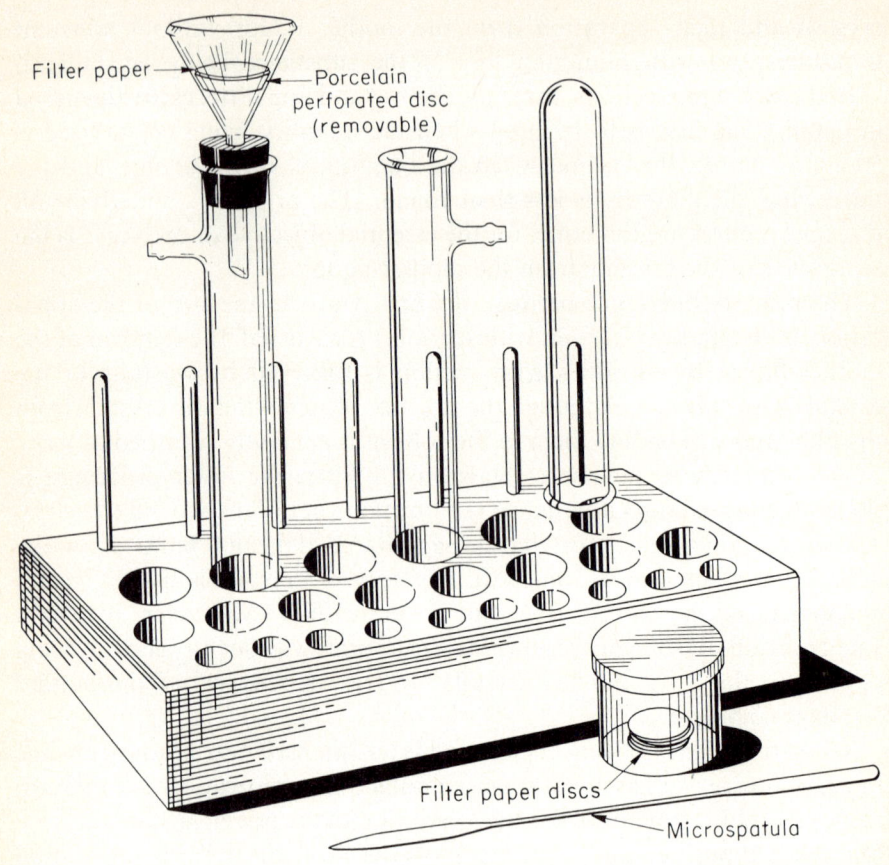

FIGURE 3.03

Apparatus for semimicro filtration (consists of a glass funnel with removable filter disc and an 8-in. tube).

and 5 mm thick, which fits inside the funnel against the ground surface. The edges of the disc are beveled so that the bottom diameter is about 15 mm (Figure 3.03). The stem of the funnel is inserted through a one-hole No. 4 rubber stopper, which is fitted into the mouth of an 8-in. tube having a side arm, and extends 10 to 12 mm below the side arm. This arm is connected to a rubber hose leading to an ordinary water aspirator (in the absence of a water aspirator, it is connected to a rubber aspirator bulb).

To prepare the funnel for filtration, the tube is placed on a test-tube rack or clamped to a small stand. The perforated porcelain disc is placed inside the funnel and arranged in place by a slight pressure of the finger. A disc of filter paper 24 to 25 mm in diameter is placed on top of the porcelain plate; by means of a pipet dropper, two drops of water are placed on two different parts of the surface of the paper and the funnel tilted slightly so that the filter paper is moistened throughout. The funnel is fitted into the mouth of the receiving tube which, in turn, is connected to the aspirator. The filter paper is sucked down into place by gentle suction. If the filter paper is properly placed, the edges are held tightly against the funnel and extend upward. If any part of the paper protrudes downward, a leak will develop during filtration; in such a case, the suction is discontinued, another drop of water is added, and the filter paper is adjusted with a spatula. After the filter paper has been made to adhere to the sides of the funnel, the suction is momentarily discontinued, 2 drops of methanol are added to the paper, and suction is applied again. The receiving tube is changed, and the funnel is ready for filtration if the solvent is an alcohol, ether, ester, or an organic acid. When a hydrocarbon (benzene, heptane, or the like) is used as a solvent, the moistening of the paper by water is followed by washing with acetone and then with 5 to 6 drops of the hydrocarbon; then, suction is applied.

The funnel, prepared as described in the preceding paragraph, is fitted into the mouth of a clean tube, and gentle suction is applied through the side arm. The tube containing the hot solution is held in one hand and its mouth is lowered over the funnel so that the solution is poured through the center of the disc. It is advisable to use a glass rod in pouring the hot solution into the funnel. The tube containing the solution is lowered so that it just touches a rod held vertically with one end touching the filter disc. The filtration is completed within a few seconds. If the amount of the solid being crystallized is very small, 0.5 ml of fresh solvent is added to the tube from which the solution was poured out and heated until all the crystalline solid adhering to the sides is dissolved; the hot solvent is then poured into the filter so as to wash down any solid adhering to it.

The funnel is removed, cleaned immediately, and set aside, upside down, to drain and be ready for the next filtration. The tube containing the filtered hot solution is immersed in a beaker or a small jar in which running tap water circulates; the solution is stirred with a glass rod from time to time. For most derivatives 5 to 10 minutes of cooling is sufficient; others require 15 minutes, or more, for complete crystallization.

If a mixed solvent, such as an alcohol (methanol or ethanol) and water, is used, it is advisable to effect solution in alcohol and, after filtration, to add warm water dropwise by means of the pipet dropper until a permanent cloudiness is obtained on shaking. The tube is heated until the cloudiness disappears and is then cooled slowly.

The crystals and mother liquor are thoroughly stirred by means of the rod so as to loosen most of the solid adhering to the walls of the tube. A clean 8-in. tube with side arm is fitted with the filter funnel prepared as described above. The tube is shaken two or three times, and the contents are poured into the funnel. The mother liquor is poured back into the crystallizing tube, and the process is repeated until practically all the adhering crystals have been transferred into the funnel. The draining is complete within a minute or two. About 0.5 to 1 ml of solvent is added to the tube in which the crystallization took place, and the tube is shaken so

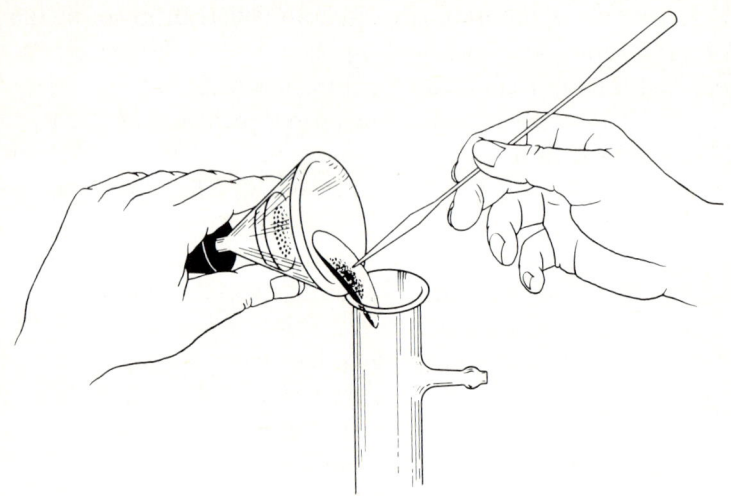

FIGURE 3.04

Transfer of filter paper and crystals directly into the solution vessel.

that washing of the adhering crystals takes place, since the tube is to be used directly for the second crystallization. The amount of solvent added should be insufficient to dissolve the crystals remaining in the tube. The suction is discontinued, and the washings are added slowly over the crystals in the filter so that the entire mass is moistened; after a minute, the washing is repeated. The filtrates are saved until the crystallization is complete, and the melting point determined. A small amount (5 to 10 mg) of the crystals is removed from the filter and dried, as directed on pages 77–79, and labeled as a sample from the first crystallization. The balance is then used in the next crystallization.

To perform a recrystallization, loosen with the spatula the filter paper containing the crystals and transfer it directly into the 8-in. tube in which the previous crystallization took place (Figure 3.04). If this manipulation is difficult for the beginner, the funnel may be placed upside down on a small piece of clean paper (Figure 3.05) and, by a slight pressure with the spatula, the porcelain disc and the crystals made to fall on the paper. The porcelain disc is pushed away with the spatula, and the crystals, together with the filter paper, are transferred into the crystallizing tube.

The amount of solvent to be added varies between 50 and 80 per cent of the amount used in the first crystallization. Assume that 7 ml of methanol were used in the first crystallization. Since a certain amount of the solid remains in the mother liquor, depending on the solubility of the derivative, it is advisable to begin with 3 to 5 ml of methanol for the second crystallization. The tube is heated until the solvent begins to boil; by means of a

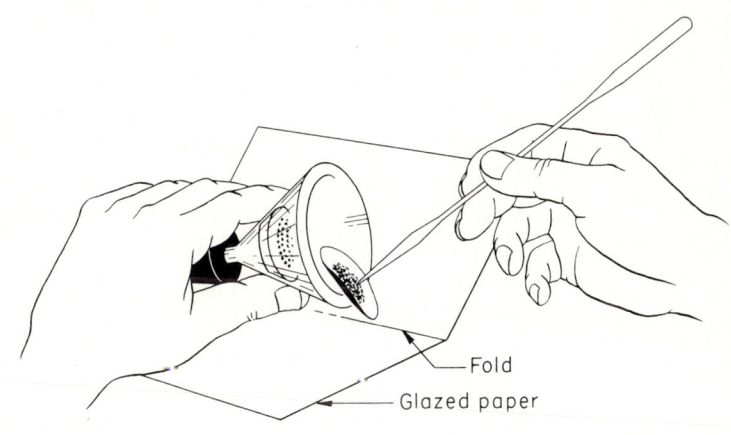

FIGURE 3.05
Transfer of filter paper and crystals on glazed paper.

spatula or a glass rod, the filter paper is pulled up on the sides about 50 mm from the bottom and the heating is resumed, so that the hot vapor condenses on the region of the filter paper and washes down any adhering solid. The filter paper is pulled out and discarded. Additional solvent is added until the solid dissolves at a temperature near the boiling point of the solvent. If proper care has been taken, the solution will be clear and without shreds of paper, so that filtration of the solution may not be necessary; if the solution is not clear, it should be filtered by the same procedure as before. The cooling of the hot solution and filtration of the crystals are accomplished in the manner already described. A sample of the crystals (5 mg) is removed, dried as directed on page 77, and properly labeled. The bulk of the crystals from the second crystallization is saved until the identification of the compound is complete.

When two crystallizations have been completed, the melting point of the dry crystals is determined. If the melting points of the crystals from two successive crystallizations are identical, or within 0.5° of each other, the derivative may be regarded as pure. If there is an appreciable difference (2° or more) between the melting points of the first two crystallizations, a third crystallization is performed. It is possible to perform three or four recrystallizations (retaining a 5-mg sample from each) in a period of one hour.

Quantities of 10 to 50 mg

A diagram of the apparatus[6] employed is shown in Figure 3.06. The tube with the side arm is 18 × 150 mm. A capillary tube about 90 mm in length, 8 mm OD and 1 mm ID, tapered to a 6-mm ground joint at one end, fits into the receiver through a rubber stopper. The funnel has a diameter of about 15 mm at the upper portion and 10 mm at the lower, which is ground to fit over the capillary; the over-all length of the funnel is about 60 to 65 mm. For filtration, a small disc of filter paper 5 mm in diameter, cut with a cork borer, is placed on the ground capillary, and the funnel is fitted over it.

The hot solution is prepared in a 3-in. tube. The funnel may be warmed by boiling some of the pure solvent and, then, lifting the tube so that the siphon tube reaches the bottom of the 3-in. tube, as shown in Figure 3.06, while mild suction is applied at the side arm. The tube with the side arm is emptied; the hot solution is brought near the boiling point, and then filtered. Crystallization is allowed to take place in the receiver and, when the crystals are ready to be filtered, the rubber stopper holding the capillary funnel is fitted over another receiver. A new siphon tube, having a downward arm of about 160 mm, is fitted over the top of the funnel and the tube containing the crystals is raised under the side arm of the siphon while mild suction is applied. A few drops of the solvent are added along the sides of the tube so as to wash the last adhering crystals and, then, the washings are passed through the siphon system. Finally, the siphon tube is removed,

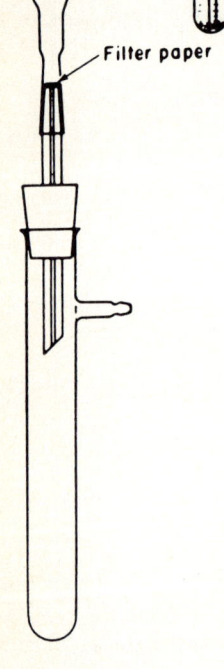

FIGURE 3.06
The Ma apparatus for microcrystallization.

[6] Micro-Ware, Inc., Vineland, N.J.

the suction is discontinued, and the crystals are washed by the addition of 0.1 to 0.2 ml of the pure solvent. After 2 to 3 minutes during which suction is applied, when the crystals have been washed and thoroughly drained, the rubber stopper is removed. If no further crystallization is to be performed, the crystals may be dried and the filter paper then removed; this procedure eliminates the necessity of scraping the filter paper fibers.

If further recrystallization of the crystals is desired, the funnel is carefully inverted over a 3-in. tube and the capillary part removed by a gentle rotary motion and placed on a clean watch glass. By means of a microspatula, the crystals are pushed into the tube. A small sample (2 to 3 mg) is saved for melting-point determination. By means of a capillary pipet, the required amount of solvent is added through the funnel so that all adhering crystals are washed down into the 3-in. tube containing the bulk of the crystals. The cycle of solution, filtration of the hot solution, crystallization, and filtration of the crystals is then repeated.

Quantities of 1 to 10 mg

An extensive discussion of microcrystallization is given in another work by one of us.[7] For the present purpose it will suffice to describe a procedure that has been found very useful in the crystallization of a few milligrams.

The apparatuses required are (a) several pipets with long capillary tips (if they are not available, they can readily be constructed as described on page 14); (b) at least two 3-in. tubes tapered at the bottom;[8] (c) a hand centrifuge.[9] It has been found that a hand centrifuge is better suited for the separation of crystals than electrically operated centrifuges that were tried. The reason for the difference appears to be that the hand centrifuge can be stopped suddenly, and the braking forces the solid particles to the bottom of the tube; in the electrically operated centrifuges there is a gradual deceleration, which results in the formation of a loose cake of solid particles at the bottom of the tube with a tendency of the crystals to rise. Whatever the explanation, there is a definite advantage in employing a hand-operated centrifuge (Figure 3.07). Aside from the fact that it costs about one half as much as the electric one, and also lasts longer, its main advantage is that a separation of clear supernate is obtained in about 10 to 20 sec, while, in the electrically operated centrifuge, the interval between placing and removing the tube is usually 45 to 60 sec. In the

---

[7] N. D. Cheronis, "Micro and Semimicro Methods," in *Technique of Organic Chemistry*, Vol. 6, A. Weissberger, ed., Interscience, New York, 1954, pp. 37–50.
[8] Corning Glass Works, Corning, N.Y.; available from most laboratory supply houses.
[9] A. H. Thomas Co., Philadelphia, Pa.

filtration of crystals, this time factor is not important; however, in the clarification of the hot solution and its transfer into another tube by means of the capillary pipet, the time interval between the removal from the flame of the tube containing the solution and the transfer of the clear supernate into another tube is of fundamental importance. This interval should be as short as possible; otherwise, crystallization will occur in the capillary pipet during transfer, and a loss of material will result.

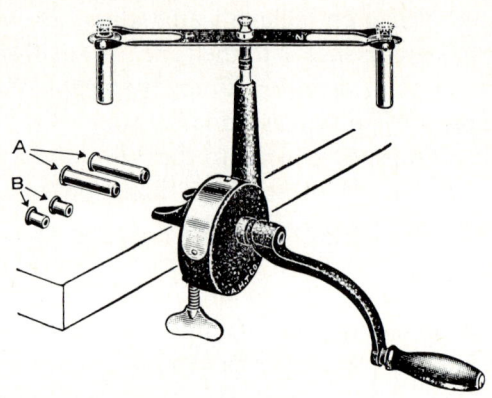

FIGURE 3.07
Hand centrifuge. (Courtesy A. H. Thomas Co.)

It should be emphasized that in all microcrystallization procedures the heat needed to effect solution should be applied slowly by means of a microflame or micro hot plate (Figure 2.13), and transfer from one vessel to another—except as directed—should be avoided. The solid to be crystallized is transferred to the vessel with glazed paper creased at one end and a microspatula. The vessel is rotated while half of the required amount of solvent is added slowly with a capillary graduated pipet whose tip is placed on the upper inner lip of the vessel. It is then stoppered (with a small cork) and centrifuged for 10 to 20 sec, after which the balance of the solvent is added, as before, and centrifuging is repeated.

Heat is applied by moving the tube cautiously back and forth over a small microflame or over a micro hot plate; a temperature near the boiling point of the solvent should be reached only after several minutes of heating. Boiling should be avoided, particularly with the 6-mm cone, owing to the danger of bumping. If solid particles are present when the solution has reached the boiling point, additional solvent is added, and the tube is reheated and re-examined. If all the solid has not dissolved, the solution is reheated almost to boiling, centrifuged for 10 to 20 sec, and the hot solution is transferred rapidly to another microtube.

If the solution is not saturated with solute, it may be transferred very easily and effectively by pouring into a second tube. If a little care is taken, the solid will remain in the first tube while practically all the solution will be transferred to the second. If the volume of the solution is 1 ml or less, the transfer is best made as shown in Figure 3.08 with a micropipet, which has been preheated over a microflame.

If the solution cannot be freed from suspended particles by centrifuging, or if it is desired to remove colored impurities, minute amounts of charcoal and Filter-cel are added; the solution is stirred and heated nearly to

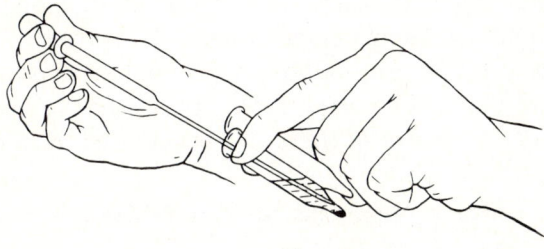

FIGURE 3.08

Transfer of hot, clear supernate in crystallization of milligram quantities.

boiling and centrifuged for about 15 sec and, then, the clear supernate is transferred rapidly to another 3-in. tube with a preheated transfer capillary pipet. The tube containing the solution is inclined at an angle, and the bulb of the pipet is pressed and, then, most of the glass part is passed over a microflame two to three times. The tip of the capillary end is inserted into the clear supernate until it reaches to about 3 mm above the compact residue at the bottom. The pressure on the bulb of the pipet is released so that the solution rises into it; then the pipet is inserted into a clean, dry 3-in. tube and the solution expelled into it. This process is repeated until all but a few droplets remain above the compact residue. The tube containing the clear solution is set aside to cool. When crystallization is complete, the crystals are gently loosened with the sharp edge of the microspatula, the tube is centrifuged, and the mother liquor is drawn off into another tube. To wash the crystals, a small amount of solvent is added, and the crystals are separated as described above; the washing is repeated twice, after which a small sample is removed and dried for a melting-point determination. The crystals are most easily dried under reduced pressure by placing the 3-in. tube into an 8-in. tube with a side arm and applying suction at room temperature. Heat can be applied by means of a water bath if the compound is stable. In the case of milligram quantities of material, the crystals should be dried without transfer, in the same tube in which the crystallization has been completed.

The crystallization of a few crystals of a solid can be done on a microscope slide.[10] To do this, a solvent should be used that is relatively nonvolatile and in which the solid has low solubility. A drop of the solvent is placed on the slide, a few crystals of the solid are added, and the mixture is stirred with a small rod to avoid spreading the drop of liquid. The slide is heated carefully with the flame of a microburner until most of the solid dissolves. More solid is added if necessary. The clear solution is drawn off with a capillary pipet, transferred to another location on the plate, and the plate is cooled to room temperature. The solvent is removed by touching the drop with a piece of filter paper, and the crystals are pressed dry with another piece of filter paper. The crystals may be examined under a microscope, or one or two of them may be removed for a melting-point determination.

Other Apparatus for Crystallization

The Cheronis Universal glass apparatus, described on page 35, permits solution, as well as filtration of the hot solution and crystals, with a minimum number of transfers (Figure 2.28).

The preparation of the nearly saturated solution is made in the arrangement shown in Figure 2.28(*A*). When the solution is ready to be filtered, the condenser is removed and the sleeve is inserted into the flask. The sintered glass funnel is fitted into an empty dry flask and a piece of filter paper is fitted on top of the sintered glass. This precaution is advisable when filtering solutions containing charcoal, since particles of carbon penetrate within the sintered glass and are difficult to remove. The funnel-flask arrangement is fitted over the sleeve, the whole apparatus is inverted as shown in Figure 2.28(*E*), and suction is applied at the side arm of the funnel. When filtration is complete, the lower flask is removed and set aside to cool while the other pieces of the arrangement are removed and immediately cleaned. After crystallization is complete, the crystals are filtered by the same procedure employed in the filtration of the hot, near-saturated solution.

The traditional filter flask-Buchner funnel arrangement may be employed for the crystallization of semimicro quantities by using small sizes of these

---

[10] N. D. Cheronis, "Micro and Semimicro Methods," in *Technique of Organic Chemistry*, Vol. 6, A. Weissberger, ed., Interscience, New York, 1954, p. 44.

apparatuses, the Buchner funnel being 15 mm in diameter at the bottom, 35 mm at the top, and 10 mm in depth[11] (Figure 3.09). In our opinion, this apparatus is not so convenient for filtering and cleaning as the arrangement with the porcelain disc, and is also more expensive.

A filter-tube arrangement described by Wexler[12] and shown at the left in Figure 3.10 is readily constructed from ordinary 6- or 8-in. tubes. Figure 3.10 (*right*) shows a filter-stick arrangement; filter sticks may be either made from glass tubing[13] or obtained commercially. A number of selected references on other types of filtration apparatus will be found at the end of the chapter. For a more complete discussion the reader is referred to the literature.[14]

The filtration process can be speeded by the use of centrifugal force, and apparatus which use this principle have been described by several investigators.[15] One of these,[15(a)] shown in Figure 3.11(*a*), can filter 10 ml of an aqueous solution in less than 1 minute through a coarse or medium porosity sintered plate at 1500 rpm, and the holdup is less than 50 mg. The apparatus in Figure 3.11(*b*) can be used for filtering hot solutions, for which the first apparatus is not suited.

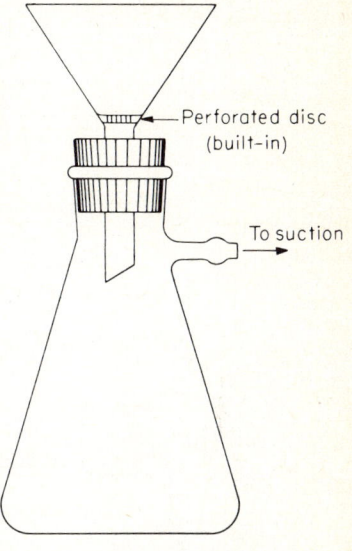

FIGURE 3.09
Filter flask with Hirsch funnel.

Drying of Crystals

To dry the crystals rapidly for a melting-point determination, the washed crystals are pushed with the edge of a paper square to form a small compact

---

[11] Manufactured by Coors Porcelain Company; available through most laboratory supply dealers.
[12] A. Wexler, *J. Chem. Educ.*, **18**, 167 (1941).
[13] A. A. Benedetti-Pichler, *Introduction to the Microtechnique of Inorganic Analysis*, Wiley, New York, 1942, pp. 202–204.
[14] F. Emich and F. L. Schneider, *Microchemical Laboratory Manual*, Wiley, New York, 1932, pp. 31–32 and 127–129; F. Pregl, *Mikrochemie*, **2**, 76 (1924); F. L. Schneider, *Organic Qualitative Microanalysis*, Wiley, New York, 1946, p. 44; N. D. Cheronis, "Micro and Semimicro Methods," in *Technique of Organic Chemistry*, Vol. 6, A. Weissberger, ed., Interscience, New York, 1954, pp. 13–50.
[15] (*a*) M. T. Bush, *Microchem. J.*, **1**, 105–108 (1957); (*b*) E. L. Skau, *J. Phys. Chem.*, **33**, 951 (1929); (*c*) L. C. Craig and O. W. Post, *Ind. Eng. Chem., Anal. Ed.*, **16**, 413 (1944); (*d*) J. English, Jr., *Ind. Eng. Chem., Anal. Ed.*, **16**, 478 (1944); and (*e*) M. T. Bush, *Ind. Eng. Chem., Anal. Ed.*, **18**, 584 (1946).

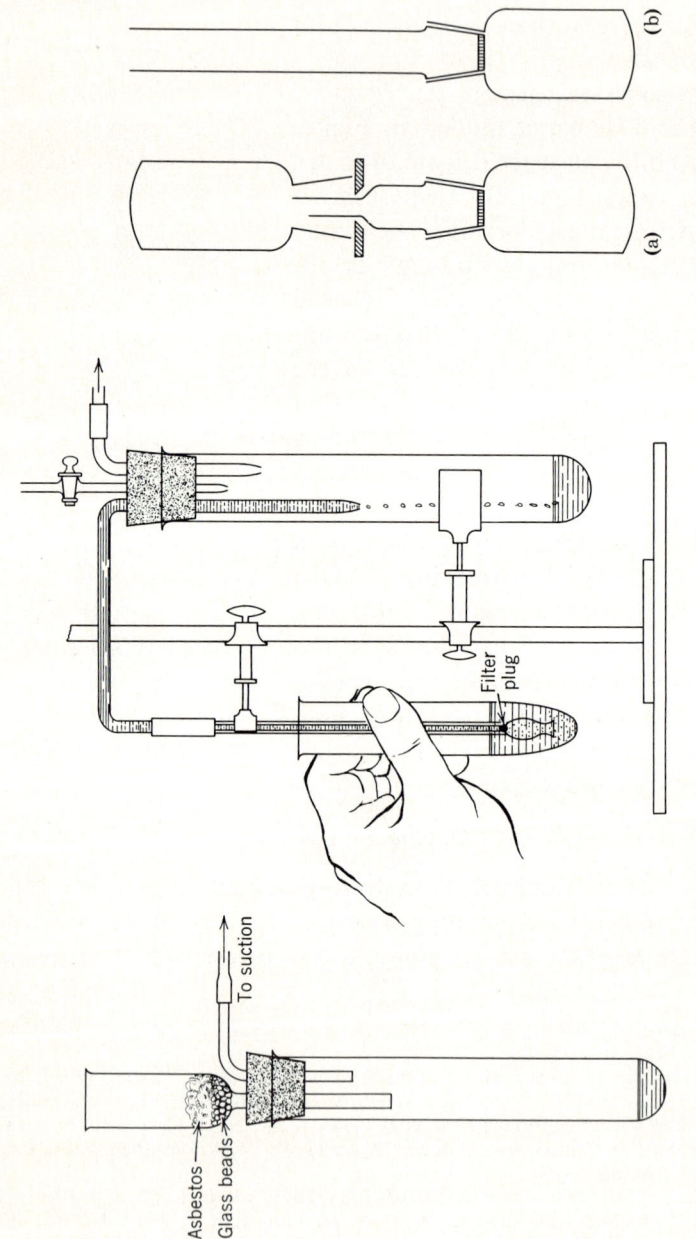

FIGURE 3.10
Left. Semimicro apparatus for suction filtration using filter tube. *Right.* Filtration by means of filter stick.

FIGURE 3.11
Apparatus for filtering with the help of centrifugal force.[15(a)]

cake; dry parts of the square are gently pressed on the cake until moisture appears to have been absorbed. The paper is examined for any adhering crystals, which are removed carefully with the microspatula.

Pieces of unglazed plates or tiles may be used very effectively for rapid drying of solids. They absorb liquids very readily and are not subject to shredding. A few milligrams of solid dry in a minute or two if the solid is pressed on the plate, scraped free, and pressed again with a spatula. The piece of porous plate or tile is never reused.

When successive crystallizations are performed with an unknown solid or in the preparation of derivatives, samples of 5 to 10 mg are retained from each crystallization for melting-point determination (see page 68). Thus, in dealing with compounds that are not affected by exposure to air, it is convenient to adopt a procedure for a rapid drying of samples of crystals from two successive crystallizations so that both their melting points can be determined at once, in order to decide whether further recrystallizations are necessary. With some planning, it is even possible to dry the crystals from the first crystallization rapidly enough that the determination of the melting point can be carried out while the clear, hot solution of the second crystallization is cooling.

Procedure and Apparatus for Air Drying

In our laboratories, the commercially available unglazed porcelain plates or tiles have been found most convenient for air drying of crystals. Paper discs may be used, but the possibility of incorporating some of the fiber with the crystals makes them less desirable.

A fresh porcelain plate or tile is broken into several pieces of varying sizes. Crystals are placed on a piece of the unglazed porcelain of the appropriate size. They are pressed firmly to the surface of the porcelain by means of a microspatula, and are then scraped gently from the surface. This process is repeated until there is no further change in the appearance of the solid. The crystals are then left to dry at room temperature for about ten minutes before a melting point is taken. If they are left on the porcelain overnight they are covered with a watch glass or a funnel. Complete dryness is indicated when none of the crystals adhere to the microspatula. The piece of porcelain is not reused, but is discarded when the compound is shown to be pure.

Drying of Crystals in Desiccators

A number of derivatives employed in the characterization of organic compounds decompose when dried in air; the osazones of sugars, for example, have 5 to 10° lower melting points when they are dried as described in the preceding section. On the other hand, samples from the same lot of the derivatives, when dried in a vacuum desiccator, melt at a higher temperature and close to the melting points given in the literature for a particular derivative.

The simplest type of hand desiccator is shown in Figure 3.12(A). It consists of a small, wide-mouth bottle of about 60-ml capacity with a

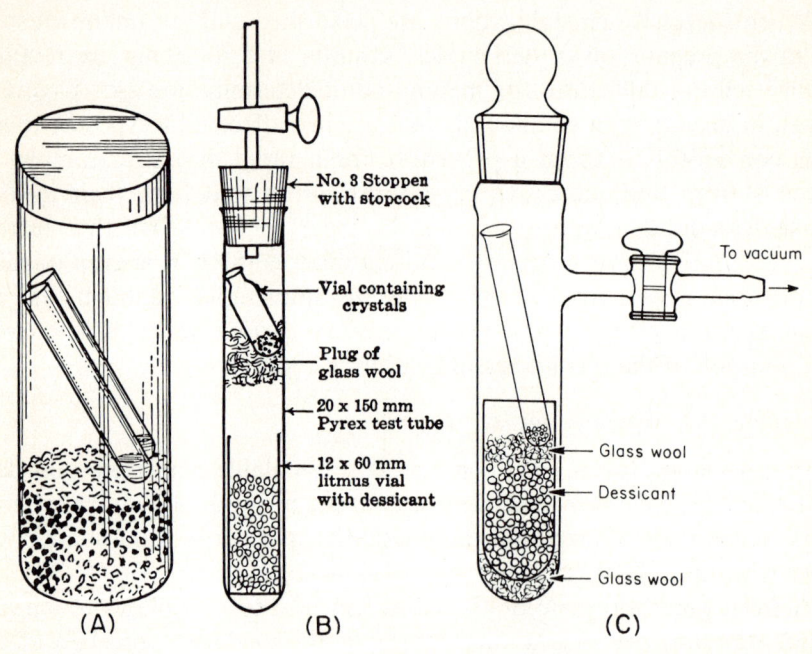

FIGURE 3.12

(A) Simple desiccator bottle. (B) Test-tube vacuum microdesiccator. (C) Vacuum microdesiccator with a ground glass stopper.

plastic screw cap. The drying agent is placed in the lower part of the bottle to a depth of about 10 to 15 mm. The sample is placed in a small tube or vial whose upper part, when placed within the bottle, comes to within 5 to 10 mm of the mouth of the bottle. The cap of the bottle is smeared with a thin coating of stopcock grease before the bottle is closed. The drying agent depends on the nature of the solvent used for crystallization. A mixture that will be satisfactory for most purposes consists of 5 g of sodium hydroxide pellets and 5 g of anhydrous calcium chloride (8 mesh). If a hydrocarbon was used as the crystallizing solvent, a few pieces of freshly cut paraffin wax are added to the bottle. When the solvent is alkaline, a small tube containing concentrated sulfuric acid is placed by the side of the tube containing the crystals. In most cases, 100 mg of crystals placed in such a desiccator will dry overnight. The tube is lifted out with forceps and held upright until its lower part has been wiped thoroughly dry; it is then tilted so that the dry crystals may be emptied out onto a clean watch glass or paper.

Figure 3.12(*B*) shows an inexpensive vacuum desiccator that is easily assembled and well adapted for the drying of most derivatives. It consists of a 20 × 150 mm test tube containing a vial into which the desiccant is placed. A plug of glass wool is placed above the mouth of the open vial, at about the middle of the test tube. The glass wool supports a small vial or tube (10 × 35 mm), which contains the crystals to be dried. A No. 3 rubber stopper holding a stopcock fits into the mouth of the test tube and serves for connection to the vacuum pump. The test tube desiccator is placed in a rack after evacuation and allowed to stand at room temperature, or it can be warmed in a bath (below the melting point of the crystals).

The desiccant to be used depends on the solvents employed for crystallization and washing of the crystals. A mixture of calcium chloride and sodium hydroxide pellets with a few small pieces of paraffin wax serves as absorbents for most of the common solvents. Other desiccants that may be used are phosphorus pentoxide, "Drierite," sulfuric acid, and soda lime.

Another type of small vacuum desiccator is shown in Figure 3.12(*C*). It consists of a 25 × 150 mm tube with a side arm that bears a stopcock. The desiccant ("Drierite," "Dehydrate," "Anhydrone," or phosphorus pentoxide) is covered with glass wool to prevent circulation of small particles. The substance to be dried is placed inside a vial or 3-in. tube, which is inserted in the drying chamber. The drying chamber can be inserted in a heating bath and heated while the system is being evacuated.

Heated Vacuum-Drying Apparatus

For precise work, the most efficient method for drying organic compounds so that both solvent and moisture are removed uses the Abderhalden type of vacuum drying apparatus. Two commercially available types, which are shown in Figures 3.13 and 3.14, may be used for drying quantities of a gram or less under reduced pressure at constant temperature.

The apparatus shown in Figure 3.13 consists of four parts connected by three standard taper ground-glass joints. The round-bottom boiling flask is connected to the heating chamber, which has a jacketed inner tube in which is placed the sample to be dried. The heating chamber is connected

80 Techniques of Organic Analysis

to a condenser and a drying chamber, which has a depression for the desiccant, and an outlet for connection to a vacuum pump. The heating chamber is maintained at constant temperature by the vapors of a liquid, which is heated in the boiling flask. Use of different liquids permits a wide range of temperatures.

The apparatus illustrated in Figure 3.14 is an improvement on that just described. It has a pear-shape vessel for the desiccant and is easier to fill. It also has a bypass for return of the condensate, which makes for greater uniformity of temperature.

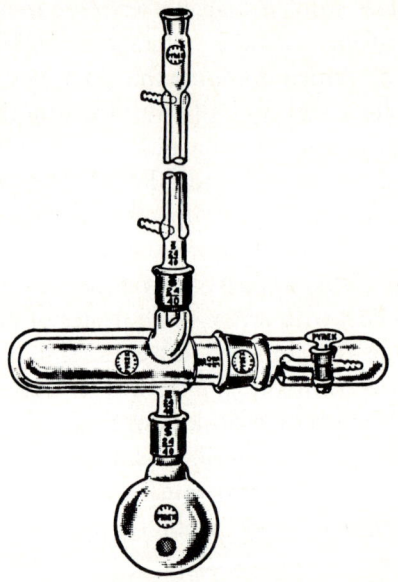

FIGURE 3.13

Pyrex vacuum drying apparatus. (Courtesy Corning Glass Works.)

By far the most versatile vacuum drying apparatus with automatic temperature control is the Ma-Schenck apparatus[16] shown in Figure 3.15. It consists of a glass pistol, fitted in an electrically heated chamber. The temperature is regulated by an automatic calibrated temperature controller. For the drying of small quantities of crystals, this instrument has been found to be the most efficient and to require least care; however, it is also the most expensive.

[16] T. S. Ma and R. T. Schenck, *Microchemie*, **40**, 242 (1942); available from Micro-Ware, Inc., Vineland, N.J.

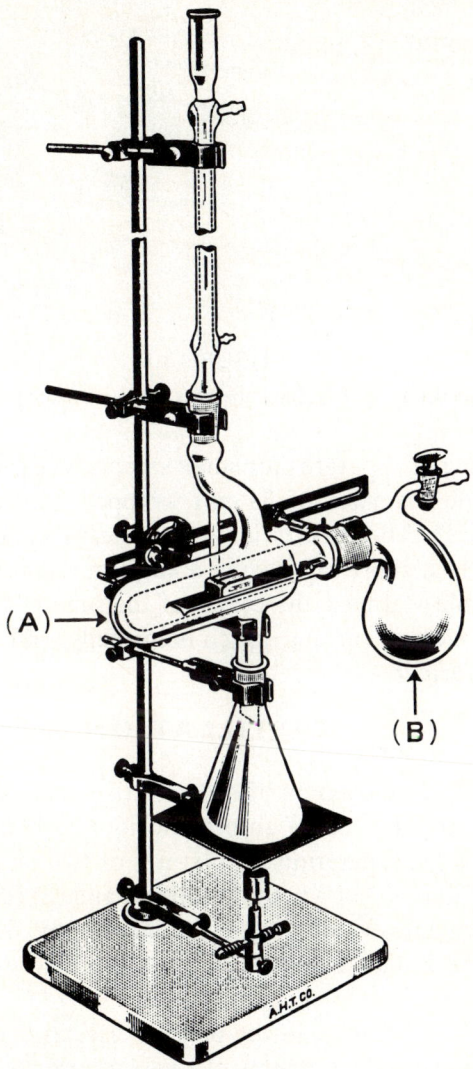

FIGURE 3.14
Abderhalden vacuum drying apparatus. (Courtesy A. H. Thomas Co.)

FIGURE 3.15

The Ma-Schenck type of drying apparatus with automatic temperature control.

A simpler drying apparatus for small samples is easily made from a tube with an outside diameter of 34 mm wrapped with asbestos paper and resistance wire.[17] The tube is joined by means of 35/40 standard-taper ground joints to a short cap having a vacuum connection. A wire gauze holds the sample vial near the center of the tube. The tube is heated by application of electric current, which is carefully controlled by means of a variable transformer.

DISTILLATION

Distillation is employed in the qualitative analysis of organic compounds for (a) purification of an organic liquid containing predominantly one component, (b) separation of mixtures of two or more organic liquid substances, (c) removal of a volatile solvent, and (d) removal by steam of a volatile component. Generally, fractionation by distillation, although employed in the initial stages of analysis, is not so extensively employed as crystallization.

No attempt will be made in the present discussion either to deal extensively with the theoretical aspects of distillation or to present a complete survey of all methods proposed for the distillation of small quantities of liquids. For these, see the literature.[18]

---

[17] J. D. Reinheimer, *J. Chem. Educ.*, **30**, 139 (1953).
[18] A. L. Glasebrook and F. E. Williams, "Ordinary Fractional Distillation. Part I. Apparatus," in *Technique of Organic Chemistry*, Vol. 4, A. Weissberger, ed., Interscience, New York, 1951, pp. 175–294; F. E. Williams, "Ordinary Fractional Distillation. Part 2. Procedure," id., pp. 295–316.

Volumes of 5 to 25 ml can be readily distilled with the arrangement of apparatus described in this section. Volumes of 2 ml can be distilled if proper *care and patience* are exercised. When the volume is less than 2 ml, fractionation is difficult (see the discussion on pages 96–99).

The most important consideration in all microdistillations, as in microcrystallizations, is avoidance of loss of material. Although it is possible for a beginner to purify 100 mg of an organic solid with fair success, the purification of 100 microliters (or 100 λ) of liquid by distillation cannot be readily accomplished, even by an experienced worker. The reason for this is to be found in the very nature of distillation. Heating of the liquid causes some decomposition; condensation of the vaporized liquid on the surface of the vessel results in an unavoidable holdup because of film formation on the glass surfaces. On the other hand, if an attempt is made to restrict the area of condensation, very little separation of the components of the liquid will result. Under these conditions, great care must be exercised in the selection of the apparatus and the technique of distillation. For small quantities of liquids whose boiling points are very close or a mixture of azeotropes, consideration should be given to separation by means of displacement adsorption as described on pages 131–145.

Simple Distillation at Atmospheric Pressure

Two assemblies for simple distillation, using the traditional distilling flask-condenser-adapter apparatus are shown in Figures 3.16 and 3.17. In the apparatus shown in Figure 3.16 the condenser is of glass and has a jacket length of 160 mm. The distilling flask may have a capacity of 5, 10, or 25 ml and either a wide opening, into which the thermometer may be inserted by means of a cork (Figure 3.16), or a narrow neck, in which case the thermometer is attached by means of a rubber sleeve. The assembly shown in Figure 3.17 is similar except that it is constructed with 19/22 interchangeable glass joints. The head adapter may be obtained either with a neck for inserting the thermometer by means of a rubber sleeve or with a ground-glass joint for a thermometer.

Figure 3.18 shows two types of assemblies[19] for simple distillation at atmospheric pressure using the finger-microcondenser system for condensation and collection of the distillate. One employs a distilling tube (5, 10, or 25 ml) and the other a pear-shape flask (5, 10, or 25 ml) for the distilling flask.

[19] Micro-Ware, Inc., Vineland, N.J.

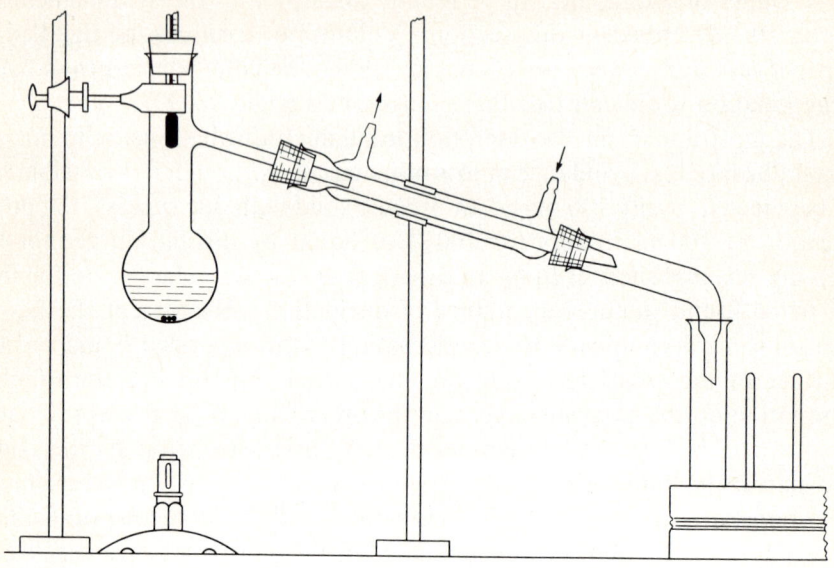

FIGURE 3.16

Semimicro distillation setup using the traditional distilling flask-condenser-adapter apparatus. Capacity of flask is 5, 10, or 25 ml; length of condenser jacket is 160 mm.

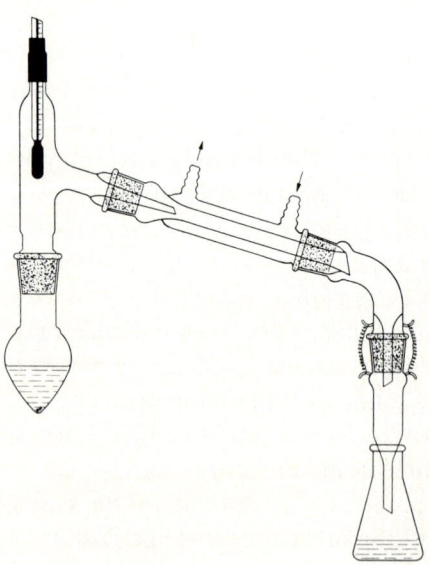

FIGURE 3.17

Semimicro distillation apparatus with standard-taper glass joints.

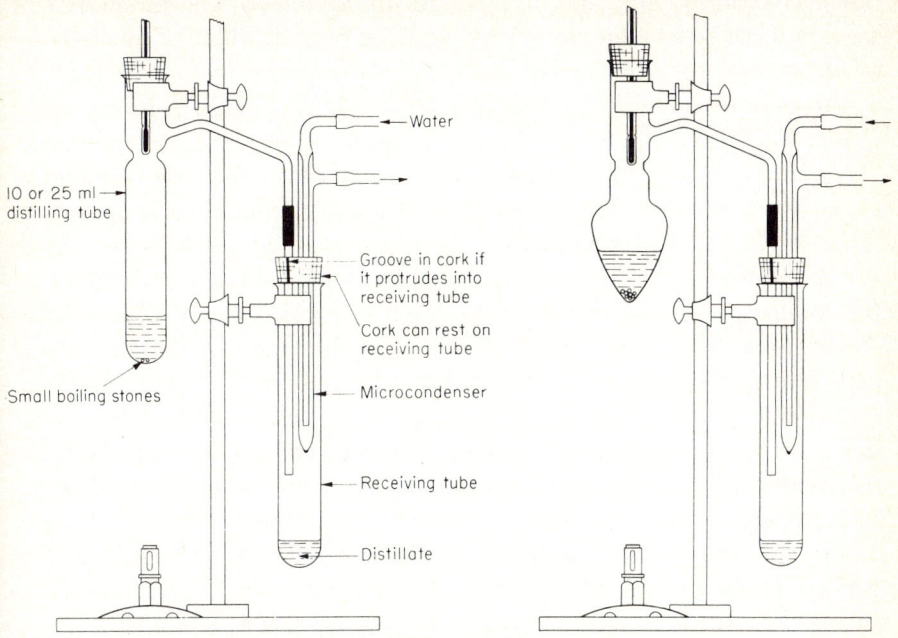

FIGURE 3.18

Left. Semimicro distillation apparatus using a distilling tube and microcondenser. *Right.* Semimicro distillation apparatus using a distilling flask and microcondenser.

The distilling vessel is fitted with a cork holding the thermometer (the flask can also be obtained with a sleeve-neck for the thermometer). The curved side arm is connected by a small piece of rubber tubing (3 to 4 mm) to a delivery tube 5 to 6 mm in diameter and 100 to 150 mm in length. Both the delivery tube and the condenser are inserted, through a cork, into an ordinary 6- or 8-in. test tube, which acts as a receiver (a test tube of any size may serve as a receiver, provided its diameter is sufficient to hold the delivery tube and the condenser). The opening in the cork for the condenser is larger than the diameter of the condenser tube to give a loose fit, and the cork rests on the receiver so that an open system results. If the cork is inserted into the receiving tube, then a groove is cut on the side of the cork as shown in the diagram.

The volume of the liquid should not be more than two thirds of the volume of the distilling tube. It is sometimes necessary to distill 25 ml of an ether or benzene solution that contains only 1 to 2 ml of the desired compound. In such cases, the liquid is distilled, a portion at a time, in a vessel of 25-ml capacity until about 5 to 8 ml remain. This liquid is then transferred, in two portions, to a 5-ml distilling vessel; that is, about 3 ml are first transferred and, after the solvent has been distilled off, the balance is added, and the 25-ml flask is washed with 1 ml of the distillate; the washings are then added to the 5-ml vessel. In this manner, losses owing to unavoidable adherence of a film on the walls of the vessel are minimized.

No matter which assembly is employed, the precautions to minimize losses are the same. The liquid to be distilled is placed in the distilling tube, and two small boiling stones about the size of grape seeds are added. All connections are inspected, and the receiving tube is so adjusted that the condenser and the delivery tube reach about two-thirds of the way to the bottom of the tube. Heat is applied with a microburner, the flame being moved to and fro. As the liquid begins to boil, the vapor condenses on the sides of the distilling vessel and returns to the boiling liquid. The flame is adjusted so that the vapors ascend very slowly and reach the thermometer bulb at least 1 minute before they pass through the side tube. In this manner the thermometer is heated to the temperature of the vapor before the first vapors reach the condenser. The vapor condenses and collects at the bottom of the tube. Thus, the progress of distillation is followed by the amount of liquid in the receiving tube.

If the distilling liquid is essentially one component, the temperature at which the liquid appears in the receiving tube is recorded. The distillation is continued slowly until the mercury column in the thermometer remains nearly stationary, when the flame is removed and the receiving tube is changed. Heating is then resumed and, as soon as distillation begins, the flame is adjusted so that the liquid in the receiving tube increases slowly. A little experience is necessary before the technique is mastered. No attempt should be made to keep the temperature indicated by the thermometer constant by varying the heating of the flask. The best procedure is to maintain a steady slow rate of distillation, and let the temperature rise as it will. In this way, we know if there is a change in the composition of the vapors about the thermometer bulb by observing the temperature change.

Toward the end of the distillation the flame, manipulated by hand, is moved to and fro in order to avoid superheating. When a minute amount of liquid is left in the distilling tube, the distillation is discontinued and the apparatus is disconnected and cleaned. If the microcondenser type of apparatus is employed, it will be found advisable to allow the cork to remain in place on the delivery tube and condenser. The delivery tube and condenser are cleaned by rinsing with a little acetone or alcohol, and then dried.

Fractional Distillation at Atmospheric Pressure

Figure 3.19 shows a simple, inexpensive, and relatively efficient assembly for semimicro fractionation by distillation.[19] It comprises a 25-ml round-bottom flask into which is fitted, by means of a cork, a glass column 160

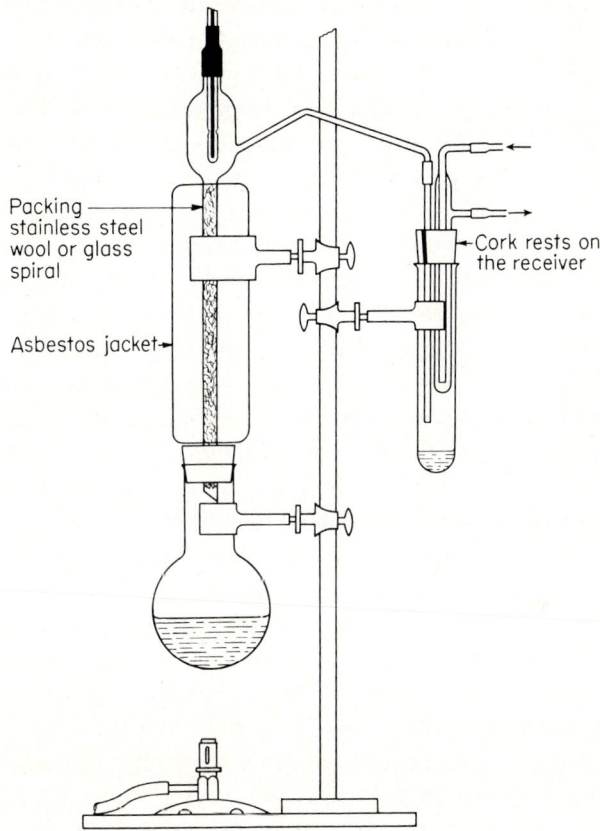

FIGURE 3.19

Apparatus for semimicro fractionation. The column may be obtained also with a wider top for insertion of a cork to hold a thermometer.

mm in length and 8 mm in diameter; the side arm of the column connects to a regular semimicro condenser assembly described in the preceding section. Several types of packing are available. The most inexpensive consists of commercially available stainless steel wool,[20] which is packed

---

[20] Stainless steel wool is available in most hardware stores.

rather loosely into the column, but not so loosely that large empty spaces are visible. Although this type of packing is attacked by some halogen compounds, it has about the same efficiency under comparable conditions (rate of distillation) as the well-known glass helix type of packing. A third type of packing consists of Nichrome wire stampings, which come in two sizes, made by the manufacturers of the Heli-Grid type of packing. This type is more efficient than either the steel wool or glass helix type.

For insulation, an asbestos- or a glass vacuum-jacket is used. The asbestos jacket is 100 to 110 mm long, and is made by winding several layers of asbestos paper, which have been dipped in a sodium silicate solution, around an 8-mm tube to a thickness of 4 to 5 mm and allowing it to dry. The vacuum jacket is made of glass and contains several metal radiation shields spaced within the evacuated section in which the pressure is about 1×10^{-5} mm. One end of the column is inserted through a cork, which is then placed in the mouth of a 10- or 25-ml round-bottom flask or a test tube. The side arm of the column is connected to the condensing-receiving system as described on page 85.

For a detailed discussion of the factors that determine the efficiency of fractionation, see the publications listed at the end of this chapter. The most important of these factors are (a) height of column, (b) nature of packing, (c) insulation, and (d) rate of distillation. Even with a column so designed as to meet the requirements in the first three categories, the efficiency of the fractionation rapidly diminishes if the withdrawal of distillate is at the rate of more than 0.3 ml per minute.

It is advisable to wet the column with some of the mixture to be fractionated before distillation is begun; this is best accomplished by adding the liquid to be distilled, slowly, through the top of the column into the boiling vessel, from which the cork has been loosened to permit the escape of air. The various connections are inspected and adjusted, and heat is slowly applied to the boiling vessel with a small flame. In the beginning, there will be a certain amount of refluxing within the column; when flooding appears on the top of the column, the flame is removed and the vessel is allowed to cool momentarily. Heating is resumed with the flame moved to and fro until the vapor begins to enter the side tube. The heating is then adjusted so that the rate of distillate withdrawal is 0.2 to 0.3 ml per minute. The receiving tubes may be calibrated at the 0.5-, 1-, 2-, and 5-ml marks with thin strips of gummed paper. If care is used, 5 to 10 ml of mixtures may be separated efficiently with a single fractionation.

The effect of the rate of distillate withdrawal on a routine fractionation of a 1 : 1 methanol-water mixture is shown, graphically, in Figure 3.20.

In order to recover the liquid retained in the column and apparatus, which often is 0.5 to 1.0 ml, a high-boiling compound or "chaser" is added to the mixture to be fractionated. The boiling point of the chaser should be, at least, 20° higher than the boiling point of the last fraction; in addition, the chaser should be relatively inert and not form azeotropes. Among the chasers commonly employed are: cymene (bp 175°); biphenyl (bp 254°); acenaphthene (bp 277°); and phenanthrene (bp 340°). The

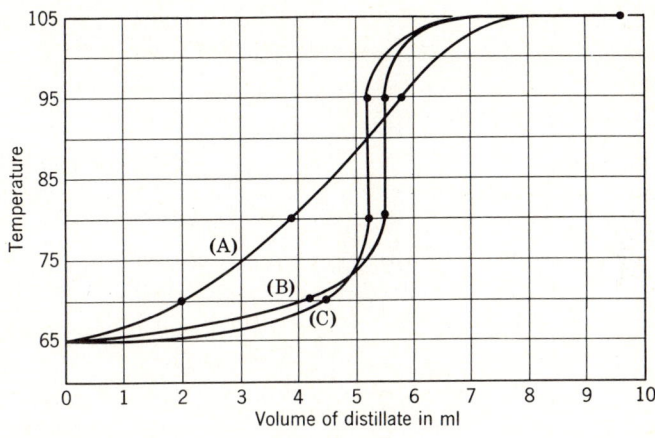

FIGURE 3.20

Effect of packing and rate of distillation on semimicro fractionation. (A) Glass wool packing, 0.5 ml/min. (B) Heli-grid packing, 0.25 ml/min. (C) Glass spiral packing, 0.25 ml/min.

amount of chaser added should be a little more than the total estimated holdup.

Figure 3.21A shows the same type of column as the one just described but with a built-in vacuum jacket and with glass-stoppered joints to the distillation pot. Figure 3.21B shows a Cheronis type of microcolumn with a Podbielniak Heli-Grid or Super-Col type of packing with a built-in vacuum jacket.[21] The distilling flask is available in either 25- or 50-ml sizes. The column is provided with a receiver changer and graduated receiving burets.

Craig[22] has described a microapparatus for fractional distillation, which gives good results, and is fairly easily constructed. With it, 1 g of a 1 : 1 by weight mixture of carbon tetrachloride, bp 76.8°, and benzene, bp 80.1°, was distilled in 143 minutes with a total recovery of 0.975 g. The

[21] Micro-Ware, Inc., Vineland, N.J.
[22] L. M. Craig, *Ind. Eng. Chem., Anal. Ed.*, **9**, 441–443 (1937).

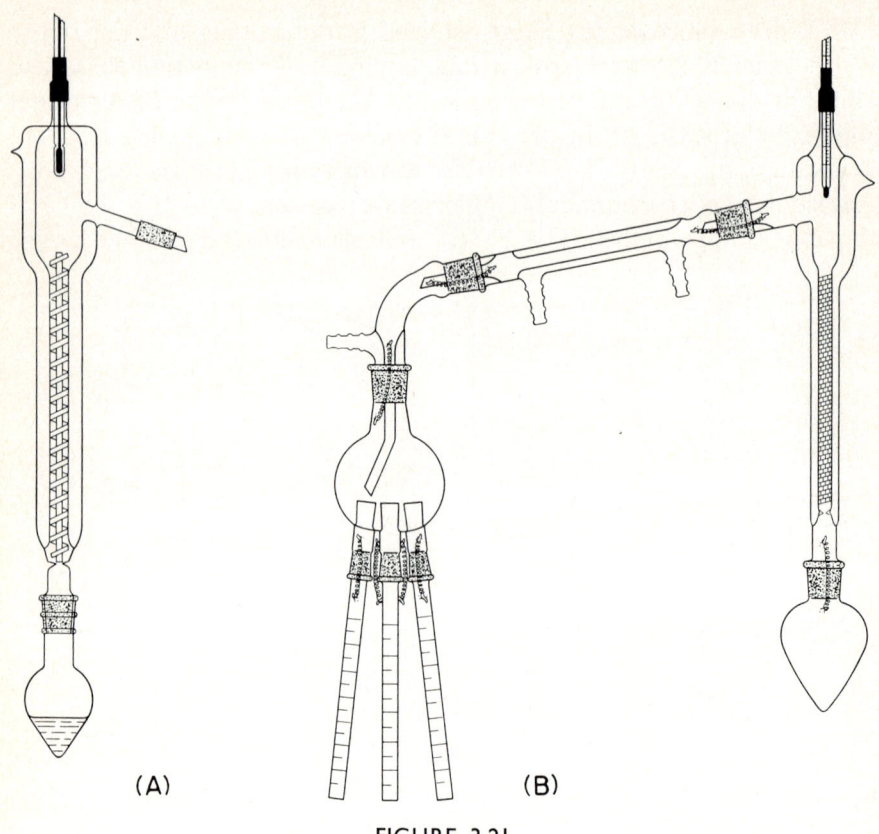

FIGURE 3.21

(A) Semimicro fractionating column with built-in vacuum jacket arm, glass spiral packing, and standard-taper glass joints. (B) Cheronis semimicro column with Podbielniak Heli-Grid packing.

first 35 per cent of the distillate consisted of 85 per cent carbon tetrachloride, while the last 25 per cent had less than 10 per cent carbon tetrachloride. The apparatus is essentially a scaled-down laboratory column with a take-off system designed to minimize loss of liquid by contact with the walls.

Distillation under Reduced Pressure

The apparatus shown in Figure 3.21B, which is provided with Podbielniak Heli-Grid packing and a receiver changer may be used for fractionation under reduced pressure. A less expensive assembly is shown in Figure 3.22. It consists of a 25-ml Claisen distilling flask, a condenser, and a receiving assembly. The capillary tube that reaches almost to the bottom of the pear-shape flask is made of 3 to 4 mm glass tubing; a piece 100 to 125 mm in length is sealed to the end of a small piece of scrap tubing. The region near the seal is heated over the flame, the tube being held at an angle of 75°. When the glass walls of the tube collapse and the inner diameter becomes small, the flame is increased to render the thick-wall section soft. The tube is removed from the flame and pulled out slowly

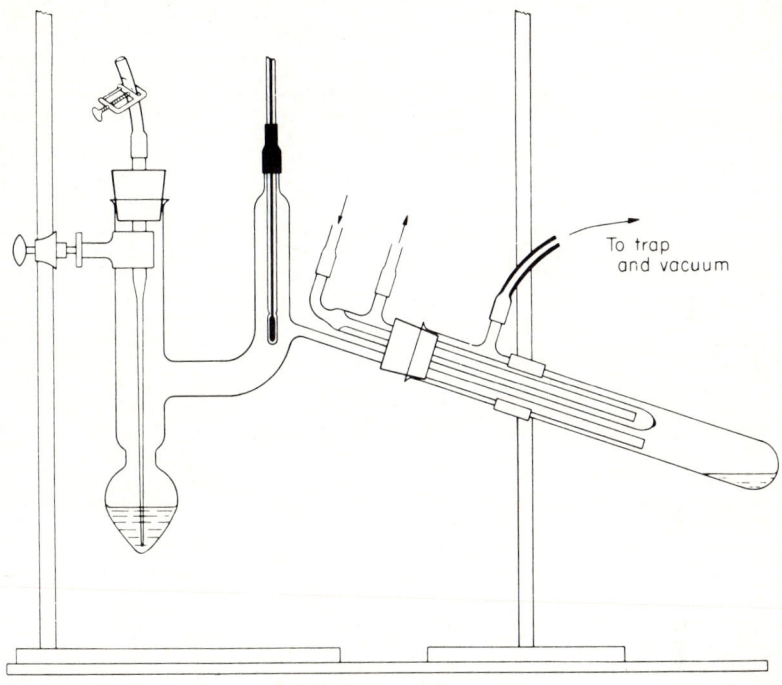

FIGURE 3.22

Simple assembly for vacuum distillation.

into a capillary. The apparatus is assembled as shown in the diagram, and the receiving assembly is connected through thick-wall rubber tubing to the trap, manometer, and suction pump.

The liquid to be distilled is placed in the boiler by removing the stopper holding the capillary. The stopper is replaced and a small oil bath is raised to the boiling vessel. The air inlet is adjusted so that fine bubbles can be counted. Heat is applied under the bath until the first sign of distillate appears, whereupon the flame is removed from under the bath. The temperature and pressure are noted, and heating is resumed until distillation begins and the temperature remains constant, with little variation in the pressure. The receiver is changed and heat is applied slowly until distillation is resumed. The temperature and pressure are recorded, and heat is applied intermittently so that the distillation proceeds evenly. When the temperature begins to rise or the amount of liquid in the boiling tube is very small, the distillation is discontinued. The residue is not rejected until all data on the distillation have been inspected.

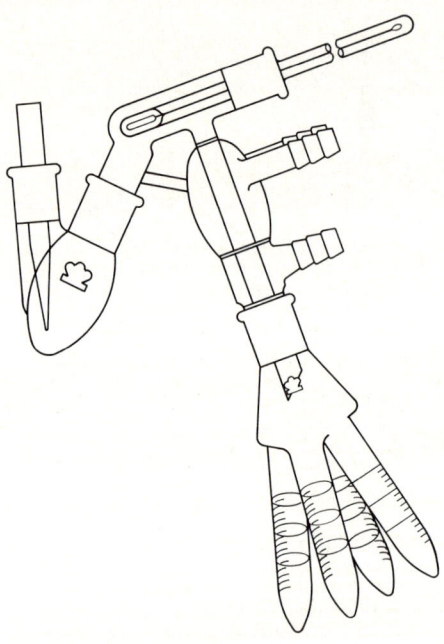

FIGURE 3.23

Short-path fractional distillation apparatus. The pear-shape flask has a capacity of 10 ml. (Courtesy Kontes Glass Company, Vineland, N.J.)

Simple, compact apparatuses are available commercially for distillation of small amounts of liquid. One of these, shown in Figure 3.23, uses the thermometer bulb and stem as column packing, and conserves product by reducing the amount of surface wetted by the liquid.

Accessory Equipment

A distillation under reduced pressure may be performed with the minimum equipment described in the last section and, in many cases, this is all that is needed. However, it is usually desirable to have also a manometer for measurement of the pressure, so that the results of the distillation can be compared with results of other distillations. For carefully controlled pressure during a fractional distillation, a manostat is required, especially if the process lasts longer than a few minutes. Some source of vacuum will be required, either a mechanical pump or a water aspirator. If a mechanical pump is used, a cold trap and a trap for acid vapors should be provided. If the water aspirator is used, a trap should be used to keep water out of the vacuum system.

Manometers may be obtained in a variety of designs, but many of these use mercury to indicate the difference between some known pressure and the pressure being measured. One of the most useful in the general organic laboratory, shown in Figure 3.24, is available commercially.[23] In this manometer, a glass tube extends downward from the standard-taper plug almost to the bottom of the container; it is closed at the top and has a small hole at the bottom. During operation of the manometer the tube holds the column of mercury. The solid standard-taper plug is notched so that the main body of the manometer can be evacuated through the side arm. To use the manometer, about 1 ml of mercury is placed in it, and the central tube is filled with mercury. The process of filling and removal of gas is easier than in the case of most dead-end manometers. The mercury-filled column can be inverted and placed in the body of the manometer.

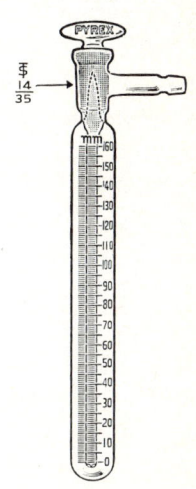

FIGURE 3.24

Pyrex manometer. Closed-end type. Also available with T-shape side arm. (Courtesy A. H. Thomas Co.)

When the pressure in the manometer is reduced by evacuation through the side arm, the level of mercury inside the central tube falls: the difference in levels of mercury indicates the pressure. The two levels are read by means of graduations on the jacket (160 mm in 1-mm divisions). By turning the standard-taper plug, the manometer is isolated from the remainder of the vacuum system.

For measuring pressures below 2 or 3 mm with more accuracy, a McLeod gauge is very desirable. Several companies have available tilting McLeod gauges with maximum readings of 1.0, 5.0, 10.0, or 15.0 mm of mercury. These must be tilted for each reading, but are sturdy and practical.

Many devices for controlling the pressure during a distillation have been

[23] Corning Glass Works (through most suppliers of scientific equipment).

described, and several of these can be obtained commercially from suppliers of scientific equipment. One of the best of these has been described by Hershberg and Huntress,[24] can be made in the laboratory fairly easily, and permits regulation of the pressure to ±0.015 mm of mercury during long usage. One side of a manometer containing sulfuric acid is connected to the system in which the pressure is to be controlled. If the pressure in the system falls too low, electric contact is made through the sulfuric acid, and a relay opens a small inlet for air. A needle valve for admission of air allows coarse adjustment of the pressure, and the automatic valve provides the fine control only. It is possible to reproduce the same pressure even if the distillation is interrupted for days.

If a rotary oil pump is used to reduce the pressure for a distillation, it should be protected from organic and acid vapors. Vapors of organic compounds are best removed by use of a cold trap, cooled with a bath of dry ice and a liquid such as alcohol, acetone, or trichloroethylene. Acid vapors should be removed by passing them through a tower of solid soda ash pellets and one with solid sodium hydroxide or potassium hydroxide pellets. These traps are so important to the maintenance of the oil pump that they should be permanently connected to any pump used for the distillation of organic compounds.

In many cases, a water aspirator is more convenient and just as good as an oil pump for distillations. A properly designed and working aspirator will lower the pressure in a closed system well below the vapor pressure of water at that temperature. However, to prevent water from entering the vacuum system during the distillation, an empty trap should be placed between the aspirator and the receiver. This trap should be arranged so the water that comes into it is sucked out when the aspirator returns to normal operation.

When liquids are vaporized in a vacuum system, the vapors occupy many times the volume that they would at atmospheric pressure and, consequently, the vapors move at high velocity. For best results, the paths traveled by the vapors should be kept as short as possible. Tubes through which they move should be as large as conveniently possible. Rubber tubing, especially old rubber tubing, may have pin holes, and so should be avoided as much as possible.

Molecular Stills

In order to distil very high-boiling materials without prolonged and excessive heating, it is necessary to take special precautions. One device

[24] E. B. Hershberg and E. H. Huntress, *Ind. Eng. Chem., Anal. Ed.*, **5**, 344–346 (1933).

for accomplishing this is the molecular still in which the distance from the heated surface to the cooled condensing surface is very short. The path, actually, is equal to or less than the average distance traveled by the molecules between collisions (mean free path). A molecular distillation is not a good method of fractionation, but it will separate nonvolatile material from slightly volatile material.

Figure 3.25 shows a molecular still in its simplest form.[25] The material to be distilled is placed in the bottom of the container, a receiver flask is joined to the right-hand tube, and the system is evacuated through the left-hand tube. Ice may be placed in the depression on top of the apparatus

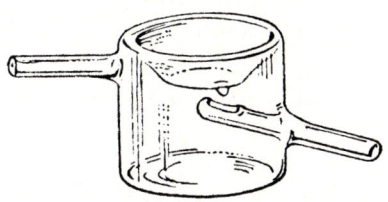

FIGURE 3.25

Hickman molecular still. (Courtesy Scientific Glass Apparatus Co.)

for cooling, and the bottom of the apparatus is heated with an oil bath, a hot plate, or a heating mantle.

Sanders and Helwig[26] have designed a compact molecular distillation apparatus that can distil from 1 mg to 5 g of high-boiling material. The apparatus is characterized by an internal heating element and a conveniently located ionization gauge for measurement of pressure. The vacuum take-off is located so that desired material is not drawn out of the apparatus.

Another molecular still having some advantages has been described by Ungnade.[27] The apparatus has a heating block, which is made to fit the pot, and a cold-finger condenser which has a clearance of 1.5 to 2.0 mm from the walls of the still and which may be cooled with water, ice, or dry ice. The distillate-collecting tube may be rotated to accept or reject a particular fraction. The apparatus can handle small analytical samples; a larger one distils samples for preparative work.

[25] K. C. D. Hickman and C. R. Sanford, *J. Phys. Chem.*, **34**, 637–653 (1930).
[26] G. R. Sanders and H. L. Helwig, *Anal. Chem.*, **31**, 484–485 (1959).
[27] H. E. Ungnade, *Anal. Chem.*, **31**, 1126–1127 (1959).

Distillation of Small Quantities at Atmospheric or Reduced Pressures

The apparatus shown in Figure 3.26 is suitable for distillation of very small quantities of material. A piece of 10-mm tubing about 90 mm in length is sealed at one end. The lower end is filled with glass wool to a height of 20 to 30 mm. A special microcondenser with a well at the lower end is fitted to the upper part of the tube. This microcondenser may be constructed from an ordinary finger condenser. The lower part is heated in a hot, pointed flame until the glass is soft and is then pushed inward and downward by means of a metal rod. The well holds 1 or 2 drops of liquid.

Two or three drops of the liquid to be fractionated are added on top of the glass packing, and the special microcondenser is inserted. The tube is heated by means of a very small flame from the microburner until the vapors of the liquid begin to condense on the sides of the tube just above the glass fibers. The flame is then adjusted so that about 5 minutes are required for it to fill the well. The distillate is removed by lifting the condenser and inserting a capillary pipet in the well. In this manner, 5 to 10 fractions may be collected from the fractionation of a few drops.

Another method for the fractionation of a few drops of liquid is the original procedure of Emich.[28] Morton and Mahoney have described[29] a refinement of Emich's method such that 30 to 70 fractions can be collected from a single drop weighing about 25 mg. The method gives good results if sufficient patience and care are exercised. The fractionation of 1 drop of an equal mixture of benzene and xylene is highly recommended as practice for beginners.

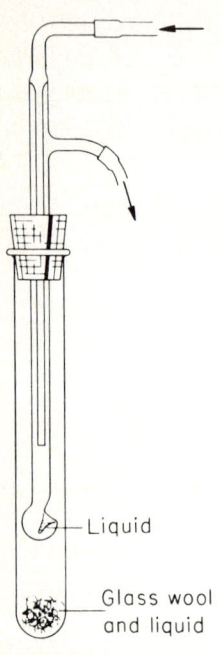

FIGURE 3.26

Assembly for fractionation of a few drops of liquid; boiler tube 10 mm in diameter and 90 mm in length.

[28] F. Emich and F. L. Schneider, *Microchemical Laboratory Manual*, Wiley, New York, 1932, p. 34; F. L. Schneider, *Organic Qualitative Microanalysis*, Wiley, New York, 1946, p. 29.
[29] A. A. Morton and J. F. Mahoney, *Ind. Eng. Chem., Anal. Ed.*, **13**, 494 (1941); N. D. Cheronis, "Micro and Semimicro Methods," in *Technique of Organic Chemistry*, Vol. 6, A. Weissberger, ed., Interscience, New York, 1954, pp. 71–73.

The distillation of quantities less than 1 ml presents serious difficulties. The best apparatus that has been described thus far is by Gould et al.,[30] which is quite expensive.[31] A very simple but less efficient arrangement, which has been described by Babcock,[32] is diagrammed in Figure 3.27. A tube about 225 mm in length and 6 mm OD is sealed at one end; then bulb A is blown and is pressed in slightly at the lower end. A rubber stopper cut to 8 mm in length is placed just below bulb A and then bulb B is blown as a lopsided blister at one side of the tube. The rubber stopper is

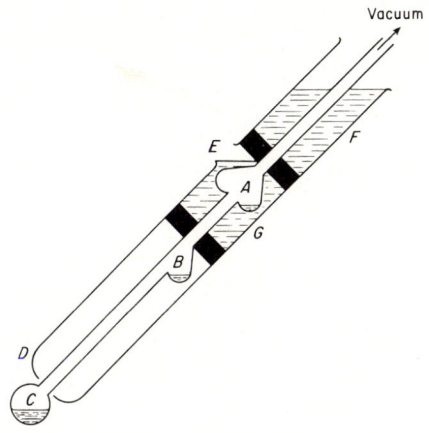

FIGURE 3.27

Apparatus for microdistillation under reduced pressure.[32]

now lowered to a position above B, and bulb C is blown. The jacket (D) is made from an 8-in. test tube by fashioning an opening at the bottom just larger than bulb C and a 5-mm hole at position E by pressing a wire through the softened test tube. The fractionating tube is clamped at an angle of 45° and, by means of a capillary pipet, 0.1 ml of water is released into bulb A; the water level is marked with paint and another mark is made at the 0.2-ml level. The setup is then assembled as shown in Figure 3.27. The sample is introduced into bulb C with a capillary pipet. Ice water is placed in cooling chamber F, and the tube is clamped at an angle of 45° so that bulb C is immersed in an oil bath with bulb B on the top side. The upper part of the distilling tube is connected to the vacuum system and the

---

[30] C. W. Gould, G. Holzman, and C. Nieman, *Anal. Chem.*, **20**, 361 (1948); N. D. Cheronis, "Micro and Semimicro Methods," in *Technique of Organic Chemistry*, Vol. 6, A. Weissberger, ed., Interscience, New York, 1954, pp. 73–75.

[31] Emil Greiner and Co., New York, N.Y.

[32] M. J. Babcock, *Anal. Chem.*, **21**, 632 (1949).

oil bath is slowly heated. When sufficient distillate has collected in bulb *A*, the lower end of the tube is raised out of the oil bath to an angle of about 30°, and heating of the oil bath is discontinued. Ice water is introduced through the opening (*E*) into cooling chamber *G* to condense the vapor in this region, and the assembly is carefully rotated on its longitudinal axis until bulb *B* is on the bottom side. The apparatus is now carefully lowered into the bath until sufficient distillate has collected in bulb *B*. The distillation is stopped by first raising the apparatus out of the bath and then

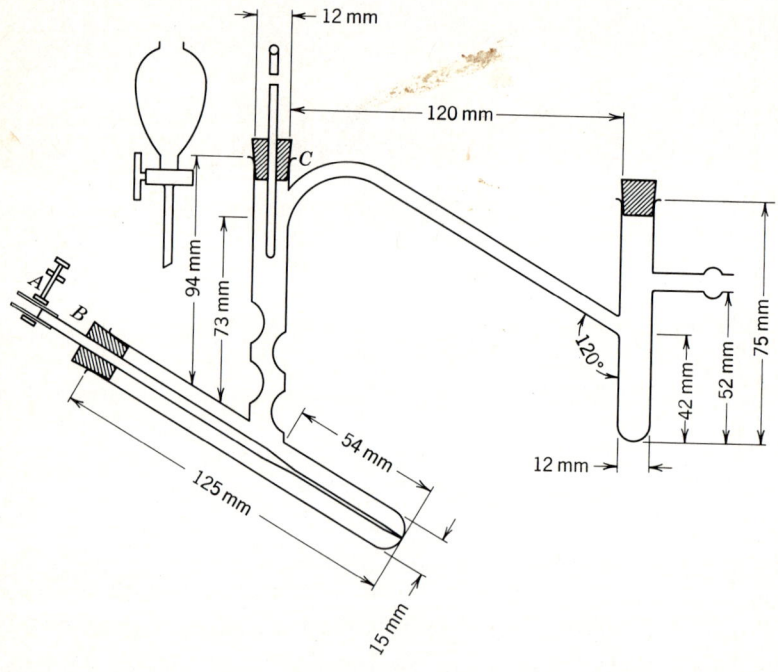

FIGURE 3.28

Apparatus for small-scale distillation at reduced pressure.[33]

releasing the vacuum. If the water in cooling chamber *G* becomes warm, it is replaced by cold water. The fractions are removed from *A*, *B*, and *C* by capillary pipets.

In another method, a tube 3 mm in diameter is drawn out to a curved capillary at one end with a 15-mm bulb at the other. The bulb is warmed and, gradually, draws up the liquid as it cools. The rate of distillation recommended is about 0.1 ml per 10 to 15 minutes. The fraction obtained may be subjected to a second distillation.

Another apparatus which is useful for distilling small quantities of material at reduced pressure can be constructed in the laboratory[33] or purchased.[34] As shown in Figure 3.28, the essential part of the apparatus is in one piece to reduce liquid holdup at joints or other connections. Ether solutions may be added slowly through a dropping funnel at *C* and flashed off to concentrate the desired product in the distillation vessel. Later a vacuum is applied, air is bled in through the capillary tube, and the product is collected in the trap at the right. The delivery arm may be cooled by application of paper or cloth which has been dipped in cold water. The vertical column gives some fractionation of the product if the distillation is conducted slowly.

An especially designed apparatus for distillation of one drop or less of liquid can be purchased.[35] It is electrically heated, and can be obtained with a vacuum take-off for distillation at reduced pressure.

Steam Distillation

The steam distillation apparatus shown in Figure 3.29 is assembled with the heating bath shown in Figure 2.15. A No. 20 cork or a No. 9 rubber stopper is fitted into the largest opening of the bath. A hole to hold an 8-in. tube, snugly, is carefully bored in the stopper. The tube is pushed gently in, far enough so that when the stopper is in place the tube reaches almost to the bottom of the bath. A No. 14 cork or No. 6 one-hole rubber stopper is fitted through the other large opening of the bath. A piece of glass tubing 4 to 5 mm in diameter and curved slightly at one end, is inserted through this stopper to serve as the steam outlet. It connects through a small piece of rubber tubing with the steam injector, which is a glass tube of the same bore, 210 to 220 mm in length, and reaches almost to the bottom of the 8-in. tube. The vapor outlet projects 40 mm above the tube before it bends to connect with the condensing system. It is advisable to make the vapor outlet of tubing 5 to 6 mm in bore, thus minimizing the danger of clogging and of carrying over into the distillate material spattered from the boiling tube. A glass tube 4 to 5 mm in bore and 300 to 350 mm in length is inserted through a stopper and placed in one of the small openings of the bath. The long glass tube reaches almost to the bottom of the bath and serves as a water and pressure gauge. The remaining opening of the bath is closed by a solid stopper.

[33] G. F. Wright, *J. Chem. Educ.*, **26**, 422–425 (1949).
[34] John's Glass Co., Toronto, Ontario, Canada.
[35] U. M. F. Science Corporation, Arcadia, Calif.

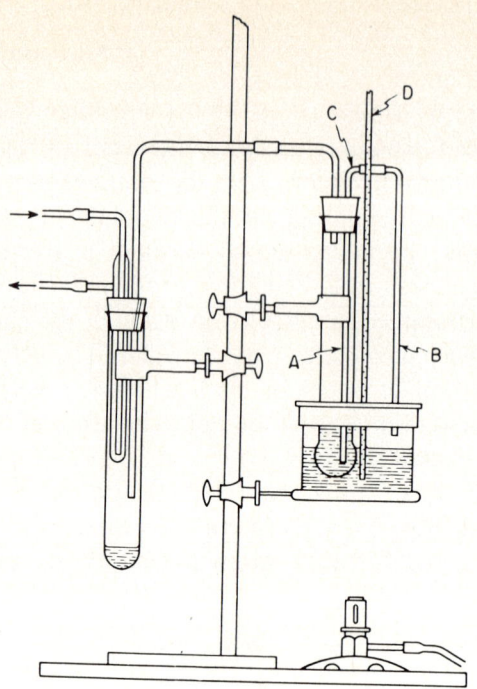

FIGURE 3.29
Semimicro apparatus for steam distillation.

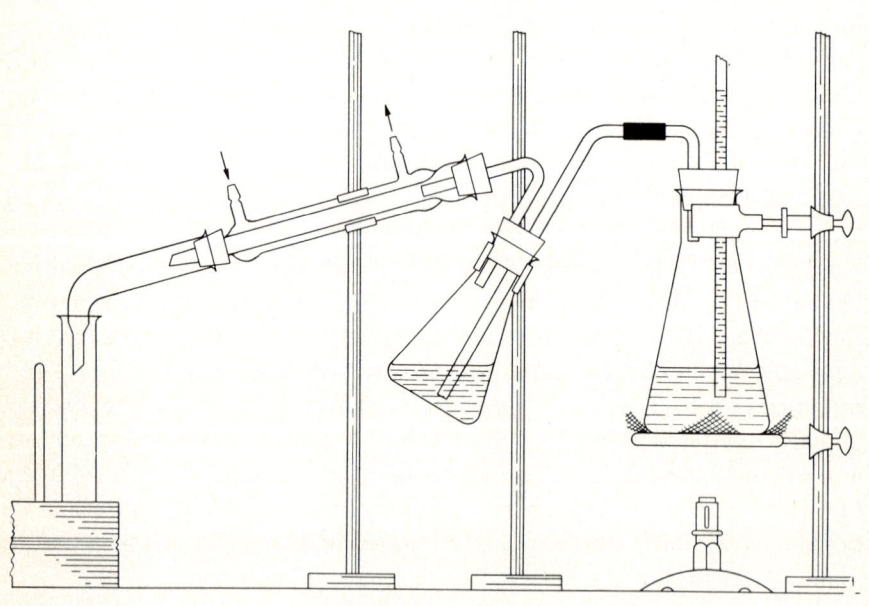

FIGURE 3.30
Semimicro steam distillation apparatus using two small Erlenmeyer flasks.

About 15 ml of the liquid to be steam-distilled is placed in the tube, and the rubber tubing that joins the steam inlet and steam injector is disconnected. The flask or bath is placed on a ring stand and clamped securely. Heat is applied until the water rises in the gauge and steam issues from the steam-outlet tube. The flame is then removed momentarily and, after about 30 sec, the steam inlet is adjusted to the injector and heating is resumed. The flame is so adjusted that the splashing does not reach the middle part of the vapor-outlet tube. An 8-in. tube is used in the receiving system and, if the microcondenser is not sufficient to cool the vapor, a beaker of cold water is raised so as to surround the lower part of the receiving tube.

With a little practice, steam distillations may be performed rapidly by use of the apparatus just described. The only difficulty that may be encountered is with foaming liquids, particularly when finely divided solids are contained in the mixture to be distilled. In such cases the volume of liquid to be steam-distilled should not exceed 5 to 7 ml. If foaming becomes troublesome, the upper part of the boiling tube, near the steam outlet, may be heated by a small flame.

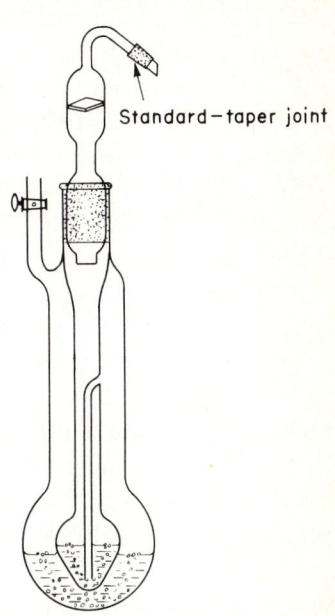

FIGURE 3.31
Compact assembly for semimicro distillation.

Another type of assembly for steam distillation, shown in the diagram of Figure 3.30, is readily arranged from apparatus available in all laboratories, since it employs two small Erlenmeyer flasks. This arrangement is useful with mixtures that froth badly with steam injection, and when the volume of the mixture is above 25 ml.

An all-glass assembly for steam distillation with 19/22 joints is shown in Figure 3.31. The capacity of the inner flask is 25 to 30 ml; the bent outlet is provided with a glass baffle trap to prevent the spray from contaminating the distillate.

SUBLIMATION

In some cases, sublimation is superior to crystallization as a method for the purification of very small quantities of solid substances, since it entails a minimum loss of material. It can be employed either for the purification of a single compound or for the separation of several components from a mixture. However, fractional sublimation is limited in application as a process of separation because of the difficulty in obtaining a series of multiple resublimations in the same vessel. Theoretically, any solid organic compound that can be distilled at atmospheric or reduced pressure without decomposition may also be sublimed. However, a solid that fails to give a sublimate at a pressure of a few microns when heated at a temperature 25 to 50° below its melting point for several hours is not considered to sublime for practical purposes.

There is no well-defined temperature at which the detachment of molecules from the sublimand begins. However, attempts have been made to define the practical subliming point[36] as the lowest temperature at which a sublimate discernible under a microscope is obtained when a substance is maintained for 20 minutes at that temperature in a specified apparatus.

The most important factors determining the efficiency of sublimation are (a) the nature of the sublimand, (b) the pressure at which the operation takes place, (c) the temperature gradient between the surfaces of the sublimand and the condenser, (d) the type of sublimation vessel, particularly the distance between the sublimand and the condensing surface, and (e) the presence of a nonvolatile impurity in the sublimand.

The nature of the sublimand or, rather, the vapor pressure of the crystals and their crystal habit determines whether a compound can be sublimed readily; if the vapor pressure of the crystals is very low, the rate of sublimation may be so slow as to make the process impractical. For example, naphthalene, benzoic acid, salicylic acid, and anthracene readily yield excellent crystalline sublimates at atmospheric pressure; other substances[37] —sucrose and lactose, for example—do not form sublimates at about 10 μ of mercury pressure. L-Leucine (m.p. 287° to 288°) sublimes slowly, phenolphthalein (m.p. 260°) very slowly, and caffeine (m.p. 236°) very rapidly, as compared with the preceding two compounds.

Evacuation of the sublimation apparatus removes most of the air molecules which interfere with the passage of the vapor detached from the sublimand. Most substances that distill at atmospheric pressure can be sublimed even if the distance between the sublimand and the condenser is not short, provided that a good vacuum is obtained. A large number of organic compounds in quantities of 1 to 10 mg can be readily sublimed at a vacuum of 1 to 5 mm. With substances like the amino acids, which cannot

---

[36] H. Hoffmann, Jr., and W. C. Johnson, *J. Assoc. Offic. Agr. Chem.*, **13**, 367 (1930).
[37] H. M. Hubacher, *Ind. Eng. Chem., Anal. Ed.*, **15**, 448 (1943); R. Eder and W. Haas, *Mikrochemie, Emich Festschr.*, **1930**, 43.

be distilled either at atmospheric pressure or under high vacuum with the usual distillation apparatus, a vacuum of 10 to 50 microns is necessary for efficient sublimation. It is advisable to evacuate the apparatus to a pressure below the vapor pressure of the substance at its melting point.

The rate of condensation depends on the temperature of the sublimand and the temperature of the condenser. If the rate of production of the vapor is high and the temperature of the condenser is low, the rate of accumulation of the sublimate is high. This, however, is not necessarily a desirable condition. Rapid evaporation of the sublimand does not favor fractionation and removal of impurities; moreover, sudden cooling and fast formation of crystals result in the accumulation of a hard micro crystalline coating on the walls of the condenser and in the formation of flocks of crystals that detach easily and fall when the condenser is removed to collect the sublimate. For compounds that melt between 40° and 100° the sublimation temperature should be at least 10° below the melting point; for compounds that melt between 100° and 200° it should be 25 to 50° below the melting point; and for compounds that melt above 200° it should be about 50 to 75° below the melting point. The temperature of the condenser may be varied between a few degrees below the melting point of the sublimate and the temperature of tap water. If the sublimate is to be used for crystallographic work, the condenser should be kept at a temperature below that of the sublimand; this condition favors the slow growth of euhedral crystals.[38] On the other hand, a hard coating of small crystals is not a disadvantage for purification or for determination of melting points, and for some purposes it is decidedly advantageous. The coating can be removed with the sharp blade of the microspatula or, better still, by washing the condenser with a small amount of appropriate solvent.

The distance between the sublimand and the condensing surface is of fundamental importance for compounds that have very low vapor pressures. In the apparatus described in this work, the distance between the bottom of the distillation pot and the condenser is 8 to 10 mm.

If the sublimand contains a nonvolatile impurity, it is possible for this layer to cover the entire surface of the sublimand so that sublimation is virtually stopped. In molecular distillations the "moving film" prevents accumulation of a nonvolatile layer. When a nonvolatile film forms and the sublimation is arrested, it is advisable to disconnect the apparatus and pulverize the sublimand further with a small agate pestle; it should then be spread in a thin layer and sublimation resumed. If this treatment does not suffice, the condenser is removed, a few drops of solvent are added,

---

[38] A. C. Shead, *Proc. Okla. Acad. Sci.*, **15**, 86 (1935); **16**, 87 (1936).

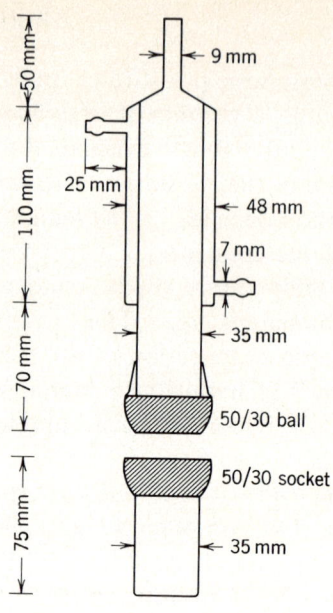

FIGURE 3.32

Sublimation apparatus with large condensing surface.[40]

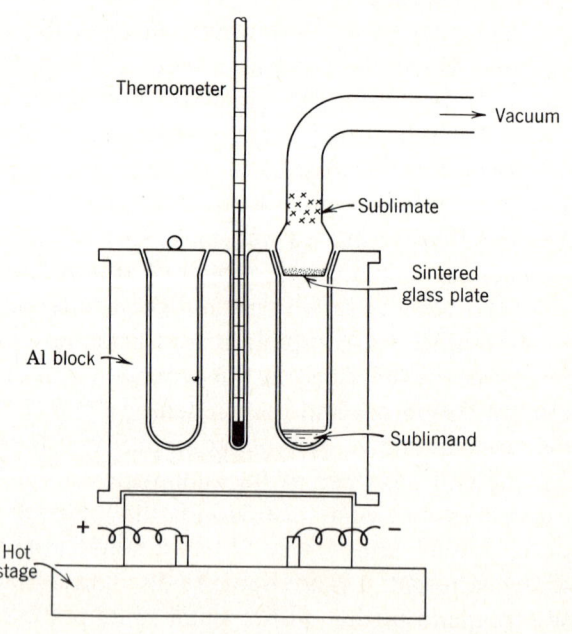

FIGURE 3.33

Apparatus for microsublimation. Sintered glass plate prevents sublimate from returning to the heated vessel.[42]

and the resulting solution is distributed so that it covers the bottom of the sublimator. As soon as practically all the solvent has evaporated, sublimation is resumed. For a more detailed discussion of sublimation of solids at atmospheric pressure by means of watch glasses or from glass slides, reference is made to the literature.[39]

Another apparatus for sublimation of solids is particularly useful because it affords a large surface for condensation of the sublimate.[40] The apparatus, shown in Figure 3.32, is essentially a condenser with a large bore and a small cup joined by a ball and socket ground joint. The cup is heated in an oil bath, water is run through the condenser jacket, and a vacuum of 0.05 mm or less is applied to the apparatus. The sublimate collects on the inner walls of the condenser about 10 to 50 mm above the oil bath level and, at the end of the experiment, is scraped out with a spatula. The apparatus may be purchased.[41]

One difficulty encountered in almost all apparatuses used for sublimation is caused by the poor adherence of the sublimed solid to the collecting surface; the sublimate often falls back into the material being sublimed. The apparatus[42] shown in Figure 3.33 prevents this very efficiently by use of a fritted glass plate. The vapors pass through the porous plate, but the solid particles cannot. Some loss of desired material, however, may occur unless the walls are cooled.

Simple sublimation may be performed on a hot stage under a microscope, using a sublimation dish and cover. The process can be observed in detail, and may be done at atmospheric pressure or at reduced pressure.[43]

Fractional Sublimation under Reduced Pressure

For an extensive survey of the various assemblies for sublimation at reduced pressure that have been described in the literature, see a work by one of us.[44] In the present section, several kinds of apparatus will be described.

---

[39] R. S. Tipson, "Sublimation," in *Technique of Organic Chemistry*, Vol. 4, A. Weissberger, ed., Interscience, New York, 1951, pp. 603–646; N. D. Cheronis, "Micro and Semimicro Methods," id., Vol. 6, pp. 84–96; A. A. Benedetti-Pichler, *Microtechnique of Inorganic Analysis*, Wiley, 1942, p. 91; and F. L. Schneider, *Organic Qualitative Microanalysis*, Wiley, 1942, p. 67.
[40] F. B. Mallory, *J. Chem. Educ.*, **39**, 261 (1962).
[41] Kontes Glass Co., Vineland, N.J., Catalog No. K-85500.
[42] W. Hausman, *Anal. Chem.*, **26**, 619–20 (1954).
[43] N. D. Cheronis, "Micro and Semimicro Methods," in *Technique of Organic Chemistry*, Vol. 6, A. Weissberger, ed., Interscience, New York, 1954, pp. 94–96.
[44] N. D. Cheronis, "Micro and Semimicro Methods," in *Technique of Organic Chemistry*, Vol. 6, A. Weissberger, ed., Interscience, New York, 1954, pp. 89–96, 380–384.

Figure 3.34 shows a microsublimator tube. The sublimator part of the apparatus is 25 × 100 mm; its bottom is rounded in the shape of a bulb with a maximum diameter of about 30 mm. The condenser is about 120 mm in length, 8 mm in diameter above the middle portion, and 24 mm in diameter at the wide part of the conelike lower portion. The upper part of the cone dips to form a well. The glass joint of the sublimator is lubricated with a light grease.[45]

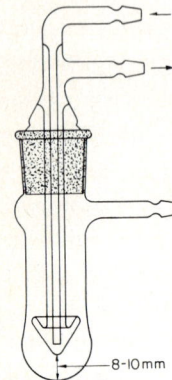

FIGURE 3.34
Microsublimator tube.

The second type of apparatus, described by Ronzio,[46] consists of a distillation pot connected by means of a semiball joint as shown in Figure 3.35. The apparatus can be obtained in two dimensions; one with a pot 80 mm in diameter and another with a 25-mm pot. The larger model, although more expensive, is more efficient because small sublimators lead to repeated "blocking" of sublimation and necessitate frequent redissolving and redistribution of the compound being purified. The specific procedure employed for purification by microsublimation depends upon whether the impurity is more or less volatile than the compound being purified. If the substance is unknown, about 10 mg of finely pulverized material is spread at the bottom of the sublimator (an alternate procedure is to place a solution of the compound to be sublimed in the kettle and remove the solvent).

The condenser joint is greased very lightly, the condenser is adjusted in place, and water is run through very slowly, heat is applied from a shallow bath into which the sublimator dips about 5 to 6 mm. For ordinary work, a pressure of 5 to 20 mm is satisfactory. The temperature is raised gradually to 45 to 50° and kept in this range for about 30 minutes. If no cloudy film forms at the lower part of the condenser, the temperature is raised, stepwise, 10 to 15° and allowed to remain for 0.5 to 1 hour at each interval until a film of sublimate is obtained. After about 1 mg of sublimate has been formed, the vacuum is released gradually, the condenser is lifted over a glass slide, and (with the sharp end of the spatula blade) a few crystals are detached from the center of the glass slide for microscopic examination, which is followed by determination of the melting point. The condenser is now washed with a thin stream of solvent and the material is recovered after evaporation of the solvent. In this manner,

[45] Among the greases used for vacuum sublimators, Celvasene Light (Distillation Products Industries, Rochester, N.Y.) is preferred.
[46] Described in N. D. Cheronis, "Micro and Semimicro Methods," in *Technique of Organic Chemistry*, Vol. 6, A. Weissberger, ed., Interscience, New York, 1954, pp. 380–384.

6 to 8 fractions are obtained, and it is possible, on the basis of the melting point data, to decide upon a plan of fractionation. An illustration of fractional sublimation for the purification of derivatives is described in some detail on pages 477–479.

If the nature of the impure sublimand is known, and if it melts below 200°, the temperature is adjusted 10 to 15° below its melting point and fractions of about 1 mg are sublimed until the sublimate gives the desired

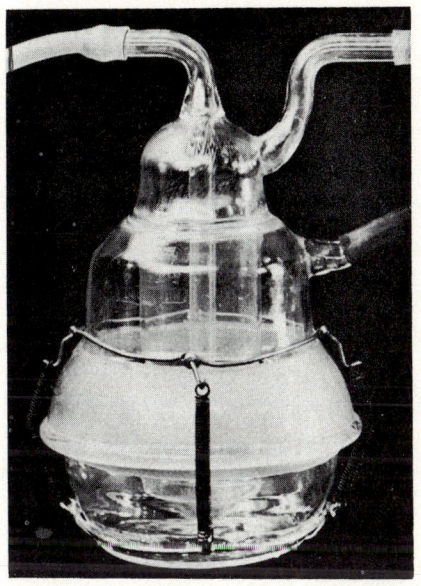

FIGURE 3.35
Sublimator with semiball joint (after Ronzio).[46]

melting point. In the purification of derivatives a rapid procedure is to fractionate one third of the sublimand and, afterward, to collect a second fraction that is usually relatively pure.

If the desired compound melts above 200° and no information is available regarding the appropriate temperature and pressure for efficient sublimation, it is best to start with a pressure of 10 to 50 microns and a temperature of 100 to 150°. If the vacuum system gives a pressure of only 1 to 10 mm, it is advisable to raise the temperature to 150 to 180° and allow the sublimation process to proceed for 6 to 12 hours or more.

An apparatus that is capable of the fractional sublimation of atmospheric pollutants has been described.[47] This apparatus, shown schematically in Figure 3.36, should prove very successful in separating small quantities of other mixtures. The mixture (10 to 20 mg) is weighed into the sample boat and placed in the sublimation tube as shown. The pressure in the tube is reduced to 0.01 to 5.0 mm of mercury and is carefully regulated throughout the sublimation. The end of the sublimation

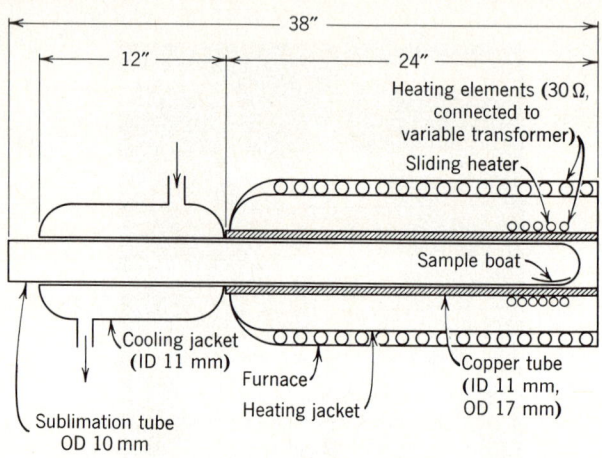

FIGURE 3.36

Apparatus for fractional sublimation. A linear temperature gradient along the sublimation tube is achieved by use of the heavy-wall copper tube and the special heater.[47]

tube holding the sample is carefully heated so as to establish a linear temperature gradient along the tube. This is done by heating one end of a thick-wall copper tube which encases the sublimation tube. The most volatile compound escapes from the boat first, moves most rapidly along the sublimation tube, and condenses first on the cold surface. The tube is then withdrawn slightly from the furnace, giving a fresh cool surface for the next fraction. The tube is cut into segments at the end of the sublimation in order to isolate each fraction. By properly manipulating the temperature, pressure, and position of the tube, it is possible to achieve remarkably good separations of high molecular weight compounds.

Another apparatus uses a linear temperature gradient to separate substances with different sublimation properties.[48] By use of a plastic sheet, the crystals can be obtained in their original state for further studies.

[47] J. F. Thomas, E. N. Sanborn, M. Mukai, and B. D. Tebbens, *Anal. Chem.*, **30**, 1954–1958 (1958).
[48] G. Schmidt, *Mikrochim. Acta.* **1959**, 406–418.

EXTRACTION

The methods for extraction of aqueous solutions by means of ether, benzene, and other solvents immiscible with water require little discussion. The apparatus for the separation of immiscible liquids is described in Chapter 2. In dealing with the extraction of unknown organic compounds, consideration should be given to the stability of the solute in the solvent, the ease of removing the solvent, and the pH of the aqueous solution. For example, it is possible with proper adjustment of pH and addition of another solute to remove a component readily from a mixture.

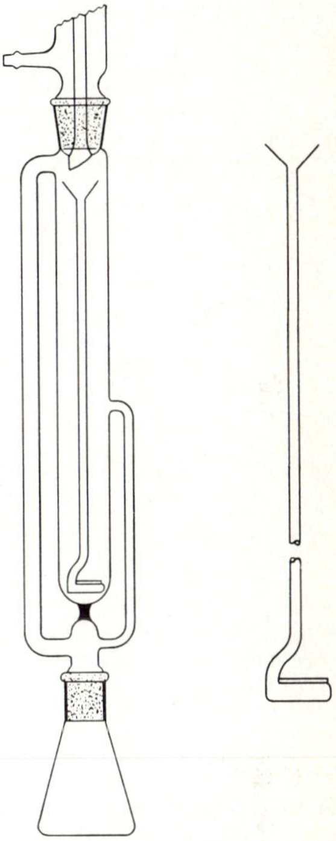

Smith and Page[49] used a chloroform solution of a long-chain tertiary amine to remove strong acids, such as hydrochloric or sulfuric, from aqueous solutions; the amine forms a salt with the acid that is preferentially soluble in chloroform; the resulting distribution ratio makes possible the removal of 98 per cent of the acid in a single extraction, provided that such cations as sodium, potassium, and the like are absent. Therefore, some preliminary trials are necessary, using 1 ml of the solution in a test tube and employing the capillary pipet technique to separate the organic solvent (page 30).

For a more extensive discussion of extraction than is given here, see the literature.[50] A thorough discussion of extraction for fractionation purposes is given by Craig and Craig.[50c]

FIGURE 3.37
Liquid-liquid extractor for use with solvents lighter than water.

Differential extraction of solid mixtures by fractionation into relatively pure substances is a standard procedure as described in Chapter 6. After exploratory trials have been made on 25 to 50 mg of the mixture and the solvent systems have been selected, it often becomes necessary to submit the major part of the unknown sample to

---

[49] E. L. Smith and J. E. Page, *J. Chem. Soc. Ind.*, **67**, 48 (1948).
[50] N. D. Cheronis, "Micro and Semimicro Methods," in *Technique of Organic Chemistry*, Vol. 6, A. Weissberger, ed., Interscience, New York, 1954, pp. 97–105; (*b*) G. H. Morrison, *Anal. Chem.*, **22**, 1388 (1950); and (*c*) L. C. Craig and D. Craig, *Laboratory Extraction and Countercurrent Distribution*, 2nd ed., Vol. 3, Part 1, A. Weissberger, ed., Interscience, New York, 1956, pp. 149–332.

differential continuous extraction by a series of solvents. It should be pointed out that many times a solvent *that does not dissolve an appreciable amount* of a component may be employed advantageously in continuous extractions of both solids and liquids.

Two assemblies for liquid-liquid continuous extraction are shown in Figures 3.37 and 3.38. The continuous extractor shown in Figure 3.37 is for use with solvents lighter than water; that shown in Figure 3.38 is for use with solvents heavier than water. Continuous liquid-liquid extractions may be used to remove organic substances from an aqueous solution or dispersion even though the distribution coefficient is relatively small, so that a large number of batch extractions would normally be necessary to effect quantitative separation. The solvent in both types of extractors is placed in the boiler flask and distilled; the vapors pass upward to the condenser, and the condensate is returned through a tube having a disperser at the end (usually fritted glass), which breaks the solvent in minute droplets that pass through the aqueous phase being extracted. In this manner the contact area between the two immiscible liquids is increased. The solvent-extract separates out and is returned by the siphon system to the boiler where the solvent is recycled; the solute remains in the flask, and when the operation is completed it is recovered either by evaporation of the solvent or stripped with another solvent in which the solute is very soluble.

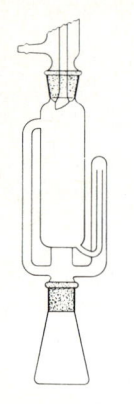

FIGURE 3.38

Liquid-liquid extractor for use with solvents heavier than water.

The effect of pH in removing small amounts of organic substances from mixtures so that they can be recovered almost quantitatively by continuous extraction can be illustrated by citing a few examples. Assume that it is desired to separate a small amount of fatty acids from a mixture that contains mostly sterols. If the mixture is shaken with dilute alkali the fatty acids are converted to their sodium salts, which are very soluble in water and thus are separated from the sterols, which now can be removed by one or two extractions with benzene. The alkaline aqueous phase is adjusted to a pH below 7.0 and subjected to continuous extraction either with chloroform in the extractor shown in Figure 3.38 or ether in the extractor shown in Figure 3.37. Similarly, a mixture of aldehydes can be separated from other organic compounds by converting them first to bisulfite addition products, which are soluble in water, and then raising the pH of the aqueous phase to decompose the addition product and subjecting it to continuous extraction. In general, it is possible by judicious selection of solvents to remove small amounts of a large number of organic compounds from their reaction mixtures by continuous extraction with the minimum losses in handling.

REFERENCES

Crystallization

A. A. Benedetti-Pichler, *Introduction to the Microtechnique of Inorganic Analysis*, Wiley, New York, 1942, pp. 202–204. Filter sticks.

A. L. Bluhm, *J. Chem. Educ.*, **35**, 200 (1958). Apparatus for semimicro low-temperature crystallizations.

M. T. Bush, *Microchem. J.*, **1**, 105 (1957). Crystallization with centrifugal filtration.

M. T. Bush, *Ind. Eng. Chem., Anal. Ed.*, **18**, 584 (1946). Small glass centrifugal filters.

N. D. Cheronis, "Micro and Semimicro Methods," in *Technique of Organic Chemistry*, Vol. 6, A. Weissberger, ed., Interscience, New York, 1954, pp. 13–50. Small scale crystallizations.

L. C. Craig, *Ind. Eng. Chem., Anal. Ed.*, **12**, 773 (1940). Micro apparatus for fractional recrystallization.

L. C. Craig and O. W. Post, *Ind. Eng. Chem., Anal. Ed.*, **16**, 413 (1944). Apparatus for small scale crystallizations.

F. Emich and F. L. Schneider, *Microchemical Laboratory Manual*, Wiley, New York, 1932, pp. 31–32 and 127–129. Small-scale crystallizations.

J. English, Jr., *Ind. Eng. Chem., Anal. Ed.*, **16**, 478 (1944). Apparatus for small-scale crystallizations.

J. P. Friedrich, *Anal. Chem.*, **33**, 974 (1961). Apparatus for small-scale crystallizations at low temperatures.

W. E. Gibbs, *Trans. Inst. Chem. Engrs. (London)*, **8**, 38 (1930). Formation and growth of crystals.

T. Kato, *J. Pharm. Soc. Japan*, **60**, 228 (1940); *C.A.*, **36**, 2182 (1942). Crystallization of small quantities.

T. S. Ma and R. T. Schenck, *Mikrochemie*, **40**, 242 (1942). Small scale drying apparatus.

M. Martin-Smith, *Lab. Practice*, **7**, 572 (1958). Apparatus for preparing and filtering hot solutions.

J. D. Piper, N. A. Kerstein, and A. G. Fleiger, *Ind. Eng. Chem., Anal. Ed.*, **14**, 738 (1942). Apparatus for repeated crystallizations.

F. Pregl, *Mikrochemie*, **2**, 76 (1924). Apparatus for small-scale crystallizations.

F. W. Quackenbush and H. Steenbock, *Ind. Eng. Chem., Anal. Ed.*, **14**, 736, 738 (1942). Crystallization at low temperature or in enclosed systems.

J. D. Reinheimer, *J. Chem. Educ.*, **30**, 139 (1953). Drying apparatus for small samples.

F. L. Schneider, *Organic Qualitative Microanalysis*, Wiley, New York, 1946, p. 44. Small-scale crystallization apparatus.

E. Sellier, *Chem. Trade J.*, **97,** 259 (1935). Principles of crystal formation and growth.

E. L. Skau, *J. Phys. Chem.*, **33,** 951 (1929). Centrifugal filtration apparatus.

E. L. Skau, J. C. Arthur, Jr., and H. Wakeham, "Determination of Melting and Freezing Temperatures" in *Technique of Organic Chemistry*, 3rd ed., Vol. 1, Part 1, A. Weissberger, ed., Interscience, New York, 1959, pp. 287–355. Melting temperatures from cooling curves, pp. 312–318.

T. B. Smith, *Analytical Processes*, Arnold, London, 1940, pp. 335–371. Supersaturation and crystallization.

R. Steinmann, *Mikrochim. Acta*, **1953,** 494. Bibliography on microcrystallization to 1949.

R. S. Tipson, "Crystallization and Recrystallization," in *Technique of Organic Chemistry*, 2nd ed., Vol. 3, Part 1, A. Weissberger, ed., Interscience, New York, 1956, pp. 395–562. Theory and practice, mostly macro scale. There are 829 references.

R. S. Tipson, *Anal. Chem.*, **22,** 628 (1950). Theory of crystallization.

A. Wexler, *J. Chem. Educ.*, **18,** 167 (1941). Small-scale crystallization.

Semimicro Fractional Distillation

R. H. Baker, C. Barkenbus, and C. A. Roswell, *Ind. Eng. Chem., Anal. Ed.*, **12,** 468 (1940). Spinning-band semimicro column.

M. Blumer, *Anal. Chem.*, **34,** 704 (1962). Efficient fractionation of milligram quantities.

J. E. Bolzan, *Chemist Analyst*, **50,** 120 (1961). Ebullition aid.

N. D. Cheronis, "Micro and Semimicro Methods," in *Technique of Organic Chemistry*, Vol. 6, A. Weissberger, ed., Interscience, New York, 1954, pp. 57–75. Small-scale fractional distillation.

E. A. Coulson and E. F. G. Herington, *Laboratory Distillation Practice*, Interscience, New York, 1958.

T. P. Carney, *Laboratory Fractional Distillation*, Macmillan, New York, 1949. Theory and practice, mostly macro.

L. C. Craig, *Ind. Eng. Chem., Anal. Ed.*, **9,** 441 (1937). Microapparatus for distillation.

A. L. Glasebrook and F. E. Williams, "Ordinary Fractional Distillation. Part 1. Apparatus," in *Technique of Organic Chemistry*, Vol. 4, A Weissberger, ed., Interscience New York, 1951, pp. 175–294. A review of apparatus, mostly macro scale. There are 198 references.

C. W. Gould, Jr., G. Holzman, and C. Nieman, *Anal. Chem.*, **20,** 361 (1948). Distillation equipment for centigram and decigram quantities.

H. M. Haendler, *Anal. Chem.*, **20,** 596 (1948). Semimicro distilling head.

R. A. Hall, *Anal. Chem.*, **31,** 437 (1959). Vapor take-off still head.

R. Handley and B. Holgate, *Chem. & Ind.* (*London*), **1959,** 1087. New packing for columns.

G. R. Lappin, *J. Chem. Educ.*, **25,** 657 (1948). Semimicro distilling tube.

H. S. Lecky and R. H. Ewell, *Ind. Eng. Chem., Anal. Ed.*, **12,** 544 (1940). Construction and packing of semimicro columns.

S. D. Lesesne and N. R. Lochte, *Ind. Eng. Chem., Anal. Ed.*, **10,** 450 (1938). Small-scale fractionating equipment.

A. A. Morton and J. F. Mahoney, *Ind. Eng. Chem., Anal. Ed.*, **13**, 494 (1941). Analysis of a single drop by micro fractionation.

E. A. Naragon and C. J. Lewis, *Ind. Eng. Chem., Anal. Ed.*, **18**, 448 (1946). Construction and packing of semimicro columns.

A. F. Nerheim, *Anal. Chem.*, **31**, 2114 (1959). Improved head and flask for small-scale distillation.

A. F. Nerheim, *Anal. Chem.*, **29**, 1546 (1957). Teflon bands for spinning-band columns.

W. J. Podbielniak and S. T. Preston, "Analytical Distillation," in *Physical Methods in Chemical Analysis*, Vol. 3, W. G. Berl, ed., Academic Press, New York, 1956, pp. 402–448. Review of theory.

E. O. Ramler and J. H. Simons, *Ind. Eng. Chem., Anal. Ed.*, **14**, 430 (1942). Small-scale fractionating equipment.

C. E. Redemann, *Ind. Eng. Chem., Anal. Ed.*, **11**, 635 (1939). Apparatus for semimicro Kjeldahl distillation.

G. S. Ross and L. J. Frolen, *J. Research Natl. Bur. Standards*, **62**, 187 (1959) RP 2951. Simple rotating molecular still.

M. L. Selker, R. E. Burk, and H. P. Lankelma, *Ind. Eng. Chem., Anal. Ed.*, **12**, 352 (1940). Construction and packing of semimicro columns.

R. Steinmann, *Mikrochim. Acta*, **1953**, 491. Bibliography on micro distillation and micro extraction to 1949.

J. T. Stock and M. A. Fill, *Mikrochim. Acta*, **1953**, 103. Alarm device for distillation.

W. Swietoslawski, *J. Chem. Educ.*, **5**, 469–472 (1928). Semimicro distillation.

F. E. Williams, "Ordinary Fractional Distillation. Part 2. Procedure," in *Technique of Organic Chemistry*, Vol. 4, A. Weissberger, ed., Interscience, New York, 1951.

Distillation under Reduced Pressure

M. J. Babcock, *Anal. Chem.*, **21**, 632 (1949). Micro vacuum distillation.

J. R. Bowman and R. S. Tipson, "Distillation under Moderate Vacuum," in *Technique of Organic Chemistry*, Vol. 4, A. Weissberger, ed., Interscience, 1951, pp. 463–494. A review with 90 references.

G. A. Dalin, *Ind. Eng. Chem., Anal. Ed.*, **15**, 731 (1943). Pressure regulation for vacuum distillation.

D. W. S. Evans, *Chem. & Ind. (London)*, **1959**, 219. Fraction collector for small-scale vacuum distillation.

W. M. Grant, *Ind. Eng. Chem., Anal. Ed.*, **18**, 729 (1946). Vacuum distillation of milliliter volumes.

K. Groves and R. R. Legault, *Anal. Chem.*, **29**, 1724 (1957).

J. C. Hecker, "Distillation under High Vacuum. Part 2. The Vacuum System," in *Technique of Organic Chemistry*, Vol. 4, A Weissberger, ed., Interscience, New York, 1951, pp. 540–602. A review with 141 references.

E. B. Hershberg and E. H. Huntress, *Ind. Eng. Chem., Anal. Ed.*, **5**, 344 (1933). Pressure regulator for distillations.

K. C. D. Hickman and C. R. Sanford, *J. Phys. Chem.*, **34**, 637 (1930). Molecular still.

R. E. Jentoff and A. A. Carlstrom, *Chemist Analyst*, **50**, 116 (1961). Automatic pressure regulator for vacuum distillations.

E. Klenk, *Z. physiol. Chem.*, **242**, 250 (1936). Fractionation under reduced pressure.

E. Knobloch, *Chem. listy*, **53**, 718 (1959). Molecular distillations.

N. B. Mehta and J. Zupicich, *Chemist Analyst*, **50**, 84 (1961). Short path tubular high vacuum distillation and sublimation.

R. G. Nester, *Rev. Sci. Instr.*, **31**, 1002 (1960). Falling film molecular still.

L. V. Peakes, *Mikrochimie*, **18**, 100 (1935). Fractionation under reduced pressure.

E. S. Perry, "Distillation under High Vacuum. Part 1. Distillation," in *Technique of Organic Chemistry*, Vol. 4, A. Weissberger, ed., Interscience, New York, 1951, pp. 495–540. A review with 72 references.

J. Radell and J. W. Connolly, *J. Chem. Educ.*, **38**, 459 (1961). Fraction collector for continuous collection of samples.

B. Riegel, J. Beiswanger, and G. Lanzl, *Ind. Eng. Chem.*, *Anal. Ed.*, **15**, 417 (1943). Vacuum sublimation and molecular still apparatus.

T. Sakuragi and F. A. Kummerow, *Anal. Chem.*, **26**, 620 (1954). Fraction cutter for high vacuum distillation.

G. R. Sanders and H. L. Helwig, *Anal. Chem.*, **31**, 484 (1959). Molecular distillation unit.

S. A. Schrader and J. E. Ritzer, *Ind. Eng. Chem.*, *Anal. Ed.*, **11**, 54 (1939). Fractionation under reduced pressure.

A. Soltys, *Mikrochim. Acta*, Melisch Festschr. **1936**, 393. Fractionation under reduced pressure.

C. Tiedcke, *Ind. Eng. Chem.*, *Anal. Ed.*, **15**, 81 (1943). Fractionation under reduced pressure.

H. E. Ungnade, *Anal. Chem.*, **31**, 1126 (1959). Molecular still.

P. R. Watt, *Chem. & Ind. (London)*, **1960**, 713, 1207. Semimicro molecular still.

P. R. Watt, *Molecular Stills*, Reinhold, New York, 1963. A compilation of all types of molecular stills.

G. F. Wright, *Can. J. Research*, **B17**, 302 (1939). Fractionation under reduced pressure.

Steam Distillation

G. S. Duboff, *Analyst*, **84**, 619 (1959). Self-cleaning steam distillation apparatus.

J. Erdos and B. Laszlo, *Mikrochim. Acta*, **3**, 304 (1938).

L. R. Fina and H. J. Sincher, *Chemist Analyst*, **48**, 83 (1959). Micro steam distillation apparatus.

J. L. Hoskins, *Analyst*, **69**, 271 (1944).

F. T. Wallenberger, W. F. O'Connor, and E. J. Moriconi, *J. Chem. Educ.*, **36**, 251 (1959). Universal apparatus for steam distillation.

Sublimation

T. H. Bates, *Chem. & Ind. (London)*, **1958**, 1319. Micro fractional vacuum sublimation and distillation apparatus.

A. A. Benedetti-Pichler, *Microtechnique of Inorganic Analysis*, Wiley, New York, 1942, p. 91.

N. D. Cheronis, "Micro and Semimicro Methods," in *Technique of Organic Chemistry*, Vol. 6, A Weissberger, ed., Interscience, New York, 1954, pp. 84–96. There are 38 references.

B. L. Clarke and H. W. Hermance, *Ind. Eng. Chem., Anal. Ed.*, **11,** 50 (1939). Micro and semimicro sublimation.

R. Eder and W. Haas, *Mikrochemie, Emich Festschr.* **1930,** 43.

R. Fischer, *Mikrochemie*, **15,** 247 (1934). Microsublimation under a microscope.

A. O. Gettler, C. J. Umbreger, and L. R. Goldbaum, *Anal. Chem.*, **22,** 601 (1950). Vertical sublimator.

W. Hausman, *Anal. Chem.*, **26,** 619 (1954) Apparatus for fractional sublimation at normal or reduced pressure.

H. Hoffmann, Jr., and W. C. Johnson, *J. Assoc. Offic. Agr. Chemists*, **13,** 367 (1930). Subliming point.

M. H. Hubacher, *Ind. Eng. Chem., Anal. Ed.*, **15,** 448 (1943). Sublimation requirements of substances.

W. Hurka, *Mikrochemie*, **30,** 193 (1942). Microsublimation.

S. Kandersteg, *Apotheker Z.*, **82,** 61, 81 (1944). Separation of physiologically active compounds from pharmaceutical preparations by sublimation.

R. Kempf, *Z. anal. Chem.*, **62,** 284 (1923). Apparatus for microsublimation.

F. B. Mallory, *J. Chem. Educ.*, **39,** 261 (1962). Sublimation apparatus with large condensing surface.

C. M. Marberg, *J. Am. Chem. Soc.*, **60,** 1509 (1938). Micro and semimicro sublimation.

W. H. Melhuis, *Nature (London)*, **184,** 1933 (1959). Fractional micro sublimation technique.

A. A. Morton, J. F. Mahoney, and G. R. Richardson, *Ind. Eng. Chem., Anal. Ed.*, **11,** 460 (1939). Microsublimator.

M. Oakley, *J. Assoc. Offic. Agr. Chemists*, **28,** 298 (1945). Separation of saccharin and benzoic acid from food by sublimation.

R. H. Petrucci and J. C. Weygandt, *Anal. Chem.*, **33,** 275 (1961). Sublimation procedure for detection and removal of impurities from organic solids.

L. Rosenthaler, *Mikrochemie*, **45,** 165 (1950). Sublimation at atmospheric pressure.

G. Schmidt, *Mikrochim. Acta*, **1959,** 406.

A. C. Shead, *Mikrochim. Acta*, **1959,** 657. Sublimation giving crystals suitable for measurement of profile angles.

A. C. Shead, *Proc. Okla. Acad. Sci.*, **15,** 86 (1935); **16,** 87 (1936).

R. Steinmann, *Mikrochim. Acta*, **1953,** 494. Bibliography on microsublimation to 1949.

J. F. Thomas, E. N. Sanborn, M. Mukai, and B. D. Tebbens, *Anal. Chem.*, **30,** 1954 (1958). Fractional sublimation apparatus.

R. S. Tipson, "Sublimation," in *Technique of Organic Chemistry*, Vol. 4, A. Weissberger, ed., Interscience, New York, 1951, pp. 603–646.

J. A. Zapotocky and L. E. Harris, *J. Am. Pharm. Assoc., Sci. Ed.*, **38,** 557 (1949). Assay of cinchona and nux vomica by microsublimation.

Extraction

L. Alders, *Liquid-Liquid Extraction. Theory and Laboratory Practice*, Elsevier, Amsterdam, 1959.

F. C. Alderweireldt, *Anal. Chem.*, **33,** 1920 (1961). Steady batchwise separation by extraction.

E. W. Berg, *Physical and Chemical Methods of Separation*, McGraw-Hill, New York, 1963, pp. 50–79.

N. D. Cheronis, "Micro and Semimicro Methods," in *Technique of Organic Chemistry*, Vol. 6, A. Weissberger, ed., Interscience, New York, 1954, pp. 97–105. There are 40 references.

L. C. Craig and D. Craig, "Laboratory Extraction and Counter-current Distribution," in *Technique of Organic Chemistry*, 2nd ed., Vol. 3, Part 1, A. Weissberger, ed., Interscience, New York, 1956, pp. 149–332. Theory and apparatus. There are 230 references.

J. Erdos, *Mikrochim. Acta*, **1961**, 515. Microapparatus for extraction and distillation.

C. H. Jarboe, *Chemist Analyst*, **50**, 120 (1961). Immersion gas-lift circulator for extraction.

P. Sosis, *Chemist Analyst*, **50**, 85 (1961). Liquid-liquid extractor.

F. A. von Metzsch, "Solvent Extraction," in *Physical Methods in Chemical Analysis*, Vol. 4, W. G. Berl, ed., Academic Press, New York, 1961, pp. 317–456. Review of theory and apparatus. Many tables.

R. Wade, C. H. Mack, and A. Mason, *Chemist Analyst*, **50**, 52 (1961). All-glass perfusion apparatus with vapor pump.

4

Fractionation Procedures

CHROMATOGRAPHY: PAPER, COLUMN, THIN-LAYER, ION-EXCHANGE, AND GAS; OTHER METHODS OF FRACTIONATION

CHROMATOGRAPHIC PROCEDURES

One of the most useful tools at the disposal of the analyst is chromatography. The advantages of this method are (a) its ability to separate complex mixtures heretofore difficult or impossible to separate by other means; (b) its ability to separate mixtures of very small quantities or at very low concentrations; and (c) the simplicity of the method and the relatively inexpensive equipment required.

The purpose of this section is to present the basic principles of the method and the more common experimental techniques of adsorption and partition chromatography. An extensive account of the theory and experimental details is beyond the scope of this work. However, the selected references at the end of this chapter will serve as a guide to a more complete understanding of the method and to recent developments in the field.

Principles of the Chromatographic Separation

The chromatographic process involves the distribution of a solute, or the "adsorptive substance," between two phases, one of which is stationary or immobile, and the other mobile. In a relative sense, this represents a countercurrent distribution. In paper chromatography the immobile phase is considered to be water molecules bound into the cellulose network of the paper, or perhaps it is a saturated solution of cellulose, whereas the mobile phase may be any one of a number of pure or mixed solvents. Distribution of the solute between the "bound" water and the mobile phase, or developing solvent, results in movement of the solute through the paper. In adsorption chromatography the immobile phase is a solid adsorbent, such as alumina, magnesia, silicic acid, calcium carbonate, or charcoal, packed in a glass column or spread in a thin layer on a plate, and, again, the mobile phase may be one of several pure or mixed solvents. In either case, the pure solvent or the mixed solvent system comprising the mobile phase is commonly called the *developer* or the *eluent*. Oftentimes, in adsorption chromatography, the term *eluent* is applied to the highly polar solvent that is used to desorb from the adsorbent column a component, which has been isolated from a mixture. For both adsorption and partition chromatography, the relative movement of the solute through the immobile phase has great significance, and is designated by a special term, the R_f value. The R_f value relates the movement of the solute to the movement of the developing solvent in the following manner.

$$R_f = \frac{\text{Rate of movement of solute}}{\text{Rate of movement of developing solvent}}$$

Obviously, if the developing solvent and the solute are started at the same time on the column of paper, the ratio can be expressed in terms of the distances moved by each:

$$R_f = \frac{\text{Distance solute moved}}{\text{Distance solvent moved}}$$

In practice, the distance moved by the solute in a given length of time is measured from the point of application of the solute to the center of its zone, whereas the distance moved by the developing solvent is measured from the same origin to the point it has reached in the same length of time.

The factors that influence the movement of the solute through the immobile phase are many and varied. In paper chromatography the partition coefficient, that is, the ratio of the concentration of solute in the immobile phase to its concentration in the mobile phase, has been shown[1] to be quantitatively related to the R_f value by the expression

$$R_f = \frac{A_1}{A_1 + \alpha A_s}$$

where α = partition coefficient
A_1 = cross-sectional area of the mobile phase
A_s = cross-sectional area of the immobile phase

[1] R. Consden, A. H. Gordon, and A. J. P. Martin, *Biochem. J.*, **38**, 224 (1944).

This equation was derived from studies in partition chromatography with columns from which the values of A_1 and A_s are more readily obtained than from paper strips or sheets. Unfortunately, the limited number of partition coefficients available, particularly for the unusual solvent systems commonly encountered in paper chromatography, render impractical the use of this equation in a preliminary calculation of the R_f value. Assuming that adsorption between the solute and cellulose is negligible (actually evidence can be presented to support or refute this assumption), we need only consider the solubility of the solute in the two liquid phases. If the solute is highly soluble in the organic phase used as the developing solvent, it will move rapidly and will exhibit a high R_f value. On the other hand, should the solute show a low solubility in organic solvents and a high water solubility, it will move slowly, if at all, and will have a low R_f value. Thus, it is seen that in paper chromatography factors affecting the solubility of the solute in the two liquid phases account for its behavior in moving through the immobile water phase.

In adsorption chromatography the picture is somewhat different. The distribution of the solute in this case takes place between the developing solvent, as before, and a solid adsorbent. In addition to the characteristics of the solvent—for example, dipole moment, dielectric constant, and hydrogen-bonding ability—we must now consider the nature of the adsorbent and the forces of adsorption by which the solute is held to it. These may be weak Van der Waal forces or they may be sufficiently strong to suggest chemical reaction. It is generally believed that in many instances adsorption takes place through the formation of relatively strong hydrogen bonds between the solute and certain active sites on the surface of the adsorbent. Such a mechanism has been suggested for the adsorption of the carboxylic acids on silica gel,[2]

$$R-C \begin{matrix} O \cdots H-O \\ \\ O-H \cdots O \end{matrix} Si-O-H$$

The process of adsorption in chromatography is often considered as a competition between the solute and the developing solvent for sites on the adsorbent surface. Thus, if the attraction between the developing solvent and the adsorbent is higher than that of the solute for the adsorbent, the solute will find itself moving along with the solvent. If, on the other hand, the attraction of the solute for the adsorbent

---

[2] A. L. Elder and R. A. Springer, *J. Phys. Chem.*, **44**, 943 (1940).

surface is higher, the solute will tend to stick to the column and will consequently exhibit low R_f values. Dielectric constants, dipole moments, hydrogen-bonding ability, and relative polarizabilities of the three components are among the factors that decide the outcome of this competition. The separation of stereoisomers even implies that the shapes of the molecules play an important role in chromatographic analysis.

Paper Chromatography

General Techniques

Paper strips are commonly employed for a preliminary investigation of a solute mixture to establish the appropriate solvent for the effective separation of such a mixture. These strips may be cut from sheets or may be obtained in rolls of 3/4-inch or 1-inch width from most of the chemical supply houses. Care should be exercised in handling these papers, since grease from the hands and rough edges produced in cutting strips from sheets can interrupt the movement of the developing solvent up the strip and give rise to erratic R_f values. A few microliters (λ) of a solution of the solute mixture are applied at a marked point near one end of the paper strip. This end of the strip is then dipped into the developing solvent, and the development is allowed to proceed until the solvent has moved a predetermined distance. The solvent may be allowed either to ascend or to descend the paper strip. In the former case, the process is somewhat slower; but conditions of equilibrium with respect to the distribution of the solute between the two phases are more nearly attained, and the isolated zones of the components of the mixture are usually more compact. If the mixture is simple and good separation is indicated, the descending method may be preferred because of the time it saves. These two methods are illustrated in Figure 4.01 (the arrangement shown in Figure 4.01(*A*) may be employed for two-dimensional chromatography).

After development is complete, the distance that the solvent has moved is noted, and the paper strip is hung to dry. If the components of the mixture are colorless, the paper is then sprayed or streaked with a color-forming reagent or, in some cases, it is dipped in a solution of such a reagent. Many investigators refer to the process of color formation by spraying, streaking, or dipping as developing the chromatogram. We prefer to designate this process as *color development* or *location of spots* by color formation or other means, and to reserve the expression *developing the chromatogram* for the process of inducing separation of the components of a mixture by the flow of solvent along the paper. Consequently, the *developing solvent*, hereafter, both in adsorption and in paper chromatography, will be referred to as the *developer*. With the appearance of color in the various zones, the distance that each solute or component has moved can be measured from the point of application of the mixture to the leading edge of the respective zones. Thus, the R_f values may be calculated. It should be noted that in a number of cases the spots are located with the aid of ultraviolet light.

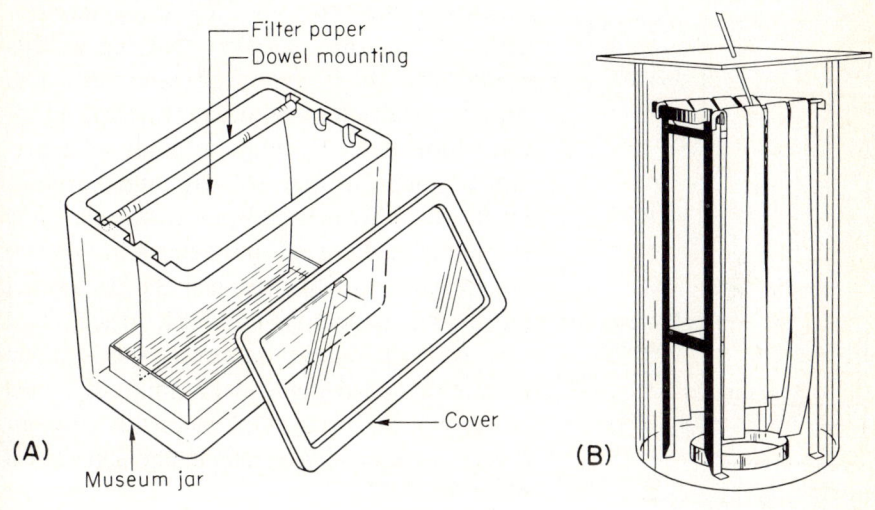

FIGURE 4.01

Apparatus for paper chromatography. (*A*) Ascending. (*B*) Descending.

To illustrate an R_f calculation, we shall assume that an organic compound is placed at position A, as shown in Figure 4.02. After the strip of paper has been developed with a mixture of 70 per cent *tert*-butyl alcohol and 30 per cent water, the solvent front has reached D. This point is marked with a pencil so that its location will not be lost on evaporation of the solvent. After the paper has been dried and sprayed with a suitable color reagent, the organic compound is found at position B. If the distance AD is 14 cm, and the distance AB is 9.5 cm, then the R_f value of this compound from *tert*-butyl alcohol-water in the percentages given above is 9.5/14, or 0.67. The organic compound has traveled along the paper at two-thirds the rate of the developing solvent.

Often the need arises for further development of a chromatogram when the solvent has reached the limit of the paper strip. By a technique described below, the investigator may take advantage of further development without the trouble of transferring partially separated components to other strips. This technique is *two-dimensional*, or two-directional chromatography. The method is to apply the solute mixture near one corner of a sheet of chromatographic paper, perhaps an inch and a half from each edge. One of these edges is dipped into the solvent and the process of development takes place as before. After the solvent has moved a reasonable distance, the paper is removed and dried. The original solvent may be replaced by a second solvent known to have particularly good developer characteristics for those components of the solute mixture that, up to this point, have not become well separated. The edge of the dried paper along which development took place with the original solvent is dipped into the new solvent, and development with the new solvent proceeds. The technique is depicted in Figure 4.01. When the solvent front has reached the prescribed limit, the paper is removed, dried, and sprayed with reagent to reveal the location of the components of the mixture. This technique has been employed with eminent success by Consden et al.[3] in the separation of amino acids.

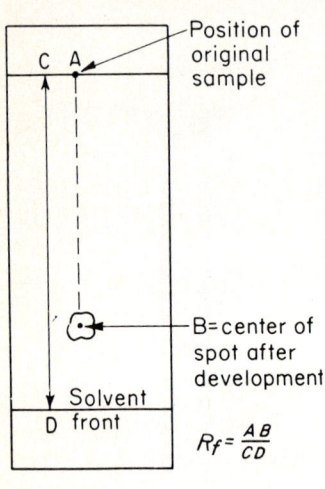

FIGURE 4.02

Calculation of R_f values on a paper chromatogram.

It is very important, in all paper chromatographic procedures, that development take place in an atmosphere of saturated solvent vapors. Otherwise, the developer reaches a point on the paper from which it evaporates more rapidly than it is replaced by capillarity, and the components of the mixture, which may have been well separated earlier, are now concentrated in the region of evaporation. Depending upon the vapor pressure of the solvent and the size of the chromatographic chamber, a few minutes to many hours are required to saturate the chamber with vapor.

In analytical and characterization work, it is impractical to duplicate the conditions of temperature, solvent purity, and paper consistency under which the reported R_f value of a particular compound was determined. Therefore, the R_f values of the unknown and the substance that has been tentatively assumed to be identical with the unknown are determined side by side on the same sheet of paper and with the same solvent system. *If under these conditions equal quantities of the two substances*

[3] R. Consden, A. H. Gordon, and A. J. P. Martin, *Biochem. J.* **38**, 224 (1944).

give substantially the same type of spot and the same R_f value, they may be tentatively presumed to be identical.

Assume, for example, that 0.05 ml of an alcohol occurs in a 10-ml aqueous solution. It is extremely difficult to separate that volume of pure alcohol to determine its boiling point, solubility, and other constants—or, for that matter, to perform many of the systematic tests outlined in this work. However, it is possible, after the presence of an alcohol has been suggested by functional group tests, to prepare directly from such a solution the 3,5-dinitrobenzoate and without isolation of the derivative, produce a paper chromatogram using 5 to 10 micrograms of the solution of the 3,5-dinitrobenzoate. Then, by examining the R_f values of the 3,5-dinitrobenzoates of alcohols, it is possible to select one that has approximately similar values. The 3,5-dinitrobenzoate of this known alcohol is prepared and chromatographed side by side with the 3,5-dinitrobenzoate of the unknown. If the spots are similar and have migrated the same distance (same R_f values), the known and unknown are tentatively assumed to be identical.

Procedure for Application of the Sample

The amount of material initially applied to the paper depends on a number of factors. First, enough of the sample must be used to produce a distinct zone that will be found when the chromatogram is subjected to color development. This amount will depend, obviously, on the sensitivity of the detection method. If radioisotopes are employed, less material is needed since a counter is generally more sensitive than a color reaction. If an insufficient amount of a mixture is applied, a component at low concentration may be missed, even though other fractions show up clearly. If too much material is applied, the spots appear too large and may obscure other components whose R_f values lie close to those of the original substance; resolution will be incomplete and overlapping of the spots will hinder identification. This problem occurs not only when too much material is applied but also when the R_f values of the components are close to each other. It has been suggested that R_f values must differ by at least 10 per cent for perfect resolution in one dimension, no matter how little material is used. If the compound is fairly insoluble in the solvent, the application of too much will cause an elongation of the spot. However, the appearance of "tails" on chromatograms does not necessarily mean that too much material is being used, for certain substances normally run in this manner. Since the actual amount of sample to be used depends on all of these factors, it is difficult to suggest any set amount as being optimum. In one-dimensional analysis, a sample of 50 to 100 micrograms is sufficient and, for the test-tube technique, an initial application of 5 to 15 micrograms is satisfactory. For two-dimensional chromatograms, 200 to 300 micrograms of substance is required.

The application of the sample should be restricted to a very small area. The material is best applied with a micropipet or microburet. A droplet of about 3 to 5 λ at the tip of the pipet is touched to the marked spot on the paper, which is spread a few inches above a hot plate. After the solution on the paper has evaporated another droplet is added on the same spot until the required quantity has been delivered. If the sample is added in large drops, the solution spreads over too large an area and a diffuse spot will result in the developed chromatogram.

The sample may be dissolved in any solvent provided it can be conveniently evaporated after application. However, since all substances that are to be chromatographed must be soluble to some extent in water, this solvent is preferred. In some cases in which the compound has a low solubility in water, it is necessary to convert it to a more soluble form by addition of an acid or a base and, after application on the paper, to reconvert it to the original compound. Thus cystine, tyrosine, histidine, arginine, and other amino acids are applied to the paper as solutions of the hydrochlorides and then neutralized by holding the paper over ammonia for a few minutes.

Selection of Solvent Systems for Development of Chromatograms

The selection of proper solvents is very important in paper chromatography. For partition on the paper, the solvent should be mixed with some water to provide water for absorption by the paper at the same time that the solvent passes over it. However, too much water is undesirable, and indiscriminate saturation of organic solvents may well lead to poor chromatograms. As a rule, the solvent should not contain more than 10 to 20 per cent of water by weight. On the other hand, there are cases of solvents completely miscible with water in which the proportion of water may be higher; 2-propanol solutions containing as much as 40 to 50 per cent water have been employed. Solvents with low vapor pressures are unsatisfactory because they may interfere with color development or allow the dissolved substances to diffuse over a large area, yielding a poor chromatogram. Collidine cannot be used, for example, with the iodoplatinate indicator for methionine, even if the chromatogram is heated at 120° for an hour, since traces that remain on the paper will decolorize the indicator. Solvents with high vapor pressures must be used with caution,

TABLE 4.01
SOLVENTS AND REAGENTS COMMONLY USED FOR PAPER CHROMATOGRAPHY OF SOME ORGANIC COMPOUNDS*

Compound	Solvent System	Reagent for Locating Spots
Organic acids	Mixtures of formic acid with: ethanol, 2-propanol, 1-butanol and other alcohols or ketones	Bromcresol green or bromphenol blue solution in ethanol
2,4-Dinitrophenyl-hydrazides of acids	Buffer pH 11.6 saturated with Me_2CO	Compounds colored; no indicator required
Hydroxy and keto acids	Toluene-acetic acid; 1-butanol-propionic acid	Ammonia and Nessler's solution; o-phenylenediamine with ultraviolet light
2,4-Dinitrophenylhydrazones of keto acids and all carbonyl compounds	1-Butanol	KOH solution (10 per cent)
Hydroxamate derivatives of organic acids	Phenol-isobutyric acid	Ferric chloride
Amino acids	Phenol; phenol-ammonia; collidine; 2,4-lutidine; α-picoline; $tert$-butyl alcohol; 1-butanol	Ninhydrin solution (0.1 to 0.2 per cent in ethanol)
Amino acids (containing S)	$tert$-Butyl alcohol	Potassium iodoplatinate solution (aqueous)
2,4-Dinitrophenylderivatives of amino acids	1-Butanol-acetic acid; collidine; phenol	Compounds colored; no indicator required
3,5-Dinitrobenzoates of alcohols	Methanol-acetone; methanol-hexane; 2-propanol-pyridine	KOH solution (5 per cent)
Amines	1-Butanol-acetic acid	Ninhydrin solution; iodine solution
2,4-Dinitrophenylhydrazones of carbonyl compounds	Ether-hexane; acetone-hexane	KOH solution (10 per cent)
Phenol derivatives; methylolphenols	1-Butanol-ammonia	p-Nitrobenzenediazonium fluoborate
Phenylazobenzenesulfonates	2-Butanol-aqueous solution Na_2CO_3	Compound colored
Sugars	1-Butanol-acetic acid; ethyl acetate-pyridine-water; 1-butanol-collidine	p-Aminodimethylaniline (tin salt); silver nitrate solution; 3,5-dinitrosalicylic acid solution; 0.3 per cent solutions of p-aminohippuric acid in ethanol.

* *Source*: N. D. Cheronis, "Micro and Semimicro Methods," in *Technique of Organic Chemistry*, Vol. 6, A. Weissberger, ed., Interscience, New York, 1954.

as they are sensitive to temperature fluctuations and are likely to distill off or condense on the paper, thus causing phase irregularities if the temperature is not carefully controlled. The solvent need not be a single substance. In the chromatography of various sugars, for example, two good solvents are butanol-acetic acid and a mixture of ethyl acetate and pyridine saturated with water. Table 4.01 lists some of the solvents and indicators used for a few types of organic compounds. The literature should be consulted for further information.

Test Tube Technique for Paper Chromatography

The most convenient test tube for a single spot is the 6-in. type shown in Figure 4.03; however, the short tube (25 × 150 mm) is more suitable when two spots are placed on the same strip (if the solvent travels rapidly the regular 8-in. tube, (25 × 200 mm, is employed). When the number of strips required is small, they may be easily cut from a pattern made from a yellow manila folder. The pattern for the 6-in. tube is 135 mm in length × 15 mm in width at the top, and 10 mm at the bottom. The pattern for the wider tube is 135 or 165 mm in length, 25 mm in width at the top, and 15 mm at the bottom. The strips are cut from three to four discs of Whatman No. 1 filter paper, 15 cm in diameter, folded in half. The pattern is placed on the paper, the outlines are marked lightly with a pencil, and the paper is cut with scissors along the inside of the pencil mark, so that the strip becomes approximately as wide as the pattern. The upper outside strips are discarded (since they have been handled); the inner ones are handled only at the extreme edges of the broad ends. The strips are now pierced at the center, about 4 to 5 mm from the top, to facilitate suspending during the drying operations. A light pencil line is drawn about 8 mm from the narrow end on which the sample is to be placed. For the preparation of a larger number of strips, sheets of Whatman No. 1 (45.7 × 56.3 cm) are employed, and the strips are cut with a paper cutter.

The sample is applied with a micropipet at the center of the pencil line,

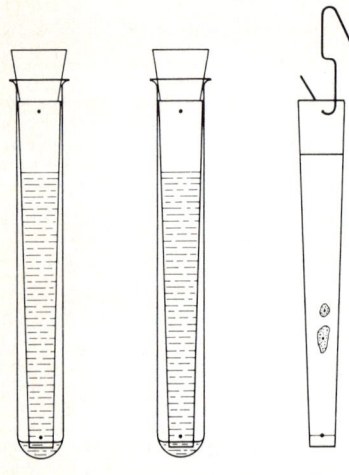

FIGURE 4.03

Test tube technique for ascending chromatography. The chromatographic strip at the right is pierced at the top and hung on a paper clip for drying.

as described in a preceding section, and allowed to dry. A test tube is provided with a well-fitting cork so that, when the tube is stoppered, the cork fails to touch the paper. About 0.4 to 0.5 ml of the appropriate solvent, saturated with water, is placed in the bottom of the tube, being released through a capillary pipet in such manner that the sides of the tube above the surface of the liquid are perfectly dry. The tube is placed on the rack, and the paper is inserted lightly so that the narrow end dips into the liquid, but the surface of the solvent is well below the penciled line. The tube is then corked and allowed to stand (Figure 4.03) until the solvent ascends near to the top of the strip or about 5 mm from the broad end. This depends on the type of solvent but, usually, takes about 1.5 to 3 hours. The tube is opened, the paper removed with a forceps, and the solvent front marked with a light pencil mark. The strip is suspended by a clip (as shown in Figure 4.03), allowed to dry and, then, sprayed with an indicator, or otherwise examined for the location of the migrated spot.

This is by far the simplest technique, since it involves no special equipment and is fairly rapid. It can be used for most identification work and, particularly, for exploratory work in determining the suitable factors for dealing with large sheets or strips.

Location of Zones on Paper Chromatograms

After the chromatogram has been developed and the solvent evaporated, a solution is applied that will react with the chromatographed compound to produce a color so that its position on the paper may be determined. A solution of the indicator (Table 4.01) is usually sprayed over the surface of the paper with an atomizer although, sometimes, brushing on the reagent gives better results; for example, chromatograms of 3,5-dinitrobenzoates. The choice of indicator is wholly dependent on the nature of the substance being chromatographed. In amino acid chromatography, for example, ninhydrin (triketohydrindene hydrate) has proved very useful. When a 0.1 to 0.2 per cent solution of ninhydrin in 1-butanol or ethanol is sprayed on a chromatogram, colored spots appear within 24 hours. This reaction may be accelerated by carefully heating the chromatogram after it has been sprayed at about 90 to 100° for 10 minutes. Colored compounds (dyes, nitro compounds, and the like) do not require the application of an indicator, although many times the paper is sprayed with dilute solutions of alkalies and complexing agents in order to intensify the spots. Compounds that fluoresce under ultraviolet light are often located by observing the chromatogram under a small ultraviolet light.

128   Techniques of Organic Analysis

Since the color of the spots often fades, the chromatogram is examined immediately after the color has been developed, and the spots are outlined in pencil. Approximate quantitative information may be obtained by measurement of the area of the spot, and comparison with the area of a standard made at the same time. For quantitative determinations of the spots by use of a photoelectric colorimeter (densitometer), the references at the end of the chapter should be consulted.

The procedure for the identification of spots on paper strip chromatograms is illustrated in the photographs of Figure 4.04. The problem was to determine the amino acids present in watermelon juice. At the left (A) is the first chromatogram obtained by applying 5 λ of the juice at the base of the strip marked by a dot, and developing for 1.5 hours with a mixture of *tert*-butyl alcohol (70 per cent) and water (30 per cent) and then spraying with 0.2 per cent ninhydrin solution in ethanol. In the center (C) is the second chromatogram with approximately the same amount of watermelon juice, which was run in a series with ten known amino acids, about 10 micrograms of each amino acid on a separate strip. The two known amino acids which gave approximately the same R_f values as the unknown spots are arginine (B) and citrulline (D). In the last chromatogram, a sample of 1 ml watermelon juice was diluted with 1 ml of a solution containing 2 mg of citrulline and 0.6 mg of arginine. From this mixture, about 5 λ were employed to make the chromatogram shown in Figure 4.04(E). Since the spots which were obtained in this chromatogram are about the same as those at (A) and (C), it is concluded that the two amino acids present in watermelon juice are arginine and citrulline, and that the latter is present in about 3 to 4 times the quantity as the former. It should be noted, however, that this identification must be regarded as tentative until two-dimensional chromatograms are made and the spots are not resolved further as the quantity of the sample placed on the paper is reduced. Rigorous identification involves, first, developing several chromatograms,

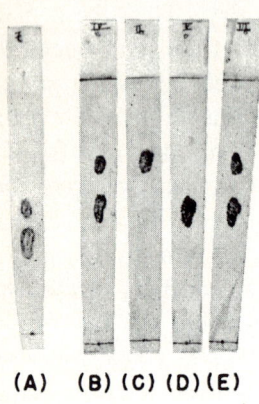

(A)   (B) (C) (D) (E)

FIGURE 4.04

(A) Chromatogram of spots obtained with test-tube technique on paper strip by application of 5 λ of watermelon juice. (B) Chromatogram of same nature as (A). (C) Chromatogram with 10 λ of arginine. (D) Chromatogram with 10 λ of citrulline. Strips (B), (C), and (D) were run side by side in separate test tubes. (E) Chromatogram of watermelon juice plus citrulline and arginine.

then cutting the zones in which the spots are located and eluting these cuttings with water. Finally, the solution is evaporated, and the amino acid is converted to the dinitrophenyl derivative (page 454).

Other Paper Chromatographic Techniques

The circular technique, developed by Rutter,[4] can be employed either by using a Petri dish or a desiccator, and is well suited for the rapid determination of purity of dyes, acid-base indicators, and other substances. Suitable procedures and equipment for small-scale two-dimensional chromatography have also been described.[5] The arrangement is shown in Figure 4.01(*A*).

An effective chromatographic tank can be prepared from two beakers of the same size.[6] Samples are placed on a line parallel to a long edge of a rectangular sheet of paper in the usual manner. The paper is rolled and the ends stapled together so that the vertical edges do not touch each other. The developing solvent is added to a beaker and the paper roll is placed inside carefully so that it does not touch the walls. The line of samples is near the lower part of the paper. The second beaker is lowered carefully over the paper, until it comes to rest on the edge of the first beaker. The edges of the beakers are then taped together to avoid loss of the solvent until the chromatogram has been developed. Beakers without pouring spouts are sealed more easily than ordinary beakers.

Another apparatus makes use of a bell jar, the bottom part of a desiccator, and a Petri dish.[7] The ground surfaces provide an effective, rapid closure for the apparatus.

A device for locating substances which absorb ultraviolet light on paper chromatograms can be made easily, according to published directions[8]. It consists of a wooden frame and holder for the chromatogram, an ultraviolet light, and a phosphor-coated glass plate. Substances which absorb the ultraviolet light appear as dark zones against a bright green background. By changing the light source to one of different wavelength and observing the response of compounds to the two sources, we can differentiate various types of compounds. Microgram quantities of material can be detected in this way. Direct exposure to ultraviolet light should be avoided, since radiation of this wavelength range is harmful to the eyes.

[4] L. Rutter, *Nature*, **161**, 435 (1948); *Analyst.*, **75**, 37 (1950).
[5] J. C. Underwood and L. B. Rockland, *Anal. Chem.*, **26**, 1253 (1955). This article is an excellent review of the small-scale chromatography of amino acids. The vessel employed is a No. 4 North American medical museum jar.
[6] D. S. Wiggins, *J. Chem. Educ.*, **34**, 536 (1957).
[7] E. C. Mathew and D. M. Das, *J. Chem. Educ.*, **32**, 352 (1955).
[8] S. Katz, *J. Chem. Educ.*, **39**, 34–35 (1962).

Laboratory Practice in Paper Chromatography

A beginner should first master the test tube technique using one or two known amino acids. Solutions of amino acids containing about 1 mg or less per ml are satisfactory for this purpose. The two most common causes for failure encountered by beginners are improper handling of strips by the fingers, which causes spots to develop, and improper application of the sample.

The strips are made as directed on page 126. The pencil dot is placed near the lower end of the strip, and the sample is applied by means of a capillary pipet with a long tip. Several of these pipets with tips of varying diameter are made ahead of time. Before applying the spot, practice by applying droplets on dots made on filter paper (of the same variety as that from which the strips are made) until the spot produced by a single droplet is 1 to 2 mm in diameter. The pipet that gives best results is selected and then three strips are prepared, placing 1 droplet on the first, 2 droplets on the second, and 3 droplets on the third, being very careful to dry each spot before applying the second droplet. The tubes are charged with 0.5 ml of a mixture of *tert*-butyl alcohol (70 per cent) and water (30 per cent) and the strips are placed in the tubes as directed on page 127 and developed. When the solvent front has ascended three fourths of the way up the strips, they are taken out, the solvent front is marked with a pencil, and the strips are then dried. A third difficulty encountered by beginners is in the application of the indicator solution, in this case a 0.2 per cent solution of ninhydrin in ethanol. The solution can be sprayed by an atomizer or painted by an artist's air brush, or the entire strip can be dipped into the solution. The first two techniques are recommended until experience has been acquired. The spraying should be even so that all the paper is uniformly wetted by the reagent solution but does not drip. The spraying should be done while the strips are hanging in the position in which they will be dried. After the spots are clearly visible the strips are examined. The distance from the dot at the lower end of the strip (where the sample was placed) to the penciled line indicating the solvent front is measured. Likewise, the distance from the original dot to the center of the spot is also measured, and in this manner the R_f value is calculated (see page 121). The difference between the R_f values of the three strips should not be larger than 0.01. Generally, the best strip is the one with the best defined spot and a minimum of diffusion outward (absence of "tails" and "beards"). Examination of the three strips made with 1, 2, and 3 droplets will reveal the best method of application to be used in future trials.

The above experiment should next be repeated, employing the conditions that were indicated as the most likely to give good results. After reproducible results are obtained, the separation of two different amino acids should be tried. After the first amino acid solution is applied on the strip, it is allowed to dry (see page 126) and, then, a droplet of the solution of the second amino acid is placed on the same spot, using a different capillary pipet. The two amino acids are selected so that there is an appreciable difference in their R_f values. For example, aspartic and glutamic acids, lysine and arginine, or arginine and alanine are good pairs for such trial runs.

When satisfactory results have been obtained in separation of two amino acids, the separation of two closely related alcohols by means of their 3,5-dinitrobenzoates, as described on page 483, can be undertaken. For example, 1 to 5 mg each of ethanol and methanol are mixed and derivatized by means of 3,5-dinitrobenzoyl chloride, and the mixture of the derivatives is then chromatographed so as to obtain spots on paper strips that are identical to the spots obtained by the pure derivatives of the alcohols. In a similar manner, two or three aldehydes or ketones can be separated as described on page 509. Only after considerable experience has been gained in separations of this type, should the composition of a protein be studied as outlined on page 457 or the composition of an aliphatic carboxylic acid mixture according to the citations given on page 439.

An integrated set of reagents and apparatus for performing simple experiments in ascending or descending paper chromatography and electrophoresis is available.[9] Included, also, is a manual with explanations of the principles of chromatography, as well as detailed directions for many experiments.

Column Chromatography

General Techniques

The characterization of an organic compound by chromatographic analysis on an adsorbent column may provide valuable supplemental evidence toward the identification of the compound. Just as in paper chromatography, the method offers the analyst a rapid means of separating simple mixtures. However, the most valuable applications of the method have come in the separation of complex mixtures.

Small columns are commonly employed for characterization experiments and for experiments to determine the feasibility of separating a

[9] Shandon Scientific Company Limited, 65 Pound Lane, Willesden, London NW10, England.

mixture. The general procedure for such an experiment is to pack the solid adsorbent (alumina, silicic acid, and the like) into a glass tube, add the sample solution to the top of the column, and wash it down the column, that is, develop the chromatogram, with the appropriate solvent. When the solvent front reaches the bottom of the column, the process is stopped, and the moist adsorbent column is extruded from the glass tube. A color-developing reagent is applied by streaking or brushing the column to reveal the location of colorless compounds. Measurement is made of the distance between the leading edge of the zone and the top of the adsorbent column, and the distance between the solvent front and the top of the column, in this case the full length of the column. The ratio of the former to the latter yields the R_f value. If the components of the mixture have been separated by such a procedure and the pure components are needed for further testing, the reagent streak may be sliced from the column and the desired regions of the column cut out. Then a polar solvent, such as acetone or methanol, is used to desorb, or elute, the pure compound from the column and, after the adsorbent is separated by filtration, the pure compound is retrieved by evaporating the solvent. An alternate procedure is to run a second column exactly as the first and, knowing the zone boundaries of the various components from observations on the first column, cut out the desired portions of the column and recover the pure compounds as before. Of course, the locations of colored compounds are always evident.

A common variation on the above procedure is to collect the zones as they pass from the column, a task made easy when the compounds are colored. When colorless materials are being chromatographed, however, fractions must be collected periodically and the content of each fraction established through characterization reactions or by some instrumental method.

Another technique, which will not be used in these characterization tests but nevertheless should be described, is that of *displacement analysis*. When somewhat larger quantities of materials are to be handled, this method has been found quite useful. Instead of developing the chromatogram, as before, in hopes of effecting a clean separation of the components of a mixture, a solvent is used that is more strongly adsorbed than any of the components of the unknown mixture. The displacing solvent actually pushes all the components of the mixture ahead of it and, in doing so, the various components line up on the column from the bottom to the top in the order of increasing adsorption affinity. Because of the nature of the method, the leading edge of each zone of material is contaminated by the trailing edge of the zone ahead of it, so that a clean separation cannot be anticipated. However, the heart of each zone will contain practically pure component, and it is this region of the zone that is sought by utilization of the displacement technique. Fractions are collected as before and, by virtue of some property such as refractive index or light absorption, the purity of a particular component in each fraction is ascertained.

Recommended Techniques for Laboratory Exercises

For most purposes, a chromatographic tube, 9 mm in diameter and 130 mm in length, is employed (Figure 4.05). Tapered tubes of approximately the same dimensions may be preferred in some cases where it is

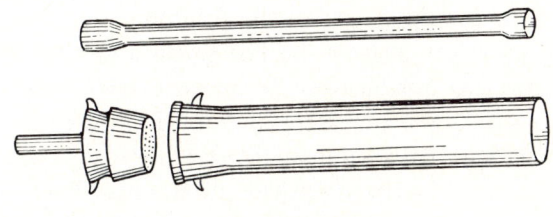

FIGURE 4.05
Chromatographic tube with glass joint and wooden stamper.

found that the moist adsorbent is difficult to extrude. Both types are available commercially.[10] An adequate, improvised chromatographic arrangement is shown in Figure 4.06. The tube is attached to a suction flask by means of a rubber stopper, and the flask, in turn, is connected to a water aspirator with the suction turned on full (Figure 4.07). A plug of cotton is then inserted and pressed down with the extruder so that it will form a support, about 3 mm thick, for the column. An extruder is easily made from a $\frac{1}{4}$-in. dowel rod (Figure 4.05). It should be about 3 in. longer than the chromatographic tube. The adsorbent powder is now poured into the tube through a powder funnel or, more conveniently, from bottles prepared in the manner shown in Figure 4.08. The small glass tube above the pinch clamp fits inside the chromatographic tube, the bottle is inverted, and the bottom of the bottle is tapped to start the flow of adsorbent into the tube. The sides of the tube are tapped gently with the extruder to settle the adsorbent, and the top of the adsorbent column is finally smoothed by very light pressure from the extruder. The adsorbent should fill approximately 5/6 of the volume of the tube. The substance to be chromatographed, dissolved in a small amount of solvent from which it is very

[10] Scientific Glass Apparatus Co., Bloomfield, N.J.; Ace Glass Co., Vineland, N.J.

strongly adsorbed, is introduced into the column. If the rate of filtration through the column is too slow, the solid adsorbent may be diluted with a filter aid. The total time required for the development of the chromatogram should not exceed 1 hour. In most cases, it is much shorter.

It is often convenient to prepare a concentrated solution of the substance in a solvent from which it is weakly adsorbed, but readily soluble, and then to dilute this solution with a solvent from which the compound is strongly adsorbed so that it forms a concentrated zone when poured on the column. The initial width of the zone of all the adsorbed substances on the column should not exceed 1/5 of the length of the column.

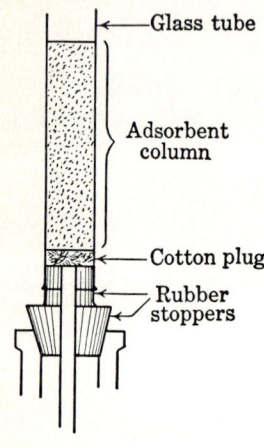

FIGURE 4.06
Improvised chromatographic assembly.

Just as the last portion of this solution is disappearing into the adsorbent, a small amount of the developer is poured on the column, followed by several smaller portions, until all of the solute has been washed into the adsorbent. The top of the column should never be allowed to become dry, since this often results in channeling and distortion of the zones.

Development is continued until the zones show the desired amount of separation or, if the zones are colorless, until the developer has reached the bottom of the column. After the development is complete, the column is allowed to drain, just to dryness, and the tube is removed from the adapter. The adsorbent column may be extruded from the tube in this manner: tap the tube against the palm of the hand several times, vigorously, in order to loosen the adsorbent from the sides of the tube. To avoid flicking the top of the column out of the tube, a cotton plug may be inserted into the top of the tube. Then, lay the tube on the table and push out the column by means of the extruder. In difficult cases, the ends of the tube may be tapped against a piece of soft wood, alternately, until it is seen that the column moves freely from one end to the other. If this technique must be resorted to, it is imperative that a cotton plug be inserted into the top of the tube. The zones are cut out and eluted separately. The adsorbed zones are often concave at their bottom edge and convex at the top edge; consequently, it is not advisable to cut directly through the column at the edge of a zone, but rather to start from the top of the column and gradually slice the adsorbent off, holding the knife at an

Fractionation Procedures   135

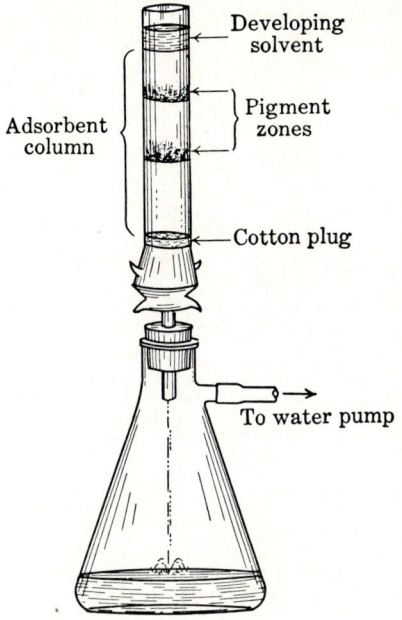

FIGURE 4.07
Standard assembly for column chromatography.

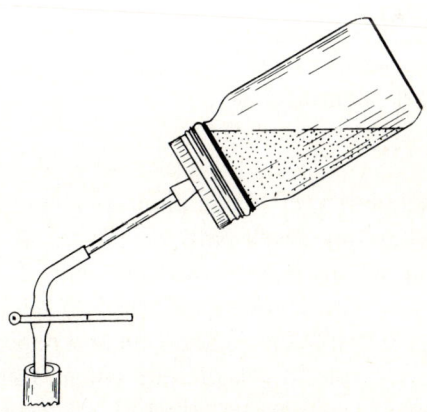

FIGURE 4.08
Bottle for charging chromatographic tubes with adsorbent powder.

angle to the long axis of the column until a zone is reached (Figure 4.09). The zone is then cautiously removed, transferred to a sintered glass funnel, and washed with eluting agent until all the adsorbed substance is removed

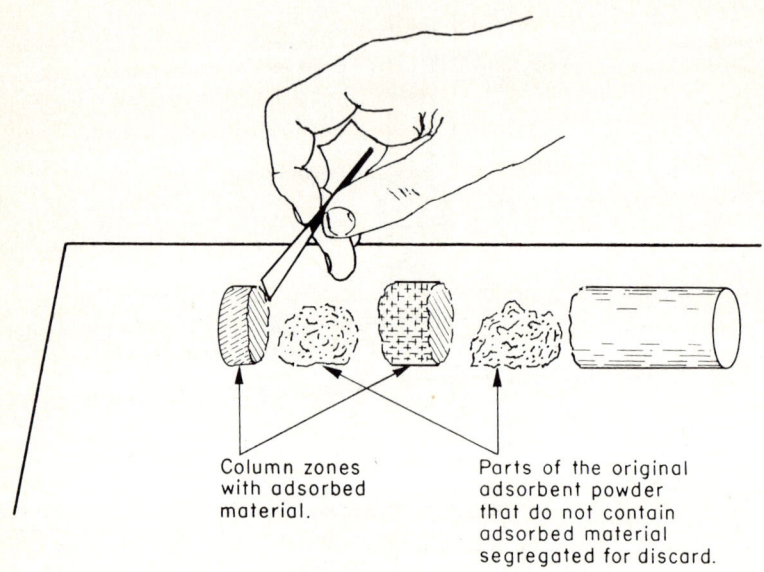

FIGURE 4.09

Separation of zones from extruded column.

from the adsorbent. The eluate is then evaporated to recover the pure compound, or it is analyzed by some means (possibly spectrophotometrically) to establish identification.

Selection of the Adsorbent

In spite of the popularity of chromatographic procedures today, the selection of adsorbents and developing solvents is at present, to a very large extent, more of an art than a science. This factor may partially account for the lack of more widespread use of adsorption chromatographic procedures. Nevertheless, adsorption and paper chromatographic procedures are valuable aids to the present schemes of organic analysis.

A suitable adsorbent should possess the following characteristics: (a) it should reversibly adsorb the substance to be chromatographed; (b) it should not cause any chemical alteration of the adsorbed compounds; and (c) its particle size should be such as to allow the developing solvent to flow at a reasonable rate, for example, 10 to 50 mm per minute. Another desirable (but certainly not essential) characteristic of the adsorbent is that it be white or neutral in color to facilitate location of the adsorbed zone.

In practice, relatively few adsorbents are found to be suited for all chromatographic analyses. Alumina and silicic acid, or silica gel, are outstanding in the number of applications to which they have been put. They possess the characteristics mentioned above, and their activity, as adsorbents, may be altered without difficulty. It has been shown that the strength with which these two adsorbents bind a solute molecule can be changed by varying the amount of water they contain. Both have

TABLE 4.02
ADSORBENTS FOR COLUMN CHROMATOGRAPHY

Adsorbent	Source
Alumina, Reagent: "Suitable for Chromatographic Purposes"	Merck & Co., Inc., Rahway, N.J
Calcium carbonate	Chemical supply houses
Calcium hydroxide	Chemical supply houses
Florisil	Floridin Company, Inc., Warren, Pa.
Magnesium oxide	Micron brand, Westvaco Chlorine Products Co., Newark, Calif.
Neutral Filtrol	Filtrol Company, Los Angeles, Calif.
Silicic acid	Merck & Co., Inc., Rahway, N.J.
Silica gel	Davison Chemical Co., Baltimore, Md.

been "activated" by drying at higher temperatures or by prewashing techniques. In either case, the result is believed to be the same: some of the water in the adsorbent is driven off. Activation by heating should be done cautiously, since too-high temperatures alter the structure of the adsorbents and result in a total loss of adsorbent power. Activation by prewashing may be accomplished using several combinations of solvents, one such combination being 1 volume of acetone, 1 volume of ether, and 2 volumes of petroleum ether or benzene. The volume mentioned here is the quantity of solvent, in milliliters, required to wet the column its entire length.

Alumina and silicic acid may be obtained in good chromatographic grades. A list of several adsorbents and convenient sources is found in Table 4.02. Another general-purpose adsorbent, although somewhat weaker, is Florisil, a synthetic magnesium silicate. Charcoal is one of the most widely used adsorbents, but its use in characterization tests is restricted because of its color. Calcium carbonate, calcium hydroxide, magnesia, sugar, and many others have found application for special purposes.

The effect of particle size on the flow rate of the developing solvent is obvious. Particles, if too large, can easily be made smaller by grinding with a mortar and pestle, or, on a larger scale, with a ball mill. When it is observed that the flow rate is too low the adsorbent may be diluted with a filter aid such as Celite or Hyflo Supercel.[11] Dilution by this means reduces the capacity of the column and sometimes causes spreading of the zones, making detection of the adsorbed compound more difficult. Although some diluents are known to possess weak adsorbent characteristics, this fact is generally neglected.

Selection of the Developing Solvent

The role of the developing solvent in chromatography has been the subject of much investigation, from which has come the conclusion that no true correlation exists between any one property of the solvent (dipole moment, dielectric constant, hydrogen-bonding ability, and the like) and its developer strength. The several physical properties of the solvent cannot be divorced from each other in explaining the action of the developing solvent in the mechanism of adsorption.

As a general rule, petroleum ether, carbon disulfide, benzene, carbon tetrachloride, and chloroform represent weak developers; ethyl ether and isopropyl ether represent moderately strong developers; and the lower-molecular-weight alcohols and ketones represent strong developers. In initial experiments, petroleum ether is a reasonable first choice of developing solvent. If it does not develop the chromatogram, a suggested order in which other solvents should be applied is: benzene, ethyl ether, ethanol. Mixed solvents have been used successfully in a number of cases. The separation of the carotenoid pigments has been effected by introducing the sample to the column dissolved in petroleum ether and, then, developing with a mixture of petroleum ether and benzene, since benzene itself is a little too strong in this case.

The strong solvents mentioned above are often used as eluents, that is, to desorb the solid sample after the extrusion of the chromatogram. Here, a solvent is needed that is more strongly adsorbed than the material already on the column and that will, consequently, displace it from the column.

Detection of Zones on Adsorption Columns

Although the expression "chromatographic analysis" implies the analysis of colored compounds, this is usually not the case. We are, therefore, concerned with the location of colorless zones on the column of adsorbent.

[11] Available from most laboratory supply firms.

Several methods are used to establish the positions of the colorless zones. The various techniques fall conveniently into two classes: *instrumental methods* and *visual methods*. The second of these is the more generally practiced in the laboratory and, consequently, is the one to which this discussion will be devoted. However, before elaborating on the visual methods, a brief description is offered of the various ways in which instrumental methods have been applied to zone detection.

Instrumental methods. Most of the instrumental methods employed in the chromatographic separation and identification of colorless organic compounds are applied after collecting fractions of the percolate as this solution flows from the column. A simple, easily constructed fraction collector consists of a circular rack of test tubes, which rotates periodically, bringing an empty tube into place beneath the column.[12] As the tube fills, its weight depresses the balance arm on which it rests, thereby removing a stop and allowing the next empty tube to move into position. The fractions collected in this manner may then be examined by any one of a variety of methods. Ultraviolet absorption spectroscopy, infrared analysis, refractive index determinations, polarographic analysis, conductance measurements, and potentiometric titrations are some of the standard instrumental methods that have been applied to the identification of colorless organic compounds after chromatographic separation.

Refractive index gradients, produced by the presence of a compound dissolved in the developer, provide another means of identifying colorless organic compounds. In this procedure, the effluate passes from the column into the cell of an interferometer and, as the solute zone enters the cell, an interference pattern is recorded from which the change in refractive index with effluate volume may be determined. From these data the identity of the adsorbed sample may be established.

The high-frequency oscillator has been applied directly to the column to detect colorless zones.[13] A small condenser of aluminum foil or copper screen is fashioned to the chromatographic column which, in turn, is connected into the circuit of the very high-frequency oscillator. Because of differences in dielectric constant between the developing solvent and the solute, the chromatographic zone passing between the plates of the small condenser alters the frequency difference between the working oscillator and the reference oscillator. In this manner, the appearance of the solute zone can easily be discerned.

[12] E. S. Sanderson, *Anal. Chem.*, **26**, 944 (1954).
[13] P. H. Monaghan, P. B. Moseley, T. S. Burkhalter, and O. A. Nance, *Anal. Chem.*, **24**, 193 (1952).

Visual methods. A common visual technique that is quite helpful in the preliminary identification of organic compounds is the use of streak reagents in conjunction with separation by means of chromatography. Streak reagents, generally, differ from the classification tests applied in test tubes and spot plates only in method of application, that is, streaking the column with a dropper pipet or brushing on the reagent. Modifications in existing test reagents are sometimes required to overcome the inherent acidity or basicity of the adsorbent column and, because of the very small amount of the compound on the surface of the adsorbent, the most sensitive reagents are necessary. The principal advantages of this technique are the possibility of a separation on the column and identification due to the fact that the R_f value exhibited by an organic compound on a particular adsorbent is characteristic. For example, homologous series of aldehydes, methyl ketones, and amines show an increase in R_f value with an increase in the length of the hydrocarbon chain. Although significant differences may not exist in the R_f values of adjacent homologs, measurable differences will generally be found where the hydrocarbon chains differ in length by two carbon atoms. Differences are more likely to be found among the low-molecular weight homologs than among those having long hydrocarbon chains.

Most of the tests that are described later have been applied to the detection of colorless zones on silicic acid columns, with benzene used as developer. Other adsorbents have been tested, and some have been found to be particularly good for certain separations, but silicic acid serves as an excellent general-purpose adsorbent. Alumina and Florisil are good alternatives. It has been suggested that standardization of the adsorbent be accomplished by prewashing successively with acetone, ether, and petroleum ether and, finally, drying 24 hours at 80°. The technique recommended in an earlier section should be employed in performing these tests. The reagent is applied to the extruded column by touching the dropper pipet to the surface of the adsorbent and, while pressing gently on the rubber bulb, drawing the pipet the length of the column. A single column may be streaked at least three times simply by rolling it until fresh surface is available.

Alkaline permanganate reacts with many oxidizable compounds and, although it is lacking in selectivity, it is very useful in detecting zones on chromatographic columns. A $0.0075M$ solution of potassium permanganate in $0.25M$ sodium hydroxide is commonly used. However, stronger alkalinity is better on alumina and calcium hydroxide columns. The zone may appear green against a purple background or light brown against a pink background.

Some of the better specific reagents follow, listed according to the classes of organic compounds for which they have been found useful. The colors reported[14] were obtained on silicic acid columns using benzene as developer.

1. *Acids.* Acids may be detected by several reagents. BDH universal indicator may be used for acids as well as for bases such as amines. The indicator develops a red color in the zone for acids, and green for bases. Another useful reagent consists of equal volumes of 1 per cent potassium bromate, 1 per cent potassium iodide, and 5 per cent starch solution, mixed just prior to use. Streaked with this reagent, acids show blue against a white background. Dimethylaminoazobenzene, a 3 per cent solution in benzene, gives red zones against a yellow background when applied to adsorbed acids. (Compare this test with Test P-7A in Chapter 10.)

2. *Alcohols.* The low-molecular-weight alcohols are difficult to detect on a column with streak reagents, possibly because of the rapid evaporation of alcohols from the surface. Alcohols of four or more carbon atoms are readily detected by streaking first with a solution of ammonium vanadate containing 200 to 300 mg of vanadium per liter and overstreaking with a 2.5 per cent solution of 8-hydroxyquinoline in 6 per cent acetic acid. The zone appears orange-brown against a dull green background. (Compare this test with Test 10.04A in Chapter 10.)

3. *Aldehydes and ketones.* A $2M$ hydrochloric acid solution, saturated with 2,4-dinitrophenylhydrazine, has been used successfully in the detection of both aldehydes and ketones. The yellow streak develops a brighter yellow color in the aldehyde or ketone zone. To differentiate between the two types of compounds, the column is turned a third of a turn and streaked with Schiff reagent. A red-violet coloration in the zone indicates an aldehyde. (Compare this test with Tests 10.10A and 10.10B in Chapter 10.)

4. *Amines.* A saturated solution of chloranil in dioxane is an excellent reagent for the aliphatic amines. A green zone against an orange background identifies this group. Aromatic amines give weak reactions, the zone usually being dull green to black in appearance. A 1 per cent aqueous solution of 4-nitrobenzenediazonium fluoroborate is very effective in the detection of aromatic amines. The zone appears yellow to red against a faint yellow background. Phenols give similar reactions. (Compare this test with Tests P-6A and 10.08-B6 in Chapter 10.)

5. *Aromatic rings.* A mixture of 10 ml of concentrated sulfuric acid and 0.2 ml of 37 per cent formaldehyde is used as a streak reagent for this class of compounds. Single rings produce a red zone, generally, whereas two

---

[14] A. L. LeRosen, R. T. Moravek, and J. K. Carlton, *Anal. Chem.*, **24**, 1335 (1952).

or more rings develop green or blue-green zones. The reagent should be freshly prepared each day and, because the reaction with benzene would obscure the presence of other aromatic compounds, the test solution is prepared and developed with heptane. (Compare this test with Test P-4B in Chapter 10.)

6. *Mercaptans.* The familiar Doctor test is used to detect the mercaptans. The reagent is prepared by adding 1.25 g of PbO to 7.5 g of NaOH in 40 ml of water. The presence of a mercaptan is indicated by a yellow color. When overstreaked with a saturated solution of sulfur in benzene, the zone turns gray to black. (Compare this test with Test 10.21E in Chapter 10.)

7. *Phenols.* Two good reagents for detecting phenols are available. These are a 1 per cent solution of vanillin in concentrated sulfuric acid and the diazonium reagent described for amines. The former gives pink to red zones against a white background, and the latter gives yellow to red zones. Substituted phenols, in general, give weaker tests than unsubstituted phenols. (Compare this test with Test 10.08-B6 in Chapter 10.)

8. *Sulfides.* Sulfides can be detected by streaking the column with a solution of the gold iodide complex which is prepared by adding 2 drops of a $2M$ gold chloride solution to a solution of 0.5 g of potassium iodide dissolved in 25 ml of water. Sulfides give a brown to black coloration against a pink background. Amines react similarly. The advantages of a chromatographic separation are distinctly evident with the application of this reagent. Amines and sulfides give the same reaction, but amines are strongly adsorbed on silicic acid where sulfides are adsorbed hardly at all. Consequently, the former class is found at the top of the column, whereas the latter is found at the bottom.

9. *Unsaturated hydrocarbons.* The column is streaked with a solution of 3 mg of *o*-safranine in 25 ml of water and overstreaked with $0.1M$ potassium iodide containing 4 or 5 drops of bromine. The original red streak of the safranine is bleached, but when unsatuated hydrocarbons, sulfones, and di- and triaryl amines are present, the red color is regenerated in the zone.

A number of complex organic compounds may be detected by virtue of the fluorescence they exhibit when irradiated with ultraviolet light.[15] Identification is generally limited to simple mixtures. A very useful procedure employing ultraviolet fluorescence is the determination of volume percentages of aromatics, olefins, and saturates in gasoline.[16] In this

[15] D. W. Criddle and R. L. LeTourneau, *Anal. Chem.*, **23**, 1620 (1951).
[16] H. Brockmann and F. Volpers, *Ber.*, **80**, 77 (1947).

method a very small amount of a dye mixture is added to the gasoline sample to be chromatographed. Displacement chromatography is the technique used and ethyl alcohol is employed as displacer. The complex dye mixture contains components which assume positions on the column at the boundaries between the various hydrocarbon fractions. Thus, as the displacer pushes the gasoline fractions through a silica gel column, the alcohol-aromatic boundary is indicated by a nonfluorescing red dye, the aromatic-olefin boundary by a blue fluorescence, and the olefin-saturate boundary by a yellow fluorescence.

Still another method of chromatographic analysis using ultraviolet light is the detection of zones by fluorescence quenching. Small amounts of certain fluorescent dyes are retained by the adsorbent when treated with a solution of the dye. Such a dye-impregnated column, when used in chromatographic separations, will fluoresce under ultraviolet light,[17] except in the location of the solute zones—assuming, of course, that the solute itself does not fluoresce. This method is nonselective but useful where the R_f value of a compound is desired to aid in its identification. Morin has been suggested as the fluorescing dye to use on alumina columns, berberine on silicic acid columns, and either morin or diphenylfluorindine sulfonic acid on magnesia or calcium hydroxide columns.

Example 4.01: Separation of Carotenoid Pigments of Plants

Fifty grams of tomato flesh, or 10 g of tomato paste, carrots, or other colored vegetable or fruit are mashed in a mortar and ground to a pulp; 50 ml of methanol are added, and the grinding is continued for a few minutes. The suspension is washed with 50 ml of methanol into a 300 ml wide-mouth glass-stoppered flask, and 100 ml of petroleum ether (BP 60 to 70°) added. The mixture is shaken for 15 minutes by hand or on a shaker. The suspension is then filtered through a Buchner funnel by suction. The filtrate is saved and the solid fibrous residue is again ground in the mortar and twice more submitted to the operations described. The final residue should be white. The amount of tomato pulp to be employed depends on the capacity of the tube. For an 18-mm tube, one third of the quantities given should be used.

The combined extracts are diluted with an equal volume of water in order to transfer the pigment to the petroleum ether, which forms a separate layer above the aqueous phase. The aqueous layer is removed in a separatory funnel and the upper (ligroin) layer is washed 8 times by

---

[17] Ultraviolet lamp manufactured by Vogal Luminescence Corp., San Francisco, Calif.

briefly shaking with water.[18] It is dried by standing over anhydrous sodium sulfate for a few minutes and then filtered. This filtrate is evaporated to about 20 ml before it is placed on the column.

A column of calcium hydroxide 230 mm long is prepared in a chromatographic tube about 33 mm in diameter, according to the directions given in the preceding section. The pigment solution is poured on the column. Just as the last part is entering the adsorbent, a little petroleum ether is poured on, and this is repeated until all the pigment has passed into the adsorbent. The top of the column should never become dry.

Next, the chromatogram is developed by ligroin until the lowest zone is about halfway down the column. The zones are located when the chromatogram is about half completed by exposing it to an ultraviolet lamp; a strongly fluorescing zone is usually observed below the lowest pigment zone. More ligroin, containing 10 per cent of acetone, is now poured on the column, and the development is continued until the lowest zone is almost at the bottom of the column. At this point, the column is allowed to drain, just to dryness. Air is allowed to enter the filter flask. The tube is removed and the adsorbent is extruded. The chromatogram usually shows a small red zone of lycoxanthine, $C_{40}H_{55}OH$, near the top of the column and, immediately below it, a broad red zone of the main pigment, lycopene, $C_{40}H_{56}$. The latter is followed by several small zones and then by a broad orange-colored β-carotene zone. The lycopene and β-carotene zones are cut out separately with a scalpel and are eluted with ligroin containing 25 per cent alcohol. The elutes may be evaporated to dryness under reduced pressure (preferably after the alcohol has been washed out) to obtain a pure pigment.[19]

Example 4.02: Chromatographic Separation of o-, m- *and* p-*Nitroaniline*

A solution of each isomer is prepared by dissolving 5 mg per ml of the respective compounds in benzene. A mixture of the three is prepared by combining 2 ml of the *o*-nitroaniline solution, 2 ml of the *p*-nitroaniline, and 4 ml of the *m*-nitroaniline solution. A sample of 1 ml of this mixture is diluted with ligroin (BP 60 to 70°) to a volume of 5 ml, just before it is poured on the column.

The adsorbent used is a mixture of 2 parts by weight of calcium hydroxide and 1 part of Celite; however, if the calcium hydroxide used allows a good rate of filtration, it need not be diluted with this filter aid. A column

---

[18] For a more efficient method of washing, see A. L. LeRosen, *Ind. Eng. Chem., Anal. Ed.*, **14**, 165 (1942).

[19] This analysis is adapted from an article by F. W. Went, A. L. LeRosen, and L. Zechmeister, *Plant Physiol.*, **17**, 91 (1942).

about 150 mm long is prepared in a chromatographic tube, approximately 200 × 18 mm.

The benzene-ligroin solution is introduced, and the chromatogram is developed with pure ligroin until the lowest zone is located about 1 cm from the bottom of the column. The column is allowed to drain dry, and is then extruded. The positions of the zones may vary with different brands of the adsorbent, but they should be located approximately the following distances (in millimeters) from the top of the column: *p*-nitroaniline, 5 to 25, bright yellow; *m*-nitroaniline, 65 to 90, yellow; and *o*-nitroaniline, 95 to 145, yellow. These zones may be cut out and eluted with benzene containing 10 per cent alcohol or with pure alcohol. Evaporation of each eluate should yield a pure compound.

Thin-Layer Chromatography

Description

Thin-layer chromatography is one of the most recently developed methods for analysis on a micro scale. Several good reviews of the subject are available, although the method is still relatively new.[20] The fractionation occurs in a bed of solid particles bound to a smooth surface, and combines desirable features of column and paper chromatography. According to the details of the procedure used, the fractionation may be considered either adsorption chromatography or partition chromatography. The separation of ionic or highly polar compounds as well as nonpolar compounds may be achieved. The method can be used for qualitative, quantitative, or preparative work within limits. The method may be used with samples as small as 1 mg or as large as 100 mg.

Among the advantages of the method are the following ones.

1. A minimum of apparatus is required. Small-scale analyses can be carried out on microscope slides or the covers of lantern slides (3 in. × 4 in.). More precise methods require larger plates and/or a spreader for obtaining a uniform layer of adsorbent.

2. The method requires a few simple operations, as will be seen in detail later.

3. The time needed for an analysis is small. The development of the chromatogram should, normally, require no more than 30 to 45 minutes.

[20] E. Demole, "Chromatostrips and Chromatoplates" in *Chromatographic Reviews*, Vol. 1, M. Lederer, ed., Elsevier Amsterdam, 1959, or *J. Chromatog.*, **1**, 24–34 (1958) (in French); E. Demole, "Recent Progress in Thin-Layer Chromatography," in *Chromatographic Reviews*, Vol. 4, M. Lederer, ed., Elsevier Amsterdam, 1962, or *J. Chromatog.*, **6**, 2–21 (1961); and E. G. Wollish, M. Schmall, and M. Hawrylyshyn, *Anal. Chem.*, **33**, 1138–1142 (1961).

4. The results can be directly transferred to the problem of preparative scale separations on columns. Since the optimum conditions for solvent, and eluent separation of a mixture can be obtained with a minimum of time and of sample, considerable savings of both occur.

5. Location of the separated compounds is facilitated by the use of corrosive chemicals, which do not harm the inorganic substrate of the layer.

6. The separation of compounds is often more effective and the spots are smaller than with paper chromatography.

7. Only small quantities of material are required; usually 0.5 micrograms is the minimum amount. However, 0.5 mg may be applied to a single spot.

8. The method can be used for micropreparative work, since a mixture having a total weight of 30 to 100 mg can be applied in separate spots to the plate. The material in each spot is separated simultaneously, and the individual compounds are scraped off the plate.

General Procedure

Plates for thin-layer chromatography are prepared by spreading carefully on a sheet of glass a slurry of the adsorbent material and drying the plates in a horizontal position in an oven or in the air. For reproducible results, the layer should be of uniform thickness, which is achievable by use of a mechanical spreader. However, for many experiments, satisfactory results can be obtained by dipping the plates in the slurry and wiping off one side of the plate. For adsorption chromatography (for separation of oil-soluble compounds), the plates are usually heated to 100 to 140° for 30 to 120 minutes in order to activate them. They are then kept from atmospheric moisture, as much as possible, until they are used.

After the plates are dry, a line is lightly drawn near one end of each plate; a solution of the sample is applied in small spots along the line with micropipets. The samples may be dissolved in small volumes of a solvent that evaporates very quickly. The plates are placed vertically in a covered jar for development by a solvent in the bottom of the jar. The liquid quickly travels by capillary action upward on the plate, carrying with it the components of the mixture. These components do not move at the same rate and, consequently, become separated along the length of the plate. Before the solvent front approaches the upper end of the plate, the latter is removed from the jar, and the leading edge of the solvent front is marked on the plate. The solvent is removed by air-drying the plate or by heating the plate in an oven.

When all the solvent has been removed, the location of the adsorbed material can be determined by a variety of methods, some of which are listed in a later section. One of the easiest and most general methods is to spray the plate with concentrated sulfuric acid, which will blacken all the spots where organic components are located. By measuring the distance traveled by each component and comparing this with the distance traveled by the solvent front, R_f values can be calculated for each compound in the mixture, as shown on page 121.

If layers of the same uniform thickness have been used, these R_f values should be the same as those obtained in other experiments with the same adsorbent and solvents. If, however, the layers are not uniform in thickness, it is desirable to apply a sample of a known compound to each plate for comparison purposes. Even under the best conditions, the R_f values, obtained here, are not as reproducible as those obtained with paper chromatography.

Adsorbents

Various solid adsorbents are used for thin-layer chromatography, but the one most used is silica gel (silicic acid). This substance can be used for the separation of either water-soluble compounds or oil-soluble compounds. Alumina is also used, but less extensively. Also used are adsorbents such as cellulose and cellulose derivatives, kieselguhr, and polyamide powder.

In general, the particles of adsorbent are smaller than those used for column chromatography. Often a binder, such as plaster of Paris or starch, is used to hold the material on the plate although, in some cases, this is not needed. Sometimes, a phosphor is added to the adsorbent to aid in the location of compounds by means of ultraviolet light. The particle size and other characteristics of the solid must be carefully controlled to achieve uniform results. For the partition-chromatographic separation of water-soluble substances, deactivated silica gel is used, together with aqueous solvents. Many types of adsorbents prepared especially for thin-layer chromatography are available both in this country and abroad; some of these suppliers and their addresses are listed.[21]

Applicators

Although plates can be prepared for thin-layer chromatography without the use of mechanical spreaders, the layers thus formed are not uniform,

[21] Brinkmann Instruments, Inc., 115 Cutter Mill Road, Great Neck, N.Y.; Excorna, Pharm. Präparate O. H. G., Mainz, Germany; Fluka S. A., Buchs, St.-Gall., Switzerland; Research Specialties Co., 200 South Garrard Blvd., Richmond, Calif.; Terra Chemicals, Inc., 500 Fifth Ave., New York 36, N.Y.; and M. Woelm, 344 Eswege, W. Germany.

and give variable results in use. Many investigators in the field report greatly improved results when devices are used to produce a layer of uniform, known thickness. Descriptions of spreaders for this use have been published.[22] Spreaders may also be purchased.[23]

Plates, already coated with any of 16 adsorbent materials, may be purchased.[24] The coatings on the plates are said to be uniformly thick and more adherent, due to a special cleaning process for the glass plates.

Developing the Chromatogram

The principles used in choosing a solvent for developing a thin-layer chromatogram are essentially those that have been discussed in the sections on paper and column chromatography. The polarity of the solvent should match that of the compounds to be separated. In the case of partition chromatography, aqueous solutions are used.

Adsorption on most solids used for thin-layer chromatography increases in this order: saturated hydrocarbon (little or none), unsaturated hydrocarbons, ethers, aldehydes and ketones, amines, alcohols, and carboxylic acids (most). The separation of compounds that have different functional groups should be achieved easily. Separation of compounds of the same class is more difficult.

The ability of solvents to elute compounds from the adsorbent varies with their polarity. This ability increases in the following series: hexane (least), carbon tetrachloride, benzene, chloroform, ether, ethyl acetate, 2-propanol, ethanol, methanol, pyridine, and water (most). Although mixtures of liquids are often used, it is generally best to keep the solvent system as simple as possible. In searching for a good eluent, it is preferable to start with a solvent that is low in polarity, and change to a more polar one as needed to achieve sufficient mobility of the adsorbed components.

Locating the Zones

One of the advantages of thin-layer chromatography over paper chromatography is the greater ease of locating the zones in which compounds are located. Since the materials used to prepare the adsorbent layer are usually inorganic and therefore relatively inert, corrosive reagents can be

[22] E. G. Wollish, M. Schmall, and M. Hawrylyshyn, *Anal. Chem.* **33**, 1138–1142 (1961); R. Wasicky, *Anal. Chem.*, **34**, 1346 (1962); E. Stahl, *Chem. Ztg.*, **82**, 323 (1958); M. Barbier, H. Jäger, H. Tobias, and E. Wyss, *Helv. Chim. Acta*, **42**, 2440 (1959); and G. Machata, *Mikrochim. Acta*, **1960**, 79.

[23] Camag, Hombergerstrasse 24, Muttenz, Switzerland or Arthur H. Thomas, Philadelphia, Pa.; C. Desaga GmbH, Hauptstrasse 60, Heidelberg, Germany; Research Specialties Co., 200 South Garrard Blvd., Richmond, Calif.; and from most laboratory supply houses.

[24] Custom Service Chemicals, New Castle, Del.

used to detect the compounds. Although this is also true for column chromatography, the flat surface of the plate is much more accessible for treatment than the column of adsorbent.

In many cases, compounds can be detected under an ultraviolet light by their fluorescence. However, the most generally used method is to spray the surface of the plate with concentrated sulfuric acid. Most organic compounds will char under this treatment and give a dark spot. The plate may be heated on a hot plate to hasten the action.

Special reagents, which will react only with specific types of compounds, may be sprayed on the surface. By using two or more plates and treating them with different reagents, the identity of many mixtures can be confirmed without further work.[25] Several reagents, which are useful for locating different classes of compounds, are listed in Table 4.03.

Example 4.03: Characterization of Essential Oils[26]

Prepare a mixture of 28.5 g of silicic acid, 1.5 g of corn starch, 0.0011 g of Rhodamine 6G, and 54 g of water. Stir the mixture while it is heating on a water bath at 85°. After the mixture thickens, remove it from the water bath and add 20 ml of water. Spread the mixture to give a layer 0.02 in. thick on glass plates 5 in. × 7 in. Dry the plates for 20 minutes in an oven at 105° and, then, for 30 minutes at a pressure of 2 mm over potassium hydroxide.

Apply 1 to 2 mg of each sample or mixture dissolved in a low-boiling hydrocarbon in small spots along a line 2 cm from a short edge of the plate. Place the plate (with the samples near the bottom edge) in a covered battery jar. The bottom of the jar should be covered well with a solution of 10 to 15% ethyl acetate in hexane. Allow the solvent front to move 12 to 14 cm from the initial line. Remove the plate from the battery jar and allow it to dry in the air.

Observe the plate under ultraviolet light, and draw with a pencil a ring about each spot found. Spray the plate with a solution of 0.4 g of 2,4-dinitrophenylhydrazine in 100 ml of $2N$ hydrochloric acid to detect carbonyl derivatives, and again mark the spots. Use ultraviolet light to locate traces of ketones. Place the plate in an oven at 105° for 10 minutes to cause heat- and acid-sensitive components to appear. Compare the locations and appearances of the spots with those caused by known compounds. If a permanent record is desired, trace the edges of the spots on a piece of transparent paper.

[25] J. M. Miller and J. G. Kirchner, *Anal. Chem.* **25,** 1107–1109 (1953).
[26] *Source.* R. H. Reitsema, *Anal. Chem.*, **26,** 960–963 (1954).

TABLE 4.03
METHODS OF DETECTING ZONES ON THIN-LAYER CHROMATOGRAMS*

Class of Compound	Reagent	Reference†
Acids	Bromocresol green	9
Aldehydes and ketones	Acidified 2,4-dinitrophenylhydrazine	16, 17
	o-Dianisidine in glacial acetic acid	7
Amino acids	Ninhydrin, AcOH, $Cu(NO_3)_2$, heat	1, 14, 15
Basic nitrogen and sulfur compounds	Potassium iodoplatinate (Munier)	5, 11, 12, 15
Essential oils and terpenes	Concentrated sulfuric acid, H_2O_2	2, 13
	5% Concentrated nitric acid in concentrated sulfuric acid	7, 8
Esters	Hydroxamic acid reaction with ferric chloride	3
Glycosides	Antimony trichloride in $CHCl_3$	18
Inert compounds	Sulfuric and nitric acids	9
Organic peroxides	KI starch, or Fe(III) rhodanide	18
Phenols	Diazotized p-nitroaniline, or ferric chloride	3
Semicarbazones	Bromophenol blue-citric acid	8
Sulfonamides	0.1% p-Dimethylaminobenzaldehyde in alcoholic HCl, Bratton-Marshall reagent	19
Terpenes	Antimony pentachloride, 20% in CCl_4	18
Unsaturated compounds	Aqueous $KMnO_4$; iodine at high temperature; 0.05% aqueous fluorescein and Br_2 vapors	3, 4, 13 6, 10, 18

* *Source.* E. G. Wollish, M. Schmall, and M. Hawrylyshyn, *Anal. Chem.*, **33**, 1138–1142 (1961).

† 1. M. Brenner and A. Niederwieser, *Experimentia*, **16**, 378 (1960).
  2. L. Bryant, *Nature*, **175**, 556 (1955).
  3. E. Demole, *Chromatog. Rev.*, **1**, 1 (1959).
  4. E. Demole, *J. Chromatog.*, **1**, 24 (1958).
  5. H. Gänshirt and A. Malzacher, *Arch. pharm.*, **293**, 925 (1960).
  6. J. G. Kirchner and J. M. Miller, *Ind. Eng. Chem.*, **44**, 318 (1952).
  7. J. G. Kirchner and J. M. Miller, *J. Agr. Food Chem.*, **1**, 512 (1953).
  8. J. G. Kirchner and J. M. Miller, *J. Agr. Food Chem.*, **5**, 283 (1957).
  9. J. G. Kirchner, J. M. Miller and G. I. Keller, *Anal. Chem.*, **23**, 420 (1951).
  10. J. G. Kirchner, J. M. Miller, and R. G. Rice, *J. Agr. Food Chem.*, **2**, 1031 (1954).
  11. M. Liebich, *Deut. Apotheker Ztg.*, **99**, 1246 (1959).
  12. M. Liebich, *Deut. Apotheker Ztg.*, **100**, 394 (1960).
  13. J. M. Miller and J. G. Kirchner, *Anal. Chem.*, **25**, 1107 (1953).
  14. E. Mutschler and H. Rochelmeyer, *Arch. Pharm.*, **292**, 449 (1959).
  15. E. Nürnberg, *Arch. Pharm.*, **292**, 610 (1959).
  16. R. H. Reitsema, *Anal. Chem.*, **26**, 960 (1954).
  17. R. H. Reitsema, *J. Am. Pharm. Assoc., Sci. Ed.*, **43**, 414 (1954).
  18. E. Stahl, *Fette, Seifen, Anstrichmittel*, **60**, 1027 (1958).
  19. E. G. Wollish, M. Schmall, and M. Hawrylyshyn, *Anal. Chem.*, **33**, 1138 (1961).

Example 4.04: Separation of Mixtures of Sugars[27]

Heat a mixture of 3 g of starch, 57 g of silica gel (Merck, No. 7729 Germany), and 115 ml of water on a water bath with stirring until the mixture reaches a temperature of 70°. Cool the mixture with stirring to 30°, and add more water as needed to allow the mixture to flow properly.

The mixture is applied to strips of single-weight window glass, 1/2 in. × 5½ in. with a spreader. The plates are allowed to dry at room temperature. When the plates are dry, marks are made 1.5 cm and 11.5 cm from one end of each plate. An aqueous solution (2 microliters in volume) containing 10 to 200 micrograms of the sugar mixture is placed on the mark which is 1.5 cm from one end.

The chromatogram is then developed with a solvent containing isopropyl alcohol, toluene, ethyl acetate, and water (10:2:5:2.5 by volume). Since polar solvents move slowly on silica gel, one to two hours are required for the solvent front to travel 10 cm to the second mark. The plates are dried and then sprayed with a naphthoresorcinol-phosphoric acid indicator (5 volumes of 0.2% naphthoresorcinol in acetone plus 1 volume of 9% phosphoric acid in water). When the plates are heated at 110° for 5 to 10 minutes, zones of different sugars become visible. Known sugars are applied to other plates, which are treated in identical fashion.

Example 4.05: Separation of Leaf Pigments[28]

Add 12 ml of water to 5 g of Biosil A-30B (TLC silicic acid) and stir vigorously for 15 sec or until well mixed. Pour about 1 ml of this dispersion on a clean microscope slide and spread it with a stirring rod. Tap the slide gently to produce an even film of silicic acid on the slide. Place the slide in the oven at 110° until it is dry.

Prepare an extract of leaves by grinding green leaves in a mortar with a few milliters of a 2:1 by volume mixture of petroleum ether and alcohol or acetone. Transfer the solution to a separatory funnel and contact with an equal volume of water. Discard the water layer, and repeat the water wash twice more. Decant the organic layer, and add 2 g of anhydrous sodium sulfate to it.

Place a drop of this extract about 1.5 cm from one end of the cool slide prepared above. Place the slide vertically in a beaker with the spot end down. The beaker should contain a 7:3 by volume mixture of benzene and acetone. Cover the beaker and let it stand at room temperature.

[27] *Source.* M. Gee, *J. Chromatog.*, **9**, 278–282 (1962).
[28] *Source.* C. Rollins, *J. Chem. Educ.* **40**, 32 (1963).

The solvent rises quickly through the thin layer of silicic acid. Just before the solvent front reaches the top end of the slide, remove the slide from the solvent, and allow it to dry. Eight colored spots should be visible, and two or three additional spots may be seen.

Example 4.06: Oxidation and Separation of Micro Quantities of Products[29]

Prepare a mixture of silicic acid, starch, and water in the proportions of 30:1.5:60 by weight. Heat the mixture on a water bath with stirring until the mixture thickens. Allow it to cool to room temperature and add water as needed to give the proper flow to the liquid. Apply the mixture with a spreader to glass strips (microscope slides). If approximate results are satisfactory, the strips may be dipped in the slurry, and one side wiped clean. Dry the slides in the air and then in an oven at 105°.

Add the material to be oxidized (10 to 200 micrograms) in 1 to 2 microliters of solution to a spot near one end of the slides, and allow the solvent to evaporate. Add one drop of a saturated solution of chromic anhydride in glacial acetic acid to the same spot. Allow the slide to stand at room temperature for a few minutes, or warm the slide carefully.

Develop the chromatogram with a solution of 15% ethyl acetate in hexane or a solution of 10% ethyl acetate in chloroform. Fortunately, inorganic salts are not removed from the original spot, and compounds such as acetic acid and ethylene glycol are so strongly adsorbed that they do not interfere with detection of the desired products.

The zones containing the separated compounds are located by viewing the slide under ultraviolet light, or by spraying with suitable reagents. Known compounds are treated in the same way for comparison.

Ion-Exchange Chromatography

In practice, the use of ion exchange materials for the separation of organic compounds is rather similar to the use of solid adsorbents in column chromatography. The ion exchange materials are usually resin-type products that have acidic or basic groups incorporated in the molecular structures. The most common of the commercially prepared resins are those resulting from the copolymerization of styrene and divinylbenzene to form the network structure of the resin. Sulfonic acid groups, carboxyl groups, amino groups, or quaternary ammonium groups must be attached to the benzene rings of the resin to make it an ion exchange material. The divinylbenzene forms the cross linkages between the chains of

---

[29] *Source.* J. M. Miller and J. G. Kirchner, *Anal. Chem.*, **25**, 1107–1109 (1953).

polymerized styrene. The pore size of the resin is controlled by the amount of divinylbenzene incorporated, since denser networks in the resin may be made by increasing the number of cross linkages.

Ion exchange materials are classified as either cation exchange resins or anion exchange resins. Cation exchange resins may contain, as the active exchange groups, either sulfonic acid groups or carboxylic acid groups. Such resins may be regarded as insoluble polyvalent anions with positively charged counter ions attached. The cation exchangers in their hydrogen form are high-molecular-weight insoluble acids. Anion exchangers may be regarded either as polyvalent cations with negatively charged counter ions attached or as high-molecular-weight insoluble bases or salts of such bases.

Ion exchange resins are presently manufactured by several companies under various trade names. They may be secured from supply houses by designating the type of resin desired, such as (a) strong-base anion resin (contains quaternary ammonium groups), (b) medium or weak-base anion resin (contains amino groups), (c) strong-acid cation resin (sulfonic acid type), or (d) weak-acid cation resin (carboxylic acid type).

The counter ions on both the anion and cation exchangers are the exchangeable ions in the ion exchange system. While much work must yet be done in order to make the use of ion exchange resins a widely useful tool for qualitative organic analysis, it is apparent that by the judicious selection of the proper resin in the properly regenerated form, we may separate many organic mixtures by these techniques. A detailed discussion of the subject is beyond the scope of this book, but we have given references on this subject at the end of this chapter, and recommend particularly the books by Samuelson and Nachod.

Techniques of Column Operation

A 50-ml glass stopcock buret is a serviceable vessel for holding the resin during an ion exchange operation. Place a small plug of glass wool in the buret and pack a column of the resin to a depth of 150 to 200 mm in the buret by pouring an aqueous slurry of the resin into the buret. The resin should be covered with water at all times, and air bubbles should not be present in the column.

The first operation in an exchange cycle is the *sorption* or *exhaustion* step. At the beginning of this step, the resin normally contains only one kind of exchangeable ions. The *influent*, which may contain one or several exchangeable ions, is passed through the column. After the column is rinsed with distilled water, the resin contains the exchangeable ions from the influent. The next step is one of *regeneration* or *elution*. The retained ions are removed by passing an excess of electrolyte solution (*elutriant* or *regenerant*) through the column. Ordinarily, the elution is performed with a solution containing only one exchangeable ion. After the column is washed with water, the cycle is finished and the exchanger is ready for a new cycle.

It is well to keep in mind that a given resin has a finite number of exchange groups, thus limiting the "capacity" of a given amount of resin to react with components of the solution that is passed through it. The capacity of various resins vary from 1 to 3 milliequivalents per milliliter of resin. For clean-cut separations, not more than 0.5 milliequivalents of reactive material should be used per milliliter of resin in the column.

Applications and Limitations

Even a brief study of the references on ion exchange given at the end of this chapter will disclose the wide variety of applications that have already been made to the separation or purification of organic compounds by various workers with ion exchange resins. These methods are most applicable when the substance to be recovered (or removed as an impurity) is acidic or basic or if it may be complexed to form an acidic or basic substance. The major applications have been in the fields of plant extracts, protein hydrolyzates, fruit juices, pharmaceuticals, alkaloids, and the like. Other applications include the removal of acids from aldehydes or ketones, removal of carbonyl compounds from alcohols, removing traces of acids or bases from alcohols, recovery of the unchanged acid in derivatization of acids to amides, and similar problems.

As a laboratory method, the ion exchange technique is restricted in use primarily by the limited capacity of the resins. With quantities of resin that are practical to use in the laboratory, only a relatively few milligrams of the adsorbable compounds can be removed by passage through the column of resin. The method is therefore useful only when it is desired to remove micro quantities of substance from semimicro quantities of a mixture. Furthermore, the general practice is to use aqueous solutions only. However, ether that is saturated with water, or 95 per cent alcohol may be used if the rate of flow of the solution through the resin is kept extremely slow. Another limitation is that the components are left in very dilute solutions from which they generally have to be recovered for identification. Hence, the ion exchange method is most applicable when the material to be examined is already a very dilute solution, particularly in aqueous solution, and where no other practical method of separation is available. Micro methods need to be used for derivatization of recovered components.

Fractionation Procedures   155

The following experiments represent types of separations that may be carried out in the laboratory in order to acquire experience in the use of ion exchange resins.

Example 4.07: Removal of Acids from Neutral Compounds

The removal of small amounts of citric acid from a dilute sugar solution may be accomplished by the following procedure. In general, acids, except those with molecules of such large size that they cannot enter the pores of the resin, may be separated from alcohols, aldehydes, ketones, sugars, and other nonelectrolytes, by this same procedure.

Pour an aqueous slurry of a strong-base anion resin into a 50 ml glass stopcock buret, which contains a thin plug of glass wool, until the column of resin measures 15 to 20 ml. Drain off the water and then allow 100 ml of a $1N$ solution of sodium carbonate to pass through the resin at a rate of 5 ml per minute. Rinse the resin with 200 ml of distilled water, allowing the water to run through as fast as the open stopcock will allow. Close the stopcock. The column is ready to receive the mixture to be separated.

Prepare a mixture of 25 mg of citric acid and 300 mg of dextrose in 100 ml of water. Allow this mixture to pass through the resin at a rate of 5 ml per minute. Rinse the resin with 100 ml of distilled water and combine this water with the effluent that contains the dextrose. The acid has been retained in the resin bed. The dextrose may be recovered by distilling under reduced pressure to a small volume and derivatizing the sugar to its osazone. The citric acid may be removed from the resin by eluting with 100 ml of a $1N$ solution of ammonium carbonate. Evaporate the effluent to a volume of 5 ml; add $2N$ sodium hydroxide solution to bring the pH to about 9, and derivatize the acid with *p*-nitrobenzyl chloride.

The column of resin that has been used in the above experiment may, of course, be regenerated for further use. If it is to be again used to remove acids, it should be washed with water, $1N$ sodium carbonate, and again with water. Leave the resin covered with water until it is to be used again.

Example 4.08: Removal of Carbonyl Compounds

Carbonyl compounds that will form reasonably stable compounds (α-oxysulfonic acids) with bisulfite ions can be separated from neutral compounds by using an ion exchanger in the bisulfite form. For example, traces of acetone may be removed from 2-propanol.

Prepare a column of a strong-base anion exchange resin as described in Example 4.07. Pass 100 ml of a $1N$ solution of sodium bisulfite through the resin at a rate of 5 ml per minute and then wash the column with 200 ml of distilled water.

Add 1 drop of acetone to a mixture of 40 ml of 2-propanol and 10 ml of water and pass this mixture through the prepared resin at a rate of 3 ml per minute. The acetone is retained in the resin bed and may be eluted off with 100 ml of $1N$ sodium chloride solution. Recovery and identification of this small amount of acetone in such a dilute solution is most difficult. By using extreme care, it may be accomplished by means of the following steps. Saturate the solution with sodium chloride and distill 10 ml. Saturate this 10 ml of distillate with sodium chloride and distill 1 ml. The acetone may then be converted to the 2,4-dinitrophenylhydrazone for identification.

Example 4.09: Multiple Separations

A mixture of acetic acid, acetaldehyde, and ethanol may be separated into the three components by combining the procedures given in Examples 4.07 and 4.08. Sufficient resin would have to be used to be capable of reaction with all of the acid and aldehyde present in the mixture. The mixture would first be passed through a strong-base anion resin as described in Example 4.07. The acetic acid would be retained on the resin and the effluent would contain both acetaldehyde and ethanol. This mixture would then be passed through a bisulfite-type resin as described in Example 4.08 to separate the acetaldehyde from the ethanol.

Example 4.10: Separation of Sugars

Borate ions react with many monosaccharides to produce anionic complexes; hence, such sugars may be separated from sugars that do not form such complexes.

Prepare a bed of a strong-base anion resin as described in Example 4.07 and wash it with 100 ml of $1N$ sodium hydroxide at a rate of 3 ml per minute. Wash the bed with distilled water until the effluent is neutral to phenolphthalein.

Prepare a solution containing 30 mg of dextrose and 30 mg of maltose in 50 ml of a 0.5 per cent solution of boric acid. Pass this mixture through the resin at a rate of 3 ml per minute. The dextrose will be retained in the resin bed and the maltose will be in the effluent. The dextrose may be removed from the resin by elution with 50 ml of $1N$ sodium chloride solution. The individual sugars may be identified by concentrating their respective solutions to 1 to 2 ml volumes and converting the sugars to their osazones.

Example 4.11: Removal of Cations

Sometimes it is necessary to remove cations from aqueous solutions of organic compounds. This may be done by passing the mixture through a bed of strong-acid cation resin.

Prepare a column of strong-acid cation resin in a buret and pass 100 ml of 10 per cent hydrochloric acid through the bed. Wash the resin with distilled water until it is neutral to methyl orange. Prepare a solution containing 10 to 15 mg of ferric chloride in 95 per cent alcohol and pass this mixture through the column at a rate of 2 to 3 ml per minute. Wash the column with 50 ml of distilled water. The iron will be retained by the resin.

Gas Chromatography

As the name implies, this method is applicable to materials that can be volatilized with heat. The stationary phase may be a solid adsorbent or a liquid, which is adsorbed on the surface of a solid. The mixture to be separated is swept along through the column by an inert gas. In this process, some compounds are adsorbed or absorbed by the stationary phase better than others, so that these are retained longer in the column. If the mixture is injected into the column quickly, each component is removed from the column at a different time and during a short period if the conditions are suitable. This time for passage through the column is characteristic for each compound under these conditions.

The compound is detected as it leaves the column by instrumental means; a strip chart recorder attached to the detector shows a peak for each compound which is separated. This series of peaks is called, in this case, the chromatogram. The size of the peak indicates the amount of that particular compound in the mixture, and its location on the chart indicates its nature.

Factors that must be considered for good separation of a mixture are: the material used for the packing of the column, the temperature of the column, the size of the sample, the rate of flow of the gas, and the nature of the compounds to be separated. Since good control of these variables is necessary, the equipment used here must be reliable. Consequently, the cost of this equipment is in the range of $1500 to $3000.

One company[30] advertises a kit by which an experimental gas chromatography apparatus can be assembled with minimum effort. A laboratory oven, a storage battery, and a 0 to 10 mv potentiometer must also be

[30] Gow-Mac Instrument Co., 100 Kings Road, Madison, N.J.

available. If these items are at hand, we can try the techniques of gas chromatography without a large investment in equipment.

Some equipment has been designed for preparative gas chromatography, and may be purchased.[31] For details of theory and practice of gas chromatography, see the books and articles listed at the end of this chapter.

Example 4.12: Separation of Isomeric Xylenes[32]

The gas chromatograph column is packed with firebrick, coated with 1-chloronaphthalene. One hundred grams of firebrick (40 to 60 mesh) are heated with 300 ml of aqua regia for 1 hour. The acid is removed, and the firebrick is washed several times with water–sodium hydroxide solution, and several more times with water. Fine particles are removed by washing the firebrick on a 60-mesh screen. The solid is dried at 110°.

A weight of 1-chloronaphthalene, equal to 15% of the weight of the firebrick to be used, is dissolved in methylene chloride, and the solution is added to the treated firebrick with stirring. The solid is warmed to dryness, and is then used to pack the column.

This column is then used to separate mixtures of *o*-, *m*-, and *p*-xylene under varying conditions of gas flow and column temperatures. The operational manual of the particular instrument used is consulted for details of procedure. Determine the best conditions for separating the isomers.

Example 4.13: Quantitative Separation of Aromatic Hydrocarbons

Benzene, toluene, and ethylbenzene may be determined quantitatively on the same column used in the previous example. Use 10 microliters of each of the above compounds separately under various conditions of flow rate and column temperature. Select a single set of conditions, which will give the best separation of the mixture.

With the same set of conditions, use a mixture of the three substances, containing a known quantity of each. Extend the lines of the peaks down to the base line to form single peaks similar to the ones obtained for the single substances. Measure the areas under the curve for each component (use paper with horizontal and vertical lines, and count the squares and parts of squares under each peak). Repeat this operation until the extent of variation is known, and the correlation of area under the peak and the amount of material is found. Use a different mixture to verify the results.

---

[31] Ask for descriptive literature from any company that markets analytical gas chromatographs.

[32] *Source.* A. Zlatkis, S. Ling, and H. R. Kaufman, *Anal. Chem.*, **31**, 945 (1959).

OTHER METHODS OF FRACTIONATION

Electrophoresis

Electrophoresis, or electrochromatography, is a fractionation process in which an electric potential is used to help separate the components of a mixture in solution. If the solution is applied to a paper strip continuously, the separation process may be carried on indefinitely and the separated compounds collected. The electric field is applied at right angles to the flow of the liquid, which is usually downward.

Electrophoresis is similar to ordinary paper or column chromatography except that migration of the components is greatly affected by the charge on the molecules. Ions are drawn toward one electrode or the other, while neutral molecules are unaffected by the electric field. Consequently, some mixtures are more easily separated by electrophoresis than by ordinary chromatography.

The usual procedure for zone electrophoresis is to add the mixture in solution to a line near the middle of a paper strip, which is wet with an electrolyte. The two ends of the paper strip are then connected to the source of electrical potential. For low molecular weight materials, a low potential gradient is used (2 to 10 volts/cm), but high molecular weight compounds require a higher potential gradient (50 to 100 volts/cm). After an appropriate time for migration of the compounds, the electric potential is removed, the paper is dried, and the zones are detected by methods already discussed for paper chromatography.

An inexpensive continuous paper electrophoresis apparatus has been described,[33] which has been used successfully to fractionate animal sera. It can be used to determine the homogeneity of various kinds of amphoteric materials and to separate inorganic elements and compounds.

A commercially available system makes possible the separation of 20 protein fractions in one column, their elution, detection, and collection.[34] The separation is by disc electrophoresis. The separating medium is a double layer of specially formulated polyacrylamide gel with upper and lower buffers. The various fractions are tightly layered in the lower "stacking gel," and move to the bottom of the column in order of their electrophoretic mobility.

Factors that should be considered in designing an experiment for an electrophoretic separation are: acidity and ionic strength of the electrolyte solution, the electrical potential gradient, the temperature, and the

---

[33] W. G. Glenn and H. A. Jaeger, *J. Chem. Educ.*, **35**, 360–361 (1958).
[34] Canal Industrial Corporation, Bethesda, Md.

mobility of the compounds to be separated. Electrophoresis can be conducted not only in paper, but in slabs of any porous materials, in columns, and in gels. For details of procedure and applications of electrophoresis to specific problems, the references at the end of the chapter should be consulted.

Dialysis

Compounds diffuse through porous membranes at different rates, which are related to their molecular weights; smaller compounds, usually, diffuse more rapidly than larger ones. Thus, separation of large and small molecules can be achieved by allowing the smaller molecules to diffuse through an appropriate membrane. The greater the difference in molecular weights, the better the separation.

Consequently, one important application of dialysis is the removal of inorganic salts from polymeric material, which is often of biological origin. In this case, complete separation is achieved. Although the method can be used for separating molecules that are more nearly alike in size, the process may need to be repeated to obtain the desired purity. With molecules that are very similar in size, the method results in only slight separation of the mixture.

Membranes, such as cellophane, collodion (nitrocellulose), parchment paper, and some plastic films, have been used. The electrokinetic charge on the membrane affects the rate of diffusion of electrolytes, particularly. Other factors are temperature, thickness of membrane, and degree of agitation of the solution. The publications listed at the end of this chapter should be consulted for details.

Molecular Sieves

Many chemical compounds may be separated from gaseous or liquid mixtures by selective adsorption on a surface. This method of purification may well supplement separation methods that are based on distillation, crystallization, and chromatography.

A solid is needed that has a very large surface area; materials such as activated carbon, activated alumina, and silica gel have long been used for this purpose. If the pores of the solid have a uniform size, a new dimension of selectivity is possible, since molecules above a certain size cannot enter the pores and, therefore, are poorly adsorbed. The molecular-sieve adsorbents have pores with uniform diameters in the range of molecular diameters and, therefore, show a high degree of selectivity in the molecules that they adsorb. These materials adsorb polar compounds very strongly.

Because of the latter property, molecular-sieve adsorbents will separate water from ethanol, ethylene from ethane, and benzene from cyclohexane. Based on their combined properties of size and polarity, many molecules may be separated cleanly and efficiently from one another. The manufacturer[35] should be consulted for the best type of molecular-sieve adsorbent to be used for specific purposes. A good review of the use of these materials has been written by Thomas and Mays.[36]

Inclusion Compounds

When compounds crystallize, they may trap other compounds in their crystal structure. Very often, this results in the inclusion of undesirable impurities but, sometimes, advantage is taken of this process to separate a compound from a mixture of others. Such inclusion compounds may be called clathrate compounds because of the actual cagelike structure of the crystals, or adducts, because of their close adherence. Generally, chemical bonds are not formed, although hydrogen bonds and van der Waal forces may be involved.

When urea crystallizes in the presence of suitable compounds, it forms a roughly cylindrical lattice with a specific diameter. The other compound may be included in this channel if its diameter is compatible in size with the core of the cylinder. The length of the second compound is not a factor in the ability of the urea to trap it, provided that it is above a minimum length. Other compounds such as thiourea show similar channellike lattices.

Some compounds crystallize with cagelike spaces, and may trap other compounds in these spaces as the crystals are formed. If the other compounds are too large, the cage will not close on them, and if they are too small they will escape from the cage. The cages are usually prepared by the interaction of a metal salt and an organic compound. The compound to be included must be present at the time of formation of the crystal. The complex is usually an insoluble solid, which can be removed by filtration, and the included compound can be obtained by decomposition of the metal complex.

Foams

Flotation has long been used for concentration of ore from less desirable materials. Usually, the process consists of wetting the ore particles with an organic liquid, and bubbling air through the mixture. The coated ore

---

[35] Linde Company, Division of Union Carbide Corporation, Tonowanda, N.Y.
[36] T. L. Thomas and R. L. Mays, "Separations with Molecular Sieves," in *Physical Methods in Chemical Analysis*, Vol. 4, W. G. Berl, ed., Academic Press, New York, 1961, pp. 45–98.

particle clings to the bubble and tends to rise toward the surface of the liquid where it is skimmed off. The selectivity of the process depends on the type of mineral surface and its bonding with the organic liquid used to coat it. Weak surface bonding usually results in poor separation of the ore particles.

This process has been applied to the separation of organic compounds with some success.[37] Its advantages should be investigated further because of the simplicity of operation. The literature referred to at the end of this chapter gives details of theory and procedure.

Zone Refining

Zone refining of inorganic materials has resulted in compounds of very high purity. A solid rod of the substance to be purified is subjected to careful heating in a small portion of the rod near one end until that part melts. The heating is continued along the rod until the other end is reached. The impurities move along the rod with the heated zone, since they remain in the liquid portion. In this way, the impurities accumulate at one end of the rod, and this portion can be removed mechanically.

The method has been applied to the purification of organic compounds when crystals of great purity are desired. Several apparatuses have been described, including some for work at the micro level. The articles at the end of this chapter should be consulted for details of the apparatus and procedure.

REFERENCES

Chromatography

General

W. G. Berl, "Chromatographic Analysis," in *Physical Methods in Chemical Analysis*, Vol. 2, W. G. Berl, ed., Academic Press, New York, 1951, pp. 591–618. General review, particularly of theory.

H. G. Cassidy, "Fundamentals of Chromatography," in *Technique of Organic Chemistry*, Vol. 9, A. Weissberger, ed., Interscience, New York, 1957.

R. Consden, A. H. Gordon, and A. J. P. Martin, *Biochem. J.*, **38**, 224 (1944).

A. L. Elder and R. A. Springer, *J. Phys. Chem.*, **44**, 943 (1940).

E. Heftmann, ed., *Chromatography*, Reinhold, New York, 1961.

E. Heftmann, *Anal. Chem.*, **34**, 13R (1962). A review of recent developments with 1313 references.

E. Lederer and M. Lederer, *Chromatography. A Review of Principles and Applications*, 2nd ed, Elsevier, Amsterdam, 1957. Principles and separation of organic and inorganic compounds.

[37] B. L. Karger and L. B. Roger, *Anal. Chem.*, **33**, 1165–68 (1961); R. Lemlich and E. Lavi, *Science*, **134**, 191 (1961).

A. G. Mistretta, *Microchem., J.* **3**, 305–314 (1959); ibid., **4**, 289–305 (1960); ibid., **5**, 305–324 (1961); ibid., **6**, 327–349 (1962). Annual reviews of developments.

I. Smith, *Chromatographic and Electrophoretic Techniques*, Vols. 1 and 2, William Heinemann Medical Books, Inc., London, 1960.

R. Stock and C. B. F. Rice, *Chromatographic Methods*, Reinhold, New York, 1963.

H. H. Strain, *Anal. Chem.*, **32**, 3R (1960). Recent fundamental developments. There are 462 references.

J. A. Thoma, *Anal. Chem.*, **35**, 214 (1963). Application and theory of unidimensional multiple chromatography.

Paper Chromatography

E. W. Berg, *Physical and Chemical Methods of Separation*, McGraw-Hill, New York, 1963, pp. 132–142 and, for discussion of the ring-oven technique, pp. 155–162.

R. J. Block, E. L. Durrum, and G. Zweig, eds., *A Manual of Paper Chromatography and Paper Electrophoresis*, 2nd ed., Academic Press, New York, 1958, pp. 3–488.

H. G. Cassidy, "Fundamentals of Chromatography," in *Technique of Organic Chemistry*, Vol. 10, A. Weissberger, ed., Interscience, New York, 1957, pp. 133–210.

R. Consden, A. H. Gordon, and A. J. P. Martin, *Biochem. J.*, **38**, 224 (1944). Separation of amino acids.

M. S. Dunn and E. A. Murphy, *Anal. Chem.*, **33**, 997 (1961). Chromatographic purity of amino acids.

K. Fink, R. E. Cline, and R. M. Fink, *Anal. Chem.*, **35**, 389–398 (1963). Paper chromatography of several classes of compounds: correlated R_f values in a variety of solvent systems.

C. S. Hanes, *Can. J. Biochem. and Physiol.*, **39**, 119 (1961).

S. Katz, *J. Chem. Educ.*, **39**, 34 (1962). Device for locating zones.

C. S. Knight, *Nature*, **188**, 739 (1960).

E. Lederer and M. Lederer, *Chromatography*, 2nd ed., Elsevier, New York, 1957, pp. 115–148.

F. W. Lima, *J. Chem. Educ.*, **31**, 153 (1954). Paper chromatography with radioactive materials.

E. C. Mathew and B. M. Das, *J. Chem. Educ.*, **32**, 352 (1955). Apparatus for two-dimensional paper chromatography.

L. C. Mitchell and P. A. Mills, *J. Assoc. Offic. Agr. Chemists*, **43**, 748 (1960).

R. L. Ory, *J. Chromatog.*, **5**, 153 (1961). Separation of glycerides of mixed fatty acid chain length by glass paper chromatography.

L. Peyron, *Bull. soc. chim. France*, **1960**, 1243. A review of radial chromatography.

F. Pocchiari and C. Rossi, *J. Chromatog.*, **5**, 377 (1961). Quantitative radio paper chromatography.

V. Prey, A. Kabil, and H. Berbalk, *Mikrochim. Acta*, **1959**, 68. Preparative micromethods in paper chromatography.

V. Prey and A. Kabil, *Mikrochim. Acta*, **1959**, 79. Qualitative organic analysis with the help of paper chromatography.

L. Rutter, *Nature*, **161**, 435 (1948); *Analyst*, **75**, 37 (1950). Circular paper chromatography.

A. T. Thomas and J. P. Phillips, *J. Chem. Educ.*, **38**, 406 (1961). Uses filter papers impregnated with ion-exchange resins.

J. C. Underwood and L. B. Rockland, *Anal. Chem.*, **26,** 1253 (1955). Small-scale filter paper chromatography of amino acids.

D. S. Wiggins, *J. Chem. Educ.*, **34,** 536 (1957). Chromatographic chamber for the student laboratory.

E. Winter, *Mikrochim. Acta*, **1961,** 816. Paper chromatography on a small scale.

Column Chromatography

E. W. Berg, *Physical and Chemical Methods of Separation*, McGraw-Hill, New York, 1963, pp. 80–106.

H. Brockmann and F. Volpers, *Ber.*, **80,** 77 (1947). Detection of fractions.

M. J. R. Cantow, R. S. Porter, and J. F. Johnson, *Nature*, **192,** 752 (1961). Polymer column fractionation.

H. G. Cassidy, "Adsorption and Chromatography," in *Technique of Organic Chemistry*, Vol. 5, A. Weissberger, ed., Interscience, New York, 1951, pp. 207–299.

H. G. Cassidy, "Fundamentals of Chromatography," in *Technique of Organic Chemistry*, Vol. 10, A. Weissberger, ed., Interscience, New York, 1957, pp. 77–132, 211–284.

D. W. Criddle and R. L. LeTourneau, *Anal. Chem.*, **23,** 1620 (1951). Detection of fractions.

P. Lebreton, *Bull. Soc. chim. France*, **1960,** 2188. Chromatographic analysis with gradient elution.

A. L. LeRosen, R. T. Moravek, and J. K. Carlton, *Anal. Chem.*, **24,** 1335 (1952). Detection of fractions.

A. L. LeRosen, *Ind. Eng. Chem., Anal. Ed.*, **14,** 165 (1942).

K. Macek, "Techniques of Liquid-Liquid Partition Chromatography," in *Chromatography*, E. Heftmann, ed., Reinhold, New York, 1961, p. 112–163.

P. H. Monaghan, P. B. Mosely, T. S. Burkhalter, and O. A. Nance, *Anal. Chem.*, **24,** 193 (1952). Detection of zones with a high-frequency oscillator.

W. Rieman, III, and R. Sargent, in *Physical Methods in Chemical Analysis*, Vol. 4, W. G. Berl, ed., Academic Press, New York, 1961, pp. 133–222.

E. S. Sanderson, *Anal. Chem.*, **26,** 944 (1954). Fraction collector.

D. P. Schwartz, O. W. Parks, and M. Keeney, *Anal. Chem.*, **34,** 669 (1962). Separation of 2,4-dinitrophenylhydrazone derivatives of monocarbonyls.

L. R. Snyder, *J. Chromatog*, **6,** 22 (1961). Compound separability with alumina.

F. W. Went, A. L. LeRosen, and L. Zechmeister, *Plant Physiol.*, **17,** 91 (1942) Separation of carotenoid pigments.

J. J. Wren, *J. Chromatog.*, **4,** 173 (1960). Chromatography of lipids on silicic acid.

Thin-Layer Chromatography

M. Barbier, H. Jager, H. Tobias, and E. Wyss, *Helv. Chim. Acta*, **42,** 2440 (1959)

I. Bekersky, *Anal. Chem.*, **35,** 261–262 (1963). Spray technique for the preparation of thin-layer chromatographic plates.

J. M. Bobbitt, *Thin Layer Chromatography*, Reinhold, New York, 1963.

M. Brenner and A. Niederwieser, *Experientia*, **16,** 378 (1960). Thin-layer chromatography of amino acids.

E. Demole, "Chromatostrips and Chromatoplates," in *Chromatographic Reviews*,

Vol. 1, M. Lederer, ed., Elsevier, Amsterdam, 1959, pp. 1–10; *J. Chromatog.*, **1**, 24–34 (1958).

E. Demole, "Recent Progress in Thin-Layer Chromatography," in *Chromatographic Reviews*, Vol. 4, M. Lederer, ed., Elsevier, Amsterdam, 1962, pp. 26–48; *J. Chromatog.*, **6**, 2–21, (1961).

M. Gee, *J. Chromatog.*, **9**, 278 (1962). Separation of sugars.

C. G. Honegger, *Helv. Chim. Acta*, **44**, 173 (1961). Separation of amines and amino acids.

J. G. Kirchner, J. M. Miller, G. I. Keller, *Anal. Chem.*, **23**, 420 (1951). Separation and identification of some terpenes.

J. G. Kirchner and J. M. Miller, *Ind. Eng. Chem.*, **44**, 318 (1952).

J. G. Kirchner and J. M. Miller, *J. Agr. Food Chem.*, **1**, 512 (1953).

J. G. Kirchner, J. M. Miller, and R. G. Rice, *J. Agr. Food Chem.*, **2**, 1031 (1954).

J. G. Kirchner and J. M. Miller, *J. Agr. Food Chem.*, **5**, 283 (1957).

J. H. Linford, *Can. J. Biochem. and Physiol.*, **34**, 1153 (1956).

H. L. MacDonnell and J. P. Williams, *Anal. Chem.*, **33**, 1552 (1961). Porous glass chromatography used for the characterization of water-soluble inks.

G. Machata, *Mikrochim. Acta*, **1960**, 79. Applicator for thin-layer chromatography.

D. C. Malins and J. C. Wekell, *J. Chem. Educ.*, **40**, 531–534 (1963). Applications.

K. H. Mangold and R. Kammereck, *Chem. & Ind.* (*London*), **1961**, 1032. Separation of hydroxylated and unsaturated acids, followed by gas chromatography.

H. K. Mangold, R. Kammereck, and D. C. Malins, "Thin-Layer Chromatography as an Analytical and Preparative Tool in Lipid Radiochemistry," in *Microchemical Techniques*, N. D. Cheronis, ed., Interscience, New York, 1962, pp. 697–714. There are 37 references.

R. S. MacDonald, *J. Am. Chem. Soc.*, **79**, 850 (1957).

J. E. Meinhard and N. F. Hall, *Anal. Chem.*, **21**, 185 (1949). Surface chromatography.

J. M. Miller and J. G. Kirchner, *Anal. Chem.*, **24**, 1480 (1952). Chromatography of terpenes.

J. M. Miller and J. G. Kirchner, *Anal. Chem.*, **25**, 1107 (1953). Chromatostrips for identifying constituents of essential oils.

J. M. Miller and J. G. Kirchner, *Anal. Chem.*, **26**, 2002 (1954). Apparatus for the preparation of chromatostrips.

V. Prey, H. Berbalk, and M. Kausz, *Mikrochim. Acta*, **1961**, 968. Separation of sugars.

K. Randerath, *Dunnschicht-Chromatographie*, Verlag Chemie, Weinheim, Germany, 1962.

K. Randerath, *Thin Layer Chromatography*, Academic Press, New York, 1963.

R. H. Reitsema, *Anal. Chem.*, **26**, 960 (1954). Characterization of essential oils by chromatography.

E. Stahl, *Chem. Ztg.*, **82**, 323–329 (1958). Thin-layer chromatography. II.

E. Stahl, *Z. anal. Chem.*, **181**, 303 (1961).

E. Stahl, *Angew. Chem.*, **73**, 646 (1961).

E. Stahl, ed., *Thin-Layer Chromatography*, Academic Press, New York, 1963.

H. Struck, *Mikrochim. Acta*, **1961**, 634. Separation of steroids.

E. V. Truter, *Thin Film Chromatography*, Wiley, New York, 1963.

R. Wasicky, *Anal. Chem.*, **34**, 1346 (1962). Use of microslides.

C. E. Weill and P. Hanke, *Anal. Chem.*, **34**, 1736 (1962). Separation of sugars.

E. G. Wollish, M. Schmall, and M. Hawrylyshyn, *Anal. Chem.*, **33**, 1138 (1961). Recent developments in equipment and applications.

E. G. Wollish, "Present States of Thin-Layer Chromatography," in *Microchemical Technique*, N. D. Cheronis, ed., Interscience, New York, 1962, pp. 687–696. There are 55 references, mostly applications.

Ion-Exchange Chromatography

E. W. Berg, *Physical and Chemical Methods of Separation*, McGraw-Hill, New York, pp. 176–216.

D. L. Buchanan and R. T. Markiw, *Anal. Chem.*, **32**, 1400 (1960). Water elution chromatography of amino acids on ion-exchange materials.

J. Cason, G. Sumrell, and R. S. Mitchell, *J. Org. Chem.*, **15**, 850 (1950). Separation of acids and amides.

H. G. Cassidy, "Adsorption and Chromatography," in *Technique of Organic Chemistry*, Vol. 5, A. Weissberger, ed., Interscience, New York, 1951, pp. 267–290. There are 50 references.

H. G. Cassidy, "Fundamentals of Chromatography," in *Technique of Organic Chemistry*, Vol. 10, A. Weissberger, ed., Interscience, New York, 1957, pp. 285–318.

M. St. C. Flett, *Physical Aids to the Organic Chemist*, American Elsevier Publishing Co., Inc., New York, 1962, pp. 31–34.

G. Gabrielson and O. Samuelson, *Acta Chem. Scand.*, **6**, 738 (1952). Separation of carbonyls and alcohols.

V. A. Haas and E. R. Stadtman, *Ind. Eng. Chem.*, **41**, 983 (1949). Components of fruit juices.

P. B. Hamilton, D. C. Bogue, and R. A. Anderson, *Anal. Chem.*, **32**, 1782 (1960). Ion-exchange chromatography of amino acids: analysis of diffusion mechanisms. Theory.

F. Helferich, *Nature*, **189**, 1001 (1961). Ligand exchange. Metal ions are held in a column of ion-exchange resin. Ligands are strongly adsorbed since they form strong complexes with the metal, and are selectively displaced.

F. Helferich, *Ion Exchange*, McGraw-Hill, New York, 1962.

J. X. Khym and L. P. Zill, *J. Am. Chem. Soc.*, **73**, 2399 (1951). Separation of sugars.

R. Kunin, *Ion Exchange Resins*, Wiley, New York, 1958.

R. Kunin and F. X. McGarvey, *Anal. Chem.*, **34**, 48R (1962). Review of recent developments with 70 references.

L. Levy and M. J. Coon, *J. Biol. Chem.*, **192**, 809 (1951).

F. C. Nachod and J. Schubert, eds., *Ion Exchange Technology*, Academic Press, New York, 1956.

G. H. Osborn, *Synthetic Ion-Exchangers*, 2nd ed., Chapman and Hall, London, 1961.

W. Rieman, III, *J. Chem. Educ.*, **38**, 338 (1961). Salting-out chromatography.

W. Rieman, III, and R. Sargent, "Ion Exchange," in *Physical Methods in Chemical Analysis*, Vol. 4, W. G. Berl, ed., Academic Press, New York, 1961, pp. 134–222. Review with 139 references.

J. E. Salmon and D. K. Hale, *Ion Exchange, A Laboratory Manual*, Butterworths, London, 1959.

O. Samuelson, *Ion Exchange Separations in Analytical Chemistry*, Wiley, New York, 1963.

J. Sherma and W. Rieman, III, *Anal. Chim. Acta*, **20**, 357–365 (1959). Solubilization chromatography, a type of partition chromatography on ion-exchange resins. Mixtures of ethers, fatty acids, and hydrocarbons were separated.

N. E. Skelly, *Anal. Chem.*, **33**, 271 (1961). Gradient elution in the separation of chlorophenols by ion exchange.

R. N. Shelley and E. J. Umberger, *Anal. Chem.*, **31**, 593 (1959). Behavior of acidic organic compounds in nonaqueous media ion-exchange resins.

S. C. Smith and S. H. Wender, *J. Am. Chem. Soc.*, **70**, 3719 (1948). Separation of xanthine and guanine.

S. R. Watkins and H. F. Walton, *Anal. Chim. Acta*, **24**, 334 (1961). Adsorption of amines on cation exchange resins.

J. C. Winters and R. Kunin, *Ind. Eng. Chem.*, **41**, 460–463 (1949). Use of ion exchangers in pharmaceuticals; separation of amino acids.

Gas Chromatography

D. Ambrose and B. A. Ambrose, *Gas Chromatography*, George Newnes, London, 1961.

E. Bayer, *Gas-Chromatographie*, Springer-Verlag, Berlin, 1962.

E. W. Berg, *Physical and Chemical Methods of Separation*, McGraw-Hill, New York, 1963, pp. 107–132.

K. J. Bombaugh, *Anal. Chem.*, **33**, 29 (1961). Separation of *o*-, *m*-, and *p*-chloronitrobenzenes.

V. J. Coates, H. J. Noebels, and I. S. Fagerson, eds., *Gas Chromatography*, Academic Press, New York, 1958.

D. H. Desty, ed., *Gas Chromatography, 1958*, Academic Press, New York, 1958.

D. H. Desty, ed., *Vapour Phase Chromatography*, Academic Press, New York, 1957.

L. S. Ettre, *Gas Chromatography*, H. J. Noebels et al., eds., Academic Press, New York, 1961.

M. St. C. Flett, *Physical Aids to the Organic Chemist*, American Elsevier Publishing Co., Inc., New York, 1962, pp. 37–58.

J. C. Giddings, *Anal. Chem.*, **32**, 1707 (1960). Optimum conditions for separation in gas chromatography.

B. J. Gudzinowicz and W. R. Smith, *Anal. Chem.*, **32**, 1767 (1960). High temperature gas chromatography.

C. J. Hardy and F. H. Pollard, "Review of Gas-Liquid Chromatography" in *Chromatography Reviews*, Vol. 2, M. Lederer, ed., Elsevier, Amsterdam, 1960, pp. 1–43.

C. Hista, J. P. Messerley, R. Reshke, D. H. Fredericks, and W. D. Cooke, *Anal. Chem.*, **32**, 88 (1960). Gas chromatography of solid organic compounds.

R. Kaiser, *Gas Chromatography*, Akad. Verlag., Leipzig, 1960. Emphasis on technique; discussion of capillary columns.

R. A. Keller, G. H. Stewart, and J. C. Giddings, *Ann. Rev. Phys. Chem.*, **11**, 347 (1960).

A. I. M. Keulemans, *Gas Chromatography*, 2nd ed., Reinhold, New York, 1959.

J. H. Knox, *Gas Chromatography*, Methuen, London; Wiley, New York, 1962. Relatively nonmathematical, general presentation. Good introduction to subject.

A. B. Littlewood, *Gas Chromatography. Principles, Techniques, and Applications*, Academic Press, New York, 1962. Both theoretical and practical aspects.

I. G. McWilliam, *Rev. Pure Appl. Chem.*, **11**, 33 (1961). Applications of gas chromatography.

S. D. Nogare and L. W. Safranski, "Gas Chromatography," in *Organic Analysis*, Vol. 4, J. Mitchell, Jr., ed., Interscience, New York, 1960, pp. 91–227.

S. D. Nogare and R. S. Juvet, Jr., *Anal. Chem.*, **34**, 35R (1962). A review of recent work with 480 references.

S. D. Nogare and R. S. Juvet, Jr., *Gas-Liquid Chromatography. Theory and Practice*, Interscience, New York, 1962. Thorough treatment of trace analysis, capillary columns, and preparative work. Lists commercial instruments available in 1961.

R. L. Pecsok, *Principles and Practice of Gas Chromatography*, Wiley, New York, 1959.

C. S. G. Phillips, "Gas Chromatography" in *Physical Methods in Chemical Analysis*, Vol. 3, W. G. Berl, ed., Academic Press, New York, 1956, pp. 1–28. An early review.

J. H. Purnell, *Gas Chromatography*, Wiley, New York, 1962. Emphasizes theoretical aspects of the subject.

R. P. W. Scott, ed., *Gas Chromatography, 1960*, Butterworth, London, 1960.

H. A. Szymanski, ed., *Lectures on Gas Chromatography, 1962*, Plenum Press, New York, 1963.

R. Teranishi, C. C. Nimmo, and J. Corse, *Anal. Chem.*, **32**, 1384 (1960). Programmed temperature control of the capillary column.

A. Zlatkis, S. Ling, and H. R. Kaufman, *Anal. Chem.*, **31**, 945 (1959).

Other Methods of Fractionation

Electrophoresis

P. Alexander and R. J. Block, eds., *A Laboratory Manual of Analytical Methods of Protein Chemistry*, Vols. 1, 2, and 3, Pergamon Press, New York, 1960.

R. Audubert and S. deMende, *The Principles of Electrophoresis*, Macmillan, New York, 1960.

E. W. Berg, *Physical and Chemical Methods of Separation*, McGraw-Hill, New York, 1963, pp. 143–154.

E. W. Bermes, Jr., and H. J. McDonald, *J. Chromatog.*, **4**, 34 (1960).

M. Bier, ed., *Electrophoresis*, Academic Press, New York, 1959.

R. J. Block, E. L. Durrum, and G. Zweig, *A Manual of Paper Chromatography and Paper Electrophoresis*, 2nd ed., Academic Press, New York, 1958.

M. A. Doran, *Anal. Chem.*, **33**, 1752 (1961).

J. E. Garvin, *J. Chem. Educ.*, **38**, 36 (1961). Student experiment with filter paper electrophoresis.

W. G. Glenn and H. A. Jaeger, *J. Chem. Educ.*, **35**, 360 (1958). An inexpensive continuous paper electrophoresis apparatus.

D. Gross, *J. Chromatog.*, **5**, 194 (1961).

B. W. Grunbaum and P. L. Kirk, *Anal. Chem.*, **32**, 564 (1960). Apparatus for micro samples.

J. B. Himes, L. D. Metcalfe, and H. Ralston, *Anal. Chem.*, **33**, 364 (1961).

A. Kolin, *Methods of Biochemical Analysis*, Vol. 6, D. Glick, ed., Interscience, New York, 1958, p. 259.

M. Lederer, *An Introduction to Paper Electrophoresis and Related Methods*, Elsevier, Amsterdam, 1957.

H. Michl, "High Voltage Electrophoresis," in *Chromatographic Reviews*, Vol. 1, M. Lederer, ed., Elsevier, Amsterdam, 1959, pp. 11–38.

D. H. Moore, "Electrophoresis," in *Technique in Organic Chemistry*, 3rd ed., Vol. 1, Part 4, A. Weissberger, ed., Interscience, New York, 1960, pp. 3113–3153. There are 116 references.

R. F. Peterson and L. W. Hanman, *J. Chromatog.*, **4**, 42 (1960).

L. P. Ribeiro, E. Mildieri, and O. R. Affonso, *Paper Electrophoresis, Review of Methods and Results*, Elsevier, Amsterdam, 1961.

R. L. Searcy and L. M. Bergquist, *Clin. Chim. Acta*, **5**, 941 (1960).

I. Smith, *Chromatographic and Electrophoretic Techniques*, Vols. 1 and 2, William Heinemann Medical Books, Inc., London, 1960.

R. D. Strickland, *Anal. Chem.*, **34**, 31R (1962). A recent review with 184 references.

M. M. Tuckerman and H. H. Strain, *Anal. Chem.*, **32**, 695 (1960).

E. J. Wawszkiewicz, *Anal. Chem.*, **33**, 252 (1961).

T. Wieland and K. Dose, "Electrochromatography (Zone Electrophoresis, Pherography)" in *Physical Methods in Chemical Analysis*, Vol. 3, W. G. Berl, ed., Academic Press, New York, 1956, pp. 29–70. Good coverage of apparatus and procedures.

C. Wunderly, *Principles and Applications of Paper Electrophoresis*. Elsevier, New York, 1961.

Dialysis

E. W. Berg, *Physical and Chemical Methods of Separation*, McGraw-Hill, New York, 1963, pp. 227–241.

C. W. Carr, "Dialysis," in *Physical Methods in Chemical Analysis*, Vol. 4, W. G. Berl, ed., Academic Press, New York, 1961, pp. 1–44. Review of theory.

L. C. Craig, *A Laboratory Manual of Analytical Methods of Protein Chemistry*, Vol. 1, P. Alexander and R. J. Block, eds., Pergamon Press, New York, 1960, pp. 103–121.

S. Hakomori and K. Takeda, *Nature*, **190**, 265 (1961). Fractionation of compound lipids by dialysation in an organic solvent against an organic solvent.

Molecular Sieves

T. L. Thomas and R. L. Mays, "Separations with Molecular Sieves," in *Physical Methods in Chemical Analysis*, Vol. 4, W. G. Berl, ed., Academic Press, New York, 1961, pp. 45–98.

Inclusion Compounds

M. Baron, "Analytical Applications of Inclusion Compounds," in *Physical Methods of Chemical Analysis*, Vol. 4, W. G. Berl, ed., Academic Press, New York, 1961, pp. 223–266.

L. W. Gamble, "The Microscopic Study of Urea and Thiourea Adducts," in *Microchemical Techniques*, N. D. Cheronis, ed., Interscience, 1962, pp. 153–164. There are 24 references.

M. M. Hagan, *Clathrate Inclusion Compounds*, Reinhold, New York, 1962.

Foams

E. W. Berg, *Physical and Chemical Methods of Separation*, McGraw-Hill, New York, 1963, pp. 310–334.

H. G. Cassidy, "Fundamentals of Chromatography," in *Technique of Organic Chemistry*, Vol. 10, A. Weissberger, ed., Interscience, New York, 1957, pp. 327–344.

B. L. Karger and L. B. Rogers, *Anal. Chem.*, **33**, 1165 (1961). Foam fractionation of organic compounds.

R. Lemlich and E. Lavi, *Science*, **134**, 191 (1961). Foam fractionation with reflux.

H. M. Schoen, ed., *Chemical Engineering Separation Techniques*, Interscience, New York, 1962. There is a chapter on foam separation in this book.

M. E. Wadsworth, "Separations with Foams," in *Physical Methods in Chemical Analysis*, Vol. 4, W. G. Berl, ed., Academic Press, New York, 1961, pp. 99–117.

Zone Refining

E. W. Berg, *Physical and Chemical Methods of Separation*, McGraw-Hill, New York, 1963, 163–175.

J. E. Benyon and R. A. Saunders, *Brit. J. Appl. Phys.*, **11**, 128 (1960). Purification of organic materials by zone refining.

M. St. C. Flett, *Physical Aids to the Organic Chemist*, American Elsevier Publishing Co., Inc., New York, 1962, pp. 59–72.

R. Handley and E. F. G. Herington, *Chem. & Ind.* (*London*), **1957**, 1184. Simplification of a semimicro zone melting apparatus.

E. F. G. Herington, *Analyst*, **84**, 680 (1959). Zone refining with some discussion of its analytical applications.

E. F. G. Herington, *Zone Melting of Organic Compounds*, Wiley, New York, 1963.

G. Hesse and H. Schildkneckt, *Angew. Chem.*, **68**, 641 (1956). Micro zone melting process for the purification of organic substances.

M. J. Joncich and D. R. Bailey, *Anal. Chem.*, **32**, 1578 (1960). Zone melting of some organic compounds.

E. T. Knypl and K. Zielenski, *J. Chem. Educ.*, **40**, 352 (1963). An automatic apparatus for refining organic substances.

J. S. Mathews and N. D. Coggeshall, *Anal. Chem.*, **31**, 1124 (1959). Concentration of impurities from organic compounds by progressive freezing.

W. G. Pfann, *Zone Melting*, Wiley, New York, 1958.

A. P. Ronald, *Anal. Chem.*, **31**, 964 (1959). Automatic multistage semimicro zone melting apparatus.

E. A. Wynne, *Microchem. J.*, **5**, 175 (1961). Application of zone refining to the purification of organic compounds.

M. Zief, H. Ruch, and C. H. Schramm, *J. Chem. Educ.*, **40**, 351–352 (1963). A low-temperature zone-refining apparatus.

5

Physical Properties of Organic Compounds

MELTING POINT—BOILING POINT—
REFRACTIVE INDEX—DENSITY—OPTICAL ROTATION—
MOLECULAR WEIGHT—MOLAR REFRACTION AND
DISPERSION

Physical properties, such as the melting point, boiling point, refractive index, density, and molecular weight, are important in establishing the identity of an organic compound. These constants alone will often indicate a great deal about the structure of a compound and, at least, will narrow the search for identification purposes to a few compounds in a class or group.

Physical properties of organic compounds may be classified in several ways, depending on the use to be made of them. For our purpose, the method of classification used by Ostwald is very useful. He described two classes of physical properties: (a) additive and (b) constitutive. Additive properties are those that depend only on the numbers and kinds of atoms in the molecule; molecular weight and vapor density are examples of this class. Constitutive properties depend not only on the numbers and kinds of atoms in the molecule, but also on their arrangement in the molecule. Two kinds of constitutive properties are recognized. Constitutive properties of the first kind are independent of molecular association, and are represented by such properties as optical rotation and molar refraction. Constitutive properties of the second kind are dependent on molecular association; properties such as melting points, boiling points, and density are representative of this class.

One other type of physical property should be mentioned in this connection. Colligative properties are those that are dependent only on the number of particles present. The lowering of the freezing point by a solute, the elevation of the boiling point by a solute, and the osmotic pressure of a solution are colligative properties. It should be noted, however, that these are properties of solutions, not of pure substances.

Several important physical properties will be discussed in the following sections. Procedures will be given for determining properties with small amounts of material, and these properties will be related to the chemical structures of the molecules.

Melting Point

The melting point is the most important single physical property of solids for qualitative analysis. Usually, the first physical property determined by the chemist, after he prepares a new solid compound or isolates a solid from a natural mixture, is the melting point. If the compound has been prepared before, he can compare the melting point of his solid with published values and, very quickly, determine if he has the desired compound. If the compound has not been previously prepared, the melting point will immediately establish one basis for comparison in case the compound is ever made again. Furthermore, the melting range of the substance is an indication of its purity. Even if the identity of the compound is unknown, the fact that the melting occurs over a range of $0.5°$ shows that the compound is fairly pure. If the range of melting is greater, we shall probably wish to recrystallize the compound (as described in Chapter 3), and redetermine the melting point. It should be noted that purifying the compound will not only shorten the melting range, but also raise the whole range to a higher value.

The melting point is an important physical property for still other reasons. (1) Very little material is required for a melting point determination—from 1 mg down to a single crystal will suffice. Thus, 5 to 20 mg of solid can be crystallized and the melting point determined several times, if necessary, in order to obtain the desired purity. (2) The apparatus required for determination of the melting point is quite simple and inexpensive. A beaker with a liquid, which can be heated, a capillary tube, and a thermometer are the minimum essentials. Melting points taken carefully with this equipment are as good as those taken with much more expensive apparatus. (3) Conditions such as the atmospheric pressure do not affect the melting point. Although application of extreme pressures will change melting points, there is no indication that small variations in pressure have a measurable effect on the melting points of solids. (4) The melting process is not subject to superheating. Although liquids may be cooled below their melting points without solidifying, the reverse is not true. Crystals, evidently, cannot be heated above their melting points without melting.

It is helpful to visualize the process of melting in order to understand the factors that influence it. Let us imagine an orderly arrangement of molecules, which we call a crystal. This arrangement is stable over a relatively wide range of conditions of temperature and mechanical stress. As the temperature is raised, the molecules must absorb energy; the higher the temperature, the more energy that must be absorbed. Some of this energy goes into the increasing of the vibrations of the molecule and some of it may go into an increase in the rotational energy of the molecule. Eventually, if the temperature continues to increase, these modes of energy absorption will be inadequate, and the molecule will burst from its lattice and acquire translational energy. At this point, it becomes a liquid in which the molecules are still close to one another but have no regular arrangement. Let us examine the factors that affect the temperature of melting.

1. The forces that operate between molecules play an important role; the stronger these forces, the more energy must be applied to break down the orderly arrangement and, consequently, the higher the temperature of melting. These forces have been classified in various ways but, for our purposes, we can list them as ionic attraction, hydrogen bonding, dipole-dipole interaction, and van der Waal forces. Some organic compounds have a great deal of ionic character, and may exist as ions; salts of amines and salts of carboxyl acids are examples. These have a great deal of attraction between particles. Molecules such as glycerol show a great attractive force between molecules, due to hydrogen bonding. A molecule such as acetone cannot form hydrogen bonds with other molecules of acetone. However, it is slightly bound to another molecule of acetone by

$$O=C\begin{matrix}CH_3\\ \\CH_3\end{matrix}$$

the mutual attraction of the dipole, which exists in each molecule. This dipole is created by the greater attraction of electrons by the oxygen atom

than by the carbon atom. It should be pointed out, however, that some acetone molecules exist in the enol form, $CH_3-C=CH_2$, which can form
$$|$$
$$OH$$
hydrogen bonds with other molecules. Even in hydrocarbons, weak forces, called van der Waal forces, operate between molecules. The nature of these forces is poorly understood, but they operate only when the molecules are very close to one another; the greater the opportunity for contact between the molecules, the greater the van der Waal forces.

2. Symmetry is one of the powerful factors governing the melting process. If a molecule is symmetrical, it can absorb more energy without disrupting the crystal lattice. Many examples of the operation of this factor can be found, and some will be discussed later.

3. Size of the molecule affects the melting point, if other factors are equal. Larger molecules, in general, melt higher, if they are otherwise similar.

4. Polymorphism is responsible for some difference in melting points. Sometimes a compound will crystallize in more than one type of lattice, and each of these two crystals will have its own melting point. It would be difficult to compare the melting points of two similar compounds, which existed in two different crystal structures. Fortunately, the phenomenon of polymorphism is not common.

Melting Points and Structures of Compounds

Although it is not possible to isolate completely each of the factors that affects melting points, we can choose examples in which one factor or another accounts for most of the difference in the melting points of two compounds. In this way, the general influence of the factor can be evaluated and, perhaps, this evaluation may be carried over to other pairs of compounds.

One of the principal factors affecting melting points is the force of attraction between the molecules or ions in the crystal lattice. A molecule of γ-aminobutyric acid, $\overset{+}{H_3N}-CH_2-CH_2-CH_2-CO_2^-$, is entirely dipolar, that is, it has a full positive charge on the nitrogen atom and a full negative charge on the oxygen atoms; butyrolactam,

$$\begin{array}{c} CH_2-CH_2 \\ | \quad\quad\quad \diagdown \\ \quad\quad\quad\quad\quad C=O \\ | \quad\quad\quad \diagup \\ CH_2-NH \end{array}$$

on the other hand, has no full electron charges. The former is a solid melting at 202° with decomposition, while the latter is a liquid at room temperature. The difference is, principally, due to the ionic attraction between the dipolar ions of γ-aminobutyric acid, and its absence in butyrolactam. In the same manner, ammonium benzoate, mp 198° d, and benzamide, mp 130°, differ in the amount of ionic attraction in the crystal lattice.

Many examples of the effect of hydrogen bonding on the melting point can be found, although it may be difficult to isolate this factor completely. Acetic acid, CH_3CO_2H, mp 16.6°, does not form hydrogen bonds as well

TABLE 5.01
MELTING POINTS OF DISUBSTITUTED BENZENE DERIVATIVES

	Ortho, °C	Meta, °C	Para, °C
$C_6H_4Cl_2$	2	−7	53
$C_6H_4(CH_3)(NO_2)$	−11	16	51
$C_6H_4(CH_3)(OH)$	30	11	36
$C_6H_4(NO_2)_2$	118	90	173

as glycolic acid, $HOCH_2CO_2H$, mp 63°. n-Butyl alcohol has a higher melting point, −89°, than does butane, −135°. 2-Chloroethanol melts higher, −69°, than chloroethane, −139°. In each case, the higher melting compound has more opportunity for molecular attraction via hydrogen bonds than the lower melting compound.

Dipole-dipole interaction constitutes another of the molecular forces, which cause higher melting points. Although acetone and isobutane have similar molecular weights and structures, and neither is able to form hydrogen bonds with itself, acetone melts at a much higher temperature, −95°, than isobutane does, −145°. Methyl sulfone, CH_3—SO_2—CH_3, has coordinate covalent bonds between the sulfur atom and the oxygen atom; this compound melts at 109°. Methyl sulfide, $(CH_3)_2S$, has a similar structure, without the two oxygen atoms, and without the coordinate covalent bonds; it melts at −83°. The difference in melting points in each case may be due to the attraction of the dipole of one molecule for the dipole of another molecule.

If two different groups are attached to the benzene ring, the relation between structure and melting point is complex. The para isomer always has the highest melting point of the three, as shown in Table 5.01. If the two groups are identical, the ortho isomer melts higher than the meta isomer. If the groups are different, but have the same effect on orientation for aromatic substitution, the ortho isomer still has the higher melting point. However, if the groups have different orientation effects, the meta isomer has the higher melting point. These effects may be explained on the basis of the dipole-dipole interaction. If the groups are identical, the dipole moment of the ortho isomer will be greater than that for the meta isomer and, consequently, its melting point will be higher. If the group moments for the two groups have different signs (leading to different

TABLE 5.02

MELTING POINTS OF SOME BENZENE DERIVATIVES (MELTING POINT OF BENZENE, 5.5°)

R	C_6H_5R °C	1,4-$C_6H_4R_2$ °C	1,2,4-$C_6H_3R_3$ °C
—CH_3	−95	13.2	−57.4
—Cl	−45	53	17
—OH	41	170.5	140.5

orientations), then the meta isomer will have a larger dipole moment, and will have a higher melting point.

Symmetry is a very powerful factor in determining melting points, other factors being equal. The molecular model of *tert*-butyl alcohol, $(CH_3)_3COH$, is almost spherical, whereas that for *n*-butyl alcohol is more linear; the former melts at 25.5° and the latter at −89.2°. Tetramethylbutane melts at 104°, although trimethylbutane melts at −25°:

$$\begin{array}{cc} CH_3 & CH_3 \\ | & | \\ CH_3-C-\!\!\!-C-CH_3 \\ | & | \\ CH_3 & CH_3 \end{array}$$
(mp 104°)

$$\begin{array}{cc} CH_3 & CH_3 \\ | & | \\ CH_3-C-\!\!\!-C-CH_3 \\ | & | \\ CH_3 & H \end{array}$$
(mp −25°)

Symmetry factors are strikingly illustrated by benzene derivatives, as shown in Table 5.02. Note that the introduction of one group on the benzene ring lowers the melting point, in all cases listed, while the introduction of a second similar group in the para position raises it again. The addition of a third group again destroys the symmetry, and the melting point drops.

In general, the larger the molecule, the higher will be its melting point. Austin[1] has suggested that the melting points of members of a particular series of compounds fit a curve represented by

$$\log M = A + BT_m$$

where M is the molecular weight, T_m the melting point, and A and B are constants for that particular series.

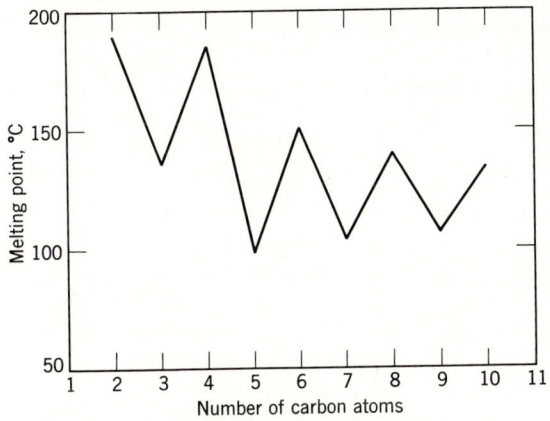

FIGURE 5.01

Melting points of the dibasic acids.

Unfortunately, the melting-point curve for members of a series is not a smooth one; some show a great deal of oscillation between high and low values. The curve for the dibasic acids, shown in Figure 5.01, is one of the best examples of oscillation. In many cases, this oscillation, or saw-tooth effect, makes it impossible to predict with any accuracy the melting point of a particular compound. However, by the use of the principles discussed here, we should be able to estimate whether the melting point of a certain compound is higher or lower than that of another, which is closely related to it structurally.

Melting Points of Mixtures

If a pure liquid A is cooled, it will produce, with time, a cooling curve shown at (*1*) in Figure 5.02(*A*). If however a liquid solution of X and Y is cooled, a cooling curve like that at (*2*), (*3*), or (*4*) in Figure 5.02(*A*) will

[1] J. B. Austin, *J. Am. Chem. Soc.*, **52**, 1049–1053 (1930).

be obtained. From these cooling curves, the phase diagram shown in Figure 5.02(B) can be derived.

From the diagram in Figure 5.02(B), it is clear that small amounts of Y in X will lower the melting point, even though Y has a higher melting point than X. A small percentage of X in Y will also lower the melting point of Y. In each case, the melting range is widened as it is lowered. Only in rare cases will a mixture of two organic compounds have a melting point no lower than those for the two pure compounds. This is the principle by which so-called "mixed" melting points are used to determine whether two substances are identical.

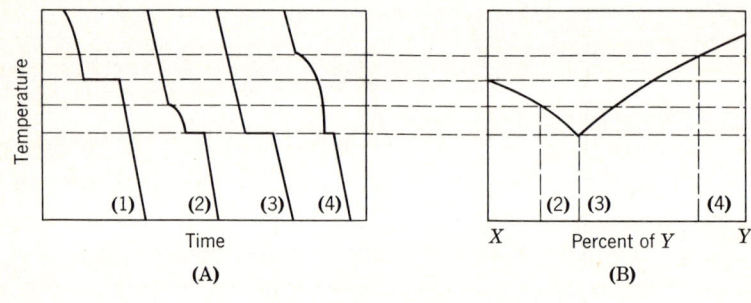

FIGURE 5.02

(A) Cooling curves. (B) Phase diagram.

Note that at the eutectic point (the lowest point on the melting point curve for the two compounds) the melting range is narrow. This mixture of substances could easily be mistaken for a pure compound. If, however, the mixture were recrystallized once more, the composition would change, and the range of melting would increase.

Apparatus

The *capillary-tube* method is commonly used for the determination of melting points. About a milligram (or less) of the solid is placed in a thin-wall glass capillary tube having a diameter close to 1 mm. The capillary tube is attached to a thermometer, then placed in a liquid bath, and heated slowly. The temperature at which the solid within the capillary tube begins to liquefy and the temperature at which the liquid is clear is recorded as the *observed melting range*. When the values are corrected they are called *corrected melting points*. It should be noted that melting points determined by this method are not *true melting points* but *capillary melting points*; the latter are slightly higher than the true melting points, which are determined by cooling or heating curves;[2] these require larger samples but give more exact information as to the purity of the compound. For most ordinary purposes, the capillary-tube method may be used. The amount of substance required for a single determination by the capillary-tube method is usually 1 to 2 mg, although a fraction of this amount may be used.

Other methods for the determination of melting temperatures of solids are *heating bars* and *heating stages*. In the former, the crystals are heated on a metal bar whose temperature is determined either by a thermometer or thermocouple; in the latter, a few crystals weighing a fraction of a microgram are placed on an electrically heated stage, and the temperature at which the crystals melt is observed. With this method, it is possible, in most cases, to observe the temperature at which liquid and fragments of crystals coexist and, for this reason, *corrected micromelting points* should be differentiated from corrected *capillary melting points*.

Several factors determine the precision and accuracy of the measurement of melting temperatures. Although it is possible to obtain a precision of 0.5° with ordinary thermometers, and with specially constructed thermometers a precision of 0.1°, the accuracy depends primarily on the calibration of the device by which the temperature is measured at the region where the crystals are situated. Since the *melting points of derivatives* are employed as a final confirmation in the systematic characterization of an unknown, see pages 197–202 for a discussion of the evaluation of melting-point data. In the following discussion the capillary method will be discussed in greater detail than the other two methods.

Construction and Filling of Melting-Point Capillaries

Glass capillaries of uniform diameter (1 to 1.2 mm) and 70 to 75 or 100 mm length are commercially available, packed in vials; we prefer capillary tubes of about 100 to 120 mm length, since they are more suitable for most melting-point apparatus. It is recommended, however, that unless suitable thin-wall capillaries can be purchased these be prepared in the laboratory. A clean soft-glass test tube 12 to 16 mm in diameter, or any thin-wall glass tube 6 to 8 mm in diameter, and 16 to 20 cm in length is heated in a flame until it is soft. It is then removed from the flame and drawn out slowly in such a manner as to insure a uniform capillary bore; this bore should be about 1 mm in diameter. The capillary

---

[2] E. L. Skau, J. C. Arthur, Jr., and H. Wakeham, "Determination of Melting and Freezing Temperatures" in *Physical Methods of Organic Chemistry*, 3rd. ed., Part 1, A. Weissberger, ed., Interscience, New York, 1959, pp. 287–356.

is heated at intervals of 180 to 240 mm, so as to seal the tube completely. They are broken at the center to form two melting-point capillaries, as needed. Care should be taken to make the sealed end oil-tight, if an oil bath is to be used. The capillaries are placed in a dry test tube, which is tightly corked to keep out moisture and other impurities. Consideration should be given to the possibility that alkali-sensitive compounds are affected by the alkalinity of the soft glass, and may give low melting temperatures.[3] In such a case, capillaries from heat resistant glass should be employed.

To load the capillary tube, a few milligrams of the crystalline material are placed on a watch glass, a piece of clean paper, or a porous plate, and crushed to a fine powder by drawing the spatula over them. The open end of the capillary tube is pressed into the fine powder; then the closed end is tapped on the desk, or the tube is lightly scratched with the flat part of a file, in order to force the sample to the bottom. The tube is filled to a height of about 1 to 2 mm, and is then attached to the thermometer so that the end of the capillary tube reaches the middle of the mercury bulb. If oil is used as a bath, the capillary tube is attached to the thermometer by means of a small rubber band cut from ordinary 3/16-in. tubing. The rubber band is placed near the top of the capillary tube well above the liquid bath. If the rubber band comes in contact with the hot bath liquid, the latter will be discolored; therefore, it is advisable to use new rubber bands for every determination. Silicone fluids do not affect rubber or metal bands. A fine copper wire, wound several times around the thermometer and capillary, has been successfully used by us. Another method, suggested in the literature, is to omit the rubber band and attach the capillary by placing a glass rod against the entire length of the thermometer above the bulb; the capillary is fitted in the groove, formed by the rod and thermometer. If sulfuric acid is used as a bath, the rubber band is unnecessary, as capillary attraction will hold the melting-point tube to the thermometer.

In some cases it is desirable to seal the capillary after filling, and even to displace the air above the crystals with an inert gas. In such cases, the capillary is filled either through a small capillary funnel or, better, by the following method. Two capillaries are constructed so that one exactly fits into the other. The thinner capillary is made 10 mm longer than the wider one. Each capillary is sealed at one end, and the thinner one is filled in

---

[3] H. A. Jones and J. W. Wood, *J. Am. Chem. Soc.*, **63**, 1760 (1941); H. A. Jones, *Ind. Eng. Chem., Anal. Ed.*, **13**, 819 (1941); L. B. Norton and R. Hansberry, *J. Am. Chem. Soc.*, **67**, 1610 (1945); and A. Georg, *Helv. Chim. Acta*, **15**, 924 (1932).

the usual manner, but without packing the sample tightly at the bottom of the tube. The filled capillary is carefully wiped off to remove any adhering powdered substances from the sides and open end. The larger capillary is fitted over the filled capillary, which is held upright and allowed to descend slowly until its closed end reaches the open end of the filled tube. Then the two capillaries are inverted rapidly, and the filled tube is raised about 10 mm above the larger tube and emptied slowly by rasping it with a file. The thinner tube is now raised another 10 mm, and its open end is slowly rubbed about the walls of the outer tube to remove any adhering particles; the capillaries are held at an angle, and the inner capillary is raised slightly and, after the process has been repeated, is finally withdrawn. In this manner it is possible to fill capillaries without any crystals adhering to the neck of the tube where they might decompose and contaminate the sample when the open end is sealed.

A method for filling capillary tubes with compounds which react with moisture or oxygen has been devised by Pinkus and Waldrop.[4] Capillary tubes with the open end enlarged are filled in a glove box in an atmosphere of nitrogen, and the enlarged end of the tube is closed with a rubber stopper. The stoppered tube is removed from the glove box and the capillary tube is sealed with a gentle flame. The same procedure can also be used in semimicro molecular weight determinations by the method of Rast.

Apparatus for Liquid Heating Baths

Figure 5.03(A) shows the well-known Thiele tube, and Figure 5.03(B) shows a modification to improve uniform heat transfer by stirring with air. The Thiele tube, even with the modification, must be heated very slowly to insure an even temperature about the sample and thermometer bulb.

Figures 5.04(A) and 5.04(B) show variations of the round-bottom flask type of apparatus. The flask shown in Figure 5.04(A) is a 25-ml Kjeldahl flask; however, a 200- or 250-ml round-bottom flask may be employed. Figure 5.04(B) shows a modification of the round-bottom flask for slow air circulation to insure uniform heat transfer. With this apparatus, also, great care must be exercised in heating, so as to have a uniform temperature about the capillary tube and thermometer bulb.

A third type of apparatus is the Markley-Hershberg type,[5] shown in

[4] A. G. Pinkus and P. G. Waldrop. *Mikrochim. Acta*, **1959**, 772–773.
[5] K. S. Markley, *Ind. Eng. Chem., Anal. Ed.*, **6**, 475 (1934); E. B. Hershberg, *Ind. Eng. Chem., Anal. Ed.*, **8**, 312 (1936); F. C. Merriam, *Ind. Eng. Chem., Anal. Ed.*, **20**, 1246 (1948); N. D. Cheronis, "Micro and Semimicro Methods" in *Technique of Organic Chemistry*, Vol. 6, A. Weissberger, ed., Interscience, New York, 1954, pp. 149–153.

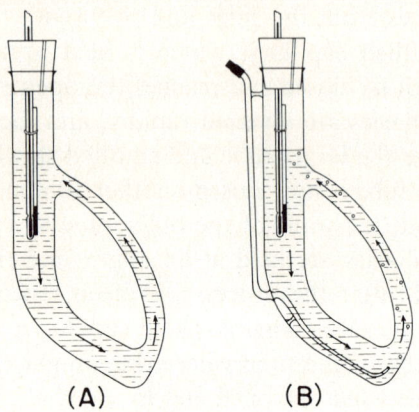

FIGURE 5.03

(*A*) Thiele tube for determination of melting points. (*B*) Modified Thiele tube for melting-point determination.

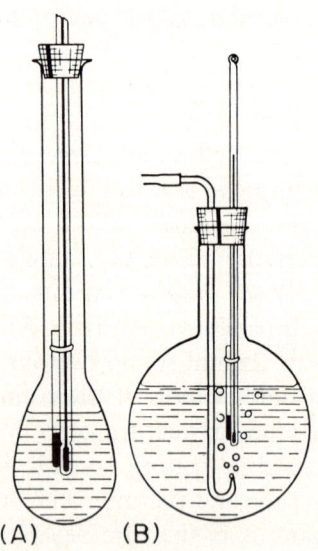

FIGURE 5.04

(*A*) Kjeldahl flask (25-ml) assembly for melting-point determinations. (*B*) Round-bottom flask assembly for melting-point determinations.

Figure 5.05, which is electrically heated; the temperature is measured with complete-immersion thermometers of the Anschütz type. Although it is possible to obtain a precision of 0.1° by the use of this apparatus, the accuracy depends on the calibration of the immersion thermometers.

An improved form of the Markley-Hershberg melting point apparatus has been described by Nickels.[6] Temperature surges in the liquid are lessened by the use of double-wall construction. Better visibility is assured by use of square-wall tubing. Rapid cooling of the liquid is achieved by circulation of air through a coil in the liquid.

A melting-point apparatus constructed by Walter[7] uses a bath of electrically heated air and no capillary tubes. A test tube is wound with resistance wire and positioned horizontally. A few crystals of the solid, whose melting point is to be determined, are placed on the bulb of the thermometer and the latter is carefully clamped inside the horizontal tube. The temperature can be raised slowly by controlling the voltage on the resistance wires with a variable transformer. Since the mass of the apparatus is small, the temperature falls rapidly when heating is discontinued. Kiplinger[8] has described a modification of the Walter apparatus, which uses a microscope and polaroid analyzer. The sample is placed on a small glass slide, which is thermally connected to the thermometer bulb by a strip of metal.

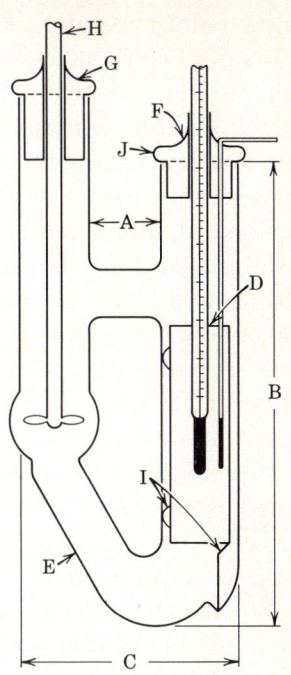

FIGURE 5.05

The Markley-Hershberg modification[5] of the Thiele apparatus. A, 28-mm Outside diameter and 25-mm inside diameter. B, 17 cm. C, 8.5 cm. D, Sleeve, 19-mm outside diameter, 17-mm inside diameter, 9 cm long; loops, No. 26 B. and S. gage platinum wire. E, 18-mm Outside diameter, wound with electrical heating element. F, Thermometer cap; thermometer tube, 7-mm inside diameter. G, Stirrer cap. H, Stirrer, 5-mm outside diameter glass tubing; ball bearings with 0.61-cm (0.25-in.) hole and 2.2-cm (0.875-in.) outside diameter, unground. I, Knobs to center sleeve. J, Lip and wedge to prevent rotation of cap.

For the determination of melting points up to 200°, a high grade of heavy petroleum (mineral) oil may be used as the liquid for the

---

[6] J. E. Nickels, *J. Chem. Educ.*, **28**, 303 (1951).
[7] J. L. Walter, *J. Chem. Educ.*, **30**, 142 (1953).
[8] C. C. Kiplinger, *J. Chem. Educ.*, **31**, 33 (1954).

melting-point apparatus. Concentrated sulfuric acid may be used for determinations up to 300°. A mixture of six parts of acid and four parts of potassium sulfate, which is solid at ordinary temperatures, may be heated up to 365°. There is *great danger* involved in heating sulfuric acid and mixtures of the acid and potassium sulfate; in addition to the danger resulting from breakage, the mixture of acid and salt may separate into two layers and, when heated, mix with explosive violence. In laboratories in which the use of sulfuric acid for heating baths is not permitted, a petroleum wax melting at 60 to 70° is used for temperatures of 250 to 350°.

Silicone fluids, although expensive, have been found to be among the better thermal conducting media for melting-point apparatus. We recommend silicone fluid 9981-LTNV-40[9] and Type-550-100 cst. These organic polysiloxanes are colorless, clear, stable to heat, and resistant to most chemical reagents; in addition, they exhibit a low rate of viscosity change over a wide temperature range, and have higher flash points than petroleum oils of equivalent viscosity. They can be used without appreciable discoloration in a Thiele tube apparatus for one year or more. Turbidity and darkening can be eliminated by filtration after shaking with a mixture of Filter-cel and charcoal.

Procedure Using Liquid-Bath Apparatus

The thermometer is arranged in the apparatus so that the lower end of the capillary is clearly visible. If a rubber band is used to hold the capillary to the thermometer, it is so adjusted that it is out of the liquid. In the Thiele apparatus, the oil level is about 10 to 15 mm above the circular side tube and the thermometer 15 mm below it, so that the latter is near the mid-point between the upper and lower side arms.

When the thermometer bearing the capillary tube has been adjusted, the tube or flask is heated rapidly to about 10 to 15° below the known melting point of the substance. If the substance is an unknown, the approximate melting point is first determined by heating fairly rapidly until the substance has melted. The bath is then allowed to cool to about 20° below the observed melting point; the thermometer is carefully removed and held until it has acquired the temperaure of the room; then a new loaded capillary tube is inserted. The thermometer is replaced and the bath heated until the temperature rises to within 10 to 15° of the melting point. The flame is removed until the temperature begins to drop. The heating is then resumed at such a rate that the temperature rises 2 to 3° per minute.

[9] Silicone fluid 9981-LTNV-40, General Electric Co., Schenectady, N.Y. Type-550-100 cst. Dow Corning Corp., Midland, Mich. Both media are carried by some dealers.

the liquid bath being stirred so that the temperature in the various parts of the apparatus will be as uniform as possible. When the temperature comes to within 2 to 4° of the melting point, a rise of 1° per minute is desirable. It should be stressed that the slower the rate of heating the greater is the precision and the rate for the last 1° should be 1° each 2 to 3 minutes. The temperature at which the substance begins to liquefy and the temperature at which the liquid is clear are noted. The interval of temperature is recorded as the melting-point range of the substance. If the compound melts without decomposition, it is suggested that a second and a third observation be made by removing the thermometer from the bath, holding it in air until the liquid in the capillary solidifies, then repeating the melting-point determination.

Procedure Using Heating Bars or Blocks

Two types of heating bars or blocks are described here in detail. A more comprehensive review will be found in the literature.[10]

The Fisher-Johns melting point apparatus,[11] shown in Figure 5.05A, consists of a small aluminum block which is electrically heated. The resistance box is equipped with a knob for temperature control. The temperature measurement is made through a direct reading thermometer graduated from 20 to 300°, imbedded immediately below the stage on which the sample is placed. To determine the melting point of a substance, a few crystals are placed between cover glasses on the well of the aluminum stage and the magnifier is adjusted over the sample. The temperature is raised rapidly at first and then at a rate of 1° per minute until the crystals coalesce and form droplets. The temperature control of the instrument is good for temperatures below 200° but somewhat difficult above this range.

There are two sources of error in the use of the instrument, both of which can be eliminated. Obviously, variations will occur unless the thermometer bulb is in contact with the hot stage, and this contact is the same at all times. To eliminate this difficulty the manufacturer furnishes a small amount of silver thread, which is wound around the thermometer bulb to give uniform contact with the stage. The second, more serious difficulty arises because the thermometer is not calibrated by the manufacturer. Since the error due to this variable may be as high as 5 to 8°, it is necessary even for routine work to calibrate the thermometer,[12] as outlined in a

[10] N. D. Cheronis, "Micro and Semimicro Methods" in *Technique of Organic Chemistry*, Vol. 6, A. Weissberger, ed., Interscience, New York, 1954, p. 158.
[11] Fisher Scientific Co., Pittsburgh, Pa., and New York, N.Y., No. 12-142.
[12] Two thermometers, one from 20 to 180° and the other from 150 to 300°, permit better spacing of the graduations.

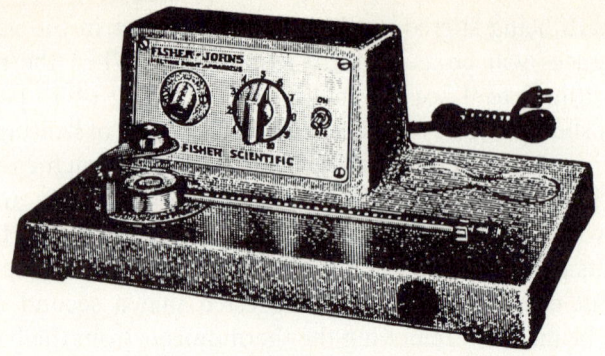

FIGURE 5.05A
Fisher-Johns melting-point apparatus. (Courtesy Fisher Scientific Co.)

later section (p. 195). When this is done, the variations become comparable to those obtained with a liquid bath apparatus provided with a calibrated thermometer.

The second type of heating block is the Ma-Schenck micro heating stage, described on page 20. As previously stated, this versatile instrument can be adapted for many purposes. Figure 5.06 shows an arrangement using the micro-heating stage for melting-point determinations. The apparatus should be calibrated with several known solids. A few crystals of the sample are placed in the round depression of the aluminum block, and a cover glass is pressed over them. The temperature is raised by means of

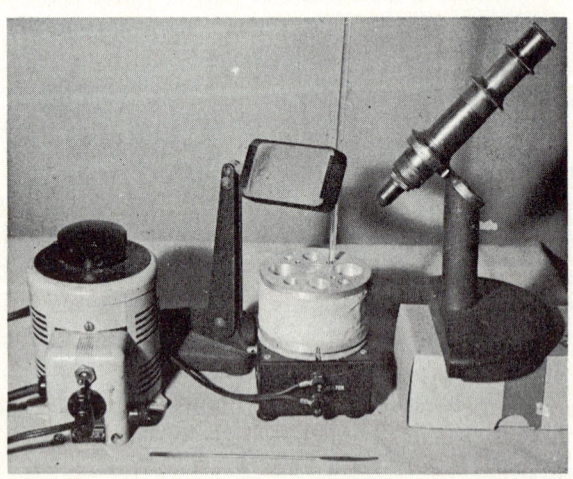

FIGURE 5.06
Ma-Schenck heating stage arranged for determination of melting points.

the regulator, and a magnifier is employed to observe the sample. With a small telescope, as shown in Figure 5.06, the apparatus is similar to a microscope hot stage.

The Dennis heating bar[13] and the Kofler hot bar[14] represent metal bars electrically heated on which a few particles of finely powdered substance are dropped and the region where the temperature is high enough to melt the sample is located. The Dennis apparatus requires a potentiometer; the Kofler bar does not. In the opinion of the authors, considering the cost, neither apparatus readily provides greater precision and accuracy than any of the apparatus described in the preceding sections in which calibrated thermometers are used.

Procedures Using a Microscope Hot Stage

The advantage of a microscope hot stage is that it permits direct observation of the changes of crystals before and during melting. The premelting phenomena are of great use to experienced analysts. An experienced worker in our laboratory has identified several hundred compounds by means of the microscope hot stage; important facts in such identifications were the crystal structure, the behavior during heating, and the rearrangement of the crystals while heating. A number of these observations can be made without a hot stage by the use of a glass slide and a very simple microscope as described by McCrone.[15] Directions are given on page 277.

Another advantage of the microscope hot stage is that it permits the determination of melting points with a single crystal. For example, a quantity of a few milligrams of an impure compound is being purified by fractional vacuum sublimation. At intervals the vacuum is discontinued and the condenser is brought over a glass slide and just touched lightly with the sharp edge of the microspatula. The few minute particles that fall on the slide are covered by a watch glass, and the melting point determined. In this manner it is possible to determine when the fraction should be removed from the condenser and collection of another begun. For example, it is possible to separate a mixture of 10 to 25 mg of a derivative in ten or more fractions; the desired derivative is collected as crystals are obtained having the melting point listed in the literature and, with care, it is possible to separate a mixture of 1 to 2 mg in ten to twenty fractions with data on the melting point of each fraction. For these

---

[13] L. M. Dennis and R. S. Shelton, *J. Am. Chem. Soc.*, **52**, 3128 (1930); Parr Instrument Co., Moline, Ill., Model MP-11.
[14] Wm. J. Hacker and Co., West Caldwell, N.J.
[15] W. C. McCrone, *Anal. Chem.*, **21**, 436 (1949); *Mikrochemie*, **38**, 476 (1951); *Fusion Methods in Chemical Microscopy*, Interscience, New York, 1956.

reasons, we believe that modern methods of fractionation and characterization should include the uses of a microscope hot stage.

Figures 5.07 and 5.08 show diagrams of the Kofler hot stage.[16] The instrument is mounted on an ordinary microscope by inserting the pins *IP* in the holes of the microscope stage provided for the clips. The electrically heated stage, *A*, supports a removable rim, *B*, and a cover glass, *C*, over the heating chamber. The apparatus is connected to the electrical power

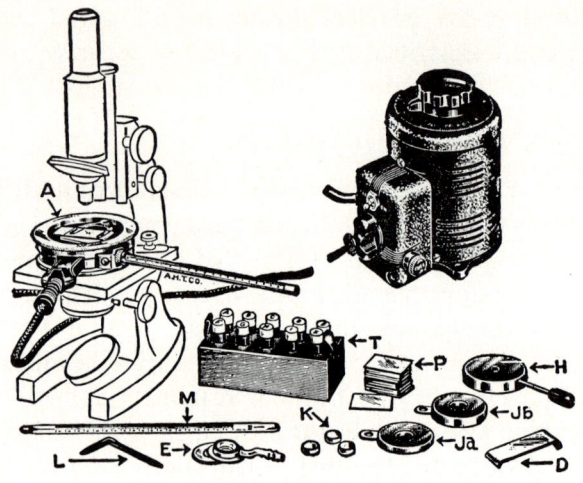

FIGURE 5.07

Kofler hot stage. (Courtesy A. H. Thomas Co.)

supply through a variable transformer. A switch in the connecting line is very useful for the control of temperatures near the melting point range. Two thermometers are supplied with the apparatus; one covers the lower range, from +30° to 230°, and the other, the high range from +60° to 350°. Both thermometers have been standardized on the actual hot stage with which they are to be used. Furthermore, the substances that were used for the calibration are furnished with the apparatus so that the thermometers may be checked at any time. The thermometer *M* (Figure 5.08) is inserted into a metal guard, which screws on the stage and is then pushed gently until it touches the wall below the center of the stage. The light from the mirror of the microscope passes through a condensing system at the base of the hot stage and through a 2-mm opening upward

[16] L. Kofler and W. Kofler, *Mikrochemie*, **34**, 374 (1949); *Thermo-Mikro-Methoden*, Weinheim/Bergstrasse, Verlag, Chemie, 1954. Stage supplied for A. H. Thomas Co., Philadelphia, Pa., or Wm. J. Hacker and Co., New York, N.Y., agents for Reichert Optische Werke, Vienna. A stage that measures temperatures up to 750° is also available from Reichert.

through the stage to illuminate the sample. The heating element is mounted inside the heating chamber. The rate of temperature rise is controlled by the variable transformer.

The transformer can be adjusted for temperatures from 50 to 350° by placing the dial in the approximate position indicated in the manufacturer's directions. If the hot stage has not been in use for some time it should be heated to about 200° without the thermometer, to drive off any moisture. The thermometer is inserted after the hot stage has again reached room temperature and the apparatus is then ready for use.

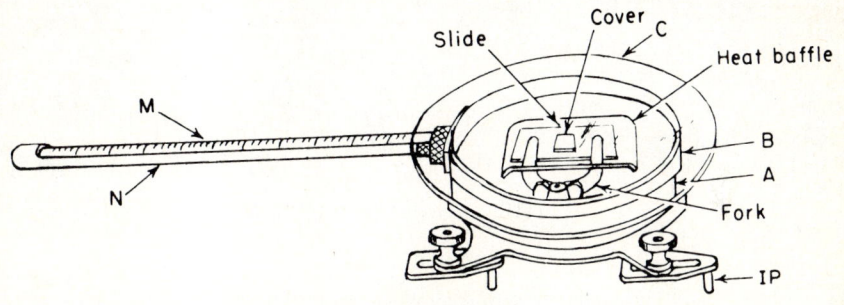

FIGURE 5.08

Kofler hot stage with assembly for melting-point determinations.

The arrangement for melting-point determinations by means of the Kofler hot stage is shown in Figure 5.08. A few crystals of the pulverized substance are placed on the glass slide with a microspatula; the cover glass is placed over the sample and pressed firmly on the slide. The slide is inserted in the stage by means of a forceps and then pushed towards the center by means of a spatula or the fork L (Figure 5.07), so that the sample is directly above the light opening. The position of the slide is adjusted, while the microscope is focused, to give the desired field which should contain several well-defined individual crystals rather than large aggregates. A field containing a few individual crystals and a small aggregate is often advantageous, since the latter gives information about premelting which is difficult to observe with single crystals. After a good field has been selected, the glass baffle is placed over the preparation to protect the microscope lens from excessive heat. Finally, the glass cover, C, is placed on the metal rim, B, the focus is readjusted, and heating is begun with the transformer adjusted to give the desired rate. For an unknown sample, the approximate melting temperature range is first obtained. The rate of heating is fairly rapid up to about 10° below the melting point of the sample. At that point heating is discontinued until the temperature begins to fall. It is resumed at a rate of about 2° per minute until a temperature 3° below the melting-point range has been reached, and from then on at a rate of 1° per 3 to 4 minutes. As the temperature approaches the melting point, the crystals often undergo rearrangement; finally, just before melting, the crystals lose their sharpness as their edges become rounded and smooth. A good example cited by Kofler is the behavior of ethyl *p*-aminobenzoate (Figure 5.09). At 89°, the crystals are distinct with sharp edges; at 90°, the sharp edges become

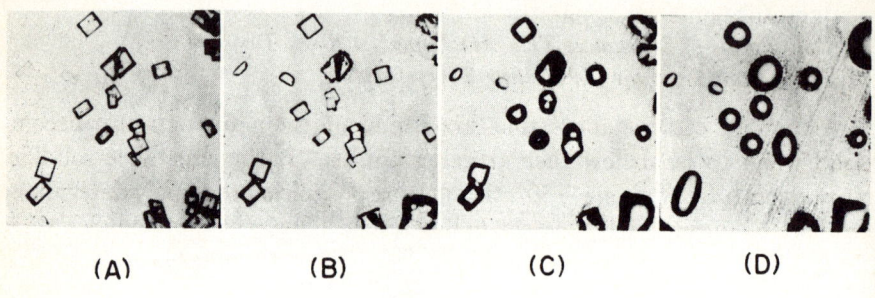

(A)   (B)   (C)   (D)

FIGURE 5.09

Melting crystals of ethyl *p*-aminobenzoate. When the melting temperature is reached, first the smallest fragments liquefy, then the larger crystals (noticeable by the rounding off of corners and edges and a gradual liquefaction). (*A*) Temperature 89°, all crystals as yet unchanged. (*B*) Temperature 90°, melting begins. (*C*) Temperature 90.5°, melting in process. (*D*) Temperature 91°, all crystals melted. (Courtesy Wm. J. Hacker and Co.)

rounded; melting begins at 90.5° and is complete at 91°. When fine globules appear around each crystal, the temperature is read. It is possible with practice and patience to hold the temperature at the point at which both globules and particles of the solid phase are in the field. When the temperature is raised a fraction of a degree, small crystals almost disappear; and when it is lowered a fraction of a degree, crystals are seen to grow. Thus the crystals can be made to grow slowly or to melt with temperature changes which are hardly detectable on the thermometer. This temperature interval is recorded as the *observed micro melting point*. If all the crystals disappear rapidly with the formation of droplets, the rate of heating has been too rapid; under these conditions, an error is introduced owing to the time required to shift the eyes from the field to the thermometer. A magnifying glass with good illumination has been found very helpful in making readings. Unless the substance melts with decomposition, it is possible—if we overshoot the melting temperature—to let the sample cool until the solid phase reappears and to repeat the observations.

It is frequently necessary to remove the slide after the determination is complete and replace it with another sample before the stage cools to room temperature. It is important to remember that the stage must not be touched, and that all manipulations must be performed either by the *lifter*, or with a microspatula and forceps. The microspatula and forceps are used to adjust the position of a newly inserted slide on a stage which is hot to give an appropriate field for observation. To cool the stage rapidly so that a new sample may be inserted 10–20° below the melting point, the metal block provided in the apparatus is cooled in tap water or an ice bath, wiped off quickly, and placed in the hot stage. The cooling block may also be used to induce rapid crystallization of a melted sample. The slides and glass baffle are cleaned with a solvent, immersed in cleaning solution for several hours, thoroughly rinsed, and dried.

For an extensive description of the application of the Kofler microscope hot stage, see the excellent treatises by the Koflers[17] and by McCrone.[18] It should be pointed out that their methods, as applied to the identification of organic compounds by *fusion techniques*, are not widely known. For example, the Koflers list the *eutectic melting points* of more than 1200 organic compounds with eight of their reference substances—such as benzanilide and phenacetin. These and other data are employed for rapid methods of characterization.

A melting-point apparatus with an elaborate optical system uses an electrically heated copper block.[19] Transmitted and reflected images of the melting point capillary tubes are projected side by side on a screen. The apparatus is particularly useful for determining melting points of materials that might explode. A description of a simple apparatus for melting-point determinations has been given by Jennings.[20] It is an electrically heated hot stage to be used with a microscope. A block of aluminum, heated with a burner, has been used for determinations of melting points by Owen and Reid,[21] who claim for it advantages over liquid baths and electrically heated hot stages.

Another new design for a melting-point apparatus, which combines simplicity, safety, speed, convenience, and versatility, has been described.[22] These advantages are achieved by the use of a metal block heated with a gas flame, adequate illumination of the samples, a good optical system for

---

[17] L. Kofler and W. Kofler, *Thermo-Mikro-Methoden*, Verlag Chemie, Weinhein/Bergstrasse, 1954.
[18] W. C. McCrone, *Fusion Methods in Chemical Microscopy*, Interscience, New York, 1957.
[19] H. E. Ungnade, E. A. Igel, and B. B. Brixner, *Anal. Chem.*, **31**, 1432–1433 (1959).
[20] W. G. Jennings, *J. Chem. Educ.*, **34**, 95 (1957).
[21] W. S. Owen and W. M. Reid, *Mikrochim. Acta*, **1956**, 1373–1376.
[22] H. J. Barber, D. P. Odell, and W. R. Wragg, *Chem. & Ind.*, (London), **1958**, 153–155.

viewing the samples, and a metal housing. This apparatus may be purchased.[22a]

Procedure for Liquids

A number of substances that are liquid at room temperature become solid when cooled below their freezing points. If the substance solidifies at temperatures of about $-50°$, the melting point is more readily determined (and with less material) than the boiling point. A simple method, described in a work by one of us,[23] is very useful in the determination of the melting points of eutectic mixtures that are liquid at room temperature.

This method uses only a drop or less of the liquid, and gives sharp melting points. Two capillary tubes are selected with diameters so that one slips inside the other. One end of the larger tube is sealed to make a melting point capillary. One end of the other tube is dipped into the liquid, which rises rapidly in it by capillary attraction. The smaller tube with the liquid is inserted into the other until it touches the bottom, and is quickly withdrawn. Some liquid remains in the bottom of the larger tube and some is smeared along its inner wall.

The melting-point apparatus is cooled to a temperature $10°$ below the estimated melting point of the compound by means of an outside bath. The capillary is attached to a thermometer, and the assembly is added to the cold melting point bath. When the solid freezes, the melting-point bath is removed from the cold bath, and allowed to warm with stirring. The solid in the capillary usually melts within $0.5°$ if it is pure. If the solid is not pure, the solid mass may not melt completely within two or three degrees of the true melting point.

An alternate method is to use about 1 ml of the liquid in a small test tube, with the bulb of a thermometer in the liquid. A wire, bent in the form of a loop large enough to encircle the thermometer bulb freely, is used as a stirrer. The test tube holding the liquid is placed in a cold bath, and the liquid is cooled without stirring until the temperature is four or five degrees below the expected freezing point. The test tube is removed from the cold bath, and the contents are stirred vigorously for several seconds. As the liquid crystallizes, the temperature rises to the freezing point of the liquid and stays there for two or three minutes. The process can be repeated with decreasing degrees of supercooling in order to obtain the true freezing point of the mixture.

[22a] Electrothermal Engineering, Ltd., London E.7, England.
[23] N. D. Cheronis, "Micro and Semimicro Methods" in *Technique of Organic Chemistry*, Vol. 6, A. Weissberger, ed., Interscience, New York, 1954, pp. 181–185.

Physical Properties of Organic Compounds

Procedure for Solids which Decompose when Heated

Some organic compounds, such as amino acids, osazones, and many of the quaternary salts and, generally, compounds that have bonds of partial ionic character, melt with decomposition. As the sample is heated, decomposition begins, and a lowering of the melting point of the substance occurs. Thus, D-glutamic acid has been reported to melt with decomposition at temperatures varying from 198 to 225°. Similarly, DL-tyrosine has been reported to decompose at 295°, 318°, and 340°; the melting point of a sample of phenyl-D-glucosazone was 210° when the liquid bath was heated at the rate of 40 to 60° per minute, and 194 to 198° when the rate of heating was 8 to 10° per minute.

The temperature range at which decomposition takes place depends on the rate of heating. It is erroneous to consider organic substances that decompose slowly on heating, such as the classes of compounds cited, as having true melting points. On the other hand, a distinction should be made for compounds that melt sharply with decomposition, such as some of the substituted malonic acids; these decompose on melting with the evolution of carbon dioxide.

The determination of melting points of compounds that decompose presents more difficulties. One method is to preheat the bath to within 10° of the expected melting point and, then, to insert the thermometer with the capillary and heat as rapidly as possible, noting the rate of the increase in temperature. The initial temperature of the bath and the rate of temperature rise should always be reported.

Procedure for Melting Points of Mixtures

The determination of the melting point of a mixture plays an important role in organic qualitative analysis; this process is often referred to as the determination of a "mixed" melting point. In an earlier discussion it was pointed out that a mixture of two substances usually has a wide range and a lower melting point than either of the two pure solids. If, therefore, an unknown material is mixed with a known solid, and the mixture melts at the same temperature as the known and the unknown substance, the two solids are probably the same chemical.

It should be remembered, however, that there are cases of unlike crystalline substances that show a higher melting point than either of the two components because of the formation of a new compound. In a number of instances, two different compounds, melting a few degrees apart, may show no depression in melting point when mixed. Thus, naphthalene picrate, mp 151°, and benzothiophene picrate, mp 149°, melt when mixed at 149°,[24] and D-dimethyl tartrate, mp 48°, and L-dimethyl tartrate, mp 43.3°, melt when mixed in equal proportions at 89.4°.[25] It is also well known that two different organic substances, when mixed in different ratios, may form two or more eutectics, which melt considerably below either component, and one or more molecular compounds, which may melt higher than either component. Therefore, the "mixed-melting-point method" should only be used in conjunction with other pertinent data. On the other hand, the use of fusion techniques eliminates most of these errors since the formation of eutectics is clearly visible under the microscope.

To illustrate the use of mixed melting points in characterization work, assume that an organic liquid under investigation boiling at 106 to 108° is provisionally identified as isobutyl alcohol. One of the derivatives prepared for the final proof of the identity is the 3,5-dinitrobenzoate; the melting range of the crystals of this derivative was 84 to 85° after the first crystallization and 85 to 86° after the second crystallization. The melting point of the 3,5-dinitrobenzoate of isobutyl alcohol is listed in Table 6A (page 714) as 87°. For a mixed melting point, the 3,5-dinitrobenzoate of a known sample (100 to 200 mg) of isobutyl alcohol is prepared, and the melting point of the crystals is determined. Approximately equal amounts of the crystals of the dinitrobenzoate derived from the "known" and the "unknown" samples are thoroughly mixed by crushing in a mortar or a watch glass and the melting point of the mixture is determined. If all three melting points are essentially the same or, if the melting point of the mixture lies between that of the two dinitrobenzoates, the unknown sample is identified as isobutyl alcohol. If the unknown one is not isobutyl alcohol, the melting point of the mixture of the two dinitrobenzoates will be at least 10° below that of the components, and the melting will not be "sharp" but will soften and melt gradually over a range of several degrees. Consult the section in Chapter 7 on fusion techniques in connection with the proof by mixed fusion as to whether two substances A and B are identical.[26]

The proof of identity is made more rapidly with the "mixed" fusion method described on page 279. If the observations indicate that the two

[24] G. Lock and G. Nottes, *Ber.*, **68**, 1200 (1935); R. Meyer and W. Meyer, *Ber.*, **52**, 1249 (1919); *Ber.*, **51**, 1571 (1918).

[25] J. H. Adriani, *Z. physik. Chem.*, **33**, 453 (1900); C. W. Gibby and W. A. Waters, *J. Chem. Soc. (London)*, **1931**, 2151.

[26] These observations can be made without the use of a microscope hot stage by using any simple microscope of 50 magnification.

substances are not identical but are isomorphous, confirmation may be obtained from the melting temperature, as determined with the hot stage, and from the eutectic melting point of each substance as determined with the same reference compound; acetanilide, benzanilide, or phenacetin may be used as the reference compound[27] (page 282).

From the above considerations it is obvious that failure to observe a lowered capillary melting point in a mixture of two derivatives, prepared from a known and an unknown sample, is not a reliable proof of identity unless all other data—solubility tests, functional group tests, and physical constants—are in agreement. With this reservation, the practice of taking "mixed" melting points, using samples of derivatives from the unknown and the compound tentatively identified as the unknown, may be resorted to whenever two successive crystallizations fail to produce a rise of more than 2° in the melting point of the derivative. This topic is discussed in greater detail in a later section.

Calibration of Thermometers

The accuracy of melting-point determinations by the capillary tube method, and by practically every method in which a thermometer is used, depends to a large extent on the calibration of the instrument by which the temperature is measured. Calibration by the National Bureau of Standards by total immersion eliminates the errors inherent in the thermometer if it is used in a total-immersion apparatus. In most apparatus the thermometer is only partially immersed and corrections must be made for the exposed stem. Aside from the inconvenience of such a procedure, the errors introduced offset the benefits of the calibration. The Anschütz thermometers that are used by total immersion are difficult to calibrate. Most of these thermometers are duplicates of the original German models and are not supplied with an ice point, which is a prerequisite for certification by the National Bureau of Standards. However, the Bureau will submit a report on each thermometer that indicates the amount of variation at certain points. If each thermometer is checked at three points, the cost of such a report for a set of Anschütz thermometers will exceed $300.00. This is extremely high when we consider that after one year a recalibration is necessary for thermometers that have been used extensively. This is true for all thermometers calibrated either by the Bureau of Standards or by direct comparison with a standard thermometer,[28] and arises from the

[27] L. Kofler, *Ber.*, **B76**, 1096 (1943); L. Kofler and W. Kofler, *Thermo-Mikro-Methoden*, Verlag Chemie, Weinhein/Bergstrasse, 1954; W. C. McCrone, *Fusion Methods in Chemical Microscopy*, Interscience, New York, 1957.
[28] T. H. Liggett, *Proc. Iowa Acad. Sci.*, **37**, 241 (1930).

fact that when thermometers are used at high temperatures, they undergo gradual irreversible changes in bulb volume. Since the bulb of an average thermometer contains mercury corresponding to about 6000° of scale length,[29] it is obvious that small changes in bulb volume will have a relatively large effect upon the reading of the thermometer.

The older methods for making corrections on observed melting temperatures involved essentially a calibration of the thermometer and applying a correction for the stem immersed in the liquid. The common thermometer has been calibrated while being totally immersed in a bath. In the melting-point apparatus described, only a part of the stem is immersed. The column of mercury above the liquid in the bath will show a lower temperature than that for which the thermometer was calibrated. Therefore, either a thermometer calibrated by partial immersion should be used, or a correction must be made for the unequal heating of the mercury in the stem of the thermometer. The correction for unequal heating of the thermometer is given by the formula

$$\text{Stem correction (degrees)} = 0.000154(t_o - t_s)N,$$

where the fraction 0.000154 represents the difference in the coefficients of expansion of glass and mercury, t_o is the temperature read, and t_s is the average temperature of the column of mercury not immersed in the substance; t_s is determined (approximately) by reading a second thermometer whose bulb is held at the midpoint of that part of the column of mercury not immersed in the substance. N is the length in degrees of the portion of the column that is not immersed. The error due to this variable is small at temperatures below 100°, but may amount to 3–6° at 200° and above.

On the basis of an extensive investigation by one of us[30] the procedure recommended is to employ a thermometer calibrated by the manufacturer by partial immersion and then calibrate the thermometer by means of reference standards *in situ*, that is, in the apparatus that is to be employed.

A thermometer is selected in which 1° is equivalent to 1 to 1.2 mm and calibrated by partial immersion. If greater precision is desired, two thermometers, one reading from 0 to 180° and the other from 150 to 320°, with subdivision in 0.5°, are employed.

The thermometers to be calibrated are first heated for 6 to 8 hours at about 300°, which is higher than the temperatures to which they are commonly exposed. The thermometers are allowed to stand at room

[29] J. M. Sturtevant, "Calorimetry," in *Physical Methods of Organic Chemistry*, 3rd ed. Part 1, A. Weissberger, ed., Interscience, New York, 1959, pp. 523–654.

[30] N. D. Cheronis, "Micro and Semimicro Methods," in *Technique of Organic Chemistry*, Vol. 6, A. Weissberger, ed., Interscience, New York, 1954, pp. 152–158.

temperature for 2 to 3 days, and are then calibrated by the reference standards listed in Table 5.03, *in the same apparatus and with the same technique that is employed for the melting point determinations.* Three determinations or more are made for each fixed point; the deviation from the average value should not exceed 0.5°. Average values are used to plot the calibration curve from which the correction to be applied to observed melting points may be read directly. Figure 5.10 shows calibration curves for three thermometers.

Since there is no great differential in the price of the thermometers

TABLE 5.03
PRIMARY STANDARDS FOR THERMOMETER CALIBRATION BY MELTING POINT*

Substance	Melting Point, °C	Substance	Melting Point, °C
Water-ice	0.0	Urea	132.8
Cyclohexanol	25.4	Salicyclic acid	158.3
Menthol	42.5	Succinic acid	182.8
Benzophenone	48.1	Anthracene	216.2
p-Nitrotoluene	51.6	Phthalimide	233.5
Naphthalene	80.2	p-Nitrobenzoic acid	241.0
Acetanilide	114.2	Phenolphthalein	265.0
Benzoic acid	122.4	Anthraquinone	286.0

*Source. N. D. Cheronis, "Micro and Semimicro Methods" in *Technique of Organic Chemistry*, Vol. 6, A. Weissberger, ed., Interscience, New York, 1954, p. 156.

calibrated by partial immersion, they are recommended with the provision that they should be calibrated. Such thermometers are marked with an etched ring at the 76-mm mark and, above it, bear the inscription "76-mm immersion." To use such thermometers, the stem is immersed so that the mark is even with the surface of the liquid bath. These thermometers are recommended in place of the common thermometers. The length of the capillary to be used with such thermometers is 100 to 120 mm.

The Use of Melting-Point Data in the Identification of Organic Substances

Melting-point data constitute one of the main criteria employed in the determination of the purity of solid organic compounds and, also, the sole criterion in the final identification of an unknown substance. The last step in any systematic scheme of characterization involves the preparation of one of two derivatives and comparison of the melting-point data of these

198 Techniques of Organic Analysis

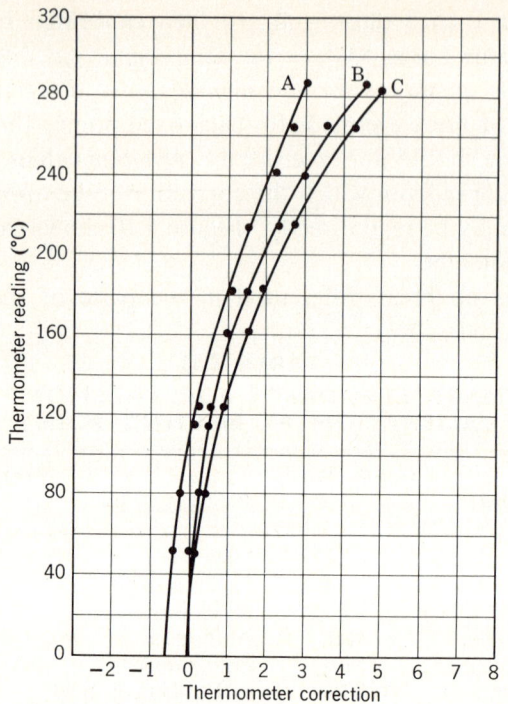

FIGURE 5.10

Calibration curves of three thermometers.

derivatives with the values listed in the literature. The values of about 35,000 derivatives are listed on Tables 1–35 on pages 667–967 of this work. Therefore, it becomes necessary to discuss at some length how these values on melting-point data were compiled and how, in our opinion melting-point data obtained in characterization work should be compared to the values listed in this, and other, standard works.

The tables of the derivatives of organic substances (pages 667–967) were originally (1940–1947) compiled with great care after an examination of the literature. More recently (1963), they were checked and expanded. In this compilation and critical evaluation of melting-point data, the following difficulties were encountered.

1. For many derivatives, several melting points were listed; the values given may vary by only 1 to 2°, or by as much as 10° or more. These variations in values are discussed in paragraphs to follow.

2. In a few cases, particularly in recent literature, the melting points of derivatives given are stated to be corrected; that is, the correction for

unequal heating of the thermometer has been applied. In most instances, however, there is no statement as to whether the correction has been applied. Obviously, however, many of the older values are uncorrected melting points; in such cases, the compounds were not prepared primarily for identification purposes but in connection with some other type of investigation and, hence, the correction of the melting point was not considered important. The correction for values of 100 to 150° may amount to 1 to 3° but, for values above 200°, the correction may be as high as 4 to 7°. Whenever it is known that a corrected melting point is given for a derivative in this text, it is noted by an asterisk.

3. Only in few instances, the following designations were found: *uncorrected capillary melting point, capillary melting point corrected, heating bar melting point, micro melting point corrected.* It is a fair assumption to consider most of the older values as uncorrected capillary melting points; also, that most of the values reported in the literature, unless otherwise indicated, represent capillary melting points. Since there is a growing tendency in reporting the melting points of new compounds to indicate the type of apparatus employed, it should be pointed out that the melting points determined by means of heating bars and hot stages are consistently higher than the corresponding capillary melting points. This difference arises because, in the capillary-tube method, an equilibrium between the solid and the liquid phases does not exist at the temperature observed as the melting point whereas, in the hot stage or heating bar method, an equilibrium condition is more nearly realized with a temperature rise of 1° per 4 to 6 minutes.

4. In some cases, the listing of two or more melting points is explained on the basis of the existence of two or more different crystalline modifications; for example, the 2,4-dinitrophenylhydrazone of acetaldehyde exists in the "stable" form, which melt at 168.5° (corrected), and the "metastable form," which melts at 157°. Under certain conditions, an equilibrium mixture of the two forms is obtained that melts at 148°. However, most discrepancies in the melting points of organic compounds listed in the literature cannot be explained on the basis of the existence of more than one crystalline form. The usual assumption is that the various investigators used compounds of varying degrees of purity. Since the presence of impurities usually lowers the melting point of a compound, the rule followed by most workers in compiling data on melting points of organic compounds is to select the highest value from those listed. There is considerable evidence in the literature to justify such practice. For example, in checking the melting point of D-camphor semicarbazone, it was found

that most modern standard works list it as 236 to 238°; an older reference was found that gave the value of 245°. Investigation of the original literature disclosed that Tiemann,[31] who first prepared this derivative, gave as its melting point the value of 236–238°. Rimini[32] pointed out that the true decomposition point is 245°. Finally, Bredt and Perkin[33] prepared the compound by several different methods and found that it melts with decomposition at 247 to 248° corrected. This value is 10° higher than the value listed in most modern works.

That such a situation is not always the case, however, is illustrated by the following two examples. The acetyl derivative of *p*-toluidine is a common derivative prepared for the identification of acetic acid, or acetic anhydride. The values listed in the literature are 147 to 148°, 148 to 149°, 151 to 152°, 152°, 153°, 155°.[34] Since this is a common derivative, it was prepared by one of us by the reaction of acetic anhydride and acetyl chloride with *p*-toluidine. After one crystallization, the derivative gave a corrected melting point of 147°, and this did not change after six additional crystallizations. Therefore, the original listed value (on page 764) of this derivative of 155° was changed to 147°.

The other example concerns the melting point of piperonyl alcohol. This has been reported as 51°, 52 to 53°, 54°, 57°, and 58°.[35] Although, in most cases, the higher melting point was selected for listing, in this particular case 52 to 53° was selected, because the compound has been extensively investigated in the laboratory of one of us.

It is not possible to explain all the discrepancies on the basis of differences in thermometer corrections, varying degrees of purity, or the existence of more than one crystalline form. It appears that present knowledge concerning the relation between crystal structure and the temperature at which there is a change from the solid to the liquid phase is incomplete. There is some evidence confirming that stresses within the crystals or crystal aggregates affect the melting point. McCrone[36] has found

---

[31] F. Tiemann, *Ber.*, **28**, 2191 (1895).

[32] E. Rimini, *Gazette*, **30**, 603 (1900).

[33] J. Bredt and W. H. Perkin, *J. Chem. Soc. (London)*, **103**, 2189 (1913).

[34] *147–148°*: W. Kelbe, *Ber*, **16**, 1200 (1883). *148–149°*: A. Hugershoff, *Ber.*, **58**, 2484 (1925). *151–152°*: E. Wedekind and E. Bruch, *Ann.*, **471**, 107 (1929). *152°*: T. Curtius, *J. prakt. Chem.* (2) **125**, 303 (1930). *153°*: I. Gazopoulos, *Ber.*, **59**, 2187 (1926). *155°*: L. Frejka and L. Cizmar, *Chem. Listy*, **31**, 460 (1937).

[35] *51°*: R. Fittig and I. Remsen, *Ann.*, **159**, 138 (1871); H. Decker and O. Koch, *Ber*, **38**, 1741 (1905); C. Mannich and O. Walther, *Arch. Pharm.*, **265**, 1 (1927). *52–53°*: G. Barger, *J. Chem. Soc. (London)*, **93**, 567 (1908); W. H. Carothers and A. Adams, *J. Am. Chem. Soc.*, **46**, 1681 (1924). *54°*: G. Vavon, *Compt. rend.*, **154**, 361 (1912); J. v. Braun and K. Wirz, *Ber.*, **60**, 102 (1927). *57°*: A. Paris, *Rec. trav. chim.*, **49**, 41 (1930). *58°*: A. M. B. Orr, R. Robinson, and M. Willams, *J. Chem. Soc. (London)*, **111**, 950 (1917).

[36] W. C. McCrone, private communication.

that octachloropropane, if crystallized in the unstrained condition, melts at 168°; if crystallized in a strained condition, it melts at 144°; this difference in the melting temperature is not accompanied by changes in crystal structure. Differences of 1 to 2° in the melting points of several *p*- and *o*-toluidides of carboxylic acids have been obtained by one of us on crystallizing the compounds from hot solutions by sudden cooling, and on crystallizing from a dilute solution by very slow cooling. It appears that, in some crystal structures, stresses apparently may develop within the crystals, which affect the melting point. It should be pointed out, however, that the factors that give rise to stresses are not known, and that it is possible that stresses may develop readily in one type of crystal structure and not in another.

It is not practical in a work of this type to list all the melting points that may be found in the literature for a particular derivative because, in many cases, more than two values are given that differ by several degrees; therefore, whenever *two or more values* have been found for the same derivative, a selection has been made in accordance with the following criteria: (a) preference has been given to values appearing in the recent literature; (b) preference has been given to values presented by investigators whose primary purpose was the determination of physical and related data for characterization work rather than to those appearing in papers dealing primarily with preparative work; and (c) in general, whenever two values have been listed, the higher has been chosen. If the difference between the values of the melting point for the same derivative was 2° or less, no attempt was made to record the variation. If, however, the difference was 3° or more, *one other value* has sometimes been recorded in parenthesis under the selected value. In some instances, a note appears at the end of the table, regarding the difference, and it is recommended that these notes be used. Moreover, the listing of two values in the present work indicates that two or more values with differences of 3° or greater are to be found in the literature; a difference of 1 to 2° in the melting points of the same derivatives, as given in standard works, is the rule rather than the exception. Consultation of the more complete listings of derivatives from the works given on pages 661–663 is highly recommended.

Recommendations for the Evaluation of Melting-Point Data of Derivatives

The practical importance of the discussion given in the preceding section becomes apparent when we prepare one or two derivatives for the final identification of a probable compound and find that they differ by 2 to 3° from the value listed in the literature; yet, the derivatives prepared do not show much alteration in the melting point on two successive crystallizations. The following practice is recommended.

Prepare one, or better, two different derivatives of the probable compound. From each derivative, save a few milligrams and recrystallize the rest. If the difference between the first and second lots of crystals is not more than 1 to 2°, check the melting point of the same derivative of the probable compound as listed in the tables. If the difference is not more than 2 to 3°, prepare the same derivative from a pure sample of the compound tentatively identified as the unknown, using the same quantities of reagents, procedures, and crystallizations as those used in the preparation of the derivative of the unknown. Mix a few milligrams of each of the two derivatives obtained from the unknown and from the pure sample of the compound tentatively identified as the unknown. If the mixture does not show a variation of more than 1° from the melting point of either component alone, the proof of identity may be considered conclusive. It is recommended that beginners prepare two different derivatives; only after considerable experience has been gained should the conclusive identification be based on the preparation of one derivative and a "mixed" melting point. The experienced worker will find the method of mixed fusion (page 279) superior to that of mixed melting points to prove whether two derivatives, *A* and *B*, are identical.

An explanation for the above practice may be in order. When semimicro quantities are used, the amount of derivative available after one or two crystallizations is usually 50 to 100 mg, if we begin with 100 to 200 mg of the unknown; but, in some cases, it may require a total of four or five crystallizations to obtain a derivative that has the melting point shown in the literature. In other cases, for reasons pointed out above, *the melting point given in the literature will not be obtained, no matter how many recrystallizations are performed.* Therefore, if all other evidence from solubility data, functional-group tests, and physical constants fits a particular probable compound, the procedure outlined in the above recommended practice, with reference to derivatives, is regarded as sound.

Boiling Points

One of the physical properties of a liquid, most important to its identification, is its boiling point. Although the boiling point does not serve as a positive identification for a liquid, a knowledge of this physical property gives valuable information about the compound, and limits consideration to a few compounds in each class.

The boiling point has some advantages and some disadvantages over the melting point for identification purposes. The determination of the boiling point is somewhat more complicated than determination of the melting point. Generally, more of the sample is needed (as much as 5 ml) for a good, accurate determination. Less can be used, but a small amount will not give a boiling range. The boiling point is affected less by traces of impurity than the melting point. Even with an adequate sample of liquid, we cannot determine the purity very exactly by means of the boiling point. The boiling point varies significantly with variations in atmospheric pressure. For accurate work, correction must be made for the pressure at which the determination is made. The boiling point determination may be in error because of superheating of the liquid if it is not done very carefully. However, the boiling point comes closer to being a constitutive property of the compound than the melting point; at least, the boiling point is more predictable than the melting point.

Think of a liquid as being a mass of closely packed molecules, which are in constant motion internally as well as externally. They are vibrating, rotating, and undergoing translational motion. Occasionally, a molecule at the surface will acquire enough energy to leave its close neighbors and take up a more solitary existence in the space above the liquid. As the temperature rises, the average energy of the molecules becomes greater, and more molecules succeed in leaving the liquid surface. When the molecules above the liquid constitute a full atmosphere of pressure we say the substance is boiling.

An important factor in determining the boiling point of a liquid is the force between the molecules. These attractive forces are the same ones that we have already discussed in the section on melting points: ionic attraction, hydrogen-bonding, dipole-dipole interaction, and van der Waal forces. The greater these forces, the more energy must be supplied to allow a molecule to escape from the surface, and the higher the boiling point.

A second factor of considerable importance is the size of the molecule. The larger the molecule, the higher the boiling point. The relationship is based partly on the fact that the molecule can absorb more energy by vibrational and rotational processes if it is large. In addition, however, the larger molecules have more opportunity to contact intimately their neighboring molecules, and the intermolecular forces operate more effectively in preventing the molecule from escaping from the surface of the liquid.

A third factor, symmetry, is not as important as in the case of the melting point and, actually, works in the opposite direction. The more symmetrical a molecule, the more it resembles a sphere, and the less is its contact with its neighboring molecules. Consequently, a spherical molecule has a lower boiling point than a long molecule of the same kind and size. This means that some nearly spherical molecules may sublime if their melting points are sufficiently high and their boiling points sufficiently low. Let us examine some specific illustrations of the operation of these principles.

Boiling Points and Structures of Compounds

Boiling points are affected strongly by forces between molecules. The result is that organic compounds with ionic bonds will, generally, decompose before they boil. Butyrolactone boils at 206°, but γ-aminobutyric acid, which exists as a dipolar ion, does not boil. This is also true for salts of amines and salts of organic acids; they decompose rather than boil. Advantage is taken of their low volatility in the separation of mixtures by distillation.

If a molecule is capable of forming hydrogen bonds with identical molecules, the compound will probably be high-boiling because of its molecular association. Ethyl alcohol, bp 78°, with one hydroxyl group boils 166° higher than ethane, bp −88°. Ethylene glycol, bp 197°, has still another hydroxyl group; it boils 109° higher than ethyl alcohol. If the —H of one of the hydroxyl groups in ethylene glycol is replaced by —CH_3, the molecule cannot form hydrogen bonds so well, and its boiling point drops to 124°. Although ethyl mercaptan, CH_3—CH_2—SH, is a heavier molecule than ethyl alcohol, it boils 62° lower than ethyl alcohol. The hydrogen atom on a sulfur atom cannot take part in hydrogen bonding, since sulfur is not as electronegative as oxygen. The reason that many other molecules boil higher than their molecular weights would indicate is because of association through hydrogen bonding.

Dipole-dipole attraction also is a factor in increasing boiling points. Nitromethane, CH_3NO_2, has a coordinate covalent bond between the nitrogen atom and the oxygen atoms; it boils at 101°. Methyl nitrite, CH_3ONO, has only simple covalent bonds; it boils at −12°. The large dipole moment of nitromethane is responsible for its high boiling point. Acetone, bp 56°, has a dipole moment, due to the unequal sharing of electrons between oxygen and carbon. Isobutane, bp −10°, has almost exactly the same molecular weight and shape, but not such a large dipole moment.

Symmetry is not as important in the case of boiling points as in the case of melting points. Benzene, bp 80°, is more symmetrical than toluene, bp 111°, or chlorobenzene, bp 132°. These boiling points seem to depend more on molecular weight than on the shape of the molecule. In some cases, a symmetrical molecule will boil lower than normal because of its symmetry. *n*-Heptane, bp 98°, boils higher than trimethylbutane, bp 80°. *n*-Octane, bp 125°, boils higher than tetramethylbutane, bp 106°. *n*-Butyl alcohol, bp 118°, boils 35° higher than *tert*-butyl alcohol.

The boiling points of the members of a given series of organic compounds are so regular that smooth curves can be drawn through them. Mathematical equations can be derived for these curves, so that the boiling points of compounds of that series can be calculated with considerable accuracy. Unfortunately, each equation is different for even closely related series of compounds, such as the 1-alkenes and the 2-alkenes.

Attempts have been made to develop the boiling point as a constitutive property. The method of Kinney[37] is one of the best, and may be referred to in the recent editions of the *Handbook of Chemistry* by N. A. Lange. Boiling point numbers are assigned to different arrangements of atoms; these are given in a table. The numbers are added, the cube root is determined, and the boiling points are calculated from this. Another table is provided to save these steps. The result is a good estimate of the boiling point if the compound is not too unusual. If we need to guess the boiling point of a new compound, this method is reliable and usable.

Variation with Pressure

The boiling point varies with atmospheric pressure or with the pressure over the liquid. Since the atmospheric pressure is continuously varying, the boiling point will vary also. If we wish to compare boiling points taken at different times and places, they should all be corrected to 760 mm pressure. For small variations in pressure from 760 mm, the data in Table 5.04 may be used to advantage. Note that the correction is different for associated and nonassociated liquids.

In regions of high altitude and low barometric pressure, the correction to be applied is of greater magnitude. However, the use of reference standards obviates the necessity of considering precise determinations of pressure and can be considered as the best method for obtaining reliable values on the boiling points of "unknowns." This method gives an accuracy of 0.5 to 0.1° for most determinations, and may be employed without complex apparatus. The boiling temperature of the liquid under investigation is determined by one of the methods described in this chapter.

---

[37] C. R. Kinney, *J. Am. Chem. Soc.*, **60**, 3032–3039 (1938); *Ind. Eng. Chem.*, **32**, 559–562 (1940).

TABLE 5.04
CORRECTION TO BE MADE IN BOILING POINTS FOR EACH 10-mm CHANGE IN PRESSURE

Boiling Point, °C	Nonassociated Liquids, °C	Associated Liquids, °C
0	0.36	0.27
50	0.42	0.31
100	0.48	0.35
150	0.53	0.39
200	0.58	0.43
250	0.63	0.46
300	0.67	0.50

Immediately afterward, a determination is made of the boiling temperature of the reference standard (see Table 5.05), which is closest in structure and boiling point to the liquid under investigation. The difference between the boiling points of the reference substance measured under standard and under experimental conditions is used to correct the boiling point of the substance under investigation.

Assume, for example, that a compound boils at 84.5°. Under the same conditions, the boiling point of benzene, which is the reference standard, is 79.5°; the boiling point of benzene at 760 mm is 80.1°. Hence, the corrected boiling point of the substance under investigation is 84.5° + 0.6° = 85.1°. For more precise work, the boiling temperature of the compound under investigation is determined at a specified pressure at which

TABLE 5.05
PRIMARY STANDARDS FOR DETERMINATION OF BOILING TEMPERATURES*

Substance	Boiling Point, °C	Substance	Boiling Point, °C
Ethyl bromide	38.4	Cyclohexanol	161.1
Acetone	56.1	Aniline	184.4
Chloroform	61.3	Methyl benzoate	199.5
Carbon tetrachloride	76.8	Nitrobenzene	210.8
Benzene	80.1	Methyl salicylate	223.0
Water	100.0	p-Nitrotoluene	238.3
Toluene	110.6	Diphenylmethane	264.4
Chlorobenzene	131.8	α-Bromonaphthalene	281.2
Bromobenzene	156.2	Benzophenone	306.1

* Source. N. D. Cheronis, "Micro and Semimicro Methods" in *Technique of Organic Chemistry*, Vol. 6, A. Weissberger, ed., Interscience, New York, 1954, p. 189.

the boiling temperature of the reference compound is known with an accuracy of 0.01° or less. A Beckmann microthermometer and a manostat are required for such determinations.

Larger pressure differences, such as between 25 and 760 mm, require more elaborate calculations. One of the best-known equations for relating vapor pressure and temperature was developed by Clausius and Clapeyron:

$$\frac{dp}{dT} = \frac{Lp}{RT^2}$$

Here, p is the pressure, T the temperature, L the molar latent heat of vaporization, and R the gas constant. The integrated form of this equation is

$$\log \frac{p_2}{p_1} = -\frac{L}{R}\left(\frac{1}{T_2} - \frac{1}{T_1}\right)$$

T_1 and T_2 are the initial and final temperatures, and p_1 and p_2 are the initial and final pressures. Unfortunately, the molar latent heat of vaporization is unknown for most organic compounds; it can be roughly approximated for nonpolar compounds by Trouton's rule:

$$L \approx 22 T_b$$

where T_b is the boiling point in degrees Kelvin. Not only is this a rough approximation for the molar heat of vaporization, but the latter does not remain constant over a wide range of temperature. Other means must be found for relating changes in pressure and temperature.

One of these, by H. B. Hass and R. F. Newton, can be found in recent editions of the *Handbook of Chemistry and Physics*, C. D. Hodgeman, ed. The compound is placed into one of seven classes, and its entropy of vaporization at 760 mm is determined from a chart. The corrected boiling point can then be calculated. If the boiling point at 760 mm is not known, it must be found by successive approximations. The equation can be used to find boiling points at reduced pressures if care is taken in substituting into the equation.

Another method[38] uses a slide rule to facilitate calculations, and another[39] uses charts for the same purpose.

Boiling Points of Solutions

If a nonvolatile compound is dissolved in a liquid, the vapor pressure of the solution will be lowered in a predictable manner. The equation

[38] F. T. Miles, *Ind. Eng. Chem.* **35**, 1052–1061 (1943).
[39] C. Bordenca, *Ind. Eng. Chem., Anal. Ed.*, **18**, 99–101 (1946).

relating the mole fraction of the solute, N_2, and vapor pressure of the solution, P, was developed by Raoult, and is as follows.

$$P = P_0 N_1 = P_0(1 - N_2)$$

where P_0 is the vapor pressure of the solvent, and N_1 is its mole fraction. Since the vapor pressure of the solution is lowered, the boiling point of the solution is higher than the boiling point of the pure solvent. This fact allows the determination of molecular weights by measuring the increase in boiling point of the solution, as will be discussed in more detail later in this chapter.

If the compound dissolved in the liquid is volatile the situation becomes

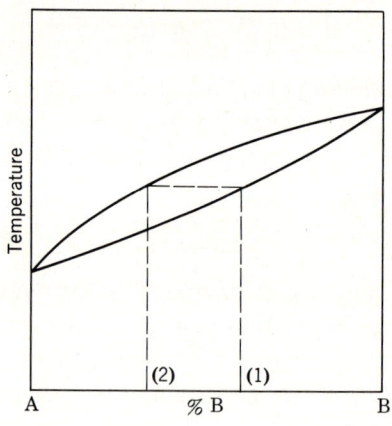

FIGURE 5.11
Vapor-liquid curves for an ideal solution.

more complicated. If the solute is completely soluble in all proportions, the solution is said to be homogeneous. Homogeneous solutions of liquids may be ideal or azeotropic. A diagram for the ideal solution of two liquids is given in Figure 5.11.

In this case, A boils lower than B. If a solution of composition represented by (*1*) is heated just to boiling, it will be in equilibrium with vapor of composition (*2*). The vapor will be richer in the lower-boiling component, A, than the original was. If this vapor is condensed, it will be in equilibrium with vapor still richer in the lower-boiling component A. If this process is repeated enough times, pure A will distill. If, however, the process is not repeated enough times, a mixture containing some B will be produced. Consequently, to separate two liquids that boil fairly close together, much patience and a good column are needed. One simple distillation will, very seldom, satisfactorily separate the two liquids.

If the homogeneous solution is not ideal, it may show a minimum- or maximum-boiling azeotrope. Diagrams for binary mixtures of these two types are shown in Figure 5.12. In the case of the pair with a minimum-boiling azeotrope [Figure 5.12(*A*)], a mixture with composition (*1*) heated just to boiling would be in equilibrium with vapor of composition (*2*). In this case, the vapor is richer in the higher boiling liquid. If the process

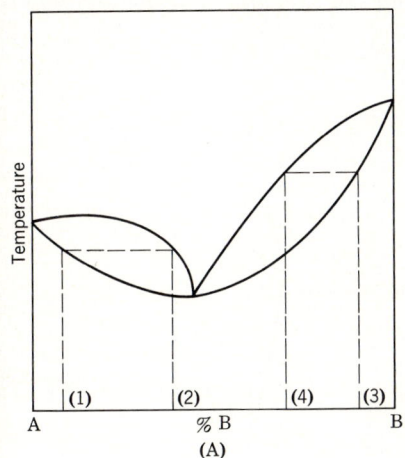
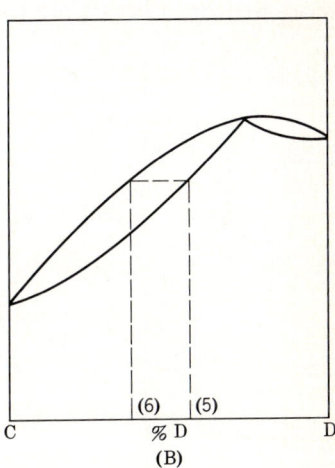

FIGURE 5.12

Homogeneous solutions of liquids with azeotropes. (*A*) Minimum-boiling. (*B*) Maximum-boiling.

is repeated enough times, the product of the distillation will be the minimum-boiling azeotrope, and pure *A* will be left in the distillation pot. If the initial composition was (*3*), the vapor would have the composition represented by (*4*), which is richer in the lower-boiling component. A large number of distillations in this case would produce the minimum-boiling azeotrope as the first product, followed by pure *B*.

Ethyl alcohol and water form a minimum boiling azeotrope. Such azeotropes are formed when more hydrogen bonds are broken than formed on mixing the liquid. For more details, see the discussion on p. 58.

Figure 5.12(*B*) is a liquid-vapor diagram for a pair of liquids that form a maximum-boiling azeotrope. Hydrochloric acid and water form a solution of this type. If a mixture with the composition represented by (*5*) is heated just to boiling, it will be in equilibrium with vapor of composition

(6). In this case, the vapor will be richer in the lower-boiling component. If the distillation is repeated enough times, pure C will distil until the composition of liquid in the distillation pot reaches that of the maximum-boiling azeotrope. From this point to the end of the distillation, the composition of the distillate will remain unchanged. The hydrochloric acid produced in this way is a primary standard for acidimetry, since its composition is so constant and well known.

Chloroform and acetone also produce a solution with a maximum-boiling azeotrope. Other liquids will form azeotropes of the same nature, if more hydrogen bonds are formed than are broken when the liquids are

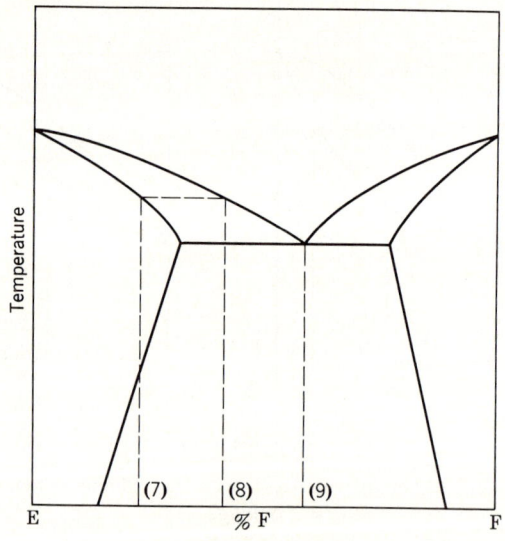

FIGURE 5.13

Boiling-point diagram of a pair of partially miscible liquids.

mixed. For more detailed discussion, see the earlier section on page 58 or the original article by Ewell et al.[40]

Liquids may not be completely miscible in all proportions; these are called heterogeneous solutions. The diagram for such a mixture is shown in Figure 5.13. Assume that a mixture of the two liquids represented by the composition (7) is heated. At first, it has two liquid phases and appears cloudy but, as the temperature rises, the two phases become one, and the solution clears. If this same solution is heated just to the boiling point, the first vapor has the composition represented by (8). As soon as the vapor

[40] R. H. Ewell, J. M. Harrison, and L. Berg, *Ind. Eng. Chem.*, **36**, 871–875 (1944).

condenses, two liquid phases appear. If the liquid condensate is redistilled, the composition of the product is represented by (9). Actually, distillation of any liquid mixture with compositions anywhere along the horizontal line will give a distillation of composition (9) in one step. This is the basis of steam-distillation.

Procedure for 0.1 ml of Liquid

The quantity of material required is 40 to 100 mg, depending on the diameter of the "boiler" capillary. This is constructed of 3 to 4 mm (OD)

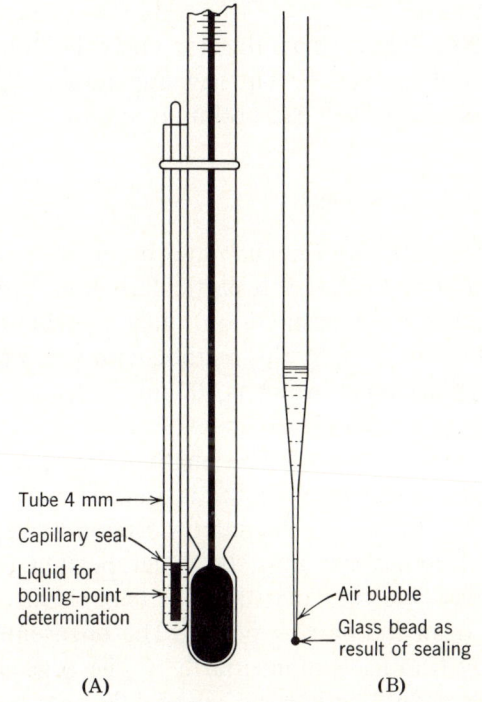

FIGURE 5.14

(A) Boiling point setup with inverted capillary. (B) Boiling point setup according to Emich.

tubing, 80 to 100 mm in length, sealed at one end. About 2 to 3 drops of liquid placed in this "boiler" with a capillary pipet will fill it to a height of 6 to 8 mm. Another capillary approximately 1 mm in diameter and 80 to 100 mm long, is sealed about 10 mm from one end and inserted into the boiler with the open end downward in the liquid. The boiler capillary is attached to a thermometer (Figure 5.14(A)), which is immersed in a liquid bath, as in melting-point determinations. Heat is applied gradually; small bubbles of air trapped in the inverted capillary force their way out as the temperature rises. When the bubbling becomes rapid, heating is discontinued, and the temperature is allowed to drop 5 to 10°. As the temperature drops, the liquid recedes into the capillary. The temperature is then raised at the rate of 1° per minute, until a steady stream of bubbles emerges from the capillary. After the temperature has been noted, heating is discontinued. When the bubbles stop emerging and the liquid begins to recede into the melting-point tube, the temperature is again noted. This interval of temperature, which is very small in the case of pure liquids, is the boiling point.

An improved Siwoloboff apparatus for micro boiling points has been devised by Karr and Childers.[41] The new apparatus, with radiant heating and with minimum heat loss, has optimum viewing, and achieves reproducibility of 0.5°.

Procedure for 0.01 ml of Liquid

The method developed by Emich,[42] and modified by Benedetti-Pichler and Schneider[43] and by Fischer,[44] is particularly useful when less than one drop is available for the determination. It is possible to make a determination of the boiling point by this method with 1–10 λ (0.001–0.01 ml) of substance. A capillary pipet, which is 100 mm in length, 0.5 mm in diameter, and tapers at the narrow point to a 0.1 mm diameter, is drawn out of 6-mm tubing [see Figure 5.14(*B*)]. The length of the narrow point should be about 10 mm. The surface of the liquid is touched by the fine point of the capillary so that the liquid fills the narrow section by capillary attraction and rises about 1 to 1.5 mm into the wider portion of the tube. The capillary is removed and the tapered end is bent slightly upward to draw the liquid away from the narrow point. The open end of the capillary point is sealed by heating it momentarily at the edge of the burner. A minute air bubble should be present, which fills most of the fine point of the capillary [as shown in Figure 5.14(*B*)], but it should not extend into the wider portion of the capillary. If the bubble is not satisfactory upon examination with a magnifying glass, the capillary is cut and resealed, so that a bubble of proper size is obtained.

The capillary tube is attached to a thermometer as a melting point tube and is heated slowly in the bath. The air bubble is observed through a

[41] C. Karr, Jr., and E. E. Childers, *Anal. Chem.*, **33**, 655 (1961).
[42] F. Emich, *Monatsh.*, **38**, 219 (1917).
[43] A. A. Benedetti-Pichler and F. Schneider, *Z. anal. Chem.*, **86**, 69 (1931).
[44] R. Fischer, *Die Chemie*, **55**, 244 (1942).

magnifying glass, with a light if necessary. When the bubble or liquid column begins to show signs of upward motion, the flame is removed. The temperature at which the bubble reaches the surface of the bath liquid is recorded as the boiling point of the sample. A slight upward movement may occur below the boiling point, but it is very slow until the boiling temperature is reached. After the bath has cooled to 10° below the recorded temperature, it is again heated slowly, and a second observation is made. A modification of this method has been proposed by Garcia.[45]

Refractive Index

Although the refractive index is not a constitutive property, it is nevertheless a very useful property of liquids for identification purposes. The refractive index is easily and quickly determined on one or two drops of liquid if a refractometer is available. The results are generally quite reproducible if proper precautions are taken to control the temperature and wavelength of light used. To some extent, the refractive index is useful for quantitative work; a standard curve is prepared, which is used to find the concentration or percentage for a particular reading. For certain specialized uses, such as determining the concentration of sucrose in water, the refractometer may be modified to read directly in the units desired. The refractometer is sometimes designed for other special purposes, such as work with fats and oils, so as to read directly in units from 0 to 100, which are easier to work with.

Theory

When light travels from one medium to another, its direction is modified. If it goes from a less dense to a more dense medium (Figure 5.15), it will be bent toward the vertical (or away from the plane of the surface). Angle a in Figure 5.15 will always be less than angle b if the second medium is more dense than the first. Since part of the wave front enters the more dense medium first and travels more slowly, the wave front changes direction as indicated. The amount of this change is characteristic for the wavelength of light, the temperature, and the two media involved.

The refractive index, n, is defined to be the ratio of the sines of the angles of incidence and of refraction:

$$n = \frac{\sin b}{\sin a}$$

[45] C. R. Garcia, *Ind. Eng. Chem., Anal. Ed.*, **15**, 648 (1943).

If the less dense medium is assumed to be a vacuum, this ratio will always be greater than 1. Since the refractive index is a ratio, it has no units. The refractive index is also the ratio of the speed of light in the two media. Light travels faster in a vacuum than in any other medium.

Factors Affecting the Refractive Index

The refractive index depends on several factors mentioned in the preceding section. If the less dense medium is always considered to be a vacuum, one of these is standardized. If the more dense medium is the liquid or solid being identified, we have another characteristic property to help in our search.

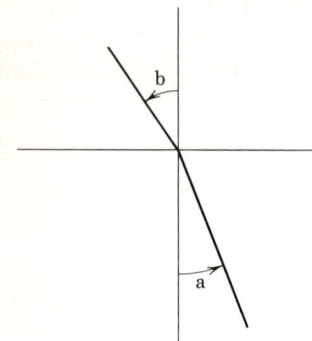

FIGURE 5.15

Refraction of light on entering a more dense medium.

The refractive index varies significantly with the temperature. The value of the refractive index decreases by 0.00035 to 0.00055 for each degree rise in temperature. If we wish to read a refractometer to 0.001 and to compare the reading with a recorded value, the sample should be within 2 or 3 degrees of the temperature at which the recorded value was taken. Consequently, for good work, the refractometer should be kept at a constant known temperature while readings are being made. The temperature should always be recorded, and indicated as a superscript; the terms n^{25} or 1.4754^{25} indicate that a temperature of 25° was used.

The refractive index varies considerably with the wavelength of the light used, being greater for shorter wavelengths than for longer ones. Therefore, this datum also should be indicated. However, it is customary now to use the value for the sodium D lines (5890 and 5896 Å) in reporting refractive indices; if no wavelength is given, it will be understood that the sodium D lines were used for the measurement. If the wavelength is designated, it appears as a subscript; n_D, 1.3241_D, n_{5890}, 1.2344_{5890}. Fortunately, most refractometers in use now can use white light (containing all visible wavelengths) and give results in terms of the sodium D line value. For those refractometers, which need monochromatic light, sodium lamps are available.

Optical Dispersion

Advantage is taken of the fact that refractive indices are different for various wavelengths in the identification of organic compounds. For some compounds, the difference in refractive indices is greater than for other compounds. This difference in the refractive indices at two wavelengths is known as optical dispersion. The optical dispersion is defined as:

$$\text{optical dispersion} = \frac{n_F - n_C}{d}$$

where n_F and n_C are the refractive indices of the material at 4861 Å and 6563 Å, respectively, and d is the density.

The optical dispersion may be determined with an Abbe refractometer with the help of a table, which is furnished with each instrument. Two readings are needed: (1) the refractive index itself and (2) the reading on the dial around the barrel of the refractometer. We, then, calculate the dispersion from the formula given with the instrument.

In general, aromatic compounds have greater optical dispersions than aliphatic compounds. Benzene, particularly in an Erlenmeyer flask, will make a brilliant display of color when viewed in the proper light. Diamond is prized, not only for its hardness and sentimental value but, also, because of its great optical dispersion. See the last section of this chapter for further information.

Refractometers and Procedures for Liquids

No matter which instrument or method is used in the determination of refractive indices, it is essential to check the instrument and procedure by means of reference standards, as described in the following section.

Several types of refractometers are used for determining the refractive indices of liquids. The Abbe refractometer is widely used because it employs only a few drops of the material and requires but a few minutes for the determination. The Fisher refractometer is simple and less versatile. The Abbe refractometer, shown in Figure 5.16, consists of (a) a pair of rotating water-jacketed prisms hinged together, (b) an observing telescope above the prisms for observing the border line of the total reflection that is formed in the prism, and (c) a sector on which the telescope is fastened. The sector is graduated from 1.300 to 1.710 and permits direct reading of the index of refraction; it is adjusted to the sodium *D* line of the spectrum.

To operate the refractometer, we adjust the thermometer in place, and connect the base of the prism enclosure with a supply of water at 20°. The left thumb is placed on the sector, and the right hand is used to open the double prism by pulling down the screwhead that is fastened on the lower prism. The prisms are wiped off carefully with a sheet of lens paper (facial tissue paper is well adapted to the same purpose). If the liquid does not evaporate rapidly, 2 drops are placed on the face of the lower prism, which is closed immediately by means of the screwhead and clamped against the upper prism. If the liquid is volatile, the prism is closed, the screw is slackened and, with a pipet dropper, a few drops of the liquid are placed into the depression of the side of the prism leading through a narrow channel into the space between the two prisms. The screw is

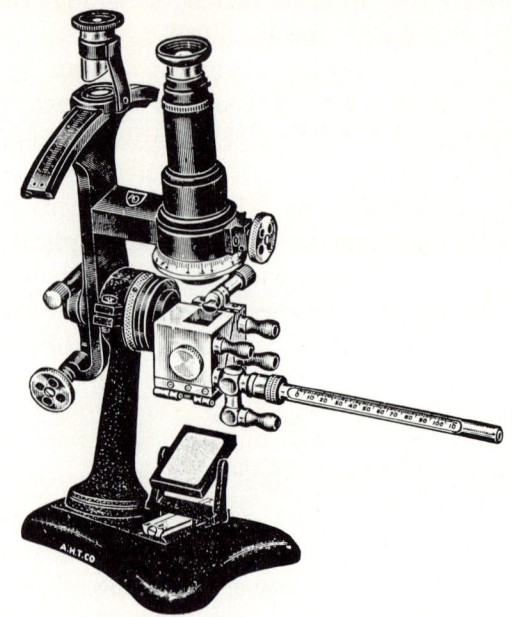

FIGURE 5.16

Abbe-Spencer refractometer, range n_D 1.300 to 1.710. Readings can be estimated to the fourth decimal. (Courtesy A. H. Thomas Co.)

tightened and the prism is rotated by moving the arm at the side of the sector. The mirror is adjusted to obtain maximum illumination, and the cross hairs are brought sharply into focus. As the prism is rotated, the illuminated field is partly darkened by a shadow moving across it. The boundary between the dark and the light field is called the *dividing line*, or *border line*. If the dividing line is not sharp but is hazy with a band of colors, the compensator wheel, which is at the lower part of the observing telescope, is moved until the dividing line becomes sharp and colorless. The arm of the prism is moved until the dividing line coincides with the intersection of the cross hairs. The refractive index is read through the small eyepiece over the sector. The pointer indicates the last three figures

Physical Properties of Organic Compounds   217

of the refractive index. The first two are read on the left side of the scale a small distance below the pointer. The double prism is opened and cleaned with a piece of lens paper and a few drops of acetone.

The Fisher refractometer is shown in Figure 5.17. A small glass slide with a beveled edge is fitted on the glass plate of the eyepiece, as shown at the right. A small clamp holds the glass slide down, so that a prism-shape well is formed between the plate of the eye piece and the glass slide. The instrument has a cord by which it is attached to an electrical outlet.

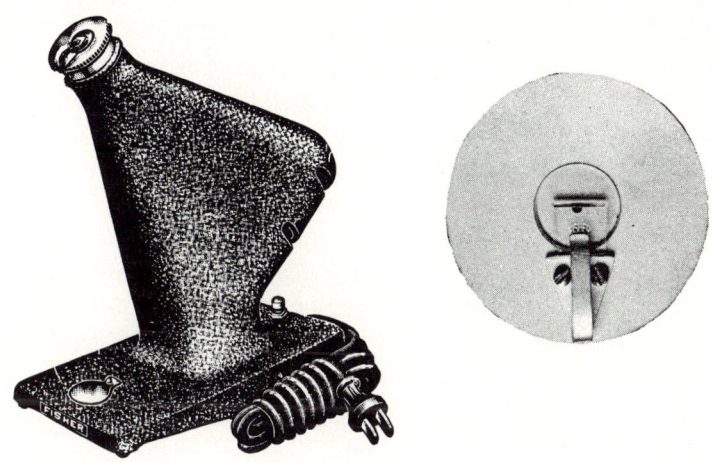

FIGURE 5.17
Fisher refractometer (*left*) and eyepiece (*right*). (Courtesy Fisher Scientific Co.)

If the push button on the base of the instrument is pressed, and observation is made through the aperture of the eyepiece, an illuminated scale appears. The graduations on the scale are divided at 1.516 by an arrow. This point corresponds to the refractive index of the glass in the eyepiece.

When a very small drop of a liquid is placed in the well formed by the glass slide and glass plate of the eyepiece, the refraction of light passing through the prism of the liquid sample produces a secondary or virtual image of the arrow on the scale. A liquid having a refractive index less than that of the glass employed will cause the light to bend downward, and the secondary image will appear above 1.516 on the scale. Conversely, if the refractive index is higher, the bending is upward, and the secondary image appears below 1.516.

For operation of the Fisher refractometer, the small glass slide is removed from the box and cleaned with lens paper. Likewise, the plate of the eyepiece is wiped clean. The glass slide is placed with the beveled edge downward over the plate so that it just covers the aperture. The cord of the instrument is connected to an electrical outlet. A very small amount (0.01 to 0.05 ml) of the liquid is added at the edge of the glass slide directly over the aperture, by means of a capillary pipet dropper. The minute droplet spreads by capillary attraction over the beveled edge and fills the well. The push button is pressed, and the second arrow either above or below 1.516 on the scale is observed. This reading is the refractive index of the liquid. If the liquid evaporates rapidly, the droplet is added with one hand while the button is pushed with the other, and the secondary image is observed immediately. When the determination is completed, the glass slide is cleaned and replaced in the proper box. If it is desired to make several determinations, the glass slide and eyepiece are cleaned with lens paper between operations.

When only a small quantity of a liquid unknown is available, it is possible to measure the refractive index using the Nichols refractometer,[46] which requires about 10 λ of the sample. Reference standards whose refractive indices are known to five significant numbers are commercially available, and a number of these should be at hand to check the accuracy of the instrument that is employed for refractometric measurements. In the absence of any reference standards the refractive index of pure water should be used. At 20°, the refractive index is 1.33299; at 25°, 1.33250.

Refractive Indices of Solids and Their Melts

The determination of the refractive index of an *isotropic crystalline solid* is far more involved than the relatively easy procedure used for the determination of the refractive index of a liquid. Two general methods are in use: (a) particles of the crystalline solid are immersed in liquids of known refractive indices until a liquid is found in which a minimum visibility of the particle is obtained—this liquid and the crystal are then considered to have the same refractive index; (b) the particle is immersed in a medium having a lower refractive index, and a second medium that has a higher value is added for dilution until minimum visibility is obtained—the refractive index of the diluted medium is then determined, and this value is taken as the refractive index of the particle. For *anisotropic crystalline solids* the method is still more involved, since the refractive index varies with the direction of transmission and the vibration of light in the specimen. See the literature given in the references at the end of this chapter.

[46] N. D. Cheronis, "Micro and Semimicro Methods" in *Technique of Organic Chemistry* Vol., 6, A. Weissberger, ed., Interscience, New York, 1954, p. 203.

Turunen[47] has described a new method for determination of refractive indices, which is applicable to the continuous study of a reaction. The method uses a two-component system, one a liquid and the other a transparent solid. The refractive index of one component is known. Light is passed through the system, and the intensity of the nonscattered light is measured. The refractive index of the other component is then calculated. The equipment is not complex, but is not particularly adapted to micro determinations. The method should work well for determination of refractive indices of solids.

There are two methods for the determination of the refractive indices of melted organic compounds. One employs the Fisher refractometer with a special eyepiece containing a heating unit and thermometer.[48] The second method employs a hot stage, the Becke line principle,[49] and a series of glass powders with stepwise variations in refractive indices by which the determinations are made.

The use of refractive indices and densities and other physical constants in calculating molar refractions and dispersion is discussed at the end of the chapter.

Density

Density is not a constitutive property taken alone but, combined with the molecular weight to give the molecular volume, it has usefulness in this sense under special conditions. However, the density is an important physical property for liquids and, often, for solids for identification purposes. The density may be obtained easily for quantities of liquid of 10 to 30 ml by means of hydrometers or the Westphal balance. For small quantities of liquid, however, specific gravity bottles and analytical balances are necessary. Both of these items should be available for work in qualitative analysis of organic compounds. The density is often needed as an adjunct to other methods of analysis, and equipment should be available in the laboratory for determining the density. In some cases, the density serves for quantitative purposes; the standard method for determination of alcohol in beverages depends on measurement of the density of a distilled sample.

[47] L. Turunen, *Anal. Chem.*, **33**, 1617–1631 (1961).
[48] H. A. Frediani, *Ind. Eng. Chem., Anal. Ed.*, **14**, 439 (1942); N. D. Cheronis, "Micro and Semimicro Methods" in *Technique of Organic Chemistry*, Vol. 6, A. Weissberger, ed., Interscience New York, 1954, pp. 204–207.
[49] E. E. Jelley, "Light Microscopy" in *Physical Methods of Organic Chemistry*," Part 2, 3rd ed., A. Weissberger, ed., Interscience New York, 1960, pp. 1347–1474.

220 Techniques of Organic Analysis

The determination of density is useful in the identification of compounds that do not form well-defined derivatives. The characterization of such substances as the liquid aliphatic hydrocarbons is usually accomplished through determination of the boiling points, refractive indices and densities. The determination of density may also be used as a general index of the relative complexity of the unknown. Compounds that have a density of less than 1.0 usually do not contain more than one functional group, whereas polyfunctional compounds have, as a rule, a density greater than 1.0.

The difficulties encountered in the accurate determination of the densities of organic liquids account for the restricted use of this important physical constant. Compared with the determination of melting points, refractive indices, and boiling points, the measurement of densities is subject to more errors because *simple, rapid,* and *reliable* instruments and techniques have not yet been developed. The macromethods that use 1 to 5 ml of liquid are reliable if proper pycnometers are used and sufficient time is available for repeated accurate weighings with an analytical balance. The micromethod described in this section gives reliable results if care is exercised. Since the amount of liquid for the determination is 0.02 to 0.03 ml, small errors in weighing or losses by evaporation will cause large discrepancies. For determination of the densities of *solids*, see the literature.[50]

Density measurements are usually expressed as grams of mass per milliliter, $d_4^t = m/V$, g/ml. This is measured by a direct comparison of the weights of equal volumes of the substance at $t°$ and of water at $4°$ (3.98°); specific gravity d_t^t is measured by direct comparison of the weights of equal volumes of the substance and water at $t°$. The temperatures most frequently used are $25°/25°$ and $20°/20°$. However, the densities and specific gravities of many organic compounds recorded in the literature were determined at temperatures ranging from 0 to 40°. With the Dreisbach[51] tables, it is possible to convert rapidly the specific gravity at $25°/25°$ to density at any temperature between 0 and 40°, if the coefficient of cubical expansion, β, is known. A value for specific gravity d_t^t, multiplied by the density d_4^t of water, gives the density d_4^t.

---

[50] E. R. Caley, *Ind. Eng. Chem., Anal. Ed.*, **2**, 177 (1930); E. W. Blank, *Ind. Eng. Chem., Anal. Ed.*, **3**, 9 (1931); E. W. Blank and M. L. Willard, *J. Chem. Educ.*, **10**, 109 (1933); N. D. Cheronis, "Micro and Semimicro Methods" in *Technique of Organic Chemistry*, Vol. 6, A. Weissberger, ed., Interscience, New York, 1954, p. 198; N. Bauer and S. Z. Lewin, "Determination of Density" in *Physical Methods of Organic Chemistry*, Part 1, A. Weissberger, ed., Interscience, New York, 1959, pp. 131–190.

[51] R. R. Dreisbach and R. A. Martin, *Ind. Eng. Chem.*, **41**, 2879 (1949).

Density and Structure

The densities of members of a homologous series fall on a smooth curve, as shown in Figure 5.18. Note that some series start low and others high, but most of them seem to approach a density of about 0.8.

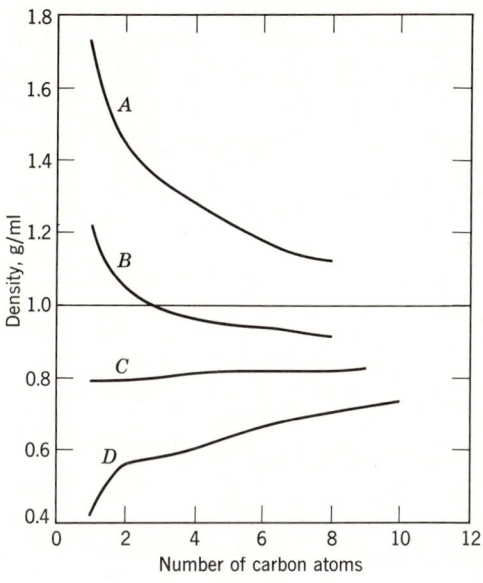

FIGURE 5.18

Densities of organic compounds. (*A*) *n*-Alkyl bromides. (*B*) Straight-chain acids. (*C*) *n*-Alcohols. (*D*) Straight-chain alkanes.

Apparatus and Procedures for 1 ml of Liquid or More

Pycnometers with a capacity of 1 to 2 ml are commercially available. The pycnometer is cleaned, dried, and then weighed. The bulb is then filled with distilled water to a point above the mark and immersed in a 25-ml beaker containing water at 20°. After 5 to 10 minutes, the level of water in the pycnometer is adjusted by means of a capillary pipet dropper. The pycnometer is then removed from the beaker, dried rapidly with a small piece of chamois, and weighed. The pycnometer is then emptied, dried, and filled with the liquid under investigation; it is adjusted at 20° as before, and weighed. The weight of the sample divided by the weight of water gives the density of the liquid at 20°. However, the density of water at 20° is not 1.0000 and, therefore, a correction must be made to express the density with reference to that of water at 4° by the following factor.

$$d_4^{20} = \frac{\text{Weight of sample}}{\text{Weight of water}} \times 0.99823$$

Apparatus and Procedures for Less than 1 ml of Liquid

Semimicro and micro determinations of density may be made with pycnometers that are easily constructed of 3 to 4 mm tubing or glass capillaries. The pycnometer shown in Figure 5.19, if made of 3-mm glass

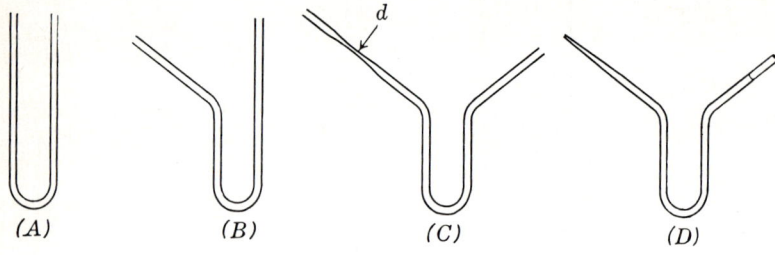

FIGURE 5.19

Construction of a micropycnometer.

tubing, has a capacity of 0.2 to 0.4 ml and, when made from a melting point capillary, 0.02 to 0.03 ml. Javes and Liddell[52] reported on the results of the use of a similar pycnometer with a capacity of 0.05 ml. The specific gravities thus obtained were well within 0.001 of the specific gravities determined on macrosamples.

Figure 5.20 shows the Fisher-Davidson gravitometer for routine work. When semimicro quantities are available, this gravitometer, based on Ciochina's[53] balanced-column method, gives rapid results with an accuracy of about 0.1 to 0.2 per cent, and requires about 0.5 ml of the sample. It gives direct readings, and can be used for the range $d_4^{20} = 0.6$ to 2.0. The scale is established by using ethylbenzene as a standard in the L tube of the instrument; the sample is placed in the Z tube, and both liquids are drawn up by the pump, which is operated by turning the center knob until the menisci of the liquid in the Z tube rest in the upper and lower arms. The fixed pressure difference between the atmosphere and the connecting lengths of the Z and L tubes varies directly with the density of the liquid in the Z tube and is measured on the graduated scale as d_4^{20}. The standard liquid for $d_4^{20} = 0.6$ to 2.0 is "certified" ethylbenzene supplied with the

[52] A. R. Javes and C. Liddell, *Anal. Chem.*, **30**, 1570–1575 (1958).
[53] I. Ciochina, *Z. anal. Chem.*, **98**, 416 (1934); **107**, 108 (1936).

instrument. For higher densities, carbon tetrachloride may be used and the readings multiplied by a conversion factor. The Z tube may be replaced with a wider (4-mm) tube for liquids of high viscosity. The amount of sample required for each determination is 0.3 to 0.7 ml, depending on the substance. Reviews and discussion on the determination of micro quantities will be found in the references. For most purposes, the

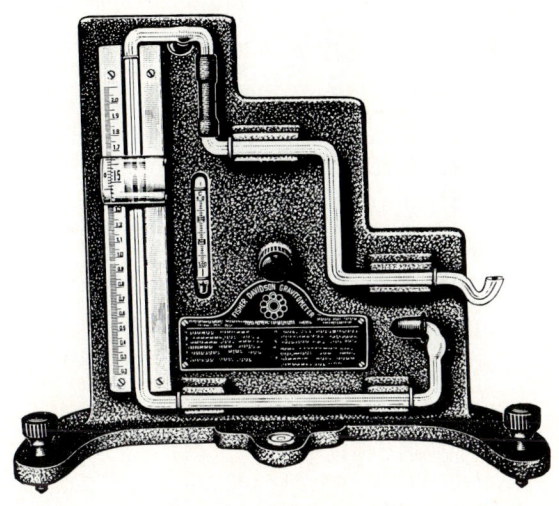

FIGURE 5.20

Fisher-Davidson gravitometer. (Courtesy Fisher Scientific Co.)

micropycnometer pipets shown in Figure 5.21 (described by Alber[54]) are suitable. Micropycnometer pipets of this type are commercially available. Figure 5.21(A) represents the decigram pipet adjusted to contain 0.1 ml of liquid at 20°. Tightly fitting ground caps are placed over each end to prevent loss by evaporation. The total weight of the pipet varies between 4 and 6 grams. With a semimicro balance (sensitivity 0.01 mg), an accuracy of 0.05 per cent is attainable; with an analytical balance (0.05 to 0.1 mg), an accuracy of 0.15 per cent is attainable.

Figure 5.21(B) shows a graduated, but not calibrated, pipet of the same type obtainable for a milligram or a centigram range. The former has a capillary bore of about 0.5 mm, and is suitable for volumes from 0.005 to 0.016 ml; the latter has a capillary bore of 1 mm, and is suitable for volumes from 0.02 to 0.08 ml. Both pipets have scales 80 mm long graduated in 1-mm divisions. The pipet should be calibrated at the first mark and

[54] H. Alber, *Z. anal. Chem.*, **90**, 87 (1932); H. Alber and M. v. Renzenberg, *Z. anal. Chem.*, **84**, 114 (1931); H. Alber, *Ind. Eng. Chem., Anal. Ed.*, **12**, 774 (1940).

four other points at 20° with distilled water or bromoform ($d_4^{20} = 2.893$) in the usual manner. After being weighed, the pipet with the rubber tubing attached to the upper end is placed in the test tube containing water, which is then immersed in a thermostatically controlled bath. After about 15 minutes, the tube is taken out of the bath and the liquid is drawn up to one of the marks by means of suction. The pipet is removed from the test tube

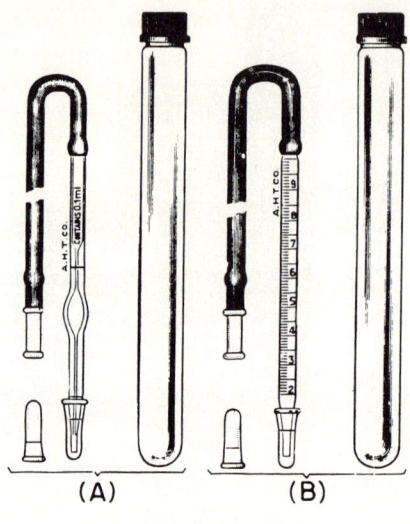

FIGURE 5.21

(*A*) Micropycnometer pipet of 0.1-ml capacity. (*B*) Micropycnometer pipet graduated but not calibrated. (Both courtesy of A. H. Thomas Co.)

and is immediately placed in a horizontal position; the free liquid is then quickly cleaned off with a piece of chamois and the meniscus is read with a low-power lens, at least to about one quarter of a division. Finally, the first cap is fitted slowly over the end of the pipet, so that the liquid column is not pressed too far in the other direction and, after removal of the rubber tubing, the second cap is adjusted. The pipet is placed on the hooks of the bow of the balance pan, and weighed after 20 minutes. The pipet should be handled with forceps or chamois fingers to avoid heat effects. This procedure is repeated for the other marks.

In a density measurement, a 2000-ml beaker filled with water a few degrees lower than the required temperature may be substituted for a thermostatically controlled water bath. A piece of cardboard with two openings, 10 mm apart, serves as a holder for two test tubes, one for the substance and pipet, and the other for 1 ml of water and a thermometer with divisions of 0.1°. The bath is heated and the pipet is filled as soon as the desired temperature is reached. Errors in microdeterminations of this type are chiefly due to inaccurate reading of the meniscus, sensitivity of the balance, and inaccurate weighing. With practice, it is possible to obtain as accurate results as with macropycnometers.

A simple densitometer for solids uses helium and a minimum of equipment.[55] Helium fills all cracks and crevices in a solid and, yet, is not absorbed on the surface. The sample is heated in a vacuum to remove any occluded gas. The volume of helium needed to fill the densitometer with and without the solid is measured by expansion into a well known volume and determination of the pressure.

Optical Rotation

The change in the direction of vibration of linearly polarized light during its passage through anisotropic substances is called *optical rotation*, and such substances are called *optically active*. For an extensive discussion of optical rotation and related phenomena, see the chapter by Heller and Fitts.[56] Although the number of organic compounds that exhibit optical activity is quite large, the determination of specific rotations for identification work is most useful in the characterization of sugars,[57] amino acids and alkaloids[58], and other naturally occurring substances. This section is restricted to a brief description of the standard method employed for the determination of specific rotation.

Optical Rotation and Chemical Structure

The property of optical rotation is shown only by molecules that are asymmetrical. If a molecule has a plane of symmetry or a point of symmetry, it will not rotate the plane of polarized light. A molecule has a plane of symmetry if an imaginary plane will divide the molecule into two similar halves, so that any atom or group of atoms in one half is duplicated by another atom or group of atoms in the other half, arranged in the same manner about the imaginary plane. If we can draw a line from each atom perpendicular to the plane and find the same kind of atom when the line is extended an equal distance on the other side of the plane, then the plane

[55] R. J. Kokes, *Anal. Chem.*, **32**, 446–447 (1960).
[56] W. Heller and D. D. Fitts, "Polarimetry" in *Physical Methods of Organic Chemistry*, 3rd ed., Part 3, A. Weissberger, ed., Interscience, New York, 1960, pp. 2147–2334.
[57] F. J. Bates et al., "Polarimetry, Saccharimetry and the Sugars," Nat. Bur. Stand. Circ. No. C 440 (1942).
[58] T. A. Henry, *Plant Alkaloids*, Blakiston, 1949.

is a plane of symmetry, and the molecule is optically inactive. Only one plane of symmetry in a molecule is necessary to rule out optical rotation.

Asymmetry of molecules arises in several ways: (1) an asymmetric carbon atom, that is, one with four different groups attached to it; (2) restricted rotation due to (a) interfering groups as in the biphenyls of the kind

[structure: biphenyl with NO$_2$, CO$_2$H on one ring and Br, CH$_3$ on the other]

(b) double bonds, as in the allenes,

[structure: allene with H$_3$C and HO$_2$C on one end, CH$_3$ and CO$_2$H on the other: C=C=C]

or (c), rings as in the cyclopropanes:

[structure: cyclopropane with H, H, CO$_2$H, CO$_2$H, H, H substituents]

More details will be found in the references at the end of this chapter.

Optical Rotatory Dispersion

The specific optical activity of an asymmetrical molecule varies with the wavelength of the light used for its determination. This variation, called optical rotatory dispersion, has recently become a powerful tool for the elucidation of the structure of molecules, particularly those having a steroid framework of carbon atoms. Measurement of the optical rotatory dispersion is done by a spectropolarimeter, and may be made automatic; commercial instruments are available. Interpretations of the data are too complicated for discussion here, but are useful in the determination of absolute configurations of steroids.[59]

[59] C. Djerassi, *Science*, **134**, 649–655 (1961); C. Djerassi, *Optical Rotatory Dispersion: Applications to Organic Chemistry*, McGraw-Hill, New York, 1960.

Procedure for Determination of Specific Rotation

The first step involves the preparation of the solution of the active compound. About 100 to 500 mg of the substance are accurately weighed and dissolved in 25 ml of the solvent in a volumetric flask. The solvents commonly used are water, methanol or ethanol, chloroform, and a mixture of ethanol and pyridine. For the rotations of hydrazones and osazones, the method recommended by Neuberg[60] requires 200 mg of the derivative dissolved in a mixture of 4 ml of pyridine and 6 ml of absolute alcohol, and a reading of the solution in a 100-mm tube.

The compound must be pure before the solution is prepared. If the solution is not clear, it should be filtered after being properly diluted. The filtrate is collected in a dry flask and is returned to the funnel until a perfectly clear filtrate passes through the stem. The funnel is placed over another dry flask and the clear filtrate is collected for the determination.

The next step is the filling of the polarimeter tube. The cup is screwed on one end and the tube is held vertically while the solution is poured in until it rises to the top end; the cover glass is placed over the end of the tube in such a manner that no bubbles appear. The cap is screwed on the tube with care. If great pressure is applied to the cover glass, the strain may produce optical activity and a serious error will be introduced in the observation.

The *zero reading* of the instrument is determined by turning the movable prism until the two halves of the field are matched so as to have a uniform illumination. Before the two fields are matched, the eyepiece of the telescope should be checked for focusing; this is accomplished by turning the eyepiece to the right or left until the line that divides the two fields is sharp. The true zero of the instrument may not coincide with the zero of the graduated circle. The main circle is usually divided into degrees and 0.25 division of 1°. The vernier outside is divided into 25 divisions, thus enabling a reading of 0.01°. Five readings are made and averaged. The polarimeter tube that contains the solution is placed in the opening between the two prisms. The movable prism is turned until uniform illumination is obtained. The reading of the circle is noted and the observation is repeated 4 or 5 times. The temperature at the time of the readings is recorded. The readings are averaged; this average rotation, less the zero reading, gives the observed notation. The specific rotation is calculated from this value, using the formula

$$\text{Specific rotation} = [\alpha]_D^t = \frac{\alpha \times 100}{L \times c}$$

[60] C. Neuberg, *Ber.*, **32**, 3384 (1899).

where α is the observed rotation of the sample in degrees, L is the length of the tube in decimeters, and c is the concentration of the dissolved active substance in grams per 100 ml of solution; t refers to the temperature of the solution, and D to the wavelength of light used.

Molecular Weights

For a more extensive discussion and description of the micromethod employed in the determination of molecular weights, the literature given in the references should be consulted. The present discussion gives details for only the cryoscopic micromethods.

All cryoscopic methods depend on the accurate determination of the melting (or freezing) point of the solvent and of a solution of the sample. The well-known Beckmann method for determining the depression, Δt, in the melting point of the solvent produced by a small quantity of solute may be adapted to micro quantities; the size of the thermometer is reduced and an apparatus suitable for small volumes is used.[61] The Beckmann microthermometer[62] may be used for semimicro quantities with a simple test-tube arrangement for the solution vessel and a magnetic stirrer. In addition to the usual difficulties encountered in macrocryoscopic measurements of this type, the method is not applicable in some cases, particularly when the observed depression is slight, owing to the small range of the thermometer. The most successful cryoscopic micromethod is based on the use of solvents with high molal depression constants, with which it is possible to obtain a depression of the melting point of 5–20° for 10 per cent solutions of the sample; the measurements are made by the capillary tube method. This microprocedure was developed by Rast[63] with camphor as the solvent, after Jouniaux[64] had noted that, in addition to its excellent solvent properties, camphor exhibited a molecular melting point depression of about 40°.

Although the Rast method is unique in its simplicity, it is subject to limitations and difficulties. The molal freezing point constant of camphor varies between 37 and 40°, depending on the source of the compound. A more serious drawback is the high mp (176 to 180°) of camphor, which limits its use to substances which are stable at these temperatures. Furthermore, it is applicable only to those compounds which are soluble in camphor and which give approximately ideal solutions. Some of these limitations

[61] R. Iwamoto, *Sci. Repts. Tôhoku Imp. Univ.*, **17**, 719 (1928).
[62] A. H. Thomas Co., Philadelphia, Pa.
[63] R. Rast, *Ber.*, **55**, 1051, 3727 (1922).
[64] A. Jouniaux, *Bull. Soc. Chim.*, **11**, 722, 993 (1912); *Compt. rend.*, **154**, 1592, 1692 (1912).

may be surmounted by the use of other solvents which have high molal freezing-point constants. For a solvent to be useful for this purpose, it should dissolve a wide variety of organic compounds; the solution should be clear at the melting point of the mixture; decomposition of the solute should not occur at the melting temperature. Furthermore, the solute should not undergo association in the solvent and the depression of the melting point of the solvent for 5 to 10 per cent solute concentrations should be large. In Table 5.06 are listed a number of solvents that have

TABLE 5.06
SOLVENTS FOR RAST MICROPROCEDURE

Solvent	Melting Point, °C	Molal Depression Constant, °C
Cyclohexanol	24.7	42.5
Camphene	49	31
Cyclopentadecanone	65.6	21.3
Perylene	276	25.7
2,4,6-Trinitrotoluene	82	11.5
Tetrabromoethane	—	86.7
Bornyl bromide	—	67.4
Bornylamine	164	40.6
Camphor	176–180	37–40
Camphoquinone	190	45.7
Borneol	202	35.8
Dicyclopentadiene	32	46.2
Dihydro-α-dicyclopentadienone	53	92.0
Tetrahydro-α-dicyclopentadiene	77	35.0
Dihydro-α-dicyclopentadien-3-ol	53	92.0

Source: N. D. Cheronis, "Micro and Semimicro Methods" in *Technique of Organic Chemistry*, Vol. 6, A. Weissberger, ed. Interscience, New York, 1954, p. 210.

been recommended for the Rast microprocedure. Of the substances that are given in the table, few are commercially available for general use. Camphene has excellent solvent properties and a low melting point. Bornylamine is useful for alkaloids, and camphoquinone can be used in place of camphor. Cyclo-pentadecanone, which is recommended for sterols, carotenoids, azo dyes, and quinones, is expensive. Cyclohexanol is preferable to camphor in our opinion, if the solubility characteristics are the same; high temperatures can be avoided, and it is possible to obtain a more uniform solution with it than with camphor.

Although cyclohexanol has good solvent characteristics and an unusually high freezing-point depression constant of 42.5° per gram molecular weight of solute per 1000 grams of the solvent, its use should be considered with caution, since Wheeler and Jones have reported that solutions of a number of organic compounds in cyclohexanol deviate from Raoult's law.

The precision of the method depends to a considerable extent on the reproducibility of the solution in the capillary tube. With experience in the preparation of a uniform dispersion of the solute, a precision of about 0.5 per cent is possible. The accuracy of the method depends largely on the nature of the solute. The molecular weight of carboxylic acids cannot be determined with camphor as the solvent; the results are high, probably because of association. Under the most favorable conditions, an accuracy of 1 to 2 per cent is attainable by the experienced worker. In the determination of the molecular weights of resins and colored substances, the accuracy is about 5 per cent.

Procedure When Camphor or Other High-Melting Solvents Are Used

The method involves the determination of the melting point of a pure solvent, usually camphor, and the melting point of a solution of the substance under investigation in camphor. The apparatus required includes capillary tubes, which can be readily constructed or obtained from dealers, and a melting-point apparatus of the type described on pages 181–184.

The melting point of the solvent is determined by the usual method (page 184). If the solvent has not been employed before in the laboratory, the molar depression constant of the particular lot must be made with the same apparatus and thermometer which will be employed in the determination. Substances recommended for standardization are acetanilide, azobenzene, chloroanthraquinone, sulfonal, and naphthalene.

The solution of the standard or unknown is made in capillary tubes about 50 to 70 mm in length with an inner diameter of 3 to 4 mm, constructed of soft glass, unless the sample is affected by traces of alkali; in that event, Pyrex glass is used. The capillary, after being thoroughly wiped, is accurately weighed. The sample (0.5 to 5.0 mg) is introduced by a smaller capillary, as described in the caption of Figure 5.22; if the quantity exceeds 1.5 mg, it is advisable to use a capillary with an inner diameter of 4 mm. Sufficient solvent to give a solute concentration of 5 to 10 per cent is introduced by the same technique. Then, after the capillary has been wiped and reweighed, it is sealed at a point about 20 mm above the level of the solid. The sealing requires some care to minimize the volatilization of the solvent; the filled part of the capillary is wrapped with wet filter paper and then rotated in an inclined position over a microflame at the point where the seal is desired until the glass collapses and forms a solid glass rod. However, the capillary may be sealed by dipping the wrapped part in water and using a fine oxygen flame. Figure 5.22 shows the capillary and the various steps by which it is filled and sealed. The capillary tube is then heated, cautiously, to melt the solvent and mix it with the solute. Errors will occur unless care is exercised to obtain an homogeneous dispersion, particularly in the case of solvents such as camphor, which crystallize rapidly below their melting points. The capillary tube is warmed in a bath until the contents melt, and is then withdrawn, and the contents are mixed by vigorous shaking; this process is repeated three or four times. An alternative procedure used in our laboratory consists in rotating the capillary between the thumb and the first two fingers of the right hand over a microflame until the solvent has melted; the left hand is placed over the capillary, which is inclined at an angle of 45°. The contents are subsequently mixed by rotating the tube rapidly between the palms above the microflame at a sufficient height so that the mixture remains liquid. Some experience is needed for this technique, since the lower part of the capillary, which contains the solution, is at a temperature above 150°. A device that facilitates thorough mixing of the melted sample and solvent without their removal from the bath has been described by Aluise.[65]

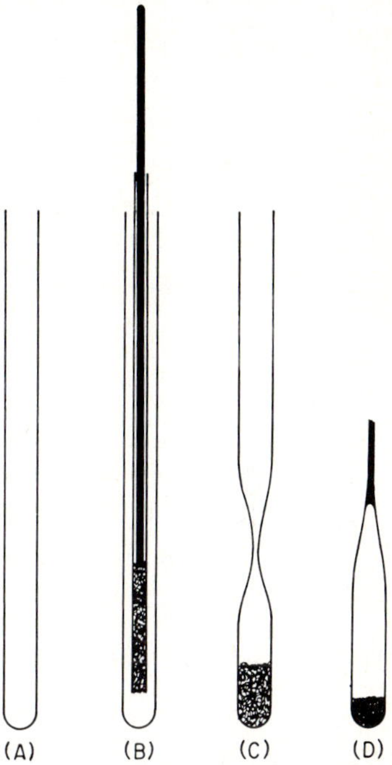

The melting points of the pure solvent (camphor, cyclohexanol, and the like) and of the solution (mixture prepared according to the steps shown in Figure 5.22) are determined several times according to the standard procedure described under melting points in this chapter. Care should be exercised in the observations for the melting point of the solution, since the melting

FIGURE 5.22

Rast method for microdetermination of molecular weight. (*A*) Capillary of 50 to 70 mm length and 3 to 4 mm diameter. (*B*) A second capillary, containing the material, is inserted into capillary (*A*), and the material is pushed out by means of a small rod. (*C*) The capillary containing solute and solvent is weighed and then sealed. (*D*) The upper part of the capillary is drawn out in the form of a rod, and the sample and solvent are melted and mixed. The tube is now ready for determination of the melting point of the mixture.

[65] V. A. Aluise, *Ind. Eng. Chem., Anal. Ed.*, **13**, 365 (1941).

point is not sharp; the temperature at which the last crystal disappears should be considered as the melting point of the mixture. For camphor and other solvents melting above 150°, a thermometer graduated in the range 140 to 230° may be employed. After each determination, the thermometer to which the capillary is fastened is removed momentarily from the heating bath and is carefully inverted to mix the contents of the capillary. The melt solidifies rapidly, and the determination of the melting point is repeated until the deviation between two consecutive measurements is about 0.2 to 0.5°. The molecular weight is calculated from the data by the equation

$$M = \frac{KS_1}{S_2 \Delta t}$$

where M is the molecular weight of the sample, K the molal depression constant of the solvent multiplied by 1000, S_1 the weight of the sample in milligrams, S_2 the weight of the solvent in milligrams, and Δt the observed difference between the melting points (averaged) of the solvent and the solution.

Procedure When Cyclohexanol Is Used

The advantage of this method is the ease with which homogeneous solutions of samples are prepared at room temperature. Purified cyclohexanol may be purchased and repurified for molecular weight determinations by fractional distillation and crystallization. For example, 200 ml of cyclohexanol are fractionally distilled and the middle fraction of 150 ml is collected, frozen, and allowed to melt until about one half is liquid, filtered rapidly and the crystals are collected and melted. The liquid is standardized, according to the directions in the preceding section. The solution vessel, either a small test tube, or an 8-mm or 6-mm tube, is weighed empty and then 5 mg of the sample are introduced by a small capillary tube in the case of solids, or by a capillary pipet in the case of liquids. About 50 to 100 mg (2 to 3 drops) of the solvent is added and the vessel is reweighed. The vessel is rotated over a microflame until a clear solution is obtained, and is then stoppered and centrifuged. The vessel is rotated and centrifuged again, the process being repeated 3 or 4 times. A sample is removed from the stock solution with a capillary tube open at both ends, and the technique used for the determination of the melting points of substances that melt below 25° is employed,[66] with a thermometer

[66] N. D. Cheronis, "Micro and Semimicro Methods" in *Technique of Organic Chemistry* Vol. 6, A. Weissberger, ed., Interscience, New York, 1954, pp. 181–184.

graduated from -30 to $+30°$ in $0.2°$ divisions. In using cyclohexanol, considerations should be given to the observation of Wheeler and Jones[67] that solutions of a number of compounds in this solvent deviate from Raoult's law.

Other Methods

The Victor Meyer method for molecular weight determination depends upon the vaporization of a known weight of liquid, and measurement of the volume of the vapor. The weight of 22.4 l of the vapor at standard conditions can then be calculated. An improved design for the determination of molecular weights by the Victor Meyer method has been described.[68] The apparatus is compact (only 33 cm long), has no danger of vapor condensation, and has durability and convenience.

One of the colligative properties of solutions is the increase of the boiling temperature of the solvent on the addition of a nonvolatile solute. This method has now been applied to microdeterminations of molecular weights. An ebulliometric apparatus has been constructed to measure the molecular weights of polymers.[69] To obtain sensitive temperature readings, a 80-junction thermopile was used; this allows the detection of a difference of 0.015 millidegrees. The remainder of the apparatus is of the Menzies-Wright type; about 15 ml of solution is needed for the determination.

A Menzies-Wright ebulliometric apparatus has been developed for the determination of molecular weights using 0.3 ml of solvent.[70] Such determinations can be carried out with good precision and accuracy (deviation of less than 2%) on samples of 1 to 3 mg. The major innovations involve two sources of the heat and better positioning of the condensation zones.

A recent development in the electrical field is the availability of thermistors. These devices, no larger than the head of a pin, change their resistance with temperature. Together with a Wheatstone bridge, these miniature probes allow the measurement of temperature within a drop of liquid. A thermistor makes it possible to reduce the amount of liquid needed for a molecular weight determination by the freezing point method.[71] The sample cup holds 0.4 ml, although a volume of liquid as small as 0.3 ml can be used.

Molecular weights in the 100 to 700 range can be obtained on samples

---

[67] C. M. Wheeler and F. S. Jones, *Anal. Chem.*, **24**, 1991 (1952).
[68] J. H. Robertson, *J. Chem. Educ.*, **37**, 152–153 (1960).
[69] C. A. Glover and R. R. Stanley, *Anal. Chem.*, **33**, 447–450 (1961).
[70] G. W. Perold and F. W. G. Schoning, *Mikrochim. Acta*, **1961**, 749–753.
[71] B. Wilkins, J. A. Knight, and Fred Sicilio, *Rev. Sci. Instr.*, **32**, 355–356 (1961); J. A. Knight, B. Wilkins, Jr., D. K. Davis, and Fred Sicilio, *Anal. Chem. Acta*, **25**, 317–332 (1961).

of 10 to 14 mg of solid with a standard deviation of 1.9% by the thermoelectric effect.[72] This method depends upon the establishment of a steady state between drops of a solvent and a solution suspended in the atmosphere of the solvent at some fixed temperature, and the accurate measurement of the temperature difference between the two drops. Thermistors are used to measure the difference in temperature. The descriptions of similar apparatus have been given.[73] A review of this method has been written by Simon and Tomlinson.[74] Instruments for determination of molecular weights based on this method are commercially available.[75]

Molar Refraction and Dispersion

The term molar refraction R is defined by the Lorentz-Lorentz equation[76]

$$R = [(n^2 - 1)/(n^2 + 2)]M/d$$

where n = refractive index, d = density, and M = molecular weight. The molar refraction can be considered (with certain restrictions) as a characteristic property of a given molecular structure. The value of R can be calculated also by adding the refractions due to atoms and atomic groupings and also to refractions due to the various types of linkages which connect the atoms and groups within the molecule. Table 5.07 gives a partial list of the values assigned to a number of atoms and atomic groupings for the sodium D lines (5890 and 5896 Å). By comparing the molar refraction calculated from atomic refractions and that obtained from measurements of the refractive index and density of pure samples of the substance, it is possible to check the structure of the compound.

Another application of the refractive index in analytical work is the measurement of the dispersion. The difference between the molar refraction of a substance measured at two wavelengths is called molar dispersion.

---

[72] C. Tomlinson, *Mikrochim. Acta*, **1961**, 457–466.

[73] A. Wilson, L. Bini, and R. Hofstader, *Anal. Chem.*, **33**, 135–137 (1961); V. J. Filipic J. A. Connelly, and C. L. Ogg, "A Study of the Thermistor Method for Molecular Weight Determination" in *Microchemical Techniques*, N. D. Cheronis, ed., Interscience, New York 1962, pp. 1039–1051; H. C. Ehrmantraut, "Use of a Vapor Pressure Osmometer in the Molecular Weight Determination on a Micro Scale" in *Microchemical Techniques*, N. D. Cheronis, ed., Interscience, New York, 1962, pp. 1063–1069; D. Wegmann, C. Tomlinson and W. Simon, "Thermoelectric Microdetermination of Molecular Weight, Part II. Routine Apparatus" in *Microchemical Techniques*, N. D. Cheronis, ed., Interscience, New York, 1962 pp. 1069–1085.

[74] W. Simon and C. Tomlinson, *Chemica.*, **14**, 301 (1960).

[75] Mechrolab, Mountain View, Calif.; Arthur H. Thomas Company, Philadelphia 5, Pa

[76] For detailed discussion of molar refractivity and dispersion, see N. Bauer, K. Fajans, and S. Z. Lewin, "Refractometry" in *Physical Methods of Organic Chemistry*, Part 2, 3rd ed A. Weissberger, ed., Interscience, New York, 1960, pp. 1139–1282.

Thus, the dispersion D may be defined as

$$D = n_2 - n_1$$

where n_1 and n_2 are indices of refraction measured at two different wavelengths as, for example, with the hydrogen alpha line, 6563 Å, and the

TABLE 5.07
ATOMIC REFRACTIONS USED IN THE CALCULATION OF MOLAR REFRACTIONS

Atom, Group, or Bond	R	Atom, Group, or Bond	R
C	2.413	CN	5.46
CH_2	4.618	Nitrogen in:	
C=C (bond)	1.733	RNH_2	2.32
C≡C (bond)	2.40	$ArNH_2$	3.21
H	1.100	R_2NH	2.49
O in OH	1.525	Ar_2NH	3.59
O in OR	1.643	R_3N	2.84
O in C=O	2.211	Ar_3N	4.36
Cl	5.967	$RCONH_2$	2.65
Br	8.865	NO_2 (group) in RNO_2	6.72
I	13.900	NO_2 (group) in $ArNO_2$	7.30
S in SH	7.69	Tricyclic structure	0.71
S in R_2S	7.97	Tetracyclic structure	0.48
S in RCNS	7.91		

Source. F. Eisenlohr, *Z. physik. Chem.*, **75**, 585 (1910); **79**, 129 (1912); C. P. Smyth, *Phil. Mag.*, **50**, 361, 715 (1925); K. Fajans and C. A. Knorr, *Ber.*, **59**, 249 (1926).

hydrogen beta line 4861 Å. For actual use, the term specific dispersion is defined as

$$S = [(n_2 - n_1)/d] \times 10^4$$

where d is the density measured at the same temperature as the indices. The factor 10^4 is employed to avoid small numbers. The value of D varies by a larger relative amount than the refractive index from closely related compounds in an homologous series. For example, inspection of the data in Table 5.08, which lists the densities, refractive indices, and dispersions, indicates that it is possible to differentiate between alkanes and cycloalkanes, on one hand, and aromatic hydrocarbons, on the other. The former have S values much below 200, while the latter have values well above 200.

An approximate method for the determination of dispersion is to employ the common refractometer provided with a calibrated ring, which turns the

TABLE 5.08
DENSITY, REFRACTIVE INDEX, AND DISPERSION OF SOME HYDROCARBONS*

Compound	Density (d^{20})	Index (n_D^{20})	Dispersion $(D^{20} \times 10^4)$	Specific Dispersion (S^{20})
Benzene	0.87910	1.50144	218.3	248.4
Toluene	0.86688	1.49682	209.3	241.4
Ethylbenzene	0.86713	1.49577	197.8	228.1
o-Xylene	0.87910	1.50449	206.4	234.8
m-Xylene	0.86410	1.49712	204.9	237.1
p-Xylene	0.86104	1.49575	205.1	238.2
n-Heptane	0.68090	1.38670	84.5	122.4
n-Hexane	0.6603	1.37563	65.1	98.6

* After R. T. Wendland, *J. Chem. Educ.*, **23**, 3 (1946).

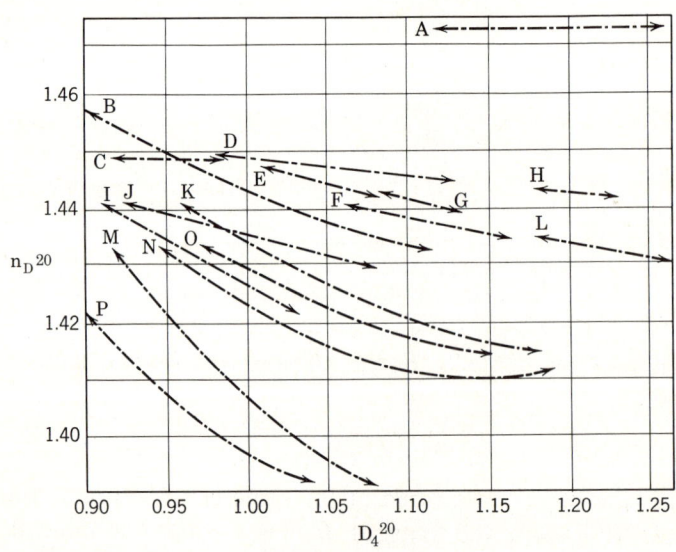

FIGURE 5.23

Densities and refractive indices of several classes of saturated aliphatic compounds.[77]
- (A) Triols
- (B) Diols
- (C) Diones
- (D) Ether diols
- (E) Keto diesters
- (F) Diether acids
- (G) Triesters
- (H) Alcohol diesters
- (I) Diether alcohols
- (J) Ketols
- (K) Ether-acids
- (L) Keto acids
- (M) Acid anhydrides
- (N) Keto esters and alcohol esters
- (O) Diesters
- (P) Ether esters

compensator and gives a reading from which the dispersion can be calculated by means of tables provided with the instrument. A more precise measurement is to use a refractometer, which has no compensator, and to determine the refractive index first with one source of monochromatic light and then with another.

Finally, the refractive index can be plotted against the density, as shown in Figure 5.23, to obtain information that will restrict the unknown into a particular group. By plotting these two values[77] for a large number of pure compounds, the graph can be divided into zones or areas, which include distinct groups of organic compounds. The zones may overlap so that it may be necessary to perform another test or to use an additional constant. However, such correlations are of great aid in the tentative identification of many compounds. See the original article by Gilmore et al.[77] which has nine charts similar to that shown in Figure 5.23.

REFERENCES

Melting Point

J. G. Aston, H. L. Fink, J. W. Tooke, and M. R. Cines, *Anal. Chem.*, **19**, 218–221 (1947).

J. B. Austin, *J. Am. Chem. Soc.*, **52**, 1049–1053 (1930).

H. J. Barber, D. P. Odell, and W. R. Wragg, *Chem. & Ind. (London)*, **1958**, 153. New design of melting point apparatus.

R. N. Chakravarti, K. N. Chaudhuri, and G. Datta, *J. Proc. Inst. Chemists (India)*, **29**, 267 (1957). Gas-heated micro melting-point apparatus.

N. D. Cheronis, "Micro and Semimicro Methods" in *Technique of Organic Chemistry*, Vol. 6, A. Weissberger, ed., Interscience, New York, 1954, pp. 145–188. There are 129 references.

G. J. Dean, *Vacuum*, **7 & 8**, 90–93 (1957–1958). Microscope hot stages. Slides coated with transparent thin layers of gold, can be operated at 350° by passing current through them.

L. M. Dennis and R. S. Shelton, *J. Am. Chem. Soc.*, **52**, 3128 (1930). Electrically heated blocks or bars.

E. Dowzard and M. Russo, *Ind. Eng. Chem., Anal. Ed.*, **15**, 219 (1943). Electrically heated liquid bath.

H. Euler and K. Guthmann, *Arch. Eisenhüttenw.*, **9**, 72 (1935). Errors due to the use of thermocouples.

C. W. Gibby and W. A. Waters, *J. Chem. Soc. (London)*, **1931**, 2151. Mixed melting points.

G. Gorbach, *Mikrochem. Acta*, **31**, 116 (1943). Electrically heated blocks or bars.

M. M. Graff, *Ind. Eng. Chem., Anal. Ed.*, **15**, 638 (1943). Liquid bath with infrared drying bulb.

---

[77] E. H. Gilmore, M. Menaul, and V. Schneider, *Anal. Chem.*, **22**, 892–896 (1950).

J. L. Hartwell, *Anal. Chem.*, **20**, 374 (1948). Thermal conducting media for melting point apparatus.

E. B. Hershberg, *Ind. Eng. Chem., Anal. Ed.*, **8**, 312 (1936). Hershberg modification of the Thiele tube; precise apparatus for melting point determinations.

E. J. Hewitt, *Chem. & Ind. (London)*, **1947**, 42. Micro melting-point apparatus.

H. Hilbck, *Mikrochemie*, **36/37**, 310 (1951). Electrically heated blocks or bars.

F. Hippenmeyer, *Mikrochemie*, **39**, 409 (1952). Electrically heated blocks or bars.

W. G. Jennings, *J. Chem. Educ.*, **34**, 95 (1957). Apparatus for the micro determination of melting points.

E. Kahame, *Mikrochemie*, **36/37**, 411 (1951). Use of mercury in heating baths.

C. C. Kiplinger, *J. Chem. Educ.*, **31**, 33 (1954). An inexpensive hot stage.

L. Kofler, *Ber.*, **B76**, 1096 (1943). Reference standards for eutectic melting points.

L. Kofler and A. Kofler, *Thermo-Mikro-Methoden*, Verlag Chemie, Weinheim, 1954. Detailed discussion of hot stage, melting points, and fusion techniques.

T. H. Liggett, *Proc. Iowa Acad. Sci.*, **37**, 241 (1930).

F. Lüdy-Tenger, *Mikrochemie*, **36/37**, 892 (1951). Electrically heated blocks or bars.

T. S. Ma and R. F. Schenck, *Mikrochemie*, **40**, 245–53 (1952). Use of heated metal blocks.

F. J. Maffei and R. Wasicky, *Anais assoc. quim. Brasil*, **7**, 111–114 (1948). Apparatus for micro and macro melting-point determination.

K. S. Markley, *Ind. Eng. Chem., Anal. Ed.*, **6**, 476 (1934). Modification of the Thiele tube.

W. C. McCrone, *Anal. Chem.*, **21**, 436 (1949); *Mikrochemie*, **38**, 476 (1951).

W. C. McCrone, *Fusion Methods in Chemical Microscopy*, Interscience, New York, 1956.

F. C. Merriam, *Ind. Eng. Chem., Anal. Ed.*, **20**, 1246 (1948). Modification of the Hershberg apparatus.

J. E. Nickels, *J. Chem. Educ.*, **28**, 303 (1951). An improved melting point apparatus.

W. S. Owen and W. M. Reid, *Mikrochim. Acta*, **1956**, 1373. Apparatus for the determination of melting points and boiling points.

A. G. Pinkus and P. G. Waldrop, *Mikrochim. Acta*, **1959**, 772. A simple method of obtaining melting points of compounds reacting with moisture or oxygen.

J. A. Ramsay, *J. Exptl. Biol.*, **26**, 57–64 (1949). Freezing-point determination for small quantities.

K. Rast, *Chem.-Ing.-Techn.*, **29**, 277 (1957). Determination of melting points to 0.1°.

E. L. Skau, J. C. Arthur, Jr., and H. Wakeham, "Determination of Melting and Freezing Temperatures," in *Physical Methods of Organic Chemistry*, Part 1, 3rd ed., A. Weissberger, ed., Interscience, 1959, pp. 287–355.

R. Steinmann, *Mikrochim. Acta*, **1953**, 490. Bibliography on melting points to 1949.

G. Talsky, *Chem. Ztg.*, **80**, 450 (1956). Melting point determination with the heating block. A review.

H. E. Ungnade, E. A. Brixner, and E. A. Igel, *Anal. Chem.*, **31**, 1432 (1959). Melting point apparatus for simultaneous observation of samples in transmitted and reflected light.

J. H. Walter, *J. Chem. Educ.*, **30**, 142 (1953). A useful melting point apparatus

G. F. Wright, *Can. J. Technol.*, **34**, 89 (1956). A useful modification of the Hershberg melting point apparatus.

Boiling Points

A. A. Benedetti-Pichler and F. Schneider, *Z. anal. Chem.*, **86**, 69 (1931). Method for less than one drop.

H. Böhme and R.-H. Böhm, *Mikrochim. Acta*, **1959**, 270–273. A micromethod for the determination of boiling points with direct temperature measurement.

W. E. Cervenansky, *Mikrochim. Acta*, **1963**, 412–415. Microdetermination of the boiling point.

N. D. Cheronis, "Micro and Semimicro Methods" in *Technique of Organic Chemistry*, Vol. 6, A. Weissberger, ed., Interscience, New York, 1954, pp. 188–194. Discussion of boiling-point techniques with 21 references.

F. Emich, *Monatsh.*, **38**, 219 (1917).

R. H. Ewell, J. M. Harrison, and L. Berg, *Ind. Eng. Chem.*, **36**, 871–875 (1944).

A. Furst and J. W. Bohner, *J. Chem. Educ.*, **22**, 531–532 (1945). Semimicro boiling-point tube.

C. R. Garcia, *Ind. Eng. Chem., Anal. Ed.*, **15**, 648 (1943). Modification of capillary tube method.

C. Karr and E. E. Childers, *Anal. Chem.*, **33**, 655–656 (1961). Improved micro boiling-point apparatus. Reproducibility of 0.5°.

C. Rosenblum, *Ind. Eng. Chem., Anal. Ed.*, **10**, 449 (1938). Microprocedures for use at different atmospheric pressures.

A. Smith and A. W. Menzies, *J. Am. Chem. Soc.*, **32**, 897 (1910). Capillary method.

W. Swietoslawski and J. R. Anderson, "Determination of Boiling and Condensation Temperatures" in *Physical Methods of Organic Chemistry*, 3rd ed., Part 1, A. Weissberger, ed., Interscience, New York, 1959, pp. 357–400.

Refractive Index

H. K. Alber and J. T. Bryant, *Ind. Eng. Chem., Anal. Ed.*, **12**, 305 (1940). Microrefractometric methods.

N. Bauer, K. Fajans, and S. Z. Levin, "Refractometry" in *Physical Methods of Organic Chemistry*, 3rd ed., Part 2, A. Weissberger, ed., Interscience, New York, 1960, pp. 1140–1281. There are 283 references.

S. G. Blohm, *Acta Chem. Scand.*, **4**, 1495 (1950); *Mikrochemie*, **36**, 321–329 (1951). Minimum amount of liquid required.

M. Brändstatter-Kuhnert and A. Martinek, *Mikrochim. Acta*, **1958**, 803–811. Kofler's glass powder method in the quantitative analysis of binary mixtures.

N. D. Cheronis, "Micro and Semimicro Methods," in *Technique of Organic Chemistry*, Vol. 6, A. Weissberger, ed., Interscience, New York, 1954, pp. 199–207. There are 22 references.

A. T. J. Dollar, *Mineralog. Mag.*, **28**, 438 (1948). Refractive index comparator for the microscope.

A. E. Edwards and C. E. Otto, *Ind. Eng. Chem., Anal. Ed.*, **12**, 459 (1940). Jelly-type microrefractometer.

H. A. Frediani, *Ind. Eng. Chem., Anal. Ed.*, **14**, 439 (1942). Methods for determining refractive indexes.

E. E. Jelley, *J. Roy. Microscop. Soc.*, **54,** 234 (1934). Jelley-type microrefractometer.

L. Kofler and A. Kofler, *Thermo-Mikro-Methoden*, Verlag Chemie, Weinheim-Bergstrasse, 1954. Use of glass powders.

R. Steinmann, *Mikrochim. Acta*, **1953,** 478–480. Bibliography to 1949.

G. Svensson, *Mikrochim. Acta*, **1956,** 645–650. Determination of absorption and index of refraction of extremely absorbent solutions on a microscale.

L. W. Tilton and J. K. Taylor, "Refractive Index Measurement," in *Physical Methods of Chemical Analysis*, 2nd ed., Vol. 1, W. G. Berl, ed., Academic Press, New York, 1960, pp. 412–462.

L. Turunen, *Anal. Chem.*, **33,** 1617–1621 (1961). New method of determination of refractive indices and study of continuous reactions.

C. L. Wilson, *Analyst*, **71,** 117–122 (1946). Microrefractometric methods.

Density

H. H. Anderson, *Anal. Chem.*, **20,** 1241–1243 (1948). Micropycnometers and micropipets.

N. Bauer and S. Z. Lewin, "Determination of Density" in *Physical Methods of Organic Chemistry*, 3rd ed., Part 1, A. Weissberger, ed., Interscience, New York, 1959, pp. 132–190. Review with 149 references.

M. Chambon, *Ann. chim. anal. chim. appl.*, **24,** 38 (1942); *Chem. Zbl.*, **1942, II,** 76; *C.A.*, **37,** 4273 (1943). Semimicro pycnometer.

B. J. Fontana and M. Calvin, *Ind. Eng. Chem., Anal. Ed.*, **14,** 185–186 (1942). Semimicro pycnometer.

H. Heller, *J. Physiol.*, **98,** 3–4 (1940). Specific gravity of small quantities of liquid.

A. A. Houghton, *Analyst*, **69,** 345–346 (1944). Micropycnometer.

A. R. Javes and C. Liddell, *Anal. Chem.*, **30,** 1570–1575 (1958). Micromethods for analysis of petroleum.

W. B. Jepson, *J. Sci. Instr.*, **36,** 319–320 (1959). Two techniques for measuring the density of a porous solid by displacement of a liquid.

R. J. Kokes, *Anal. Chem.*, **32,** 446–447 (1960). A simple helium densitometer.

J. G. Reynolds, *Chem. & Ind.* (*London*), **1947,** 176–177. Pycnometer for small volumes of liquid.

R. Steinmann, *Mikrochim. Acta*, **1953,** 477. Bibliography to 1949.

C. L. Wilson, *Metallurgia*, **33,** 157 (1946). Use of micropycnometer.

Optical Rotation

M. v. Bekesy, *Biochem. Z.*, **312,** 103 (1942). Micropolarization.

B. Carroll and I. Blei, *Science*, **142,** 200–208 (1963). New approaches to measurement of optical activity.

C. Djerassi, *Optical Rotatory Dispersion: Applications to Organic Chemistry*, McGraw-Hill, New York, 1960.

C. Djerassi, *Science*, **134,** 649–655 (1961). Application of optical rotatory dispersion to studies of configuration of steroids.

H. Kacser and A. R. Ubbelohde, *J. Soc. Chem. Ind.* (*London*), **68,** 135 (1949). Thermostatically controlled micropolarimeter tube.

A. P. Marion, *Ind. Eng. Chem., Anal. Ed.*, **12,** 777 (1940). Adapting polarizing microscope for use as a polarimeter.

D. Smith and S. A. Ehrhardt, *Ind. Eng. Chem., Anal. Ed.*, **18**, 81 (1946). A rapid-filling capillary polarimeter tube.

R. Steinmann, *Mikrochim. Acta*, **1953**, 473. Bibliography to 1949.

Molecular Weight

V. A. Aluise, *Ind. Eng. Chem., Anal. Ed.*, **13**, 365 (1941). Microdetermination of molecular weights.

H. H. Anderson and L. D. Shubin, *Anal. Chem.*, **29**, 852 (1957). An improved Dumas method.

R. Belcher and M. Sobotka, *J. Chem. Soc. (London)*, **1961**, 480. Micro and semimicro method.

G. Cauquil, *Compt. rend.*, **180**, 1207 (1925). Solvents for Rast procedure.

N. D. Cheronis, "Micro and Semimicro Methods," in *Technique of Organic Chemistry*, Vol. 6, A. Weissberger, ed., Interscience, New York, 1954, pp. 207–229. Discussion of determination of molecular weight with 77 references.

A. F. Colson, *Analyst*, **83**, 169 (1958). The ebullioscopic microdetermination of molecular weight. An improved form of the Menzies-Wright ebulliometer.

E. J. Cowles and M. T. Pike, *J. Chem. Educ.*, **40**, 422–423 (1963). Semimicro modification of the Rast method.

P. Csokan, *Magyar Chem. Folyoirat*, **48**, 56–61 (1942); *C.A.*, **38**, 2854 (1944). Microdetermination of molecular weights.

G. W. Dailey, R. B. Huff, J. Kang, L. D. Queen, and C. S. Patterson, *J. Chem. Educ.*, **38**, 28 (1961). Isopiestic vapor pressure apparatus.

M. Dimbat and F. H. Stross, *Anal. Chem.*, **29**, 1517–1520 (1957). Microebulliometer for determination of molecular weight.

H. C. Ermantraut, "Use of a Vapor Pressure Osmometer in the Molecular Weight Determination on a Micro Scale" in *Microchemical Techniques*, N. D. Cheronis, ed., Interscience, New York, 1962, pp. 1063–1069.

V. J. Filipic, J. A. Connelly, and C. L. Ogg, "A Study of the Thermistor Method for Molecular Weight Determination," in *Microchemical Techniques*, N. D. Cheronis, ed., Interscience, New York, 1962, pp. 1039–1051.

D. T. Gibson and J. Currie, *Mikrochim. Acta*, **1957**, 644–646. Molecular weight determination on a hot stage microscope.

C. A. Glover and R. R. Stanley, *Anal. Chem.*, **33**, 447–450 (1961). Ebulliometric apparatus for studying number-average molecular weights of polymers.

H. Gysel, W. Padowetz, and K. Hamberger, *Mikrochim. Acta*, **1960**, 192–196. The microdetermination of molecular weights of dissolved substances by vapor pressure measurements.

C. Heitler, *Analyst*, **83**, 223 (1958). An improved ebulliometer.

E. B. Hershberg, *Ind. Eng. Chem., Anal. Ed.*, **8**, 312 (1936).

H. Hoyer, *Angew. Chem.*, **73**, 465 (1961). Micromolecular weight determination by isothermal distillation. Apparatus.

H. Keller and H. v. Halban, *Helv. Chim. Acta*, **27**, 1439 (1944). Solvents for Rast procedure.

D. Kennedy, et al., *Anal. Chem.*, **31**, 1884 (1959). Ebullioscopic method.

J. A. Knight, B. Wilkens, Jr., D. K. Davis, and F. Sicilio, *Anal. Chim. Acta*, **25**,

317–321 (1961). Semimicro cryoscopic molecular weight determination with a thermistor thermometer.

L. Kofler and M. Brandstätter, *Mikrochemie*, **33**, 20 (1947). Microscopic determination.

N. Kornblum and R. J. Clutter, *J. Am. Chem. Soc.*, **76**, 4494 (1954).

J. J. Neumayer, *Anal. Chim. Acta*, **20**, 519 (1959). Determination of molecular weights using thermistors.

C. L. Ogg, *J. Assoc. Offic. Agr. Chemists*, **41**, 294 (1958). Microdetermination of molecular weights.

I. A. Pastak, *Bull. soc. chim.*, **39**, 82 (1926). Solvents for Rast procedure.

J. G. v. Pelt, *Phillips Tech. Rev.*, **20**, 357 (1958–1959). Thermistors in the ebullioscopic determination of molecular weight.

G. W. Perold and F. W. G. Schoning, *Mikrochim. Acta*, **1961**, 749–753. A Menzies-Wright ebulliometric apparatus using 0.3 ml of solvent.

J. Pirsch, *Ber.* **65**, 862, 1227, 1839 (1932); **66**, 349, 506, 815, 1694 (1933); **67**, 101, 115, 1303 (1934); **68**, 67 (1935); and *Z. angew. Chem.*, **51**, 73 (1938). Solvents for Rast procedure.

K. Rast, *Ber.*, **55**, 1051, 3727 (1922). Cryoscopic micromethod using capillary tube.

J. H. Robertson, *J. Chem. Educ.*, **37**, 152–153 (1960). Improved Victor Meyer vapor density apparatus.

M. Schnitzer and J. G. Desjardins, *Chemist Analyst*, **50**, 117 (1961). Cryoscopic molecular weight apparatus.

W. Simon and C. Tomlinson, *Chimica*, **14**, 301 (1960). Thermoelectric microdetermination of molecular weight. A review.

W. M. Smit, J. H. Ruyter, and H. F. van Wijk, *Anal. Chim. Acta*, **22**, 8 (1960). A new cryoscopic micromethod for the determination of molecular weights.

J. H. C. Smith and W. G. Young, *J. Biol. Chem.*, **75**, 289–298 (1927). Refinements of Rast microprocedure.

W. T. Smith, Jr., and R. L. Shriner, *Examination of New Organic Compounds*, Wiley, New York, 1956, pp. 67–70.

M. Sobotka, *Mikrochemie*, **32**, 49 (1944); **36/37**, 408 (1951). Problems in determinations of physical constants by micromethods.

R. Steinmann, *Mikrochim. Acta*, **1953**, 482. Bibliography to 1949.

F. Toffoli and M. Boccacci, *Ann. Chim. (Rome)*, **3**, 574 (1959). Further studies of a semimicro method for determining molecular weights.

C. Tomlinson, *Mikrochim. Acta*, **1961**, 457–466. Thermoelectric microdetermination of molecular weight.

S. Uchida and K. Shimoyama, *J. Soc. Chem. Ind., Japan*, **36**, 388 (1932). Solvents for Rast procedure.

D. Wegmann, C. Tomlinson, and W. Simon, "Thermoelectric Microdetermination of Molecular Weight, Part II. Routine Apparatus," in *Microchemical Techniques* N. D. Cheronis, ed., Interscience, New York, 1962, pp. 1069–1085.

G. Wendt, *Ber.*, **B75**, 425–429 (1942). Solvents for Rast procedure.

K. B. Wiberg, *Laboratory Technique in Organic Chemistry*, McGraw-Hill, New York, 1960, p. 92.

B. Wilkens, J. A. Knight, and F. Sicilio, *Rev. Sci. Instr.*, **32**, 355–356 (1961).

Determination of molecular weight with a small volume by lowering of the freezing point.

C. L. Wilson, *Analyst*, **73**, 585–596 (1948). Micromolecular weight methods.

H. N. Wilson and A. E. Heron, *J. Chem. Soc. Ind.* (*London*), **60**, 168–171 (1941). Cyclohexanol as a solvent for the Rast procedure.

A. Wilson, L. Beni, and R. Hofstader, *Anal. Chem.*, **33**, 135–137 (1961). Thermistor micromethod for molecular weight.

6

The Separation of Mixtures

An organic material that is presented for investigation or for identification may be (a) a relatively pure substance, (b) an impure organic compound (that is, one containing only small amounts of impurities), or (c) a mixture from which it is desired to identify each of the components. A pure organic substance is one that, on repeated fractionation, yields fractions that have essentially identical physical constants and identical chemical properties. The difference between an *impure compound* and a *mixture* is a matter of degree. In general practice, the term *mixture* is applied to materials in which no single component constitutes more than 90 per cent of the total. Furthermore, only one component is normally to be identified from an *impure* compound, whereas two or more purified compounds are expected to be recovered from a *mixture*.

An adequate preliminary examination will allow us to classify a material as being essentially pure, of "practical grade" purity, or as a mixture. If the sample is a solid, microscopic examination is helpful. A very thin layer of the sample is spread on a slide, and the crystal structure, refractivity, and color of the individual particles are noted. In this manner it is possible to determine homogeneity or nonhomogeneity in most cases. A polarizing microscope, or an ordinary microscope provided with a polarizing unit, may be used to examine the material. Differences in the appearance of some crystals may indicate that the sample is a mixture when, in fact, the substance is polymorphic.

In all cases, however, the melting point of the solid sample before and after a single crystallization gives reliable information. If both samples melt sharply at substantially the same temperature (within 0.5 to 1.0°), the unknown is a relatively pure substance. If the first sample softens and melts over a range of several degrees and the second sample melts within a range of 1 to 2°, the unknown contains a small amount of impurities. If both the initial and recrystallized sample melt over a wide range, the unknown is a mixture of two or more substances. The experienced analyst may find it advantageous to approach the matter of the purity of solids through fusion techniques, as described in Chapter 7.

If the sample is a liquid, the boiling range of about 2 to 5 ml is first determined (page 82) and two fractions are collected. If the boiling ranges of the two fractions are within 1 to 2° and the refractive indices (page 213) are not appreciably different, the unknown liquid is a relatively pure substance. If the liquid distills over a wide range and the refractive indices of the various fractions are appreciably different, the liquid is a mixture, and must be fractionated. If the liquid sample distills with rising temperature until about 10 per cent has distilled and then the temperature remains relatively constant, the receiver is changed and two fractions boiling within 1 to 2° range are collected. If the combined volume of the two fractions is about 80 per cent of the original and their refractive indices are substantially the same, the liquid sample is a commercial or impure grade of the compound.

In the case of very small samples, one of the chromatographic procedures, electrophoresis, or sublimation should be used to evaluate the probable purity of the substance under investigation.

MIXTURES

The purification of impure organic compounds and the separation of relatively simple mixtures were discussed in the fractionation procedures covered in Chapters 3 and 4 to which reference is made. The present chapter deals with systematic procedures that may be employed in the fractionation of a mixture of unknown composition so that its various components may be separated in essentially pure form, suitable for identification. Even small amounts of an impurity may cause an experimenter to reach false conclusions when the substance is being tested for elemental analysis, acid-base character, or chemical reactivity. Hence, the chapters which follow and which provide methods leading to detection and final identification of specific compounds presuppose that the substance being tested is essentially pure. Procedures for the separation of mixtures must, therefore, receive consideration here, even though the methods used to separate mixtures employ material that will be found in later chapters.

General Principles

The final separation of a mixture is based on differences in the physical properties of the components at the time of separation. The two physical properties that are most useful are vapor pressure (volatility) and solubility. Sufficient differences may exist among the components of a mixture with regard to one or the other of these properties to make possible the separation of the mixture directly by some form of distillation or extraction by inert solvents, or by fractional crystallization. However, *it is usually necessary to produce the requisite differences in physical properties of the components of the mixture by the use of one or more chemical agents* that change the chemical nature of one or more of the components, thus producing the necessary changes in physical properties (volatility or solubility). Chapters 8 and 9, which discuss the relationship of molecular structure and polarity to volatility and solubility, should be thoroughly considered at this time. The relationship of polarity and of other properties to selective adsorbability on solids should also be reviewed; the relationships are discussed in the section of Chapter 4 that deals with chromatography.

The more important methods for separating mixtures are the following ones.

1. Extraction by solvents that produce chemical change.
2. Extraction by solvents without chemical change.
3. Fractional crystallization.
4. Fractional distillation.
    (a) Direct distillation at atmospheric pressure.
    (b) Distillation under reduced pressure.
    (c) Steam-distillation.
    (d) Sublimation.
5. Chromatographic techniques and ion exchangers.
6. Various combinations of the above five methods.

Preliminary Tests for a General Mixture

If the mixture consists of more than one liquid phase, or of a solid phase in a liquid, these phases should be separated, and treated individually. In such mixtures, it is probable that the same compounds will exist in more than one of the phases. It is important that the preliminary tests be run on representative samples of the mixture, so that all the components will be tested. A complete record should be kept of all the tests made, including the results of the tests and the deductions that are made. It is wise, also, to keep a record of classes that are eliminated as possibilities because of the results of the tests. All samples that are to be set aside for later examination or use should be adequately labeled.

The following tests should be made, together with any others that the nature or behavior of the mixture may suggest.

Composition

The tests suggested in Chapter 7 are recommended. Care should be taken, in making the analysis for the elements, to insure that all components of the mixture are present in the sample taken for the fusion with sodium (test for water; dehydrate before fusing with sodium). In the preparation of the scheme for separating the mixture, it is extremely important to know what elements are present.

Solubility

Solubility determinations should be made on well-mixed samples, utilizing all of the solubility classification solvents, as discussed in Chapter 8. It should be recalled that one or more of the components of the mixture may dissolve in any one solvent, and also that the same compound may partially dissolve in more than one solvent. The solubility in ether should be determined even if the material is not soluble in water. Other solvents that may be used to advantage include methanol, ethanol, carbon tetrachloride, and chloroform. In cases where it is difficult to determine whether or not the solvent has dissolved appreciable amounts of the mixture, the solvent should be separated from the residue and distilled. Exceptions to this technique would be solutions in sodium hydroxide or sulfuric acid. An alkaline extract should be tested by acidifying it and extracting with ether. In the case of concentrated sulfuric acid as a solvent, some classes of compounds that dissolve in it may be recovered by pouring the acid onto an excess of cracked ice.

In evaluating the data from the solubility tests, it is essential to remember the elements that were found in the mixture.

Acid-Base Character

The mixture should be examined with the indicators used in Chapter 9. The information thus obtained is helpful but not completely reliable as regards possible components of the mixture. The use of the indicator solutions on various fractions as separated is also recommended.

Distillation

Chapter 3 should be consulted for procedures for fractional distillation of small quantities.

Any evidence of thermal decomposition of any of the compounds during the distillation should be noted and, if decomposition is evident, the distillation should be abandoned. In other cases, it should be observed whether a solid residue remains after distillation is complete; if such a residue exists, it should be steam-distilled. The various fractions obtained by distillation should be examined. Distillation under reduced pressure should be considered if normal distillation is unsatisfactory.

Solid Mixtures

In the case of solid mixtures that did not appear to be separable by cold solvents, hot solvents should be used in an attempt to separate the mixture by fractional crystallization. Steam distillation may often be used to advantage on solids.

Chemical Tests

Selected tests from Chapter 10 should be applied to the original mixture, or to fractions that have been separated from it by the preliminary testing methods. Every ascertainable fact about the presence or absence of the various chemical classes in the mixture will aid in devising a scheme of separation that will have maximum effectiveness with the minimum number of operations.

A brief discussion of ways to separate some binary mixtures will illustrate the application of general principles to methods of separation.

Example 6.01: Salicylic Acid and Malonic Acid

Although both salicylic acid and malonic acid are polar and both are soluble in water, malonic acid is not appreciably soluble in ether because of its two carboxyl groups (highly polar). Therefore, the salicylic acid may be extracted from the malonic acid by ether.

Example 6.02: 2-Nitrophenol and 4-Nitrophenol

The solubilities of 2- and 4-nitrophenol are too similar to allow these isomers to be effectively separated by extraction or fractional crystallization. Moreover, any chemical agent that is added to react with one of them will react with both. However, since 2-nitrophenol is *chelated* (therefore exists as a monomer), whereas the 4-nitrophenol is intermolecularly associated, the *ortho* isomer can be steam-distilled while the *para* isomer can not. Both of these phenols are sufficiently soluble in water that the loss of material is considerable, if we merely separate the phenols from the two aqueous mixtures (distillate and residue) by filtration. This loss may be largely avoided by making each of the aqueous filtrates alkaline with sodium carbonate and then evaporating the solutions nearly to dryness, acidifying with 50 per cent sulfuric acid, and extracting the recovered phenols with benzene. The phenols may then be separated from the benzene by distillation.

Feigl[1] has pointed out that these isomeric phenols may be separated by careful sublimation at 120°.

Example 6.03: A Mixture of Aniline and Nitrobenzene

The boiling points of aniline and nitrobenzene are too close together to make fractional distillation a practical method of separating these compounds in pure enough condition to get reliable physical constants for each component. Neither compound is very soluble in water, and both are soluble in ether. The addition of dilute hydrochloric acid to the mixture will not affect the nitrobenzene but it will convert the aniline into a highly polar salt (anilinium chloride, which is soluble in water but not in ether). The nitrobenzene may then be separated from the aqueous layer, washed with dilute hydrochloric acid to get out any remaining aniline, washed with water and very dilute sodium hydroxide to remove traces of the hydrochloric acid, dried with an anhydrous salt, and finally distilled. The aniline may be recovered from the solution of its hydrochloride salt by making the solution distinctly alkaline, extracting the aniline with ether, and separating the ether-aniline mixture by fractional distillation.

The separation of a neutral compound from a base is not always successfully accomplished by extracting the neutral compound from an acidified solution of the base. For example, a mixture of acetanilide and nitroaniline will yield the base in sufficiently pure form for identification, but repeated extractions fail to yield pure acetanilide.

Example 6.04: Benzene and Cyclohexylamine

Benzene and cyclohexylamine could be separated by the same method as used in the preceding example. However, since the boiling points of these two compounds are more than 50° apart, it would be possible to separate them by direct distillation using a good fractionating column.

---

[1] F. Feigl, *Spot Tests in Organic Chemistry*, 6th ed., Elsevier, Amsterdam, 1960, p. 412.

Example 6.05: Benzaldehyde and Benzoic Acid

A dilute alkaline solution will convert benzoic acid into a salt and the benzaldehyde may then be separated by extraction with ether or by steam distillation. In the case of steam distillation the acid could be recovered by making the aqueous solution acidic with sulfuric or phosphoric acid and then extracting with ether. The mixture cannot be separated by steam distillation directly because both compounds will steam-distill.

Example 6.06: Separation of Water-Insoluble Alkylamines from Arylamines

Most alkylamines will form water-soluble salts in an aqueous mixture at pH 5.5. Arylamines are not protonated at this pH and, hence, may be separated from the mixture by extraction.

Example 6.07: Separation of Very Weak Bases from Neutral Compounds

Concentrated sulfuric acid is capable of protonating most ethers (weak Lewis bases); hence, for example, a mixture of butyl ether and chlorobutane may be separated by treatment with concentrated sulfuric acid at room temperature. The chlorobutane fails to dissolve. The ether may be recovered by pouring the acid solution onto crushed ice.

Example 6.08: Separation of Acids Based on their Relative Acidities

By extracting aqueous solutions or suspensions of acidic substances at various pH levels, the substances may be separated into weak, intermediate, and relatively strong acids. Extraction by ether or benzene at decreasing pH levels of about 8, 4, and 1 is recommended.

Summary

All pertinent factors should be considered before these various methods are applied to the separation of mixtures. For emphasis, some of the more important factors are listed below.

1. Methods that are quite satisfactory when large quantities of the mixture are available may not be suitable for semimicro quantities.
2. Safety factors should be considered. In particular, in working with ether (or benzene), extreme care must be used to avoid igniting it with nearby open flames or exposed, hot wires (cone-type electric heaters).
3. Some types of compounds, such as low–molecular-weight acid chlorides and anhydrides, are hydrolyzed by water. It is particularly important to remember this fact when carrying out steam distillations.

The Separation of Mixtures   251

4. Several classes of compounds are hydrolyzed in hot alkaline solutions or hot acidic solutions. If steam distillation is carried out while a mixture that contains such compounds is definitely alkaline or acidic, hydrolysis may occur, at least to a sufficient extent to complicate the separation of the mixture and the later identification of the original components.

5. Any method that involves the introduction of considerable quantities of water into the mixture is to be avoided if possible, especially if it is believed that a water-soluble component is present because it may prove very difficult to recover the compound from the water (for example, sugars, glycols, amino acids, sulfonic acids, low-molecular-weight alcohols, carbonyls, and amines and, in general, compounds that form azeotropic mixtures with water).

6. Attempts to use fractional distillation with high-boiling mixtures (except under reduced pressure) may cause (a) thermal decomposition of some components, (b) a chemical reaction between components that would not appreciably react at normal temperatures, or (c) oxidation of one or more of the components.

7. Separation of liquids by fractional distillation is limited by the fact that a great number of liquid mixtures form azeotropic solutions.

8. Extraction of a substance from water to ether or from ether to water is not as complete as would be predicted from the solubilities of the substance in each pure solvent. This is due to the fact that ether and water are moderately soluble in each other and the two phases are a saturated solution of ether in water and a saturated solution of water in ether, respectively.

A General Procedure for the Separation of Mixtures

The following general scheme may be applied to the separation of a mixture into several fractions, many of which correspond to the usual Solubility Classes of Chapter 8. The scheme is offered with the belief that it will serve as a general guide, and not with the expectation that it is applicable to all problems of separation. The scheme should be modified in accordance with the observed facts for any one mixture. The later sections of this chapter take up the problems of separating mixtures that are present within single fractions as separated by this scheme.

The procedures outlined in the following eight steps are summarized in the flow sheet on page 256. *The Roman numerals in the flow sheet refer to the numbers of the steps.*

If the mixture is a solid, or if the preliminary distillation test showed that there was no distillate below 100°, Step I should be omitted. If the preliminary distillation showed the presence of a low-boiling amine, the distillate in Step I should be absorbed in $3N$ HCl.

Step I. Place 3 to 10 ml of the liquid mixture in a 25-ml distilling flask or distilling tube. Using a well-cooled receiver, distill the mixture to remove all the components that distill below 100°. The distillate may contain low-molecular-weight members of practically all the nonaromatic classes of compounds. Generally speaking, the molecules will contain 5 or fewer carbon atoms. A few saturated cyclic hydrocarbons, a few heterocyclic compounds, and benzene also boil below 100°. Most of these volatile compounds are soluble in both water and ether but the hydrocarbons and their halogen derivatives are not soluble in water. By noting the boiling range of the distillate, a good estimate may be made as to whether or not the distillate is a mixture. If the distillate is a mixture, chemical separation and solvent extraction will be possible in a few cases but very careful fractional distillation will be required for most of these mixtures. Test the distillate for the elements and make appropriate classification tests.

It should be noted that chemical reactions may occur between components of the original mixture during this period of heating, even if the compounds did not react in the cold mixture. For example, if a mixture of aniline hydrochloride, sodium benzoate, and ethanol is heated to distill the ethanol, ethyl benzoate is formed in good yield. Examination of a sample of the original mixture in comparison with the final results of the analysis will usually detect any such change in composition of the mixture during its separation.

Failure to obtain a distillate from a mixture heated to 100° does not *prove* the absence of some lower boiling components, considering the facts of the vapor pressures of solutions. Furthermore, there is the possibility of low-boiling or high-boiling azeotropic mixtures causing confusion.

Test the residue from the distillation for water. If it is present, it must be removed before proceeding to the next step.

Step II. The residue from Step I should be shaken with ether, using 5 ml of ether for each gram of the mixture. Allow the ether to remain in contact with the mixture for 3 minutes (shake occasionally). Treat any undissolved residue by Step III and save the ether solution for Step V.

Step III. Warm the ether-insoluble residue to drive off the ether. Add 5 ml of water for each gram of residue and shake the mixture vigorously. Remove the aqueous solution. Again extract the residue with water, using 10 ml of water for each gram of residue. The water will remove Solubility

Division S_2 compounds (see page 324). The two aqueous solutions may be combined, or they may be examined separately. Owing to marked differences in solubility among various components in this group, it is entirely possible that the two aqueous solutions represent a fair separation of Division S_2 compounds. Examine the aqueous solution by evaporating the water out of a 5-ml sample. If the residue is extremely small, the Division S_2 compounds are not represented in the mixture. If a residue exists after evaporating the water, test other samples of the aqueous solution for acidity, for carbohydrates, and for other likely types of compounds of the Division.

If carbohydrates are present, the water may be removed by vacuum distillation or by azeotropic distillation. The compound or compounds introduced for such purpose should be soluble in ether so that any of these liquids remaining after all the water has been distilled may be removed by ether extraction. Water-insoluble acids that are present in the original mixture as their soluble salts may be separated from the aqueous solution by making the solutions acidic with mineral acids and then distilling or extracting with ether. The salts of amines may be decomposed by sodium hydroxide and the amines removed by distillation or ether extraction.

Step IV. The ether-insoluble, water-insoluble residue from Step III should be shaken with a volume of cool methanol equal to 5 times the weight of the residue. The alcoholic solution may be separated from any insoluble residue by filtration or decantation. The alcohol should then be distilled. Thus, an alcohol-soluble and an alcohol-insoluble fraction may be obtained. Examine these fractions for homogeneity. If either fraction appears to be a mixture, extract such a mixture with $1.2N$ hydrochloric acid and with $2.5N$ sodium hydroxide in an attempt to separate the components. If these extractions fail, fractional crystallization from various solvents should be tried.

Unfortunately, exact and complete data are lacking on the solubilities of most organic compounds in various solvents, including the common solvents. The attempt to use solvents in the separation of mixtures is further complicated by the fact that in many cases isomers of the same compound do not have similar solubilities. However, incomplete lists of some of the types of compounds that may be expected in the two fractions resulting from the methanol extraction are given below.

Some compounds insoluble in alcohol, ether, and water:
    Many dinitro derivatives of the aromatic hydrocarbons and their amino, hydroxy, and acid derivatives.
    Many trinitro compounds of the above types.

Several dihalo derivatives of anthracene.
Several amino-substituted sulfonic acids; a few amides and imides.
Benzyl and benzoyl ureas; several derivatives of anthraquinone.

Some compounds soluble in alcohol, but insoluble in ether and in water:
Some dibromo- and dinitrobenzoic acids and a few other aromatic acids.
Several polyhydroxy- and polyaminoquinones and quinolines.
A few aminophenols; a few amides and anilides; a very few amines.

Step V. Pour the ether solution from Step II into a distilling flask or distilling tube the capacity of which is twice the volume of the ether and distill the ether. Cool the residue and then extract it twice with water, using 3 ml of water per gram of residue for the first extraction and 7 ml of water per gram of residue for the second extraction. These aqueous solutions will contain the Solubility Division S_1 compounds (see page 324). Examine these two aqueous solutions separately, since many compounds of Division S_1 are highly soluble in water, whereas others are only moderately soluble.

Since many compounds that are slightly soluble in water do not belong to Division S_1, the aqueous extract may be given some color or odor by such compounds. Extract the aqueous component with 5 ml of ether and discard the ether. To determine whether or not the water has removed a Division S_1 fraction, test the solution with litmus. Also distill a 5-ml portion, noting the boiling range, the properties of the distillate, and the residue. If it is concluded that one or more components have been removed by the water, saturate the aqueous solution with potassium carbonate. Any acids originally present will be converted to salts and most of the other compounds will separate from the salt solution. Shake the solution with half its volume of ether. Separate the ether layer and distill the ether. The residue will be the Division S_1 compounds, with the exception of the acids. When acids have been detected in the aqueous extract by the litmus test, the potassium carbonate solution should be neutralized with dilute sulfuric acid to the yellow end-point of bromothymol blue, and the solution extracted with half its volume of ether. It is best to remove drops of the solution and add them to bromothymol blue test-paper, rather than add the indicator to the solution. The ether extraction will remove most of the phenols or amides that were present in the aqueous solution. The aqueous solution should now be made definitely acidic with dilute sulfuric acid and distilled to remove the volatile acids. If some acid is precipitated in the water when it is acidified, it may be removed by ether extraction.

Step VI. The residue that was insoluble in water at the beginning of Step V should be dissolved in ether. (If nitrogen was absent in the residue, omit the remainder of this step and proceed to the next one.) Place the ether solution in a separatory funnel and shake it thoroughly with one fourth its volume of 0.6N hydrochloric acid. Separate the two layers. Again extract the ether with one half its volume of 1.2N hydrochloric acid. The acid will remove Solubility Division **B** compounds. The two acidic solutions should be examined separately on the chance that some separation of the amine components may have been accomplished. Make the solutions slightly alkaline with 1N sodium hydroxide. Extract them twice with ether and combine the ether solutions. Dry the ether solution with anhydrous sodium carbonate and distill the ether to obtain the Division **B** fraction.

It should be recalled that many amines are not extractable by dilute acids and will be found in the Division **M** fraction.

Step VII. Shake the ether solution remaining after the Division **B** compounds have been removed in Step VI with half its volume of 2.5N sodium hydroxide solution. Extract the ether again with half as much 2.5N sodium hydroxide solution. Combine the two alkaline extracts and warm the mixture to drive off the dissolved ether. Neutralize the alkaline solution to the yellow end point of bromothymol blue by adding dilute hydrochloric acid dropwise, while vigorously stirring the solution. To test the solution for the proper pH, remove a drop of the mixture from time to time and apply it to a strip of bromothymol blue test-paper (if the indicator is added directly to the solution, it will be extracted by ether and cause confusing colorations). Now extract the aqueous solution twice with ether to remove the Solubility Division A_2 compounds. Dry the ether with anhydrous sodium sulfate. Decant the ether and distill it, leaving the Division A_2 compounds as the residue. Not all the phenols will be extracted at the pH used and the later acid fraction should be tested for phenols.

The aqueous solution from which most of the phenols and other weakly acidic compounds have been extracted should be further acidified to the end point of methyl orange. Ether extraction at this pH will remove most of the negatively substituted phenols and the intermediate acids. This fraction represents the less acidic members of Division A_1 compounds.

Concentrate the aqueous solution by evaporation to about half its original volume. Cool the solution and acidify it to the red end point of thymol blue. Extract the solution with ether to get out the most acidic compounds of Division A_1.

Step VIII. The ether solution from which the acids have been removed should be washed twice with 5-ml portions of water to remove any remaining sodium hydroxide. Dry the ether with an anhydrous salt and decant

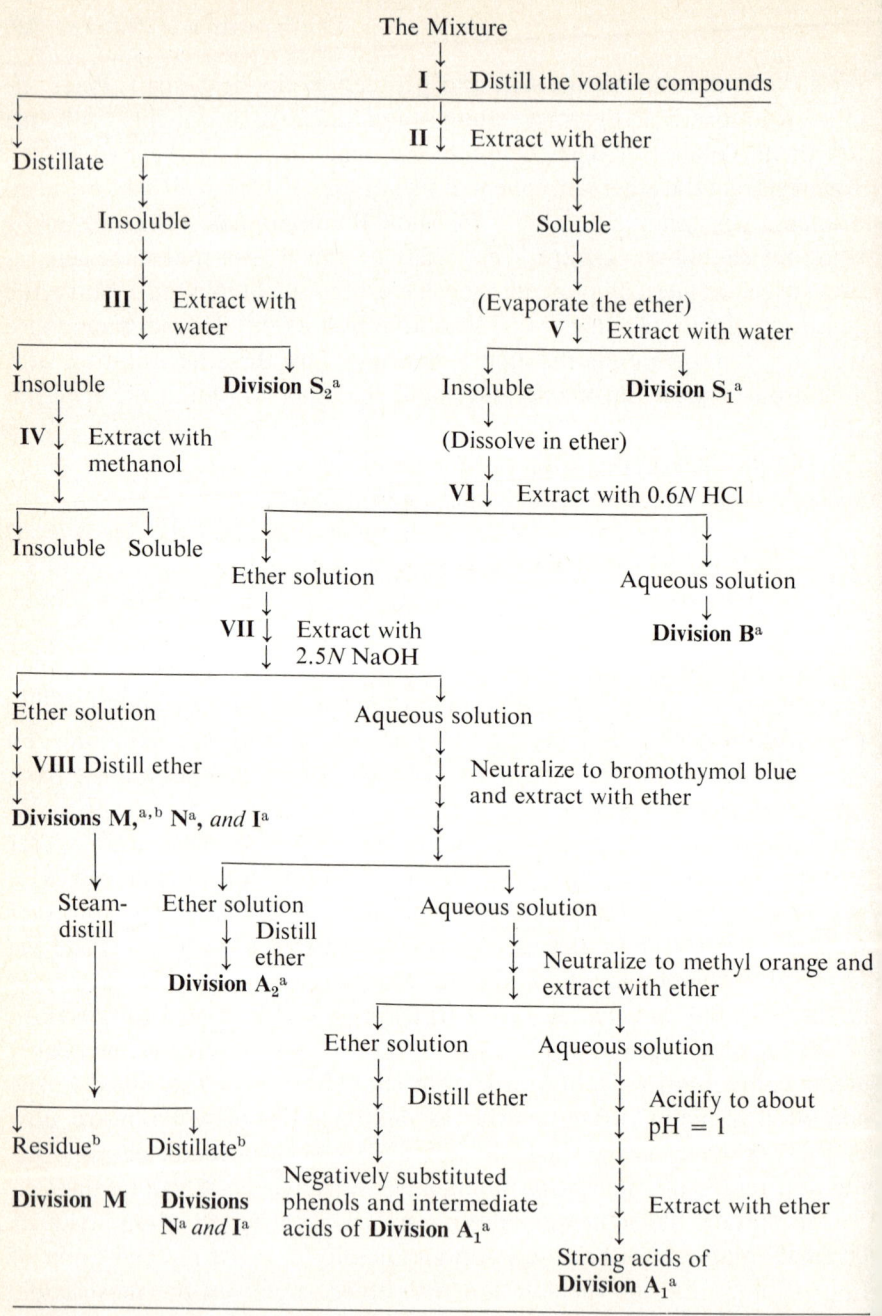

[a] Refer to the Divisional Solubility Classifications for the classes and subclasses of compounds that may be present (pages 324–325).
[b] Division **M** compounds are not present unless nitrogen and/or sulfur was found present on elemental analysis. Distilling with steam does not give a "clean-cut" separation in these cases, since some Division **M** compounds do distill with the steam, and because some Division **N** and **I** compounds do not distill, appreciably, with the steam.

and distill it. The residue will contain compounds that are in Solubility Divisions **M**, **N**, and **I**. Suggestions for further separation of this mixture are given on pages 259–260.

Suggestions for Separating Intraclass Mixtures

Assuming that the mixture was treated by the scheme suggested, it has been separated into a maximum of ten fractions. It is improbable, however, that any one mixture will contain compounds that would separate in all ten of these fractions. Because of overlapping solubilities, it is entirely possible that some of the compounds have been partially separated in two or more fractions. This fact should be kept in mind when the individual fractions are purified, and tests are being performed on them.

In connection with attempts at purification of the individual fractions, it may be discovered that the fraction represents a mixture of two or more compounds, not counting the impurities due to imperfect separation. No simple set of directions can be given for the separation of such intraclass mixtures. The usual methods of distillation, fractional extraction, and fractional precipitation are often useful. Hot solvents, the less commonly used solvents, and mixed solvents should also be tried. Occasionally, resort may have to be made to chemical reactions that will make separation possible. Benzene may be separated from cyclohexane by nitrating or sulfonating the benzene. A mixture of an ester and an ether that cannot be fractionated may be separated by saponifying the ester.

Mixtures of Division S_2 Compounds

Aqueous mixtures of Division S_2 compounds should be tested for carbohydrates, amine salts, metallic salts, and ammonium salts. If carbohydrates are absent, such mixtures may be distilled to remove the water but, if carbohydrates are present, it is best to distill under reduced pressure. If amine salts are present, make the solution alkaline. The method to be used for the recovery of the free amine will depend on the characteristic of the amine. Volatile amines should be distilled into a dilute solution of hydrochloric acid. Less volatile amines may often be steam-distilled. In many cases, the amine may be recovered from the alkaline solution by extraction. The salts of acids could, of course, be converted to free acids by adding a mineral acid. The acids thus liberated may or may not be extractable by ether, or be capable of being steam-distilled. In general, molecules having two or more polar groups cannot be steam-distilled.

Hot alcohol is a convenient solvent for separating mixtures of Division S_2 compounds after the water has been removed from the mixture. Sugars do not dissolve in the hot alcohol (levulose is reasonably soluble). Most of the carboxylic acids will dissolve in hot alcohol but will crystallize out on cooling. Many of the other compounds of this class remain in solution in the alcohol and may be recovered by distilling the alcohol.

The hydrogen atom in chloroform is an *acceptor* in hydrogen-bonding. Hence, compounds having functional groups that act as *donors* will dissolve in chloroform, even if they do not dissolve in carbon tetrachloride. Chloroform will extract some types of compounds from nonaqueous mixtures of this class.

Mixtures of Division S_1 Compounds

If the aqueous mixture of Division S_1 compounds is either acidic or basic, neutralize the solution. Steam distillation will separate the volatile components from the salts, the polyhydroxy phenols, and other nonvolatile compounds. The nonvolatile residue may often be separated by fractional crystallization from hot water. Ether and chloroform are good solvents for extracting the residue after the water has been removed.

The volatile compounds, which would be present in the distillate, will include the alcohols, esters, aldehydes, and ketones. If a test for aldehydes and ketones is positive, these classes may be separated from the alcohols and esters by conversion to the sodium bisulfite complexes or to the phenylhydrazones. The alcohols and esters may often be separated by fractional distillation. Another method is to "salt out" the alcohols and esters by saturating the solution with potassium carbonate, separating the alcohol-ester fraction by means of a separatory funnel or pipette, and then adding a few grams of calcium chloride to the alcohol-ester fraction. After a few minutes, add just enough water to dissolve the salt. The alcohol will remain in solution with the calcium chloride, whereas the ester will separate. To recover the alcohol, saturate the salt solution with sodium sulfate and extract with ether.

Mixtures of Division **B** Compounds

Many, but not all, of the amines of this class are volatile with steam. Hence, steam distillation is sometimes helpful in separating such mixtures. Fractional crystallization and, less often, fractional distillation may be used. Of course, benzenesulfonyl chloride or *p*-toluenesulfonyl chloride will react with primary and secondary amines, but not with tertiary amines.

It is occasionally advisable to treat the amine mixture with one of these reagents and then extract the tertiary amine with 1.2N hydrochloric acid. The derivatives of the primary and secondary amines may be separated by the difference in solubility of their substituted amides in alkaline solution.

Aromatic amines may be separated from many impurities by converting them into picrates in alcohol solution. The amines may be regenerated from the picrates by treatment with ammonia.

Mixtures of Divisions **M**, **N**, *and* **I** *Compounds*

The scheme of separating mixtures proposed in this chapter places, in one residual group, all of the compounds that are soluble in ether but insoluble in water and were not extracted by hydrochloric acid or sodium hydroxide. There is, therefore, considerable probability that this residue will be a mixture of two or more compounds. If neither nitrogen nor sulfur are present in this residue, the Division **M** compounds are absent. *In general*, the Division **N** and Division **I** compounds are volatile with steam, whereas only a few Division **M** compounds are volatile with steam. Hence, steam distillation will *usually* separate the Division **M** compounds from the other two classes.

Although sulfuric acid and phosphoric acid are not usually satisfactory for the separation of mixtures, the use of these acids is recommended for small samples to help in determining what types of compounds are present.

Division **I** compounds are insoluble in concentrated sulfuric acid. Of the compounds that dissolve in sulfuric acid, only the lower-molecular weight ones will dissolve in 85 per cent phosphoric acid. It should be recalled that the members of these classes were removed by water extraction if they did not contain more than 4 to 5 carbon atoms per molecule. The phosphoric acid will dissolve members of these classes if they do not contain more than 8 to 9 carbon atoms per molecule.

Mixtures of Division **M** compounds may be best separated by fractional extraction or fractional crystallization. Mixed solvents are frequently useful. Several of the more common members of this class may be extracted by hot water, from which they will separate when the solution is cooled. Carbon tetrachloride will dissolve many of the compounds of Division **M** but it fails to dissolve many of the dinitro and polynitro compounds, anilides, amides, sulfones, and other compounds of similar structure. Chloroform will dissolve most of those compounds that are insoluble in carbon tetrachloride, especially if they contain active *donor* groups for hydrogen-bonding. Chloroform is not a good solvent for the sulfonamides. Methanol is useful in fractionating the mixture that is insoluble in carbon tetrachloride; it dissolves the anilides and amides but not the nitro compounds or sulfones. As previously mentioned, some few of the Division **M** compounds can be steam-distilled.

The mixtures of Division **N** and Division **I** compounds may frequently be fractionally distilled, either at atmospheric pressure or under vacuum. If aldehydes or ketones are present, the mixture may be dissolved in ether and extracted with a saturated solution of sodium bisulfite. For solid mixtures of these divisions, fractional crystallization from hot solvents is often the best method. Aromatic hydrocarbons may usually be separated from nonaromatic hydrocarbons by chromatographic methods using a silica gel column and eluting with alcohol.

Alternate Methods for the Separation of Mixtures

It cannot be emphasized too strongly that no one schematic procedure for the separation of mixtures is equally applicable to all types of mixtures. Two of the many possible schemes are outlined below. Details are omitted since, in most cases, the separation of the compound or compounds from each fraction obtained by these methods may be accomplished by regular methods or by methods suggested in the general procedure and discussed in more detail in the preceding sections.

Solubility classification may be related to *probable* volatility with steam as follows. Most Solubility Division S_1 compounds are volatile with steam whereas Division S_2 compounds are not; most Division A_1 and A_2 compounds are not volatile, but there are several exceptions; many Division **B** compounds are volatile; some Division **M** compounds are volatile; most of the Division **N** and **I** compounds are volatile with steam. When steam distillation is used, the matter of possible hydrolysis of compounds must not be overlooked.

The classification of a given mixture as water-soluble or water-insoluble presents difficulties. Some of the components of the mixture may be very soluble in water and the others relatively insoluble. In such a case, the mixture could be separated by water into two mixtures, one of which would be treated as a water-soluble mixture and the other as a water-insoluble mixture. More often, however, various components of the mixture partially dissolve in water but fail to be completely dissolved in the quantity of water that may be used. In that case, the aqueous solution may be considered as a water-soluble unknown mixture and the undissolved material approached as a water-insoluble unknown. Note that the presence of water-soluble organic solvents in the mixture may take into solution in the aqueous phase many compounds that would not dissolve in water alone. The fact that a given mixture completely dissolves in water does not necessarily prove that there are no water-insoluble substances present.

The outline on page 262 suggests one way to separate a water-soluble mixture. Solubility Division **M** compounds are usually not very soluble in aqueous mixtures; however, if such compounds should be present they are generally not volatile with steam and hence would be left in the residue with Division S_2 compounds. Division **I** compounds would not be present.

The outline on page 263 represents one way to separate a mixture that is not soluble in water. Benzene could be substituted for the ether as a solvent in this procedure.

THE USE OF OTHER METHODS OF SEPARATION

The methods for the systematic separation of mixtures presented thus far in this chapter represent what may be called *classical* methods. In general, these methods require a few hundred milligrams of a solid or a milliliter or more of a liquid as starting material. It is assumed that the separated components will be in sufficient quantity to allow the determination of certain physical constants to make the necessary tests for the elements present and for the chemical classification and, finally, to allow the conversion of each component to a solid derivative. When the quantity of substance available is adequate, this approach to the separation of a mixture has many advantages, particularly when the mixture contains several chemical types and when the nature of these chemical types is unknown or unsuspected by the analyst. Only simple laboratory equipment and chemicals are required for these methods. Very extensive data on the physical constants of compounds and on derivatives of these compounds are available for use in identifying the components that have been separated.

However, the methods previously described do have disadvantages — they require relatively large samples and they take considerable time. The methods of fractionation discussed in Chapter 4 and other instrumental methods have been developed to overcome one or both of the disadvantages of the classical methods. Most of the new methods allow the successful use of very small samples; hence, they may be used in situations where the classical methods are completely useless. In many cases, the time required is only a small fraction of that required by the older method. However, these techniques also have their disadvantages. The cost of the equipment used for each of these methods may well be from 10 to 100

262   Techniques of Organic Analysis

ALTERNATE METHOD I
MIXTURES OF WATER-SOLUBLE COMPOUNDS

Mixture (or aqueous solution)
↓
Distill to near dryness[a]
↓

Residue I
(Division S_2)

Distillate I
(Divisions S_1,[b])
↓
(a) Acidify with H_3PO_4
(b) Distill to near dryness

Residue II
(Amines as phosphates)
↓
(a) Make alkaline (NaOH)
(b) Distill
(c) Collect distillate in HCl

—Residue = Na_3PO_4

Distillate III
Evaporate to near dryness
↓
Amine hydrochlorides

Distillate II
(Neutral and acidic compounds)
↓
(a) Make alkaline
(b) Distill to near dryness

Residue III
(Salts of acids)
↓
(a) Evaporate to small volume
(b) Saturate with CO_2
(c) Extract with ether
↓

Aqueous solution   Ether solution
of salts   (Phenols)
↓
(a) Acidify with H_2SO_4
(b) Distill or extract
    with ether
↓
(Acids)

Distillate IV
(Neutral compounds)
↓
(a) Saturate with K_2CO_3 and separate any second liquid phase.
(b) Distill 20–30% of volume of salt solution
(Separate phase or distillate: water soluble materials)

---

[a] Interrupt the distillation at 130 to 140° (unless the mixture has solidified, already), change receivers, and introduce steam into the distilling flask to see if any components can be distilled by steam.

[b] Division S_1 contains neutral, acidic, and basic compounds.

The Separation of Mixtures

ALTERNATE METHOD II
MIXTURES OF WATER-INSOLUBLE COMPOUNDS

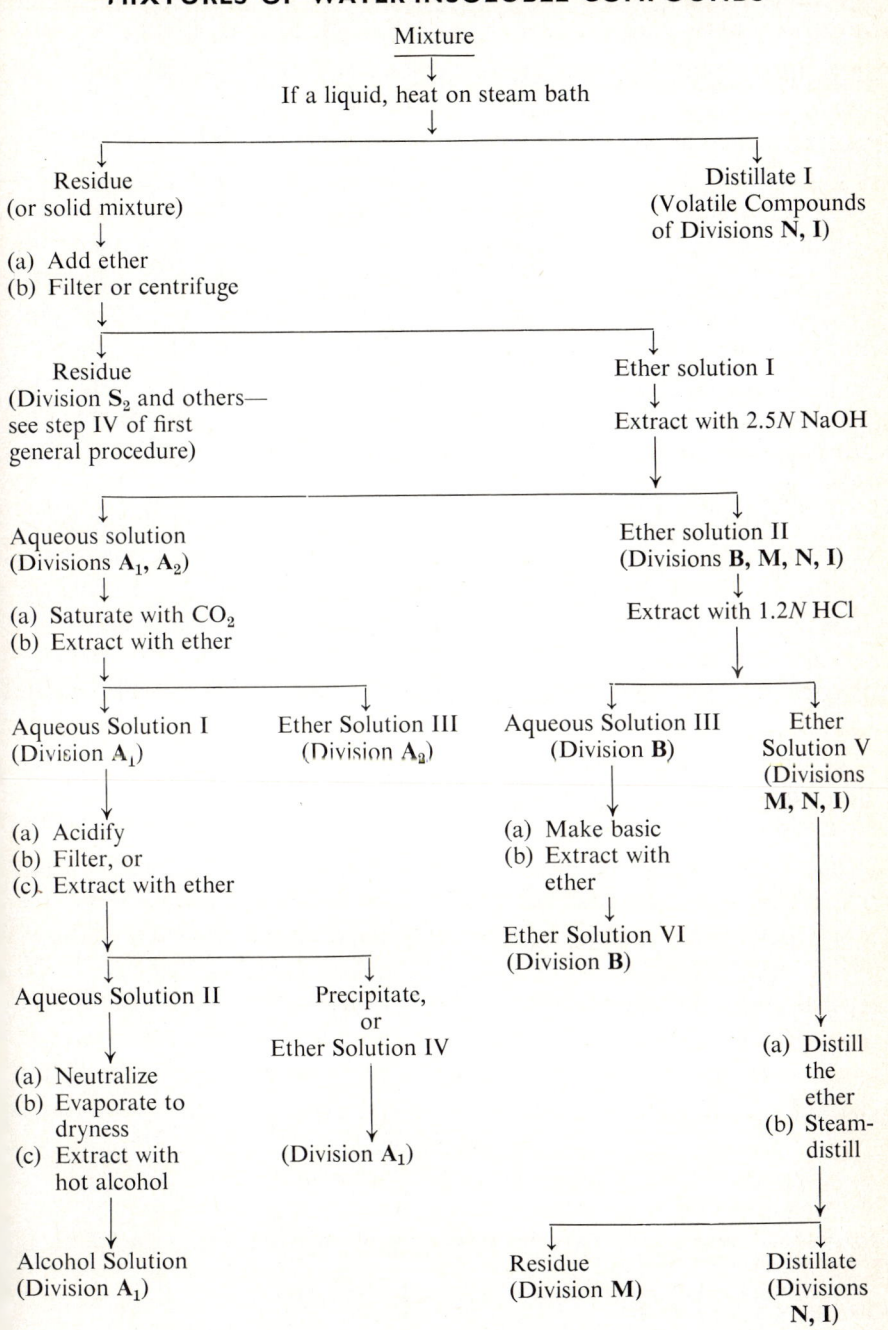

times the cost of equipment for the classical method. Identification of the separated components depends on the availability of information regarding the "response" of a known compound that has been treated in the same instrument under very similar conditions. The lack of sufficient data on "knowns" has been the major disadvantage in the use of instrumental techniques for the separation and identification of "unknowns." Currently, the usefulness and reliability of instrumental methods for identification work is rapidly improving in at least three ways: (1) information is being accumulated and tabulated regarding the "response" of an ever increasing number of "known" compounds when examined by various instrumental methods; (2) certain instruments, particularly gas-liquid chromatographs, are commercially available which automatically fractionate larger samples into separate samples of components that allow chemical derivatives to be prepared (the so-called "preparative chromatographs"); and (3) methods are being developed for the chemical derivatization of microgram quantities.

Walsh and Merritt[2] have published an article that shows how the qualitative functional group detection may be made on the effluent gas from a regular gas-liquid chromatographic apparatus. Procedures are given for detecting twelve chemical classes in the 20 to 100 microgram range. Graphs are given to show that, by plotting the "log retention volume" against the number of carbon atoms per molecule for each of the homologous series, the individual compounds may very often be identified after the class has been detected by the reagents used with the effluents.

Hoff and Feit[3] have shown that a gaseous mixture that contains several chemical classes of compounds may be separated into classes before being introduced into a gas chromatographic apparatus by placing the sample in a hypodermic syringe and introducing classification reagents.

The applications of the various chromatographic techniques to the separation of organic mixtures is almost limitless. The scope of current practice in this field may be obtained by consulting the selected references to the literature that may be found in Chapter 4 and at the end of this chapter.

REFERENCES

Attention is called to the extensive lists of references given at the end of Chapters 3 and 4 that relate to the various methods of fractionation. Many of these references

[2] J. T. Walsh and C. Merritt, Jr., *Anal. Chem.*, **32,** 1378 (1960).
[3] J. E. Hoff and E. D. Feit, *Anal. Chem.*, **35,** 1298 (1963).

deal with practical problems involved in the separation of mixtures. Selected additional references are given here. Papers that deal primarily with a single chromatographic method have been listed in separate sections.

General Works

K. E. Almin, *Acta Chem. Scand.*, **13**, 1263, 1274, 1278, 1287, and 1293 (1959). Fractionation of polymers by countercurrent distribution.

Azeotropic Data, published as No. 6 of *Advances in Chemistry Series*, American Chemical Society, Washington, D.C., 1952.

O. L. Baril et al. "Separation of Hydrocarbons by Azeotropic Distillation with Acetone, 1,4-Dioxane, Ethanediol, and 2-Ethoxyethanol," *Petroleum Research Fund Reports:* No. **4** (1959), p. 80; No. **5** (1960), p. 72; No. **6** (1961), p. 42; and No. **7** (1962), p. 15.

D. S. Binnington and W. F. Geddes, *Ind. Eng. Chem., Anal. Ed.*, **6**, 461 (1934). Addition of alcohol during steam-distillation to prevent solids forming in the condenser.

M. T. Bush, *Microchem. J.*, **3**, 315 (1959); **4**, 289 (1960); and **5**, 325 (1961). Progress reports on distillation, sublimation, and crystallization methods.

C. Karr, Jr. et al., *Anal. Chem.*, **32**, 463 (1960). Countercurrent distribution.

K. A. Kobe and L. R. Reinhart, *J. Chem. Educ.*, **36**, 300 (1959). Urea and thiourea complexes.

W. H. Melhuish, *Nature*, **184**, 1933 (1959). Fractional microsublimation of hydrocarbons.

J. Porath, *Nature*, **183**, 1657 (1962). Gel filtration.

L. D. Metcalfe, *Anal. Chem.*, **33**, 1559 (1961). A review of methods of separation, with 71 references.

S. Siggia and J. G. Hanna, *Anal. Chem.*, **33**, 896 (1961). Analysis by differential reaction rates: alcohols and carbonyls.

W. Seaman et al., *J. Am. Chem. Soc.*, **67**, 1571 (1945). Separation of alkaryl amines.

R. S. Tipson, *Anal. Chem.*, **22**, 628 (1950). Theory and methods of recrystallization.

A. Tiselius, J. Porath, and P. Albertson, *Science*, **141**, 13 (1963). Separation and fractionation of macromolecules.

F. T. Wallenberg et al., *J. Chem. Educ.*, **36**, 251 (1959). Universal apparatus for steam distillation.

B. R. Warner and L. Z. Raptis, *Anal. Chem.*, **27**, 1783 (1955). Azeotropic distillation using chloroform to separate formic acid from other acids.

F. T. Weiss, A. E. O'Donnell, R. J. Shreve, and E. D. Peters, *Anal. Chem.*, **27**, 198 (1955). Schematic analysis of RAr-sulfonate detergents.

J. C. Winters and R. A. Dinerstein, *Anal. Chem.*, **27**, 546 (1955). Distillation in miniature columns.

General Chromatography

H. G. Cassidy, *Technique of Organic Chemistry*, Vols. V and X, Interscience, New York, 1951 and 1957.

Handbook of Analytical Chemistry, L. Meites, ed., McGraw-Hill, New York, 1963. H. H. Strain, Tables 10-10 to 10-13: Adsorbents and solvents for chromatographic techniques for various classes of organic compounds.

E. Heftmann, ed., *Chromatography*, Reinhold, New York, 1961.

I. M. Kolthoff and P. J. Elving, ed., *Treatise on Analytical Chemistry*, Interscience, New York, 1961, Part 1, Vol. 3, pp. 1411–1723. All types of chromatography with hundreds of references.

L. Lederer, *Chromatography in Organic and Biological Chemistry*, Masson, Paris, Vol. I, 1959, and Vol. II, 1960.

S. Patton in *Microchemical Techniques*, Interscience, New York, 1962, pp. 757–769. The use of 2,4-dinitrophenylhydrazine with carbonyls; with 78 references.

R. Sargent and W. Rieman III, *J. Org. Chem.*, **21,** 594 (1956). New technique for the separation of organic compounds.

Column Chromatography

R. Bassette and C. H. Whitnah, *Anal. Chem.*, **32,** 1098 (1960). Chemical reactions.
D. C. Bogue, *Anal. Chem.*, **32,** 1779 (1960). Theory of phase diffusion.
J. Gasparić, J. Petranek, and J. Borecký, *J. Chromat.*, **5,** 408 (1961). Separation of alkylated phenols.
A. Hallén, *Acta Chem. Scand.*, **14,** 2249 (1960). Separation of sugars as borate ions.
E. Honkanen, *Acta Chem. Scand.*, **15,** 449 (1961). Separation of 2,4-dinitrophenylhydrazones on a cellulose column impregnated with dimethylformamide.
J. R. Howe, *J. Chromat.*, **3,** 389 (1960). Correlation of chromatographic behavior and structure of 111 organic acids.
D. P. Johnson and J. B. Johnson, *Anal. Chem.*, **31,** 1373 (1959). 3,5-Dinitrobenzamides.
D. F. Keummel, *Anal. Chem.*, **34,** 1003 (1962). Separation of the mercury derivatives of esters.
J. G. Kirchner and A. J. Haagen-Smit, *Ind. Eng. Chem., Anal. Ed.*, **18,** 31 (1946). Acids.
H. M. Koehler and E. G. Feldman, *Anal. Chem.*, **32,** 28 (1960). Local anaesthetics.
A. Meister and P. Abendschein, *Anal. Chem.*, **28,** 172 (1956). α-Ketoacids.
K. J. Monty, *Anal. Chem.*, **30,** 1350 (1958). Partition and spectrophotometry of the 2,4-dinitrophenylhydrazones of aliphatic carbonyls.
D. P. Schwartz et al., *Anal. Chem.*, **34,** 664 (1962). 2,4-Dinitrophenylhydrazones.
J. W. Spanyer and J. P. Phillips, *Anal. Chem.*, **28,** 253 (1956). Xanthates of alcohols.
H. H. Strain, *Anal. Chem.*, **21,** 79 (1949). Adsorbents and solvents for column and paper chromatography.
H. H. Strain, *Anal. Chem.*, **31,** 818 (1959). Differential migration requirements.
H. M. Tenny and F. E. Sturgis, *Anal. Chem.*, **26,** 946 (1954). Hydrocarbons.
J. W. White, Jr., *Anal. Chem.*, **20,** 726 and 853 (1948). 2,4-Dinitrophenylhydrazones (from alcohols).
E. O. Woolfolk, F. Beoch, and S. P. McPherson, *J. Org. Chem.*, **20,** 391 (1955). *p*-Phenylazobenzoates of alcohols.
E. O. Woolfolk and J. M. Taylor, *J. Org. Chem.*, **22,** 827 (1957). *p*-Phenylazobenzoates of phenols.
E. O. Woolfolk, W. E. Reynolds, and J. L. Hudson, *J. Org. Chem.*, **24,** 1445 (1959). Separation of sulfonamides.
M. L. Wolfrom and G. P. Arsenault, *Anal. Chem.*, **32,** 693 (1960). Separation of the 2,4-dinitrophenylhydrazones of highly oxygenated carbonyl compounds.
V. Zbinovsky, *Anal. Chem.*, **27,** 764 (1955). Carboxylic acids.

Paper Chromatography

J. A. Attaway et al., *Anal. Chem.*, **34**, 671 (1962). Urethanes of alcohols.

J. A. Attaway, R. W. Welford, G. E. Alberding, and G. J. Edwards, *Anal. Chem.*, **35**, 234 (1963). Terpene alcohols.

R. C. Bean and G. G. Porter, *Anal. Chem.*, **31**, 1929 (1959). Sugars.

G. Biserte, J. W. Halleman, J. Holleman-dehove, and P. Sautière, *J. Chromatog.*, **2**, 225 (1959). Dinitrophenylamino acids.

R. J. Block, E. L. Durrum, and G. Zweig, *Paper Chromatography*, Academic Press, New York, 1958.

M. A. Buchanan, *Anal. Chem.*, **31**, 1616 (1959). Fatty acids.

D. P. Burma, *Anal. Chem.*, **25**, 549 (1953). Partition mechanisms.

J. W. Chittum, T. A. Gustin, R. L. McGuire, and J. T. Sweeney, *Anal. Chem.*, **30**, 1213 (1958). Halogenated acetic and propionic acids.

F. Cramer, *Paper Chromatography*, St. Martin's Press, New York, 1954.

E. A. Day and S. Patton, *Microchem. J.*, **3**, 137 (1959). 2,4-Dinitrophenyl sulfide derivatives of thiols and mercaptoacids.

B. A. Dehority, *J. Chromatog.*, **2**, 384 (1959). Antioxidants.

F. W. Denison, Jr., and E. F. Phares, *Anal. Chem.*, **24**, 1628 (1952). Acids.

J. Franc and G. Čělikovská, *Coll. Czeck Chem. Comm.*, **26**, 667 (1961). R_f values for aldehydes and ketones.

A. M. Gaddis and R. Elleis, *Anal. Chem.*, **31**, 870, 1997 (1959). 2,4-Dinitrophenylhydrazones.

J. Gasparič, *J. Chromatog.*, **4**, 75 (1960). Aminoanthraquinones.

J. Gasparič and J. Borecký, *J. Chromatog.*, **5**, 466 (1961). 3,5-Dinitrobenzoates of alcohols, phenols, thiols and amines.

W. G. Gennings, *Anal. Chem.*, **31**, 1117 (1959). 2,4-Dinitrophenylhydrazones.

H. O. Heisey, *Chemist-Analyst*, **52**, 86 (1963). Technique for applying large volumes in paper chromatography.

J. B. Himes, L. D. Metcalfe, and H. Ralston, *Anal. Chem.*, **33**, 1364 (1961). Sugars.

W. Huber, *Mikrochim. Acta*, **1960**, 44. Olefins.

D. P. Johnson and J. B. Johnson, *Anal. Chem.*, **31**, 1373 (1959). 3,5-Dinitrobenzamides.

M. Jureček, J. Čhuracek, and V. Červinka, *Mikrochim. Acta*, **1960**, 102. Fatty acids.

P. Kabasakalian and A. Basch, *Anal. Chem.*, **32**, 458 (1960). Steroids.

A. J. Landua, R. Fuerst, and J. Awapara, *Anal. Chem.*, **23**, 162 (1951). Amino acids.

W. N. Martin and R. M. Husband, *Anal. Chem.*, **33**, 840 (1961). Phenols on paper impregnated with polyamides.

H. J. McDonald, L. P. Roberto, and L. J. Banaszak, *Anal. Chem.*, **31**, 825 (1959). Centrifugal force in paper chromatography and electrophoresis.

E. F. McFarren, *Anal. Chem.*, **23**, 168 (1951). Amino acids.

N. A. Milas and I. Belic, *J. Am. Chem. Soc.*, **81**, 3358 (1959). Organic peroxides.

L. C. Mitchell, *J. Assoc. Offic. Agr. Chemists*, **43**, 810 (1960). Insectisides.

E. D. Moffat, and R. I. Lytle, *Anal. Chem.*, **31**, 926 (1959). Amino acids.

R. Neu, *Mikrochim. Acta*, **1957**, 196; **1958**, 267. Spot testing.

S. D. Nogare, *Anal. Chem.*, **28**, 903 (1956). Amine hydrochlorides.

M. Nonaka, E. L. Pippen, and G. F. Bailey, *Anal. Chem.*, **31**, 875 (1959). Chromatography and spectrophotometric examination of 2,4-dinitrophenylhydrazones.

J. L. Occolowitz, *J. Chromat.*, **5**, 373 (1961). Dibasic acids.

L. R. Ory, W. G. Bickford, and J. W. Dieckert, *Anal. Chem.*, **31**, 1449 (1959). Fatty acids.

J. B. Pridham, *J. Chromatog.*, **2**, 605 (1959). Phenols by chromatography and electrophoresis.

R. J. Rice, G. J. Keller, and J. G. Kirchner, *Anal. Chem.*, **23**, 194 (1951). 2,4-Dinitrophenylhydrazones of aldehydes and ketones and 3,5-dinitrobenzoates of alcohols.

D. P. Schwartz in *Handbook of Analytical Chemistry*, L. Meites, ed., McGraw-Hill, New York, 1963, Section 10, pp. 69–100.

J. B. Stark, A. E. Godban, and H. S. Owens, *Anal. Chem.*, **23**, 413 (1951). Acids.

E. Sundt and M. Winter, *Anal. Chem.*, **29**, 851 (1957). 3,5-Dinitrobenzoates of alcohols. Gives R_f data.

J. C. Underwood and L. B. Rockland, *Anal. Chem.*, **26**, 1553 (1954). Amino acids.

J. Van Espen, *J. Pharm. Belg.*, **11**, 45 (1960). Sulfonamides by chromatography and electrophoresis.

O. F. Wiegand and A. R. Schrank, *Anal. Chem.*, **28**, 259 (1956). Solute concentrations.

Ion-Exchange

B. Alfredsson, S. Bergdahl, and O. Samuelson, *Anal. Chim. Acta*, **28**, 371 (1963). Hydroxyacids.

D. L. Buchanan, *Anal. Chem.*, **32**, 1400 (1960). Amino acids.

C. Calmon and T. R. E. Kressman, *Ion Exchangers in Organic and Biochemistry*, Interscience, New York, 1957.

R. Djurfeldt and O. Samuelson, *Acta Chem. Scand.*, **4**, 165 (1950). Theory and uses.

R. Kunin, *Anal. Chem.*, **21**, 87 (1949). A review.

R. Kunin and F. X. McGravey, *Anal. Chem.*, **26**, 106 (1954). Review of progress.

G. H. Osborne, *Synthetic Ion-Exchangers*, Chapman and Hall, London, 1955.

O. Samuelson, *Ion Exchangers in Analytical Chemistry*, Wiley, New York, 1953.

O. Samuelson and B. Swanson, *Anal. Chim. Acta*, **28**, 426 (1963). Sugars.

D. P. Schwartz, A. R. Johnson, and O. W. Parks, *Microchem. J.*, **6**, 37 (1962). 2,4-Dinitrophenylhydrazones.

E. R. Tompkins, *Anal. Chem.*, **22**, 1352 (1950). Theory and procedures.

Gas Chromatography

V. E. Cates and C. E. Meloan, *Anal. Chem.*, **35**, 658 (1963). Sulfoxides.

D. H. Desty, *Vapor Phase Chromatography*, Butterworths, London, 1959.

M. Dimbat, P. E. Porter, and F. H. Stross, *Anal. Chem.*, **28**, 290 (1956). Apparatus and methods.

L. R. Durrett, L. M. Taylor, C. F. Wantland, and I. Doretzay, *Anal. Chem.*, **35**, 637 (1963). Hydrocarbons.

F. T. Eggertsen, H. S. Knight, and S. Groennings, *Anal. Chem.*, **28**, 303 (1956). Hydrocarbons.

L. S. Ettre and W. Averill in *Microchemical Techniques*, N. D. Cheronis, ed., Interscience, New York, 1962, pp. 715–732 with 28 references. Gas chromatography in microanalysis.

E. M. Fredericks and F. R. Brooks, *Anal. Chem.*, **28**, 297 (1956). Hydrocarbons.

B. J. Gudzinowicz and W. R. Smith, *Anal. Chem.*, **32**, 1767 (1960). High temperature, gas-liquid chromatography.

C. Hishta, J. P. Messerley, R. Reshke, D. H. Fredericks, and W. D. Cooke, *Anal. Chem.*, **32**, 88 (1960). Solid organic compounds.

J. E. Hoff and E. D. Feit, *Anal. Chem.*, **35**, 1298 (1963). Functional group analysis by pretreatment of sample in a hypodermic syringe.

E. C. Horning, W. J. A. Vanden Heŭvel, and B. G. Greech, in *Methods of Biochemical Analysis*, D. Glick, ed., Vol. 2, Interscience, New York, 1963. Steroids.

W. B. Innes, W. E. Bambrick, and A. J. Andreatch, *Anal. Chem.*, **35**, 1198 (1963). Hydrocarbon analysis using differential absorption.

A. T. James in *Methods of Biochemical Analysis*, Vol. VIII, D. Glick, ed., Interscience, New York, 1960. Fatty acids.

C. Karr, Jr., P. M. Brown, and P. A. Estep, *Anal. Chem.*, **31**, 1413 (1959). Phenols.

A. I. M. Keulmans, *Gas Chromatography*, Reinhold, New York, 1957.

S. D. Nogare, *Anal. Chem.*, **32**, 19R (1960). A review with 244 references.

J. G. O'Conner et al., *Anal. Chem.*, **32**, 710 (1960). Molecular sieves for hydrocarbons.

K. D. Parker and P. L. Kirk, *Anal. Chem.*, **33**, 1378 (1961). Barbiturates.

K. D. Parker et al., *Anal. Chem.*, **34**, 757 (1962). Tranquilizers.

R. L. Pecsok in *Handbook of Analytical Chemistry*, L. Meites, ed., McGraw-Hill, New York, 1963, Section 10, pp. 101–125. Retention volumes and retention times.

R. Rowan, Jr., *Anal. Chem.*, **33**, 658 (1961). Hydrocarbons.

B. Smith, *Acta Chem. Scand.*, **13**, 480 (1959). Use of aqueous mixtures.

C. F. Spencer, F. Baumann, and J. F. Johnson, *Anal. Chem.*, **30**, 1473 (1958). Thiols and disulfides.

J. T. Walsh and C. Merritt, Jr., *Anal. Chem.*, **32**, 1378 (1960). Qualitative functional group analysis of gas chromatographic effluents.

Thin-Layer Chromatography

E. F. L. J. Anet, *J. Chromatog.*, **9**, 291 (1962). 2,4-Dinitrophenylhydrazine derivatives.

M. Beroza and T. P. McGovern, *Chemist-Analyst*, **52**, 82 (1963). A transfer tool.

J. M. Bobbitt, *Thin-Layer Chromatography*, Reinhold, New York, 1963.

J. Davidek, *J. Chromatog.*, **9**, 363 (1962). Gallic acid esters.

M. Gee, *J. Chromatog.*, **9**, 278 (1962). Sugars.

G. Machata, *Mikrochim. Acta*, **1960**, 79. Spot testing.

H. K. Mangold, R. Kammereck, and D. C. Malins in *Microchemical Techniques*, N. D. Cheronis, ed., Interscience, New York, 1962, pp. 697–714 with 37 references.

W. N. Martin and R. M. Husband, *Anal. Chem.*, **33**, 841 (1961). Phenols on paper impregnated with polyamides.

K. Randerath, *Thin-Layer Chromatography*, Academic Press, New York, 1963.

J. Rosmus and Z. Deyl, *J. Chromatog.*, **6**, 2, 187 (1961). 2,4-Dinitrophenylhydrazones.

J. E. Spikner and J. C. Towne, *Chemist-Analyst*, **52**, 50 (1963). Technic for the separation and solution of zones.

E. Stahl, ed., *Thin-Layer Chromatography*, Academic Press, New York, 1963.

M. W. Subbarao, *J. Chromatog.*, **9**, 295 (1962). Oxygenated fatty compounds.

E. G. Woolish in *Microchemical Techniques*, N. D. Cheronis, ed., Interscience, New York, 1962, pp. 687–696, with 55 references.

Electrophoresis

M. Bier, *Electrophoresis—Theory, Methods, and Applications*, Academic Press, New York, 1959.
S. Hjertén, *Arkiv. Kemi*, **13**, 151 (1958).
V. Jakl and V. O. Sukupová-Kolková, *Cesk. Farm.*, **10**, 197 (1961). Separation and characterization of local anaesthetics.
B. Lindberg and B. Swan, *Acta Chem. Scand.*, **14**, 1043 (1960). Electrophoresis of carbohydrates.
J. Porath, E. B. Lindner, and S. Jerstedt, *Nature*, **182**, 744 (1958).
J. Porath and K. Störiko, *J. Chromat.*, **7**, 385 (1962).
J. Porath and S. Hjertén in *Methods of Biochemical Analysis*, Vol. IX, D. Glick, ed., Interscience, New York, 1962.
W. Reuter, *Biochem. Z.*, **1959**, 331, 337. Preparative zone electrophoresis.
L. A. Williams et al., *Anal. Chem.*, **32**, 1883 (1960). Alkaloids.

Zone Refining

E. F. G. Herrington, *Zone Melting of Organic Compounds*, W. Heffer and Sons, Cambridge, England, 1963.
E. T. Knypl and K. Ziélenski, *J. Chem. Educ.*, **40**, 352 (1963).
N. L. Parr, *Zone Refining and Allied Techniques*, G. Newness, Ltd., London, 1960.
W. G. Pfann, *Zone Melting*, Wiley, New York, 1958.
W. G. Pfann, *Science*, **135**, 1101 (1962).
E. Stahl and U. Kaltenbach, *J. Chromat.*, **5**, 351, 358 (1961).
R. Tschesche, F. Lampert, and G. Snatzke, *J. Chromat.*, **5**, 217 (1961).
E. G. Wollish, M. Schmall, and M. Hawrylyshyn, *Anal. Chem.*, **33**, 1138 (1961).
M. Zief, H. Ruch, and C. H. Schramm, *J. Chem. Educ.*, **40**, 351 (1963).

Extraction

L. C. Craig, *Techniques of Organic Chemistry*, 2nd ed., Vol. 3, Part 1, pp. 149–332, Interscience, New York, 1956.
W. G. Batt, and H. K. Alber *Ind. Eng. Chem., Anal. Ed.*, **13**, 127 (1941). Microextraction.
M. T. Bush and P. M. Densen, *Anal. Chem.*, **20**, 121 (1948). Systematic multiple fractional extraction procedures.
M. T. Bush, *Microchem. J.*, **5**, 73 (1961). Extraction of drugs and related compounds.
C. Karr, Jr., *Anal. Chem.*, **32**, 463 (1960). Countercurrent distribution.
P. J. Lloyd and A. D. Carr, *Analyst*, **86**, 335 (1961). Extraction of amines as the cobalt-thiocyanate complex in a pentanol-kerosene mixture.
F. A. V. Metzsch, *Angew. Chemie*, **65**, 586 (1953). Nearly 400 nearly immiscible solvent systems for extraction by countercurrent distribution.
H. A. Pagel and F. W. McLafferty, *Anal. Chem.*, **20**, 272 (1948). Use of tributyl phosphate for extracting organic acids.
R. H. Petrucci and J. C. Weygand, *Anal. Chem.*, **33**, 275 (1961). Microscopic-sublimation procedure for the detection and removal of impurities from organic solids.

G. Rudstam, *Anal. Chem.*, **32**, 1664 (1960). Partition extraction of labeled compounds.
J. E. Spikner, V. F. Ward, and J. C. Towne, *Chemist-Analyst*, **52**, 50 (1963). New extraction apparatus.
F. Will, III, *Anal. Chem.*, **33**, 647 (1961). New solvent system for the separation of fatty acids, C_{10} to C_{18} by countercurrent distribution.
M. Wayman and G. F. Wright, *Ind. Eng. Chem., Anal. Ed.*, **12**, 91 (1940). Extraction by a 3:1 mixture of petroleum ether and acetone from a saturated solution of NaCl.

PART TWO

Procedures for Tentative Identification of an Unknown

7

Preliminary Examination of the Pure Compound

Once the fraction of the unknown substance into pure compounds has been accomplished, the systematic identification of the pure sample can be undertaken. However, in the examination of the original unknown substance and subfractionation procedures, one or two constants of the pure sample have been determined. These data are recorded, since they will be of use in later steps of the identification.

The first step in the systematic procedure for the identification of an organic substance is to obtain some general information as to the nature of the unknown and then to determine, by qualitative tests, the elements present. A variety of procedures and tests may be employed for obtaining general information that may give valuable clues as to the probable nature of the unknown. For example, the observations made, such as color, odor, and appearance under the microscope are of value. Phenolic compounds, aliphatic and aromatic hydrocarbons, most amines, the lower aliphatic acids, and a large number of carbonyl compounds have characteristic odors which can be easily identified by an experienced person. Nitro compounds, quinones, azo compounds, and many derivatives of triphenylmethane and anthraquinone are colored. Color as a result of impurities usually diminishes or disappears on purification.

It is advisable to proceed *systematically through all the steps* of the characterization without attempting to use curtailed procedures and short-cut methods. The experienced worker may be able to obtain, through the fusion techniques outlined in subsequent pages of this chapter, sufficient information about the nature of the unknown to permit the curtailment of several steps in the identification. Indeed, the experienced analyst may be able, through the fusion techniques and one or two functional group tests, to arrive at a tentative identification and confirm it by derivatization. However, use of such a procedure by a beginner usually results in confusion and requires more time than the systematic procedure. As experience is gained, one may begin to utilize curtailed procedures.

Ignition and Preliminary Tests with Reagents

Ignition Test

After the color, odor, and crystal structure of the sample have been noted, a small amount of material is heated until it burns, and its behavior on combustion is noted. If the substance is a solid, observations are made during the first stages of heating as to whether it melts or sublimes, sputters and explodes, or gives off vapors.

Organic compounds containing metals, such as the salts of carboxylic acids, sulfonic acids, and the like, leave a residue consisting mainly of the carbonate of the metal. Aromatic compounds burn with a smoky flame, whereas the lower aliphatic ones give an almost nonsmoky flame. Compounds containing oxygen burn with a bluish flame. Sugars and proteins burn with characteristic odors. Halogen compounds burn with a smoky flame; polyhalogen compounds, however, as a rule do not ignite until the flame is applied directly to the substance, which then momentarily renders the flame of the burner smoky.

The test may be performed with 1 to 2 mg of the substance, using a crucible cover or small evaporating dish or with a single crystal or droplet (0.01 mg) on a platinum or iron microspatula. About 1 to 2 mg of the substance are placed near the center of the porcelain crucible cover or small evaporating dish, and heat is applied directly by means of a small flame. From time to time the flame is applied directly to the top of the substance so that it will ignite before it volatilizes. If the substance carbonizes, the flame is increased and, finally, the substance is strongly heated. If a residue remains, it should be nearly white. If gases are given off in the initial stage of heating—that is, before ignition—a test should be made by means of litmus or pH paper to determine whether the gas has acidic or basic properties.

If the test is performed on a wire microspatula, a crystal or droplet of the substance is placed on the flat end. The wire is then heated about 8 to 10 mm from the flattened end in the colorless flame of a microburner. The spatula is gradually moved so that the sample is brought into the flame. The residue, if any, is dissolved in a drop of water placed on a microscope slide. The pH is tested by placing a droplet on a piece of pH paper.

Test for Presence of Water

The presence of water in organic liquids is usually detected in the purification of the unknown; traces of water, however, may escape detection. It is important that appreciable amounts of water should not be present in several tests, such as those for sodium fusion and detection of hydroxyl groups. If water is found to be present, it should be removed before further tests are made.

Several reagents may be employed to detect the presence of water in organic substances. The most sensitive is tetraisopropyl titanate; about 0.01 to 0.05 ml (one drop or less) of the reagent is added to a few drops of the liquid (or a solution of the compound in anhydrous methanol). Water hydrolyzes the reagent to produce hydrated titanium oxides, which precipitate as chalky solids. The amount of precipitate is a measure of the amount of water in the sample.

The sensitivity of the test is shown by adding 1 drop of water in 10 ml of anhydrous ether; a cloudy appearance is obtained in testing a few drops of the ether. A few per cent of water will give a voluminous precipitate, or will cause a gel to form. Commercial methanol gives only a very faint test; when methanol is dried over calcium hydride, a negative test is obtained.

Other reagents for the detection of water are aluminum ethoxide or isopropoxide and anhydrous copper sulfate. A drop of the liquid or solution is added to a few crystals of the aluminum alkoxide; the presence of water is indicated by the appearance of gelatinous aluminum hydroxide. The disadvantage of both aluminum alkoxides is that once the bottle is opened it is difficult to keep the moisture in air from attacking the reagent. For the detection of traces of moisture with anhydrous copper sulfate, 25 to 50 mg of the colorless salt are placed on a watch glass and a drop of the liquid to be tested is placed on the copper salt. If water is present, it will hydrate the salt to the blue pentahydrate.

Observations during Fusion and Cooling

Fusion Techniques

Fusion techniques (developed primarily by the Koflers[1] and their co-workers in Europe and by McCrone[2] and his collaborators in this country),

---

[1] A. Kofler and L. Kofler, *Thermo-Mikro-Methoden*, Weinheim/Bergstrasse, Verlag Chemie 1954).

[2] W. C. McCrone, *Fusion Techniques in Chemical Microscopy*, Interscience, New York, 956; W. C. McCrone et al., *Ind. Eng. Chem., Anal. Ed.*, **18**, 578 (1946); W. C. McCrone, *Anal. Chem.*, **21**, 436 (1949); W. C. McCrone, *Mikrochemie*, **38**, 476 (1951).

include observations made on a few milligrams of the solid material during heating, during melting and solidification of the melt, and during cooling. "Mixed" fusion with a reference substance may be included. For a detailed discussion of these methods, see the works of the Koflers[1] and McCrone.[2] In the present section, a brief summary of the methods will be given together with a detailed description of the mixed fusion technique for establishing identity or nonidentity of two solid substances.

If a microscope hot stage is available, the sample is examined as described on page 187. If a hot stage is not available, a regular glass slide is employed and, after the cover glass is placed on the sample, the slide is held at one edge and carefully passed back and forth above a microflame not more than 5 mm in height. When the first change in the crystals is noted (either by direct observation or by means of a lens) the slide is examined immediately under the microscope. It is then heated again until melting begins and observation under the microscope is repeated. Finally the slide is heated until all the crystals have fused and the liquid has distributed itself on most of the area under the cover glass; during the cooling period the sample is examined periodically until it reaches room temperature.

In general, low magnification (20 to 50 ×) with crossed Nicol crystals is most useful. An old microscope with a 32 or 40 mm objective is suitable, since an objective of shorter working distance would be injured by the heat from the preparation. When the slide is hot, the condenser is lowered; as the preparation cools, the condenser can be raised and higher power objectives can be inserted for examination at higher magnifications or for conoscopic observation.

As the temperature of the sample approaches the melting point, the ease of sublimation, the nature of the sublimate, and the tendency to decompose or to dehydrate can be observed; dehydration shows up as characteristic gas spaces as the melt is formed. Many organic compounds (hexamethylenetetramine, benzoic acid, carbazole, hydroquinone, and 3,5-dinitrobenzoic acid)[3] give sublimates on which most of the geometric and optical properties of the crystals can be determined.[4]

With a hot stage the same phenomena are observed while they are taking place and, in addition, the characteristic premelting rearrangement of the crystals may give valuable information to the experienced worker. For example, although ascorbic acid decomposes at 185 to 192°, it can be identified by the characteristic rearrangement of crystals above 175° McCrone[4] has pointed out that in one research program on high explosive

[3] W. C. McCrone, *Anal. Chem.*, loc. cit.
[4] W. C. McCrone, *Mikrochemie*, loc. cit.

the analyst had no difficulty remembering the key properties of nearly fifty explosives, boosters, and mixtures thereof. As a result any substance in that group could be recognized unequivocally in a few minutes, which included preparation of the sample.

Other observations, useful in characterization work, that may be made during the heating period are: (a) the temperature at which the first discernible polymorphic transformation takes place; (b) the temperature at which loss of solvent of crystallization begins; (c) the temperature at which decomposition of crystals first appears; and (d) the temperature at which eutectic melts appear. The "eutectic melting point" developed by the Koflers is particularly useful for identification of those compounds that decompose or sublime before melting. Finally, after the sample has melted, the refractive index of the melt can be measured.

Observations during cooling and crystallization give information about the crystal habit, the rate of crystal growth, and the form of crystal front.[6] Of importance are the presence and relative amount of residual melt in the crystallizing sample because they give information concerning the purity of the compound. If a compound is pure and does not decompose on melting, the crystalline film resulting from the melt will not show even traces of a residual eutectic melt. If an impurity is present, a residual melt will appear after solidification of the principal component. The relative size of the fused zone indicates the extent of the impurity. However, in a few cases in which the impurity is isomorphous with the principal component, no eutectic melt will appear. This case is discussed in the following section.

"Mixed" Fusion

The technique of mixed fusion[7] is very useful for the study of the comparative purity of different samples of the same compound and particularly for determining whether two solid compounds are identical. In the identification of organic compounds, it is often more convenient for experienced workers to determine whether two derivatives are identical by means of mixed fusion (without the use of a hot stage) than by means of mixed melting points.

---

[6] W. C. McCrone, op. cit.; E. M. Chamot and C. W. Mason, *Handbook of Chemical Microscopy*, Wiley, 1940, New York, Vol. II; J. Mitchell, *Anal. Chem.*, **21**, 449 (1950).

[7] For detailed discussion of this technique, see A. Kofler and L. Kofler, op. cit.; W. C. McCrone, *Fusion Techniques in Chemical Microscopy*, Interscience, New York, 1956; N. Goetz-Luthy, *J. Chem. Ed.*, **26**, 159 (1949); V. Gilpin, *Anal. Chem.*, **23**, 365 (1951); C. J. Arceneaux, *Anal. Chem.*, **23**, 906 (1951); and D. E. Lakowski et al., *Anal. Chem.*, **25**, 1400 (1953).

The steps involved in the mixed fusion technique are illustrated in Figure 7.01. About 1 to 2 mg of substance A are placed under the cover glass, melted carefully over a microflame, and allowed to solidify. The other substance, B, is then placed at the edge of the cover glass and heated so that it melts and runs under the cover glass into contact with A. The slide, which is now reheated so that all of substance B and some of substance A melt, is examined under the microscope, preferably with crossed Nicols.

If the two compounds A and B are identical, there will be no zone of mixing as the sample solidifies and the crystals will grow throughout with

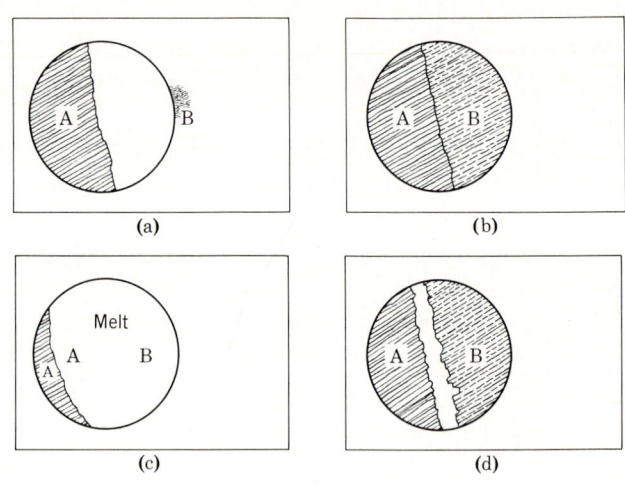

FIGURE 7.01

Steps in mixed fusion. (a) Substance A melted and allowed to crystallize, then substance B placed near the cover glass. (b) Substance B melted and allowed to run under the cover glass and crystallize. (c) All of B and part of A remelted to form a zone of mixing. (d) Substances A and B recrystallized to zone of mixing. (After McCrone.)

no discontinuity either in the rate of growth or the form of the crystals. If the two are identical and one is more pure than the other, there may be a change in the rate of growth and perhaps crystal size as the crystal front progresses through the zone of mixing. If A and B are not identical, there will be a discontinuity in rate at the zone of mixing. In the case of isomorphous compounds a solid solution will be formed; there will be a gradually increasing rate of growth up to the interface, where the rate will decrease and crystal growth will continue slowly through the zone of mixing; then it will increase again as the second substance solidifies.

When *A* and *B* are different and not isomorphous, crystals of the pure substances will grow from either side until the interface is reached where the rate of crystal growth will decrease and finally cease, leaving a thin region of the melt, which is either an addition product or a eutectic that may or may not solidify. The following pairs that form eutectics are useful for practice in the technique of mixed fusion: acetanilide and phenacetin; acetanilide and biphenyl; coumarin and vanillin; naphthalene and azobenzene. The mixed fusion of acetanilide and phenacetin is shown in Figure 7.02. The photograph was taken using crossed Nicols after the

FIGURE 7.02
Mixed fusion acetanilide and phenacetin.

sample had completely solidified; the eutectic region is well defined. When such a preparation is placed on a hot stage and heated gradually, the eutectic will appear at 90° as a dark band when viewed with crossed Nicols; the eutectic of acetanilide and biphenyl appears at 64.5°.

The eutectic region can often be recognized with the naked eye. In the example cited above (acetanilide and phenacetin), after the pure compounds have crystallized separately, the eutectic zone remains as a band of liquid for at least thirty seconds to one minute, depending on the rate of cooling. Similarly, if such a slide is heated very slowly by passing it over a minute microflame, the eutectic zone melts first and appears as a miniature stream. The eutectic melting point has been employed by the Koflers and by McCrone in the same manner as the melting point of derivatives for the characterization of organic compounds. The Koflers determined the eutectic melting points of about 1200 organic compounds with each of two reference substances as listed in Table 7.01.

TABLE 7.01

REFERENCE SUBSTANCES EMPLOYED IN THE KOFLER METHOD OF CHARACTERIZATION OF ABOUT 1200 COMPOUNDS BY MEANS OF THEIR EUTECTIC MELTING POINTS*

Melting Point of Compound To Be Identified, °C	Component Employed in Mixed Fusion for Determination of Eutectic Melting Point
20–100	Azobenzene, benzil
100–120	Benzil, acetanilide
120–140	Acetanilide, phenacetin
140–170	Phenacetin, benzanilide
170–190	Benzanilide, salophen
190–240	Salophen, dicyandiamide
240–340	Phenolphthalein

* Tables of the eutectic melting points, refractive indices of the melts, and other properties for these 1200 organic compounds will be found in the previously cited works by the Koflers and by McCrone. The eutectic melting points are particularly valuable for substances that decompose near their melting temperatures.

ANALYSIS FOR ELEMENTS

Detection of individual elements involves the decomposition of the organic molecules and the conversion of the elements present to ions or simple molecules for which known methods of identification are available. Not any single method of decomposing the organic substance serves for the detection of all types of elements; hence, several procedures are given here, and still others may be found in the references given at the end of this chapter.

Detection of Carbon and Hydrogen

Carbon is usually detected in the preliminary ignition test. With rare exceptions, such as compounds in which halogen has completely replaced hydrogen, all compounds that contain carbon also contain hydrogen. A simple test for the presence of carbon may be made by modifying the Feigl test for microgram samples as given in the next section by using a 3-in. test tube and larger quantities of material. If it becomes necessary to test for both carbon and hydrogen, the first procedure in which the sample is heated with cupric oxide may be employed for samples as small as 0.1 mg. For the detection of a few micrograms of carbon in a sample, one of the procedures described after the cupric oxide test is selected.

When a test sample containing carbon and hydrogen is heated with cupric oxide, the presence of carbon is shown by the evolution of carbon dioxide, and the presence of hydrogen by the evolution of water vapor. Carbon dioxide is detected by passing the evolved gases through a solution containing barium hydroxide; the water vapor is condensed over anhydrous copper sulfate, which changes from nearly colorless to the blue pentahydrate. The test, as described, can be applied to quantities of 0.1 to 1 mg of the sample.

Apparatus and Procedure for Milligram Quantities of Carbon and Hydrogen

Select a piece of glass tubing 6 mm OD and 60 to 70 mm long. Seal one end by heating it in a flame with continuous rotation until the glass is soft. Remove the tube from the flame and, by means of forceps, draw out a short piece of glass so as to seal the end. If forceps are not available, seal another piece of glass tube to the soft part and use it as a handle. Rotate the sealed end of the tube over the hot flame until a small mass of red-hot glass accumulates; then remove it from the flame and blow gently into the tube so as to form a bulb about 8 mm in diameter, as shown in Figure 7.03(*A*).

Heat about 50 to 100 mg of copper oxide powder in a small tube for a few minutes to drive off the moisture. Stopper the tube with a cork and allow to cool. In a watch glass, mix about 0.1 to 1 mg of the substance with the fine copper oxide and transfer to the bulb of the ignition tube. If the substance is a liquid, the copper oxide is first placed in the ignition tube. After a droplet is added to the copper oxide by means of a capillary pipet, the contents are mixed by shaking. Insert a plug of loose glass wool above the bulb, so that it covers a length of about 10 mm, then insert copper oxide wire to fill a length of 15 to 20 mm of the tube. Insert another plug of glass wool extending 15 to 20 mm above the copper oxide wire, and compress the upper part of the plug slightly. Add a small amount of anhydrous copper sulfate on top of the glass wool. Heat the tube with continuous rotation over the flame, just beyond the portion containing the copper salt. When the glass is soft enough, remove the tube from the

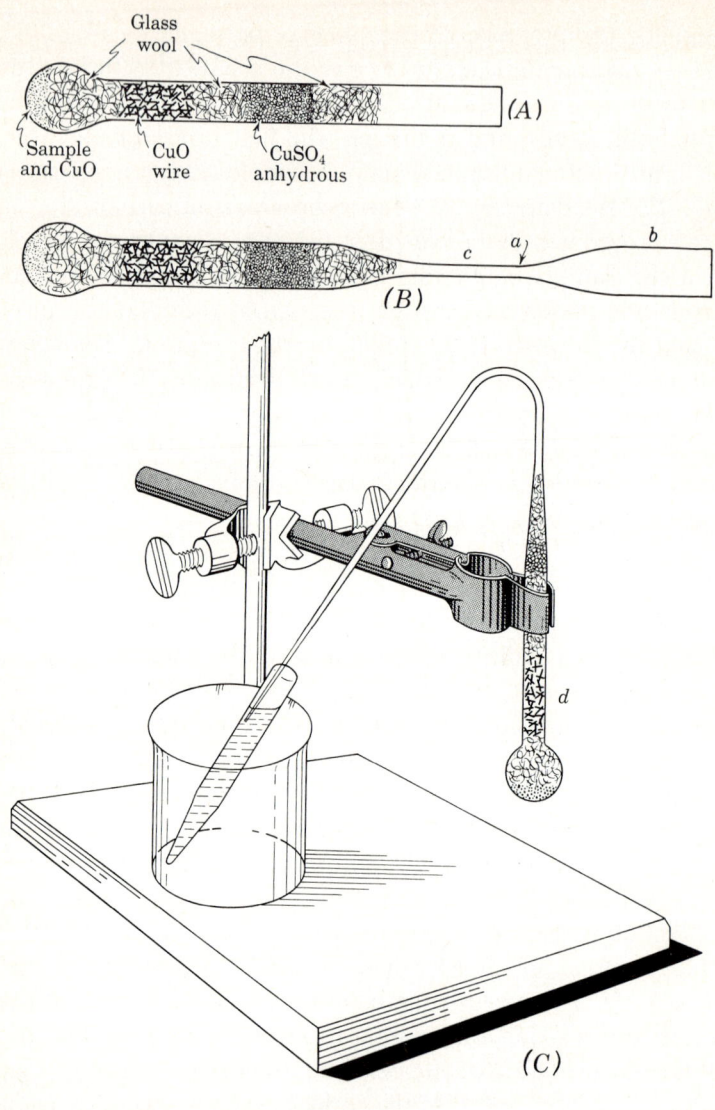

FIGURE 7.03

Microdetection of carbon and hydrogen. (A) and (B) Construction of ignition tube. (C) Ignition-tube assembly.

flame and draw out to a total length of 110 to 130 mm, as shown in Figure 7.03(B). Allow the tube to cool, and cut it at section a. Heat, momentarily, the tapered portion at point c, and bend as shown in Figure 7.03(C). The tube is then clamped so that the end of the capillary dips under the surface of 0.5 ml of barium hydroxide solution placed in the small tube, obtained by sealing the tapered end of section *b* of the tube shown in Figure 7.03(*B*). Heat the tube at location *d*, Figure 7.03(*C*); then, slowly, move the flame downward until all the copper oxide wire has been thoroughly heated. When this region is red hot, heat the bulb. The sample volatilizes, and the presence of carbon is indicated by the reduction of the copper oxide wire to copper and the appearance of carbonate in the upper layer of the alkaline hydroxide. The presence of hydrogen is indicated by the change of the anhydrous copper salt to the blue hydrate. The test can be performed with a 3-in. tube by employing larger quantities of sample and reagents and then placing a one-hole stopper at the mouth of the tube, through which is inserted a delivery tube.

Apparatus and Procedure for Microgram Quantities of Carbon

Feigl[8] has described several microdetection tests for carbon. In one of these, a small quantity of the test substance is placed in a tube about 6 mm in diameter and 70 to 80 mm in length. The tube is then half-filled with powdered molybdenum trioxide and connected with a pump by means of suction tubing. After the air has been removed, the tube, clamped at an angle, is heated for 1 to 2 minutes by means of a small flame, first at the upper zone of the oxide and then at the lower in which the sample is located. If there is carbon in the sample, the molybdenum oxide (MoO_3) is reduced to lower oxides (Mo_2O_5 and the like), which have a blue color; hence, a blue zone appears at the lower end, whose size and color intensity varies according to the carbon content of the sample. Although this test is usually reliable as indicating the presence of carbon, ammonium salts, sulfites, and other substances that readily reduce molybdenum oxide also produce the blue color.

The Emich[9] microdetection test can be applied to quantities less than 100 micrograms. A Pyrex capillary tube, about 1 mm in diameter and sealed at one end, is used. If the sample is a solid, it is introduced with a glass thread, and if a liquid, with a capillary pipet. The capillary is then sealed at the open end and heated just above the sample. When this region is red hot, the sample is heated so that it vaporizes and passes through the hot region. If carbon is present it forms a shining mirror. The tube is cooled and then cut into two pieces near the mirror. If the section containing the mirror is heated, the mirror volatilizes and ignites.

Microgram quantities of carbon can be detected by a wet oxidation

---

[8] F. Feigl and D. Goldstein, *Mikrochim. Acta*, **1956**, 1317; F. Feigl, *Spot Tests in Organic Analysis*, Elsevier, Amsterdam, 1960, p. 78.
[9] F. Emich and F. Schneider, *Microchemical Laboratory Manuals*, Wiley, New York, 1932, p. 112.

method,[10] which involves heating the sample at about 100° for 3 to 5 minutes with a mixture of iodic, chromic, sulfuric, and phosphoric acids in a capillary tube. The carbon dioxide formed by the oxidation of the organic compound is detected by precipitation as barium carbonate.

Another proposal[11] for the detection of carbon by spot-test analysis involves the dry heating of the sample with mercuric oxide and ammonium chloride to produce hydrogen cyanide, which may be detected by the copper acetate–benzidine reagent.

Decomposition of Organic Compounds

Before the individual elements may be detected and identified, an organic compound must be decomposed with the production of ions or small molecules that are capable of identification. Selection of the method to be used for the decomposition generally is based on (1) the nature of the substance to be examined, (2) the elements to be detected, (3) the quantity of the available sample, (4) the facilities available, and (5) the personal preferences and experience of the analyst. Only semimicro methods are covered in this text. A great variety of procedures, which generally combine chemical concentration methods with instrumental procedures, are available for specialized purposes and for microquantities, but reference must be made to the literature for details on these methods.

To illustrate the influence of the nature of the substance to be examined and the elements that it is desired to detect, assume that the substance is a natural product (vegetable or animal) and that it is desired to detect metals. Fusion with an active metal or combustion in an oxygen flask are completely impractical. Depending on the circumstances, the material may be ashed to produce the oxides and carbonates of the metals, or it may be "wet-ashed" in a mixture of concentrated sulfuric acid and some oxidizing agent such as nitric or perchloric acids. Another technique of wet-ashing is to cover the sample with concentrated sulfuric acid and add, dropwise with careful agitation, 30 to 50 per cent hydrogen peroxide. It is claimed that this method will completely decompose vegetable and animal tissues in a very few minutes.

Flask Methods

In 1955, Schöniger[12] published his first paper on the use of an oxygen-filled flask for the combustion of organic compounds so that such elements

[10] E. L. Bennett, C. W. Gould, E. H. Swift, and C. Neimmann, *Anal. Chem.*, **19**, 1035 (1947); ibid., **21**, 1582 (1949).
[11] A. Caldas and V. Gentil, *Talanta*, **2**, 220 (1959).
[12] W. Schöniger, *Mikrochim. Acta*, **1955**, 123; ibid., **1956**, 869; ibid., **1959**, 670.

as the halogens, sulfur, nitrogen, and phosphorous could be qualitatively detected and quantitatively determined. Since that time, nearly 200 papers have been published that are related to the usefulness of this method. A few selected references are given at the end of this chapter. Recently, Corliss and Rush[13] proposed that the halogens, nitrogen, sulfur, and phosphorous may be qualitatively detected by the flask method, using only an air-filled flask instead of the oxygen-filled flask. Because only air is used, very simple equipment is adequate: a 500-ml Erlenmeyer flask and a rubber stopper to which a hypodermic needle is attached to serve as a spear for the small piece of paper that holds the sample. The work cited should be consulted for details of the procedure. Experimentation with this method in the laboratory of one of us indicates that inexperienced workers may get satisfactory results for the halogens at least. Without doubt, the oxygen-flask method, and perhaps the air-flask method serve a useful purpose in the qualitative organic analysis laboratory. However, it is not believed that the more classical methods of decomposing organic compounds have been replaced, and these methods will be described in this chapter.

Sodium Fusion

The classical Lassaigne method of decomposing organic compounds depends on heating the compound with an alkali metal—usually sodium; hence, the method is designated as the *sodium-fusion method*. When an organic compound is heated with metallic sodium, extensive decomposition occurs, free carbon is deposited, and some carbon dioxide and monoxide are formed. Nitrogen, in the presence of carbon and sodium, forms sodium cyanide; sulfur, if present, forms sodium sulfide; and the halogens form the corresponding halides. The residue obtained in the fusion is extracted with water, and portions of the filtered alkaline solution are used for the detection of the cyanide, sulfide, and halide ions.

Procedure. Add a *small* drop of the liquid to be tested, or a *few* milligrams (if it is a solid) to a clean, dry 4-in. test tube. Now, add a piece of sodium (about a 3-mm cube) to the test tube. Warm the bottom of the tube slightly, and allow the mixture to stand for 2 to 3 minutes. Clamp the tube in a vertical position and, gradually, apply heat to the bottom of the tube until the sodium has melted. Being careful not to hit the sides of the hot tube with the sample, drop a *few* milligrams of the substance being tested directly onto the melted sodium. Heat the bottom inch of the tube

---

[13] J. M. Corliss and C. A. Rush, in *Microchemical Techniques*, N. D. Cheronis, ed., Interscience, New York, 1962, pp. 407–416.

until the glass is red hot, and continue heating for 2 minutes. Allow the tube to cool to room temperature and add 5 drops of methanol. If the residue is a globular mass, break it up with a clean glass rod to allow contact between the alcohol and any excess sodium metal. If gas bubbles are produced, wait for the reaction to be completed. Then add 3 ml of distilled water. Boil the mixture, and filter it. Instead of filtering, the mixture may be centrifuged and the clear centrifugate decanted or removed by means of a pipet. If, at this point, the solution is dark-colored so as to obscure further tests, it is probable that the amount of sample was too large or that the fused mass was not heated to a sufficiently high temperature; the fusion procedure should be repeated. The solution obtained by the above procedure is to be used in the tests for nitrogen, sulfur, the halogens and, perhaps, other elements. For the procedures, see pp. 291–296.

The procedure described has been used successfully for the detection of nitrogen in such nitro compounds as picric acid, trinitrotoluene, *p*-nitrochlorobenzene, and *p*-nitrophenol and in proteins, amino acids, and aminonaphthalenesulfonic acids; with care it may be used to detect nitrogen in organic nitrites and nitrates. In a comparison of the sodium fusion and the zinc-alkali carbonate fusion tests for the detection of nitrogen in 20 mg of ten polynitro compounds and in butyl nitrite and ethyl nitrate, the sodium-fusion test gave consistently better results, as judged by the development of a distinct blue color. With a number of pyrrole derivatives potassium metal gives better results.[14]

When the quantity of the unknown is small, the test may be performed with 100 to 200 micrograms. The sample is placed in the bulb of a small fusion tube approximately 80 mm in length, prepared from 5 to 6 mm of glass tubing as described in the test for carbon and hydrogen. Metallic sodium (3 to 5 mg), pressed between filter paper to form a slender rod, is pushed into the tube with a wire so that it comes in contact with the sample. After standing for 3 to 5 minutes, the fusion tube is heated, while clamped, from the top to the bottom; the bulb is finally heated strongly and the tube is allowed to cool. When the tube is perfectly cold, 2 drops of methanol are added with a capillary pipet. The tube is allowed to stand and warmed if necessary until any unreacted sodium has decomposed. After 0.5 to 0.7 ml of water has been added, the tube is heated cautiously almost to boiling over a microflame, which is directed toward the upper part of the tube and not toward the bulb, to avoid sudden bumping. After heating, the tube is centrifuged or the suspended material is allowed to settle. The supernate is then withdrawn with a capillary pipet and diluted to 1 ml.

[14] G. Kainz and A. Resch, *Microchemie*, **39**, 75 (1952).

Portions of the solution are used to test for nitrogen, sulfur, and halogens.

Sodium fusion, properly executed, is highly efficient, and provides a filtrate that may be used to test for all four halogens, nitrogen, and sulfur. Care must be exercised if the sample is quite volatile. Nitrogen is probably the most difficult element to detect satisfactorily. For this reason, a number of methods have been suggested to replace sodium fusion in testing for nitrogen. Most of these methods involve fusion of the sample with a mixture of an active metal and an alkali carbonate.

Magnesium-sodium carbonate fusion. About 50 mg of the sample to be tested are mixed with 100 mg of a mixture of equal parts of magnesium and anhydrous sodium carbonate. This mixture is placed in a 4-in. Pyrex test tube arranged as for the sodium fusion. Then about 100 mg of the magnesium mixture is added on top of the sample. The fusion mixture is heated, beginning at the end that is farthest from the sample and, when it begins to burn, the lower end of the tube is heated until it is red hot. The tube is cooled, 5 ml of water are added, and the mixture is boiled and filtered. The filtrate is tested for cyanide ion as directed in a later section.

Zinc-sodium carbonate fusion. A small fusion tube about 50 to 60 mm in length is prepared from 6-mm glass tubing, as directed on page 283 for testing the presence of carbon and hydrogen. About 30 to 50 mg of the compound are mixed with about 150 mg of an intimate mixture of equal parts zinc dust and anhydrous sodium carbonate, and this mixture is introduced into the ignition tube. If the compound is a liquid, 2 drops are added to the empty ignition tube and then 150 mg of the zinc-carbonate mixture are added. More of the zinc-carbonate mixture is added on top of the sample; after it is tapped down, the tube is filled to a height of 25 mm above the bulb. The ignition tube is clamped at an angle with the open end of the tube near the clamp. By means of a flame, the region of the tube beyond the column is heated, and then gradually the flame is moved downward so that by the time the bulb is reached, a portion of the column is red hot. The bulb is heated strongly for 2 to 3 minutes, the flame being moved upward so that the column remains red hot. By means of tongs, or a test-tube holder, the ignition tube is removed and dropped in a 5-in. test tube containing 5 to 8 ml of distilled water. If the tube does not break, it is removed, broken in a mortar, and the contents transferred to the test tube. The mixture is boiled for a minute and filtered. The filtrate should be colorless. About 1 ml of this alkaline filtrate is tested for cyanide ion (see page 292).

If it is desired to test for sulfide ion in the mixture resulting from the zinc–sodium carbonate fusion, it is necessary to employ the residue left after the fusion mixture has been extracted with water. In such fusions, any sulfur that may have been present is converted to zinc sulfide, which is sparingly soluble. Therefore, a portion of the residue remaining after the fusion mixture has been extracted with water is introduced into a 3-in. tube and acidified with 2 to 3 drops of acetic acid. The mouth of the tube is covered with a small disc of filter paper moistened with lead acetate as described in a following section. The tube is then warmed and observations are made on the lead acetate paper.

Gansel[15] recommends the use of lithium carbonate for microfusions of the compound with zinc.

Calcium-oxide fusion. About 25 to 50 mg of the compound are mixed with 100–200 mg of pulverized soda-lime. The mixture is placed in a small evaporating dish or crucible and then covered with a watch glass, on the under surface of which is laid a strip of moist red litmus paper or pH paper. The dish is heated gently to avoid spattering. If the litmus paper turns blue or the pH paper indicates a pH of 10 to 11, ammonia was evolved. The dish is allowed to cool, then a portion of the mixture is tested for sulfide ion.

For small quantities an ignition tube is prepared as described under the test for carbon, and hydrogen is employed. The charge consists of 1 mg or less of the sample and 2 to 5 mg of pulverized soda-lime. Wet the pH paper and then place it on the mouth of the ignition tube, which has been thoroughly cleaned and tested for absence of soda-lime.

Brown and Hoffpauir[16] recommend calcium oxide fusion for the detection of micro amounts of nitrogen. They detect the evolved ammonia by 8-quinolinol-zinc.

Oxidation by manganese dioxide. Feigl[17] has published a test for nitrogen in organic compounds based on the oxidation of the organic compound by manganese dioxide, which has been previously heated at 600°C for 15 minutes to decompose all nitrates. The treated manganese dioxide should be freshly prepared, and it is imperative that a blank test be made on this reagent. The nitrous acid that is evolved during the test is detected by the Greiss reagent, which is prepared by mixing equal volumes of a 1 per cent solution of sulfanilic acid in 30 per cent acetic acid and a 1 per cent solution of 1-naphthylamine in 30 per cent acetic acid. The following procedure has been found to give good results, particularly with solid samples.

[15] E. E. Gansel, *Microchem. J.*, **3**, 91 (1959).
[16] L. E. Brown and C. L. Hoffpauir, *Anal. Chem.*, **23**, 1035 (1951).
[17] F. Feigl, *Anal. Chem.*, **30**, 1148 (1958).

Insert a 3-in. test tube through a 6 × 6 in. sheet of asbestos. Place 30 mg of the compound to be tested in the tube, and add 0.2 g of recently prepared manganese dioxide. Cover the mouth of the test tube with a filter paper that has been moistened with Greiss reagent. Heat the bottom of the test tube strongly for 1 to 2 minutes. The formation of a medium-to-dark rose color on the paper is a positive test.

Detection of the Elements in the Sodium Fusion Filtrate

Detection of Sulfur

(A) If sulfur is present in the unknown substance, it will be converted to sodium sulfide by the sodium fusion test. The presence of sulfide ion is detected by the formation of lead sulfide on addition of lead ion. About 0.2 ml of the alkaline filtrate or centrifugate from the sodium fusion is placed in a 3-in. tube; 1 drop of lead acetate solution is added, followed by dilute acetic acid until the solution is acidic. A brown-to-black precipitate or coloration indicates the formation of lead sulfide.

For detection of small quantities (when the total filtrate or centrifugate from the sodium fusion is 1 ml) about 0.1 ml of the test solution in a microtube is acidified with dilute acetic acid. A small disc of filter paper (20 to 25 mm) is placed over the mouth of the tube and moistened with 1 to 2 drops of lead acetate solution. The paper is pressed on the sides of the tube and the solution is boiled gently for a few seconds; appearance of a dark coloration on the paper indicates the presence of sulfur. The test can be performed on 1 drop of test solution by placing it in the well of a microscope slide having a depression. A cover slip is provided with a very small circle of filter paper (2 mm) moistened with a droplet of lead acetate solution. The cover slip is placed over the well so that the adhering test paper is underside. If the depression is shallow then a small glass ring is placed around it, and the cover slip is fitted on the glass ring. The cover slip is removed momentarily, and 1 to 2 drops of $6N$ acetic or perchloric acid are added. A dark color forms within 5 minutes if sulfide is present.

(B) Place 0.2 ml of the filtrate from sodium fusion in a 3-in. test tube and add 1 drop of a 0.1 per cent solution of sodium nitroprusside. A deep red or purple color indicates the presence of the sulfide ion.

(C) Feigl[18] has described a microdetection procedure, based on the sulfide-catalyzed reaction between sodium azide and iodine to liberate small bubbles of nitrogen gas.

---

[18] F. Feigl, *Spot Test in Organic Analysis*, 6th ed., Elsevier, Amsterdam, 1960, p. 242.

Detection of Nitrogen

If sulfur has been found present, a slight modification of the procedures for nitrogen is desirable, since ferrous sulfide is only slightly soluble. Hence, to be sure that sufficient ferrous ions are available for both the sulfide and cyanide ions, the amount of ferrous salt used should be increased by 50 per cent. Furthermore, when the sulfuric acid is added to dissolve the oxides of iron, a residue of ferrous sulfide may remain in the bottom of the tube. This residue will not interfere with the test for nitrogen.

(A) Place 0.8 ml of the sodium fusion filtrate (or filtrate from one of the metal-alkali carbonate fusions) in a 4-in. test tube and add 1 drop of 30 per cent potassium fluoride, 1 drop of 10 per cent sodium hydroxide and 15 to 20 mg of ferrous sulfate crystals or ferrous ammonium sulfate crystals. Boil the mixture, gently, for 1 minute. If a bluish precipitate does not exist at this point, add 2 drops more of the 10 per cent sodium hydroxide and reboil the mixture. Add a small droplet of 1 per cent ferric chloride and, again, bring the mixture to boiling. Add $6N$ sulfuric acid (dropwise and with shaking) until the oxides of iron have just dissolved. Allow the tube to stand for 2 to 5 minutes. A blue color or a blue precipitate is a positive test for nitrogen (ferri-ferrocyanide). A green or greenish-blue coloration indicates a weak test for nitrogen (perhaps due to poor fusion). The addition of a drop of dilute phosphoric acid to such a greenish mixture will, usually, cause the blue color to show up, since the yellow color is due to ferric chloride, and the ferric ions are largely removed by the phosphoric acid. An alternate procedure is to filter the greenish solution; if a blue color shows up on the filter paper, the presence of nitrogen in the sample may be assumed. A yellow-colored solution, after acidification by the sulfuric acid, indicates a negative test for nitrogen, if the fusion and test procedure were properly performed.

(B) In the copper acetate-benzidine test[19] for cyanide ions, about 0.1 to 0.2 ml of the filtrate or centrifugate is acidified with a drop of 10 per cent acetic acid and then 1 to 4 drops of the reagent are added carefully by means of a capillary pipet so that there is no appreciable mixing of the two solutions. If cyanide ion is present, a blue ring develops; if sulfide ion is present, it is removed by adding 1 drop of lead acetate solution and centrifuging. The copper acetate-benzidine reagent is prepared from

---

[19] A. Sieverts and A. Hermsdorf, *Z. angew. Chem.*, **34**, 3 (1921); K. N. Campbell and B. K. Campbell, *J. Chem. Educ.*, **27**, 261 (1950); G. Kainz and F. Schoeller, *Mikrochim. Acta*, **1954**, 333; and F. Feigl, 183 (cited in Reference 18).

two stock solutions, one containing 150 mg of benzidine in 100 ml of water and 1 ml of acetic acid, and the other 286 mg of copper acetate in 100 ml of water. The two solutions are kept separately (in dark bottles) and the reagent is prepared by mixing equal volumes of each just before use.

(C) Baker[20] et al. have proposed a method for the detection of micro quantities of cyanide in the presence of an excess of sulfide. They use bromine to oxidize sulfide to sulfate and cyanide to cyanogen bromide, which is detected by pyridine-benzidine.

Detection of Halogens

The presence of chlorine, bromine, and iodine in organic compounds can be readily detected by the simple copper halide (Beilstein) test. However, since a number of other organic compounds that do not contain halogen (particularly those containing cyano and thiocyanate functional groups) give positive reactions with this test, it is always necessary to confirm the presence of halogens by means of one of the fusion tests. The alkaline filtrate or centrifugate from the fusion mixture is first treated so as to eliminate the cyanide and sulfide ions (if these are present) and then tested by various procedures so as to detect the presence of chloride, bromide, iodide or fluoride.

(A) *Beilstein test for halogens.* A small loop is made on a piece of copper wire 120 mm in length. The end of the wire having the loop is heated until no green color is detected, dipped (while hot) in some copper oxide, and reheated until the oxide adheres to the loop. A minute amount of the compound to be tested is placed on the copper oxide and heated in the nonluminous flame of a burner—first in the inner, then in the outer, zone near the lower edge. A blue-green flame indicates the presence of chlorine, bromine, or iodine; the color is due to the vapor of copper halide. Copper fluoride is not volatile and hence the test is not suitable for the detection of fluorine. The copper oxide can be omitted if the copper wire is first heated to redness (a film of oxide is formed), but the procedure with the oxide is more sensitive. A number of compounds that contain no halogen, such as urea, thiourea, hydoxyquinolines, pyrimidines, pyridines, and some carboxylic acids, have been reported[21] to give strong positive tests, probably owing to the formation of volatile copper salts, such as copper cyanide.

[20] M. O. Baker, R. A. Foster, B. G. Post, and T. A. Hiett, *Anal. Chem.*, **27**, 448 (1955).
[21] Hans Meyer, *Analyse und Konstitutionsermittlung organischer Verbindungen*, 6th ed., Springer, Berlin, 1938, p. 166; H. Gilman and J. E. Kirby, *J. Am. Chem. Soc.*, **51**, 1571 (1929).

Modifications of the test have been described by Ruigh,[22] Stenger,[23] and Hayman[24] for testing gases and volatile liquids. In general, for liquids, solids, or gases, the test—if carefully executed as described—is sensitive to a few micrograms. It is best to confirm a positive test with the fusion test.

(B) *Silver halide test.* The sodium fusion filtrate may contain one or more of the following halide ions: fluoride, chloride, bromide, and iodide. Silver fluoride is highly soluble in water; hence, the presence of fluoride ions cannot be detected by the addition of silver ions. A separate test for fluorides is given later in this chapter. Silver cyanide and silver sulfide are both sparingly soluble in water; hence, if either nitrogen or sulfur has been found present by previous tests, the cyanide or sulfide ions must be removed before adding silver ions to test for the halides. To remove cyanide or sulfide ions, add 1 ml of distilled water to 0.3 to 0.5 ml of the fusion filtrate in a 4-in. test tube and acidify the solution by adding, dropwise, concentrated nitric acid. Hold the tube under the hood and boil the mixture until the volume has been reduced by one half. Cool the mixture, and add 3 drops of 5 per cent silver nitrate. A white or slightly yellow precipitate indicates the presence of chloride, bromide, or iodide ions. If neither nitrogen nor sulfur has been found present, the fusion filtrate may be acidified with nitric acid and tested without boiling.

If the amount of test solution available is less than 0.1 ml, the tests can be made in capillaries instead of microtubes with as little as 25 to 50 λ of solution. Glass capillaries of 1.5 to 2 mm OD are employed; the solution and reagents are delivered by capillary pipets with long tips that fit within the capillary tubes. After the delivery of each sample or reagent, the capillary tube is centrifuged; then the contents are mixed with a fine glass rod and observed by means of a lens.

Differentiation between Chloride, Bromide, and Iodide, in the Fusion Extract

Numerous methods have been suggested for the detection of iodine, bromine, and chlorine, either alone or in mixtures, based on the differences in the ease of oxidizability of their ions to the elements. The following procedure by Hanson[25] is simple, and has proved reliable.

Place 0.5 ml of the sodium fusion filtrate in a 4-in. test tube. Add 0.5 ml of carbon tetrachloride and 3 drops of concentrated nitric acid. Shake the tube and allow the liquids to stratify. The presence of iodine is indicated by a violet color in the carbon tetrachloride. If iodine is present,

---

[22] W. L. Ruigh, *Ind. Eng. Chem., Anal. Ed.*, **11**, 250 (1939).
[23] V. A. Stenger, *Ind. Eng. Chem., Anal. Ed.*, **11**, 121 (1939).
[24] D. F. Hayman, *Ind. Eng. Chem., Anal. Ed.*, **11**, 470 (1939).
[25] M. W. Hanson, *J. Chem. Educ.*, **38**, 412 (1961).

remove the carbon tetrachloride layer by a pipet and add 0.5 ml of carbon tetrachloride to the original test solution. Add 1 drop of concentrated nitric acid, shake the mixture, and allow it to stratify. If the violet color still shows, remove the lower layer by means of a pipet and, without adding more nitric acid, repeat the extraction of the iodine by carbon tetrachloride until the lower layer remains essentially colorless. Now, add 2 ml of concentrated nitric acid to the mixture being tested. Shake the solution and allow it to stratify. The presence of bromine is indicated by a tan or tan-red color in the carbon tetrachloride. Repeat the extraction of the bromine, if present, until the lower layer becomes colorless. Add 3 drops of 5 per cent silver nitrate to the aqueous layer. The immediate formation of a white precipitate indicates chlorine.

The following microprocedure, when more than one halogen is present, has been proposed by Wilson.[26] About 0.2 ml of the filtrate or centrifugate liquid is placed in a microtube with 1 to 2 drops of chloroform and 1 drop of dilute sulfuric acid. A drop of freshly prepared chlorine solution or 2 to 3 drops of 3 per cent hydrogen peroxide are added and the mixture is shaken. The chloroform layer becomes almost colorless if the halogen is a chloride, brown or tan if it is a bromide, and violet if it is an iodide. The iodide test may be made on a drop of the filtrate or centrifugate placed on a glass slide (over a piece of white paper) with a droplet of dilute sulfuric acid. The end of a fine rod is dipped in starch solution, shaken, and used to stir the test drop. A small crystal of sodium nitrite is added to the drop, stirred, and observed for the development of a blue color.

The bromine test may be made on a drop of the filtrate or centrifugate by placing it in the depression of a glass slide, adding 1 drop of dilute sulfuric acid, 1 drop of potassium sulfate, and then 1 drop of a 1:1000 solution of platinum sulfate. The test drop is allowed to evaporate. Red-to-brown crystals of potassium bromoplatinate indicate the presence of bromine.

Filler[27] recommends the use of N-chlorosuccinimide instead of chlorine water to convert bromide and iodide ions to the atoms.

Detection of Fluoride Ion in the Fusion Extract

(A) Place 15 to 20 mg of lanthanum chloranilate (2,5-dichloro-3,6-dihydroxy-*p*-benzoquinone, lanthanum salt) in a 3-in. test tube. Add 1 to 2 ml of distilled water, 5 drops of 6M acetic acid, and 5 to 10 drops of

---

[26] D. W. Wilson and C. L. Wilson, *J. Chem. Soc.* (*London*), **1939**, 1956. See also A. J. Llacer, *C.A.*, **42**, 838–839 (1948).

[27] R. Filler, *J. Chem. Educ.*, **35**, 407 (1958).

the sodium fusion filtrate. The development of a light violet or pink color within 10 minutes is a positive test for fluorine.

(B) It is reported[28] that fluoride ions form a lilac-blue double complex with the red-colored cerium(III) chelate of 1,2-dihydroxyanthraquinonyl-3-methylamine-N,N-diacetic acid. Add dilute acetic acid to a few drops of the sodium fusion filtrate to bring the pH to 4.5. To one drop of this test solution on a spot plate, add, in sequence, 1 drop of a $0.001M$ solution of the reagent lightly buffered to pH 4.5 with sodium acetate-acetic acid, followed by 1 drop of $0.001M$ cerous nitrate. Stir immediately. A blank should be run, using a drop of water instead of the test solution. A positive test for fluoride involves the formation of a lilac-blue color within 1 minute.

Detection of Oxygen

There is no simple and specific test for oxygen in organic compounds. In cases where a direct method is desired to prove the presence of oxygen in organic compounds, the procedure developed by Schütze and improved by Unterzaucher[29] may be employed. The sample is vaporized at 950 to 1000° in a stream of pure nitrogen and the products carried over pure carbon heated to 1100 to 1150°, whereby the oxygen is converted to carbon monoxide, which is then oxidized to carbon dioxide by iodine pentoxide.

Ferrox Test

Ferrox,[30] ferric hexathiocyanatoferriate, an intensely colored salt, is not soluble in hydrocarbons or their halogen derivatives but it is soluble in compounds that contain oxygen or sulfur and in most compounds that contain nitrogen.

(A) *Solid reagent.* Prepare the ferric hexathiocyanatoferriate by placing a small crystal of ferric ammonium sulfate and one of potassium thiocyanate in a dry 3-inch test tube and grinding the crystals with a glass stirring rod until the red or green mass adheres to the stirring rod. Withdraw the rod and insert it into a dry 3-in. test tube. Pour 2 to 4 drops of the liquid to be tested down the rod and stir the mixture. The liquid will dissolve the salt and show red to reddish-purple colorations if the compound contains oxygen or sulfur and, generally, if it contains nitrogen. The salt does not dissolve in hydrocarbons or halogenated hydrocarbons,

---

[28] R. Belcher, M. H. Leonard, and T. S. West, *Talanta*, **2**, 92 (1959).
[29] J. Unterzaucher, *Ber.*, **B73**, 391 (1940); P. J. Elving and W. B. Ligett, *Chem. Rev.*, **34**, 129 (1944); A. D. Kirchenbaum and A. G. Streng, *Anal. Chem.*, **25**, 638 (1953).
[30] D. Davidson, *Ind. Eng. Chem., Anal. Ed.*, **12**, 40 (1940); D. Davidson and D. Perlman, *A Guide to Qualitative Organic Analysis*, Brooklyn College Bookstore, 2nd Ed., Brooklyn, N.Y. (1958).

hence these liquids retain their original color or are only faintly tinted. Solids may be tested by using a saturated solution of the compound in warm benzene or purified carbon tetrachloride

NOTE. Solids that are not sufficiently soluble in benzene or carbon tetrachloride give negative tests even if they contain oxygen. Some high-molecular-weight compounds such as diphenyl ether, alkyl naphthyl ethers, and triphenylcarbinol do not give positive tests. Some esters, such as dimethyl oxalate, do not give the test. Most nitrogen containing compounds that have been tested gave positive tests except the alkyl amines which generally do not dissolve the salt.

This test is most useful if applied to compounds that fall in Solubility Division N so as to distinguish the hydrocarbons that are soluble in concentrated sulfuric acid from the oxygen-containing compounds. Since ethers are hard to detect by functional group tests, the Ferrox test is useful in distinguishing most of them from hydrocarbons.

Ferric chloride may be substituted for the ferric ammonium sulfate in the preparation of the reagent.

(B) *Paper method.* The above test may be modified so as to use a test-paper rather than the solid salt. Dissolve 0.3 g of ferric ammonium sulfate in 5 ml of water and add 0.5 g of potassium thiocyanate. Stir the solution and soak filter paper in the solution for 5 minutes. Remove the paper and dry it. Cut the paper into strips about 0.5 by 1 cm and store in a brown bottle. For testing, insert a strip of the dyed paper in a dry 3-inch test tube and add the compound to be tested. The solid salt seems to give more reliable results than the test-paper method, but the paper method is generally satisfactory.

Detection of Additional Elements

The number and variety of elements that may be found in organic compounds is quite large, and the present trend indicates that the number will increase rapidly. The scope of this book does not permit a discussion of methods for the detection of many of these additional elements. A few selected methods will be mentioned, and additional references included at the end of this chapter.

(A) *The detection of phosphorus.* Ketcham and de Low-Beer[31] have suggested a very simple test for phosphorus in organic compounds. During the usual sodium fusion procedure, a piece of filter paper that has previously been moistened with silver nitrate solution is placed over the mouth of the ignition tube immediately after the sample has been dropped on the hot sodium. The *immediate* production of a jet-black color of the paper is a positive test for phosphorus. For highly oxidized compounds, such as phosphate esters, sucrose should be added to the substance before

[31] R. Ketcham and Anne de G. Low-Beer, *J. Chem. Educ.*, **38**, 414 (1961).

fusion. Compounds that contain nitrogen, sulfur, or iodine may cause the paper to turn gray or brown, but these colors should be considered as a negative test.

Organophosphorus compounds may be degraded by either hot concentrated sulfuric acid or by fusion with sodium peroxide. The phosphate ions thus produced may be detected by conventional methods, some of which are indicated in the references at the end of the chapter.

Tests for arsenic and phosphorus are performed after a small amount of the organic compound has been fused with a mixture of potassium nitrate and anhydrous sodium carbonate. The melt is allowed to cool, is then dissolved in water and the solution is tested by the usual methods.

(B) *Detection of metals.* After ignition for 15 to 20 minutes, the crucible is allowed to cool and the residue is moistened with a few drops of concentrated sulfuric acid or aqua regia. Excess acid is evaporated by means of a small flame (hood) and the residue ignited again. If a carbon residue persists, the crucible is allowed to cool and a few drops of concentrated nitric acid are added and carefully evaporated before ignition. The white or yellow residue is then analyzed by the usual scheme for cation analysis.

If a white residue is obtained in the ignition test, it is likely to be due to a metallic carbonate or oxide. About 200 to 400 mg of the sample are mixed with 500 mg of ammonium nitrate and placed in a clean crucible or evaporation dish and strongly heated.

Mercury, if present, is volatilized during ignition. To test for mercury, the organic compound is oxidized by heating 200 mg of the substance nearly to boiling with 10 ml of concentrated solution of potassium chlorate until the mixture is colorless. After evaporation, the residue is dissolved. A small amount of the solution is placed in a tube, together with a clean piece of copper wire. An amalgam indicates the presence of mercury.

A rapid method for the detection of mercury in organic compounds utilizes the reaction of the volatilized mercury with cuprous iodide. About 5 to 10 mg of the compound to be tested are mixed with about 500 mg of anhydrous sodium carbonate and placed in a test tube. A glass rod is inserted through a slotted one-hole cork and fitted into the tube so that the end is about 15 mm above the surface of the sodium carbonate mixture. The end of the rod is coated with a mixture of equal parts of powdered cuprous iodide and water. With the tube held at an angle of 45°, its lower end is heated gently over a small flame. In the presence of mercury the cuprous iodide paste turns quickly from white to a salmon or pink color

This method is stated to be sensitive to 0.02 mg of mercury. Evolution of alkaline vapors or hydrogen sulfide interfere with the sensitivity of the test. The interference due to alkaline vapors can be prevented by carefully placing a 2-mm layer of potassium pyrosulfate over the alkaline fusion mixture; interference due to sulfide can be avoided by using a 2-mm layer of litharge.

A systematic qualitative and approximately quantitative procedure of elementary analysis of organic compounds with a sample of 20 to 40 mg was developed by Swift and Niemann and first published under the title *A System for the Ultimate Analysis of Chemical Warfare Agents*.[32] The procedures are applicable to the elementary analysis of organic compounds in general; a brief description of the method has appeared in *Analytical Chemistry*.[33] The basis of the procedures is fusion of the sample in a Parr microbomb with sodium peroxide. Treatment of the fusion mass with boiling water results in a residue consisting of those basic elements forming insoluble oxides and carbonates (Fe, Ti, Mn, Ni, Cd, Mg, Ca, Sr, Ba, Sn, Te, Sb, Cu, Pb). A portion of the fusion solution is analyzed for the predominantly amphoteric elements (Se, Te, Cu, Pb, Cd, Zn, As, Sb, Sn) and another portion for the acidic elements (F, Cl, Br, I, S, Cr, Te, As, P, Se, S, B); separate small samples of the solution are tested by specific reactions for F, Si, I, Br, Te, Se, As, P, and N. It is claimed that any element of the system that constitutes 1 per cent of the sample should be detected and, with certain exceptions, the accuracy of the estimation should be within ± 0.3 mg. The procedures are applicable to samples that are solid or, if liquid, have boiling points greater than 40°. With lower-boiling liquids, or gases, the fusion is made with a known weight of the sample in a sealed glass ampoule. The apparatus required, besides the microbomb, are a centrifuge, microburets, a photoelectric colorimeter, and several readily fabricated pieces of special glassware.

REFERENCES

Detection of Elements—Reviews

R. Belcher et al., "Analytical Chemistry," *Ann. Rep. Chem. Soc. London*, **51**, 346 (1954). Qualitative and quantitative.

P. Gouverneur, *Chem. Weekblad.*, **57**, 313 (1961).

A. M. G. MacDonald, *Analyst*, **86**, 3 (1961). Quantitative.

[32] Published by Chemical Warfare Service, U.S.A., and deposited with the American Documentation Institute, 1719 N Street N. W., Washington, D.C.

[33] E. H. Swift and C. Niemann, *Anal. Chem.*, **26**, 538 (1954).

W. Schöniger and H. Lieb, *Farnaco Ed. Sci.*, **2,** 81 (1961). Quantitative.
C. L. Wilson, *Ann. Rep. Chem. Soc. London*, **48,** 328 (1951).
R. Steinmann, *Mikrochim. Acta*, **1953,** 534. Bibliography on microdetection.

Schemes for the Detection of Several Elements

Advances in Analytical Chemistry and Instrumentation, C. N. Reilley, ed., Interscience, New York, 1960, pp. 199–241.
C. Barkenbus and R. H. Baker, *Ind. Eng. Chem., Anal. Ed.*, **9,** 135 (1937). Detection of N, S, and halogens.
R. Bennewitz, *Mikrochim. Acta*, **1960,** 54. Modification of the Schöniger method.
M. Boetuus et al., *Mikrochim. Acta*, **1958,** 321. Schöniger method.
D. G. Foulke and F. Schneider, *Ind. Eng. Chem., Anal. Ed.*, **10,** 104 (1938). Detection of N, S, and halogens.
L. Mázor, L. Erdey, and T. Meisel, *Mikrochim. Acta*, **1960,** 412, 417. Fusion with potassium in contact with glass beads.
C. L. Ogg et al., in *Microchemical Techniques*, N. D. Cheronis, ed., Interscience, New York, 1962. pp. 427–433. A safe oxygen-filled flask method.
J. Patrick and F. Schneider, *Mikrochim. Acta*, **1960,** 970. Detection of N and S.
W. Schöniger, *Mikrochim. Acta*, **1959,** 670. Includes 262 references.
E. H. Swift and C. Niemann, *Anal. Chem.*, **26,** 538 (1954). Semimicro scheme for the detection of 32 elements.
F. Vojtĕch, *Chem. průmsyl*, **10,** 135 (1960). Quick, automatic combustion apparatus —used for several elements.
G. Widmark, *Acta Chem. Scand.*, **7,** 1395 (1953). Microdetection of C, S, P, As, and halogens.
C. L. Wilson, *Analyst*, **63,** 332 (1938). Detection of N, S, and halogens.

Detection of Carbon

E. L. Bennett, C. W. Gould, E. H. Swift, and C. Niemann, *Anal. Chem.*, **19,** 1035 (1947); **21,** 1582 (1949). Microgram quantities of carbon by wet combustion.

Detection of Oxygen

D. Davidson, *Ind. Eng. Chem., Anal. Ed.*, **12,** 40 (1940). Ferrox test.
J. Goerdeler and H. Domgörgen, *Mikrochemie*, **41,** 212 (1952). Qualitative micro adaptation of the Unterzaucher method.
J. Goerdeler and H. Domgörgen, *Mikrochemie ver Mikrochim. Acta*, **40,** 212 (1953). Micro ferrox test.
A. D. Kirchenbaum and A. Streng, *Anal. Chem.*, **25,** 638 (1953). Direct oxygen method.
J. Unterzaucher, *Ber.*, **B73,** 391 (1940). Direct method for oxygen.

Detection of Nitrogen

A. Acosta, *Quimica (Mexico)*, **3,** 135 (1945). *C.A.*, **40,** 3359 (1946).
K. N. Campbell and B. K. Campbell, *J. Chem. Educ.*, **27,** 261 (1950). Method involves change in pH.

F. Feigl and J. R. Amaral, *Anal. Chem.*, **30,** 1148 (1958). Oxidizes nitrogen to nitrous acid.
G. Kainz and F. Schöller, *Mikrochim. Acta,* **1954,** 327. Use of benzidine to detect cyanide.
E. Rathenasinkam, *J. Proc. Inst. Chemists (India),* **18,** 151 (1946). Detection of small quantities of cyanogen compounds.
A. Spěvák, V. Kratochvil, and M. Večeřa, *Coll. Czech. Chem. Comm.,* **26,** 887 (1961). Converts cyanide to cyanogen chloride for detection as a polymethine dye.
L. Velluz, *Ann. pharm. franç.,* **4,** 12 (1946). Micromethod.

Detection of Sulphur

R. J. Bertolacini et al., *Anal. Chem.,* **30,** 202 (1958); ibid., **29,** 281 (1957).
G. Bussmann, *Helv. Chim. Acta,* **32,** 235 (1949).
F. L. Hahn, *Ind. Eng. Chem., Anal. Ed.,* **17,** 199 (1945). Micro detection of sulfur.
R. W. Klipp et al., *Anal. Chem.,* **31,** 596 (1959).
G. E. Mapstone, *Anal. Chem.,* **18,** 498 (1946). Elemental sulfur.
E. A. Wynne et al., *Anal. Chem.,* **33,** 807 (1961).

Detection of the Halogens

R. Belcher et al., *Mikrochim. Acta,* **1954,** 104. Microanalysis for fluorine.
R. Belcher, M. A. Leonard, and T. S. West, *Talanta,* **2,** 92 (1959). Fluorides.
R. Belcher, M. A. Leonard, and T. S. West, *J. Chem. Soc. (London),* **1960,** 4477. Fluorides.
R. J. Bertolacini and J. E. Barney II, *Anal. Chem.,* **30,** 202 (1958); ibid., **29,** 1187 (1957). Fluorine and chlorine by salts of chloranilic acid.
R. Filler, *J. Chem. Educ.,* **35,** 407 (1958). Use of *N*-chlorosuccinimide instead of chlorine to liberate bromine and iodine from their ions.
L. H. Fine and E. A. Wynne, *Microchem. J.,* **3,** 515 (1959). Fluoride by lanthanum chloranilate.
F. J. Frere, *Anal. Chem.,* **33,** 644 (1961). Comparison of fluoride methods.
R. Greenalgh and J. P. Riley, *Anal. Chim. Acta,* **25,** 179 (1961). Fluorides in water.
F. A. Gunther, R. C. Blinn, and D. E. Ott, *Anal. Chem.,* **34,** 302 (1962).
A. L. Hensley and J. E. Barney II, *Anal. Chem.,* **32,** 828 (1960). Fluoride by thorium chloranilate.
A. J. Llacer, *Abales asoc. quím. Argentina,* **34,** 43–73 (1946); *C.A.,* **42,** 839 (1948). A systematic separation scheme.
T. S. Ma, *Anal. Chem.,* **30,** 1557 (1958). Quantitative determination of fluorine.
T. S. Ma, *Microchem. J.,* **2,** 91 (1958). Methods for the halogens.
R. L. Menville and W. W. Parker, *Anal. Chem.,* **31,** 1901 (1959). Potentiometric method.
N. R. Rao and K. H. Shah, *Mikrochim. Acta,* **1953,** 254. Microanalysis for chlorine and bromine.
J. G. Sharefkin and H. E. Schwerz, *Anal. Chem.,* **32,** 996 (1960). Detection of iodine without sodium fusion.
For modifications of the Beilstein test for volatile compounds, see *Ind. Eng. Chem., Anal. Ed.,* **11,** 121, 250, and 470 (1939).

Procedures for Tentative Identification of an Unknown

Detection of Phosphorus

K. D. Fleischer et al., *Anal. Chem.*, **30,** 152 (1958). Schöniger flask method.

K. Hayashi et al., *Talanta*, **4,** 244 (1960). Use of lanthanum chloranilate.

J. W. Robinson and P. W. West, *Microchem. J.*, **1,** 93 (1957). Selective test for orthophosphate.

W. I. Stephen, *Ind. Chemist*, **37,** 86 (1961).

C. M. Welch and P. W. West, *Anal. Chem.*, **29,** 874 (1957). Microdetermination, using *o*-dianisidine molybdate.

H. H. Willard, L. L. Merritt, and J. A. Dean, *Instrumental Methods of Analysis*, Van Nostrand, New York, 1958, p. 17. Color test for phosphate ion.

Detection of Boron

T. S. Burkhalter and D. W. Peacock, *Anal. Chem.*, **28,** 1186 (1956). Spot test for boron using sorbitol.

R. Ruggieri, *Anal. Chim. Acta*, **25,** 145 (1961). 1,8-Dihydroxy-anthraquinone as a reagent for boric acid.

Detection of Metals

R. Belcher et al., *Mikrochimie*, **40,** 76 (1952). A review on the microdetermination of metals with 179 references.

R. Belcher and J. W. Robinson, *Mikrochim. Acta*, **1954,** 49. Microdetermination of potassium by phosphomolybdate.

I. Stone, *Ind. Eng. Chem., Anal. Ed.*, **5,** 220 (1933). Microtest for mercury.

8

Classification by Solubility

The determination of the solubility of organic compounds in selected solvents and the use of the solubility behavior in classifying such compounds for identification purposes has been common practice for many years. Classification by solubility, then, is one of the important steps for the systematic identification of an unknown substance. These steps are (1) determination of the physical constants (Chapter 5) and the elements present (Chapter 7), (2) the solubility behavior, (3) the acid-base character (Chapter 9) and the detection of the major functional groups that are present in the molecules (Chapter 10), and (5) the preparation of one or more derivatives (Chapters 12 to 21). By considering a combination of the facts learned from elemental analysis, the solubility, and the acid-base character of the unknown, decisions may be made as to a logical sequence for making the tests for the functional groups that may be present.

The process of one substance dissolving in another is not as simple a phenomenon as that often implied in elementary chemistry courses. All liquids and solids tend to retain their physical identity and composition because of interparticle forces acting among the ions or molecules of the substance. Hence, if one substance is to *dissolve* in another, the interparticle forces that are involved in maintaining the physical identity of both the *solute* and the *solvent* must be broken before the individual particles of these two substances may admix. The energy required to break the interparticle bonding forces would come, if a solution is to result, from the release of energy due to the formation of new interparticle bonding forces between units of the solute and the solvent. Based on past research work, the type and quantitative energy value of the interparticle bonding forces of many molecular and ionic structures have been determined. The classification of compounds into Solubility Divisions, as proposed in this chapter, is based on practical observations as to the extent that various kinds of organic molecules dissolve in various solvents. If the structure of a compound is known and the types of bonds involved in that structure have been established, the solubility behavior of that compound in solvents of known structure and bond types can, in general, be predicted. When we have established the types of molecules that do or do not dissolve in each selected solvent, it is obvious that if we know the solubility behavior of an unknown substance we may make logical deductions about its structure and, thus, move toward its identification.

A detailed discussion of the theoretical explanations of solubility is unnecessary for the purposes of this book. It is assumed that the reader is at least moderately familiar with the various types of both intramolecular and intermolecular bonds. If further study in this field is desired, reference should be made to a modern text in organic chemistry and to the books listed at the end of this chapter. However, a brief discussion of bond types is included here to lay a groundwork for understanding the action of the solvents used for classification on various kinds of solutes. The nature of the ionic bond is too well known to need discussion. Within individual molecules, it is common practice to speak of nonpolar covalent bonds and polar covalent bonds, the latter being also referred to as a covalent bond with partial ionic character. In molecular orbital terminology, covalent bonds may be either *sigma* bonds or a combination of *sigma* and *pi* bonds. With regard to solubility classifications, the main importance of these intramolecular bonds is the effect that they have on the bonding potential of the molecule as a whole with other molecules, either molecules of the same substance or of a different substance.

In addition to the interaction of oppositely charged ions, three types of attractive forces are of importance concerning the possibility of solvation of a substance. These forces are (1) the van der Waals forces, (2) dipole-dipole interactions, and (3) ion-dipole interactions. The hydrogen bond is considered to be a special case of dipole-dipole interaction. Some combination of these forces is involved when any solute dissolves in any solvent. The applications will be discussed later in this chapter, in connection with the individual solvents; however, some basic principles and facts are appropriate here.

When molecules approach each other without forming an actual bond, there is a slight attraction between them, due to the mutual distortion of their electron clouds. Such molecules will be weakly bonded to each other by so-called van der Waals forces. Since London was the first to interpret these cohesion forces in terms of the quantum theory, they are often called the van der Waals–London forces, or dispersion forces. The energy involved in such intermolecular attractions is of the order 0.5 to 5 kcal/mole. In molecules that show little or no external dipole, rapid fluctuations occur in the distribution of charge of the electronic cloud, thus causing the molecules to have dipoles for very short periods of time and with continuous change in both direction and magnitude.

The covalent bond between two atoms involves a pair of electrons that belong to the charge-cloud of both partners. The distribution of the electrons around the two atoms is never uniform, except when the joined atoms are identical. If the molecule is so constructed that the centers of the positive and negative charges do not coincide, one or more internal dipoles will be created. In molecules of certain composition and structures, these internal dipoles will cancel each other; for example, molecules of carbon tetrachloride, 1,4-dichlorobenzene, and the like. Such molecules will not possess permanent molecular dipoles. The direction of the dipole moment is determined by the positions of the positive and negative charge centers, and the magnitude of the dipole moment by the product of the charge and the distance between the centers. Molecules with dipoles, like nonpolar molecules, possess polarizability which may be defined as the measure of the ease with which intramolecular displacements of the charge occur under the influence of an electrical field. The dielectric constant, or specific inductive capacity, is a measure of the ratio of electrical displacement to the electrical field intensity. Molecules with permanent dipoles exert forces on one another and associate with each other by dipole-dipole interaction. A most important application of polarity to solubility is the possibility of forming *hydrogen bonds*. The hydrogen bond results from a single proton sharing two sources of electrons rather than being satisfied by one. Usually, the proton serves as a bridge between two strongly electronegative atoms, being attached to one atom by a covalent bond and attached to the other atom by an electrostatic force, the strength of which is only 5 to 10 per cent of that of a covalent single bond. The energy of formation of a hydrogen bond is 5 to 10 kcal/mole. When covalently bonded hydrogen is the positive end of a dipole, such as —O—H or —N—H, it may form a hydrogen bond with atoms of oxygen, nitrogen, or fluorine in the same molecule (intramolecular hydrogen bonding, or *chelation*) or in another molecule (intermolecular bonding or *association*). Only atoms of these elements are strong enough in electronegativity or basicity to form an electrostatic attachment to a proton that is already covalently bound to another atom.

In representing the differences in the bonds on the proton in a hydrogen bond, it is customary to use a full line to show the covalent bond and a dotted line to show the electrostatic bond, thus

$$O—H--O, \quad O—H--F, \quad N—H--O, \quad \text{and the like.}$$

The ion-dipole bond is another force of attraction that is important in solubility, particularly if the solvent is water. Since the water molecule has hydrogen as the positive end of its dipole and oxygen as the negative end of the dipole, it may bond through hydrogen to anions or through oxygen to cations. The hydration of ions is a very important reason for their solubility in water. Of course, other molecules besides water are capable of ion-dipole interactions as, for example, amines to cations. The strength of ion-dipole bonds is much greater than that of hydrogen bonds described above, and may approach the strength of a covalent bond.

The Classification Solvents

The particular solvents used in this text are water, ethyl ether, $1.2N$ hydrochloric acid, $2.5N$ sodium hydroxide, $1.1N$ sodium bicarbonate, and concentrated sulfuric acid. The usefulness of each of these solvents will be discussed separately. It is to be noted that water is the actual solvent when the dilute solutions of hydrochloric acid, sodium hydroxide, and sodium bicarbonate are used. From the acid-base viewpoint, three acids are used to detect bases of varying degrees of basicity, namely, water, the hydronium ions in the hydrochloric acid solution, and concentrated sulfuric acid. Likewise, three bases are used: water, the hydroxide ion, and the bicarbonate ion. Ether, although it is a very weak base, is not used as a base in this scheme. Some workers may question the inclusion of concentrated sulfuric acid as a strong acid but, in this case, it should be pointed out that while the hydronium ion concentration is low, the tendency to transfer protons is very large and can be measured by Hammett's acidity function, H_0.

The classification of compounds into Solubility Divisions is based on the practical observations as to how well various kinds of substance dissolve in one or more of these solvents. For the purposes of this text, a compound is considered soluble in a solvent if 30 mg of it dissolve in 1 m of the solvent at room temperature. Obviously, a different classification would result if some other criterion of solubility were chosen. Furthermore, a different classification could be made by using different solvents, either instead of, or in addition to, the ones specified. For example, benzene may be substituted for the ethyl ether without much change in the classification results. On the other hand, the replacement of either ether or water by such a solvent as methanol, acetone, acetonitrile, or dimethylformamide would result in quite a different classification. Unfortunately, adequate data are not available for most of these solvents to justify a prediction of what the classifications by solubility would be. The best guide to their probable solvent action would be their respective dielectric constants. Phosphoric acid may be used in addition to sulfuric acid in an attempt to subclassify the water-insoluble neutral compounds. The 85 per cent phosphoric acid dissolves most of the oxygen-containing neutrals that contain less than ten carbon atoms.

Concentrated sulfuric acid (96 to 98 per cent) dissolves the alkenes and alkynes and the aromatic hydrocarbons that are easily sulfonated. The indications are that 80 per cent sulfuric acid will dissolve the alkenes and alkynes but not the aromatics. Berg and Parker[1] report that a mixture of one part glacial acetic acid to two parts 15 per cent fuming sulfuric acid will dissolve most aromatic hydrocarbons as well as the alkenes and alkynes. It should be stated that, for purposes of classification, a compound is considered "soluble" in sulfuric acid if it obviously reacts with the acid, even though the product of the reaction may be insoluble (a polymerized product from an alkene, for example).

Comparison of Water and Ether as Solvents

The process of producing a solution involves the properties of both the solute and the solvent. As was mentioned earlier, bonds must be broken in both the solute and the solvent and new bonds formed between the solute and the solvent if a solution is to result. Since water and ether are the primary classification solvents used, their differences in properties are important. There are three major differences between water and ether that affect their solvent action: (1) the great difference in their dielectric constants which, measured at 20°C, are 80.4 for water and 4.34 for ether; (2) water molecules are highly associated because of hydrogen bonding, whereas ether is unassociated since a hydrogen bond does not form between atoms of carbon and oxygen; and (3) water may act as both an acid and a base while ether is only a very weak base and is not an acid. In summary,

[1] C. Berg and F. D. Parker, *Anal. Chem.*, **20,** 456 (1948).

we may say that water is a strongly polar, highly associated solvent that has the potential of acting as either an acid or a base. Ether is essentially nonpolar, unassociated, and neutral. It is these differences that make it possible to classify, tentatively, organic compounds according to composition and structural types on the basis of their solubility in these solvents.

Solubility in Water

Electrolytes as solutes. In general, organic compounds that are highly ionized are soluble in water. Two oppositely charged ions separated by a certain distance will have only one eightieth the attraction for each other when they are in solution in water (dielectric constant equals 80) that they would have in a vacuum. However, since salts are much more soluble in water than they are in hydrogen cyanide (dielectric constant equals 115), it is evident that the dielectric constant alone is not the major cause of the solubility of ionic substances in water. The ionic species become hydrated due to the ion-dipole interactions between water molecules and the ions, since the oxygen atom of water, being the negative end of the water dipole, can interact with positive ions and the hydrogen atoms of water, being the positive ends of the dipole, can interact with negative ions. Thus, a cation in aqueous solution will be surrounded by oxygen atoms of water molecules, the number of which will be determined by the balance between the attractive forces and the steric forces that are involved. Anions would be surrounded by water molecules bonded through one of the hydrogen atoms. The water molecules that are directly bonded to the ions will, of course, retain some ability to become hydrogen bonded to other water molecules that are adjacent to them in the solvent. Since hydrogen atoms attached to carbon atoms do not normally become involved in hydrogen bonding, hydrogen cyanide, even though it has a higher dielectric constant than water, does not solvate ions appreciably and hence does not dissolve them to the extent that water does. As previously mentioned, the energy released by the formation of an ion-dipole bond is relatively high. The energy released by ion-hydrate bond formation is adequate to supply the energy necessary to separate the ion-ion bonds of most electrolytes and to break the hydrogen bonds of the associated water complex.

Acids and bases as solutes. The water molecule is both an acid and a base; hence, it may ionize an amine by donating a proton to the amine or it may ionize an acid by accepting a proton from the acid. These reactions may be illustrated thus:

$$B: + HOH \rightarrow B:H^+ + OH^-$$
$$RCOOH + H_2O \rightarrow RCOO^- + H_3O^+$$

The number of acids and bases that can be ionized by water alone is limited, and it is probable that most acids and bases that dissolve in water do so more by hydrogen bonding than as a result of ionization and the hydration of the ions thus produced. However, water-insoluble acids and bases may be converted into soluble ions by the reaction solvents discussed later in this chapter. Compounds that are only slightly soluble in water but that hydrolyze readily to give water-soluble products will, of course, dissolve.

Polar compounds as solutes. Nonionic substances do not dissolve in water unless they are capable of forming hydrogen bonds with water molecules. It should be recalled that hydrogen bonds of significant strength are only formed when the proton is sandwiched between two atoms that are strongly electronegative (strong Lewis bases) and that for practical purposes only the elements fluorine, oxygen, and nitrogen are involved. Therefore, hydrocarbons, the halogen derivatives of hydrocarbons, and the thiols (mercaptans) are very sparingly soluble in water, even considering the low molecular-weight members of these classes. Polar organic molecules contain hydrocarbon groups of varying magnitude. Whether a given polar compound will be soluble in water will depend on the effect on solubility of the polar group or groups that are present in the molecule relative to the effect of the nonpolar hydrocarbon group or groups that are also present. Considering our arbitrary basis for calling a substance "soluble," namely, 30 mg/ml, it has been found that one polar group capable of forming hydrogen bonds with water will make a molecule soluble in water if the hydrocarbon part of that molecule does not exceed four or five carbon atoms in a "normal" chain or five or six carbon atoms in a "branched" chain. The strength of the dipole of the potential solute has an influence on how many carbon atoms may be present. If more than one polar group is present in the solute molecule, the allowed ratio of carbon atoms is usually three or four to each polar group for water-soluble substances. For solubility purposes, the phenyl group* has approximately the effect of a normal butyl group. These "rules" are very useful observations but they should not be taken as infallible truths.

The effect of structure. While isomerism within the hydrocarbon group affects solubility, isomerism with respect to the location of the polar group within the molecule has a much greater effect. For example, 2-methyl-1-propanol is 20 per cent more soluble in water than 1-butanol, but 2-butanol is 50 per cent more soluble than 1-butanol, and 2-methyl-2-propanol is miscible in all proportions. The boiling point of a liquid is related to

---

* Since the word "radical" more properly implies "free radical," the word "group" will be used to avoid confusion.

the bonding energies among its molecules. Since these bonding energies must be broken in order for a liquid to be dispersed in another liquid, it is reasonable that, other things being equal, liquids with lower boiling points should be more soluble in a solvent. The solubilities of the isomeric pentanols in water illustrate the influence of these factors, as shown in Table 8.01.

It will be noted that for the primary alcohols, isomerism within the alkyl radical changes the solubility about 50 per cent of the normal pentanol

TABLE 8.01
SOLUBILITY OF THE PENTANOLS IN WATER[a]

Alcohol	BP	Solubility[b]
PRIMARY		
1-Pentanol	138	2.36
3-Methyl-1-butanol	134	2.85
2-Methyl-1-butanol	129	3.18
2,2-Dimethyl-1-propanol	113	3.74
SECONDARY		
2-Pentanol	120	4.86
3-Pentanol	116	5.61
3-Methyl-2-butanol	114	6.07
TERTIARY		
2-Methyl-2-butanol	102	12.15

[a] *Source.* P. M. Ginnings and R. Baum, *J. Am. Chem. Soc.*, **59**, 1111 (1937).
[b] Solubility is expressed in g/100 g water at 20°.

value, whereas shifting the hydroxyl group from the first to the second carbon in the normal chain more than doubles the solubility, and the 3-pentanol is even more soluble. Where the effects of both types of isomerism are combined, the solubility is much greater. The tertiary alcohol is more than five times as soluble in water as the normal primary alcohol. The inductive effect of the groups attached to the *alpha* carbon atom may be the most important factor in the solubility differences mentioned. Palit[2] considers the solubility differences to be due to the effect of the hydroxyl group on the electron displacements in the different alkyl radicals.

[2] S. R. Palit, *J. Phys. & Colloid Chem.*, **51**, 837 (1947).

Prediction of probable solubility based on the number of carbon atoms present in the radicals is more reliable when applied to liquids than to solids of equal carbon content, because the solid state involves greater intermolecular forces that must be overcome before solution may occur. If two solids have approximately the same heats of fusion, the one with the lower melting point will be the more soluble in any given solvent. Similarly, of two solids having approximately the same melting points, the one with the lower heat of fusion will have the greater solubility. Table 8.02

TABLE 8.02
DIBASIC ACIDS

Number of Carbon Atoms	Systematic Name	Common Name	MP	Solubility (g/100 g water at 20°)
2	Ethanedioic	Oxalic	189	9.5
3	Propanedioic	Malonic	135	73.5
4	Butanedioic	Succinic	188	6.8
5	Pentanedioic	Glutaric	97	64
6	Hexanedioic	Adipic	153	2
7	Heptanedioic	Pimelic	105	5
8	Octanedioic	Suberic	144	0.16
9	Nonanedioic	Azelaic	107	0.24
10	Decanedioic	Sebacic	133	0.10
4	2-Butene-1,4-dioic (*trans*)	Fumaric	300	0,7
4	2-Butene-1,4-dioic (*cis*)	Maleic	130	75

shows the relationship of both carbon content and melting point to solubility in the case of some dibasic acids in water.

It will be noted that the dibasic acids with an even number of carbon atoms have higher melting points and lower solubilities than the next higher homologue. Between *cis-trans* isomers, the *cis* form usually has the lower melting point and the higher solubility.

Jordan[3] has published very extensive data on the vapor pressures of organic compounds at various temperatures. Interesting studies may be made regarding water solubility of members of a given homologous series by noting which members of the series have vapor pressures of at least 5 mm at 20°C.

[3] T. E. Jordan, *Vapor Pressure of Organic Compounds*, Interscience, New York, 1954.

Solubility in Ethyl Ether

Nonpolar and slightly polar substances will, in general, dissolve in ether because, like ether itself, they are largely unassociated. Ionic compounds, such as salts, are not soluble in ether because the cation-anion attraction is too great to be overcome by the weak dipole of the ether and therefore the ions do not separate from their lattice structure. Whether or not a polar compound will dissolve in ether will depend on the influence of the polar group or groups relative to the influence of the nonpolar radical that is present. In general, compounds that have only one polar group per molecule will dissolve in ether unless they are very highly associated or of extreme polarity, as for example, the sulfonic acids.

Most organic compounds that are not soluble in water are soluble in ether. Hence, solubility in ether is not a useful criterion for classification by solubility except for those substances that are also soluble in water. Since water has a high dielectric constant and is associated, whereas ether has a low dielectric constant and is unassociated, the kinetics of their action as solvents will tend to be opposite. Some generalizations may be deduced from the foregoing discussions. If a compound is soluble in both water and ether, it most likely (1) is nonionic, (2) contains five or less carbon atoms, (3) has a functional group that is polar and capable of forming hydrogen bonds, and (4) does not contain more than one strongly polar group. If a compound is soluble in water but not in ether, it may (1) be ionic (salt), or (2) contain two or more polar groups with not more than four carbon atoms per polar group. As with all generalizations, there are exceptions to the statements just made.

Commercial ether contains small percentages of alcohol and water. These components increase the observed solubility above expectations for water-soluble and alcohol-soluble compounds that are sparingly soluble in ether. In the process of extracting a solute from water by ether or from ether by water, the two phases are actually not water and ether, but a saturated solution of ether in water and a saturated solution of water in ether. In terms of the arbitrary standards adopted for calling a compound "soluble," ether is soluble in water and water is soluble in ether. If, for example, benzoic acid is extracted from water by ether, much less of the acid will be recovered than would be expected according to the distribution law based on the solubility of the acid in pure water and in pure ether. Benzoic acid is much more soluble in water saturated with ether than it is in pure water, and less soluble in ether saturated with water than it is in pure ether.

Benzene may be substituted for ethyl ether in testing for solubility in case there is doubt as to whether the compound is soluble in ether. It may be assumed that the classification will be the same as if ether had been used but, as in all generalizations, there are exceptions.

Solubility in Dilute Hydrochloric Acid

Most water-insoluble compounds which are soluble in dilute hydrochloric acid have a basic nitrogen atom; generally, these will be amines. Hence, if nitrogen is not found during analysis for the elements, it is very unlikely that a substance will dissolve in the acidic solution any more than it does in water. It is true that some water-insoluble salts are dissolved by the acidic solution. A few rare-type molecules, such as the pyrones, do contain oxygen atoms that are basic enough to form salts with dilute hydrochloric acid. In the case of most water-insoluble oxygen-containing compounds the oxygen is so weakly basic that such a strong proton donor as concentrated sulfuric acid is required to produce ions by protonation. Such cases will be discussed later.

The basicity of an amine is due to the presence of an unshared pair of electrons on the nitrogen atom. Instrumental measurements show that in alkylamines, like in ammonia, the nitrogen uses sp^3 orbitals which are directed to the corners of a tetrahedron. Three of these orbitals overlap orbitals with hydrogen or carbon, while the fourth orbital contains an unshared pair of electrons. The strength of the amine as a base will be determined by the availability of this electron pair for reaction with an acid. Since amines will have one, two, or three alkyl or aryl groups attached to the nitrogen atom, and since these groups have varying degrees of electronegativity, it is obvious that amines will vary greatly with regard to their basicity. Alkyl groups have slightly less attraction for electrons than does hydrogen, whereas aryl groups have much more attraction for electrons than does hydrogen. The aliphatic amine is more basic than ammonia because the electron-releasing alkyl groups tend to disperse the positive charge on the substituted ammonium ion and therefore stabilize it as compared to the ammonium ion. From another point of view, the greater basicity of the amine could be explained by saying that the alkyl groups push electrons toward the nitrogen atom, and thus make the fourth electron pair more available for reaction with the incoming proton from the acid. The basicity constant, K_b, of the alkylamines is of the order of 10^{-4}, as compared with ammonia, K_b, 10^{-5}. The lower weight alkylamines are soluble in water while the others dissolve in dilute hydrochloric acid. These amines are stronger bases than water and weaker bases than the hydroxide ion:

$$RNH_2 + H_3O^+ \rightarrow RNH_3^+ + H_2O$$

$$RNH_3^+ + OH^- \rightarrow RNH_2 + H_2O$$

Whereas alkylamines are slightly stronger bases than ammonia, the arylamines are very much weaker, with basicity constants, K_b, of 10^{-10} or less. The low basicity of the arylamines is due to the fact that the amine is stabilized by resonance more than is the ion, which it can form by combining with a proton. The more stable the molecule is, relative to the ion, the less basic is the amine. Petrarca[4] has reported that an aqueous solution of acetic acid buffered with sodium acetate to pH 5.5 will dissolve alkylamines but not arylamines.

A much more detailed discussion of the basicity of amines will be found in the next chapter, which deals with acid-base character and the classification of compounds on this basis. Not all arylamines will dissolve in dilute hydrochloric acid to the extent of 30 mg/ml. Two or more aryl groups on the same nitrogen, or strong electron-withdrawing groups on the ring weakens the basic property of the amine to such an extent that it does not dissolve in $1.2N$ hydrochloric acid.

Thus, compounds like diphenylamine, most nitro- and polyhaloarylamines are not soluble in dilute hydrochloric acid and are classed in the miscellaneous Solubility Division **M**. Water-insoluble amides do not dissolve in $1.2N$ hydrochloric acid, but some of the mono- or dialkyl-substituted amides do dissolve in $3N$ or $6N$ acid. A few amines that would be predicted to be soluble fail to form solutions because of the low solubility of their chloride salts; 1-naphthylamine is a good example. Hydrazines, except such compounds as 2,4-dinitrophenylhydrazine, are soluble in the dilute acid. The amino derivatives of carboxylic acids aminophenols, aminothiophenols, aminosulfonamides, and ketoximes with less than eight carbon atoms, are amphoteric and, if water-insoluble will be soluble in the dilute acid. Aminosulfonic acids are not soluble in water or in dilute acid but are soluble in bases.

Using general formulas (alkyl=R; aryl=Ar), the more common types of nitrogen compounds may be summarized with regard to their probable solubility as follows.

---

[4] A. E. Petrarca, *J. Org. Chem.*, **24**, 1171 (1959).

Intermediate bases. RNH_2, R_2NH, R_3N, and $Ar(CH_2)_nNH_2$. Low-molecular weight amines with only one basic group are soluble in both water and in ether. The higher weight compounds are soluble in dilute acid.

Weak bases. $ArNH_2$, $ArNHR$, $ArNR_2$, some $RCONR_2$, and amphoteric compounds. These types are generally not soluble in water but are generally soluble in $1.2N$ hydrochloric acid unless the aryl group is substituted by strong electronegative groups. (Aniline is soluble in water on the basis of the criterion set.)

Almost neutral. Ar_2NH, Ar_2NR, Ar_3N, $RCONH_2$, $RCONHR$, many $RCONR_2$, $ArCONH_2$, $ArCONHR$, and $ArCONR_2$. These are not soluble in $1.2N$ hydrochloric acid; some are soluble in $3N$ or $6N$ hydrochloric acid.

Solubility in Dilute Sodium Hydroxide

Water-insoluble compounds that are capable of donating a proton to a base in aqueous solution may form water-soluble products. The stronger the acid, the weaker the base may be that is required to cause a reaction. Three bases are used in this scheme for classification by solubility, namely, water, the bicarbonate ion, and the hydroxyl ion. The use of water as a base has been discussed. The usefulness of the bicarbonate ion will be covered in a later section, which compares solubilities in $1.1N$ sodium bicarbonate solution with solubilities in $2.5N$ sodium hydroxide solution. For water-insoluble compounds, aqueous sodium hydroxide may be considered as a detecting solvent and aqueous sodium bicarbonate as a subclassifying solvent. Substances that are soluble in both the sodium hydroxide and the sodium bicarbonate solutions are classified in Solubility Division A_1 while substances that are soluble in the sodium hydroxide solution but not in the more weakly basic bicarbonate solution are classified in Solubility Division A_2 (see Table 8.03, pages 324–325). A more detailed discussion of the relative strength of acids is given in the next chapter.

Acids, usually, owe their acidic character to the presence in the molecule of —OH, —SH, or =NH groups that are attracted to an electronegative group or groups. The relative electronegativity of the attached group or groups determines, in most cases, the strength of the substance as an acid and, thus, determines whether the substance will be soluble in both sodium hydroxide and sodium bicarbonate solutions or only in the former. The order of electronegativity of the common groups is

$$\text{sulfonyl} > \text{aroyl} > \text{acyl} > \text{aryl} > \text{alkyl}$$

Substituents for hydrogen atoms on any of these groups will increase or decrease the electronegativity of the group as a whole, dependent on whether the substituent is electron-withdrawing or electron-contributing as compared with hydrogen. Classes of compounds in which the proton is removed from a hydroxyl group include the sulfonic, sulfinic, and carboxylic acids; phenols, oximes, enols, hydroxamic acids, and the *aci* forms of primary and secondary nitro compounds. Thiophenols and mercaptans represent classes in which the proton leaves a sulfur atom (—SH group). Classes in which the proton may be removed from nitrogen by the hydroxyl ion include the sulfonamides and N-monoalkyl sulfonamides, the imides of both aliphatic and aromatic acids, and some N-monomethyl aromatic amides. It will be noted that one sulfonyl radical, when substituted for hydrogen in ammonia, is sufficient to make the hydrogen on nitrogen ionizable, whereas two acyl or aroyl radicals (or, in some cases one aroyl radical and a methyl radical) on nitrogen are necessary to make the remaining hydrogen atom ionizable.

The carboxylic acids, phenols, thiophenols, and mercaptans, if not soluble in water, dissolve in dilute sodium hydroxide because of the production of water-soluble salts of conventional types. The sulfonamides ionize to yield ions of the general type $ArSO_2N(H)^-$. Hydroxamic acids and oximes produce the $RCON(H)O^-$ and $RC(H)=NO^-$ or $R_2C=NO^-$ ions respectively. The ionization of imides may be explained by two different mechanisms: either direct ionization of hydrogen from nitrogen, or enolization. Thus, the ion of succinimide might be written as either of these formulas:

$$\begin{array}{cc}
H_2C-C-O^- & H_2C-C=O \\
| \quad \searrow & | \quad \searrow \\
| \quad \quad N & | \quad \quad N^- \\
| \quad \swarrow & | \quad \swarrow \\
H_2C-C=O & H_2C-C=O
\end{array}$$

Primary and secondary nitro compounds usually dissolve in dilute sodium hydroxide. One explanation for their solubility is based on the formation of the *aci* isomer, thus:

$$CH_3N\overset{O}{\nearrow}=O \rightleftharpoons CH_2=N\overset{O}{\nearrow}-OH \rightleftharpoons CH_2=N\overset{O}{\nearrow}-O^- + H^+$$

$$(CH_3)_2\overset{H}{\underset{}{C}}-N\overset{O}{\nearrow}=O \rightleftharpoons (CH_3)_2C=N\overset{O}{\nearrow}-OH \rightleftharpoons (CH_3)_2C=N\overset{O}{\nearrow}-O^- + H^+$$

Another explanation for the solubility of these nitroparaffins is based on the resonance stabilization of the ion form in alkaline solution:

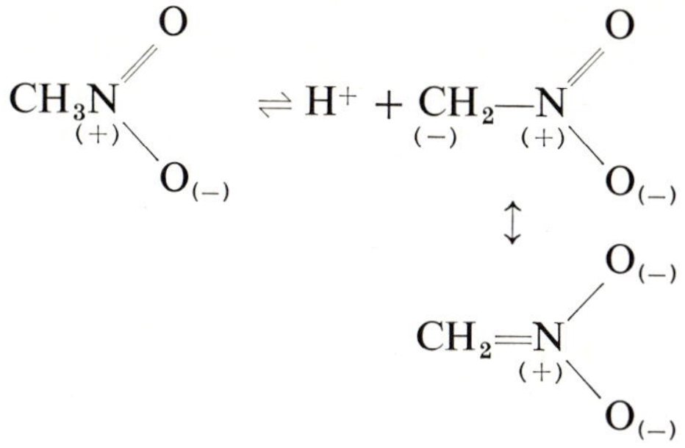

$$CH_3N^{(+)}(=O)(O^{(-)}) \rightleftharpoons H^+ + CH_2^{(-)}-N^{(+)}(=O)(O^{(-)}) \leftrightarrow CH_2=N^{(+)}(O^{(-)})(O^{(-)})$$

The enolic forms of triacylmethanes, 1,3-diketones, and beta-ketoesters contain the —C(OH)=C— grouping that is characteristic of the phenols. Hence, enols and phenols have many reactions in common, including the reaction with alkaline solutions to form water-soluble salts. The K_a of 2,4-pentanedione is 5.8×10^{-9}. The K_a's of the nitroparaffins are of the order of 10^{-8} to 10^{-11}.

Many of the amphoteric substances are soluble in water, but if they are not water-soluble, they dissolve in both dilute hydrochloric acid and dilute sodium hydroxide. Amino-substituted carboxylic acids, aminophenols, aminothiols, and the oximes of low-molecular-weight ketones constitute the major classes of amphoteric compounds. Amino-substituted sulfonic acids are not amphoteric; they are soluble in alkali but not in dilute acids.

Chelated phenols do not dissolve in dilute sodium hydroxide solution as readily as unchelated isomers of the same compounds. Thus, ortho nitrophenol and salicylaldehyde dissolve much more slowly and to somewhat less extent than their meta and para isomers. These results are owing to the partial inactivation of the compound as an acid due to intramolecular hydrogen bonding:

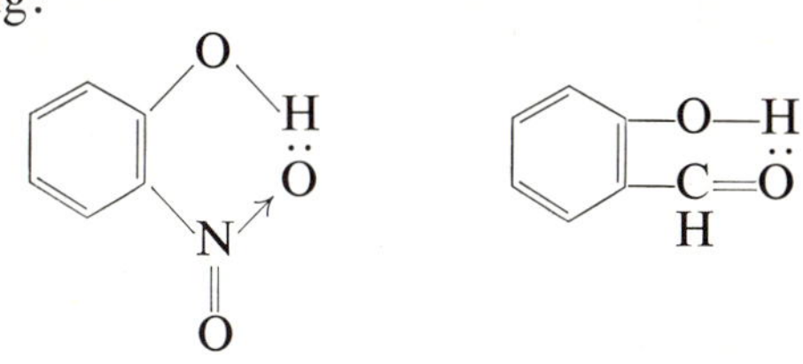

Solubility in Dilute Sodium Bicarbonate

The use of sodium bicarbonate solution only serves to distinguish relatively strong acids from weak acids. Indicators (Chapter 9) serve the same purpose and should be used, particularly if the solubility difference is inconclusive. The most common classes of compounds that are soluble in both 2.5N sodium hydroxide and 1.1N sodium bicarbonate are the carboxylic and sulfonic acids, acid anhydrides, and acid halides. Polynitrophenols, amino-substituted aromatic acids, and N,N-diarylamino acids are soluble in the bicarbonate solution if only one amino group is present in the molecule. Carboxylic acids that contain ten or more carbon atoms form colloidal dispersions (soaps) in sodium hydroxide, and disperse so slowly in the bicarbonate solution that they are often not properly classified as soluble unless the mixture is shaken for several minutes. Carboxylic acids have K_a's in the range of 10^{-3} to 10^{-5} and the sulfonic acids are still more acidic. Phenols, imides, sulfonamides, and the other classes of weak acids that are soluble in sodium hydroxide but not in sodium bicarbonate have K_a's of 10^{-9} or less in most cases. These compounds are stronger acids than water but are less strong than carbonic acid (K_a of about 10^{-7}), therefore these acids are not converted to their conjugate bases by the bicarbonate ion and hence do not dissolve.

Solubility in Concentrated Sulfuric Acid

Concentrated sulfuric acid is a very effective proton donor and thus is capable of protonating even very weak bases. Also, because of the equilibrium

$$2H_2SO_4 \rightleftharpoons H_3O^+ + HSO_4^- + SO_3$$

the strong Lewis acid, SO_3, is present and can convert easily sulfonated aromatic hydrocarbons to the ionic sulfonic acids. Furthermore, concentrated sulfuric acid has a very high dielectric constant (estimated as being 110 or higher) and thus favors the electrolytic dissociation of ionic or ion-dipole products produced because of its other reactions with solutes. The solubility of a substance is not tested in sulfuric acid until it has been established that it is not soluble in water, or in the dilute aqueous solution of hydrochloric acid and sodium hydroxide. Since most compounds that contain either nitrogen or sulfur, and that have failed to dissolve in the previously used solvents, dissolve in concentrated sulfuric acid because they form electrolytes with this acid, these compounds are arbitrarily placed in a miscellaneous Solubility Division **M**. Compounds that do not contain nitrogen or sulfur and that are soluble in concentrated sulfuri acid are classed in Solubility Division N. Compounds that obviously react chemically with the sulfuric acid, even if the product is not soluble in the acid, are classed as "soluble" for purposes of classification.

Three major types of compounds are classified in Division N, namely (1) oxygen-containing compounds (except diaryl ethers and perfluoro compounds that contain oxygen), (2) alkenes and alkynes, and (3) aromatic hydrocarbons that are quite easily sulfonated, such as the meta isomers of di- and trialkyl-substituted benzenes and hydrocarbons that have three or more aromatic rings.

The oxygen-containing compounds classed in Solubility Division N are all weak Lewis bases that react with the concentrated sulfuric acid to form either ionic products or ion-dipole products which, in either case, dissolve because of the high dielectric constant of this acid. In the case of alcohols, it seems probable that the mechanism is (1) protonation of the alcohol (ROH_2^+), (2) dehydration of the protonated alcohol to leave the carbonium ion (R^+), a Lewis acid, and then (3) the reaction of the carbonium ion with the HSO_4^- to form an alkylsulfuric acid which is soluble in the excess acid. Instead of reacting with the bisulfate ion to form the alkylsulfuric acids, the carbonium ions may rearrange to produce alkenes, which could also react with the acid or, in certain cases, they would polymerize. For purposes of classification, substances that react with the sulfuric acid are classed as "soluble" in it.

Ether molecules are protonated by the sulfuric acid to the conjugate acids (R_2OH^+). The fact that the proton is actually transferred to the ether is indicated by the fact that the freezing point of 100 per cent sulfuric acid is lowered approximately twice the theoretical amount when ether is dissolved in it. Since water is a stronger base than ether, if the solution of an ether in concentrated sulfuric acid is diluted sufficiently with water (the ether-concentrated sulfuric acid mixture should be poured over chipped ice to cause the dilution), the ether may be recovered. These reactions may be illustrated as

$$R_2O + H_2SO_4 \rightleftharpoons R_2OH^+ + HSO_4^-$$

$$R_2OH^+ + H_2O \rightleftharpoons R_2O + H_3O^+$$

Aldehydes, ketones, esters, and other oxygen-containing compounds are also protonated by the acid with the proton being attracted to the negative end of the dipole of the carbonyl group. A few carbonyl compounds polymerize in concentrated sulfuric acid.

The alkenes and alkynes are also weak Lewis bases, due to the extra pair (or pairs) of electrons in the pi cloud. The addition of the proton converts the hydrocarbon to a carbonium ion, which is the conjugate acid of the hydrocarbon base. For example:

$$(C_6H_5)_2C=CH_2 + H_2SO_4 \rightleftharpoons (C_6H_5)_2\overset{+}{C}-CH_3 + HSO_4^-$$

Relatively few aromatic hydrocarbons dissolve in concentrated sulfuric acid at room temperature; hence, most aromatic hydrocarbons fall in Solubility Division **I** (insolubles) together with the halogen-substituted hydrocarbons, the diaryl ethers, and most of the perfluoro compounds.

Consult Table 8.03 for listings of the types of compounds that may be expected in each of the Solubility Divisions.

Solubility Determinations

For the purposes of this text, a compound is considered soluble in a solvent if 30 mg of it dissolve in 1 ml of the solvent at room temperature. In doubtful cases, the mixture should be shaken for 2 minutes before a decision is reached as to whether or not the compound has dissolved. The use of a narrow, 3-in. test tube is recommended for the solubility determinations.

The solvents to be used for the solubility classifications are, in the order in which they should be used, water, ether,* 1.2N aqueous hydrochloric acid, 2.5N aqueous sodium hydroxide, 1.1N aqueous sodium bicarbonate, and concentrated sulfuric acid. All solubilities are to be determined at room temperature. It is permissible to warm the mixture *slightly* during the test but the mixture should be cooled before a decision is reached as to its solubility.

It is not necessary to test every compound in all of the solvents; rather, the solvents should be used in the order listed above and the first solubility class found should be accepted. If the compound is soluble in water, its solubility may be tested in ether, but in no other solvent. If the compound is not soluble in water, it need not be tested with ether. Water-insoluble compounds that are soluble in sodium hydroxide should also be tested with sodium bicarbonate to distinguish between relatively strong acids and the weakly acidic compounds. If an obvious chemical reaction occurs in any solvent, the compound may be considered as soluble in that solvent.

* Benzene may be used instead of ether with about the same results as regards classification.

A few of the amine hydrochlorides are slightly soluble in 1.2N hydrochloric acid; hence, in these cases the original amine will react when it is first added to the aqueous acid to form the amine hydrochloride and will then precipitate as that salt. To determine whether the insoluble material is the original amine or a salt that has been formed, a "mixed" melting-point determination should be made, using the *washed and dried* solid from the acid solution with some of the original sample.

In making the solubility tests, bear in mind the elements that were found present in the compound. Unless nitrogen is present, there is no need to test for solubility in hydrochloric acid. If the compound contains nitrogen or sulfur and is insoluble in water, hydrochloric acid, and sodium hydroxide, it is arbitrarily placed in a special "miscellaneous" division without a test for its solubility in concentrated sulfuric acid.

Other concentrations of the HCl, NaOH, and $NaHCO_3$ can sometimes be used to advantage. If it appears that any of these reagents has reacted with the original compound to form a slightly soluble salt, more water may be added to dissolve the salt. In some cases, additional information may be gained by using solutions of these reaction solvents that are twice as concentrated as the regular reagents. For example, amines of low basicity, such as the nitroanilines, may not be classified as "soluble" in 1.2N HCl but they do react with 3N HCl. One point to keep in mind, however, in using more concentrated HCl is that many substituted amides, such as the N-alkyl acetanilides, are basic enough to dissolve in the stronger acid but not in the 1.2N acid solution.

If there is doubt as to whether or not an appreciable amount of a substance has dissolved in the HCl, NaOH, or $NaHCO_3$ reagents, the clear solution should be separated from the excess of the compound and neutralized to see if the original compound will separate.

The quantity of the compound to be used for solubility determinations may be conserved in several ways when necessary. For the determination of solubility in water or ether, a very few milligrams of the substance may be added to 2 to 3 drops of the solvent. The substance may be recovered from the ether by evaporation of the solvent. If the substance is not soluble in water, a few drops of 2N HCl may be added to the aqueous suspension to test the solubility in dilute acid. If the compound is insoluble in the acid solution, 6N NaOH may be added, dropwise, to change the acidic solvent to a basic solvent. Such methods are not equivalent to the regular methods as regards concentration but they are adequate for detecting lack of solubility and at least approximate the regular classification solvents for situations in which the quantity of compound to be examined is very limited.

If the amount of substance available for examination is very small, the solubility may be determined by the following procedure. Carefully weigh a clean, dry capillary tube (3 to 4 mm in diameter and 75 to 100 mm long, sealed at one end) on an analytical balance. Introduce 2 to 4 mg of the substance and again weigh the tube. By means of a capillary pipet, add 0.1 ml of the solvent for each 3 mg of the substance being tested, taking care to wash the substance down the sides to the tube while adding the solvent. A 1-mm glass rod may be used as a stirring rod.

Designation for the Solubility Divisions

Division S_1. This division includes the compounds that are soluble in both water and ether.

Division S_2. This division includes the compounds that are soluble in water but insoluble in ether.

Division B. Compounds that are insoluble in water but soluble in 1.2N hydrochloric acid belong in this division. They all contain nitrogen. Not all the amines will dissolve in dilute acid, however, and many of them fall in Division **M**.

Division A_1. The compounds in this division are insoluble in water but soluble in both 1.1N sodium bicarbonate solution and 2.5N sodium hydroxide solution.

Division A_2. This division includes the compounds that are insoluble in water and insoluble in a 1.1N sodium bicarbonate solution, but which are soluble in a 2.5N sodium hydroxide solution.

Division M. Compounds that contain nitrogen or sulfur, and which have been found insoluble in all of the solvents used thus far, are placed in this miscellaneous division. The list of compounds that fall in this division is very long. Only the most common chemical classes are included in the solubility tables of this chapter. Halogens may, of course, be present in the compounds of this division.

Division N. Compounds that are soluble in concentrated sulfuric acid but are insoluble in the other solvents used belong in this division. Nitrogen and sulfur are both absent, since their presence would classify the compound in Division **M**.

Division I. Compounds that are insoluble in all the classification solvents and do not contain nitrogen or sulfur belong in this division.

An outline of the solubility classification procedures is provided.

AN OUTLINE FOR SOLUBILITY CLASSIFICATION

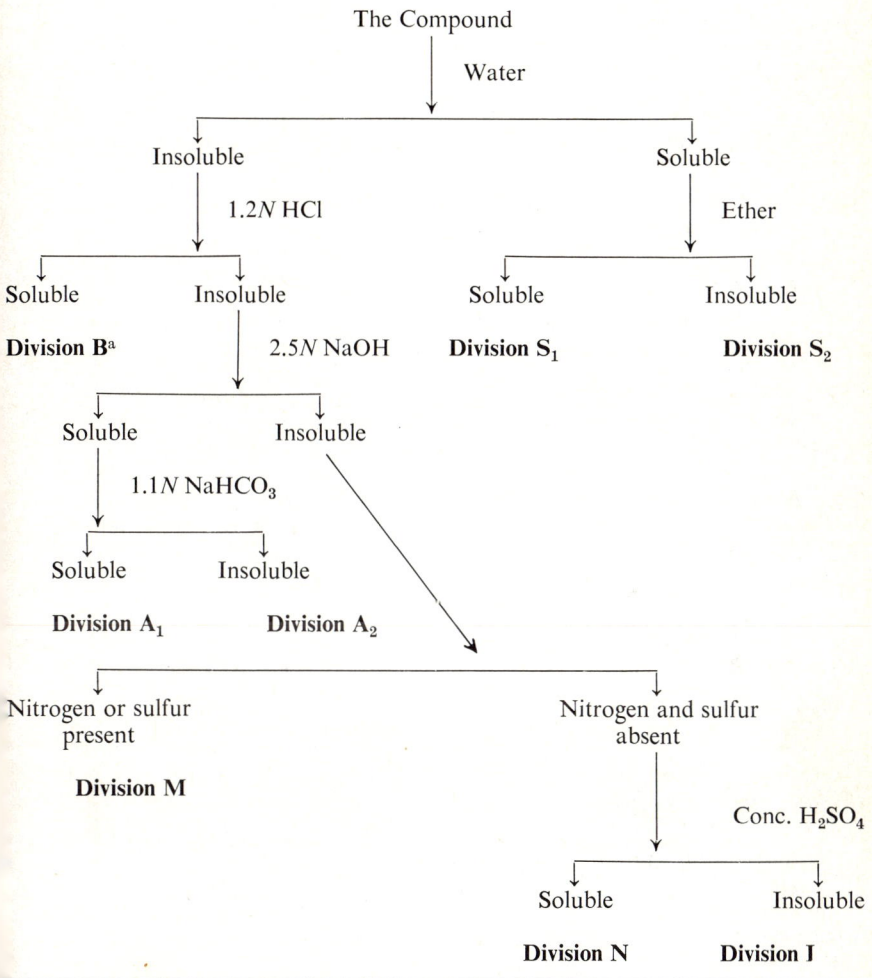

[a] If the water-insoluble compound is soluble in HCl, determine its solubility in NaOH to detect amphoteric compounds.

TABLE 8.03
DIVISIONAL SOLUBILITY CLASSIFICATIONS[a,d]

Division S_2[b]

1. Only C, H, and O present:
   DIBASIC AND POLYBASIC ACIDS[d]
   HYDROXY ACIDS
   POLYHYDROXY ALCOHOLS
   POLYHYDROXY PHENOLS
   Simple carbohydrates
2. Metals present:
   SALTS OF ACIDS AND PHENOLS
   Miscellaneous metallic compounds
3. Nitrogen present:
   AMINE SALTS OF ORGANIC ACIDS
   AMINO ACIDS
   AMMONIUM SALTS
   Amides
   Amines
   Amino alcohols
   Semicarbazides
   Semicarbazones
   Ureas
4. Halogen present:
   HALO ACIDS
   Acyl halides (by hydrolysis)
   Halo alcohols, aldehydes, etc.
5. Sulfur present:
   SULFONIC ACIDS
   Alkyl sulfuric acids
   Sulfinic acids
6. Nitrogen and halogen present:
   Amine salts of halogen acids
7. Nitrogen and sulfur present:
   AMINO SULFONIC ACIDS
   BISULFATES OF WEAK BASES
   Cyano sulfonic acids
   Nitro sulfonic acids

Division S_1[c]

1. Only C, H, and O present:
   ALCOHOLS
   ALDEHYDES AND KETONES
   CARBOXYLIC ACIDS
   Acetals
   Anhydrides
   Esters
   Ethers
   Some glycols
   Lactones
   Polyhydroxy phenols
2. Nitrogen present:
   AMIDES
   AMINES
   Amino heterocyclics
   Nitriles
   Nitro paraffins
   Oximes
3. Halogen present:
   Halogen-substituted compounds of 1 above
4. Sulfur present:
   Hydroxy heterocyclic sulfur compounds
   Mercapto acids
   Thio acids
5. Nitrogen and halogen present:
   Halogenated amines, amides, and nitriles
6. Nitrogen and sulfur present:
   Amino heterocyclic sulfur compounds

Division B

AMINES[e]
Amino acids
Amphoteric compounds (e.g., amino phenols, amino thiophenols, amino sulfonamides)
Aryl substituted hydrazines
N, N-Dialkyl amides
Some salts

Division A_1

1. Only C, H, and O present:
   ACIDS[f] and ANHYDRIDES
2. Nitrogen present:
   AMINO ACIDS
   NITRO ACIDS
   Cyano acids
   Heterocyclic nitrogen carboxylic acids
   Imides
   Polynitro phenols
3. Halogens present:
   HALO ACIDS
   ACID HALIDES
   Polyhalo phenols

---

[a] Nitrogen, halogens, and sulfur are absent unless specified.
[b] Moderate-weight compounds with two or more polar groups, except for the sulfonic and sulfinic acids when only one polar group is necessary.
[c] Generally, mono functional compounds with 5 carbons or less.
[d] In this table, the more common classes are printed in SMALL CAPITAL letters.
[e] Amines with sufficiently strong negative substituents as well as diaryl and triarylamines fall in **Division M**.
[f] Generally with 10 carbons or less; many form colloidal soap solutions.

TABLE 8.03 (Continued)
DIVISIONAL SOLUBILITY CLASSIFICATIONS

4. Sulfur present:
   SULFONIC ACIDS
   Sulfinic acids
5. Nitrogen and sulfur present:
   Amino sulfonic acids
   Nitro thiophenols
   Sulfates of weak bases
   Sulfonamides
6. Sulfur and halogens present:
   SULFONYLHALIDES

Division A_2

1. Only C, H, and O present:
   ACIDS[g]
   ANHYDRIDES
   PHENOLS, including esters of phenolic acids
   Enols
2. Nitrogen present:
   AMINO ACIDS
   NITRO PHENOLS
   Amides[h]
   Amino phenols
   Amphoteric compounds
   Cyano phenols
   Imides
   N-monoalkyl aromatic amides
   N-substituted hydroxylamines
   Oximes
   p- and s-Nitroparaffins
   Trinitro aromatic hydrocarbons
   Ureides
3. Halogens present:
   HALO PHENOLS
4. Sulfur present:
   Mercaptans (thiols)
   Thiophenols
5. Nitrogen and halogen present:
   Polynitro halogenated aromatic hydrocarbons
   Substituted phenols
6. Nitrogen and sulfur present:
   Amino sulfonamides
   Amino sulfonic acids
   Amino thiophenols
   Sulfonamides
   Thioamides

Division M[i]

1. Nitrogen present:
   ANILIDES AND TOLUIDIDES
   AMIDES AND IMIDES
   NITRO ARYLAMINES
   NITRO HYDROCARBONS
   Amino phenols
   Azo, hydrazo, and azoxy compounds
   Di- and triarylamines
   Dinitro phenylhydrazines
   Nitrates
   Nitriles
2. Sulfur present:
   N-dialkyl sulfonamides
   Sulfates; sulfonates
   Sulfides; disulfides
   Sulfones
   Thio esters
   Thiourea derivatives
3. Nitrogen and sulfur present:
   Sulfonamides
   Thiocyanates
4. Nitrogen and halogen present:
   Halogenated amines, amides, nitriles, and nitro compounds

Division N[j]

ALCOHOLS
ALDEHYDES AND KETONES
ESTERS
ETHERS
UNSATURATED HYDROCARBONS[k]
Acetals
Anhydrides
Lactones
Polysaccharides[l]

Division I

HYDROCARBONS[m]
Halogen derivatives of hydrocarbons
Diaryl ethers
All perfluoro esters, ethers, aldehydes, and ketones

---

[g] High-molecular-weight acids form colloidal soaps.
[h] Including N-monoalkyl amides.
[i] Only the most common classes are listed.
[j] Halogens may be present as substituents.
[k] Noncyclic unsaturated hydrocarbons, and those unsaturated cyclics that are easily sulfonated, such as di- or polyalkyl-substituted benzenes.
[l] Char in the acid.
[m] Including most of the cyclic hydrocarbons, and all of the saturated, noncyclic hydrocarbons.

REFERENCES

Selected Books

L. F. Audrieth and J. K. Kleinberg, *Nonaqueous Solvents*, Wiley, New York, 1953, Chapter 9.

G. I. Brown, *An Introduction to Electronic Theories of Organic Chemistry*, Longmans Green, New York, 1958; Chapters 4, 5, and 8.

C. A. Coulson, *Valence*, Oxford Univ. Press, Oxford, 1952; Chapter 12.

F. Feigl, *Spot Tests in Organic Analysis*, Elsevier, Amsterdam, 1960; pp. 148–155.

L. N. Ferguson, *Electronic Structures of Organic Molecules*, Prentice-Hall, New York, 1952; Chapters 2 and 6.

E. S. Gould, *Mechanism and Structure in Organic Chemistry*, Henry Holt, New York, 1959; Chapters 1, 2, and 3.

D. Hadži, *Hydrogen Bonding*, Pergamon Press, New York, 1959; particularly pp. 339 ff., 423 ff., and 443 ff.

L. P. Hammett, *Physical Organic Chemistry*, McGraw-Hill, New York, 1940.

P. H. Hermans, *Introduction to Theoretical Organic Chemistry*, Elsevier, Amsterdam, 1954; Chapters 6 and 10.

C. K. Ingold, *Structure and Mechanism of Organic Chemistry*, Cornell Univ. Press, Ithaca, N.Y., 1953; Chapters 1, 3, and 7.

J. A. A. Ketelaar, *Chemical Constitution*, Elsevier, Amsterdam, 1953; Chapter 5.

E. A. Maelwyn-Hughes, *Physical Chemistry*, Pergamon Press, New York, 1957; Chapters 7 and 17.

L. Pauling, *Nature of the Chemical Bond*, 3rd ed., Cornell Univ. Press, Ithaca, N.Y., 1960; Chapter 12.

G. C. Pimental and A. L. McClellan, *The Hydrogen Bond*, Freeman Co., San Francisco, 1960; a comprehensive text with 2241 references.

A. E. Remick, *Electronic Interpretation of Organic Chemistry*, 2nd ed., Wiley, New York, 1949; Chapter 5.

A. Seidell, *Solubility of Organic Compounds*, 3rd ed., Van Nostrand, New York, 1941, and supplement to 3rd edition, 1952 (with W. F. Linke).

C. P. Smythe, *Dielectric Behavior and Structure*, McGraw-Hill, New York, 1955; pp. 202–392.

Y. K. Syrkin and M. E. Dyatkina, *Structure of Molecules and the Chemical Bond*, Interscience, New York, 1950; Chapters 6 and 12.

W. A. Waters, *Physical Aspects of Organic Chemistry*, Routledge and Sons, London, 1950; Chapters 4 and 11.

G. W. Wheland, *Advanced Organic Chemistry*, 3rd ed., Wiley, New York, 1960; Chapters 1, 4, and 11.

Selected Journal References

I. Fischer-Hjalmans and R. Grahn, *Acta Chem. Scand.*, **12**, 584 (1958); hydrogen bonding.

J. Hildebrand, *Science*, **83**, 21 (1936); hydrogen bonding.

J. Hildebrand, *Chem. Rev.*, **44**, 37 (1949); theory of solubility of nonelectrolytes.

M. L. Huggins, *J. Org. Chem.*, **1**, 407 (1936); hydrogen bonding.

T. Kiba et al., *Bull. Chem. Soc. Japan*, **30,** 482 (1957); phosphoric acid as a classifying solvent.

S. Ohashi and H. Sugatani, *Bull. Chem. Soc. Japan*, **30,** 864 (1957); phosphoric acid as a classifying solvent.

M. A. Paul and F. A. Long, *Chem. Rev.*, **57,** 1 (1957); H_0 and related indicator acidity functions.

H. O. Pritchard and H. A. Skinner, *Chem. Rev.*, **55,** 745 (1955); electronegativity.

L. Rosenthaler, *Pharm. Ztg.*, **100,** 475 (1955); trichloroacetic acid as a solvent.

S. Saloway and P. Rosen, *Anal. Chem.*, **29,** 1820 (1957); classification by salvochromic and thermochromic methods in connection with solubility and acid-base methods.

F. Schneider and D. G. Foulke, *Anal. Chem.*, **10,** 104, 445 (1938); capillary method.

F. Schneider et al., *Mikrochim. Acta*, **1960,** 967; **1959,** 801; determination of solubility by a vapor method.

E. L. Skau and R. E. Boucher, *J. Phys. Chem.*, **58,** 460 (1954); calculation of the solubilities of missing members of a series.

O. E. Schultz and R. Gmelin, *Arch. Pharm.*, **1954,** 287, 344; trichloroacetic acid as a solvent.

9

Classification by the Indicator Method

The classification of a compound by its solubility behavior and the elements present in it is usually adequate as a basis for proceeding with functional-group tests. However, particularly with water-soluble compounds and compounds that show borderline solubilities, further aid may be obtained for classification purposes by determining the relative acidity or basicity of the compound. Most low-molecular weight compounds are soluble in water, hence no chemical classification is deducible from their solubility determination. High-molecular weight acids or amines may fail to dissolve sufficiently in the classification solvents to be properly classified on that basis, and yet many of these compounds may be successfully detected by the indicator method. For example, stearic acid is frequently classified as "insoluble" in dilute sodium hydroxide; it is properly classified as an intermediate acid by the indicator system. Furthermore, there is some advantage in knowing that an acid or a base is strong, intermediate, or weak as determined by indicators. In general, the solubility classification does not make such distinctions. Hence, it is felt that the indicator method of classification is a useful supplement to the method of classification based on solubility and on the elements present. The method used in this chapter is an adaptation of the work of Davidson.[1]

[1] D. Davidson, *J. Chem. Educ.*, **19**, 221 (1942); reprinted as "More Acids and Bases," *J. Chem. Educ.*, Easton, Pa., 1944; also D. Davidson and D. Perlman, *A Guide to Qualitative Organic Analysis*, Brooklyn College Bookstore, Brooklyn, N.Y., 1952.

The subject of acids and bases among organic compounds was covered in a general way in Chapter 8. Practical use of these facts was made in the classification of compounds on the basis of their solubility. The present chapter extends somewhat the considerations previously given. For a detailed coverage of the subject of acids and bases, reference should be made to the literature. A number of suggested references are given at the end of this chapter.

We shall consider an acid as a proton donor and a base as a proton acceptor. When an acid donates a proton, the residue is a base, called the conjugate base of that acid. Similarly, when a base accepts a proton, an acid is produced, which is called the conjugate acid of that base. An acid may be uncharged, positively charged, or negatively charged, but it is always true that the acid has one unit more positive charge than does its conjugate base. For example:

Acids	Bases	
$C_6H_5COOH \rightleftharpoons$	$C_6H_5COO^- + H^+$	(1)
$C_6H_5NH_3^+ \rightleftharpoons$	$C_6H_5NH_2 + H^+$	(2)
$HOOCCH_2COO^- \rightleftharpoons$	$^-OOCCH_2COO^- + H^+$	(3)

We may write such equilibria as

$$HA \rightleftharpoons H^+ + A^- \tag{4}$$

where HA designates the acid and A^- its conjugate base without, in either case, designating the electrical charge that is actually present. The acidity constant, K_a for the acid HA may then be written as

$$K_a = \frac{[H^+][A^-]}{[HA]} \quad \text{or} \quad [H^+] = K_a \cdot \frac{[HA]}{[A^-]} \tag{5}$$

The usual units for K_a are moles/liter. The values of K_a for organic acids are very small numbers; hence, it is often more desirable to use the negative logarithm of K_a, designated pK_a:

$$pK_a = -\log K_a \tag{6}$$

Since $pH = -\log [H^+]$, it is true that

$$pH = pK_a + \log \frac{[A^-]}{[HA]} \tag{7}$$

From Equation 7 we see that, if the concentration of the acid and its conjugate base are equal in a solution, the pH of the solution is equal to the pK_a of the acid. Even if the ratio of the concentration of the acid to its conjugate base varies by a hundredfold, it will still be true that the $pH = pK_a \pm 1$.

It is to be noted that in Equation 4 the terms HA and A$^-$ designate the acid and its conjugate base without reference to ionic charges, if any. Hence, if the terms of Equation 2 are substituted in Equation 5, the K_a obtained is the acidity constant of the anilinium ion:

$$K_a = \frac{[H^+][C_6H_5NH_2]}{[C_6H_5NH_3^+]} \tag{8}$$

It is becoming common practice to express the basicity constant of bases, K_b, as the K_a of their conjugate acids. However, if it is desired to express the basicity constant, it is obvious that since

$$H^+ + A^- \rightleftharpoons HA, \quad \text{then} \quad \frac{1}{K_a} = \frac{[HA]}{[H^+][A^-]} \tag{9}$$

Comparing the acidity constant of the anilinium ion as given by Equation 8, we see that the basicity constant of aniline is proportional to $\frac{1}{K_a}$.

$$\frac{1}{K_a} = \frac{[C_6H_5NH_3^+]}{[H^+][C_6H_5NH_2]} = \frac{[C_6H_5NH_3^+][OH^-]}{K_w[C_6H_5NH_2]} \tag{10}$$

If water is the solvent, Equation 10 conventionally would be written as

$$\frac{K_w}{K_a} = K_b = \frac{[C_6H_5NH_3^+][OH^-]}{[C_6H_5NH_2]} \tag{11}$$

Analogous to the case of acidity constants, the K_b may be expressed as pK_b:

$$pK_b = -\log K_b \tag{12}$$

Since K and pK values vary inversely, the stronger acids and bases will have the smaller pK_a or pK_b values.

The effect of solvents. In the above discussion, no consideration was given to what happens to the hydrogen ion in Equations 1, 2, and 3, nor to the source of the hydrogen ions in Equation 9. Obviously these are important matters in determining the ratio of the conjugated acids and bases that are involved. The strength of acids may be measured only in terms of equilibria involving at least two acids. Likewise, measurement of the strength of bases must involve equilibria with at least two bases present. Commonly, the solvent acts as one of the acids or bases. Individual solvents may be acids (protogenic), bases (protophilic), amphoteric (amphiprotic), or essentially neutral (aprotic). The nature of the solvent is one very important factor in determining the ratio of an acid to its conjugate base (or base to its conjugate acid) in a solution of that substance in the given solvent. If the solvent acts as a base, B, in Equation 4, we have

$$HA + B \rightleftharpoons BH^+ + A^- \qquad (13)$$

The stronger the base is, the more the equilibrium will shift to the right and the larger the K_a for the acid HA will be. For example, water is a stronger base than methanol. The K_a for benzoic acid in water as a solvent is 12 times as large as it is in a 50 per cent methanol-water solvent. If pyridine were used as the solvent, the K_a of benzoic acid would be much greater than in water. The weaker the acid is, the stronger the base needs to be to detect or measure the acid.

Basic substances, such as amines, must receive a proton from an acid in order to be converted to their conjugate acids, thus allowing the determination of their basicity constant, K_b. For a given solvent acting as an acid, the relative basicities of two amines may be compared. For example, using water as the solvent, it is determined that the K_b for an alkyl amine is about one million times the K_b for aniline. Water is too weak an acid to be useful in detecting very weak bases and some stronger acid, such as acetic acid, must be used. As was mentioned in the previous chapter, the very strong proton-donor, concentrated sulfuric acid, is needed to make the very weak bases such as the ethers and esters accept protons and thus be converted into their respective conjugate acids.

The effect of substituents on acidity. The ability of an acid to donate a proton to a base depends on the total structure of the acid. Substitution of electron-attracting or electron-releasing groups for hydrogen in the hydrocarbon group of an organic acid affects the acidity of the acid. This is true both as regards the identity of the substituent and its location in the molecule. Several other factors also affect the acidity constant, such as resonance, solvation, and steric effects.

Tables 9.01 and 9.02 point out the effects of different groups and also the effects of identical groups in different positions in aliphatic acids. In these cases the effects are primarily inductive effects. Groups that are more powerful electron attractors than hydrogen exhibit negative inductive effects ($-I$), whereas those that have less attraction for electrons than hydrogen have positive inductive effects ($+I$). Alkyl groups have $+I$ effects. Propionic acid is weaker than acetic acid, which is a weaker acid than formic acid. Aside from the alkyl groups, all the other substituents included in the tables have $-I$ effects, hence they strengthen the acids.

Inductive effects decrease rapidly with distance from the active center; this is to be noted in the case of the chlorobutyric acids.

In contrast with the aliphatic acids, in which induction is the major concern related to the variations among the acidity constants, the aromatic

TABLE 9.01
EFFECT OF SUBSTITUENTS ON pK_a*

Position on the Chain		Number of Substituents	
Butyric	4.824	Acetic	4.757
2-Chlorobutyric	2.857	Chloroacetic	2.866
3-Chlorobutyric	4.056	Dichloroacetic	1.300
4-Chlorobutyric	4.523	Trichloroacetic	0.89

* *Source.* Adapted from Brown, page 78.[2]

TABLE 9.02*
EFFECT ON THE pK_a OF DIFFERENT SUBSTITUENTS AS X IN XCH_2COOH

H	4.757	Br	2.903
$H_2C=CH$	4.342	Cl	2.866
C_6H_5	4.312	F	2.586
CH_3O	3.571	CN	2.469
I	3.175	NH_3^+ (protonated cation)(pK_1)	2.350
C_6H_5O	3.171		

* *Source.* Bower and Bates.[3]

acids and phenols are affected by inductive effects, resonance effects and, in some cases, by steric effects. For purposes of this text, it is not important to go into detail as to the relative importance of induction and resonance in particular acids. Gould[4] discusses the probable relationships of these effects and also comments on the terms used by different authors in this field. Although overly simplified, it may be said that (1) resonance predominates over induction when the group is located para to the reaction center (since the inductive effect is small when separated from the reaction site by four carbons); (2) resonance effect is minimized when the substituent is in the *meta* position due to the alternating nature of the atoms

[2] G. I. Brown, *An Introduction to Electronic Theories of Organic Chemistry*, Longmans Green, New York, and Wiley, New York, 1958.

[3] V. E. Bower and R. G. Bates in *Handbook of Analytical Chemistry*, L. Meites, ed., McGraw-Hill, New York, 1963, pp. 1–21 to 1–27.

[4] E. S. Gould, *Mechanism and Structure in Organic Chemistry*, Holt, New York, 1959; Chapters 4 and 7.

in the conjugated system, hence, inductive effects tend to predominate in meta substitutions; and (3) both effects act strongly in the case of ortho substitutions.

Carboxylic acids yield resonance-stabilized anions, with the ion more stabilized than the molecular acid. The phenoxide ion is also more stabilized by resonance than the phenol. Any substituent that stabilizes the ion more than its conjugate acid will make the acid stronger and, of course, any substituent that makes the ion less stable lessens the acidity of the aromatic acid or phenol. For example, 2,6-dihydroxybenzoic acid

TABLE 9.03*

EFFECT ON THE pK_a OF DIFFERENT SUBSTITUENTS AS X IN XC_6H_4COOH

X-Group	Ortho	Meta	Para
H	4.212	4.212	4.212
CH_3	3.908	4.272	4.373
C_2H_5	3.793	—	4.353
OH	2.996	4.080	4.582
CH_3O	4.094	4.088	4.492
CH_3CO	4.126	3.825	3.700
F	3.267	3.865	4.141
Br	2.854	3.809	4.002
Cl	2.943	3.824	3.986
I	2.86	3.85	3.93
NO_2	2.170	3.450	3.442

* *Source.* Bower and Bates.[5]

is about 100 times as strong an acid as 2,6-dimethoxybenzoic acid because hydrogen-bonding helps stabilize the ion in the hydroxy acid. Groups that withdraw electrons stabilize the anion and thus produce stronger acids. Groups that release electrons to the ring destabilize the ion and thus reduce the acidity constant of the acid or the phenol. Tables 9.03 and 9.04 give data on the effect of different groups and the same group in different positions on the acidity constants of selected benzoic acids and phenols that illustrate the general principles mentioned.

The effect of substituents on basicity. An alkylamine is more basic than ammonia because the electron-releasing alkyl groups tend to disperse the positive charge of the substituted ammonium ion and thus stabilizes it more than is possible for the simple ammonium ion. Looked at another

[5] V. E. Bower and R. G. Bates in *Handbook of Analytical Chemistry*, L. Meites, ed., McGraw-Hill, New York, 1963.

TABLE 9.04*
EFFECT OF SUBSTITUENTS ON THE pK_a OF PHENOL ($pK_a = 9.998$)

Group	Ortho	Meta	Para
CH_3	10.29	10.09	10.26
NO_2	7.234	8.399	7.149
F	8.8	9.3	9.95
Cl	8.477	9.023	9.378
Br	9.0	9.4	9.8
I	9.0	9.4	9.7

2,4-dinitrophenol, 4.11; 2,6-dinitrophenol, 3.706; 2,5-dinitrophenol, 5.216; 3,4-dinitrophenol, 5.424, 2,4,6-trinitrophenol (picric acid), 0.29.

* Data for fluoro-, bromo-, and iodophenols was adapted from Badger.[6] Other data is from Bower and Bates.[5]

way, the alkyl groups push their bonding electrons with nitrogen toward the nitrogen, thus making the fourth pair of electrons more readily available for sharing with the proton of an acid. As would be expected, dialkylamines are more basic than monoalkylamines. However, trialkylamines are generally intermediate, probably due to steric and solvation factors. Table 9.05 illustrates the relationships with data adapted from Morrison and Boyd.[7]

Aromatic amines are much less basic than ammonia. This is due to the fact that the nitrogen tends to share its fourth pair of electrons with the ring, thus making the nitrogen atom have a partial positive charge. Resonance stabilizes the molecular amine more than it stabilizes its conjugate cation acid. As regards substituents on the aromatic ring, those that release electrons to the ring (such as $—NH_2$, $—OCH_3$, and $—CH_3$) increase the basicity as compared to the unsubstituted amine. Those

TABLE 9.05
$10^4 K_b$ FOR SELECTED ALKYLAMINES (AMMONIA = 0.18)

Methylamine	4.4	Ethylamine	4.7	Propylamine	3.8
Dimethylamine	5.1	Diethylamine	9.5	Dipropylamine	8.1
Trimethylamine	0.6	Triethylamine	5.5	Tripropylamine	4.5

[6] G. M. Badger, *The Structure and Reactions of Aromatic Compounds*, Cambridge Univ. Press, New York, 1954, p. 197; Bower and Bates.[5]

[7] R. T. Morrison and R. N. Boyd, *Organic Chemistry*, Allyn and Bacon, Boston, 1959, p. 520.

groups that withdraw electrons from the ring (such as —NO_2, —COOH, and halogens) reduce the basicity of the amine. A given substituent affects the basicity of an amine and the acidity of a carboxylic acid in opposite ways.

Table 9.06 gives the approximate values for some substituted anilines as representative of the effect of substituents on basicity constants. In

TABLE 9.06*
EFFECT OF SUBSTITUENTS ON THE BASICITY OF ANILINE

Compound	pK_b			pK_a of Conjugate Acid		
Aniline	9.4			4.6		
	Ortho	Meta	Para	Ortho	Meta	Para
Toluidine	9.6	9.3	8.9	4.4	4.7	5.1
Chloroaniline	11.4	10.7	10.0	2.6	3.3	4.0
Fluoroaniline	—	10.6	9.5	—	3.4	4.5
Nitroaniline	—	11.5	13.0	−0.3	2.5	1.0

* Source. Bower and Bates.[8]

general, groups that strengthen acids will weaken bases, but there are exceptions. It may be noted that nitroaniline is much weaker than aniline and that, as is the case with acids, the methyl radical does not have much effect (toluidines compared to aniline). The pK_a values given above are for the conjugate acids of the compounds listed. Meta-directing groups are decidedly base-weakening when situated ortho or para to an amino group, but have much less effect when located meta to an amino group.

The product of K_a and K_b for an amphiprotic solvent is the autoprotolysis constant, K_{auto}.

$$pK_{auto} = pK_a + pK_b \tag{14}$$

At 25°C, the K_{auto} for water is 14.0, and that for methanol is 16.7. With water as the solvent, the pK_b of aniline is 9.4; therefore, the pK_a of the anilinium ion, $C_6H_5NH_3^+$ is 4.6.

Indicator Method of Classifying Acids and Bases

An indicator is a conjugate acid-base system in which the colors of the acid and base are different. The acid form of the indicator may be either a

[8] V. E. Bower and R. G. Bates in *Handbook of Analytical Chemistry*, L. Meites, ed., McGraw-Hill, New York, 1963.

molecule or a cation, thus making the conjugate base an anion or a molecule. (In rare cases, the acid form may be an anion and the conjugate base a higher valent anion.)

The value of the pK_a of an acid differs with different solvents. For example, the pK_a of acetic acid is 4.8 in water but is 9.7 in methanol. To distinguish clearly weak acids from feeble acids or nonacids, a solvent that is even weaker as an acid than methanol must be used. A mixture of methanol and pyridine is satisfactory for that purpose. Similarly, properly to classify weak bases, an acidic solvent, such as acetic acid, is used. By the use of mixed indicators in appropriate solvents (see Table 9.07), acids

TABLE 9.07
COMPOSITIONS OF INDICATOR REAGENTS FOR DAVIDSON'S SYSTEM OF CLASSIFICATION

Component		Solvent	Composition of Reagent, ml			
			A-I	A-II	B-I	B-II
Alizarin yellow-R	0.1%	Methanol	25	—	—	—
Bromothymol blue	0.1%	Methanol	25	25	—	25
Bromocresol purple	0.1%	Methanol	—	37.5	—	37.5
Thymol blue	0.1%	Methanol[a]	—	25	—	25
Benzeneazodiphenylamine	0.1%	Acetic acid	—	—	25	—
Methylene blue	0.1%	Acetic acid	—	—	10	—
Potassium hydroxide	2M	Methanol	25	25	—	—
Hydrochloric acid	Conc.		—	—	4.5	4.5
Methanol			425	887.5	—	933
Pyridine			500	—	—	—
Acetic acid (glacial)			—	—	960.5	—

[a] Contains 0.3 ml of 2M KOH (in methanol) per 100 ml.

and bases may each be divided into four subclasses: strong, intermediate, weak, and feeble; designated, respectively, as A_s, A_i, A_w, A_f, B_s, B_i, B_w, and B_f. The possible acid-base conjugates would therefore have the following probable combinations: $A_s\text{-}B_f$, $A_i\text{-}B_w$, $A_w\text{-}B_w$, $A_w\text{-}B_i$, $A_f\text{-}B_s$. Ampholytes, that is, compounds that may be both proton-donors and proton-acceptors, would have one of the following combinations: A_iB_w, A_wB_w, or A_wB_i. They are usually designated by the symbol A_m. Substances that are, for all practical purposes, neutral would classify as $A_f\text{-}B_f$ and are called *neutrals*. Neutrals are designated by the symbol N.

In Davidson's system of classification, the indicator reagent also serves as the solvent system for the compound that is to be examined. The reagents should be prepared fresh every few months. They should be stored in brown, tightly stoppered bottles, and dispensed from dropper-bottles to avoid contamination. To prevent the possible use of contaminated or deteriorated reagents, the color of the side-shelf reagents should always be noted before they are used for a test. The normal colors are: A-I, light purple; A-II, blue-violet; B-I, light purple; B-II, yellow; and inverted A-I, yellow. Colored compounds, or compounds that give abnormal colors with the reagents, should be tested in the same solvent mixtures that are used in preparing the indicator reagents, but without the indicators being present. The colors produced in such cases must be considered in evaluating the color changes that occur with the indicator reagents. If a solid is to be tested, it should be finely powdered before introducing it into the reagent and any solid that remains undissolved should be ground against the side of the tube by means of a clean glass rod.

The flow-sheet on page 340 shows the sequence to be used in testing a compound with these indicator reagents. Compounds that do not change the colors of either A-I or B-I are neutral. Table 9.08 lists the common types of molecules as classified by these indicators.

Procedures for Testing

To test a compound by the use of the classification indicators, place 0.5 ml of the proper indicator in a clean, dry 3-in. test tube and add 15 to 30 mg of the compound that is being examined. Record the color change, if any, that occurs because of the compound. The following steps are recommended.

I. Test the compound in reagents A-I and B-I. Record the results.
   Deductions. If the color of A-I does not change, the compound is neutral or basic. If the color of B-I does not change, the compound is neutral, amphoteric, or acidic. Therefore, if the compound did not change *either* A-I or B-I, it is a neutral substance, N, and no further testing with indicators is required.

II. If there is a color change in A-I but not in B-I, proceed to Step V. If there is a color change in B-I but no change in A-I, proceed to Step III.

Basic Compounds

III. Test the compound in B-II and record the results.
   Deductions. No change in B-II indicates a weak base, B_w. If the indicator solution becomes dichromatic, the substance is a weak or intermediate base, B_w or B_i. If the color is blue-violet, proceed to Step IV.

TABLE 9.08
INDICATOR CLASSIFICATIONS

ACIDS AND AMPHOLYTES[a]
Acids

Strong (A_s)	Intermediate (A_i)	Weak (A_w)	Ampholytes (A_m)
1. Only C, H, and O present: Acid anhydrides Carboxylic acids 2. Halogen present: Acid halides 3. Nitrogen present: Nitro acids Polynitro phenols 4. Sulfur present: Alkyl acid sulfates Bisulfates of bases Sulfate salts of weak bases Sulfonic acids 5. Nitrogen and sulfur present: Sulfonamides Sulfonimides 6. Sulfur and halogen present: Sulfonyl halides	1. Only C, H, and O present: Acid anhydrides β-Keto lactones Carboxylic acids Cyclic β-diketones Esters (easily hydrolyzed) Phenolic aldehydes, ketones and esters Tri acyl methanes 2. Halogen present: α-Chloro esters Halo acids Polyhalo phenols Salts of weak bases 3. Nitrogen present: Nitramines Nitro- and cyanophenols Nitrogenous acids Salts of weak bases Ureides 4. Sulfur present: Dialkyl sulfates Thiophenols 5. Nitrogen and sulfur present: Aromatic aminosulfonic acids Salts of weak bases Sulfonamides 6. Sulfur and halogen present: Sulfonyl halides	1. Only C, H, and O present: Alkyl formates α-Hydroxy aldehydes and ketones α-Keto aldehydes β-Diketones β-Keto esters Cyclic-α-diketones Esters of phenols and weak acids Phenols Quinones 2. Nitrogen present: Acylamino phenols Aliphatic hydroxamic acids Amides Imides Oximes (except aliphatic ketoximes) p- and sec-Nitro compounds Salts of intermediate bases or ampholytes Ureides 3. Sulfur present: Thiols (mercaptans) 4. Nitrogen and sulfur present: Aliphatic amino sulfonic acids Aryl isothiocyanates Aryl thioureas Salts of intermediate bases or ampholytes Sulfonamides Thioamides Thioureides	Aminoacids Aminophenols Aromatic aminosulfonamides Primary salts of diamines, dibasic acids, dihydric phenols, and phenolic acids Sulfadiazines

[a] Halogens may occur as extra elements in all divisions. The listing of a class in a division does not imply that all homologs of that class will necessarily fall in that division. The word "salt" in these tables refers to compounds of amines with acids.

Classification by the Indicator Method 339

TABLE 9.08 (Continued)
INDICATOR CLASSIFICATIONS

Bases	BASES AND NEUTRALS[b]		
	N, S, and P Absent		N, S, or P Present

Weak (B_w)
Aliphatic ketoximes
α-Oxides
Aryl alkyl amines
Aryl dialkyl amines
Aryl hydrazines
Azomethines
Beatines
Heterocyclic amines
  (benzenoid)
Primary aryl amines
Salts of intermediate
  acids or ampholytes
 	
Intermediate (B_i)
Aliphatic amines (*p-*,
  sec-, *tert-*)
Aliphatic hydrazines
Amidines
Aryl guanidines
Imidazoles
Salts of weak acids and
  ampholytes
 	
Strong (B_s)
Alkyl guanidines
Quaternary ammonium
  compounds
	Positive to Ferrox test:[c]
	Acetals
	Alcohols
	Aldehydes
	Alkyl esters (not of
	  phenols)
	Carbohydrates
	Ethers (not diaryl)
	Ketones
	Lactones
		Negative to Ferrox test:[c]
		Aryl alkyl ethers (a
		  few)
		Diaryl ethers
		Furanes
		Halogenated hydro-
		  carbons
		Hydrocarbons
		Triaryl methanols
			1. Nitrogen present:
			  Alkyl nitrates and
			    nitrites
			  Amides
			  Azo, azoxy, and
			    hydrazo com-
			    pounds
			  Di- and triaryl
			    amines
			  Hydrazones
			  Nitriles
			  Nitro and nitroso
			    compounds
			  Polyhalogenated
			    aromatic amines
			  Guanidinium and
			    quaternary am-
			    monium salts of
			    strong acids
			  Ureas and ure-
			    thanes
			2. Sulfur present:
			  Alkyl sulfites and
			    sulfates
			  Aryl sulfonates
			  Salts of alkyl sul-
			    fates and sul-
			    fonic acids
			  Sulfides and di-
			    sulfides
			  Sulfoxides and sul-
			    fones
			  Thio esters
			  Thiophenes
			3. Nitrogen and sul-
			  fur present:
			  Alkyl isothiocya-
			    nates
			  Alkyl thioureas
			  N,N-Dialkyl sul-
			    fonamides
			  Guanidinium and
			    quaternary am-
			    monium sul-
			    fates
			  Thiocyanates
			4. Phosphorus
			  present:
			  Phosphates
			  Phosphites

[b] Halogens may be present as substituents.
[c] Chapter 7, p. 296.

Flow-Sheet for Use of Indicators

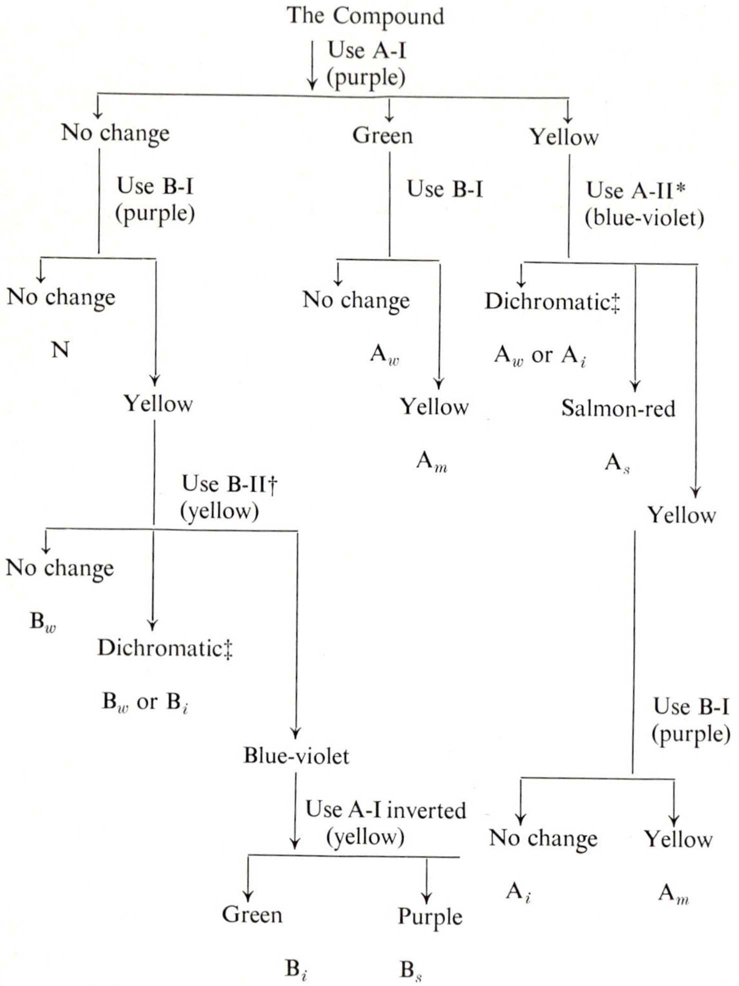

The Compound
| Use A-I
↓ (purple)

No change → Use B-I (purple)
- No change → N
- Yellow → Use B-II† (yellow)
  - No change → B_w
  - Dichromatic‡ → B_w or B_i → Blue-violet → Use A-I inverted (yellow)
    - Green → B_i
    - Purple → B_s

Green → Use B-I
- No change → A_w
- Yellow → A_m

Yellow → Use A-II* (blue-violet)
- Dichromatic‡ → A_w or A_i
- Salmon-red → A_s → Yellow → Use B-I (purple)
  - No change → A_i
  - Yellow → A_m

---

* With A-II, acids between A_w and A_i may produce a green color.

† With B-II bases between B_f and B_w give various colors between purple and yellow.

‡ *Note.* In their dichromatic zones the indicator reagents A-II and B-II appear blue when viewed through thin sections of the solution (detected by tilting the tube), and red when viewed through the full depth. With reagent A-II, the dichromatic character indicates intermediate uncharged acids or weak cation acids. With reagent B-II, the dichromatic character indicates intermediate uncharged bases or weak anion bases.

IV. Test the compound in A-I *which has been inverted* by adding microdrops of $1N$ hydrochloric acid to the regular A-I reagent until the color *just* changes from purple to yellow. Record the results.

Deductions. A green color indicates an intermediate base, B_i, and a purple color indicates a strong base.

Acidic Compounds

V(A). If the color of A-I changes to green, recall the color produced in B-I.

Deductions. If the color in B-I is yellow, the compound is amphoteric, A_m. If the compound did not change B-I, the compound is a weak acid A_w.

V(B). If the color of A-I changes to yellow, proceed to Step VI.

VI. Test the substance in reagent A-II and record the results.

Deductions. If A-II becomes dichromatic, the compound is a weak or intermediate acid, A_w or A_i. If A-II changes to salmon-red, the compound is a strong acid, A_s. If the color of A-II is yellow, recall the color produced in B-I. If B-I was yellow, the compound is amphoteric, A_m, whereas if B-I remained purple, the compound is an intermediate acid, A_i.

REFERENCES

L. F. Audrieth and J. K. Klienberg, *Nonaqueous Solvents*, Wiley, New York, 1953. Chapters 1 and 2.

R. P. Bell, *Acids and Bases: Their Quantitative Behavior*, Wiley, New York, 1952.

R. P. Bell, *The Proton in Chemistry*, Cornell Univ. Press, Ithaca, 1959. Chapters 1–7.

L. N. Ferguson, *Electronic Structures in Organic Molecules*, Prentice-Hall, New York, 1952.

L. P. Hammett, *Physical Organic Chemistry*, McGraw-Hill, New York, 1940.

P. H. Hermans, *Introduction to Theoretical Organic Chemistry*, Elsevier, Amsterdam, 1954. Chapters 11 and 23.

J. Hine, *Physical Organic Chemistry*, 2nd ed., McGraw-Hill, New York, 1962. Chapter 2.

C. K. Ingold, *Structure and Mechanism of Organic Chemistry*, Cornell Univ. Press, Ithaca, 1953. Chapters 2 and 13.

I. M. Kolthoff and P. J. Elving, *Treatise on Analytical Chemistry*, Interscience, New York, 1959. Part I, Vol. I, Chapters 11–13.

W. F. Luder, and S. Zuffanti *The Electronic Theory of Acids and Bases*, Wiley, New York, 1946.

M. S. Neuman, ed., *Steric Effects in Organic Chemistry*, Wiley, New York, 1956.

Y. K. Syrkin and M. E. Dyatkina, *Structure of Molecules and the Chemical Bond*, Interscience, New York, 1950.

W. A. Waters, *Physical Aspects of Organic Chemistry*, 4th ed., van Nostrand, New York, 1950. Chapter 10.

G. W. Wheland, *Resonance in Organic Chemistry*, Wiley, New York, 1955. Chapter 7.

G. W. Wheland, *Advanced Organic Chemistry*, 3rd ed., Wiley, New York, 1960. Chapter 5.

Selected Articles

Hammett Acidity Functions.

Hammett et al., *J. Am. Chem. Soc.*, **54**, 2721 (1932); **56**, 827 (1934).

H. H. Jaffé, *Chem. Rev.*, **53**, 191 (1953); M. A. Paul and F. A. Long, *Chem. Rev.*, **57**, 1 (1957).

Grunwald Acidity Scale.

Grunwald et al., *J. Am. Chem. Soc.*, **73**, 4934, 4939 (1951); ibid., **75**, 559, 565 (1953).

Relative Acidities in Different Solvents.

R. A. Benkeser and H. R. Krysiak, *J. Am. Chem. Soc.*, **75**, 2423 (1953).

H. C. Brown and A. Cahn, *J. Am. Chem. Soc.*, **72**, 2939 (1950).

N. Bjerrum, *Chem. Rev.*, **16**, 287 (1935).

R. E. Dessy et al., *J. Am. Chem. Soc.*, **84**, 2899 (1962).

T. Higuchi et al., *Anal. Chem.*, **34**, 400 (1962).

R. R. Miron and D. M. Hercules, *Anal. Chem.*, **33**, 1770 (1961).

S. R. Palit and P. Ghosh in *Microchemical Techniques*, N. D. Cheronis, ed., Interscience, New York, 1962, pp. 663–676.

C. A. Streuli et al., *Anal. Chem.*, **30**, 1978 (1958); ibid., **31**, 1652 (1959); ibid., **32**, 407 (1960).

Use of Dissociation Constants in Qualitative Organic Chemistry.

T. V. Parke and W. W. Davis, *Anal. Chem.*, **26**, 642 (1954). Many tables.

Review Articles.

J. A. Riddick, *Anal. Chem.*, **24**, 41 (1952); ibid., **26**, 77 (1954).

10

Tests for the Classification of an Unknown

Until recently, the only methods available for the identification of functional groups and, thus, able to provide information about the chemical class or classes to which an unknown belonged, involved chemical reactions that are characteristic of the individual groups. However, within the past 20 years, numerous instruments have become available that provide quite definite information regarding many functional groups and other structural features of organic compounds. Most of these instrumental methods make use of some type of spectroscopy. Instruments tend to give precise information, using very small samples, and require little operating time. Positive identification of an *unknown* by instrumental methods alone requires that a *known* that gives the same instrumental responses be available for comparison. Chemical and instrumental methods largely compliment each other. Organic chemists must be familiar with both approaches to identification and must recognize the usefulness and the limitations of each method. Each instrumental method is becoming highly specialized, and several books and journal articles are available, which discuss the theory back of the various methods and the techniques for using each type of instrument. (Brief introductions to instrumental methods applicable to qualitative organic analysis may be found in Chapter 22 of this text, together with selected references to the more detailed literature.) The objective of this book is to present chemical methods for the identification of an unknown organic compound.

A List of the Tests

Test No.	Class Detected and Reagents	Page
P-1	Gross Observation of Substance	347
P-2	Ignition Test	348
P-3	Test for Salts	349
P-4	Tests for Aromatic Structures	
	A. Chloroform and Aluminum Chloride	350
	B. Formaldehyde and Sulfuric Acid	351
	C. 2,4,7-Trinitrofluorenone	352
P-5	Tests for Active Unsaturation	
	A. Bromine in Carbon tetrachloride	353
	B. Permanganate	354
P-6	Tests for Oxidizable Compounds	
	A. Chloranil	356
	B. Ceric Nitrate	356
	C. Ferricyanide	357
	D. Iodic Acid	357
	E. Nitrochromic Acid	358
	F. Tollens' Reagent	358
P-7	Tests for Acidic Substances	
	A. Iodate-Iodide Reagent	359
	B. Rhodamine B-Uranyl Acetate	359
	C. Liberation of Nitrous Acid	360
P-8	Test for Compounds that contain Nitrogen and Oxygen in the Same Group	360
P-9	Iodoform Formation Test	361
10-01	Acids	
	A. Carboxylic Acids	363
	B. Amino Acids	365
	C. Sulfonic Acids	365
10.02	Acid Anhydrides	366
10.03	Acid Halides	
	A. Amide Formation	366
	B. Ferric Hydroxamate Test	366

A List of the Tests (Continued)

Test No.	Class Detected and Reagents	Page
10.04	Alcohols	
	A. Vanadium-Oxine	368
	B. Ferric Hydroxamate	368
	C. Xanthate	368
	D. N-Bromosuccinimide	369
	E. Lucas	370
	F. Chromic Acid	371
10.05	Alkyl and Aryl Halides	
	A. Alcoholic Silver Nitrate	372
	B. Hydrolysis, followed by Silver Nitrate	372
	C. Formaldehyde-Sulfuric Acid	372
10.06	Amides, Unsubstituted	
	A. Ammonia Liberation	373
	B. Ferric Hydroxamate	373
	C. Distinguishing Aliphatic from Aromatic	374
	D. Sulfonamides	375
	E. Ureas	375
10.07	Amides, N-Substituted	
	A. N-Alkyl Substituted	376
	B. Anilides	377
	C. Other Amides	377
10.08	Amines	
	A. Tetraphenyl Borate	379
	B. Copper Ions	379
	C. Fluorescein Chloride	379
	D. Basicity Test	380
	E. Quinhydrone	380
	F. N-Halosuccinimide	380
	G. 3,3',5,5'-Tetrabromophenolphthalein	381
	H. Hinsberg's Test	382
	I. Diazotization	384
	J. Lignin	384
	K. Chloranil	385
	L. 2,4-Dinitrofluorobenzene	385
	M. Tests for Primary Amines	385
	N. Tests for Secondary Amines	387
	O. Tests for Tertiary Amines	387
10.09	Carbohydrates	
	A. Anthrone	388
	B. *p*-Toluidine	389
	C. Molisch's Test	390
	D. Resorcinol	390
	E. Oxidation by Copper Ions	390
	F. Test for Ketoses	391
	G. Test for Pentoses	392

A List of the Tests (Continued)

Test No.	Class Detected and Reagents	Page
10.10	Carbonyl Compounds	
	A. 2,4-Dinitrophenylhydrazine	392
	B. 3,5-Dinitrobenzoic Acid	393
	C. N-Hydroxybenzenesulfonamide	394
	D. Schiff's Test	394
	E. Methone	395
	F. Benedict's and Tollens' Tests	395
	G. 2-Hydrazinobenzothiazole	396
	H. Tests for Ketones	396
10.11	Esters	397
10.12	Ethers	398
10.13	Hydrazines	399
10.14	Hydrocarbons	400
10.15	Nitrates and Nitrites	401
10.16	Nitriles	402
10.17	Nitro Compounds	
	A. Ferrous Hydroxide	403
	B. Diphenylamine	403
	C. Di- and Trinitro- Hydrocarbons	404
	D. Nitroparaffins	404
	E. Nitrophenols	405
10.18	Nitroso Compounds	405
10.19	Oximes, Hydrazones, and Semicarbazones	406
10.20	Phenols	
	A. Ferric Chloride	406
	B. Coupling with a Diazonium Salt	407
	C. Indicator Formation	407
	D. Indophenol Formation	408
	E. Millon's Test	408
	F. 4-Aminoantipyrine	409
10.21	Sulfides, Disulfides, and Sulfones	410
10.22	Thiols	
	A. Iodine-Azide	411
	B. Nitroprusside	411
	C. Isatin	412
	D. Lead Ions	412

Tests for the Classification of an Unknown 347

This chapter consists of two parts. Part I contains introductory material and preliminary tests, and Part II consists of specific tests for certain functional groups that characterize the various classes of organic compounds.

PART I: AN INVENTORY AND A FORWARD LOOK

In order that further work may be taken in proper sequence, an inventory should now be made of all of the facts that have been learned about the compound under examination. These facts should be recorded in logical order and, as further examination of the substance progresses, a complete record should be kept of the results of each test made, together with any deductions that may be logically drawn from these results. A test or observation that eliminates a class is significant.

It is logical to suspect a class that is commonly used in laboratories and industry before considering an uncommon one. When one is considering the probable classes of a compound on the basis of the elements present, the most common combinations should be considered first.

Up to this point, the compound, purified if necessary, has been examined to establish the elements present, the solubility behavior, and the apparent acidity, basicity, or neutrality of the substance. On the basis of these facts one may restrict the substance to a relatively few possible classes by examining the tables on pages 324–325. In some cases, the information available at this point may definitely suggest a certain classification of compounds. In some cases it is wise to proceed directly to tests for that specific class as given in Part II of this chapter. However, in many cases, it is advantageous to use one or more of the following preliminary tests, which are not specific for any one class but which serve as guides for a logical selection for tests for a specific class.

P-1 Gross Observations

Note the color and odor of the substance and try to recall what class of compounds has similar properties. Many phenols and aryl amines develop color in storage due to oxidation. Colors due to impurities usually diminish or disappear when the substance is purified. Low–molecular weight acids, many phenols, amines, alcohols, esters, hydrocarbons, and carbonyl compounds have characteristic odors. Regularly, students of organic chemistry should carefully note the color, odor, and other observable characteristics of each compound that they use and try to relate its properties to the chemical class represented.

The great majority of carbon compounds are colorless. Therefore, if the pure compound is colored, certain groups may be suspected, and the selection of further tests to be applied would take those facts into account. Although the exact cause of color in molecules is still somewhat uncertain, several authors have logically summarized the present theories on the subject. Among compounds containing carbon, hydrogen, oxygen, halogens, or sulfur, there are a few colored substances, mainly quinones, aromatic ketones that have unsaturated side-chains, and a very few alkanediones.

If nitrogen is present, with or without the additional presence of halogens or sulfur, the most common colored compounds are substituted anilines, toluidines, polycyclic amines, or hydrazines; nitro-, nitroso-, or amino-phenols; polynitro- or polyaminohydrocarbons; nitro- or aminoquinones; azo or diazo compounds; picrates; hydrazones or osazones. Aside from the azo compounds, which are generally orange or red, compounds that contain only chromophoric groups, such as —N=N—, =C=S, =C=O, —N=O, —NO$_2$, and the o- or p-quinoid structures are usually yellow. If, however, the compound also contains auxochromic groups, such as —NH$_2$, —NHR, —OH or —SH, the color is deepened and intensified. Some of the halogenated nitrohydrocarbons are colored. The color tends to deepen as the halogen is changed from chlorine to bromine and from bromine to iodine.

Unsubstituted anilides are colorless. Many compounds tend to darken on exposure to light or air, and hence a purified sample should always be used for color estimation.

P-2 The Ignition Test

Much may be learned about a substance by an ignition test. Place 1 or 2 drops, or a few milligrams of a solid, on the inverted lid of a crucible or in a shallow evaporating dish. Heat the vessel with a small flame. From time to time, contact the substance directly with the flame so that it may ignite before it volatilizes. If gases are given off by the heated substance before it ignites, a pH paper that has been moistened with distilled water should be held in the fumes to detect any acidic or basic properties. If the substance ignites, note the odor and the general appearance of the flame. If the substance carbonizes, apply a hot flame to the vessel until the carbon is all oxidized. The residue, if any, is most likely the oxide or carbonate of some metal, usually white but sometimes colored. Such a residue should be tested for metals by the semimicro methods for cation analysis as published in texts on that subject.

Most aromatic compounds, particularly if they do not contain oxygen, burn with a smoky flame. Low–molecular weight nonaromatic compounds burn with a nonsmoky flame which tends to be bluish if oxygen is present in the organic compound. Halogen compounds are difficult to ignite, but burn with a smoky flame when ignited. Sugars and proteins tend to carbonize and produce characteristic odors.

P-3 Tests for Salts

Commonly encountered organic salts may be divided into three classes: ammonium salts of organic acids, metallic salts of acids or phenols, and salts of amines with acids. Such compounds will generally be suspected as a result of the ignition test (test P-2), the elemental analysis, and the solubility determinations.

A. Ammonium Salts

1. Place 25 to 50 mg of the suspected salt in a 4-in. test tube and add 0.5 ml of 10 per cent sodium hydroxide. Place a small filter paper over the top of the tube and crush it down around the tube. Add 2 drops of 10 per cent copper sulfate on the filter paper covering the mouth of the test tube. Heat the tube with a small flame to boil the mixture. Ammonium salts liberate ammonia gas which will react with the copper ions on the paper to develop a blue color. This test is more reliable than the "sniff test" or the litmus paper test for ammonia arising from the alkaline hydrolysis of an ammonium salt.

2. A very sensitive test for ammonia is based on the oxidation of ammonia to nitrous acid which reacts with phenol to produce nitrosophenol. The equilibrium isomer of nitrosophenol is the monoxime of quinone. It may be assumed that the color produced in the test is the result of a reaction between phenol and nitrosophenol.

Mix 1 ml of a 4 per cent solution of phenol in water with 1 ml of a 5 per cent aqueous solution of sodium hypochlorite in a 4-in. test tube. Add a few milligrams of a solid that is to be tested or a few drops of a solution of the compound in water. Warm the mixture. A blue color develops if ammonia or ammonium ions are present (or substances that yield ammonia or ammonium ions under the conditions of the experiment).

NOTE. This is a very sensitive test. It has been found that many amides also give a blue color when used in this test. This reaction may be accounted for either by assuming that some ammonium salts are present in the amide as impurities or on the basis of some hydrolysis of the amide in the warm alkaline solution. The test has not been investigated sufficiently to establish certainty of what additional interferences may exist in this method of detecting ammonia and ammonium salts. Nessler's reagent, as used in inorganic chemistry, may also be used to detect ammonia.

B. Metallic Salts

The presence of metals is indicated by a white or light-colored residue on ignition of a sample (test P-2). The identity of the metal can be determined by treating this oxide or carbonate by the conventional methods of inorganic analysis.

C. Salts of Amines

Salts formed by reacting amines with acids are generally ionic and soluble in water. Most commonly, the anion present is the chloride or sulfate ion. Less commonly, the anion will be phosphate, acetate, or some other ion. Detection methods are given in Chapter 7 for the halide, sulfate, and phosphate ions.

NOTE. Salts of amines that are not sufficiently soluble in water for direct testing may be decomposed by treatment with 5 per cent sodium hydroxide. The amine may be extracted by ether, the aqueous layer acidified with acetic acid and then tested for the anions.

The detection of the acetate ion in amine acetates is much more involved. One method is as follows: Place the salt in a test tube and acidify it with phosphoric acid. Distill the evolved acetic acid into 0.5 ml of water in a 4-in. test tube. Neutralize the distillate with sodium bicarbonate and evaporate it to dryness. Add a few milligrams of dry sodium formate to the dry sodium acetate in the test tube and cover the mouth of the test tube with a filter paper to which has been added 1 drop of a 20 per cent solution of morpholine and 1 drop of a 5 per cent solution of sodium nitroprusside. Heat the salt in the test tube with a small flame. The development of a blue color on the test paper is a positive test for acetaldehyde which is formed by the reaction of sodium formate with sodium acetate.

P-4 Tests for Aromatic Structure

It is frequently an advantage to know as soon as possible if the unknown has aromatic structural characteristics. The following two tests, although not completely reliable or specific, are very useful.

A. The Chloroform–Aluminum Chloride Test

Compounds that have an aromatic structure usually react with chloroform in the presence of aluminum chloride to produce colored products.

Place 100 mg of aluminum chloride in a dry 4-in. test tube and heat it in a strong flame to sublime aluminum chloride up onto the sides of the tube. Allow the tube to cool. Prepare a solution of 10 to 20 mg of the compound in 5 to 8 drops of chloroform and run this solution down the side of the test tube containing the sublimed aluminum chloride. Note any color produced by contact of the solution with the salt.

NOTE. Stock samples of aluminum chloride generally have absorbed and reacted with water vapor. Freshly sublimed salt is a much more efficient catalyst for this reaction. If this test is carefully performed it is useful but not completely reliable for detecting aromatic type structures. It is most useful for distinguishing aromatic hydrocarbons or their chlorine compounds from nonaromatic hydrocarbons and their chlorine compounds. Many nonaromatic compounds that contain bromine produce yellow colors, and many nonaromatic compounds that contain iodine produce violet colorations. As a rule, nonaromatic compounds fail to produce a color on the aluminum chloride, whereas monocyclic aromatic compounds give rise to a yellow-orange or red color; bicyclic aromatics give blue or purple, and more complex aromatics produce green colorations on the salt.

B. *The Formaldehyde–Sulfuric Acid Test*[1]

The formaldehyde–sulfuric acid test is particularly useful in distinguishing aromatic compounds from nonaromatic compounds for those substances that fall in Solubility Division I, that is, compounds that are insoluble in concentrated sulfuric acid.

Feigl[2] has found that phenols with a free *para* position give colors (usually red) with the formaldehyde-sulfuric acid reagent. He also reports that naphtholsulfonic acids produce colors with this reagent, whereas naphthylaminesulfonic acids do not.

Prepare a solution of about 30 mg of the compound to be tested in 1 ml of a nonaromatic solvent (hexane, cyclohexane, or carbon tetrachloride). Add 1 or 2 drops of this solution to 1 ml of the reagent, which is prepared at the time of use by adding 1 drop of formalin (37 to 40 per cent formaldehyde) to 1 ml of concentrated sulfuric acid and shaking the solution slightly. Note the color of the surface layer of the reagent after the test solution has been added and the color of the reagent after the tube has been shaken.

NOTE. With time, colors for most of the compounds that give positive tests change to various shades of brown or black. It is advisable to run a blank test on the solvent before it is used, since it may contain aromatic contaminants. A few milligrams of trioxane may be substituted for the formalin in preparing the reagent. Typical colors

[1] H. E. Morris, R. B. Stiles, and W. H. Lane, *Ind. Eng. Chem., Anal. Ed.*, **18**, 294 (1946); A. L. Le Rosen, R. T. Moravek, and J. K. Carlton, *Anal. Chem.*, **24**, 1335 (1952); L. Silverman and W. Bradshaw, *Anal. Chem.*, **27**, 96 (1955); M. J. Rosen, *Anal. Chem.*, **27**, 111 (1955).

[2] F. Feigl, *Spot Tests in Organic Analysis*, Elsevier, Amsterdam, 1960, pp. 135–138

produced in this test are:[3] benzene, toluene, and *n*-butylbenzene give red; *sec*-butylbenzene gives pink; *tert*-butylbenzene and mesitylene produce orange; diphenyl- and triphenylbenzene give blue or greenish blue; naphthalene and phenanthrene produce blue-green to green; aryl halides produce pink to purple colors; naphthyl ethers give purple colors. Alkanes, cycloalkanes, and their halogen derivatives produce either no color or a pale yellow. Often, but not always, precipitates form.

The formaldehyde-sulfuric acid reagent will also produce colors with many other classes of aromatic compounds besides the classes that are insoluble in concentrated sulfuric acid, but in those classes the color production is not consistent enough to be very useful in classification work. Unsaturated, noncyclic hydrocarbons react with the reagent and usually produce brown precipitates. It is reported[4] that alcohols with more than four carbon atoms produce yellow, brown or red-brown colors with the reagent.

C. Adduct Formation with 2,4,7-Trinitrofluorenone

Using the fusion technique, Laskowski and McCrone[5] have determined many types of aromatic structures that will form adducts with 2,4,7-trinitrofluorenone; also types that fail to react with this reagent. These workers conclude that substituents on benzene may be divided as follows: (1) monosubstitution by —NH$_2$, —OH, —C$_6$H$_5$, —Br, —I, —OR, or the vinyl group will allow adduct formation; (2) adduct formation does not take place with monosubstitution by —NO$_2$, —COOH, —CN, —Cl, —CHO, —COOR, —C=O, —CH$_3$, and some other groups. The cited
$\quad\quad\quad\quad\quad\quad\quad\quad\quad\quad\quad\quad$R
article should be consulted for the results with di- and polysubstituted benzenes.

Gordon and Huraux[6] have discussed the use of 2,4,7-trinitrofluorenone as a reagent for spot testing for aromatic structures. The cited article should be consulted for details and for methods of interpretation. These authors give several tables, which include the colors of the adducts with various types of molecules.

P-5 Tests for Active Unsaturation

Both test *A* and test *B*, described below, should be performed and the results compared, as shown in Table 10.01, before reaching tentative conclusions regarding the nature of the substance being examined.

---

[3] A. L. Le Rosen, R. T. Moravek and J. K. Carlton, cited in Reference 1.
[4] L. Rosenthaler, *Pharm. Acta Helv.*, **32**, 440 (1957).
[5] D. E. Laskowski and W. C. McCrone, *Anal. Chem.*, **26**, 1497 (1954).
[6] H. T. Gordon and M. J. Huraux, *Anal. Chem.*, **31**, 302 (1959).

TABLE 10.01

COMPARISON OF RESULTS OF PERMANGANATE ION AND BROMINE TESTS

Types of Compounds	Permanganate	Bromine	
		Addition	Substitution
Most alkenes	Positive	Positive	
Nonconjugated alkadienes	Positive	Positive	
Conjugated alkadienes	Negative	Positive	
Many ArC=CAr	Positive	Negative	
Phenols; many aryl amines	Positive		Positive
Many aldehydes	Positive		Positive
Primary and secondary alcohols (pure)	Negative		Negative
Thiophenols	Positive		Positive
Thiols; sulfides	Positive		Negative
Some ketones	Negative		Positive
Some alkynes	Positive	Negative	

NOTE. The term *alkene* refers, in this case, to molecules that have a nonaromatic double bond between carbon atoms and is not restricted to hydrocarbons. If electronegative groups are attached to each of the unsaturated atoms, bromine will react very slowly and the test may appear negative. Alkynes with at least one hydrogen attached to the unsaturated carbon apparently react with bromine by a free-radical mechanism. Ketones that are highly enolized will be oxidized by permanganate. Methyl ketones react with bromine more than other ketones. Formaldehyde and benzaldehyde do not react appreciably with bromine. Secondary alcohols are more readily oxidized by permanganate than primary ones. High–molecular weight alcohols react very slowly so that the test may appear negative. Molecules of the $Ar_2C=CAr_2$ type generally give negative tests for unsaturation by both methods. Authors give different opinions concerning the reactivity of different type molecules with bromine and with permanganate ions. For example, see the article by Daniels and Bauer, cited in Reference 7, and J. G. Sharefkin and H. E. Shwerz, *Anal. Chem.*, **33**, 635 (1961); J. S. Swinehart, *J. Chem. Educ.*, **41**, 392 (1964). For various methods of detection and determination of unsaturation, see the article by F. A. Leisey and W. P. Cropper in *Handbook of Analytical Chemistry*, L. Meites, ed., McGraw-Hill, New York, 1963, pp. 12-50 to 12-69.

A. Bromine (in CCl_4) Test

Bromine will form addition compounds with most actively unsaturated compounds. It will also react to substitute bromine for hydrogen and liberate hydrogen bromide with compounds that are easily brominated. Typical equations are

$$\underset{H}{\overset{H}{HC}}=\underset{}{\overset{H}{C}}-CH_2OH + Br_2 \rightarrow \underset{Br}{\overset{H}{HC}}-\underset{Br}{\overset{H}{C}}-CH_2OH$$

Dissolve 50 to 100 mg of the compound to be tested in 1 to 2 ml of carbon tetrachloride and add, dropwise, a 2 per cent solution of bromine in carbon tetrachloride. If more than 2 drops of the bromine solution are required to cause the bromine color to remain for at least 1 minute, a reaction by either addition or substitution is indicated. See the *Notes*, including those that follow Table 10.01.

NOTE. Most compounds that possess alkene structures add bromine quite rapidly, with the exception of molecules that have electronegative groups attached to both of the carbon atoms that are involved in the unsaturation. A compound that readily reacts with bromine without the evolution of hydrogen bromide may be presumed to be unsaturated, but this conclusion should be confirmed by the permanganate test (Test B).

Discharge of the color of more than a drop or two of the bromine solution, accompanied by the liberation of hydrogen bromide gas, indicates a phenol, amine, enol, aldehyde, ketone, or some other compound containing an active methylene group. Amines do not evolve hydrogen bromide after the first substitution by bromine because the amine reacts with the first hydrogen bromide produced to form a salt. Not all members of these classes react with bromine under the specified conditions. The fact that hydrogen bromide is being evolved may usually be determined by blowing the breath across the top of the reaction tube and noting the fog that is produced.

Alkynes react very slowly with bromine, and may not be expected to give a positive test with bromine.[7] Both nonconjugated and conjugated alkadienes react with bromine to give a positive test for unsaturation. Most cyclodienes react very slowly with bromine.

B. Permanganate Ion Test

The results of the permanganate ion test, when compared with the results of the bromine test, may be used to detect active unsaturated linkages or to indicate certain other type structures. The test is based on the fact that easily oxidized compounds will reduce the permanganate ion, thus causing the disappearance of the purple color and the appearance of the brown-colored hydrated oxides of manganese. Applied to hydrocarbons, the test is positive for the alkenes and alkynes (Baeyer test).

$$3R_2C\!\!=\!\!CR_2 + 2MnO_4^- + 4H_2O \rightarrow 3R_2C\!\!-\!\!CR_2 + 2MnO_2 + 2OH^-$$
$$\underset{H\ \ \ H}{\underset{O\ \ \ O}{|\ \ \ |}}$$

Dissolve 25 to 30 mg of the compound in 2 ml of water or acetone (free of alcohols) in a 4-in. test tube. Add a 1 per cent aqueous solution of potassium permanganate drop by drop, with vigorous shaking of the tube.

[7] R. Daniels and L. Bauer, *J. Chem. Educ.*, **35**, 444 (1958).

If more than 1 drop of the permanganate is reduced, an unsaturated hydrocarbon or some other easily oxidized compound is indicated.

NOTE. Recall that bromine reacts by addition with many unsaturated compounds and by substitution with many others. The permanganate ion, in the presence of water, oxidizes alkene and alkyne types of bonds to glycols which are cleaved and oxidized to the corresponding acids by more vigorous treatment with permanganate. However, the permanganate ion will also oxidize several other types of compounds. Table 10.01 summarizes the results of the bromine test and the permanganate test on different types of structures.

P-6 Tests by Selected Oxidizing Agents

Several types of organic molecules may be oxidized. Since different oxidizing agents are somewhat selective concerning the structures that they will react with, the performance of at least some of the following preliminary tests may serve as a guide for further testing by specific class tests.

TABLE 10.02
SUMMARY OF THE RESULTS OF SIX TESTS

Chemical Type	A	B	C	D	E	F
Acids, hydroxy		+			+	+
Acids, aromatic			−	−		P
Alcohols, p- and s-		+	±	±	+	
Alcohols, polyhydroxy		+	±	±		
Aldehydes			+	+	−	+
Amines, alkyl	+	P	+	−		
Amines, aryl	+	P	+	+		±
Carbohydrates			−	±	+	+
Phenols	+	±	+	+	−	±
Phenols, polyhydroxy			+		−	±
Thiols	+		+			P

In making use of these general tests, the selection should be made in the light of what is already known about the substance being investigated, such as the elements present, the solubility classification, the acid-base character, and the like.

Six tests have been selected from the large number available. Table 10.02 summarizes in a very incomplete way the results that may be expected.

The individual tests, designated A through F, should be consulted for more details as to types of compounds that will give positive results as well as structures that fail to react. The plus and minus symbols in the table indicate that some members of the class give positive results, while others do not. The letter *P* indicates a precipitate. Blank spaces indicate lack of information. The letters at the top of the columns refer to the tests that follow.

A. *Chloranil Test*

Chloranil (tetrachloro-*p*-benzoquinone) reacts with most primary and secondary amines to produce compounds that are blue, red, green, or reddish-brown. It reacts with many phenols to produce colored products that are generally yellow, yellow-orange, or reddish-orange. Chloranil also reacts with condensed-ring hydrocarbons to produce adducts, many of which are red. Divalent sulfur compounds react with this reagent to yield amber or orange products.

Place 3 drops of a saturated solution of chloranil in dioxane on a spot plate and add 30 mg of the compound under examination to the reagent. Color formation is rapid in a positive test.

NOTE. The structures responsible for the colors have not been established. Amino acids do not give the amine test with this reagent. Amine salts, such as aniline acetate and the naphthylamine hydrochlorides, produce a gray color with the reagent. Many anilides produce yellowish-green or reddish-orange colors. In general, amides do not react.

B. *Ceric Nitrate Test*

The ceric nitrate reagent produces amber or red colorations with solutions of most hydroxy compounds that are soluble in water. Dioxane that is alcohol-free may be used rather than water as the solvent for compounds that contain up to about ten carbon atoms. Some aromatic amines and compounds that are easily oxidized to chromophoric groups also produce colors or precipitates with the reagent. Many phenols give brown or green colors or precipitates.

Dissove 25 to 30 mg of the compound in 2 ml of water (or dioxane) in a 4-in. test tube. Add 0.5 ml of the ceric nitrate reagent to the solution and shake the tube.

NOTE. It has been found[8] that alcohols, glycols, hydroxy acids, hydroxy esters, and hydroxy aldehydes and ketones that do not contain more than ten carbon atoms cause the reagent to change from a bright yellow color to amber, yellow-orange, or red. Amino alcohols do not give the test and usually cause ceric hydroxide to precipitate. Many phenols produce a green-brown or brown precipitate in aqueous solution and a red-brown or brown precipitate in dioxane. Aromatic amines and thiophenes[9] and

[8] F. R. Duke and G. F. Smith, *Ind. Eng. Chem., Anal. Ed.*, **12**, 201 (1940).
[9] W. D. Hartough, *Anal. Chem.*, **20**, 860 (1948).

easily oxidized compounds that produce colored products often produce various colors with the reagent.

For preparation of the ceric nitrate reagent, see the Appendix.

C. Ferricyanide Test

Compounds that are easily oxidized will reduce ferricyanide ions to ferrocyanide ions, which may be detected by ferric ions. Primary and secondary amines, most phenols, and thiols give positive tests. Some aldehydes, a few ketones, some alcohols (particularly secondary alcohols), and a very few acids are oxidized. Most acids, esters, and amides are not oxidized appreciably.

Place 30 to 50 mg of the compound on a spot plate and add 1 drop of 2.5 per cent aqueous potassium ferricyanide, followed by 1 drop of 2.5 per cent aqueous ferric chloride. A green or blue color, developing within 1 minute, is a positive test.

D. Iodic Acid Test

It has been found[10] that iodic acid is a selective oxidizing agent. The following are oxidized by iodic acid, under the conditions specified: simple alcohols up to heptanol (except methanol), aldehydes, methyl ketones, phenols, and aniline derivatives. The following are not oxidized: polyhydroxy alcohols (except 1,2- and 1,3-propandiol), acids, and sugars (except fructose and sucrose).

If ethanol is the compound used, the equations for this test are:

$$3C_2H_5OH + HIO_3 \rightarrow 3CH_3CHO + HI + 3HOH$$

$$3CH_3CHO + HIO_3 \rightarrow 3CH_3COOH + HI$$

$$5HI + HIO_3 \rightarrow 3I_2 + 3HOH$$

The reagent is prepared by carefully adding 2.5 ml of concentrated sulfuric acid to 8 ml of water, cooling the mixture to room temperature, and adding 100 mg of potassium iodate. To this reagent, add 50 mg of the compound and keep the tube immersed in boiling water for 1 hour, unless oxidation occurs more quickly. A brown color as a result of suspended iodine is a positive test.

NOTE. Periodic acid may be used effectively to detect polyhydroxy alcohols with vicinal hydroxyl groups, many carbohydrates, and such di- or polyhydroxy acids as tartaric acid. The periodic acid is reduced to iodic acid in these reactions, and formaldehyde and formic acid are evolved. Since the presence of the iodate ion may be

[10] R. J. Williams and M. A. Woods, *J. Am. Chem. Soc.*, **59,** 1408 (1937).

detected by adding silver ions and, since very sensitive tests are available for both formaldehyde and formic acid, periodic acid may be used to supplement the iodic acid test for oxidizable organic compounds.

E. Nitrochromic Acid Test[11]

A mixture of nitric acid and potassium dichromate will oxidize most primary and secondary alcohols, saccharides, formaldehyde, and lactic and tartaric acids.

Add 5 drops of a 5 per cent solution of potassium dichromate to 5 ml of cold $7.5N$ nitric acid in a 4-in. test tube. Add 1 ml of a 10 per cent aqueous solution of the compound to be tested. The development of a distinctly blue color within 5 minutes is a positive test.

NOTE. If the compound is not soluble in water, 50 to 100 mg of it may be added directly to the nitrochromic acid; thoroughly shake this mixture.

This test gives negative results with tertiary alcohols (unless other alcohols are present as impurities), aldehydes (except formaldehyde), ketones, alkanoic acids, oxalic and citric acids, phenols, and amino compounds.

In the case of sugars, the test may be modified as follows. Add 10 to 15 mg of the compound to 1 ml of concentrated nitric acid, followed by 4 to 5 drops of a 5 per cent solution of potassium dichromate. A blue color will develop in the cold mixture within 1 minute.

F. Tollens' Test[12]

The silver ions in a solution containing silver-ammonia complex ions are reduced to metallic silver by most aldehydes, readily oxidized sugars, polyhydroxy phenols, aminophenols, hydroxylamines, and other reducing agents.

Add 30 to 50 mg of the compound to be tested to 2 ml of freshly prepared reagent. Shake the tube and allow it to stand for 10 minutes. If no reaction occurs in this time, place the tube in a beaker of water at about 35° for 5 minutes. A precipitate of silver is a positive test.

The reagent is prepared as follows: Add 2 drops of 5 per cent sodium hydroxide to 2 ml of 5 per cent aqueous silver nitrate. Shake the tube and add $2N$ ammonium hydroxide dropwise and with shaking until the precipitated silver hydroxide just dissolves.

NOTE. The test tube to be used in this test must be clean. It is best to clean the tube by boiling a 10 per cent solution of sodium hydroxide in it and then discarding this solution. The silver produced in this test will generally precipitate as a "mirror" on the glass tube if it is thoroughly clean but the formation of black metallic silver in the mixture is also a positive test.

[11] W. R. Fearon and D. M. Mitchell, *Analyst*, **57**, 372 (1932).
[12] B. Tollens, *Ber.*, **15**, 1635, 1828 (1882); G. T. Morgan and F. M. G. Micklewait, *J. Soc Chem. Ind.*, **21**, 1375 (1902).

(CAUTION. *Silver fulminate, which is very explosive when dry, may be present in the residues from the use of Tollens' solution. Hence, as soon as the test is completed the contents of the tube should be poured down the sink and washed through the trap with water. Also, rinse out the test tube with dilute nitric acid.*)

P-7 Detection of Acidic Substances

Normally, acidic substances are detected during the classification of the substance by solubility and by indicators (Chapters 8 and 9). However, it is sometimes useful to use other methods, including tests with commercially available mixed indicator solutions or test papers. The following three procedures have proven useful in detecting acidic substances even when they are too weakly acidic to respond to indicators.

A. The Iodate-Iodide Test[13]

About 5 mg of the substance to be tested (or a saturated solution of it in 2 drops of neutral alcohol) is placed in a 3-inch test tube. Add 2 drops of a 2 per cent solution of potassium iodide and 2 drops of a 4 per cent solution of potassium iodate. Stopper the test tube and hold it in boiling water for 1 minute. Cool the tube and add 1 to 4 drops of a freshly prepared 0.1 per cent solution of starch. If the substance is an acid, a blue color appears

$$5I^- + IO_3^- + 6H^+ \rightarrow 3H_2O + 3I_2$$

NOTE. In some cases it is necessary to add more starch solution to cause the characteristic blue iodine-starch complex to appear.

Solid acids may also be detected by grinding together a few milligrams of the acid with a few milligrams of the dry potassium iodide and potassium iodate. The development of a brown color due to free iodine is a positive test. In case of doubt, add 5 drops of water and 2 to 4 drops of starch solution.

B. The Rhodamine B-Uranyl Acetate Test[14]

This test is particularly useful for acidic substances that are soluble in benzene but not appreciably soluble in water. Acids, negatively substituted phenols, many oximes, salicylamides, and the like all give a positive test.

Dissolve a few milligrams of the substance in 1 or 2 drops of benzene and add 5 drops of a saturated solution of rhodamine B in benzene and 1 drop of a 1 per cent aqueous solution of uranyl acetate. Shake or stir the

---

[13] F. Feigl, *Spot Tests in Organic Analysis*, Elsevier, Amsterdam, 1960, p. 117.
[14] F. Feigl, pp. 120, 514 (cited in Reference 13).

mixture. A positive test is indicated by a pink to red color in visible light which fluoresces orange in ultraviolet light.

C. Test by Liberation of Nitrous Acid[15]

Acids release nitrous acid from sodium nitrite which diazotizes sulfanilic acid which then couples with 1-naphthylamine to produce a red or orange color within a few minutes (Griess test for nitrous acid).

Place a few milligrams of the substance on a spot plate and add 1 drop each of: (1) 0.48 g of sodium sulfanilate and 0.12 g of sodium nitrite in 5 ml of water; (2) 0.18 g of 1-naphthylamine in 4 ml of ethanol or dioxane.

P-8 Test for Compounds that have Nitrogen and Oxygen in the same Group

Feigl[16] has developed a method for testing compounds that have nitrogen linked to oxygen in the same group. Such compounds include nitro and nitroso compounds, oximes, nitrates and nitrites, nitramines, azoxy compounds, amine oxides, and hydroxamic acids. The basis of the test is the premise that such compounds decompose when dry-heated to liberate nitrous acid which may be detected by the Griess reagent. Alkali salts interfere with the test; hence, if they are suspected of being present, the substance should be evaporated to dryness with a drop of concentrated hydrochloric acid before making the test. Because volatile substances distill before they are thermally decomposed, they may not be tested by this procedure. The Griess reagent is prepared by mixing equal volumes (just before use) of a 1 per cent solution of sulfanilic acid in 30 per cent acetic acid with a 0.1 per cent solution of 1-naphthylamine in 30 per cent acetic acid.

Insert a 3-in. test tube through a hole in an asbestos sheet. Add a few milligrams of the substance to be tested to the test tube. Cover the top of the test tube with a filter paper that has been moistened with Griess reagent. Heat the test tube, beginning above the sample, to thermally decompose the substance—usually, 1 or 2 minutes is adequate. A pink-to-red coloration on the filter paper is a positive test. The nitrous acid diazotizes the sulfanilic acid, which couples with the naphthylamine.

This preliminary test is particularly useful as a guide to what class tests should be made for compounds that contain nitrogen.

[15] F. Feigl, p. 119 (cited in Reference 13); Y. Nomura, *Bull. Chem. Soc. Japan*, **32**, 536 (1959).

[16] F. Feigl, *Spot Tests in Organic Analysis*, Elsevier, Amsterdam, 1960, pp. 162–164; F Feigl and J. R. Amaral, *Mikrochim. Acta*, **1958**, 337.

P-9 Iodoform Formation Test

Except for acetic acid and its derivatives, most compounds that contain the $CH_3C{=}O$ group or that can be readily oxidized to produce this group will yield iodoform when the substance is warmed in an alkaline solution of sodium hypoiodite.[17] Stodola[18] pointed out that even traces of iodoform may be detected by treating with potassium hydroxide and resorcinol. Gillis[19] has shown that compounds that can be deaminated or hydrolyzed to produce the $CH_3C{=}O$ or $CH_3C(H)OH$ structure will also produce iodoform. Seelye and Turney[20] have discussed the probable mechanism for the iodoform formation. They also propose an alternate method using cyanide ions in the alkaline-iodine solution, and claim that this technique allows differentiation between substances that already contain the CH_3CO group and other structures that also give positive tests.

According to Gillis,[19] isopropylamine, 2-aminopropionic acid, 2-acetoxypropionic acid, and the like, all yield iodoform.

Dissolve 100 mg of the compound being tested in 1 ml of water (use dioxane if the compound is insoluble in water). Add 3 ml of 10 per cent sodium hydroxide solution and then add dropwise a 10 per cent solution of iodine in a 20 per cent solution of potassium iodide in water, until a slight excess of iodine exists in the solution. Place the tube in a beaker of 60° water. Add more iodine until the iodine color persists for 2 minutes and then add drops of 10 per cent sodium hydroxide solution until the brown iodine color just disappears. Remove the tube from the warm water and add 10 ml of water. Iodoform precipitates as a yellow solid, which melts at 120°.

NOTE. Acetaldehyde is the only alkanal that yields iodoform. Acrolein, furfural, and aldehydes of similar structure also form iodoform. Ethanol is the only primary alcohol that gives a positive test. Tertiary alcohols give negative results. Secondary alcohols (2-alkanols) that can be oxidized to methyl ketones by sodium hypoiodite yield iodoform as do the methyl ketones (2-alkanones). Compounds that contain the CH_3CO grouping but which, on hydrolysis, produce acetic acid do not give the iodoform test (for example, acetoacetic acid and its esters, acetanilide, and the like). If dioxane is used as a solvent, it should be previously tested for impurities that might give a positive iodoform test.

---

[17] R. C. Fuson and C. W. Tullock, *J. Am. Chem. Soc.*, **56**, 1638 (1934); R. C. Fuson and B. A. Bull, *Chem. Rev.*, **15**, 275 (1934); R. Bright and P. R. Johnson, *J. Am. Chem. Soc.*, **63**, 1558 (1941).
[18] F. H. Stodola, *Ind. Eng. Chem., Anal. Ed.*, **15**, 72 (1943).
[19] B. T. Gillis, *J. Org. Chem.*, **24**, 1027 (1959).
[20] R. N. Seelye and T. A. Turney, *J. Chem. Educ.*, **36**, 572 (1959).

PART II: TESTS FOR SPECIAL CLASSES

The specific classes of organic compounds are generally recognized by the detection of the main functional group that is present in the molecule. For example, detection of the carbonyl group indicates an aldehyde or a ketone; the presence of nitrogen with basic properties indicates an amine; detection of hydroxyl groups, depending on the properties exhibited, indicates either alcohols or phenols. In the case of polyfunctional compounds, two or more functional groups may be detected during the course of the investigation, and a decision must be made as to which functional group dominates the properties of the compound so that the compound may be found in the proper table that gives its physical constants and those of suitable derivatives. For example, a cyano-substituted acid should be classified as an acid rather than as a nitrile, and an aminophenol should be sought in the table of amines since it would be derivatized as an amine rather than as a phenol.

In every case, an attempt should be made to determine the major functional group first. For instance, a compound of Solubility Division A_2 that proves to be weakly acidic and contains nitrogen should be tested for phenols before an attempt is made to determine the nature of the nitrogen substituent. Probably, if it proves to be a phenol, the compound can be identified by its physical constant and the physical constants of its derivatives as a phenol without the necessity of testing specifically for the nature of the nitrogen-containing group. Similarly, since acids and phenols with halogen substituents are much more common than acid halides, a compound that falls in Solubility Division A_1 and contains halogen should be tested for acids and phenols before acid halides.

When testing for functional groups in organic compounds, it must be recognized that the group is not an individual identity such as an ion would be in inorganic analysis, but rather it is a component of the molecule, the total structure of which affects the properties of all functional groups at least to some extent. In some cases, this influence may be so great that the functional group for which the particular test is being made may fail to give a positive reaction. For this reason, it is advisable to use more than one method of identifying the groups whenever possible.

Students should perform the tests given in this chapter using known compounds of the appropriate classes to gain experience in the procedures and to observe the reactions which are characteristic of each class. Also it is often worthwhile when working with an unknown to simultaneously test a known compound that is assumed to be similar to the unknown.

Methods for the preparation of special reagents used in the tests described in this chapter will be found in the Appendix.

10.01 Acids

A. Carboxylic Acids

Three classes of derivatives of carboxylic acids may be detected by conversion to hydroxamic acids which may then be reacted with ferric ions to produce the highly colored ferric hydroxamates. Acids may be converted into one of these classes—acid chlorides, anhydrides, or esters—and thus detected indirectly. Obviously, it is necessary to establish that the unknown substance is not an anhydride (test 10.02), an acid halide (test 10.03), or an ester (test 10.11) before proceeding to test for acids. Furthermore, since ferric ions produce colored products with other classes, particularly with phenols, the unknown should be tested with a solution of ferric chloride (test 10.20A) before making any test for a hydroxamic acid. Procedures for converting acids to derivatives suitable for further testing follow.

1. *Conversion to acid chlorides.* (a) Place 100 mg (or 3 drops) of the compound in a 6-in. test tube and add 8 drops of thionyl chloride. Insert a microcondenser and gently reflux the material for 10 to 20 minutes.

$$RCOOH + SOCl_2 \rightarrow RCOCl + SO_2 + HCl$$

Test for the acid halide by test 10.03.

NOTE. Thionyl chloride may convert dibasic acids to the anhydride instead of the acid chloride (particularly, in case of $HOOC(CH_2)_nCOOH$, where n is 2 or 3, or to the cyclic ketone where n is 4 or 5). Anhydrides respond to the same hydroxamate test as do acid halides. α-Hydroxyacids react with thionyl chloride to produce formic acid and an aldehyde or ketone. Amino acids, di- and trichloroacetic acid, and dicarboxylic acids do not convert to the acid chloride successfully with thionyl chloride.

Phosphorus pentachloride may be used instead of thionyl chloride for the conversion of acids to acid chlorides.

(b) Merliss and Weinheimer[20] have proposed a method for converting acids in water solution to acid chlorides which may then be detected by the ferric hydroxamate test. It is claimed that this method can be used for low molecular weight acids that may be lost by vaporization by other methods and for acids whose acid chlorides are unstable, such as the α-hydroxy acids. This procedure should be carried out under the hood,

---

[20] F. E. Merliss and A. J. Weinheimer, unpublished paper given at the Southwest Regional Meeting of the ACS, Dallas, Texas, December, 1962.

and a safety mask should be worn over the face since experience in our laboratory shows that there is some danger that decidedly acidic or basic liquids may be thrown out of the tube by excessive reactivity unless care is taken. The method is particularly useful when the substance being examined is a relatively dilute aqueous solution.

Place 2 ml of the aqueous solution (or 50 to 75 mg of the acid in 2 ml of water) in an 8-in. test tube and insert the tube in an ice bath. Add, in small portions, 2 ml of thionyl chloride to the tube. Occasionally shake the tube gently and keep it in the ice bath until the mixture appears homogeneous (less than 5 minutes, generally). While keeping the tube in the ice bath, add, in portions, 3 ml of a saturated solution of hydroxylammonium chloride in ethanol. After all of this reagent has been added, remove the tube from the ice bath and allow it to stand at room temperature for 3 to 5 minutes. Because of the large excess of thionyl chloride, a sufficient amount of the acid will be converted to the acid chloride to allow its detection by the ferric hydroxamate test. To make this test, insert a glass stirring rod into the test tube and, with stirring, add, dropwise, $6N$ aqueous potassium hydroxide (addition may be made more rapidly if the tube is put back in the ice bath to prevent overheating). Continue the addition of the potassium hydroxide until the pH of the mixture reaches 10 to 11 as shown by an indicator paper. As the base is added, copious quantities of potassium chloride will precipitate. This precipitated salt will not interfere with the final test and, therefore, may be neglected (it may be dissolved by adding more water). Now add 1 drop of 10 per cent ferric chloride. The solution will turn yellow, and a yellow-to-brown precipitate may form. Add, dropwise, $12M$ hydrochloric acid with stirring until the yellow color is discharged and the solution becomes essentially colorless. Finally, add 1 to 3 drops of 10 per cent ferric chloride. A bluish-red color is a positive test.

2. *Conversion to esters.* If the original substance is an acid that has been converted to an acid chloride by procedure (1) above, it may be converted to an ester by the following method.

Add 0.5 ml of an alcohol (for example, butanol) to the tube in which the acid halide has been prepared and gently reflux the mixture for 2 minutes. If a precipitate exists, add more alcohol dropwise and continue heating until the precipitate dissolves. Cool the tube and add 1 ml of water to hydrolyze any excess thionyl chloride. Remove the water-insoluble layer by means of a pipet and test it for esters by test 10.11.

NOTE. Another method for converting an acid to an ester for detection by the hydroxamate test is as follows: Dissolve or suspend 30 to 40 mg of the acid in 0.5 m of ethylene glycol. Add 1 drop of concentrated sulfuric acid and gently reflux the mixture for 5 minutes. Cool, and apply the ester test.

3. *Neutralization equivalent.* Acids may be tentatively identified by determining the neutralization equivalent; see the Appendix.

B. Amino Acids

Ninhydrin[21] (triketohydrindene hydrate) reacts with α-amino acids and hydrolysis products of proteins to produce violet, blue, or purple products. The final colored products are not single compounds, but are condensation products, some of which depend on the amino acids originally present; but one product that has been observed in all cases is the violet bis-1,3-diketoindyl.[22]

Dissolve 1 to 5 mg of the substance to be tested in 1 to 3 ml of water in a 4-in. test tube. Add 1 ml of a 0.1 per cent solution of ninhydrin in water. Heat the contents of the tube to boiling for 1 to 2 minutes. The production of a violet, blue, or purple color is a positive test.

NOTE. Proteins, peptones, and peptides give positive tests with ninhydrin after hydrolysis, so that at least some free amino acids are present. Proline and hydroxyproline do not give the characteristic blue colored products, but produce yellow-red colors instead. Feigl[23] states that β-amino acids and primary and secondary alkyl amines give positive tests with ninhydrin in some cases.

Bergmann and Bentov[24] have proposed a new reagent for amino acids, namely, 2,4-dinitro-5-fluoroaniline. This reagent is said to have two advantages: (1) high yields of derivative, and (2) the arylamino group may be diazotized and coupled, thus making possible the detection of very small amounts of an amino acid.

C. Sulfonic Acids

Sulfonic acids are usually detected by the elemental analysis coupled with their solubility behavior and decided acidity. If further testing is desired, the sulfonic acid may be converted to the acid chloride by using thionyl chloride and then treating with concentrated ammonium hydroxide to prepare the sulfonamide which can be detected by test 10.06D. Feigl[25] suggests detecting sulfonic acids by converting them to ferric acetyhdroxamate.

[21] S. Ruhemann, *J. Chem. Soc.*, **97**, 1438, 2025 (1910); **99**, 792, 1486 (1911).
[22] R. Moubacher and M. Ibrahim, *J. Chem. Soc. (London)*, **1949**, 702; A. Schoenberg and R. Moubacher, *Chem. Rev.*, **50**, 272 (1952).
[23] F. Feigl, *Spot Tests in Organic Analysis*, Elsevier, Amsterdam, 1960, p. 293.
[24] E. D. Bergmann and M. Bentov, *J. Org. Chem.*, **26**, 1480 (1961).
[25] F. Feigl, p. 265 (cited in Reference 23).

10.02 Acid Anhydrides

Most anhydrides can be converted to hydroxamic acids by the method given in test 10.03, since both anhydrides and acid halides may be converted to hydroxamic acids by this procedure. If an anhydride is suspected even though the hydroxamate test is negative, heat 50 to 75 mg of the compound with 3 drops of butanol for 3 to 5 minutes and test for an ester (test 10.11).

NOTE. Containers of "anhydrides" that have been frequently opened and thus exposed to moist air will often be found to contain that acid rather than the anhydride. Samples of anhydrides frequently fail to give the expected physical constants because of partial or complete hydrolysis.

10.03 Acid Halides

A. Amide Formation

Acid halides may be reacted with an amine such as benzylamine or aniline to form substituted amides that are only slightly soluble in cold water.

Add 2 drops of the compound to 3 drops of benzylamine in a 4-in. test tube. After the reaction has subsided, add 2 ml of cold water and shake the tube vigorously. The white benzylamide may be recrystallized from water.

NOTE. Benzylamine reacts more readily with acid halides, particularly high-molecular-weight aroyl halides or sulfonyl halides, than does aniline. If it is suspected that the acid halide is a sulfonyl halide, 5 drops of pyridine should be added to the test tube and the mixture gently heated for 3 minutes before the water is added. Sulfonyl halides react with pyridine to form a sulfonpyridinium cation which is a good acylating agent. Acid anhydrides will not react appreciably to form amides under the conditions specified. Acids may react to form salts but not amides.

B. The Ferric Hydroxamate Test[26]

Hydroxamic acids reacts with ferric chloride in acidic solutions to form soluble ferric hydroxamates that are highly colored. The most common color for these salts is bluish red (magenta) but a deep red color is often noted, particularly if the concentration is high. Many classes of organic compounds may be converted into hydroxamic acids and hence may be tested for either directly or indirectly by the ferric hydroxamate test.

[26] R. E. Buckles and C. J. Thelen, *Anal. Chem.*, **22**, 676 (1950); D. Davidson, *J. Chem. Educ.*, **17**, 81 (1940); F. Feigl et al., *Mikrochemie*, **15**, 9, 23 (1934); F. Feigl, *Spot Tests in Organic Analysis*, Elsevier, Amsterdam, 1960, p. 249.

It must be recalled that ferric chloride reacts directly with several classes of organic compounds, particularly with phenols, to produce colored products, some of which have colors very similar to the colors produced with hydroxamic acids. Hence, *it is advisable to test the compound with ferric chloride alone before applying any hydroxamate test.* This may be done as follows:

Dissolve 30 mg of the compound in 1 ml of ethanol in a 4-in. test tube and add 1 ml of $1N$ hydrochloric acid, followed by 1 drop of 10 per cent ferric chloride. If the mixture has any color except yellow, it is very doubtful that reliable conclusions could be drawn from performing a hydroxamate test.

Procedure for acid halides and anhydrides. Add 30 to 40 mg of the compound being examined to 0.5 ml of a $1N$ solution of hydroxylammonium chloride in alcohol. Add 2 drops of $6M$ hydrochloric acid to the mixture; warm it slightly for 2 minutes and then boil it for a few seconds. Cool the solution and add 1 drop of 10 per cent ferric chloride. The formation of a reddish blue or bluish red color is a positive test. If the color is more red than blue, adjust the pH of the solution to a pH of 2 to 3 by adding dropwise $2N$ hydrochloric acid. The color should shift toward purple (see the *Note* that follows). Typical reactions are:

$$RCOX + NH_2OH \rightarrow RCO(NHOH) + HX$$

$$(RCO)_2O + NH_2OH \rightarrow RCO(NHOH) + RCOOH$$

$$RCO(NHOH) + Fe^{+++} \rightarrow Fe(RCONHO)^{++} + H^+$$

NOTE. It has been determined[27] that the color of the soluble complex formed between ferric ions and the hydroxamate ions varies with the pH of the solution. The probable predominating species are: $Fe(RCONHO)^{++}$ in strongly acidic solution, $Fe(RCONHO)_2^+$ in weakly acidic solution, and $Fe(RCONHO)_3$ in neutral and weakly alkaline solution.

For additional procedures for the hydroxamate test, see the test for esters (test 10.11) and the test for amides (test 10.06B).

10.04 Alcohols

Tests for alcohols and certain other classes were given in test P-6 (page 356). Particularly, attention is called to test P-6B for water-soluble alcohols and to test P-6D for detecting di- and polyhydroxy alcohols. The following tests are more specific for alcohols.

[27] G. Aksnes, *Acta Chem. Scand.*, **11**, 719 (1957).

A. Vanadium-Oxine Test[28]

Alcohols, even when they are present in mixtures with other classes, such as hydrocarbons, ethers, ketones, and halogenated hydrocarbons, react with a mixture of vanadate ions and 8-hydroxyquinoline(oxine). Many esters, phenols, amines and some acids also give colors with the reagent but the color is not red unless an alcohol is present. Weak tests for an alcohol may be due to the presence of an alcohol as an impurity in some other solvent.

Place about 30 mg of the compound to be tested in a 3-in. test tube and add 0.5 ml of a solution of ammonium vanadate (30 mg in 100 ml of water). Now add 1 or 2 drops of a 2.5 per cent solution of 8-hydroxyquinoline in 6 per cent acetic acid. A green precipitate of the vanadium-oxine complex forms which reacts with alcohols to produce a complex that is red and which is soluble in aromatic hydrocarbons. Add 0.5 to 0.7 ml of benzene or toluene and shake the tube. A pink to red color in the hydrocarbon layer denotes an alcohol.

NOTE. Aqueous solutions of an alcohol may also be used to test for alcohols by this method providing the amount of hydrocarbon used is sglihtly greater than the volume of the aqueous solution. Additional references for the use of the vanadium-oxine test for alcohols may be found at the end of this chapter.

B. Hydroxamate Test

Alcohols may be converted to esters by reaction with acetyl chloride. Tertiary alcohols are partly converted to alkyl chlorides when they are treated with acetyl chloride, owing to the reaction of the liberated hydrogen chloride on another molecule of the alcohol. This result can be avoided by introducing dimethylaniline into the mixture to react preferentially with the hydrogen chloride. The dimethylaniline may be omitted if the alcohol is known not to be a tertiary alcohol.

Mix 0.1 ml of acetyl chloride with 0.1 ml of dimethylaniline and add 0.2 ml of the alcohol. Shake the mixture frequently for 5 minutes. Add about 1 g of ice; or add, dropwise and with shaking, 1 ml of cold water to decompose the remaining acetyl chloride. Pour the mixture into a small test tube so that the stratified layers may be easily distinguished. Remove 2 or 3 drops of the upper layer and test for esters by test 10.11.

C. Xanthate Test

1. It has been shown[29] that the alcohols may be qualitatively detected and quantitatively estimated by reacting the potassium alkoxides with

---

[28] F. Buscarons, J. L. Marin, and J. Claver, *Anal. Chim. Acta*, **3**, 310, 417 (1949); *Anales real soc. espān. fis y quím*, **49B**, 367 (1953); *C.A.*, **48**, 80 (1954); and A. J. Blair and D. A. Pantony, *Anal. Chim. Acta*, **13**, 1 (1955).

[29] W. F. Whitmore and E. Lieber, *Ind. Eng. Chem., Anal. Ed.*, **7**, 127 (1935).

carbon disulfide to form the potassium alkyl xanthates. The test is also satisfactory for the cellosolves (monoalkyl ethers of glycol), but the carbitols (monoalkyl ethers of diethyleneglycol) yield heavy red oils rather than the customary light yellow precipitates. The xanthates of tertiary alcohols are rather easily hydrolyzed, but enough precipitate is generally formed to give a positive test.

$$ROH + KOH \rightleftharpoons ROK + H_2O$$
$$ROK + CS_2 \rightarrow ROCSSK$$

Add 1 pellet of solid potassium hydroxide to 0.5 ml of the alcohol *in a dry test tube* and heat until the KOH dissolves (very volatile alcohols require a reflux condenser). Cool the tube and add 1 ml of ether. Add, dropwise, carbon disulfide until a pale yellow precipitate forms. If a precipitate does not form by the time 0.5 ml of carbon disulfide has been added, the test should be considered negative.

NOTE. Ketones that can enolize effectively give a positive xanthate test.

2. Feigl[30] considers that the xanthate test is applicable only to primary and secondary alcohols. Also, he confirms the presence of an alkali alkyl xanthate by using the molybdate ion. A summary of his method follows.

Add 1 drop of the suspected alcohol to a few drops of ether in a 3-in. test tube and add 1 or 2 drops of carbon disulfide and 200 mg of powdered sodium hydroxide. Shake the mixture for 5 minutes and then add 1 or 2 drops of 1 per cent ammonium molybdate. Carefully acidify the mixture with $1M$ sulfuric acid. Add 2 drops of chloroform, shake the mixture and then allow it to stratify. A violet color in the chloroform is considered a positive test for a primary or secondary alcohol, according to Feigl.

D. *Tests to Distinguish Primary, Secondary, and Tertiary Alcohols*

1. Kruse, Grist, and McCoy[31] recommend the use of N-bromosuccinimide for distinguishing the three subclasses of alcohols.

Dissolve 50 to 75 mg of the alcohol in 1 to 2 ml of a 0.01 per cent (by weight) solution of bromine in carbon tetrachloride in a 4-in. test tube. Add 20 to 30 mg of N-bromosuccinimide and place the test tube in a water bath at 78 to 80°. At this temperature, the solution will boil gently. Any color change from the initial pale yellow will occur within 13 minutes (usually within 5 minutes). Primary alcohols give a permanent orange color. Secondary alcohols produce a transitory orange color which fades

---

[30] F. Feigl, pp. 186, 249 (cited in Reference 23).
[31] P. F. Kruse, K. L. Grist, and T. A. McCoy, *Anal. Chem.*, **26**, 1319 (1954).

(usually very rapidly), leaving the solution colorless after continued boiling. After the tube is cooled, an orange precipitate is often apparent in the case of a secondary alcohol. Tertiary alcohols generally do not produce color changes with the reagent.

NOTE. Allyl, benzyl, and *tert*-amyl alcohols all give false results in that all three give the test for secondary alcohols. Primary alcohols up to octadecanol react regularly. The monoalkyl ethers of ethylene glycol give the test for primary alcohols. The cyclohexanols react regularly (secondary alcohols).

Although the chemistry involved in this test has not been established, Kruse, Grist, and McCoy, basing their deductions on other studies,[32] suggest that the primary and secondary alcohols react to form hypobromites which split out hydrogen bromide to yield aldehydes and ketones. The hydrogen bromide reacts with unchanged N-bromosuccinimide to yield succinimide and bromine which is probably the cause of the color. The fading of the color in the test for secondary alcohols would then be explained by the greater reaction of the hydrogen bromide with secondary alcohols, thus allowing less bromine formation, or to the greater reaction of bromine with the ketones than with the aldehydes.

2. A distinction may be made among the three subclasses of alcohols that contain up to six carbon atoms by noting the rates of formation of the alkyl chlorides when the alcohol is reacted with a saturated solution of anhydrous zinc chloride in concentrated hydrochloric acid (Lucas[33] reagent).

Add 3 or 4 drops of the alcohol to 2 ml of Lucas reagent in a 3-in. test tube and shake the mixture vigorously. Allow the mixture to stand at room temperature. A reaction is detected by the clouding of the solution, owing to the formation of an insoluble alkyl chloride. Tertiary alcohols react immediately; secondary alcohols react within 2 to 3 minutes; primary alcohols require a much longer time.

If there is any doubt as to whether the alcohol is secondary or tertiary, mix 2 to 4 drops of the alcohol with 2 ml of concentrated hydrochloric acid. Secondary alcohols are not converted to alkyl chlorides under these conditions, whereas tertiary alcohols yield the alkyl chloride within 10 minutes.

NOTE. Alcohols with six or more carbon atoms are not soluble in the reagent, hence, they emulsify on shaking with the reagent and appear to give a test for a tertiary alcohol. The cloudiness of the solutions produced by secondary and tertiary alcohols is due to an emulsion of the alkyl chloride in the reagent. The mixture will clear on standing and a separate layer of the alkyl chloride will form. If a separate layer does not form, it is likely that the alcohol detected was only an impurity in a less reactive

[32] L. F. Feiser and S. Rajagopalan, *J. Am. Chem. Soc.*, **71**, 3935, 3938 (1949); M. Z. Barakat and G. M. Mousa, *J. Pharm. and Pharmacol.*, **4**, 115, (1952).

[33] H. J. Lucas, *J. Am. Chem. Soc.*, **52**, 802 (1930).

alcohol. Allyl, benzyl, and cinnamyl alcohols behave irregularly in that they give the test for a tertiary alcohol.

3. A number of workers have proposed tests for alcohols based on selective oxidation and some of these methods are referred to in the list of references that follow this chapter. Of these oxidation methods, one that has proven useful in our laboratory is that given by Davidson and Perlman.[34] In this case, tertiary alcohols are not detected but both primary and secondary alcohols are not only detected but are converted to solid derivatives which may be used for characterization.

Chromic acid test. (a) Place 0.5 ml of formic acid in a 6-in. test tube and add 100 to 150 mg of the alcohol followed by 1 ml of $2M$ chromic acid in acetic acid. Maintain the temperature at 60° until the chromic acid is reduced (violet to green color). Cool the mixture and add 5 ml of a saturated solution of 2,4-dinitrophenylhydrazine in $2M$ hydrochloric acid. A precipitate should form if a carbonyl compound has been produced by oxidation. Filter the precipitate, wash it with a few drops of water, and test a few mg of it for a 2,4-dinitrophenylhydrazone by adding it to 2 ml of $2M$ potassium hydroxide in methanol. A red or purple color in this test confirms the hydrazone.

NOTE. The test for secondary alcohols is more evident than that for primary alcohols because some of the aldehyde that is produced by oxidizing the primary alcohol is further oxidized to an acid. The purpose of using the formic acid is to reduce the excess chromic acid before the 2,4-dinitrophenylhydrazine is added. Some phenols, aryl ethers, and a very few aromatic hydrocarbons may be oxidized to quinones by this procedure. In the confirmatory test, the 2,4-dinitrophenylhydrazones of quinones produce a blue color rather than a red or purple color.

(b) Bordwell and Wellman[35] use the chromic acid to detect tertiary alcohols.

Dissolve 1 g of chromic acid in 1 ml of concentrated sulfuric acid and then very carefully dilute this mixture with 3 ml of water. Add 1 drop of this reagent to a solution of 15 to 30 mg of the alcohol dissolved in 1 ml of alcohol-free acetone. The orange color of the reagent is not affected by a tertiary alcohol but primary or secondary alcohols react within 10 seconds to discharge the orange color and to produce an opaque, blue-green suspension.

NOTE. Easily oxidized compounds, such as aldehydes, phenols, many amines, and the like, react with this chromic acid reagent to discharge the orange color. Hence, this test is not specific for tertiary alcohols unless it has been established that the compound is an alcohol.

[34] D. Davidson and D. Perlman, *A Guide to Qualitative Organic Analysis*, 2nd ed., Brooklyn College Bookstore, Brooklyn, N.Y., 1958, p. 59.
[35] F. G. Bordwell and K. M. Wellman, *J. Chem. Educ.*, **39**, 308 (1962).

10.05 Alkyl and Aryl Halides

The halogens may occur as substituents in all classes of organic compounds. However, with the exception of several classes of perfluoro compounds, only the halogenated hydrocarbons fall in Solubility Division I. The following tests are useful in distinguishing alkyl halides from aryl halides with the limitations noted.

A. Alcoholic Silver Nitrate

1. Add 30 to 40 mg of the compound to 0.5 ml of a saturated solution of silver nitrate in ethanol. Do not heat the solution. Note any precipitate that forms within 2 minutes.

Alkyl bromides and iodides and tertiary alkyl chlorides precipitate silver halide by this test. Alicyclic bromides and iodides, allyl halides, and 1,2-dibromoalkanes also give a positive test at room temperature. Aryl halides and primary and secondary chlorides do *not* form precipitates by this method.

2. If no precipitate forms within 2 minutes at room temperature, heat the solution from (1) above and boil it for 30 seconds. Silver chloride will precipitate from primary and secondary alkyl halides. Aryl halides, vinyl halides, and compounds such as chloroform do not cause precipitation.

B. Hydrolysis

The alkyl halides hydrolyze much more readily than the aryl halides. In the following test, all of the alkyl halides (except fluorides) will cause the precipitation of the silver halide, but the aryl halides produce no more than a light cloudiness.

Mix 100 mg of the halogen compound with 5 ml of 5 per cent alcoholic solution of potassium hydroxide and reflux the mixture for 5 minutes. Cool the mixture and add 10 ml of distilled water. Acidify the solution with dilute nitric acid. Unless the solution is clear, filter it. Add 2 drops of 5 per cent silver nitrate solution.

C. Formaldehyde–Sulfuric Acid Test

According to a limited study,[36] the following test produces pink, red, or bluish red colors with aryl halides, whereas alkyl halides produce yellow, amber, or brown colors.

Add 1 drop of the compound to be tested to 1 ml of hexane or carbon

---

[36] B. Berry, *Proc. of La. Acad. Sci.*, **18**, 92 (1955).

tetrachloride. Add 1 or 2 drops of this solution to 1 ml of a reagent which is prepared at time of use by adding 1 drop of formalin (37 per cent formaldehyde) to 1 ml of concentrated sulfuric acid. Shake the tube gently and note the color.

10.06 Unsubstituted Amides

Procedures for the detection of amides that are not N-substituted will be given in this section. Methods for detecting N-substituted amides will be given in 10.07.

A. Liberation of Ammonia

Amides may be hydrolyzed by treatment with alkali to form the salt of the acid and liberate ammonia, a compound that may be easily detected. Ammonium salts also may liberate ammonia by the same treatment; however, ammonium salts do not give the hydroxamate test (test B below). Ammonia may be liberated from unsubstituted amides by (*a*) boiling a mixture of 50 mg of the substance in 2 ml of 20 per cent sodium hydroxide for 1 minute, or (*b*) by dry fusion of a mixture of 50 mg of the compound with 200 to 300 mg of pulverized sodium hydroxide. In either case, the vaporized ammonia may be detected by test P-3A. N-Substituted amides in which the alkyl groups have only a few carbon atoms will also give the blue coloration to copper sulfate by this method, since low-weight amines as well as ammonia form blue complexes with copper ions (test P-3A1). They do not give positive results by the phenol-hypochlorite test (test P-3A2). See test 10.08A to establish the presence of amines.

Ammonium salts may be differentiated from amides by the fact that these salts will liberate ammonia when heated in a suspension of magnesium oxide in water. Amides are not hydrolyzed by this mixture except after long heating.

B. Hydroxamate Test

Amides that are not N-substituted may be detected by the hydroxamate test,[37] provided it is known that the substance does not belong to some other class that will also yield an hydroxamic acid under the test conditions.

Add 30 mg of the substance to 2 ml of $1N$ hydroxylammonium chloride in propylene glycol in a 4-in. test tube. Boil the mixture for 2 minutes,

---

[37] S. Soloway and A. Lipschitz, *Anal. Chem.*, **24**, 898 (1952).

cool, and add 0.5 to 1 ml of a 5 per cent ferric chloride. A red-to-violet color is a positive test.

Davidson and Perlman[38] have proposed a modification of the test, which they claim is more general for amides and ureas but which does not give a positive test with esters and the like.

Mix 50 to 60 mg of the amide with 30 mg of solid hydroxylammonium chloride in a 3-in. test tube. Insert this tube in an oil bath (for example, a Thiele tube), which has been preheated to 170°. Heat the mixture at 170° until foaming ceases; cool, and dissolve the melt in 1 ml of methanol. Finally, add 1 drop of 10 per cent ferric chloride in methanol.

C. Distinguishing Aliphatic Amides from Aromatic Amides

1. Most aromatic amides are converted directly to the hydroxamic acid by hydrogen peroxide, whereas the aliphatic amides fail to form the hydroxamic acid in this way. Most aliphatic amides are converted to hydroxamic acid by hydroxylamine in aqueous or ethanolic solution; aromatic amides react much less readily.

$$ArCONH_2 + H_2O_2 \rightarrow ArCONHOH + H_2O$$
$$RCONH_2 + H_2NOH \cdot HCl \rightarrow RCONHOH + NH_4Cl$$

(*a*) *Aliphatic amides.* Add 50 mg of the amide to 1 ml of $1N$ hydroxylammonium chloride in ethanol and boil the mixture for 3 minutes. Cool the tube and add 1 or 2 drops of 5 per cent ferric chloride. A bluish red color is a positive test.

(*b*) *Aromatic amides.* Suspend 50 mg of the amide in 2 to 3 ml of water. Stopper the tube and shake it vigorously for a few seconds. Add 4 or 5 drops of 6 per cent hydrogen peroxide and heat the mixture to near boiling. If the amide does not completely dissolve add a few drops more of hydrogen peroxide. Cool the solution and add 1 drop of 5 per cent ferric chloride. If a bluish red color does not develop within 1 minute, warm the tube gently but do not boil the solution. In most cases the reaction eventually goes past the hydroxamate stage and a brown color develops, gradually settling out as a brown precipitate.

It has been observed that the addition of a few milliliters of 10 per cent sodium hydroxide to the final mixtures obtained in this test produces a clear, deep reddish-brown solution.

NOTE. Any time ferric chloride is to be used as one of the reagents in a test, the reader must remember that ferric chloride reacts directly with several classes of

---

[38] See D. Davidson and D. Perlman, p. 65 (cited in Reference 34).

compounds to give colored products (see test 10.20A). For example, salicylamide is a phenol as well as an amide; hence, the hydroxamate test on this compound is not significant because of the interfering color produced by the interaction of ferric chloride with the phenolic group.

2. The amides of aliphatic amides may be distinguished from the amides of aromatic acids by heating a mixture of the amide, dimethyl oxalate, and thiobarbituric acid for 2 to 3 minutes at a temperature of 130 to 160°. The aliphatic amides react with the dimethyl oxalate to produce oxamide which, in turn, reacts with thiobarbituric acid to form red condensation products[39] which are soluble in water or alcohol. Aromatic amides may produce a yellow color by this procedure. Substances that release ammonia when they are heated interfere with this test for aliphatic amides, since ammonia reacts with dimethyl oxalate to form oxamide.

D. Sulfonamides

To prepare the test paper for this text, mix equal volumes of a 1 per cent solution of N,N-dimethyl-α-naphthylamine in methanol and a 1 per cent aqueous solution of sodium nitrite. Dip a piece of filter paper in the mixture and allow it to dry in the dark.

In making this test[40] for sulfonamides, place 1 or 2 drops of a solution or a suspension of the compound in water on the test paper and touch the spot with a drop of 0.2 to 0.5 per cent hydrochloric acid. A red or dark rose color develops quickly if a sulfonamide is present.

NOTE. This is an extremely sensitive test for sulfonamides. It may be used when the concentration of the sulfonamide is of the order of 5 mg per cent in the solution being tested. The test may be made on blood by treating the blood with an equal volume of 10 per cent trichloroacetic acid before applying it to the test paper.

Amides other than sulfonamides do not give this test but an orange-red ring of color may appear around the spot where the compound was placed on the paper.

E. Urea, Thiourea, and Substituted Ureas

When urea, N-substituted urea, or N,N'-symmetrically substituted urea is heated with an excess of phenylhydrazine at about 200°, diphenylcarbazide is formed and ammonia or an amine is liberated. For example

$$H_2NCONHR + 2H_2NNHC_6H_5 \rightarrow (C_6H_5NHNH)_2CO + NH_3 + RNH_2$$

Thiourea gives an analogous reaction and forms diphenylthiocarbazide. The diphenylcarbazide reacts with nickel ions to form violet inner complex salts that are soluble in chloroform.

[39] F. Feigl, *Spot Tests in Organic Analysis*, Elsevier, Amsterdam, 1960, p. 299.
[40] Chr. Hackmann, *Deut. med. Wechschr.*, **72**, 71 (1947); *C.A.*, **41**, 4824 (1947).

Place 10 to 20 mg of the compound in a 3-in. test tube and add 2 or 3 drops of phenylhydrazine. Heat the tube in an oil bath at about 195° for 5 minutes. After cooling the tube, add 6 drops of concentrated ammonium hydroxide and 6 drops of a 10 per cent solution of nickel sulfate. Shake the tube vigorously and allow it to stand for 3 minutes. Extract the mixture with 10 drops of chloroform. A red-violet or violet color in the chloroform layer constitutes a positive test.

NOTE. At the temperature used, part of the urea will be converted to biuret, but this also reacts with phenylhydrazine to produce diphenylcarbazide. Many urethanes also give a positive test with this procedure. The test is not as sensitive with thiourea as with urea. The presence of thiourea may be detected by heating the dry sample of the original compound to about 200° and noting the evolution of hydrogen sulfide.

10.07 Substituted Amides

Substituted amides vary greatly in their compositions and in the methods by which they may be detected. In fact, no specific test for all types of amides is on record. Most methods for detecting N-substituted amides depend on first hydrolyzing the amide and then, by one method or another, identifying the amine that is liberated. Two major problems present themselves. First, many amides, particularly those with nitro and halogen substituents, are resistant to hydrolysis. Second, the recovery of the amine, produced by hydrolysis, is sometimes difficult. The following tests cover the more common types of substituted amides.

A. N-Alkyl-Substituted Amines

Low-weight alkyl amines as well as ammonia can be detected by the formation of blue complexes with copper ions (see test 10.08-A2). To differentiate ammonia from primary or secondary amines, proceed as follows:

Place 200 to 300 mg of the amide in a 6-in. test tube and add 5 ml of 10 per cent sodium hydroxide. Insert a microcondenser into the tube and gently reflux the mixture for 15 minutes. After cooling the test tube, remove the condenser and arrange to distill from the test tube into another 6-in. test tube which contains 5 ml of water to which 2 drops of concentrated hydrochloric acid have been added. Have the delivery tube extend to within 3 to 4 mm of the surface of the acidified water. *Slowly* distill the hydrolyzed mixture until about 0.5 ml has passed into the receiving tube. Neutralize the distillate by adding, dropwise, 1.1N sodium bicarbonate solution. Divide the distillate into two equal portions and apply test 10.08C for primary alkyl amines and test 10.08D for secondary alkyl amines. Tertiary amines would not be produced by hydrolyzing an amide.

B. Anilides

The most common N-substituted amides are anilides; they may be detected by the following two methods with the limitations noted.

1. Anilides that do not have substituents on the ring produce a rose color when treated at room temperature with sulfuric acid and potassium dichromate.

Add 100 mg of the compound to 3 ml of concentrated sulfuric acid. Stopper the tube and shake it vigorously. Add 50 mg of finely powdered potassium dichromate. A bluish pink color is a positive test.

2. A modification of the aniline acetate test for carbohydrates (compare test 10.09A2) is quite satisfactory for determining anilides, both simple anilides and anilides with certain substituents (see the *Note* below).

Mix 100 mg of the compound with one pellet of crushed sodium hydroxide. Add the mixture to a dry 4-in. test tube and wrap a piece of filter paper over the mouth of the tube. Moisten the paper with 1 or 2 drops of 4M acetic acid. Heat the tube and fuse the mixture. Aniline and certain substituted anilines will vaporize and react with the acetic acid on the paper. Place 50 mg of a sugar in a dry 4-in. test tube and transfer the filter paper that had covered the fusion mixture over to the tube which contains the sugar. Heat the sugar until it chars. Decomposition of the sugar produces furfural, which reacts with the aniline acetate to give a pink color on the paper.

NOTE. The procedure produces positive results even when halogens, methyl, or hydroxy groups are present on the aniline ring. Obviously, only those anilines which are at least moderately volatile will give this test.

C. Other Substituted Amides

Proof that the compound is an amide involves hydrolysis and identification of the amine and the acid thus produced. Since hydrolysis of these amides is difficult, it is desirable to detect the products by derivatizing them in the same operation, thus detecting the classes and determining the individual compounds produced at the same time. The sections of this book relating to the preparation of derivatives of amides, amines, and acids should be consulted before selecting a procedure for hydrolyzing the amide.

Hydrolysis of the simpler substituted amides is effected by boiling the compound for a few hours with either $6N$ hydrochloric acid or with 20 per cent sodium hydroxide. Alkaline hydrolysis is usually faster than acid hydrolysis. If alkaline hydrolysis is used, the freed amine may be recovered by filtration (if it is a solid), by steam distillation, or by extraction by ether. The aqueous solution of the sodium salt of the acid should be reserved for recovery of the acid. If acid hydrolysis is used, the organic acid may be separated by filtration, distillation, or extraction. The acid solution of the amine salt may be made alkaline and the amine recovered as above.

Other alkaline-hydrolyzing media include (a) a saturated solution of potassium hydroxide in ethanol, (b) a 5 per cent solution of sodium ethoxide in ethanol, and (c) a 5 to 20 per cent solution of potassium hydroxide in propylene glycol or glycerol. Amides that contain halogens or nitro groups in the amine moiety are quite resistant to hydrolysis. In such cases, 100 per cent phosphoric acid, made by mixing the 85 per cent acid with phosphorus pentoxide, is recommended. N-Alkyl–substituted sulfonamides are very resistant to hydrolysis. It is reported that they require 10 to 40 hours of refluxing with dilute hydrochloric acid. A mixture of 48 per cent hydrobromic acid and phenol has been recommended[41] as the best medium for hydrolyzing the substituted sulfonamides.

10.08 Amines

Simple amines are easily detected by the fact that they contain nitrogen, give basic reactions, and are soluble in acidic solutions if they are not soluble in water. Many of the substituted aromatic amines fail to dissolve in dilute acids and fail to give positive tests with reagents that are effective on most amines. Since no specific test works perfectly for all amines of a given class, it is wise to test any nitrogen-containing compound that is not easily classified by at least two of the tests for amines. A study has been made of the dependability of a large number of tests that have been proposed for amines. The following tests seem to be the most dependable. It may be noted that the tests are divided into sections: *A*, general tests for amines; *B*, a number of tests suitable for distinguishing primary, secondary and tertiary amines; and, finally, sections *C*, *D*, and *E*, which give tests that are relatively specific for the subclasses.

---

[41] H. R. Snyder et al., *J. Am. Chem. Soc.*, **74**, 2006, 4864 (1952).

A. General Tests

1. *Tetraphenylborate test.*[42] Amines that have a pK_a of 11 or less may be detected by the formation of a white precipitate when the protonated amine is treated with a solution of tetraphenylborate. Two procedures that have been found useful are presented.

(a) Place 5 to 15 mg of the substance in a 4-in. test tube and add 3 drops of $3M$ hydrochloric acid, followed by 5 ml of water. Shake the mixture and add 1 drop of a 5 per cent aqueous solution of tetraphenylborate. Immediate formation of a dense white precipitate at the top of the solution is a positive test.

(b) Place 5 to 15 mg of the substance on a spot-plate (black background) and dissolve the sample in a few drops of methanol. Add one drop of $1M$ hydrochloric acid and 1 drop of a 5 per cent aqueous solution of tetraphenyl borate. A *dense* white precipitate is a positive test. Slight precipitation may be caused by other classes than amines. Dimethylformamide does precipitate the reagent but other amides do not.

2. *Copper ion test.* Low-molecular weight amines that are soluble in water, including amino alcohols, may be detected by adding 30 mg of the substance to 1 ml of a 10 per cent solution of copper sulfate. A blue or blue-green coloration or precipitate indicates an amine. This test may also be made by allowing the vapors of the suspected amine to come in contact with a filter paper that has been treated with a few drops of a copper sulfate solution.

3. *Fluorescein chloride test.* Feigl[43] recommends fusion of the amine salt with fluorescein chloride as a method of detecting amines through the formation of red water-soluble rhodamine dyes. Hydrazo compounds and some N-aryl amides also give this test. Feigl also discusses the additional evidence that may be obtained from fluorescence studies.

Dissolve or suspend a few milligrams of the substance in water, acidify with hydrochloric acid, and evaporate to dryness in a small test tube. Add 30 to 40 mg of fluorescein chloride and 60 to 80 mg of powdered, anhydrous zinc chloride. Fuse the mixture at 250 to 260° until all of the zinc chloride has melted. Cool the tube and dissolve the melt in water. A red color is a positive test.

B. This section includes nine tests that serve to subclassify amines but which are not specific for any one subclass in most cases. Hence, after

---

[42] F. E. Crane, Jr., *Anal. Chem.*, **28**, 1794 (1956); and **30**, 1426 (1958); A. J. Barnard, Jr. and W. W. Wendlandt, *Revista de la sociedad de Mexico*, **3**, 269 (1959).

[43] F. Feigl, *Spot Tests in Organic Analysis*, Elsevier, Amsterdam, 1960, p. 275.

an amine has been tentatively identified as to subclass, the appropriate tests given in sections C, D, or E should be tried for confirmation.

1. *Basicity test.* Differentiation of alkyl amines from aryl amines may generally be made on the basis of their difference in basicity. It is reported[44] that an aqueous solution buffered at pH 5.5 will dissolve alkyl amines but not aryl amines. While this method is not completely reliable, it has proven useful.

NOTE. The buffer solution is made by adding 24 g of glacial acetic acid to 164 g of anhydrous sodium acetate and diluting to 1 liter.

2. *Quinhydrone test.* Quinhydrone has been found to be an excellent reagent for detecting the subclass to which an amine belongs.[45] The following method has[46] proven satisfactory in our laboratory.

Place 6 ml of water in a 6-in. test tube and add 15 to 20 mg (or 1 drop) of the amine. Shake the mixture. If the amine dissolves in the water, add 6 ml more of water. If the substance did not dissolve in the first 6 ml of water, add 6 ml of ethanol. Shake the mixture. To the tube, now containing 12 ml of solvent, add 1 drop of a 2.5 per cent solution of quinhydrone in methanol. Shake the mixture and allow it to stand for 2 minutes. Aliphatic amines produce colors with only 1 drop of the reagent: primary amines produce a violet color; secondary amines a rose color; and tertiary amines a yellow color. Aryl amines generally do not produce any color with only 1 drop of the reagent, hence, if no color develops from 1 drop, add 5 more drops of the quinhydrone reagent. Shake the mixture and allow it to stand for 2 minutes. Aryl amines produce the following colors: primary, rose; secondary, amber; and tertiary, yellow.

NOTE. Nitro substituted aromatic amines can not be identified by this test. Phenylenediamines give false colors. In using this test for the first time, known amines of each subclass should be tested and the colors carefully noted.

3. N-*Halosuccinimide test.*[47] N-Bromosuccinimide and N-iodosuccinimide have been shown to be good detecting and differentiating reagents for primary, secondary, and tertiary amines. The N-bromosuccinimide may also be used to differentiate primary, secondary, and tertiary alcohols (test 10.04D). Confirmatory tests should be run for amines and alcohols when using this reagent.

[44] A. E. Petrarca, *J. Org. Chem.*, **24**, 1171 (1959).
[45] E. Kröller, *Süddeut. Apoth. Ztg.*, **90**, 724 (1950); *C.A.*, **45**, 1464 (1951).
[46] W. F. Meek and J. B. Entrikin, *J. Chem. Educ.*, **41**, 420 (1964).
[47] P. F. Kruse, K. L. Grist, and T. A. McCoy, *Anal. Chem.*, **26**, 1319 (1954).

Add 100 mg of the compound to 1 ml of carbon tetrachloride in a 4-in. test tube. Add 30 mg of N-iodosuccinimide and a few mg of benzoyl peroxide. Wash the sides of the test tube with 1 ml of carbon tetrachloride. Place the test tube in a water bath at 80° and keep the tube in the bath at that temperature for 10 minutes. Note the formation of a brown color and note whether the color remains for the full 10 minutes or fades to a yellow-tan or to no color within a few minutes in the water bath. If a brown color develops and remains for the full time, dissolve 30 mg of the original compound being tested in 1 ml of carbon tetrachloride in a 4-in. test tube and add 30 mg of N-bromosuccinimide. Look for the immediate. appearance of an orange precipitate. Primary and tertiary amines react with N-iodosuccinimide to give brown colors that do not fade during heating. Secondary amines produce brown colors that do fade to yellow-tan or colorless within a few minutes in the bath. Tertiary amines, but not primary amines, produce an orange precipitate when contacted with N-bromosuccinimide.

NOTE. The original reference cited covers only the application of these reagents to alkyl amines. However, experience has shown that the tests are also useful for many aromatic amines. The chemistry involved in this test is not well established, but suggestions regarding the probable mechanisms are given in the original article. The test for tertiary amines is particularly useful as a confirmatory test where other tests are used for detecting amines.

4. *Differentiation test, using 3,3',5,5'-tetrabromophenolphthalein ethyl ester with aliphatic amines.* The use of this ester as an acid-base indicator was discussed by Davis[48] and co-workers. The present use of this ester to subclassify primary, secondary and tertiary amines (aryl amines do not react) was developed by Valentine.[49] The ester is very sensitive to water; hence, all glassware used should be dry. Test tubes that have previously contained amines must be very thoroughly cleaned before reuse. Cork stoppers can only be used once, since the reagent is extremely sensitive to alkyl amines. See the following *Note* for the method of preparing the reagent and for preparing standards for comparison of colors. Special tests for butylamine and isobutylamine are also given in the note.

Place 2 to 3 ml of the reagent (3,3',5,5'-tetrabromophenolphthalein ethyl ester dissolved in carbon tetrachloride) in a clean, dry 4-in. test tube. Add 1 *small drop* of a liquid amine or, if the amine is a solid, use 1 drop of a solution of 10 mg of the amine in 1 ml of methanol. Stopper the tube

---

[48] Marion M. Davis and Priscilla J. Schuhmann, *J. Research, Nat. Bur. Standards*, **39**, 221 (1947); Marion M. Davis and Hannah B. Hetzer, ibid., **46**, 496 (1951); **48**, 381 (1952).
[49] J. Valentine, J. B. Entrikin, and W. M. Hanson, *J. Chem. Educ.*, in press.

with a clean cork and shake it vigorously. Observe the color change, if any, immediately. The straw-yellow color of the indicator solution is changed to purple by primary amines, to blue by secondary amines, and to red by tertiary amines.

NOTE. The reagent is prepared as follows: Weigh out 0.05 grams of 3,3',5,5'-tetrabromophenolphthalein ethyl ester and add it to a clean, dry 500 ml volumetric flask. Add approximately 100 ml of carbon tetrachloride (reagent grade, free of chloroform). Shake the flask a few minutes and then dilute to 500 ml with additional carbon tetrachloride. If any of the ester remains undissolved, continue to shake the flask. The concentration of the reagent must be between 1.5×10^{-4} to 1.6×10^{-4} M for the proper colors to be produced. The straw-colored indicator solution has been found to be stable for at least 3 months if kept well stoppered and out of direct sunlight. Always add the amine to indicator solution, not the indicator solution to the amine.

Since the indicator produces "iridescent" colors, a standard method of identifying colors must be employed. Hold all solutions for which a color is to be detected up to some light source. Before using this indicator method, we should become familiar with the colors produced by each subclass of alkyl amines. This may be done by preparing four tubes each of which contains 2 to 3 ml of the indicator solution in carbon tetrachloride. Add a small drop of a primary alkyl amine to the first tube and, similarly, add a secondary, a tertiary, and an aryl amine to the other three tubes.

Butylamine and isobutylamine may each be distinguished from all other primary amines by the following facts. If the original color change is from yellow to purple (test for a primary amine), set the tube aside and note the color at the end of 10 minutes. If during this time the color changes from the original purple to light blue, the amine is butylamine. If the color is still purple at the end of 10 minutes, allow the tube to sit for an additional 30 minutes. A light blue color at this time indicates isobutylamine.

Of the many amines tested, only benzylamine and triethanolamine gave irregular colors—both gave a blue color by the regular test.

5. *Hinsberg test.*[41] This test is based on the fact that primary and secondary amines react with sulfonyl halides to form N-substituted sulfonamides, whereas tertiary amines do not react with the reagent. Furthermore, since the sulfonamides from primary amines are weak acids, they are soluble in alkaline solution and may be separated from the sulfonamides from secondary amines which are not acidic and therefore not soluble in dilute basic solutions. Many modifications of the Hinsberg method have been proposed; two are presented here.

NOTE. Paul E. Fanta and C. S. Wang, in *J. Chem. Educ.*, **41**, 280 (1964), point out that the benzenesulfonamides prepared from tertiary alkyl amines, high-molecular weight alkyl amines, and cycloalkyl amines are not appreciably soluble in 10 per cent sodium hydroxide and hence falsely appear to be secondary amines by test (b).

(a) In this test, it is very important that the measurements be made accurately. Add 2 drops of the amine to 2 ml of pyridine; then add 0.8

[41] O. Hinsberg, *Ber.* **23**, 2962 (1890).

ml of *freshly prepared* (free of carbonates) 2 per cent aqueous sodium hydroxide. Shake the mixture thoroughly and add 1 drop of benzenesulfonyl chloride. Shake the mixture again. A yellow color indicates a primary amine; an orange color indicates a secondary amine; and a deep red or purple color indicates a tertiary amine. This test appears reliable for alkyl amines and for many aryl amines that have only one ring.

(b) A more involved procedure, but one that allows for the formation of a solid derivative for primary and secondary amines, is as follows.

To an 8-in. test tube, add 2 ml of methanol, 100 mg of the amine, and 400 mg of *p*-toluenesulfonyl chloride (or benzenesulfonyl chloride). By heating over a microburner, bring the mixture just to boiling, then cool, and add 8 ml of 10 per cent sodium hydroxide. Shake quite frequently for 5 minutes and then allow the tube to stand, with occasional shaking, for 10 minutes. Now, warm the mixture to hydrolyze any excess acid chloride. Cool the mixture and acidify it by adding $6M$ hydrochloric acid, dropwise, and with stirring. The sulfonamides of both primary and secondary amines are insoluble in acidic solution; hence, if a precipitate is present at this point, either a primary or a secondary amine is indicated. A precipitate should be removed by filtration and the precipitate washed with $2M$ hydrochloric acid to remove any adsorbed tertiary amine. Reserve the precipitate, if any, and combine the filtrate and the acidic washings. The acidic solution will contain any tertiary amine that was originally present. After making this filtrate alkaline by adding a sodium hydroxide solution, the tertiary amine may be recovered by ether extraction or by distillation (steam distillation for high-weight amines).

If a precipitate formed when hydrochloric acid was added to the original mixture (as reserved above), wash it with 3 ml of water and transfer it to a 6-in. test tube. Add 5 ml of 5 per cent sodium hydroxide and warm the mixture to 50°. Place a stopper in the tube and shake it vigorously for 2 minutes. If the precipitate dissolves, the original amine was a primary amine; if it fails to dissolve, a secondary amine is indicated (see the *Note* that follows). The sulfonamide of a primary amine may be recovered by acidifying the alkaline solution.

NOTE. Primary amines may form disulfonyl derivatives, and such derivatives are often not soluble in dilute alkaline solutions. Hence, they would be confused with derivatives for a secondary amine. The disulfonyl derivatives may be hydrolyzed to the monosulfonyl derivative by refluxing for 15 to 20 minutes in a solution of sodium methoxide prepared by dissolving 0.5 g of sodium in 10 ml of methanol.

The sulfonamides from primary alicyclic amines and high-molecular weight alkyl amines from sodium salts that are not appreciably soluble in water; hence, such amines give false results by this method.

6. *Diazonium salt test*. Most aryl amines and phenols (see test 10.20B) undergo coupling reactions with diazonium salts to form colored azo compounds. Since amines and phenols fall in different solubility divisions and give other characteristic tests, these two classes need not be confused by the fact that they both undergo azo compound formation. Alkyl amines do react with diazonium salts, but the color (usually yellow) is much less intense than with aryl amines. While several diazonium salts may be used, experience in the author's laboratories indicates that 4-nitrobenzenediazonium fluoroborate[50] is the most satisfactory that has been tried.

Place 25 mg of the compound in a 3-in. test tube and add, dropwise, while slightly warming the tube, 1.2N hydrochloric acid until the substance dissolves. Cool the tube and add, dropwise, 10 per cent ammonium hydroxide until the solution *just* begins to become cloudy. Coupling with amines is optimum in a slightly acidic solution. Add 3 to 5 drops of a freshly prepared 1 per cent aqueous solution of 4-nitrobenzenediazonium fluoroborate reagent. A decidedly yellow, orange, or red coloration or precipitate is a positive test.

NOTE. A precipitate will form in a 1 per cent aqueous solution of the reagent if the solution is allowed to stand for a few days. Experience has shown, however, that the precipitates that form may be removed by filtration and still leave a reagent of sufficient strength to give positive tests.

The reagent gives the positive tests with virtually all aryl amines and phenols that will undergo coupling reactions with diazonium compounds. Of course, compounds that have both of the ortho positions and the para positions substituted do not react. Compounds such as 4-aminobenzoic acid and 4-aminoacetophenone produce orange-red precipitates; anilides do not give the test. Alkyl amines generally give very weak or negative tests. However, this test is not considered reliable as a method of distinguishing alkyl amines from aryl amines.

Amines couple with diazonium salts most effectively at pH 3.5 to 7. Since the aryl amines are not appreciably soluble in water, the addition of acid is necessary to put them in solution. The solution should not be too acidic—hence, the addition of ammonium hydroxide or sodium acetate to reduce the acidity.

7. *Lignin test*. Webster[51] has proposed a very simple test that has proved quite reliable for primary and secondary amines—especially so if they are aryl amines. The test depends on the action of lignin in newsprint paper, but the chemistry of the reaction is not known.

Dissolve 10 to 20 mg of the compound in a few drops of ethanol and moisten a small area of newsprint paper with the solution. Place 2 drops

---

[50] A. L. LeRosen, P. H. Monahan, C. A. Rivet, E. D. Smith, and H. A. Suter, *Anal. Chem.*, **22**, 809 (1950); C. Hanot, *Bull. soc. chim. Belges*, **66**, 76 (1957).

[51] V. S. Webster, *Proc. S. Dakota Acad. Sci.*, **24**, 85 (1944).

of 6N hydrochloric acid on the moistened spot. The immediate development of a yellow or orange color is a positive test for a primary or secondary aryl amine. If the test is negative, repeat it, using a hot solution of the amine in ethanol and hot hydrochloric acid. Primary and secondary alkyl and alicyclic amines do not give the yellow or orange colors at room temperature but do give them when hot solutions are used. Tertiary amines, aliphatic amino acids, and amides do not give the test. Negatively substituted aryl amines that are too weakly basic to show definite basic properties will give this test (even aminosulfonic acids in most cases).

NOTE. Moerke[52] has discussed the lignin color reactions with amino compounds. He concluded that the color formation is due to some quinonelike material reacting with amino compounds—not necessarily aryl amines.

8. *Chloranil test.* Primary and secondary aryl amines react with chloranil to produce blue, red, or brown products; See preliminary test P-6A.

9. *2,4-Dinitrofluorobenzene test.*[53] Moisten a piece of filter paper with a saturated solution of 2,4-dinitrofluorobenzene in ethanol and add a drop of a solution of the amine in water. An intense yellow color strongly indicates a primary alkyl amine. Other amines may produce orange, red, or brown colors. Ammonia does not give this test, nor do the amino acids.

C. *Tests for Primary Amines*

1. *Primary alkyl amines.* Assuming that it has been established by previous tests that the substance is an alkyl amine, the Rimini test may be used to detect that it is a primary amine. Place 5 ml of a very dilute solution of the amine (or 1 or 2 drops of the amine in 5 ml of water) in a 4-in. test tube and add 1 ml of acetone. Add 1 drop of a 1 per cent solution of nitroprusside. The development of a definite violet-red color within 2 minutes is a positive test for a primary alkyl amine.

NOTE. The amine must be fairly soluble in water to give satisfactory results. This test for primary alkyl amines is positive, even in the presence of secondary alkyl amines. The acetone used must be free of acetaldehyde.

2. *Primary aryl amines;* (a) *Diazotization test.* Primary aryl amines are converted to diazonium salts by nitrous acid. At low temperatures,

---

[52] G. A. Moerke, *J. Org. Chem.*, **10**, 42 (1945).
[53] F. J. Smith and E. Jones, *A Scheme of Qualitative Organic Analysis*, Blackie and Son, London, 1948, p. 110; F. Feigl, *Spot Tests in Organic Analysis*, Elsevier, Amsterdam, 1960, p. 276. F. C. McIntire et al., *Anal. Chem.*, **25**, 1757 (1953); and D. T. Dubin, *J. Biol. Chem.*, **235**, 783 (1960).

these salts are stable and will couple in the alpha position with the sodium salt of 2-naphthol to form a red coloration or precipitate.

Prepare about 100 ml of a mixture of crushed ice, salt, and water to be used as a chilling bath. In one 4-in. test tube, mix 30 to 50 mg of the amine with 1 ml of water and 4 drops of concentrated sulfuric acid. In a second tube, place 1 ml of 10 per cent sodium nitrite. In a third tube, dissolve 100 mg of 2-naphthol in 2 ml of 10 per cent sodium hydroxide. Chill all three solutions in the ice bath. After the solutions are thoroughly chilled, add the sodium nitrite solution, dropwise and with shaking, to the acidified amine solution. Now add, dropwise, the sodium naphthoxide solution. A red color or precipitate indicates a primary aryl amine.

NOTE. Some electronegatively substituted amines, such as 2,4-dinitroaniline, do not diazotize by this method. *o*-Diamines do not diazotize, but form dark-colored azoimides. *m*-Diamines diazotize, but react with undiazotized amine to form brown dyes. *p*-Diamines diazotize and couple with the naphthol in the normal way.

It must be remembered that nitrous acid also reacts with secondary amines with the formation of N-nitroso products. These products are usually yellow or red oils that are insoluble in water. Nitrous acid reacts with N,N-dialkyl amines to form *p*-nitroso compounds. It also reacts with phenols and some other classes of compounds, but none of these reactions yield diazonium salts that will couple with naphthol.

Veibel[54] states that amines that do not diazotize by the regular method may be diazotized by adding the amine to 1 to 2 ml of concentrated sulfuric acid which contains crystals of sodium nitrite, shaking the mixture for 2 minutes, and then pouring the mixture into 20 ml of water.

(*b*). *Isocyanide test.* Most primary amines will react with chloroform to form isocyanides which have nauseating odors. The test is very *sensitive* and the reaction will be given by primary amines at low concentrations even if secondary amines are present in large amounts. *This test is more useful with aromatic amines than with others.*

$$C_6H_5NH_2 + CHCl_3 + 3KOH \rightarrow C_6H_5NC + 3KCl + 3HOH$$

Mix 50 mg of the primary arylamine with 2 drops of chloroform and 1 ml of $2N$ potassium hydroxide in methanol. Warm the mixture slightly and note the odor.

NOTE. Hydrazines and hydrazo compounds are reported to give a positive test by this procedure. The odor of the isocyanide product should be destroyed by pouring the test sample into some concentrated hydrochloric acid before pouring the sample into the sink.

[54] S. Viebel, *The Identification of Organic Compounds*, G.E.C. Gad, Copenhagen, 1954, p. 215.

D. Tests for Secondary Amines

In addition to the tests already discussed, particularly those in section B, tests 2 to 5, which give rather specific information on secondary amine detection, the following tests are useful in confirming secondary amines.

1. The following test[55] is very selective for secondary alkyl amines. Add 1 ml of a freshly prepared 5 per cent solution of acetaldehyde to 5 ml of a dilute aqueous solution of the amine (for example, 1 or 2 drops of amine in 5 ml of water). Add 1 or 2 drops of 10 per cent sodium nitroprusside and 2 drops $1.1N$ $NaHCO_3$. A blue color will develop within 3 minutes. On standing, the solution may change color to green and finally to yellow. Primary amines produce a purple color in this test.

2. *Nickel-dithiocarbamate test.* The original article by Duke[56a] gives separate tests for primary and secondary amines; the test for secondary amines is particularly useful. The reagent is prepared by dissolving 0.5 g of nickel chloride hexahydrate in 100 ml of water and adding enough carbon disulfide to saturate the solution and leave a small globule of carbon disulfide.

Add 30 to 50 mg of the amine to 5 ml of water in a 4-in. test tube. If the amine does not dissolve, add a drop of hydrochloric acid and shake the solution. In another 4-in. test tube, place 1 ml of the nickel chloride–carbon disulfide reagent and add 0.5 to 1 ml of concentrated ammonium hydroxide, followed by 0.5 to 1 ml of the amine solution. Secondary alkyl amines usually give a greenish yellow precipitate, and secondary aryl amines give a white or tan precipitate.

NOTE. Feigl[56b] recommends the formation of copper dithiocarbamates as a selective test for secondary alkyl amines.

Primary amine impurities may be blocked before making the test for secondary amines. See the series of papers by Critchfield and Johnson listed in the references at the end of this chapter.

E. Tests for Tertiary Amines

1. *Aconitic anhydride test.*[57] The reagent is prepared by dissolving 0.25 g of *cis*-aconitic anhydride in 40 ml of acetic anhydride and diluting to 100 ml with toluene.

---

[55] F. Feigl and V. Anger, *Mikrochim. Acta,* **1,** 138 (1937); C. F. Cullis and D. J. Waddington, *Anal. Chim. Acta,* **15,** 158 (1956).
[56a] F. R. Duke, *Ind. Eng. Chem., Anal. Ed,* **17,** 196 (1945).
[56b] F. Fiegl, *Spot Tests in Organic Analysis,* Elsevier, Amsterdam, 1960, p. 274.
[57] S. Sass, J. J. Kaufman, A. A. Cardenas, and J. J. Martin, *Anal. Chem.* **30,** 529 (1958); M. Palumbo, *Farm sci e tec (Pavia),* **3,** 675 (1948); and B. T. Cromwell, *Biochem. J.* **46,** 578 (1950).

Dissolve the amine in toluene and add 1 drop of this solution to a 4-in. test tube. Add 1 ml of the aconitic anhydride reagent and place the tube in a bath of boiling water for 15 seconds. Remove the tube from the hot water and allow it to stand for 15 minutes. Add 5 ml of toluene and again let the tube stand for 15 minutes. A red to red-purple color remaining at the end of the 30 minutes is a positive test for a tertiary amine.

NOTE. It has been noted that secondary amines sometimes give a red color shortly after mixing the components but this red color disappears before the 30 minute time limit for this test. Triethanolamine and dimethylaniline failed to give a positive test in our laboratory.

2. *Citric acid-acetic anhydride test.*[58] A reagent that seems very selective for tertiary amines of both the aliphatic and aromatic types has been proposed by Ohkuma.[58] The reagent consists of a solution prepared by heating 2 g of citric acid in a 100 ml of acetic anhydride.

Add 2 or 3 drops of the amine or its solution in alcohol to 3 drops of the reagent in a small test tube. Place the test tube in a bath of boiling water or heat briefly over a small flame. The development of a red to purple color within 1 to 2 minutes is a positive test.

10.09 Carbohydrates

A. General Tests for Carbohydrates

1. *Anthrone test.*[59] Anthrone dissolved in concentrated sulfuric acid will react with all but the most complex carbohydrates to produce a green color. This test is relatively specific for carbohydrates if the reaction mixture is not heated, and if only a green color is considered positive. See the *Note* that follows for other uses of this reagent.

Place 0.5 ml of water containing 1 to 5 mg of the compound in a 3-in. test tube. While holding the tube at an angle, pour 1 ml of a 0.2 per cent solution of anthrone in 95 per cent sulfuric acid down the side of the tube so as to stratify the liquids. Allow the tube to stand at room temperature for one minute and, if a green zone has not appeared at the interface, shake the tube gently and observe it for 3 more minutes. Carbohydrates produce a green color which changes to blue-green.

This test and others that require the stratification of liquids of different densities (such as the Molisch test, B(1) below) may be carried out using very

---

[58] S. Ohkuma, *J. Pharm. Soc. Japan*, **75**, 1124 (1955); F. Feigl, p. 281, cited in Reference 56).
[59] R. O. Dreywood, *Ind. Eng. Chem., Anal. Ed.*, **18**, 499 (1946); L. Sattler and F. W. Zerban, *Science*, **108**, 207 (1948); L. Sattler and F. W. Zerban, *J. Am. Chem. Soc.*, **72**, 3814 (1950); and L. H. Koehler, *Anal. Chem.*, **24**, 1576 (1952).

small volumes of reactants by the following procedure. Seal one end of a 6-mm piece of glass tubing for use as a "test tube." Place a few drops of the denser solution in this tube. Partially fill using a capillary tube (melting-point tube, open at both ends), with the less dense liquid and, holding a finger over the open end so as to retain the liquid, insert the capillary tube into the first "test tube." When the finger is withdrawn from the capillary tube, the denser liquid will rise into the capillary tube. A colored zone, which forms at the interface of the liquids in the capillary, shows up very clearly, particularly if the tube is viewed against a white background.

NOTE. The solution of anthrone in sulfuric acid should be prepared fresh every few days. The final mixture must be at least 50 per cent with respect to sulfuric acid to hold the anthrone in solution. Mono-, di-, and polysaccharides and their acetates, dextrins, dextrans, gums, glucosides, and starches give a positive test with anthrone. Furfural produces a transitory green color which rapidly changes to brown.

Alcohols, aldehydes, ketones, ethers (perhaps due to the presence of alcohols as impurities), and some aryl amines and proteins produce a red color with the anthrone reagent.[60] In a few cases, the red color develops without heat but, in most cases, heating in a boiling water bath for 3 to 5 minutes is necessary to develop the red color.

Easily dehydrated compounds may produce a tan, brown, or black color with this reagent. This reagent has been found very useful in the quantitative determination of very small amounts of carbohydrates.[61]

2. p-*Toluidine acetate test.* Carbohydrates are decomposed by heat or by hot phosphoric acid to yield furfural or furfural derivatives that produce a red color with *p*-toluidine acetate.

Place 5 to 10 mg of the compound in a microcrucible (or evaporate a few drops of a solution of the compound to dryness in a crucible). Cover the crucible with a filter paper that has been moistened with 2 drops of a 10 per cent solution of *p*-toluidine in 10 per cent acetic acid. Heat the bottom of the crucible with a microburner for 1 minute.

If a pink to red color does not form on the paper, repeat the experiment, adding a drop of syrupy phosphoric acid to the compound before heating it.

NOTE. Furfural and furfural derivatives may be detected by this test by heating an aqueous solution of the aldehyde and allowing the vapors to contact the *p*-toluidine treated paper. Tested dry, carbohydrates, including agar, starch, many gums, and alkyl- and acyl-substituted cellulose, give a positive test, especially if the phosphoric acid is used.

[60] G. Correa, *Colegio farm. Santiago, Chile*, **10**, 121 (1953); W. F. Meek, unpublished studies.
[61] D. L. Morris, *Science*, **107**, 254 (1948); F. J. Viles and L. Silverman, *Anal. Chem.*, **21**, 951 (1949); and J. A. Kowald and R. D. McCormack, *Anal. Chem.*, **21**, 1383 (1949); **24**, 1576 (1952).

Aniline or 4-bromoaniline may be substituted for the *p*-toluidine in this test but it has been found that the *p*-toluidine gives a deeper pink to red color and that the color persists much longer on the paper. The addition of a microdrop of 10 per cent tin(II) chloride in acetic acid to the test paper before heating the crucible tends to maintain the red color longer.

B. Test for Water-Soluble Carbohydrates

1. *Molisch test*.[62] Dissolve 20 mg of the compound in 1 ml of water and add 2 drops of a 5 per cent solution of 1-naphthol in methanol. Place 1 ml of concentrated sulfuric acid in a 4-in. test tube and, while holding the tube at an angle, slowly introduce the solution to be tested by means of a pipet so that it stratifies on top of the acid. The development of a violet-purple color at the interface is a positive test.

NOTE. Pentoses and their disaccharides are decomposed by concentrated sulfuric acid to form furfural. Hexoses and their disaccharides analogously produce hydroxymethylfurfurals. These furfurals produce colored condensation products with 1-naphthol. Some more complex carbohydrates give faintly positive tests.

2. *Resorcinol test*.[63] Add 20 mg of the compound to 1 ml of a 0.1 per cent solution of resorcinol in water. Stratify this mixture on top of 2 ml of concentrated sulfuric acid. An orange to red zone at the interface is a positive test.

C. Distinguishing Monosaccharides from Disaccharides

The monosaccharides are more easily oxidized than the disaccharides. Barfoed's solution will oxidize monosaccharides within 2 minutes, but it will not oxidize disaccharides unless heated for several minutes. Benedict's solution will oxidize all of the common sugars except sucrose. Methods of preparation of the reagents are given in the Appendix.

Place 2 ml of Barfoed's reagent or Benedict's reagent in a 4-in. test tube and add 10 to 20 mg of the carbohydrate (or 1 ml of a dilute solution of it in water) to the reagent. Place the tube in a bath of boiling water for 3 minutes. Remove the tube from the bath and allow it to cool. A yellow-orange or orange-red precipitate is a positive test. It should be remembered that a yellow suspension in a blue solution appears green.

NOTE. These reagents are not specific tests for sugars since they also oxidize other easily oxidized compounds, particularly α-hydroxyaldehydes and ketones and α-ketoaldehydes. Their use in this test is to distinguish monosaccharides from disaccharides and both of these types from more complex carbohydrates.

---

[62] H. Molisch, *Monatsh.* **7**, 198 (1886).
[63] R. Morgenstern, *Centr. Zuckerind*, **50**, 226 (1942); *C.A.*, **38**, 530 (1944).

D. *Ketoses and Aldoses*

1. The Seliwanoff[64] test for ketoses is based on the conversion of the ketose to hydroxymethylfurfural and its subsequent condensation with resorcinol to form colored complexes.

Mix 1 ml of Seliwanoff's reagent with 1 ml of about a 5 per cent solution of the sugar in water. Heat the mixture to boiling. A red color develops within 2 minutes if the sugar is a ketose. Long standing, or prolonged heating, will develop the color with aldoses. See the Appendix for the preparation of the reagent.

2. Tauber[65] has used aminoguanidine and dichromate ions in sulfuric acid to distinguish ketohexoses from aldohexoses (or sugars that contain these hexose units).

Place 0.5 ml of a 2 per cent aqueous solution of the sugar in a 4-in. test tube and add 0.5 ml of a 2.5 per cent aqueous solution of aminoguanidine sulfate (monohydrate). Shake the mixture and then add 1 ml of a 1 per cent solution of potassium dichromate in concentrated sulfuric acid. Ketohexoses give rise to an immediate wine-red color. Aldohexoses produce a deep blue color within 1 minute and pentoses give a yellow color within 1 minute.

3. Providing the quantities of reagents are kept very low, the use of diphenylamine in methanol-hydrochloric acid solution is a good test for ketoses.[66]

Mix 2 ml of concentrated hydrochloric acid with 2 ml of methanol in a $\frac{1}{2}$-in. test tube. Place the tube in a bath of boiling water and add 3 to 5 mg of the sugar and 3 to 5 mg of diphenylamine. Levulose and sucrose will produce a red coloration within 1.5 minutes and this color will change to blue within 3 minutes. Dextrose, maltose, and lactose fail to produce colors at these concentrations. However, if larger quantities of the sugar and diphenylamine are used (for example, 10 to 15 mg of each), dextrose and lactose will produce a light red color within 2.5 minutes with a change to light blue within 4 minutes. Even at these higher concentrations, maltose does not produce a color.

4. *Toluene-3,4-dithiol-zinc complex test.* Clark and Neville[67] have reported a new test for keto sugars. The cited article should be consulted for the procedure.

NOTE. Although levulose and dextrose produce the same osazone with phenylhydrazine, asymmetrical methylphenylhydrazine yields different osazones with these two sugars.

[64] Th. Seliwanoff, *Chem. Berichte*, **20**, 181 (1887).
[65] H. Tauber, *Anal. Chem.*, **25**, 826 (1953).
[66] Chemical Society (London), *Annual Reports*, **1960**, 329.
[67] R. Clark and R. G. Neville, *J. Org. Chem.*, **24**, 110 (1959).

E. Pentoses

1. Tollens' test for pentoses is based on the reaction of the pentose with hydrochloric acid to form furfural, which is then condensed with phloroglucinol to yield red complexes. Other sugars may produce yellow, orange, or brown colors.

Dissolve about 10 mg of the sugar in 5 ml of $6N$ hydrochloric acid and add about 10 mg of phloroglucinol. Boil the mixture for 1 minute. A red coloration indicates a pentose.

> NOTE. Hexuronic acid also produces a red color with phloroglucinol. It is reported that pentoses but not hexuronic acid produce a green color and a blue-green precipitate in a test when orcinol is substituted for the phloroglucinol.

2. Thomas[68] has proposed a test which, it is claimed, will distinguish pentoses from hexoses and from hexuronic acids.

Stratify 1 ml of an aqueous solution containing 1 mg of the carbohydrate on 3 to 4 ml of a 0.3 per cent solution of 2-naphthol in concentrated sulfuric acid. The development of a blue ring at the interface identifies a pentose. Hexoses produce a green-yellow to brown color and hexuronic acids a red-brown colored zone.

F. Osazone Formation

Methods for forming osazones and a discussion of them are given in Chapter 15.

10.10 Carbonyl Compounds

A. Aldehydes and Ketones

1. Most aldehydes and ketones react alike with all reagents that condense with the carbonyl group. Of the several useful carbonyl reagents, 2,4-dinitrophenylhydrazine has proven to be very effective in detecting carbonyl compounds. However, it is possible for erroneous deductions to be drawn from the formation of a precipitate when using this reagent, particularly if it has been used for a preliminary "spot test" without having properly classified the unknown by solubility. Like many dinitro compounds, 2,4-dinitrophenylhydrazine can form slightly soluble adducts with phenols and with many hydrocarbons, aryl halides, and ethers. Hence, if there is any doubt about the validity of the test as given, repeat the test substituting 4-nitrophenylhydrazine for the 2,4-dinitrophenylhydrazine.

---

[68] P. Thomas, *Bull. soc chim. biol.*, **7**, 102 (1925).

Although this reagent is not as selective a carbonyl reagent as the 2,4-dinitrophenylhydrazine, it does not form adducts with other classes.

$$R_2CO + H_2NNHC_6H_3(NO_2)_2 \rightarrow R_2C{=}NNHC_6H_3(NO_2)_2 + H_2O$$
$$\underset{H}{RCO} + H_2NNHC_6H_3(NO_2)_2 \rightarrow \underset{H}{RC}{=}NNHC_6H_3(NO_2)_2 + H_2O$$

Place 5 ml of a saturated solution of 2,4-dinitrophenylhydrazine in $2N$ hydrochloric acid in a 6-in. test tube. Add a solution of 30 to 40 mg of the compound in 0.5 ml of methanol. Stopper the test tube and shake it vigorously. If a precipitate does not form, heat the mixture to boiling for 30 seconds and shake it again. The formation of a precipitate represents a positive test. Most aromatic compounds produce red products, whereas aliphatic compounds tend to produce yellow products. No definite conclusions may be drawn as to the type of the carbonyl compound if the hydrazone is orange in color.

NOTE. Aqueous solutions of aldehydes or ketones may also be tested by this method. If the carbonyl compound is only slightly soluble in water, it may be necessary to add more alcohol than the 0.5 ml recommended.

In the 2,4-dinitrophenylhydrazine test, an excess of the carbonyl compound is to be avoided since the phenylhydrazone is more soluble in the carbonyl compound.

Lohman[69] states that 2,4-dinitrophenylhydrazones may be quantitatively separated from excess 2,4-dinitrophenylhydrazine by extracting the hydrazone with hexane.

Identification of a precipitate as being a 2,4-dinitrophenylhydrazone may be made by adding a very small sample of the precipitate to 2 ml of $2M$ potassium hydroxide in methanol. A deep red, purple, or blue color indicates a positive test.

2. A modified method for preparing 2,4-dinitrophenylhydrazones has been proposed by Shine,[70] who also discussed the role of acid-catalysis in the reaction of carbonyl compounds with the hydrazine. He prepared the reagent by dissolving 1 g of 2,4-dinitrophenylhydrazine in 30 ml of warm di-(methoxyethyl) ether ("diglyme").

Dissolve 100 mg of the carbonyl compound in 1 ml of ethanol or di-(methoxyethyl) ether in a 4-in. test tube. Add 5 ml of the 2,4-dinitrophenylhydrazine reagent. Shake the mixture and add 3 drops of concentrated hydrochloric acid. Generally, the hydrazone precipitates immediately; if not, add water dropwise.

3. *3,5-Dinitrobenzoic acid test for carbonyl compounds.* It has been found that the sodium hydroxide-acetone test for dinitro compounds (Test 10.17C) may be adapted for use in the detection of low-molecular weight aldehydes and ketones, even if they exist in aqueous solution.

[69] F. H. Lohman, *Anal. Chem.*, **30**, 972 (1958).
[70] H. J. Shine, *J. Org. Chem.*, **24**, 252 (1959); *J. Chem. Educ.* **36**, 575 (1959).

Dissolve 50 to 75 mg of 3,5-dinitrobenzoic acid in 2 ml of methanol in a 4-in. test tube. Add 2 ml of water that contains 2 to 4 drops of the compound to be tested. Now add 1 ml of 10 per cent sodium hydroxide and shake the mixture. The formation of a red-purple color within 3 minutes is a positive test for a low-molecular-weight carbonyl compound.

NOTE. Formaldehyde does not give satisfactory results, but all other aldehydes and ketones that contain five carbons or less give definite positive tests. If the substance to be tested is an aqueous solution of a suspected aldehyde or ketone, it is advisable to increase the amount of the 3,5-dinitrobenzoic acid used and to use a higher concentration of the sodium hydroxide.

The test may also be made by mixing the substance with the alcoholic solution of the 3,5-dinitrobenzoic acid and adding a solid pellet of sodium hydroxide. In this case, the color forms at the surface of the pellet.

B. *Aldehydes*

The following tests are given by aldehydes but not by ketones, except where noted.

1. N-*Hydroxybenzenesulfonamide test.* Dissolve a few milligrams of N-hydroxybenzenesulfonamide in 0.5 ml of methanol in a 4-in. test tube and add 30 mg of the compound to be tested. Now, add 0.5 ml of $2N$ potassium hydroxide in methanol. Heat the mixture just to boiling. Cool the tube, acidify the mixture with dilute hydrochloric acid, and add 1 drop of 10 per cent ferric chloride. A bluish red color (ferric hydroxamate) is a positive test.

NOTE. The exact chemistry of this reaction is not known, but one explanation is shown by the following equations.

$$C_6H_5SO_2NHOH + 2KOH \rightarrow C_6H_5SO_2K + KNO + 2H_2O$$
$$KNO + HCl + RCHO \rightarrow RCONHOH + KCl$$
$$3RCONHOH + FeCl_3 \rightarrow (RCONHO)_3Fe + 3HCl$$

It is reported that some nitro- and hydroxyaromatic aldehydes do not give this test. Benzyl ketones do give a positive test.

2. *Schiff's test.* Nauman et al.[71] have discussed the chemistry involved in the production and reaction of this reagent in some detail. It is their conclusion that the final (colored) product is *p*-rosanilinemethylsulfonic acid.

Add 3 drops of an aldehyde to 2 ml of colorless Schiff's reagent (see the Appendix for preparation). Do not warm the mixture. A wine-purple coloration will develop within 10 minutes. A few ketones give faint

---

[71] R. V. Nauman, P. W. West, F. Tron, and G. C. Gaeke, Jr., *Anal. Chem.*, **32**, 1307 (1960).

colorations with this test. There are compounds other than aldehydes that will give light-pink colorations, but these colors lack the blue cast characteristic of aldehydes.

3. *Methone test.* The compound 5,5-dimethyl-1,3-cyclohexanedione is often called methone. It is recommended[72] as a reagent for aldehydes and does not give the test with ketones. A milky suspension forms immediately when the reagent is added to very small amounts of aldehydes.

Add 50 mg of the aldehyde to 1 ml of water and then add 3 drops of a 5 per cent solution of methone in ethanol. Shake the mixture. The formation of a milky suspension within 2 minutes is a positive test for aldehydes.

4. *Oxidation tests*; (*a*) *Benedict's reagent.* It has long been believed that Benedict's reagent would oxidize all aliphatic aldehydes but not aromatic aldehydes or any of the ketones except the α-hydroxyketones. Daniels, Rush, and Bauer[73] have recently claimed that the precipitate formed by the reaction of Benedict's reagent on common aliphatic aldehydes is not cuprous oxide and that, therefore, this reagent really gives a negative test for these compounds. It is their contention that the reagent only gives a positive test of the α-hydroxyaldehydes, α-ketoaldehydes, and the α-hydroxyketones. Regardless of the composition of the precipitate, it is true that aliphatic aldehydes in general do produce a yellow- to orange-colored precipitate or suspension when heated with Benedict's reagent. A yellow suspension in a blue solution appears green. Other classes of compounds are also oxidizable by Benedict's reagent, and the test is mentioned here only as a means of helping to distinguish between aliphatic aldehydes and aromatic aldehydes and ketones. Tollens' reagent, discussed below, gives a positive test for both aliphatic and aromatic aldehydes but a negative test for ketones, except those substituted with hydroxyl, alkoxy, or dialkylamino groups on the α-carbon.

(*b*) *Tollens' reagent.* The silver ions in a solution containing silver–ammonia complex ions are reduced to metallic silver by most aldehydes, readily oxidized sugars, polyhydroxyphenols, aminophenols, hydroxylamines, and other reducing agents.

Add 30 to 50 mg of the compound to be tested to 2 ml of freshly prepared reagent. Shake the tube and allow it to stand for 10 minutes. If no reaction occurs in this time, place the tube in a beaker of water at about 35° for 5 minutes. A precipitate of silver is a positive test.

The reagent is prepared as follows: Add 2 drops of 5 per cent sodium

---

[72] W. Weinberger, *Ind. Eng. Chem. Anal. Ed.*, **3**, 365 (1931); D. Vorlander, *Z. anal. Chem.*, **77**, 241 (1929); and E. C. Horning and M. G. Horning, *J. Org. Chem.*, **11**, 95 (1946).

[73] R. Daniels, C. C. Rush, and L. Bauer, *J. Chem., Educ.*, **37**, 205 (1960).

hydroxide to 1 ml of 5 per cent aqueous silver nitrate. Shake the tube and add $2N$ ammonium hydroxide dropwise and with shaking until the precipitated silver hydroxide just dissolves.

NOTE. The test tube to be used in this test must be clean. It is best to clean the tube by boiling in it a 10 per cent solution of sodium hydroxide and then discarding this solution. The silver produced in this test will generally precipitate as a "mirror" on the glass tube if it is thoroughly clean but the formation of black metallic silver in the mixture is also a positive test.

(CAUTION. *Silver fulminate, which is very explosive when dry, may be present in the residues from the use of Tollens' solution. Hence, as soon as the test is completed the contents of the tube should be poured down the sink and washed through the trap with water. Also, rinse out the test tube with dilute nitric acid.*)

5. *Detection of aliphatic aldehydes with 2-hydrazinobenzothiazole.* A spot test for detecting aliphatic aldehydes has been suggested by Sawicki and Hauser.[74] The following is a slight modification of one of the methods suggested by these authors.

Place 1 drop or 30 mg of the aldehyde on a spot plate. Add 2 or 3 drops of dimethylformamide. Add a few milligrams of 2-hydrazinobenzothiazole to the mixture and allow to stand for 1 to 2 minutes. Add 1 drop of 1 per cent aqueous potassium ferricyanide solution. Let the mixture stand for 2 to 3 minutes and then add 2 or 3 drops of 20 per cent potassium hydroxide solution. A deep blue color develops within 5 minutes if aliphatic aldehydes are present.

NOTE. Higher molecular weight aldehydes require more than 5 minutes (up to 30 minutes) for the development of the blue color. The aromatic aldehydes do not develop the blue color, but those which have been tested give an orange or brown-green coloration. Ketones failed to produce a color. A mechanism for the chemistry involved in this test is suggested in the reference cited, in which the authors give quantitative data and directions for additional methods for using this reagent.

Drucker and Rosen[75] distinguish ketones from saturated aliphatic aldehydes by oxidation with peroxytrifluoroacetic acid, which produces esters or lactones only with ketones. The esters or lactones are detected by the ferric hydroxamate test (see test 10.11).

C. *Ketones*

1. According to Feigl,[76] aldehydes do not give a color test by the following procedure whereas ketones that have a methyl or methylene group attached to the carbonyl produce red or blue colors.

[74] E. Sawicki and T. R. Hauser, *Anal. Chem.*, **32**, 1434 (1960).
[75] R. Drucker and M. J. Rosen, *Anal. Chem.*, **33**, 273 (1961).
[76] F. Feigl, *Spot Tests in Organic Analysis*, Elsevier, Amsterdam, 1960, p. 236.

Mix 1 drop of an aqueous or alcoholic solution of the compound with 1 drop of 5 per cent sodium nitroprusside and 1 drop of 30 per cent sodium hydroxide. A slight red or yellow color generally develops within a few minutes. Add 1 or 2 drops of glacial acetic acid. A red or blue color on the addition of the acetic acid is a positive test.

2. As pointed out in preliminary test P-9, methyl ketones, along with several other classes, yield iodoform. Kamet[77] has suggested a way to detect the 2-alkanones. These ketones react with 2-nitrobenzaldehyde to form indigo.

Dissolve or suspend 100 mg of the compound in 5 ml of a 5 per cent solution of 2-nitrobenzaldehyde in ethanol. Add 1 ml of 10 per cent sodium hydroxide dropwise. After 1 minute, place a few drops of the mixture on a piece of filter paper and allow the liquid to absorb. Wash the treated area with water. A blue stain on the paper is a positive test.

10.11 Esters

The best test for esters is the ferric hydroxamate test. Unlike the acid anhydrides and acid halides (compare with test 10.03B), esters react with hydroxylamine to form hydroxamic acids only when the reaction is carried out in alkaline solution. Lactones act like esters in this test.

Place 0.5 ml of $1N$ hydroxylammonium chloride in methanol in a 4-inch test tube and add 30 mg of the compound to be tested. Now add, dropwise, a $2N$ solution of potassium hydroxide in methanol until the mixture is alkaline to litmus and then add 4 drops more of the potassium hydroxide solution. Heat the mixture just to boiling, then cool it, and add, dropwise and with shaking, $2N$ hydrochloric acid until the pH of the mixture is approximately 3 (use Hydrion paper or a similar indicator). Add 1 drop of 10 per cent ferric chloride and note the color. A bluish red or reddish blue color is a positive test. See the following *Note* for exceptions.

NOTE. It is, of course, true that acid anhydrides or acid halides and some imides would give a positive response by this method but, since these classes fall in a different Solubility Division from esters, it is not likely that they would cause confusion. If there is any doubt as to whether the compound under consideration is an anhydride or an acid halide as compared with an ester, warm a few milligrams of the substance with a few drops of 10 per cent sodium hydroxide. This would convert the anhydride or acid chloride to a salt, which does not give a positive test.

West and Qureshi[78] have pointed out that a number of substances, particularly low-molecular weight aldehydes and amides and trichloromethyl compounds, produce

[77] J. Kamet, *Ind. Eng. Chem. Anal. Ed.*, **16**, 362 (1944).
[78] P. W. West and M. Qureshi, *Anal. chim. Acta*, **26**, 506 (1962).

ferric hydroxamate by this procedure. These authors recommend an enzymatic method for the detection of esters. In view of their findings, consideration should be given to the desirability of testing the compound for aldehydes or amides.

Goldenberg and Spoerri[79] have discussed the variation in time required for *maximum* formation of hydroxamic acids from esters. They found that for most esters the time required was less than 2 minutes but for some it was more than 1 hour.

A white precipitate of potassium chloride often forms during the performance of this test. This salt does not interfere with the test but, if we desire, it may be dissolved by adding a few drops of water.

A few esters, particularly the esters of carbonic, carbamic, and sulfonic acids, do not give the test.

In general, the ferric chloride test for phenols (test 10.20A) does not interfere with the ferric hydroximate test for esters and the like because, in the acid solution employed in this test, the ferric phenolate ion is largely converted to the phenol.

Methods are given in the Appendix for determining the saponification equivalent, saponification number, and the iodine number. These values are frequently useful in identifying esters.

10.12 Ethers

Ethers are quite unreactive to all chemical reagents. Since they are the least easily detected class in Solubility Division N, tests for all of the other classes should be made first. The tests given below are not specific for ethers and the results may be misinterpreted unless it has been established that other classes of similar solubility are not present.

NOTE. Ether peroxides present a hazard, particularly when sizeable volumes of ether are evaporated to dryness. The presence of peroxides in ether may be detected by shaking the ether with a solution of ferrous ammonium sulfate and potassium thiocyanate. A deep red ferric thiocyanate complex is formed if peroxides are present. Peroxides may be removed from ether by shaking the ether with a solution of ferrous ammonium sulfate. Concerning the hazards, see *J. Chem. Educ.*, **41**, A575 (1964).

A. Esterification[80]

A great many ethers may be hydrolyzed and converted into acetate esters by heating a mixture of the ether, acetic acid, and concentrated sulfuric acid. The reaction is not complete, but sufficient ester is formed to give the hydroxamate test for esters (test 10.11). The presence of unchanged ether does not interfere with this test.

$$2CH_3COOH + R_2O + (H_2SO_4) \rightarrow H_2O + 2CH_3COOR$$

Mix 0.5 ml of the ether with 2 ml of glacial acid and 0.5 ml of concentrated sulfuric acid. Reflux the mixture for 5 minutes and distill 1 drop.

---

[79] V. Goldenberg and P. E. Spoerri, *Anal. Chem.*, **30**, 1327 (1958); **31**, 1735 (1959).

[80] D. Davidson and D. Perlman, *A Guide to Qualitative Organic Analysis*, 2nd ed., Brooklyn College Bookstore, Brooklyn, N.Y., 1958, p. 61.

Test this drop for esters by test 10.11 (be sure that enough potassium hydroxide solution is used to make the mixture alkaline). If the drop of distillate does not give a positive test for esters, cool the mixture that was refluxed and add 5 ml of ice water to it. If a separate liquid phase appears test it for esters. It is sometimes advisable to extract the mixture with 0.5 ml of benzene and test the benzene extract for esters.

B. Additional Tests

Owing to their lack of activity, ethers may be confused with hydrocarbons. Alkyl ethers may be distinguished from hydrocarbons by the Ferrox test (page 296).

Aryl ethers may be distinguished from alkyl ethers by the formaldehyde-sulfuric acid reagent (P-4B).

Alkyl ethers are generally soluble in concentrated hydrochloric acid, whereas aryl ethers and alkyl aryl ethers are not soluble.

C. Epoxides

A general qualitative test for epoxides has been proposed.[81]

Add exactly 2 drops of concentrated nitric acid to 2 ml of a 0.5 per cent solution of periodic acid and then add 1 or 2 drops of the compound to be tested. Compounds that are not soluble in water may be dissolved in acetic acid. Shake the mixture and add 1 or 2 drops of 5 per cent silver nitrate. A positive test is the appearance of a white precipitate of silver iodate. Simple alcohols, aldehydes, and ketones do not interfere. A blank should be run in all cases.

10.13 Hydrazines

Hydrazines may be detected by condensing them with some aldehyde or ketone.

Suspend 25 to 50 mg of the compound in 1 ml of water and add enough acetic acid to dissolve the hydrazine. Add a few drops of a 5 per cent solution of acetaldehyde. The formation of a precipitate strongly indicates a hydrazine.

NOTE. In case the hydrazine does not readily dissolve in acetic acid, use dilute hydrochloric acid instead. Hydrazines liberate ammonia when heated with 10 per cent sodium hydroxide (test P-3A). They are readily oxidized by Tollens' (test 10.10B4) and Benedict's (test 10.10B4) reagents.

[81] R. Fucks, R. C. Warters, and C. A. Vanderwerf, *Anal. Chem.*, **24**, 1514 (1952).

10.14 Hydrocarbons

Most hydrocarbons are detected more by their lack of reactivity than by their reactions. Knowledge of the solubility behavior of the substance is a definite aid, as is information regarding the elements that are present. Hydrocarbons that are soluble in concentrated sulfuric acid are either actively unsaturated or easily sulfonated compounds. The Ferrox test (page 296) will usually distinguish all hydrocarbons (negative reaction) from other classes (positive reaction) that fall in Solubility Division N. Confirmatory tests may be made on compounds that do not give positive reactions for functional groups (esters, alcohols, aldehydes, ketones, and the like) of this solubility division. Such tests would include the tests for active unsaturation (tests P-5) and the formaldehyde–sulfuric acid test (test P-4B).

A recently proposed test for olefins[82] employs the Friedel-Crafts acetylation to produce a ketone which is then detected by addition of 2,4-dinitrophenylhydrazine, followed by the addition of a base to produce a deep red color.

Another procedure[83] detects olefins by converting them to ozonides which, in turn, are converted to the 2,4-dinitrophenylhydrazones.

Acetylene and monosubstituted acetylenes react with Nessler's solution to precipitate mercuric acetylides. The mercuric acetylides are not as explosive as the silver and copper acetylides. Hydrocarbons that are not soluble in concentrated sulfuric acid include the alkanes, cycloalkanes, and most of the aromatic compounds. These Solubility Division I compounds may be distinguished from the alkyl and aryl halides by elemental analysis. The formaldehyde–sulfuric acid reagent will distinguish the alkanes from the aromatic hydrocarbons. Test P-4B is particularly useful for distinguishing aromatic hydrocarbons from cycloalkanes and alkanes. Ethers of aryl radicals frequently are not soluble in concentrated sulfuric acid. Unfortunately, many of these ethers do not give positive results with the formaldehyde–sulfuric acid reagent; hence, no direct way is known to distinguish them from aromatic hydrocarbons, except by physical methods such as infrared spectroscopy.

Benzal or piperonal chloride in trifluoroacetic acid has been suggested[84] as detecting agents for many aromatic hydrocarbons.

[82] J. G. Sharefkin and T. Sulzberg, *Anal. Chem.*, **32**, 993 (1960).
[83] J. G. Sharefkin and A. Riber, *J. Chem. Educ.*, **37**, 296 (1960).
[84] E. Sawicki, R. Miller, T. Stanley, and T. Hauser, *Anal. Chem.*, **30**, 1130 (1958).

Chloranil reacts with many types of organic compounds to produce colored products. It may, however, be useful in distinguishing bi- and polycyclic aromatic hydrocarbons or polyalkylated benzenes from alkanes or the more simple aromatic hydrocarbons. Colored products form rather quickly if a small quantity of the reactive hydrocarbons is added to a few drops of a saturated solution of chloranil in benzene or chloroform.[85] See test P-6A.

10.15 Nitrates and Nitrites

A. Diphenylamine Test

Both nitrites and nitrates act as oxidizing agents on diphenylamine in sulfuric acid. Since the test is very sensitive, enough nitrate radicals are liberated from organic nitrates by sulfuric acid to give the test. Diphenylamine is first oxidized to diphenylbenzidine, which is further oxidized to the quinonoid form.

$$2(C_6H_5)_2NH \rightarrow (C_6H_5NHC_6H_4)_2 \rightarrow C_6H_5N{=}C_6H_4{=}C_6H_4{=}NC_6H_5$$

Prepare the reagent by suspending 1 mg of diphenylamine in 1 ml of concentrated sulfuric acid and then add water, dropwise, to just dissolve the amine. Cool the solution and add 9 ml more concentrated sulfuric acid. To test for nitrates, nitrites or nitramines, add a few milligrams of the substances (or 1 to 3 drops of a solution) to 1 ml of the amine reagent. A blue color is a positive test.

Since the first step in the test for nitrates and nitrites with diphenylamine involves its oxidation to diphenylbenzidine, this later compound may be substituted for the diphenylamine and thus make the test more sensitive.

NOTE. Whitman and Fauth[86] report that nitrates and nitrites, but not nitramines, give the blue color if a 0.2 per cent solution of diphenylamine in 85 per cent phosphoric acid is used as the reagent instead of the amine-sulfuric acid mixture.

Clarke[87] has proposed a test to detect nitrates, even in the presence of nitrites. Antazoline reagent produces a red color with nitrates and a yellow color with nitrites.

Both nitrates and nitrites oxidize iron(II) hydroxide to iron(III) hydroxide by test 10.17A.

---

[85] F. Feigl, V. Gentil, and C. Stark-Mayer, *Mikrochim. Acta*, **1957**, 350.
[86] C. L. Whitman and M. I. Fauth, *Anal. Chem.*, **30**, 1672 (1958).
[87] E. G. C. Clarke, *Analyst*, **84**, 662 (1959).

B. Brucine Test

It is reported that as little as 6×10^{-8} g of nitric acid may be detected by brucine in sulfuric acid. A red color develops when a few milligrams of brucine are dissolved in 1 ml of concentrated sulfuric acid and nitrate ions are added.

C. Alkyl Nitrites

Care should be taken in handling the alkyl nitrites as they have a pronounced action on the heart. They may be detected by the fact that they will react with 2-phenylindole to precipitate 3-isonitroso-2-phenylindole.

$$\text{(indole)}-C_6H_5 + HONO \rightarrow \text{(indole)}=NOH\ -C_6H_5 + HOH$$

Dissolve 100 mg of 2-phenylindole in boiling ethanol and add 100 mg of the nitrite. On cooling, the 3-isonitroso-2-phenylindole will precipitate. It may be recrystallized from amyl acetate as yellow needles and has a melting point of 280°.

10.16 Nitriles

A nitrogen-containing compound that does not give positive results with the hydroxamate tests for esters (for example, nitro-substituted esters) or amides (including anilides and the like) may be tentatively identified as a nitrile if it reacts positively to the following test.[88] Confirmation may be made by hydrolysis.

Add 30 mg of the compound to 2 ml of $1N$ hydroxylammonium chloride in propylene glycol. Then add 1 ml of $1N$ potassium hydroxide in propylene glycol and boil the mixture for 2 minutes. Cool the test tube and add 0.5 to 1 ml of 5 per cent ferric chloride. A red to violet color is a positive test.

NOTE. Owing to the high boiling point of propylene glycol, these conditions are the most rigorous used to convert a compound into a hydroxamic acid. Hence, it must be established that the substance does not belong to any more readily converted class before this test is significant. Anilides and similar substituted amides, like the nitriles, are not converted to hydroxamic acids by the less drastic methods; hence, they may be confused with nitriles by this test and must be distinguished from them by other tests (see test 10.07B).

10.17 Nitro Compounds

Nitro groups are frequently present as substituents in several classes of aromatic compounds, such as acids, aldehydes, ketones, amines, azo

[88] S. Soloway and A. Lipschitz, *Anal. Chem.* **24**, 898 (1952).

compounds, and ethers, as well as hydrocarbons. Less commonly, nitro groups are found in nonaromatic acids, alcohols, and the like. In these cases, except for the nitro hydrocarbons and possibly the nitro ethers, the class to which the substance belongs would generally be established and the compound identified by solubility data, acid-base character, functional group tests for the groups that are more chemically active than the nitro group and derivatization of this more active group. The presence of the nitro group or groups would, therefore, generally not have to be proved directly, but would be indicated by the physical constants of the compound and of its derivatives. Nitro compounds can be reduced and the resulting amine identified. See pages 623–626 for methods of reducing nitro compounds. The presence of one or more nitro groups in a compound may be detected by proving the absence of other classes of compounds that also act as oxidizing agents on ferrous hydroxide. Nitro compounds give positive reactions by test P-8. Nitro groups may be detected by infrared spectroscopy.

A. Ferrous Hydroxide Test

It has been shown[89] that organic compounds that are oxidizing agents will oxidize ferrous hydroxide to ferric hydroxide with a change of color from blue to red-brown. The most common organic compounds that are oxidizing agents are the nitro compounds; less common classes are the nitroso compounds, quinones, hydroxylamines, nitrates, and nitrites.

$$C_6H_5NO_2 + 4H_2O + 6Fe(OH)_2 \rightarrow C_6H_5NH_2 + 6Fe(OH)_3$$

In a 3-in. test tube, mix about 20 mg of the compound with 1.5 ml of freshly prepared 5 per cent solution of ferrous ammonium sulfate. Add 1 drop of $3N$ sulfuric acid and 1 ml of $2N$ potassium hydroxide in methanol. Stopper the tube quickly and shake it. A positive test is indicated by the precipitate turning red-brown within 1 minute.

NOTE. The use of a small tube is required so that very little air will come in contact with the ferrous hydroxide.

B. Diphenylamine Test

Colored molecular compounds are formed between molten diphenylamine and nitro compounds.[90] The color appears on heating the mixture to 100° or less and fades when the melt is cooled but reappears on reheating

---

[89] W. M. Hearon and R. T. Gustavson, *Ind. Eng. Chem., Anal. Ed.*, **9**, 352 (1937).
[90] F. Feigl, *Spot Tests in Organic Analysis*, Elsevier, Amsterdam, 1960, p. 172.

the mixture. The color is usually orange-red, which is unfortunate, since many nitro compounds are similarly colored. This test is effective for mono- as well as for di- and polynitro compounds.

Place 1 or 2 drops of an ether or benzene solution of the compound in a small conical tube. Add 1 or 2 drops of a 5 per cent solution of diphenylamine in benzene. Hold the tube in boiling water to evaporate off the solvents and to melt the mixture.

C. Dinitro and Trinitro Hydrocarbons

These nitro derivatives of benzene and its homologs may usually be classified by the following test.[91]

Add 50 mg of the compound to 5 ml of acetone in a test tube and then add, while shaking the tube, 2 ml of 5 per cent sodium hydroxide. Mononitro compounds do not produce colors, but colors develop quickly for dinitro compounds (purplish blue) and trinitro compounds (deep red).

NOTE. The chemistry involved in this test has not been proven. Two dinitrobenzenes give irregular colors: 1,2-dinitrobenzene (no marked color formation) and 1,4-dinitrobenzene (greenish yellow). The presence of amino, alkylamino, acylamino, hydroxy, or acylated hydroxy groups on the benzene nucleus interfere with the test. Most dinitro and trinitro phenols tested gave yellow, yellow-orange, or greenish yellow colors. 2,4-Dinitroaniline gives a red color that is confusing with the trinitro hydrocarbons.

It is stated[92] that m-dinitro compounds may be detected, even in the presence of other dinitro compounds, by heating a few milligrams of the compound with a drop of 10 per cent aqueous potassium cyanide solution in a microcrucible. A red or violet color appears on heating and this color is not affected by the addition of a few drops of $2N$ hydrochloric acid.

Trinitro compounds may produce a red color in alkaline solution without adding the carbonyl compound.[93]

D. Nitroparaffins

Nitrous acid test. The test for the nitrocompounds is given by some but not all of the nitroparaffins. For example, nitromethane and 2-nitropropane give positive results, but nitroethane and 1-nitropropane fail to oxidize the ferrous hydroxide under the conditions of this test. By vigorous reduction the nitroparaffins may be converted to primary amines.

The action of nitrous acid on the nitroparaffins in alkaline solution may be used to distinguish primary, secondary, and tertiary nitroparaffins. Under the conditions given in the test below, primary nitroparaffins

[91] R. W. Bost and F. Nicholson, *Ind. Eng. Chem., Anal. Ed.*, **7**, 190 (1935); F. L. English *Anal. Chem.*, **20**, 745 (1948).
[92] F. Feigl, p. 174 (cited in Reference 90).
[93] J. J. Carr, *Anal. Chem.*, **25**, 1859 (1953).

produce a reddish amber color, secondary nitroparaffins a sky-blue color, and tertiary nitroparaffins do not produce any color.

Nitrous acid reacts with a primary nitroparaffin to yield a nitrolic acid. The salts of the nitrolic acids are red in solution (these salts are explosive when dry). Nitrous acid reacts with a secondary nitroparaffin to yield a pseudonitrole. These compounds are blue in solution. Nitrous acid fails to react with tertiary nitroparaffins.

Add 5 drops of the nitroparaffin to 2 ml of 10 per cent sodium hydroxide. Allow the mixture to stand for 3 minutes. Add 1 ml of 10 per cent sodium nitrite solution and then add dropwise 10 per cent sulfuric acid, but do not add enough acid to neutralize the mixture completely.

NOTE. Ferric chloride was suggested as a reagent for nitroparaffins by Scott and Treon.[93] Jones and Riddick[94] recommend resorcinol in 66 per cent sulfuric acid as a colorimetric reagent.

E. *Nitrophenols*

Nitrophenols give a rather intense yellow or yellow-orange color immediately when they are dissolved in alkaline solutions. See test 10.20 for additional tests.

Add 30 mg of the compound to 2 ml of a 10 per cent aqueous solution of sodium hydroxide. A yellow color is given by most 4-nitrophenols, whereas 2-nitrophenols usually give an orange color.

NOTE. Many compounds besides nitrophenols produce some color in sodium hydroxide, but the intensity and rapidity of formation of the yellow color in this test is quite distinctive for nitrophenols. Nitrosophenols tend to produce a yellow-green color. Compounds like 2-nitrophenol, that do not give the color test with ferric chloride, will give a good positive test by this method.

10.18 Nitroso Compounds

A. The following procedure gives positive tests for most common nitroso compounds.

Dissolve 30 to 40 mg of the compound in 2 ml of concentrated sulfuric acid and add 50 mg of phenol. Shake the tube and warm it slightly. The development of a blue or green color, which changes to red when water is added dropwise to the mixture, constitutes a positive test.

NOTE. This is the same as the Liebermann test for phenols by using nitrous acid, except that the organic nitroso compound is used instead of sodium nitrite. Nitroso compounds may be detected by test 10.17A.

[93] E. W. Scott and J. F. Treon, *Ind. Eng. Chem., Anal. Ed.*, **12**, 189 (1940).
[94] L. R. Jones and J. A. Riddick, *Anal. Chem.*, **24**, 1533 (1952).

It is claimed that C-nitroso compounds liberate iodine immediately when added to an acidified solution of potassium iodide, whereas N-nitroso compounds do not. C-nitroso compounds are usually yellow-green in color. They dissolve in ether to form colorless solutions which turn blue on warming. N-Nitroso compounds are soluble in ether and produce a green color in solution.

B. It has been reported[95] that all true nitroso compounds (but not isonitroso compounds) give positive results by the following procedure.

Prepare a mixture of 7 ml concentrated sulfuric acid in 3 ml of water. After cooling the acid, add 10 mg of N,N'-diphenylbenzidine. Add 0.5 ml of this solution to a test tube containing a few milligrams of the substance to be tested. True nitroso compounds develop a blue color immediately or after brief warming in a boiling water bath.

10.19 Oximes, Hydrazones, and Semicarbazones

All three classes of oximes, hydrazones, and semicarbazones may be hydrolyzed by concentrated hydrochloric acid and thus converted into the hydrochloride salts of hydroxylamine, the hydrazine, and semicarbazide, respectively. These classes may be detected and identified by reactions used to prepare derivatives of carbonyl compounds.

10.20 Phenols

Phenols are to be suspected when a compound falls in Solubility Division A_2 and proves to be a weak or intermediate acid in the acid-base classification. Generally, more than one test for phenols must be made before a conclusion can be drawn, since the nature and location of substituent groups on a phenol markedly affect the reactions that it will undergo.

A. Ferric Chloride Test

Most phenols, enols, hydroxamic acids, many hydroxy acids, some oximes, and enolizable compounds in which the enolic structure is present to the extent of at least 5 per cent of the compound, react with ferric chloride to produce colored complexes. The colors produced by a large number of the common phenols have been published by Wesp and Brode[96] and by Soloway and Wilen.[97] The colors vary somewhat depending on the solvent used, the concentration of the reactants, and the elapse of time between the reaction and observation.

[95] V. Anger, *Mikrochim. Acta*, **1960**, 58.
[96] E. F. Wesp and W. R. Brode, *J. Am. Chem. Soc.*, **56**, 1037 (1934).
[97] S. Soloway and S. H. Wilen, *Anal. Chem.*, **24**, 979 (1952).

Dissolve 30 to 50 mg of the compound in 1 to 2 ml of water, or a mixture of water and alcohol, and add up to 3 drops of a 2.5 per cent aqueous solution of ferric chloride. Note any change in color or the formation of a precipitate.

NOTE. Most phenols produce red, blue, purple, or green colorations with ferric chloride. Soloway and Wilen have shown that the use of an anhydrous solvent (chloroform) and a weak base (pyridine) causes the test to be much more sensitive and to allow the detection of a large number of phenols that give negative results when tested in water. The function of the pyridine is apparently that of a proton acceptor, thus increasing the concentration of the phenolate ion. The composition of the colored complexes has not been established. Most nitrophenols, hydroquinone, guaiacol, 3- and 4-hydroxybenzoic acids and their esters, and 2,6-di-*tert*-butyl-*p*-cresol give negative tests.

Aliphatic hydroxy acids produce distinctly yellow solutions with ferric chloride. Many aromatic acids yield tan precipitates (gallic acid gives a black precipitate). Enols usually produce tan, red, or red-violet colorations. Oximes, if they give positive tests usually give red colors, as do the sulfinic acids. Hydroxypyridines and hydroxyquinolines give red, blue, or green colors.

B. Coupling with a Diazonium Salt

As was pointed out in test 10.08B6, phenols as well as amines couple with diazonium salts to form colored azo compounds. The reaction apparently involves the diazonium ion, an electrophilic agent, substituting in the para (preferably) or ortho positions on the ring. Since the oxide ion is more of an electron-releasing group than the hydroxyl group, the phenoxide ions are more readily reacted with the diazonium ion than are the phenol molecules. The test should be carried out in slightly basic solution.

Dissolve 25 mg of the compound in a few drops of 2 per cent sodium hydroxide. If necessary, warm the mixture to form the solution, but cool it before proceeding to add 3 or 4 drops of a 1 per cent solution of 4-nitrobenzenediazonium fluoroborate. A decided coloration or a precipitate that is red, orange, yellow-green, or blue is a positive test. Read the note and the end of test 10.08B6.

C. Indicator Formation

Most phenols condense with phthalic anhydride to form indicators that have blue, purple, red, or green colors in alkaline solution.

Place about 200 mg of anhydrous zinc chloride in a 4-in. test tube and heat it to be sure that it is anhydrous. Add 300 mg of phthalic anhydride and 50 mg of the compound to be tested. Heat the mixture sufficiently to fuse it and then cool the tube. Add 1 ml of 2 per cent sodium hydroxide and stir the mixture with a strong glass rod to break up the fused mass. Add more sodium hydroxide until the mixture is alkaline. Note the color.

NOTE. It is important not to add too much excess sodium hydroxide, since many indicators lose their characteristic colors in excess alkali. The fact that the compound is an indicator may be established by successively making the mixture alkaline and acidic.

D. Nitrous Acid Test

This test is given by many phenols which yield *p*-nitroso derivatives which, in turn, react with excess phenol to form indophenols. The indophenols are acid-base indicators.

$$C_6H_5OH \xrightarrow{HONO} ONC_6H_4OH \rightleftharpoons O=C_6H_4=NOH$$

$$O=C_6H_4NOH + C_6H_5OH \xrightarrow{H_2SO_4} O=C_6H_4=NC_6H_4-OH$$

$$(HOC_6H_4N=C_6H_4=OH^+)SO_4^-H \xleftarrow{H_2SO_4} O=C_6H_4=NC_6H_4OH \xrightarrow{NaOH}$$
$$(O=C_6H_4=NC_6H_4O)^-Na^+$$

Add about 50 mg of the compound to 1 ml of concentrated sulfuric acid in a 4-inch test tube and then add about 20 mg of sodium nitrite. Shake the tube and warm it slightly. A positive test is indicated by a green, blue, or purple color. Cautiously pour the mixture into 5 ml of water. The color will generally change to red or blue-red. Make the solution alkaline by adding 20 per cent sodium hydroxide. The color in alkaline solution is generally blue or green.

NOTE. Nitrophenols and para-substituted phenols do not give this test. It is reported[98] that —CHO, —COOH, and —COCH$_3$ groups on the ring also prevent the reaction.

Feigl[99] suggests using a nitroso phenol (for example, 1 per cent 5-nitroso-8-hydroxyquinoline) in concentrated sulfuric acid as the reagent to test for phenols, thus making it unnecessary to first convert part of the phenol to a nitroso phenol.

Many aromatic oximes react with phenol and sulfuric acid to yield indophenols.

E. Millon's Test

Millon's test is for monohydroxy phenols that have at least one ortho position open. It is also given by tyrosine, tyrosine-containing proteins,

---

[98] T. A. Turney, *J. Org. Chem.*, **22**, 1692 (1957).
[99] F. Feigl, *Spot Tests in Organic Analysis*, Elsevier, Amsterdam, 1960, p. 196.

phenolic acids, and other compounds that have one phenolic group with an ortho position open. Add 50 mg of the phenol to 1 ml of Millon's reagent. Place the tube in a beaker of water and heat it to boiling. A red color will develop. The chemistry involved in this color formation is not clear. For the preparation of Millon's reagent, see the Appendix.

F. *Aminoantipyrine Test*

4-Aminoantipyrine reacts with many phenols in the presence of ferricyanide ions and in alkaline solution to produce quinoid-type compounds. The ferricyanide ions act as oxidizing agents in the reaction. The amine group of the aminoantipyrine is converted, in the case of phenol, to the $-N=C_6H_4=O$ group.

A number of variations have been proposed for carrying out this test, some of which are given in the references cited.[100]

Inglett and Lodge[101] have made a comparative study on the spot tests for phenols by four reagents: 4-aminoantipyrine, diazotized sulfanilic acid, formaldehyde-sulfuric acid, and Millon's reagent. They give the limits of detection for some thirty phenols by each of these reagents.

The following procedure has proven satisfactory in the laboratory of one of us. The substances should be added in the order specified and the concentrations should be exact.

Place 20 to 30 mg of the compound in a 4-in. test tube and add 1 ml of water. If the substance is not dissolved by the water, add 1 ml of methanol to dissolve the compound. Add 2 drops of a 2 per cent solution of 4-aminoantipyrine, followed by 2 drops of an 8 per cent solution of potassium ferricyanide. Finally, add 4 drops of 1 per cent sodium carbonate. A red (usual), violet, or orange color represents a positive test; yellow is negative. 2-Naphthol is an exception and gives a green color (1-naphthol gives a dark red).

NOTE. It is important that the mixture be basic before the addition of the ferricyanide solution, since other classes of compounds produce colorations in neutral or acidic solutions. A pH of 10 seems optimum.

This test gives weak or negative results with phenols that have the para position substituted by any of the following radicals: alkyl, aryl, nitro, nitroso, benzoyl, or aldehydic groups. The following groups do not prevent the test even if they are present in the para position: hydroxyl, halogen, carboxyl, sulfonic acid, and methoxy.

[100] E. Emerson, *J. Org. Chem.*, **8**, 1692 (1943); S. Gottlieb and P. B. Marsh, *Ind. Eng. Chem., Anal. Ed.*, **18**, 16 (1946); R. W. Martin, *Anal. Chem.*, **21**, 1419 (1949); M. B. Ettinger, C. C. Ruckhoft, and R. J. Lishka, *Anal. Chem.*, **23**, 1783 (1951); and R. J. Lacoste, S. H. Venable, and J. C. Stone, *Anal. Chem.*, **31**, 1246 (1959).
[101] G. E. Inglett and J. P. Lodge, *Anal. Chem.*, **31**, 248 (1959).

G. Additional Tests

Nitrophenols dissolve in alkaline solution to produce yellow or orange colors (see test 10.17E).

Numerous tests have been proposed to identify and differentiate the dihydroxy- and the trihydroxy- aromatic hydrocarbons. Veibel[102] discusses tests for these classes and for hydroxy hetercyclic compounds. Feigl[103] proposes fusion at 160° with oxalic acid as a method for detecting polyhydroxybenzenes. Many of these compounds are very easily oxidized by Tollens' reagent, or even by air when they are in alkaline solution.

10.21 Sulfides, Disulfides, and Sulfones

These three classes of compounds undergo decomposition during fusion with sodium hydroxide. The sulfides and disulfides produce sodium sulfide, while the sulfones produce sodium sulfite. After acidification of the fused mass, hydrogen sulfide or sulfur dioxide may be detected by its odor or by conventional chemical methods.

A. Sulfides

The chemical tests for sulfides and thiols (mercaptans) are sufficiently similar as to be confusing. However, since these two classes fall in different solubility divisions and have quite different odors, they are not usually confused. If sulfides are used instead of thiols in test 10.22E, the first precipitate is a very light yellow, rather than a golden yellow. The addition of free sulfur gives an orange color, but unless the compound is hydrolyzed it does not turn black. If alkyl sulfides are substituted for thiols in test 10.22B, the color is red rather than bluish red and tends to become yellow.

B. Disulfides

Disulfides may be reduced easily to the corresponding thiols which may be detected by test 10.22C or other test for thiols.

Add 30 mg of the compound to 1 ml of $1N$ hydroxylammonium chloride in methanol. Add a few mg of zinc dust and shake the mixture for 1 minute. Decant the liquid after the excess zinc has settled and test the solution for thiols.

---

[102] S. Veibel, *The Identification of Organic Compounds*, G.E.C. Gad, Copenhagen, 1954, pp. 69–72.

[103] F. Feigl, 205 (cited in Reference 99).

10.22 Thiols (Mercaptans and Thiophenols)

The low-molecular weight mercaptans and thiophenols are only slightly soluble in water but dissolve in sodium hydroxide to form salts. Both have penetrating, objectionable odors. The thiophenols are not very common. They may be detected by the test for aromatic structures and by their ease of nitration or bromination. The tests given below are designed particularly for mercaptans.

A. Iodine-azide Test

Feigl[104] has made extensive use of the catalytic effect of certain types of sulfur compounds on the reaction:

$$2NaN_3 + I_2 \rightarrow 2NaI + 3N_2$$

He has found that thiols and thioketones act as catalysts, whereas thioethers, disulfides, sulfones, sulfinic, and sulfonic acids do not materially increase the rate of this reaction which proceeds very slowly in the absence of a catalyst. The reagent is prepared by dissolving 1 gram of sodium azide in 33 ml of $0.1N$ iodine solution.

Place 1 drop (10 to 15 mg of a solid) of the sulfur containing compound on a spot plate. Add 2 drops of the iodine-azide reagent. An immediate and close observation is necessary to see if small bubbles of nitrogen are evolved. Evolution of nitrogen within a few seconds may be taken as a positive test for thiols or thioketones.

NOTE. Inorganic sulfides, thiosulfates, and thiocyanates do catalyze this reaction; hence, they must be absent if the test is valid for thiols or thioketones. Thioketones may be differentiated from thiols by the fact that thioketones are not reactive with an iodine solution, whereas thiols are oxidized to disulfides that are not catalysts for this reaction. Oxidation of a thiol may be accomplished by warming the thiol with a solution of iodine in alcohol in the presence of a small amount of sodium acetate.

B. Nitroprusside

Mercaptans, in a slightly alkaline solution of sodium nitroprusside, give about the same deep wine color as is given by hydrogen sulfide. The exact composition of the color complex is not known, but it is believed to involve a union of the sulfur with the nitroso group of the nitroprusside. Alkyl sulfides also react with sodium nitroprusside, but the color is more red than blue. Thiophenols also give this test if ammonium hydroxide is substituted for the sodium hydroxide.

Add 1 drop of the mercaptan to 2 ml of a 1 per cent solution of sodium

[104] F. Feigl; (cited in Reference 99) p. 242; *Mikrochemie*, **15**, 1 (1934).

nitroprusside and then add 3 drops of 10 per cent sodium hydroxide. A deep wine color forms. The color changes to yellow if the solution is acidified with hydrochloric acid. Aryl sulfides do not give this test.

NOTE. An alkaline solution of sodium nitroprusside produces colors with some nonsulfur-containing classes—for example, methyl ketones.

Fretag[105] carefully considered ten recommended tests for mercapto compounds and recommended the following procedure. Place 2 ml of a solution containing the sulfur compound in a test tube and add 0.5 ml of 1 per cent sodium nitroprusside, some crystals of sodium chloride, and 3 drops of ammonium hydroxide. A blue-red color develops in a positive test.

C. Isatin Test

Mercaptans give a green color of unknown cause in this test. Alkyl sulfides and hydrogen sulfide do not interfere, since they do not give such colorations.

Add 3 drops of a dilute solution of the mercaptan in ethanol to 2 ml of a 1 per cent solution of isatin in concentrated sulfuric acid.

D. Lead Mercaptides

Mercaptans react with lead or mercuric salts of weak acids to form lead mercaptides or mercuric mercaptides. Lead acetate or mercuric cyanide are generally used for these tests.

$$2HOH + 2RSH + Pb^{+2} \rightarrow Pb(SR)_2 + 2H_3O^+$$

Add 2 drops of the mercaptan to 5 ml of a saturated solution of lead acetate in ethanol. Lead mercaptide (yellow) precipitates.

E. Lead Sulfide

The petroleum industry makes extensive use of a so-called *doctor test* for *sour* distillates, that is, those containing mercaptans. With mercaptans, the reagent (sodium plumbite) first forms the yellow lead mercaptides, which are converted by sulfur to the black lead sulfide and the alkyl disulfides. The chemical changes in this test are variable, but one pair of equations may be written as follows.

$$Pb(OH)_2 + 2RSH \rightarrow Pb(SR)_2 + HOH$$
$$Pb(SR)_2 + S \rightarrow PbS + RSSR$$

Add 1 drop of a mercaptan to 2 ml of the sodium plumbite solution and shake the mixture vigorously. A yellow precipitate forms. Now add about

---

[105] Fretag, Z. anal. Chem., **138**, 259 (1953); C.A., **47**, 7374 (1953).

50 mg of finely powdered sulfur. The color may first change to orange, but will become black within a few minutes.

Coordination of Data

Thus far, in a systematic examination of an unknown, small samples have been subjected to: (a) a preliminary examination; (b) the determination of the elements present; (c) solubility tests and reaction to indicators thus leading to the classification of the unknown as probably belonging to a restricted number of chemical classes; and (d) specific class tests to determine the nature of functional groups thus indicating, with considerable assurance, a certain chemical class. In addition, the melting point or boiling point and perhaps other physical constants would have been determined. At this point, all of these data should be organized and compared with information about known compounds listed in the literature.

Extensive tables are provided in later sections of this book, in which the major classes of organic compounds are arranged alphabetically. Separate tables are provided for liquid and solid members of each class. Within each table, the compounds are listed in the order of their boiling points or melting points. The table of appropriate chemical compounds, indicated by the solubility, indicator, and specific class tests, is consulted and a list is made of all the compounds in that class for which the melting point or the boiling point is within 3° of the value found for the unknown. At this point it can be assumed as probable that the unknown is one of the compounds listed, provided its boiling point or its melting point has been accurately determined and also provided that it has been properly classified. It is, of course, possible that the unknown is not one of the compounds in the first list of possibilities.

In some cases, it is possible to use "spot tests" for specific compounds or specific structures as a definite aid in the tenative identification of the compound at this point. Feigl, Sawicki, and others have published extensively in this field. Attention is called to the references at the end of this chapter, since many of these refer to selective tests that are not given in this book. Texts by other authors, such as those listed in Chapter 1 (p. 6), may profitably be consulted for additional tests. Also, the Indexes of *Chemical Abstracts* are most useful in locating tests for the identification of individual compounds.

After an unknown has been sufficiently tested to allow tenative identification, it may be apparent that one of the physical methods could profitably be used. For example, if it is believed that the unknown is an amino acid, the chromatographic determination of the R_f values of the unknown and the amino acid that it is believed to be would be indicated. Likewise, certain spectrographic or microscopic procedures may either confirm the conclusions already reached or point toward more likely possibilities.

REFERENCES

General References

F. Amelink, *Rapid Microchemical Identification Methods in Pharmacy and Toxicology*, Netherlands Univ. Press and Wiley, New York, 1962. Detailed methods for many sulfonamides and sulfones.

L. R. Axebrod and J. E. Pulliam, *Anal. Chem.*, **32,** 1200 (1960). Detection of functional groups on steroid nucleus.

M. L. Bender, *Chem. Rev.*, **60,** 53 (1960). Catalysis of carboxylic acid derivatives.

N. D. Cheronis, *Microchem. J.*, **2,** 43 (1958). Microidentification of organic compounds.

N. D. Cheronis, *Microchem. J.*, **5,** 525 (1961); **3,** 433 (1959). Progress reports in Qualitative Organic Analysis, with references.

N. D. Cheronis, *Microchem. J.*, **4,** 555 (1960). Proof of identification of organic compounds.

N. D. Cheronis, ed., *Microchemical Techniques*, Interscience, New York, 1962, pp. 117–149. Characterization of organic compounds in microgram and submicrogram ranges, with 139 references.

P. J. Elving et al., *Anal. Chem.*, **22,** 376 (1950). Determination of functionality in organic compounds.

D. G. Faulke and F. Schneider, *Ind. Eng. Chem., Anal. Ed.*, **12,** 554 (1940); **14,** 94 (1942). Micro technique in qualitative organic analysis.

M. St. C. Flett, *Physical Aids to the Organic Chemist*, American Elsevier, New York, 1962. A good book on instrumentation.

D. Glick, ed., *Methods of Biochemical Analysis*, Interscience, New York, 1955. New color reactions for sugars and polysaccharides.

W. I. Hassid and R. M. McCreaody, *Ind. Eng. Chem., Anal. Ed.*, **14,** 683 (1942). Sugars.

G. M. Kline, ed., *Analytical Chemistry of Polymers: Part III*, Interscience, New York, 1962. Systematic classification by solubility, ignition tests, color tests, and the like.

J. W. Ladburg and C. F. Cullis, *Chem. Rev.*, **58,** 403 (1958). Kinetics and mechanism of oxidation by permanganate.

F. A. Long and M. A. Paul, *Chem. Rev.*, **57,** 953 (1957). Acid catalysis.

T. S. Ma, *Microchem. J.*, **4,** 373 (1959). A review.

J. C. Martin, *J. Chem. Educ.*, **38,** 287 (1961). NMR spectroscopy.

R. L. Peck, *Anal. Chem.*, **23,** 97 (1951). Review.

R. L. Peck, *Anal. Chem.*, **24,** 116 (1952). Review.

M. Perez and M. Legrand, *Bull. Soc. Chim. France*, **1960** (3), 453. Review with 80 references.

P. V. Peurifoy and M. Nager, *Anal. Chem.*, **32**, 1135 (1960). Detection of nitrogen compounds.

J. A. Quense and W. M. Dehn, *Ind. Eng. Chem., Anal. Ed.*, **12**, 556 (1940). Sugars and polyhydroxy alcohols.

L. Schubert and I. May in *Treatise on Analytical Chemistry*, I. M. Kolthoff and P. J. Elving, eds., Interscience, New York, 1961. Reagents in organic analysis.

G. Sosnovsky, *Chemist-Analyst*, **52**, 81 (1963). Detection of acylating agents.

D. E. Taxbell and T. Huang, *J. Org. Chem.*, **24**, 887 (1959). Tetracyanoethylene as a reagent.

B. S. Thyagarajan, *Chem. Rev.*, **58**, 439 (1958). Oxidation by ferricyanide.

S. Veibel, *Anal. Chem.*, **23**, 665 (1951). Determination of functionality in organic compounds.

Acids

R. Antoszewski and L. T. Antoszewska, *Anal. Chim. Acta*, **20**, 595 (1959). The detection of micro amounts of amino acids.

R. Circo and B. A. Freeman, *Anal. Chem.*, **35**, 262 (1963). Diethylamine-ninhydrin reagent for amino acids in chromatography.

R. E. Dunbar and F. J. Ferrin, *Microchem. J.*, **4**, 59 (1960). Photomicrographs and melting points of 20 amino acids as dibenzofuran-2-sulfonates.

M. S. Dunn and W. Drell, *J. Chem. Educ.*, **28**, 480 (1951). Qualitative identification of amino acids.

O. A. Guagnini and E. E. Vonesch, *Mikrochim. Acta*, **1959**, 372. A study of the hydroxamic acids.

C. D. Hurd and D. G. Batteron, *J. Org. Chem.*, **11**, 207 (1946).

R. F. Keeler, *Science*, **129**, 1618 (1959). Color reactions of amino acids.

T. S. Ma and R. Breyer, *Microchem. J.*, **4**, 481 (1960). Micro determination of amino acids.

D. J. McCaldin, *Chem. Rev.*, **60**, 39 (1960). The chemistry of ninhydrin.

Ping-Yuan Yeh et al., *Anal. Chem.*, **34**, 990 (1962). Differentiation of organic acids and phenols.

A. Said and D. Fleita, *Chemist-Analyst*, **52**, 79 (1963). 1,2,3-Phenalenetrione as an amino acid reagent in chromatography.

A. Saifer and I. Oreskes, *Anal. Chem.*, **28**, 501 (1956). Amino acid detection.

K. Sataka et al., *J. Biochem. Tokio*, **47**, 654 (1960). Tests for amino acids.

R. W. Storherr, *Anal. Chem.*, **31**, 268 (1961). Tests for individual amino acids.

R. L. Sublett and J. P. Jewell, *Anal. Chem.*, **32**, 1841 (1960). A specific test for glycine, pyridine, or alkyl chloroformates.

R. T. Wendland and D. H. Wheeler, *Anal. Chem.*, **26**, 1469 (1954). Characterization of common organic acids.

M. Yamagishi, *J. Pharm. Soc. Japan*, **74**, 1001 and 1233 (1954). Ninhydrin reactions.

Alcohols

G. M. Christensen, *Anal. Chem.*, **34**, 1030 (1962). Identification of alcohols.

F. E. Critchfield and J. A. Hutchinson, *Anal. Chem.*, **32**, 862 (1960). Secondary alcohols.

B. T. Dewey and N. F. Witt, *Ind. Eng. Chem., Anal. Ed.*, **14**, 648 (1942); ibid., **12**, 459 (1940). Identification of alcohols.

R. E. Dunbar and F. J. Ferrin, *Microchem. J.*, **3**, 65 (1959). Microscopy of the 3,5-dinitrobenzoates of hydroxy compounds.

H. C. Heim and C. F. Poe, *J. Org. Chem.*, **9**, 299 (1944). Glycols.

E. Henry-Basch and P. Freon, *Comp. rend.*, **248**, 2597 (1959). Periodic acid oxidation.

J. P. Johnson and F. E. Critchfield, *Anal. Chem.*, **32**, 865 (1960). Detection of low concentrations of alcohols.

Ichiro Kudo and Ichiro Aoki, *Buneski Kagaku*, **6**, 791 (1957). Vanadium-oxine.

D. E. Laskowski and O. W. Adams, *Anal. Chem.*, **31**, 148 (1959). Identification of alcohols.

L. Maros and E. Schulek, *Acta Chim. Hung.*, **20**, 359 (1959). Periodic acid oxidation.

S. Maruta and F. Iwana, *J. Chem. Soc. Japan* (*Pure Chem. Sect.*), **80**, 1131 (1959). Vanadium-oxine reagent.

V. Pandu Ranga Rao et al., *Chemist-Analyst*, **52**, 68 (1963). Vanadium-2,4-pentandione reagent.

M. Perez and J. Bartos, *Talanta*, **5**, 216 (1960). Identification of alcohols.

F. O. Ritter, *J. Chem. Educ.*, **30**, 395 (1953). Permanganate oxidation.

L. Schotte and S. Veibel, *Acta Chem. Scand.*, **7**, 1357 (1953). Tertiary alcohols.

J. C. Speck, Jr., and A. A. Forist, *Anal. Chem.*, **26**, 1942 (1954). 1,2-Glycols.

Alkyl and Aryl Halides

P. M. G. Bavin, *Anal. Chem.*, **32**, 554 (1960). Characterization of alkyl halides.

R. N. Boyd and M. Meodou, *Anal. Chem.*, **32**, 551 (1960). Characterization of alkyl halides.

H. P. Burchfield and P. Schuldt, *Journal of Agr. and Food Chem.*, **6**, 106 (1958). Analysis of pesticides containing active halogen atoms.

R. E. Dunbar and W. M. King, *Microchem. J.*, **3**, 143 (1959). Aryl halide derivatives.

F. Feigl et al., *Talanta*, **1**, 80 (1958). Differentiation of aliphatic and aromatic bound halogen by fusion with succinic acid.

R. J. Gritter, *J. Chem. Educ.*, **35**, 475 (1958). Free radicals.

A. E. Pavlath and J. Leffler, *Aromatic Fluorine Compounds*, Reinhold, New York, 1962.

L. Schotte and S. Veibel, *Acta Chem. Scand.*, **7**, 1357 (1953). Preparation of S-alkylthiuronium picrates.

E. Wilson, *Chem. Rev.*, **16**, 149 (1935). The effect of structure on the reactivity of some organic halogen compounds.

Amides

J. A. Calamari, R. Hubata, and P. B. Roth, *Ind. Eng. Chem., Anal. Ed.*, **14**, 534 (1942). Differentiation and identification of sulfadrugs.

F. Feigl, *Anal. Chem.*, **27**, 1315 (1955). Detection of amides, ureas, and urethanes.

F. Feigl, *Spot Tests in Organic Analysis*, Elsevier, Amsterdam, 1960, pp. 301–302. Differentiation of anilides by fusion with guanidine carbonate and with hydrated manganese sulfate.

P. L. De Reeder, *Anal. Chim. Acta*, **7**, 42, 417 and 428 (1952); ibid., **8**, 6 (1953); ibid., **11**, 68 (1954). Sulfonamides.

A. B. Sample, *Anal. Chem.*, **17**, 151 (1945). Identification of sulfonamides.

E. Sawicki, *Chemist-Analyst*, **47**, 9 (1958). Piperonal chloride test for aromatic acylamines.

S. Trofimenko and J. W. Sease, *Anal. Chem.*, **30**, 1432 (1958). Detection of nitriles and amides.

Amines

A. M. Albrecht, W. L. Scher, Jr., and H. J. Vogel, *Anal. Chem.*, **34**, 398 (1962). Test for aliphatic amines.

J. Bartos and J. F. Burtin, *Am. Pharm. Franc.*, **17**, 144 (1959). Condensation of primary amines with *o*-aminobenzaldehyde.

F. E. Critchfield and J. B. Johnson, *Anal. Chem.*, **29**, 1174 (1957); ibid., **28**, 430 (1956).

F. Feigl and E. Jungreis, *Chemist-Analyst*, **47**, 64 (1958). Spot test for determining the purity of N,N-dialkyl anilines.

F. Feigl and D. Goldstein, *Anal. Chem.*, **32**, 861 (1960). Spot test for diphenylamine and its derivatives.

F. Feigl, *Chemist-Analyst*, **50**, 15 and 18 (1961). Spot test for piperidine in the presence of piperazine.

R. G. Frieser and P. A. Scardaville, *Anal. Chem.*, **32**, 196 (1960). Identification of the isomers of phenylenediamine.

R. G. Gillis, *J. Chem. Educ.*, **31**, 344, 625 (1954). Mechanism of diazotization.

W. E. Hearn and R. Kinghorn, *Analyst*, **85**, 766 (1960). Test for aromatic amines using ceric ammonium nitrate.

A. Hirsch et al., *Chemist-Analyst*, **50**, 7 (1961). Spot tests for phenylenediamines.

F. H. Lohman and W. E. Norteman, Jr., *Anal. Chem.*, **35**, 707 (1963). Spectrophotometric determination of p- and s-aliphatic amines.

G. A. Lugg, *Anal. Chem.*, **35**, 899 (1963). Table of colors of 15 amines with 13 diazonium salts.

T. S. Ma and H. Moss, *Mikrochim. Acta*, **1962**, 111. Amino alcohols and diamines.

A. J. Milun and J. P. Nelson, *Anal. Chem.*, **31**, 1655 (1959). Secondary amines as impurities.

F. W. Neumann and C. W. Gould, *Anal. Chem.*, **25**, 755 (1953). Oxidation of N-alkylarylamines.

E. D. Olleman, *Anal. Chem.*, **24**, 1425 (1952). Long review with 190 references.

Hch. Zollinger, *Chem. Rev.*, **51**, 347 (1952). Kinetics of the diazo coupling reaction.

I. A. Pearl and P. McCoy, *Anal. Chem.*, **32**, 132 and 1407 (1960). Study of 20 amines with 27 commercial dyes.

M. Perez and J. Bartos, *Anal. Chim. Acta*, **20**, 187 (1959). Primary aromatic amines.

M. Perez, J. Bartos, and J. F. Burtin, *Talanta*, **5**, 213 (1960). Microgram scale determination of aromatic amines.

M. Perez and J. Bartos, *Talanta*, **5**, 216 (1960). Colorimetric determination of primary aliphatic amines.

J. H. Ridd, *Quarterly Reviews* (The Chem. Soc., London), **15**, 418 (1961). Theory of diazotization.

S. Sass et al., *Anal. Chem.*, **30**, 529 (1958). Use of aconitic anhydride.

E. Sawicki et al., *Chemist-Analyst*, **48**, 30 (1959). Diazotization.

E. Sawicki et al., *Anal. Chem.*, **33**, 93 (1961); ibid., **33**, 722 (1961). Aromatic amines and heteroaromatic compounds.

R. M. Silverstein, *Anal. Chem.*, **35**, 154 (1963). Spectrophotometric determination of primary, secondary, and tertiary fatty amines.

G. W. Stevenson and S. H. Biers, *Anal. Chem.*, **31**, 2095 (1959). Cyanoethylation.

A. Streitwieser, Jr., *J. Org. Chem.*, **22**, 861 (1957). Reaction of aliphatic primary amines with nitrous acid.

E. O. Woolfolk and E. H. Roberts, *J. Org. Chem.*, **21**, 436 (1956). Use of *p*-phenylazobenzoyl chloride for the identification and chromatographic separation of amines.

E. O. Woolfolk, W. E. Reynolds, and J. L. Mason, *J. Org. Chem.*, **24**, 1445 (1959). Separation and identification of amines with *p*-phenylazobenzenesulfonyl chloride.

Carbohydrates

S. S. Cohen, *J. Biol. Chem.*, **201**, 71 (1953). Ketose test.

M. J. Crumpton, *Biochem. J.*, **72**, 479 (1959). Amino sugars.

W. M. Dehn et al., *Ind. Eng. Chem., Anal. Ed.*, **4**, 413 (1932). Results of 30 reagents on 21 carbohydrates.

Z. Dische and E. Borenfreund, *J. Biol. Chem.*, **192**, 583 (1951). Ketose test.

W. J. Frearon and J. A. Drum, *Analyst*, **74**, 56 (1950). Urea reagent for ketohexoses.

F. L. Green, *Anal. Chem.*, **30**, 1164 (1958). Specific tests for heptoses.

W. K. Hall and T. S. Decker, *Anal. Chem.*, **31**, 1746 (1959). Improved phenylhydrazine reagent for osazone formation.

J. R. Helbert and K. D. Brown, *Anal. Chem.*, **28**, 1098 (1956). Anthrone reagent.

L. E. Hessler, *Anal. Chem.*, **31**, 1234 (1959). *p*-Anisidine and 3,3-dimethoxybenzidine as detecting reagents.

R. Johanson, *Nature*, **172**, 956 (1953). Anthrone reagent for paper chromatography.

R. Johanson, *Anal. Chem.*, **26**, 1331 (1954). Pentose interference in hexose test.

B. Klein and M. Weissman, *Anal. Chem.*, **25**, 771 (1953). Detection of hexoses in the presence of pentoses.

L. H. Koehler, *Anal. Chem.*, **26**, 1914 (1954) and **24**, 1576 (1952). Anthrone reagent.

A. P. MacLennan, H. M. Randall, and D. W. Smith, *Anal. Chem.*, **31**, 2020 (1959). Detection and identification of deoxysugars by chromatography.

E. B. Mano and L. Cunha Lima, *Anal. Chem.*, **32**, 1772 (1960). Detection of cellulose and its derivatives.

Methods in Carbohydrate Chemistry, R. L. Whistler and M. L. Wolfrom, eds., Academic Press, New York, 1962, Vol. I, pp. 477–514: Color reactions of carbohydrates.

G. L. Miller, R. H. Golder, and E. E. Miller, *Anal. Chem.*, **23**, 903 (1951). Pentoses.

J. B. Pridham, *Anal. Chem.*, **28**, 1967 (1956). *p*-Anisidine as a detector.

P. S. Rao and R. M. Beri, *Proc. Indian Acad. Sci.*, **33A**, 368 (1951); ibid., **34A**, 236 (1951). R_f values for sugars in paper chromatography.

T. A. Scott and E. H. Melvin, *Anal. Chem.*, **25**, 1650 (1953). Anthrone reagent.

C. D. Steinecker and M. S. Rheims, *Am. J. of Med. Tech.*, **25**, 377 (1959). Anthrone as a detector of carbohydrates.

H. Tauber, *Anal. Chem.*, **25**, 826 (1953). Aminoguanidine as a reagent for sugars.

T. H. Whitehead and W. C. Bradbury, *Anal. Chem.*, **22**, 651 (1950). Fructose.

Carbonyls

V. Anger, *Mikrochim. Acta*, **1959**, 386. Tests for quinones.

V. Anger and G. Fischer, *Mikrochim. Acta*, **1960**, 592. Detection of aldehydes.

V. Anger and G. Fischer, *Mikrochim. Acta*, **1960**, 592. Indole reagent.

L. I. Braddock, *Anal. Chem.*, **25**, 301 (1953). Theory of separation of the 2,4-Dinitrophenylhydrazones.

R. A. Braun and W. A. Mosher, *J. Am. Chem. Soc.*, **80**, 3048 (1958). A new reagent: 2-Diphenylacetyl-1,3-indandione-1-hydrazone.

D. M. Coulson, *Anal. Chim. Acta*, **19**, 284 (1958). Polarographic determination of semicarbazones.

E. P. Crowell, W. A. Powell, and C. J. Varsel, *Anal. Chem.*, **35**, 184 and 189 (1963). Characterization and determination of aldehydes by UV spectral changes resulting from acetal formation.

R. Drucker and M. J. Rosen, *Anal. Chem.*, **33**, 273 (1961). Detection of ketones.

M. Feldstein and N. C. Klendshoj, *Anal. Chem.*, **26**, 932 (1954). Test for formaldehyde.

G. D. Johnson, *J. Am. Chem. Soc.*, **75**, 2720 (1953). Correlation of color and constitution of 2,4-dinitrophenylhydrazones.

L. A. Jones, J. C. Holmes, and R. B. Seligman, *Anal. Chem.*, **28**, 191 (1956). Spectrophotometric studies of some 2,4-dinitrophenylhydrazones.

L. A. Jones and C. K. Hancock, *J. Org. Chem.*, **25**, 226 (1960). Spectrometric studies of some 2,4-dinitrophenylhydrazones.

E. Jungreis, *Chemist-Analyst*, **49**, 14 (1960). Spot test for formaldehyde.

M. Keeney, *Anal. Chem.*, **29**, 1489 (1957). Regeneration of carbonyls from their 2,4-dinitrophenylhydrazones.

D. E. Kramm and C. L. Kolb, *Anal. Chem.*, **27**, 1076 (1955). Schiff reagent.

T. Kwon and B. M. Watts, *Anal. Chem.*, **35**, 733 (1963). Anthrone reagent.

G. R. Lappin and L. C. Clark, *Anal. Chem.*, **23**, 541 (1951). Colorimetric determination of traces of carbonyl compounds.

D. E. Laskowski, *Anal. Chem.*, **32**, 1171 (1960). Investigation of the interaction between quinones and hydrocarbons.

V. E. Levine and M. Taterka, *Anal. Chim. Acta*, **15**, 237 (1956). Color reactions of aldehydes and ketones with vanillin.

F. H. Lohman, *Anal. Chem.*, **30**, 972 (1958). Determination of carbonyl oxygen.

F. W. Neumann and C. W. Gould, *Anal. Chem.*, **25**, 755 (1953). Identification of volatile carbonyl compounds.

J. H. Pomeroy and C. B. Pollard, *Quart. J. Florida Academy of Sci.*, **10**, 13 (1948). Sensitivity of aldehyde reagents.

V. V. Rachinskij and E. J. Knyazyatova, *Doklady Akad. Naut.*, *SSSR*, **85**, 1119 (1952); *C.A.*, **47**, 448 (1953). R_f values in partition chromatography.

E. Sawicki, T. W. Stanley, and T. R. Hauser, *Chemist-Analyst*, **47**, 87 (1958). Differentiation of aromatic aldehydes and ketones.

E. Sawicki, T. W. Stanley, and T. R. Hauser, *Chemist-Analyst*, **47**, 31 (1958). Detection of aromatic aldehydes.

E. Sawicki, *Chemist-Analyst*, **48**, 4 (1959). Detection of acetonyl grouping in organic compounds.

E. Sawicki and T. W. Stanley, *Anal. Chem.*, **31**, 122 (1959). Test for aliphatic ketones.

E. Sawicki and T. R. Hauser, *Anal. Chem.*, **32**, 1434 (1960). Spot test detection and colorimetric determination of aliphatic aldehydes.

E. Sawicki, J. Noe, and T. W. Stanley, *Mikrochim. Acta*, **1960**, 286. Test for arylalkyl and dialkyl ketones.

E. Sawicki and W. Elbert, *Anal. Chim. Acta*, **22**, 448 (1960). Specific test for inner-ring *o*-quinones.

E. Sawicki and T. W. Stanley, *Chemist-Analyst*, **49**, 107 (1960). Fluorescence spot test for glyoxal, pyruvaldehyde, salicylaldehyde, and some other aromatic aldehydes.

E. Sawicki and W. Elbert, *Anal. Chim. Acta*, **23**, 205 (1960). Detection and characterization of 1,4-naphthoquinones.

E. Sawicki and T. W. Stanley, *Mikrochim. Acta*, **1960**, 510. Test for aliphatic, aromatic, and heterocyclic aldehydes.

E. Sawicki, T. W. Stanley, T. R. Hauser, and W. Elbert, *Anal. Chem.*, **33**, 93 (1961). Detection and determination of aliphatic aldehydes.

E. Sawicki, *Microchemical Techniques* (Microchem. J. Symposium Series, Vol. 2), Wiley, New York, 1962, pp. 59–106. Spot-test detection and spectrophotometric determination of microgram amounts of aldehydes.

L. Segal, *Anal. Chem.*, **23**, 1449 (1951). Determination of formaldehyde.

H. Shechter, M. J. Collis, R. Dessy, Y. Okuzumi, and A. Chen, *J. Am. Chem. Soc.*, **84**, 2905 (1962). The effects of ring size on the rates of acid and base catalyzed enolization of homologous cycloalkanones and cycloalkyl phenyl ketones.

H. J. Shine, *J. Chem. Educ.*, **36**, 575 (1959). Acid catalysis for 2,4-dinitrophenylhydrazones.

H. J. Shine, *J. Org. Chem.*, **24**, 252 (1959). 2,4-Dinitrophenylhydrazones.

H. Siegel and F. T. Weiss, *Anal. Chem.*, **26**, 917 (1954). Determination of aldehydes in presence of ketones.

S. Siggia and E. Segal, *Anal. Chem.*, **25**, 640 (1953). Determination of aldehydes in presence of acids, ketones, acetals, and vinyl ethers.

T. W. Stanley, *Chemist-Analyst*, **47**, 91 (1958). Spot test for methyl ketones.

T. W. Stanley, *Chemist-Analyst*, **49**, 47 (1960). Detection of polynuclear diaryl ketones.

A. Stoll and A. Ruegger, *Helv. physiol. Acta*, **10**, 385 (1952). R_f values.

H. H. Stroth, *Chem. Ber.*, **90**, 352 (1957); **91**, 2645–2663 (1958). Substituted hydrazines.

M. Tannenbaum and C. E. Bricker, *Anal. Chem.*, **23**, 354 (1951). Determination of formaldehyde.

R. B. Wearn, W. M. Murry, Jr., M. Ramsay, and N. Chandler, *Anal. Chem.*, **20**, 922 (1948). Determination of some *alpha, beta* unsaturated aldehydes.

P. W. West and B. Sen, *Anal. Chem.*, **27**, 1460 (1955). Spot test for formaldehyde.

Esters

G. F. Baumann and S. Steingiser, *J. Appl. Polymer Sci.*, **1**, 251 (1959). Differentiation of polyester and polyether based urethane polymers.

F. Feigl et al., *Mikrochemie*, **15**, 9 and 23 (1934). The ferric hydroxomate test for esters.

F. Feigl and E. Jungreis, *Anal. Chem.*, **31**, 2101 (1959). Phenyl esters.

F. Feigl and V. Anger, *Chemist-Analyst*, **49**, 13 (1960). Spot test for esters of noncarboxylic acids.

V. Goldenberg and P. E. Spoerri, *Anal. Chem.*, **30**, 1327 (1958). Reaction times and absorbance for several esters; hydroxomate test.

H. Y. Yale, *Chem. Rev.*, **33**, 209 (1943). The hydroxamic acids and their reactions with iron (III) ions.

Ethers and Epoxides

R. L. Barwell, Jr., *Anal. Chem.*, **26**, 615 (1954). Cleavage of ethers.

W. Dasler and C. D. Bauer, *Ind. Eng. Chem., Anal. Ed.*, **18**, 52 (1946). Removal of peroxides from organic solvents.

F. Feigl and E. Silva, *Analyst*, **82**, 582 (1957). Detection of methoxy and ethoxy groups.

F. Feigl and E. Jungreis, *Anal. Chem.*, **31**, 2099 and 2101 (1959). Phenyl ethers are decomposed by fusion with potassium iodide and hydrated oxalic acid. The liberated phenol is detected by the Gibbs reagent.

F. Feigl, R. J. Amaral, and D. Haguenauer-Castro, *Mikrochim. Acta*, **1960**, 821. Detection of ethers as their peroxides with copper acetate-benzidine.

D. Peters and N. Kharasch, *J. Org. Chem.*, **21**, 592 (1956). Reaction of epoxides.

M. J. Rosen, *Anal. Chem.*, **27**, 787 (1955). Pyrolysis of polyoxyethylene and polyoxypropylene with phosphoric acid.

J. D. Swan, *Anal. Chem.*, **26**, 878 (1954). Determination of epoxides.

W. C. Tobie, *Ind. Eng. Chem., Anal. Ed.*, **15**, 433 (1943). Qualitative test for alkoxy groups.

A. N. Wrigley, A. J. Stirton, and E. Howard, Jr., *J. Org. Chem.*, **25**, 439 (1960). Higher alkyl monoethers of the glycols.

Hydrocarbons

A. P. Alysuhller and I. R. Cohen, *Anal. Chem.*, **32**, 1843 (1960). Spectrophotometric methods for olefins.

R. Daniels and L. Bauer, *J. Chem. Educ.*, **35**, 444 (1958). Reactivity of acetylenes and olefins toward bromine.

L. W. Gamble in *Microchemical Techniques*, Wiley, New York, 1962. Study of urea and thiourea adducts.

H. T. Gordon and M. J. Huraux, *Anal. Chem.*, **31**, 302 (1959). Trinitrofluorenone reagent.

T. R. Hauser, *Chemist-Analyst*, **48**, 96 (1959). Detection of anthracene compounds.

D. E. Laskowski, D. G. Grabar, and W. C. McCrone, *Anal. Chem.*, **25**, 1400 (1953). Use of 2,4,7-trinitrofluorenone in fusion analysis.

D. E. Laskowski and W. C. McCrone, *Anal. Chem.*, **30**, 542 (1958). Use of 2,4,7-trinitrofluorenone; ibid., **26**, 1497 (1954).

R. E. Merrifield and W. D. Phillips, *J. Am. Chem. Soc.*, **80**, 2778 (1958). Tetracyanoethylene.

M. Orchin, L. Reggel, and E. Woolfolk, *J. Am. Chem. Soc.*, **69**, 1225 (1947). 2,4,7-Trinitrofluorenone.

P. V. Peurifoy, S. C. Slaymaker, and M. Nager, *Anal. Chem.*, **31**, 1740 (1959). Test for aromatic hydrocarbons.

E. Sawicki, R. Miller, T. W. Stanley, and T. R. Hauser, *Anal. Chem.*, **30**, 109 and 1130 (1958). Polynuclear hydrocarbons.

E. Sawicki, T. W. Stanley, and T. R. Hauser, *Chemist-Analyst*, **47**, 69 (1958). Detection of aromatics in air.

E. Sawicki, T. W. Stanley, and J. Noe, *Anal. Chem.*, **32**, 816 (1960). Test for compounds containing cyclopentadiene group.

E. Sawicki and T. W. Stanley, *Chemist-Analyst*, **49**, 77 (1960). Test for hydrocarbons.

J. G. Sharefkin and T. Sulzberg, *Anal. Chem.*, **32**, 993 (1960). Test and derivatives for olefins.

J. G. Sharefkin and E. M. Boghosian, *Anal. Chem.*, **33**, 640 (1961). Detection and characterization of acetylenes.

J. G. Sharefkin and H. E. Shwerz, *Anal. Chem.*, **33**, 635 (1961). Detection of olefins.

G. H. Schenk and M. Ozolins, *Talanta*, **8**, 109 (1961). Tetracyanoethylene.

G. H. Schenk et al., *Anal. Chem.*, **35**, 167 (1963). Pi complexes with tetracyanoethylene.

D. S. Tarbell and T. Huang, *J. Org. Chem.*, **24**, 887 (1959). Tetracyanoethylene.

Nitro Compounds and Nitrites

P. G. Butts et al., *Anal. Chem.*, **20**, 947 (1948). Reduction by titanous chloride.

B. H. Dolin, *Anal. Chem.*, **15**, 242 (1943). Color tests to distinguish the di-nitro derivatives from benzene toluene, and the xylenes.

F. Feigl and V. Gentil, *Anal. Chem.*, **27**, 432 (1955). Rhodamine-B reagent for enolizable polynitro compounds.

F. Feigl and D. Goldstein, *Anal. Chem.*, **29**, 1521 and 1522 (1957). Spot tests for nitromethane and nitroethane.

F. Feigl, *Chemist-Analyst*, **52**, 47 (1963). Test for nitrocellulose.

H. B. Hass, M. L. Bender, and E. J. Berry, *J. Am. Chem. Soc.*, **71**, 2290 (1949). *p*-Nitrobenzyl derivatives for nitroparaffins.

J. P. Hestis and J. W. Cavett, *Anal. Chem.*, **31**, 1977 (1959). *m*-Dinitro compounds.

W. B. Koniecki and A. L. Linch, *Anal. Chem.*, **30**, 1134 (1958). Reagent for reducing nitro compounds.

W. C. McCrone, *Microchem. J.*, **3**, 479 (1959). High explosives (nitros).

D. E. Pearson and A. Morrissey, *Microchem. J.*, **6**, 175 (1962). Test for mononitro compounds.

E. Sawicki and T. W. Stanley, *Anal. Chem. Acta*, **25**, 166 (1961). Test for nitrites.

E. Sawicki and T. W. Stanley, *Anal. Chim. Acta*, **23**, 551 (1960). Color tests for di- and polynitro compounds.

E. Sawicki and J. L. Noe, *Anal. Chim. Acta*, **25**, 166 (1961). Test for nitrites.

E. Sawicki et al., *Anal. Chem.*, **34**, 297 (1962). Spot test for nitrites using *p*-phenylazoaniline.

R. W. Shellman, *J. Org. Chem.*, **22**, 818 (1957). Color produced in acetone by nitro compounds.

M. H. Swann, *Anal. Chem.*, **29**, 1504 (1957). Acetone-sodium hydroxide test for nitrocellulose in lacquers.

G. N. Smith, *Anal. Chem.*, **32**, 32 and 978 (1960). Use of methylamine and dimethylformamide to test for *m*-dinitro aromatic amides.

R. P. Zimmerman and E. Lieber, *Anal. Chem.*, **22**, 1151 (1950). Reduction by titanous chloride in buffered solution at room temperature.

Phenols

R. D. Blasco, A. Dimenza, and L. N. Pizzon, *Rev. Farm.*, **101**, 105 (1959). *o*-Diphenols.

F. Feigl and E. Jungreis, *Anal. Chem.*, **31**, 2099 and 2101 (1959). Gibbs' reagent.

F. Feigl and D. Haguenauer, *Chemist-Analyst*, **49**, 43 (1960). Spot tests for *o*- and *p*-nitrophenols.

F. Feigl, V. Gentil, and D. Haguenauer-Castro, *Microchem. J.*, **4**, 445 (1960). Spot test for phenoxy compounds.

G. E. Inglett and J. P. Lodge, *Anal. Chem.*, **31**, 248 (1959). Tabular comparative results of reacting 34 phenols with four different reagents.

G. A. Lugg, *Anal. Chem.*, **35**, 899 (1963). Table of colors of 14 phenols with 17 diazonium salts.

T. S. Ma and A. Hirsch, *Chemist-Analyst*, **50**, 12 (1961). Spot tests for phenols.

E. Sawicki et al., *Anal. Chem.*, **39**, 1130 (1958). Phenols and polynuclear hydrocarbons.

E. Sawicki et al., *Anal. Chem.*, **31**, 1664 (1959). Tests for phenols.

I. Takanobu and S. Kamuja, *Bunsoki Kajaku*, **7**, 616 (1958). Spot tests for phenols.

G. R. Tallon and R. D. Hepner, *Anal. Chem.*, **30**, 1521 (1958). *p*-Substituted phenols.

E. O. Woolfolk and J. M. Taylor, *J. Org. Chem.*, **22**, 827 (1957). Identification and chromatographic separation of phenols.

Sulfur Compounds

S. Akerfeldt, *Acta Chem. Scand.*, **13**, 627 (1959). 6-Chloromercuri-2-nitrophenol as a reagent for thiols. Said to be specific for thiols at pH of 4.

I. W. Grote, *J. Biochem.*, **93**, 25 (1931). Color reactions for thiols and sulfides.

C. Karr, *Anal. Chem.*, **26**, 528 (1954). Twenty tests for various sulfur compounds.

PART THREE

Procedures for Final Characterization of an Unknown

INDEX TO PREPARATION OF DERIVATIVES

Melting points of derivatives, evaluation of, 436
Selection of derivatives, 430
Small quantities of derivatives, preparation of, 434
  General procedure, 434
  Special procedure, 435

Derivatives*

Acetals, 514 (667)
Acid anhydrides, 459 (668)
Acid halides, 458 (670)
Acids,
  Amino, 450 (672)
  Carboxylic, 437 (684)
  Sulfonic, 635 (959)
Alcohols, 465 (714)
Aldehydes, 496 (728)
Alkanes, 572 (862)
Alkenes, 574 (870)
Alkyl halides, 548 (830)
Alkynes, 574 (870)
Amides, 607 (744)
Amines, 589 (750)(799)
Amino acids, 450 (672)
Aromatic hydrocarbons, 577 (877)
Aryl halides, 558 (842)
Azo compounds, 615 (909)
Azoxy compounds, 617 (911)
Carbohydrates, 519 (809)
Carboxylic acids, 437 (684)
Cycloalkanes, 572 (862)
Cycloalkenes, 574 (870)
Cycloalkyl halides, 548 (830)
Dienes, 574 (870)
Esters, 532 (814)

Ethers, 541 (823)
Fluorine compounds, 564 (857)
Halides, 549 (830)(842)
Hydrazines, 607 (750)
Hydrazo compounds, 617 (911)
Hydrocarbons, 571 (862)(877)
Imides, 607 (744)
Isocyanates, 617 (912)
Isocyanides, 617 (912)
Isothiocyanates, 640 (961)
Ketones, 510 (892)
Mercaptans, 641 (966)
Nitriles, 617 (913)
Nitro compounds, 623 (916)
Nitrogen functions, 590
Nitroso compounds, 628 (923)
Phenols, 484 (924)
Sulfonamides, 631 (948)
Sulfonyl chlorides, 634 (953)
Sulfonic acids, 635 (959)
Sulfur functions, 632
Thiocyanates, 640 (961)
Thioesters, 643
Thiols, 641 (966)
Thiophenols, 641 (966)
Ureas, 607 (744)

* Page numbers of tables of derivatives are in parentheses.

11

Problems in the Derivatization of Organic Compounds

The preparation of suitable derivatives is the last and conclusive step in the identification of an organic compound. The identification is regarded as rigorous and conclusive if (a) all experimental data fit the properties of the compound selected from the literature search, (b) the derivatives prepared from the unknown melt within 1 to 2° of the melting points given in the literature for the same derivatives of the probable compound, and (c) a mixture of the same derivative prepared from the known and unknown shows identity by either the fusion or the melting point method.

A number of problems must be considered in the preparation of the derivatives. First, the most suitable derivative must be selected from among the several listed in the tables of derivatives (pages 659–967). Second, consideration should be given to the preparative procedure that will be employed; in dealing with small quantities it is imperative to minimize side reactions so as to obtain as pure a derivative as possible after one crystallization and to avoid, insofar as possible, losses in transfer.

It can not be emphasized too strongly that the preparation, purification, and identification of a derivative when one uses only 50–100 mg of starting material requires more planning and better techniques of operation than is demanded when 1 to 3 g of the compound is used.

Selection of Derivatives

Requirements for Useful Derivatives

Most organic compounds undergo a number of reactions that give rise to other compounds that may be used as *derivatives*. For example, an aldehyde may be oxidized to an acid, reduced to an alcohol, or converted into an oxime, phenylhydrazone, or semicarbazone. Although theoretically all these reactions may be used to prepare derivatives for the characterization of an aldehyde, practical considerations demand that a derivative must fulfill certain requirements in order to be suitable for characterization purposes. These requirements are as follows.

1. The derivative should be a solid, melting, if possible, above 50° and below 250°. If the derivative is an oil, it cannot be purified in small quantities; as a rule organic crystalline compounds melting below 50° do not crystallize well and in such cases they have a tendency to separate as oils. A derivative melting between 100 to 200° is preferable (other factors being equal) to a derivative melting much above 200°; the determination of melting points much above 200° is more difficult and requires considerable care in ascertaining the thermometer correction to be applied.

2. The derivative should have a melting point that is quite different from that of the original compound from which it is prepared; further, the melting point of the derivative should differ by more than 5° from the derivatives of closely related compounds. For example, let us assume that a sample of an unknown liquid, on the basis of tests, is tentatively identified as propionic acid. In the selection of the derivatives to be prepared, reference is made to Table 5A (page 684), and it is found that the anilide of propionic acid melts at 106°, whereas the anilide of isobutyric acid, a closely related acid and one of the possibilities, melts at 105°. On the other hand, the *p*-toluidide of propionic acid melts at 126°, whereas that of isobutyric melts at 107°; therefore the *p*-toluidide in this particular case is a much more suitable derivative than the anilide.

3. The reaction by which the derivative is made should be complete within 30 minutes; should be subject to few, if any, side reactions; and should afford a good yield.

4. The reagents used in the preparation of a derivative should be readily available.

5. The derivative should be readily purified; it should be slightly soluble in some common solvent in the cold and somewhat soluble at the boiling

point of the solvent. More specifically, the ratio of the solubility at room temperature to that at the boiling point of the solvent should be more than 1:5.

Recommended Derivatives

The number of derivatives described in the literature as suitable may be rather large for some groups of compounds. For example, as shown in

TABLE 11.01

PARTIAL LIST OF DERIVATIVES DESCRIBED IN THE LITERATURE FOR IDENTIFICATION OF CARBONYL COMPOUNDS AND ALCOHOLS

Carbonyl Compounds	*Alcohols*
Benzothiazoles	Alkyl hydrogen tetrachlorophthalates
Benzothiazolines	Aryloxyacetic acids
Benzylideneaminomorpholines	Allophanates
p-Carboxyphenylhydrazones	S-Benzylthiuronium derivatives
o- and p-Chlorobenzohydrazones	p-Bromophenylurethans
Dibromomethone derivatives	p-Chlorophenylurethans
2,4-Dinitrophenylhydrazones	3,5-Dinitrobenzoates
3,5-Dinitrophenylsemicarbazones	3,5-Dinitro-4-methylphenylurethans
Diphenylhydrazones	3,5-Dinitrophenylurethans
Hydantoins	2,4-Dinitrophenyl ethers
Methone derivatives	Hydrogen 3-nitrophthalates
α- and β-Naphthylsemicarbazones	Hydrogen phthalates
3-Nitroguanylhydrazones	4-Iodobiphenylurethans
o-, m-, and p-Nitrobenzenesulfonhydrazones	p-Iodophenylurethans
p-Nitrophenylhydrazones	α-Naphthylurethans
Oximes	p-Nitrobenzoates
Phenylhydrazones	m-Nitrophenylurethans
Semicarbazones	o-Nitrophenylurethans
Thiosemicarbazones	p-Nitrophenylurethans
o-, m-, and p-Tolylsemicarbazones	Phenylurethans
Xenylsemicarbazones	Trityl ethers
	Xanthates
	p-Xenylurethans

Table 11.01, a considerable number of derivatives are available for the identification of carbonyl compounds and alcohols.

The same is true of carboxylic acids (see page 438) and many other groups of compounds. The selection of the appropriate derivative is even more important when small quantities of the unknown are available. For example, when semimicro quantities are used, the quantity of the derivative to be prepared seldom exceeds 100 to 200 mg.

We have selected *recommended derivatives* on the basis of actual tests.

In the tables in Chapters 12 to 21, these recommended derivatives are listed first, and each is preceded by an asterisk; in the tables in Part Four they appear under the column head "Recommended." The preparation of the recommended derivatives is described in detail in Chapters 12 to 21. Since semimicro quantities are used, specific directions are indispensable. In addition, a general method described for one member of a particular group of compounds often requires radical alterations for different members of the same group. The rates of a particular reaction vary greatly among members of a group, and in some cases the reaction may proceed in such a manner that the preparation of a derivative may be impractical; for example treatment of formaldehyde with semicarbazide does not give a semicarbazone, but a polycondensation product. Under the discussion of the general method, brief notes are included about the changes in procedure necessary for the preparation of the same type of derivative from other members of a given group or subgroup of compounds.

Detailed Procedures for Recommended Derivatives

Most methods described in original papers on the preparation of derivatives for identification work are based on macro quantities, usually 1 to 5 g of the compound. The amount of the derivative given by macro quantities is so great that even a careless worker can perform two or three crystallizations and still have 100 mg of the derivative left. In the transition from macro to semimicro quantities, the amounts of reagents are reduced tenfold or more; consequently conditions must be chosen that will insure: (a) completion of the reaction; (b) a minimum number of crystallizations to obtain a pure product; (c) exact quantities of solvent and conditions of crystallization; and (d) a sufficient yield of pure product for several melting-point determinations.

To meet some of these conditions the material in Chapters 12 to 21 is arranged according to the following plan. For each major group or class of organic compounds—amines, carbonyl compounds, and the like—a section is first devoted to the general discussion of (a) the most important derivatives that have been proposed in the literature and the reactions by which each type of derivative is prepared; (b) the limitations of each reaction; (c) side reactions and impurities that may separate along with the derivative; and (d) recommendations of the derivatives that should be tried first. This discussion is followed by a description of the general procedure for the preparation of important derivatives and by several examples of the derivatives of representative compounds of the series; in some cases, such as alcohols, eight to ten derivatives are described. For each type of recommended derivative, an attempt has been made to select a representative member from both the aliphatic and the aromatic series. The reactions by which derivatives are made should be considered as general reactions exhibited by all members of the group, but at varying rates. The rate of reaction may be so slow that the preparation of the derivative is impractical; since in any reaction of organic compounds there may be several equilibria involved and hence some members of the group will not give the expected derivative. As mentioned, treatment of formaldehyde with semicarbazide yields not a semicarbazone, but a polycondensation product.

Generally, there is a difference in reactivity between acyclic and cyclic derivatives, particularly between aliphatic and aromatic compounds. In some cases the aromatic compounds react faster than the open-chain compounds, although no general guiding rule can be given. However, with reference to the solubility of derivatives, such a rule can be formulated. Derivatives of aromatic compounds are far less soluble in water and methanol (the pair of solvents most widely used in this work) than the same derivatives from open-chain compounds. Therefore, if a derivative is described for an open-chain compound with 4 to 6 carbon atoms, and it is desired to prepare the same type of derivative from a simple aromatic compound, the quantitity of methanol used in the purification of the derivative should be increased. In most cases it is possible to start with about 50 mg or less of the aromatic compound, prepare a derivative, perform two or more crystallizations, and have 20 to 30 mg of the pure derivative left for determination of the melting point.

Other Derivatives

The tables of derivatives (Part Four) and the discussion in Chapters 12 to 21 include other types of derivatives in addition to those recommended. No attempt has been made to include all the derivatives that have been described in the literature for any particular group, although in many cases (for example, derivatives of carboxylic acids, alcohols, and carbonyl compounds) the list is believed to be exhaustive. However, the general procedures for the preparation of a number of derivatives other than the recommended ones are described in Chapters 12 to 21.

The selection of "*other derivatives*" was made after careful consideration of all the factors involved. Often such a derivative is suitable for some particular members of the group of compounds under which it is described. It should be emphasized that if a derivative is not listed as being recommended, *this does not mean* that it is not useful for some or all members of the series. Therefore, if the recommended derivatives do not prove suitable, a selection should be made from the other derivatives listed. For this selection the literature should be consulted; most of the derivatives listed in the tables in Part Four are covered in the references at the ends of Chapters 12 to 21. The first factor to be considered in choosing derivatives other than those recommended is the availability of the reagent. If the reagent in question is not commercially available and not easily prepared, it is best, in most cases, to eliminate such a derivative from consideration.

In working with quantities of a few milligrams the following factors must be considered: (a) the purity of the derivatizing agent; (b) the ratio of the reactants; (c) the effect of the purification procedure. These factors have been discussed in great detail by Cheronis and Vavoulis.[1]

General Procedure for Preparation of Small Quantities of Derivatives

A large number of examples in the preparation of derivatives using small quantities is given in Chapters 12 to 21. However, it is desirable to consider in some detail what alterations must be made in macro procedures described in the literature in order to adapt them to small-scale experimentation. In addition, the special precautions necessary for work with small quantities will be stressed.

In general, once one becomes familiar with the common microprocedures, it is relatively easy to adapt a given preparation from a method designed to give a yield of 2.5 to 10.0 g of product to a scale of 50 to 500 mg. This adaptation usually involves reduction in the size of the vessels and a few changes in the methods of purification.

When the scale of preparation is a few milligrams or less, *exact procedures* must be employed. Such procedures involve quantitative methods in weighing or measuring the reactants and special methods for the isolation and purification of the desired product. The following problem illustrates the necessity for exact procedures when it became necessary to prepare a derivative of a compound of which only a few milligrams were available. It was obvious that reduction in the size of reaction vessels and modification of preparative procedures, such as crystallization and filtration, and the like, would not be suitable for handling a few milligrams

---

[1] N. D. Cheronis and A. Vavoulis, *Mikrochemie*, **38**, 428 (1952); N. D. Cheronis, *Micro and Semimicro Methods*, Vol. VI of Weissberger (ed.), *Technique of Organic Chemistry*, Interscience, New York, 1954, pp. 481–489.

of material. Therefore the necessity arose for developing a procedure by which the reaction, isolation of the crude product, and purification would be made in one vessel with little or no transfer until the final product was obtained. Therefore the reaction was carried in a modified microcone, the end of which was broken off and placed in a microsublimator, and the product was purified by fractional microsublimation.

In the present work very little space is devoted to exact procedures in the preparation of derivatives using a few milligrams or less. In this connection, see a work by one of us[2] dealing with such procedures. The scale of most procedures described in Chapters 12 to 21 is 100–200 mg; however, in many instances the scale of experimentation can be reduced to 25 to 50 mg, since all the derivatives are solids and not appreciably soluble.

Precautions When Small Quantities of Reactants Are Used

The following precautions are recommended for beginners when they undertake to prepare 50 to 200 mg of a derivative.

1. The procedure should involve as few transfer operations as possible, and yield a product that requires the minimum number of purification steps.
2. The reaction vessel should be as small as possible, since the larger the vessel the greater is the loss from adsorption on the walls of the vessel.
3. The method for the isolation of the crude product and its subsequent purification should be carefully selected.
4. The purity of the derivatizing agent should be considered.
5. When there is doubt about the procedure a "trial run" should be made, using a sample of *the pure compound tentatively identified as the unknown*. This will indicate changes that should be made in the procedure and will also give the analyst practice that will be helpful in dealing with the preparation of the same derivative from the unknown.

When the scale of preparation is reduced below 50 mg, the description of exact procedures should be consulted.[2]

Example 13.09 (Chapter 13) illustrates exact procedures. For more detailed treatment of the characterization of milligram quantities or less, the references at the end of each chapter should be consulted.

[2] N. D. Cheronis, op. cit. pp. 479–571.
[3] N. D. Cheronis, pp. 479–571 (cited in Reference 1).

Evaluation of the Melting Points of Derivatives

After the crude derivative is prepared, it is purified by crystallization. In the description of the general procedures and the illustrative examples the number of crystallizations usually required for a pure product is usually given. However, when such information is not available the general procedure for recrystallizations and checking of melting points at the same time, as described on pages 55–70, should be followed. The problems involved in the evaluation of the melting-point data of derivatives are discussed in detail in the sections entitled *The Use of Melting Point Data in the Identification of Organic Substances* and *Recommendations for the Evaluation of Melting Point Data of Derivatives* of Chapter 5 (pages 197–201). It should be stressed again that *at times the melting point given in the literature for a particular derivative will not be obtained* no matter how many recrystallizations are performed. Therefore, it is necessary to correlate the melting point and other data in arriving at conclusions.

12

Derivatives of Carboxylic Acids and Acid Derivatives

CARBOXYLIC ACIDS—AMINO ACIDS—
ACID HALIDES—ACID ANHYDRIDES

CARBOXYLIC ACIDS

Table 12.01 lists a large number of derivatives that have been described in the literature for the derivatization of the carboxylic function. The number of references at the end of this chapter further indicates the tremendous interest in the characterization of carboxylic acids.

Generally, the derivatives listed in Table 12.01 fall into four classes: (a) amides and substituted amides, such as anilides, *p*-bromoanilides, and *p*-toluidides, which are formed by reaction of the carboxyl group with ammonia or amines; (b) esters, such as the *p*-nitrobenzyl, phenacyl, and *p*-bromophenacyl, formed by the reaction of the arylalkyl halides and the sodium salts of the acids; (c) salts of the carboxylic acids with such bases as phenylhydrazines, 2,4-dinitrophenylhydrazine, benzylamine, piperazine; and (d) a variety of derivatives formed by reaction of the carboxyl group with various functional groups; for example, reaction of the carboxyl group with the two amino groups of 1,2-diaminobenzene (*o*-phenylenediamine) yields a 2-alkylbenzimidazole (see page 450). Similarly, the ureides and thioureides formed by reaction of the carboxyl group with substituted ureas can be regarded as substituted amides.

TABLE 12.01
DERIVATIVES FOR THE IDENTIFICATION OF CARBOXYLIC ACIDS

Acid esters of dibasic acids	Hydrazides
N-Acyl-p-aminoazobenzenes	Hydroxamic acids
N-Acylanthranilic acids	p-Hydroxyanilides
N-Acylcarbazoles	Lactone derivatives
N-Acyl-2-nitro-p-toluidides	Menthyl esters
N-Acylphenothiazines	N-Methylamides
N-Acylsaccharins	2-Methyl-5-isopropylanilides
N-Acyl-2-acylcarbazoles	Monoureides
p-(N-Acylamino)benzoic acids	Monothioureides
3-Acylaminodibenzylfurans	*α- and β-Naphthylamides
2-Alkylbenzimidazole picrates	S-(α-Naphthylmethyl)thiuronium salts
2-Alkylbenzimidazoles	o-, m-, and p-Nitroanilides
*Amides	*p-Nitrobenzyl esters
*Anilides	S-(p-Nitrobenzyl)thiuronium salts
N-Benzylammonium salts	Octadecylamides
*S-Benzylthiuronium salts	Octadecylammonium salts
Bis(p-dimethylaminophenyl)ureides	*Phenacyl derivatives
*p-Bromoanilides	o- and p-Phenetides
p-Bromobenzylthiuronium salts	4-Phenylazophenacyl esters
*p-Bromophenacyl esters	α-Phenylethylammonium salts
α-Bromo-β-naphthylamides	Phenylhydrazides
o-Bromo-p-toluidides	Phenylhydrazonium salts
Carbodi-imide derivatives	N-Phenylimides of dibasic acids
p-Chlorobenzylthiuronium salts	p-Phenylphenacyl esters
p-Chlorophenacyl derivatives	Phenylmercuric salts
2,8-Diacylcarbazoles	Piperazonium salts
Diazomethane derivatives	Tetraphenylstilbonium salts
N-Diethanolamides	Thiophenylamides
Dimethylamides	o-Toluidides
*2,4-Dinitrophenylhydrazides	*p-Toluidides
2,4-Dinitrophenylhydrazones of the p-phenylphenacyl esters	p-Tolylmercuric salts
	2,4,6-Tribromoanilides
Diphenylamides	Triphenyllead salts
Dodecylamides	p-Xenylamides
Dodecylammonium salts	m-Xylidides
N-Ethanolamides	

* Recommended derivative.

The discussion that follows deals mostly with the recommended derivatives listed in Table 12.01. In the section titled "*Salts and Other Derivatives of Carboxylic Acids*" (page 449) a few of the other derivatives are briefly described, but reference to the cited articles is suggested for a more extensive treatment. In general, if the carboxylic compound is in the anhydrous condition or can be readily obtained in this form (for example, by evaporation of the neutralized aqueous solution of the acid), the preparation of the *p*-toluidide, *p*-bromoanilide, or anilide is recommended. If the carboxylic compound is in aqueous solution or if the water cannot be readily removed, derivatization by means of one of the sparingly soluble esters (*p*-nitrobenzyl and the like) is recommended. For the identification of microgram quantities of carboxylic acids by chromatographic methods, the 2,4-dinitrophenylhydrazides are employed.

The microscopic identification of carboxylic acids based on the crystalline properties of their derivatives has been described by several investigators. Deniges[1] has described the use of cholesterol and other compounds for the identification of organic acids; Bryant and Mitchell[2] have described the use of the optical properties of the *p*-bromoanilides of lower fatty acids; Behrens-Kley, Klein and Wenzl, and Steenhauer[3] have described the use of various salts for the microscopic identification of carboxylic acids. For derivatization of microgram quantities of carboxylic acid the reader is referred elsewhere.[4]

The chromatographic separation of mixtures of fatty acids has been reviewed by Markley and Ralston.[5] A method for the chromatographic separation of the 2,4-dinitrophenylhydrazides has been described by Cheronis.[6]

Amides, *p*-Toluidides, Anilides, *p*-Bromoanilides, and Other Substituted Amides of Carboxylic Acids

Discussion

The formation of amides, anilides, *p*-bromoanilides, *p*-toluidides, and the diamides of carboxylic acids is illustrated in the equations on page 440. For the first three derivatives the acid is first converted to the acid chloride by treatment with thionyl chloride and then treated with ammonia or the amine. The direct reaction of the carboxylic acid with the amine is possible in some cases, as shown in the fourth equation, and for the toluidides of the lower fatty acids the method is described in Example 12.03.

[1] G. Deniges, *Compt. Rend.*, **196**, 1504 (1933); *Bull. Trav. Soc. Pharm. Bordeaux*, **76**, 173 (1938).
[2] W. M. D. Bryant and J. Mitchell, *J. Am. Chem. Soc.*, **60**, 2748 (1938).
[3] H. Behrens and C. Kley, *Organische Microchemische Analyse*, Voss, Leipzig, 1922, p. 311; G. Klein and H. Wenzl, *Mikrochemie*, **10**, 70 (1931/1932); **11**, 73 (1932); A. J. Steenhauer, *Pharm. Weekblad*, **72**, 667 (1935).
[4] N. D. Cheronis, "Micro and Semimicro Methods" in *Technique of Organic Chemistry*, Vol. 6, A. Weissberger, ed., Interscience, New York, 1964, p. 541.
[5] K. S. Markley, *Fatty Acids*, Interscience, New York, 1947; A. W. Ralston, *Fatty Acids and Their Derivatives*, Wiley, New York, 1948.
[6] N. D. Cheronis, *Mikrochim. Acta*, **1956**, 925; N. D. Cheronis, p. 543 (cited in Reference 4).

$$CH_3(CH_2)_4COOH \xrightarrow{SOCl_2} CH_3(CH_2)_4COCl + 2NH_3 \longrightarrow$$
n-Caproyl chloride

$$CH_3(CH_2)_4CONH_2 + NH_4Cl$$
n-Caproamide

$$CH_3(CH_2)_2COOH \xrightarrow{SOCl_2} CH_3(CH_2)_2COCl + 2C_6H_5NH_2 \longrightarrow$$
Aniline

$$CH_3(CH_2)_2CONHC_6H_5 + C_6H_5\overset{+}{N}H_3\overset{-}{Cl}$$
n-Butyranilide
(N-Phenyl-n-butyramide)

$$CH_3(CH_2)_2COCl + 2BrC_6H_4NH_2 \longrightarrow$$

$$CH_3(CH_2)_2CONHC_6H_4Br + BrC_6H_4\overset{+}{N}H_3\overset{-}{Cl}$$
n-Butyro-p-bromoanilide
(N-p-Bromophenyl-n-butyramide)

$$CH_3CH_2COOH + CH_3C_6H_4NH_2 \xrightarrow{180°} CH_3CH_2CONHC_6H_4CH_3 + H_2O$$
p-Toluidine   Propiono-p-toluidide
(N-p-Tolylpropionamide)

$$2CH_3COOH + CH_2(C_6H_4NH_2)_2 \longrightarrow CH_2(C_6H_4NHCOCH_3)_2 + 2H_2O$$
4,4′-Diaminodiphenylmethane   4,4′-Methylenebisacetanilide

General Method for the Preparation of Amides

The preparation of aliphatic amides (equation 1) is not recommended for semimicro quantities. The amides are usually obtained by first preparing the acid chloride and treating the latter with an excess of aqueous ammonia at low temperatures. One difficulty arises from the fact that the acid chloride undergoes hydrolysis and ammonolysis at the same time, thus reducing considerably the yield of the amide. If the temperature is kept low and the acid halide is not very reactive, as in the case of the higher fatty acids and aromatic acids, the degree of hydrolysis is kept at a minimum. Another difficulty arises from the significant solubilities of amides in alcohol-water mixtures, which render it difficult to crystallize 100 to 200 mg of an amide several times and leave a sufficient amount of the pure derivative for a melting-point determination.

About 1 to 1.5 millimoles of acid chloride is prepared according to the procedure described for benzoyl chloride in Example 12.05. If the carboxylic acid has less than 6 carbon atoms, it is advisable to dissolve the acid chloride in 10 ml of dry benzene and then pass in gaseous ammonia, while the mixture is cooled until an excess of ammonia is present. The solution is filtered from the ammonium chloride, the filtrate is evaporated to dryness, and the amide is crystallized from the minimum amount of petroleum ether. In the case of higher aliphatic acids or any aromatic carboxylic acids the halide is treated with excess of ice-cold aqueous ammonia and the mixture is stirred until the reaction is complete. The amide, which separates out, is filtered and crystallized. If the yield is low an additional amount may be obtained by extraction of the aqueous filtrate with ether followed by evaporation of the solvent.

Example 12.01: Preparation of n-*Butyramide*

To 10 ml of dry benzene saturated with ammonia contained in an 8-inch test tube, add the butyryl chloride prepared from 0.2 ml of *n*-butyric acid according to the procedure described in Example 12.05. Cork the tube and allow it to stand in the cold for 15 minutes. Pour the mixture through a fluted filter paper and wash the residue of ammonium chloride with 2 to 3 ml of dry benzene. Cautiously evaporate the benzene from a water bath. About 125 to 150 mg of the amide are obtained; the crystals melt at 115°.

Example 12.02: Preparation of Cinnamamide

Prepare the acid chloride from 0.2 g of cinnamic acid and 0.8 ml of thionyl chloride as described in Example 12.05. Remove the tube or flask and cool in an ice-water mixture (hood). Add 5 ml of ice-cold concentrated aqueous ammonia. Stir by means of a glass rod and allow the tube to stand for 5 minutes in the cold, shaking from time to time. Filter the crude amide and dissolve in alcohol, filter, and add water. The yield is 70 mg, melting at 147 to 148°.

General Method for the Preparation of p-*Toluidides, Anilides, and* p-*Bromoanilides*

Of the substituted amides the *p*-toluidides are recommended. They are easily prepared and have desirable crystallizing properties. If the carboxylic acid contains less than 8 carbon atoms, the *p*-toluidide may be conveniently prepared by heating the acid with an excess of *p*-toluidine at 180 to 200°. The reaction mixture is extracted with dilute hydrochloric acid to remove the excess of the base, and with dilute sodium hydroxide to remove traces of unreacted acid, and is then crystallized from an alcohol-water mixture. The toluidides of the higher fatty acids and of the aromatic carboxylic acids are prepared by first converting the acid to the acid chloride and then adding an excess of *p*-toluidine. The same method is used for the preparation of anilides and *p*-bromoanilides.

The acid chlorides are conveniently prepared by heating under reflux 1 millimole of the acid with 1.1 millimoles of pure thionyl chloride at about 70 to 78° for about 30 minutes. If the acid contains water, it is neutralized with sodium carbonate and evaporated to a small volume and then placed in the reaction tube and evaporated to dryness. In such cases the amount of thionyl chloride is slightly increased (1.3 millimoles/millimole of acid). If the acid is unsaturated, there is danger of the reaction between the hydrogen chloride evolved from the thionyl chloride and the unsaturated linkages. This tendency can be kept at a minimum if the acid is first dissolved in dry benzene and then the thionyl chloride is added. Another limitation of the use of thionyl chloride is the formation of anhydrides with dicarboxylic acids having the carboxyl groups separated by two or three carbon atoms; these anhydrides will react readily with the amines to form the desired derivative, but the yield will be lowered.

After the formation of the acid chloride is completed, dry benzene and an excess of the amine are added and the mixture is refluxed. The solution is then extracted with dilute hydrochloric acid to remove the excess of amine; the benzene solution is evaporated; and the crude amide (anilide, *p*-toluidide, or *p*-bromoanilide) is purified by crystallization.

Example 12.03: Direct Preparation of the p-*Toluidide of a Monocarboxylic Acid: Propiono-*p-*toluidide*

Place in an 8-in. test tube 0.3 ml of propionic acid and 1.2 g of *p*-toluidine. Immerse the tube in an oil bath and raise the temperature slowly over a period of 10 minutes to 190°. Keep the temperature at 190 to 200° for 30 minutes; then remove the tube and allow to cool. Add 7 ml of 5 per cent hydrochloric acid and wash down the inner walls of the tube, by means of a pipet dropper, with 1 ml of methanol. Heat nearly to boiling and set in cold water to cool for 10 to 15 minutes, shaking the tube from time to time. Filter; wash the solid with a mixture of 2 ml of 5 per cent hydrochloric acid and 3 ml of water, then with 5 ml of water, followed by 5 ml of 2 per cent sodium hydroxide solution, and finally twice with 3 ml of water. By means of the spatula remove the solid, together with the filter paper, and place in an 8-in. test tube. Add 3 ml of methanol and heat nearly to boiling until the crystals are completely dissolved. Add as much charcoal as can be placed on one half of the spatula blade and heat for a few seconds; then set aside while a small apparatus for suction filtration is made ready. Heat the solution of the derivative to boiling and pour slowly into the filter. The filtrate should be clear. If a minute amount of charcoal has passed through the filter paper, it may be overlooked, since the compound is to be recrystallized. About 1 ml of methanol is poured dropwise down the sides of the tube that was used to prepare the solution and, after rotating and heating for a few seconds, the washings are added into the filter. The suction is discontinued and a few drops of water are added by means of the pipet dropper to the filtrate until cloudiness appears. The tube is heated until the cloudiness disappears and is then cooled for 15 to 20 minutes. The crystals are filtered and washed three times with 3 to 4 ml of water. The yield of dry crystals is 75 to 100 mg; the melting point is 122 to 123°. The crystals are recrystallized immediately after filtration from a 2:1 mixture of methanol and water. The yield of pure propiono-*p*-toluidide is 50 to 70 mg, melting at 124 to 125°.

NOTE. The solubility of *p*-toluidides decreases with an increase of the molecular weight of the acid. Thus, by using 0.3 ml of *n*-caproic acid with the procedure described, and the same number of crystallizations, about 110 to 130 mg of *n*-capro-*p*-toluidide are obtained, melting at 74 to 75°. However, as the molecular weight increases, there is a tendency for the *p*-toluidides to separate as oils. The *p*-toluidide of *n*-caprylic acid illustrates this tendency. When the reaction mixture of 0.3 ml of acid and 1.2 g of *p*-toluidine is heated with dilute hydrochloric acid, an oil separates that does not crystallize on cooling. If, during cooling, the tube is shaken frequently and the inner sides of the tube are scratched with a glass rod, the oil crystallizes into a dark mass. After washing as described, it is dissolved in 5 ml of hot ethanol, treated with charcoal, and filtered; the filtrate is then diluted with 2 ml of water. The crystals that separate on cooling are crystallized again from 5 ml of ethanol and 3 ml of water. The yield is about 200 to 220 mg of *n*-caprylo-*p*-toluidide, melting at 70°.

The preparation of the derivatives of fatty acids having 5 to 8 carbon atoms may be accomplished by using 0.1 to 0.2 ml of the sample. The dry sodium salt of the carboxylic acid may be used instead of the free acid. In such cases equal amounts of sodium salt and concentrated hydrochloric acid are placed in the tube. The toluidine is added and the mixture is heated gradually until a temperature of 140° is reached; it is kept at this temperature for 15 minutes. The temperature is then raised to 190 to 200° and maintained for 30 minutes. This method is effective when aqueous solutions of the lower carboxylic acids are involved. Since it is difficult to extract small amounts, it is more appropriate to neutralize the solution carefully with sodium carbonate and evaporate the salt solution to dryness, using this salt mixture for the preparation of the derivative.

*Example 12.04: Direct Preparation of the Toluidide of a Dicarboxylic Acid: Adipobis-*p*-toluidide*

The procedure described in Example 12.03 is followed, using 200 mg of adipic acid and 1 g of *p*-toluidine. Add to the cooled reaction mixture 7 ml of 5 per cent hydrochloric acid and heat nearly to boiling. Cool slightly and pour the contents of the tube into a mortar. Grind the mass of crystals until no lumps remain. Replace the mixture in the 8-inch tube and heat. Filter and wash as described in Example 12.03. Dissolve in ethanol and, after treating with charcoal, filter. Add a few drops of water and cool. The solid that separates is crystallized after filtration from ethanol. About 150 mg of the pure crystals, melting at 239°, are obtained.

Note. Dicarboxylic acids may react with only one mole of the base to give the mono-*p*-toluidide. This tendency is particularly manifested when the dicarboxylic acid forms an anhydride upon heating. For example, phthalic and maleic acids give the mono-*p*-toluidide when they are heated with *p*-toluidine. The literature value for the melting point of this bis-*p*-toluidide is 241°; however, the authors have not been able to obtain values above 239° (see discussion in Chapter 5 on the use of melting-point data).

Example 12.05: Preparation of a p-*Toluidide through the Preparation of the Acid Chloride:* Benzo-p-*toluidide*

Place in a 25-ml tube or flask arranged as shown in Figure 13.01 (page 472) 180 mg of benzoic acid. Lift the cork holding the microcondenser and add, by means of the pipet dropper, 1.2 ml of thionyl chloride. Heat in a water bath at 75 to 80° for 30 minutes. Lift the cork and add a solution of 500 mg of *p*-toluidine in 25 ml of dry benzene. (CAUTION. *Flames should not be in the vicinity.*)

Replace the microcondenser and reflux gently for 15 minutes. Cool, add 5 ml of water, and transfer the mixture into a small separatory funnel or into an 8-in. tube provided with separatory stopper. Wash the distilling tube with 1 ml of ethanol and add to the mixture in the separatory vessel. Shake the benzene-water mixture gently to insure thorough mixing but to avoid formation of an emulsion. Remove the aqueous layer and wash the benzene solution successively with 5 ml of 5 per cent hydrochloric acid, 5 ml of 5 per cent sodium hydroxide, and 5 ml of water. Allow sufficient time after shaking for complete separation of the two immiscible layers. Pour the benzene solution into an evaporating dish and evaporate cautiously over a water bath. Wash the separatory vessel with 1 to 2 ml of ethanol and add the washings to the evaporating dish. Add to the residue a few milliliters of ethanol and warm until the solid is completely dissolved. Add charcoal and filter by suction. Wash the dish, by means of the pipet dropper, with 1 to 2 ml of ethanol and pour the washings through the filter. Add water slowly to the filtrate until cloudiness appears and heat the tube until the cloudiness disappears. Cool and filter the crystals; wash the crystals twice with water, and dissolve them directly in 5 ml or less of hot ethanol. Filter and add 2 ml of water. Cool and filter the toluidide. About 180 mg of pure benzo-*p*-toluidide, melting at 157 to 158°, are obtained.

Note. This method illustrates the preparation of *p*-toluidides, anilides, *p*-bromoanilides, and other substituted amides by reaction of the amine with the acid chloride. The amount of carboxylic acid can be reduced to 0.5 millimole or less (about 50 to 60 mg for most common acids) with proportionate amounts of other reagents with fairly good results.

Example 12.06: Preparation of an Anilide through the Acid Chloride: n-Butyranilide

Place in an 8-in. tube 150 mg of butyric acid and use the procedure as described in Example 12.05, except that a solution of 0.5 ml of aniline dissolved in 25 ml of dry benzene is substituted for *p*-toluidine. The residue, after evaporation of benzene, is dissolved in 4 ml of methanol and 2 ml of water. The crystals that separate after the first crystallization are dissolved in hot methanol and filtered; to the hot filtered solution water is added. The yield of pure *n*-butyranilide, melting at 94 to 95° is about 70 to 75 mg.

NOTE. The anilides of aryl carboxylic acids are much less soluble than those of the aliphatic acids; hence the same variation in the amount of alcohol and water for crystallization is observed as in the *p*-toluidides. For the preparation of *p*-bromoanilides, the same procedure is followed as for the preparation of benzo-*p*-toluidide (Example 12.05). The preparation of *p*-bromoanilides of the lower carboxylic acids is preferred over the anilides if the quantities of the carboxylic acid available are small.

As stated on page 443, the sodium salt of the carboxylic acid may be used instead of the free acid for the preparation of the substituted amides. In cases where it is required to prepare the acid chloride and then react it with the arylamine, the dry salt is treated with the thionyl chloride as described under the preparation of benzo-*p*-toluidide (Example 12.05), using the same quantities of salt instead of the acid. Dicarboxylic acids, such as succinic, glutaric, and maleic, which form anhydrides easily, are likely to give the anhydride rather than the acid chloride by reaction of the salt with thionyl chloride; in such cases the resulting anhydride may form the monosubstituted amide instead of the diamide. For example, maleic acid may yield under such conditions the mono-*p*-toluidide instead of the di-*p*-toluidide. On the other hand, aromatic acids that contain a negative substituent in the position para to the carboxyl group (*p*-chlorobenzoic, *p*-bromobenzoic, *p*-hydroxybenzoic, and the like), do not react easily with thionyl chloride. In such cases it is often possible to convert the carboxylic acid to the acid chloride by the use of phosphorus pentachloride.

General Methods for the Preparation of Other Substituted Amides

The preparation of other substituted amides, such as α-naphthylamides, β-naphthylamides, diphenylamides, *o*-bromo-*p*-toluidides and the like, is similar to the procedures described for the *p*-toluidides and anilides. When the boiling points of the carboxylic acid and the amine are high, direct heating may be employed instead of converting the acid to the halide and then reacting it with the amine. This is illustrated by the reaction of stearic acid with 4,4'-diaminodiphenylmethane to give the substituted diamide (or methylenebisanilide).

Example 12.07: Preparation of 4,4'-Methylenebisstearanilide

Place in an 8-in. test tube 80 mg of 4,4'-diaminodiphenylmethane and 250 mg of stearic acid. Clamp the tube to a stand and heat, by means of a microburner, for about 5 minutes. The temperature within the tube should be such that the water vapor formed rises to the mouth of the tube. Cool and add 5 ml of water and 2 ml of ethanol. Heat to boiling and add a drop of phenolphthalein and, by means of a pipet dropper, sufficient 5 per cent sodium hydroxide to render the solution alkaline. Filter and wash the crude diamide twice with 4 to 5 ml of water. Transfer the crystals into a test tube and add 4 ml of methanol and 5 ml of benzene. Heat, filter, and cool the solution; after cooling for 15 minutes, filter and dry the crystals.

Example 12.08: Preparation of the β-Naphthylamide of Palmitic Acid: N-β-Naphthylpalmitamide

Place in an 8-in. test tube, fitted with a microcondenser, 1 millimole (256 mg) of palmitic acid. Lift the cork holding the microcondenser and add, by means of a pipet dropper, 0.8 ml of thionyl chloride. Reflux in an oil or water bath at 85 to 90° for 30 minutes. While the test tube is still in the bath the condenser is removed and the unreacted thionyl chloride is removed with the aid of a vacuum (or by increasing the temperature and driving off the excess thionyl chloride). Lift the cork and add a solution of 5 millimoles (572 mg) of β-naphthylamine dissolved in 25 ml dry benzene (CAUTION. *Benzene is flammable*). Replace the microcondenser and reflux gently for another 30 minutes. Cool, filter and to the solution add 5 ml of water, and transfer the mixture to a small separatory funnel. Shake the benzene-water mixture gently to insure thorough mixing but to avoid formation of an emulsion. Remove the aqueous layer (the separation may also be done with an ordinary capillary pipet from a test tube) and wash the benzene solution successively with 5 ml of 5 per cent hydrochloric acid, 5 ml of 5 per cent sodium hydroxide solution, and 5 ml water. Allow sufficient time after shaking for the complete separation of the two immiscible layers. If crystallization does not occur, pour the benzene solution into an evaporating dish and cautiously evaporate on a water (steam) bath. Wash the separatory funnel with 1 to 2 ml of ethanol and add the washings to the evaporating dish. Add to the residue 5 ml of ethanol and warm until the solid is completely dissolved. Add charcoal and filter by suction. Wash the dish, by means of the pipet dropper, with 1 to 2 ml of ethanol and pour the washings through the filter. Add water slowly to the filtrate until cloudiness occurs, and heat the tube until cloudiness disappears. Cool and filter; wash the crystals twice with 3 ml water, and dissolve directly in hot ethanol. Filter and add 2 ml of water. Cool and filter. The yield of crystals is about 80 per cent of the theoretical. The crystals melt at 109 to 110°.

Solid Esters of Carboxylic Acids

General Method

The following equations show the preparation of *p*-nitrobenzyl and phenacyl esters of carboxylic acids that may be used for characterization.

$$CH_3COONa + O_2N{-}C_6H_4{-}CH_2Cl \rightarrow CH_3COOCH_2C_6H_4NO_2 + NaCl$$
p-Nitrobenzyl chloride   *p*-Nitrobenzyl acetate

$$CH_3COONa + C_6H_5{-}COCH_2Cl \rightarrow CH_3COOCH_2COC_6H_5 + NaCl$$
Phenacyl chloride   Phenacyl acetate

$$CH_3COONa + Br{-}C_6H_4{-}COCH_2Br \rightarrow CH_3COOCH_2COC_6H_4Br + NaBr$$
p-Bromophenacyl bromide   *p*-Bromophenacyl acetate

In addition to the *p*-nitrobenzyl, phenacyl, and *p*-bromophenacyl, the *p*-chlorophenacyl and *p*-phenylphenacyl esters may be prepared by similar reactions. A great amount of caution, however, should be exercised in the use of these reactions. All of the phenacyl halides have lachrymatory properties and, further, cause blisters on contact with the skin. Handling of the crystalline esters may cause an irritation between the fingers, probably due to the small amounts of halide adhering to the crystals of the esters. Nevertheless, the preparation of the *p*-nitrobenzyl and the phenacyl esters as derivatives is very valuable when the acids cannot be easily separated from aqueous solutions, as is often the case in the hydrolysis of the esters of lower carboxylic acids.

For the preparation of the *p*-nitrobenzyl and phenacyl esters, 1 millimole of the carboxylic acid is converted to the sodium salt by neutralization with dilute sodium hydroxide (5 per cent) or sodium carbonate (5 per cent) and is then heated for an hour or more with an alcoholic solution of 0.9 millimole of the halide in 5 to 8 ml of alcohol. It is essential to avoid the use of an excess of the halide, since it cannot be easily removed from the ester. The esters are recrystallized from alcohol; their separation from solution is rather slow and sufficient time should be allowed for the oils that often separate to change to the crystalline state. The formation of *p*-nitrobenzyl and substituted phenacyl esters is slow and refluxing of semimicro quantities should be continued for 1.5 to 2 hours or more to insure complete reaction. For example, by using *p*-nitrobenzyl chloride in quantities as outlined above and refluxing for one hour, the following results were obtained—in each case the temperature listed after the name of the acid is the melting point of the *p*-nitrobenzyl ester after one crystallization; the temperature in the parentheses is the melting point listed in the literature: acetic, 72° (78°); succinic, 71° (88°); citric, 74° (102°); benzoic, 65° (89°). Since pure *p*-nitrobenzyl chloride melts at 71°, it is evident that after one hour the reaction was only partially completed. For these reasons the preparation of *p*-nitrobenzyl and substituted phenacyl esters is to be undertaken only if other derivatives are not suitable.

The 2,4-dinitrophenylhydrazones of the *p*-phenylphenacyl esters of fatty acids serve as a second derivative for these acids. Greater differences exist between the melting points of these derivatives than between the melting points of the original esters; the hydrazone derivatives, which are colored, can be separated by chromatography on silicic acid–nitromethane columns.[7] Melting points and R_f values are given for derivatives of 18 acids.

Example 12.09: *Preparation of* p-*Nitrobenzyl Salicylate*

Place in an 8-in. test tube 200 mg of salicylic acid and then add a drop of phenolphthalein and 2 or 3 drops of 5 per cent sodium carbonate. Warm the tube over a free flame and continue the addition of carbonate dropwise until the acid has been neutralized and the color of the solution is just pink. Warm the solution in order to be certain that all the acid has reacted. Add 2 drops of 5 per cent hydrochloric acid so that the pink color of the solution is discharged. Add 250 mg of *p*-nitrobenzyl chloride or *p*-nitrobenzyl bromide, 8 ml of alcohol, and a small boiling stone. (CAUTION: *Handle* p-*nitrobenzyl halides with care. Replace the stopper of the bottle immediately and avoid contact with the skin.*) Arrange tube for reflux (page 34) and boil gently for 1.5 hours. Cool and add 1 ml of water and scratch the sides of the tube. After 20 minutes filter the ester and wash first with 4 ml of 5 per cent sodium carbonate and then twice with 4 ml of water. Dissolve the crystals in hot alcohol, filter, and add water to the filtrate dropwise until a cloudiness appears. Heat the tube until cloudiness disappears and then cool, scratching the sides of the tube with a glass rod. Cool for 15 minutes; then filter the crystals. Wash with 1 to 2 ml of 50 per cent methanol and dry on a clay plate. Avoid handling or contact of the crystals with the skin. The yield is 110 mg; the melting point, 97 to 98°.

NOTE. *p*-Bromophenacyl esters are prepared in the same manner. From 180 mg of benzoic acid and 400 mg of *p*-bromophenacyl bromide (2.5 hours heating) about 150 to 180 mg of the derivative (mp 118 to 119°) were obtained.

[7] N. Hawkins, A. D. Webb, and R. E. Kepner, *Anal. Chem.*, **28**, 1975–1977 (1956).

Salts and Other Derivatives of Carboxylic Acids

General Method

The preparation of salts of carboxylic acids with various organic bases that may be used for characterization is represented by the following equations.

$$CH_3COOH + C_6H_5NHNH_2 \rightarrow C_6H_5NH\overset{+}{N}H_3[\overset{-}{O}COCH_3]$$
Phenylhydrazine   Phenylhydrazonium acetate

$$CH_3COOH + C_6H_5CH_2NH_2 \rightarrow C_6H_5CH_2\overset{+}{N}H_3[\overset{-}{O}COCH_3]$$
Benzylamine   Benzylammonium acetate

In addition to phenylhydrazine and benzylamine, other bases reported in the literature for the same purpose are phenethylamine and piperidine. Although phenylhydrazine is a common reagent and the preparation of the salt appears relatively simple, the preparation of this type of derivatives is not recommended for semimicro quantities except with acids that are not easily soluble in water. The phenylhydrazine salts should be purified immediately after preparation and dried as rapidly as possible, since they change slowly when exposed to air. The use of benzylamine with heating gives N-benzylamides; the cited article[7a] should be consulted for details and the melting points of the N-benzylamides.

Formation of the S-benzylthiuronium derivative of a carboxylic acid is represented by the following reactions.

$$C_6H_5CH_2SC(NH_2)_2\overset{+}{C}\overset{-}{l} + CH_3COONa \rightarrow$$
S-Benzylthiuronium chloride

$$C_6H_5CH_2SC(NH_2)_2\overset{+}{[}\overset{-}{O}COCH_3] + NaCl$$
S-Benzylthiuronium acetate

The reagent, S-benzylthiuronium chloride (see page 635), prepared by heating thiourea with benzyl chloride, forms the crystalline derivative easily on mixing with a solution of the alkaline salt of the organic acid. The resulting thiuronium salts should be recrystallized from anhydrous solvents (alcohol, dioxane) to avoid hydrolysis.

Tetraphenylstilbonium sulfate may also be used as a reagent for derivatizing organic acids. The organic acid salts of the tetraphenylstilbonium ion are prepared by adding a $0.05M$ aqueous solution of the sulfate to aqueous solutions of the acid. If the resulting solution is slightly acid, the salt comes out of solution as small needle crystals in good yields. The salts may be recrystallized from *n*-hexane, water, or water-alcohol mixtures.

[7a] O. C. Dermer and J. King, *J. Org. Chem.*, **8**, 168–173 (1943); R. W. Stafford, R. J. Francel, and J. F. Shay, *Anal. Chem.*, **21**, 1454–57 (1949).

The melting points are sharp and spread over the useful range. Derivatives of 15 acids have been reported.[8]

The following equation represents the formation of a 2-alkylbenzimidazole by the reaction of the carboxylic acid with *o*-phenylenediamine.

$$CH_3COOH + \underset{o\text{-Phenylenediamine}}{\begin{array}{c}\diagup\hspace{-6pt}\diagdown\\ NH_2 \\ NH_2\end{array}} \rightarrow \underset{2\text{-Methylbenzimidazole}}{\begin{array}{c}NH\\ \diagup\hspace{-6pt}\diagdown\hspace{-6pt}CCH_3\\ N\end{array}} + 2H_2O$$

Equal amounts of the amine and the carboxylic acid are refluxed with a small amount of dilute hydrochloric acid for 15 to 20 minutes; after the solution is cooled, aqueous ammonia is added until the alkylbenzimidazole separates. The derivative is crystallized from alcohol or converted to the picrate by dissolving it in the minimum amount of alcohol and adding the solution to a saturated alcoholic solution of picric acid.

For the procedures employed in the preparation of the 2,4-dinitrophenylhydrazides of carboxylic acids and their chromatographic separation, see work by one of us.[9]

AMINO ACIDS

The amino acids listed in Table 4 (page 672) are, for the most part, α-aminocarboxylic acids derived from proteins by hydrolysis. The three isomeric aminobenzoic acids that are included in the table are characterized by similar derivatives, although their properties do not bear close relation to those of the α-aminocarboxylic acids.

The α-aminocarboxylic acids are nonvolatile compounds and generally melt above 200° with extensive decomposition; this property is assumed to be due to the dipolar structure of the solid state. Since the melting of the crystalline solid is accompanied by decomposition, there is no fixed temperature at which the solid and liquid phase coexist, but *a temperature range at which decomposition takes place*; further, this decomposition range depends on the *rate of heating*. As a consequence there exists a considerable confusion in the literature as to the so-called "melting points" of amino acids. A few examples will serve as illustrations: D-glutamic acid has been reported to melt with decomposition at various temperatures from 198° to 225°. L-Tyrosine decomposes according to Fischer with

---

[8] H. E. Affsprung and H. E. May, *Anal. Chem.*, **32**, 1164–1166 (1960).
[9] N. D. Cheronis, "Micro and Semimicro Methods", in *Technique of Organic Chemistry*, Vol. 6, A. Weissberger, ed., Interscience, New York, 1954, p. 543.

rapid heating at 314 to 318° (corr.) and with slow heating at 290 to 295°; the same compound has been reported to decompose at 344°. Similarly, DL-tyrosine has been reported to decompose at 295°, 318°, and 340°. The nature of the difficulty is shown by the decomposition point of thyroxine; if heated at 10° per minute, it melts at 250°, and if heated at 3° per minute, it melts at 230 to 235°. In most cases, if the compound is quickly heated it decomposes at a higher temperature than when heated slowly; the difference may be as high as 40 to 50°. This brief discussion shows that since the decomposition points are very unreliable, they cannot be used except

TABLE 12.02
DERIVATIVES FOR THE IDENTIFICATION OF AMINO ACIDS

Acetamides and formamides	p-Nitrobenzyloxycarbonyl derivatives
Benzamides	*Phenylureas and phenylhydantoins
Carbobenzoxy derivatives	Phosphotungstates
Copper salts	Phosphomolybdates
*2,4-Dinitrophenyl derivatives	Phthaloyl derivatives
*3,5-Dinitrobenzamides	Picrates and flavianates
β-Naphthalene sulfonates	*Picronolates
β-Naphthalenesulfonamides	*p-Toluenesulfonamides

*Recommended derivative.

to indicate a range of possibilities. Therefore it is advisable, in the determination of the melting point of a substance suspected to be an α-aminocarboxylic acid, to make determinations both with rapid and slow heating of the bath.

The reactions used to prepare solid derivatives of the amino acids are those that characterize the amino group. The most important derivatives are listed in Table 12.02 and discussed below in three sections: (a) N-acyl and aroyl derivatives; (b) N-ureido derivatives; and (c) salts of complex acids. The recommended derivatives are the 3,5-dinitrobenzoyl, p-toluenesulfonyl, and 2,4-dinitrophenyl derivatives, the phenylureas and phenylhydantoins, and the picronolates, which are preferred for microscopic examination.[10] Procedures for the microdetermination of amino acids based on physical constants have been published by Lacourt[11] and based on crystallographic properties using 2-nitroindan-1,3-dione as a derivatizing agent by Larsen et al.[12] The chromatographic procedures are discussed on pages 457–458.

[10] P. A. Levene and D. D. Van Slyke, *J. Biol. Chem.*, **12**, 127 (1912).
[11] A. Lacourt et al., *Nature*, **172**, 906 (1953).
[12] J. Larsen et al., *Mikrochemie*, **34**, 1–14 (1948).

N-Acyl and N-Aroyl Derivatives of Amino Acids

Discussion

The reaction of the amino acid with acyl and aroyl chlorides yields the substituted amides:

$$\underset{\text{p-Toluenesulfonyl chloride}}{CH_3C_6H_4SO_2Cl} + \underset{\text{Glycine}}{H_2NCH_2COOH} \rightarrow \underset{\text{N-p-Toluenesulfonylglycine}}{CH_3C_6H_4SO_2NHCH_2COOH} + HCl$$

The following chlorides have been used for derivatization: acetyl, formyl, benzoyl, 3,5-dinitrobenzoyl, benzenesulfonyl, p-toluenesulfonyl, β-naphthalenesulfonyl, and 4-nitrotoluene-2-sulfonyl. The usual procedure is to shake an alkaline solution of the amino acid with an equivalent amount of the acid chloride. The reaction in most cases takes place very slowly, often requiring 3 to 4 hours of mechanical shaking. This slowness is probably due to the inability of the amino group to react in the charged (dipolar) form; the addition of alkali increases the concentration of the form having a free amino group:

$$\overset{+}{N}H_3CH_2CO\overset{-}{O} + NaOH \rightarrow NH_2CH_2CO\overset{-}{O} + \overset{+}{Na} + HOH$$

$$H_2NCH_2CO\overset{-}{O} + RCOCl \rightarrow RCONHCH_2CO\overset{-}{O} + HCl$$

Attention should be given to the possible racemization of the optically active amino acids by the acylating agents—a reaction that has been shown to occur.[13] For preparative purposes, acylation under nonracemizing conditions[14] is desirable since acylated amino acids are often employed in peptide synthesis. Acylation of the active forms has become important as a result of the development of asymmetric enzymatic hydrolysis of the acylated derivatives, whereby appreciable quantities of synthetic D- and L-amino acids have been made available.[15] As a first trial the use of 3,5-dinitrobenzoyl chloride is recommended wherever possible, since the reaction requires only a few minutes of shaking.

The 2,4-dinitrophenyl derivatives are formed by the action of 2,4-dinitrochlorobenzene or 2,4-dinitrofluorobenzene and are preferred for chromatographic work. The melting points of about sixteen 2,4-dinitrophenyl

---

[13] M. Bergmann and L. Zervas, *Z. Physiol. Chem.*, **203**, 280 (1928); W. M. Cahill and I. F. Burton, *J. Biol. Chem.*, **132**, 161 (1940).

[14] V. du Vigneaud and C. E. Meyer, *J. Biol. Chem.*, **98**, 295 (1932); **99**, 143 (1932); H. D. DeWitt and A. W. Ingersoll, *J. Am. Chem. Soc.*, **73**, 3359 (1951); H. Wolff and A. Berger, *J. Am. Chem. Soc.*, **73**, 3533 (1951).

[15] J. P. Greenstein et al., *J. Biol. Chem.*, **182**, 451 (1950).

derivatives have been reported. A number of these (phenylalanine, isoleucine, tyrosine, cystine, threonine, etc.) melt below 200°. It should be noted that other groups (besides the amino), such as the hydroxyl, thiol, and imidazole, form similar derivatives. Thus tyrosine may form either the O- or the N- or the O,N-bis-2,4-dinitrophenyl derivatives.

General Method for the Preparation of Benzoyl and 3,5-Dinitrobenzoyl Derivatives

One millimole of the amino acid is dissolved in 2.5 to 3 ml of $1N$ sodium hydroxide solution in a 6-in. tube. One millimole of benzoyl or 3,5-dinitrobenzoyl chloride is added in the case of a monoamino acid, and 2 millimoles in the case of a diamino acid. The tube is stoppered and shaken vigorously for 2 minutes and then at intervals for 15 to 30 minutes. The reaction mixture is acidified with dilute hydrochloric acid to pH 4 to 5, using congo red or Universal indicator. The crystals are filtered and washed with 25 per cent methanol. The derivative can be crystallized by being dissolved in hot alcohol, and adding water cautiously until permanent cloudiness results.

The diamino acids usually react with 2 moles of the chloride forming bis(benzoyl or 3,5-dinitrobenzoyl) derivatives. In general, the dicarboxylic acids react much slower than the monoamino monocarboxylic acids. A few amino acids, for example, tyrosine, form benzoyl derivatives but not 3,5-dinitrobenzoates.

NOTE. If excess of aroyl halide is used or if the reaction is incomplete, on acidification free benzoic or 3,5-dinitrobenzoic acid will precipitate and contaminate the derivative. Therefore, if the melting point of the derivative after one crystallization is several degrees below the expected value, the derivative should be treated with sodium carbonate solution as described in Example 13.01. In some cases better results are obtained if 3,5-dinitrobenzoyl chloride is dissolved in benzene before addition to the amino acid solution.

General Method for the Preparation of p-Toluenesulfonyl Derivatives

One millimole of the amino acid, 3 ml of $1N$ sodium hydroxide solution, and 250 mg of *p*-toluenesulfonyl chloride dissolved in 2 ml of ether are placed in a small bottle with a ground glass stopper, the top of which is flat so that it may be fastened with wire. The stopper is lightly greased, fitted in the bottle, and wired across the top to make a perfectly tight joint. The bottle is shaken, manually for a few seconds every 5 to 10 minutes over a period of 5 hours, or mechanically for 2 to 3 hours. The ethereal layer is separated, and the aqueous layer is acidified to pH 4 to 5, using congo red or Universal indicator; the solution is cooled for 1 hour. The derivative is separated and crystallized from an alcohol-water mixture.

NOTE. The cited article by McChesney and Swann[15a] should be consulted for the preparation of the derivatives of tyrosine, alanine and dicarboxylic acids. The following amino acids yield oils that do not crystallize: glutamic, aspartic, arginine, lysine, tryptophane, and proline.

General Method for the Preparation of Acetyl Derivatives

The addition of the anhydride and alkali in two portions and control of the pH and temperature as described below are necessary to prevent racemization. Two millimoles of the amino acid (about 200 mg) are suspended in 1 ml of water in a 6-in. microtube provided with a solid rubber stopper. The tube is cooled to about 10° and 5.0 millimoles of acetic anhydride are added in two portions over a period of 5 minutes, together with 2.0 ml of $6N$ sodium hydroxide solution, also added in two portions immediately after the anhydride. The tube is shaken vigorously after each addition; the temperature is kept between 10° and 15°, and the pH between 8.0 and 10.0; a drop of Universal indicator is added to the solution before addition of the reagents. The total amount of added alkali should be 12 millimoles. The mixture is shaken at frequent intervals during the introduction of the reagents and for a few minutes longer. It is then acidified to pH 3 to 4 with $6N$ hydrochloric acid and placed in a cold bath for 4 to 5 hours or in a refrigerator overnight. The tube is centrifuged and the supernatant liquid is transferred to another tube; the crystals are washed twice with 0.4 ml of ice water and the washings are combined with the supernate. The crystals are dried at 100°, while the combined supernates are evaporated to dryness by passing warm air over the solution. This residue is extracted with 1 to 2 ml of hot water, and upon cooling an additional amount of the acetyl derivative is obtained. The combined yield is about 1.5 to 1.8 millimoles.

An alternative method employs the sodium salt of the amino acid and 5 to 6 millimoles of alkali.

General Method for the Preparation of 2,4-Dinitrophenyl Derivatives

The reagent used for the preparation of the 2,4-dinitrophenyl derivatives is 2,4-dinitrofluorobenzene, which reacts at room temperature. 2,4-Dinitrochlorobenzene may be used in some cases if the reaction mixture can be heated to boiling.

The use of 2,4-dinitrofluorobenzene is illustrated by the preparation of

[15a] E. W. McChesney and W. K. Siwann, Jr., *J. Am. Chem. Soc.* **59**, 1116–1118 (1937).

the derivative from phenylalanine. A solution of 200 mg of phenylalanine is prepared in 5 ml of water and 400 mg of sodium bicarbonate in a small flask. To this is added a solution of 400 mg (0.28) ml of 2,4-dinitrofluorobenzene in 10 ml of ethanol. The flask is stoppered and shaken at room temperature for 2 hours. The ethanol is removed by concentration under reduced pressure and the aqueous solution is extracted with ether to remove the excess reagent. The mixture is then acidified, and the oil that separates soon solidifies. The derivative is crystallized twice from a methanol-water mixture. The yield is 270 mg of crystals melting at 186°.

NOTE. When 2,4-dinitrochlorobenzene is used, the mixture is refluxed for 4 hours, then treated as described above. When the oily derivatives do not crystallize on standing, they are extracted with chloroform, dried with anhydrous sodium sulfate, and the solvent removed.

α-Naphthylureido Derivatives and Hydantoins of Amino Acids

Discussion

Amino acids react with isocyanates, in the same manner as amines, to form N-substituted ureas:

$$NH_2CHRCOO\overline{} + C_6H_5NCO \rightarrow C_6H_5NHCONHCHRCOO\overline{}$$
Phenylurea derivative

The phenyl- or α-naphthylureido derivatives are prepared by shaking a solution of the amino acid in sodium or potassium hydroxide with phenyl or α-naphthyl isocyanate until the odor of the isocyanate disappears; the derivatives, called hydantoin acids, separate mostly as gelatinous precipitates, which may be converted to crystalline form through careful crystallization from water-alcohol mixtures. The hydantoin acid, when boiled with 10 per cent hydrochloric acid solution for a short time (3 to 5 minutes), undergoes ring closure by elimination of water to form the hydantoin as shown by the following equations.

$$C_6H_5NCO + \underset{\underset{\alpha\text{-Aminobutyric acid}}{HOOC-CHCH_2CH_3}}{H_2N} \rightarrow \underset{HO-\underset{\parallel}{C}-CHCH_2CH_3}{\overset{C_6H_5N-CO-NH}{\overset{|H|}{}}}$$

$$\underset{HO-CO-CHCH_2CH_3}{\overset{C_6H_5N-CO-NH}{\overset{|H}{}}} \rightarrow \underset{COCHCH_2CH_3}{\overset{C_6H_5N-CO-NH}{||}}$$
3-Phenyl-5-ethylhydantoin

The hydantoins are easily crystallized from alcohol solutions and, when their formation is possible, they are preferable to the corresponding hydantoin acid. Reference to Table 4 (page 672) shows that the majority of the substituted ureas and hydantoins are phenyl derivatives. This fact is due to the earlier use of phenyl isocyanate as a reagent for the amino group; since there is less danger of toxicity in the use of α-naphthyl isocyanate, we believe that the α-naphthylurea derivatives and the corresponding hydantoins are more suitable for characterization work.

General Method for the Preparation of α-Naphthylurea Derivatives of Amino Acids

Use an 8-in. tube provided with a solid rubber stopper. Place 3 millimoles of the amino acid dissolved in 3 ml of $1N$ sodium hydroxide solution and 5 ml water. Add 0.6 ml of α-naphthyl isocyanate, stopper the tube, and shake for 2 to 3 minutes; then allow to stand for 30 to 45 minutes with occasional shaking. Filter the insoluble α-naphthylurea and acidify the filtrate to pH 4 to 5 (use congo red or Universal indicator). The α-naphthylhydantoin acid separates out on cooling for an hour. Dissolve the derivative in hot alcohol, filter, and add cautiously a few drops of water until a permanent cloudiness results; then cool.

NOTE. For cystine it is advisable to use very dilute potassium hydroxide (250 mg cystine, 2 ml $1N$ potassium hydroxide, and 20 ml water), since the sodium salt is difficultly soluble.

General Method for the Preparation of Phenylhydantoins of Amino Acids

Dissolve 200 mg of the amino acid in 2 ml of $1N$ potassium hydroxide solution and 3 ml of water. Add 200 mg of phenyl isocyanate. (CAUTION. *Use care since phenyl isocyanate is toxic.*) Stopper the tube with a solid rubber stopper and shake until the odor of isocyanate disappears. Filter with suction and neutralize with dilute hydrochloric acid solution to pH 4 to 5 (use congo red or Universal indicator). A gelatinous precipitate separates on cooling. Filter the precipitate and transfer it by means of a microspatula to an 8-in. tube. Add 5 ml of 10 per cent hydrochloric acid solution in such a manner as to wash the hydantoin acid down into the tube. Boil gently for about 3 minutes and then cool; on standing needles of the hydantoin separate out. Crystallize from alcohol. The yield is 200 to 250 mg.

Salts of Amino Acids

The salts of amino acids are employed mostly for microscopic detection and characterization. Among the polynitro acids used are: picric acid, flavianic acid (1-naphthol-2,4-dinitro-7-sulfonic acid), and picrolonic acid (1-*p*-nitrophenyl-3-methyl-4-nitro-5-pyrazolone). The picrolonates are claimed to have sharper melting points than the picrates. Other salts are the phosphotungstates and phosphomolybdates, the copper salts, and the silver salts.

The picronolates, which are the most useful of all these salts, are prepared by warming a few crystals of the solid amino acid in a drop of water on a microscope slide to produce a concentrated solution; a small drop of picrolonic acid is then added and the slide is cooled; on standing the crystals of the salt separate. For crystallographic data the cited papers by Kirk and co-workers[15b] should be consulted.

Chromatographic Detection of Amino Acids

The number of papers that has been published on the chromatographic detection and estimation of amino acids, peptides, and other hydrolytic products of proteins totals over a thousand and no attempt will be made to discuss them even cursorily. The reader is referred to recent works on chromatography cited at the end of Chapter 4 and in the References of this chapter. The present section deals with a simple strip chromatography of amino acids in protein hydrolysates. For rapid micro methods of hydrolyzing 25 to 50 mg of protein and more details on the detection of amino acids, see a work by one of us.[16] For practice with known amino acids, consult page 130.

For the determination of a single amino acid (for example, cystine or methionine) the test tube method is the simplest. Either 6- or 8-in. tubes are thoroughly cleaned and dried. The strips of paper are prepared as described on page 126. About 0.5 ml of an 80:20 *tert*-butyl alcohol-water mixture is added to each of six tubes on a rack by a capillary pipet without wetting the walls of the tube; the tubes are then stoppered. Six strips are prepared by placing 5, 10, and 15 λ, respectively, of the hydrolyzate on three strips and the same amount of a standard on the last three strips. The application of the sample is made according to the directions given on page 123. Two types of standards have been used by the authors. One is a hydrolyzate prepared in the manner described from a sample of casein, the amino acid content of which has been determined microbiologically. The other type of standard is a synthetic amino acid mixture. The

---

[15b] P. L. Kirk et al., *Mikrochemie*, **18**, 137 (1935); **21**, 245 (1936); *Ind. Eng. Chem., Anal. Ed.*, **13**, 587 (1941).

[16] N. D. Cheronis, "Micro and Semimicro Methods," in *Technique of Organic Chemistry*, Vol. 6, A. Weissberger, ed., Interscience, New York, 1954, pp. 526–529.

amount of standard solution placed on the three strips is such that the amount of cystine or methionine is in the range of 10, 20, and 40 micrograms. The amount of sample applied is marked lightly with pencil on the upper part of each strip; they are then placed in the tubes in such a manner that the sample spot is 15 to 20 mm above the surface of the liquid. The tubes are stoppered and the rack is allowed to stand at room temperature, preferably away from air currents and light. After the solvent front has advanced almost to the top (1 to 3 hours), the strips are removed and hung by clips on a wire in the hood, where they are left for several hours until all the solvent has evaporated. The strips are sprayed with a 0.2 per cent solution of ninhydrin in 95 per cent ethanol. For methionine and other amino acids containing sulfur, the strips are painted with a solution of 0.1 per cent iodoplatinate reagent prepared by mixing equimolar quantities of chloroplatinic acid and potassium iodide.

ACID HALIDES

Consult the text on carboxylic acids (pages 437–450) if the substance under investigation has been tentatively identified as an acid halide. From the discussion of the derivatization of carboxylic acids it will be seen that, since carboxylic acids are usually converted to the acid chlorides in order to prepare amides, anilides, *p*-bromoanilides, and toluidides, these same derivatives may be used to advantage for the characterization of acid halides. About 100 to 200 mg or less of the acyl halide are dissolved in 10 ml of dry benzene and 500 mg of the arylamine (aniline, *p*-toluidine, *p*-bromoaniline) is added; the mixture is refluxed for 10 minutes and then treated as described in Example 12.05.

Acid chlorides react with alcohols or phenols to form esters. If the ester is solid, it may be used as a useful derivative for identification (see page 470). If the hydrolysis of the chloride gives an acid that is solid and only slightly soluble in water, it may be used for identification. In such cases, 100 to 200 mg of acid chloride are boiled with 5 ml of 5 per cent sodium carbonate solution for 20 minutes. If the acyl halide is reactive, the reflux time may be shortened to a few minutes. The solution is cooled, extracted with 5 ml of ether, and the aqueous layer is separated and acidified with dilute sulfuric acid to liberate the carboxylic acid.

The selection of the appropriate derivative for acyl halides should be based on the solubility and the melting point of the derivative. For example, assume that tests of the compound under investigation indicate that it is a nitrobenzoyl chloride. The anilides and *p*-toluidides of the nitrobenzoic acids melt above 200°, whereas their esters with the lower alcohols melt between 70 to 150°; in this case it will be more appropriate to react the acid chloride with methanol and prepare the ester as directed on page 470, rather than to prepare any of the substituted amides.

ACID ANHYDRIDES

The discussion given for the acyl halides (page 458) applies with few modifications to the acid anhydrides. The anhydrides most commonly encountered are: acetic, succinic, maleic, and phthalic. Hydrolysis of the anhydrides yields acids; reaction with amines yields amides or substituted amides. Anhydrides of dibasic acids, on reacting with arylamines, may give the monoamide or the imide:

$$\underset{\substack{\text{}}}{\text{C}_6\text{H}_4(\text{CO})_2\text{O}} \xrightarrow{\underset{p\text{-Toluidine}}{\text{CH}_3\text{C}_6\text{H}_4\text{NH}_2}} \begin{cases} \text{C}_6\text{H}_4(\text{COOH})(\text{CONHC}_6\text{H}_4\text{CH}_3) \\ \text{Phthalic mono-}p\text{-toluidide (mp 160°)} \\ \\ \text{C}_6\text{H}_4(\text{CO})_2\text{NC}_6\text{H}_4\text{CH}_3 \\ \text{N-}p\text{-Tolylphthalimide mp 204°} \end{cases}$$

In reacting with alcohols, the anhydrides of dibasic acids yield the acid esters; thus phthalic anhydride and 3-nitrophthalic anhydride yield, with alcohols, sparingly soluble acid esters that may be used as derivatives (see page 473).

The preparation of aceto-*p*-toluidide illustrates the derivatization of an acid anhydride.

Example 12.10: Preparation of Aceto-p-toluidide

Place in a test tube 100 mg of acetic anhydride and 50 mg of *p*-toluidine. Join a reflux condenser to the tube and heat over a very small flame for 10 minutes so that the mixture just boils; when the mixture has cooled, add 2 ml of methanol and heat until the solid has disintegrated. Add 6 ml of water and filter the crude toluidide. Wash with water and recrystallize from 50 per cent methanol (page 474). The yield is 40 to 50 mg, melting at 147°.

REFERENCES*

Identification of Carboxylic Acids

Amides and Substituted Amides

N-Acyl-*p*-aminoazobenzenes: Escher, *Helv. Chim. Acta*, **12**, 27 (1929).
Amides: Assano, *J. Pharm. Soc. Japan*, **480**, 97 (1929); DeConno, *Gazz. chim. ital.*,

* The references to this and the following chapters have been arranged according to subject matter. Authors' initials have been omitted in order to increase the ease of reading.

47, I, 93 (1917); Mitchell and Reid, *J. Am. Chem. Soc.,* **53,** 1879 (1931); Swera et al., *J. Am. Chem. Soc.,* **71,** 2215, 3017 (1949).

Anilides: Barnicoat, *J. Chem. Soc. (London),* **1927,** 2927; Blodinger and Anderson, *J. Am. Chem. Soc.,* **74,** 5514 (1952); Carre and Libermann, *Compt. rend.,* **194,** 2218 (1932); *Bull. Soc. Chim.,* **53,** 293 (1953); DeConno, *Gazz. chim. ital.,* **47,** I, 93 (1917); Hann and Jamieson, *J. Am. Chem. Soc.,* **50,** 1442 (1928); Hardy, *J. Chem. Soc. (London),* **1936,** 398; Robertson, *J. Chem. Soc.,* **93,** 1033 (1908); **115,** 1210 (1919); Shah and Deshpande, *J. Univ. Bombay* (2 *Pt.*), **2,** 125 (1933); *C.A.,* **28,** 6127 (1934).

Benzidine diamides: Buu-Hoi, *Bull. soc. chim.,* **12,** 587 (1945).

N-Benzylamides: Dermer and King, *J. Org. Chem.,* **8,** 168 (1943); Stafford et al., *Anal. Chem.,* **21,** 1454 (1949).

p-Bromoanilides: Bryant, *J. Am. Chem. Soc.,* **60,** 1394 (1938); Bryant and Mitchell, *J. Am. Chem. Soc.,* **60,** 2748 (1938); Houston, *J. Am. Chem. Soc.,* **62,** 1303 (1940); Kuehn and McElvain, *J. Am. Chem. Soc.,* **53,** 1173 (1931); Robertson, *J. Chem. Soc.,* **93,** 1033 (1908); **115,** 1210 (1919).

α-Bromo-β-naphthylamides: Robertson, *J. Chem. Soc.,* **93,** 1033 (1908); **115,** 1210 (1919).

o-Bromo-*p*-toluidides: Robertson, *J. Chem. Soc.,* **93,** 1033 (1908); **115,** 1210 (1919).

Diamides of *p,p'*-diaminodiphenylmethane: Ralston and McCorkle, *J. Am. Chem. Soc.,* **61,** 1604 (1935).

N-Diethanolamides: D'Alelio and Reid, *J. Am. Chem. Soc.,* **59,** 109 (1937).

Dimethylamides: Kirsavov and Zolotov, *Zhur. Obshchei Khim.* (*J. Gen. Chem.*) *U.S.S.R.,* **21,** 1166 (1951); Mitchell and Reid, *J. Am. Chem. Soc.,* **53,** 1879 (1931); Prelog, *Collection Czechoslov. Chem. Commun.,* **2,** 712 (1930).

Dodecylamides, octadecylamides: Hunter, *Iowa State Coll. J. Sci.,* **15,** 223 (1941).

p-Hydroxyanilides: DeConno, *Gazz. chim. ital.,* **47,** I, 93 (1917).

2-Methyl-5-isopropylanilides: Hann and Jamieson, *J. Am. Chem. Soc.,* **50,** 1442 (1928).

α- and β-Naphthylamides: DeConno, *Gazz. chim. ital.,* **47,** I, 93 (1917); Robertson, *J. Chem. Soc.,* **93,** 1033 (1908); **115,** 1210 (1919); Shah and Deshpande, *J. Univ. Bombay* (2 *Pt.*), **2,** 125 (1933); *C.A.,* **28,** 6127 (1934).

o-, *m*-, and *p*-Nitroanilides: Shah and Deshpande, *J. Univ. Bombay* (2 *Pt.*), **2,** 125 (1933); *C.A.,* **28,** 6127 (1934).

o- and *p*-Phenetides: DeConno, *Gazz. chim. ital.,* **47,** I, 93 (1917).

o-Phenyldiamides: Shah and Deshpande, *J. Univ. Bombay* (2 *Pt.*), **2,** 125 (1933); *C.A.,* **28,** 6127 (1934).

Thiophenylamides: Shah and Deshpande, *J. Univ. Bombay* (2 *Pt.*), **2,** 125 (1933); *C.A.,* **28,** 6127 (1934).

o- and *p*-Toluidides: DeConno, *Gazz. chim. ital.,* **47,** I, 93 (1917); Robertson, *J. Chem. Soc.,* **93,** 1033 (1908); **115,** 1210 (1919).

2,4,6-Tribromoanilides: Robertson, *J. Chem. Soc.,* **93,** 1033 (1908); **115,** 1210 (1919).

p-Xenylamides: Kimura and Nihayashi, *Ber.,* **68B,** 2028 (1935).

m-Xylidides: DeConno, *Gazz. chim. ital.,* **47,** I, 93 (1917).

Esters

Acid esters of dibasic acids: Cazeneuve, *Bull. Soc. Chim.,* **9,** 90 (1893); Veibel and Lillelund, *Acta Chem. Scand.,* **8,** 1954; Walker, *J. Chem. Soc.,* **61,** 1088 (1892).

Lactone derivatives: Meyer, *Monatsh.*, **20**, 717 (1899).
Methyl esters using diazomethane: Herzig and Wenzel, *Monatsh.*, **229**, 22 (1901).
Menthyl esters: Brauns, *J. Am. Chem. Soc.*, **42**, 1478 (1920); Tchugaeff, *Ber.*, **31**, 360 (1898).
p-Nitrobenzyl esters: Blike and Smith, *J. Am. Chem. Soc.*, **51**, 1947 (1929); Kelly and Segura, *J. Am. Chem. Soc.*, **56**, 2497 (1934); Reid et al., *J. Am. Chem. Soc.*, **39**, 124, 701, 1727 (1917); **43**, 629 (1921).
Phenacyl, *p*-chlorophenacyl, and *p*-bromophenacyl esters: Chen and Shih, *Trans. Sci. Soc. China*, **7**, 81 (1931); Erickson et al., *J. Am. Chem. Soc.*, **73**, 5301 (1951); Hann et al., *J. Am. Chem. Soc.*, **52**, 818 (1930); Harmon and Marvel, *J. Am. Chem. Soc.*, **54**, 2515 (1932); Kelly et al., *J. Am. Chem. Soc.*, **54**, 4444 (1932); Kimura, *J. Soc. Chem. Ind. Japan*, **35**, Sb, 221 (1932); Lundquist, *J. Am. Chem. Soc.*, **60**, 2000 (1938); Price and Griffith, *J. Am. Chem. Soc.*, **64**, 2884 (1942); Reid et al., *J. Am. Chem. Soc.*, **41**, 4175 (1919); **42**, 1043 (1920); **52**, 818 (1930); **53**, 1172 (1931); **54**, 2101 (1932); Wilson, *J. Am. Chem. Soc.*, **67**, 2161 (1945).
4-Phenylazophenacyl esters: Masuyama, *J. Chem. Soc. Japan, Pure Chem. Sect.*, **71**, 402 (1950); Mowrey and Frode, *J. Am. Chem. Soc.*, **63**, 2281 (1941).
p-Phenylphenacyl esters: Drake et al., *J. Am. Chem. Soc.*, **52**, 3717 (1930); **54**, 2059 (1932); **58**, 1502 (1936); Erickson et al., *J. Am. Chem. Soc.*, **73**, 5301 (1951); Ford, *Iowa State Coll. J. Sci.*, **52**, 818 (1930); Kass et al., *J. Am. Chem. Soc.*, **64**, 1061 (1942); Kelly et al., *J. Am. Chem. Soc.*, **58**, 1502 (1936).
2,4-Dinitrophenylhydrazones of the *p*-phenylphenacyl esters: N. Hawkins, A. D. Webb, and R. E. Kopner, *Anal. Chem.*, **28**, 1975–1977 (1956).

Salts and Other Derivatives

N-Acyl-2-acylcarbazoles and 2,8-diacylcarbazoles: Ford, *Iowa State Coll. J. Sci.*, **12**, 121 (1937); Gilman and Ford, *Iowa State Coll. J. Sci.*, **13**, 135 (1939).
2-Alkylbenzimidazoles: Pool, Harwood, and Ralston, *J. Am. Chem. Soc.*, **59**, 178 (1937); Seka and Mueller, *Monatsh.*, **57**, 95 (1931).
2-Alkylbenzimidazole picrates: Brown and Campbell, *J. Chem. Soc. (London)*, **1937**, 1699.
N-Benzylammonium and α-phenylethylammonium salts: Boudet, *Bull. soc. chim.*, **390**, (1948); Buehler, Carson, and Edds, *J. Am. Chem. Soc.*, **57**, 2181 (1935).
S-Benzylthiuronium salts: Bolliger, *Helv. Chim. Acta*, **34**, 916 (1951); Chambers and Sherer, *Ind. Eng. Chem.*, **16**, 1272 (1924); Donleavy, *J. Am. Chem. Soc.*, **58**, 1004 (1936); Friediger and Pedersen, *Acta Chem. Scand.*, **9**, 1425 (1955); Kass et al., *J. Am. Chem. Soc.*, **64**, 1061 (1942); *Am. J. Biol. Chem.*, **192**, 301 (1951); Viebel and Lillelund, *Bull. soc. chim.* (5), **5**, 1153 (1938); Veibel and Ottung, *Bull. soc. chim.* (5), **6**, 1434 (1939).
S-Benzylthiuronium salts (melting points): Berger, *Acta Chem. Scand.*, **8**, 427 (1954); Walker, *J. Chem. Soc. (London)*, **1949**, 1999.
S-(*p*-Bromobenzyl)- and S-(*p*-chlorobenzyl)thiuronium salts: Dewey and Shasky, *J. Am. Chem. Soc.*, **63**, 3526 (1941); Dewey and Sperry, *J. Am. Chem. Soc.*, **61**, 3251 (1939).
Bis(*p*-dimethylaminophenyl)ureides: Breusch and Ulusoy, *Arch. Biochem.*, **11**, 489 (1946).

2,4-Dinitrophenylhydrazides: Cherezo and Olay, *Anales soc. espan. fis. y quim.*, **32**, 1090 (1934); Gilman and Ford, *Iowa State Coll. J. Sci.*, **13**, 135 (1939).

Hydrazides: Hanus and Vorishek, *Collection Czech. Chem. Commun.*, **1**, 223 (1929); Kyame et al., *J. Am. Oil Chemists' Soc.*, **24**, 332 (1947); Pajari, *Fette u. Seifen*, **51**, 347 (1944); Paschke and Wheeler, *J. Oil Chemists' Soc.*, **26**, 637 (1949); Sah, *Rec. trav. chim.*, **59**, 1046 (1940) (in English).

Hydrazides and phenylhydrazonium salts: Stempel and Schaffel, *J. Am. Chem. Soc.*, **64**, 470 (1942).

Hydroxamic acids: Inoue, Noda et al., *J. Agr. Chem. Soc. Japan*, **23**, 294, 368 (1950); Inoue and Yukawa, *J. Agr. Chem. Soc. Japan*, **16**, 504, 510 (1940); **17**, 411, 491, 771 (1941); **18**, 415, 875 (1942); *Bull. Agr. Chem. Soc. Japan*, **16**, 100 (1940); **17**, 44, 89 (1941) (in English); **17**, 59 (1941); **18**, 33, 72 (1942) (in English).

Monoureides and monothioureides: Jacobson, *J. Am. Chem. Soc.*, **58**, 1984 (1936); Stendahl, *Compt. rend.*, **196**, 1810 (1933).

S-(α-Naphthylmethyl)thiuronium salts: Bonner, *J. Am. Chem. Soc.*, **70**, 3508 (1948).

S-(*p*-Nitrobenzyl)thiuronium salts: Rupe and Zweidler, *Helv. Chim. Acta*, **23**, 1025 (1940).

Octadecyl- and dodecylammonium salts: Hunter, *Iowa State Coll. J. Sci.*, **15**, 223 (1941).

Phenylhydrazides: Brauns, *J. Am. Chem. Soc.*, **42**, 1478 (1920); Cheronis and Cohn, *Mikrochim. Acta*, **1956**, 925; Shah and Deshpande, *J. Univ. Bombay* (2 Pt), **2**, 125 (1933); Van Alpen, *Rec. trav. chim.*, **44**, 1064 (1925); Veseky and Haas, *Chem. Listy*, **21**, 351 (1927).

Phenylmercuric and *p*-tolylmercuric salts: Ford, *Iowa State Coll. J. Sci.*, **12**, 121 (1937); Gilman and Ford, *Iowa State Coll. J. Sci.*, **13**, 135 (1939).

Piperazonium salts: Pollard and Adelson, *J. Am. Chem. Soc.*, **56**, 1759 (1934).

Tetraphenylstilbonium salts: H. E. Affsprung and H. E. May, *Anal. Chem.*, **32**, 1164–1166 (1960).

Triphenyl lead salts: Ford, *Iowa State Coll. J. Sci.*, **12**, 121 (1937); Gilman and Ford, *Iowa State Coll. J. Sci.*, **13**, 135 (1939).

Ureides: Schmidt et al., *Ber.*, **71**, 1933 (1938); **73**, 286 (1940); Zetzsche et al., *Ber.*, **71B**, 1088, 1516, 2095 (1938); **72**, 1599 (1939); **72B**, 1735, 2095 (1939); **73B**, 465, 1114 (1940); **74B**, 183 (1941); **75B**, 100 (1942).

Chromatographic Procedures

Airan et al., *Anal. Chem.*, **25**, 659 (1953). For dicarboxylic acids.

Asselineau, *Bull. Soc. Chim. France*, **1952**, 884.

Boldingh, *Rec. Trav. Chem.*, **69**, 247 (1950). Hydroxamic acids and hydrazides.

Brown, *Biochem. J.*, **47**, 598 (1950); *Nature*, **166**, 66 (1950).

Buch et al., *Anal. Chem.*, **24**, 489 (1952).

Cheronis and Cohn, *Mikrochim. Acta*, **1956**, 925.

Denison and Phares, *Anal. Chem.*, **24**, 1628 (1952).

Fink and Fink, *Proc. Soc. Exptl. Biol. Med.*, **70**, 654 (1949). Hydroxamic acids and hydrazides.

Hiscox and Berridge, *Nature*, **166**, 522 (1950).

Italman, *Progress in the Chemistry of Fats*, Vol. I, Academic Press, New York 1952, p. 104. Review.

Kennedy and Barker, *Anal. Chem.*, **23**, 1033 (1951).
Lugg and Overell, *Australian J. Sci. Res.*, **1**, 98 (1948).
Reid and Lederer, *Biochem. J.*, **50**, 60 (1951).
Sataki and Seki, *J. Jap. Chem.*, **4**, 557 (1950). Hydroxamic acids and hydrazides.
Thompson, *Australian J. Sci. Res.*, **B4**, 180 (1951). Hydroxamic acids and hydrazides.

Microscopical Identification

Deniges, *Bull. Trav. Soc. Pharm. Bordeaux*, **76**, 173 (1938). Microcrystalloscopic identification of organic acids.
Feigl et al., *Mikrochim. Acta*, **1**, 127 (1937). Microchemical detection of organic compounds.
Steenhauer, *Pharmal. Weekblt.*, **72**, 667 (1935). Photomicrographic characterization of carboxylic acids by means of Zwikker's reagent.
Whitmore and Croons, *J. Am. Chem. Soc.*, **60**, 2078 (1938). Differentiation between primary, secondary, and tertiary aliphatic carboxylic acids.

Identification of Amino Acids

N-Acyl and Aroyl Derivatives

Acetyl and formyl derivatives: Bergman and Zervas, *Biochem. Z.*, **203**, 288 (1928); Chattaway, *J. Chem. Soc. (London)*, **1931**, 2405; Curtius, *Ber.*, **17**, 1665 (1884); Fischer, *Ber.*, **39**, 2330 (1906); Herbst and Shemin, *Organic Syntheses*, Vol. 19, Wiley, New York, 1939, p. 4; Kraut and Hartmann, *Ann.*, **133**, 105 (1865).

Benzoyl derivatives: Fischer, *Ber.*, **32**, 2451 (1899); Fischer and Bergell, *Ber.*, **35**, 3779, 3784 (1902); **39**, 597 (1906).

Carbobenzoxy derivatives: Bergman, Zervas, and Ross, *J. Biol. Chem.*, **111**, 245 (1945); Newberger and Sanger, *Biochem. J.*, **37**, 515 (1943).

3,5-Dinitrobenzoyl derivatives: Saunders, *Biochem. J.*, **28**, 580 (1934); *J. Chem. Soc. (London)*, **1938**, 1397; Saunders et al., *Biochem. J.*, **36**, 368 (1942); Town, *Biochem. J.*, **35**, 578 (1941).

2,4-Dinitrophenyl derivatives: Sanger, *Biochem. J.*, **39**, 507 (1945).

β-Naphthalenesulfonyl derivatives: Bergman and Stein, *J. Biol. Chem.*, **129**, 609 (1939); Fischer and Bergell, *Ber.*, **35**, 3779, 3784 (1902); **39**, 597 (1906).

p-Nitrobenzyloxycarbonyl derivatives: Gish and Carpenter, *J. Am. Chem. Soc.*, **75**, 950 (1953).

Phthaloyl derivatives: Billman and Hartig, *J. Am. Chem. Soc.*, **70**, 1473 (1948).

p-Toluenesulfonyl derivatives: Fischer and Bergell, *Ber.*, **35**, 3779, 3784 (1902); **39**, 597 (1906); McChesney and Swann, *J. Am. Chem. Soc.*, **59**, 1116 (1937).

α-Naphthylureido Derivatives and Hydantoins

α-Naphthylureas: Neuberg and Rosenberg, *Biol. Z.*, **5**, 456 (1907).
Phenylureas and phenylhydantoins: Patten, *Z. physiol. Chem.*, **39**, 350 (1903).

Salts of Amino Acids

Copper salts: Cunningham, McIntyre, and Kirk, *Mikrochemie*, **21**, 245 (1936).
Phosphotungstates and phosphomolybdates: Bullock and Kirk, *Mikrochemie*, **18**, 129 (1935).

Picrates: Kirk et al., *Mikrochemie*, **18**, 137 (1935); **21**, 245 (1936); *Ind. Eng. Chem., Anal. Ed.*, **13**, 587 (1941); Levene, *J. Biol. Chem.*, **1**, 413 (1906).

Picrates and picronolates: Levene and Van Slyke, *J. Biol. Chem.*, **12**, 127, 285 (1912).

Others: Gish and Carpenter, *J. Am. Chem. Soc.*, **75**, 950 (1953); Larsen, *Mikrochemie*, **34**, 351 (1949).

Chromatographic Detection

Block, LeStrange, and Zweig, *Paper Chromatography*, Academic Press, New York, 1952.

Boissonnas, *Helv. Chim. Acta*, **33**, 1966 (1950).

Cassidy, *Adsorption and Chromatography in Technique of Organic Chemistry*, Vol. 5, A. Weissberger, ed., Interscience, New York, 1951.

Consden et al., *Biochem. J.*, **38**, 224 (1944); **40**, 590 (1947); **46**, 8 (1950).

Cifonelli and Smith, *Anal. Chem.*, **27**, 1501 (1955). Detection of amino acids on paper chromatograms.

Decker and Riffart, *Chem.-Ztg.*, **74**, 261 (1950). R_f values.

Dent, *Biochem. J.*, **43**, 169 (1948).

Lederer and Lederer, *Chromatographic Separation of Amino Acids*, Elsevier, New York, 1953.

Martin and Synge, *Biochem. J.*, **35**, 1358 (1941).

Peck and Gale, *Anal. Chem.*, **24**, 118 (1952).

Rockland et al., *Anal. Chem.*, **23**, 1142 (1951); *Science*, **109**, 539 (1949).

Underwood and Rockland, *Anal. Chem.*, **26**, 1553 (1954).

13

Derivatives of Alcohols and Phenols (Monohydric and Polyhydric)

Table 13.01 summarizes the most important derivatives that have been described in the literature for the derivatization of compounds having the hydroxyl group. These compounds include alcohols, glycols, polyhydroxy compounds, and phenols. The last-named, however, although they are derivatized by many of the reagents described in this section, are discussed separately, since the presence of aromatic structures imparts a number of additional reaction properties by which they may be derivatized.

The derivatives marked with an asterisk in Table 13.01 are those that are recommended for a first trial. It will be noted that the preparation of suitable derivatives for alcohols is based to a large extent on the following two reactions: (a) formation of sparingly soluble esters with aromatic acids, preferably nitro aromatic acid chlorides:

$$\underset{\text{Alcohol}}{\text{ROH}} + \underset{\substack{\text{Nitroaromatic}\\\text{acid chloride}}}{\text{Ar(NO}_2)_x\text{COCl}} \rightarrow \underset{\substack{\text{Nitroaromatic}\\\text{ester}}}{\text{Ar(NO}_2)_x\text{COOR}} + \text{HCl}$$

and (b) reaction of the hydroxyl compound with an isocyanate to form a urethan, which can be called also a carbamate or an ester of carbamic acid, NH_2COOR. Thus, when ethanol reacts with α-naphthyl isocyanate, the product can be called either a urethan or a derivative of carbamic acid:

$$C_2H_5OH + C_{10}H_7NCO \rightarrow C_{10}H_7NHCOOC_2H_5$$

Ethanol α-Naphthyl Ethyl α-naphthylurethan
isocyanate or
Ethyl N-α-naphthylcarbamate
or
Ethyl N-1-naphthalenecarbamate

The discussion in the following section deals with: (a) 3,5-dinitrobenzoates and p-nitrobenzoates; (b) 3-nitrophthalates; (c) α-naphthylurethans; (d) phenylurethans and substituted phenylurethans; (e) derivatives for glycols and polyhydric alcohols; and (f) other derivatives.

TABLE 13.01
DERIVATIVES FOR THE IDENTIFICATION OF ALCOHOLS, GLYCOLS, AND POLYHYDROXY COMPOUNDS

Acetates	*α-Naphthylurethans
Allophanates	*p-Nitrobenzoates
p-Anisylurethans	m-Nitrophenylurethans
Aryloxyacetic acids	o-Nitrophenylurethans
*Benzoates	p-Nitrophenylurethans
S-Benzylthiuronium derivatives	*3-Nitrophthalates
β-Bromopropionylcarbamates	p-Nitrophenylacetates
p-Chlorophenylurethans	p-Phenylazobenzoates
*3,5-Dinitrobenzoates	4-Phenylazophenylurethans
3,5-Dinitro-4-methylphenylurethans	Phenylurethans
2,4-Dinitrophenylurethans	Pseudosaccharin ethers
2,4-Dinitrophenyl ethers	Tetrachlorophthalates
2,4-Dinitrophenylhydrazones (by oxidation to aldehydes)	Trityl ethers
	Xanthates
Hydrogen phthalates	p-Xenylurethans
p-Iodophenylurethans	

* Recommended derivative. Literature references to all derivatives listed in this table are given on pages 492–495.

The two derivatives recommended for small quantities are the *3,5-dinitrobenzoates* and *α-naphthylurethans*. The reagents (3,5-dinitrobenzoyl chloride and α-naphthyl isocyanate) for the preparation of these esters are commercially available. If it is necessary to use other substituted esters of aromatic acids, the p-nitrobenzoates and 3-nitrophthalates should be tried first. For the preparation of other urethans the commercially available isocyanates—p-bromophenyl, p-nitrophenyl, and β-naphthyl—should be tried first. Phenyl isocyanate is not recommended, since it reacts slowly and the urethans that are formed are somewhat soluble. If it is necessary to prepare a substituted urethan for which the isocyanate is not commercially available, the substituted carbamyl chloride or the substituted azide should be prepared according to the directions given in the articles cited in the selected references for this chapter. For polyhydroxy compounds, benzoates and *p*-nitrobenzoates and other derivatives are suitable, and these are discussed in the section *Other Derivatives* (page 480).

3,5-Dinitrobenzoates and *p*-Nitrobenzoates

Discussion

The following equations illustrate the formation of *p*-nitrobenzoates and 3,5-dinitrobenzoates:

$$NO_2C_6H_4COCl + CH_3OH \rightarrow NO_2C_6H_4COOCH_3 + HCl$$
p-Nitrobenzoyl chloride — Methyl *p*-nitrobenzoate

$$(NO_2)_2C_6H_3COCl + C_2H_5OH \rightarrow (NO_2)_2C_6H_3COOC_2H_5 + HCl$$
3,5-Dinitrobenzoyl chloride — Ethyl 3,5-dinitrobenzoate

The alcohol is heated for several minutes with the acid chloride in a dry vessel and then water is added to separate the solid ester, which is filtered and purified. In many macro procedures an excess of alcohol is employed in order to convert all of the acid chloride to the ester. Therefore, it is assumed that no chloride will be left to form nitrobenzoic acid when water is added after the reaction is complete. It has been shown[1] that a large excess of alcohol heated with 3,5-dinitrobenzoyl chloride does not prevent the formation of impurities, which are essentially 3,5-dinitrobenzoic acid and its anhydride. Therefore, only a small excess of alcohol is employed and the product is thoroughly pulverized and extracted with dilute sodium carbonate solution to remove the acid and anhydride formed; the solid is recrystallized. The *p*-nitrobenzoates are formed and purified in the same manner. When the amount of hydroxy compound is less than 100 mg, it is advisable to heat it with the acid chloride in the presence of an inert solvent such as isopropyl ether, dry benzene, or pyridine, as described in Example 13.02. In general, this method should be used whenever the rate of formation of the ester is slow, as in the case of tertiary alcohols. For example, when 30 mg each of *n*-butyl, isobutyl, *sec*-butyl, and *tert*-butyl alcohols and 50 mg of 3,5-dinitrobenzoyl chloride were heated for 30 minutes, the yields of the dinitrobenzoates in milligrams were, respectively, 16, 20, 18, and 0.5. It is obvious that the yield of the ester from the tertiary alcohol is insufficient for the ordinary methods of purification and melting-point

[1] N. D. Cheronis and A. Vavoulis, *Mikrochemie*, **38**, 428 (1952); N. D. Cheronis, "Micro and Semimicro Methods," in *Technique of Organic Chemistry*, Vol. 6, A. Weissberger, ed., Interscience, New York, 1954, p. 483.

determination. In such cases it is advisable to boil the alcohol for an hour or two with the dinitrobenzoyl chloride and pyridine dissolved in isopropyl ether (Example 13.02). In the absence of a proton acceptor there is a tendency for tertiary alcohols to form the halide, and an olefin.

The 3,5-dinitrobenzoates form crystalline complexes with aromatic amines. The complexes with α-naphthylamine are particularly useful because they have characteristic colors from orange to red according to the radical in the alcohol. Further, they form well-defined crystals, having definite melting points and may be used to purify those 3,5-dinitrobenzoates that are not easily crystallized.

When the 2,4,6-trinitrobenzoates of alcohols are mixed with naphthalene or phenanthrene and heated carefully, addition compounds are formed. It is possible to determine the melting points of the two eutectic mixtures, the ester, and the molecular addition compound on a microscope hot stage. From these data the identity of the starting compound can be ascertained.[2] Data for 29 alcohols are given.

Although the presence of water is a disadvantage in the preparation of *p*-nitrobenzoates, 3,5-dinitrobenzoates, and other nitro aryl esters, it is possible to prepare these derivatives from dilute aqueous solutions of alcohols. The acid chloride is dissolved in purified ligroin and then is shaken with a 5 per cent aqueous solution of the alcohol in the presence of sodium acetate. It is possible by this method to prepare derivatives from 500 mg of alcohol when it is admixed with 10 ml of water.

Chemical microscopy may be applied to the identification of alcohols and phenols by means of their 3,5-dinitrobenzoates.[3] There is a total of 40 alcohols, glycols, phenols, and other hydroxy compounds, which have been converted to their 3,5-dinitrobenzoates, and microphotographs made of the resulting crystals. Mixtures of alcohols give two types of crystals which are clearly distinguishable.

General Method for Preparation

If the alcohol has less than 5 per cent water, 1.5 to 2 millimoles are heated in a small test tube with 1 millimole of the reagent. Both *p*-nitrobenzoyl chloride and 3,5-dinitrobenzoyl chloride are susceptible to hydrolysis during storage by atmospheric moisture. Their melting points should be checked;[4] if the melting point is more than 1 to 2° lower than

---

[2] D. E. Laskowski and O. W. Adams, *Anal. Chem.* **31**, 148–152 (1959).
[3] R. E. Dunbar and F. J. Ferrin, *Microchem. J.*, **3**, 65–82 (1959).
[4] In derivatizing a few milligrams of alcohol, check the purity of the reagent. See N. D. Cheronis, "Micro and Semimicro Methods" in *Technique of Organic Chemistry*, Vol. 6, A. Weissberger, ed., Interscience, New York, 1954, p. 481.

that recorded in the literature, the acid chloride should be crystallized from carbon tetrachloride. The preparation of the acid chlorides is described in Example 13.05.

The mixture of alcohol and reagent is heated cautiously over a small flame at the lowest temperature at which the mixture remains liquid. The lower alcohols are heated for 3 to 5 minutes and the higher alcohols for 10 to 15 minutes. The melt is allowed to solidify, after which the crystalline mass is broken up thoroughly with a microspatula; it is then shaken with 2 per cent sodium carbonate at 50 to 60° for about 10 to 30 seconds and filtered. Prolonged treatment with sodium carbonate solution tends to hydrolyze the ester. The crystalline mass is then washed thoroughly with water and crystallized by dissolving in hot methanol or ethanol and adding water to cloudiness. Usually only one crystallization is required. However, where the extraction with sodium carbonate is faulty, two or three crystallizations may be necessary. From 1 millimole of the reagent the yield of pure derivative after one crystallization is 150 to 200 mg.

When the alcohol is not reactive or when a polyhydric alcohol is derivatized, the use of a solvent is advantageous. The procedure is described in one of the examples given in this section. Isopropyl ether or n-butyl ether may be used as the solvent. The mixture of reagent and alcohol is refluxed for 0.5 to 1 hour and then washed with sodium carbonate solution; the ethereal layer is separated and dried; then the solvent is evaporated to obtain the crude ester. This method, which entails several transfers, cannot normally be employed for quantities less than 50 mg. With tertiary alcohols the pyridine alone is employed as the solvent. One millimole of reagent is mixed with 1.5 millimoles of the alcohol in a tube containing 2 ml of pyridine. The mixture is refluxed 0.5 to 1 hour, cooled, and extracted with 5 ml of 1 per cent sulfuric acid, which removes the pyridine and separates a crystalline mass of the crude ester.

When the alcohol contains water, it is diluted so that 250 to 500 mg are contained in a volume of about 5 to 6 ml. The mixture is cooled to 0° and then shaken with a solution of 500 mg of the acid chloride in 2 ml of specially purified hexane (ligroin or petroleum ether) and 3 ml of benzene. The hexane is specially purified by washing, first, with concentrated sulfuric acid and then with water, drying with anhydrous calcium chloride or calcium sulfate, and distilling. The mixture is well shaken, sodium acetate is added, and the mixture is kept below 5° with frequent shaking for 15 to 30 minutes. Alcohol-free ether is then added, the upper layer separated and washed, first with dilute sodium hydroxide, then with dilute hydrochloric acid, then with water. The solvent is then evaporated.

The following examples illustrate the preparation of several 3,5-dinitrobenzoates and *p*-nitrobenzoates. If the derivatizing reagent is not available, it is prepared according to the directions given in Example 13.05.

Example 13.01: Derivatization of Ethanol: Ethyl 3,5-Dinitrobenzoate

Place in a 6-in. tube 400 mg of pure 3,5-dinitrobenzoyl chloride and 0.12 ml of 95 per cent ethanol. Heat for about 5 minutes by means of a microflame so that the melt at the bottom of the tube does not solidify. Avoid a hot flame. If there is much evidence of condensation on the sides of the tube 20 to 30 mm above the reaction mixture, the flame should be reduced or the tube raised.

Allow the melt to solidify. By means of a glass rod or a microspatula break and thoroughly pulverize the crystalline mass, so that *no lumps* remain. Add to the tube 5 ml of 2 per cent sodium carbonate and continue the grinding of crystals against the walls of the tube. Heat the mixture gradually to 50 to 60°, place a solid rubber stopper on the mouth of the tube, and shake for about 15 seconds. Filter the mixture (page 65) and wash 3 times with 3 to 4 ml of water.

Place the crystals in the tube in which the derivative was prepared, add 15 ml of ethanol or methanol, and heat until solution is effected. Filter and add water to the filtrate until cloudiness appears; reheat until the cloudiness disappears. If the cloudiness persists near the boiling point of the solution, add alcohol dropwise. Cool for 10 to 15 minutes and filter. Wash twice with 3-ml portions of equal parts of alcohol and water, and dry on a clay plate. About 250 to 300 mg of pure ethyl 3,5-dinitrobenzoate, melting at 93°, are obtained.

NOTE. Good results are obtained with 100 mg of 3,5-dinitrobenzoyl chloride and 0.05 ml of the alcohol. For example, 40 mg of pure *n*-propyl 3,5-dinitrobenzoate melting at 73.5 to 74° were obtained after one crystallization from 0.05 ml of 1-propanol. In the case of alcohols having 6 or more carbon atoms, the heating of the reaction mixture should be prolonged to 10 minutes and, if the results are poor, the procedure described for the preparation of β-naphthyl 3,5-dinitrobenzoate (page 486) should be used. When the quantity of the alcohol available is 1 drop or less, the procedure described for isobutyl 3,5-dinitrobenzoate or cyclohexyl 3,5-dinitrobenzoate is used. If the melting point of the dinitrobenzoate after crystallization is more than 2 to 4° below that recorded in the literature, proceed as follows: Dissolve the dinitrobenzoate (it need not be dry) in 5 ml of ethyl or isopropyl ether and wash the ethereal solution first with 3 ml of 2 per cent sodium hydroxide solution and then with 3 ml of water. Evaporate the ether and crystallize the residue once from an alcohol-water mixture. If this method is applied to a crude sample of ethyl 3,5-dinitrobenzoate melting at 84 to 86°, a product is obtained melting at 92 to 93° without further crystallization. The 3,5-dinitrobenzoyl chloride should be pure if it is to be used in semimicro work.

Derivatives of Alcohols and Phenols   471

The method of preparation and purification is given on page 472. Commercially available chloride should be recrystallized from carbon tetrachloride unless the purity is specified. The stopper of the bottle in which the chloride is kept should be sealed with paraffin wax and exposure to air should be kept at a minimum.

If experience in derivatizing milligram quantities of an alcohol is desired, cyclohexanol is useful. A microcone is constructed and charged (according to the directions given in Examples 13.09 and 13.12) with 15 mg of finely pulverized 3,5-dinitrobenzoyl chloride and 0.01 ml of cyclohexanol. The contents of the tube are heated as directed (Example 13.09), extracted with sodium carbonate solution and washed (Example 13.12), and then repeatedly extracted with 0.5-ml portions of hot methanol, removing the hot solution in a 3-in. tube. Water is added dropwise until cloudiness appears and the solution is cautiously heated until clear, then cooled. The crystals are separated and washed by centrifugation (page 71). About 10 mg of crystals, melting at 112°, are obtained.

Example 13.02: Derivatization of Isobutyl Alcohol: Isobutyl 3,5-Dinitrobenzoate

Place in a 6-in. tube 1 drop of isobutyl alcohol, 40 mg of 3,5-dinitrobenzoyl chloride, 5 ml of isopropyl ether (free from alcohol), and 1 drop of pyridine. Place a microcondenser in the tube so as to permit refluxing and heat in a beaker containing water for 1 hour; adjust the flame of the microburner so that the isopropyl ether boils gently (56°). Remove the tube from the water bath and cool in running water. Add 0.5 to 1 ml of dilute sulfuric acid and 4 ml of water. Stopper the tube with a solid rubber stopper and shake to remove the pyridine. Transfer the ether layer into another tube and wash it once with 1 ml of 10 per cent sodium hydroxide solution and twice with 4 ml of water to remove the dinitrobenzoic acid. Transfer the ether layer to a small casserole or evaporating dish; wash the vessel from which the ether solution was transferred with 1 ml of fresh isopropyl ether; and add the washings to the dish. Evaporate the ether carefully over a water bath; add to the residue 0.5 ml of alcohol and then 2 ml of water; and transfer the liquid into a small test tube. Cool for about 5 minutes and then scrape the sides of the tube with a glass rod. Filter the crystals and then wash them with 0.5 ml of water. The yield is 5 to 10 mg of crystals, melting at 85 to 86°.

NOTE. This procedure may be used whenever the quantity of hydroxy compound is small or the hydroxy compound is not very reactive. For example, tertiary alcohols give good yields of dinitrobenzoates from 50 to 100 mg of the hydroxy compound.

Example 13.03: Preparation of a p-Nitrobenzoate: Methyl p-Nitrobenzoate

Place 0.2 ml of methanol and 100 mg of pure *p*-nitrobenzoyl chloride in a 6-in. tube and proceed as described in Example 13.01. Recrystallize the crude product by dissolving in hot methanol and adding water to the hot filtered solution. The yield is 90 mg melting at 95 to 96°.

Example 13.04: Derivatization of Alcohols in Aqueous Solution

Use 5 to 10 ml of the aqueous solution containing 250 to 500 mg of the alcohol. Dissolve separately 1 g of 3,5-dinitrobenzoyl chloride or p-nitrobenzoyl chloride in 2 ml of specially purified hexane or ligroin (washed with sulfuric acid, then with water, dried and distilled), and 8 ml of dry benzene. Place the alcoholic solution in an 8-in. tube, cool to 0°, and add 5 ml of the acid chloride solution and 500 mg of sodium acetate. Close the tube with a stopper, and shake for 2 minutes with cooling (ice bath) so that the temperature remains below 5°. Place the tube in the ice bath for 30 minutes and shake it occasionally. Add 25 ml of ether, shake well, and separate the ether layer. Wash first with water, then with 5 ml of 5 per cent sodium hydroxide, then with 5 ml of 5 per cent hydrochloric acid solution and, finally, with water again. Evaporate the ether solution and crystallize as in the Note on page 470.

An alternative procedure[5] is to dissolve the acid chloride in 5 ml of ethanol-free ether and then to add 1 drop of 0.1 per cent phenolphthalein in benzene and shake it with the aqueous alcohol, adding dilute sodium hydroxide solution dropwise until the aqueous layer assumes a pink color. The mixture is shaken frequently for 10 to 15 minutes, then separated and treated as directed above.

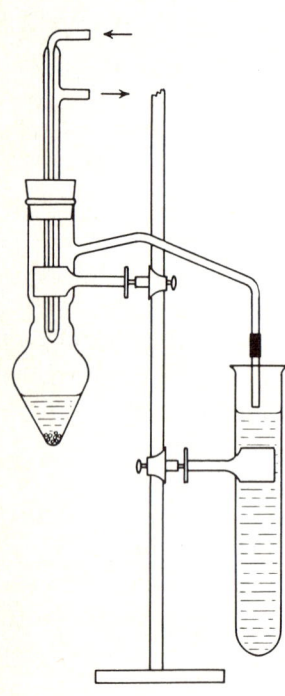

FIGURE 13.01

Arrangement of apparatus for preparation of acid halides. The flask may be replaced by an 8-in. distilling tube.

Example 13.05: Preparation of 3,5-Dinitrobenzoyl Chloride

Arrange an 8-in. distilling tube or a 25-ml distilling flask as shown in Figure 13.01. The side arm of the distilling tube is connected with a receiving tube partially filled with water; the short delivery tube reaches just above the water level. (CAUTION. *Use the hood and wear goggles*

---

[5] H. Henstock, *J. Chem. Soc. (London),* **1933**, 216.

throughout.) Remove the stopper and introduce into the tube 1 g of 3,5-dinitrobenzoic acid and 1.3 g of phosphorus pentachloride. Replace the stopper and heat the tube with a microburner until a vigorous reaction begins; remove the flame momentarily until the reaction subsides and then adjust it so that slow refluxing takes place. Heat for 15 minutes. Cool to 50° and insert a thermometer in place of the condenser; also insert a dry receiving tube. Heat the tube (moving the flame up and down) until all the oxychloride has distilled and the thermometer indicates a temperature of 120°. Remove the flame and add 5 ml of *dry* carbon tetrachloride. Cool in an ice-salt bath for 5 minutes; then filter the crystals with suction and wash them with 1 ml of solvent. Return the filtrates to the distilling tube and distil until 2 ml of solution remain. Disconnect and cool the distilling tube as before. Remove the mass of crystals by means of a rod directly to a clay plate and dry in air for about 10 minutes. The second crop of crystals (350 to 400 mg) is pure, melting at 74°; the first crop (600 to 700 mg) melts at 67 to 69°; upon crystallization from 2 ml of solvent the yield is about 400 mg, melting at 72 to 73°. The first lot is suitable for the preparation of 3,5-dinitrobenzoates. The acid chloride should be stored in a dry tube or small bottle and the stopper sealed with wax. The *p*-nitrobenzoyl chloride is prepared by the same method, using *p*-nitrobenzoic acid.

3-Nitrophthalates

Discussion

The use of esters of phthalic and tetrachlorophthalic acid to form with alcohols the monoalkyl esters is not suitable for small quantities; however, with the acid esters of 3-nitrophthalic acid good results are obtained from 200 to 300 mg of alcohol.

The reaction of 3-nitrophthalic anhydride with alcohols gives chiefly 2-monoalkyl esters of 3-nitrophthalic acid and only small quantities of the isomeric 1-monoalkyl esters as shown in the following equations:

$$\underset{NO_2}{\underset{}{\bigcirc}}\overset{CO}{\underset{CO}{>}}O + ROH \rightarrow \underset{\underset{\text{Main product}}{NO_2}}{\bigcirc}\overset{COOH}{\underset{COOR}{}} + \underset{\underset{\text{Small amount}}{NO_2}}{\bigcirc}\overset{COOR}{\underset{COOH}{}}$$

The reaction with the lower alcohols takes place readily if the mixture of 3-nitrophthalic anhydride and alcohol is heated in a water bath until a homogeneous melt is obtained. For higher alcohols, the mixture is dissolved in dry toluene and refluxed. One disadvantage of the 3-nitrophthalates for semimicro work is their tendency to separate as oils, which renders their separation from the reaction mixture and subsequent crystallization difficult. Nevertheless, their use is advisable whenever the 3,5-dinitrobenzoates and urethans are not suitable.

The general method for the preparation of the 3-nitrophthalates is illustrated by the preparation of *n*-butyl 3-nitrophthalate. For alcohols with more than 4 carbon atoms it is advisable to employ the method outlined under the preparation of isobutyl 3,5-dinitrobenzoate, using 2 millimoles of the reagent and 3 millimoles of the alcohol with 5 ml of *n*-butyl ether and 0.5 ml of pyridine.

Example 13.06: Preparation of n-*Butyl 3-Nitrophthalate*

Place 200 mg of 3-nitrophthalic anhydride and 0.2 ml of *n*-butyl alcohol into an 8-in. test tube. Add 2 boiling stones and boil gently under reflux for 10 to 15 minutes. Pour 5 ml of water into the test tube and heat nearly to boiling, stirring the oil with a glass rod. Cool and pour off the aqueous layer from the oil adhering to the tube. Pour 10 ml of ethanol and 30 ml of water into the tube and heat to boiling. Decant the hot solution from any undissolved oil. Allow to cool overnight. Filter and wash with 20 per cent ethanol. The yield is 60 to 90 mg, melting at 145 to 146°.

NOTE. The nitrophthalates of methanol, ethanol, and the propanols form with greater ease and crystallize more readily. The melting points of the nitrophthalates, after one crystallization, are sometimes 5 to 10° below the melting points given in the literature. This fact is due to incomplete reaction and to the presence of the isomeric ester. As many as four crystallizations are sometimes required.

α-Naphthylurethans

Discussion

As it was pointed out in the introduction to the derivatization of alcohols, a urethan or carbamate is formed by the reaction of an isocyanate and a hydroxy compound:

$$\underset{\text{Isocyanate}}{RNCO} + \underset{\substack{\text{Hydroxy}\\\text{compound}}}{R'OH} \rightarrow \underset{\substack{\text{Urethan } or \text{ Ester of}\\\text{substituted carbamic acid}}}{RNHCOR'}$$

$$\underset{\text{α-Naphthyl isocyanate}}{C_{10}H_7NCO} + \underset{\text{1-Propanol}}{C_3H_7OH} \rightarrow \underset{n\text{-Propyl α-naphthylurethan}}{C_{10}H_7NHCOOC_3H_7}$$

Instead of the isocyanate, either a substituted carbamyl chloride or an azide may be used:[6]

$$NO_2C_6H_4NHCOCl \rightarrow NO_2C_6H_4NCO + HCl$$
p-Nitrophenylcarbamyl chloride   p-Nitrophenyl isocyanate

$$NO_2C_6H_4NCO + CH_3OH \rightarrow NO_2C_6H_4NHCOOCH_3$$
Methyl p-nitrophenylurethan

$$BrC_6H_4CON_3 \rightarrow BrC_6H_4NCO + N_2$$
p-Bromobenzazide   p-Bromophenyl isocyanate

$$BrC_6H_4NCO + C_2H_5OH \rightarrow BrC_6H_4NHCOOC_2H_5$$
Ethyl p-bromophenylurethan

However, with the exception of diphenylcarbamyl chloride, substituted carbamyl chlorides or azides are not commercially available. If it becomes necessary to use another urethan and the isocyanate is not commercially available, it is preferable to prepare the azide instead of the isocyanate; the preparation of the azide is from the carboxylic acid whereas the aryl isocyanate involves the use of phosgene. The recommended reagent is α-naphthyl isocyanate, which is commercially available. Further discussion appears in the section "Other Substituted Urethans."

General Method for Preparation

A ratio of 1.25 millimoles of α-naphthyl isocyanate to 1 millimole of alcohol is used. If the amount of alcohol to be derivatized is less than 10 mg, use fractional microsublimation, as illustrated in Example 13.09. The anhydrous alcohol and isocyanate are mixed in an appropriate 3-in. or 6-in. test tube and heated in a water bath at 60 to 70° for 10 to 15 minutes. The crude urethan solidifies on cooling and is pulverized in the reaction vessel by means of the microspatula. It is then extracted with a minimum amount of petroleum ether to remove the soluble impurities.[7] The first extract, which contains a considerable amount of the derivative, is set aside and the residue is extracted with a fresh amount of the solvent. The second extraction yields the pure derivative. The residue contains di-α-naphthylurea formed by the reaction of moisture present in the alcohol and the reaction vessel.

Attention should be paid to the filtration of the petroleum ether extract of the reaction mixture. These and other details are described in the following examples.

---

[6] P. P. T. Sah et al., *Rec. trav. chim.*, **58**, 453, 582, 591, 595, 1012 (1939); **59**, 238, 357 (1940).
[7] N. D. Cheronis, pp. 485–489, 497–499 (cited in Reference 4).

Example 13.07: n-*Propyl α-Naphthylurethan*

Heat a 6-in. test tube over a flame until all moisture has been driven off; then cork and allow to cool. Put rapidly into the tube, by means of a pipet dropper, 0.2 ml of 1-propanol and 0.25 ml of α-naphthyl isocyanate; cork the tube immediately. Place the tube in a water bath at 60 to 70° for 5 minutes. Remove the tube and add 4 ml of petroleum ether (bp 90 to 110°) or 4 ml of commercial heptane. Heat to boiling and filter through a funnel prepared as follows: Insert the perforated disc inside of the funnel and place upon it a disc of filter paper; add 2 drops of water and apply light suction; then adjust filter paper. Add 3 drops of methanol and apply light suction again. Finally, add 4 drops of petroleum ether and repeat the application of suction. Fit the filter into the mouth of a clean and dry 8-in. test tube with a side arm, apply suction, and add the hot solution. Set the filtrate aside and transfer the residue with the filter paper back into the reaction tube and add 8 ml of the solvent. Heat the tube cautiously and as the solvent boils lift the filter paper from the bottom of the tube by means of the microspatula so that it adheres to the sides of the tube about 5 mm above the liquid. In this manner as the vapors condense, the material adhering to the paper is washed down. Prepare the funnel for a second filtration as described. Chill the second filtrate for a few minutes and scratch the sides of the tube by means of a glass rod. After 15 minutes, filter the crystals and crystallize from 5 to 6 ml of hydrocarbon. The yield is 80 to 90 mg of crystals, melting at 79 to 80°.

NOTE. It is necessary that the alcohol be free from water; otherwise a considerable amount of dinaphthylurea forms, with a corresponding decrease in the yield of the desired urethan. It is clear, therefore, that the preparation of the urethan should not be attempted unless it is known that the hydroxy compound is anhydrous. For secondary alcohols the duration of heating should be increased to 10 minutes. For example, from 0.2 ml of 2-butanol or 2-pentanol, 80 to 90 mg of the pure urethan are obtained. For smaller quantities of the hydroxy compound and for higher alcohols, the procedure described below is recommended.

Example 13.08: Cyclohexyl α-Napthylurethan

Put 100 mg of cyclohexanol and 125 mg of α-naphthyl isocyanate into an 8-in. test tube dried as described in the preparation of *n*-propylurethan. Fit a two-hole stopper into the test tube holding a microcondenser and a calcium chloride tube. Add 6 ml of petroleum ether and heat in a water bath at about 90° for 30 minutes. Prepare a funnel as described in Example 13.07 above, and filter the hot solution from the dinaphthylurea. Cool the solution for 10 to 15 minutes and then filter the crystals. Dissolve these in the minimum amount of hot petroleum ether and crystallize. The yield is 40 to 50 mg, melting at 128 to 129°.

NOTE. Tertiary alcohols react very slowly and give small yields of urethans; dehydration of the alcohol occurs and results in increased formation of the dinaphthylurea. For example, 200 mg of *tert*-amyl alcohol with 0.25 ml of the isocyanate, treated as described in Example 13.08, gave 1.5 mg of a gummy substance that could not be purified for melting-point determination.

Example 13.09: Derivatization of 1 Milligram of an Alcohol

This is an example of the exact procedures by which it is possible to prepare a derivative from a few milligrams of a substance and purify the derivative without loss in transfer. The preparation requires the use of a quantitative balance, a microscope heating stage for the determination of melting points using a few crystals of the derivative (see pages 187–191), and a vacuum fractional sublimator as described in Chapter 3.

Prepare a cone from a piece of clean 8-mm tubing as shown in Figure 13.02. The glass is first heated at point *a* and drawn very slightly so as to

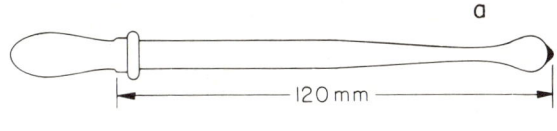

FIGURE 13.02

Microcone for preparation of derivatives on a milligram scale.

constrict its bore to about one half. Then the tube is heated about 20 mm from the constriction and drawn in the form of a cone; the lower end should not be thickened by allowing the glass to melt because the part below the constriction is to be crushed in the microsublimator by means of a sturdy rod. However, it should not be so thin that it will break in handling. The tube now is cut so that the total length is about 100 to 120 mm. The open end is then fire-polished and when cool it is fitted with a bulb from a medicine dropper, which serves as a stopper.

The tube is charged as follows. Take a regular melting-point capillary that is open at both ends and weigh it accurately on an analytical balance and record the weight. Hold the capillary upright and open momentarily the bottle of the reagent (α-naphthyl isocyanate) and touch the lower end of the capillary at the surface of the liquid so that the liquid rises to a height of about 2 mm. Now tilt the capillary so that the liquid flows to about the middle and wipe the tip carefully with a soft tissue and lay it flat on the balance pan and weigh it. If the sample is more than the desired amount the capillary is tilted carefully so that the liquid flows toward one end and then momentarily touched to a small piece of clean filter paper. With some practice it is possible to weigh 2.5 mg of the reagent rapidly. Place one end of the capillary against the inner wall of the cone and, by means of a rubber bulb, blow air through it several times to transfer practically all the liquid into the cone. Stopper the microcone by means of the rubber bulb and centrifuge it for 1 minute to force the liquid to the bottom. If it is desired to determine the exact amount of the reagent transferred into the cone the capillary is reweighed after air is blown through it. About 0.2 to 0.3 mg should remain in the capillary since the required amount of reagent is 2.3 mg. In the same manner 1.0 mg of cyclohexanol is transferred into the microcone, which is stoppered and well centrifuged. The stoppered cone is immersed in a water bath at 70 to 75°, then centrifuged and reheated again for 10 minutes, and then allowed to stand at room temperature for 30 minutes.

By means of the sharp end of a wire microspatula pulverize well the crystalline mass within the microcone. By means of a sharp file cut the lower end of the microcone about 1 mm above the crystalline mass and place the closed end in the microsublimation apparatus (page 106) and carefully break the glass into small pieces. Before beginning the purification read the section on sublimation in Chapter 3, with particular attention to the lubrication of the glass joint and general procedure to be followed. Place the microsublimator in a small beaker and add enough heavy paraffin oil that it rises just above the lower part of the condenser. It is essential that good visibility of the lower part of the condenser be maintained. Connect the outlet of the microsublimator to a good vacuum system so that a pressure of less than 5 mm is obtained. Apply heat to the bath by means of a small flame so that the temperature is raised gradually to 90 to 100°. After about 30 minutes, or when about a third of the material has sublimed, the pressure is released with care and the condenser of the sublimator is removed. The sublimate is washed with methanol into a small crucible as shown in Figure 13.03. The solvent is evaporated slowly by placing it over a warm plate, being careful that no creeping of the solid occurs. The condenser is replaced and another fraction representing one third of the original material is sublimed. The condenser is removed and a few micrograms of the sublimate are scraped from the condenser to a glass slide for melting-point determination; the balance is transferred to a watch glass by means of a fine stream of solvent from the wash bottle (Figure 13.03), and then evaporated to recover the second sublimate, the

Derivatives of Alcohols and Phenols

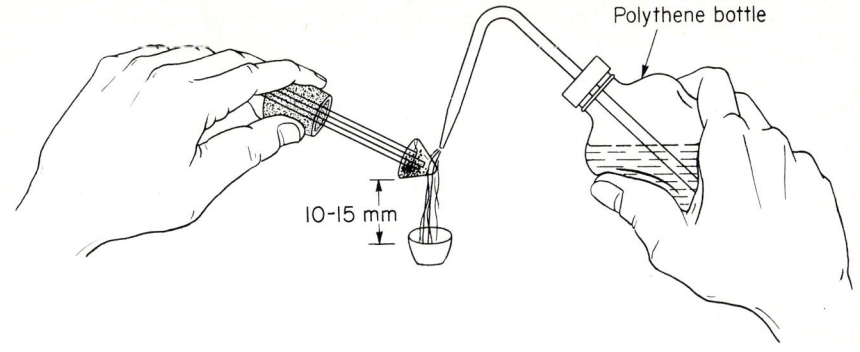

FIGURE 13.03
Washing sublimate from condenser.

desired fraction. About 0.3 to 0.4 mg of crystals, melting at 128 to 129° are obtained.

Other Substituted Urethans

Besides α-naphthyl isocyanate the following are commercially available: phenyl, β-naphthyl, p-bromophenyl, o-nitrophenyl, m-nitrophenyl, and p-nitrophenyl isocyanates. Phenyl isocyanate has appreciable lachrymatory properties and yields urethans that are somewhat soluble; hence, it is not recommended. Of the nitro-substituted reagents the p-nitrobenzyl isocyanate may be employed in the same manner as α-naphthyl isocyanate.

β-Bromopropionyl isocyanate reacts with many alcohols to give solid derivatives; melting points of the solid carbamates thus formed from at least 50 alcohols have been published.[8] Even long-chain alcohols form solid derivatives easily, although secondary alcohols tend to form oils. The reagent is easily prepared from N-bromosuccinimide, which can be stored without decomposition. Details of the procedure for the formation of the reagent and derivatization of an alcohol are given at the end of this section.

When it becomes necessary to employ an azide, either the p-bromobenzazide or 3,5-dinitrobenzazide is prepared according to methods described in the literature listed at the end of this chapter. The urethan is prepared from 1 millimole of the alcohol and 1.2 millimoles of the azide, by the procedure described in Example 13.08, except that the mixture is refluxed for 1 to 2 hours.

[8] H. W. Johnson, Jr., H. A. Kreyssler, and H. L. Needles, *J. Org. Chem.*, **25**, 279–281 (1960).

The degree of ease with which urethans are formed depends on the character of the radical attached to the hydroxyl group. Primary alcohols react with ease; secondary alcohols react slowly; and tertiary alcohols react with difficulty. As the rate of reaction with the isocyanate diminishes, the rates of the side reactions increase; thus, with the tertiary alcohols, olefin formation through dehydration may become the dominant reaction. The water produced by the dehydration of the alcohol reacts with the isocyanate in the manner indicated, and consequently the chief product may be a substituted urea instead of a urethan. Enols, phenols, hydroxy acids, and esters form urethans; however, with the α- and β-hydroxy acids and esters other side reactions occur, leading to the formation of cyclic compounds.[9] In the case of phenols, the introduction of electron-attracting substituents retards the rate of reaction. Thus phenol, *m*-cresol, and thymol yield urethans readily, whereas *p*-nitrophenol reacts with difficulty and trinitrophenol (picric acid) does not react.

Example 13.10: n-*Propyl β-Bromopropionylcarbamate*

Make a solution of N-bromosuccinimide (0.5 g) in 5 ml of chloroform that has been dried over calcium chloride, and add about 0.5 ml of allyl chloride and a trace of benzoyl peroxide. Heat the mixture under reflux until the solid N-bromosuccinimide dissolves and then for 30 minutes more. After the solution cools to room temperature, add 0.2 to 0.3 ml of *n*-propyl alcohol. A slight excess of the isocyanate appears to help in the recrystallization. Cool the solution and remove and dry the crystals which appear. If no crystals separate on cooling of the solution, evaporate some of the solvent and cool the solution again. Generally, better results are obtained if the preparation is completed within a few hours rather than left overnight.

Other Derivatives

Of the derivatives listed in Table 13.01 a number will be found useful in the derivatization of tertiary alcohols, glycols, and polyhydroxy compounds. For example, if the tertiary alcohol yields dinitrobenzoates or urethans with difficulty, it can be converted to the alkyl chloride at room temperature by shaking with excess of concentrated hydrochloric acid, and then the halide is derivatized as the S-alkylthiuronium picrate (see page 550).

The hydrogen phthalates and tetrachlorophthalates do not offer any particular advantage over the 3-nitrophthalates, which were described in this chapter. The benzoates are useful in the derivatization of glycols and

---

[9] R. F. Rekker et al., *Rec. trav. chim.*, **70**, 113 (1951).

polyhydroxy compounds, and an example of the procedure employed is described in this section. The trityl ethers are formed readily by primary alcohol groups but not by secondary or tertiary alcohols:

$$RCH_2OH + (C_6H_5)_3CCl \rightarrow RCH_2OC(C_6H_5)_3 + HCl$$
$$\text{Trityl chloride} \qquad\qquad \text{Trityl ether}$$

This reaction can be employed for introducing different radicals in polyhydroxy compounds. An example of this is the conversion of glucose to 6-trityl-β-D-glucose tetra-acetate.[10] Similarly, glycerol can be converted to ditritylglycerol and then reacted with *p*-nitrobenzoyl chloride to form the 2-*p*-nitrobenzoate; the trityl groups are then removed by means of hydrobromic acid to yield glycerol 2-*p*-nitrobenzoate.[11] The trityl ethers of glycols, Carbitols, and Cellosolves are useful for identification.[11]

The pseudosaccharin ethers are useful derivatives for primary and secondary alcohols, glycols, and glycol derivatives:

[Pseudosaccharin chloride] + RCH$_2$OH → [Pseudosaccharin ether] + HCl

The acetates and *p*-nitrophenylacetates are formed from the respective chlorides and the alcohols. However, many of the derivatives are liquid at room temperature.

p-Phenylazobenzoates are highly crystalline solids, which are easily prepared and purified. Since these derivatives are colored, they can be used to advantage in the chromatographic separation of alcohols or phenols. Melting points of 23 alcohol derivatives have been reported.[12]

The oxidation of alcohols with potassium permanganate in the presence of $2N$ sulfuric acid to the corresponding carbonyl compound and the derivatization of the latter by means of the 2,4-dinitrophenylhydrazones should be considered only if the quantity available is more than 500 mg.

Example 13.11: Preparation of Glycerol Tribenzoate

Place in an 8-in. test tube 100 mg (3 drops) of glycerol and add 0.5 ml of benzoyl chloride. Select a solid rubber stopper that fits securely in the mouth of the tube. Add 5 ml of 10 per cent sodium hydroxide solution

---

[10] B. Helferich et al., *Ber.*, **58**, 877 (1955); *Z. phys. Chem.*, **170**, 31 (1927); **175**, 311 (1928); N. D. Cheronis, pp. 301–302 (cited in Reference 4).
[11] M. K. Seikel and E. H. Huntress, *J. Am. Chem. Soc.*, **63**, 593 (1941).
[12] E. O. Woolfolk and J. M. Taylor, *J. Org. Chem.*, **20**, 391 (1955).

and shake vigorously for 1 minute and then intermittently for 5 minutes or until the solid derivative separates out. Allow to stand in a cold bath for 30 minutes, shaking the tube at intervals so that the lumps forming at the beginning break up into small granules; use a rod for this purpose if necessary. Add 5 ml of water, shake vigorously for a minute, and then filter. Wash twice with 5 ml of water and place on a drying plate.

The recrystallization of the tribenzoate of glycerol and of the dibenzoate of glycols entails considerable difficulty, owing to the tendency of these esters (glycerides and glycol esters) to separate as oils. The following procedure gives fairly good results. Retain about 5 mg of the crystals and dissolve the rest in hot ethanol; then filter. Add water until cloudiness appears and reheat until the cloudiness disappears. Cool and add a few crystals from the lot set aside for seeding. Allow the tube to stand in the cold bath for 1 to 2 hours, and then filter; wash twice with 2 to 3 ml of water and set aside to dry. The yield is 200 to 225 mg of crystals, melting at 75 to 76°. This derivative has also been reported to melt at 71 to 72°.

Example 13.12: Preparation of Benzyl Pseudosaccharin Ether

Place in a dry test tube provided with a reflux microcondenser and calcium-chloride tube 100 mg of benzyl alcohol, 0.2 ml pyridine, 5 ml of dry chloroform, and 210 mg of pseudosaccharin chloride. Reflux for 30 minutes, then cool. If crystals do not separate wash the chloroform, once with 2 ml of $1N$ sulfuric or hydrochloric acid and once with water; then pour the solution into a small dish and evaporate it to a small volume. Filter the crystals, wash with dilute acid and water, and crystallize the derivative from a solvent mixture of ether-chloroform. The pure derivative melts at 100°.

Chromatographic Detection of Small Amounts of Alcohols

For the separation and identification of alcohols by paper chromatography the 3,5-dinitrobenzoates are recommended. These esters should be prepared as outlined in this section. For the use of xanthates and dinitrophthalates for this purpose, see the literature cited at the end of this chapter. The procedure described in Example 13.12 serves as a trial run. A difference in R_f values of 0.08 is needed to differentiate well between alcohols. Once experience is obtained, it is possible to apply the same procedure to the identification of alcohols present in small quantities in various extracts and fluids from natural products, such as fruit juices.[13]

[13] A. C. Rice et al., *Anal. Chem.*, **23**, 195 (1951).

Chromatographic separation of alcohols is facilitated by the use of *p*-phenylazobenzoyl chloride reagent, since the resulting esters are crystalline, colored solids. A 1:1 mixture of alumina and Celite or a 2:1 mixture of silicic acid and Celite is used as the adsorbent. One of these adsorbents will sometimes separate mixture which the other cannot.[14]

Example 13.13: Identification of Small Amounts of Methanol and Ethanol in a Mixture by Chromatography

Charge a 3-ml tapered tube (Figure 3.01E) with 20 mg of finely pulverized 3,5-dinitrobenzoyl chloride. The reagent is weighed rapidly on a piece of glazed paper creased at one end and introduced in the 3-ml tube, which has been previously heated to remove the moisture and then stoppered while it is cooling. The stopper is removed momentarily to introduce the reagent. About 1.0 mg of methanol and a similar quantity of ethanol are weighed in separate capillaries and introduced into the tube following every detail of the procedure described in Example 13.09. The tube is well stoppered and immersed in hot water for 15 minutes and then cooled.

Pulverize thoroughly the crystalline mass at the bottom of the cone with the sharp edge of the microspatula and add 1 ml of 2 per cent solution of sodium carbonate and continue the grinding of the crystals. Immerse the tube for 2 minutes in a water bath at 50 to 60° and continue grinding the crystals until no lumps remain. Centrifuge the tube and withdraw the supernate by means of a capillary pipet. Wash the crystals three times with 0.5-ml portions of water (see page 71). Spread the residue on the sides of the tube by means of the microspatula and heat at 50 to 60° in a water bath and apply suction by inserting a one-hole rubber stopper having a glass tube that is connected to a vacuum system. The small amount of water rapidly evaporates within a few minutes. However, it is not necessary to remove all the moisture from the crystals. Cool the tube and add 0.5 ml of chloroform by means of a graduated pipet. Stir with the microspatula until the residue is completely dissolved and stopper the tube. If the conversion of the 2-mg mixture of methanol and ethanol to their respective dinitrobenzoates is nearly quantitative, the maximum yield of the derivatives is about 10 to 12 mg, and the maximum concentration of dinitrobenzoate in the chloroform solution is about 10 micrograms per microliter ($10\gamma/\lambda$). Therefore the maximum amount of solution to be placed on the paper strip is 10 to 20 λ, which would be 100 to 200 micrograms. If this amount produces large blurred spots the sample applied on the strip is reduced to 5 λ or the solution is further diluted.

[14] E. O. Woolfolk, F. E. Beach, and S. P. McPherson, *J. Org. Chem.*, **20**, 391 (1955).

Use the technique described on pages 123–129 of Chapter 4 and apply on the pencil spot made on the paper strip about 10 λ of the solution and not more than 1 to 2 λ each time, allowing the solvent to evaporate before the next application. Prepare two strips with the mixed dinitrobenzoates, two with the pure 3,5-dinitrobenzoate of methanol, and two with the pure 3,5-dinitrobenzoate of ethanol. The reference derivatives are prepared separately using 2 mg of pure alcohol instead of a mixture. As soon as each strip is prepared it is placed in a labeled, dry test tube and then stoppered. Prepare in separate test tubes two types of solvents. The first solvent consists of 2 ml of dioxane and 8 ml of water and the second of a mixture of 3 ml of pyridine and 7 ml of water. Observing all the precautions given in Chapter 4 in the description of the test-tube technique add about 0.5 ml of the first solvent to one test tube containing the mixed dinitrobenzoates and to one of each from the known pair of tubes containing the pure derivatives. To each of the other three tubes add 0.5 ml of the second solvent. Allow the strips to develop for 1 to 2 hours until the front has reached the pencil line near the top. The front is located by playing a beam of a flashlight on the strip from the side. Remove the strips and after the solvent has evaporated, paint them lightly (using artist's air brush) with a 0.5 per cent solution of 1-naphthylamine. When dry, apply lightly a 10 per cent solution of potassium hydroxide; this treatment is merely to intensify the spots, which appear as red-orange colorations. Compare the R_f values of the spots on the strip containing the mixed dinitrobenzoates with the spots of the chromatograms of the pure dinitrobenzoates, and select the best solvent system.

Other solvent systems that may be suitable for the separation of the 3,5-dinitrobenzoates of alcohol are (a) 4 ml of 2-propanol and 6 ml of water, and (b) 9.8 ml of petroleum ether (bp 20 to 40°) and 0.2 ml of methanol. The last solvent is more suitable for the separation of the derivatives of the higher alcohols. It should be remembered that the results of chromatographic procedures improve with experience.

PHENOLS

Table 13.02 lists *only the most important* derivatives for the characterization of phenols. For a complete list, see Table 13.01 (page 466), which gives the derivatives of alcohols because, in general, most of the derivatives used for the characterization of alcohols may also be used for phenols. The selection of the most suitable derivative depends on the nature of the phenol. With monocyclic phenols the 3,5-dinitrobenzoate, or *p*-nitrobenzoate, should be considered first if the melting point of the derivative is not above 200°. If the phenol is dicyclic or polycyclic and has a melting point of 100° or above, acetylation should be considered. Although the

TABLE 13.02
DERIVATIVES FOR THE IDENTIFICATION OF PHENOLS

Acetates, benzoates, p-nitrobenzoates, and other esters	*α-Naphthylurethans
	Nitro derivatives
*Aryloxyacetic acids	p-Phenylazobenzoates
*Bromo derivatives	Picrates
*3,5-Dinitrobenzoates	Pseudosaccharin ethers
2,4-Dinitrophenyl ethers	Sulfonic acid esters
Diphenyl- and other urethans	

* Recommended derivative.

formation of urethans is slow, it should be considered wherever possible, for the reaction can be catalyzed. The preparation of aryloxyacetic acid, and 2,4-dinitrophenyl ethers is based on the greater reactivity of the phenolic function as compared to the alcoholic. The preparation of these derivatives is recommended if the quantity of material available is over 200 mg. Finally, a number of typical "aromatic" reactions, such as bromination and, to a less extent, nitration and oxidation, may at times be employed for the derivatization of phenols. In the case of nitrophenols, the corresponding amino compounds obtained by reduction should be considered because the formation of esters and urethans is slow and in some cases not feasible. For example, p-nitrophenol can be readily derivatized, even in 10 mg quantities, by catalytic reduction to p-aminophenol, as described on page 491.

The discussion here is restricted to the recommended derivatives in Table 13.02. Literature references appear at the end of the chapter.

3,5-Dinitrobenzoates, p-Nitrobenzoates, Benzoates, and Acetates

General Method for Preparation of 3,5-Dinitrobenzoates

The following equation represents the reaction of 3,5-dinitrobenzoyl chloride with phenol:

$$C_6H_3(NO_2)_2COCl + C_6H_5OH \rightarrow C_6H_3(NO_2)_2COOC_6H_5 + HCl$$

3,5-Dinitrobenzoyl chloride   Phenyl 3,5-dinitrobenzoate

The preparation of 3,5-dinitrobenzoates is accomplished by employing the same general procedures as described on pages 468–473 for the preparation of the 3,5-dinitrobenzoates of alcohols. Since the reaction is slow, the pyridine method is used as illustrated in the derivatization of β-naphthol.

The 3,5-dinitrobenzoates of some phenols separate as oils upon addition of the dilute acid. In such cases the aqueous layer is poured off and 4 to 5 ml of ethyl or isopropyl ether is added, after which the ether solution is washed successively with water, 2 per cent sodium hydroxide solution, and again with water. The ether is then evaporated, and the residue is crystallized from methanol or ethanol.

p-Phenylazobenzoates of phenols are highly crystalline solids, which can be prepared from the acid chloride and the phenol. They are easily purified, and are colored. A mixture of *p*-phenylazobenzoyl chloride (approximately 0.1 g), phenol (0.0003M excess), and 3 to 6 ml of pyridine are refluxed gently for 4 hours. Derivatives of 29 phenols have been prepared, and their melting points reported.[15]

Example 13.14: Preparation of β-Naphthyl 3,5-Dinitrobenzoate

Place in a 6-in. tube 100 mg of β-naphthol, 110 mg of 3,5-dinitrobenzoyl chloride, 2 ml of pyridine, and 2 boiling stones. Arrange for reflux and boil gently for 1 hour. Cool, add 1 ml of 5 per cent sulfuric acid and 5 ml of water; shake well and filter. Replace the crystals, together with the filter paper, into the test tube, add 5 ml of two per cent sodium hydroxide, shake well to remove the 3,5-dinitrobenzoic acid, and filter; then wash twice with 2 ml of water. Suspend the crystals in 5 ml of methanol, heat almost to boiling, and filter. The crystals *remaining on the filter* are used for the melting-point determination. The yield is 80 to 100 mg, melting at 209 to 210°. A small crop of crystals may also be obtained from the filtered alcoholic solution.

NOTE. The solubility of the 3,5-dinitrobenzoates of naphthols in alcohol is small as compared with the like derivatives of phenols and cresols. For this reason the derivative is purified by removing soluble impurities, which in this case are mainly traces of dinitrobenzoic acid and unreacted naphthol. When the derivative separates as an oil after washing with acid, the aqueous layer is poured out and 5 to 8 ml of ethyl ether or isopropyl ether are added, after which the ether solution is washed successively with water, 2 per cent sodium hydroxide solution, and finally with water. The ether is then evaporated and the residue is crystallized from methanol or ethanol.

General Method for Preparation of Benzoates and Acetates

Methods for the preparation of benzoates have been described on page 481. For the preparation of acetates the usual method with acetic anhydride is employed and a drop of sulfuric acid is added to catalyze the

[15] E. O. Woolfolk and J. M. Taylor, *J. Org. Chem.*, **22**, 827–829 (1957).

reaction. If this method does not yield good results, 1 to 2 millimoles of the phenol in 5 ml of dry benzene are refluxed for about one hour with 1.1 millimoles of acetyl chloride and 100 mg of magnesium powder. The reaction mixture is diluted with 5 ml of ether and added to a mixture of 5 ml of 5 per cent sodium carbonate solution and 5 g of ice. The mixture is stirred well, transferred into a separatory funnel and the aqueous layers removed. The benzene-ether solution is washed with water and then evaporated. The crude acetate is crystallized from aqueous methanol.

Example 13.15: Preparation of Phenyl Benzoate

Place in an 8-in. test tube 100 mg of phenol, 0.2 ml of benzoyl chloride, and 4 ml of 10 per cent sodium hydroxide in the order given. Close the tube with a solid rubber stopper and shake vigorously for 1 minute and then at intervals for 10 minutes. Add 6 ml of water and filter the solid ester. Dissolve the crude phenyl benzoate in 5 ml of methanol by heating, filter the hot solution, and add water dropwise until a permanent cloudiness occurs. Warm the solution until it becomes clear; then cool. Filter the solid and wash with 25 per cent methanol. The yield is 150 to 180 mg of crystals, which melt at 69°.

Example 13.16: Acetylation of β-Naphthol

Place in an 8-in. test tube 100 mg of β-naphthol, 0.4 ml of acetic anhydride, and 1 drop of sulfuric acid. Heat gently for 3 minutes and then allow to cool. Add 3 to 4 ml of water and stir the oil by means of a glass rod until it solidifies. Filter, wash the solid with cold water, and crystallize from a methanol-water mixture. The yield is 100 to 120 mg of crystals, which melt at 71 to 72°.

Urethans

General Method for Preparation

The general methods are the same as those for alcohols (page 474). If the phenol is relatively reactive the isocyanate and the phenol are mixed in a perfectly dry vessel and heated for about 2 to 5 minutes. If the phenol is not reactive 1 ml of pyridine and a drop of 10 per cent solution of trimethylamine in hexane or heptane are added and the mixture is heated for 20 to 30 minutes. The mixture is cooled and if the urethan does not separate, 1 ml of 5 per cent sulfuric acid is added. The crude urethan is purified by crystallization from petroleum ether. Generally the introduction of electron-attracting substituents in the phenol causes a retardation in the formation of urethans. Thus, phenol, *m*-cresol, and thymol yield α-naphthylurethans readily but *p*-nitrophenol does so with difficulty, even in the presence of catalysts; picric acid does not react.

Diphenylcarbamyl chloride is commercially available and can be employed for the preparation of the diphenylurethans, which are suitable derivatives for the cresols, *o*-nitrophenol, and resorcinol. The pyridine method as described above is used, but after heating, the mixture is cooled and 2 ml of a 5 per cent solution of sodium carbonate are added and the mixture is cooled in an ice bath. The crystalline solid is washed twice with 1 ml of water containing a few drops of bicarbonate solution and twice with water. Then it is recrystallized either from 95 per cent ethanol or petroleum ether.

Other substituted urethans that may be employed for characterization are: phenyl, β-naphthyl, *o*-nitrophenyl, *p*-nitrophenyl, *m*-nitrophenyl, 3,5-dinitrophenyl, *p*-bromophenyl, *p*-xenyl, *p*-iodobiphenyl, and 3,4-dimethoxyphenyl. A more extensive list of urethans is given in Table 13.01.

Example 13.17: Preparation of Thymol α-Naphthylurethan

Dry a test tube as directed on page 476. While the tube is cooling, fit a cork (of a size equal to that with which the tube is stoppered) with a calcium chloride tube. Place rapidly in the dry test tube 200 mg of thymol and 0.25 ml of α-naphthyl isocyanate and stopper it with the cork holding the calcium chloride tube. Clamp the tube on a stand and heat it by means of a small direct flame so that the mixture boils gently for 2 minutes. Allow to cool for 3 minutes and then rub the mixture with a glass rod until it sets into a crystalline mass. Add 10 ml of petroleum ether, cool, and filter as described under the preparation of *n*-propylurethan in Example 13.07. Extract the residue with another portion of 8 to 10 ml of boiling petroleum ether. Filter the crystals and wash them with 1 ml of petroleum ether. The yield is 120 to 140 mg of crystals, melting at 159°. On crystallizing from 7 ml of petroleum ether, 75 to 90 mg of pure urethan are obtained, melting at 160°.

NOTE. Nitrophenols do not yield urethans with ease. *p*-Nitrophenol, heated for 1 hour with the isocyanate, gives a product melting 30° below the temperature recorded in the literature as the melting point of the derivative. Naphthols react slowly, and therefore it is advisable to heat, under reflux, the naphthol and isocyanate dissolved in 5 to 10 ml of petroleum ether for 0.5 to 1 hour. Tertiary amines catalyze the reaction; a drop of 10 per cent solution of trimethyl-, triethyl-, or tributylamine in petroleum ether accelerates formation of urethans. The tertiary amine may be also used to induce crystallization in case the reaction mixture of phenol and isocyanate forms a viscous oil after heating.

Derivatives of Alcohols and Phenols

Example 13.18: Preparation of p-tert-*Butylphenyl Phenylurethan*

An 8-in. dry test tube is provided with a microcondenser and arranged for reflux. About 1 ml of a petroleum distillate (bp 160 to 180°), obtained by fractionating 10 ml of kerosene, is placed in the tube together with 100 mg each of *p-tert*-butylphenol and phenyl isocyanate. Care is used in handling the isocyanate. Two boiling stones are added; the reaction mixture is gently refluxed for 2 hours and then cooled. The crystals that separate out are filtered and purified by crystallization from petroleum ether (bp 90 to 110°) as described on page 476. The yield is 80 to 90 mg, melting at 148 to 149°.

Aryloxyacetic Acids

Phenols react with chloroacetic acid in the presence of sodium hydroxide to yield aryloxyacetic acids:

$$C_6H_5ONa + \underset{\text{Chloroacetic acid}}{ClCH_2COOH} \rightarrow \underset{\text{Phenoxyacetic acid}}{C_6H_5OCH_2COOH} + NaCl$$

The aryloxyacetic acids are crystalline solids having well-defined melting points; in addition, the determination of their neutralization equivalents may be used as a confirmatory test. For semimicro work it is necessary to have available at least 200 mg of the phenol; otherwise the yield is not sufficient.

General Method for Preparation

For the preparation of the derivative 0.5 ml of 50 per cent aqueous chloroacetic acid are added to 200 mg of the phenol and 1 ml of a $6N$ solution of sodium hydroxide in a small test tube. More water should be added if the phenol salt does not dissolve completely. The test tube is provided with a microcondenser and heated in a water bath at 90 to 100° for 1 hour. The solution is then cooled, two volumes of water added, and then acidified to congo red with dilute hydrochloric acid. The derivative is extracted with two 4-ml portions of ether. The ether extract is washed with 2 ml of water and then extracted with 5 per cent sodium carbonate solution. The sodium carbonate extract is next acidified with dilute hydrochloric acid to precipitate the aryloxyacetic acid. The acid is recrystallized from water. The melting points of about fifty of these derivatives of phenols have been reported in the literature.

Example 13.19: Preparation of an Aryloxyacetic Acid from m-*Cresol*

Place in an 8-in. test tube 200 mg of *m*-cresol, 1 ml of $6N$ sodium hydroxide solution, and 0.5 ml of a 50 per cent solution of chloroacetic acid.

Provide the tube with a microcondenser arranged for reflux and heat in a water bath at 90 to 100° for about 1 hour. Cool and add 3 ml of water and 1 ml of 6N hydrochloric acid. Extract with two 4-ml portions of ether; wash the combined ethereal solutions first with 2 ml of water and then with 5 ml of 10 per cent sodium carbonate solution, which removes the aryloxyacetic acid. Transfer the sodium carbonate solution to a beaker and slowly add dilute hydrochloric acid until the solution is distinctly acid. Cool and filter the crystals. The yield is 45 to 55 mg of *m*-tolyloxyacetic acid.

Other Derivatives

Bromo Derivatives

Phenols react rapidly with bromine to give bromo-substituted phenols, which in many cases are useful derivatives. Thus, phenol forms 2,4,6-tribromophenol, while *o*-cresol and *m*-cresol yield, respectively, dibromo and tribromo derivatives.

$$C_6H_5OH + 3Br_2 \rightarrow C_6H_2Br_3OH + 3HBr$$
$$\text{2,4,6-Tribromophenol}$$

The following example illustrates a general method for bromination of phenols.[16] Since bromination of phenols is relatively easy, the preparation of the bromo derivative should be considered.

Example 13.20: Preparation of Tribromophenol

Place in an 8-in. tube 0.8 g of potassium bromide and add 5 ml of water. Shake the tube until the salt dissolves; add *carefully* 0.5 g of bromine. Place in a 6-in. test tube 100 mg of phenol, 1 ml of methanol, and 1 ml of water. Add about 1.5 ml of the prepared bromine solution and shake the tube; continue the addition of bromine solution until the mixture retains a yellow color after shaking. Add 3 to 4 ml of water and shake vigorously. Filter the bromophenol and wash well with water. Dissolve the crystals in hot methanol and filter; add water dropwise to the methanol solution until a permanent cloudiness results. The yield is 180 to 200 mg of crystals, which melt at 95°.

2,4-Dinitrophenyl Ethers

The reaction of a phenol with 2,4-dinitrochlorobenzene produces a 2,4-dinitrophenyl ether:

$$ArONa + ClC_6H_3(NO_2)_2 \rightarrow ArOC_6H_3(NO_2)_2 + NaCl$$
$$\text{2,4-Dinitrochlorobenzene} \qquad \text{2,4-Dinitrophenyl ether}$$

---

[16] For a more extensive discussion of bromination of small quantities, see N. D. Cheronis, "Micro and Semimicro Methods," in *Technique of Organic Chemistry*, Vol. 6, A. Weissberger, ed., Interscience, New York, 1954, pp. 286–289.

The phenol is dissolved in sodium hydroxide solution and is mixed with an alcoholic solution of 2,4-dinitrochlorobenzene. The mixture is refluxed for about one half hour and diluted with water; the precipitated dinitrophenyl ether is filtered and crystallized from alcohol.

The melting points of about thirty 2,4-dinitrophenyl ethers have been reported. About 5 millimoles of the phenol are dissolved in 0.8 ml of $6N$ sodium hydroxide solution and added to 5 millimoles of 1-chloro-2,4-dinitrobenzene dissolved in 15 ml of 95 per cent ethyl alcohol. In some cases more alcohol must be added to keep the substance in solution. The precipitate sometimes formed is an addition product and is not to be confused with the salt, which forms only after heating. The solution, which is always highly colored, is refluxed for about 30 minutes on a steam bath until most of the color disappears and a precipitate of sodium chloride appears. The 2,4-dinitrophenyl ether is precipitated by the addition of an equal amount of water, filtered, washed with water, and recrystallized from alcohol.

Reduction of Nitrophenols to Aminophenols

Nitrophenols are readily converted to aminophenols by catalytic hydrogenation. Thus 100 mg of *p*-nitrophenol can be converted to *p*-aminophenol within 20 to 30 minutes by catalytic hydrogenation at atmospheric pressure by the general method described in the note of Example 20.05. The yield is almost quantitative and one crystallization yields a product that melts within 1° of the value listed in the literature for the melting point of *p*-aminophenol. Similarly, *o*-nitrophenol yields *o*-aminophenol and ethyl-*p*-nitrophenol the corresponding crystalline amino derivative. In general, where the corresponding amino derivative is a crystalline compound melting between 60° and 220° the derivatization of the nitrophenol is readily accomplished even with quantities of a few milligrams.

Chromatography of Phenols

The general method described for the preparation and chromatography of the 3,5-dinitrobenzoates of the alcohols is followed (see Example 13.13). The solvents employed are butanol saturated with $5N$ ammonium hydroxide,[17] butanol-pyridine and saturated ammonium chloride,[18] and butanol-acetic acid.[19] For most exploratory work 4 parts of butanol,

---

[17] E. Lederer, *Australian J. Sci.*, **11**, 208 (1949).
[18] R. A. Evans et al., *Nature*, **164**, 674 (1949); **170**, 249 (1952).
[19] S. Rydel and M. Macheboeuf, *Bull. soc. chim. biol.*, **31**, 1265 (1949).

1 part of acetic acid, and 5 parts of water are shaken together and after separation the upper organic layer is removed and employed as a solvent.

The spots on the strips are located by means of several indicators, the most useful of which are diazotized sulfanilic acid,[20] ammoniacal silver nitrate,[21] and ferric chloride.[22] Other indicators are phosphomolybdic acid[23] and diazotized *p*-nitroaniline.[20]

The *p*-phenylazobenzoates of phenols have an advantage in the separation and identification of phenols by chromatography since they are colored. These esters are formed from *p*-phenylazobenzoyl chloride and the phenol by refluxing in pyridine for four hours. Chromatographic separations are made on columns of 1:1 mixtures of alumina and Celite or of 2:1 mixtures of silicic acid and Celite. Some mixtures of phenols are more easily separated on one adsorbent than on the other.[24]

REFERENCES

For Alcohols

3,5-Dinitrobenzoates and Other Nitrobenzoates

3,5-Dinitrobenzoates: Adamson and Kenner, *J. Chem. Soc. (London)*, **1935**, 287; Bryant, *J. Am. Chem. Soc.*, **54**, 3758 (1932); King, *J. Am. Chem. Soc.*, **61**, 2383 (1939); Lipscomb, Malone, and Reid, *J. Am. Chem. Soc.*, **51**, 3424 (1929); Reichstein, *Helv. Chim. Acta*, **9**, 799 (1926); Dunbar and Ferrin, *Microchem. J.*, **3**, 65–82 (1959).

3,5-Dinitrobenzoates of hexanols: Sutter, *Helv. Chim. Acta*, **21**, 1266 (1938).

o-Nitrobenzoates: Lowe, *J. Am. Chem. Soc.*, **74**, 841 (1952).

p-Nitrobenzoates: Adamson and Kenner, *J. Chem. Soc. (London)*, **1935**, 287; Armstrong and Copenhaver, *J. Am. Chem. Soc.*, **65**, 2252 (1943); Henstock, *J. Chem. Soc. (London)*, **1933**, 216; King, *J. Am. Chem. Soc.*, **61**, 2383 (1939); Meisenheimer and Schmidt, *Ann.*, **475**, 157 (1929).

Urethan Derivatives

p-Anisylurethans: Brunner and Wohrl, *Monatsh.*, **63**, 374 (1933).

3,5-Dinitrophenylurethans: Hoeke, *Rec. trav. chim.*, **54**, 505 (1935).

α-Naphthylurethans: Bickel and French, *J. Am. Chem. Soc.*, **48**, 747 (1926); French and Wirtel, *J. Am. Chem. Soc.*, **48**, 1736 (1926); Neuberg and Kansky, *Biochem. Z.*, **20**, 445 (1909).

m-Nitrophenylurethans: Hoeke, *Rec. trav. chim.*, **54**, 505 (1935); Veibel et al., *Dansk Tidsskr. Farm.*, **14**, 241 (1940); **17**, 187 (1943).

[20] R. L. Hossfeld, *J. Am. Chem. Soc.*, **73**, 852 (1951); R. L. Hossfeld et al., *J. Am. Chem. Soc.*, **74**, 5766 (1952).
[21] H. G. Bray et al., *Biochem. J.* **46**, 271 (1950); R. S. Asquith, *Nature*, **168**, 738 (1951)
[22] E. A. H. Roberts, *Biochem. J.*, **49**, 414 (1951).
[23] R. F. Riley, *J. Am. Chem. Soc.*, **72**, 5782 (1950).
[24] E. O. Woolfolk and J. M. Taylor, *J. Org. Chem.*, **22**, 827–829 (1957).

p-Nitrophenylurethans: Shriner and Cox, *J. Am. Chem. Soc.*, **53**, 1601, 3186 (1931); Van Hoogstraten, *Rec. trav. chim.*, **51**, 414 (1932).

4-Phenylazophenylurethans: Masuyama and Hamada, *J. Chem. Soc. Japan, Pure Chem. Sect.*, **70**, 198 (1949).

Phenylurethans: Dewey and Witt, *Ind. Eng. Chem., Anal. Ed.*, **12**, 459 (1940); **14**, 648 (1942); Lanbling, *Bull. soc. chim.* (*3*), **19**, 771 (1898); McKinley et al., *Ind. Eng. Chem., Anal. Ed.*, **16**, 304 (1944); Witten and Reed, *J. Am. Chem. Soc.*, **69**, 2470 (1947).

Substituted azides and urethans: Sah et al., *Science Repts., Natl. Tsing Hua Univ.* (*A*), **3**, 109 (1935); *J. Chin. Chem. Soc.*, **2**, 229 (1934); *Rec. trav. chim.*, **58**, 453, 582, 591, 595, 1013 (1939); **59**, 238, 357 (1940).

p-Xenylurethans: Morgan and Hardy, *Chemistry and Industry*, **1933**, 519; Morgan and Pettet, *J. Chem. Soc.* (*London*), **1931**, 1124.

β-Bromopropionylcarbamates: Johnson, Kreyssler, and Needles, *J. Org. Chem.*, **25**, 279–281 (1960).

Other Derivatives

Identification and detection of volatile alcohols and acids in biological materials: Friedemann and Brook, *J. Biol. Chem.*, **123**, 161 (1938).

Oxidation of alcohols to aldehydes and preparation of the 2,4-dinitrophenylhydrazones: Duke and Witman, *Anal. Chem.*, **20**, 490 (1948).

Acetates of tertiary alcohols: Spassow, *Ber.*, **70**, 1926 (1937); **75**, 779 (1942).

N-(Alkoxymethyl)phthalimides: Maxera and Lemberger, *J. Org. Chem.*, **15**, 1253 (1953).

Allophanates: Behal, *Compt. rend.* **168**, 945 (1919); Lane, *J. Chem. Soc.* (*London*), **1951**, 2764.

Aryloxyacetic acids: Hayes and Branch, *J. Am. Chem. Soc.*, **65**, 1555 (1943); Koelsch, *J. Am. Chem. Soc.*, **53**, 305 (1931).

S-Benzylthiuronium derivatives: Bair and Suter, *J. Am. Chem. Soc.*, **64**, 1978 (1942).

2,4-Dinitrophenyl ethers: Bost and Nicholson, *J. Am. Chem. Soc.*, **57**, 2368 (1935).

2,4-Dinitrophenylurethans from, 1-(2,4-dinitrophenyl)-3-methyl-3-nitrourea: Van Ginkel, *Rec. trav. chim.*, **61**, 149 (1942).

Hydrogen phthalates: Fessler and Shriner, *J. Am. Chem. Soc.*, **58**, 1384 (1936); Goggans and Copenhaver, *J. Am. Chem. Soc.*, **61**, 2909 (1939); Reid, *J. Am. Chem. Soc.*, **39**, 1250 (1917).

Substituted hydrogen phthalates: Lawlor, *J. Ind. Eng. Chem.*, **39**, 1419, 1424 (1947).

p-Nitrophenylacetyl derivatives: Ward and Jenkins, *J. Org. Chem.*, **10**, 371 (1945).

3-Nitrophthalates: De Graef and Pierret, *Bull. soc. chim. Belg.*, **57**, 307 (1948); Dickinson, *J. Am. Chem. Soc.*, **59**, 1094 (1937); Dickinson, Crosson, and Copenhaver, *J. Am. Chem. Soc.*, **59**, 1094 (1937); Nicolet and Sacks, *J. Am. Chem. Soc.*, **47**, 2348 (1925); Veraguth and Diehl, *J. Am. Chem. Soc.*, **62**, 233 (1940).

p-Phenylazobenzoates: Woolfolk, Beach, and McPherson, *J. Org. Chem.*, **20**, 391 (1955).

2,4,6-Trinitrobenzoates: Laskowski and Adams, *Anal. Chem.*, **31**, 148–152 (1959); Pseudosaccharin ethers: Böhme and Opper, *Z. anal. Chem.*, **139**, 255 (1953). Meadow and Reid, *J. Am. Chem. Soc.*, **65**, 457 (1943).

Trityl ethers: Seikel and Huntress, *J. Am. Chem. Soc.*, **25**, 495 (1942); Sobetay, *Compt. rend.*, **203**, 1164 (1936).

Xanthates: Shupe, *J. Assoc. Official Agr. Chem.*, **25**, 495 (1942); Whitmore and Lieber, *Ind. Eng. Chem., Anal. Ed.*, **7**, 127 (1935).

Miscellaneous: Naphthylamine addition products of 3,5-dinitrobenzoates, Benfey et al., *J. Org. Chem.*, **20**, 1777 (1955); Sutter, *Helv. Chim. Acta*, **21**, 1266 (1938). Alkyl 3,5-dihydroxybenzoates, Suter and Weston, *J. Am. Chem. Soc.*, **61**, 531 (1939). Fluorenyl, triphenylmethyl, and iodobiphenyl substituted urethans, Witten and Reid, *J. Am. Chem. Soc.*, **69**, 2470 (1947); Kawai and Tamura, *J. Chem. Soc. Japan*, **52**, 77 (1931). S-Benzylthiuronium deriv., Schotte and Veibel, *Acta Chem. Scand.*, **7**, 1357 (1953). Derivs. of glycols, Nason and Manning, *J. Am. Chem. Soc.*, **62**, 1635, 3136 (1940). Tri-iodobenzoates, O'Donnell et al., *J. Am. Chem. Soc.*, **68**, 1865 (1946); **70**, 1657 (1948).

Chromatographic Procedures

Cerbulis, *Anal. Chem.*, **27**, 1400 (1955). Chromatography of sugar alcohols and glycosides.

Hough, *Nature*, **165**, 400 (1950). The separations of polyhydroxy compounds by paper chromatography and column chromatography.

Kariyone and Hashimoto, *Nature*, **168**, 511 (1951). Potassium xanthates.

Lederer and Lederer, *Chromatography*, Elsevier, New York, 1953, p. 106. The separation of polyhydroxy compounds by paper chromatography and column chromatography.

Masuyama and Hamada, *J. Chem. Soc. Japan, Pure Chem. Sect.*, **70**, 198 (1949). Chromatographic separation of aliphatic 4-phenylazophenylurethans.

Meigh, *Nature*, **169**, 706 (1952). Microdetection of 3,5-dinitrobenzoates of alcohols by paper chromatography.

Momosa and Yamada, *J. Pharm. Soc. Japan*, **71**, 980 (1951). 3,6-Dinitrophthalates.

Rice et al., *Anal. Chem.*, **23**, 195 (1951). Microdetection of 3,5-dinitrobenzoates of alcohols by paper chromatography.

Spayner and Phillips, *Anal. Chem.*, **28**, 253 (1956). Separation as xanthates.

Sundt and Winter, *Anal. Chem.*, **29**, 851–852 (1957). Chromatographic separation of 3,5-dinitrobenzoates of alcohols.

Woolfolk, Beach, and McPherson, *J. Org. Chem.*, **20**, 391 (1955). *p*-Phenylazobenzoates.

For Phenols

Microscopic identification of dinitrophenols: Van Zijp, *Pharm. Weekblad*, **78**, 745 (1941).

Acetates: Chattaway, *J. Chem. Soc. (London)*, **1931**, 2495.

Aryloxyacetamides: Namstkin et al., *Zhur. Anal. Khim.*, **5**, 7 (1950); *C.A.*, **44**, 4375 (1950).

Aryloxyacetic acids: Hayes and Branch, *J. Am. Chem. Soc.*, **65**, 1555 (1943); Koelsch, *J. Am. Chem. Soc.*, **53**, 304 (1931).

Benzoates: Baumann, *Ber.*, **19**, 3218 (1886); Schotten, *Ber.*, **17**, 2544 (1884).

p-Bromo-, *p*-chloro-, *β*-naphthyl-, *p*-nitro-, and 3,5-dinitro-4-methylurethans: Sah et al., *Rec. trav. chim.*, **58**, 453, 582, 591, 595 (1939).

3,5-Dinitrobenzoates: Brown and Kremers, *J. Am. Pharm. Assoc.*, **11**, 607 (1922); Phillips and Keenan, *J. Am. Chem. Soc.*, **53**, 1924 (1931); Reichstein, *Helv. Chim. Acta*, **9**, 799 (1926).

2,4-Dinitrophenyl ethers: Bost and Nicholson, *J. Am. Chem. Soc.*, **57**, 2368 (1935).

3,5-Dinitrophenylurethans: Hoeke, *Rec. trav. chim.*, **54**, 514 (1935); Sah and Ma, *J. Chinese Chem. Soc.*, **2**, 229 (1934); Veibel and Lillelund, *Dansk Tidsk. Farm.*, **14**, 241 (1940); Veibel et al., *Dansk Tidsk. Farm.*, **17**, 187 (1943).

Diphenylurethans: Herzog, *Ber.*, **40**, 1831 (1907).

α-Naphthylurethans: French and Wirtel, *J. Am. Chem. Soc.*, **48**, 1736 (1926); Sah et al., *Rec. trav. chim.*, **58**, 453, 582, 591, 595 (1939).

p-Nitrobenzoates: Adamsen and Kenner, *J. Chem. Soc. (London)*, **1935**, 287; Armstrong and Copenhaver, *J. Am. Chem. Soc.*, **65**, 2252 (1943); Henstock, *J. Chem. Soc. (London)*, **1933**, 216; King, *J. Am. Chem. Soc.*, **61**, 2383 (1939); Meisenheimer and Schmidt, *Ann.*, **475**, 157 (1929).

p-Nitrobenzyl ethers: Lyman and Reid, *J. Am. Chem. Soc.*, **42**, 615 (1920); Reid, *J. Am. Chem. Soc.*, **39**, 304 (1917).

p-Nitrophenylacetates: Ward and Jenkins, *J. Org. Chem.*, **10**, 371 (1945).

p-Phenylazobenzoates: Woolfolk and Taylor, *J. Org. Chem.*, **22**, 827–829 (1957).

Phenylurethans: McKinley et al., *Ind. Eng. Chem., Anal. Ed.*, **16**, 304 (1944).

Pseudosaccharin ethers: Meadow and Reid, *J. Am. Chem. Soc.*, **65**, 457 (1943).

Picrates: Baril and Hauber, *J. Am. Chem. Soc.*, **53**, 1087 (1931).

Sulfonic acid esters: Hazlet, *J. Am. Chem. Soc.*, **60**, 399 (1941); Sekera, *J. Am. Chem. Soc.*, **55**, 421 (1933).

Use of substituted azides for urethans: Hoeke, *Rec. trav. chim.*, **54**, 514 (1935); Sah et al., *J. Chinese Chem. Soc.*, **2**, 229 (1934); *Rec. trav. chim.*, **58**, 453, 582, 591, 595, 1013 (1939); **59**, 238, 357 (1940); *Science Repts., Natl. Tsing Hua Univ. (A)*, **3**, 109 (1935); Veibel et al., *Dansk Tidsskr. Farm.*, **14**, 241 (1940).

14

Derivatives of Aldehydes, Ketones, and Acetals

The principal discussion in this chapter pertains to the preparation of derivatives suitable for the characterization of aldehydes; however, a large number of the derivatives and the procedures for their preparation apply equally to the characterization of ketones, ketoesters, ketoacids, and many of the polyfunctional compounds that have a carbonyl group. For this reason the discussion on the preparation of derivatives of ketones is brief.

Table 14.01 lists many of the compounds proposed in the literature for the derivatization of carbonyl compounds. Those marked with an asterisk are the ones that should be first considered and a detailed discussion of their preparation and purification is given in the following sections. A brief discussion of a few of the other derivatives appears in a later section. All factors should be considered in selecting the particular derivative to be prepared. For example, the recommended derivatives for glyoxal are the semicarbazone (mp 270°), 2,4-dinitrophenylhydrazone (mp 328°), phenylhydrazone (mp 180°), and methone (mp 228°). As a first trial the phenylhydrazone should be selected since it usually forms readily and its melting point is below 200°, but if it is found unsuitable owing to decomposition on purification (that is, if the melting point consistently differs by 5 to 10° from the value listed in the literature), the methone derivative should be selected.

Derivatives of Aldehydes, Ketones, and Acetals 497

TABLE 14.01
DERIVATIVES FOR THE IDENTIFICATION OF ALDEHYDES

4-Aminoantipyrine derivatives	Hydrazinobenzoic acid derivatives
Aminomorpholines	1,2-Bis(*p*-methoxybenzylamino)ethane
Benzothiazoles	derivatives
Benzothiazolines	2-Naphthylhydrazones
Bromobenzohydrazides (*o*-, *m*-, *p*-)	1-Naphthylsemicarbazones
p-Carboxyphenylhydrazones	2-Naphthylsemicarbazones
m-Chlorobenzohydrazides	Nitrobenzenesulfonhydrazones
p-Chlorobenzohydrazides	3-Nitrobenzohydrazides
1,3-Cyclohexanedione derivatives	Nitroguanylhydrazones
Dibromomethone derivatives	**p*-Nitrophenylhydrazones
*Dimethone or Dimedone derivatives	*Oximes
Diphenylhydrazones	*Phenylhydrazones
*2,4-Dinitrophenylhydrazones	5-(α-Phenylethyl)semioxamazide
α-(2,4-Dinitrophenyl)-α-methylhydrazones	derivatives
3,5-Dinitrophenylsemicarbazones	Phenylsemicarbazones
2-Diphenylacetyl-1,3-indandione	*Semicarbazones
1-hydrazone derivatives	Solid alcohols obtained on reduction
p-Dimethylaminoanils	Solid carboxylic acids obtained on
Hydantoins	oxidation
	*Thiosemicarbazones
	Tolylsemicarbazones (*o*-, *m*-, *p*-)
	Xenylsemicarbazones

* Recommended derivative. Bibliography to all derivatizations will be found on pages 515 to 518.

Phenylhydrazones

Discussion

The reaction of phenylhydrazine with an aldehyde yields a phenylhydrazone:

$$\underset{}{RC\!\!=\!\!O} + \underset{\text{Phenylhydrazine}}{H_2NNHC_6H_5} \rightarrow \underset{\text{Phenylhydrazone}}{\overset{H}{RC\!\!=\!\!NNHC_6H_5}} + H_2O$$

Since phenylhydrazine is easily available and the derivatives form readily, the use of phenylhydrazones, particularly for the aryl carbonyl compounds, is advisable. For the preparation of the derivatives, the carbonyl compound is dissolved in methanol or ethanol and heated with phenylhydrazine base and a small amount of acetic acid; the phenylhydrazone separates even while the solution is hot. After filtration the derivative should be dried rapidly and the melting point determined at once, since phenylhydrazones as a rule undergo slow decomposition when dried in air.

In general, when derivatives of phenylhydrazine are involved, we recommend that the product be crystallized immediately and dried as rapidly as possible for melting-point determination.

A number of carbonyl compounds fail to give stable phenylhydrazones even when the derivatives are prepared from pure reagents and with the utmost care. For example, cyclohexanone and acetophenone yield derivatives that melt 5 to 10° below the melting point of the pure compound; even after crystallization the derivatives undergo change in the melting point when dried in a desiccator.

General Method for Preparation

About 1 millimole of the carbonyl compound is dissolved in 4 to 5 ml of methanol or ethanol, and 1.05 millimoles of phenylhydrazine base are added, followed by an equivalent amount of glacial acetic acid. The mixture is refluxed for 3 to 15 minutes, depending on the reactivity of the carbonyl group. The mixture is cooled and water is added dropwise until a cloudiness indicates the separation of the derivative. The mixture is then chilled until precipitation is complete and the phenylhydrazone is filtered, washed, and purified as usual. The actual yield of the pure derivative varies from 40 to 50 per cent of the theoretical. The following example illustrates the preparation of the phenylhydrazone from an aromatic aldehyde.

Example 14.01: Phenylhydrazone of an Aryl Aldehyde:
Piperonal Phenylhydrazone

Place in an 8-in. tube 150 mg of piperonal and 5 ml of methanol. Heat for a few seconds to effect solution; then add 0.1 ml of phenylhydrazine. (CAUTION. *Use care in handling the reagent; some workers are sensitive to it and develop skin rashes.*) Boil the mixture for 1 minute and add 1 drop of glacial acetic acid and boil gently for 3 minutes. Add dropwise 1.5 ml of water until a permanent cloudiness results. Cool, filter the crystals, and wash with 1 ml of water containing 1 drop of acetic acid. Recrystallize the product immediately by dissolving it in 3 ml of hot methanol. Add 0.5 ml of water to the hot solution, cool, and scratch the sides of the tube if crystals do not separate readily. Filter, wash with a few drops of 50 per cent methanol, and dry rapidly by pressing the crystals on a porous clay plate. Determine the melting point as soon as the crystals are dry. The yield is about 80 to 100 mg of the product, melting at 99 to 100°.

NOTE. Other aromatic aldehydes, such as vanillin and *m*-nitrobenzaldehyde, yield phenylhydrazones readily by the same method as described for piperonal.

2,4-Dinitrophenylhydrazones and p-Nitrophenylhydrazones

Discussion

The equations below represent the formation of p-nitrophenylhydrazones and 2,4-dinitrophenylhydrazones:

$$\underset{p\text{-Nitrophenylhydrazine}}{RC\overset{H}{=}O + H_2NNHC_6H_4(NO_2)} \rightarrow \underset{p\text{-Nitrophenylhydrazone}}{RCH=NNHC_6H_4(NO_2)} + H_2O$$

$$\underset{\text{2,4-Dinitrophenylhydrazine}}{CH_3CHO + H_2NNHC_6H_3(NO_2)_2} \rightarrow \underset{\text{Ethanal 2,4-dinitrophenylhydrazone}}{CH_3CH=NNHC_6H_3(NO_2)_2} + H_2O$$

The preparation of nitrosubstituted phenylhydrazones, particularly the 2,4-dinitrophenylhydrazones, is advisable for semimicro quantities. In most cases it is possible to start with as little as 20 to 30 mg of the carbonyl compound and obtain a sufficient quantity of the pure derivative for several determinations of the melting point. For the preparation of the 2,4-dinitrophenylhydrazones, the carbonyl compound and a small amount of hydrochloric acid are added to a hot saturated alcoholic solution of 2,4-dinitrophenylhydrazine. The derivative that separates out on cooling in many cases does not require further purification. In some cases the use of nitrosubstituted phenylhydrazones may be limited by the high melting points of the derivatives; in such instances the use of phenylhydrazones may be found suitable. For example, the 2,4-dinitrophenylhydrazone of piperonal melts at 266°, and the semicarbazone of the same aldehyde melts at 234°; in this case the derivative with phenylhydrazine melting at 102° is preferable, since it may be prepared easily and is relatively stable. Another limitation of the 2,4-dinitrophenylhydrazones is that they are not very satisfactory for α-hydroxy aldehydes, ketones, and sugars, and a number of substituted aldehydes and ketones. In such cases the initial products often undergo secondary reactions; for example, the derivative of an α-hydroxy aldehyde may slit off a molecule of water. The evaluation of the melting points of the 2,4-dinitrophenylhydrazones often leads to difficulties on account of the tendency of this group to polymorphism, and also because of the possibility of geometrical isomers from the C=N linkage.

More than one modification has been shown to exist for the 2,4-dinitrophenylhydrazones of acetaldehyde,[1] furfuraldehyde,[2] propanal, butanal, and acetone.[3] The differences in the melting points of the modifications

[1] W. M. D. Bryant, *J. Am. Chem. Soc.*, **54**, 3758 (1932); **55**, 3201 (1933); **58**, 2335 (1936); **60**, 2815 (1938).
[2] H. Bredereck, *Ber.*, **65**, 1833 (1932).
[3] G. L. Clark, W. I. Kaye, and T. D. Parks, *Ind. Eng. Chem., Anal. Ed.*, **18**, 311 (1946).

may be 1 to 2°, as for the polymorphs of propanal and butanal, or as much as 24°, as in the case of furfural. When relatively pure compounds are used no difficulty will be experienced with the methods outlined in the following sections, but when mixtures are used the literature[4] should be consulted on melting points and on the use of crystallographic data and other optical properties as supplementary information. In general it is advised to determine the melting point of the 2,4-dinitrophenylhydrazone, allow the melt to solidify and then redetermine the melting point.

General Method for Preparation

The ratio of the reactants is 0.5 to 0.6 millimole of the carbonyl compound to 0.4 millimole of the nitrophenylhydrazine, which is about 40 mg. The nitrophenylhydrazine is placed in a tube containing 4 ml of methanol and 0.1 ml of $6N$ hydrochloric acid, and dissolved by heating the tube in a water bath. Another method of preparing a stable solution is to pass hydrogen chloride gas through 50 ml of methanol until 3.5 g have been absorbed; about 2 g of 2,4-dinitrophenylhydrazine are added and the mixture is shaken until the solid dissolves. The solution contains 40 mg/ml.

The carbonyl compound (0.5 to 0.6 millimole) dissolved in 1 ml of methanol is added to an amount of the reagent containing 40 mg and the mixture is heated in the water bath (50 to 60°) for 1 to 2 minutes and then allowed to stand for 15 to 30 minutes, depending on the reactivity of the carbonyl compound. The derivatives often separate out on cooling, but it is advisable to add water dropwise to cloudiness. The derivatives are purified by crystallization from a solvent pair, usually alcohol and water. When the solubility in boiling methanol or ethanol is low, dioxane, ethyl acetate, toluene, or xylene may be used.

Several acids (sulfuric, phosphoric, and perchloric) has been proposed to effect solution of the 2,4-dinitrophenylhydrazine in ethanol or methanol. It is claimed that the derivative separates more rapidly from such solutions than from the hydrochloride outlined in the preceding section. For example, 2 g of 2,4-dinitrophenylhydrazine are dissolved in 25 ml of 85 per cent phosphoric acid by heating the mixture on a steam bath and diluted after cooling to 50 ml with ethanol. The reagent solution contains 40 mg/ml and is employed as directed in the preceding section. In many cases the derivative separates at once without heating.

[4] H. A. Iddles and C. E. Jackson, *Ind. Eng. Chem., Anal. Ed.*, **6**, 454 (1934); N. R. Campbell, *Analyst* **61**, 391 (1936); C. F. H. Allen and J. H. Richmond, *J. Org. Chem.*, **2**, 222 (1937); M. Brandstatter, *Mikrochemie*, **32**, 33 (1944); G. Matthiessen and H. Hagedorn, *Mikrochemie*, **29**, 55 (1941); L. I. Braddock et al., *Anal. Chem.*, **25**, 301–306 (1953).

One of the difficulties in the use of 2,4-dinitrophenylhydrazine for the formation of derivatives is its low solubility in most solvents. This difficulty can be overcome by the use of the dimethyl ether of diethylene glycol (diglyme) as a solvent.[5] Solutions of 2,4-dinitrophenylhydrazine in this solvent are stable at room temperature, and may be stored until used. 2,4-Dinitrophenylhydrazones are prepared from this solution as follows.

Prepare a solution of 2,4-dinitrophenylhydrazine (1 g) in 30 ml of the dimethyl ether of diethylene glycol. To 5 ml of this solution, add 0.1 g of the carbonyl compound in 1 ml of 95% ethyl alcohol and 3 drops of concentrated hydrochloric acid. In some cases the hydrazone precipitates immediately; in others, it precipitates when water is added. Recrystallization is usually unnecessary.

For the preparation of *p*-nitrophenylhydrazones, the same general method is employed as for the phenylhydrazones (page 498). Following are examples of the preparation of *p*-nitrophenylhydrazones and 2,4-dinitrophenylhydrazones.

Example 14.02: Preparation of Benzaldehyde p-*Nitrophenylhydrazone*

Place in an 8-in. tube 8 ml of methanol and 50 mg of *p*-nitrophenylhydrazine; heat until solution is complete. Add 0.1 ml of benzaldehyde and boil for 1 minute; add 1 drop of glacial acetic acid and boil gently for 4 to 5 minutes. Then add water dropwise until a faint cloudiness results. Heat until the solution becomes clear and then cool. Filter the crystals and wash with 3 to 4 ml of water containing 2 drops of acetic acid. Crystallize the product by dissolving it in 7 to 10 ml of methanol and adding 1 to 1.5 ml of water to the hot solution. The yield is 80 to 100 mg of crystals, melting at 190°.

Example 14.03: Preparation of Formaldehyde 2,4-Dinitrophenylhydrazone

Place in an 8-in. tube 4 ml of methanol, 40 mg of 2,4-dinitrophenylhydrazine, and 0.1 ml of 6N HCl. Boil for a minute to effect solution. If all the solid does not dissolve, remove the flame and, after 1 minute, pour the clear solution into another test tube. Add to the clear solution 0.4 ml of 40 per cent aqueous formaldehyde solution and heat to boiling, then allow to stand for 5 minutes. If no precipitate appears, add water drop by drop to cloudiness. Filter the crystals and recrystallize from a mixture of 3 ml of methanol and 1 ml of water. The yield is about 20 mg of crystals, melting at 166 to 167°.

---

[5] H. J. Shine, *J. Org. Chem.*, **24**, 252–253 (1959).

Note. For aromatic carbonyl compounds, the amount of the materials (carbonyl compound, solvent, and reagent) may be reduced to one half. When 50 mg of benzaldehyde are used, 20 mg of the pure derivative are obtained after crystallization from 4 ml of methanol; o-chlorobenzaldehyde yields 25 mg of the pure derivative. With piperonal it is possible to obtain a pure derivative with 25 micrograms.

Semicarbazones and Thiosemicarbazones

Discussion

The formation of semicarbazones and thiosemicarbazones from carbonyl compounds is represented by:

$$C_6H_5CHO + H_2NNHCONH_2 \rightarrow C_6H_5CH{=}NNHCONH_2 + H_2O$$
Benzaldehyde — Semicarbazide — Benzaldehyde semicarbazone

$$CH_3(CH_2)_3CHO + H_2NNHCSNH_2 \rightarrow$$
Pentanal — Thiosemicarbazide
$$CH_3(CH_2)_3CH{=}NNHCSNH_2 + H_2O$$
Pentanal thiosemicarbazone

Generally, carbonyl compounds react rapidly with semicarbazide to yield crystalline derivatives, so that in many cases the separation of crystals begins upon warming of the reaction mixture. However, this fact should not be taken as evidence that all carbonyl compounds react readily with semicarbazide for, with the lower aldehydes, the time required for complete reaction is several days. For example, a mixture of formaldehyde and semicarbazide does not yield a crystalline derivative even after 10 days; if sodium acetate is omitted from the reaction mixture, an amorphous polycondensation product of formaldehyde and semicarbazide is formed. Acetaldehyde produces slowly a soluble semicarbazone that is not easily isolated. Complications may also rise on prolonged standing or heating in order to complete the reaction, such as the formation of acetylsemicarbazones and hydrazodicarbonamides. Therefore, it is advisable not to attempt the preparation of semicarbazones of aldehydes with less than 5 carbon atoms.

The semicarbazones may be used for identification by microscopy of aldehydes and ketones.[6] One advantage over some other methods is that individual members of a mixture may be identified without the necessity of separating the mixture into its components.

Thiosemicarbazide is a valuable reagent for the preparation of derivatives of semimicro quantities of the lower aldehydes and ketones, particularly in the presence of alcohols. Since it is approximately ten times as

[6] R. E. Dunbar and A. E. Aaland, *Microchem. J.*, **2**, 113–127 (1958).

costly as semicarbazide, its use is indicated only when other derivatives are found unsuitable.

Both semicarbazones and thiosemicarbazones from unsymmetrical carbonyl compounds are capable of existing in two stereoisomeric forms. Thus a pure isomer may undergo a rearrangement under the effect of heating and be converted to a mixture that melts at a lower temperature than either of the pure isomers. Hence it is advisable, when the melting point of the derivative is lower or higher than that listed in the literature, that the same derivative be prepared from a pure sample of the compound thought to be the unknown and its melting point determined in the same apparatus using the same rate of heating.

General Method for Preparation

The semicarbazones are formed when the carbonyl compounds are heated with aqueous solutions of the semicarbazide hydrochloride buffered with sodium acetate. The ratio employed is 3 millimoles of semicarbazide hydrochloride in 2 ml of water buffered with about 4 millimoles of sodium acetate to 2.75 millimoles of the carbonyl compound. The mixture is heated for 10 to 15 minutes and then allowed to stand. In the cases of the higher carbonyl compounds the separation of the derivative begins immediately, but in most cases the reaction is complete after 1 hour. The crude derivative is filtered and crystallized from the minimum amount of water. When the solubility of the semicarbazone in water is low (as in the case of the semicarbazones of the aromatic carbonyl compounds), the derivative is crystallized from methanol or ethanol and water. When the semicarbazone is sparingly soluble in hot alcohol, purification may be effected by extracting the derivative with a small amount of the hot solvent.

The same general procedure is followed for the preparation of the thiosemicarbazones. The period of heating required for the initial separation of the derivative and completion of the reaction is much less than that required for the semicarbazones. For example, 10 mg of acetone yield a thiosemicarbazone in 10 to 15 minutes, but require 1 hour to yield the semicarbazone.

Example 14.04: Preparation of the Semicarbazone of an Aldehyde: Benzaldehyde Semicarbazone

Place in an 8-in. tube 100 mg of semicarbazide hydrochloride, 150 mg of sodium acetate, 1 ml of water, and 1 ml of alcohol. Add 0.1 ml of benzaldehyde and heat the tube in a water bath at 70° for 10 minutes. Add

2 ml of water and cool. Filter the crystals and wash with two 1-ml portions of water. Recrystallize from a 3:1 mixture of methanol and water. The yield is 80 mg of solid, melting at 221 to 222°.

NOTE. The semicarbazones of aromatic aldehydes have melting points usually above 200° and care must be taken to apply proper thermometer corrections.

Dimethone Derivatives

Discussion

The reaction of an aldehyde with 5,5-dimethyl-1,3-cyclohexanedione, commonly known as "methone" reagent (also called in the literature dimethyldihydroresorcinol), is represented by the equation:

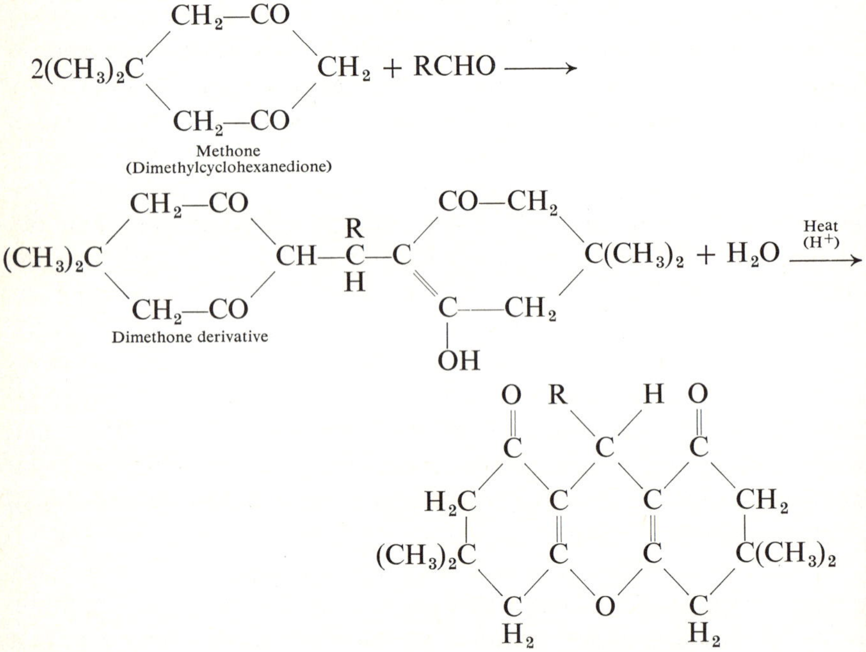

$$2(CH_3)_2C\begin{matrix}CH_2-CO\\ \\CH_2-CO\end{matrix}CH_2 + RCHO \longrightarrow$$

Methone
(Dimethylcyclohexanedione)

$$(CH_3)_2C\begin{matrix}CH_2-CO\\ \\CH_2-CO\end{matrix}CH-\underset{H}{\underset{|}{C}}-C\begin{matrix}CO-CH_2\\ \\C-CH_2\end{matrix}C(CH_3)_2 + H_2O \xrightarrow{\text{Heat} \atop (H^+)}$$

Dimethone derivative

Octahydroxanthene derivative

One mole of aldehyde condenses with 2 moles of the reagent, and therefore derivatives are often named with the prefix of the aldehyde and the ending dimethone or "dimedone," for example, formaldimethone and acetaldimethone. The reaction is not given by ketones and is helpful for detecting traces of the aldehydes. The bromo-substituted (dibromomethone) reagent is very useful in spot-plate tests for the detection of micro quantities of aldehydes. The dimethones are more suitable for the lower aldehydes than are the 2,4-dinitrophenylhydrazones.

The methone derivatives of most aldehydes can be made to undergo cyclization to give octahydroxanthenes. The usual method is to heat the methone derivative with acetic anhydride or with alcohol containing a small amount of hydrochloric acid. The cyclization usually requires 5 minutes, and the yield is nearly quantitative. In most cases the melting points of the new derivatives (octahydroxanthenes) differ by more than 15° from those of the methones; therefore it is possible to prepare two different derivatives through the reaction. The dimethone is first prepared, and after the determination of the melting point it is then converted by cyclization to the xanthene derivative.

General Method for Preparation

A ratio of 2.1 millimoles of the reagent to 1 millimole of the aldehyde is employed. The reagent is dissolved in 3 to 4 ml of 50 per cent aqueous methanol or ethanol; the carbonyl compound is added and the mixture is heated to boiling for about 30 seconds and then allowed to crystallize for 3 to 4 hours. Prolonged heating is avoided in order to prevent the cyclization to the xanthene derivative. The crystals are filtered and crystallized by dissolving in the minimum amount of alcohol and then adding water until cloudiness appears.

Example 14.05: Preparation of Dimethone of Butanal

Place in a test tube 300 mg of methone, 3 ml of a 50 per cent methanol-water mixture, 50 mg of butanal and 1 drop of piperidine. Heat in a water bath under reflux for 10 minutes. If the solution is clear at this point, add water dropwise until a cloudiness appears, then cool the mixture. Filter and wash the crystals with two 1-ml portions of 30 per cent methanol. The crystals melt at 134 to 135°; the yield is 120 to 130 mg. If crystallization is necessary, the solid is dissolved in 1 ml of methanol and water is added dropwise until a permanent cloudiness results; the solution is warmed until clear and then cooled. About 60 mg of the methone derivative are obtained and additional amounts of crystals separate out from the filtrate on standing.

NOTE. The preparation of methone derivatives is recommended when only a small amount of the aldehyde is available. For example, 1 drop of butanal is added to a solution of 50 mg of methone, dissolved in 0.5 ml of methanol, in a small test tube and allowed to stand for 4 hours. A sufficient amount of the derivative is obtained for several determinations of the melting point. The methone derivatives separate out slowly from solutions; a cloudy filtrate is an indication that crystallization is incomplete; in such cases the mixture is corked and allowed to stand in the cold overnight. For the cyclization of the methone derivative, 50 to 100 mg of the crystals are dissolved in 2 to 4 ml of a hot 80 per cent methanol-water mixture; 1 drop of concentrated hydrochloric acid solution is added and the solution is heated under reflux for 5 minutes. Water is added dropwise until a cloudiness appears. The xanthene derivative separates on cooling. The values of the melting points of the xanthene derivatives do not appear in the tables of this text, but can be found by consulting the original article cited in the bibliography section.

Other Derivatives

N,N-Dimethyl-*p*-phenylenediamine is said to react with aldehydes, but not with ketones.[7] Since the resulting *p*-dimethylaminoanils are soluble in 20 to 30% acetic acid they can be separated from ketones and other neutral compounds which may be water-insoluble. These derivatives of aromatic aldehydes crystallize well, and serve not only for separation of aldehydes but also for their characterization. A typical procedure for forming this derivative follows.

Example 14.06: Preparation of the p-*Dimethylaminoanil of* o-*Chlorobenzaldehyde*

To a solution of 2 g of N,N-dimethyl-*p*-phenylenediamine in 85 ml of alcohol, add a solution of 1 g of *o*-chlorobenzaldehyde in 25 ml of alcohol. Cool the resulting solution for 2 hours, during which time yellow crystals will separate. Crystallize the resulting anil from absolute alcohol. Dry the crystals and determine their melting point, which should be 60 to 61° for this derivative.

2-Diphenylacetyl-1,3-indandione 1-hydrazone is a valuable reagent for identifying and characterizing carbonyl compounds, both aldehydes and ketones, and has some advantages over commonly used carbonyl reagents. Over 50 different carbonyl compounds containing a variety of functional groups in addition to the carbonyl group have been reacted with this reagent and the melting points of their derivatives reported.[8] A mixture of 5 millimoles of the carbonyl compound, 4.5 millimoles of 2-diphenylacetyl-1,3-indandione 1-hydrazone (1.58 g), 20 to 30 ml of chloroform, and two drops of concentrated hydrochloric acid is heated at reflux until a nearly clear, deeply colored solution is obtained (5 to 10 minutes). The solution is filtered while hot to remove unreacted reagent, and is diluted with methanol to precipitate the product. The product may be recrystallized once from a mixture of chloroform and methanol. The derivatives are strongly fluorescent in ultraviolet light and intensely colored. The yields are 95%.

[7] von G. E. Utzinger and F. A. Regenass, *Helv. Chim. Acta*, **37**, 1901 (1954).
[8] R. A. Braum and W. A. Mosher, *J. Am. Chem. Soc.*, **80**, 3048–3050 (1958).

or better. The melting points of the solids are well spread and in the useful range. Over 60 aldehydes and ketones have been characterized by this method. The derivatives should be useful in the chromatographic separation of aldehydes and ketones. The equation for formation of the derivative follows.

$$\text{(indanedione-aminohydrazone)} + R-\overset{O}{\underset{\|}{C}}-R' + H^+ \longrightarrow$$

(structure with N—N=C(R)(R') azine product)

Many cyclic aldehydes react even in dilute solutions with 4-aminoantipyrine to give Schiff bases.[9] The crystalline forms of the derivatives of various aldehydes are so distinctive that they can be used for microscopic identification. These aldehydes can be quantitatively determined on a semimicro scale by this method.

The oximes that are marked with asterisks in Table 14.01 are described later in this chapter because they are more suitable for identification of ketones. In general, oximation of aldehydes is not recommended for semimicro quantities unless there are directions in the literature for the preparation and purification of the oxime from 300 to 500 mg of the aldehyde. See the discussion of the derivatives of ketones for the general method of their preparation.

Many of the derivatives listed in Table 14.01 are substituted hydrazones and semicarbazones. These are prepared by the same general methods described in Examples 14.01 to 14.05. However, of the derivatizing reagents required for these derivatives only diphenylhydrazine and 2-naphthylhydrazine are commercially available. For other substituted hydrazones and semicarbazones the derivatizing reagent must be prepared by consulting the literature at the end of the chapter.

[9] O. Manns and S. Pfeifer, *Mikrochem. Acta*, **1958**, 630–637.

The conversion of aldehydes to carboxylic acids or alcohols is recommended only when these derivatives are solid and readily isolated and purified. The oxidation of *m*-nitrobenzaldehyde to *m*-nitrobenzoic acid by alkaline permanganate described in this section illustrates the first type of reaction; the catalytic reduction of salicylaldehyde illustrates the second type. In general, these methods for derivatization are undertaken only if other types of derivatives have been found unsuitable. Catalytic reductive methods can be applied successfully to quantities of material as low as 10 mg or less. The oxidative methods require 400 to 500 mg in most cases; although the oxidation of ketones to carboxylic acids is involved, it is possible in the case of methyl ketones to remove the methyl group by means of oxidation with sodium hypochlorite (haloform reaction). This reaction is particularly useful in the case of unsaturated ketones, since the double bond is not attacked.

Example 14.07: Oxidation of m-*Nitrobenzaldehyde to* m-*Nitrobenzoic Acid*

Place 500 mg of *m*-nitrobenzaldehyde, 5 ml of water, and 6 drops of $6N$ sodium hydroxide solution in a 125-ml Erlenmeyer flask. Place in another 125-ml flask 1 g of potassium permanganate with 25 ml of water and heat nearly to boiling. Add the permanganate solution to the mixture in small (5-ml) portions, shaking the flask constantly by hand. When about 20 ml of permanganate solution have been added, shake well and wash the sides of the flask with a small amount of water. Allow the mixture to stand for 2 minutes, heat nearly to boiling, and test for excess of permanganate by dipping a rod into the flask and transferring a drop of the liquid into a test tube containing 10 ml of water. If the purple color of the permanganate does not disappear, oxidation is complete; otherwise, more permanganate is added to the flask until the color persists. Filter the solution from the manganese dioxide, and change the filter paper after about 10 ml of solution have been filtered to facilitate rapid filtration. Return the colored solution to the flask; add 2 ml of dilute sulfuric acid and about 500 mg of sodium bisulfite. If the solution does not decolorize, add a small amount of solid bisulfite until the permanganate and adhering film of manganese dioxide have been changed to the manganous state. Cool and filter off the crystals of *m*-nitrobenzoic acid; wash with water and recrystallize from methanol. After the hot alcoholic solution of the acid has been filtered, add water dropwise until a permanent cloudiness results. Cool and filter. The yield is 450 to 500 mg. The filtration of manganese dioxide may be omitted and an excess of bisulfite added to convert all the manganese dioxide to the manganous salt. It has been found, however, that the purity of the product is higher if the manganese dioxide is removed by filtration.

Example 14.08: Catalytic Reduction of Salicylaldehyde to Salicyl Alcohol

Place 100 mg of 5 per cent palladium carbon in the hydrogenating tube (Figure 2.34) and add about 25 ml of ethanol. Pass hydrogen for 2 minutes and then add about 200 mg (0.25 ml) of salicylaldehyde. Heat the bath at 55 to 60° and pass hydrogen for 15 to 20 minutes following the procedure described on pages 40 to 45. Filter and wash the residue with two 5-ml portions of alcohol. Evaporate the filtrates in a dish over a water bath until a small amount of solvent remains, then conduct the evaporation slowly. Cool the dish in a freezing mixture and rub the oily residue, with a spatula, against the walls of the vessel until it solidifies. Scrape the crude mass and transfer it into a tube. Add 6 to 8 ml of heptane or petroleum ether, boil to effect solution, and filter by suction into an 8-in. test tube having a side arm. Extract the oily mass remaining in the solution tube, which is crude salicyl alcohol, using the filtrates from the first crystallization. Cool the filtered solution and scratch the inner surface of the tube with a glass rod. Allow to stand 10 minutes and filter. Conduct successive extractions of the crude salicyl alcohol, using the filtrates from the previous crystallization, until the crude residue is exhausted. Collect the crystals on the suction funnel and wash with a few milliliters of the pure solvent. Place the crystals on a clay plate. The yield is about 175 to 200 mg of crystals that melt at 86°.

Chromatographic Identification and Separation of Aldehydes

For the detection of microgram quantities of aldehydes in mixtures the carbonyl compounds are first derivatized,[10] then chromatographed.

The same general procedure is followed as for the chromatographic detection of small amounts of alcohols described in Chapter 13 (p. 482), except that the 2,4-dinitrophenylhydrazones are prepared instead of the 3,5-dinitrobenzoates. For solvents two mixtures may be used: (a) petroleum ether boiling at 65 to 110° (9.5 ml) and ethyl ether (0.5 ml); or (b) acetone (3 ml), water (7 ml) and petroleum ether boiling at 20 to 40° (0.1 ml). The spots are made visible by painting with a 10 per cent solution of potassium hydroxide in water using an artist's airbrush. While still wet, the paper is placed against a white background and the spots are outlined with a pencil because they fade. The literature references on the chromatographic procedures for carbonyl compounds given at the end of the chapter should be consulted.

[10] N. D. Cheronis, "Micro and Semimicro Methods," in *Technique of Organic Chemistry*, Vol. 6, A Weissberger, ed., Interscience New York, 1954, p. 508.

For the preparation of dinitrophenylhydrazones in milligram quantities the procedures described in Example 14.03 can be modified to a milligram scale according to methods described in Examples 13.09 and 13.13. In working with milligram or microgram quantities the concentrations of the reactants should be kept not lower than $0.1 M$.

KETONES

The derivatives listed in Table 14.01 (page 497) for the characterization of aldehydes are also suitable for the characterization of ketones. Of the recommended derivatives—2,4-dinitrophenylhydrazones, *p*-nitrophenylhydrazones, phenylhydrazones, semicarbazones, thiosemicarbazones, dimethones, and oximes—only the *dimethones cannot* be prepared from the ketones. For all the rest the same general procedures described for aldehyde derivatives also apply to the preparation of derivatives from ketones. The modifications that must be introduced into the general method are essentially an increase in the time of heating the reaction mixture and an increase in the amount of solvent employed for the crystallization of the derivative. For example, in the preparation of piperonal phenylhydrazone the reaction mixture (carbonyl compound, phenylhydrazine, and acetic acid) is heated for 3 to 4 minutes (Example 14.01), while in the preparation of the same derivative of benzophenone (Example 14.09) the reaction mixture is heated for 15 minutes. In general, the carbonyl group of a ketone reacts at a slower rate than that of an aldehyde. Similarly, the amount of solvent employed for the crystallization of the same quantity of phenylhydrazone from a ketone is much greater than that employed for an aldehyde of the same molecular type.

Substituted Hydrazones

The description of the general methods for substituted hydrazones, semicarbazones, and other derivatives for aldehydes applies as well for ketones. For example, it was noted that the 2,4-dinitrophenylhydrazone is not well suited for α-hydroxyaldehydes, α-hydroxyketones, sugars, and a number of other substituted aldehydes and ketones. Just as an α-hydroxyaldehyde in reacting with dinitrophenylhydrazine may split off a molecule of water, so the derivative of α-ketoesters may undergo ring closure to yield cyclic pyrazolones. For example, ethyl acetoacetate yields a derivative that readily undergoes cyclization:

$$\begin{array}{c}\text{CH}_3\text{CO} + \text{H}_2\text{NNHC}_6\text{H}_3(\text{NO}_2)_2 \\ |\\ \text{CH}_2\text{COOC}_2\text{H}_5\end{array} \rightarrow \begin{array}{c}\text{CH}_3-\text{C}=\text{NNHC}_6\text{H}_3(\text{NO}_2)_2 \\ |\\ \text{CH}_2\text{COOC}_2\text{H}_5\end{array}$$

$$\downarrow$$

$$\begin{array}{c}\text{CH}_3-\text{C}=\!=\!=\text{N} \\ | \qquad\qquad\quad \diagdown \\ \qquad\qquad\qquad \text{NC}_6\text{H}_3(\text{NO}_2)_2 \\ \text{CH}_2-\text{CO} \diagup \end{array}$$

Dinitrophenylmethylpyrazolone

Example 14.09: Preparation of the Phenylhydrazone of an Aryl Ketone: Benzophenone Phenylhydrazone

Place in a 8-in. tube 150 mg of benzophenone and 5 ml of methanol. Heat for a few seconds to effect solution; then add 0.1 ml of phenylhydrazine and 1 drop of glacial acetic acid. Then heat to boiling under reflux for about 15 minutes. Add water dropwise until a permanent clouding results. Cool, filter the crystals, and wash with 1 ml of water containing 1 drop of acetic acid. Crystallize the product from 10 ml of methanol. It may require two crystallizations for complete purification. The yield of phenylhydrazone is about 90 to 100 mg, melting at 136 to 137°.

NOTE. A number of ketones, such as acetophenone and cyclohexanone, react anomalously with phenylhydrazine. For example, following the same method as outlined, acetophenone yields a product that shows an initial melting point of 98°, which is 7° below the melting point of the pure derivative. Even after crystallization, the product undergoes decomposition when dried in air.

Example 14.10: Preparation of Acetone 2,4-Dinitrophenylhydrazone

Proceed by the same general method as described in Example 14.03 (page 501), and use 0.2 ml of acetone. The yield is about 50 mg of solid, melting at 126°. If it is desired to recrystallize the derivative, use methanol and add water to the hot alcohol solution until a cloudiness results.

Semicarbazones and Thiosemicarbazones

Most ketones yield semicarbazones and thiosemicarbazones quite readily on mixing the reagents and slight warming. However, acetone requires one hour of heating to form the semicarbazone and 15 to 20 minutes to form the thiosemicarbazone.

512   Procedures of Final Characterization of an Unknown

Example 14.11: Preparation of the Semicarbazone of a Ketone: Butanone Semicarbazone

Place in an 8-in. tube 200 mg of semicarbazide hydrochloride, 300 mg sodium acetate, and 2 ml of water. Warm for a few seconds over a small flame to effect solution. Add 0.2 ml of butanone with a pipet or a dropper. Stopper with a cork provided with a reflux condenser, place the tube in a beaker containing water at 70 to 75°, and heat at this temperature for 10 minutes. Allow the tube to remain in the water bath for 10 additional minutes. Filter and wash the crystals with 5 drops of cold water. Keep about 5 mg of the crystals and recrystallize the main portion from water. The yield is 60 to 70 mg, melting at 135 to 136°.

NOTE. For acetone, it is advisable to heat at 50° for about 1 hour; otherwise the yield is poor. For the higher aliphatic ketones, use 2 ml of methanol and 2 ml of water as a solvent for the reaction mixture. Since the solubilities of the semicarbazones decrease with increase in the complexity of the molecule, good results are obtained by using 100 mg of the carbonyl compound. Thus, 2-heptanone (methyl *n*-pentyl ketone) and cyclohexanone give about 100 mg of pure derivative from 0.1 ml of the compound. Aromatic ketones react more slowly, and a longer period of heating the reaction mixture is recommended.

Example 14.12: Preparation of the Thiosemicarbazone of a Ketone: Acetone Thiosemicarbazone

In a 6-in. test tube place 50 mg of thiosemicarbazide and add 2 drops of acetone (about 50 mg). Place in a separate tube 100 mg of sodium acetate and 1 ml of water; heat until the sodium acetate dissolves and add the solution to the mixture of acetone and thiosemicarbazide. Warm for a minute and set aside. The thiosemicarbazone begins to separate immediately. After 15 minutes cool and filter with suction, washing with 0.5 ml of water. About 25 to 30 mg of fine crystals are obtained, which melt at 177 to 178°.

NOTE. For the preparation of the thiosemicarbazones of citral and citronellal and other carbonyl compounds not readily soluble in water a small amount of alcohol is added, and the tube is then heated in a water bath for 15 to 30 minutes at 60 to 70°.

Oximes

As noted on page 507, oxime formation is a better derivatizing procedure for ketones than for aldehydes (with aldehydes the reaction requires several days for completion). The following equation represents the formation of an oxime from a ketone:

$$(CH_3)_2C{=}O + H_2NOH \rightarrow (CH_3)_2C{=}NOH + H_2O$$

Acetone Hydroxylamine Acetoxime

Derivatives of Aldehydes, Ketones, and Acetals

In general, the reaction between carbonyl compounds and hydroxylamine is slow, often requiring hours or days for completion. Further, the melting points of a large number of oximes are below 70° and as a result crystallization, isolation, and purification are difficult. In the preparation of oximes it is advisable to use 500 mg of the carbonyl compounds since, for the reasons just given, the yields are usually poor. Furthermore, there is always the danger of the rearrangement of the oximes into N-substituted amides.

General Method for Preparation

Place in an 8-in. test tube 500 mg of the carbonyl compound, 500 mg of hydroxylamine hydrochloride, 3 ml of pyridine, and 3 ml of absolute alcohol. Arrange for reflux and boil the mixture gently for 2 hours on a steam bath. Pour the mixture into an evaporating dish and remove the solvent in a current of air under a hood. Scrape the residue by means of the microspatula and grind with 3 ml of cold water; filter and recrystallize the oxime from methanol, or methanol-water mixture.

NOTE. In the preparation of some oximes, the pyridine may be replaced with 4 ml of $1N$ sodium hydroxide solution, and absolute alcohol with ordinary methanol. The mixture is heated for 10 to 20 minutes and cooled in an ice-salt mixture.

When there is a danger of rearrangement the mixture is not refluxed but warmed and then allowed to stand for 24 hours. If on dilution with water and cooling the derivative does not separate, the mixture is heated for a longer period of time and then cooled again. The crystals that separate are crystallized from the minimum amount of aqueous methanol.

Other Derivatives

The hydantoins, which are formed by the action of sodium cyanide and ammonium carbonate on ketones, have been recommended as suitable derivatives for a number of ketones:

$$R_2CO + CN^- + NH_4^+ + CO_2 \rightarrow \underset{\underset{\text{Hydantoin}}{NHCONH}}{R_2C\text{——}CO} + H_2O$$

The action of hypochlorite on methyl ketones including unsaturated methyl ketones should be considered, since this oxidizing agent acts selectively on the methyl group (haloform reaction) yielding a carboxylic acid without attacking the unsaturated linkage:

$$RCH\text{=}CHCOCH_3 + 3NaOCl \rightarrow$$
$$CHCl_3 + 2NaOH + RCH\text{=}CHCOONa$$

The literature given at the end of this section should be consulted for the application of these two reactions for the preparation of suitable derivatives.

ACETALS

The preparation of derivatives of acetals is based on their hydrolysis to aldehydes and alcohols:

$$RCH(OR')_2 + H_2O \rightarrow 2R'OH + RCHO$$

The alcohol is then identified by the preparation of the 3,5-dinitrobenzoate or other suitable derivative. For the aldehyde, either the semicarbazone or the 2,4-dinitrophenylhydrazone is prepared. A more detailed discussion of these derivatives is given in earlier sections.

(CAUTION. *Where appreciable amounts of acetals are involved, it should be kept in mind that these compounds form peroxides like the ethers and therefore the precautions outlined on page 541 should be followed.*)

Hydrolysis of Acetals

The hydrolysis of acetals is best accomplished by heating with dilute acids; the lower acetals are usually hydrolyzed by boiling with 3 to 5 per cent hydrochloric or sulfuric acid; for the hydrolysis of the higher acetals an organic solvent miscible with water and heating for 20 to 30 minutes or longer is necessary. One method is to use a 50 per cent solution of dioxane and water and, after the hydrolysis is complete, to neutralize the hydrolyzate and divide in half; one portion is used for the characterization of the aldehyde and the second portion for the derivatization of the alcohol. For semimicro quantities it is often more convenient to hydrolyze 100 to 200 mg quantities of the acetal separately for each characterization. For example, in the characterization of benzaldehyde dimethyl acetal (page 515) 100 mg of acetal were required for the carbonyl component and 250 mg for the alcoholic part. In general, however, the hydrolysis of the acetal and subsequent preparation of the derivatives for the alcohol and the aldehyde depend on the nature of the compound and the hydrolysis products. Therefore the discussions of preparation of derivatives of alcohols (pages 465 to 484) and aldehydes (pages 496 to 510) should be consulted. The following general procedure is based upon the derivatization of the aldehyde as a semicarbazone and the alcohol as the 3,5-dinitrobenzoate.

General Method for the Identification of the Aldehyde

Place in an 8-in. tube 100 mg (4 or 5 drops) of the acetal, 0.5 ml (10 drops) of $6N$ sulfuric acid, 2 ml of water, and 2 ml of ethanol. Arrange for reflux and boil the mixture gently for 5 minutes. Add 100 mg of semicarbazide hydrochloride and a drop of Universal indicator and then dropwise 6N sodium hydroxide solution to pH 5; add 200 mg of sodium acetate and shake to dissolve the solid. The semicarbazone separates immediately if the aldehyde is aromatic or aliphatic with 3 or more carbon atoms. Cool for a few minutes and then filter; wash the solid with water and then transfer it back to the 8-in. tube from which it was removed. Add 2.5 to 3 ml of methanol, heat to boiling, and filter; add to the hot filtrate 0.5 ml of water and cool for 5 minutes. Filter the crystals of the semicarbazone and dry. If the semicarbazone does not form readily the methone derivative is prepared according to the procedure given on page 505.

General Method for the Identification of the Alcohol

Place in an 8-in. distilling tube 200 to 300 mg of the acetal, 0.5 ml of 6N hydrochloric acid solution, 2.5 ml of water, and 2.5 ml of dioxane. Insert in the mouth of the tube a cork holding a reflux condenser that protrudes into the tube at least 25 mm below the side arm. Connect the end of the side arm with a delivery tube into an arrangement for regular distillation. Add a boiling stone and reflux for 15 minutes. Pull the condenser upwards until the tip is just above the side-arm opening and continue the gentle boiling so that distillation proceeds slowly. When 3 ml of distillate have been collected, discontinue the distillation.

Since the alcohol is present in an aqueous solution, the 3,5-dinitrobenzoate must be prepared under conditions that will insure the minimum hydrolysis of the 3,5-dinitrobenzoyl chloride and removal of the hydrolytic products. Example 13.04 (page 472) is employed using proportionate amounts. For example, with 200 to 300 mg of acetal having a molecular weight of about 100 to 120 the maximum amount of a lower alcohol is 120 to 180 mg. Hence the quantities of reagents employed should be about one half of those given in Example 13.04.

NOTE. From 100 mg of benzaldehyde dimethyl acetal the yield of benzaldehyde semicarbazone is about 70 to 80 mg. The yield of methyl 3,5-dinitrobenzoate from 250 mg of the acetal is 60 to 90 mg.

REFERENCES

Aldehydes

2,4-Dinitrophenylhydrazones

Derivatives of aminocarbonyl compounds: Johnson, *J. Am. Chem. Soc.*, **73**, 5888 (1951).

X-ray identification and crystallography of aldehydes and ketones as 2,4-dinitrophenylhydrazones: Clark, Kaye, and Parks, *Ind. Eng. Chem., Anal. Ed.*, **18**, 310 (1946).

2,4-Dinitrophenylhydrazones: Allen, *J. Am. Chem. Soc.*, **52**, 2955 (1930); Allen and Richmond, *J. Org. Chem.*, **2**, 222 (1937); Brady, *Analyst*, **51**, 77 (1926); *J. Chem. Soc. (London)*, **1931**, 756; Campbell, *Analyst*, **61**, 391 (1936); Castillo, *Farm. mod.*, **47**, 640 (1936); Iddles, Low, Rosen, and Hart, *Ind. Eng. Chem., Anal. Ed.*, **11**, 102 (1939); Johnson, *J. Am. Chem. Soc.*, **75**, 2720 (1953); Neuberg and Grauer, *Anal. Chim. Acta*, **7**, 238 (1952); Perkins and Edwards, *Am. J. Pharm.*, **107**, 208 (1935); Purgotti, *Gazz. Chem. Ital.*, **24**, I, 555 (1894); Strain, *J. Am. Chem. Soc.*, **57**, 758 (1935); Shine, *J. Org. Chem.*, **24**, 252–253 (1959).

Microscopical identification and polymorphism: Allen and Richmond, *J. Org. Chem.*, **2**, 222 (1937); Braddock et al., *Anal. Chem.*, **25**, 301 (1953); Brandstatter, *Microchim. Acta*, **32**, 33 (1944); Marovic, *Mikrochim. Acta*, **32**, 6 (1944); Sandulesco, *Helv. Chim. Acta*, **19**, 1095 (1936).

Other Substituted Hydrazones

*o-, m-, p-*Bromobenzohydrazides: Kao, *J. Chinese Chem. Soc.*, **4**, 69 (1936); *Science Repts. Natl. Tsing Hua Univ.*, **4**, 62 (1936); Wang, *Science Repts., Natl. Tsing Hua Univ.* (A), **3**, 279 (1935).

p-Carboxyphenylhydrazones: Veibel et al., *Dansk Tidsk. Farm.*, **14**, 184 (1940); *Acta Chem. Scand.*, **1**, 54 (1947); **2**, 545 (1948); *C.A.*, **42**, 434 (1948); **43**, 5764 (1949).

o- and *m*-Chlorobenzohydrazides: Sah and Wu, *Science Repts., Natl. Tsing Hua Univ.*, **3**, 443 (1936); Sun and Sah, *Science Repts., Natl. Tsing Hua Univ.* (A), **2**, 359 (1934).

p-Chlorobenzohydrazides: Sah and Wang, *J. Chinese Chem. Soc.*, **14**, 39 (1946); Shih and Sah, *Science Repts., Natl. Tsing Hua Univ.* (A), **2**, 353 (1934).

α-(2,4-Dinitrophenyl)-α-methylhydrazones: Blanksma and Wackers, *Rec. trav. chim.*, **55**, 655 (1936); Vis, *Rec. trav. chim.*, **58**, 387 (1939).

Diphenylhydrazones: Maurenbrecher, *Ber.*, **39**, 3583 (1906).

β-Naphthylhydrazones: Chen and Sah, *Science Repts., Natl. Tsing Hua Univ.*, **4**, 62 (1936); Lei, Sah, and Kao, *Science Repts., Natl. Tsing Hua Univ.* (A), **2**, 335 (1934).

o-, m-, and *p*-Nitrobenzenesulfonhydrazones: Cameron and Storrie, *J. Chem. Soc. (London)*, **1934**, 1330.

3-Nitrobenzohydrazides: Chen, *J. Chinese Chem. Soc.*, **3**, 251 (1935); Meng and Sah, *Science Repts., Natl. Tsing Hua Univ.* (A), **2**, 347 (1934); Strain, *J. Am. Chem. Soc.*, **57**, 758 (1935).

Nitroguanylhydrazones: Whitmore, Revukas, and Smith: *J. Am. Chem. Soc.*, **57**, 706 (1935).

p-Nitrophenylhydrazones: Bamberger, *Ber.*, **32**, 1806 (1899); **34**, 546 (1901); Petit, *Bull. Soc. Chim. France*, **1948**, 141.

Semicarbazones

3,5-Dinitrophenylsemicarbazones: Sah and Tao, *J. Chinese Chem. Soc.*, **4**, 506 (1936).

α-Naphthylsemicarbazones: Sah and Chiang, *J. Chinese Chem. Soc.*, **4**, 496 (1936). Phenylsemicarbazones: Sah and Ma, *J. Chinese Chem. Soc.*, **2**, 32 (1934).

Semicarbazones: Angla, *Ann. chim. anal. chim. appl.*, **22**, 10 (1940); Michael, *J. Am. Chem. Soc.*, **41**, 417 (1919); Shriner and Turner, *J. Am. Chem. Soc.*, **52**, 1267 (1930); Thiele and Stange, *Ber.*, **27**, 31 (1894); Zelinsky, *Ber.*, **30**, 1541 (1897).

Semicarbazones (reliability of melting points): Veibel, *Bull. Soc. Chim.* (*4*), **41**, 1410 (1947).
4-Phenylthiosemicarbazones: Tisler, *Z. anal. chem.*, **149**, 164 (1956).
4-(*p*-Tolyl)thiosemicarbazones: Tisler, *Z. anal. chem.*, **150**, 345 (1956).
4-Arylthiosemicarbazones: Tisler, *Z. anal. chem.*, **151**, 187 (1956); **155**, 186 (1957).
Thiosemicarbazones: Busch, *J. pract. Chem.*, **124**, 301 (1930); Freund and Schander, *Ber.*, **29**, 2501 (1896); **35**, 2602 (1902); Kitamura, *J. Pharm. Soc. Japan*, **57**, 51 (1937).
m-Tolylsemicarbazones: Sah, Wang, and Kao, *J. Chinese Chem. Soc.*, **4**, 187 (1936).
o-Tolylsemicarbazones: Lei, Sah, and Shih, *J. Chinese Chem. Soc.*, **3**, 246 (1935).
p-Tolylsemicarbazones: Sah and Lei, *J. Chinese Chem. Soc.*, **2**, 167 (1934).
p-Xenylsemicarbazones: Sah and Kao, *Rec. trav. chim.*, **58**, 459 (1939).
Microscopy: Dunbar and Aaland, *Microchem. J.*, **2**, 113–127 (1958).

Methones

Dibromomethones: Voitila, *Suomen Kemistilehti.*, **10B**, 14 (1937).
Methone derivatives: Klein and Linser, *Mikrochemie, Pregl Festscht.*, **1929**, 204; Volander, *Z. anal. Chem.*, **77**, 245 (1929); *Z. angew Chem.*, **42**, 46 (1929); Weinberger, *Ind. Eng. Chem., Anal. Ed.*, **3**, 365 (1931).
Methone derivatives (and their cyclization to xanthenes): Horning and Horning, *J. Org. Chem.*, **11**, 95 (1946).

Other Derivatives

Aldehydes in cigarette smoke: Touey, *Anal. Chem.*, **27**, 1788 (1955).
Benzothiazoles and benzothiazolines: Lankelma and Sharnoff, *J. Am. Chem. Soc.*, **53**, 2654 (1931).
Benzylidineaminomorpholines: Dugan and Handler, *J. Am. Chem. Soc.*, **64**, 552 (1942).
1,3-Cyclohexanedione derivatives: King and Felton, *J. Chem. Soc. (London)*, **1948**, 1371.
1,2-Bis(*p*-methoxybenzylamino)ethane derivatives: Billman et al., *J. Org. Chem.*, **17**, 1375 (1952).
Mercaptals: Ritter and Lover, *J. Am. Chem. Soc.*, **74**, 5576 (1952).
5-(1-Phenylethyl)semioxamazide derivatives: Leonard and Boyer, *J. Org. Chem.*, **15**, 42 (1950).
Miscellaneous: Deriv. of bisulfite addition products, Von Wacek and Kratzl, *Ber.*, **76**, 1209 (1943); Adams and Gerber, *J. Am. Chem. Soc.*, **71**, 522 (1949). *p*-Dimethylaminoanils, Utzinger and Regenass, *Helv. Chim. Acta*, **37**, 1901 (1954). Carboethoxy-, carbomethoxy-, and carboxyphenylhydrazones, Rabjohn and Barnstorff, *J. Am. Chem. Soc.*, **75**, 2259 (1953); Zellner, *Monatsh.*, **80**, 330 (1951). Iodo-, methyl-, and nitro-substituted benzohydrazides, Sah et al., *J. Chinese Chem. Soc.*, **14**, 24. 31, 45 (1946); *Rec. trav. chim.*, **59**, 349 (1949); Gaudemaris and Dubois, *Bull. Soc. Chim. France*, **1950**, 63. 2,4-Dinitrophenyl-, α-, and β-naphthylsemicarbazones, McVeigh and Rose, *J. Chem. Soc. (London)*, **1945**, 713; Sah and Tao, *J. Chinese Chem. Soc.*, **4**, 501 (1936). Thiosemicarbazones and L-menthylsemicarbazones, Sah and Daniels, *Rec. trav. chim.*, **69**, 1545 (1950); Woodward et al., *J. Am. Chem. Soc.*, **63**, 120 (1941). α- and β-Naphthyl- and 5-phenylaminosemioxamazides, Sah et al., *J. Chinese Chem. Soc.*, **14**, 39, 101 (1946). Substituted hydantoins, Henze and Speer, *J. Am. Chem. Soc.*, **64**, 522

(1942). N-Methyl-β-carbohydrazidopyridinium *p*-toluenesulfonate deriv., Allen and Gates, *J. Org. Chem.*, **6**, 596 (1941).

2-Diphenylacetyl-1,3-indandione 1-hydrazone derivatives: Braun and Mosher, *J. Am. Chem. Soc.*, **80**, 3048–3050 (1958).

4-Aminoantipyrine derivatives: Manns and Pfeifer, *Mikrochem. Acta*, **1958**, 630–637.

p-Dimethylaminoanils: von Utzinger and Regenass, *Helv. Chim. Acta*, **37**, 1901 (1954).

Optical Properties of Derivatives

Grammaticakis, *Bull. Soc. Chim.*, **7**, 527 (1940); **8**, 38, 101, 427 (1941). Spectral studies of oximes, phenylhydrazones and semicarbazones.

Chromatographic Procedures

Kirchner and Keller, *J. Am. Chem. Soc.*, **72**, 1867 (1950).
Meigh, *Nature*, **170**, 579 (1952).
Rice et al., *Anal. Chem.*, **23**, 195 (1951).

Ketones

Action of hypochlorite on ketones: Hurd and Thomas, *J. Am. Chem. Soc.*, **55**, 1646 (1933).

Improved hydroxylamine method for the determination of aldehydes and ketones: Bryant and Smith, *J. Am. Chem. Soc.*, **57**, 57 (1935).

Optical properties of derivatives of ketones: Grammaticakis, *Bull. Soc. Chim.*, **8**, 28 (1941); **7**, 527 (1940).

Reaction of acetophenone derivatives with sodium hypochlorite: Van Arendonk and Cupery, *J. Am. Chem. Soc.*, **53**, 3184 (1931).

Oximes: Bachmann and Barton, *J. Org. Chem.*, **3**, 307 (1938); Bachmann and Boatner, *J. Am. Chem. Soc.*, **58**, 2099 (1936); Buck and Ide, *J. Am. Chem. Soc.*, **53**, 1541 (1931).

Salts of oximes: Grammaticakis, *Compt. rend.*, **224**, 1568 (1947).

Substituted hydantoins: Henze and Speer, *J. Am. Chem. Soc.*, **64**, 2502 (1942).

Chromatography of metal β-diketone chelates: Berg and Strassner, *Anal. Chem.*, **27**, 1131 (1955).

For other references on chromatographic identification of ketones, see aldehyde references on this page.

Acetals

Bibliographic references on the derivatization of alcohols appear on pages 492 to 495 and on the derivatization of aldehydes on pages 515 to 518. However, the general methods for the derivatization of alcohols and aldehydes should be reviewed before any literature search is made.

15

Derivatives of Carbohydrates

Table 15.01 summarizes the various types of derivatives that are most commonly employed for the derivatization of carbohydrates. It should be

TABLE 15.01

DERIVATIVES FOR THE IDENTIFICATION OF CARBOHYDRATES

Acetates and benzoates	*Substituted phenylhydrazones
Azoates	Thiobenzhydrazides
Benzimidazole derivatives	Tosyl esters
*Osazones	Trityl ethers
*Osotriazoles	

* Recommended derivative.

pointed out that for the most part these derivatives are applicable to relatively simple sugars—that is, monosaccharides—and to a less degree to disaccharides. The discussion is divided into: (a) recommended derivatives, which include the substituted phenylhydrazones, osazones and osotriazoles; and (b) other derivatives listed in Table 15.01.

In general, although the sugars undergo a great variety of reactions, their derivatization is a rather difficult matter, particularly when the sample contains traces of related compounds; the present discussion, therefore, is limited to the derivatization of pure sugars. Reference should be made to standard works on sugars for their separation and identification in mixtures. The specific rotation of sugars and their derivatives is discussed on pages 225–228.

Substituted Phenylhydrazones, Osazones, and Osotriazoles

Discussion

The formation of phenylhydrazones and substituted phenylhydrazones is represented in the following equation:

$$\begin{array}{c} \text{H} \\ \text{C}=\text{O} \\ | \\ \text{HC}-\text{OH} \\ | \\ \text{R} \end{array} + \text{H}_2\text{NNHC}_6\text{H}_5 \rightarrow \begin{array}{c} \text{H} \\ \text{C}=\text{NNHC}_6\text{H}_5 \\ | \\ \text{HC}-\text{OH} \\ | \\ \text{R} \end{array} + \text{H}_2\text{O}$$

Hexose   Phenylhydrazine   Hexose Phenylhydrazone

The formation of hydrazones is accomplished by treatment of the sugar with a little more than an equimolecular quantity of phenylhydrazine. The sugar is dissolved in a small quantity of water (100 mg/ml), and the required amount of phenylhydrazine in an equal volume of 50 per cent acetic acid is added; the mixture is allowed to stand for 24 hours in the cold. In the case of substituted phenylhydrazines a small amount of alcohol is used in the reaction mixture; for example, 100 mg of the sugar dissolved in 1 ml of water are mixed with 100 mg of *p*-nitrophenylhydrazine hydrochloride suspended in 1 ml of methanol. After standing, the hydrazones are filtered, washed with water, and crystallized by first dissolving in the minimum amount of hot methanol or ethanol and then precipitating by the cautious addition of water.

The hydrazones may be reconverted to the original sugar by reacting them with benzaldehyde or formaldehyde; the hydrazone of the aldehyde separates, leaving the sugar in solution. This property and the fact that the hydrazones of the various sugars separate with varying speeds make them useful in the separation of mixtures; the hydrazones are filtered off as they are formed and then are converted to the original sugar by treatment of the hydrazone with an aldehyde.

Among the most important substituted phenylhydrazines used for the preparation of derivatives are *p*-nitrophenylhydrazine (also the *m*- and *o*-isomers), *p*-bromophenylhydrazine, methylphenylhydrazine, diphenylhydrazine, and β-naphthylhydrazine.

The osazones are formed by reacting sugars with excess of phenylhydrazine:

$$\begin{array}{c} \text{H} \\ \text{C=O} \\ | \\ \text{HC-OH} \\ | \\ \text{R} \end{array} + 3\text{C}_6\text{H}_5\text{NHNH}_2 \rightarrow \begin{array}{c} \text{H} \\ \text{C=NNHC}_6\text{H}_5 \\ | \\ \text{C=NNHC}_6\text{H}_5 \\ | \\ \text{R} \end{array} + \text{NH}_3 + \text{C}_6\text{H}_5\text{NH}_2 + 2\text{H}_2\text{O}$$

Hexose   Phenylhydrazine   Hexose osazone

The reaction takes place rapidly, as compared with the formation of hydrazones, when a solution of the sugar is warmed with excess of the hydrazine. The mechanism of the reaction is assumed to be, first, the formation of the hydrazone, followed by oxidation of an adjacent carbon atom to the carbonyl stage and subsequent reaction with the hydrazine to produce the osazone. In the preparation of osazones the sugar solution is mixed with a solution of phenylhydrazine acetate, or phenylhydrazine hydrochloride and sodium acetate, in a tube and then heated in boiling water for 30 minutes. The time required for the formation of the osazone may be used as additional evidence in the characterization of the unknown sugar, *provided the sample is a pure substance* and not contaminated with small amounts of other sugars. The exact time required for the appearance of the osazone, after the tube is immersed in boiling water, depends on several factors, such as the amount of sugar, reagent, pH of the solution, and the amount of solvent. Generally, however, the following decreasing ease of formation of phenylosazones is observed: fructose, sorbose, glucose, xylose, rhamnose, arabinose, galactose; sucrose undergoes hydrolysis and slowly forms (after about 20 minutes) a small amount of the glucosazone. The osazones of maltose and lactose are soluble in the hot solution and separate only on cooling. It should be again noted that the time of formation of osazone is of value only in the case of pure sugars. The presence of impurities of other osazones greatly influences the rate of crystallization.

The purification of osazones must be undertaken immediately after they have been filtered and washed with cold water. A small amount is kept for the determination of the melting point, and the balance is dissolved in the minimum amount of hot methanol or ethanol and then precipitated cautiously by addition of water. Osazones should not be dried in air. D-Glucosazone does not show appreciable change when dried in air, but a number of other osazones show extensive lowering of the melting point. For example, in one experiment the osazone from 100 mg of lactose was divided into 3 portions; the sample dried in a watch glass over a water bath gave a melting point of 188 to 190°; the sample dried in air melted at 191 to 192°; while the sample dried in a vacuum desiccator melted at 209 to 210°, which is the value listed in the literature for the pure derivative.

The melting points of the osazones are not to be considered with the same regard as the melting points of other derivatives because the melting points are really decomposition points (see page 193), which vary greatly, depending on the rate of heating. For example, the melting point of D-glucosazone listed in the literature is 210°. For a given sample of the pure derivative the melting point of 210° will be observed if the rate of heating is 40 to 60° per minute. If the temperature is raised 8 to 10° per minute (which is regarded as very rapid near the melting point of a substance in the usual practice), the observed melting point will be below 200° and usually between 194 to 198°. It is obvious that under these conditions reproducibility of observations requires great care. The osazone of the sugar suspected to be the unknown, should always be prepared for comparison.

Another limiting factor in the use of osazones for the characterization of sugars is the fact that a number of isomeric sugars give the same osazone. The following serve as examples: D-glucose, D-mannose, and D-fructose; D-arabinose and D-ribose; D-xylose and D-lyxose. In addition, the corresponding sugars of the L-series, which have the same configuration beyond the second carbon atom, yield the same osazone, differing from the D-osazone only in the direction of the rotation.

Methylphenylhydrazine is useful for differentiating between aldoses and ketoses, which yield the same osazone. For example, D-fructose reacts readily with methylphenylhydrazine to form a characteristic osazone, whereas D-glucose and D-mannose do not.

The osazone may be converted to the osotriazole:[1]

$$\begin{array}{c} H \\ C=NNHC_6H_5 \\ | \\ C=NNHC_6H_5 \\ | \\ (HCOH)_3 \\ | \\ CH_2OH \end{array} \xrightarrow{Cu^{++}} \begin{array}{c} HC=N \\ | \quad \diagdown \\ \quad \quad NC_6H_5 \\ C=N \\ | \\ (HCOH)_3 \\ | \\ CH_2OH \end{array} + C_6H_5NH_2$$

Phenyl-D-glucosazone   Phenyl-D-glucosotriazole

In this reaction, one of the phenylhydrazine groups is reduced to aniline with formation of a ring containing three nitrogen (triazo) atoms. The

---

[1] C. S. Hudson et al., *J. Am. Chem. Soc.*, **66**, 735 (1944); **67**, 939 (1945); **68**, 1769 (1946); **69**, 1050, 1461 (1947).

phenylosotriazoles differ from the osazones by their sharp melting points, and hence they are recommended for confirmation of the identity of phenylosazones. The preparation of the phenylosotriazoles from 100 mg of the osazone is feasible.

General Method for Preparation of Substituted Hydrazones of Sugars

The ease of preparation of the hydrazone varies; for example, at room temperature fructose yields an impure phenylhydrazone but a relatively pure methylphenylhydrazone, whereas the action of glucose is the reverse. Similarly, the phenylhydrazone of mannose and the *o*-tolylhydrazone of galactose form readily and may be used for the characterization of these sugars. About 1 millimole of sugar and 2 millimoles of solid sodium acetate, dissolved in 2 ml of water, are mixed with 0.95 millimole of the substituted hydrazine hydrochloride, dissolved in 2 ml of methanol. The mixture is allowed to stand 24 to 48 hours, then cooled; the crystals of the hydrazone are filtered and crystallized from 95 per cent ethanol.

NOTE. In preparing phenylhydrazones the amount of the base should be reduced to 0.9 millimole per millimole of sugar. For pentoses benzylphenylhydrazine hydrochloride is useful. The benzylphenylhydrazone of arabinose (mp 174) and xylose (mp 99) form readily.

If crystals of the hydrazone do not separate readily, water is added dropwise to turbidity and the mixture is cooled.

Example 15.01: Glucose p-*Nitrophenylhydrazone*

Place in a 6-in. tube 100 mg of *p*-nitrophenylhydrazine hydrochloride and 1 ml of methanol; shake for a few seconds and then add 100 mg of glucose, 150 mg of powdered sodium acetate, and 1 ml of water. Cork the tube and shake gently so as to mix its contents; allow to stand overnight. Add 2 ml of water and filter the crystals; wash with water and crystallize from 95 per cent ethanol. The yield is 125 mg of crystals, which melt at 189 to 190°.

General Method for Preparation of Osazones and Conversion to Osotriazoles

Osazones are formed with excess of phenylhydrazine. The rate of formation of osazones and their purification and variation of melting points are discussed on page 521. The general methods for the preparation and purification of osazones and osotriazoles are illustrated in the examples given below.

Example 15.02: Rate of Osazone Formation

Place 100 mg of the sugar, 100 mg of sodium acetate, and 2 ml of water in a 6-in. tube. Add 5 or 6 drops (0.2 g) of phenylhydrazine and 6 or 7 drops (0.12 g) of glacial acetic acid. Close the tube loosely with a cork and set it in a 600-ml beaker half-filled with water that is already boiling. Note the time required for the appearance of the osazone. When the sample of sugar is pure, the following intervals of time (in minutes) are observed as measured from the moment the tube is immersed in the boiling water to the appearance of the osazone: mannose, 0.5 to 1; fructose, 1 to 2; glucose, 4 to 5; xylose, 6 to 8; arabinose, 9 to 10; galactose, 14 to 16; sucrose, 20 to 30 (by hydrolysis); and lactose and maltose, on cooling.

NOTE. Phenylhydrazine hydrochloride may be used in place of the base; in such a case, for 100 mg of sugar use 200 mg of phenylhydrazine hydrochloride, 300 mg of sodium acetate, and 2 ml of water. On standing the salt undergoes decomposition and darkens. It may be purified by crystallization from hot water; the salt is dissolved in the minimum amount of boiling water; the tarry impurities remain undissolved. A small amount of charcoal is added, and the hot solution is filtered rapidly. The solution is cooled, and concentrated hydrochloric acid is added so that the volume of the solution increases by one third. After an hour the cold mixture is filtered, and the crystals are washed with ice-water and dried.

Example 15.03: Preparation of D-Glucosazone

Place in an 8-in. tube 100 mg of glucose, 100 mg of sodium acetate, 6 drops (0.21 g) of phenylhydrazine, 2 ml of water, and 7 drops (0.13 g) of glacial acetic acid. Immerse the tube in boiling water for 30 minutes. Add 5 ml of water and cool. Filter with suction and wash the tube and crystals —first with 2 ml of water to which 2 drops of acetic acid have been added and then twice with 3 ml of water. Remove a small amount of the osazone and dry in a vacuum desiccator. Transfer the rest of the crystals to the reaction tube and add 20 ml of methanol. Heat to boiling, adding more alcohol until practically all the osazone has dissolved. Filter and add 2 to 3 ml of water to the filtrate and cool in an ice-water mixture. Filter the crystals and wash twice with 2 ml of 25 per cent methanol. Dry in a vacuum desiccator. The yield is 90 to 100 mg. The melting point of the crystals, when determined by the capillary method in an oil bath with a temperature rise of 40 to 60° per minute, is 209 to 210° (corrected). The unrecrystallized material usually melts at 207 to 208°. The identity of phenyl-D-glucosazone is confirmed by conversion to the glucosotriazole as directed in Example 15.05.

NOTE. The formation of an osazone of an aldose (such as glucose) causes the loss of asymmetry in the α carbon atom (to the carbonyl), and hence it yields the same osazone as the related ketose, D-fructose. The formation of the osazone from fructose takes place with greater ease and a slightly better yield. Some authors recommend the use of pyridine in conjunction with alcohol for crystallization of the osazone. No advantage has been found for this solvent pair. When the osazone is required for determination of the optical rotation, it should be washed with acetone before crystallization. This operation is best accomplished just before the osazone is removed from the filter funnel for crystallization. About 2 to 3 ml of acetone are added to the crystals, and after a minute suction is applied and the washing repeated.

Many of the osazones can be identified by microscopical examination of the crystals. The original article cited in the bibliography section should be consulted.

Example 15.04: Preparation of Lactosazone

Use the same quantities as in the preparation of glucosazone. Immerse the tube for 20 minutes in boiling water and then add 2 ml of water and cool in an ice-water mixture for 30 minutes. Filter and wash first with 2 ml of water containing 1 or 2 drops of acetic acid and then twice with 2 ml of cold water. Remove a small portion to a watch glass and place immediately in a vacuum desiccator. Recrystallize the balance from a mixture of 2 ml of water and 1 ml of methanol. After the mixture is brought to boiling, a small amount of charcoal is added and the mixture filtered by suction. The hot solution is cooled in an ice-water mixture for 1 hour, and then the crystals are filtered, washed with two 0.5-ml portions of water, and placed at once in a vacuum desiccator. The yield is 40 to 45 mg of crystals, which have a melting point of 210 to 211°.

Example 15.05: Preparation of D-*Glucosotriazole*

Place 100 to 110 mg of glucosazone in an 8-in. tube; add 9 ml of water, 2 drops of $6N$ sulfuric acid, 300 mg of copper sulfate (pentahydrate), 6 ml of isopropyl alcohol, and 2 boiling stones. The mixture is boiled for 1 hour under reflux. The yellowish-green solution is poured into an evaporating dish and concentrated over a water bath to a volume of 3 to 4 ml. Cool the dish in ice-water and filter the granular crystals; dissolve the crude material in 12 to 14 ml of boiling water; add charcoal and filter. Cool the solution overnight in an icebox. Filter the derivative and wash twice with 1 ml of water. The yield is 16 to 18 mg of crystals, melting at 193 to 194°. To crystallize the glucosotriazole, place the crystals in a 6-in. tube; add 1 ml of 95 per cent ethanol and heat to boiling. Add 1 ml of water to the clear solution and cool for 2 hours in an ice-salt mixture. Filter the crystals and wash with 0.5 ml of water. The yield is 13 to 14 mg of crystals, melting at 195 to 196°.

Other Derivatives

Acetates and Benzoates

The conversion of sugars to acetates, although relatively easy, is not recommended for the preparation of derivatives for identification work.

Aside from difficulties encountered in crystallization, α and β forms are possible in most cases. For example, D-glucose treated with acetic anhydride in the presence of anhydrous zinc chloride forms the α-penta-acetate and in the presence of sodium acetate yields the β-form:

$$\begin{array}{c} H \\ C=O \\ | \\ (HCOH)_4 \\ | \\ CH_2OH \\ \text{Glucose} \end{array} + 5(CH_3CO)_2O \rightarrow \begin{array}{c} HC-OCOCH_3 \\ | \\ (HC-OCOCH_3)_3 \\ | \\ HC-O \\ | \\ H_2C-OCOCH_3 \\ \text{α-Glucose penta-acetate} \end{array} + 5CH_3COOH$$

The benzoyl derivatives are prepared by the action of the chloride on the sugars in presence of pyridine or quinoline as proton acceptors. Their usefulness is restricted by the fact that both α- and β-forms are usually obtained.

Azoates

The preparation of azoates[2] is an important method for the experienced worker and useful in the identification and separation of sugars. The basis of the method is the preparation of colored esters of the sugars and their separation by chromatographic adsorption. The reagent, *p*-phenylazo-benzoyl chloride, commonly called azoyl chloride, is prepared by reacting *p*-aminobenzoic acid and nitrosobenzene and then converting the acid to the chloride:

$$C_6H_5NO + H_2NC_6H_4COOH \rightarrow C_6H_5N=NC_6H_4COOH + H_2O$$
$$C_6H_5N=NC_6H_4COOH + SOCl_2 \rightarrow$$
$$C_6H_5N=NC_6H_4COCl + SO_2 + HCl$$
$$\text{Azoyl chloride}$$

The azoyl chloride reacts slowly in pyridine solution with the hydroxyl groups of the sugar to form the azoates:

$$\begin{array}{c} HC-OH \\ | \\ (HC-OH)_3 \\ | \\ HC-O \\ | \\ H_2C-OH \\ \text{α-D-Glucose} \end{array} \xrightarrow{\text{Azoyl chloride}} \begin{array}{c} HC-OAz \\ | \\ (HC-OAz)_3 \\ | \\ HC-O \\ | \\ H_2C-OAz \\ \text{α-Penta-azoyl-D-glucose} \\ \text{(α-D-Glucose azoate)} \end{array}$$

[2] Coleman et. al., *J. Am. Chem. Soc.*, **64**, 1501 (1942); **65**, 1588 (1943); Reich, *Biochem. J.* **33**, 1000 (1939).

The azoates are colored from orange to dark red and are adsorbed from their solutions in organic solvents by many of the adsorbents commonly employed for chromatographic adsorption, such as silicic acid, Magnesol, and Diccalite (page 137).

The preparation of azoates is very useful in the systematic work of sugars, although it may not be widely applicable to routine work, since the length of time required for the reaction of azoyl chloride with sugars is 8 to 10 days.

Miscellaneous Derivatives

When aldoses are oxidized by means of potassium iodate, aldonic acids are formed that, on reaction with *o*-phenylenediamine, yield characteristic benzimidazoles.[3] The method is applicable to quantities of about 10 mM and above.

When carbohydrates are reacted with *p*-toluenesulfonyl chloride the resulting esters are known as tosyl derivatives. The method has been employed to esterify preferentially the primary (CH_2OH) group of many carbohydrates and does not yield good results with small quantities. Similarly, the primary alcohol group can be preferentially derivatized by means of trityl chloride to form trityl ethers (see page 481). See the original papers listed in the References.

The use of thiobenzhydrazide[4] has been proposed as a reagent for the derivatization of both aldoses and ketoses.

Of greater application is the oxidation of some carbohydrates to specific carboxylic acids. For example, the oxidation of galactose to mucic acid may be used for characterization of this sugar and also for lactose and polysaccharides, which yield galactose as one of the hydrolytic products. Mucic acid, $HOOC(CHOH)_4COOH$, is sparingly soluble in water and may be readily identified by its melting point.

The polysaccharides, such as starches and celluloses, are characterized, first, by means of color reactions and physical constants and, second, through the products that they yield on hydrolysis. This fact also applies to the ever-increasing number of cellulose derivatives, such as cellulose acetate, ethylcellulose, and other esters and ethers of cellulose.

Example 15.06: Preparation of Mucic Acid from Galactose and Lactose

On a watch glass, which is placed on a steam bath, add 100 mg of lactose or galactose, 1 ml of water, and 0.5 ml (15 drops) of concentrated nitric

---

[3] K. P. Link et al., *J. Biol. Chem.*, **133**, 293 (1940); **150**, 345 (1943); *J. Org. Chem.*, **5**, 639 (1940).

[4] B. Holmberg, *Arkiv Kemi*, **4**, 33 (1952).

acid; evaporate the mixture to dryness. By means of a microspatula scrape the residue and transfer into a test tube; wash the watch glass with 2 ml of water and add the washings to the residue in the tube. Heat the mixture to boiling and then cool in an ice-water mixture. Filter the crystalline mass and wash three times with 1 ml of water. Dry the crystals on a watch glass over a steam bath. The yield is 60 to 70 mg.

NOTE. The melting point of mucic acid listed in the literature is 214°. Since the compound melts with decomposition, values may be obtained that vary between 210° and 224°, depending on the rate of heating. Values near 214° are obtained when the heating is very slow.

Specific Rotation of Carbohydrates and Their Derivatives

The specific rotation of sugars and their osazones, hydrazones, and azoates is often used as a means of identification, since it consists simply in the determination of the rotation of a solution of known concentration. However, the sample must be pure. The rotation must be measured under specified conditions as to quantity of material and nature of solvent. For osazones a mixture of 40 per cent pyridine and 60 per cent alcohol is commonly used as a solvent; for azoates, alcohol-free chloroform is employed. The derivative must be of high purity. Special methods of purification are necessary when the original sample is an impure sugar. For example, glucose, for a long time, was considered a ketohexose sugar, that yielded a phenylosazone melting at 163 to 165°. It is now known[5] to be a fructose anhydride mixture. The phenylosazone, which melts at 163 to 165°, is a mixture of glucosazone and of the osazone of methylglyoxal and may be recrystallized from alcohol and other osazone solvents without change in its melting point. When it is treated with dry acetone the osazone of methylglyoxal dissolves, leaving the pure glucosazone. Therefore, it is recommended[6] that osazones prepared from impure sugars be washed with acetone before final purification. The procedure for determination of optical rotation is described on page 227.

Chromatography of Sugars

For detailed procedures employed in the separation of sugars the original literature, cited in the References, should be consulted. The

[5] L. Sattler and F. W. Zerban, *Sugar*, **39**, 12 (1944).
[6] L. Sattler, private communication.

present procedure deals with simple directions for the identification of a number of common sugars. The technique described on pages 123 to 129 is employed and the amount of solution applied is 1 to 10 λ so that about 100 to 500 γ are placed on each spot. The spot is placed 2 cm above the surface of the solvent to reduce streaking. The mixtures most commonly used as solvents are: (a) water 42 per cent, ethyl acetate 42 per cent, and acetic acid 16 per cent; (b) equal parts of water, 1-butanol, and collidine; (c) water 20 per cent, pyridine 20 per cent, and ethyl acetate 60 per cent; and (d) water 37 per cent, collidine 36 per cent, and ethanol 27 per cent. The components of each mixture are shaken together and allowed to stand for 24 hours, and if any separation occurs the organic phase is separated and used as the solvent. After development with the solvent, the chromatogram is dried and sprayed with the indicator solution. Various methods have been described in the literature,[7] but for most routine work ammoniacal silver nitrate is employed (0.1N silver nitrate and an equal volume of 5N ammonium hydroxide). After spraying, the chromatogram is heated in an oven at 105° for 5 to 10 minutes. The reducing sugar appears as brown spots. Of the other indicators the authors have found the salts of *p*-aminodimethylaniline and the alkaline 3,5-dinitrosalicylate to be the most useful for general work. The latter is prepared by dissolving 500 mg of 3,5-dinitrosalicylic acid in 100 ml of 0.1N sodium hydroxide solution. The spots produced on heating the sprayed chromatogram are brown against a pale yellow background.

For the preparation of *p*-aminodimethylaniline, 1.9 g of *p*-nitrosodimethylaniline (prepared from dimethylaniline according to the standard procedures) are added, in small portions and with shaking, to a solution of 5 g of stannous chloride in 6 ml of concentrated hydrochloric acid. If the reaction does not start, the mixture is warmed, but care should be used to cool the mixture if the reaction proceeds vigorously. At the end of the reaction the bright yellow solution is evaporated to a small volume and the crystals that separate are filtered after cooling the mixture. The compound is crystallized from 95 per cent ethanol and the crystals dried in air. The developed strip is sprayed with a 0.3 per cent ethanolic solution of the reagent and then heated for 10 minutes at 120°. The position of the spots is indicated by red to carmine red colors against a white background. Sorbose and fructose give bright red and glucose and mannose, golden orange, while xylose, arabinose, sucrose, raffinose, and maltose give red colors.

---

[7] See the references on chromatography of sugars (p. 531).

REFERENCES

Armstrong and Armstrong, *The Carbohydrates*, Longmans Green, New York, 1934.

Bates et al., *Polarimetry and Saccharimetry and the Sugars*, Bur. Standards Circ. C440, 1942.

Bomer, Zuckenack, and Tillmans, *Untersuchungsmethoden*, 2nd Part, Springer, Berlin, 1935.

Browne and Zerban, *Physical and Chemical Methods of Sugar Analysis*, 3rd ed., Wiley, New York, 1941. The most serviceable book in English on carbohydrates; contains a chapter on derivatives of sugars.

Jackson and Dehn, *Ind. Eng. Chem., Anal. Ed.*, **6**, 382 (1934). Glycosides.

Micheel, *Chemie der Zucker und Polysaccharide*, Akademische Verlagsgesellschaft, Leipzig, 1939.

Tollens and Elsner, *Kurzes Handbuch der Kohlenhydrate*, 4th ed., Barth, Leipzig, 1935.

Van der Haar, *Anleitung zum Nachweis zur Trennung und Bestimmung der Monosaccharide und Aldehydsäuren,* Gebr. Borntraeger, Berlin, 1920.

Vogel and Georg, *Tabellen der Zucker und ihrer Derivate*, Springer, Berlin, 1931.

Carbohydrate Tests

Devor, *J. Am. Chem. Soc.*, **72**, 2008 (1950); *Anal. Chem.*, **24**, 1626 (1951).

Feigl, *Spot Tests*, Vol. II, Elsevier, New York, 1953.

Klein and Weissman, *Anal. Chem.*, **25**, 771 (1953).

Steinmann, *Mikrochim. Acta*, **1953/5**, 537. Bibliography on micro tests.

Hydrazones

Thiobenzhydrazides: Holmberg, *Arkiv Kemi*, **4**, 33 (1952).

o-Tolylhydrazones of galactose, mannose: Fowweather, *Biochem. J.*, **55**, 718 (1953).

Other Derivatives

Acetates: Hudson, *Ind. Eng. Chem.*, **8**, 380 (1916).

Benzimidazoles: Link et al., *J. Biol. Chem.*, **150**, 345 (1943); **133**, 293 (1940); *J. Org. Chem.*, **5**, 639 (1940).

Azoates: Coleman et al., *J. Am. Chem. Soc.*, **64**, 1501 (1942); **65**, 1588 (1943); Reich, *Biochem. J.*, **33**, 1000 (1939).

Benzoates: Fischer and Noth, *Ber.*, **51**, 321 (1918); Levene and Meyer, *J. Biol. Chem.*, **76**, 513 (1928).

Osotriazoles: Hann and Hudson, *J. Am. Chem. Soc.*, **66**, 735 (1944); **67**, 939 (1945); **68**, 1766 (1946); **69**, 1050, 1461 (1947).

Tosyl esters: Compton, *J. Am. Chem. Soc.*, **60**, 395 (1938); Freudenberg et al., *Ber.*, **55**, 929, 3233 (1922); **58**, 294 (1925); **59**, 714 (1926).

Trityl ethers: Helferich et al., *Ber.*, **56**, 766 (1923); **58**, 872 (1925); Reynolds and Ewans, *J. Am. Chem. Soc.*, **60**, 2559 (1938).

Derivatives of sugars: Pittenger, *Sugar Research Foundation Sci. Rept. Ser.*, **5**, 51 (1947).

Derivatives of Carbohydrates

Microdetection of Sugars by Paper Chromatography

Bersin and Müller, *Helv. Chim. Acta*, **35**, 475 (1952).
Boggs et al., *Nature*, **166**, 520 (1950).
Hirst et al., *Nature*, **163**, 177 (1949).
Hough et al., *Nature*, **161**, 720 (1948); **164**, 1107 (1949); *J. Chem. Soc.* (London), **1949**, 928.
Jeanes et al., *Anal. Chem.*, **23**, 415 (1951).
Jermyn and Isherwood, *Biochem. J.*, **44**, 402 (1949); **48**, 515 (1951).
Johanson, *Nature*, **172**, 956 (1953).
Lemieux and Bauer, *Anal. Chem.*, **26**, 920 (1954).
Partridge et al., *Nature*, **158**, 270 (1946); **164**, 443, 479 (1949); *Biochem. J.*, **42**, 238, 251 (1948).
Rafique and Smith, *J. Am. Chem. Soc.*, **72**, 4634 (1950).
Sattler and Zerban, *Anal. Chem.*, **24**, 826 (1952).
Schneider and Erlemann, *Zucker-Beihefte* No. **3**, 40 (1951); *C.A.*, **46**, 381 (1952).
Wiggins and Williams, *Nature*, **170**, 279 (1952).

Microscopical Identification

Dehn, Jackson, and Ballard, *Ind. Eng. Chem., Anal. Ed.*, **4**, 413 (1932).
Hassid and McGready, *Ind. Eng. Chem., Anal. Ed.*, **14**, 683 (1942).
Quense and Dehn, *Ind. Eng. Chem., Anal. Ed.*, **11**, 555 (1939); **12**, 556 (1940).
Secor and White, *Anal. Chem.*, **27**, 1998 (1955).
White and Secor, *Anal. Chem.*, **27**, 1016 (1955).

16

Derivatives of Esters and Ethers

DERIVATIVES OF ESTERS

Table 16.01 lists the derivatives that have been proposed for the characterization of esters. There are two general approaches to the preparation

TABLE 16.01
DERIVATIVES FOR THE IDENTIFICATION OF ESTERS

*N-(β-Aminoethyl)morpholides	Hydrazides
*Anilides	*Hydrolysis followed by derivatization of acidic
*N-Benzylamides	and alcoholic components
*3,5-Dinitrobenzoates	p-Toluidides

* Recommended derivative

of derivatives for the characterization of esters. One is to derivatize the ester directly, as, for example, react the ester directly with 3,5-dinitrobenzoyl chloride so as to obtain a derivative of the alcohol function or prepare the anilide or p-toluidide by reacting the ester with a Grignard reagent so as to obtain a derivative of the acid function. The other is to hydrolyze the ester and derivatize the resulting acid and alcohol. However, as will be discussed in the following sections, a judicious selection of the proper type of derivative and detailed procedure must be made with each case depending on the nature of the ester.

In order to prepare derivatives for the final step of the identification of esters, it is necessary to hydrolyze the esters to their acidic and hydroxy components. The hydrolysis of semimicro quantities and isolation of 50 to 100 mg of an alcohol or an acid for derivatization require the utmost care. Even when the amount of ester used for hydrolysis is 2 to 3 g, the isolation of the hydrolytic products involves difficulties. The information obtained from the preliminary tests, boiling- or melting-point and refractive-index determinations, is of value in determining the probable nature of the ester; the type of hydroxy and acidic components present in the ester determines to a large extent the best method of procedure for the preparation of derivatives.

The most common esters are those of the lower alcohols having 1 to 4 carbon atoms. In the usual hydrolytic methods by aqueous alkali, the alcohol that is distilled after completion of hydrolysis contains a considerable amount of water and cannot be used for the preparation of nitrobenzoates or urethans. The preparation of benzoates through the Schotten-Baumann reaction is of little value for the lower alcohols, since these derivatives are chiefly liquids. On the other hand, the separation of anhydrous, or nearly so, methanol or ethanol from 5 to 10 ml of aqueous distillate is an extremely difficult operation requiring special apparatus. The general procedure for the separation of an alcohol is to saturate the distillate with potassium carbonate and then extract with ether (free from alcohol). This procedure gives fair results when the amount of ester is 5 to 10 g but is not satisfactory for the esters of lower alcohols, when the amount of ester hydrolyzed is less than 2 g.

Difficulties are also encountered in the identification of the acidic group of the esters. The common procedure is to evaporate the alkaline residue after distillation or extraction of the hydroxy compound and to use the sodium salt for the preparation of the p-toluidide, anilide, or other suitable derivative of the acid derived from the ester. The presence of excess alkali, however, complicates the preparation of such derivatives; as a consequence, even when one starts with 1 g of ester, little or no derivative of the acid is obtained with the use of such methods. This is particularly true in the case of the esters of the lower aliphatic carboxylic acids.

Consideration must be given to the effect of the excess alkali used in the hydrolysis upon labile functional groups such as are encountered in the β-keto esters, and in the esters of halogen acids; for example, alkali hydrolysis of ethyl acetoacetate will cause cleavage with the formation of acetone; similarly, alkaline hydrolysis of either α-chloro or α-bromobutyrates will give rise to both crotonic and α-hydroxybutyric acids.

Finally, esters hydrolyze at vastly different rates. Most esters of alcohols with less than 4 carbon atoms hydrolyze when they are treated for 30

534 Procedures of Final Characterization of an Unknown

minutes or less with hot 6N sodium or potassium hydroxide solution. Esters boiling above 200° require from 1 to 2 hours for complete hydrolysis. The disappearance of the ester layer cannot be used as a criterion for completion of hydrolysis of compounds that are slightly miscible with water; as the hydrolysis proceeds, the hydroxy compound formed rises to the top and therefore, even at the completion of hydrolysis, there remains an immiscible layer.

From this brief discussion it is evident that the exact procedure to be followed in the hydrolysis of esters and identification of the hydroxy and acyl radicals depends on the nature of the ester. The saponification equivalent of the ester given in the Appendix (pages 975 to 976) is often an aid in the identification.

Derivatization of the Acidic Part of an Ester

There are three methods by which the acidic component of the ester may be derivatized: (a) by reaction of the ester with ammonia, benzylamine, hydrazine, or morpholine to obtain solid derivatives; (b) by reaction of the ester with ethylmagnesium bromide and an arylamine such as *p*-toluidine or aniline to obtain the *p*-toluidide or anilide; and (c) by hydrolysis of the ester and isolating the acid itself if it is a solid or converting it into a solid derivative. Each of these three methods will be discussed and illustrated.

Conversion to Amides, Hydrazides, N-Benzylamides, and Morpholides

The acidic component of the ester may be identified by reaction with aqueous ammonia, benzylamine, hydrazine hydrate, or N-(β-aminoethyl)-morpholine, as shown by the following equations:

$$RCOOR' + NH_3 \rightarrow \underset{\text{Amide}}{RCONH_2} + R'OH$$

$$RCOOR' + C_6H_5CH_2NH_2 \rightarrow \underset{\text{N-Benzylamide}}{RCONHCH_2C_6H_5} + R'OH$$

$$RCOOR' + HNHNH_2 \rightarrow \underset{\text{Hydrazide}}{RCONHNH_2} + R'OH$$

$$RCOOR' + H_2NCH_2CH_2N\underset{\text{N-(β-Aminoethyl)morpholine}}{\overset{CH_2CH_2}{\underset{CH_2CH_2}{\diagup\!\!\!\diagdown}}}O \rightarrow$$

$$RCONHCH_2CH_2N\underset{\text{Morpholide}}{\overset{CH_2CH_2}{\underset{CH_2CH_2}{\diagup\!\!\!\diagdown}}}O + R'OH$$

The high solubility of the amides renders the reaction with ammonia unsuitable in most cases. The preparation of benzylamides is effected by boiling benzylamine with the ester. Halogenated esters yield N-benzylamines and benzylammonium halides by this treatment:

$$XCH_2COOR + 2C_6H_5CH_2NH_2 \rightarrow C_6H_5CH_2NHCH_2COOR + C_6H_5CH_2NH_3^+X^-$$

The reaction of esters with benzylamine has been applied to the identification of about sixty-five acyl groups in esters using about 1 ml of the liquid ester. The esters that do not give good results with benzylamine owing to the nature of the hydroxy group can be converted by alcoholysis to the methyl esters. In general the N-benzylamides should be tried first, then the hydrazides and morpholides; the amides are suitable when the acidic group is complex.

General Method for Preparation of N-Benzylamides from Esters

Place in a test tube arranged with a micro condenser for reflux 100 mg of powdered ammonium chloride, 1 ml of the ester (or 1 g if the ester is solid), 3 ml of benzylamine, and a small boiling stone. Boil gently for 1 hour and after cooling wash the reaction mixture with water to remove the soluble salts and then adjust the pH to 5 to 6 with benzylamine. In most cases this produces crystallization of the benzylamide. If there is a considerable amount of unreacted ester the benzylamide does not crystallize. In such cases the contents of the tube are placed in an evaporating dish and the tube washed with 1 ml of water; the washings are added to the dish. The mixture is boiled for a few minutes in order to volatilize the unreacted ester and then chilled. The crude benzylamide is extracted with 1 ml of ligroin and filtered then recrystallized from aqueous alcohol or acetone.

For the alcoholysis of esters with more than three carbon atoms in the alcohol radical of the ester, 1 ml of the ester is refluxed for 30 minutes with 5 ml of absolute methanol containing a trace of sodium methoxide (obtained by adding 100 mg of sodium metal to the 5 ml of absolute methanol). After refluxing, the excess of methanol is distilled off or evaporated, and the residue is reacted with benzylamine.

General Method for Preparation of Hydrazides from Esters

Higher esters are converted to methyl esters by alcoholysis. About 1 ml of the ester and 1 ml of 85 to 90 per cent hydrazine hydrate are refluxed in a tube as described in the preceding paragraph for 10 to 15 minutes, the heating is discontinued, and absolute ethanol or methanol is added dropwise until a clear solution results. The mixture is refluxed for 1 to 2 hours.

The mixture is poured into a small evaporating dish, and the alcohol is evaporated. On cooling the residue, the crude hydrazide is obtained, which is recrystallized from aqueous alcohol.

General Method for Preparation of Morpholides from Esters

About 10 millimoles of the ester are mixed with 10 millimoles of N-(β-aminoethyl)morpholine and refluxed for 2 to 3 hours. The reaction mixture is chilled, and if the morpholide does not separate, a few milliliters of petroleum ether are added and the mixture is stirred by a rod scratching the inner walls of the vessel to induce nucleation. The crude morpholide is crystallized from aqueous ethanol if the esters contain more than 11 carbon atoms and from ligroin if fewer.

The derivatives of low-boiling esters of formic, acetic, and propionic acids are best prepared by refluxing the mixture of ester and reagent with 3 ml of ethylene glycol for 3 hours followed by removal of the glycol under reduced pressure. For esters of dibasic acids a ratio of 10 millimoles of the ester to 20 millimoles of the reagent is employed. In the preparation of morpholides from aryl esters the reaction mixture is first heated to effect the reaction; then the cold mixture is treated with 5 ml of ether to facilitate precipitation of the derivative. The morpholides yield quaternary methiodides.

Conversion to a Toluidide or Anilide

If the acid derived from the ester is a liquid with fewer than 6 carbon atoms, the *p*-toluidide or the anilide may be prepared directly from the ester by means of the Grignard reaction. Ethylmagnesium bromide is prepared, reacted with *p*-toluidine or aniline, and then boiled with the ester:

$$RMgX + CH_3C_6H_4NH_2 \rightarrow CH_3C_6H_4NHMgX + RH$$
$$\text{\textit{p}-Toluidine}$$

$$\underset{\text{Ester}}{R'COOR''} + 2CH_3C_6H_4NHMgX \rightarrow R'\underset{OMgX}{C(NHC_6H_4CH_3)_2} + R''OMgX$$

$$R'\underset{OMgX}{C(NHC_6H_4CH_3)_2} + 2HCl \rightarrow R'CONHC_6H_4CH_3 + CH_3C_6H_4NH_3Cl$$
$$\text{\textit{p}-Toluidide}$$
$$+ MgXCl$$

The procedure is illustrated by Example 16.01, which describes the preparation of aceto-*p*-toluidide from ethyl acetate. However, if the acid is a liquid and is insoluble in water (such as the higher fatty acids) and the

Derivatives of Esters and Ethers 537

preparation of the *p*-toluidide through the Grignard reagent does not yield satisfactory results, the best procedure is to hydrolyze the ester. If aqueous potassium hydroxide is used, the solution is acidified and the residue is extracted with benzene. A slight excess of thionyl chloride is added to convert the carboxylic compound to the acid chloride, and the *p*-toluidide or anilide is prepared by the methods described on page 441. If potassium hydroxide dissolved in diethylene glycol is used to effect the hydrolysis of the ester, the residue is treated according to the last paragraph of Example 16.02. The details of the conversion of an ester to a toluidide are given in Example 16.01.

*Example 16.01: Preparation of Aceto-*p*-toluidide from Ethyl Acetate by Means of the Grignard Reagent*

Prepare the Grignard reagent according to the directions given in the general procedure on page 45, and observe all the precautions as to removal of moisture from the apparatus and reagents. Use 120 mg of magnesium, 5 ml of absolute ether, and 550 mg of ethyl bromide. When practically all the magnesium has dissolved, cool and add slowly a solution of 500 mg of *p*-toluidine in 4 ml of ether; after a minute add 0.2 ml of ethyl acetate and reflux the mixtue for 5 to 10 minutes. Cool and hydrolyze with 5 ml of dilute hydrochloric acid. Separate the ether layer and wash it first with dilute hydrochloric acid and then with water. Evaporate the ethereal solution and dissolve the residue in boiling ethanol. Filter and add 1 ml of water to precipitate the aceto-*p*-toluidide. The yield is 30 to 35 mg of crystals, melting at 146 to 147°.

Derivatization of the Acidic and Alcoholic Components after Hydrolysis

Discussion

If the acid is a solid it can be separated after hydrolysis. The ester is hydrolyzed by saponification with potassium or sodium hydroxide and, if the alcohol is volatile, the reaction mixture is distilled and the distillate is used for the derivatization of the alcoholic component. If the acid is a solid the residue is acidified, the solid acid is isolated and purified, and its melting point is determined. This procedure is illustrated in Example 16.02, which deals with derivatization of the acidic and alcoholic components of butyl phthalate. If the acidic component is a liquid the residue can be derivatized directly by reaction with *p*-nitrobenzyl chloride to obtain the *p*-nitrobenzyl ester, or it may be evaporated and the sodium salt converted first to the acid chloride by means of thionyl chloride and then to the anilide or *p*-toluidide by reaction of the acid chloride with an arylamine. This is illustrated in the hydrolysis of isopropyl acetate (Example 16.03).

Example 16.02: Identification of Butyl Phthalate

Place in an 8-in. distilling tube 1.5 ml of diethylene glycol, 0.3 g (1 pellet) of potassium hydroxide, and 5 drops of water. Heat by means of a small flame until the pellet dissolves. Cool by means of tap water to room temperature and add 0.5 ml of the ester. Arrange the tube with a reflux condenser and connect the outlet with a condensing setup. Heat to gentle boiling, shaking the tube from time to time; when the ester layer disappears (3 to 5 minutes), the reflux condenser is removed and 2 ml of dry pyridine are added. The tube is carefully heated until 2.2 ml of distillate have been collected. The distillate is used to prepare the 3,5-dinitrobenzoate, as described in Example 13.04. The residue is diluted with 5 ml of water and then acidified with $6N$ sulfuric acid. The crystals that separate are filtered and the melting point determined.

If a derivative of the acid is desired, the residue left in the distilling flask is diluted with 5 ml of water and 5 ml of ethanol and then neutralized to phenolphthalein with $6N$ sulfuric acid; it is then set aside to permit separation of potassium sulfate. The mixture is filtered and the clear filtrate is used for the preparation of *p*-nitrobenzyl ester as described in Example 12.09.

Derivatization of the Alcoholic Component

There are two general methods for the derivatization of the alcoholic component. One is to hydrolyze the ester, separate the alcohol (or phenol) and derivatize it by the methods discussed in Chapter 13. The other method is to react the ester directly with 3,5-dinitrobenzoyl chloride to obtain the 3,5-dinitrobenzoate of the alcohol, as described in Example 16.03.

Examples of Complete Characterization of Esters

Example 16.03: Identification of Isopropyl Acetate

Place 300 mg of isopropyl acetate, 500 mg of 3,5-dinitrobenzoyl chloride, and 3 ml of pyridine in an 8-in. tube, provided with a condenser arranged for reflux. Add 2 small boiling stones and heat for 1.5 to 2 hours over a small flame so that the mixture boils gently. Cool and add a mixture of 1 ml of 6N sulfuric acid and 9 ml of water (10 ml of 3 per cent acid). Cool and shake the mixture vigorously; extract the mixture with 5 ml of isopropyl ether or 5 ml of ethyl ether (which has been washed with water, placed over calcium chloride and sodium, and then filtered to remove all traces of ethanol). Separate the ether layer and wash it, first with 5 ml of 2 to 3 per cent sulfuric acid, then with 4 ml of 2 per cent sodium hydroxide solution, and finally with 3 ml of water. Evaporate the ether layer from a small dish. Dissolve the residue in hot methanol, add a minute amount of charcoal, filter, and add water to the filtrate. About 60 to 70 mg of crystals separate out, melting at 116 to 118°. Dissolve the crystals in methanol and precipitate with water. Filter and dry the 3,5-dinitrobenzoyl ester of isopropyl alcohol. The yield is 30 to 45 mg of crystals, melting at 121 to 122°.

For the identification of the acidic part of the ester, a new portion of the ester is hydrolyzed. Use 300 mg of the ester, 0.3 g (1 pellet) of potassium hydroxide, 1 ml of water, and 1 boiling stone. Boil for 15 minutes and then distill off most of the water and finally dry the residue within the tube either in an oven or on the water bath. By means of a spatula or a glass rod pulverize the residue and convert the salt to the halide and then to the *p*-toluidide according to Example 12.03 (page 442). The yield is 50 to 60 mg melting at 146 to 147°.

NOTE. The acidic component of the ester can be converted to the *p*-nitrobenzyl ester instead of the *p*-toluidide as follows. After boiling to hydrolyze the ester dilute to 2 ml with water; add 1 ml of 6N hydrochloric acid and 1 drop of phenolphthalein; if the solution is acid, add sodium hydroxide solution (5 to 10 per cent, until the color of the solution is just pink). Add 2 drops of 5 per cent hydrochloric acid so that the pink color is discharged. If the original solution is alkaline, add dilute hydrochloric acid (5 to 10 per cent) until the color of phenolphthalein just fades. Add 200 mg of *p*-nitrobenzyl bromide or chloride and 8 ml of methanol and a small boiling stone. (CAUTION. *Be careful in handling* p-*nitrobenzyl halides*.) Arrange a reflux condenser for the tube and boil gently for 1.5 hours. Cool and add 2 to 3 ml of water; by means of a glass rod scratch the inner sides of the tube. After 20 minutes filter the ester and wash first with 4 ml of 5 per cent sodium carbonate and then twice with 2 ml of water. Crystallize according to the method described on page 448. The yield is 40 mg, melting at 77 to 78°.

Example 16.04: *Characterization of Triphenyl Phosphate*

Place in an 8-in. tube 0.3 g (1 pellet) of potassium hydroxide, 1 ml of water, 200 mg of the solid ester, and 1 boiling stone. Boil gently for 15 minutes; dilute to 4 ml with water and cool. Remove 1 ml into an 8-in. tube and add 2 ml of water, 1 ml of 6N nitric acid, and 1 ml of ammonium molybdate solution, in the order given. Allow to stand for 5 minutes; a yellow precipitate indicates the presence of phosphate.

Acidify the remaining 3 ml of the hydrolyzed solution by means of dilute hydrochloric acid. Brominate as described on page 490 to identify the phenol.

Derivatization of Lactones and Esters of Inorganic Acids

Lactones

These compounds are listed under esters because they are formed by the reaction of the carboxyl and hydroxyl groups of the same molecule; it is quite probable, however, that the lactones will be detected as acidic compounds by the functional group tests, since they undergo hydrolysis to yield hydroxycarboxylic acids. The most common compounds of this group are ascorbic acid, a lactone of a sugar unsaturated acid (see page 668), and γ-butyrolactone, which is easily derivatized by ammonolysis forming γ-hydroxybutyramide.

Attention should be drawn to the formation of lactones from γ- and δ-hydroxy acids by intramolecular esterification. In dealing with aqueous solutions of lactones it should be remembered that an equilibrium exists between the hydroxy acid and the lactone. The solution is slightly acidic, but on addition of sufficient sodium hydroxide solution to change the color of phenolphthalein the lactone ring opens and after some time the pink color is discharged. The addition of another small amount of sodium hydroxide causes the color to reappear; after some time it is discharged again. After several small additions the lactone has been completely converted to the sodium salt of the hydroxy acid. This process of a gradual opening of the lactone ring is called "lactone titration."

The simplest derivative for lactones is the phenylhydrazide. One millimole of lactone is heated to 100° with 1.5 millimoles of phenylhydrazine for a few minutes; after cooling 1 ml of ether is added to precipitate the phenylhydrazide, which is filtered and crystallized from chloroform.

Alkyl Nitrites, Nitrates, and Sulfates

A few of the esters of the lower alcohols and inorganic acids, which are used occasionally in the organic laboratory, are listed in the tables on esters. (CAUTION. *Exercise care in handling these compounds; most of them have toxic effects.*) The nitrites hydrolyze very readily and may be detected by the ease with which they diazotize aniline in the presence of glacial acetic acid, or by the addition of 2-phenylindole, which forms a precipitate of 3-isonitroso-2-phenylindole. The three common nitrites are ethyl, *n*-butyl, and isoamyl; the alcohol is readily identified after hydrolysis by the usual methods. The sulfates are characterized either by the preparation of phenyl ethers or by conversion to the thiuronium derivatives, as outlined on page 636.

ETHERS

Table 16.02 gives a summary of the few derivatives that may be prepared for characterization of ethers. Generally, ethers are relatively inert com-

TABLE 16.02
DERIVATIVES FOR THE IDENTIFICATION OF ETHERS

Alkyl halides	*3,5-Dinitrobenzoates
*Bromo derivatives	Picrates
Carbonyl compounds	*Sulfonamides

* Recommended derivative.

pounds; the carbon-oxygen bond in ethers cannot be easily split and hence the preparation of derivatives from semimicro quantities involves some difficulties. In the case of aromatic ethers, it is possible by means of bromination, chlorosulfonylation, or other substitution reactions to prepare suitable solid derivatives that may be used for characterization. For the derivatization of aliphatic ethers, however, it is necessary to cleave the ether linkage in order to prepare a derivative of the resulting hydroxy compound. Aliphatic and aromatic ethers are, therefore, discussed separately.
(CAUTION. *In working with ethers bear in mind that they form peroxides easily, particularly when exposed to light and air. The peroxides detonate when heated, and hence the distillation of an appreciable amount of ether, which contains peroxide, involves the danger of an explosion if the distillation is allowed to proceed to dryness.* The presence of peroxides in ethers is detected by means of starch iodide paper that has been moistened with dilute hydrochloric acid. The peroxides are removed by washing the ether with water containing a small amount of ferrous sulfate and dilute sulfuric acid. For small quantities of ethers, as, for example, 10 ml, washing with 2 to 3 ml of water containing 5 drops of 10 per cent solution of ferrous sulfate and 1 drop of sulfuric acid is sufficient.)

Derivatives from Aliphatic Ethers

Discussion

Two general methods have been proposed for the splitting of ethers but neither is well suited for small quantities. The first method employs heating in the presence of anhydrous zinc chloride and 3,5-dinitrobenzoyl chloride. The ether is cleaved to the hydroxy compound, which reacts with the acid chloride to give the 3,5-dinitrobenzoate. The first step in the cleavage is assumed to be the formation of an alcohol and an olefin; the alcohol thus formed reacts with 3,5-dinitrobenzoyl chloride to give the ester and hydrogen chloride; the latter converts some of the alcohol into an alkyl chloride.

$$CH_3CH_2OCH_2CH_3 \xrightarrow{ZnCl_2} CH_3CH_2OH + CH_2=CH_2$$
$$CH_3CH_2OH + C_6H_3(NO_2)_2COCl \longrightarrow C_6H_3(NO_2)_2COOC_2H_5 + HCl$$
$$CH_3CH_2OH + HCl \longrightarrow CH_3CH_2Cl + H_2O$$

Aside from the reactions that diminish the amount of 3,5-dinitrobenzoate, the chief difficulty of the method as applied to micro and semimicro quantities is that a number of aliphatic ethers boil at low temperatures and prolonged heating results in considerable losses of the compounds before cleavage is effected. The amount of 3,5-dinitrobenzoate obtained from a series of experiments with 0.5 ml of ethyl ether and isopropyl ether was often less than 10 mg, while in many runs no derivative at all was obtained. It has been observed that, if the zinc chloride is freshly fused and precautions are taken to dry thoroughly the tube and microcondenser, better yields are obtained. By the use of a sealed tube and heating under pressure, it is possible to prevent the loss of the ether and obtain a sufficient amount of the derivative.

The second method of cleavage utilizes vapor-phase pyrolysis; when an ether is pyrolyzed in the vapor phase, the chief products are a hydrocarbon and an aldehyde or ketone, as shown by the following equations.

$$2RCH_2OCH_2R' \xrightarrow{500°} RCH_3 + R'CH_3 + RCHO + R'CHO$$
$$2R_2CHOCHR_2' \xrightarrow{500°} R_2CH_2 + R_2'CH_2 + R_2CO + R_2'CO$$
$$R_2CHOCHR_2 \xrightarrow{500°} RCH_2R + RCOR$$
$$RCH_2OCH_2R \xrightarrow{500°} RCH_3 + RCHO$$

The first two equations illustrate the pyrolysis of unsymmetrical ethers and the last two, the pyrolysis of symmetrical ethers. In general, pyrolysis should not be undertaken with small quantities; even in the case of symmetrical ethers in which only one carbonyl compound is produced, at least 0.1 ml of distillate should be obtained for which 200 to 300 mg of ether must be pyrolyzed. Since unsymmetrical ethers produce a mixture of aldehydes and ketones and these must be separated by fractionation, a sufficient amount of ether must be pyrolyzed to yield at least 1 ml of distillate. The carbonyl compounds are readily derivatized as semicarbazones or 2,4-dinitrophenylhydrazones.

The cleavage of ethers by hydriodic acid is not feasible for semimicro quantities:

$$R_2O + 2HI \rightarrow 2RI + H_2O$$

$$ArOR + HI \rightarrow ArOH + RI$$

In order to obtain good results by cleavage of alkyl or arylalkyl ethers it is necessary to use a sample of at least 4 to 5 g. The alkyl iodide and phenols are derivatized as described in Chapters 17 and 13, respectively.

General Method for Preparation of Alkyl 3,5-Dinitrobenzoates

The zinc chloride employed for cleavage is freshly fused and finely powdered. About 2 g of powdered commercial zinc chloride are fused in an iron crucible and then stirred with an iron rod while the melt cools. When the consistency becomes doughlike, the mass is pulverized with a small pestle to give a freely running fine powder that is still hot. The powder is transferred immediately to a dry bottle that has been previously warmed and tightly closed with a Bakelite screw cap. The small amount of iron oxide with which the zinc chloride is contaminated does not interfere with its activity.

A 6-in. tube is dried by being heated over a free flame and stoppered while hot. About 400 to 500 mg of zinc chloride are introduced, followed by 0.5 ml of ether and 250 mg of 3,5-dinitrobenzoyl chloride; the stopper is raised momentarily for the addition. A microcondenser is inserted, after it has been thoroughly wiped to remove any adhering moisture. The tube is immersed in a bath and heated at such a temperature that the ether does not condense much above the end of the microcondenser, which is about 10 to 15 mm from the bottom of the tube. The mixture is heated for 2 hours. The microcondenser is removed; the residue is heated in the bath until it is dry, and is then carefully pulverized with a rod. About 5 ml of 10 per cent sodium carbonate are heated separately at 60 to 70° and then added to the powdered mixture. A solid rubber stopper is inserted in the mouth of the tube and the contents are shaken for about 1 minute; then the remaining solid is stirred and crushed on the sides of the tube until no small lumps can be detected. The mixture is again heated to 60 to 70° and filtered by suction; the residue in the funnel is washed twice with 2 ml of sodium carbonate solution and twice with 2 to 3 ml of water. At this point the residue remaining in the filter should be a fine powder. If any lumps are present the treatment with warm sodium carbonate solution is repeated. The residue is transferred along with the filter paper to a dry tube and extracted with 1.5 to 2 ml of boiling ethanol. The hot solution is filtered, the tube and residue are washed with an additional 1 ml of hot ethanol, and the filtrate is combined with the first extract. If no solid separates, the filtrate is evaporated until the volume is about 2 ml, and water is added dropwise until the solution becomes cloudy. The tube is reheated until the solution is clear and is allowed to cool slowly. The crystals are filtered and dried and the melting point is determined. It is not uncommon for the melting point to be 5 to 10° below that of the pure derivative if the extraction is faulty. The yield varies from 5 to 50 mg. With low-boiling ethers, a yield of only 1 mg or less of the ester is not unusual. In such cases a closed tube (ignition tube) is used and the mixture is heated under pressure at 100° for 1 hour.

Derivatives from Aromatic Ethers

Discussion

Several reactions can be employed to form suitable derivatives of aryl ethers. For the bromination of the aromatic ether the compound is dissolved in glacial acetic acid, alcohol, or chloroform and the required amount of bromine is added dropwise:

$$CH_3C_6H_4OCH_3 + Br_2 \rightarrow CH_3C_6H_3BrOCH_3 + HBr$$

o-Cresyl methyl ether   Monobromo-*o*-cresyl methyl ether

The bromo-substituted ether is obtained either by addition of water or evaporation of the solvent. The extent of bromination depends on the groups already present: *o*-cresyl methyl ether forms a monobromo, whereas guaiacol (2-methoxyphenol) forms a tribromo derivative. Ethers that have an unsaturated linkage undergo addition and substitution; thus anethole (*p*-propenylphenyl methyl ether) forms a monobromo dibromide.

The following equation represents the formation of a molecular compound with picric acid:

$$C_6H_5OCH_3 + C_6H_2(NO_2)_3OH \rightarrow C_6H_5OCH_3 \cdot C_6H_2(NO_2)_3OH$$

Anisole   Picric acid   Anisole picrate

The picrates form readily by mixing equimolecular amounts of the ether and picric acid dissolved in the minimum amount of warm chloroform, and then allowing the mixture to stand for a short time. Although a number of the picrates are unstable on exposure to air, their preparation offers a relatively convenient method for identification of a number of aromatic ethers. Since halides, when treated with β-naphthol, form naphthyl ethers readily, which may be identified by the picrates, the method may be employed for the derivatization of a number of alkyl halides.

An aromatic ether is converted to a substituted benzenesulfonyl chloride through reaction with chlorosulfonic acid:

$$C_6H_5OCH_3 \xrightarrow{ClSO_2OH} CH_3OC_6H_4SO_2Cl$$

Anisole   p-Methoxybenzenesulfonyl chloride

The sulfonyl chloride is converted by ammonolysis to the substituted benzenesulfonamide, which may be identified through its melting point:

$$CH_3OC_6H_4SO_2Cl \xrightarrow{NH_3} CH_3OC_6H_4SO_2NH_2 + NH_4Cl$$

p-Methoxybenzenesulfonamide

The method is feasible when the preparation of a bromo derivative or a picrate does not afford satisfactory results.

Other reactions of aromatic ethers that may be used for the preparation of derivatives are nitration and oxidation. Oxidation is used in the case of ethers having side chains—as, for example, the cresyl ethers; these, by oxidation, yield alkoxybenzoic acids. With a few exceptions both nitration and oxidation give poor yields of derivatives.

General Method for Preparation of Bromo Derivatives

Quantities as low as 1 to 5 mg can be brominated. The ether is dissolved in glacial acetic acid or chloroform, and a slight excess of bromine is added slowly while the reaction mixture is cooled. A solution of bromine in glacial acetic acid is convenient to handle. If glacial acetic acid is used as the solvent, the bromo compound is separated by the addition of water; if the bromo compound tends to separate as an oil, the solvent is evaporated and the crude solid is crystallized from alcohol, petroleum ether, or isopropyl ether. The bromination of some ethers (such as anethole) is best accomplished in isopropyl ether. About 50 mg of the compound are added to 1 ml of isopropyl ether (or absolute ethyl ether) and cooled in an ice bath or tap water. To this is added over a period of 5 minutes a solution of 120 mg of bromine in 1 ml of ether. The mixture is allowed to stand in the cold for 5 minutes; then the crystals are filtered and crystallized from petroleum ether. The yield is about 75 to 90 mg of crystals melting at 107°.

For bromination in glacial acetic acid the ether is added to a cold 1 per cent solution of bromine in glacial acetic acid. The solution should contain slightly more bromine than the theoretical amount. For example, in the bromination of 100 mg of 2-naphthyl methyl ether (0.7 mM) 6 ml of 1 per cent bromine solution is used since the monobromo derivative is formed. The bromine solution is first cooled in an ice bath and the ether is added. After a few minutes the tube containing the mixture is removed from the ice bath and allowed to stand at room temperature for 10 minutes. The derivative is precipitated by the addition of water and the crude crystals are separated and crystallized from an alcohol-water mixture. The yield from 100 mg of ether varies from 100 to 150 mg of the pure derivative.

General Method for Chlorosulfonylation of Ethers and Preparation of Sulfonamides

About 0.25 ml of the aromatic ether is dissolved in 2 ml of dry chloroform (in a clean tube) and cooled in an ice bath. About 1 g of chlorosulfonic acid is added dropwise over a period of 3 to 5 minutes. The tube is removed from the ice bath and allowed to stand at room temperature for 20 to 30 minutes. The contents of the tube are poured slowly into a small separatory funnel containing 5 ml of ice water. The tube is washed with 1 ml of chloroform and the washings are added to the funnel. The mixture is shaken gently and the chloroform layer is run into an 8-in. distilling tube. The chloroform is distilled off and, while the tube is still immersed in the water bath, 1 g of powdered ammonium carbonate and 5 ml of a concentrated solution of aqueous ammonia are added. The stopper of the distilling tube is replaced and the mixture is heated at 60° for 15 minutes and then at 80 to 90° for 10 minutes. The receiving tube is charged with water to absorb the ammonia given off during heating. The hot solution is filtered to remove any solid that has separated out; this may be crude sulfonamide and is crystallized separately. The clear filtrate is evaporated to dryness on the water bath; the residue is dissolved in 5 to 10 ml of boiling water; the solution is treated with charcoal, filtered, and cooled. Another method of purification consists in dissolving the crude sulfonamide in an alcohol-water mixture. If the melting point of the derivative is 5 to 10° below the value recorded in the literature, the sulfonamide may be contaminated with products of side reactions (sulfones and chlorinated compounds). The sulfonamide is dissolved by being heated gently in 5 ml of 5 per cent sodium hydroxide and any undissolved material is filtered off. The filtrate is acidified with dilute hydrochloric acid, and the sulfonamide is filtered and purified. The yield varies from 10 to 20 mg.

General Method for Preparation of Picrates of Phenolic Ethers

A number of phenolic ethers form picrates, the molecular compounds consisting of 1 molecule of ether to 1 to 2 molecules of picric acid. About 5 mM of the ether are dissolved in 5 to 6 ml of boiling chloroform and 5 mM of the picric acid are dissolved in 3 ml of boiling chloroform. The solutions are mixed, shaken, and then allowed to cool. The crystals are filtered and dried rapidly between filter papers; the melting point is determined immediately. The picrates should not be recrystallized or allowed to dry in air as a number of these decompose.

REFERENCES

Identification of Esters

Rapid saponification of esters by potassium hydroxide in diethylene glycol: Redemann and Lucas, *Ind. Eng. Chem., Anal. Ed.*, **9**, 521 (1937).

Removal of acyl groups: Baltzly and Buck, *J. Am. Chem. Soc.*, **63**, 2022 (1941).

Use of N-(β-aminoethyl)morpholine: Bost and Mullen, *J. Am. Chem. Soc.*, **73**, 1967 (1951).

Anilides from esters: Hardy, *J. Chem. Soc. (London)*, **1936**, 398.

N-Benzylamides: Buehler and Mackenzie, *J. Am. Chem. Soc.*, **59**, 421 (1937); Dermer and King, *J. Org. Chem.*, **8**, 168 (1943).

3,5-Dinitrobenzoates: Renfrow and Chaney, *J. Am. Chem. Soc.*, **68**, 150 (1946).

Hydrazides: Sah, *Rec. trav. chim.*, **59**, 1036 (1940).

p-Toluidides from esters: Hardy, *J. Chem. Soc. (London)*, **1936**, 398; Koelsch and Tannenbaum, *J. Am. Chem. Soc.*, **55**, 3049 (1933).

Identification of Ethers

Chlorosulfonylation of ethers: Huntress and Carten, *J. Am. Chem. Soc.*, **62**, 511 (1940).

Semimicro identification of ethers by catalytic conversion to carbonyl compounds: Sah, *Rec. trav. chim.*, **58**, 758 (1939) (in English).

3,5-Dinitrobenzoates: Underwood, Baril, and Toone, *J. Am. Chem. Soc.*, **52**, 4087 (1930).

Picrates: Andersen, *Acta Chem. Scand.*, **8**, 157 (1954); Baril and Megrdichian, *J. Am. Chem. Soc.*, **58**, 1415 (1936).

Picrates of naphthyl alkyl ethers: Dermer and Dermer, *J. Org. Chem.*, **3**, 289 (1938).

Sulfonamides from aryl ethers: Huntress and Carten, *J. Am. Chem. Soc.*, **62**, 603 (1940).

Miscellaneous derivatives: Naphthylamine adducts of 3,5-dinitrobenzoates from esters, Benfrey et al., *J. Org. Chem.*, **20**, 1777 (1955). Use of ethanolamine for esters, Rauscher et al., *Ind. Eng. Chem., Anal. Ed.*, **22**, 923 (1940); *J. Am. Chem. Soc.*, **70**, 438 (1948). *p*-Nitrobenzoates from ethers, Ward and Jenkins, *J. Org. Chem.*, **10**, 371 (1945). S-Alkylthiuronium picrates from alkoxyl compounds, Kratzle and Osterberger, *Monatsh.*, **81**, 998 (1950). Hydrazides and phenylhydrazides from lactones, Darapsky et al., *J. pract. Chem.*, (2) **147**, 150 (1936); Pummerer et al., *Ber.*, **68**, 371 (1945); Seib, *Ber.*, **60**, 1399 (1935).

17

Derivatives of Halogen Compounds

ALKYL AND CYCLOALKYL HALIDES—ARYL
HALIDES—FLUORINE COMPOUNDS

Table 17.01 summarizes the derivatives that have been proposed for the characterization of alkyl, cycloalkyl, and aryl halides. The fluorine compounds (fluorocarbons and perfluoro compounds) are discussed on pages 564–569.

The discussion will include derivatives suitable for (a) alkyl and cycloalkyl halides, and (b) aryl halides. The sulfonamides and nitro compounds are suitable for aryl halides. The majority of the other derivatives are only for alkyl and cycloalkyl halogen compounds.

The polyhalides are not discussed separately. Except in the few instances in which derivatization is possible, the alkyl and cycloalkyl polyhalides are characterized by means of physical constants and functional group reactions. In general, aryl polyhalides are derivatized by the same procedures as the monohalides.

ALKYL AND CYCLOALKYL HALIDES

The majority of the derivatives listed in Table 17.01 have been proposed for the characterization of alkyl and cycloalkyl halides. The selection of the proper type of derivative is difficult. Of those recommended, the S-alkylthiuronium picrates should be tried first if inspection of Table 14 (page 830) discloses that the melting points of these derivatives are listed. Preparation of the acyl anilides, p-toluidides, α-naphthalides, and alkylmercuric halides, which are recommended as a choice when the S-alkylthiuronium picrates are not listed, involves conversion of the halide

TABLE 17.01
DERIVATIVES FOR THE IDENTIFICATION OF HALOGEN COMPOUNDS

9-Alkylfluorene-9-carboxylates	*3,5-Dinitrobenzoates
S-Alkyl-2-mercapto-4,5-dihydro-glyoxalinium picrates	Ethers of 2,4-dinitrothiophenol
	Ethers of p-hydroxybenzoic acid
*Alkylmercuric halides	Ethers of p-hydroxydiphenylamine
N-Alkyltetrachlorophthalimides	Iodoso derivatives
N-Alkyl-3-nitrophthalimides	*Nitro derivatives
N-Alkylphthalimides	6-Nitro-2-mercaptobenzothiazole derivatives
N-Alkyl-p-bromobenzenesulfon-p-anisidides	
	*Picrates of β-naphthyl ethers
N-Alkyl-p-toluenesulfontoluidides	Piperidyl derivatives
*S-Alkylthiuronium picrates	*Sulfonamides
*Anilides and p-toluidides	Tri-iodophenyl ethers

* Recommended derivative.

into a Grignard reagent. Consideration also should be given to the preparation of α-naphthyl ethers, tri-iodophenyl ethers, N-alkylphthalimides, nitrophthalimides, and N-alkylsaccharin derivatives.

S-Alkylthiuronium Picrates

Discussion

The following equation represents the formation of S-alkylthiuronium picrates:

$$C_2H_5Br + 2NH_2CSNH_2 \rightarrow NH_2C(NH)SC_2H_5 + NH_2CSNH_3{}^+Br^-$$
Thiourea   S-Ethylisothiourea

$$NH_2C(NH)SC_2H_5 + C_6H_2(OH)(NO_2)_3 \rightarrow$$
Picric acid

$$NH_2C(NH)SC_2H_5 \cdot C_6H_2(OH)(NO_2)_3$$
S-Ethylthiuronium picrate

These molecular compounds, whenever they can be prepared, should be among the first derivatives to be tried. They are easily obtained in good yields by heating a mixture of the halide and thiourea. The mixture, dissolved in ethylene glycol, is refluxed for 30 minutes; then a saturated alcoholic solution of picric acid is added and the vessel cooled. The S-alkylthiuronium picrate separates out. The addition compound is recrystallized after filtration. The number of halides from which S-alkylthiuronium picrates have been obtained is about 50; however, among these are the most common alkyl halides—that is, up to 10 carbon atoms.

General Method of Preparation

For the preparation of S-alkylthiuronium picrates, 1 mM of the alkyl halide and 2 mM of thiourea are dissolved in ethylene glycol and heated in an oil bath for 30 minutes. Most primary and secondary alkyl bromides and chlorides react at an oil bath temperature of 117°. Most primary alkyl iodides react at an oil bath temperature of 65°, and secondary alkyl iodides react at 117°. Tertiary alkyl halides do not react. In most cases, two reaction mixtures should be heated simultaneously; one with the oil bath at 117° and the other with the oil bath at 65°. After the alkyl halide and the thiourea have been refluxed for 30 minutes, 1 ml of a saturated alcoholic solution of picric acid is added and the heating is continued for 15 more minutes. The solution is then cooled, 5 ml of cold water is added, and the solution is placed in an ice bath and shaken periodically for 15 minutes. The mixture is filtered and the solid is recrystallized from methanol.

Due to steric effects, tertiary alkyl halides do not react to form the picrates. It has been suggested by earlier workers that the reaction could be facilitated by adding some potassium iodide to the reaction mixture, thus converting the alkyl halide to the alkyl iodide. This addition of potassium iodide has been found to give unreliable results. Therefore, if the compound has been established as an alkyl halide and the above procedure does not produce a satisfactory derivative, it can be assumed with reasonable certainty that the alkyl halide is tertiary, and positive identification must be made with another method.

Example 17.01: Preparation of S-n-Butylthiuronium Picrate[1]

Place 300 mg of thiourea, 5 ml of ethylene glycol, two boiling stones, and 6 drops of *n*-butyl bromide in an 8-in. test tube. Attach a finger condenser and insert the test tube in an oil bath that has been preheated to 117°.

[1] H. M. Crosby and J. B. Entrikin, *J. Chem. Ed.*, **41**, 360 (1964).

Keep the test tube in the oil bath at that temperature for 30 minutes. Add 1 ml of a saturated solution of picric acid in ethanol and keep the tube in the oil bath at 117° for an additional 15 minutes. Cool the test tube and add 5 ml of cold water. Shake the tube and then place it in an ice bath and shake it periodically for 15 minutes. Filter the mixture, and recrystallize the derivative from methanol.

NOTE. The crystallization of the alkylthiuronium picrates may be accomplished from hot solutions.

The yield of the picrate from 0.2 to 0.25 ml of alkyl halides varies from 40 to 200 mg. Generally the higher primary alkyl halides give better yields than the lower halides or the secondary halides. For example, from 200 mg of the halide the following yields of the pure picrate were obtained: methyl iodide, 50 mg; *n*-propyl iodide, 60 mg; *n*-amyl bromide, 150 mg; and *sec*-amyl bromide, 50 mg.

Picrates of β-Naphthyl Ethers

An alkyl halide is converted to an alkyl β-naphthyl ether by reaction with β-naphthol:

$$CH_3CH_2Br + \underset{\beta\text{-Naphthol}}{\text{C}_{10}H_7\text{OH}} \xrightarrow{NaOH} \underset{\text{Ethyl }\beta\text{-naphthyl ether}}{\text{C}_{10}H_7\text{OC}_2H_5} + NaBr + H_2O$$

The alkyl β-naphthyl ether is converted to the picrate as described on page 547 for the derivatization of aryl alkyl ethers.

For the preparation of the aryl alkyl ether, 5 mM of the alkyl halide and slightly less than 5 mM (700 mg) of β-naphthol and 2 ml of 10 per cent sodium hydroxide are warmed for 30 minutes on a steam bath; about 4 ml of water are added and the tube is chilled in an ice bath. The crystals are removed by centrifugation or filtration, dissolved in the minimum amount of alcohol and precipitated by addition of water, and then separated and converted to the picrate as outlined on page 547. However, if the melting point of the solid alkyl β-naphthyl ether is sharp, the conversion to the picrate is unnecessary.

Anilides, *p*-Toluidides, α-Naphthalides, and Alkylmercuric Salts

Discussion

The reactions represented by the following equations are based on the conversion of the halide to a Grignard reagent, which can react either with an isocyanate to form a substituted amide or with a mercuric halide to form an alkylmercuric halide:

$$CH_3I \xrightarrow[(C_2H_5)_2O]{Mg} CH_3MgI \begin{bmatrix} + C_{10}H_7NCO \rightarrow CH_3CONHC_{10}H_7 \\ \text{α-Naphthyl} \quad\quad\quad \text{Aceto-α-naphthalide} \\ \text{isocyanate} \quad\quad\quad \text{(N-α-napthylacetamide)} \\ \\ + C_6H_5NCO \rightarrow CH_3CONHC_6H_5 \\ \text{Phenyl} \quad\quad\quad\quad \text{Acetanilide} \\ \text{isocyanate} \\ \\ + C_6H_4(CH_3)NCO \rightarrow CH_3CONHC_6H_4CH_3 \\ \text{p-Tolyl isocyanate} \quad\quad\quad \text{Aceto-p-toluidide} \end{bmatrix}$$

$$CH_3I \xrightarrow[(C_2H_5)_2O]{Mg} CH_3MgI + HgI_2 \rightarrow CH_3HgI + MgI_2$$
$$\text{Methylmercuric iodide}$$

The preparation of the Grignard reagent requires some care (see pages 45–48). If the reagents are pure, the semimicro preparation should be complete within one-half hour. Once prepared, the Grignard reagent should be used immediately, either with the isocyanate or with the inorganic mercuric halide.

The selection of the isocyanate depends upon the availability of the reagent and the probable nature of the halide. For the lower halides α-naphthyl isocyanate is preferable because of the lower solubilities of the naphthalides as compared with the corresponding anilides. For the higher halides either phenyl isocyanate or *p*-tolyl isocyanate may be used.

The side reactions of the isocyanates have been discussed in connection with the identification of alcohols (page 475). The formation of ureas through the reaction of isocyanates with moisture present in the vessel and reagent necessitates extraction of the substituted amide with petroleum ether. In general, two crystallizations are necessary for purification of the product obtained by extraction of the crude mixture with petroleum ether; however, in some cases three or four crystallizations may be required. The use of alcohol as a recrystallization solvent, although it is given in the literature, is not recommended for semimicro work. The solubility of the substituted ureas formed by side reactions is greater in methanol or ethanol than in petroleum ether; hence their elimination by crystallization, when alcohol is used as a medium, is more difficult. When 90 per cent methanol is used to recrystallize the crude naphthalide obtained from the reaction of methylmagnesium iodide and α-naphthyl isocyanate, four or five crystallizations are required to obtain a pure product, whereas when petroleum ether is employed, one or two crystallizations are sufficient.

The Grignard reagent from the alkyl halide may be converted to an alkylmercuric halide by addition of the corresponding mercuric halide; if, for example, an organic bromide was employed to prepare the Grignard reagent, mercuric bromide is used. Although in some cases it is possible to obtain good yields with as little as 0.1 ml of the halide, it is advisable in semimicro work to use 0.3 to 0.5 ml of the halide for the preparation of the alkylmercuric halides. The crystallization of some of the organomercuric compounds is slow and purification difficult at times. Other precautions are discussed in the experimental section dealing with the preparation of alkylmercuric halides.

The general methods for the preparation of derivatives by converting the alkyl or cycloalkyl halide to the Grignard reagent are illustrated by the following three examples.

Example 17.02: Preparation of N-(α-Naphthyl)valeramide from n-Butyl Bromide

Prepare the Grignard reagent according to directions on page 45. Use 300 mg of *n*-butyl bromide, 5 ml of anhydrous ether, and 50 mg of clean magnesium turnings. After the initial reaction has subsided, heat by means of a very small flame of the microburner or by a water bath for 20 to 30 minutes. Then immerse the tube in cold water. Mix 5 ml of anhydrous ether and 0.22 ml or 260 mg (2 mM) of α-naphthyl isocyanate in a dry test tube; add this mixture in small portions to the Grignard reagent and allow it to stand for 15 minutes. Decompose the intermediate by slowly pouring the reaction mixture into a tube containing 5 ml of water and 5 ml of dilute hydrochloric acid so that any unreacted magnesium adheres to the side of the reaction tube. Use 2 ml of ordinary ether to wash the reaction tube and add the washings to the mixture being hydrolyzed. Stir the mixture cautiously to insure complete hydrolysis. Transfer it to a separatory funnel and draw off the aqueous layer. If a solid separates and remains suspended in the aqueous layer, filter it off and add it to the ether solution. Transfer the ether layer to a small evaporating dish and evaporate to dryness on a steam bath. Pulverize the residue by means of the microspatula and transfer it to a 6- or 8-in. tube. Extract successively with two 5-ml portions of hot petroleum ether. Filter the solution by suction from the undissolved impurities after first preparing the filter funnel according to directions given under the preparation of urethans (page 476). Cool the petroleum ether extracts. Filter the crystals and save the filtrates, since a further quantity of the derivative may be obtained by their evaporation. Crystallize the solid from 6 to 7 ml of hot petroleum ether. The yield of pure crystals, melting at 111 to 112°, is 45 to 55 mg.

NOTE. The amount of halide employed for the preparation of the Grignard reagent may be reduced to 0.1 ml, but precautions to prevent losses should be taken; for example, it is possible to start with 0.1 ml of methyl iodide and obtain 40 to 50 mg of the pure aceto-α-naphthalide if the circulation of cold water in the microcondenser (during the preparation of methylmagnesium iodide) is rapid. Since the success of the preparation depends on the completeness of conversion of the halide to the Grignard reagent, proper precautions should be taken to insure that the vessel and the reagents are dry and that the magnesium turnings are clean.

The amount of isocyanate used is in excess of that required, and therefore some substituted naphthylurea is admixed with the naphthalide; petroleum ether does not dissolve the urea; hence it is a better crystallizing medium than methanol or ethanol.

Example 17.03: Preparation of N-Phenylcapramide from n-*Amyl Chloride*

Prepare the Grignard reagent in an 8-in. tube from 50 mg of magnesium turnings, 5 ml of dry ether, and 0.25 ml of *n*-amyl chloride. Add a few crystals of iodine at the beginning to activate the magnesium metal; after the initial reaction has subsided, heat by means of a water bath for 45 minutes. Follow the procedure in Example 17.02 using 0.2 ml of phenyl isocyanate in place of α-naphthyl isocyanate. (CAUTION. *Care should be exercised in the handling of phenyl isocyanate, owing to its lachramatory properties.*) The yield of the pure anilide is 35 to 40 mg.

Example 17.04: Preparation of n-*Butylmercuric Bromide from* n-*Butyl Bromide*

Prepare the Grignard reagent in an 8-in. tube from 100 mg of magnesium turnings, 5 ml of dry ether, and 0.3 ml of *n*-butyl bromide (see page 553). The mixture is refluxed for 30 minutes. Filter the Grignard reagent into an 8-in. tube containing 1.5 g of mercuric bromide. Use a small funnel with a glass wool plug for the filtration. Stopper the tube and shake the mixture vigorously; warm the tube cautiously by momentary immersion in the steam bath and then repeat the vigorous shaking. Place the tube in a water bath and evaporate to dryness. Add 7 ml of ethanol and heat the tube in a water bath until the alcohol boils; then filter; add 3.5 ml of water to the solution and place in an ice bath. After 30 minutes filter off the crystals, wash, and crystallize from 7 to 8 ml of 60 per cent ethanol. The yield of the pure compound, melting at 134 to 135°, is 200 to 220 mg.

NOTE. The above procedure may be used for the preparation of a number of alkylmercuric bromides and iodides and a few alkyl chlorides. The mercuric halide used to react with the Grignard reagent should have the same kind of halogen atom as the organic halide; thus, when an organic chloride is used, mercuric chloride is added to the Grignard reagent; otherwise mixtures of the organomercuric halides will result.

The number of crystallizations required varies with the compounds. Generally, the lower alkyl halides, having 1 to 3 carbon atoms, require a greater number of crystallizations to obtain pure derivative than the higher halides. For example, in comparative runs of ethyl bromide and *n*-butyl bromide, the former required three crystallizations to raise the melting point from 185° to 192°, whereas in the case of *n*-butyl bromide, only one crystallization was necessary. This fact would indicate that the lower halides, being more reactive, undergo side reactions.

3,5-Dinitrobenzoates

A sparingly soluble ester is formed by reaction of the silver salt of 3,5-dinitrobenzoic acid and an alkyl iodide:

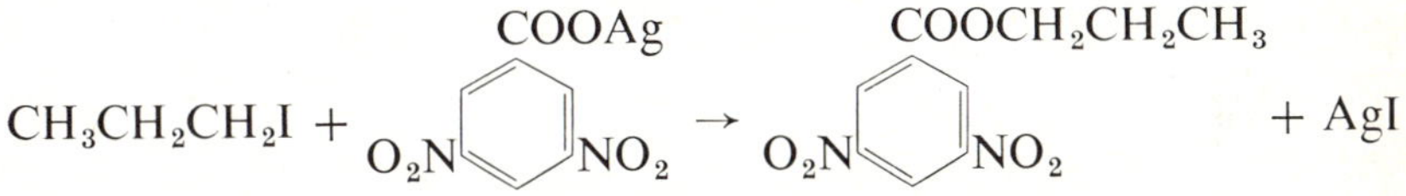

$$CH_3CH_2CH_2I + \underset{\text{Silver 3,5-dinitrobenzoate}}{\underset{O_2NNO_2}{C_6H_3(COOAg)}} \rightarrow \underset{\text{n-Propyl 3,5-dinitrobenzoate}}{\underset{O_2NNO_2}{C_6H_3(COOCH_2CH_2CH_3)}} + AgI$$

The method is applicable to a number of alkyl iodides even in very small quantities but does not give satisfactory results with the corresponding chlorides or bromides. The alkyl iodide, dissolved in a small amount of alcohol, is heated with a slight excess of dry, finely powdered silver 3,5-dinitrobenzoate. The reaction mixture is evaporated to dryness and extracted with ether to remove the ester. The crude ester is then purified as described under the preparation of 3,5-dinitrobenzoates of alcohols (page 468). The melting points of the derivatives are listed in Tables 6A and 6B.

Other Derivatives of Alkyl and Cycloalkyl Halides

Aromatic Cyclic Imides

The use of aromatic cyclic imides for the preparation of derivatives from halides is represented by the following reactions:

$$C_4H_9Br \; \text{(n-Butyl bromide)} \begin{bmatrix} + C_6H_4(CO)_2NK \rightarrow C_6H_4(CO)_2NC_4H_9 + KBr \\ \text{Potassium phthalimide} \quad\quad \text{N-}n\text{-butyl-phthalimide} \\ + C_6H_4(CO)(SO_2)NNa \rightarrow C_6H_4(CO)(SO_2)NC_4H_9 + NaBr \\ \text{Sodium saccharin} \quad\quad \text{N-}n\text{-Butyl-}o\text{-sulfobenzimide or N-}n\text{-butylsaccharin} \end{bmatrix}$$

Potassium phthalimide, potassium tetrachlorophthalimide, potassium 3-nitrophthalimide, and sodium saccharin (sodium *o*-sulfobenzimide) yield well-defined crystalline derivatives when boiled with alkyl halides. The time required for the reaction is usually 1 to 2 hours; the yields of the pure derivatives when semimicro quantities are employed are not of sufficient magnitude to warrant their preparation before other derivatives are tried. Since sodium saccharin is commercially available and at a low price, its use is preferable if this type of derivative is considered.

About 28 derivatives from sodium saccharin have been described and are listed in Table 14. The derivatives are prepared by refluxing for 30 minutes 5 mM of the halide with 6 mM of the sodium saccharin in 10 ml of 2-(2-butoxyethoxy)ethanol and 2 ml of water together with 1 gram of potassium iodide if a bromide or chloride is employed. The derivatives crystallize out on cooling. When oils result they are separated, heated cautiously to vaporize the unreacted halide, and dissolved in a small amount of alcohol; then water is added until a slight cloudiness results. On cooling the material crystallizes. The crude derivative is crystallized from aqueous alcohol.

Substituted Phenolic Ethers

When tri-iodophenol or *p*-hydroxybenzoic acid is refluxed with an alkyl halide in the presence of a base, an ether is formed. For example, ethyl iodide heated with tri-iodophenol in the presence of sodium ethoxide yields tri-iodophenyl ethyl ether.

Substituted phenolic ethers may be prepared by using *p*-hydroxydiphenylamine as the derivatizing agent. However, of the ten derivatives described five melt below 50°.

The methods described in the literature (see page 569) for the preparation of these ethers are in the scale of 20 to 30 mM. If the derivatizing agent is readily available their preparation can be tried by reducing the scale of preparation to 5 to 10 mM of halide.

2,4-Dinitrophenyl Sulfides and Sulfones

Primary and secondary alkyl iodides, bromides, and chlorides, as well as halohydrins, halonitriles, and haloesters, react readily with 2,4-dinitrothiophenol in alkaline media to form crystalline 2,4-dinitrophenyl thioethers:

$$2,4\text{-}(NO_2)_2C_6H_3SH + CH_3I \xrightarrow{KOH} 2,4\text{-}(NO_2)_2C_6H_3SCH_3 + KI + H_2O$$

2,4-Dinitrothiophenol $\qquad\qquad\qquad$ Methyl 2,4-dinitrophenyl sulfide

The melting points of about fifty 2,4-dinitrophenyl thioethers have been listed by Bost et al.[2] Most of these are included in Tables 14 and 29.

[2] R. W. Bost et al., *J. Am. Chem. Soc.*, **73**, 1968 (1951).

The general method for the preparation of the thioethers from alkyl bromides and iodides is to mix 5 millimoles of the 2,4-dinitrophenol in 10 ml of 2-(2-butoxyethoxy)ethanol with 1 ml of 28 per cent potassium hydroxide and then add 5 mM of the halide. In many cases there is an immediate reaction; otherwise the mixture is heated at about 70° on the steam bath for about 15 to 30 minutes, or allowed to stand at room temperature overnight. After the reaction is complete, the cooled mixture is diluted with 30 ml of ice-water to precipitate the alkyl sulfide, which is filtered and dissolved in the minimum amount of ethanol-dioxane or acetone solution by heating and then diluted while hot with water to turbidity. For compounds melting above 120°, 1-butanol is used for purification.

In the preparation of sulfides from chlorides 600 mg of potassium iodide are added to the reaction mixture to increase the rate of reaction. The 2,4-dinitrophenyl sulfides may be oxidized to sulfones by the permanganate method.[3]

2,2'-(Alkylenedithio)bis(6-nitrobenzothiazoles) from Alkyl Dihalides

Both alkyl mono- and dihalides react with 6-nitro-2-mercaptobenzothiazole. However, since the monohalides can be derivatized more readily by other reagents, this type of derivative is considered more suitable for dihalides. The melting points of the derivatives of about fourteen dihalides have been described.

Five millimoles of the reagent with 2.5 mM of the halide in 15 ml of 2-(2-butoxyethoxy)ethanol and 2.5 ml of $2N$ sodium hydroxide are refluxed for 1 hour. The mixture is cooled and poured into 200 ml of ice water. The crude derivative is filtered, washed with dilute ($1N$) sodium hydroxide until the filtrate becomes colorless or slightly yellow, then with water, and finally with ethanol. The crude product is dried and recrystallized from acetic anhydride, using decolorizing carbon, and dried in a vacuum desiccator over solid sodium hydroxide. Five millimoles of potassium iodide are used and the time of refluxing is increased to: 4 hours for 1,2-dibromobutane and 1,3-dichlorobutane; 5 hours for 1,1-dibromoethane, and 8 hours for 1,2-dichlorobutane.

9-Alkylfluorene-9-carboxylates

Alkyl halides may be derivatized by their reaction with methyl fluorene-9-carboxylate to give methyl 9-alkylfluorene-9-carboxylates. Many of these derivatives are crystalline solids whose melting points may be used for

[3] R. W. Bost, J. O. Turner, and R. D. Norton, *J. Am. Chem. Soc.*, **54**, 1985 (1932).

identification purposes. However, the esters may also be quantitatively saponified to obtain an equivalent weight which helps to identify the alkyl group further.[4]

S-Alkyl-2-mercapto-4,5-dihydroglyoxalinium Picrates

Ethylenethiourea (2-imidazolidinethione) has been used as a reagent for the identification of alkyl halides. The S-alkyl-2-mercapto-4,5-dihydroglyoxalinium salts, the corresponding free bases, and the picrates of the free bases of 61 alkyl halides have been prepared and their melting points determined. The picrates seem most suitable for identification purposes. The equivalent weight of the free bases and of the picrates can be determined to give further proof of the identity of the alkyl group of the alkyl halide.[5]

ARYL HALIDES

The relatively slow rates of reaction of halogen atoms when they are attached to carbon atoms of an aromatic ring render most of the reactions discussed in the preceding section inapplicable to aryl halides. Whenever it is feasible to prepare a Grignard reagent from the aromatic halide, the anilide, *p*-toluidide, or naphthalide may be prepared. However, there are several typical aromatic reactions by which aryl halides may be derivatized.

Nitration should be the first reaction to be considered. Good results are obtained from about 50 mg of the halogen compound and in some cases the sample can be as low as 5 to 10 mg. Several (three to five) recrystallizations may be necessary for purification of the crude derivative.

Chlorosulfonylation of aromatic halides and subsequent ammonolysis to produce sulfonamides should be considered if the reaction proceeds without ring chlorination. Suitable derivatives may be prepared by other reactions as follows.

(a) Reactive bromo and iodo aromatic compounds may be converted into Grignard reagents and these reacted with isocyanates. Bromobenzene, for example, may be derivatized by converting it to phenylmagnesium bromide and then to the naphthalide.

(b) Side chains, particularly methyl groups, are oxidized to the carboxylic acids. Thus the chloro-, bromo-, and iodotoluenes may be readily oxidized to the corresponding halobenzoic acids.

[4] P. M. G. Bavin, *Anal. Chem.*, **32**, 554–556 (1960).
[5] R. N. Boyd and M. Meadow, *Anal. Chem.*, **32**, 551–554 (1960).

(c) A number of polycyclic compounds, as, for example, the α-chloro- and β-chloronaphthalenes, may be derivatized by preparation of the picrates.

(d) Aryl halides in which the halogen atom is activated by nitro groups form derivatives with piperidine.

(e) Aryl iodides react with chlorine (addition) to form dichlorides which may serve as derivatives.

The brief discussion above indicates that whenever nitration and chlorosulfonylation is not feasible, a judicious selection of the reaction to be employed for derivatization depends on the probable nature of the aromatic halide. A search of the literature will, in most cases, suggest the proper derivative to be prepared. For example, in the derivatization of 1,2-dichloro-4-nitrobenzene, one of the halogen atoms may be ammonolyzed to form a chloronitroaniline (mp 104 to 105°). The ammonolysis requires heating of the aromatic halide with alcoholic ammonia at 210°— that is, under pressure. An easier method for derivatization of this compound, which contains a halogen atom activated by a para nitro group, is to react it with morpholine.[6] The halogen compound is refluxed for 2 to 3 hours with 2 to 3 mM of morpholine; the reaction mixture is treated with an excess of dilute hydrochloric acid, and the insoluble product is separated and crystallized from aqueous alcohol. The morpholine derivative of 1,2-dichloro-4-nitrobenzene melts at 127°. Other aromatic halides that give derivatives with morpholine are o-nitrochlorobenzene, p-nitrochlorobenzene, 2,4-dinitrochlorobenzene, and 3,5-dinitro-2-chlorobenzoic acid.

Nitro Derivatives of Aryl Halides

Discussion

For the preparation of mononitro derivatives the halide is dissolved or dispersed in concentrated sulfuric acid and then treated with an equal volume of concentrated nitric acid. The mixture is frequently shaken and kept at 45 to 55° for 5 minutes (or more) and then diluted with water.

An alternate method that is particularly applicable to the mononitration of micro quantities is to treat the nitro compound with specially prepared 100 per cent nitric acid. About 50 mg of the compound are mixed with 75 to 300 mg of the acid and kept at room temperature, or heated at 45 to 50°, for 5 to 15 minutes. The mixture is then diluted with water, and the product separated and crystallized. When two or three nitro groups are to be introduced, one of the following mixtures is used: (a) fuming nitric

---

[6] R. H. Harradence and F. Lions, *J. Proc. Roy. Soc. N. S. Wales*, **70**, 406 (1937).

acid and concentrated sulfuric acid; (b) specially prepared 100 per cent nitric acid and concentrated sulfuric acid; or (c) fuming sulfuric acid and nitric acid described under (a) or (b).

The selection of the proper procedure is important in obtaining nitro derivatives. For example, consider the nitration of chlorobenzene. A search in the literature discloses the following nitro derivatives: 2-nitro, mp 32°; 4-nitro, mp 83°; 3-nitro, mp 44°; and 2,4-dinitro, mp 52°. The desirable derivatives are 4-nitrochlorobenzene and 2,4-dinitrochlorobenzene. Since there are no precise directions in the literature for obtaining these derivatives starting with about 100 mg of chlorobenzene, the experimental conditions and procedures can be easily determined by a series of test-tube experiments. In the following trials, 50 mg of chlorobenzene were treated with the quantity (mg) of acid appearing outside the parentheses, while the numbers within the parentheses represent the melting point of the crude product before crystallization: (a) 150 concentrated nitric acid and 200 concentrated sulfuric acid for 15 minutes at 25° (oil); (b) 150 fuming nitric acid for 15 minutes at 25° (oil); (c) 300 nitric acid 100 per cent for 5 minutes at 25° (81–82); (d) 150 fuming nitric acid and 300 concentrated sulfuric acid for 15 minutes at 50 to 55° (30–34); (e) same as in (d), but heated at 60° for 30 minutes (40–44); (f) 150 fuming nitric acid and 300 fuming sulfuric acid for 30 minutes at 80 to 90° (48–50). From these results it is possible to select procedure (c) to obtain the mononitro derivative melting at 83° and procedure (e) or (f) to obtain the dinitro derivative melting at 52°. It is also clear from these results that several crystallizations would be required to remove the undesirable nitro derivatives from the crude nitration product. In most cases of nitration the formation of more than one nitro compound is the rule rather than the exception; therefore several crystallizations are necessary for purification of nitro derivatives.

General Method for Nitration

For the introduction of one nitro group in compounds that undergo nitration *with ease*, proceed as follows. Place 200 mg of the compound to be nitrated in an 8-in. test tube; add first 2 ml of concentrated sulfuric acid; immerse the tube in a beaker containing cold water and add slowly 2 ml of concentrated nitric acid, shaking the tube from time to time so as to prevent a rise in temperature. Place the tube in a water bath at 50 to 55° and heat for 15 to 20 minutes. The tube is shaken frequently to insure better contact of the halide with the nitrating mixture. After heating, the tube is cooled for a minute or two, and then its contents are poured into another tube, which contains 8 to 10 ml of cold water. The diluted mixture is returned to the original tube and cooled. The nitro compound that separates is filtered and washed twice with 3 to 4 ml of water; it is then transferred to the reaction tube and dissolved in the minimum amount of boiling methanol or ethanol. The hot solution is filtered, and water is added cautiously dropwise until a permanent cloudiness appears. The tube is heated until the solution is clear, and then it is cooled. The crystallization is repeated until the crystals from two successive crystallizations differ by no more than 0.5 to 1° in the melting point.

For the introduction of two nitro groups or one nitro group in *unreactive halides*, the same procedure is used as described in the preceding paragraph, but fuming nitric acid is used in place of concentrated nitric acid. (CAUTION. *Care should be exercised in handling fuming nitric acid; its addition to the mixture of sulfuric acid and halide should be at such rate that no great amount of brown fumes appear.*) The mixture is then heated at 45 to 50°. If the melting point of the nitration product after one crystallization shows a difference of 10° or more from the value listed in the literature, the nitration is repeated and the temperature is raised first to 80°; if the result is not satisfactory, the nitration is repeated at 90 to 100°. The alternative procedure is to prepare 100 per cent nitric acid as described below, since the fuming nitric acid is seldom above 88 per cent.[7]

For mononitration 50 mg of the compound are placed in a 3- or 4-in. tube and 0.1 to 0.2 ml of the prepared nitric acid is added. If the compound is very reactive, the amount used is 0.1 ml and the tube is allowed to stand at room temperature for 15 minutes; about 2 ml of water is added and the tube is chilled. The crystals are filtered and crystallized from alcohol. If the melting point differs more than 10° from that of the desired derivative, a second trial is made, using 50 mg of the compound, and 0.2 ml of the nitric acid, and heating the tube for 15 minutes at 50°. When upon addition of nitric acid to the compound a reaction occurs in which oxides of nitrogen are evolved, oxidation is likely to have occurred; in such cases it is advisable to cool the reagent and perform the nitration at 0 to 5°.

Preparation of 100 Per Cent Nitric Acid

The acid is not commercially available, but it can be readily prepared as required. A microretort is constructed as in Figure 7.03 (repeated here for convenience). The retort is cooled, then drawn out as shown in *B*; however, the length of section *a–c* is 3 times that shown in the diagram. The

---

[7] White fuming nitric acid 96 to 98 per cent is commercially produced in connection with work on missiles, but its availability is restricted.

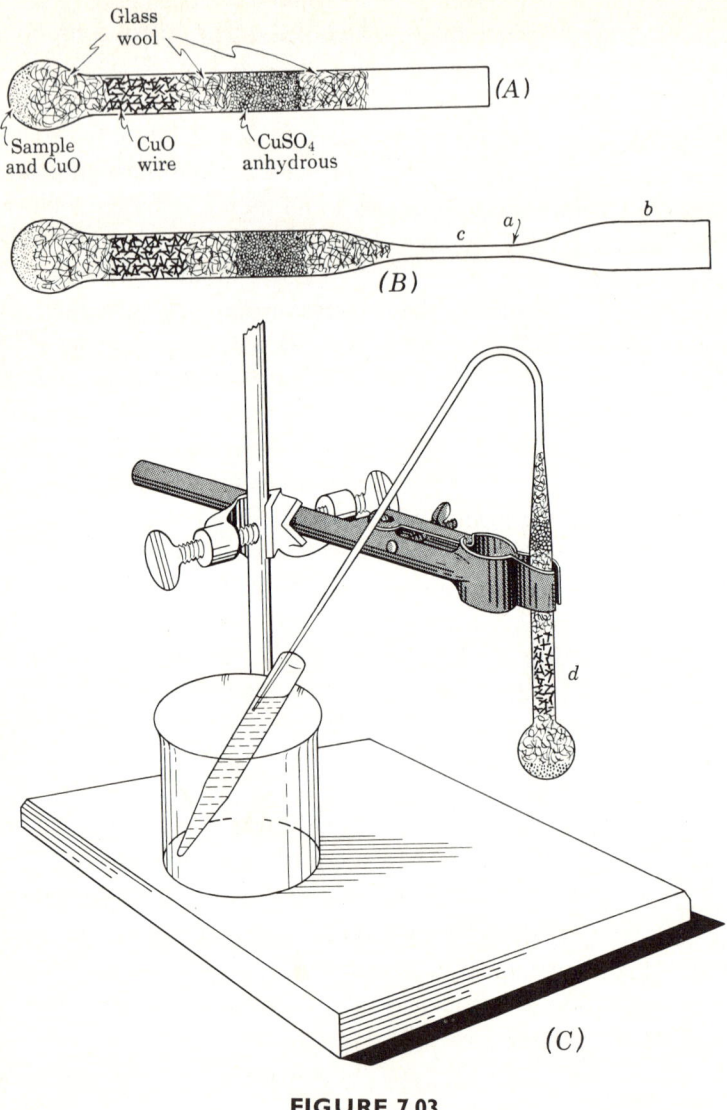

FIGURE 7.03

Microdetection of carbon and hydrogen. (*A*) and (*B*) Construction of ignition tube. (*C*) Ignition-tube assembly.

retort is cooled and the charge of 0.7 ml of concentrated nitric acid and 1 ml of concentrated sulfuric acid is introduced with a capillary pipet. Section *b* is cut at point *a* and sealed to serve as a receiver. The retort is held upright, heated cautiously at point *c*, and rapidly bent as shown in Figure 7.03C, which represents the complete setup. The receiver is marked with a pencil to a height corresponding to a volume of 0.3 ml if concentrated nitric acid is used in the charge, and to 0.5 ml if fuming nitric acid is used. The receiver fits into a 3-in. tube, which in turn rests in a 50- or 100-ml beaker containing cold water. The end of the delivery tube of the microretort is just above the mark on the receiver. The microretort, which contains 0.7 ml of concentrated nitric acid and 1 ml of concentrated sulfuric acid, is heated cautiously with a microflame until 0.3 ml has been collected. If fuming nitric acid was used in the charge, the amount of distillate collected is 0.5 ml. The distillate is almost colorless or faintly yellow, and contains 99.5 to 100 per cent nitric acid. The concentration of acid drops below 95 per cent if a greater volume of distillate is collected. The specially prepared acid may be used for dinitration and trinitration by admixing it with concentrated or fuming sulfuric acid.

Example 17.05: Mononitration of 4-Bromochlorobenzene

In a 3-in. microtube place 50 mg of the halide and 0.1 ml of 100 per cent nitric acid; allow to stand at room temperature for 15 minutes. Add 2 ml of water and chill the tube. After 5 minutes centrifuge the tube and remove the supernate by means of a capillary pipet. Wash with 0.5 ml of water twice (page 73) and then add 1 to 1.5 ml of methanol and heat the tube cautiously until all the crude derivative is dissolved. Transfer the hot solution to another tube (page 73) by means of a preheated capillary pipet. Add 2 or 3 drops of water to the solution and chill. Separate the crystals and wash. Remove a few crystals for determination of melting point and crystallize the remainder, using the same procedure, if the value is 2° below 71 to 72°. About 25 to 30 mg of crystals are obtained, melting at 71 to 72°.

NOTE. *p*-Dichlorobenzene and *p*-dibromobenzene nitrate easily; when the above procedure and quantities are employed, one crystallization is required to obtain the pure nitro compound. *p*-Dichlorobenzene yields 130 to 150 mg, and *p*-dibromobenzene 100 to 120 mg, of the nitro compound. It should be noted that *p*-dichlorobenzene yields, under the conditions described, a mononitro compound (2-position), whereas *p*-dibromobenzene yields a dinitro compound (2,5-positions). Unless the compound is reactive, the introduction of two nitro groups requires treatment with fuming nitric acid. Instead of 100 per cent nitric acid a mixture of 1 ml of nitric acid and 1 ml of sulfuric acid may be used, but the mixture should be heated at 45 to 50° for 15 minutes.

Example 17.06: Nitration of Chlorobenzene

Place in an 8-in. tube 2 ml of concentrated sulfuric acid and 200 mg of chlorobenzene. Cool and add slowly 2 ml of fuming nitric acid. Heat the tube in a bath at 90 to 100° for 30 minutes and mix the contents by shaking frequently. Cool and add the mixture to 10 ml of water and crystallize as described on page 65. Two or three crystallizations are required to obtain a product that melts at 52°. The yield is 100 to 120 mg.

NOTE. When fuming nitric acid is used, care should be taken if oxidizable groups are present. If a considerable amount of brown fumes are evolved during nitration, it is advisable to use a lower temperature.

For the preparation of 4-chloronitrobenzene, mix in a 4-in. test tube 100 mg of the halide and 0.3 ml of 100 per cent nitric acid, and allow to stand at room temperature for 20 minutes. Add 2 ml of water and, after cooling, extract with 2 ml of ether. Withdraw the ether layer by means of a capillary pipet and place in another tube. Add one pellet of sodium hydroxide and shake gently for a few minutes. Separate and evaporate the ether from a small dish and dissolve the residue in 2 ml of hot methanol. Filter with suction and add water dropwise until a permanent cloudiness results. Cool and filter the crystals. About 50 to 70 mg of 4-chloronitrobenzene is obtained; the compound melts at 82 to 83°.

Sulfonamides of Aryl Halides

As indicated earlier, aryl halides undergo chlorosulfonylation by reaction with chlorosulfonic acid. The resulting sulfonyl chloride may then be converted by ammonolysis to a sulfonamide:

$$\text{ArX} \xrightarrow{ClSO_2OH} \text{ArX-}SO_2Cl \xrightarrow{2NH_3} \text{ArX-}SO_2NH_2$$

In actual practice the product from the main reaction for a number of aryl halides is different from that indicated in the above equation. In a number of cases ring chlorination takes place. For example, p-di-iodobenzene yields 1,2,4,5-tetrachlorobenzene. In general, aryl iodides do not give good results. In the cases of fluorobenzene, iodobenzene, o-dichlorobenzene, and o-dibromobenzene the chief product is a sulfone instead of a sulfonyl chloride. Although the sulfones can be purified and employed as derivatives, more suitable compounds can be obtained by nitration procedures. The general method for the preparation of sulfonyl chlorides and their conversion to sulfonamides is described on pages 638 and 639.

FLUORINE COMPOUNDS*

Table 17 (pages 857–861) lists about 160 highly fluorinated compounds from the relatively new field of fluorocarbon chemistry. Few functional

* This section is largely based on the contribution of Elliott Bergman to the second edition of this work.

derivatives of highly fluorinated alkanes had been prepared prior to World War II. Since that time, however, the field of fluorocarbon chemistry has expanded at a startling rate. Perhaps the most significant contribution to this rapid expansion was the invention of Simons' process[8] for the preparation of perfluorinated[9] functional derivatives, such as perfluoroacids, perfluoro tertiary amines and perfluoroethers.

The perfluoroacids serve as excellent starting materials for the preparation of perfluoroaldehydes, esters, acid halides, alcohols, alkyl halides, and nitriles by means of usual procedures. Several of the common reactions are illustrated by the following equations.

$$CH_3CO_2H + HF \xrightarrow{\text{5-7 Volts}} CF_3COF \xrightarrow{H_2O} CF_3CO_2H$$

$$CF_3CO_2H + C_2H_5OH \xrightarrow{H_2SO_4} CF_3CO_2C_2H_5$$

$$CF_3CO_2H + LiAlH_4 \longrightarrow CF_3CHO + CF_3CH_2OH$$

$$CF_3CO_2C_2H_5 + NH_3 \longrightarrow CF_3CONH_2 \xrightarrow{P_2O_5} CF_3CN$$

$$CF_3CO_2H + Ag_2O \longrightarrow CF_3CO_2Ag \xrightarrow{I_2} CF_3I$$

$$2CF_3CO_2H + P_2O_5 \longrightarrow (CF_3CO)_2O$$

$$CF_3COCl + C_6H_6 \xrightarrow{AlCl_3} CF_3COC_6H_5$$

Identification and Characterization of Fluorocarbons

Since the physical and chemical properties of fluorinated compounds differ in many respects (for example, volatility and reactivity) from the corresponding nonfluorinated analogues, slightly different techniques for the identification of these compounds must be employed. The present section presents a summary of the modifications of the standard procedures.

Gases. Gases are best treated in solution (for example, solutions in alcohol or acetone).

Liquids and solids. Liquids and solids may be treated as usual.

Test for unsaturation. The most reliable test for the unsaturated linkage in highly fluorinated systems is the potassium permanganate test carried out in acetone. The bromine test is poor because fluorinated olefins do not brominate under ionic conditions.

Ignition test. Highly fluorinated compounds fail to burn with a self-sustaining flame.

[8] J. H. Simons, *J. Electrochem. Soc.*, **95**, No. 2, 47 (1949).
[9] The term *perfluoro* indicates that all hydrogen atoms, except those on the functional group, are replaced by fluorine.

Sodium fusion test. The sodium fusion can be run as usual with liquids and solids. Gases are fused by passing them through a U-tube containing molten sodium.

Detection of fluoride. Fluoride is best detected by lanthanum chloranilate as described on page 295.

Detection of fully fluorinated compounds. Compounds such as polymers of fluoro-olefins are characterized by their stability to boiling fuming nitric acid. All other organic matter will be oxidized to carbon dioxide while the fluorocarbon polymer will be left as a clear oil or white powder.

Solubility tests. Major changes in the solubility classification of fluorinated functional compounds are to be noted. Fluorinated acids are water-soluble and are very strong acids ($pK_a \sim 1$). Higher acids act as soaps due to this property and may be detected by their pK_a and their lowering of surface tension (soap action). The indicator method (page 328) is useful in showing these acids to be strong. The lower acids are ether-soluble and hence fall in Division S_1. Solubility data for the higher acids have not been reported in detail. Dicarboxylic acids form addition compounds with ether that are ether soluble; hence these acids also fall into Division S_1.

Fluorinated amines of the type $CF_3(CF_2)_nCH_2NH_2$ are weak bases ($pK_b \sim 9$) and presumably will fall in Division **B**. Perfluoro tertiary amines are so weakly basic that they are in Division **M**.

Fluorinated alcohols of the type $CF_3(CF_2)_nCH_2OH$ are weakly acidic ($pK_a \sim 12$), while alcohols of the type $[CF_3(CF_2)_n]_2CHOH$ and $[CF_3(CF_2)_n]_3COH$ are slightly stronger acids (pK_a 11 and 10) and should fall in Division A_1.

All other functions—esters, ethers, anhydrides, aldehydes, ketones, and halides—fall into Division **I** by virtue of their sulfuric acid insolubility (note that even the oxygen-containing functions fail to dissolve in this reagent since the inductive effect of the fluoroalkyl group lowers the basicity of the oxygen atoms in the functional group). Exceptions to this rule will be noted with those compounds that dissolve in water as a result of reaction, for example lower aldehydes or ketones, which form water-soluble hydrates and which also dissolve in ether and hence fall in Division S_1.

Highly fluorinated compounds, such as perfluoroalkanes, exhibit characteristic insolubility in water. They are difficultly soluble in ether, benzene, and other organic solvents. These compounds are in Division **I** but may be characterized further by their low solubility in organic solvents.

Specific Class Tests for Fluorine Compounds

Acid halides, acid anhydrides, and acids (via the anhydride or halide) may be detected by their reaction with aniline to yield base-soluble anilides by the procedure described on page 459.

$$CF_3COCl + C_6H_5NH_2 \rightarrow CF_3CONHC_6H_5 + HCl$$
$$CF_3CONHC_6H_5 + NaOH \rightarrow [CF_3CONC_6H_5]^-Na^+ + H_2O$$

After the precipitated anilide is washed with water, it is tested for base solubility by suspending 30 mg in 2 ml of $1N$ sodium hydroxide.

The ferric hydroxamate test gives a fading to negative test with a variety of fluorinated esters. The ester is best identified by its characteristic fruity odor and its hydrolysis products.

Derivatization of Fluorine Compounds

The derivatization or perfluoro acids, carbonyl compounds, amines, and other perfluoro compounds is carried out according to the procedures described under the various groups except that provision is made to account for either the increased or decreased reactivity of the function as a result of the perfluorination. For example, the preparation of 2,4-dinitrophenylhydrazones and semicarbazones of the perfluoro carbonyl compounds requires a longer reaction time if good yields are to be obtained. On the other hand, many perfluoroketones readily form adducts with ammonia and water (hydrates), which are suitable derivatives. In many cases suitable derivatives have not been described for characterization. Perfluoro amines, alcohols, ethers, halides, and fluorocarbons are characterized best by their solubility behavior and their physical characteristics.

Most aldehydes and ketones form ordinary carbonyl derivatives but at a much slower rate than do the hydrocarbon analogues. If a 2,4-dinitrophenylhydrazone forms very slowly, its presence can be detected as follows: A solution of 2 drops of aldehyde or ketone is allowed to stand overnight in 1 ml of $0.05M$ reagent prepared as in Example 14.03 (page 501). After this period one drop of the solution is added to 1 ml of $2M$ methanolic potassium hydroxide. An intense red color indicates that a dinitrophenylhydrazone is present. For some of the less reactive ketones the solution must be refluxed for a few days before this color test is positive.

$$CH_3COC_3F_7 + 2,4\text{-}(NO_2)_2C_6H_3NHNH_2 \rightarrow$$
$$2,4\text{-}(NO_2)_2C_6H_3NHN\!\!=\!\!C(CH_3)(C_3F_7)$$
$$2,4\text{-}(NO_2)_2C_6H_3NHN\!\!=\!\!C(CH_3)(C_3F_7) + KOH \rightarrow$$
$$[2,4\text{-}(NO_2)_2C_6H_3N\!\!-\!\!N\!\!=\!\!C(CH_3)(C_3F_7)]^-K^+ + H_2O$$
$$\text{red}$$

Mercury derivatives of fluorocarbon carboxylic acid amides are useful in the identification of these compounds.[10] A sample of the amide is heated with an excess of red mercuric oxide to 150 to 170° for about 5 minutes; the mixture is heated with alcohol to extract the mercury derivative. The alcohol is evaporated to give the white crystalline solid.

A very reliable test for perfluoro aldehydes and ketones consists of their cleavage by bases. A mixture of 0.1 g of compound and 5 ml of 5N sodium hydroxide is warmed to reflux for 5 minutes. The evolution of a gaseous or volatile monohydrogen perfluoroalkane indicates the presence of a perfluoroalkyl carbonyl compound. Careful acidification of the mixture gives a volatile or a solid acid as the case may be. Aromatic fluoroalkyl ketones cleave to yield the aromatic acid.

$$C_6H_5COCF_3 + NaOH \rightarrow C_6H_5CO_2Na + CHF_3$$
$$C_6H_5CO_2Na + HCl \rightarrow C_6H_5CO_2H + NaCl$$

The following three examples illustrate the modification of standard procedures for the derivatization of perfluoro compounds.

Example 17.07: Preparation of 2,4-Dinitrophenylhydrazones[11]

Dissolve 500 mg of 2,4-dinitrophenylhydrazine in 10 ml of 50 per cent sulfuric acid and 10 ml of 95 per cent ethanol. Add 500 mg of ketone. Let stand for one week, or until sufficient derivative has formed. The derivative that precipitates is then filtered and recrystallized from ethanol and water.

Example 17.08: Preparation of Semicarbazones

Place in an 8-in. test tube, fitted with a microcondenser, 500 mg of phenyl perfluoro-n-propyl ketone, 500 mg of semicarbazide hydrochloride, and 5 ml of 95 per cent ethanol. Add water until the solution is just turbid, and reflux for 4 hours. Evaporate the solvent in vacuo and extract the residue with ether. About 550 mg of crystals are obtained that, on recrystallization from ether, melt at 143.8 to 144.8°.

Example 17.09: Preparation of Urethans from Perfluoroalcohols: Phenyl Perfluoro-n-propyl Carbinol

Place in a test tube 520 mg of the carbinol and 1 g of phenyl isocyanate and 0.1 ml of pyridine in the order given. The reaction takes place immediately after the addition of pyridine or on slight warming. Cool and

---

[10] R. H. Patton and J. H. Simons, *J. Org. Chem.*, **21**, 1199–1200 (1956).
[11] J. H. Simons et al., *J. Am. Chem. Soc.*, **75**, 5621 (1953).

add about 5 ml of carbon tetrachloride and wash the solution with 2 ml of 1N hydrochloric acid and then with 2 ml of water; filter. Concentrate the filtrate by evaporation to 1 ml and filter the crystals. About 500 mg of the derivative are obtained melting at 96 to 97°.

REFERENCES

Identification of Alkyl and Aryl Halides

Conversion of esters to sulfonamides: Huntress and Carten, *J. Am. Chem. Soc.*, **62**, 511 (1940).

Oxidation of thioethers to sulfones: Bost et al., *J. Am. Chem. Soc.*, **54**, 1985 (1932); **73**, 1967 (1951).

Use of iodoso chlorides: Nicol and Sandin, *J. Am. Chem. Soc.*, **67**, 1307 (1945).

N-Alkyl-*p*-bromobenzenesulfon-*p*-anisidides: Gillespie, *J. Am. Chem. Soc.*, **56**, 2740 (1934).

S-Alkylthiuronium picrates from tertiary alkyl halides: Veibel and Lillelund, *Bull. Soc. Chim.* (5), **5**, 1153 (1938).

S-Alkylthiuronium picrates: Brown and Campbell, *J. Chem. Soc. (London)*, **1937**, 1699; Levy and Campbell, *J. Chem. Soc. (London)*, **1939**, 1442; Schotte, *Arkiv Kemi*, **5**, 11 (1952); *Acta Chem. Scand.*, **7**, 1357 (1953); Crosby and Entrikin, *J. Chem. Ed.*, **41**, 360 (1964).

Alkylmercuric halides: Hill, *J. Am. Chem. Soc.*, **50**, 167 (1928); Marvel, Gauerke, and Hill, *J. Am. Chem. Soc.*, **47**, 3009 (1925); Slotta and Jacobi, *J. prakt. Chem.*, **120**, 249 (1929).

Alkylsaccharins: Merritt, Levey, and Cutter, *J. Am. Chem. Soc.*, **61**, 15 (1939).

N-Alkyltetrachlorophthalimides: Allen and Nicholls, *J. Am. Chem. Soc.*, **56**, 1409 (1934).

N-Alkyl-*p*-toluenesulfontoluidides: Young, *J. Am. Chem. Soc.*, **56**, 2167, 2783 (1934); **57**, 773 (1935).

Anilides, *p*-toluidides, and α-naphthalides: Gilman and Furry, *J. Am. Chem. Soc.*, **50**, 1214 (1928); Schwartz and Johnson, *J. Am. Chem. Soc.*, **53**, 1063 (1931); Underwood and Gale, *J. Am. Chem. Soc.*, **56**, 2117 (1934).

3,5-Dinitrobenzoates from alkyl halides, Furter, *Helv. Chim. Acta*, **21**, 872 (1938); Benfey et al., *J. Org. Chem.*, **20**, 1782 (1955).

Ethers of 2,4-dinitrothiophenol: Bost et al., *J. Am. Chem. Soc.*, **73**, 1967 (1951).

Ethers of *p*-hydroxybenzoic acids: Lauer et al., *J. Am. Chem. Soc.*, **61**, 3050 (1939).

Ethers of *p*-hydroxydiphenylamine: Houston, *J. Am. Chem. Soc.*, **71**, 395 (1949).

6-Nitro-2-mercaptothiazole derivatives: Cutter et al., *J. Am. Chem. Soc.*, **69**, 831 1947); *Anal. Chem.*, **25**, 198 (1953).

3-Nitrophthalimides: Sah and Ma, *Ber.*, **65B**, 1930 (1932); *Science Repts., Natl. Tsing Hua Univ.*, **2**, 147 (1933).

4-Nitrophthalimides and their use for saponification equivalents: Billman and Cash, *. Am. Chem. Soc.*, **75**, 2499 (1953).

Picrates of alkyl β-naphthyl ethers: Baril and Megrdichian, *J. Am. Chem. Soc.*, **58**, 415 (1936); Dermer and Dermer, *J. Org. Chem.*, **3**, 289 (1938).

Piperidyl derivatives of aromatic halogenonitro compounds: Seikel, *J. Am. Chem. Soc.*, **62,** 750 (1940).

Tri-iodophenyl ethers: Drew and Sturtevant, *J. Am. Chem. Soc.*, **61,** 2666 (1939).

S-Alkyl-2-mercapto-4,5-dihydroglyoxalinium picrates: Boyd and Meadows, *Anal. Chem.*, **32,** 551–554 (1960).

9-Alkylfluorene-9-carboxylates: Bavin, *Anal. Chem.*, **32,** 554–556 (1960).

Identification of Fluorine Compounds

Analysis of fluorine-containing compounds: Simons, *Fluorine Chemistry*, Vol. II, Academic Press, New York, 1954.

Detection of fluoride: Steiger, *J. Am. Chem. Soc.*, **30,** 219 (1908).

Amides: Husted and Ahlbrecht, *J. Am. Chem. Soc.*, **75,** 1605 (1953).

Amides: Patton and Simons, *J. Org. Chem.*, **21,** 1199–1200 (1956).

2,4-Dinitrophenylhydrazones: Husted and Ahlbrecht, *J. Am. Chem. Soc.*, **74,** 5422 (1952); Simons, Black, and Clark, *J. Am. Chem. Soc.*, **75,** 5621 (1953).

Hydrates and amine adducts of aldehydes and ketones: Hauptschein and Braun, *J. Am. Chem. Soc.*, **77,** 4930 (1955); Husted and Ahlbrecht, *J. Am. Chem. Soc.*, **74,** 5422 (1952).

p-Toluenesulfonates: Tiers, Brown, and Reid, *J. Am. Chem. Soc.*, **75,** 5978 (1953).

Chromatographic Procedures

Moynihan and O'Colla, *Chem. & Ind.*, **1951,** 407.

Schmeiser and Jerchel, *Angew. Chem.*, **65,** 366 (1953).

Winterringham et al., *Nature*, **166,** 999 (1950).

18

Derivatives of Hydrocarbons

ALKANES AND CYCLOALKANES—ALKENES AND CYCLOALKENES—ALKYNES—DIENES—AROMATIC HYDROCARBONS

Table 18.01 summarizes the various types of derivatives that have been proposed for the characterization of hydrocarbons. It will be noted that there are no suitable derivatives for the characterization of alkanes and cycloalkanes, only a few for alkenes, alkynes, and cycloalkenes, and several for aromatic hydrocarbons.

TABLE 18.01
DERIVATIVES FOR THE IDENTIFICATION OF HYDROCARBONS

Alkanes and Cycloalkanes:

None

Alkenes and Cycloalkenes, Alkynes, Dienes:

*Bromine addition products
*Dinitrophenyl sulfides
Dithiocyanates
Hydration to ketones and derivatization of latter
*Mercuric salts
Nitrosates, nitrosites, and nitrosyl chlorides

Aromatic Hydrocarbons:

*Acetamido derivatives
Aroylbenzoic acids
2,4-Dinitrophenyl sulfides
*Nitro derivatives
Oxidation of side chains
Picrates
*Sulfonamides
*Trinitrofluorenone adducts

* Recommended derivative.

CHARACTERIZATION OF ALKANES AND CYCLOALKANES

Alkanes (paraffins) and cycloalkanes (cycloparaffins) do not form derivatives useful for characterization work. Although most of these hydrocarbons undergo a variety of reactions under appropriate conditions, the large number of isomers formed and the difficulties involved in the separation of these isomers render such reactions useless for derivatization. For instance, *n*-hexane can be easily brominated; the three monobromo derivatives cannot be easily separated in a state of purity so that they may be used further for the preparation of solid derivatives. A hydrocarbon with a tertiary hydrogen atom may yield a greater proportion of a particular monohalide, but even in such instances other isomers are formed.

Halogenation and oxidation may be used in the characterization of certain cycloparaffins. For example, cyclohexane may be oxidized by hot nitric acid to adipic acid; the chlorination or bromination of cyclohexane may be controlled so as to produce mostly the monohalide, which in turn may be derivatized.

Cyclohexane and substituted cyclohexanes can be distinguished from cyclopentanes and paraffins by catalytic dehydrogenation. The cyclohexanes are converted to aromatic hydrocarbons, which are readily characterized by methods described in this chapter. For identification work it is possible to pass a mixture of a few milliliters of methylcyclopentane and methylcyclohexane through the semimicro dehydrogenation apparatus described by Cheronis and Savoy[1] and obtain the methylcyclopentane unchanged while the methylcyclohexane has been converted quantitatively to toluene. In general, cyclohexane systems can be characterized by conversion to aromatic compounds either by catalytic dehydrogenation in the vapor phase or through dehydrogenation using sulfur or selenium in the liquid phase. The literature cited at the end of the section should be consulted for details of the application of this general method.

Characterization by Means of Physical Constants

From the above considerations it is evident that the characterization of paraffins and cycloparaffins must be based almost exclusively on the physical constants, particularly on the boiling point, density, and refractive index (see the references at the end of this chapter). These constants lead

---

[1] N. D. Cheronis, "Micro and Semimicro Methods," in *Technique of Organic Chemistry* Vol. 6, A. Weissberger, ed., Interscience, New York, 1954, p. 262.

to fairly accurate results with pure compounds; impurities, particularly those due to isomers, render accurate characterization extremely difficult. In recent years other physical measurements, such as absorption and nuclear magnetic resonance spectroscopy, have been used for the identification of hydrocarbons. See the literature cited for a description of the optical methods.

The work published by Rossini and co-workers[2] gives a detailed discussion of the methods employed for the separation and characterization of pure hydrocarbons from petroleum. Hydrocarbons can be separated by gas chromatography and the individual compounds tentatively identified by retention times.

Oxidation of a Cycloalkane to a Dicarboxylic Acid

Discussion

A number of cycloalkanes are readily oxidized by hot nitric acid, undergoing rupture of the ring to form a mixture of dicarboxylic acids. Thus cyclohexane heated with nitric acid gives a mixture of adipic and glutaric acids and smaller amounts of succinic acid and nitro compounds. However, conditions can be chosen under which the oxidation gives mostly adipic acid and only small amounts of the other products. The method consists in heating nitric acid to boiling and adding the hydrocarbon 1 drop at a time and waiting until it has reacted *completely* before adding another drop. The reaction is very vigorous and the method is not recommended for the oxidation of more than 0.5 ml of the hydrocarbon added in portions of 0.05 ml. About 0.2 to 0.3 ml or less of the hydrocarbon yields a sufficient amount of the dicarboxylic acid for determination of its melting point after one or two crystallizations.

Example 18.01: Preparation of Adipic Acid from Cyclohexane

Place 2 ml of concentrated nitric acid in an 8-in. test tube and add 1 or 2 boiling stones. Clamp the tube to a stand in the hood and heat the acid to boiling. Reduce the flame of the microburner and, by means of a dropper, add cautiously 1 drop of cyclohexane; shake the tube and, when the vigorous reaction has subsided, repeat the addition. Add a total of 0.2 ml (8 to 10 drops) of hydrocarbon over a period of 10 minutes. Boil gently for 1 minute and cool. Filter the crystals of adipic acid. Recrystallize from boiling water. The yield is 70 to 80 mg, melting at 152 to 153°.

[2] F. D. Rossini et al., *Anal. Chem.*, **20**, 110 (1948).

ALKENES AND CYCLOALKENES, ALKYNES, DIENES

Although unsaturated hydrocarbons undergo a large number of reactions, there is no simple general method for converting them into solid derivatives suitable for characterization work. A number of unsaturated hydrocarbons, particularly terpenes, may be characterized by the addition of nitrosyl chloride, nitrogen trioxide, and nitrogen tetroxide.

$$RCH=CHR' + NOCl \rightarrow \underset{\text{Nitrosochloride}}{RCH(NO)CHClR'}$$

$$RCH=CHR' + N_2O_4 \rightarrow \underset{\text{Nitrosate}}{RCH(NO)CH(ONO_2)R'}$$

$$RCH=CHR' + N_2O_3 \rightarrow \underset{\text{Nitrosite}}{RCH(NO)CH(ONO)R'}$$

The nitrosochlorides are generally prepared by addition of concentrated hydrochloric acid to a mixture of the hydrocarbon and ethyl nitrite dissolved in acetic acid. See the literature for details on the preparation of these derivatives for the characterization of terpenes.

Unsaturated compounds on oxidation yield carboxylic acids:

$$RCH=CHR \xrightarrow{[O]} 2RCOOH$$

$$RC \equiv CR \xrightarrow{[O]} 2RCOOH$$

A number of symmetrical olefins or acetylenes (or 1-alkenes and 1-alkynes) that yield predominantly one type of solid carboxylic acid may be identified by this method. The olefin is either shaken or refluxed with alkaline permanganate. The manganese dioxide is filtered and the alkaline solution is concentrated, acidified, and extracted with ether. If difficulty is encountered in the filtration of manganese dioxide, the solution is concentrated and then sodium bisulfite and dilute hydrochloric acid are added until the solution is clear; it is then cooled and extracted with ether. The difficulty in the application of this method to semimicro quantities arises from the fact that in most cases the oxidation proceeds further than represented by the equations and gives rise to small amounts of other carboxylic acids. For this reason several crystallizations are required to obtain a pure product, and, since the yield of the expected carboxylic acid is usually 50 to 70 per cent of the theoretical, at least 500 mg of the hydrocarbon should be used.

Olefins may be characterized by epoxidation and rearrangement of the epoxides to carbonyl compounds.[3] Epoxidation is accomplished with 40%

---

[3] J. G. Sharefkin and H. E. Schwerz, *Anal. Chem.*, **33**, 635–639 (1961).

peroxyacetic acid, and the rearrangement with boron trifluoride-diethyl ether complex. The carbonyl compounds may then be derivatized with 2,4-dinitrophenylhydrazine. About 1 g of the olefin is needed to give enough of the final derivative for identification.

The addition of thiocyanogen to form *dithiocyanates* has been used for the derivatization of ethylene, cyclohexene, 3-methylcyclohexene, and styrene. Since three of these hydrocarbons may be readily characterized by other derivatives and the handling of thiocyanogen and dithiocyanates involves some hazard, this method is not recommended.

The additions of mercaptans (thiols), thiophenols, and thioacids have been proposed as means of identification of olefins. The addition to unsymmetrical olefins may take place in two ways, depending on the presence or absence of peroxides and other catalysts; the first equation shows addition according to Markownikoff's rule; most aliphatic mercaptans and thiophenols add to unsymmetrical olefins according to the second equation:

$$CH_3CH=CH_2 + C_2H_5SH \longrightarrow \underset{\text{Ethyl isopropyl sulfide}}{CH_3CH(SC_2H_5)CH_3}$$
(Ethyl mercaptan)

$$CH_3CH=CH_2 + C_2H_5SH \xrightarrow{\text{Peroxide}} \underset{\text{Ethyl } n\text{-propyl sulfide}}{CH_3CH_2CH_2SC_2H_5}$$

This method suffers from two disadvantages: first, if the reaction is run at room temperature, several weeks are required for completion; or the reaction may be run at 180° by heating in a bomb for 10 hours. Second, most of the thioethers obtained by such reactions are liquids and must be derivatized either by oxidation to sulfones or by preparation of addition compounds with mercuric chloride or palladous chloride.

The following two types of derivatizations should be explored first: (a) addition of bromine and, if the bromide is liquid, conversion to a 2,4-dinitrophenyl thioether by reaction with 2,4-dinitrothiophenol;[4] and (b) reaction with 2,4-dinitrobenzenesulfenyl chloride to yield the corresponding sulfides (Kharasch and co-workers have described derivatives from alkenes). The latter method is discussed further on page 577.

The monosubstituted acetylenes, $RC{\equiv}CH$, react with an alkaline solution of mercuric iodide or cyanide to give simple mercuric salts:

$$2RC{\equiv}CH + K_2HgI_4 + 2KOH \rightarrow (RC{\equiv}C)_2Hg + 4KI + 2H_2O$$

The acetylene salts are prepared by adding a dilute solution of alkaline mercuric iodide to an alcoholic solution of the acetylene; the acetylene

---

[4] R. W. Bost et al., *J. Am. Chem. Soc.*, **73**, 1968 (1951).

mercuric salt separates immediately. The product is filtered, washed with alcohol, and then crystallized from alcohol or benzene.

Disubstituted acetylenes and some monoalkylacetylenes, such as 1-hexyne, 1-heptyne, and 1-octyne, may be converted to the corresponding ketones by catalytic hydration:

$$\underset{\substack{\text{1-Phenyl-1-butyne}\\\text{(Ethylphenylacetylene)}}}{CH_3CH_2C\equiv CC_6H_5} + H_2O \xrightarrow[\text{HgSO}_4]{\text{H}_2\text{SO}_4} \underset{n\text{-Butyrophenone}}{CH_3CH_2CH_2COC_6H_5}$$

$$\underset{\text{1-Hexyne}}{CH_3(CH_2)_3C\equiv CH} + H_2O \xrightarrow[\text{HgSO}_4]{\text{H}_2\text{SO}_4} \underset{\text{2-Hexanone}}{CH_3(CH_2)_3COCH_3}$$

An alkyne may also be converted to a ketone by the following method. About 1 g of the acetylene is dissolved in methanol containing a suspension of a catalyst prepared from red mercuric oxide, trichloroacetic acid, methanol, and boron trifluoride etherate.[5] The reaction may be warmed to complete the reaction, cooled and diluted with water. The ketones are separated by filtration, fractional distillation, or extraction. They are derivatized by 2,4-dinitrophenylhydrazine for identification.

Dienes with conjugated double bonds may be derivatized by condensing the hydrocarbon with maleic anhydride or naphthoquinone.

The characterization of terpenes is discussed in the work of Simonsen (see References).

Example 18.02: Addition of Bromine to an Alkene: Styrene Dibromide

Dissolve 0.2 ml of styrene in 1 ml of dry carbon tetrachloride. Add 10 drops of bromine (use care in handling bromine) and then cool the tube. Add to the solid mass of crystals that separates 5 ml of methanol, heat until the solid dissolves, filter, and cool the filtrate. The yield of *styrene dibromide*, melting at 71 to 72°, is about 300 mg.

Example 18.03: Oxidation of a Cycloalkene to a Dicarboxylic Acid: Adipic Acid from Cyclohexene

In a 125-ml Erlenmeyer flask, place 1.5 g of potassium permanganate, 25 ml of water, and 1 ml of 6N sodium hydroxide solution. Warm to effect solution of the permanganate; add 0.3 ml of cyclohexene and, after stoppering the flask with a solid rubber stopper, shake at intervals for 10 to 15 minutes or until the odor of cyclohexene has completely disappeared. Filter with suction and evaporate the filtrate to dryness; add 2 ml of 6N hydrochloric acid solution and 3 ml of water. Extract three times with 5 ml

[5] J. G. Sharefkin and E. M. Boghosian, *Anal. Chem.*, **33**, 640–644 (1961).

of ether. Evaporate the ether and crystallize the residue from hot water. The yield is 50 to 60 mg of adipic acid, melting at 149 to 151°.

Preparation of Adducts of Olefins with 2,4-Dinitrobenzenesulfenyl Chloride[6]

2,4-Dinitrobenzenesulfenyl chloride (SC reagent) promises to be a versatile reagent for the characterization of a number of classes of organic compounds. With symmetrical olefins it yields readily a crystalline sulfide:

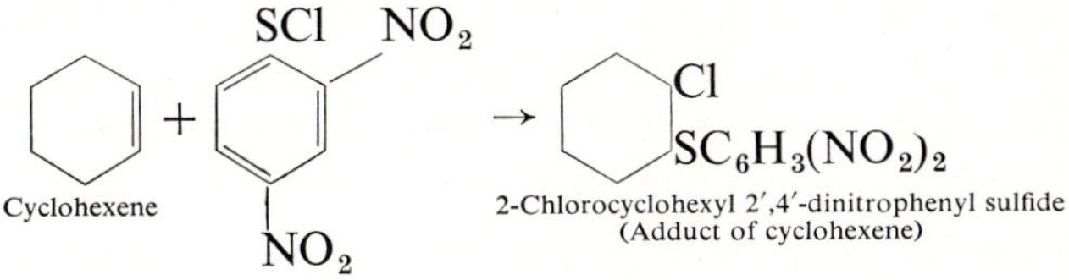

Cyclohexene + 2,4-dinitrobenzenesulfenyl chloride → 2-Chlorocyclohexyl 2',4'-dinitrophenyl sulfide (Adduct of cyclohexene)

With symmetrical olefins or negatively substituted olefins, such as styrene, only a single adduct is formed. Other olefins, such as propene and 2-pentene, yield mixtures of isomeric adducts. The reagent also reacts with acetylene, 2-butyne, 3-hexyne, and diphenylacetylene to give derivatives with sharp melting points.

The general method for the preparation of derivatives for the characterization of olefins is to mix 200 mg of the reagent and 200 to 300 mg of the olefin in 5 ml of glacial acetic acid and heat in the steam bath for 15 minutes or until the potassium iodide test shows that there is no unreacted reagent present. A drop of the reaction solution is added to a drop of potassium iodide on a spot plate. The presence of unreacted reagent is revealed by liberation of iodine: $2RSCl + 2I^- \rightarrow RSSR + I_2 + 2Cl^-$.

The reaction mixture is chilled. If crystals are formed they are removed by filtration or centrifugation. If an inadequate amount (or no product at all) is obtained in this manner the reaction mixture or filtrate is poured onto 5 to 10 g of crushed ice. The resulting solid or oil is crystallized from ethanol.

AROMATIC HYDROCARBONS

Aromatic hydrocarbons can be derivatized by: (a) nitration; (b) nitration followed by reduction of the nitro compound and acetylation of the amine to form the acetamido derivative; (c) chlorosulfonylation, followed by ammonolysis of the sulfonyl chloride to the sulfonamide; (d) condensation with phthalic anhydride in the presence of aluminum chloride to form the *o*-aroylbenzoic acids; (e) addition products with picric acid and 2,4,7-trinitrofluorenone; (f) oxidation of the side chains to carboxylic acids (this is applicable to toluene and the other alkylbenzenes); and (g)

[6] N. Kharasch et al., *J. Am. Chem. Soc.*, **71**, 2724 (1949); **74**, 3422 (1952); **75**, 1081 (1953).

reaction of a number of aromatic hydrocarbons with 2,4-dinitrobenzenesulfenyl chloride in the presence of catalysts whereby aryl 2,4-dinitrophenyl sulfides are formed.

The first two methods should be tried first; in nitration the formation of dinitro and trinitro derivatives is preferable, as these are less likely to be contaminated by isomers than the mononitro compounds. The monoalkyl benzenes do not yield solid nitro derivatives readily but can be derivatized by formation of the mononitro compound and conversion to the 4-acetamido derivative; their preparation can be carried out with 10 to 20 mg of the hydrocarbon. A few monoalkyl benzenes, for example, the monobutylbenzenes, may be readily characterized by formation of the aryl 2,4-dinitrophenyl sulfides; at least 200 mg of the hydrocarbon is required. The picrates of the hydrocarbons form very readily, but they are not useful in most cases, with the exception of anthracene and naphthalene, because they are not stable enough to be isolated in the pure form. The 2,4,7-trinitrofluorenone addition products are stable and are therefore useful for the characterization of polynuclear hydrocarbons. For the conversion of aromatic hydrocarbons to sulfonamides at least 250 mg of the hydrocarbon is required, and for the preparation of aroyl benzoic acid, at least 400 mg.

Nitration of Aromatic Hydrocarbons

Discussion

Micronitration of aromatic hydrocarbons in most cases even with milligram quantities of material gives fairly good yields because of the low solubility and high melting point of most nitro compounds. The usual nitrating agents may be used with slight modifications; however, particular care must be taken to select conditions which will give the minimum amount of undesirable nitro products. For example, the yield of the mononitro derivative from 10 g of a hydrocarbon may be more than 5 g even after four or five crystallizations, despite the formation of about 20 per cent of the dinitro compound. However, the nitration of 10 mg under similar conditions will result in the loss of most of the mononitro derivative in the purification step.

The optimum conditions can best be ascertained by several trials in which the nitrating mixture, temperature, and time of contact between reactants are varied. For example, trials[7] with the three xylenes, toluene, and isopropylbenzene showed that it is not feasible to prepare a solid derivative with *o*-xylene and isopropylbenzene, but that a dinitro derivative

[7] N. D. Cheronis, p. 318 (cited in Reference 1).

can be obtained readily from toluene and a trinitro from *m-* or *p-*xylene. For the characterization of alkylbenzenes it is possible to prepare the mononitro compounds and then to convert them without separation to the 4-acetamido derivatives.[8]

The common nitrating agents are: (a) ordinary nitric acid (density 1.40 to 1.41), containing about 58 per cent acid, mixed either with concentrated or fuming sulfuric acid; (b) fuming nitric acid, which usually contains about 88 to 90 per cent acid; (c) ordinary or fuming nitric acid in acetic acid and acetic anhydride; and (d) alkali nitrates and sulfuric acid. In addition, the following have also been used: organic nitrates, such as benzoyl and acetyl nitrates; mixtures of nitric, acetic, and sulfuric acids; mixtures of nitric and sulfuric acids in chloroform, acetone, or ethers; and nitrogen oxides. For micronitrations it has been found convenient to replace the commercially available fuming nitric acid with 100 per cent nitric acid, since the commercial acid does not have a constant composition—the density seldom indicates the nitric acid content. The preparation described on page 561 requires a simple microretort, which is easily constructed from glass tubing and gives 99.5 to 100 per cent nitric acid containing only traces of nitrogen oxides. This acid can be used for either mono- or polynitration. For example, when *p-*bromochlorobenzene is nitrated with the usual acid mixture, it does not readily yield the mononitro derivative; however, when 50 mg of the aryl halide are treated with 0.1 ml of 100 per cent acid at room temperature for 15 minutes, the yield is 25 mg of the pure mononitro derivative after one crystallization.

When difficulties are encountered because of migration of labile groups[9] during nitration, it is advisable to work with quantities larger than 500 mg. Although it is possible to vary the amount of the acid or the solvent, the effect of nitric acid alone should be tried first in such cases.

Example 18.04: Nitration of Toluene

Place in an 8-in. test tube 1.5 ml of concentrated sulfuric acid and 0.25 ml of toluene. Cool and add slowly 1.5 ml of fuming nitric acid. Heat in a water bath for 15 minutes. Remove the tube from the bath every few minutes and shake in order to mix the two layers. Cool and then add 7 to

---

[8] V. N. Ipatieff and L. Schmerling, *J. Am. Chem. Soc.*, **59**, 1056 (1937); **60**, 1476 (1938); L. I. Smith et al., *J. Am. Chem. Soc.*, **57**, 1289 (1935); **61**, 989 (1939); **62**, 2635 (1940); C. A. Mann et al., *Ind. Eng. Chem.*, **28**, 598 (1936); K. A. Kobe and T. F. Doumani, *Ind. Eng. Chem.*, **31**, 257 (1939); A. Newton, *J. Am. Chem. Soc.*, **65**, 2434 (1943).
[9] L. C. Raiford et al., *Am. Chem. J.*, **43**, 393 (1910); **44**, 209 (1911); *J. Am. Chem. Soc.*, **55**, 2125 (1933); **66**, 1872 (1944); H. N. Hodgson et al., *J. Chem Soc. (London)*, **1931**, 1500, 2268; **1932**, 273.

8 ml of cold water or the same amount of ice. Filter the solid and wash twice with 2 to 3 ml of water. Dissolve in hot methanol, filter with suction, and add 1 or 2 drops of water to the filtrate. Cool for 10 minutes, filter the solid, and repeat the crystallization. The yield is 110 to 140 mg of 2,4-dinitrotoluene, melting at 69 to 70°.

NOTE. As stated under the discussion of aromatic hydrocarbons, the procedure outlined above will not be successful for the nitration of the three xylenes. It is recommended that the entire discussion on nitration (pages 559–561) be read before nitration of an aromatic hydrocarbon is attempted. By the experimental approach outlined on page 560 the best conditions for dinitration of mesitylene were found to be as follows: a mixture of 500 mg of mesitylene, 4 ml of chloroform, and 2 ml of sulfuric acid in a tube is cooled to −5°, and 0.4 ml of fuming nitric acid is added dropwise keeping the temperature below 0°. After 20 minutes the chloroform layer is separated, washed successively with 1-ml portions of water, 5 per cent sodium carbonate, and water. The solvent is evaporated and the crude product crystallized from ethanol-water. About 450 mg of crystals are obtained, melting at 86 to 88°.

Example 18.05: Nitration of m-Xylene

Mix in an 8-in. test tube 3 ml of concentrated sulfuric acid and 1.5 ml of fuming nitric acid. Add 0.25 ml of *m*-xylene. Stopper the tube with a solid rubber stopper and shake for 2 to 3 minutes; at first heat is generated by the reaction, but the tube should not be immersed in cold water. Let it stand for 2 minutes and then heat in a water bath at 70 to 80° for 15 minutes. Add 8 ml of cold water and allow the oil that first separates out to crystallize. Filter the solid, wash with water, and recrystallize from 5 to 6 ml of methanol; repeat the crystallization twice, using slightly less solvent for the second and third crystallizations. The yield is 100 to 110 mg of 2,4,6-trinitro-*m*-xylene, melting at 181 to 182°.

NOTE. For the preparation of the mononitro derivatives of the alkylbenzenes, see the original article by Ipatieff and Schmerling; the method given on page 581 will also be found useful.

Example 18.06: Nitration of p-Xylene

Mix 2 ml of concentrated sulfuric acid and 2 ml of fuming nitric acid in an 8-in. test tube. Add 0.2 ml of *p*-xylene and immerse the tube for 30 minutes in a water bath at 90 to 95°, shaking the tube frequently. Cool and add 20 ml of water. Filter the solid and recrystallize twice from methanol. After the hot alcoholic solution has been filtered, water is added with shaking. The filtrate should be cooled for 10 minutes before the crystals are removed. The yield is 100 to 120 mg of 2,3,5-trinitro-*p*-xylene, melting at 136°.

Acetamido Derivatives of Aromatic Hydrocarbons

Discussion

Monoalkylbenzenes do not yield solid nitro derivatives with ease. For example, if a mixture of 200 mg of isopropylbenzene, 3 ml of concentrated sulfuric acid, and 1.5 ml of fuming nitric acid is heated at 70 to 80° for 45 minutes, the product is an oil; this oil contains a considerable proportion of the dinitro derivative and may be used for characterization by reduction and subsequent conversion to the diacetamido derivative. The method developed by Ipatiff and his collaborators is applicable to the identification of monoalkylbenzenes. The nitration is effected by the methods developed as described on page 560. Under the condition described in trial (a), relatively pure 4-nitroalkylbenzenes are formed with very little of the ortho-substituted isomers. With the more vigorous nitrating conditions described in trial (d), 2,4-dinitroalkylbenzenes are formed. The nitro derivatives are extracted with ether and are reduced by means of tin and hydrochloric acid to the corresponding amines, which are acetylated to obtain the 4-acetamido- and 2,4-diacetamidoalkylbenzenes. The first are more suitable for the characterization of the pure hydrocarbons; the second for the components of aromatic mixtures. The general method for the preparation of the 4-acetamido derivatives is illustrated by the derivatization of isopropylbenzene.

Example 18.07: Preparation of p-*Acetamido Derivative of Isopropylbenzene*

Place in an 8-in. test tube 3 ml of nitrating mixture, consisting of equal volumes of concentrated sulfuric and nitric acids. Add 0.25 ml of isopropylbenzene and shake the mixture in the tube for 5 minutes. Add 8 ml of water and extract with 5 ml of ether. Separate the ether layer and place in an evaporating dish. When the solvent has evaporated, transfer the remaining oil, by means of 1.5 ml of methanol, to an 8-in. tube. Add 2 ml of concentrated hydrochloric acid and 2 g of tin. Boil gently until practically all the tin dissolves. Cool and dilute with 5 ml of water; add 8 ml of $6N$ sodium hydroxide solution and then extract with ether. Add to the ether solution 0.5 ml of acetic anhydride and then evaporate the solvent; heat the residue over a small free flame for 1 to 2 minutes at 100°. Add 2 ml of water and neutralize the solution by addition of sodium hydroxide solution. Cool and filter the *p*-acetamido derivative; recrystallize from methanol by adding water to the hot filtrate. It may require two or three crystallizations to obtain a pure *p*-acetamido derivative melting at 105 to 106°. The yield is about 30 to 50 mg.

Note. When the diacetamido derivative is desired, the nitration mixture consists of 3 ml of concentrated sulfuric acid and 1.5 ml of concentrated or fuming nitric acid. All other directions are the same.

Picrates and 2,4,7-Trinitrofluorenone Adducts of Aromatic Hydrocarbons

Discussion

A large number of hydrocarbons react with picric acid and other trinitro compounds to form addition products; however, only a few of these addition compounds may be isolated in the pure form for determination of melting points. In order that a picrate may be useful in characterization work, it should form when the hydrocarbon is added to an alcoholic or benzene solution of picric acid and be sufficiently stable to be filtered and dried rapidly in air. When these conditions are used, the following results are obtained (the first figure represents the melting point obtained, and the second, in parentheses, the melting point listed in the literature): naphthalene, 148 (149); anthracene, 135 (138); phenanthrene, 97 to 99 (144); β-methylnaphthalene, 91 to 96 (116); α-methylnaphthalene, 101 (140). In general, it is recommended that only the picrates of naphthalene and anthracene be used for characterization work.

Of much greater use is the application of 2,4,7-trinitrofluorenone to the characterization of not only complex aromatic hydrocarbons but also many polynuclear substituted aromatic compounds. The melting points of 650 aromatic hydrocarbon adducts have been reported since the initial introduction of this reagent by Orchin[10] and collaborators in 1946. More recently McCrone and Laskowski and co-workers[11] developed a rapid micro method for the identification of polynuclear aromatic compounds using this reagent. By means of fusion techniques (page 277) it is possible to determine with a few milligrams four physical properties of the unknown: the melting point of the unknown, the melting point of the derivative of the unknown and the reagent, the melting point of the eutectic between the unknown and the derivative, and the melting point of the eutectic between the reagent and the derivative.[12] A microscope heating stage is employed.[13]

[10] M. Orchin et al., *J. Am. Chem. Soc.*, **68**, 1727 (1946); **69**, 1225 (1947).
[11] D. E. Laskowski, D. G. Grabar, and W. L. McCrone, *Anal. Chem.*, **25**, 1400 (1953); D. E. Laskowski and W. C. McCrone, *Anal. chem.*, **30**, 542–544 (1958).
[12] D. E. Laskowski and W. C. McCrone, *Anal. Chem.*, **26**, 1497 (1954).
[13] N. D. Cheronis, p. 541 (cited in Reference 1).

The preparation of adducts with 2,4,7-trinitrofluoronene is recommended in preference to those with picric acid. The melting points of about 150 T.N.F. adducts have been reported. The general methods are illustrated by the following examples. References to the literature appear on page 586.

Example 18.08: Preparation of the Picrate of Naphthalene

Dissolve 100 mg of naphthalene in 6 ml of hot methanol; cool the solution and add 1.5 ml of a saturated solution of picric acid in methanol. Filter the solid with suction and wash with 0.5 ml of alcohol. Dry by pressing the crystals between filter paper and determine the melting point at once. The yield is 40 to 50 mg of crystals, melting at 148°.

NOTE. An alternative procedure is to use dry benzene as a solvent. Only a few hydrocarbons yield pure stable picrates. Even when benzene is used and the picrate is filtered immediately after mixing the reagents, results are only slightly better than when methanol or ethanol is used as a solvent.

Example 18.09: Preparation of the 2,4,7-Trinitrofluorenone (T.N.F.) Adduct of Anthracene

Dissolve 100 mg of T.N.F. (2,4,7-trinitrofluorenone) in a mixture of 10 ml of absolute methanol or ethanol and 2 ml benzene. Boil for a few seconds and add a solution of 60 mg anthracene in 3.5 ml methanol and 1.5 ml of benzene. Heat for 30 seconds and cool. Filter the red flocculent crystals, wash with 1 ml of methanol, and dry. The yield is 50 to 60 mg of solid melting at 192 to 193°. The complex may be recrystallized from absolute alcohol or alcohol-benzene.

Aroylbenzoic Acids from Aromatic Hydrocarbons

The preparation of aroylbenzoic acids by the condensation of phthalic anhydride with the hydrocarbon in presence of aluminum trichloride (Friedel-Crafts reaction) has been applied to the characterization of about twenty aromatic hydrocarbons:

$$ArH + C_6H_4\begin{matrix}CO\\ \\CO\end{matrix}\!\!\!\!\!>\!O \xrightarrow{AlCl_3} C_6H_4\begin{matrix}COAr\\ \\COOH\end{matrix}$$

o-Aroylbenzoic acid

This method is recommended for the hydrocarbons that do not give suitable derivatives through nitration or nitration followed by reduction and acetylation or through addition of trinitrofluorenone.

Use the apparatus shown on page 472. Place in the tube a mixture of 10 ml of carbon disulfide, 400 mg of phthalic anhydride, 800 mg of anhydrous aluminum chloride, and 400 mg of the hydrocarbon to be identified. Heat the mixture on a water bath until no more hydrogen chloride is evolved. Remove the test tube and cool in tap water. If the mixture separates into two layers, the upper carbon disulfide layer is decanted; if it does not, 10 ml of $6N$ hydrochloric acid are added, drop by drop at first, and later in 1-ml portions with frequent stirring. The resulting product, if solid, is separated by filtration and washed with two 5-ml portions of cold water. If a viscous liquid separates, the mixture is cooled in an ice bath until it solidifies; the aqueous layer is decanted and the residue is washed in the test tube with two 5-ml portions of cold water. The product in either case is transferred to a small beaker and boiled for 1 minute with a mixture of 10 ml of $6N$ ammonium hydroxide, 20 ml of water, and about 100 mg of decolorizing carbon. The solution is filtered by suction while hot, and the filtrate is poured onto about 25 g of crushed ice contained in a small beaker, and then acidified with $6N$ hydrochloric acid with stirring. After standing for 10 minutes, the precipitate of aroylbenzoic acid is filtered, washed with small portions of water until free from acid, and dried in air. The product is recrystallized by dissolving in 3 to 4 ml of alcohol, filtering while hot, and then adding about 7 ml of water and cooling the filtrate.

NOTE. For *n*-propylbenzene, *n*-butylbenzene, cumene, and cymene the general procedure is modified as follows. Use 2.5 times the quantities of reagents except carbon disulfide, of which 10 ml is sufficient. At the end of the reaction, the tube is cooled and the carbon disulfide layer removed; then 15 ml of $6N$ hydrochloric acid are added with the precautions previously mentioned. After cooling the mixture in an ice bath, the aqueous layer is decanted, and the oily product is washed with two 5-ml portions of cold water. To remove the unconverted hydrocarbon, a current of steam is passed through this mixture until the odor of the former is no longer perceptible. The aqueous layer is then decanted and the residue extracted under reflux with three 20-ml portions of ligroin (bp 90 to 120°). The combined ligroin extractions are cooled in an ice bath, and the gummy precipitate of aroylbenzoic acid that appears is allowed to stand with occasional stirring until it becomes granular. The solid is collected on a filter, dried in air, and finally recrystallized.

For ethylbenzene and the three xylenes tetrachlorophthalic anhydride is used. It is more convenient, however, to derivatize the three xylenes by nitration, and ethylbenzene through the *p*-acetamido derivative.

Preparation of Derivatives of Aromatic Hydrocarbons with 2,4-Dinitrobenzenesulfenyl Chloride[14]

Benzene, alkylbenzenes, biphenyl, bibenzyl, phenanthrene, and other aromatic hydrocarbons react with the 2,4-dinitrobenzenesulfenyl chloride (page 577) to yield sulfides:

$$C_6H_5H + ClSC_6H_3(NO_2)_2 \xrightarrow{AlCl_3} C_6H_5-S-C_6H_3(NO_2)_2 + HCl$$

2,4-Dinitrobenzenesulfenyl chloride → Phenyl 2,4-dinitrophenyl sulfide

The general method for derivatizing aromatic hydrocarbons is to add 500 mg of the reagent to 7 ml of redistilled 1,2-dichloroethane. The mixture is chilled in an ice bath to 0 to 5° and 500 mg of anhydrous aluminum chloride are added. To the resulting red mixture add about 1 ml of the aromatic hydrocarbon. If the compound is a solid, dissolve it in the minimum amount of 1,2-dichloroethane prior to adding the reagent. The temperature rises and hydrogen chloride is evolved.

The vessel is rotated for 1 to 2 minutes in the ice bath. About 2.5 ml of ethanol are then added dropwise with shaking so that the temperature does not exceed 25 to 30° at any time. The solution is then washed twice with 1N hydrochloric acid to remove the aluminum compounds. The organic layer is removed and concentrated to about 1.5 to 2 ml in a small evaporating dish over a steam bath. About 2 ml of absolute methanol or ethanol are added and the mixture is transferred to a tube and chilled. The derivative is filtered and crystallized from absolute alcohol or alcohol-benzene mixture.

Oxidation of Side Chains

Aromatic hydrocarbons with side chains are oxidized by either alkaline permanganate or chromic acid solution to aryl carboxylic acids, which in many cases can be used as derivatives. The general methods for the oxidation of side chains are described on pages 626 to 627. It is obvious, however, that the number of hydrocarbons that can be directly derivatized by oxidation of the side chains to yield either monocarboxylic or dicarboxylic acids is small. Toluene and all monoalkylbenzenes yield benzoic acid; however, the three isomeric xylenes give different dicarboxylic acids, and mesitylene a tricarboxylic acid.

[14] C. M. Buess and N. Kharasch, *J. Am. Chem. Soc.*, **72**, 3529 (1950).

REFERENCES

Addition compounds of dienes: Diels and Alder, *Ber.*, **62**, 2081, 2337 (1939).

Addition products of aromatic hydrocarbons with picric acid: Baril and Hauber, *J. Am. Chem. Soc.*, **53**, 1087 (1931).

Addition of mercaptans to unsaturated hydrocarbons: Jones and Reid, *J. Am. Chem. Soc.*, **60**, 2452 (1938).

Characterization of olefins with 2,4-dinitrobenzenesulfenyl chloride: Grani, *J. Am. Chem. Soc.*, **71**, 3883 (1949); N. Kharasch et al., *J. Am. Chem. Soc.*, **71**, 2724 (1949); **74**, 3422 (1952); **75**, 1081 (1953).

Characterization of terpenes by addition products: Simonsen, *The Terpenes*, Cambridge Univ. Press, New York, 1931.

Chlorination (semimicro) of hydrocarbons using chlorine: Cheronis, *J. Chem. Educ.*, **20**, 611 (1943).

Chlorination of hydrocarbons using sulfuryl chloride: Cutter and Brown, *J. Chem. Educ.*, **21**, 443 (1944).

Conversion of aromatic hydrocarbons to sulfonamides: Huntress and Autenrieth, *J. Am. Chem. Soc.*, **63**, 3446 (1941).

Epoxidation and rearrangement of olefins: Sharefkin and Schwerz, *Anal. Chem.*, **33**, 635–639 (1961).

Hydration of acetylenes: Sharefkin and Boghosian, *Anal. Chem.*, **33**, 640–644 (1961).

Hydration of alkylacetylenes: Thomas, Campbell, and Hennion, *J. Am. Chem. Soc.*, **60**, 718 (1938).

Hydration of disubstituted acetylenes: Johnson, Swartz, and Jacobs, *J. Am. Chem. Soc.*, **60**, 1883 (1938).

Identification of alkylbenzenes by means of the acetamido derivatives: Ipatieff and Schmerling, *J. Am. Chem. Soc.*, **59**, 1056 (1937); **60**, 1476 (1938); **65**, 2470 (1943).

Identification of aromatic hydrocarbons: Levy and Campbell, *J. Chem. Soc. (London)*, **1939**, 141–142.

Identification of aromatic hydrocarbons by means of substituted *o*-aroylbenzoic acids: Lewenz, *J. Am. Chem. Soc.*, **75**, 4087 (1953); Underwood and Walsh, *J. Am. Chem. Soc.*, **57**, 940 (1935).

Identification of monosubstituted acetylenes by mercuric salts: Johnson and McEwen, *J. Am. Chem. Soc.*, **48**, 469 (1926).

Identification of olefins as dithiocyanates: Dermer and Dysinger, *J. Am. Chem. Soc.*, **61**, 750 (1939).

Nitrosochlorides and nitrolamines of unsaturated hydrocarbons: Perrot, *Compt. rend.*, **203**, 329 (1936).

Reaction of aliphatic olefins with thiophenol: Ipatieff, Pines, and Friedman, *J. Am. Chem. Soc.*, **60**, 2731 (1938).

Reaction of thiol compounds with olefins: Ipatieff and Friedman, *J. Am. Chem. Soc.*, **61**, 70 (1939).

2,4,7-Trinitrofluorenone adducts of polynuclear hydrocarbons: Beaton and Tucker, *J. Chem. Soc. (London)*, **1952**, 3870; Bergmann and Orchin, *J. Am. Chem. Soc.*, **71**, 1917 (1949); Campbell and Kidd, *J. Chem. Soc. (London)*, **1954**, 2155; Cason and Philips, *J. Org. Chem.*, **17**, 298 (1952); Chu et al., *J. Phys. Chem.*, **57**, 504 (1953);

Descamps and Martin, *Bull. Soc. Chim. Belges* **60**, 223 (1952); Gross and Lankelma, *J. Am. Chem. Soc.*, **73**, 3429 (1952); Herran et al., *J. Org. Chem.*, **16**, 899 (1951); Hofer and Peebles, *Anal. Chem.*, **24**, 822 (1952); Huisgen and Sorge, *Ann.*, **566**, 162, (1950); King and King, *J. Chem. Soc. (London)*, **1954**, 1375; Klemm and Sprague, *J. Org. Chem.*, **19**, 1464 (1954); Kloetzel and Metel, *J. Am. Chem. Soc.*, **72**, 4786 (1950); Lambert and Martin, *Bull. Soc. Chim. Belges*, **61**, 224 (1953); Laskowski, Grabar, and McCrone, *Anal. Chem.*, **25**, 1400 (1953); Laskowski and McCrone, *Anal. Chem.*, **26**, 1497 (1954); Newman and Hart, *J. Am. Chem. Soc.*, **69**, 298 (1947); Newman and Kosak, *J. Org. Chem.*, **14**, 375 (1949); Newman and Wheatley, *J. Am. Chem. Soc.*, **70**, 1915 (1948); Newman and Whitehouse, *J. Am. Chem. Soc.*, **71**, 3664 (1949); Orchin et al., *J. Am. Chem. Soc.*, **68**, 1727 (1946); **69**, 505, 1225 (1947); **70**, 1745 (1948); **71**, 3002 (1949); **73**, 436, 1877 (1951); *J. Org. Chem.*, **18**, 609 (1953); Price and Halpern, *J. Am. Chem. Soc.*, **70**, 1915 (1948); Riegel et al., *J. Am. Chem. Soc.*, **70**, 1073 (1948); Riegel, Siegel, and Kritchevsky, *J. Am. Chem. Soc.*, **70**, 2950 (1948); Soffer and Stewart, *J. Am. Chem. Soc.*, **74**, 567 (1952); Stubbs and Tucker, *J. Chem. Soc. (London)*, **1950**, 3288; **1954**, 231; Takemura, Cameron, and Newman, *J. Am. Chem. Soc.*, **75**, 3280 (1953); Tucker, *J. Chem. Soc. (London)*, **1949**, 2182; Tucker, Forrest, and Whalley, *J. Chem. Soc. (London)*, **1949**, 3194; Tucker and Whalley, *J. Chem. Soc. (London)*, **1949**, 632, 3213; Woolfolk, Orchin, and Storch, *Fuel*, **26**, 78 (1947); Laskowski and McCrone, *Anal. Chem.*, **30**, 542–544 (1958).

Use of Physical Methods for Characterization of Hydrocarbons

Berl, *Physical Methods in Chemical Analysis*, Vol. I, Academic Press, New York, 1960.

Coggeshall, "Spectra," in *Organic Analysis*, Vol. I, Interscience, New York, 1953.

Doss, *Physical Constants of the Principal Hydrocarbons*, The Texas Co., New York, 1943.

Egloff, *Physical Constants of Hydrocarbons*, Reinhold, New York, 1939.

Esafoy, *J. Applied Chem. (U.S.S.R.)*, **14**, 140 (1941), translated in *Foreign Petroleum Tech.*, **9**, 344 (1941). Detection and determination of diene hydrocarbons with a conjugated system of double bonds.

Forziati et al., *J. Research Natl. Bur. Standards*, **36**, 129 (1946). Purifications and properties of 29 paraffin, 4 alkylcyclopentane, 10 alkylcyclohexane, and 8 alkylbenzene hydrocarbons.

Ferris, *Handbook of Hydrocarbons*, Academic Press, New York, 1955.

Gibbon et al., *J. Am. Chem. Soc.*, **68**, 1130 (1946). Purification and physical constants of aromatic hydrocarbons.

Gilman-Miller, *Applications of Infrared and Ultraviolet Spectra to Organic Chemistry*, Wiley, New York, 1953.

Gooding, Adams, and Rall, *Ind. Eng. Chem., Anal. Ed.*, **18**, 2 (1946). Determination of aromatics, naphthenes, and paraffins by refractometer methods.

Grosse and Linn, *J. Am. Chem. Soc.*, **61**, 151 (1939). Refraction data on propane hydrocarbons.

International Critical Tables, McGraw-Hill, New York, 1932.

Jacob, *Chimie et Industrie*, Special No., 341 (September, 1926). Identification of hydrocarbons by magnetic rotatory power.

Javelle, *Chimie et Industrie*, 264 (April, 1928). Identification of hydrocarbons by magnetic rotatory power.

Mousseron and Winternitz, *Bull. soc. chim. France*, **12**, 70 (1945). Constants of hydrocarbons of the cyclohexadiene series.

Moutte, *Chimie et Industrie*, Special No., 202 (April, 1928). Analysis of hydrocarbons by means of their refractive dispersions.

Randall, Fowler, Fuson, and Dangl, *Infrared Determination of Organic Structure*, Van Nostrand, New York, 1949.

Miscellaneous methods: Identification of hydrocarbons by gas-liquid partition chromatography, Harvey and Chalkley, *Fuel*, **34**, 191 (1955); James and Phillips, *J. Chem. Soc. (London)*, **1953**, 1900; Lichtenfels, Fleck, and Burow, *Anal. Chem.*, **27**, 1510 (1955); Phillips, *Discussions Faraday Soc.*, **7**, 241 (1949); Ray, *J. Appl. Chem.*, **4**, 21, 82 (1954), Separation of normal paraffins from branched by urea, Zimmerschied et al., *J. Am. Chem. Soc.*, **71**, 2947 (1949). Derivatives of saturated hydrocarbons with desoxycholic acid, Huntress and Phillips, *J. Am. Chem. Soc.*, **71**, 458 (1949). Addition of nitriles to unsaturated hydrocarbons, Ritter et al., *J. Am. Chem. Soc.*, **70**, 4045, 4048 (1948); **71**, 4128, 4130 (1949). Characterization of aromatic hydrocarbons with 2,4-dinitrobenzenesulfenyl chloride, Buess and N. Kharasch, *J. Am. Chem. Soc.*, **72** 3529 (1950).

19

Derivatives of Amino-Nitrogen Functions

AMINES—HYDRAZINES—AMIDES, IMIDES, AND UREAS

AMINES

Table 19.01 summarizes the derivatives proposed in the literature for the characterization of primary, secondary, and tertiary amines. The majority of these are: (a) substituted amides formed by the reaction of amines with acyl or aroyl halides; (b) substituted thioureas and ureas formed with isothiocyanates and isocyanates; and (c) salts with acids.

The recommended derivatives listed and marked with an asterisk at the beginning of Table 19.01 are discussed in detail in the following sections. Bibliographical references to all the derivatives will be found on pages 611 to 613. The most suitable derivatives for tertiary amines will be found on pages 612 to 613 under Salts.

Acetamides and Benzamides of Amines

Discussion

The use of acetic anhydride for the acetylation of amines is represented by the reaction with *p*-bromoaniline:

$$BrC_6H_4NH_2 + (CH_3CO)_2O \rightarrow CH_3CONHC_6H_4Br + CH_3COOH$$

p-Bromoaniline   Acetic anhydride   N-(*p*-Bromophenyl)acetamide
(*p*-Bromoacetanilide)

TABLE 19.01
DERIVATIVES FOR THE IDENTIFICATION OF AMINES

*Acetamides and benzamides
*Benzenesulfonamides
*3,5-Dinitrobenzoates
*α-Naphthylthioureas
*α-Naphthylureas
*Phenylthioureas
*Picrates
*Quaternary ammonium salts

Substituted Ureas and Thioureas:[a]

p-Chlorophenylureas
3,5-Dinitro-4-methylphenylureas
3,5-Dinitrophenylureas
p-Iodo-, p-bromo-, m-bromophenylureas
β-Naphthylureas
p-Nitrophenylureas
N-Nitro-N'-2,4-dinitrophenylureas
o-Tolylthioureas
Xenylthioureas

Substituted Amides:

α-Toluenesulfonamides
p-Bromo- and m-nitrobenzenesulfonamides

2,4-Dinitro- and o-nitrobenzenesulfenamides
3-Nitrophthalimides
Methanesulfonamides
p-Phenylazobenzenesulfonamides
p-Phenylazobenzamides
Sulfonylbis(N-alkylacetamides)

Salts:

Acetates and benzoates
Arylsulfonates
2,4-Dinitrobenzoates
2-Nitro-1,3-indandiones
p-Nitrophenylacetates
Picramides
β-Resorcylates

Miscellaneous Derivatives:[a]

Alkanolamine hydrochlorides
2-Isonitrosocyclohexane derivatives
N-(2-Naphthyl)nitroamines
Nitrophenacyl derivatives
N-Substituted dinitroanilines
N-Substituted trinitroanilines
2,4,5-Trinitrotoluene derivatives

* Recommended derivative.
[a] See also References (pages 611 to 613).

Acetylation is useful with arylamines, but the acetamides from alkylamines often have low melting points. The usual method of acetylation is to heat the amine with a slight excess of acetic anhydride and then to separate the substituted amide by adding water. In a few cases, acetic anhydride may be used for acylations under similar conditions as benzoyl chloride—namely, in cold alkaline solution. Difficulties are encountered in the acetylation of substituted arylamines, particularly when the substituent is in the ortho position to the amino group. The substituent may exert a retarding effect on the acetylation or an accelerating effect, which gives rise to considerable amounts of diacetyl derivatives. The retarding effect is illustrated by a comparison of the acetylation of o-nitroaniline and p-nitroaniline; 100 mg of the base, 300 mg of acetic anhydride, and 2 ml of pyridine were boiled for 60 minutes. About 100 mg of the acetyl derivative of p-nitroaniline were isolated; from the acetylated mixture of

o-nitroaniline a product was isolated melting 4° below the melting point of the pure amine, indicating that very little acetylation took place. This behavior may be ascribed to chelation and to electron attraction by the nitro group, which lowers the basicity of the amine; the effect is greater in o-nitroaniline, as shown by its basic ionization constant, 10^{-14}, as compared with 10^{-12} for p-nitroaniline. The accelerating effect in acylation is illustrated by the acetylation of α-naphthylamine. Heating of 100 mg of α-naphthylamine with 300 mg of acetic anhydride for 3 to 5 minutes gives a product that melts, after one crystallization, at 142°, whereas heating for 10 minutes gives a product melting at 128°. Although the diacetyl derivative may be removed by hydrolysis, such a procedure is not recommended unless another suitable derivative cannot be prepared easily.

A substituted benzamide is formed by reaction of the amine with benzoyl chloride:

$$C_2H_5NH_2 + C_6H_5COCl \rightarrow C_6H_5CONHC_2H_5 + HCl$$

Ethylamine   Benzoyl chloride   N-Ethylbenzamide

The best procedure for benzoylation is to suspend the amine in an aqueous alkaline solution and add the aroyl chloride in small amounts with vigorous shaking, keeping the mixture cold; this method is often called the *Schotten-Baumann method* or *reaction*. On the whole, the preparation of substituted benzamides is satisfactory; it is recommended that the preparation of substituted benzamides be considered if their melting points are above 60°. p-Nitrobenzoyl chloride and 3,5-dinitrobenzoyl chloride react in the same manner as benzoyl choride; the melting points of the nitro-substituted benzamides are, as a rule, above 200° and hence are not recommended as a first choice. Furthermore, since the chlorides are solids, heating in the presence of pyridine is necessary in most cases. The N-substituted 3,5-dinitrobenzamides should not be confused with the 3,5-dinitrobenzoates; the latter are salts formed by the reaction of 3,5-dinitrobenzoic acid with the amines.

p-Phenylazobenzoyl chloride reacts with primary and secondary amines to give highly crystalline solids which are colored. The reaction gives good yields, and the products are easily purified. Because of the color of the products, the p-phenylazobenzamides may be used to advantage in the chromatographic separation and purification of primary and secondary amines. The melting points of the derivatives for 38 amines have been reported.[1]

[1] E. O. Woolfolk and L. H. Roberts, *J. Org. Chem.*, **21**, 436 (1956).

General Method for Preparation of Acetamides

About 1 millimole of the amine mixed with 2 to 4 millimoles of acetic anhydride is refluxed 3 to 5 minutes and then cooled, and the reaction mixture is diluted by the addition of 3 to 4 ml of water. The acetyl derivative usually separates as a crystalline mass or as an oil that slowly changes to a solid. The crystals are filtered and washed thoroughly with dilute hydrochloric acid to remove any unchanged amine and then crystallized from an alcohol-water mixture. The acetyl derivatives may at times separate as oils when water is first added to the reaction mixture; the oils usually solidify, however, when the mixture is vigorously stirred and rubbed against the side of the tube. When stirring and scratching fail to induce crystallization, dilute sodium hydroxide solution is cautiously added to neutralize part of the acetic acid formed by the reaction of the anhydride with the amine.

When the acetyl derivative is formed slowly, as in the case of amines that have low basicity, the reaction mixture is heated for 0.5 to 1 hour in the presence of anhydrous benzene or pyridine. Since the latter solvent is a proton-acceptor, it is more suitable; this is illustrated by the acetylation of *p*-nitroaniline. Acetyl chloride may be used as an acetylating agent for compounds that react slowly with acetic anhydride. Another method is to add 0.1 g of zinc chloride to the mixture of acetic anhydride and amine and heat for 10 to 15 minutes.

As previously mentioned, diacetylation occurs to some extent with primary amines. The formation of the diacetyl derivatives can be minimized by using a ratio of 1 millimole of the amine to 1.1 millimoles of the anhydride. In general, it is advisable to undertake the preparation of other derivatives if acetylation proves unsatisfactory. References to the literature regarding diacetylation of amines are given in the References at the end of the chapter.

Example 19.01: Acetylation of o-*Bromoaniline:* N-(o-*Bromophenyl)acetamide or* o-*Bromoacetanilide*

Place in an 8-in. tube 100 mg of *o*-bromoaniline and 0.3 ml of acetic anhydride. Provide the tube with a micro reflux condenser and heat for 5 minutes by means of a small flame so that the mixture boils gently. The condenser may be omitted if the flame is so adjusted that the vapors of the mixture rise to about the middle of the tube. Cool and add 2 ml of water. Heat for a few minutes to dislodge the mass of crystals from the bottom of the tube and then cool and filter the crystals; wash once with 3 ml of 10 per cent hydrochloric acid solution and twice with 2 ml of water. Recrystallize by dissolving the solid in 8 ml of methanol and adding 3 ml of water to the hot, filtered solution. Cool, filter the solid, and wash twice with 3 ml of water. The yield is 100 to 120 mg of crystals, melting at 98 to 99°.

Example 19.02: Acetylation of p-Nitroaniline: N-(p-Nitrophenyl)acetamide or p-Nitroacetanilide

Place in an 8-in. tube 100 mg of *p*-nitroaniline, 0.3 ml of acetic anhydride, and 2 ml of pyridine. Provide the tube with a micro reflux condenser and adjust the flame so that the mixture boils gently. Heat for 0.5 hour and then cool; add 10 ml of 2 per cent sulfuric acid solution and shake the tube so as to mix the contents thoroughly. Cool in running water for 10 minutes. Filter the solid and wash twice with 2 ml of 2 per cent sulfuric acid. Dissolve the crystals in 6 ml of methanol, filter, and add 3 ml of water. Cool for 10 to 15 minute and then filter; wash twice with 1 ml of 50 per cent methanol. The yield is 110 to 130 mg of crystals, melting at 214 to 215°.

General Method for Preparation of Benzamides

About 1 mM of the amine is suspended in an aqueous solution of sodium hydroxide (0.6 to 1.0 ml of 10 per cent solution), and 2 to 3 mM of benzoyl chloride are added in small amounts, with vigorous shaking, while the mixture is kept cold. After about 5 to 10 minutes of shaking the reaction mixture is carefully neutralized to about pH 8; this insures the separation of the derivatives of the primary amines, $RNHCOC_6H_5$ which are somewhat soluble in strong alkaline media, owing to the presence of an amino hydrogen atom. The N-substituted amides separate (usually in lumpy or granular masses) and are filtered, washed thoroughly with water, and crystallized from an alcohol-water mixture.

For the preparation of *p*-nitrobenzoyl or 3,5-dinitrobenzoyl derivatives the aroyl chloride is dissolved in 2 to 3 ml of dry benzene, and after addition of the amine the reaction mixture is shaken with 1 ml of 10 per cent sodium hydroxide for 10 to 15 minutes. The crude derivative is separated by evaporating the benzene solution. If this procedure is not satisfactory, the amine and aroyl chloride are mixed with 2 ml of pyridine and then boiled for 30 minutes. The N-substituted *p*-nitrobenzamides and 3,5-dinitrobenzamides have melting points above 200° and, therefore, are not recommended as derivatives unless other more suitable ones cannot be prepared.

Example 19.03: Benzoylation of Ethylamine

In an 8-in. tube, provided with a solid rubber stopper, place 0.4 ml of an aqueous (33 per cent) solution of ethylamine; add, by means of a pipet dropper, 0.6 ml of benzoyl chloride and then 6 ml of 10 per cent solution of sodium hydroxide. Stopper the tube and shake for 1 minute, then at intervals over a period of about 5 minutes. After each shaking, carefully release the rubber stopper (*preferably in a hood, since the vapors of benzoyl chloride have lacrymatory properties*). The oil that separates at first soon crystallizes in shiny plates. Cool and filter the crystals; wash twice with water. Neutralize the filtrate cautiously to about pH 7 to 8 in order to precipitate an additional amount of the derivative dissolved by the excess of alkali. Dissolve the combined solid in boiling methanol and filter; wash the tube with 0.5 ml methanol and add the washings to the filtrate. To the filtrate add 3 ml of water and cool for 30 minutes. If crystals do not separate within 5 minutes, scratch the inner side of the tube by means of a glass rod. Filter and wash twice with 1 ml of water. The yield is 40 to 50 mg, melting at 70 to 71°.

NOTE. The benzoyl derivatives of the primary amines have a tendency to dissolve in an excess of alkali; it is advisable, therefore, to neutralize the filtrate with dilute hydrochloric acid. It is evident, however, that the precipitated derivative from the filtrate will contain a small amount of benzoic acid resulting from the hydrolysis of the reagent; the amount of benzoic acid which is coprecipitated can be kept to a minimum if on neutralizing with dilute acid the pH is adjusted (with the aid of Universal indicator or Hydrion paper) to about 8.

Example 19.04: Benzoylation of o-Toluidine

Use 0.1 ml of *o*-toluidine, 0.5 ml of benzoyl chloride, and 6 ml of 10 per cent sodium hydroxide solution and proceed as in Example 19.03. Dissolve the derivative in 10 ml of hot methanol and add 1 ml of water to the filtered hot solution. The yield is about 120 mg of solid, melting at 142 to 143°.

NOTE. Ethyl *p*-aminobenzoate and anthranilic acid are benzoylated easily by the same method. The reaction mixture, however, should be cautiously neutralized with dilute hydrochloric acid in order to precipitate completely the derivative. From 100 mg of ethyl *p*-aminobenzoate, 120 to 130 mg of pure benzoyl derivative are obtained, having a melting point of 147 to 148°; similarly, 100 mg of anthranilic acid yield 110 to 120 mg of the benzoyl derivative, melting at 180 to 181°.

Sulfonamides and Sulfenamides from Amines

Discussion

The formation of various substituted *sulfonamides* is represented in the following equations:

$$\left.\begin{array}{l}\underset{\text{Aniline}}{C_6H_5NH_2} \\ \underset{\text{N-Methylaniline}}{C_6H_5NHCH_3}\end{array}\right\} + \underset{\text{Benzenesulfonyl chloride}}{2C_6H_5SO_2Cl} \rightarrow \left[\begin{array}{l}\rightarrow \underset{\text{N-Phenylbenzenesulfonamide}}{C_6H_5SO_2NHC_6H_5} + HCl \\ \rightarrow \underset{\text{N-Phenyl-N-methylbenzenesulfonamide}}{C_6H_5SO_2N(CH_3)C_6H_5} + HCl\end{array}\right.$$

$$\underset{\text{Dimethylamine}}{(CH_3)_2NH} + \underset{p\text{-Toluenesulfonyl chloride}}{CH_3C_6H_4SO_2Cl} \rightarrow \underset{\text{N,N-Dimethyl-}p\text{-toluenesulfonamide}}{CH_3C_6H_4SO_2N(CH_3)_2} + HCl$$

Many substituted benzenesulfonamides have been described in the literature as suitable derivatives for amines; the following partial list is arranged in decreasing order of utility for semimicro work: *p*-phenylazobenzenesulfonamides, *p*-toluenesulfonamides, *p*-bromo-α-toluenesulfonamides, *m*-nitrobenzenesulfonamides, and methanesulfonamides.

p-Phenylazobenzenesulfonyl chloride forms solid derivatives with many primary aliphatic, secondary aliphatic, and primary aromatic amines which are suitable for identification purposes.[2] Melting points for derivatives of 45 amines are given. These sulfonamides are crystalline, colored solids, which are also useful for the chromatographic separation of amines.

A reagent related to the arene sulfonyl chlorides for the derivatization of amines is the arene sulfenyl chloride, ArSCl, which on reaction with amines forms amides of the sulfenic acid:

$$ArSCl + 2R_2NH \rightarrow \underset{\text{Sulfenamide}}{ArSNR_2} + R_2NH \cdot HCl$$

The advantage of these reagents is that the reaction takes place rapidly at room temperature. The *o*-nitrobenzenesulfenyl chloride is prepared by the action of chlorine on *o,o'*-dinitrodiphenyl disulfide,[3] whereas the 2,4-dinitrobenzenesulfenyl chloride is prepared by a similar method from 2,2',4,4'-tetranitrodiphenyl disulfide;[4] the sulfenyl chloride, sometimes called "SC-Reagent." is commercially available.[5]

General Method for Preparation of Sulfonamides

The benzenesulfonamides and *p*-toluenesulfonamides are prepared by mixing 1 mM of the amine, 1.2 to 1.5 mM of the sulfonyl chloride, and

[2] E. O. Woolfolk, W. E. Reynolds, and J. L. Mason, *J. Org. Chem.*, **24**, 1445–1450 (1959).
[3] M. H. Hubacher, *Organic Synthesis*, **15**, 45 (1935) and Coll. Vol. **2**, 455 (1943); J. H. Billman and E. O'Mahoney, *J. Am. Chem. Soc.*, **61**, 2340 (1939).
[4] J. H. Billman et al., *J. Am. Chem. Soc.*, **63**, 1920 (1941).
[5] Versatile Chemicals Inc., Glendale, Calif.

4 to 5 ml of 10 per cent sodium hydroxide solution and shaking the mixture intermittently for 3 to 5 minutes. If the crystals do not separate at once, the reaction mixture is carefully acidified with $6N$ hydrochloric acid to pH 6. The granular mass of crystals is filtered, washed thoroughly with water, and crystallized from an alcohol-water mixture.

Most of the other substituted sulfonamides (p-bromobenzene-, p-nitrobenzene-, m-nitrobenzene-, α-naphthyl-, α-toluene-, and methanesulfonamide), are prepared by heating for 5 to 10 minutes a mixture of 1 mM of the chloride, 2.1 mM of the amine, and 4 ml of dry benzene, and allowing the mixture to cool. The amine hydrochloride that separates is removed by filtration or centrifugation and the clear solution is evaporated to obtain the crude sulfonamide, which is purified by crystallization from alcohol. The benzene may be eliminated in many cases as described in the examples of this section.

Example 19.05: Preparation of the Benzenesulfonyl Derivatives of Aniline

Place in an 8-in. test tube 0.1 ml of aniline, 0.2 ml of benzenesulfonyl chloride, and 5 ml of 10 per cent sodium hydroxide solution. Close the tube by means of a solid rubber stopper and shake at frequent intervals for 3 minutes. Remove the stopper and warm the tube; then shake again for 1 minute. Cool in running water and carefully add dilute hydrochloric acid to neutralize the excess of sodium hydroxide. Filter the mixture and wash the crystals twice with 3 ml of water. Dissolve the derivative in hot methanol, filter, and add 3 ml of water to the filtrate. The product melts at 109 to 110° and requires an additional recrystallization to give a melting point of 111 to 112°. The yield of the pure sulfonamide is about 140 to 150 mg.

Example 19.06: Derivatization and Separation of the Sulfonamides of a Primary and a Secondary Amine

Place in an 8-in. test tube 0.1 ml of aniline, 0.1 ml of N-methylaniline, 0.4 ml of benzenesulfonyl chloride, and 2 ml of methanol. Heat for 30 seconds and add 8 ml of 10 per cent sodium hydroxide solution and shake at frequent intervals over a period of 3 to 4 minutes. Allow to stand for 10 minutes with occasional shaking. Warm to hydrolyze any excess of sulfonyl chloride; then cool. Filter and wash crystals with 3 ml of water. Place the filtrate aside and transfer the crystals to a test tube; add 5 ml of 2 per cent sodium hydroxide solution and warm to about 50°. Place a solid rubber stopper in the mouth of the tube and then shake the contents vigorously for 1 minute. Allow the mixture to stand for 2 to 3 minutes, but at intervals shake the tube vigorously. Cool and filter the mixture. Combine the filtrates with those previously obtained which contain the benzenesulfonyl derivative of aniline in solution. Wash the crystals of the methylanilide derivative with 2 to 3 ml of water. Transfer the solid to a test tube and dissolve it in hot methanol; filter and add 1 ml of water to the hot solution. Cool and filter the crystals. The yield is 150 mg of crystals, melting at 78 to 79°.

The combined filtrates containing the derivative of aniline in solution are carefully neutralized with dilute hydrochloric acid. The precipitated crystals are filtered and washed 3 times with 3 ml of water, then transferred to a test tube and suspended in 5 ml of water. Sodium hydroxide solution (10 per cent) is added dropwise, the tube being shaken after the addition of 2 or 3 drops, until an opalescent solution results. A minute amount of charcoal is added and the solution filtered. The substituted sulfonamide is precipitated by the addition of dilute hydrochloric acid, filtered, and then crystallized as in Example 19.05 above. The yield is 90 to 100 mg of crystals, melting at 111°.

NOTE. The above procedure (often called the *Hinsberg method*) may be used for the separation of primary, secondary, and tertiary amines. If any tertiary amine is present, the reaction mixture is acidified with dilute acid and the precipitated sulfonamides are separated and washed with a 10 per cent solution of hydrochloric acid. Any tertiary amine originally present is removed. The mixed sulfonamides are then separated by treatment with dilute alkali. The arenesulfonyl derivatives of primary and secondary amines are usually contaminated with disulfonyl derivatives formed by such reactions. Such disulfonyl derivatives are often insoluble in dilute alkali and are a source of contamination to the arenesulfonyl derivatives of secondary amines. When the melting point of such an arenesulfonyl derivative varies more than 5° from the melting point listed in the literature, the crystals are refluxed for 10 to 15 minutes with a solution of sodium methoxide, prepared by dissolving 0.5 g of sodium in 10 ml of methanol. By this treatment the disulfonyl derivative is hydrolyzed to the mono-derivative. The mixture is evaporated almost to dryness and then diluted with 10 ml of water; the crystals are filtered and purified.

The separation of the benzenesulfonyl derivative by means of its solubility in alkali due to the presence of an amino hydrogen, is not applicable to all primary amines. As the solubility of the alkylamines decreases, the benzenesulfonyl derivatives $[RN(Na)(SO_2C_6H_5)]$ become insoluble; therefore, the method is not satisfactory with alicyclic amines and the higher alkylamines. See the exceptions noted by Fanta and Wang, *J. Chem. Educ.*, **41**, 280 (1964).

Example 19.07: Preparation of p-*Bromobenzenesulfonamide and Other Substituted Sulfonamides:* N-n-*Butyl-*p-*bromobenzenesulfonamide*

Place in an 8-in. test tube 100 mg of *p*-bromobenzenesulfonyl chloride and 70 mg (3 drops) of *n*-butylamine. Warm the tube by means of a small flame until the solid chloride melts and then adjust the flame so that the mixture boils gently. After 10 minutes remove the flame and add 3 ml of water and 2 ml of 6N hydrochloric acid. Cool and then filter the reaction mixture; wash the solid twice with 2 to 3 ml of 10 per cent hydrochloric acid and twice with the same amount of water. Recrystallize as in Example 19.05 (page 596). The yield is 50 to 70 mg of crystals, melting at 86 to 87°.

NOTE. The *p*-bromobenzenesulfonamides of benzylamine, ethylaniline, and piperidine are prepared by the same procedure as described for butylamine, although the amount of alcohol required for crystallization is somewhat greater; the yield of derivatives from 100 mg of the base is about 90 to 110 mg.

The *p*-bromobenzenesulfonamides of methylamine, dimethylamine, and piperidine are prepared by a similar method as that employed for the benzenesulfonamides.

m-Nitrobenzenesulfonyl chloride may be used to advantage in the preparation of derivatives of diethylamine and di-*n*-butylamine. The procedure is the same as that for the preparation of the *p*-bromobenzenesulfonamides.

In the preparation of the bromo- and nitrosubstituted sulfonamides, it is important to wash the crude derivative thoroughly with dilute hydrochloric acid since there is always a small amount of unreacted amine present.

General Method for Preparation of 2,4-Dinitrobenzenesulfenamides

About 1 mM of the reagent (2,4-dinitrobenzenesulfenyl chloride) dissolved in 3 to 4 ml of dry ether is added dropwise to a solution of 2 mM of the amine in an equal volume of dry ether. The reaction is complete when the addition of the reagent fails to produce further precipitation of the amine hydrochloride. The amine salt is filtered and washed with dry ether. The clear solution is evaporated and the crude sulfenamide is crystallized from methanol or ethanol. Two to three crystallizations are required for purification.

When the amine is present in 25 to 33 per cent aqueous solutions enough ether is added to the amine sample to form a homogeneous solution, and then the reagent is added dropwise.

Substituted Ureas and Thioureas of Amines

Discussion

Substituted ureas and thioureas are formed by reaction of amines with isocyanates and isothiocyanates:

$$CH_3(CH_2)_4NH_2 + C_6H_5NCS \rightarrow C_6H_5NHCSNH(CH_2)_4CH_3$$
n-Amylamine   Phenyl isothiocyanate   N-*n*-Amyl-N'-phenylthiourea

$$(C_4H_9)_2NH + C_6H_5NCS \rightarrow C_6H_5NHCSN(C_4H_9)_2$$
Di-*n*-butylamine   N-Phenyl-N',N'-di-*n*-butylthiourea

$$C_6H_5NH_2 + C_{10}H_7NCO \rightarrow C_6H_5NHCONHC_{10}H_7$$
Aniline   α-Naphthyl isocyanate   N-Phenyl-N'-α-naphthylurea

The preparation of thioureas is preferable to that of ureas because the isothiocyanates do not react so easily with moisture as the isocyanates do. Because of the lower rates of reaction, isothiocyanates may even be used in presence of alcohols. The mixture of amine and isothiocyanate dissolved in alcohol is heated for a short time; on cooling and careful addition of water, the thiourea separates, usually as an oil, which on standing crystallizes as a well-defined solid. The tendency of thioureas to separate as oils from concentrated alcoholic or aqueous-alcoholic solutions is particularly marked in the case of alkylamines. However, when all factors are considered, the thioureas are often the most suitable derivatives for many of the alkylamines. α- or β-Naphthyl isothiocyanate may be used in place of phenyl isothiocyanate if a thiourea with a higher melting point is desirable. Nitrosubstituted amines either fail to react with phenyl isothiocyanate or do so incompletely.

Among other isothiocyanates that have been proposed in the literature for the preparation of thioureas are: o- and p-tolyl, p-chlorophenyl, p-xenyl, m-nitrophenyl, and α- and β-naphthyl. The following isocyanates and substituted azides may be used for the preparation of substituted ureas: α- and β-naphthyl, m- and p-nitrophenyl, 3,5-dinitrophenyl, m- and p-chlorophenyl, and m- and p-bromophenyl. A complete list is given in Table 19.01 and in the References at the end of this chapter.

General Method for Preparation

For the preparation of thioureas, a solution of 1 mM of the isothiocyanate in 1 to 2 ml of alcohol is mixed with 1.1 to 1.3 mM of the amine and the mixture is gently refluxed for 5 to 10 minutes. To the warm mixture, water is added dropwise until a permanent cloudiness occurs, and the mixture is chilled. The derivative usually separates as an oil, which solidifies when it is carefully rubbed by means of a glass rod against the wall of the vessel. The solid is dried rapidly by pressing between filter papers and extracted with 1 to 2 ml of petroleum ether (commercial hexane or heptane). A minute amount of the crude product is saved for "seeding" and the rest is crystallized by dissolving in hot alcohol, adding water to cloudiness, seeding with a crystalline aggregate of the crude material, and scratching the inner wall of the vessel.

For the preparation of substituted ureas 1.1 mM of anhydrous amine are placed in a perfectly dry tube (previously heated to 150° and stoppered while warm) and 1 mM of the isocyanate is added. The tube is stoppered and placed in a bath at 40 to 50° for 30 minutes. About 2 ml of alcohol are added and the tube is warmed to disintegrate the crystalline mass and then 1 ml of water is added dropwise. The mixture is chilled and filtered, and the crude product is crystallized from alcohol.

Example 19.08: Preparation of a Substituted Phenylthiourea from Methylamine: N-Methyl-N'-phenylthiourea

Place 0.15 ml (7 drops) of phenyl isothiocyanate in an 8-in. test tube; add 2 drops of methylamine solution (33 per cent) and 1 ml of methanol so that a clear solution results. Heat for 10 minutes at 60 to 70°; if the tube is partially immersed in the water bath, the alcohol that boils off from the mixture condenses on the sides of the tube. If the boiling is brisk, either a reflux condenser is used or an additional 1 ml of methanol is added after about 5 minutes of heating.

Add to the reaction mixture, while it is still hot, 1.5 ml of water and cool the tube in tap water; by means of a glass rod scratch the inner wall of the tube until the oil that separates at first begins to crystallize. Allow the tube to stand in the cold bath for 10 minutes; then filter. Wash first with 2 ml of water to which 1 drop of $6N$ hydrochloric acid has been added, then with 1 ml of plain water. Transfer the solid into the 8-in. reaction tube and dissolve in about 1.5 ml of hot methanol. Filter and add to the hot solution 0.7 ml of water. Stir and scratch the inner wall of the tube until crystallization begins. Filter and wash the crystals with 1 ml of 25 per cent methanol. The yield is 110 mg of crystals, melting at 112 to 113°.

NOTE. As stated in the discussion of the general method, the main disadvantage of the thioureas is that these derivatives often separate as oils from the reaction mixture and are, at times, slow in crystallizing. The derivatives of the lower alkylamines are quite soluble in alcohol and hence water must be added for separation. The addition of water invariably causes the separation of the thiourea as an oil. Therefore, in the purification of the crude thiourea, it is advisable to save a small amount of crystals so that a minute amount may be used to seed the oily mixture that separates from the filtered solution of the derivative upon cooling.

Example 19.09: Preparation of a Substituted α-Naphthylthiourea from n-Butylamine: N-Butyl-N'-α-naphthylthiourea

Place in an 8-in. tube, provided with a condenser, 100 mg of α-naphthyl isothiocyanate, 0.1 ml of *n*-butylamine, and 1 ml of ethanol; heat for 30 minutes in a water bath at 60 to 70°, or over a small free flame of the microburner, so that the alcohol boils gently. Cool and, if the solid thiourea does not separate out, add 1 or 2 drops of water and rub the oil that separates against the side of the tube by means of a glass rod until it solidifies. Add 1 to 2 ml of 50 per cent ethanol to facilitate the transfer of the crystals and filter by suction; wash once with 1 ml of 50 per cent ethanol and once with 1 to 2 ml of 90 per cent ethanol. Transfer the solid to a test tube and add 2 ml of petroleum ether; heat to boiling and then cool and filter by suction. Transfer the solid remaining on the filter to a test tube and dissolve in 3 ml of ethanol; filter by suction and add 0.5 ml of water to the hot filtrate. By means of a glass rod rub the oily emulsion against the wall of the tube until crystals begin to separate; then cool for 15 minutes. Filter by suction and wash with 1 to 2 ml of 50 per cent ethanol. The melting point of the crystals is 107 to 108°; if a lower melting point is obtained, it indicates that the unreacted isothiocyanate and the small amount of dinaphthylthiourea were not completely removed. If the melting point is in the vicinity of 100°, repeat the process of washing with 90 per cent ethanol and with petroleum ether and then recrystallize from alcohol and water. The yield of the pure thiourea varies between 75 to 150 mg.

NOTE. If an impure product is obtained, 4 to 5 crystallizations will be required for purification. A slight excess of amine should be used so that very little isothiocyanate remains unreacted; the proportions should be about 1.3 millimoles of the amine to 1 millimole of the isothiocyanate. Alcohols react slowly with isothiocyanates and are therefore used as solvents; the alcohol may be omitted in the case of the lower amines, which react readily. The amine and isothiocyanate are heated for 1 to 2 minutes over a small flame of the microburner until a homogeneous solution is obtained. The mixture is then cooled while stirring with a rod; the solid mass that separates is extracted with 90 per cent ethanol and then with petroleum ether and crystallized as directed.

The preparation of thioureas is preferred to that of the corresponding ureas by reaction of the amine with α-naphthyl isocyanate. Comparative runs with the same amine indicate that far greater precautions are necessary to obtain a pure product with α-naphthyl isocyanate than with the corresponding isothiocyanate. The chief difficulty is the reaction of the isocyanate with the moisture present in the tube and in the reagent to yield dinaphthylurea, which is difficult to separate from the alkyl or aryl-naphthylurea. Since isocyanates react with alcohols to form urethans, alcohol should not be used as a solvent to form ureas.

Quaternary Ammonium Salts of Amines

Discussion

A number of amine salts of primary, secondary, and tertiary amines are readily formed if the amines are basic enough. For example, the hydrochlorides can be prepared readily by passing dry hydrogen chloride gas into a solution of the amine in ether:

$$\left.\begin{array}{l} RNH_2 \\ R_2NH \\ R_3N \end{array}\right\} + HCl \rightarrow \begin{array}{l} [RNH_3]^+Cl^- \\ [R_2NH_2]^+Cl^- \\ [R_3NH]^+Cl^- \end{array}$$

The salts are filtered, washed with ether, and dried in a desiccator. However, the melting points of hydrobromides and hydrochlorides are not particularly suitable since these salts melt with decomposition at temperatures that often depend on the rate of heating.

Inspection of Table 19.01 shows that among the recommended derivatives are the 3,5-dinitrobenzoates, picrates, and quaternary ammonium salts.

The preparation of the 3,5-dinitrobenzoates described in Example 19.10 illustrates the general procedure employed for the preparation of salts from primary and secondary amines.

$$(NO_2)_2C_6H_3COOH + CH_3NH_2 \rightarrow (NO_2)_2C_6H_3COONH_3CH_3$$

3,5-Dinitrobenzoic acid　　　　　　　　　　Methylammonium 3,5-dinitrobenzoate

However, the salts are important mainly as derivatives of tertiary amines. These latter compounds do not have amino hydrogen atoms; hence the reactions by which derivatives may be made are restricted to the formation of salts. Among the most useful of the salts are the picrates. These are easily formed by boiling the amine with a saturated solution of picric acid in methanol. The picrates may be purified by crystallization without appreciable decomposition.

Another type of salt suitable for tertiary amines, particularly the N-heterocyclics, are the compounds formed with β-resorcylic acid (2,4-dihydroxybenzoic acid). The acid is dissolved in ether and the calculated amount of amine dissolved in a small amount of ether is added. The salt separates as a solid or as an oil, which crystallizes on scratching the wall with a spatula. Since β-resorcylic acid does not form salts with aromatic amines, it can be employed to separate a number of amine mixtures—for example, quinoline from aniline.[6]

Another type of derivative is formed by tertiary amines by the acceptance of an organic group to form quaternary ammonium salts:

$$R_3N: + R'^+X^- \rightarrow [R_3NR']^+X^-$$
$$C_6H_5N(CH_3)_2 + CH_3I \rightarrow [C_6H_5N(CH_3)_3]^+I^-$$

N,N-Dimethylaniline　　　　　　　　　　Phenyltrimethylammonium iodide

The most useful types of quaternary salt derivatives are those formed with methyl iodide, methyl *p*-toluenesulfonate, and benzyl chloride. The quaternary methyl *p*-toluenesulfonates are useful in the case of nitrogen cyclic compounds and are less useful with other types of amines.

General Method for Preparation of Dinitrobenzoates of Amines

The general procedure for the preparation of 2,4- and 3,5-dinitrobenzoates of the amines is to add 1 millimole of the amine to an alcoholic solution of

[6] K. W. Wilson, F. E. Anderson, and R. W. Donohoe, *Anal. Chem.*, **23**, 1032 (1951).

Derivatives of Amino-Nitrogen Functions

1 millimole of the aromatic nitro acid and evaporate the solution; the crude salt is then crystallized. The details are illustrated in Example 19.10.

Example 19.10: Preparation of β-Naphthylammonium 3,5-Dinitrobenzoate

Prepare a solution of 3,5-dinitrobenzoic acid containing 212 mg of the acid dissolved in 20 ml of methanol. Weigh accurately 143 mg of β-naphthylamine and place in a small evaporating dish; add the alcoholic solution of the nitrobenzoic acid to the amine and evaporate the mixture on a water bath. Scrape the crystals and transfer to a 6-in. test tube. Add 2.5 ml of methanol, heat to boiling for a second, and then, if solution is not complete, add more alcohol dropwise until a clear solution is obtained. Cool rapidly and allow to stand for 5 minutes. Filter and wash the crystals with 0.5 ml of methanol. The yield is 100 to 200 mg, melting at 195 to 196°. The crude, unrecrystallized derivative melts at 189 to 191°.

NOTE. This method, although effective in the identification of some aromatic amines, fails to give easily crystallizable products from such tertiary amines as dimethylaniline. Other amine salts described in the literature for the identification of amines are the 2,4-dinitrobenzoates. The original literature should be consulted; references are given on page 613.

General Method for Preparation of β-Resorcylates

About 100 mg of β-resorcylic acid are dissolved in 2 ml of dry ether and the calculated amount of amine dissolved in 1 ml of ether is added by means of a pipet dropper. In most cases the salt precipitates immediately; if an oil is produced, it can be made to crystallize by vigorous scratching with a spatula. When no visible reaction occurs, the tube is stoppered and allowed to stand overnight. The salt is removed by filtration or centrifugation and washed with dry ether then dried in air. The salt is crystallized from ethyl acetate. The yields are about 60 to 80 per cent of the theoretical.

General Method for Preparation of Picrates of Tertiary Amines

One millimole of the tertiary amine dissolved in 5 ml of alcohol is mixed with 2 to 3 ml of a saturated solution of picric acid in methanol, and is then allowed to cool. The crystals that separate are filtered, washed with a small amount of methanol, and dried. If crystals do not separate on cooling the solution is concentrated to about 4 ml (see page 31). The picrates may be recrystallized from methanol or ethanol; however, in many cases no purification is necessary.

Example 19.11: Preparation of the Picrate of Dimethylaniline

Place into an 8-in. test tube 3 ml of a saturated solution of picric acid in methanol and 0.1 ml of N,N-dimethylaniline. Add 5 ml of methanol, boil for a few minutes using a reflux condenser, and allow to cool. Filter the crystals and wash twice with 1 ml of methanol. The yield is 180 to 200 mg of crystals, melting at 162 to 163°.

Preparation of Quaternary Salts of Tertiary Amines

The most suitable quaternary salts are the methiodides and methyl *p*-toluenesulfonates. The general method is to mix equimolar quantities of the tertiary amine and the reagent and heat for a few minutes; the use of isopropyl ether or benzene facilitates the handling of small quantities of material.[6a] The following examples illustrate details of the general procedure.

*Example 19.12: Preparation of the Methiodide of Tri-*n*-butylamine: Methyltri-*n*-butylammonium Iodide*

Place 0.1 ml of tri-*n*-butylamine and 0.1 ml of methyl iodide in a test tube and add 1 ml of isopropyl ether. Heat under reflux for 5 minutes; then cool. Filter and wash the crystals with 1 ml of isopropyl ether. The yield is 180 mg of crystals, melting at 179 to 180°.

Example 19.13: Preparation of N-Methylpyridinium p-*Toluenesulfonate*

Pour into a 6-in. test tube 0.2 ml of pyridine, 300 mg of methyl *p*-toluenesulfonate, and 1 ml of isopropyl ether. Boil (under reflux) in a water bath for 20 minutes; then cool. Pour off the ether from the crystals and add 1 ml of methanol. Heat until a solution results and then add 5 ml of ethyl acetate. Cool and filter the product; wash with 1 ml of ethyl acetate and dry a small amount of crystals by pressing between filter paper: dry in the air for 2 to 3 minutes and then determine the melting point. If the melting point is below 139°, the product is recrystallized. The yield is 175 to 200 mg of crystals.

NOTE. The preparation of the methyl *p*-toluenesulfonate derivative is undertaken if the picrate is found unsuitable. In the case of pyridine, the picrate is easily prepared and purified.

Other Derivatives of Amines

Table 19.01 (page 590) lists a large number of substituted amides, thioureas, ureas, salts, and miscellaneous derivatives that may be employed for the derivatization of amines. Among these a few will be briefly mentioned, and the reader is referred to the literature cited in the References for further details.

[6a] Marvel et al., *J. Am. Chem. Soc.*, **51**, 3638 (1929).

Substituted 3-Nitrophthalimides

Primary amines form N-substituted 3-nitrophthalamic acids, which, when heated to 145°, are converted to substituted 3-nitrophthalimides; secondary amines form N,N-disubstituted nitrophthalamic acids, which are stable to heat:

$$RNH_2 \atop R_2NH \Bigg\} + \text{3-Nitrophthalic anhydride} \longrightarrow \begin{cases} \text{N-Substituted 3-Nitrophthalamic acid} \xrightarrow{-H_2O} \text{3-Nitrophthalimide} \\ \text{N,N-Disubstituted 3-Nitrophthalamic acid} \end{cases}$$

A mixture containing derivatives from primary and secondary amines may be separated by the addition of sodium bicarbonate solution, which dissolves the latter but not the former. The use of 3-nitrophthalic anhydride is indicated for amines, such as o-nitroaniline, which fail to give derivatives by the usual methods. The use of 3-nitrophthalic anhydride also permits in many cases, a differentiation between primary, secondary, and tertiary amines.[7]

Salts with Acids

In addition to those salts already described, the amine salts of p-toluenesulfonic acid, m-nitrobenzoic acid, anthraquinonesulfonic acid, and other aryl carboxylic and sulfonic acids have been employed. The preparation of a number of arylamine salts of sulfonic acid is described on page 637.

Miscellaneous Derivatives

The following reagents are among the most promising that have been described to form suitable derivatives of primary and secondary amines: p-nitrophenylacetyl chloride,[8] N-(2-isonitrosocyclohexyl)pyridinium chloride,[9] 2,4-dinitrobenzaldehyde,[10] and 2-nitro-1,3-indanedione.[11] The

---

[7] J. W. Alexander and S. M. McElvain, *J. Am. Chem. Soc.*, **60**, 2285 (1938).

[8] H. P. Ward and E. F. Jenkins, *J. Org. Chem.*, **10**, 371 (1945).

[9] A. J. Birch, *J. Chem. Soc.* (*London*), **1944**, 314.

[10] A. Lowry et al., *J. Am. Chem. Soc.*, **42**, 849 (1920); **43**, 346 (1921); **45**, 1060 (1923); G. M. Bennett and G. N. Willis, *J. Chem. Soc.* (*London*), **1928**, 1962; G. M. Bennett and W. L. C. Pratt, *J. Chem. Soc.* (*London*), **1929**, 1465.

[11] B. E. Christensen et al., *Anal. Chem.*, **21**, 1573 (1949); G. Wanag et al., *Ber.*, **69B**, 1006 (1936); **70B**, 547 (1937).

last-named reagent is a relatively strong acid and forms well-defined salts having sharp melting points with a variety of amino compounds including simple and substituted amines, alkaloids, nitrogen heterocyclics, amino acids, amines, and nitroamines. The derivatives of about 100 amino compounds have been described.

Primary and secondary amines react with 2,4-dinitrochlorobenzene to yield N-substituted 2,4-dinitroanilines. Similarly, picryl chloride yields 2,4,6-trinitroanilines. A number of primary arylamines readily yield derivatives by bromination, as illustrated by the bromination of aniline. For example, when 0.15 ml of bromine dissolved in 1 ml of glacial acetic acid is mixed with 0.1 ml of aniline and 1 ml of water is added, crystals of 2,4,6-tribromoaniline separate. A single crystallization of the crude product from 50 per cent methanol yields crystals that melt at 119°.

Chromatographic Detection of Amines

If the amine or mixture of amines is volatile, the same general procedure is followed as described for the chromatography of the 3,5-dinitrobenzoates of alcohols. A suitable derivative is formed and this is subjected to chromatographic separation in the usual manner.

Since the *p*-phenylazobenzamides are easily formed and are colored they are particularly useful for this purpose.[12] The *p*-phenylazobenzenesulfonamides are also suitable for the chromatographic detection of amines.[13]

If the amine is a nonvolatile liquid or a solid, it can be chromatographed directly. If the amine is not soluble in water it is dissolved in chloroform and several strips are made with varying quantities of samples (5, 25, 50, and 100 micrograms). However, better results are obtained if the sample is placed in the form of aqueous solutions of the amine salts. For example, the amine is dissolved in the minimum amount of dilute hydrochloric acid. The solvent used for development is a mixture of 4 ml of 1-butanol, 5 ml of water, and 1 ml of acetic acid. If a slower development is desired 1-butanol saturated with water is employed (butanol is shaken with water and the alcohol layer is separated and used). The indicator used for development of the spots is a 0.2 per cent ethanolic solution of ninhydrin applied in the same manner as described for the chromatography of amino acids. For more detailed information, consult the references at the end of this Chapter.

[12] E. O. Woolfolk and E. H. Roberts, *J. Org. Chem.*, **21**, 436 (1956).
[13] E. O. Woolfolk, W. E. Reynolds, and J. L. Mason, *J. Org. Chem.*, **24**, 1445–1450 (1959).

HYDRAZINES

Phenylhydrazine, $C_6H_5NHNH_2$, and its substituted derivatives are the most common hydrazines; these compounds, for the most part, are weak or intermediate bases, dissolve in dilute hydrochloric acid to form salts, and reduce Fehling solution; they are also easily reduced to ammonia and a primary arylamine, as, for example, in the formation of the osazones of the carbohydrates. The derivatives most suitable for their characterization are the hydrazones with the carbonyl compounds; these derivatives are not listed in the table on amines but can be found in the tables of carbonyl compounds. For example, the derivatives listed for phenylhydrazine are benzenesulfonamide, acetamide, benzamide, and the like; however, it is more convenient to identify phenylhydrazine by the condensation with an aldehyde, as, for example, benzaldehyde, to obtain the phenylhydrazone.

AMIDES, IMIDES, AND UREAS

This group (amides, imides, and ureas) includes unsubstituted amides such as acetamide and benzamide; substituted amides, such as anilides, o- and p-toluidides; imides, such as succinimide, phthalimide, and substituted imides; ureas and their many derivatives, such as the mono- and disubstituted ureas and ureides. Table 19.02 summarizes the few types of derivatives that may be employed for their characterization.

TABLE 19.02
DERIVATIVES FOR THE IDENTIFICATION OF AMIDES, IMIDES, AND UREAS

N-Acylphthalimides	p-Nitrobenzyl derivatives
*Hydrolysis and derivatization of the hydrolytic products	Oxalates
	*N-Xanthylamides
*Mercury derivatives	

* Recommended derivative.

By far the best method of preparing derivatives for this group of compounds is hydrolysis and identification of the hydrolytic products. For example, the identification of the amides, $RCONH_2$, and substituted amides, $RCONHR'$, such as anilides, toluidides, and the like, is best accomplished by hydrolysis:

$$RCONH_2 + H_2O + HCl \rightarrow RCOOH + NH_4Cl$$
$$RCONHR' + NaOH \rightarrow RCOONa + R'NH_2$$

The hydrolytic products are then characterized by methods suitable for identification of carboxylic acids and amines.

The formation of xanthylamides by reaction with xanthydrol can be applied to the direct derivatization of about 25 to 30 amides including urea and several ureides and substituted ureas. The formation of mercury salts has been applied to the derivatization of 15 amides and the oxalates to about 6 lower aliphatic amides.

Characterization by Hydrolysis

Discussion

The hydrolysis of the simpler amides and substituted amides is effected by boiling the compound with $6N$ hydrochloric acid or with a 10 to 20 per cent solution of sodium hydroxide. Alkaline hydrolysis is usually faster than acid hydrolysis. Amides and substituted amides that are resistant to hydrolysis when boiled with aqueous solutions of acids or bases may be hydrolyzed by heating with 100 per cent phosphoric acid. Another method for the hydrolysis of resistant amides is to heat the compound at 200° in a 20 per cent solution of potassium hydroxide in glycerol.

If acid hydrolysis is used, the reaction mixture is made alkaline in order to separate the amine. If the original compound is an unsubstituted amide, the ammonia is identified by its odor or it may be distilled into a receiver containing dilute hydrochloric acid and then tested with Nessler solution or chloroplatinic acid. In the case of substituted amides, the amine that separates may be isolated either by filtration if it is a solid, or by extraction with ether or by distillation. Two or three extractions with ether will remove the amine from the reaction mixture. The ether extract is shaken with a small amount of hydrochloric acid. The amine, through salt formation, passes into the aqueous layer, which is then separated, carefully neutralized, and treated with an acid chloride as described in Examples 19.04–19.09, pages 594–601.

After removal of the ammonia or amine the remaining alkaline solution is evaporated to a small volume and carefully neutralized. This solution is used for the preparation of the *p*-nitrobenzyl ester, as described in Example 12.09 (page 448). If the acid is a solid, the concentrated alkaline solution is acidified and cooled, and the carboxylic acid that separates is removed by filtration.

The general method for the hydrolysis of amides and derivatization of the hydrolytic products is illustrated in the following example describing the characterization of acetanilide.

Example 19.14: Hydrolysis and Characterization of Acetanilide

Place in a small distilling tube, arranged as in Figure 13.01 (page 472), 5 ml of 10 per cent sodium hydroxide solution and 300 mg of acetanilide. The receiving tube contains 2 ml of 10 per cent hydrochloric acid, and the delivery tube reaches to about 20 mm above the surface of the acid. Add 2 boiling stones to the distilling tube and boil gently for 10 to 15 minutes. Remove the reflux condenser from the distilling tube and place it into the receiving tube (Figure 3.19, page 87). Distill the reaction mixture until about 3 ml of distillate have been collected in the receiving tube containing the acid. Save the residue in the distilling tube for identification of the acid.

Add to the tube containing the distillate 5 ml of 10 per cent sodium hydroxide and 0.2 ml of benzenesulfonyl chloride and proceed according to the directions given in Example 19.05 (page 596).

Pour the alkaline residue remaining in the distilling tube into an 8-in. tube. Wash the distilling vessel with 1 to 2 ml of water and add the washings to the mixture in the 8-in. tube. Add a drop of phenolphthalein and then cautiously add dilute hydrochloric acid until the color is just discharged and prepare the *p*-nitrobenzyl acetate as described in the *Note* under Example 16.03 (page 538).

NOTE. Certain acyl-substituted amides, particularly those containing nitro and halogen groups, are resistant to hydrolysis. In such cases use is made of 100 per cent phosphoric acid. About 300 to 400 mg of the amide are placed in a tube containing a mixture of 1.0 g of 85 per cent phosphoric acid and 400 mg of phosphorus pentoxide. The mixture is boiled gently with a very small flame for 1 hour under reflux to effect hydrolysis. The hydrolytic products are treated in the manner outlined.

Xanthyl Derivatives of Amides

Discussion

The reaction of amides with xanthydrol takes place according to the following equation.

Xanthydrol + $RCONH_2$ → N-Xanthylamide + H_2O

The N-xanthylamides (or 9-acylamidoxanthenes) crystallize readily when xanthydrol is heated with the amide in the presence of acetic acid. The derivatives are purified by crystallization from aqueous dioxane or alcohol mixtures. The N-substituted amides and a number of other amides (oxamides, trichloroacetamides, salicylamides) do not form derivatives. Urea forms a derivative readily, whereas the substituted ureas and ureides (barbiturates) react more slowly so that heating for 30 to 45 minutes may be necessary.

The general method for the formation of xanthyl derivatives is to dissolve 2 mM (400 mg) of xanthydrol in 5 ml of glacial acetic acid and, after adding 1 to 1.5 mM of the amide, warm the mixture for 10 to 30 minutes; upon cooling the xanthylamide separates. The crystals are filtered and crystallized from a mixture of 70 per cent dioxane or alcohol and 30 per cent water.

When the amide is not soluble in acetic acid it is first dissolved in 2 ml of ethanol and this is added to the acetic acid. After the mixture is heated, 1 ml of water is added and the mixture is cooled.

Example 19.15: Xanthyl Derivative of Acetamide

Place 400 mg of xanthydrol and 4 ml of glacial acetic acid in a 6-in. tube. Shake at intervals for 5 minutes until the xanthydrol dissolves. If an oil separates, decant the clear supernatant solution into another tube. Add to the xanthydrol solution 100 mg of acetamide and place the tube in a beaker of water heated to 80 to 85° for 15 to 20 minutes, then cool for 30 minutes. Filter off the solid and recrystallize it from a mixture of 70 per cent dioxane and 30 per cent water. The yield is 100 to 120 mg of crystals, melting at 238 to 240° uncorrected.

Mercuric Salts and Other Derivatives of Amides

Mercuric Salts

The formation of a salt through the weak acidic properties of the amides is represented by the equation:

$$2RCONH_2 + HgO \rightarrow (RCONH)_2Hg + H_2O$$
$$\text{Mercuric salt}$$

The mercury salts are formed by heating a mixture of mercuric oxide and the amide in alcohol; an alternate method consists in heating a mixture of the oxide and the amide to the melting point of the amide. The melting points of 15 mercuric derivatives have been described; most of these are above 200°.

The general method is illustrated by the derivatization of benzamide described in Example 19.16.

Example 19.16: Mercuric Derivative of Benzamide

Place in an 8-in. tube arranged for reflux 300 mg of benzamide, 400 mg of finely powdered mercuric oxide, and 4 ml of methanol. Add a boiling stone and boil gently for 30 minutes. Filter the hot mixture with suction. Cool the filtrate in an ice-water mixture for 10 to 15 minutes. Filter off the solid and wash with 1 to 2 ml of cold methanol. The yield is 90 to 100 mg of crystals, melting at 222° uncorrected.

NOTE. An alternative method for the preparation of the mercury derivatives is to place the mixture of amide and mercuric oxide in an 8-in. tube and then heat by means of a small flame until the amide melts and the reaction begins. If the yellow color is not discharged, more amide is added in small portions until the color is removed completely. The mixture is heated for a few minutes and then allowed to cool. About 3 to 4 ml of alcohol are added and the mixture is heated until the solid dissolves; the solution is allowed to crystallize.

Since the derivatives of aliphatic amides are soluble in cold ethanol and methanol, the amount of solvent should be restricted. On the other hand, the derivatives of aromatic amides have low solubilities and in certain cases purification is effected by leaching out the unreacted amide with boiling alcohol.

Oxalates and N-Acylphthalimides

The formation of an amide oxalate is an example of salt formation through the weak, basic properties of the amides:

$$RCONH_2 + (COOH)_2 \rightarrow RCONH_2 \cdot (COOH)_2$$
Amide oxalate

The reaction is effected by heating anhydrous oxalic acid and the amide in the presence of ethyl acetate. The derivatization of six lower aliphatic amides by this method has been described.

Phthalyl chloride reacts with amides to form N-acylphthalimides:

$$\text{Phthalyl chloride} + RCONH_2 \rightarrow \text{N-Acylphthalimide} + 2HCl$$

See the literature for details on these derivatives.

REFERENCES

Derivatives of Amines

Substituted Ureas

4-Biphenylylthioureas and β-naphthylthioureas: Brown and Campbell, *J. Chem. Soc.* (*London*), **1937**, 1699.

m-Bromophenylureas: Sah et al., *J. Chinese Chem. Soc.*, **4**, 513 (1936).

m-Bromophenylureas by use of *m*-bromobenzazide: Sah and Chang, *Rec. trav. chim.*, **58**, 8 (1939).

p-Bromophenylureas: Sah et al., *J. Chinese Chem. Soc.*, **4**, 193 (1936).

p-Chlorophenylureas: Sah et al., *J. Chinese Chem. Soc.*, **3**, 137 (1935); **4**, 513 (1936).

3,5-Dinitro-4-methylphenylureas from 3,5-dinitro-4-methylbenzazide: Sah et al., *Rec. trav. chim.*, **58**, 1008 (1939).

1-(2,4-Dinitrophenyl)-3-alkyl(or -aryl)ureas: McVeigh and Rose, *J. Chem. Soc. (London)*, **1945**, 621.

3,5-Dinitrophenylureas: Sah et al., *J. Chinese Chem. Soc.*, **2**, 159 (1934); **4**, 75 (1936).

p-Iodophenylureas: Sah and Wang, *Rec. trav. chim.*, **59**, 364 (1940).

α-Naphthylthioureas: Suter and Moffett, *J. Am. Chem. Soc.*, **55**, 2496 (1933).

α-Naphthylureas: French and Wirtel, *J. Am. Chem. Soc.*, **48**, 1736 (1926).

β-Naphthylureas: Sah et al., *J. Chinese Chem. Soc.*, **5**, 100 (1937).

m-Nitrophenylureas: Sah and Meng, *J. Chinese Chem. Soc.*, **4**, 75 (1936).

p-Nitrophenylureas by use of *p*-nitrobenzazide and *p*-nitrophenyl isocyanate: Sah, *Rec. trav. chim.*, **59**, 231 (1940).

Phenylthioureas and *o*-tolylthioureas: Brown and Campbell, *J. Chem. Soc. (London)*, **1937**, 1699; Fry, *J. Am. Chem. Soc.*, **35**, 1544 (1913); Whitmore and Otterbacher, *J. Am. Chem. Soc.*, **51**, 1909 (1929).

p-Xenylthioureas: Brewster and Honer, *Trans. Kansas Acad. Sci.*, **40**, 101 (1937).

p-Chlorobenzoylthioureas: Tisler, *Z. anal. chem.*, **165**, 272 (1959).

Substituted Amides

Acetylation of amines: Chattaway, *J. Chem. Soc. (London)*, **1931**, 2495; Kaufmann, *Ber.*, **42**, 3480 (1909); Raiford et al., *J. Am. Chem. Soc.*, **46**, 205 (1924).

Benzoylation of amines: Henstock, *J. Chem. Soc. (London)*, **1933**, 216; Menalda, *Rec. trav. chim.*, **49**, 967 (1930).

p-Bromobenzenesulfonamides and *m*-nitrobenzenesulfonamides: Marvel et al., *J. Am. Chem. Soc.*, **45**, 2696 (1923); **47**, 166 (1925).

2,4-Dinitrobenzenesulfenamides: Billman et al., *J. Am. Chem. Soc.*, **63**, 1920 (1941).

Methanesulfonamides: Marvel et al., *J. Am. Chem. Soc.*, **51**, 1272 (1929).

o-Nitrobenzenesulfenamides: Billman and O'Mahoney, *J. Am. Chem. Soc.*, **61**, 2340 (1939).

p-Phenylazobenzamides: Woolfolk and Roberts, *J. Org. Chem.*, **21**, 436 (1956).

p-Phenylazobenzenesulfonamides: Woolfolk, Reynolds, and Mason, *J. Org. Chem.*, **24**, 1445–1450 (1959).

Sulfonylbis(*N*-alkylacetamides): Alden and Houston, *J. Am. Chem. Soc.*, **56**, 413 (1934).

α-Toluenesulfonamides: Marvel and Gillespie, *J. Am. Chem. Soc.*, **48**, 2943 (1926).

Salts

Amine salts of anthraquinonesulfonic acids: Perkin and Sewell, *J. Soc. Chem. Ind.*, **42**, 27T (1923).

Amine salts of arylsulfonic acids: Foster and Keyworth, *J. Soc. Chem. Ind.*, **43**, 165T (1924); **43**, 299T (1924); Keyworth, *J. Soc. Chem. Ind.*, **43**, 341T (1924); **46**, 20T (1927); **46**, 397T (1927); Noller and Liang, *J. Am. Chem. Soc.*, **54**, 670 (1932).

Amine salts of 3,5-dinitrobenzoic and 2,4-dinitrobenzoic acids: Buehler et al., *Ind. Eng. Chem., Anal. Ed.*, **5**, 277 (1933); **6**, 351 (1934).

β-Resorcylates: Wilson et al., *Anal. Chem.*, **23**, 1032 (1951).

Other Derivatives

Separation of amines by the Hinsberg reaction: Bell, *J. Chem. Soc. (London)*, **1929**, 2787; **1930**, 1072; Herzog and Hancu, *Ber.*, **41**, 636 (1908); Hinsberg, *Ber.*, **23**, 2962 (1890); Ssolonina, *Centralblatt II*, 848 (1897); *II*, 867 (1898).

Alkanolamine hydrochlorides: Jones, *J. Assoc. Official Agr. Chem.*, **27**, 467 (1944).

N-(Arylaminomethyl)phthalimides: Winstead and Heine, *J. Am. Chem. Soc.*, **77**, 1913 (1955).

Diliturates (nitrobarbiturates). Optical properties of: Plein and Dewey, *Ind. Eng. Chem., Anal. Ed.*, **15**, 534 (1943); **18**, 575 (1946).

N-(2-Naphthyl)nitroamines using 2-naphthol: Borodkin and Burmistrov: *J. Gen. Chem. (U.S.S.R.)*, **17**, 63 (1947).

2-Nitro-1,3-indandione: Christensen et al., *Anal. Chem.*, **21**, 1573 (1949); Wanag and Dombrowski, *Ber.*, **75B**, 82 (1942); Wanag and Lode, *Ber.*, **70**, 547 (1937); **69B**, 1066 (1936).

p-Nitrophenylacetamides of aromatic amines and color phenomena: Smirnov, *J. Gen. Chem. (U.S.S.R.)*, **20**, 733 (1950).

p-Nitrophenylacetyl chloride: Ward and Jenkins, *J. Org. Chem.*, **10**, 371 (1945).

3-Nitrophthalimides: Alexander and McElvain, *J. Am. Chem. Soc.*, **60**, 2285 (1938).

Picramides: Linke et al., *Ber.*, **65**, 1282 (1932).

Substituted 2-isonitrosocyclohexane derivatives: Birch, *J. Chem. Soc. (London)*, **1944**, 314.

N-Substituted 2,4-dinitroanilines using 2,4-dinitrochlorobenzene: Van Der Kam, *Rec. trav. chim.*, **45**, 722 (1926).

N-Substituted 2,4,6-trinitroanilines using picryl chloride: Mulder, *Rec. trav. chim.*, **25**, 108 (1906); Van Romburgh, *Rec. trav. chim.*, **2**, 103 (1883); **4**, 189 (1885).

Chromatographic Procedures

Baker et al., *J. Chem. Soc. (London)*, **1952**, 3215.

Bremmer and Kenten, *Biochem. J.*, **49**, 651 (1951).

Burmistrov, *Zhur. Anal. Khim.*, **5**, 39 (1950).

Cuthbertson and Ireland, *Biochem. J.*, **34**, 52 (1952).

Ekman, *Acta Chem. Scand.*, **2**, 383 (1948).

James et al., *Biochem. J.*, **52**, 238, 242 (1952). Gas-liquid chromatographic separation of volatile amines and pyridines.

Kariyoni and Hashiomot, *Nature*, **168**, 739 (1951).

Roche and Lafon, *Bull. soc. chim. biol.*, **33**, 1437 (1951).

Walker et al., *Australian J. Sci.*, **3**, 84 (1950).

Wickström and Salvesen, *J. Pharm. Pharmacol*, **4**, 631 (1952).

Miscellaneous derivatives: Reineckates for tertiary amines, Aycock et al., *J. Am. Chem. Soc.*, **73**, 1351 (1951). Quaternary salts of tertiary amines, Marvel et al., *J. Am. Chem. Soc.*, **51**, 3638 (1929). Compounds of amines with phenols and *p*-nitrobenzyl halides, Buehler et al., *J. Am. Chem. Soc.*, **54,** 2398 (1932); Lyons, *J. Am. Pharm. Assoc.*, **21,** 224 (1932). Ethyl alkylaminomethylenemalonates, Lappin, *J. Chem. Educ.*, **28,** 126 (1951). Halogenated quinone derivatives, Buu-Hoi et al., *Rec. trav. chim.*, **71,** 1059 (1952).

Derivatives of Amides, Imides, and Ureas

Barbiturates derivatized directly: Castle and Poe, *J. Am. Chem. Soc.*, **66,** 1440 (1944).

Hydrolysis of amides with phosphoric acid: Dehn and Jackson, *J. Am. Chem. Soc.*, **55,** 4285 (1933).

Hydrolysis of arylamides without affecting alkoxyl groups: MacGregor and Wilson, *J. Soc. Dyers Colourists*, **55,** 449 (1939).

Reaction of potassium salts of imides with alkyl halides to form N-alkylimides: Gabriel and co-workers, *Ber.*, **20,** 2224 (1887); **21,** 566 (1888); **22,** 2220 (1889); **25,** 3056 (1892); **26,** 2197 (1893); **35,** 3805 (1902).

Saponification of amides and nitriles: Olivier, *Rec. trav. chim.*, **46,** 600 (1927).

Urethans and isocyanates converted to sulfamic acids: Bieber, *J. Am. Chem. Soc.*, **75,** 1405 (1953).

N-Acylphthalimides: Evans and Dehn, *J. Am. Chem. Soc.*, **51,** 3651 (1929).

m-Bromobenzazides for identification of amides: Sah and Chang, *Rec. trav. chim.*, **58,** 8 (1939).

Hydrazides for amides and ureas: Sah, *Rec. trav. chim.*, **59,** 1036 (1940).

Mercury derivatives of amides: Williams et al., *J. Am. Chem. Soc.*, **64,** 1738 (1942).

Oxalates of amides: McKenzie and Rawles, *Ind. Eng. Chem., Anal. Ed.*, **12,** 737 (1940).

N-Xanthylamides: Adriani, *Rec. trav. chim.*, **35,** 180 (1916); Kny-Jones and Ward, *Analyst*, **54,** 574 (1929); Phillips and Frank, *J. Org. Chem.*, **9,** 9 (1944); Phillips and Pitt, *J. Am. Chem. Soc.*, **65,** 1355 (1943).

Xanthyl derivatives of ureas, monosubstituted ureas, and ureides; *p*-Nitrobenzyl chloride or bromide for derivatives of barbituric acids: Jespersen et al., *Dansk Tidsk. Farm.*, **8,** 212 (1935); *C.A.*, **29,** 3459 (1935).

Sulfonylamidines: King, *J. Org. Chem.*, **25,** 352–356 (1960).

20

Derivatives of Other Nitrogen Functions

AZO COMPOUNDS—AZOXY AND HYDRAZO COMPOUNDS—ISOCYANATES AND ISOCYANIDES—NITRILES—NITRO COMPOUNDS—NITROSO COMPOUNDS

DERIVATIVES OF AZO COMPOUNDS

The identification of azo compounds is important because more than 1000 azo dyes are sold commercially, for the most part in impure form. Although the systematic identification of azo dyes is beyond the scope of this work (see the specialized treatises), the reaction by which azo compounds may be derivatized will be briefly discussed.

The first of the two equations that follow shows the conversion of azobenzene to hydrazobenzene by mild reduction, and the second represents oxidation by hydrogen peroxide to azoxybenzene:

$$C_6H_5N{=}NC_6H_5 + 2[H] \rightarrow C_6H_5NHNHC_6H_5$$
Azobenzene   Hydrazobenzene

$$C_6H_5N{=}NC_6H_5 + H_2O_2 \rightarrow C_6H_5N{=}NC_6H_5 + H_2O$$
$$\underset{\text{Azoxybenzene}}{\overset{\downarrow}{O}}$$

Although both of these methods may be used for a number of azo compounds, the general procedures that are recommended are illustrated in the equations below. If the azo compound contains an amino or hydroxy group, as is the case with most azo dyes, acetylation or benzoylation readily yields a crystalline derivative:

$$C_6H_5N{=}NC_{10}H_6OH + (CH_3CO)_2O \rightarrow$$
Benzeneazo-β-naphthol

$$C_6H_5N{=}NC_{10}H_6OCOCH_3 + CH_3COOH$$
Acetyl benzeneazo-β-naphthol

A disadvantage of this method arises from the fact that few acyl derivatives of the azo compounds are described in the literature. Therefore the most commonly used method is energetic reduction:

$$C_6H_5N{=}NC_{10}H_6OH + 2Zn + 6HCl \rightarrow$$
$$C_6H_5NH_3Cl + HOC_{10}H_6NH_3Cl + 2ZnCl_2$$

The two amino compounds produced represent (a) the original amine from which the azo compound was formed by diazotization and (b) the coupling agent, which contains an amino group in the position previously occupied by the azo group. In this example the original amine is aniline and the coupling agent β-naphthol. The hydrochlorides of the two amines may be separated by differences in solubilities; aniline hydrochloride is soluble while the aminonaphthol salt separates and is filtered. The separation may also be accomplished by steam distillation after the reduction mixture has been made alkaline; aniline, being more volatile, distils with steam.

A general method for the reduction of the azo compounds is to treat 500 mg of the purified substance with 5 ml of a 25 per cent solution of stannous chloride in concentrated hydrochloric acid. The subsequent method of separation of the two amines depends on the nature of the azo compound reduced. Usually the reduction mixture is cooled, and separated by filtration from any precipitate that separates out. The mixture is then made alkaline and steam distilled. The residue in the distilling flask is cooled and extracted with ether. The steam distillate and ether extract are used to prepare derivatives of the two amines.

Azo compounds containing naphthylamine (or naphthylaminesulfonic acids) as the coupling component yield, by energetic reduction, 1,2-diaminonaphthalene. For example, congo red on reduction yields benzidine and 1,2-diaminonaphthalene-4-sulfonic acid. The 1,2-diaminonaphthalene derivatives, when heated with phenanthrenequinone, yield characteristic colored compounds (derivatives of quinoxaline).

DERIVATIVES OF AZOXY AND HYDRAZO COMPOUNDS

Only a few azoxy and hydrazo compounds are readily available. The azoxy compounds are easily converted by reduction to hydrazo or azo compounds.

$$C_6H_5N=NC_6H_5 \atop \quad\downarrow \atop \quad O \xrightarrow[C_2H_5OH]{Zn,\ NaOH} \underset{\text{Azobenzene}}{C_6H_5N=NC_6H_5}$$

Azoxybenzene

$$CH_3C_6H_4N=NC_6H_4CH_3 \atop \qquad\quad\downarrow \atop \qquad\quad O \xrightarrow[C_2H_5OH]{Zn,\ NaOH} \underset{\textit{o}\text{-Hydrazotoluene}}{CH_3C_6H_4NHNHC_6H_4CH_3}$$

o-Azoxytoluene

The hydrazo compounds may be easily oxidized to azo compounds or often rearranged in acid media to diamino compounds.

$$\underset{\text{Hydrazobenzene}}{C_6H_5NHNHC_6H_5} \xrightarrow{[O]} \underset{\text{Azobenzene}}{C_6H_5N=NC_6H_5}$$

$$\underset{\text{Hydrazobenzene}}{C_6H_5NHNHC_6H_5} \xrightarrow{[H^+]} \underset{\text{Benzidine}}{H_2NC_6H_4\cdot C_6H_4NH_2}$$

In some cases if the hydrazo compound is stable, it may be derivatized by acylation.

DERIVATIVES OF ISOCYANATES AND ISOCYANIDES

A few of the isocyanates like phenyl isocyanate and α-naphthyl isocyanate, are among the most common reagents used for the characterization of alcohols, phenols, and amines. Isocyanides are not very common. The alkyl isocyanates and isocyanides are readily hydrolyzed to the corresponding alkylamines; for example, methyl isocyanate is hydrolyzed when boiled with sodium hydroxide solution to methylamine, and ethyl isocyanide reacts vigorously at room temperature with concentrated hydrochloric acid to give ethylamine. The hydrolysis of the aryl isocyanates to substituted ureas was discussed in the section dealing with the preparation of urethans from alcohols.

DERIVATIVES OF NITRILES

Table 20.01 gives a summary of the derivatives that have been described for the characterization of nitriles. The preparation of these derivatives involves (a) hydrolysis of the nitriles to carboxylic acids or amides and subsequent characterization of these products; (b) reduction to amines and characterization of these by conversion to substituted thioureas;

TABLE 20.01
DERIVATIVES FOR THE IDENTIFICATION OF NITRILES

*Amides	*α-Iminomercaptoacetic acid hydrochlorides
*Amines and substituted thioureas	
*Carboxylic acids	Ketones
Hydrazides	

* Recommended derivative.

(c) reaction of the nitriles with Grignard reagents to form ketones; and (d) condensation with mercaptoacetic acid to form α-iminoalkylmercaptoacetic acid hydrochloride.

Hydrolysis of Nitriles to Carboxylic Acids and Amides

Discussion

Hydrolysis of the nitriles to carboxylic acids may be effected by heating either with acids

$$RCN + H_2SO_4 + 2H_2O \rightarrow RCOOH + NH_4HSO_4$$

or with alkalies

$$RCN + NaOH + H_2O \rightarrow RCOONa + NH_3$$

The most common procedure for acid hydrolysis is to heat the nitrile with 75 per cent sulfuric acid at 160 to 190°. For semimicro quantities, a mixture of four parts phosphoric acid (85 per cent) to one part of sulfuric acid (75 per cent) gives better results. If the acid is volatile, or distils below 200° without decomposition, a small amount of water is added after the hydrolysis, and the mixture is distilled; the distillate is used for the preparation of the *p*-nitrobenzyl ester. If the carboxylic acid formed in the hydrolysis is solid, the reaction mixture is diluted with water and carefully treated with sodium hydroxide solution in order to reduce the excess acidity; the solution is then extracted with ether to obtain the solid carboxylic acid.

For alkaline hydrolysis the nitrile is heated with a solution of potassium hydroxide in diethylene glycol or glycerol; if an aqueous solution is used, the heating period should be doubled. After hydrolysis the reaction mixture is carefully acidified, and the carboxylic acid is either distilled or extracted with ether and derivatized either as the *p*-toluidide, *p*-bromobenzyl ester, or S-benzylthiuronium salt.

A number of aryl nitriles and hydroxy substituted alkyl nitriles, when treated with a small amount of concentrated sulfuric acid at temperatures ranging from 20 to 80°, undergo partial hydrolysis to form amides:

$$RCN + H_2O(H_2SO_4) \rightarrow RCONH_2$$

Since the amides are solids, it is easy to determine whether the nitrile under consideration undergoes such hydration; a few drops of the nitrile are placed in a tube and 2 drops of 95 per cent sulfuric acid are added. The mixture is stirred with a thermometer against the side of the tube; if there is an immediate reaction, the mixture solidifies instantly. If there is no immediate reaction, the mixture is warmed to 50 to 60° and, after a few minutes, cooled. The reaction mixture is treated with water rendered slightly alkaline with dilute sodium carbonate, and then filtered to obtain the amide.

The following three examples illustrate the general methods employed for acid hydrolysis and alkaline hydrolysis of nitriles and the conversion of *nitriles* to *amides*.

Example 20.01: Acid Hydrolysis of Acetonitrile

Place in an 8-in. distilling tube 4 ml of phosphoric acid (85 per cent), 2 ml of sulfuric acid (75 per cent), and 0.5 ml of acetonitrile. Add 2 boiling stones and attach to the tube a reflux condenser; boil gently for 1 hour. Add 3 ml of water and distil until 3.5 ml of distillate have been collected in an 8-in. tube, which serves as a receiver. Add 1 drop of phenolphthalein and sufficient sodium hydroxide solution (10 per cent) to develop a pink color; then make the solution just acid to phenolphthalein with 2 or 3 drops of dilute hydrochloric acid. Add 200 mg of *p*-nitrobenzyl bromide and 12 to 15 ml of methanol so that, when the solution is barely refluxing, it is homogeneous. Reflux for 2 hours. Cool for about 30 minutes and filter the crystalline mass that separates out. Recrystallize from alcohol (see page 448). The yield is 70 to 90 mg.

NOTE. When the carboxylic acid is not volatile with steam and boils above 200°, it is separated from the phosphoric-sulfuric acid mixture by extraction; the reaction mixture is first diluted with 3 ml of water, cooled, and partially neutralized with 6*N* sodium hydroxide solution. It is then extracted with three 5-ml portions of ether. If the acid is solid (from aryl cyanides), the ether is evaporated; otherwise it is used for the preparation of the *p*-toluidide or anilide. Addition of 100 mg of solid sodium chloride to the sulfuric acid mixture before boiling increases the rate of hydrolysis.

Since hydrochloric acid is more effective for many nitriles than sulfuric acid,[1] it can be employed for nitriles which are resistant to hydrolysis.

An alternative method for the characterization of the carboxylic acid is to neutralize the hydrolytic mixture and prepare the S-benzylthiuronium salt directly.

Example 20.02: Alkaline Hydrolysis of Benzonitrile

Place 4 g of glycerol, 2 g of potassium hydroxide pellets, and 0.2 ml of benzonitrile in an 8-in. test tube.

(CAUTION. *Care should be used in handling benzonitrile as it is a lachrymator.*)

Attach a condenser to the tube and boil gently for 1 hour. Dilute with 1 ml of water, cool, and add 2 ml of ether. Shake gently and allow the immiscible layers to separate. Pour off the ether from the aqueous viscous layer. Cool the tube and make the solution just acid by slow addition of 6N hydrochloric acid solution. Extract three times with 4 to 5 ml portions of ether. Distill the ether or evaporate cautiously in a dish over the steam bath and crystallize the crude benzoic acid by dissolving it in the minimum amount of 90 per cent hot methanol and adding water to the hot solution until a permanent cloudiness results. The yield is 50 to 70 mg.

NOTE. The yield of the carboxylic acid from 200 mg of benzonitrile or phenylacetonitrile is usually about 150 mg, but the product melts 2° or more below the melting point of the pure compound. When this method is used with the lower aliphatic nitriles, it is advisable to begin with 400 to 500 mg of substance.

When the carboxylic acid is not a solid the ether is transferred to a distilling tube and after removal of the solvent the procedure described in Example 12.05 (page 444) is followed to convert the acid to the *p*-toluidide.

Example 20.03: Hydrolysis of Benzonitrile to Benzamide

Place 100 mg of benzonitrile in a 6-in. tube.

(CAUTION. *Use the hood in handling benzonitrile as it has lachrymatory properties.*)

Rotate the tube in order to distribute the liquid over the lower interior surface. Use a thermometer as a stirring rod; add 5 drops of concentrated sulfuric acid and mix it with the nitrile by means of the thermometer spreading the mixture over the glass surface. The temperature rises to 40 to 50°, and the mixture sets into a solid mass. Place the tube in a water bath at 60 to 70° and allow it to stand for 2 to 3 minutes with occasional stirring. Then add 1 ml of water and cool. Remove the thermometer, add 2 ml of sodium carbonate solution (10 per cent), and shake the mixture

---

[1] M. L. Kilpatrick, *J. Am. Chem. Soc.*, **69**, 40–46 (1947).

vigorously so as to dissolve any benzoic acid that may have formed. Filter the crystals of benzamide and wash twice with water, then dry. The yield is 40 to 50 mg of crystals that melt at 127 to 128°.

Reduction of Nitriles to Amines and Their Characterization by Preparation of Substituted Thioureas

Discussion

The reduction of the nitrile to a primary amine is represented by the following equation.

$$RCN + 4(H) \rightarrow RCH_2NH_2$$

The most common method for reduction is to employ sodium in the presence of alcohol as the reducing agent. The reaction mixture is distilled to remove the alcohol. The amine is distilled and derivatized by means of phenyl isothiocyanate to yield a substituted thiourea.

The yield in the reduction of nitriles to amines is not very good since side reactions occur and hence it is necessary to use 1 to 1.5 g of the nitrile. When the quantity available is 500 mg or less, catalytic hydrogenation of the nitrile in 90 per cent ethanol containing 10 per cent of ammonium carbamate should be tried.

General Method for Reduction of Nitriles

An 8-in. distilling tube provided with a reflux condenser is charged with 10 ml of absolute ethanol and 0.5 g of an aliphatic nitrile or 0.75 g of an aromatic nitrile. About 0.75 g of finely cut sodium is added at such a rate that the reaction proceeds vigorously under control. To add the sodium, momentarily lift the cork if a microcondenser is employed or insert it through the top of the condenser if a Liebig-type condenser is used.

After the reaction is complete (15 minutes) the mixture is cooled and 5 ml of concentrated hydrochloric acid are added dropwise with stirring of the contents of the tube. Test the reaction mixture and if it is not distinctly acid add a small amount more of hydrochloric acid. The tube is arranged for distillation, two boiling stones are added, and the mixture is distilled until 10 ml of distillate (ethanol) have been collected, which are set aside. The flask is cooled by an ice-salt mixture and 7 ml of 40 per cent sodium hydroxide are added in small amounts. The temperature of the mixture should not be allowed to rise above 30°. The tube is arranged for distillation; the receiving 8-in. tube is charged with 2 ml of water and 1 ml of 5N hydrochloric acid; the delivery tube is arranged so that it just dips into the dilute hydrochloric acid. The distillation of the amine-water mixture is continued until the volume of the distillate is 7 to 8 ml.

The distillate in the tube is cautiously neutralized by dropwise addition of sodium hydroxide solution to pH 8, and 0.25 to 0.30 ml (12 to 14 drops) of phenyl isothiocyanate is added. The amine is converted to the substituted phenylthiourea according to the procedure described on page 600.

Other Derivatives of Nitriles

α-Iminoalkylmercaptoacetic Acid Hydrochlorides

In the presence of hydrogen chloride, a nitrile undergoes thioalcoholysis to form a compound analogous to the one obtained by treatment of the nitrile with alcohol and hydrogen chloride (imino ester hydrochloride):

$$RCN + HSCH_2COOH + HCl \rightarrow RC\begin{matrix}NH \cdot HCl \\ \diagdown \\ SCH_2COOH\end{matrix}$$

$$\text{Thioglycolic acid} \quad\quad\quad\quad\quad \text{α-Iminoalkylmercaptoacetic}$$
$$\text{(Mercaptoacetic acid)} \quad\quad\quad\quad\quad \text{acid hydrochloride}$$

The nitrile with twice its weight of thioglycolic acid (mercaptoacetic acid) is dissolved in a small amount of dry ether and saturated in the cold with dry hydrogen chloride. The crystals of the derivative that separate on standing are filtered and dried in a vacuum. The decomposition point of the derivative is determined by the standard method used for melting points.

The general method for the preparation of the derivative is to place 0.5 g of the nitrile with 1.0 g of thioglycolic acid and 7 ml of absolute ether in a 6-in. tube and saturate the mixture with *dry* hydrogen chloride while the tube is immersed in an ice-salt mixture. After complete saturation the tube is stoppered and kept in the ice bath until crystals separate out, which may take from 1 hour with aliphatic nitriles to 24 hours with aromatic nitriles. The crystals are removed by filtration or centrifugation, washed several times with absolute ether, and dried in a vacuum desiccator (Figure 3.13). The melting points are really decomposition points; however, additional confirmation may be obtained by determining the neutralization equivalent by titration with $0.1N$ sodium hydroxide using thymol blue as an indicator.

Ketones from Nitriles by Use of the Grignard Reagent

The addition of a nitrile to a Grignard reagent is followed by decomposition and hydrolysis to form a ketone, which may be converted to the semicarbazone, as shown, or to other carbonyl derivatives:

$$\text{RCN} + \text{R}'\text{MgX} \rightarrow \underset{\underset{\text{R}'}{|}}{\text{RC}}{=}\text{NMgX}$$

$$\underset{\underset{\text{R}'}{|}}{\text{RC}}{=}\text{NMgX} + 2\text{HX} + \text{H}_2\text{O} \rightarrow \text{RCOR}' + \text{NH}_4\text{X} + \text{MgX}_2$$

$$\text{RCOR}' + \text{H}_2\text{NNHCONH}_2 \rightarrow \underset{\text{Semicarbazone}}{\text{RR}'\text{C}{=}\text{NNHCONH}_2} + \text{H}_2\text{O}$$

The Grignard reagent recommended in the literature for this reaction is phenylmagnesium bromide in the ratio of 4 moles of the Grignard reagent to 1 mole of the nitrile. The original paper should be consulted for details.[2]

Ketones from Nitriles Followed by Condensation

By means of phloroglucinol in the presence of a condensing agent (Hoesch synthesis), the nitrile may be converted to a ketone,

$$\text{C}_6\text{H}_2(\text{OH})_3\text{COR}.$$

From the point of view of derivatization, this reaction is analogous to the Grignard method of converting the nitrile, RCN, to the ketone, RCOR. The nitrile and phloroglucinol are dissolved in dry ether; anhydrous zinc chloride is added, and dry hydrogen chloride is passed in for about 30 minutes. The oil that separates from the reaction mixture is taken up with water and the aqueous layer is concentrated by heating. The alkyl trihydroxyphenyl ketone separates on cooling. The method is applicable to quantities of about 500 mg of the nitrile; however, only six such derivatives have been described. The original literature should be consulted for details (see the References at the end of this chapter).

DERIVATIVES OF NITRO COMPOUNDS

The reactions by which nitro compounds may be derivatized are: (a) reduction to amines which are then converted to N-substituted benzamides, arenesulfonamides, or substituted phenylthioureas; or (b) further nitration, or introduction of other substituents. For example, *p*-nitrotoluene may be reduced to *p*-toluidine, nitrated to 2,4-dinitrotoluene, or oxidized to *p*-nitrobenzoic acid. In general, most aromatic mononitro compounds may be converted to dinitro or trinitro derivatives; in addition,

[2] R. L. Shriner and T. A. Turner, *J. Am. Chem. Soc.*, **52**, 1267 (1930).

other substituents already present may be altered; for example, the aryl nitro compound may be brominated or an alkyl side chain may be oxidized. The selection of the derivative should be the result of a judicious consideration of all the factors involved. The example of p-nitrotoluene may be further considered as an illustration; if p-nitrotoluene is reduced to p-toluidine, the melting point of the amine is low (45°) and must be acylated for identification. The second alternative is to oxidize p-nitrotoluene to p-nitrobenzoic acid; the disadvantage of this method lies in the high melting point of the acid (242°). The third alternative, which is selected as a trial, is the preparation of 2,4-dinitrotoluene. The ease of nitration and purification of the dinitro compound as compared with the preparation of other possible derivatives are the factors that suggest this selection.

The derivatization of dinitro, trinitro, and in general polynitro compounds must be considered individually for each compound. Addition compounds of polynitro derivatives often prove desirable derivatives. At least two nitro groups on each benzene ring are required for the formation of addition compounds. The relative position of the nitro groups and the nature of other substituents present in the ring also have an influence on the formation of the addition compound. Nitro groups *ortho* to each other and methyl groups situated between nitro groups appear to hinder addition compound formation. α-Naphthol is stated in the literature as having a greater tendency to form addition compounds than naphthalene; the latter, however, is the reagent recommended because it is easily available in the pure form.

Nitro compounds are readily separated by column adsorbents.[3] As has been discussed under the derivatives of alcohols, carbonyl compounds, and carboxylic acids, the chromatographic separation of their polynitro derivatives is feasible, since the spots can be readily detected without the application of indicators. Therefore the separation and characterization of micro quantities of nitro compounds can be undertaken after nitration to give dinitro and trinitro compounds as described in this section. For example, a known mixture of a few milligrams of the three xylenes can be nitrated by means of 100 per cent nitric acid and 100 per cent sulfuric acid for a sufficient length of time to insure the formation of the trinitro derivatives of the m- and p-xylenes and the dinitro derivative of the o-xylene. A few micrograms of the nitrated mixture is chromatographed alongside a sample of a mixture containing known amounts of xylenes nitrated in the same manner. The developed chromatogram gives information about the

---

[3] E. Lederer and M. Lederer, *Chromatography*, Second edition, Elsevier, New York, 1957 p. 200.

Derivatives of Other Nitrogen Functions 625

isomers present and the approximate amount of each. References to the derivatives of nitro compounds are given on page 630.

Derivatization of Nitro Compounds by Reduction to Amines

Discussion

The reduction of nitro compounds to amines may be accomplished by: (a) metals, such as tin or zinc, in acid media; (b) catalytic hydrogenation; or (c) ions, such as S^{-2} and $S_2O_4^{-2}$.

Tin and hydrochloric acid are most commonly used for reduction in acid media. After the reduction of the nitro compound the solution is rendered alkaline and the amine is extracted with ether. If the product is a lower alkylamine, the alkaline solution is distilled, and the distillate is collected in a small amount of dilute acid. The distillate is used directly for the preparation of derivatives by means of acylation. In some cases it is possible to reduce the nitro compound to the substituted hydroxylamine, which can be identified by its melting point or by derivatization.

Catalytic hydrogenation is advisable when the quantity of the nitro compound available is less than 100 mg. The general method is given on pages 40–45, and several examples of its application are described in this section.

The details of all methods of reduction are described in the following examples.

Example 20.04: Reduction by Metals in Acid Media: Reduction of 1 Nitropropane to n-Propylamine

Place in an 8-in. test tube 200 mg of 1-nitropropane and 3 ml of 6N hydrochloric acid solution. Add 500 mg of tin in two portions over a period of 10 minutes, warming at first to start the reaction. Boil the mixture gently under reflux for 30 minutes or until the odor of the nitro compound has disappeared. Cool the mixture by immersion of the tube in running tap water and add slowly 6 ml of 6N sodium hydroxide. Transfer the contents of the tube to an 8-in. distilling tube; wash the vessel with 1 to 2 ml of water and unite the washings with the alkaline mixture. Add 2 boiling stones and distil the alkaline solution until 4 ml of distillate have been collected in a receiving tube containing 2 ml of 6N hydrochloric acid and 1 drop of aqueous methyl red or methyl orange solution. If the distillate becomes alkaline, a small amount of additional hydrochloric acid is added. Add to the distillate 0.4 ml of benzoyl chloride and then, while the tube is cooled in tap water, 8 ml of 6N sodium hydroxide solution. The tube is closed with a solid rubber stopper and shaken vigorously at intervals for 10 minutes. An oil separates that, on cooling and shaking, solidifies. Then proceed to prepare the benzoyl derivative as directed in Example 19.03 (page 594). The yield is 30 to 35 mg of the pure derivative, melting at 84°.

NOTE. If the amine boils much above 100°, it is best to extract the alkaline solution with ether. In such a case the acid solution that contains the amine salt is made alkaline, care being taken not to use a great excess of alkali; it is then extracted with 3 portions of 4 to 5 ml of ether (which is free from alcohol), and this extract is used directly for the preparation of the derivative.

Example 20.05: Catalytic Reduction: Reduction of Nitrobenzene to Aniline

The general method described on pages 40–45 is employed. About 200 mg of nitrobenzene in 15 to 20 ml 95 per cent ethanol and 100 mg of 5 per cent palladium carbon are used. The reduction is complete in 10 to 15 minutes. The alcohol is poured into a dish and neutralized with glacial acetic acid. After the alcohol has been evaporated, a mixture of 1 ml of acetic anhydride and 2 ml of glacial acetic acid is added to the residue. The mixture is boiled gently over a small direct flame for about 3 minutes. Water (5 ml) is added, and then, dropwise, $6N$ sodium hydroxide solution, until the reaction is neutral or only slightly acidic. After cooling, the crystals of acetanilide are filtered off. No further purification is required.

NOTE. *m*-Dinitrobenzene is readily reduced to the diamine by the same method; after reduction the solution is neutralized with hydrochloric acid to convert the amine to the hydrochloride and evaporated, then converted to the solid amine with a few drops of ammonium hydroxide. Similarly, *o*- and *p*-nitrophenol are reduced to the aminophenols by using the same procedure and quantities as in the reduction of nitrobenzene. After the reduction has passed the midpoint, there is a tendency for the disperser to clog (see page 43).

Other Derivatives of Nitro Compounds

Oxidation of Side Chains

Acid dichromate or alkaline permanganate solution may be used for the oxidation of the side chains of aromatic hydrocarbons to the carboxylic stage. Generally, permanganate oxidation is preferred for the more resistant side chains—that is, in the presence of nitro groups, or whenever extensive degradation by oxidation is necessary. The dichromate and permanganate procedures are illustrated by the following examples.

Derivatives of Other Nitrogen Functions

Example 20.06: Permanganate Oxidation of o-*Nitrotoluene to* o-*Nitrobenzoic Acid*

Place 1.5 g of solid potassium permanganate, 25 ml of water, 0.5 ml (5 to 8 drops) of 6N sodium hydroxide, and 2 boiling stones in an 8-in. tube arranged for heating under reflux. Lastly, add 400 to 500 mg of o-nitrotoluene and boil gently for 1 hour or longer until the purple color of the permanganate has disappeared. Cool the reaction mixture and carefully acidify with dilute sulfuric acid; then heat to boiling. If there is an appreciable amount of manganese dioxide present, add a small amount of solid sodium bisulfite. Cool and filter the acid; recrystallize from 4 to 5 ml of hot alcohol. Filter the hot alcoholic solution and add water dropwise until a permanent cloudiness results. Cool, and filter off the crystals. The yield is 300 to 400 mg.

NOTE. *m*- and *p*-Nitrotoluene are oxidized to the respective nitrobenzoic acids by the same method. Similarly, *o*-, *m*-, and *p*-xylene are oxidized, respectively, to phthalic, isophthalic, and terephthalic acids. In the case of xylenes, since there are two side chains to be oxidized, the amount of all reagents except the hydrocarbon is doubled. A 125-ml Erlenmeyer flask is used for the boiling vessel, and the mixture is heated for 1.5 to 2 hours.

Example 20.07: Dichromate Oxidation of p-*Nitrotoluene to* p-*Nitrobenzoic Acid*

In an 8-in. tube dissolve 1 g of sodium dichromate in 3 ml of water and add 2 ml of concentrated sulfuric acid. Add 200 to 250 mg of *p*-nitrotoluene and 2 boiling stones. Boil for 20 to 30 minutes. Cool and add 2 to 3 ml of water; then filter. Wash three times with water. Recrystallize from hot methanol. Filter the hot methanol solution and add water to the filtrate until a permanent cloudiness results. Cool, and filter off the crystals. The yield is 180 to 230 mg.

Preparation of Derivatives of Mononitro Compounds by Further Nitration

The discussions of nitration on pages 559–564 and 578–580 should be reviewed. The derivatization of *o*- and *p*-nitrotoluenes by further nitration is based on the fact that both yield the same derivative, 2,4-dinitrotoluene.

Example 20.08: Nitration of o-*Nitrotoluene or* p-*Nitrotoluene to* 2,4-*Dinitrotoluene*

The general method on page 560 is employed. Use 200 mg of the mononitro compound. In the case of *p*-nitrotoluene, after one crystallization 50 to 70 mg of the pure dinitro compound (mp 70°) are obtained. In the case of *o*-nitrotoluene, two crystallizations are required, and the yield of the pure derivative is 40 to 50 mg.

A number of polynitro compounds with substituents other than nitro groups may be derivatized by alteration of their substituents. For example 2,4-dinitrochlorobenzene may be converted by hydrolysis to 2,4-dinitrophenol; similarly, 2,4,6-trinitroanisole and 2,4,6-trinitrophenetole may be hydrolyzed to give picric acid. Oxidation of methyl groups is feasible, although such oxidation should be done **with care even with semimicro quantities;** for example, 2,4,6-trinitrotoluene may be converted by oxidation to 2,4,6-trinitrobenzoic acid, but this is not recommended unless precautions are used.

The polynitro compounds may be derivatized by preparation of adducts with α-naphthol or naphthalene. To prepare such adducts with naphthalene, equimolecular amounts of the polynitro compound and the reagent are heated cautiously until a homogeneous melt is obtained. The melt is cooled, recrystallized from alcohol, dried rapidly, and the melting point determined.

DERIVATIVES OF NITROSO COMPOUNDS

The derivatives to which nitroso compounds may be converted for characterization are: (a) amines, (b) hydroxylamines, (c) azo compounds, and (d) hydrazines. The first three are suitable for the C-nitroso compounds, while the hydrazines are derived from N-nitroso compounds.

Nitroso compounds are converted to amines by reduction:

$$ArNO + 4[H] \rightarrow ArNH_2 + H_2O$$
Arylamine

The reduction may be accomplished either by tin and hydrochloric acid or by catalytic hydrogenation. The latter is to be preferred when the amount available is very small.

Lithium aluminum hydride reduces nitroso compounds to azo compounds, and if the reagent is available it is recommended for reduction of small quantities of material. Otherwise the method by which azo compounds are derived from nitroso is by reaction with an arylamine:

$$ArNO + Ar'NH_2 \rightarrow ArN{=}NAr' + H_2O$$
Azo compound

Thus, by reaction of aniline and nitrosobenzene, orange-red crystals of azobenzene (mp 68°) are formed. The rate at which the azo compound is formed varies. For example, *p*-iodonitrosobenzene requires several days to react with *p*-aminobenzoic acid to give *p*-iodophenylazobenzoic acid. *p*-Bromoaniline is the most suitable arylamine for reaction with nitroso compounds, giving substituted *p*-bromoazobenzenes.

A nitroso compound may be reduced to the *hydroxylamine* by means of zinc dust in the presence of ammonium chloride or calcium chloride:

$$ArNO + 2[H] \rightarrow ArNHOH$$
Arylhydroxylamine

The nitroso compound is dissolved or suspended in a mixture of alcohol and water, and zinc dust is added, keeping the temperature at 40 to 50°. The hydroxylamine separates on cooling the filtrate of the reaction mixture.

The N-nitrosoamines, such as R_2NNO, Ar_2NNO, and ArRNNO, are derivatized by reduction to hydrazines, which in turn are converted to hydrazones through reaction with carbonyl compounds:

$$R_2NNO + 4[H] \rightarrow R_2NNH_2 + H_2O$$

The yields of hydrazines by reduction of the nitrosoamines using semimicro quantities are extremely poor, and hence, if the hydrazine is not obtained on the first trial, it is advisable to use energetic reduction by means of zinc and hydrochloric acid to convert the nitrosoamine to the corresponding secondary amine, which is then derivatized. For example, N-nitrosodiphenylamine, $(C_6H_5)_2NNO$, and N-nitrosodi-*n*-butylamine fail to give appreciable quantities of the corresponding hydrazines when reduced by means of zinc dust and acetic acid; a further difficulty arises in that many of the derivatives (hydrazones) of the substituted hydrazines are not described in the literature. For these reasons it seems preferable to reduce the nitrosoamine to the corresponding secondary amine and derivatize the latter.

For reduction of the nitroso compounds to amines by catalytic hydrogenation or by the action of acids and metals the same general procedures described for nitro compounds on pages 40–45 should be followed, including the derivatizing of the amines.

For reduction to azo compounds by lithium aluminum hydride the citations given in the references should be consulted.

REFERENCES

Identification of Azo, Azoxy, and Hydrazo Compounds

Identification of reduction products of azo dyes: Ueno, *Bull. Inst. Phys. Chem. Research (Tokyo)*, **7**, 49–88 (1928); *C.A.*, **22**, 3399 (1928).

Reduction of amidine by iodine: Stephen, *Anal. Chem.*, **24**, 180 (1952).

Reduction of azoxy compounds to azo with lithium aluminum hydride : Nelson and Laskowski, *Anal. Chem.*, **23**, 1495 (1951).

Separation of azo compounds by paper chromatography: Emery et al., *Stain Technol*, **27**, 21 (1952); *C.A.*, **46**, 5650 (1952); Graichen. *J. Assoc. Offic. Agr. Chem.*, **34**, 795 (1951); Hough et al., *J. Chem. Soc. (London)*, **1949**, 2511; Lederer, *Science*, **112**, 504 (1950); Zalokar, *J. Am. Chem. Soc.*, **74**, 4213 (1952).

Substituted benzotriazoles of *o*- and *p*-semidines: Witt and Schmidt, *Ber.*, **25**, 1017 (1892).

Identification of Nitriles

Addition compounds of nitriles with mercaptoacetic acid: Condo, Hinkel, Fassero, and Shriner, *J. Am. Chem. Soc.*, **59**, 230 (1937).

Hydrolysis of nitriles with potassium hydroxide in diethylene glycol or glycerol: Hovira and Palfray, *Compt. rend.*, **211**, 396 (1940).

Reduction to amines: Cutter and Taras, *Ind. Eng. Chem., Anal. Ed.*, **13**, 830 (1941).

Alkyl 2,4,6-trihydroxyphenyl ketones: Howells and Little, *J. Am. Chem. Soc.*, **54**, 2451 (1932).

Alkyl phenyl ketones: Shriner and Turner, *J. Am. Chem. Soc.*, **52**, 1267 (1930).

Identification of Nitro and Nitroso Compounds

Identification of nitro compounds by catalytic hydrogenation at atmospheric pressure: Cheronis and Koeck, *J. Chem. Educ.*, **20**, 488 (1943); Cheronis and Levin, *Chem. Educ.*, **21**, 603 (1944).

Identification of polynitro compounds as addition compounds: Asahina and Shinomiya, *J. Chem. Soc. Japan*, **59**, 341 (1938); Dermer and Smith, *J. Am. Chem. Soc.*, **61**, 748 (1939); Shinomiya, *Bull. Soc. Japan*, **15**, 92 (1940).

Derivatives of nitroparaffins: Dermer and Hutcheson, *Proc. Okla. Acad. Science*, **23**, 60 (1943); *C.A.*, **38**, 2008 (1944).

Miscellaneous methods and derivatives: Reduction of nitriles, nitro, nitroso, and azo compounds by means of lithium aluminum hydride, Brown in Adams (ed.), *Organic Reactions*, Vol. VI, Wiley, New York, 1951, pp. 469–509; Cheronis, *Micro and Semimicro Methods*, Vol. VI of Weissberger (ed.), *Technique of Organic Chemistry*, Interscience, New York, 1954, pp. 252–256. Phenols from azoxy compounds, Gaudry and Kirstad, *Can. J. Research*, **27**, 897 (1949). Identification of nitroso compounds by use of *p*-bromoaniline, Levy and Campbell, *J. Chem. Soc. (London)*, **1939**, 1442. Conversion of nitriles to methylenebisamides, Margat et al., *J. Am. Chem. Soc.*, **73**, 1028 (1951).

21

Derivatives of Sulfur Functions

SULFONAMIDES—SULFONYL CHLORIDES—SULFONIC ACIDS—THIOCYANATES AND ISOTHIOCYANATES—THIOETHERS—THIOLS

Table 21.01 lists the derivatives that have been described in the literature for the characterization of sulfonamides, sulfonyl chlorides, sulfonic acids, isothiocyanates, thioethers, and thiols. For the microchemical detection and characterization of sulfonamides and the microscopic and chromatographic identification of many sulfonic acids the literature listed in the references at the end of this chapter should be consulted.

DERIVATIVES OF SULFONAMIDES

The hydrolysis of a sulfonamide will yield a sulfonic acid and ammonia or an amine:

$$RSO_2NH_2 + HOH \xrightarrow{HCl} RSO_2OH + NH_4Cl$$

$$RSO_2NHR' + HOH \xrightarrow{HCl} RSO_2OH + R'NH_3Cl$$

The hydrolysis is effected by heating with 25 per cent hydrochloric acid, 80 per cent sulfuric acid, or a mixture of 85 per cent phosphoric acid and 80 per cent sulfuric acid. In the case of a substituted sulfonamide, the amine may be separated by making the solution alkaline and distilling, if

TABLE 21.01
DERIVATIVES FOR THE IDENTIFICATION OF SULFUR FUNCTIONS

Sulfonamides:

*Hydrolysis and characterization of the amine and sulfonic acid
Monoacetyl and diacetyl derivatives
*N-Sulfonylphthalimides
*N-Xanthylsulfonamides

Sulfonyl chlorides:

*Sulfonamides
*Sulfonanilides
*Sulfontoluidides

Sulfonic Acids:

*S-Benzylthiuronium salts
Chlorosulfonanilides and chloroarylsulfonanilides
N-(p-Nitrobenzyl)pyridinium salts
*Phenylhydrazine and pyridine salts

*Sulfonamides and sulfon-α-naphthylamides
*p-Toluidine and other arylamine salts
Xanthydrol derivatives

Isothiocyanates:

*Substituted thioureas
Thiosemicarbazides

Thioethers (Sulfides):

*Sulfones

Thiols (Mercaptans and Thiophenols):

Anthraquinonyl thioethers
*2,4-Dinitrophenyl thioethers
*3,5-Dinitrobenzoyl thioesters
Nitrosylmercaptides
Saccharin derivatives

* Recommended derivative.

the amine is volatile, or extracting with an appropriate solvent. Thus it is possible to identify both the amine and the sulfonic acid.

Unsubstituted sulfonamides, RSO_2NH_2, may be reacted either with phthaloyl chloride to give N-sulfonylphthalimides or with xanthydrol to form N-xanthylsulfonamides.

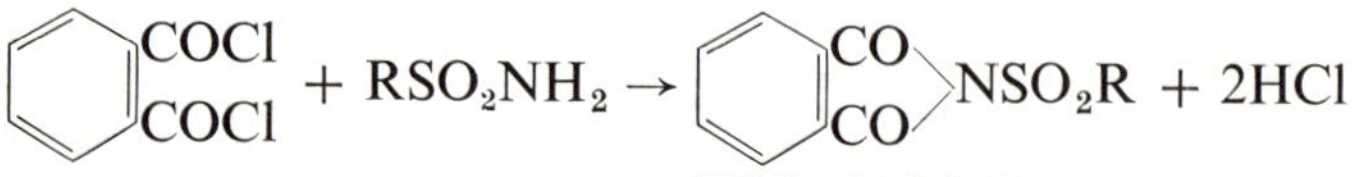

$$C_6H_4(COCl)_2 + RSO_2NH_2 \rightarrow C_6H_4(CO)_2NSO_2R + 2HCl$$
N-Sulfonylphthalimide

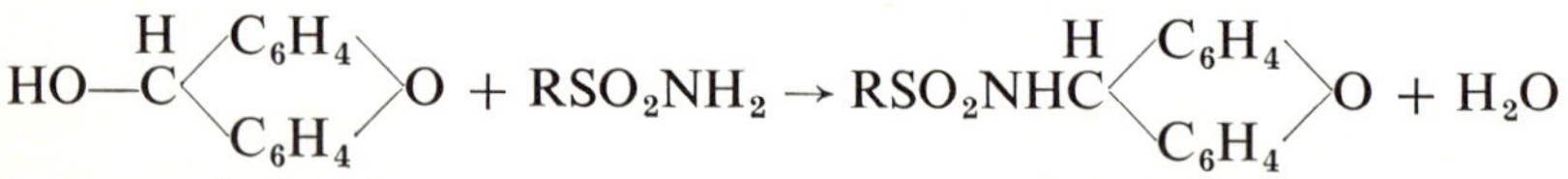

$$HO-CH(C_6H_4)_2O + RSO_2NH_2 \rightarrow RSO_2NHCH(C_6H_4)_2O + H_2O$$
Xanthydrol   N-Xanthylsulfonamide

The preparation of N-xanthylsulfonamides may be applied successfully, using semimicro quantities; however, only about a dozen derivatives have been reported, and the method is not successful for benzenoid amides that contain branched alkyl groups on the ring. The alkylation of sulfonamides that have amino hydrogen atoms has been used to prepare derivatives. Either alkyl halides, such as methyl iodide and ethyl bromide, or alkyl sulfates may be used for the alkylation.

The chlorination and the preparation of N-acetyl and diacetyl derivatives[1] should be considered for the characterization of small quantities of sulfonamides. For example 1 to 2 mg of sulfanilamide can be derivatized rapidly on a glass slide and its melting point determined.[2] Chlorination and acylation are illustrated in Examples 21.03 and 21.04.

The general methods for hydrolysis and preparation of N-xanthyl-sulfonamides and acetyl derivatives are illustrated by the following examples.

Example 21.01: Hydrolysis of Sulfonamide

Place in a distilling tube 1 ml of concentrated sulfuric acid (sp. g. 1.84) and add cautiously 5 drops of water and 1 ml of 85 per cent phosphoric acid in the order given. Add 500 to 800 mg of the sulfonamide, place a thermometer in the tube, and heat gradually until the temperature reaches 160°. Keep the temperature at 155 to 165° for about 5 to 10 minutes or until the sulfonamide has completely dissolved. Cool the dark viscous solution and add to it 6 ml of water. While the mixture is being cooled, add slowly 25 to 30 per cent sodium hydroxide solution until the solution is distinctly alkaline.

If the amine resulting from the hydrolysis of the sulfonamide is volatile, the tube is arranged for distillation, and, after addition of boiling stones, the alkaline mixture is distilled until the volume is reduced to one half; after 8 ml of distillate have been collected, a drop is collected separately and tested with litmus or pH paper; if it is alkaline, the distillation is continued; otherwise it is discontinued. The distillate may be used directly for benzoylation as described on page 593, or, if the free amine is desired, it may be extracted with ether; the extract, after drying with a few pellets of sodium hydroxide, is distilled to remove the solvent.

If the amine resulting from the hydrolysis of the sulfonamide has a low volatility, it may be extracted with ether directly from the cold alkaline solution. The ether extract is contaminated with hydrolytic decomposition products; therefore, after the evaporation of ether, the tarry mass is boiled with 3 ml of water, 1 ml of 6*N* hydrochloric acid solution, and a small amount of charcoal and filtered. The filtrate is made alkaline, and the amine is extracted or derivatized. The residue remaining in the distilling tube is poured into an evaporating dish, treated with 100 mg of

---

[1] M. L. Crossly, E. H. Northey, and M. E. Hultguist, *J. Am. Chem. Soc.*, **61**, 2950–2955 (1939).

[2] N. D. Cheronis, *Micro and Semimicro Methods*, in *Technique of Organic Chemistry*, Vol. 6, A. Weissberger, ed., Interscience, New York, 1954, p. 305.

charcoal, evaporated to about 4 to 5 ml, and filtered while hot. The solution is carefully neutralized and then used for the preparation of the arylamine salt or S-benzylthiuronium derivative (Example 21.06).

Example 21.02: Preparation of N-Xanthylsulfonamides

Place in an 8-in. tube 10 ml of glacial acetic acid, 200 mg of xanthydrol, and 200 mg of the sulfonamide. Insert a clean solid rubber stopper, and shake the mixture for 2 to 3 minutes. Filter the solid and crystallize from dioxane-water (3:1 mixture). One crystallization is usually sufficient.

NOTE. The following twelve sulfonamides have been derivatized, using xanthydrol as a reagent [the number following the abbreviation of the sulfonamide is the melting point (uncorrected)]: Benzenesulfonamide, 200–200.5; 2-Me-, 182–183.5; 4-Me-, 197–197.5; 4-Et-, 196; 4-n-Pr-, 200; 4-n-Bu-, 186; 4-n-Am-, 165; 3,4-di-Me-, 190; 2,4-di-Me-, 188; 2,5-di-Me-, 176; 2,4,6-tri-Me-, 204; 4-NH_2-, 208; Saccharin-, 199.

Example 21.03: Chlorination of Sulfanilamide

Place in an 8-in. test tube 430 mg (2.5 mM) of sulfanilamide and 15 ml of 6N hydrochloric acid and, after complete solution, add 2 ml of 30 per cent hydrogen peroxide (20 mM) and let the tube stand at room temperature for 30 minutes. The solid 3,5-dichloro-4-aminobenzenesulfonamide is filtered, washed, and crystallized from 125 ml of boiling water. The crystals melt at 205 to 205.5°.

Example 21.04: Acetylation of Sulfanilamide

Place 50 mg of sulfanilamide in a 3-in. tapered tube, add 5 drops of acetic anhydride, and let stand at room temperature for 5 minutes. Add 1 ml of water and a few drops of 1N sodium hydroxide solution to about pH 5 to 6 and then centrifuge the tube and remove the supernate by means of a capillary pipet. Recrystallize according to the procedure described on page 71 from an alcohol-water mixture. The crystals of the monoacetyl derivative melt at 214°.

DERIVATIVES OF SULFONYL CHLORIDES

The sulfonyl chlorides are easily converted to the sulfonamides. They may be also derivatized by preparing the sulfonanilides or sulfon-p-toluidides. The procedures are described in a following section (pages 637–640).

DERIVATIVES OF SULFONIC ACIDS

The three most important derivatives for the characterization of sulfonic acids are: (a) the S-benzylthiuronium derivatives prepared by reacting the alkali sulfonate with S-benzylthiuronium chloride; (b) salts formed by the reaction of the sulfonic acid and an arylamine such as aniline or one of the toluidines; and (c) sulfonamides or N-α-naphthylsulfonamides formed by conversion of the sulfonic acid to the sulfonyl chloride followed by reaction with ammonia or naphthylamine.

Other derivatives that have been proposed for the characterization of sulfonic acids are listed in Table 21.01.

S-Benzylthiuronium Derivatives of Sulfonic Acids

Discussion

The equation below represents the formation of the S-benzylthiuronium derivative:

$$C_6H_5CH_2SC(NH_2)_2Cl + RSO_2ONa \rightarrow$$
$$\text{S-Benzylthiuronium chloride}$$

$$C_6H_5CH_2SC(NH_2)_2OSO_2R + NaCl$$
$$\text{S-Benzylthiuronium derivative}$$

A number of sulfonic acid derivatives suitable for characterization work are prepared by reacting the alkali sulfonate with S-benzylthiuronium chloride. The reagent is easily prepared by refluxing an alcoholic solution of benzyl chloride and thiourea. A concentrated neutral solution of the sodium or potassium salt of the sulfonic acid to be derivatized is added with stirring to a slight excess of the reagent dissolved in water; the method is satisfactory for mono- and disulfonic acids if other functional groups are absent. The presence of hydroxy or amino groups is disadvantageous.

A number of important substituted naphthalene sulfonic acids may be identified in micro quantities by microscopic examination of the benzoyl or S-benzylthiuronium derivatives.

The preparation of the reagent and of the derivatives is described in the following examples.

Example 21.05: Preparation of S-Benzylthiuronium Chloride

If S-benzylthiuronium chloride is not available for the preparation of the derivatives it may be prepared conveniently by heating for 20 to 30 minutes under reflux 2 g of benzyl chloride, 1.2 g of thiourea, and 3 ml of methanol. The pale yellow solution is cooled in an ice-water mixture and the mass of crystals is filtered by suction, then washed twice with 1-ml portions of ethyl acetate. The product is dried rapidly by pressing between filter papers and placed in a stoppered tube. The yield is 2.5 to 3.0 g.

Example 21.06: Preparation of S-Benzylthiuronium Derivatives of Sulfonic Acids

Dissolve 200 mg of the sodium or potassium salt of the sulfonic acid in the minimum amount of water; in the case of the free sulfonic acid, dissolve 200 mg in dilute sodium hydroxide solution (0.5 ml of 10 per cent sodium hydroxide solution and 1 ml or more of water). Add a drop of phenolphthalein and neutralize the excess sodium hydroxide by addition of dilute hydrochloric acid solution. Prepare separately in a test tube a water solution of 250 mg of S-benzylthiuronium chloride for each acidic group present in the sulfonic acid molecule of the 200-mg sample (for example, use 250 mg of the reagent if the acid taken is 1-naphthalenesulfonic acid, but use 500 mg of the reagent if the acid taken is naphthalene-2,7-disulfonic acid). Cool both solutions and mix by adding the sulfonic salt solution to the reagent slowly with shaking. If this procedure fails to give the derivative, dissolve the benzylthiuronium chloride in sufficient hot alcohol to give a 15 per cent solution and add to this the sulfonic acid salt solution. The derivative is filtered, washed with water, and recrystallized by dissolving in the minimum amount of hot alcohol; add water dropwise until a permanent cloudiness results. Dry the crystals rapidly by pressing between filter papers or by placing in a vacuum desiccator. The derivatives often develop an offensive odor owing to the formation of α-toluenethiol (mercaptan) by decomposition of the benzylthiuronium chloride.

Arylamine Salts of Sulfonic Acids

Discussion

The formation of an arylamine salt of the sulfonic acid is represented by the following equation.

$$\underset{\text{Sodium sulfonate}}{RSO_2ONa} + \underset{\text{Arylammonium salt}}{[ArNH_3^+]Cl^-} \rightarrow \underset{\text{Arylammonium sulfonate}}{[ArNH_3^+]OSO_2R^-} + NaCl$$

The arylamines that have been proposed in the literature for the identification of sulfonic acids are aniline, *o*-toluidine, *p*-toluidine, pyridine, and phenylhydrazine. The first three arylamines are recommended. The salts are easily prepared by heating together an aqueous solution of the free acid or of the alkali salt, a slight excess of the amine, hydrochloric acid, and enough water to bring all of the material into solution at the boiling point. The salt separates out on cooling and, after filtration, it is recrystallized from 1 per cent acetic acid to minimize hydrolysis. Aromatic aminosulfonic acids must first be acetylated in order to form the arylamine salt. An alternative method is to remove the amino group; the amino sulfonic acid is diazotized, and the diazo group is replaced by chlorine through the Sandmeyer reaction; this last method is suitable for about 1 gram of material and should not be tried with less than 500 mg.

Example 21.07: Preparation of Arylammonium Salts of Sulfonic Acids

The following directions apply to the preparation of the sulfonates of aniline, *o*-toluidine, and *p*-toluidine.

Dissolve about 200 mg of the sodium salt of the sulfonic acid in water in an 8-in. tube; if the free sulfonic acid is available, use the same amount and dissolve it in the minimum amount of water or dilute sodium hydroxide. For the barium salt of the sulfonic acid, use 300 mg and boil it with 2 ml of water and 1 ml of $6N$ sulfuric acid; add a minute amount of charcoal or filter-aid and filter the hot solution to remove the barium sulfate.

To the solution of the alkali sulfonate or free sulfonic acid, add 300 mg of the arylamine (aniline, *o*-toluidine, or *p*-toluidine), 1 to 2 ml of $6N$ hydrochloric acid, and enough water to bring all the material into solution at the boiling point. Add about 50 to 100 mg of charcoal, filter the hot solution, and cool. Filter the arylammonium sulfonate and recrystallize to constant melting point from 1 per cent acetic acid.

NOTE. The arylammonium salt should be thoroughly dried before the melting point is determined. When the salt melts above 180°, the sample may be dried by pressing the material on a filter paper and then filling the capillary; however, the capillary should be placed in the bath when the temperature is below 100° to insure proper drying while the temperature of the bath rises.

For the arylamine it is best to use either aniline or *p*-toluidine, since these are commonly available in the laboratory in a greater state of purity than *o*-toluidine.

Preparation of Sulfonyl Chlorides from Sulfonic Acids and Conversion to Sulfonamides and N-α-Naphthylsulfonamides

Discussion

The equations below represent the method most commonly employed in the literature for the derivatization of sulfonic acids, which involves the conversion of the alkali sulfonate to the sulfonyl chloride by treatment with phosphorus pentachloride:

$$RSO_2ONa + PCl_5 \rightarrow \underset{\text{Sulfonyl chloride}}{RSO_2Cl} + POCl_3 + NaCl$$

$$RSO_2Cl + 2NH_3 \rightarrow \underset{\text{Sulfonamide}}{RSO_2NH_2} + NH_4Cl$$

The sulfonyl chloride is ammonolyzed to the sulfonamide or converted to the anilide. The disadvantage of this method is that it requires 1 to 2 g of dry salt and considerably more time and work than the other methods; in addition, the method is not applicable to compounds containing groups that react with phosphorus halides. It is therefore recommended as the third choice. Instead of ammonia an amine such as α-naphthylamine may be used to prepare a naphthylsulfonamide, RSO_2NH-α-$C_{10}H_7$.

Example 21.08: Preparation of Sulfonyl Chlorides from Sulfonic Acids

Arrange a distilling tube as shown on page 472, but omit the reflux condenser. Place in the tube 1 g of finely pulverized phosphorus pentachloride—the material is rapidly pulverized in a hood, using a mortar and pestle, and transferred immediately into the distilling tube. Add to the phosphorus halide 500 mg of dry and finely pulverized acid or its alkali salt. Raise an oil bath under the distilling tube and heat to 100 to 108° for about 30 minutes; then raise the temperature to 140° in order to distill over the phosphorus oxychloride. For this purpose the receiving tube containing the water is raised so that the water level is just below the outlet of the side arm. The oil bath may be omitted if a good microburner with an adjustable flame is available. In such a case a thermometer is inserted through the rubber stopper so that it reaches the bottom of the distilling tube. The flame may be adjusted so that the initial and final heating is done at the temperatures indicated. The tube is cooled, and 2 to 3 ml of ice water are added; the mixture is stirred by means of a glass rod to wash the sulfonyl chloride and remove the phosphorus halides. Decant the aqueous layer, being careful that the oily sulfonyl chloride adheres to the sides of the tube and stirring rod. Repeat the washing with ice water and decant the washings. The crude product is ammonolyzed to the sulfonamide as directed in Example 21.09.

If a pure sample of the sulfonyl chloride is desired, one should start with 1 to 2 g of the sulfonic acid and double its weight of phosphorus pentachloride. The procedure is the same as described until the removal of phosphorus oxychloride. About 10 ml of benzene are then added, and the mixture is transferred into a separatory funnel; the distilling tube is washed with 5 ml of the organic solvent, and the washings are united with the main portion of the solution. The benzene solution is washed twice with 10 ml of ice water and then transferred into a small flask and dried with anhydrous calcium chloride. The dry benzene solution is transferred into a distilling tube, and the solvent is removed by distillation until the volume of the solution is about 4 to 5 ml. At this point it may be transferred to a small distilling tube arranged for vacuum distillation, and the crude sulfonyl chloride is distilled at a pressure of 10 to 20 mm.

If the sulfonyl chloride can be crystallized from carbon tetrachloride, chloroform, or petroleum ether, the vacuum distillation may be omitted, and the crude product purified by crystallization. In such a case the benzene is entirely removed and 3 to 4 ml of the solvent added and heated to boiling. If any appreciable amount of oil remains undissolved at the sides or bottom of the distilling tube, the amount of solvent is cautiously increased until complete solution is effected at or near the boiling point of the solvent. The hot solution is filtered and cooled in an ice-salt mixture. The solid mass that separates is filtered rapidly, washed with a small amount of the pure solvent, and dried in a vacuum desiccator.

Example 21.09: Ammonolysis of Sulfonyl Chlorides to Sulfonamides

Use the crude sulfonyl chloride prepared as described in the preceding example from 500 mg of the sulfonic acid or its salt. Add to the distilling tube containing the crude sulfonyl chloride 10 ml of concentrated solution of aqueous ammonia and 2 g of powdered ammonium carbonate. Stir the mixture for a few minutes by means of the stirring rod that was used in washing the chloride and set aside for 5 minutes. Warm at 60° for 15 minutes, then to 80 to 90° for 10 minutes, and cool. Filter any solid that separates out at this point and evaporate the filtrate to dryness on the water bath. The residue of crude sulfonamide is crystallized separately from any amount that separated from the initial solution. For crystallization, the sulfonamide is dissolved in 10 to 25 ml of boiling water; the solution is treated with charcoal, filtered, and cooled.

Example 21.10: Conversion of a Sulfonic Acid to an N-α-*Naphthylsulfonamide:* N-α-*Naphthyl-*m*-nitrobenzenesulfonamide*

Place in an 8-in. distilling tube, fitted with a microcondenser, 1 mM (186 mg) of *m*-nitrobenzenesulfonic acid. Lift the cork holding the microcondenser and add, by means of a pipet dropper, 0.8 ml of thionyl chloride. Reflux in an oil or water bath at 85° for 30 minutes. While the test tube is still in the bath the condenser is removed and the unreacted thionyl chloride is removed with the aid of a vacuum (or by increasing the temperature and driving off the excess thionyl chloride). Lift the cork and add a solution of 5 mM (570 mg) of α-naphthylamine dissolved in 25 ml of dry benzene (CAUTION. *Flames should not be in the vicinity.*) Replace the microcondenser and reflux gently for another 30 minutes. Cool, filter, and to the solution add 5 ml water, and transfer the mixture into a small separatory funnel. Shake the benzene-water mixture gently to insure thorough mixing but to avoid formation of an emulsion. Remove the aqueous layer (the separation may also be effected with a capillary pipet in a test tube). Wash the benzene solution successively with 5 ml of 5 per cent hydrochloric acid, 5 ml of 5 per cent sodium hydroxide, and 5 ml of water. Allow sufficient time after shaking for complete separation of the two immiscible layers. If crystallization does not occur, pour the benzene solution into an evaporating dish. Add to the residue 5 ml of ethanol and warm until it is completely dissolved. Add charcoal and filter by suction. Wash the dish, by means of the pipet dropper, with 1 to 2 ml of ethanol and pour the washings through the filter. Add 2 to 3 ml of water slowly to the filtrate and heat the tube until cloudiness disappears. Cool and filter the crystals; wash the crystals twice with 3 ml of water, and dissolve directly in hot ethanol. Filter and add 2 ml of water. Cool and filter. The yield of reddish crystals, melting point 166 to 167°, is about 90 per cent of the theoretical.

DERIVATIVES OF THIOCYANATES AND ISOTHIOCYANATES

Isothiocyanates (mustard oils) are very readily derivatized by reaction with primary or secondary amines to give thioureas:

$$\text{ArNH}_2 + \text{C}_6\text{H}_5\text{NCS} \rightarrow \text{C}_6\text{H}_5\text{NHCSNHAr}$$

Arylamine   Phenyl isothiocyanate   N-Aryl-N'-phenylthiourea

The general procedures, together with several examples, are discussed under the derivatization of amines (pages 598–601).

The preparation of thiosemicarbazides, by reaction of the isothiocyanate with phenylhydrazine or other substituted hydrazines, is another method of derivatizing the isothiocyanates:

$$\text{RNCS} + \text{C}_6\text{H}_5\text{NHNH}_2 \rightarrow \text{C}_6\text{H}_5\text{NHNHCSNHR}$$

Substituted phenylthiosemicarbazide

The use of *p*-carboxyphenylhydrazine as a derivatizing reagent for isothiocyanates has been described by Veibel.[3]

[3] S. Veibel, *Dansk Tids. Farm.*, **17**, 42 (1943).

DERIVATIVES OF THIOETHERS

The most convenient method for the preparation of derivatives of thioethers is to oxidize them to sulfones:

$$O_2N\text{-}C_6H_3(NO_2)\text{-}S\text{-}R \xrightarrow{KMnO_4} O_2N\text{-}C_6H_3(NO_2)\text{-}SO_2\text{-}R$$

2,4-Dinitrophenyl thioether   2,4-Dinitrophenyl sulfone

The general procedure is described in the two following examples.

Example 21.11: Oxidation of 2,4-Dinitrophenyl Thioethers to Sulfones

Dissolve 3 mM of the thioether in the minimum quantity of glacial acetic acid and treat it with 0.7 g of potassium permanganate dissolved in 25 ml of water. Add the permanganate solution in portions of 2 to 3 ml, shaking after each addition until the color is discharged. Continue the addition of permanganate until the color persists after shaking for several minutes. Remove the excess of permanganate by careful addition of sodium bisulfite solution; the sulfone precipitates at this point on cooling by addition of 25 to 30 g of ice. Filter the solid and dry by pressing the solid between filter papers. Purify the sulfone by crystallization from methanol.

Example 21.12: Preparation of the Sulfone of Methionine

To prepare the sulfone dissolve 135 mg (1 mM) of methionine in 1 ml of water and 1.25 ml of $1M$ perchloric acid. To this solution add 2 drops (0.1 ml) of $0.5M$ ammonium molybdate and then 6 drops (0.3 ml) of 30 per cent hydrogen peroxide. The white precipitate that forms on the addition of the molybdate solution dissolves when the peroxide is added, and a yellow solution results. Immerse the tube containing the solution in a water bath at 20° for 2 hours; then add equal volumes of methanol and butylamine or amylamine to pH 9. Add about 25 ml of acetone and, after 10 minutes, wash the precipitated sulfone several times with acetone by decantation, then filter by suction, and wash with acetone and finally with ether. Dry by continuous suction and then place the powder in a watch glass and heat for 10 minutes at about 100°. The yield is 140 to 150 mg.

DERIVATIVES OF THIOLS (MERCAPTANS AND THIOPHENOLS)

The derivatives for the characterization of open-chain thiols (mercaptans) and thiophenols are listed in Table 21.01 (page 632). The most suitable are the 2,4-dinitrophenyl ethers and their sulfones and the 3,5-dinitrobenzoyl thioesters.

Thioethers

Discussion

A 2,4-dinitrophenyl thioether may be formed by reacting the mercaptan with 2,4-dinitrochlorobenzene:

$$RSH \xrightarrow{NaOH} RSNa + Cl\text{-}C_6H_3(NO_2)_2 \longrightarrow (NO_2)_2C_6H_3\text{-}S\text{-}R + NaCl$$

  2,4-Dinitrochlorobenzene   2,4-Dinitrophenyl thioether

The reaction takes place with ease when a sodium mercaptide solution is added to an alcoholic solution of the aromatic nitrohalide. The thioether separates on cooling; for further identification the thioether may be oxidized to the corresponding sulfone. The preparation of these derivatives is convenient for the characterization of many mercaptans.

Another type of solid thioether may be prepared by reacting the thiol with sodium anthraquinone-α-sulfonate; the thioether may be oxidized to the corresponding sulfone; another anthraquinone reagent proposed for the same purpose is disodium 1,5-anthraquinonedisulfonate. One disadvantage of this method is that the reaction takes place slowly, requiring several hours of heating.

$$\alpha\text{-}C_{14}H_7O_2SO_3Na \xrightarrow{RSH} \alpha\text{-}C_{14}H_7O_2SR + NaHSO_3$$

  Sodium anthraquinone-   Anthraquinonyl
  α-sulfonate   thioether

The general procedure for the preparation of 2,4-dinitrophenyl ethers is described in Example 21.13.

Example 21.13: Preparation of 2,4-Dinitrophenyl Ethers from Thiols

Place in an 8-in. tube 8 ml of methanol, 3 mM of the mercaptan, and 3 mM of sodium hydroxide (9 or 10 drops of $6N$ sodium hydroxide solution). Add the sodium mercaptide solution to a tube containing 600 mg of 2,4-dinitrochlorobenzene dissolved in 4 ml of methanol. Add a boiling stone and arrange for reflux. Boil the mixture gently for 5 to 10 minutes and filter the solution rapidly while hot. Cool for 10 minutes and filter the solid thioether. Recrystallize once or twice from methanol.

NOTE. If a red coloration results when the sodium hydroxide solution is added to the alcoholic solution of the mercaptan, a slight excess of the latter is used in order to remove the color caused by excess of alkali.

For further identification of the thioether, it may be converted by oxidation to a sulfone, as outlined on page 641.

Thioesters

Discussion

The equations below represent the formation of thioesters by reacting the mercaptan with 3,5-dinitrobenzoyl chloride or 3-nitrophthalic anhydride:

[3,5-Dinitrobenzoyl chloride] + RSH $\xrightarrow{\text{NaOH}}$ [3,5-Dinitrobenzoyl thioester] + NaCl + H$_2$O

[3-Nitrophthalic anhydride] + RSH $\rightarrow$ [3-Nitrophthalic thioester with COSR and COOH groups]

The preparation of the 3,5-dinitrobenzoyl thioester should be a second choice. The method for the preparation of the esters is similar to that used for the alcohols. About 200 mg of the acid chloride or the anhydride are heated with 5 or 6 drops of the thiol until a uniform melt has been obtained; a few drops of pyridine may be added in the reaction of the thiol with the acid chloride to aid the removal of hydrogen chloride. After addition of water, the solid derivative is filtered and purified by crystallization.

The substituted thiocarbamic esters formed by reaction of thiols with azides may serve as derivatives. For example, *m*-nitrobenzazide yields with thiols *m*-nitrophenylurethans:

$$\underset{\textit{m-Nitrobenzazide}}{NO_2C_6H_4CON_3} + \underset{\text{Thiol}}{RSH} \rightarrow \underset{\textit{m-Nitrophenylurethan}}{NO_2C_6H_4NHCOSR} + N_2$$

The general procedure for the preparation of thioesters is given in Example 21.14.

Example 21.14: Preparation of 3,5-Dinitrothiobenzoates from Thiols

Dry an 8-in. tube by heating it over a flame and then close it with a cork and allow to cool. Place in it 200 mg of 3,5-dinitrobenzoyl chloride and arrange for reflux. Add to the solid chloride 5 or 6 drops of the thiol and 1 drop of pyridine. Adjust the microburner so that the reaction mixture melts into a homogeneous mass and heat in this manner for about 10 minutes. If at this point a strong odor of the thiol persists, add 25 to

50 mg of the chloride and heat for an additional 5 minutes. Add 2 ml of water, cool, and stir by means of a glass rod until the oily mass solidifies. Filter with suction and wash with water. To remove the small amount of 3,5-dinitro-benzoic acid and to crystallize the derivative, follow the directions given in Example 13.01 (page 470).

REFERENCES

Sulfinic Acids

Aryl- and alkylmercuric chlorides from sulfinates: Coffey, *J. Chem. Soc.* (*London*), **1926**, 637; Kharasch and Chalkey, *J. Am. Chem. Soc.*, **43**, 607 (1921); Marvel et al., *J. Am. Chem. Soc.*, **68**, 2735 (1946); Peters, *Ber.*, **38**, 2567 (1905); Whitmore et al., *J. Am. Chem. Soc.*, **45**, 1066 (1923).

1,2-Dialkylsulfonylethanes: Allen, *J. Org. Chem.*, **7**, 23 (1942).

Sulfonamides

Microcharacterization: Chiarino et al., *Anales Asoc. Quim. Arg.* **31**, 72, 233 (1943); *C.A.*, **38**, 530 (1944); **39**, 255 (1945); Vonesch, *Anal. farm. y bioquim.* (*Buenos Aires*), **14**, 81 (1943); *C.A.*, **39**, 1430 (1945).

Paper chromatography: Bray et al., *Biochem. J.*, **46**, 271 (1950); Longenecker, *Anal. Chem.*, **21**, 1042 (1949); de Reeder, *Anal. Chim. Acta*, **8**, 325 (1953); Robinson, *Nature*, **168**, 512 (1951); San and Ultee, *Nature*, **169**, 586 (1952); Steel, *Nature*, **168**, 877 (1951).

N-Xanthylsulfonamides: Phillips and Frank, *J. Org. Chem.*, **9**, 9 (1944).

Sulfonic Acids

Identification of aromatic sulfonic acids containing an amino group: Allen et al., *J. Org. Chem.*, **7**, 15 (1942); **10**, 1 (1945).

Identification of sulfobenzoic acids: Suter and Campaigne, *J. Am. Chem. Soc.*, **67**, 1860 (1945).

Microscopic identification of some important substituted naphthalenesulfonic acids: Chambers and Scherer, *Ind. Eng. Chem.*, **16**, 1272 (1924); Garner, *J. Soc. Dyers Colourists*, **43**, 12 (1927); **52**, 302 (1936); Hann and Keenan, *J. Phys. Chem.*, **31**, 1082 (1927); Whitmore and Gebhart, *Ind. Eng. Chem., Anal. Ed.*, **10**, 654 (1938).

Micro identification of naphthalenesulfonic acids by means of benzylisothiourea: Garner, *J. Soc. Dyers and Col.*, **52**, 302 (1936).

Aniline salts of aromatic sulfonic acids: Dermer and Dermer, *J. Org. Chem.*, **7**, 581 (1942).

N-(p-Nitrobenzyl)pyridinium salts of aromatic sulfonic acids: Huntress and Foote, *J. Am. Chem. Soc.*, **64**, 1017 (1942).

N-Sulfonylphthalimides: Evans and Dehn, *J. Am. Chem. Soc.*, **51**, 3651 (1929).

Phenylhydrazine salts of aliphatic sulfonic acids: Latimer and Bost, *J. Am. Chem. Soc.*, **59**, 2501 (1937).

Pyridine salts of acetylated aminosulfonic acids: Chen and Gross, *J. Soc. Dyers Colourists*, **59**, 144 (1943).

S-Benzylthiuronium salts of sulfonic acid: Chambers and Scherer, *Ind. Eng. Chem.*, **16,** 1272 (1924); Chambers and Watt, *J. Org. Chem.*, **6,** 376 (1941); Campaigne and Suter, *J. Am. Chem. Soc.*, **64,** 3040 (1942); Donleavy, *J. Am. Chem. Soc.*, **58,** 1005 (1936); Hann, *J. Am. Chem. Soc.*, **57,** 2166 (1935); Veibel, *J. Am. Chem. Soc.*, **67,** 1867 (1945); Veibel and Lillelund, *Bull. Soc. Chim. France* (5), **5,** 1153 (1939).

o- and *p*-Toluidine salts of aromatic sulfonic acids: Dermer and Dermer, *J. Org. Chem.*, **7,** 581 (1942).

p-Toluidine salts of monoarylsulfates: Barton and Young, *J. Am. Chem. Soc.*, **65,** 294 (1943).

p-Toluidine salts of sulfonic acids: Feiser, *J. Am. Chem. Soc.*, **51,** 2463 (1929).

Xanthydrol derivatives of sulfonic acids: Phillips and Frank, *J. Org. Chem.*, **9,** 9 (1944).

Thiocyanates and Isothiocyanates

Thiosemicarbazides from reaction with *p*-carboxyphenylhydrazine: Veibel, *Dansk Tidsk. Farm.*, **17,** 42 (1943).

Substituted thioureas from reaction with ammonia or primary amines: Weller et al., *J. Am. Chem. Soc.*, **74,** 1104 (1952).

Thioethers and Thiols

Anthraquinonyl thioethers: Reid and Ellis, *J. Am. Chem. Soc.*, **54,** 1674 (1932); Reid and Hoffman, *J. Am. Chem. Soc.*, **45,** 1831 (1923); Reid, Mackall, and Miller, *J. Am. Chem. Soc.*, **43,** 2104 (1921).

Nitrosylmercaptides from reaction with nitrous acid: Rheinboldt, *Ber.*, **59,** 1311 (1926); **60,** 184 (1927); Tasker and Jones, *J. Chem. Soc.*, **95,** 1917 (1909); Vorlander and Mittag, *Ber.*, **52,** 422 (1919).

Use of 3,5-dinitrobenzoyl chloride and 3-nitrophthalic anhydride: Wertheim, *J. Am. Chem. Soc.*, **61,** 3660 (1939).

Use of 2,4-dinitrochlorobenzene for thioethers: Bost, Turner, and Norton, *J. Am. Chem. Soc.*, **54,** 1985 (1932); Bost, Turner, and Conn, *J. Am. Chem. Soc.*, **55,** 4956 (1933).

Use of saccharin chloride: Meadow and Cavagnol, *J. Org. Chem.*, **17,** 488 (1952).

Chromatographic Procedures

Schmeiser and Jerchel, *Angew. Chem.*, **65,** 366 (1953). Paper chromatography of sulfur compounds.

22

Instrumental Methods

ULTRAVIOLET AND VISIBLE—INFRARED—NUCLEAR
MAGNETIC RESONANCE—MASS SPECTROSCOPY

The instrumental methods which are most useful for the qualitative analysis of organic compounds are ultraviolet and visible absorption spectroscopy, infrared absorption spectroscopy, nuclear magnetic resonance spectroscopy, and mass spectroscopy. These methods supplement those described in this book previously, and should be applied whenever their use would speed the identification of an organic compound.

Unfortunately all these methods require the use of complex instruments which are expensive, costing from about one thousand dollars to a hundred thousand dollars, depending on the instrument. Maintenance of the more complex instruments requires considerable skill and time of the user. A skilled, well-trained operator is needed to work with the more complicated instruments and to properly evaluate the results in terms of structures of compounds. Only a brief summary can be given here of the scope of these instrumental methods, but extensive bibliographies, articles, and collections of data are given at the end of this chapter.

Advantages of the Instrumental Methods

Most of these instrumental methods require only a few minutes for each sample. The infrared spectrum of a compound can be obtained in about the same time required for determination of a melting point.

The amount of material required for most of these methods is usually in the order of a few milligrams—scarcely more than the quantity needed for a melting point determination. Even this small amount of compound is not necessarily destroyed. Ultraviolet and infrared absorption spectra can both be determined on less than 1 mg of material and then the same material can be used in the mass spectrometer.

Each of the instrumental methods gives fundamental information concerning the nature of the compound or material. The results of one determination often allow a pure compound to be placed in a particular structural class, or to be identified positively as a specific member of a class.

Types of Information Obtained

Absorption of ultraviolet and visible radiation is associated with the resonance capabilities of compounds. Saturated aliphatic hydrocarbons do not absorb in the ultraviolet or visible regions, but aromatic hydrocarbons do. Ketones and aldehydes have very definite absorption bands in the ultraviolet, but aliphatic alcohols do not. In general, each class of compound which absorbs in the visible or ultraviolet region has its maximum absorption in a characteristic region. This wavelength of greatest absorption may be shifted somewhat by the presence in the molecule of other functional groups. The bands are often broad, but some are sharp and therefore may be more useful in identification. Tentative identification of some functional groups or of particular compounds may be made in this way.

Functional groups in a compound are more easily identified by means of infrared spectroscopy. The bands tend to be sharper and more characteristic for each class of compound. Position isomers of a compound may be distinguished by their different absorption bands. Ortho, meta, and para derivatives, for example, may absorb at slightly different wavelengths. Many other isomers (except optical isomers) may be distinguished from one another.

Nuclear magnetic resonance spectroscopy can be used to study only compounds having an element with an odd atomic number, such as hydrogen, fluorine, or phosphorus. Functional groups containing one or more of these atoms may be identified. If a high-resolution instrument is used, the structural environment of these atoms may be determined. The position of an absorption maximum gives information concerning the chemical nature of a hydrogen atom, and the size and number of peaks are related to the number of hydrogen atoms on neighboring carbon atoms. Consequently, position isomers may be positively identified if the number of hydrogen atoms on adjacent carbon atoms are different. The areas under the peaks are proportional to the number of hydrogen atoms of that particular species.

When a sample is injected into a mass spectrometer the sample is not only ionized but often degraded into smaller particles which also have a charge. By considering the masses of the particles produced one can often determine the structure of the original component. Since organic compounds break into well-defined segments in a uniform way, the pattern of particle weight distribution can be used to derive a probable structure of the original compound. In addition, some information concerning the molecular formula of the compound may be derived from the ratio of the masses of the ions.

Sample Handling

Ultraviolet and Visible Absorption Spectroscopy

Ultraviolet and visible absorption spectra are usually determined for solutions of liquid or solid compounds, although the gas phase can be used. The most common cells for liquids are 1 cm square and require 2 to 3 ml of solution. The concentration of solute may be varied in order to examine different regions of the spectrum in an optimum fashion. Normally the sample required is from 0.1 to 100 mg.

The most common solvents are cyclohexane, ethyl alcohol, and dioxane; other solvents such as water, chloroform, ether, acetonitrile, methanol, and isooctane are used successfully. These must be carefully freed of impurities before they can be used.

Infrared Absorption Spectroscopy

Infrared spectra are most often determined for solutions of compounds; however, solids, gases, and pure liquids may also be used. Solutions or pure liquids are placed in cells made of rock salt plates spaced from 0.03 to 1 mm apart. The concentration of solute varies from 0.1 to perhaps 5 per cent, depending on the cell and the compound. Usually at least 1 mg of the compound is needed, and 15 mg should be the maximum. Solvents are most often carbon tetrachloride, carbon disulfide, and chloroform. Other solvents sometimes used are bromoform, tetrachloroethylene, and dichloromethane.

Solids may be mulled with a heavy mineral oil, Fluorolube S, or hexachlorobutadiene. The suspension is viewed by placing it between two rock salt plates. Solids are also examined in the form of a disc made by grinding the solid with dry potassium bromide and pressing the mixture under pressures of 20,000 to 50,000 lb per square inch.

Gases are examined in special cells, which may have effective lengths of several meters.

Nuclear Magnetic Resonance Spectroscopy

For determination of the nuclear magnetic resonance spectrum the sample must be in the liquid state either as a solution or as the pure liquid. Normally the sample is placed in a glass tube with an outside diameter of almost 5 mm. About 0.4 ml of the liquid or solution is required. As little as 10 to 50 mg of the sample may be used if it is dissolved in 0.4 ml of solvent.

Solvents commonly used for determination of proton spectra are carbon tetrachloride and carbon disulfide, which have no hydrogen atoms and therefore no absorption peaks. Acetone, acetonitrile, benzene, chloroform, cyclohexane, dimethyl sulfoxide, and water have only one peak each. Fully deuterated solvents such as chloroform-d, benzene-d_6 and acetone-d_6 have no absorption peaks.

Mass Spectroscopy

Gases, liquids, and solids may be used in determination of mass spectra. However, samples must be slightly volatile or possess reproducible breakdown patterns under conditions inside the mass spectrometer. Ordinarily, a vapor pressure of 0.01 to 0.05 mm under these conditions is necessary, but a compound with a vapor pressure of only 10^{-6} mm may be used under special conditions.

Samples may vary in size from several milligrams to less than one microgram. Solvents are not normally needed or used.

Special Procedures for Reducing the Size of the Sample

For most of these instrumental methods the sample size can be decreased by a more efficient arrangement of the sample. One method suggested for reducing the volume of solution needed for absorption studies in the ultraviolet or visible region with the usual 1-cm glass or quartz cell is to add clean glass beads to the lower part of the cell.[1] Another method involves the use of a Teflon insert which fits directly in the 1-cm cell and allows spectra determinations with 1 ml of solution.[2] A Teflon or Kel-F insert for a Lowry and Bessey microliter absorption

---
[1] K. M. Dubowski, *Chemist Analyst*, **41**, No. 4, 96 (1954).
[2] L. H. Sharpe, *Anal. Chem.*, **26**, 1528 (1954).

cell[3] permits the use of solutions of 5 microliters or less[4] with the Beckman Model DU spectrophotometer.

A spectrometer which allows absorption studies on as little as 10 microliters uses a light beam only 0.4 mm in diameter. The cuvette is a capillary tube 1 mm in diameter.[5] A low magnification microscope attached to a Cary model 14 spectrophotometer has been used to record absorption spectra in the range 3000 to 7000 Å and in fields of 0.1 to 1 mm in diameter. With a better microscope the range is 3500 to 6500 Å, and the diameter of the field may be 4 microns.[6] A scanning microspectrophotometer can be used to study behavior of animal pigments *in situ* and in solution.[7]

For infrared absorption determination cells are commercially available which have volumes of 0.6 microliters and a path of 0.05 mm. An absorption spectrum can thus be obtained on a sample of a few micrograms. Another infrared microcell has a path 10 mm in length, but requires only 0.5 ml of solution.[8] Many organic compounds can be detected at a level of 1 to 10 ppm. A cell with diamond or sapphire windows has been used to measure infrared spectra of solids and corrosive liquids with sample sizes as small as 4 micrograms.[9] The visible and ultraviolet regions can also be studied with the same cell. Infrared spectral methods useful for solids are compared by these authors and many references to the original literature are given.

More efficient use of the solid sample can be obtained by use of a normal potassium bromide disc with a 3-mm hole in the center. The sample mixed with more potassium bromide is pressed into this hole.[10]

The usual double-beam infrared spectrometer may be converted for use with very small samples. Small crystals and microgram quantities of organic compounds in solution give good spectral data.[11] Full-scale spectra have been obtained with 1 to 5 microgram samples by use of a double-beam instrument and reflecting-type 6X beam condensing optics. With the help also of electronic ordinate expansion, spectra can be obtained on samples as small as 0.05 microgram.[12]

[3] O. H. Lowry and O. A. Bessey, *J. Biol. Chem.*, **163**, 633 (1946).
[4] D. Glock and B. W. Grunbaum, *Anal. Chem.*, **29**, 1243–1244 (1957).
[5] D. F. H. Wallach and D. M. Surgenor, *Anal. Chem.*, **30**, 1879–1882 (1958).
[6] P. K. Brown, *J. Opt. Soc. Am.*, **51**, 1000–1008 (1961).
[7] P. A. Liebman, *Biophys. J.*, **2**, 161–178 (1962).
[8] D. S. Erley, *Appl. Spectroscopy*, **15**, 80–81 (1961).
[9] E. R. Lippincott, F. E. Welsh, and C. E. Weir, *Anal. Chem.*, **33**, 137–143 (1961).
[10] G. Fabbri, *Ann. chim.* (Rome), **47**, 394–401 (1957).
[11] E. R. Blout and M. J. Abbate, *J. Opt. Soc. Am.*, **45**, 1028–1030 (1955).
[12] M. Sparagana and W. B. Mason, *Anal. Chem.*, **34**, 242–247 (1962).

For nuclear magnetic resonance studies with small quantities of material the volume of the solutions needed can be reduced to about 0.2 ml with the help of special plugs in the sample holder. A specially constructed sample holder is said[13] to use as little as 25 microliters of solution.

REFERENCES

Ultraviolet and Visible Absorption Spectroscopy

General

R. P. Bauman, *Absorption Spectroscopy*, John Wiley & Sons, New York, 1962. Theory, equipment, and applications.

G. H. Beavan, E. A. Johnson, H. A. Willis and R. G. J. Miller, *Molecular Spectroscopy*, Macmillan, New York, 1961. Methods and applications in chemistry.

E. A. Braude, "Ultraviolet and Visible Light Absorption" in *Determination of Organic Structures by Physical Methods*, Vol. I, E. A. Braude and F. C. Nachod, eds., Academic Press, New York, 1955, pp. 131–194. There are 156 references.

A. B. F. Duncan and F. A. Matsen, "Electronic Spectra in the Visible and Ultraviolet" in *Technique of Organic Chemistry*, Vol. 9, A. Weissberger, ed., Interscience, New York, 1956, pp. 581–706.

M. St. C. Flett, *Physical Aids to the Organic Chemist*, Elsevier, New York, 1962, pp. 73–136.

A. E. Gillam and E. S. Stern, *An Introduction to Electronic Absorption Spectroscopy in Organic Chemistry*, Second Edition, Edward Arnold, London, 1958. Applications and many references.

H. H. Jaffe and M. Orchin, *Theory and Applications of Ultraviolet Spectroscopy*, John Wiley & Sons, New York, 1962.

F. A. Matsen, "Applications of the Theory of Electronic Spectra," in *Technique of Organic Chemistry*, Vol. 9, A. Weissberger, ed., Interscience, New York, 1956, pp. 629–706. There are 144 references.

W. C. Price, "Spectroscopy in the Vacuum Ultra-Violet" in *Advances in Spectroscopy*, Vol. 1, H. W. Thompson, ed., Interscience, New York, 1959, pp. 56–75.

D. A. Ramsey, "Electronic Spectra of Polyatomic Molecules and the Configurations of Molecules in Excited Electronic States" in *Determination of Organic Structures by Physical Methods*, Vol. 2, F. C. Nachod and W. D. Phillips, eds., Academic Press, New York, 1962, pp. 245–338. There are 200 references. Theoretical.

C. N. R. Rao, *Ultra-Violet and Visible Spectroscopy: Chemical Applications*, Butterworth, London, 1961.

R. M. Silverstein and G. C. Bassler, *Spectrometric Identification of Organic Compounds*, John Wiley & Sons, New York, 1963, pp. 90–103.

D. W. Turner, "Far and Vacuum Ultraviolet Spectroscopy" in *Determination of Organic Structures by Physical Methods*, Vol. 2, F. C. Nachod and W. D. Phillips, eds., Academic Press, New York, 1962, pp. 339–400. There are 37 references. Applications to structure determination.

[13] R. M. Silverstein and G. C. Bassler, *Spectrometric Identification of Organic Compounds*, Wiley, 1963, p. 74.

H. H. Willard, L. L. Merritt, Jr., and J. A. Dean, *Instrumental Methods of Analysis*, Third Edition, Van Nostrand, Princeton, N.J., 1958, pp. 96–138.

Bibliographies and Collections of Spectra

American Petroleum Institute Research Project and the Manufacturing Chemists Association. Texas A. and M. College, College Station. Spectra are on looseleaf sheets.

R. A. Friedel and M. Orchin, *Ultraviolet Spectra of Aromatic Compounds*, John Wiley & Sons, New York, 1951. There are 579 spectra given.

H. M. Hershenson, *Ultraviolet and Visible Absorption Spectra, Index for 1930–1954*, Academic Press, New York, 1956. References to the original publications.

H. M. Hershenson, *Ultraviolet and Visible Absorption Spectra, Index for 1955–59*, Academic Press, New York, 1961.

M. J. Kamlet, ed., *Organic Electronic Spectral Data*, Vol. 1, Interscience, New York, 1960. Covers the literature from 1946 to 1952.

C. Karr, Jr., *Appl. Spectroscopy*, **13**, 15–25, 40–45 (1959); *ibid.*, **14**, 146–153 (1960). A tabulation of absorption bands for 612 condensed aromatic hydrocarbons and polynuclear heterocyclic aromatics.

L. Lang, *Absorption Spectra in the Ultraviolet and Visible Region*, Vols. 1 and 2, Hungarian Academy of Sciences, Budapest. There are 349 spectra given.

J. P. Phillips and F. C. Nachod, *Organic Electronic Spectral Data*, Vol. 4, John Wiley & Sons, New York, 1963. Covers the literature from 1958 to 1959.

Sadtler Research Laboratories, Philadelphia, Pa. Collection of 6000 ultraviolet spectra.

H. G. Ungnade, *Organic Electronic Spectral Data*, Vol. 2, Interscience, New York, 1960. Covers the literature from 1953 to 1955.

O. H. Wheeler and L. Kaplan, *Organic Electronic Spectral Data*, Vol. 3, John Wiley & Sons, New York, 1963. Covers the literature from 1956 to 1957.

Infrared Absorption Spectroscopy

General

H. C. Allen, Jr., and P. C. Cross, *Molecular Vib-Rotors: The Theory and Interpretation of High Resolution Infrared Spectra*, John Wiley & Sons, New York, 1963. Theory.

D. H. Anderson, N. B. Woodall, and W. West, "Infrared Spectroscopy" in *Physical Methods of Organic Chemistry*, Third Edition, Part 3, A. Weissberger, ed., Interscience, New York, 1960, pp. 1959–2019. Short review, primarily instrumentation. There are 102 references.

R. B. Barnes, R. C. Gore, U. Liddel, and V. Z. Williams, *Infrared Spectroscopy*, Reinhold, New York, 1944. Industrial applications and bibliography.

G. M. Barrow, *Introduction to Molecular Spectra*, McGraw-Hill, New York, 1962. Theory.

R. P. Bauman, *Absorption Spectroscopy*, John Wiley & Sons, New York, 1962. Theory, equipment, and qualitative analysis.

G. H. Beavan, E. A. Johnson, H. A. Willis, and R. G. J. Miller, *Molecular Spectroscopy*, Macmillan, New York, 1961. Methods and applications in chemistry.

L. J. Bellamy, *The Infrared Spectra of Complex Molecules*, Second Edition, John Wiley & Sons, New York, 1958. Interpretations in terms of structure.

W. Brugel, *An Introduction to Infrared Spectroscopy*, John Wiley & Sons, New York, 1962. Theory, equipment, and applications. There are over 800 references.

A. D. Cross, *Introduction to Practical Infra-red Spectroscopy*, Butterworth, London, 1960. Description of equipment and tables of frequencies of absorption bands.

M. Davies, *Infra-red Spectroscopy and Molecular Structure*, Elsevier, New York, 1963. An outline of the principles.

M. St. C. Flett, *Physical Aids to the Organic Chemist*, Elsevier, New York, 1962, pp. 138–218.

A. Elliott, "The Infra-Red Spectra of Polymers" in *Advances in Spectroscopy*, Vol. 1, H. W. Thompson, ed., Interscience, New York, 1959, pp. 214–287.

R. C. Gore, "Infrared Light Absorption" in *Determination of Organic Structures by Physical Methods*, Vol. 1, E. A. Braude and F. C. Nachod, eds., Academic Press, New York, 1955, pp. 195–230. There are 67 references.

H. L. Hackforth, *Infrared Radiation*, McGraw-Hill, New York, 1960. Basic principles, equipment, and applications.

G. Herzberg, *Infrared and Raman Spectrum of Polyatomic Molecules*, Second Edition, Van Nostrand, New York, 1954. Theory.

R. N. Jones and C. Sandorfy, "The Application of Infrared and Raman Spectrometry to the Elucidation of Molecular Structure" in *Technique of Organic Chemistry*, Vol 9, A. Weissberger, ed., Interscience, New York, 1956, pp. 247–580. There are 979 references.

C. E. Meloan, *Elementary Infrared Spectroscopy*, Macmillan, New York, 1963.

K. Nakanishi, *Infrared Absorption Spectroscopy*, *Practical*, Holden-Day, San Francisco, 1962. Tables of characteristic frequencies and practical problems in structure determination. Some data on NMR spectral shifts.

H. H. Nielsen and R. A. Oetjen, "Infrared Spectroscopy" in *Physical Methods of Chemical Analysis*, Second Edition, Vol. 1, W. G. Berl, ed., Academic Press, New York, 1960, pp. 327–388. Largely theory and equipment. There are 85 references.

W. J. Potts, *Chemical Infrared Spectroscopy*, Vol. 1, "Techniques," John Wiley & Sons, New York, 1963. Theory and techniques.

H. M. Randall, R. G. Fowler, N. Fuson, and J. R. Dangl, *Infrared Determination of Organic Structures*, Van Nostrand, New York, 1949. Practical applications, interpretations, and techniques.

N. Sheppard, "Rotational Isomerism about C—C Bonds in Saturated Molecules as Studied by Vibrational Spectroscopy" in *Advances in Spectroscopy*, Vol. 1, H. W. Thompson, ed., Interscience, New York, 1959, pp. 288–353.

R. M. Silverstein and G. C. Bassler, *Spectrometric Identification of Organic Compounds*, John Wiley & Sons, New York, 1963, pp. 49–70.

H. A. Szymanski, *Interpreted Infrared Spectra*, Plenum Press, New York, 1963.

H. A. Szymanski, *Progress in Infrared Spectroscopy*, Vol. 1, Plenum Press, New York, 1962.

H. A. Szymanski, *Progress in Infrared Spectroscopy*, Vol. 2, Plenum Press, New York, 1963.

H. A. Szymanski and N. Alpert, *IR-Theory and Practice of Infrared Spectroscopy*, Plenum Press, New York, 1963.

H. H. Willard, L. L. Merritt, Jr., and J. A. Dean, *Instrumental Methods of Analysis*, Third Edition, Van Nostrand, Princeton, N.J., 1958, pp. 139–178.

E. B. Wilson, J. C. Decius, and P. C. Cross, *Molecular Vibrations—The Theory of Infrared and Raman Vibrations*, McGraw-Hill, New York, 1955.

M. K. Wilson, "Infrared and Raman Spectroscopy" in *Determination of Organic Structures by Physical Methods*, Vol. 2, F. C. Nachod and W. D. Phillips, eds., Academic Press, New York, 1962, pp. 181–243. There are 125 references.

Bibliographies and Collections of Spectra

American Petroleum Institute Research Project 44 and the Manufacturing Chemists Association. Texas A. and M. College, College Station, Texas. Spectra are on looseleaf sheets.

K. Dobriner, E. R. Katzenellenbogen, and R. N. Jones, *Infrared Absorption Spectra of Steroids, An Atlas*, Vol. 1, Interscience, New York, 1953. See also, G. Roberts et al., for Vol. 2.

H. H. Hershenson, *Infrared Absorption Spectra: Index for 1945–1957*, Academic Press, New York, 1959. References to the original literature.

Ministry of Aviation (Great Britain). *An Index of Published Infra-Red Spectra*, Vol. I and II, British Information Service, 45 Rockefeller Plaza, New York, N.Y., 1960. Complete to 1957. Gives references to original literature.

National Research Council—National Bureau of Standards. Spectra are on edge-punched cards.

G. Roberts, B. S. Gallagher, and R. N. Jones, *Infrared Absorption Spectra of Steroids, An Atlas*, Vol. 2, Interscience, New York, 1958. See also K. Dobriner et al., Vol. 1. More than 700 spectra are given in the two volumes.

Samuel P. Sadtler and Sons, Inc., Philadelphia, Pa. Largest collection of spectra available.

H. A. Szymanski, *Infrared Band Handbook*, Plenum Press, New York, 1963. Annual supplements are available.

Documentation of Molecular Spectroscopy (DMS), Butterworths, London, and Verlag Chemie GMBH, Weinheim/Bergstrasse, West Germany, in cooperation with the Infrared Absorption Data Joint Committee, London, and the Institut für Spectrochemie und Angewandte Spectroskopie, Dortmund. Spectra are on punch cards.

Nuclear Magnetic Resonance Spectroscopy

General

E. R. Andrews, *Nuclear Magnetic Resonance*, Cambridge University Press, New York, 1955.

H. Conroy, "Nuclear Magnetic Resonance in Organic Structural Elucidation" in *Advances in Organic Chemistry*, Vol. 2, R. A. Raphael, E. C. Taylor, and H. Wynberg, eds., Interscience, New York, 1960.

M. St. C. Flett, *Physical Aids to the Organic Chemist*, Elsevier, New York, 1962, pp. 255–308.

H. Foster, "Application of Nuclear Magnetic Resonance Spectroscopy to Organic Analysis" in *Organic Analysis*, Vol. 4, Interscience, New York, 1960, pp. 229–291.

H. S. Gutowsky, "Analytical Applications of Nuclear Magnetic Resonance" in *Physical Methods in Chemical Analysis*, Vol. 3, Academic Press, New York, 1956, pp. 304–383.

H. S. Gutowsky, "Nuclear Magnetic Resonance" in *Physical Methods of Organic Chemistry*, Third Edition, Part 4, A. Weissberger, ed., Interscience, New York, 1960, pp. 2663–2799. There are 256 references.

L. M. Jackman, *Applications of Nuclear Magnetic Resonance Spectroscopy in Organic Chemistry*, Pergamon Press, New York, 1959.

P. C. Lauterbur, "Nuclear Magnetic Resonance Spectra of Elements Other than Hydrogen and Fluorine" in *Determination of Organic Structures by Physical Methods*, Vol. 2, F. C. Nachod and W. D. Phillips, eds., Academic Press, New York, 1962, pp. 465–536. There are 125 references. Applications.

J. C. Martin, *J. Chem. Educ.*, **38**, 286–291 (1961). NMR spectroscopy as an analytical tool in organic chemistry.

D. E. McGreer and M. M. Mocek, *J. Chem. Educ.*, **40**, 358–361 (1963). A qualitative approach to the study of complex NMR spectra.

C. T. O'Konski, "Nuclear Quadrupole Resonance Spectroscopy" in *Determination of Organic Structures by Physical Methods*, Vol. 2, F. C. Nachod and W. D. Phillips, eds., Academic Press, New York, 1962, pp. 661–726. There are 226 references. Theory and applications.

W. D. Phillips, "High Resolution H^1 and F^{19} Magnetic Resonance Spectra of Organic Molecules" in *Determination of Organic Structures by Physical Methods*, Vol. 2, F. C. Nachod and W. D. Phillips, eds., Academic Press, New York, 1962, pp. 401–463. There are 140 references. Applications to structure determination.

J. A. Pople, W. G. Schneider, and H. J. Bernstein, *High-Resolution Nuclear Magnetic Resonance*, McGraw-Hill, New York, 1959.

R. E. Richards, "Nuclear Magnetic Resonance Spectra of Organic Solids" in *Determination of Organic Structures by Physical Methods*, Vol. 2, F. C. Nachod and W. D. Phillips, eds., Academic Press, New York, 1962, pp. 537–562. There are 47 references.

J. D. Roberts, *Nuclear Magnetic Resonance: Applications to Organic Chemistry*, McGraw-Hill, New York, 1959.

J. D. Roberts, *J. Chem. Educ.*, **38**, 581 (1961).

J. D. Roberts, *An Introduction to the Analysis of Spin-Spin Splitting in High-Resolution Nuclear Magnetic Resonance Spectra*, W. A. Benjamin, New York, 1961. Theory.

R. M. Silverstein and G. C. Bassler, *Spectrometric Identification of Organic Compounds*, John Wiley & Sons, New York, 1963, pp. 71–89.

Varian Associates, *NMR and EPR Spectroscopy*, Macmillan, New York, 1960. Collection of papers.

Bibliographies and Collections of Spectra

N. S. Bhacca, L. F. Johnson, and J. N. Shoolery, *NMR Spectra Catalog*, Varian Associates, Palo Alto, Calif., 1962. High resolution proton spectra of 368 representative organic compounds.

K. B. Wiberg and B. J. Nist, *Interpretation of NMR Spectra*, W. A. Benjamin, New York, 1962. Spectra calculated with aid of computer.

Mass Spectroscopy

General

J. H. Beynon, *Mass Spectrometry and Its Applications to Organic Chemistry*, Elsevier, Amsterdam, 1960. Equipment, correlation of molecular structure and mass spectra.

K. Biemann, *Mass Spectrometry, Organic Chemical Applications*, McGraw-Hill, New York, 1962. Equipment and applications.

J. Blears, *Applied Mass Spectrometry*, The Institute of Petroleum, London, England, 1954. Report of a conference: bibliography for 1950–1953.

V. H. Dibeler, "Analytical Mass Spectrometry" in *Organic Analysis*, J. Mitchell, ed., Vol. III, Interscience, New York, 1956, pp. 387–441.

H. E. Duckworth, *Mass Spectroscopy*, University Press, Cambridge, England, 1958. Equipment.

R. M. Elliott, ed., *Advances in Mass Spectrometry*, Macmillan, New York, 1963. Proceedings of a conference. There are 44 papers.

M. St. C. Flett, *Physical Aids to the Organic Chemist*, Elsevier, New York, 1962, pp. 309–350.

F. W. McLafferty, "Mass Spectrometry" in *Determination of Organic Structure by Physical Methods*, Vol. II, F. C. Nachod and W. D. Phillips, eds., Academic Press, New York, 1962. There are 139 references.

F. W. McLafferty, *Mass Spectrometry of Organic Ions*, Academic Press, New York, 1963.

R. I. Reed, "Mass Spectrometry as a Structural Tool" in *Advances in Organic Chemistry: Methods and Results*, Vol. 3, R. A. Raphael, E. C. Taylor, and H. Wynberg, eds., Interscience, New York, 1963. There are 114 references.

R. M. Silverstein and G. C. Bassler, *J. Chem. Educ.*, **39**, 546–553 (1962). Application of IR-, UV-, NMR and mass spectrometry to the study of organic compounds.

R. M. Silverstein and G. C. Bassler, *Spectrometric Identification of Organic Compounds*, John Wiley & Sons, New York, 1963, pp. 4–48.

D. W. Stewart, "Mass Spectrometry," in *Technique of Organic Chemistry*, A. Weissberger, ed., Vol. I, Part IV, Third Edition, Interscience, New York, 1960, pp. 3449–3539.

J. D. Waldron, ed., *Advances in Mass Spectrometry*, Pergamon Press, London, 1959. Collection of papers—includes interpretation of spectra in terms of structure of organic compounds.

H. W. Washburn in *Physical Methods in Chemical Analysis*, W. G. Berl, ed., Academic Press, New York, 1950, p. 592.

Bibliographies and Collections of Spectra

American Petroleum Institute Research Project 44. Texas A. and M. College, College Station, Texas. Catalog of mass spectra data.

Associated Electrical Industries, Ltd., Instrumentation Division, *Bibliography on Mass Spectrometry, 1938–1957* Inclusive, Pergamon Press, London, 1961.

F. W. McLafferty, *Mass Spectral Correlations*, Advances in Chemistry Series No. 40, American Chemical Society, Washington, D.C., 1963. Tabular correlations of mass spectral data.

PART FOUR

Tables of Organic Compounds with their Constants and Derivatives

Preface to the Tables

More than 7550 compounds are listed in the 35 tables that follow. A definite effort has been made to include as many members of each class as possible, provided that these conditions are met: (1) the compound is available commercially or is of definite interest in research; (2) it has a well-defined boiling point or melting point at atmospheric pressure; and (3) the melting point of one or more derivatives is available or sufficient physical constant data could be presented to make identification possible. It is regrettable that hundreds of commercially available compounds could not be included because, although the boiling point or melting point is known, no chemical derivative has been prepared and its melting point has not been published. As Wiberley and Drake[1] have said, "There never will be sufficient up-to-date tables of physical properties of compounds. The problem grows every year... A tremendous amount of physical data is to be found in technical periodicals, but it is indexed by compound and not by properties. The analyst, who most needs tables, wants to find the compound from its properties, not the properties from the compound."

Nomenclature

Much thought has been given during the preparation of these tables to the problem of nomenclature. It must be granted that in common usage the naming of organic compounds is far from being consistent or systematic. Our real desire is to have each name used in the tables easily recognized as referring to a specific structure. For many compounds, the trivial names are so well established that they have been retained even though systematic names might be used. For example, few chemists say *ethanoic* acid when referring to *acetic* acid, or ethyl *ethanoate* for ethyl *acetate*. On the other hand, it is believed that the name *methoxybenzene* is more descriptive of methyl phenyl ether than the name *anisole*, even though the latter is a widely used name. *Chemical Abstracts* uses *ortho*, *meta*, and *para* to designate the location of substituents for disubstituted derivatives of benzene. However, some of the most modern and extensive catalogs of organic compounds use *numbers* instead of *o-*, *m-*, and *p-* for most of the disubstituted derivatives as well as for the more complex structures. This is the practice that is being used in these tables even though the *o-*, *m-*, and *p* system is used in Chapters 11 to 21. Likewise, numbers are used instead

[1] J. S. Wiberley and H. W. Drake, *Microchem J.*, **4**, 283 (1960).

of Greek letters to designate the position of substituents for nonaromatic compounds; for example, 2-aminopropionic acid rather than α-aminopropionic acid. The positions on naphthalene are numbered instead of using Greek letters. The prefix *n-* (normal) is not used, and is to be assumed as applying unless another prefix is used. In general, as far as seems practical, the IUPAC nomenclature system is used. Pertinent comments related to the nomenclature for certain classes of compounds follows, and should be noted for each class.

Table 1: Acetals and ketals. These compounds are named as dialkoxy compounds. Thus, acetal (acetaldehyde diethyl acetal) is named 1,1-diethoxyethane.

Tables 2, 3, and 5: Acid anhydrides, acid halides, and carboxylic acids. The names formic, acetic, propionic and butyric have been retained. The higher alkanoic and alkenoic acids have been named systematically. For purposes of reference, certain trivial names are arranged alphabetically together with the systematic names used in these tables on page 661.

Table 8: Amides. In general, the amides are named to correspond with the names used for acids in Table 5.

Table 9: Primary and secondary amines. As is true for other aromatic compounds, numbers are used to designate positions for substituents. In this table, xylidenes are named as dimethylanilines, aminophenols as hydroxyanilines, and substituted toluidines as substituted methylanilines to avoid the confusion that exists in some literature because some systems, in naming the toluidines, consider the methyl group to be in position 1 while others designate the amino group as being in position 1.

Table 10: Tertiary amines. Picolines are named as methylpyridines, lutidines as dimethylpyridines, and collidines as trimethylpyridines.

Table 12: Esters. More trivial names have been retained in the table for esters than is true for the acids. Admittedly, this is not being consistent, but it is done in the belief that the names are more commonly used.

Table 13: Ethers. Most of the compounds are named as ethers. However, other names are used when it is believed that the name is more descriptive or more simple. For example, "ethylene glycol mono butyl ether" is named *butoxyethanol*, and *ethoxybenzene* is used instead of *phenetole*.

Table 14. In accordance with current usage, *alkyl halides* are named as halogen substituted hydrocarbons.

Table 29: Phenols. The name *cresol* is retained, but the substituted cresols are named as methylphenols to avoid confusion as to whether the methyl group or the hydroxyl group is to be considered as occupying position 1. The xylenols are named as dimethylphenols.

It is not possible to include in the tables all types of derivatives that have been prepared for any one of the common classes of organic compounds. As a supplement to certain of the tables, selected references to recent literature have been added to these tables. Attention is also called to the references given at

NAMES OF ACIDS

Trivial Names of Acids	Systematic Names Used in this Text
Acrylic	Propenoic
Adipic	Hexanedioic
Anisic	4-Methoxybenzoic
Arachidic	Eicosanic
Azelaic	Nonanedioic
Behenic	Docasanic
Capric	Decanoic
Caproic	Hexanoic
Caprylic	Octanoic
Cinnamic	3-Phenylpropenoic
Crotonic	2-Butenoic
Glutaric	Pentanedioic
Glycollic	Hydroxyacetic
Hydrocinnamic	3-Phenylpropionic
Isocaproic	4-Methylpentanoic
Isovaleric	3-Methylbutanoic
Lauric	Dodecanoic
Lignoceric	Tetracosanoic
Mandelic	2-Hydroxy-2-phenylacetic
Margaric	Heptadecanoic
Melissic	Tricontanoic
Myristic	Tetradecanoic
Palmitic	Hexadecanoic
Pivalic	Trimethylacetic
Pyromucic	2-Furoic
Pyruvic	2-Ketopropionic
Sebacic	Decanedioic
Stearic	Octadecanoic
Suberic	Octanedioic
Succinic	Butanedioic
Valeric	Pentanoic

the end of the chapters dealing with the preparation of derivatives, since many of these articles also give data on derivatives of a number of members of certain classes. In addition to the tables of physical constants for pure compounds and their derivatives found in textbooks on qualitative organic analysis, such as those listed at the end of Chapter 1, the following specialized works are very valuable sources of data on physical constants.

Selected General Reference Works

Amelink, Felix, *Rapid Microchemical Identification Methods in Pharmacy and Toxicology*, Interscience, New York, 1962.

Beilstein, F., *Handbuch der Organischen Chemie*, 4th ed., F. Richter, ed., Springer, Berlin, 1918–1963. This impressive work consists of four series of publications.

The main series, volumes 1–27, covers the literature to 1910. The first supplement covers from 1910–1919; the second from 1920–1929; and the third, which is now in the course of publication, covers the literature from 1930–1950 This work surveys the literature for all organic compounds of known structure.

Egloff, G., *Physical Constants of Hydrocarbons*, Reinhold, New York, 1939–1947, 4 vols.

Heilbron, I., *Dictionary of Organic Compounds*, 2nd ed., Oxford, New York, 1953, 4 vols. This is the most extensive catalog of organic compounds in the English language that has been published in complete form. Contains the formula, physical data, and characteristic reactions, together with the melting point of identification derivatives and many references. In addition to the commonly encountered organic compounds, Heilbron contains data on compounds such as amino, diamino, halo, and halo-amino combinations of anthraquinones, quinolines, pyridines, pyrimidines, xylenes, xylidines, piperidines, and pyrenes. Also, many hydroxy, nitro, methyl, and methoxy combinations of the above mentioned compounds may be found in these works.

Hodgman, C. D., ed., *Tables for the Identification of Organic Compounds* Chemical Rubber Co., Cleveland, Ohio, 1960.

International Critical Tables of Numerical Data, Physics, Chemistry, and Technology, McGraw-Hill, New York, 1926–1933, 7 vols. and index.

Kempf, R., and Kutter, F., *Schmelzpunkt Tabellen zur Organischen Molekular-Analyse*, Edwards Bros., Ann Arbor, Mich., 1944.

Lange, N. A. ed., *Handbook of Chemistry*, 10th ed., McGraw-Hill, New York, 1961.

Markley, K. S., *Fatty Acids*, Interscience, New York, 1960. Two volumes published, more to follow. Extensive tables on the physical constants of acids and their derivatives.

McCrone, W. C., *Fusion Methods in Chemical Microscopy*, Interscience, New York, 1957. Numerous melting points of pure compounds and eutectics.

Mulliken, S. P., *The Identification of Pure Organic Compounds*, Wiley, New York, 1904–1922, 4 vols. Volume I revised by C. H. Huntress, Wiley, New York, 1941.

Radt, F., ed., *Elseviers Encyclopedia of Organic Chemistry*, Elsevier, Houston, 1946–1956. Two volumes in 16 parts containing bicyclic compounds, compounds of napthalene, tricyclic, tetracyclic and higher compounds.

Rausch, D. A., and Postelnek, W., and Lovelace, A. M., *Aliphatic Fluorine Compounds*, Reimhold, New York, 1958.

Reid, E. E., *Organic Chemistry of Bivalent Sulfur*, Chemical Publishing Co., New York, 1958–1963. Five volumes.

Rodd, E. H., *Chemistry of Carbon Compounds*, Elsevier, Houston, 1952–1962. Five volumes in 10 parts including a general index. Vol. I, Aliphatics; Vol. II, Alicyclics; Vol. III, Aromatics; Vol. IV, Aromatics; Vol. V, Miscellaneous and index.

Timmermans, J., *Physico-Chemical Constants of Pure Organic Compounds*, Elsevier, Houston, 1950.

Utermark, W., and Schicke, W., *Melting Point Tables of Organic Compounds*, Interscience, New York, 1963.

Weissberger, A., consulting ed., *The Chemistry of Heterocyclic Compounds*, Interscience, New York, 1963. A series of monographs. Sixteen titles have been published to

1963. For example: H. D. Hartough, *Thiophene and its Derivatives* (1952, being Vol. 3 of the series); E. Kingsberg, *Pyridine and its Derivatives* (in 4 parts, 1960–1962, being Vol. 14; D. J. Brown, *The Pyrimidines* (1962, being Vol. 16).

General Comments Regarding the Tables

Compounds melting above 20° are classed as solids and are arranged according to increasing *melting points*. Compounds melting below 20° are arranged according to increasing *boiling points*. Compounds melting between 20° and 30° are usually listed in both the solid and liquid sections of a table. All temperatures are degrees centigrade; density is given at 20°/4°, and refractive index as n_D^{20}, unless otherwise indicated.

An asterisk following the melting point or boiling point of a derivative indicates that the value is corrected. The degree sign is usually omitted after the figures denoting *mp* or *bp* (or *m* or *b*).

Those compounds containing more than one functional group may generally be found in each of the tables corresponding to those functional groups. In some cases, derivatives may be obtained for more than one of the functional groups, in which case reference is made to the compound of interest in another table. Polyfunctional compounds for which derivatives are given for only one functional group are generally listed in that table only. The melting points (or boiling points) of some additional derivatives may be found in the notes as indicated. Any notes accompanying a compound should be consulted for pertinent information.

When a derivative exists in two or more forms, each having a different melting point, the values are connected by the word *and*; thus, *118* and *145* indicates that the particular derivative exists in two forms, one that melts at 118° and another at 145°. On the other hand, there are many derivatives for which several melting points are listed in the literature; in some cases, besides the value selected for the tables in this book, one or two additional values have been listed in parentheses under the selected value. For a more elaborate discussion of this topic, see pages 197–201.

Consult pages 429–434 with reference to the selection of the derivative to be prepared.

In case the available data indicate a particular compound but the prepared derivative does not give the melting point listed in the appropriate table, consult pages 201–202.

For easier reference and also for identification in the index, numbers appear in sequence (1, 2, 3, etc.) to the left of the names of the compounds listed in the left column of each table. Whenever a table continues across two facing pages, the names of the compounds are omitted on the second page, but the numbers are repeated.

The melting points of hundreds of compounds may be found in these tables even though they are not listed under the specific class name and, hence, the names of these compounds are not included in the index. For example, there are no "tables" of such compounds as oximes or phenylhydrazones, but the melting point of all of the common oximes and phenylhydrazones may be found by consulting the appropriate columns of derivatives in the tables for aldehydes and ketones.

Commercial products often melt over a range of at least two degrees. The catalogs of commercial suppliers of organic chemicals usually indicate the purity of their products by giving the ranges over which the chemicals boil or melt.

For possible changes and additions—particularly for corrections in the physical constants of the compounds listed, new additions, and new derivatives—we should appreciate suggestions from all who are interested in this undertaking. Such communications, which we earnestly solicit, may be sent to us in care of the publishers.

Abbreviations

As far as possible, abbreviations conform to those used in the *Chemical Abstracts*. Listed below are a few abbreviations used in the tables that may be less familiar.

N (after the boiling point, or melting point): Indicates a note to be found at the end of the group of compounds, in which this particular compound appears. The reference number for the note will be found in the column headed Note.

r.h.: rapid heating   *d:* decomposes
s.h.: slow heating   *v.s.:* volatile with steam
s.t.: sealed tube   *s:* sublimes

List of Tables

The following *List of Tables* by class of compound is given in order that the desired table may be located quickly.

LIST OF THE TABLES

Class of Compound	Table Number[a]	Page
Acetals and Ketals	1	667
Acid Anhydrides	2	668
Acids, amino	4	672
Acids, carboxylic	5A	684
	5B	688
Acids, sulfonic	32	959
Acid Halides, acyl and aroyl	3	670
Acid Halides, sulfonyl	31	953
Alcohols	6A	714
	6B	722
Aldehydes	7A	728
	7B	734
Amides and Ureas	8	744
Amides, sulfonamides	30	948
Amines, primary and secondary	9A	750
	9B	762
Amines, tertiary	10A	799
	10B	804
Azo compounds	22	909
Azoxy compounds	23	911
Carbohydrates	11	809
Esters	12A	814
	12B	821
Ethers	13A	823
	13B	827
Halogen compounds, haloalkanes	14	830
Halogen compounds, aromatic	15A	842
	15B	846

[a] Many of the tables are divided into two sections, A for liquids and B for solids.

LIST OF THE TABLES (continued)

Class of Compound	Table Number[a]	Page
Halogen compounds, di- and poly-, nonaromatic	16A	852
	16B	856
Halogen compounds, perfluoro-	17	857
Hydrazo compounds	24	911
Hydrocarbons, alkanes and cycloalkanes	18	862
Hydrocarbons, alkenes, alkynes	19	870
Hydrocarbons, arenes(aromatics)	20A	877
	20B	886
Isocyanates and Isocyanides	25	912
Isothiocyanates and Thiocyanates	33	961
Ketones	21A	892
	21B	899
Nitriles	26A	913
	26B	914
Nitro compounds	27A	916
	27B	917
Nitroso compounds	28	923
Phenols	29A	924
	29B	926
Thioethers (sulfides)	34	962
Thiols (mercaptans and thiophenols)	35A	966
	35B	967

[a] Many of the tables are divided into two sections, A for liquids, and B for solids.

TABLE I
ACETALS AND KETALS

	Name of Compound	BP	Products of Hydrolysis	
			Alcohol	Carbonyl Compound
1	Dimethoxymethane	43	Methanol	Methanal
2	1,1-Dimethoxyethane	64	Methanol	Ethanal
3	2-Methyl-1,3-dioxolane	82	Ethanediol	Ethanal
4	2,2-Dimethoxypropane	83	Methanol	Propanone
5	Diethoxymethane	88	Ethanol	Methanal
6	1,1-Diethoxyethane	103	Ethanol	Ethanal
7	1,3-Dioxane	106	1,3-Propanediol	Methanal
8	2-Methyl-1,3-dioxane	110	1,3-Propanediol	Ethanal
9	1,1-Di(2-propoxy)ethane	114	2-Propanol	Ethanal
10	2,2-Diethoxypropane	114	Ethanol	Propanone
11	1,1-Dimethoxybutane	114	Methanol	Butanal
12	Diisopropoxymethane	121	2-Propanol	Methanal
13	1,1-Diethoxypropane	124	Ethanol	Propanal
14	1,1-Diethoxy-2-propene	126	Ethanol	2-Propenal
15	1,1-Diethoxy-2-methylpropane	137	Ethanol	2-Methylpropanal
16	Dipropoxymethane	137	Propanol	Methanal
17	1,1-Dimethoxy-2-chloroethane	140	Methanol	2-Chloroethanal
18	1,1-Diethoxybutane	143	Ethanol	Butanal
19	1,1-Dimethoxy-2-bromoethane	145	Methanol	2-Bromoethanal
20	1,1-Dipropoxyethane	147	Propanol	Ethanal
21	1,1-Diisopropoxypropane	147	2-Propanol	Propanal
22	3,3-Diethoxypentane	154	Ethanol	3-Pentanone
23	1,1-Diethoxy-2-chloroethane	157	Ethanol	2-Chloroethanal
24	Di(2-methylpropoxy)methane	164	2-Methylpropanol	Methanal
25	1,1-Diethoxy-2-bromoethane	170	Ethanol	2-Bromoethanal
26	1,1-Di(2-butoxy)ethane	171	2-Butanol	Ethanal
27	1,1-Di(2-methylpropoxy)ethane	176	2-Methylpropanol	Ethanal
28	Dibutoxymethane	181	Butanol	Methanal
29	1,1-Dipropoxybutane	182	Propanol	Butanal
30	1,1-Diethoxy-2,2-dichloroethane	184	Ethanol	Dichloroethanal
31	1,1-Dibutoxyethane	186	Butanol	Ethanal
32	Benzaldehyde dimethyl acetal	207(199)	Methanol	Benzaldehyde
33	1,1-Diethoxyheptane	209	Ethanol	Heptanal
34	1,1-Dibutoxybutane	213	Butanol	Butanal
35	Di(2-chloroethoxy)methane	218	2-Chloroethanol	Methanal
36	Dipentoxymethane	219	Pentanol	Methanal
37	1,1-Dimethoxy-2-phenylethane	220	Methanol	2-Phenylethanal
38	Dipentoxyethane	222	Pentanol	Ethanal
39	Benzaldehyde diethyl acetal	222	Ethanol	Benzaldehyde
40	Dihexoxymethane	255	Hexanol	Methanal
41	Benzaldehyde dibutyl acetal	262	Butanol	Benzaldehyde

TABLE 2
ACID ANHYDRIDES

	Name of Compound	Note	BP	MP	Acid BP	Acid MP	Amide	Anilide	p-Toluidide
1	Trifluoroacetic		39		72		75	88	
2	Acetic		140		118	16	82	114	147(153)
3	Aceticpropionic		154						
4	Propionic		167		141		81	106	126
5	Isobutyric		182		154		128	105	107
6	Trimethylacetic		190		164	35	154	129	120
7	Butyric		198		162.5		115	96	75
8	γ-Butyrolactone	1	206						
9	γ-Valerolactone		207						
10	Citraconic		214			92d	185-7d(di)	175(di)	
11	3-Methylbutanoic		215		176.5		135(137)		107
12	Dichloroacetic		216d		194		98s	118	153
13	Pentanoic		218		186.3		106	63	74
14	Crotonic		248		189	72	161(158)	118(115)	132
15	Hexanoic		254-7 (245)		205.1		100	95(92)	75(73)
16	Heptanoic		258	17	223		96	70(65)	81
17	Octanoic		280-5		239.3	16	110(106)	57(55)	70
18	Oleic			22	216/5 mm	16	76	41	43
19	Decanoic		285	24	268-70	31.2	108(98)	70	78
20	Undecanoic			37	284	30	103(99)	71	80
21	o-Toluic			39		104-5	143	[125]	144
22	Bromoacetic			41-2	208	50	91	131	
23	Dodecanoic			42	299	44	110(102)	78	87
24	Benzoic		360	42		122	130	163	158
25	Chloroacetic			46	189	63	121	134	162
26	Tridecanoic			50	312	44	100	80	88
27	Maleic		200 (82/14 mm)	52-4		130	181	173-5 (mono) 187(di)	142(di)
28	Tetradecanoic			54	202/16 mm	54.1	107(103)	84	93
29	Glutaric			56	200/20 mm	98	175-6(di)	224	218
30	Suberic (dimer)			56-7		144(141)	127(mono) 217(di)	128(mono) 186(di)	218(di)
31	Hexadecanoic			64	222/16 mm	63	106-7	90	98
32	Heptadecanoic			67	231/16 mm	61	108		
33	Itaconic	2		67-8		165	192(di)	190(185)	
34	Sebacic (dimer)			68	243/15 mm	133s	210(di) 170(mono)	201(di) 122(mono)	201
35	Octadecanoic			70		70	109	95	102
36	m-Toluic			71		112	94	126	118
37	Phenylacetic			72		76.5	156	118	136
38	1-Cyclohexene-1,2-dicarboxylic anhydride			74		126			155d
39	Aconitic(cis)	4		74		125		170(mono)	
40	Dodecandioic			76-8					
41	Eicosanic			77.5	204/1 mm	75.2	108-9	92	96
42	2-Chlorobenzoic			79		142(140)	142	114(118)	131
43	Docosanic			82		80.2			
44	3-Chlorobenzoic			95		158(155)	134	122	
45	p-Toluic			95		179-80	160	145	160
46	4-Methoxybenzoic			99		184-6	167(163)	169-71	186
47	4-Ethoxybenzoic			108		198	202	170	
48	3,5-Dinitrobenzoic			109		204-5	183	234	

TABLE 2 (Continued)
ACID ANHYDRIDES

	Name of Compound	Note	BP	MP	Acid BP	Acid MP	MP of Recommended Derivatives Amide	MP of Recommended Derivatives Anilide	MP of Recommended Derivatives p-Toluidide
49	4-Nitrophthalic			119		165	200d	192	172(mono)
50	Butanedioic		261	119–20	235d	186–8	157(mono) 260(di)	148(mono) 230(di)	180(mono) 255(di)
	(131/10 mm)								
51	Dichloromaleic			120		116	175d (mono)	170 and 193	
52	Nicotinic			123		237–8s	128	85	150
53	o-Sulfobenzoic			129		68–9		194–5(di)	
54	Phthalic		295	131.6		206d	220(di) 149(mono)	253–5 (170)	201(150)
55	Aconitic(trans)	4		134		195	250	189(di)	
56	2-Nitrobenzoic			135		146	176	155	
57	Cinnamic			136 (130)		133	148(142)	151(153)	168
58	1,2-Cyclohexane dicarboxylic (trans)			140		221			
59	1-Naphthoic			146		162	202	163	
60	2,4-Dinitro-benzoic			160		183	203		
61	3-Nitrobenzoic			160		140	143	154	162
62	3-Nitrophthalic			162		218	201d(di)	234(di)	226(di)
63	1,2-Naphthalic			169					
64	4-Nitrobenzoic			189		241	201(198)	211(204)	204(192)
65	4-Chlorobenzoic			194		240	179(170)	194	
66	Diphenic			217		229	212(di)	230	
67	4-Bromobenzoic			218		251	189	197	
68	d-Camphoric (dl, mp 221–3)			221		188	177(mono) 193(di)	204(mono) 226(di)	α-212–4 β-190–6
69	2,3-Naphthalic			246		239–41			
70	Tetrachloro-phthalic			256		250d			
71	1,8-Naphthalic	3		274					
72	Tetrabromo-phthalic			275		266			
73	Tetraiodo-phthalic			329–31		327			

NOTES ON ACID ANHYDRIDES

1. On ammonolysis with concentrated ammonia gives amide, mp 99(87).
2. Not volatile with steam (differentiation from citraconic). Anilide obtained only by boiling excess of amine with acid.
3. Naphthalimide, mp 300, is formed by heating compound with excess aqueous NH_3; derivative is purified by boiling with Na_2CO_3 solution. The compound boiled with aniline forms N-phenylnaphthalimide, mp 202.
4. The triamide of the *trans* isomer turns brown at about 250 and sinters without melting at about 260. The dianilides of both isomers melt at 190 and 200.

TABLE 3
ACID HALIDES

	Name of Compound	MP	BP	MP of Recommended Derivatives	
				Amide	Anilide
	FLUORIDES				
1	Acetyl fluoride		20.5	82	114
2	Propionyl fluoride		44–6	81	106
3	Fluoroacetyl fluoride		50.5–51.0	108	
4	Trichloroacetyl fluoride		66–8	141	97(94)
5	Butyryl fluoride		67	115	96
6	Chloroacetyl fluoride		73–5	120	137(134)
	CHLORIDES				
7	Acetyl chloride		51–2	82	114
8	Oxalyl chloride		64	419d	246
9	Mesaconyl dichloride		64–5(di)	177(di)	186(di)
10	Fluoroacetyl chloride		71.5–73	108	
11	Propenoyl chloride		76	85	105
12	Propionyl chloride		80	81	106
13	Isobutyryl chloride		92	128	105
14	Ethyl chloroformate		94–5	49	52
15	Butyryl chloride		101–2	115	96
16	Trimethylacetyl chloride		105–6	154	129
17	Chloroacetyl chloride		105–6	120	137(134)
18	Dichloroacetyl chloride		108	98s	118
19	2-Chloropropionyl chloride		109–10	80	92
20	Methoxyacetyl chloride		113	97	58
21	3-Methylbutanoyl chloride		115	135(137)	109
22	*dl*-2-Methylbutanoyl chloride		115–6	112(121)	
23	Trichloroacetyl chloride		118	141	97(94)
24	Pentanoyl chloride		126	106	63
25	Crotonyl chloride		126	161(158)	118(115)
26	Ethyloxalyl chloride		131–2		66–7
27	Bromoacetyl chloride		134	91	131
28	2-Ethylbutanoyl chloride		137–9	112(107)	127
29	4-Methylpentanoyl chloride		147	121	112(110)
30	Hexanoyl chloride		153	100	95
31	Fumaryl chloride		160–2	266(di)	
32	Heptanoyl chloride		175	96	70
33	Phenyl chloroformate		187d	141	
34	Butanedioyl chloride		190d	157(mono) 273(di)	148(mono) 230(di)
35	Octanoyl chloride		196	110(106)	57(55)
36	Benzoyl chloride		197	130	163
37	Diethylmalonyl dichloride		197	146(mono)	
38	Phenylacetyl chloride		210	156	118
39	Nonanoyl chloride		215.3	99	57
40	2,6-Dimethylbenzoyl chloride		217	139	
41	Pentanedioyl chloride		218	175–6(di)	224
42	2-Ethylbenzoyl chloride		219	151–3	
43	4-Chlorobenzoyl chloride	16	222	179(170)	194
44	3-Chlorobenzoyl chloride		225	134	122
45	Decanoyl chloride		232	108(98)	70
46	2-Chlorobenzoyl chloride		238	142	114(118)
47	3,4-Dichlorobenzoyl chloride		242	133	

TABLE 3 (Continued)
ACID HALIDES

	Name of Compound	MP	BP	MP of Recommended Derivative	
				Amide	Anilide
48	3-Bromobenzoyl chloride		243(239)	155	136
49	3-Methoxybenzoyl chloride (benzylamine salt of the acid, mp 113-4)		244		
50	2-Bromobenzoyl chloride		245	156	141
51	2-Methoxybenzoyl chloride		254	129	131
52	Phthaloyl chloride(di)		281	149(mono)	170(mono)
		15–6	(276)	220(di)	253(di)
53	Dodecanoyl chloride	−17	145/18 mm	110(102)	78
54	Tetradecanoyl chloride	1–3	174/16 mm	107(103)	84
55	Hexadecanoyl chloride	11–2	194/17 mm	106–7	90
56	Salicylyl chloride	19–20	92/15 mm	142	136
57	2-Nitrobenzoyl chloride	20	148/9 mm	176	155
58	Octadecanoyl chloride	23	202–3/6 mm	109	94
59	4-Methoxybenzoyl chloride	24	145/14 mm	162	169
60	3-Nitrobenzoyl chloride	35	278	143	154
61	Cinnamoyl chloride	36	257.5	149	153
62	4-Bromobenzoyl chloride	42	245–7d	190	197
63	Picolinyl Chloride	46		106.5	76
64	2,4-Dinitrobenzoyl chloride	46		203	
65	4-Nitrophenylacetyl chloride	48		198	198
66	3,5-Dinitrobenzoyl chloride	68–9(74)		183	234
67	4-Nitrobenzoyl chloride	75	150–2/15 mm	201(198)	211(204)
68	3-Nitrophthaloyl chloride	77		201d(di)	234(di)
69	Picryl chloride (see Table 27)	83			
70	Diphenylcarbamyl chloride	86		189	
71	4-Phenylazobenzoyl chloride	93–4		224–5	
72	4-Phenylbenzoyl chloride	114–5		223	
	BROMIDES				
73	Acetyl bromide		81	82	114
74	Propionyl bromide		103	81	106
75	2-Methylpropionyl bromide		114–6	130	105
76	Chloroacetyl bromide		127	120	137(134)
77	Butyryl bromide		128	115	96
78	3-Methylbutanoyl bromide		138–40	135(137)	
79	Bromoacetyl bromide		149	91	131
80	2-Bromopropionyl bromide		153	123	99(110)
81	Benzoyl bromide		218	130	163
82	4-Iodobenzoyl bromide	55		217	210
83	3,5-Dinitrobenzoyl bromide	60		183	234
84	4-Nitrobenzoyl bromide	64		201(198)	211(204)
	IODIDES				
85	Acetyl iodide		108	82	114
86	Propionyl iodide		127	81	106
87	Butyryl iodide		146–8	115	96

REFERENCE FOR TABLE 3

Sonntag, N. O. V., *Chem. Rev.* **52**, 260 (1953). The reactions of aliphatic chlorides. Lists the melting points of the amides derived from a number of acid chlorides.

TABLE 4
AMINO ACIDS[a]

	Name of Compound	MP	Recommended		Others		
			p-Toluene-sulfonyl	Phenyl-urea	Dinitro-phenyl	Benzoyl	Acetyl
1	Histamine	83–4					
2	m-Aminohydrocinnamic acid	84–5					162
3	L-(+)-Valine	93–6 (315s.t.)	147			127	156
4	N-Methyl-β-alanine	99–100					
5	Aminomalonic acid	109				61	
6	2-Aminophenylacetic acid	119				179	158
7	2-Amino-1-naphthoic acid	126					195–6
8	N-Phenylglycine	127		195		63	194
9	p-Aminohydrocinnamic acid	132				194–5	124(hyd.) 143(anh.)
10	L-Ornithine	140N		190		240(mono) 189(di)	
11	Anthranilic acid (2-Aminobenzoic) (see Table 9B)	147 (145)	217	181		182	185
12	3-Aminophenyl acetic acid	151					
13	D-(−)-Valine	156–7 (239s.t.)					156
14	5-Aminopentanoic acid	157				105 (94)	
15	o-Aminocinnamic acid	158–9				191–3	250–1 (mono) 158(di)
16	2-Chloro-3-amino-benzoic acid	160–1					207
17	DL-3-Aminopentanoic acid	160–5				145–6	
18	2-Amino-p-toluic acid	165					279–81
19	3-Aminobenzoic acid (see Table 9B)	174		270		248	250
20	3-(4-Aminophenyl)-propenoic acid	175				274	259–60
21	ω-Aminotridecyclic acid	177				111 (105)	
22	4-Amino-1-naphthoic acid	177					189
23	3-Amino-p-toluic acid	177					184
24	m-Aminocinnamic acid (trans)	181				229	237
25	3-Amino-1-naphthoic acid	181–2					254–5
26	N-Ethylglycine	182					

[a] The melting points of amino acids are strictly decomposition points, which vary widely depending on the rate of heating and the temperature at which the crystals were dried. Miscellaneous derivatives are given in the notes at the end of the table that are numbered to correspond to the number of the acid in the table.

TABLE 4 (Continued)
AMINO ACIDS[a]

| | | | Melting Point of Derivatives | | | | |
| | | | Others (continued) | | | | |
	Formyl	Picrate	Picrolonate	α-Naphthyl-urea	3,5-Dinitro-Benzoyl	β-Naphthalene-sulfonyl	Phenyl-hydantoin
1		160–2(mono) 238–42(di)	262–4				
2							
3			170–80		157–8		131–3
4							
5	48						
6	110						
7							
8	125						
9							
10		208	220–1(mono) 235–6(di)			189	
11	169	104			278		
12							
13	156						
14							
15							
16							
17						134	
18							
19					270		
20							
21							
22							
23	186						
24							
25							
26							

(Continued)

TABLE 4 (Continued)
AMINO ACIDS[a]

	Name of Compound	MP	Melting Point of Derivatives				
			Recommended		Others		
			p-Toluene-sulfonyl	Phenyl-urea	Dinitro-phenyl	Benzoyl	Acetyl
27	L-Canavanine $(\alpha)_D^{20}$, +7.90 $(C_5H_{12}O_3N_4)$	184				86d(tri)	
28	L-Glutamine	185					
29	Hippuric acid	187					
30	4-Aminobenzoic acid (see Table 9B)	188		300		278	252
31	DL-Canaline	190–5				158–60	
32	DL-β-Aminobutyric acid	194				154	
33	m-Aminocinnamic acid (cis)	195d					
34	DL-Glutamic acid	199	117			153 (155–7)	185–7
35	4-Aminophenylacetic acid	199–200d				205–6	168–70
36	β-Alanine	200 (196)		168 (174)		120	
37	D,L-Isoserine	200d				107–9*	
38	6-Aminohexanoic acid	202–3				75–8	
39	4-Aminobutyric acid	203d					
40	DL-Proline (mono hyd. mp 191)	203 (205)		170			
41	3-Aminooctanoic acid	204–6d					
42	4-Aminohexanoic acid	205–7d				150–2	
43	6-Amino-1-naphthoic acid	205–6					170–2 (252–3)
44	1-Amino-2-naphthoic acid	205					
45	L-Arginine $(\alpha)_D^{20}$ 11.37	207d			252	298(mono) 235(di)	
46	L-Glutamic acid	211–3N	131				
47	5-Amino-1-naphthoic acid	212					296
48	Sarcosine (N-methylglycine)	212–3 (210)		102		104	135
49	L-3,5,-Diiodotyrosine	213					
50	L-α-Asparagine	213–15					
51	DL-4-Aminopentanoic acid	214*				132	
52	L-Canaline	214d				99(di)	
53	3-Amino-2-naphthoic acid	214					238
54	2-Chloro-4-amino benzoic acid	214–5					206
55	5-Chloro-3-amino-benzoic acid	216					265–7
56	4-Chloro-3-amino-benzoic acid	216–7					265
57	2-Amino-2-methyl butyric acid	217					
58	β-Hydroxyvaline	218d		182		153(mono)	
59	L-(−)-Proline	222d	130–3	170N	137	156(mono)	

TABLE 4 (Continued)
AMINO ACIDS[a]

	Melting Point of Derivatives						
				Others (continued)			
	Formyl	Picrate	Picrolonate	α-Naphthyl-urea	3,5-Dinitro-Benzoyl	β-Naphthalene-sulfonyl	Phenyl-hydantoin
27		163–4					
28							
29							
30	268				290		
31							
32							
33							
34	182		184d	236			165*
35							
36				231–3	202.5		
37							
38							
39							
40	135–7				217		118*
41							
42							
43							
44							
45		217(mono) 190d(di)			150	87–9	
46							
47							
48					153–5		
49							
50							
51							
52		192–3d					
53							
54							
55							
56							
57							
58							
59		154				138(anh.) 134(hyd.)	144

(Continued)

TABLE 4 (Continued)
AMINO ACIDS[a]

	Name of Compound	MP	Melting Point of Derivatives				
			Recommended		Others		
			p-Toluene-sulfonyl	Phenyl-urea	Dinitro-phenyl	Benzoyl	Acetyl
60	L-Citrulline	222 (226)					
61	7-Amino-1-naphthoic acid	223–4					229
62	D- or L-Lysine	224–5d		184		149–50(di)N	
63	6-Amino-2-naphthoic acid	225					230–2
64	β-L-Asparagine (β-aspartamide)	227N (234–5)	175	164		189	
65	DL-Threonine (α-amino-β-hydroxybutyric acid)	227–9N		177–8	152	145N	
66	L-Serine	228d					
67	Glycine	228–32N (262d)	147 (150)	197*	195	185–7	206
68	β-Hydroxynorvaline	230–1d		156			170–1(mono)
69	DL-Thyroxine	231–3N				210–5d	
70	DL-3-Amino-3-phenylpropionic acid	231				199	161–2
71	D- or L-3-Amino-3-phenylpropionic acid	234–5					
72	3-Aminosalicylic acid	235d				189	215
73	DL-Allothreonine	237–9N				175–6	
74	DL-Arginine	238				230(di, anh.) 176(di, hyd.)	
75	4-Dimethylaminobenzoic acid	242					
76	3-Amino-3-phenylisobutyric acid	243				205	
77	DL-Serine	246d,N	213	169	199	149–50	
78	DL-Isoserine (β-aminolactic acid)	248d,N		184		151*	
79	4-Hydroxyphenylglycine	248 (200d)				117	203(mono) 175(di)
80	Isoaspartic acid	250					
81	D- or L-Threonine	251–2				147–8N	
82	DL-α-Aminophenylacetic acid	256(subl.)				178	198
83	3-Chloro-5-aminosalicylic acid	259–60					258(mono) 218(di)
84	L-Cystine	260d	204–5*	160	109	147–8N 181(di)N	
85	DL-2-Amino-2-phenylpropanoic acid	260(subl.)					
86	L-Glycylglycine (diglycine)	260–2 (262–4)	178	176		208	187–9
87	2-Aminodecanoic acid	264				136	
88	DL-Phenylalanine	264d,N (272)	134–5	182	186	188*	
89	DL-Aspartic acid	270d			196	119(hyd.) (177, anh.)	
90	D- or L-Aspartic acid	270–1	140	162		185*	

TABLE 4 (Continued)
AMINO ACIDS[a]

	Melting Point of Derivatives						
	Others (continued)						
	Formyl	Picrate	Picrolonate	α-Naphthyl-urea	3,5-Dinitro-Benzoyl	β-Naphthalene-sulfonyl	Phenyl-hydantoin
60		206					
61							
62		266d	246–52				183–4*
63							
64		180d		199	196 (199)		
65							164–5
66							
67	153–4	190	214–5d	191	179	159*	
68							154–5
69							
70	128–9						
71	142–3						
72							
73							
74		200–1(mono) 196(di)	248 (231)				
75							
76							
77			265d	192	95	214*	168–9
78							
79							
80							
81							
82	180						
83							
84					180d	226–30	117
85							
86				217	210	180–2	
87							
88	168–9	173	238N	143–4	93		173–4
89			130				
90				115d		153	

(Continued)

TABLE 4 (Continued)
AMINO ACIDS[a]

	Name of Compound	MP	Melting Point of Derivatives				
			Recommended		Others		
			p-Toluene-sulfonyl	Phenyl-urea	Dinitro-phenyl	Benzoyl	Acetyl
91	2-Amino-octanoic acid	270 (264)				128	
92	DL-2-Amino-nonanoic acid	273				128	
93	L-Hydroxyproline	274N	153	175		100(mono)	
94	DL-Tryptophane	275–82	176				
95	L-(+)-Alloisoleucine	278d,N		151			
96	2-Aminoisobutyric acid	280(subl).				199 (202)	
97	DL-2-Aminoheptanoic acid	281				135	
98	DL-Methionine	281 (272)	105		117	145	114
99	L-Methionine	283d,N					98–9
100	DL-Lanthionine	283				195–8 (di)	
101	5-Aminosalicylic acid	283 (280d)				252	184(di) 218(mono)
102	D- or L-Phenyl-alanine	283r.h. (320)N	164–5 (161)	181*		146	
103	L-(+)-Isoleucine	283d,N	130–2	121		117	
104	L-Histidine	288N (253)	202–4d			230d (mono)	
105	L-Tryptophane	290N	176	166	175	183	
106	DL-Isoleucine	292N	141*	120	166	118	
107	DL-N-Methyl-α-alanine	292					
108	L-(+)-2-Amino butyric acid	292d (303s.t.)				121	
109	DL-Leucine	293–5d,s.t. (332)		165		137–41	157–8
110	D or L-Leucine	293–5r.h. (337)	124*	115		105–7 (anh.)	189–90
111	DL-2-Amino-2-methyl pentanoic acid	295s.t.					
112	DL-Alanine	293–5	139	174dN		166*	137
113	D- or L-2-Amino-2-phenylpropionic acid	295 (subl.)					
114	D- or L-Alanine	297d	133 (92–4)	175d (168)		151	116
115	DL-Norleucine	297–300 (275s.t.)	124				
116	DL-Valine	298d,s.t. (282)(292)	110	164		132	
117	L-(+) or D-(—)-Norleucine (s.275)	301		53			
118	Creatine	303					165(di)
119	DL-Norvaline	303s.t.		117			
120	DL-2-Aminobutyric acid	304		170		147	
121	Lanthionine (meso)	304				198–200	
122	Creatinine	305d (260)					

TABLE 4 (Continued)
AMINO ACIDS[a]

	Melting Point of Derivatives						
				Others (continued)			
	Formyl	Picrate	Picrolonate	α-Naphthyl-urea	3,5-Dinitro-Benzoyl	β-Naphthalene-sulfonyl	Phenyl-hydantoin
91							
92							
93							123–4
94					240		
95	126			166–8			
96							
97							
98	99–100		179–80				
99				188	94–5(hyd.) 150(anh.)		
100							
101							
102	167		208d	155	93		
103	156			178–9d			
104		86	232(mono) 265d(di)		189	149–50	
105	195–6		203–4	159–60	233d	185	
106	121			178			
107							
108	126			195		148	120–7d
109			150r.h. 180s.h.	163	187	145–6	
110	141–4		150		187	68*	163.5
111							
112			216d	198	177	152–3*	
113	194–5						
114			217d(mono) 145(di)	202d (198)		122–3	
115	114.5						
116	140–5		150r.h. 250s.h.	204	158		125
117	115–6					149	
118		218–20					
119	132						103
120					194		126
121							
122		220					

(Continued)

TABLE 4 (Continued)
AMINO ACIDS[a]

	Name of Compound	MP	Melting Point of Derivatives				
			Recommended		Others		
			p-Toluene-sulfonyl	Phenyl-urea	Dinitro-phenyl	Benzoyl	Acetyl
123	Betaine	305					
124	L-α-Aminophenylacetic acid	305–10					191
125	Diaminosuccinic acid (meso)	306					235(di)
126	[L-(+) or D-(−)-] Norvaline	307d				64	137
127	[L-(−)-] Tyrosine	314–8d,r.h. 290–5d,s.h.	N-188* 114(di)	104 194		N-166–7 211–12(di)	N-148 172(di)
128	L-(+)-Valine	315	147			127	
129	DL-Tyrosine	340r.h. 290–5s.h.	224–6			N-197	
130	Djenkolic acid	300–50d (250)				166(mono) 85(di)	
131	DL-Ornithine		188(mono)	192		4-N-285–88 188(di)	
132	DL-Lysine			196		249(mono)N 146(di)	
133	DL-Norvaline	303s.t.,N		117			
134	L-Cysteine	None					

NOTES ON AMINO ACIDS

The number in parenthesis refers to the compound as listed in Table 4.

(1) Hydrochloride, mp 244–6.
(2) Hydrochloride, mp 191.
(3) N-Benzenesulfonyl, mp 153; Carbobenzoxy, mp 64–65.
(4) Hydrochloride, mp 105.
(5) Heat converts aminomalonic acid to glycine.
(10) No melting point is given in any of the standard works. D-Ornithine is listed as a syrup. The melting point of 140 listed was obtained from Vickery and Cook, J. Biol. Chem. **94**, 398 (1931). Flavinate, mp 234–5; Hydrochloride, mp 233.
(11) N-Benzenesulfonyl, mp 214; N-Chloroacetyl, mp 187–8.
(12) Amide, mp 164–6; N-Chloroacetyl, mp 187–8.
(20) Ethyl ester, mp 68–69.
(21) N-Benzenesulfonyl, mp 120;
(22) Amide, mp 175d; Nitrile, mp 174.
(26) Hydrochloride, melting point about 180.
(28) Carbobenzoxy, mp 137.
(29) p-Nitrobenzyl ester, mp 136; p-bromophenacyl ester, mp 151.
(31) γ-Carbobenzoxy, mp 208–10.
(32) The D and L forms both melt at 220d. Hydrochloride, mp 110 (very hydroscopic).
(34) N-Chloroacetyl, mp 123; hydrochloride, mp 202(193).
(35) Amide, mp 161–2; N-chloroacetyl, mp 158–60.
(36) Carbobenzoxy, mp 106.
(37) N-Chloroacetyl, mp 143.
(38) Amide, mp 50–1 (hydroscopic). Heating produces the lactam, 2-keto-hexamethyleneimime, which forms an N-benzoyl derivative melting at 45–7.
(39) Hydrochloride, mp 135–6. Changes on long standing in 20% alcohol to melting point of 202. Heating above the melting point converts the compound to pyrrolidone.
(40) DL-Proline forms an anilide, mp 170. Amide, mp 93.
(41) Phenylurethane, mp 114;
(42) Hydrochloride, mp 120–1; on heating the lactam is formed (5-ethyl-2-pyrrolidone), mp 22; bp 256–7.
(45) The melting point of D-arginine, 207, is often confused with the melting point of DL-arginine, 238. For the picrolonate of D-arginine see Heyl, J. Am. Chem. Soc., **41**, 681 (1919). Flavinate, mp 258 (mono); 220 (di).
(46) L-Glutamic acid melts at 211–3 on rapid heating; on slow heating it melts at 197–8d. Its HCl salt melts at 202 on slow heating but at 213 on rapid heating. N-Chloroacetyl, mp 143; N-benzenesulfonyl, mp 129–32.

TABLE 4 (Continued)
AMINO ACIDS[a]

				Melting Point of Derivatives			
				Others (continued)			
	Formyl	Picrate	Picrolonate	α-Naphthyl-urea	3,5-Dinitro-Benzoyl	β-Naphthalene-sulfonyl	Phenyl-hydantoin
123		183					
124	190						
125					212(di) (hyd.)		
126							
127	N-171–4d 147(di)		260d (mono)	205–6 (mono)		102N	
128					157–8		
129			260d	205–6	252–4(di)		
130							
131			203(di)	221d(mono) 236(di)			
132		225d(mono)					
133	132*						103
134							

NOTES ON AMINO ACIDS (Continued)

(47) Nitrile, mp 139; ethyl ester, mp 92.
(48) Hydrochloride, mp 168–70.
(50) Carbobenzoxy, mp 164.
(52) Flavinate, mp 211d, hydrochloride, mp 166d.
(53) Amide, mp 234–6; benzyl ester, mp 105–6.
(57) Hydrochloride, mp 120.
(58) β-Hydroxyvaline forms a phenylurethane, mp 162, and a 2-naphthalenesulfonamide, mp 261.
(59) The L-proline phenylurea derivative may precipitate as a resin, and should be converted to the hydantoin, which melts at 118(144). The reineckate derivative has mp 199d.
(60) Flavinate, mp 218d; Hydrochloride, mp 185d.
(62) The dibenzoyl derivative of DL-lysine melts at 145–6. L-lysine forms a 5-benzoyl derivative (mono) melting at 235. L-Lysine forms a monohydrochloride, mp 235–6, and a dihydrochloride, mp 193. The DL-lysine forms a monohydrochloride, mp 235–6, and a dihydrochloride, mp 188–90.
(64) Heated rapidly, the compound melts at 234–5; heated slowly (sealed tube), it melts at 226–7. The D- and L- forms have the same melting point.

(65) DL-Allothreonine melts at 237–9. D- and L-threonine melt at 251–2. The dibenzoyl derivative melts at 174, the monobenzoyl at 176, and the eutectic of the two at 145.
(66) The methyl ester of L-serine forms a hydrochloride, mp 167.
(67) Glycine turns brown above 220 and decomposes at 262. A commercial source lists the melting point at 245. Glycine picrate contains 2 mole of glycine; it softens at 199–200 and decomposes at 202. The earlier figure of 190 is stated to be erroneous. The barium salt crystallizes even from dilute solutions and is suitable for separation from other amino acids. Glycine forms an anilide, mp 62, and a p-toluidide, mp 107. The hydrochloride melts at 185.
(69) Thyroxine melts at 250 when heated 10°/minute and at 230–5 when heated 3°/minute; iodine is evolved in all cases. DL-Thyroxine yields a methyl ester, mp 156. The N-Chloroacetyl melts at 201–2d.
(70) Hydrochloride, mp 218.
(72) N-Benzenesulfonyl, mp 194.
(73) See note 65.
(74) DL-Argine nitrate has a melting point of 230.

NOTES ON AMINO ACIDS (Continued)

(75) Amide, mp 206; anilide, mp 182–3.
(77) L-serine melts at 228d. DL-Serine yields the following derivatives: phenylurethan, mp 159; ethyl ester, mp 256d; and methyl ester hydrochloride, mp 114. The N-Chloroacetyl melts at 122–3.
(78) D- and L-Isoserine melt at 199–201. DL-Isoserine forms a phenylurethan, mp 183–4.
(79) The methyl ester has a melting point of 97–8. The amide melts at 135–6.
(80) Diamide, mp 200–1. Heat converts the acid to alanine.
(81) See note 65.
(84) The phenylurea derivative gives hydantoin, mp 117. The benzoyl (di) derivative of DL-cystine melts at 170. L-Cystine diethyl ester dihydrochloride melts at 177; DL-cystine diethyl ester dihydrochloride melts at 185. Mesocystine melts at 200–18d.
(85) N-Carbethoxy, mp 191.
(86) N-Chloroacetyl, mp 178–80.
(88) Variable values are given in the literature for the decomposition point of the pure compound and of the picrolonate. The N-Chloroacetyl melts at 130–1 and the amide has a melting point of 138–44.
(89) The DL-α-ethyl ester melts at 165 and the β-ethyl ester melts at 200d.
(90) Forms a dihydrazide, mp 135; N-Chloroacetyl, mp 142; amide, mp 131 (di).
(91) Amide, mp 158.
(93) 4-Hydroxyproline exists in two isomeric forms. Natural 4-hydroxyproline has $(\alpha)_D^{20}$ −81 and forms a picrate that melts at 188.
  I. mp 274; $(\alpha)_D^{21}$ 75.2 in water
     mp 274; $(\alpha)_D^{26}$ −74.6
     D,L mp, 261
  II. mp 237–41; $(\alpha)_D^{18}$ 58.6
      mp 238–41; $(\alpha)_D^{18}$ 58.1
      D,L, mp 250
(94) N-Benzenesulfonyl, mp 185d.
(95) L-(+)-Isoleucine melts at 283–4d and D-(−)-isoleucine melts at 285–6d; their $(\alpha)_D^{20}$ is 10.7°. They give the same derivatives. L-(+)-alloisoleucine melts at 278d and D-(−)-alloisoleucine melts at 274–5d; their $(\alpha)_D^{20}$ is 14. They give the same derivatives. DL-Isoleucine and DL-alloleucine have been reported as melting at 275 in a sealed tube. The N-chloroacetyl melts at 72–4 and the N-benzenesulfonyl melts at 147–8.
(96) Amide, mp 127; α-naphthylisocyanate, mp 198.
(99) D- and L-Methionine shrink and darken above 278 and melt at 283 with decomposition.
(102) The decomposition point is variously given as 283, 275–80 and 320.
(103) See note 95.
(104) The melting point of L-histidine is variously reported as 253, 272, 277, and 288. Flavinate, mp 224–6d.
(105) The melting point is variously reported as 252, 278, 282, and 290. N-Chloroacetyl, mp 159; N-benzenesulfonyl, mp 185d.
(106) See note 97. Phthalyl derivative, mp 120–1.
(107) Methyl amide, mp 43; hydrochloride, mp 110.
(108) N-Chloroacetyl, mp 119.
(109) Forms a p-nitrobenzyl ester, mp 184–5. Amide, mp 106–7; Phthalyl derivative, mp 140–1.
(110) Phthalyl derivative, mp 115–6.
(111) α-Naphthylurethane, mp 196.
(112) The phenyl urea derivative has been reported as melting at 150, 168, 174, and 190; forms as p-nitrobenzyl ester, mp 229; amide, mp 62; phthalyl derivative, mp 160–1.
(114) Amide, mp 72.
(115) N-Chloroacetyl, mp 104–7; phthalyl derivative, mp 111.5–112.5.
(116) Amide, mp 78–80; phthalyl derivative, mp 101.5–102.0.
(119) Ethyl ester, mp 65; hydrochloride, mp 188d; phthalyl derivative, mp 103–4.
(120) N-Chloroacetyl, mp 130; phthalyl derivative, mp 95.5–96.5.
(123) Hydrochloride, mp 236–7.
(126) N-Chloroacetyl, mp 107.
(127) The di-1-napthalenesulfonyl derivative of L-form gives at 100–2 a viscous oil. It is very slightly soluble in hot water and very soluble in hot alcohol. DL-Tyrosine quickly heated decomposes at 340 but, in a preheated bath, at 295. Forms both di- and mono- derivatives. See McChewey and Swann, *J. Am. Chem. Soc.*, **59**, 1117. The melting point of L-tyrosine has been confused in some works with the melting point of DL-form. The decomposition point of the L-form with rapid heating is 314–8, and with slow heating 290–5 (Fischer, *Ber.* **32**, 3641). N-Chloroacetyl, mp 155–6; amide, mp 153–4.
(130) Dihydantoin derivative, mp 200.
(131) See note 10. The oxalate salt has a melting point of 218; hydrochloride, mp 215d (mono).
(132) The melting point of the 1-N-benzoyl derivative is reported at 235–49; the benzoyl derivative is reported at 254–68. The 1-N-benzoyl-5-N-p-toluenesulfonyl, mp 140; 5-N-benzoyl-1-N-p-toluenesulfonyl, mp 199; hydrochloride, mp 235–6d (mono).
(133) The D- and L-acids sinter at 307, ethyl ester, mp 65.
(134) Hydrochloride, mp 175–8d. Cysteine may be oxidized to cystine, mp 260, q.v.

REFERENCES FOR TABLE 4

Dunbar, R. E., and Ferrin, F. J., *Microchem.*, **4**, 59 (1960). Amino Acids and their Dibenzofuran-2-Sulfonates in Qualitative Organic Analysis. Lists the melting points and photomicrographs of 20 α-amino acids.

Greenstein, J. P., and Winitz, M., *Chemistry of the Amino Acids*, Wiley, New York, 1960. Three volumes. Hundreds of tables and figures (including infrared spectra) listing the chemical and physical properties of thousands of amino acids, peptides, and derivatives thereof.

Inouye, K., Sanderlin, R., and Kirk, P. L., *Anal. Chem.* **13**, 587 (1941). Amino Acids: Microscopy of the Amino Acids and Their Compounds.

Lacourt, A., and Delande, N., *Microchemical Techniques*, Wiley, New York, 1962. Identification of Amino Acids by Thermomicro Methods: Eutectic Melting Temperature. Many melting points are given.

Levey, G. B., in *Handbook of Analytical Chemistry*, L. Meites, ed., McGraw-Hill, New York, 1963. Optical Rotation of Natural (α-L-) Amino Acids and Related Compounds and Table 6–46, pp. 6–246 to 6–251.

Mathieu, J. P., Roche, J., and Desnuelle, P., *Optical Rotatory Power III. Amino Acids*. Pergamon Press, New York, 1959.

Ronwin, E. J., *J. Org. Chem.*, **22**, 1180 (1957). Optical Activity and the Direct Method of Acylation. Gives the melting points of several nitrobenzoylated amino acids. See Table II of this article.

TABLE 5A
CARBOXYLIC ACIDS-LIQUID

	Name of Compound	Note	BP	Melting Point of Derivatives		
				Recommended		Others
				p-Tolui-dide	Anilide	2-Naph-thyl-amide
1	Trifluoroacetic		72		88	
2	Thioacetic		93	130	76	
3	Thiopropionic acid		—		67	
4	Formic (mp 8.4)	3	101	53	50	129
5	Acetic (mp 16)	2, 4	118	147*	114.2*	132
6	Difluoroacetic	5	134–5N		52	
7	Propenoic(acrylic; mp 13)		140	141	105	
8	Propionic	2	141	126	106	
9	2-Cyanopropionic	7	142–5			
10	Propynoic	8	144d		87	
11	Isobutyric	1, 2	154.6	107	105	
12	2-Methylpropenoic		161			
13	Fluoroacetic		164			
14	Butyric	2	164	75	96	125
15	3-Butenoic	6	164 (169)		58	
16	2-Ketopropionic (mp 13)	10	165d	109 (130)	104s	
17	2-Butenoic (*cis*) (mp 15)		169	132	102	
18	3-Methylbutanoic	2	176.5	107	110	138.5
19	*dl*-2-Methylbutanoic		176	93	110	
20	2-Ketopentanoic (see Table 21-A)		179			
21	3,3-Dimethylbutanoic		184	134	132	
22	*dl*-2-Chloropropionic		186	124	92	
23	Cyclopropanecarboxylic		186			
24	Pentanoic	2. 11	186.3	74	63	112
25	2,2-Dimethylbutanoic		187 (190)	83	92	
26	3-Pentenoic		188–9			
27	Cyclopropylacetic		190			
28	*dl*-2,3-Dimethylbutanoic		192	113	78	

TABLE 5A (Continued)
CARBOXYLIC ACIDS-LIQUID

Melting Point of Derivatives
Others (Continued)

	4-Nitro-benzyl Ester	4-Phenyl-phenacyl Ester	4-Chloro-phenacyl Ester	4-Bromo-phenacyl Ester	Amide	4-Bromo-anilide	Methyl-enebis-anilide	Benz-imid-azole	Phenyl-hydra-zide
1					75				
2					108				
3					43				
4	31	74	128	101		119		177	143d
5	78	110N	67	85	82	168	228	177	129
6					52				
7					85				
8	31	103N	98	59 (63)	81	148	213	175	157
9					97 (s.t.)				
10					61–2				
11		89N		77	130	155		224	140
12					106	116			
13					107				
14	35	82N	55	63	115	115	198	157	102
15				60.5	73				
16					124	168			
17				81	102				
18		79N		68	136	128		187	
19		71		55	112	122			
20					108s				
21					132				
22					80				95
23					125				
24		69N	98	64 (75)	106				
25		87			132 (103)				
26					94				
27		83							
28		74			132				

(Continued)

TABLE 5A (Continued)
CARBOXYLIC ACIDS-LIQUID

	Name of Compound	Note	BP	Melting Point of Derivatives		
				Recommended		Others
				p-Tolui-dide	Anilide	2-Naph-thyl-amide
29	Dichloroacetic		194	153	118	
30	2-Ethylbutanoic	1	195	116	127	
31	Cyclobutanecarboxylic		195			
32	dl-2-Methylpentanoic		196	81	95	
33	dl-3-Methylpentanoic		197	75	87	
34	4-Methylpentanoic	2	199.4	63	112	
35	Methoxyacetic		204		58	
36	2-Ethyl-2-methylbutanoic		204			
37	dl-2-Bromopropionic (mp 26) (see Table 5B)		205d			
38	Hexanoic	2	205.1	75	95	107
39	Ethoxyacetic		207	32		
40	2-Ethylpentanoic		209	129	94	
41	2-Methylhexanoic		210	85	98	
42	2-Bromobutyric	1	217d	92	98	
43	4-Methylhexanoic		218		77	
44	2,2-Dimethylhexanoic		218			
45	Heptanoic	2	223	81	70	101
46	2-Ethylhexanoic	1	228			
47	Cyclohexylacetic (mp 31) (see Table 5B)		237			
48	Octanoic (mp 16)	2	239.3	70	57	103
49	Nonanoic (mp 12.3)	2	255	84	57	103
50	d-Citronellic		257			
51	2-Phenylpropionic		265			
52	Undecylenic (mp 24.5) (see Table 5B)	12	275			
53	5-Ketohexanoic (mp 13–4)	13	275d	123		
54	Undecanoic (mp 30) (see Table 5B)		284			
55	Oleic (mp 16)		216/5 mm	43	41	169
56	dl-Lactic (mp 17)		226/10 mm	107	59	137.5

TABLE 5A (Continued)
CARBOXYLIC ACIDS-LIQUID

Melting Point of Derivatives
Others (Continued)

	4-Nitro-benzyl Ester	4-Phenyl-phenacyl Ester	4-Chloro-phenacyl Ester	4-Bromo-phenacyl Ester	Amide	4-Bromo-anilide	Methyl-enebis-anilide	Benz-imid-azole	Phenyl-hydra-zide
29			93	99	98s				
30		77			114				
31					152–3				
32		64			80				
33		47			125			159	
34		70N		77	121				144
35					97	85		136	
36					78				
37									
38		70N	62	72	101	105	186	163	98
39			94	105	82				
40					105	148			
41					72	114			
42	49				112 (108)				
43					98				
44					89				
45		65N	65	72	96	98	184	138	103
46		54			104				
47									
48		68N	63	67	107	103	183	145	106
49		71N	59	68	99	100	177	140	
50					84–5				
51					92				
52									
53					114				
54									
55		61	40	40 (46)	76				73
56		145		113	76			179	117

TABLE 5B
CARBOXYLIC ACIDS—SOLID

	Name of Compound	Note	MP	Melting Point of Derivatives		
				Recommended		Others
				p-Toluidide	Anilide	2-Naphthyl-amide
1	2-Methylpropenoic		16			
2	3-Bromobutyric		20			
3	3-Ethyl-2-methylacrylic(*trans*)		24			
4	2-Ethoxybenzoic		25			
5	Undecylenic (*bp* 275)	12	24.5	68	67	
6	DL-2-Bromopropionic (bp 205*d*)		26	125	99	
7	Undecanoic (bp 284)	2, 14	29*N*	80	71	
8	2-Chloro-2-methylpropanoic		31		70	
9	2-Keto-3-methylbutanoic (bp 171) (see Table 21B)		31			
10	Fluoroacetic (bp 179)		31–2			
11	Cyclohexanecarboxylic	1	31			
12	Decanoic (bp 168–70)	1, 2	31.2	78	70	104
13	2-Hexenoic		32		110	
14	2-Acetobutyric (see Table 21B)	15	32			
15	2-Ketobutyric (see Table 21B)	16	32			
16	1,3-Pentadiene-1-carboxylic (sorbic)(*cis-trans*)	1	32–5			
17	10-Docosenoic		33–4			
18	Levulinic	17	33–5	109	102	
19	13-Docosenoic(*cis*)		34	58	55	87
20	Trimethylacetic (bp 164)		35	[120]	129 (133)	
21	2-Methyl-3-phenylpropionic		36.6	130		
22	Acetoacetic (see Table 21B)	18	36–7*N*	95	86	
23	4-Ketohexanoic (see Table 21B)		40			
24	3-Chloropropionic		42			
25	2-Phenylbutyric		42			
26	4-Diethylaminobenzoic		43			
27	Dodecanoic	2	43.9	87	78	106
28	Tridecanoic	2	44	88	80	
29	2-Bromo-3-methylbutyric		44	124	116	
30	Elaidic (oleic-*trans*)		44–5 (52)			
31	4-Cyanobutyric (bp 245)		45			
32	Angelic (bp 185*v.s.*)		45–6		126	
33	3-Phenylpropionic	19	48	135		135
34	Dibromoacetic		48			
35	4-Phenyl-2-ketobutyric (see Table 21B)		48–50			
36	2-Bromoisobutyric (bp 198–200)		48–9	93	83	135
37	3-Cyanopropionic		48–50			
38	Bromoacetic (bp 208)	1	50	91	131	134
39	3,4-Dibromobutyric		50			
40	2-Pentynoic		50			
41	3-Isopropylbenzoic		51–2		118	
42	4-Phenylbutyric (bp 290)		52			
43	Pentadecanoic		52	93	78	

TABLE 5B (Continued)
CARBOXYLIC ACIDS—SOLID

Melting Point of Derivatives (Continued)

Others (Continued)

	4-Nitro-benzyl-Ester	4-Phenyl-phenacyl Ester	4-Chloro-phenacyl Ester	4-Bromo-phenacyl Ester	Amide	4-Bromo-anilide	Methyl-enebis-anilide	Benz-imid-azole	Phenyl-hydra-zide
1					106	116			
2					92				
3				91	80				
4					132				
5					87				
6					123				
7		80N	60 (62)	68	103	102	176	114	110
8									
9					110				
10					108				
11					172				
12			62	67	108	102	179	127	105
13									
14									
15					116–7				
16					113–4				
17		72.5	54.5	60.5					
18	61			84	108d				
19		76	56	61	84 (66)				83
20		113–4		78	178 (154)				
21		73			109				
22									
23									
24					101				
25					86				
26					137				
27		86 (84)	70	76	110 (102)	107 (104)	175	107	106
28		87N	67	75	100		173	109	
29					133				
30		73	56	65	93–4				
31					69–70s.t.				
32					127–8				
33	36	95		104	105			186	
34					156				
35					180				
36					148				
37					97s.t.				
38	88				91				
39					86				
40					146				
41				96					
42					84				
43	40	92	74		102		168	99	

(*Continued*)

TABLE 5B (Continued)
CARBOXYLIC ACIDS—SOLID

	Name of Compound	Note	MP	Melting Point of Derivatives		
				Recommended		Others
				p-Toluidide	Anilide	2-Naphthyl-amide
44	4-Carboxy-1,3-dithiolane		53			
45	2-Carboxytetramethylene sulfide		53			
46	Tetradecanoic	2	54.1	93	84	108
47	Trichloroacetic (bp 197)		57–8	113	97 (94)	
48	2-Ethylbutane-1,1-dicarboxylic		58		219(di)	
49	Transbrassidic		60		78	
50	5-Phenylpentanoic		60		90	
51	2-Chloroisocrotonic (bp 195)		61		108	
52	Heptadecanoic		61			
53	3-Bromopropionic		62.5			174
54	Hexadecanoic	2	62.8	98	90	109
55	Chloroacetic (bp 189)	20	63N	162	137	117–8
56	2-Isopropylbenzoic		63–5		138	
57	2-Methyl-2-butenoic-cis (tiglic)(bp 199)		64	71 (76)	77	96
58	3-Benzoylpropenoic (see Table 21B)	21	64			
59	Benzoylformic (see Table 21B)		66			
60	Cyanoacetic		66		198	
61	2,3-Dibromopropionic	22	67N			
62	3-Methyl-2-butensic		67			
63	2-Furylacetic		67		85	
64	2-Chloro-2-butenoic(cis)	24	67			
65	2-Ethylbenzoic		68			
66	d-Chaulmoogric		68.5	100	89	96–8
67	Nonadecanoic	23	68.6			
68	Octadecanoic	2	69.9	102	95	112
69	2-Hydroxydecanoic		70		79	
70	2-Butenoic(trans) (bp 189v.s.)		72	132	118 (115)	
71	2-Dimethylaminobenzoic		72			
72	4-Bromo-2-butenoic		74			
73	Eicosanic		75.2	96	92	112
74	Phenylacetic	1	76.5	136	118	159
75	3-(3-Chlorophenyl)propionic		78		86–7	
76	2-Hydroxy-2-methylpropionic		79	133	136	157.5
77	Tricosanoic	26	79.1			
78	Docosanoic		79.9		101–2	
79	Hydroxyacetic (glycollic)	25	80	143	97	138
80	2-Benzylidenepropionic		81			
81	2-Benzoylpropionic	27	82–3		137–8	
82	2-Iodopropionic		82 (85)			
83	Iodoacetic		83		143–4	

TABLE 5B (Continued)
CARBOXYLIC ACIDS—SOLID

Melting Point of Derivatives (Continued)

Others (Continued)

	4-Nitro-benzyl Ester	4-Phenyl-phenacyl Ester	4-Chloro-phenacyl Ester	4-Bromo-phenacyl Ester	Amide	4-Bromo-anilide	Methyl-enebis-anilide	Benz-imid-azole	Phenyl-hydra-zide
44					90				
45					132				
46		90N	76	81	103	110 (107)	171	105	108
47	80				141				123
48									
49		86	69	74	94				98
50					109				
51					110				
52	49	96	79	83 (78)	108		165	94	
53					111				
54	42	94N	82	86 (82)	106–7	113	168	97	111
55		116	93.8	104	118				111
56				54–5					
57	64			68	76				
58									
59					91				
60					120				
61					130 (133)				
62				101	108				
63									
64									
65					151–3				
66					106				
67									
68		97N	86	90	109	115	165	94	115 (110)
69				93					
70	67			96	161				
71					140				
72					101				
73		86	86	89	108–9				
74	65	63d (88)		89	156			187	175
75									
76	80				98				
77									
78					111				
79	107			138	120			172	
80					128				
81					145–6				
82					101 (142)				
83					95				

(Continued)

TABLE 5B (Continued)
CARBOXYLIC ACIDS—SOLID

	Name of Compound	Note	MP	Melting Point of Derivatives		
				Recommended		Others
				p-Toluidide	Anilide	2-Naphthyl-amide
84	4-Chloro-2-butenoic		83			
85	Pentacosanoic	28	83.5			
86	Tetracosanoic		84			
87	2-Benzoylbutyric		85–7			
88	dl-2-Bromo-2-phenylacetic		86 (84)			
89	4-Methoxyphenylacetic		87 (84)			
90	Hexacosanoic		87.7			
91	3-Benzylidenepropionic		88			
92	Dibenzylacetic	1	89	175	155	
93	o-Tolylacetic		90 (88)			
94	Octacosanic	29	90			
95	Nonacosanic	30	90.3			
96	2-Benzoylbenzoic (anh. m 127)	1	90(hyd) (93)		195	
97	Citraconic		92d	170(mono)	153(mono) 175.5(di)	
98	Triacontanoic	31	93.6			
99	1,3-Cyclohexadiene-1-carboxylic		94.5			
100	2-Chlorophenylacetic		95	170	138.5	
101	p-Tolylacetic		95 (91)			
102	Pyrrole-1-carboxylic	32	95			
103	3,4-Dimethoxyphenylacetic	33	95–9			
104	2-Hydroxy-3-phenylpropionic		97			
105	Pentanedioic (glutaric)		98	218	224	
106	3-Phenoxypropionic		98			
107	2-Aldehydobenzoic (see Table 7B)		98–9			
108	Phenoxyacetic		99		99	
109	2-Hydroxymalonic	34	100			
110	Citric(monohyd.)(anh. m 153)	35	100	189(tri)	199(tri) (192)	
111	2-Methoxybenzoic	1	100–1	[62]	131	
112	2-Hydroxybutanedioic (l-Malic)	118	100–1	206–7(di)	197(di)	
113	Oxalic(dihydrate)	36	101	169(mono) 268(di)	149(mono) 254(di)	
114	2,4-Diketopentanoic		101			
115	Butylmalonic		101		193(di)	
116	2-Cyano-3-phenylpropionic		101–2			
117	6-Chloro-2-methylbenzoic		102			
118	2-Bromophenylacetic		103–4		107–8	
119	Benzoylacetic	37	103–4d		108	
120	o-Toluic		105 (107–8)	144	[125]	

TABLE 5B (Continued)
CARBOXYLIC ACIDS—SOLID

Melting Point of Derivatives (Continued)

Others (Continued)

	4-Nitro-benzyl-Ester	4-Phenyl-phenacyl Ester	4-Chloro-phenacyl Ester	4-Bromo-phenacyl Ester	Amide	4-Bromo-anilide	Methyl-enebis-anilide	Benz-imid-azole	Phenyl-hydra-zide
84					130–2				
85									
86			100	91					
87					149				
88					148				
					(144)				
89					189				
90					105–7				
91					130				
92					128				
93					161				
94									
95									
96	100				165				
97	71	109			185–7(di)				
98									
99					105				
100					175				
101					185				
102					166				
103					145–7				
104					112				
105	69	152		137	175–6(di)				
106					119				
107									
108				148	101				
109									
110	102	146		148	210d(tri)				
					(215)N				
111	113	131		113	129				
112	87.2 (mono) 124.5(di)	106		179	156.5–8				
113	204(di)	166d		242	210(mono) 419d(di)				
114					131–2				
115					200(di)				
116					130				
117					167				
118					113				
119					113				
120	91	94.5		57	143				

(Continued)

TABLE 5B (Continued)
CARBOXYLIC ACIDS—SOLID

	Name of Compound	Note	MP	Melting Point of Derivatives		
				Recommended		Others
				p-Toluidide	Anilide	2-Naphthyl-amide
121	Heptanedioic		105	206	109(mono) 155(di)	
122	2-Phenylpropenoic		106–7		134	
123	Nonanedioic		107	201	108(mono) 186(di)	
124	2,5-Dichlorophenylacetic		107			
125	dl-2-(2-Naphthoxy)propionic		107-8			
126	Dehydracetic		109		115	
127	3-Methoxybenzoic	38	110			
128	Allo-2-chlorocinnamic(cis)	39	111		138–9	
129	Cyclohexanol-2-carboxylic		111	155–6		
130	Ethylmalonic		111		150	
131	m-Toluic		112	118	126	
132	4-Carboxypentamethylene sulfide		112.5			
133	2-Phenylbenzoic		113			
134	2-Phenoxybenzoic		113			
135	Bromomalonic		113d	217(di)		
136	4-Bromophenylacetic		114s			
137	2-Benzylbenzoic		114			
138	2-Acetobenzoic (see Table 21B)		114–5			
139	Pyrotartaric (Methylsuccinic)	40	115	164	200(di) 159(mono)N	
140	2-Phenoxypropionic		115–6	115	118–9	117
141	3-Benzoylpropionic (see Table 21B)		116		150	
142	Dichloromaleic	41	116(120)		170N	
143	2,6-Dimethylbenzoic		116			
144	d- or l-α-(2-Naphthoxy)-propionic		117			
145	4-Isopropylbenzoic		117–8			
146	Benzylmalonic		117d		217(di)	
147	2-Ketopentanedioic (see Table 21B)		117–8			
148	dl-3-Hydroxy-2-phenyl-propionic (d, m 130)		117–8			
149	dl-2-Hydroxy-2-phenylacetic (d m 133; l m 134)		118	172	152	189
150	l-2,3,4,5-Tetrahydroxy-pentanoic		118–9	200	204	
151	2-Chloro-2,2-diphenylacetic		118–9d		88	
152	Cyclopentane-1,3-dicarboxylic(cis)		120			
153	2-Hydroxy-2-(4-chlorophenyl)acetic		119–22			
154	3-Nitrophenylacetic		120			
155	Anilinomalonic		120		162(di)	
156	(4-Chloro-2-methyl-phenoxy)-acetic	43	120(anh)			
157	Picrolonic	44	120–1			

TABLE 5B (Continued)
CARBOXYLIC ACIDS—SOLID

Melting Point of Derivatives (Continued)

Others (Continued)

	4-Nitro-benzyl Ester	4-Phenyl-phenacyl Ester	4-Chloro-phenacyl Ester	4-Bromo-phenacyl Ester	Amide	4-Bromo-anilide	Methyl-enebis-anilide	Benz-imid-azole	Phenyl-hydra-zide
121		148d		137	175(di)				
122					121–2				
123	44	141		131(di)	95(mono)	215(di)			
					172(di)				
124					157				
125					169				
126									
127									
128					134				
129									
130	75				214(di)				
131	87		136.5	108	95				
132					184.5				
133					177				
134					131				
135					181(di)				
136					192–4				
137					163				
138					116.5				
139					225(di)				
140					132–3				
					(130)				
141									
142					175(mono)				
143					139				
144					197				
145					133				
146	119(di)				228(di)				
147									
148					169				
149	124			113	132			202	
150					136				
151					115				
152					226(di)				
153					122–3				
154					110				
155					156(di)				
156					149–50				
157									

(Continued)

TABLE 5B (Continued)
CARBOXYLIC ACIDS—SOLID

Name of Compound	Note	MP	Melting Point of Derivatives		
			Recommended		Others
			p-Toluidide	Anilide	2-Naphthyl-amide
158 3-(2-Nitrophenyl)-2-keto-propionic (see Table 21B)	42	121			
159 1-Cyclopentenecarboxylic		121	122	126	
160 2-Bromo-3-phenyl-propenoic(cis)		121			
161 Furan-3-carboxylic		121–2		130	
162 Benzoic (v.s.)		122.36*	158	163	
163 Picric (see Table 29B)		122.5			
164 3-Fluorobenzoic		124			
165 Trichlorolactic		124		164	
166 2-Acetopropenoic (see Table 21B)		125			
167 3-Nitrosalicylic(hyd).		125			
168 Diethylmalonic		125			
169 Aconitic(cis)	46	125		170d(mono) 200(di)N	
170 1-Cyclohexene-1,2-dicarboxylic	47	126		155(mono)	
171 d- or l-α-(1-Naphthoxy)-propionic		126		205	
172 2,4-Dimethylbenzoic(anh) (hyd., m 90)		127		141	
173 3-(2-Chlorophenyl)-propenoic(cis)		127			
174 2-Benzoylbenzoic		127		195	
175 3-Benzoylbutyric (see Table 21B)		127			
176 2-Fluorobenzoic		127			
177 Dodecanedioic		128	165(di)	191(di)	
178 Thiophene-2-carboxylic	48	129		140	
179 Maleic (cis-butenedioic)	49	130N	142(di)	187(di)	
180 Tribromoacetic (bp 235d)		131			
181 1-Naphthaleneacetic	1	132		155 (160)	
182 2,5-Dimethylbenzoic		132		140	
183 3-(3-Chlorophenyl)-propenoic	50	132	142	134.5	
184 Indole-3-propionic	51	132–3			
185 2-Furoic		133–4	107.5	123.5	
186 3-Phenylpropenoic (Cinnamic)(trans)		133	168	153	
187 Decanedioic		134.5	201	122(mono) 201(di)	
188 1,3-Pentadiene-1-carboxylic (Sorbic)(trans, trans)	1	134.5		153	
189 Malonic		135	86(mono) 253(di)	132(mono) 230(di)	
190 Acetylsalicylic		135(r.h.)		136	
191 Acetone-1,3-dicarboxylic	52	135d		155(di)	

TABLE 5B (Continued)
CARBOXYLIC ACIDS—SOLID

Melting Point of Derivatives (Continued)

Others (Continued)

	4-Nitro-benzyl- Ester	4-Phenyl-phenacyl Ester	4-Chloro-phenacyl Ester	4-Bromo-phenacyl Ester	Amide	4-Bromo-anilide	Methyl-enebis-anilide	Benz-imid-azole	Phenyl-hydra-zide
158					165–6				
159									
160					129				
161					169				
162	89	167	119	119	130				168
163					190				
164					130				
165					96				123
166									
167					145				
168	91(di)				146(mono)				
					224(di)				
169									
170									
171					192				
172					179–81				
173					112				
174	100				165				
175									
176					116				
177					185(di)				
178					180(174)				
179	91	168		168 (190)	181(mono)				
					266(di)				
180					122				
181					180				
182					186				
183					76				
184									
185	133.5	86		138.5	143				
186	117	182		146	148				
187	73.5(di)	140		147	170(mono)				194
					210(di)				
188					168				162–3
189	86	175			50(mono)				194
					170(di)				
190	90.5				138				
191									

(Continued)

TABLE 5B (Continued)
CARBOXYLIC ACIDS—SOLID

	Name of Compound	Note	MP	Melting Point of Derivatives		
				Recommended		Others
				p-Toluidide	Anilide	2-Naphthyl-amide
192	Pyridine-2-carboxylic (picolinic)		136–7	104	76	
193	Glutaconic(cis) and (trans)	53	136–8		228(di) 167(mono)	
194	3-Ethoxybenzoic		137			
195	Phenylpropynoic		137s	142	126	
196	2-Chloro-3-phenyl-propenoic(trans)	54	137–8	116	118	
197	Thiophene-3-carboxylic		138			
198	Methylmalonic	55	138d	145d(mono) 228(di)	182	
199	5-Chloro-2-nitrobenzoic		139		164	
200	3-Nitrobenzoic		140	162	154	
201	2-Chloro-4-nitrobenzoic		140		168	
202	meso-Tartaric		140			
203	2-Bromo-4-methylbenzoic		140			
204	2,4-Dichlorophenoxyacetic		141			
205	3-Sulfobenzoic(anh.)		141			
206	2-Nitrophenylacetic		141			
207	4-Chloro-2-nitrobenzoic		142			
208	2-Naphthaleneacetic		142			
209	2-Chlorobenzoic	1	142 (140)	131	118	
210	Octanedioic		142 (144)	218(di)	128(mono) 186(di)	
211	2-Bromophenoxyacetic		142.5			
212	2,4,5-Trimethoxybenzoic		144		154.5	
213	2,6-Dichlorobenzoic		144 (139)			
214	2-Chlorophenoxyacetic		145–6		121	
215	2-Nitrobenzoic		146		155	
216	4'-Methylbenzophenone-2-carboxylic		146			
217	3-Hydroxy-2-methylbenzoic	56	146			
218	Phthalonic(anh.)		146		176(mono) 208(di)	
219	2-Hydroxyphenylacetic		147 (141)			
220	2-Aminobenzoic (see Tables 4 and 9B)		147 (145)	151	131	
221	4'-Chlorobenzophenone-2-carboxylic	57	148			
222	Diglycolic (oxydiethanoic)		148 (142)	148(mono)	118(mono) 152(di)	
223	Diphenylacetic		148	[173]	180	191–2
224	3-Nitrosalicylic		148–9			
225	Oxanilic		148–9		154(di)	
226	4-Hydroxyphenylacetic		148–50			
227	dl-Thiomalic	58	149–50			

TABLE 5B (Continued)
CARBOXYLIC ACIDS—SOLID

Melting Point of Derivatives (Continued)

Others (Continued)

	4-Nitro- benzyl Ester	4-Phenyl- phenacyl Ester	4-Chloro- phenacyl Ester	4-Bromo- phenacyl Ester	Amide	4-Bromo- anilide	Methyl- enebis- anilide	Benz- imid- azole	Phenyl- hydra- zide
192					106.5				
193									
194					139				
195	83				100 (109)				
196					121–2				
197				130	180				
198	75				217				
199					154				
200	141	153		132	143				
201					172				
202	93				187(di)				
203					137s				
204					130				
205					170(di)				
206					161				
207					172				
208					200				
209	106	123		106	142				
210	85(di)	151(di)		144(di)	127(mono) 217(di)				
211					151				
212					184.5				
213					202				
214					149.5				
215	112	140		107	176				
216					175–6				
217									
218					α-179d β-155d				
219					118				
220	105				109				
221									
222					135(mono)				
223		111		112	168				
224					155 (145)				
225					228				
226					175				
227					103(mono)				

(Continued)

TABLE 5B (Continued)
CARBOXYLIC ACIDS—SOLID

	Name of Compound	Note	MP	Melting Point of Derivatives		
				Recommended		Others
				p-Toluidide	Anilide	2-Naphthyl-amide
228	2-Bromobenzoic		150		141	
229	Benzilic	1	150	[190]	175	
230	2,6-Dibromobenzoic	59	150.5			
231	2-Hydroxy-5-methylbenzoic	1	153			
232	Citric(anh.)		153	189(tri)	199(tri)	
	(monohyd. m 100)				(192)	
233	4-Nitrophenylacetic		153	210	198	
					(212)	
234	dl-α-(1-Naphthoxy)-propionic		153		173	
235	2,5-Dichlorobenzoic		153			
236	Hexanedioic (adipic)		153	239	151–3(mono)	
					241(di)	
237	Phenylmalonic		153			
238	3-Bromobenzoic		155		136	
239	2,4,6-Trimethylbenzoic		155			
240	2-Chloro-4-methylbenzoic		155–6			
241	2-Naphthoxyacetic		156		145	
242	3-Phenyl-2-ketopropionic		157		126	
			(154)			
243	2-Hydroxy-1-naphthoic	62	157			
244	2,5-Dibromobenzoic	60	157			
245	2-Nitrophenylpropynoic		157			
246	4-Chlorophthalic	61	157			
247	Quinoline-2-carboxylic		157			
248	Cyclobutane-1,1-dicarboxylic		157		214–5(di)	
249	3-Chlorobenzoic	1	158		122	
250	4-Chlorophenoxyacetic		158		125	
			(156)			
251	Salicylic(v.s.) (see Table 29B)		158.3	156	136	188.9*
252	3-Hydroxyphthalic		161d			
	(see Table 29B)					
253	1-Naphthoic	1	162		163	
254	2-Iodobenzoic	1	162		141	
255	5-Chloro-2-hydroxy-3-nitrobenzoic		163			
256	Cholanic		163			
257	2,4,6-Trichlorobenzoic		164			
258	Thiosalicylic	63	164			
259	2,4-Dichlorobenzoic	64	164			
260	2,3-Dichlorobenzoic	65	164			
261	3,4-Dinitrobenzoic		165		188–9	
262	4-Nitrophthalic	66	165	172	192	
263	Propene-2,3-dicarboxylic		165		190	
	(itaconic)					
264	5-Bromo-2-hydroxybenzoic	68	165		122	
			(167)			
265	2-Chloro-5-nitrobenzoic	67	165			
266	2-Mercaptobenzoic		165	230	237	168
267	3-(4-Isopropylphenyl)propenoic		165			
			(157–9)			
268	Propane-1,2,3-tricarboxylic		166			252(tri)

TABLE 5B (Continued)
CARBOXYLIC ACIDS—SOLID

Melting Point of Derivatives (Continued)

Others (Continued)

	4-Nitro-benzyl-Ester	4-Phenyl-phenacyl Ester	4-Chloro-phenacyl Ester	4-Bromo-phenacyl Ester	Amide	4-Bromo-anilide	Methyl-enebis-anilide	Benz-imid-azole	Phenyl-hydra-zide
228	110	98		102	155				
229	99.5	122		152	155				
230					208.5				
231	147				178				
232	102	146		148	210–5(tri)				
233				207	198				
234					152				
235					155				
236	106	148		154.5	224(di) 125–30 (mono)				209
237					233				
238	105	155		126	155				
239					188				
240					182				
241					147				
242					190				
243									
244									
245					159				
246									
247					133				
248					275–7(di)				
249	107	154		116	134				
250				136	133				
251	98	148		140	142				
252									
253				135.5	202				
254	111	143		110	184				
255					199				
256					75				
257					181				
258									
259					194				
260									
261					165–6				
262		120			200d				
263	91			117	192(di)				
264					232				
265					178				
266									
267					185.6				
268			126(tri)	138(tri)	205–7d(tri)				

(Continued)

TABLE 5B (Continued)
CARBOXYLIC ACIDS—SOLID

	Name of Compound	Note	MP	Melting Point of Derivatives		
				Recommended		Others
				p-Toluidide	Anilide	2-Naphthyl-amide
269	3,4-Dimethylbenzoic		166		104	
270	2-Hydroxy-3-methylbenzoic		166		83	
271	3,5-Dimethylbenzoic		166			
272	dl-Phenylsuccinic (2-phenylbutanedioic)	70	167	α175 β169	175α 171β 222(di)	
273	3-Hydroxyphthalic (see Table 29B)	69	166–7rh			
274	2-Hydroxy-6-methylbenzoic	71	168			
275	8-Hydroxy-1-naphthoic	72	169			
276	d-Tartaric		169–71		180d(mono) 264d(di)	
277	8-Chloro-1-naphthoic		171–2			
278	4-Bromo-2,5-dimethylbenzoic		172			
279	5-Chloro-2-hydroxybenzoic		172			
280	4-Chloro-2-methylbenzoic		172			
281	2,4-Dibromobenzoic		174			
282	3-Aminobenzoic (see Tables 4 and 9B)	1	174		140	
283	3,5-Dinitrosalicylic(hyd.) (see Table 29B)		174			
284	4-Hydroxy-3-methylbenzoic	73	174–5			
285	3-Aldehydobenzoic (see Table 7B)	75	175			
286	Naphthalene-1,2-dicarboxylic	74	175			
287	8-Bromo-1-naphthoic		178		151	
288	p-Toluic		179–80	160 (165)	145 (148)	
289	Acetylenedicarboxylic		179			
290	2-Bromo-5-nitrobenzoic		180		166	
291	Ureidoacetic	76	180			
292	5-Bromo-2,4-dimethylbenzoic		180–1			
293	4-Nitrophenylpropynoic	77	181			
294	N-Benzoylanthranilic		181		279	
295	3,4-Dimethoxybenzoic(anh.)		181		154	
296	4-Chloro-3-nitrobenzoic		181–2		131	
297	N-Methylanthranilic	78	182			
298	2,4-Dinitrobenzoic		183			
299	4-Fluorobenzoic		182.6			
300	4-Hydroxy-2-naphthoic (see Table 29B)		183			
301	4-Hydroxy-1-naphthoic	79	184d			
302	5-Hydroxy-2-methylbenzoic	80	184			
303	3-Bromosalicylic		184			
304	4-Methoxybenzoic (anisic)		184–6	186	169–71	
305	Acetylanthranilic		185		167	
306	2,4-Dinitrophenylacetic	81	185d		181	
307	2-Naphthoic		185.5	192	171	

TABLE 5B (Continued)
CARBOXYLIC ACIDS—SOLID

Melting Point of Derivatives (Continued)

Others (Continued)

	4-Nitro-benzyl Ester	4-Phenyl-phenacyl Ester	4-Chloro-phenacyl Ester	4-Bromo-phenacyl Ester	Amide	4-Bromo-anilide	Methyl-enebis-anilide	Benz-imid-azole	Phenyl-hydra-zide
269					130				
270	98.5				112				
271					133				
272					159α				
					145β				
					211(di)				
273									
274									
275									
276	163	203–4		204(di) (216)	196d				240
277					207.5				
278					209–10				
279					226–7				
280					183				
281					198				
282	201				111 (114)				
283					181				
284									
285					190d				
286					265(di)				
287					179–80				
288	104.5	165		153	160				
289					294d(di)				
290					197–8				
291					180(204)				
292					197.5–8				
293									
294					219				
295					164				
296					156				
297					159–60				
298	142			158	203				
299					154.5				
300									
301									
302									
303					165				
304	132	160		152	167 (163)				
305					177				
306					180				
307					192 (195)				

(Continued)

TABLE 5B (Continued)
CARBOXYLIC ACIDS—SOLID

	Name of Compound	Note	MP	Melting Point of Derivatives		
				Recommended		Others
				p-Toluidide	Anilide	2-Naphthyl-amide
308	3-Chlorophthalic	82	186			
309	Hippuric		187		208	
310	4-Methoxy-3-nitrobenzoic	1	187		163	
311	3-(2-Methoxypheny)-propenoic(trans)		187			
312	3-Iodobenzoic	1	187			
313	4-Bromo-2-methylbenzoic		187			
314	Butanedioic (succinic)		188 (185)	180(mono) 255(di)	148(mono) 230(di)	
315	4-Nitrophenoxyacetic	83	188		170	
316	d-Camphoric (l,m 187)	84	188	α212–4 β190–6	204(mono) 226(di)	
317	4-Aminobenzoic (see Tables 4 and 9B)		188			
318	2-(2-Naphthyl)benzoic		190			
319	N-Methylacetylanthranilic		192–3			
320	Dimethylmalonic		193s			
321	1-Naphthoxyacetic		193.5		144	
322	Aconitic(trans)	85	194–5d, N		189(di)	
323	1-Hydroxy-2-naphthoic		195		154	
324	Pyridine-2,3-dicarboxylic	86	195			
325	3-(3,4-Dihydroxyphenyl)-propenoic (3,4-dihydroxycinnamic)	87	195d			
326	Thiazole-4-carboxylic	88	196			
327	4-Ethoxybenzoic		198		170	
328	dl-Glutamic (l,m 211–3) (see Table 4)	89	199d			
329	4-Aminophenylacetic		199–200			
330	3,4-Dihydroxybenzoic		200d		166	
331	Fumaric		200s, N (mp 300r.h.)		314	
332	3-Hydroxybenzoic (see Table 29B)	1	200	163	157	
333	2,5-Dihydroxybenzoic (see Table 29B)	90	200			
334	3,5-Dichloro-4-methoxybenzoic		202		154	
335	2,6-Dinitrobenzoic	91	202			
336	6-Bromo-3-nitro-4-methylbenzoic,		203			
337	3,4,5-Trichlorobenzoic	92	203			
338	4-Bromo-3-nitrobenzoic		203–4		156	
339	2,3-Dihydroxybenzoic	93	204			
340	dl-Tartaric		204		236	
341	Apocamphoric		204		212(mono)	
342	4-Hydroxyphthalic (see Table 29B)	96	205			
343	3-(3-Nitrophenyl)-propenoic	1	205(199)			
344	4-Hydroxy-3,5-dimethoxy-benzoic (see Table 29B)	95	205			

TABLE 5B (Continued)
CARBOXYLIC ACIDS—SOLID

Melting Point of Derivatives (Continued)

Others (Continued)

	4-Nitro-benzyl-Ester	4-Phenyl-phenacyl Ester	4-Chloro-phenacyl Ester	4-Bromo-phenacyl Ester	Amide	4-Bromo-anilide	Methyl-enebis-anilide	Benz-imid-azole	Phenyl-hydra-zide
308									
309	136	163		151	183				
310									
311					195 (192)				
312	121	147		128	186				
313					180				
314					157(mono) 260(di)				
315					156–8				
316	65.5			142.5	177(mono) 193(di)				
317					183				
318					170				
319				155					
320	83.6				269(di)				
321					155				
322	70			186	250N				
323					202				
324					209N				
325									
326					150				
327	110			141	202				
328									
329					161–2				
330	188				212				
331	151			256–7d	266(di)				
332	106–8			176	170				
333									
334									
335									
336					191				
337					176				
338					156				
339									
340	148				226				
341									
342									
343	174			178 (173)	196				
344									

(Continued)

TABLE 5B (Continued)
CARBOXYLIC ACIDS—SOLID

	Name of Compound	Note	MP	Melting Point of Derivatives		
				Recommended		Others
				p-Toluidide	Anilide	2-Naphthyl-amide
345	Acetylglycine	97	206			
346	4-Coumaric		206			
347	Phthalic	98	206–8	201(di)	253(di)	
348	3,5-Dinitrobenzoic	1	207	145–7	234	
349	Vanillic		207 (210)			
350	3,4-Dichlorobenzoic	1	208–9 (201–2)			
351	3-Nitroanthranilic (see Table 9B)		208			
352	Acetophenone-4-carboxylic (see Table 21B)		208			
353	Anthracene-9-carboxylic	99	208			
354	5-Chloro-2-hydroxy-4-methylbenzoic (see Table 29B)		208		222	
355	4-Acetobenzoic (see Table 21B)		208			
356	2-Coumaric(*trans*)		208d			
357	6-Hydroxy-1-naphthoic	100	209		193–4	
358	Oxamic		210		148–9	
359	4-Chloro-1-naphthoic		210 (224)			
360	5-Hydroxy-2-naphthoic	101	211		163–4	
361	2-Amino-5-chlorobenzoic	102	211–12			
362	3-(2-Chlorophenyl)-propenoic(*trans*)		212		176	
363	*l*-Glutamic (see Table 4)		211–13			
364	2-Bromo-3,5-dinitrobenzoic		213			
365	Mucic	103	214d, N			
366	3-(2-Hydroxylphenyl)-propenoic (see Table 29B)		214			
367	3-Chloro-4-methoxybenzoic		214–5			
368	4-Hydroxybenzoic		215	204	198	
369	Tetrahydrophthalic	104	215r.h.			
370	Piperic		216			
371	2,4-Dihydroxybenzoic (β-resorcylic)		216d (213)		126–7	
372	3-Chloro-2-naphthoic		216–7			
373	6-Chloro-1-naphthoic		217		197	
374	3-Nitrophthalic		218(215)	226(di)	234(di)	
375	Thiazole-5-carboxylic	105	218			
376	4-Cyanobenzoic		219 (214)		179	
377	3,5-Dibromobenzoic		219			
378	4-Mercaptobenzoic		220		264	283
379	3,6-Dihydroxyphthalic	106	220			
380	4-Aminophenoxyacetic (see Table 9B)		220		104–5	
381	3,5-Dichlorosalicylic		220			
382	Acenaphthalene-5-carboxylic		220			

TABLE 5B (Continued)
CARBOXYLIC ACIDS—SOLID

Melting Point of Derivatives (Continued)
Others (Continued)

	4-Nitro-benzyl Ester	4-Phenyl-phenacyl Ester	4-Chloro-phenacyl Ester	4-Bromo-phenacyl Ester	Amide	4-Bromo-anilide	Methyl enebis-anilide	Benz-imid-azole	Phenyl-hydra-zide
345					137				
346	152				197				
347	155	167		153	220(di)				161(di)
348	157	154		159	183				
349	140d								
350			123		133				
351					234				
352									
353									
354					239–40				
355									
356	152.5				209d				
357									
358					419d				
359				131	236				
360									
361					172				
362					168				
363					165				
364					216				
365	310	149d		225	192d(mono) 220(di)				
366									
367					193				
368	180–2	240		191.5	162(hyd.)				
369									
370	145								
371	189				222				
372					237				
373		143			216				
374	189	149		166	201d(di)				
375					186				
376	189				223				
377					187				
378									
379									
380					128				
381					209				
382					198				

(Continued)

TABLE 5B (Continued)
CARBOXYLIC ACIDS—SOLID

	Name of Compound	Note	MP	Melting Point of Derivatives		
				Recommended		Others
				p-Toluidide	Anilide	2-Naphthyl-amide
383	3-Hydroxy-2-naphthoic		223*	222	244 (249)	
384	4-Hydroxy-2-naphthoic		225–6*	206		
385	4-Methyl-2,6-dinitrobenzoic		226			
386	4-Bromo-6-nitro-2-methylbenzoic		226			
387	Methyliminodiacetic		227d			
388	2,4,6-Trinitrobenzoic		228			
389	Biphenyl-4-carboxylic	107	228			
390	3,5-Dibromosalicylic	108	228			
391	2,2'-Diphenic		229		230(di) 181–3(mono)	
392	Piperonylic		229			
393	8-Hydroxy-2-naphthoic	109	229		239	
394	5-Nitrosalicylic		229–30		224	
395	4-Chloro-3-hydroxy-2-naphthoic	110	231			
396	4-Bromo-3-hydroxy-2-naphthoic		233–5		161–2	
397	4-Chloro-1-hydroxy-2-naphthoic		234	143–4	180–1	
398	3-Aminosalicylic (see Table 4)		235			
399	5-Hydroxy-1-naphthoic (see Table 29B)	111	235			
400	3-Chloro-2-nitrobenzoic		235		186	
401	5-Bromo-2-hydroxy-3-methylbenzoic		236		125	
402	7-Bromo-1-naphthoic		237		202	
403	Nicotinic	112	237–8s	150	132N	
404	2-Hydroxyisophthalic		239 (244)			
405	3-(2-Nitrophenyl)-propenoic(trans)		240			
406	Mesaconic	94	240.5	212(di)	186(di)	
407	4-Nitrobenzoic		241	204	211	
408	Naphthalene-2,3-dicarboxylic	113	241			
409	4-Chlorobenzoic	1	242		194	
410	7-Chloro-1-naphthoic		243		185	
411	5-Chloro-1-naphthoic		245 (242)			
412	4-Dimethylaminobenzoic		245		182–3	
413	6-Hydroxy-2-naphthoic	114	245		197–8	
414	Anthracene-1-carboxylic		245 (252)			
415	Barbituric		245d			
416	3-Hydroxy-1-naphthoic (see Table 29B)		248		112–3	
417	Tetrachlorophthalic		250d			
418	3-(4-Bromophenyl)-propenoic		251–3		183	
419	7-Hydroxy-1-naphthoic		254		209–10	
420	Gallic	115	254dN		207	
421	4-Aldehydobenzoic (see Table 7B)		256			

TABLE 5B (Continued)
CARBOXYLIC ACIDS—SOLID

Melting Point of Derivatives (Continued)

Others (Continued)

	4-Nitro-benzyl-Ester	4-Phenyl-phenacyl Ester	4-Chloro-phenacyl Ester	4-Bromo-phenacyl Ester	Amide	4-Bromo-anilide	Methyl-enebis-anilide	Benz-imid-azole	Phenyl-hydra-zide
383					218*				
384			109–10		217–8				
385					255–7				
386					235				
387					169				
388					264d				
389					223N				
390					183(170)				
391	187				193(mono) 212(di)				
392					169				
393									
394					225				
395					225				
396									
397									
398									
399									
400									
401					75–8				
402					247				
403					128				
404				109 (mono) 162(di)	185(di) 245(mono)				
405	132	146		141	185				
406	134				177 and 222				
407	168	182		137	201				
408									
409	129	160		126	179				
410					237				
411					239				
412					206				
413									
414					260				
415									
416					209–11				
417									
418									
419									
420	141	160		134	189				
421									

(Continued)

TABLE 5B (Continued)
CARBOXYLIC ACIDS—SOLID

	Name of Compound	Note	MP	Melting Point of Derivatives Recommended		Others
				p-Toluidide	Anilide	2-Naphthyl-amide
422	4-Bromobenzoic	1	258 (254)		197	
423	Chelidonic		262			
424	β-Phenylalanine (see Table 4)		264 (273)			
425	4-Iodobenzoic		270		210	
426	5-Chloro-2-naphthoic		270		202.5	
427	7-Hydroxy-2-naphthoic		270		219–20	
428	Imidazole-4-carboxylic		275d			
429	5-Aminosalicylic (see Table 4)		283			
430	4-Amino-3-nitrobenzoic	116	284 (290)		215–6	
431	3-(4-Nitrophenyl)-propenoic(trans)	1	285			
432	Muconic		289d			
433	Anthraquinone-2-carboxylic		290–2		258–60	
434	9,10-Anthraquinone-1-carboxylic		293–4		288–9	
435	Bromoterephthalic		299			
436	Terephthalic		>300s		334–7	
437	Pyridine-4-carboxylic	117	306s			
438	3-Hydroxy-triphenylene-2-carboxylic		309–10	308	270	
439	4-Hydroxyisophthalic		310			
440	1,4-Diphenyl-triphenylene-2,3-dicarboxylic		330–5	341	358	
441	Triphenylene-2-carboxylic		336–8			
442	Isophthalic		348s			
443	Trimesic		380			118

NOTES FOR TABLES 5A AND 5B (ACIDS)

1. The melting points of the S-benzylthiuronium salts of the numbered acids are listed. A method for the preparation of these salts may be found under the heading of the derivatives for sulfonic acids. Table 5A: (15) 143; (36) 133–4; (48) 146–7; (52) 129–30. Table 5B: (11) 165–6; (12) 148–9; (16) 137; (38) 144.5; (74) 165–6; (92) 138–9; (96) 177–8; (111) 163–4; (181) 162–3; (188) 169–70; (209) 165–6; (229) 140; (231) 185 (249) 164–5; (253) 147–8; (254) 162–3; (282) 160–1; (309) 158–9; (312) 178–9; (332) 161–2; (343) 158–9; (348) 178; (350) 168–9; (409) 190; (422) 195–6; (431) 208–9.
2. The melting points of the 2,4-dinitrophenylhydrazones of the p-phenylphenacyl esters of the numbered compounds follows (*Anal. Chem.* **28,** 1976 (1956). Table 5A: (7) 184; (11) 139; (15) 156; (19) 143; (23) 150; (29) 150; (40) 145–6; (44) 136–7; (51) 121–2; (54) 114; (55) 104–5. Table 5B: (7) 101–2; (12) 100; (27) 103; (28) 101–2; (46) 101–2; (54) 101–2; (68) 101–2.
3. Decomposed by concentrated sulfuric acid. Readily reduced to formaldehyde by magnesium and hydrochloric acid. Hydrazide, mp 54.
4. Hydrazide, mp 67.
5. Only one reference was found that gave a melting point for this acid (40.35).
6. The dibromide melts at 50.
7. The methyl ester, bp 215.
8. Can be reduced by sodium amalgam to propionic acid.
9. Ureide, mp 143.
10. 2,4-Dinitrophenylhydrazone, mp 218.
11. The 4-phenylphenacyl ester melts at 63.5 when purified by the usual methods; the melting point of 69 was obtained on

TABLE 5B (Continued)
CARBOXYLIC ACIDS—SOLID

Melting Point of Derivatives (Continued)

Others (Continued)

	4-Nitro-benzyl Ester	4-Phenyl-phenacyl Ester	4-Chloro-phenacyl Ester	4-Bromo-phenacyl Ester	Amide	4-Bromo-anilide	Methyl-enebis-anilide	Benz-imid-azole	Phenyl-hydra-zide
422	180	193			190				
423		198d			245				
424	222				140				
425	141	171		146	217				
426					186–7				
427									
428					215				
429					194–7d				
430					227				
431	186	192		191	204 (217)				
432					240d(di)				
433					280				
434					280				
435					270(di)				
436	263(di)			225(di)					
437					156				
438									
439					250(di)				
440									
441					292 3				
442	202	280		179	280(di)				
443				197(s.t.)	365				

NOTES FOR TABLES 5A AND 5B (ACIDS) (Continued)

samples purified chromatographically.
12. Forms a copper salt, mp 232–4, and a lead salt, mp 80.
13. Oxime, mp 104–5; semicarbazone(hydrated) 175d.
14. One form of the acid is reported to melt at 14–16.
15. The methyl ester boils at 200, and yields a phenylhydrazone, mp 88, and a 4-nitrophenylhydrazone, mp 147.
16. The amide forms an oxime, mp 133–5.
17. Oxime, mp 96; phenylhydrazone, mp 93; 2,4-dinitrophenylhydrazone, mp 206.
18. One reference lists a melting point for acetoacetic acid as 36–7 and states that it decomposes below 100, no other melting point data were found. It is reported that it forms an amide, mp 54, that yields a phenylhydrazone, mp 128. Additional derivatives: 2-chloroanilide, mp 105; o-toluidide. mp 104.

19. Treated with $AlCl_3$, it ring-closes to form 1-indanone, mp 42 [*Bull. soc. chim.* (4) **41**, 942 (1927)].
20. Heilbron lists three forms, melting at 61.3, 56.2, and 52.5.
21. Dibromide, mp 148.
22. A labile form melts at 51 but the stable form melts at 67.
23. Methyl ester, mp 3819; ethyl ester, mp 36.1.
24. By heating with pyridine, the *cis* isomer is converted to the *trans* isomer, mp 99–100. Reacts with Cl_2 in CS_2 to yield trichlorobutyric acid, mp 60.
25. Long heating at 100 produces the anhydride, mp 128–30.
26. Methyl ester, mp 54; ethyl ester mp 52–3.
27. Phenylhydrazone, mp 100–4.
28. Methyl ester, mp 61–2; ethyl ester, mp 58–9.
29. Ethyl ester, mp 65.

NOTES FOR TABLES 5A AND 5B (ACIDS) (Continued)

30. Methyl ester, mp 68.8; ethyl ester, mp 66.6.
31. Methyl ester, mp 71.5; ethyl ester, mp 70.5.
32. Decomposes at the melting point to pyrrole and CO_2; ethyl ester boils at 180.
33. Phenacyl ester, mp 67; hydrazide, mp 115–6.
34. Phenacyl ester, mp 106; bisphenylhydrazide, mp 220.
35. Monoanilide, mp 164; dianilide, mp 179. The anhydrous acid, mp 153, may be obtained by continued boiling of the aqueous solution of the acid. Warmed with acetic anhydride and pyridine, it produces a carmine color.
36. Melts at 101 when heated rapidly; anhydrous acid, mp 189.5.
37. Decomposes to acetophenone on being strongly heated; the ethyl ester, bp 265–70, yields a semicarbazone, mp 125d, and a 2,4-dinitrophenylhydrazone, mp 247.
38. Benzylamine salt, mp 113–4; methyl ester, bp 236–8.
39. The *cis* isomer on being heated at 155, converts to the *trans* isomer, mp 138.
40. The monoanilide (from ethyl acetate), mp 159; the monoanilide from chloroform, mp 123.
41. The dianilide exists in two forms: yellow, mp 170, and white, 193; imide, mp 179; anil, mp 203s.
42. Methyl ester, mp 83–7; ethyl ester (from ligroin), 46–7.
43. The hydrate of the acid, mp 86; ammonium salt, mp 113
44. Used as a reagent to derivatize amino acids. Picrolonates of: glycine, mp 214–5; DL-phenylalanine, mp 208d.
46. The dianilide has two forms: mp 200 and 188–9. The *cis* form of the acid converts to the *trans* form on being heated.
47. Imide, mp 169–70; anhydride, mp 74.
48. Hydrazide, mp 136.
49. Ordinarily, this acid contains about 3% fumaric acid and melts at 130, but the pure acid melts at 137; the anhydride mp 60, is formed by heating the acid to 160; hydrazide, mp 292.
50. Transforms to the *trans* isomer, mp 142, when heated at 170.
51. Hydrazide, mp 129–30; picrate, mp 141–3; methyl ester, mp 79–80.
52. Oxime, mp 53–4; decomposes by hot water, acids, or alkalis to acetone and CO_2.
53. The labile *cis* isomer melts at 136, but changes on melting into the *trans* isomer, mp 138. It is dehydrated by acetic anhydride at 40 to yield 6-hydroxy-α-pyrone, mp 88.
54. 2-Naphthylamide, mp 139.
55. Dihydrazide, mp 172–3.
56. Also named 3-hydroxy-o-toluic acid; methyl ester, mp 75; ethyl ester, mp 69; acetyl derivative, mp 145.
57. Methyl ester, mp 109–10; ethyl ester, mp 88; hot, concentrated sulfuric acid converts it to 2-chloroanthraquinone.
58. The D-acid, and the L-acid, mp 152–3 form a mono amide, mp 125; S-acetyl derivative, mp 125–6; S-benzoyl derivative, mp 181.
59. Hydrazide, mp 204; methyl ester, mp 83;
60. 4-Nitrophenyl ester, mp 183–4; 3-nitrophenyl ester, mp 165.
61. Anil, mp 174; imide, mp 210–11; anhydride, mp 99.
62. Methyl ester, mp 80 (76); ethyl ester, mp 55; acetyl derivative, mp 131.
63. Acetyl derivative, mp 125; S-ethyl derivative, mp 134–5.
64. 3,5-Dinitro-derivative, mp 212.
65. Distilled with soda lime it yields 1,2-dichlorobenzene.
66. Imide, mp 199–200.
67. Methyl ester, mp 73; N-methylamide, mp 174.
68. Acetyl derivative, mp 168 ("5-bromoaspirin").
69. Imide, mp 255–6; methyl ester, mp 173–4; dimethyl ester, mp 73–4.
70. The d- and l- acids melt at 173–4, and form an anhydride, mp 84, whereas the anhydride from the dl-acid melts at 54.
71. Acetyl derivative, mp 131 (from benzene).
72. Lactone, mp 108.
73. Ethyl ester, mp 98–9; acetyl derivative, mp 75–6.
74. Imide, mp 224; dimethyl ester, mp 80.
75. Anil, mp 156.
76. Also named hydantoic acid; ethyl ester mp 139; butyl ester, mp 119. Melting points for the acid have been reported at 156 and 163.
77. Also named 4-nitrophenylpropiolic acid; ethyl ester, mp 126.
78. Salt with HCl, mp 141; ethyl ester, mp 39; see Table 9B for other data.
79. Methyl ester, mp 178; ethyl ester, mp 134; acetyl derivative, mp 178–9.
80. Methyl ester, mp 75; ethyl ester, mp 67.
81. Hydrazide, mp 136–7; methyl ester, mp 82; ethyl ester, mp 35.
82. Imide, mp 118–20 in sealed tube; acid heated above melting point forms the anhydride, mp 124–5.
83. Methyl ester, mp 100; ethyl ester, mp 75–6; N-methylamide, mp 166.
84. The dl-acid melts at 202 (208). The oxime mp 117–9.
85. Cis-aconitic acid melts at 125. The melting point of the *trans* isomer may vary widely due to the partial transformation to itaconic acid. The triamide turns brown at about 250, and sinters without melting at about 260.

NOTES FOR TABLE 5A AND 5B (ACIDS) (Continued)

86. Diamide melts with decomposition at 209 (some decomposition starts as low as 190); imide, mp 233 (230); dihydrazide, mp 224.
87. Diacetyl derivative, mp 198; dibenzoyl derivative, mp 204–6.
88. Hydrazide, mp 144.
89. The d-acid, mp 224–5, and the l-acid mp 213 ($r.h.$).
90. Diacetyl derivative, mp 62–4 in sealed tube; methyl ester, mp 88.
91. Methyl ester, mp 147; ethyl ester, mp 75.5; acid chloride, mp 98.
92. Ethyl ester, mp 86; acid chloride, mp 36.
93. Methyl ester, mp 130.5; diacetyl derivative, mp 148–50.
94. Also named methylfumaric acid. Mono-p-toluidide, mp 196; α-monoanilide, mp 202; β-monoanilide, mp 163; α-monoamide, mp 222; β-monoamide, mp 174.
95. 4-Acetyl derivative, mp 191 (187); 4-benzoyl derivative, mp 229–32.
96. Imide, mp 290; dimethyl ester, mp 107–8.
97. Hydrazide, mp 115.
98. Melts at 191 in sealed tube, and at 230 by rapid heating; mono-p-toluidide melts at 150 (slow) and 160–5 (rapid heat); monoanilide, mp 170; monoamide, mp 149; imide, mp 235–7.
99. Methyl ester, mp 111.
100. Methyl ester, mp 112–3; ethyl ester, m.p. 105–7.
101. Ethyl ester, mp 150–1.
102. N-Acetyl derivative, mp 173; methyl ester, mp 76.
103. Melts at 214 (slow heat) and varies from 225 to 235 (rapid heat).
104. Also named 2-cyclohexene-1,2-dicarboxylic acid; imide, mp 172–3; anhydride, mp 78–9 (from ether).
105. Hydrazide, mp 157–9.
106. Diacetyl derivative mp 165–6; dimethyl ester, mp 141–2; anhydride, mp 261.
107. The amide, mp 223, was recrystallized from acetic acid; methyl ester, mp 118.
108. 2-Acetyl derivative, mp 163.
109. Acetyl derivative, mp 177; ethyl ester, mp 135–7.
110. Acetyl derivative, mp 186; acid chloride, mp 128.
111. Methyl ester, mp 130–1; ethyl ester, mp 73; acetyl deriv., mp 202–3.
112. Also named pyridine-3-carboxylic acid; the anilide melts at 132 when crystallized from benzene or chloroform; when crystallized from water it has been reported at 85 (a hydrate) and at 265; hydrazide, mp 158–9; piperidine salt, mp 122; hydrochloride salt, mp 273–5.
113. Imide, mp 275; phenylimide, mp 277–8; anhydride, mp 246.
114. Ethyl ester, mp 111–2; acetyl derivative, 223.
115. The melting point varies widely, often as low as 220–40d.
116. N-Acetyl derivative, mp 220–1.
117. Hydrazide, mp 163 (171–4); also named isonicotinic acid.
118. dl-Malic acid, mp 130–1, is reported to yield an amide, mp 173–5.

REFERENCES FOR TABLES 5A AND 5B

The references for Tables 5 and 6 are grouped together under Table 6 on p. 726. References for acids and alcohol are to be found at the end of Table 6.

TABLE 6A
ALCOHOLS—LIQUID

	Name of Compound	Note	BP	Melting Point of Derivatives Recommended	
				α-Naphthyl-urethan	3,5-Dinitro-benzoate
1	Methanol	81	64.65	124	108*
2	Ethanol	81	78.32	79	93
3	2-Propanol	81	82.2	106	123
4	2-Methyl-2-propanol (mp 25.5; see table 6B)	81	82.5		
5	dl-3-Buten-2-ol	1	94–6		
6	Difluoroethanol	2	96		
7	2-Propen-1-ol		97.1	108	49
8	Propanol	81	97.15	105	74
9	2-Butanol	81	99.5	97	76
10	2-Methyl-2-butanol		102.35	72	116
11	2-Methyl-3-butyn-2-ol		103.6		112
12	2-Fluoroethanol		105	128	
13	2-Methyl-1-propanol		108.1	104 (98)	87
14	3-Penten-2-ol		112		
15	3-Methyl-3-buten-2-ol		112–3		
16	3-Buten-1-ol		113		
17	dl-3-Methyl-2-butanol	3	114	109	76
18	2-Propyn-1-ol		114	63	
19	1-Penten-3-ol		114–6		
20	3-Pentanol		116.1	95	101 (97)
21	Butanol	81	117.7	71	64 (62.5)
22	2,3-Dimethyl-2-butanol		118–9	101	111
23	dl-2-Pentanol		119.85	74.5	62
24	dl-3,3-Dimethyl-2-butanol		120.4		107
25	2-Methyl-2-pentanol		121		72
26	3-Methyl-1-pentyn-3-ol		121.4		72–3
27	3-Methyl-3-pentanol		123	104	96.5 (62.5)
28	2-Methoxyethanol	4	124.2	113	
29	1-Chloro-2-propanol	76	127		77
30	2-Methyl-3-pentanol		127.5		85
31	2-Methyl-1-butanol		128	82	70
32	2-Methoxy-1-propanol		130	60	97
33	2-Chloroethanol	76	131	101	95
34	4-Methyl-2-pentanol		132	88	65
35	3-Methyl-1-butanol		132	68	61
36	2,4-Dimethyl-2-pentanol		132–3		
37	3,4-Dimethyl-1-pentyn-3-ol		133		
38	dl-2-Chloro-1-propanol	5	133–4		76
39	2-Dimethylaminoethanol (see Table 10A)		133–5		
40	3-Methyl-2-pentanol		134	72	43.5
41	1-Hexen-3-ol		134		
42	2-Ethoxyethanol	4	135	67	75
43	1-Methylcyclopentanol		135–6		115.5
44	3-Hexanol		136	72	77
45	3-Ethyl-1-pentyn-3-ol		135–7		

TABLE 6A (Continued)
ALCOHOLS—LIQUID

Melting Point of Derivatives (Continued)

Others

	Phenyl-urethan	4-Nitro-phenyl-urethan	p-Xenyl-urethan	3,5-Dinitro-phenyl-urethan	4-Nitro-benzoate	Hydrogen 3-nitro-phthalate	Hydrogen phthalate	Pseudo-saccharin Derivative	Allo-phanate
1	47	179.5	127	127	96	153*	82.5*	182*	208(212)
2	52	129	119	83	57	158*	48	219*	188(198)
3	88	116	138	112	110	154*		137*	180(185)
	(75)								
4									190
5					43–4		52–3		151–2
6									
7	70	108		114	28	124			
8	57	115	129	97	35	145.5*	54.4*	124.5*	167(175)
		(110)							
9	64.5	75	105.5	120	26	131*	60	65.5*	159.5
10	42				85				151–2
11					126–7				
12									
13	86	80		119	69	180.5*	65	100*	174
	(82)						(68)		
14									158–9
15									150–1
16	23.4–24.5								
17	68					127	39		163(170)
18									
19									151–2
20	49				17	121			172(179)
21	61	95.5	109	70	36	147*	73.5	96*	149–50
				(64)					
22	66	100.5			82				
23			94.5		24–5	103	61	38*	154(158)
24	78						86		
25									126(128)
26					72		95–7		
27	43.5				69.5				152–4
28		111			51	129			163
29									
30	50					150.7	70		179(186)
31	31					158			150
32									
33	51					98			179–82
34	143		95.5		26				161–3
35	57	97.5			21	163.3*		64*	150
36									132
37									144
38									
39									
40									
41					60–2		62–3		140
42		80				118			
43					83				157
44						127–8	76		185.5
45									130–1

(Continued)

TABLE 6A (Continued)
ALCOHOLS—LIQUID

	Name of Compound	Note	BP	Melting Point of Derivatives Recommended	
				α-Naphthyl-urethan	3,5-Dinitro-benzoate
46	2,2-Dimethylbutanol		136.7	81	51
47	3-Methyl-1-hexyn-3-ol		137		
48	2,4-Dimethyl-3-pentanol		138		
49	1-Pentanol	81	138*	68	46.4
50	2-Hexanol		138–9	60.5	38.5
51	4-Penten-1-ol		138–9		
52	3-Methyl-5-hexen-3-ol		139		
53	3-Methyl-3-hexanol		140		
54	2,2-Dimethyl-3-pentanol	6	140	95	
55	Cyclopentanol		140.8	118	115
56	2-Penten-1-ol		141–2		
57	Cyclobutylmethanol		142	112	99
58	3-Ethyl-3-pentanol	6	142		
59	2,3-Dimethyl-1-butanol		145		51.5
60	3-Hydroxy-2-butanone (see Table 21A)		145		
61	1-Hydroxy-2-propanone (see Table 21A)		146		
62	5-Methyl-3-hexanol		147–8		
63	2-Methyl-1-pentanol		148	76	50.5
64	2-Ethyl-1-butanol		148	60–1	51.5
65	2-Bromoethanol		149d	86	
66	4-Methyl-3-hexanol		149–50		
67	Trichloroethanol		151	120	142.5
68	4-Methyl-1-pentanol		151–2	58–9	72
69	3-Methyl-1-pentanol		152–3	58	38
70	dl-1-Hepten-3-ol		153–4		
71	3-Methoxy-2-methyl-1-propanol		155		64
72	2-Ethyl-2-methyl-1-butanol		155		
73	2,2-Dimethyl-3-hexanol		155–7	113–4	
74	4-Hexen-1-ol		155–8		
75	3-Heptanol		156		
76	4-Heptanol		156	80	64
77	1-Hexanol	81	157.5	59 (62)	58.4
78	2,3-Dimethyl-3-hexanol		158		
79	dl-2-Heptanol		158.5	54	49
80	2,3-Dimethyl-2-hexanol		159		
81	2,4-Dimethyl-1-pentanol		159.8		
82	1-Amino-2-propanol (see Table 9A)		159–60		
83	3-Ethyl-3-hexanol		161		
84	Cyclohexanol (mp 25) (see Table 6B)	7, 81	161.1		
85	3-Chloro-1-propanol		161–2	76	77
86	Cyclopentylmethanol		162	85–6	90
87	2-Diethylaminoethanol		163		
88	3-Methyl-3-heptanol		164		
89	2-Methyl-1-hexanol		165		
90	dl-4-Methyl-1-hexanol		165	50	
91	2-Methylcyclohexanol(cis)	8	165.5	154.5	99
92	2-Ethylpentanol		166		
93	4-Hydroxy-4-methyl-2-pentanone (see Table 21A)		166		55
94	2,3-Epoxy-1-propanol		167d	102	

TABLE 6A (Continued)
ALCOHOLS—LIQUIDS

Melting Point of Derivatives (Continued)

Others

	Phenyl-urethan	p-Nitro-phenyl-urethan	p-Xenyl-urethan	3,5-Dinitro-phenyl-urethan	4-Nitro-benzoate	Hydrogen 3-nitro-phthalate	Hydrogen phthalate	Pseudo-saccharin Derivative	Allo-phanate
46	66					154.5	69		
47									
48									133
49	46	86	99	58	11	136*	75.5	62*	174–5
50				40		62–3			159–60
51									170
52									147–8
53									132–3
54	96–9				155				148–9
55	132				62				180
56									157.5
57	65–6								
58									173
59	29								
60									
61									
62									185–7
63			99			145(141)	51		151(156)
64						147	54		158(161)
65	76					172			
66									172–4
67	87				71				183
68	48					140	48		
69						152			
70					24–5		56–7		156–7
71									
72									102
73	70–1								
74									161.5
75									185
76					35		60		206
77	42	104	98	75	5	124*	25	60*	165(160)
78	90.5								
79							57.5		
80	74								
81			75			155			117
82									
83									149
84									
85	38								166
86	105–6								173
87		60							138
88									130
89			88.5			132			
90						144			
91	103				56		104		177
92			77.5			128			
93					48				
94	60				56				

(Continued)

TABLE 6A (Continued)
ALCOHOLS—LIQUID

	Name of Compound	Note	BP	Melting Point of Derivatives Recommended	
				α-Naphthyl-urethan	3,5-Dinitro-benzoate
95	2-Methylcyclohexanol (mp 21)(*trans*)	8	167	155	115
96	2-Ethylaminoethanol (see Table 9A)		167.1		
97	1-Amino-2-butanol (see Table 9A)		168–70		
98	2-Methylaminoethanol (see Table 9A)		169–70	125	
99	2-Aminoethanol (see Table 9A)		171		
100	2-Butoxyethanol	4	171 (743 mm)		
101	3,5-Dimethyl-4-heptanol		171		
102	Furfuryl alcohol	9	172	130	81
103	4-Amino-2-butanol (see Table 9A)		172		
104	3-Methylcyclohexanol (*cis* or β)		173–4	128–9	91–2
105	4-Methylcyclohexanol (*cis* or β)		173–4 (750 mm)	160	134
106	4-Methylcyclohexanol (*trans* or α)		173–5 (745 mm)		140
107	dl-2-Amino-1-propanol (see Table 9A)		173–6		
108	3-Methylcyclohexanol (*trans* or α)		174–5	122 (118)	97–8
109	2,6-Dimethyl-4-heptanol		175		
110	2,4-Dimethylcyclohexanol(*trans*)		175		
111	1,3-Dichloro-2-propanol		176	115	
112	3-Bromo-1-propanol		176	73	
113	1-Heptanol		176.8	62	47
114	2-Hydroxypentan-4-one (see Table 21A)		177		
115	Tetrahydrofurfuryl alcohol		177–8 (743 mm)	90	83–4
116	2-Methyl-1,2-propanediol		178		
117	3,3-Dimethyl-4-heptanol		178		73
118	2,2-Dibromoethanol	10	179–81		
119	2-Octanol		179	63–4	32
120	2,3-Dichloro-1-propanol	13	182	93	
121	2,3-Butanediol	11	182.5		
122	Cyclohexylmethanol	12	182	108–9	95–6
123	4-Methyl-1-heptanol		182.7		
124	2-Ethyl-1-hexanol		184.6	61	
125	3,3-Dimethylcyclohexanol	14	185		
126	1,2-Propanediol	15	187.4		
127	6-Methyl-1-heptanol		187.6		
128	3-Amino-1-propanol (see Table 9A)		188		
129	3,4-Dimethylcyclohexanol		189		
130	3-Diethylaminopropanol (see Table 10A)		190		
131	2-Nitroethanol	16	194		

TABLE 6A (Continued)
ALCOHOLS—LIQUID

Melting Point of Derivatives (Continued)

Others

	Phenyl- urethan	p-Nitro- phenyl- urethan	p-Xenyl- urethan	3,5-Dinitro- phenyl- urethan	4-Nitro- benzoate	Hydrogen 3-nitro- phthalate	Hydrogen phthalate	Pseudo- saccharin Derivative	Allo- phanate
95	105				65		124	54	
96									
97									
98									
99									
100	62	59				120.5			118
101									163
102	45				76		85	55	167.5
103									
104	87–8				65		82–3		141
105	118–9				94		72–3		
106	125				67		120		
107									
108	93–4				58		93–4		178.8
109	154		118		118				156
110	96								
111	73								182
112									
113	60 (65)	102 (105)		61	10	127*	17.5	55*	160
114									
115	61				46–8				161
116	140.5								
117									
118									
119	Oil	Oil			28		55		155
120	73				37–8				
121	201(di)								
122	83								
123						133		34*	
124	34		80			108		53.5*	124–5
125					83				
126	153 (bis)				127(di)				
127	81								
128									
129	119								
130									
131					100				

(*Continued*)

TABLE 6A (Continued)
ALCOHOLS—LIQUID

	Name of Compound	Note	BP	Melting Point of Derivatives Recommended	
				α-Naphthyl-urethan	3,5-Dinitro-benzoate
132	2-(2-Methoxyethoxy)-ethanol		194(191)		
133	5-Nonanol		194		
134	1-Octanol	81	195	67	61
135	1-Bromo-3-chloro-2-propanol	17	197		
136	1,2-Ethanediol (ethylene glycol)	77	197.85	176 (bis)	169 (di)
137	2-Nonanol		198.2	55.5	42.8*
138	*l*-Linalool		199	53	
139	1-Phenyl-1-ethanol		202	106	95
140	2-(2-Ethoxyethoxy)-ethanol	79	202(196)		Oil
141	Benzyl alcohol	81	205.5	134	113
142	1,3-Butanediol	18	207.5	184	
143	2-Decanol		211	69	44
144	1-Nonanol	19,81	213.5 (215)	65.5	52
145	1,3-Propanediol	20	214.7	164 (di)	178 (di)
146	3-Methylbenzyl alcohol		217	116	
147	1-*p*-Tolyl-1-ethanol		219		
148	2,3-Dibromo-1-propanol	21	219d		
149	1-Phenyl-1-propanol		219	102	
150	2-Phenylethanol	81	219.8	119	108
151	Citronellol	22	222		
152	2-Methyl-1-phenyl-1-propanol	23	222–4	116–7	
153	2,4-Heptanediol		224.9		
154	*dl*-2-Undecanol		228–9		
155	2-(2-Butoxyethoxy)-ethanol	80	228–30		
156	Geraniol	78	230	48	63
157	1,4-Butanediol (mp 19.5)		230 (235)	199(bis)	
158	1-Decanol	81	231	73	57
159	3-Phenyl-1-propanol		237.4 (235)		92
160	1,5-Pentanediol		242 (238)	147(di)	
161	Undecanol		243	73	55
162	2,2′-Dihydroxyethyl ether	24	244.5	149	151*
163	2-Phenoxyethanol	25	245 (237)		
164	2-Methoxybenzyl alcohol	26	247	136	
165	Dodecanol (mp 24)		259	80	60
166	4-Methoxybenzyl alcohol (mp 24–5)	27	259		
167	2,2′-Iminodiethanol (see Table 9B)(mp 28)		270		
168	1-Phenyl-1-heptanol		275		
169	1,2-Bis-(2-hydroxyethoxy)-ethane	28	285		
170	Glycerol	29	290d	192(tri)	
171	1-(4-Methoxyphenyl)-1-ethanol	30	310d		
172	9-Octadecen-1-ol(*cis*)	31	335		
173	2,2′,2″-Nitrilotriethanol	32	360		

TABLE 6A (Continued)
ALCOHOLS—LIQUID

Melting Point of Derivatives (Continued)

Others

	Phenyl-urethan	p-Nitro-phenyl-urethan	p-Xenyl-urethan	3,5-Dinitro-phenyl-urethan	4-Nitro-benzoate	Hydrogen 3-nitro-phthalate	Hydrogen Phthalate	Pseudo-saccharin Derivative	Allo-phanate
132		73.5 (76)			92	91(anh.)			
133							45		158
134	74	111		69	12	128*	22	46*	156–7
135	73								
136	157 (bis)	135.5			140				
137							42–4		
138	66				70				
139	92				43		108		182(186)
140		66			Oil	Oil			102–4
141	77	157	156	181	85	176	106	130*	182–3
142	122–3								
143							69		
144	60N	104		60 (66)	10	125*	42.5	49*	158
145	137 (di)				119(di)				
146									
147	96								
148	84								
149	60			71	59–60				150
150	78 (80)	135		139	62*	123	189		156
151									
152									
153	101(di)								
154							50		
155		55							
156					35	117	47		124.5
157	183.5 180(bis)				175(di)				
158	60	117		70	30	122.8*	38*	47.5*	159
159	45	104			47	117			
160	176(di)				104.5(di)				
161	62	99.5				123.2*	44	58.5*	155–8
162									
163						113			
164									180
165	74	117		81	45 (42)	124*	50.3*	54*	
166	92								
167									
168	75								
169	108								
170	180(tri)	216			188(tri)				
171	83								
172		85–91							135 and 129
173									

TABLE 6B
ALCOHOLS—SOLID

	Name of Compound	Note	MP	Melting Point of Derivatives Recommended	
				α-Naphthyl-urethan	3,5-Dinitro-benzoate
1	2,2',2"-Nitrilotriethanol	32	20–22		
2	1-Phenyl-1-ethanol (bp 202) (see Table 6A)		20		
3	2-Methylcyclohexanol(*trans*) (bp 167) (see Table 6A)	8	21		
4	Dodecanol (see Table 6A) (bp 259)		24		
5	1-Phenyl-2-methyl-2-propanol		24		
6	4-Methoxybenzyl alcohol (bp 259)	27	24–5		
7	Cyclohexanol (bp 161.1)		25	129	113
8	2-Methyl-2-propanol		25.45	101	142
9	3-Nitrobenzyl alcohol	33	27		
10	2,2'-Iminodiethanol (see Table 9B)		28		
11	1-Ethynyl-1-cyclohexanol	34	30–1		
12	Tridecanol		30.63		
13	2,4-Hexadien-1-ol		31		
14	Cinnamyl alcohol		33	114	121
15	2,3-Butanediol		34		
16	*dl*-α-Terpineol	35	35	152	79
17	Pinacol	36	35(43)		
18	2-Methylbenzyl alcohol		36		
19	9-Octadecen-1-ol(*trans*)	31	36.7 (34)	71N	
20	2-Phenyl-2-propanol		37		
21	Tetradecanol	37	39	82	67
22	*dl*-Fenchyl alcohol	38	41.5 (38–9)	149	104
23	1,6-Hexanediol	39	42		
24	7-Tridecanol		42	51	
25	Pentadecanol		44	72	
26	*l*-Menthol	40	44N	119	153
27	1-Phenyl-1-butanol		49	99	
28	1-Hydroxydecalin(*trans*)(2)		49		
29	Hexadecanol		49	82	66
30	1,1,1-Trifluoro-2-propanol		52		
31	2,2-Dimethyl-1-propanol		52–3	100	
32	Piperonyl alcohol	42	52–3		
33	Heptadecanol		54	88.5	
34	1-Hydroxydecalin(*cis*)(2)		55		
35	2-Butyne-1,4-diol	41	55		189–91 (di)
36	Octadecanol	43	58.5	89	77 (66)
37	4-Methylbenzyl alcohol		59–60		117–8
38	Nonadecanol		62		
39	1-Hydroxydecalin(*trans*)(1)		63		
40	1-Hexadecoxyglycerol		63–4		58–9 (di)
41	Eicosanol		65		
42	1-(1-Naphthyl)-1-ethanol		66		
43	1,2-Diphenylethanol		67		
44	1-Phenyl-1,2-ethanediol	44	67–8		
45	Diphenylcarbinol	45	68	136 (139)	141
46	Docosanol	46	71 (74)		

TABLE 6B (Continued)
ALCOHOLS—SOLID

Melting Point of Derivatives (Continued)

Others

	Phenyl-urethan	p-Nitro-phenyl-urethan	p-Xenyl-urethan	3,5-Dinitro-phenyl-urethan	4-Nitro-benzoate	Hydrogen 3-nitro-phthalate	Hydrogen Phthalate	Pseudo-saccharin Derivative	Allo-phanate
1									
2									
3									
4									
5									
6	94								
7	82		166		50	160	99		
8	136			166d	116				
9									
10									
11									
12					37.4*	124*	52.5	66*	
13	78–9								
14	90–1				78				
15	200								
16	113				(97) 139		118		
17									
18	79								
19	56								
20									124–6d
21	74				51.2*	123.5*	60*	62*	
22	104				109	95	169		
23									
24									
25	72				45.8*	122.5*	60.5	72*	
26	112				62		122N		215
27					58		91		
28	134				116		121		
29	73	118		86	58.4*	122*	67	69.5*	159.7d
30									
31	144								
32	102.5								176.5
33					53.8*	121.5*	66.5*	76*	
34	80–1				86		142		
35	130–2 (di)								
36	79–80	115		88	64.3*	119*	72.5*	74.5*	
37	79								
38					58.9*		71	80.5	
39	114				86		168		
40	100 (di)								
41					69.4*		77		
42							132		
43							131*		
44									
45	140				132		165		
46	86								

(Continued)

TABLE 6B (Continued)
ALCOHOLS—SOLID

Name of Compound	Note	MP	Melting Point of Derivatives Recommended	
			α-Naphthyl-urethan	3,5-Dinitro-benzoate
47 Erythritol	47	72		
48 Dihydroxyacetone	48	72		
49 2-Nitrobenzyl alcohol	49	74		
50 2-Chlorobenzyl alcohol	50	74		
51 4-Chlorobenzyl alcohol	51	75		
52 Hexacosanol	52	79		
53 Tribromoethanol	53	80		
54 2-Hydroxybenzyl alcohol	54	86–7		
55 Phenacyl alcohol	55	86		
56 2,5-Dimethyl-2,5-hexanediol	56	87.5–89		
57 d-Sorbitol(anh.)	57	89–93 (112)		
58 3-Nitrophenacyl alcohol	58	93		
59 1-Hydroxydecalin(cis)(1)		93		
60 4-Nitrobenzyl alcohol	59	93		
61 Terpin hydrate	60	117		
62 meso-Erythritol	61	121		
63 4-Nitrophenacyl alcohol	62	121		
64 4-Hydroxybenzyl alcohol	63	125		
65 1,2-Diphenyl-1,2-ethanediol		134		
66 dl-Benzoin	64	137	140	
67 β-Sitosterol	65	137		202–4
68 1,4-Cyclohexanediol(trans)	66	143		
69 l-Cholesterol(anh.)	67	148.5	176	
70 Triphenylmethanol	68	162		
71 Ergosterol(anh.)	69	165	202	
72 d-Mannitol	70	166		
73 Dulcitol	71	188.5		
74 d-Borneol	72	208	132 (127)	154–5
75 dl-Isoborneol		212s.t.		
76 meso-Inositol	73	225		86
77 d-Quercitol	74	232 (234)		
78 Pentaerythritol	75	262 (253)		

NOTES FOR TABLES 6A AND 6B (ALCOHOLS)

1. Forms a constant boiling mixture with water, bp 80 (21.76% water).
2. Acetyl derivative, bp 106; oxidizes to the acid, bp 134.
3. The melting points of the derivatives are for the dl-form; the hydrogen phthalate of the d- or l-forms melt at 34.
4. The lower monoalkyl ethers of ethylene glycol (1,2-ethanediol) are commonly known as "cellosolves"; for urethane derivatives of cellosolves, see J. Am. Chem. Soc., 62, 3136 (1940), and for phenylphenacyl esters, see J. Am. Chem. Soc., 62, 1635 (1940).
5. Density 1.103 and refractive index, 1.436.
6. Camphorlike odor.
7. Oxidation with chromic acid gives cyclohexanone; may be oxidized by nitric acid to adipic acid.
8. To get the desired derivatives, one must start with pure compounds, either cis or trans. The 2-methylcyclohexanol obtained by the complete hydrogenation of o-cresol consists of a mixture of two stereoisomers, each of which may be resolved into two optically active forms. The derivatives listed in this table are for the dl-cis (or β-form) and the dl-trans (or α-form).
9. Diphenylurethane, mp 81.
10. Urethane, mp 90–1.
11. The commercial product is a mixture, if obtained by fermentation, of the meso and dl-forms. The derivatives listed are for the

TABLE 6B (Continued)
ALCOHOLS—SOLID

Melting Point of Derivatives (Continued)

	Phenyl-urethan	p-Nitro-phenyl-urethan	p-Xenyl-urethan	3,5-Dinitro-phenyl-urethan	4-Nitro-benzoate	Hydrogen 3-nitro-phthalate	Hydrogen Phthalate	Pseudo-saccharin Derivative	Allo-phanate
47									
48									
49									
50					93–4				
51									
52									
53									
54									
55					128.6				
56									
57									
58									
59	118				83		176		
60									
61									
62									
63									
64									
65									280(di)
66	165	183		220d	123				
67									
68									
69	168	205		198	185 (190–3)		161		
70									
71	185								
72	303								
73									
74	138–9				137 (153)				
75					129		167		
76									
77									
78									

NOTES FOR TABLES 6A AND 6B (ALCOHOLS) (Continued)

meso isomer; other derivatives are: dibenzoate, mp 76, and dibromobenzoate, mp 139.5; the *d*-isomer melts at 25, and the *l*- at 19.
12. Density, 0.9280 and refractive index, 1.4649.
13. β-Naphthylurethane, mp 99.
14. 2-Nitrobenzoate, mp 62.
15. Distilled from anhydrous zinc chloride, it yields propionaldehyde; di-4-chlorobenzoate, mp 104; di-3-chlorobenzoate, bp 198.
16. Benzoyl derivative mp 42–44.
17. Oxidizes to bromochloroacetone, mp 35.
18. The boiling point of the *dl*-form is also reported at 204; the phenylurethane of the *d*-form melts at 115–6; the diphenyl- urethane of the *l*-form melts at 127–8.
19. The melting point of the phenylurethane has also been reported at 62–4 and 69.
20. Dibenzoate, mp 59; di-4-chlorobenzoate, mp 101; di-3-chlorobenzoate, mp 67.
21. 3,5-Dinitrophenylurethane, mp 71.
22. Isomeric with rhodinol; roselike odor; can be oxidized to β-methyladipic acid, mp 89.
23. Oxidizes with chromic acid-sulfuric acid mixture to isopropyl phenyl ketone.
24. Di-2-naphthyl ether, mp 122; di-4-chlorobenzoate, mp 139.
25. Benzoate, mp 64; 4-toluenesulfonate, mp 80.
26. Benzoate, mp 59.
27. Also named *p*-anisyl alcohol; acetate, mp 184.

NOTES FOR TABLES 6A AND 6B (ALCOHOLS) (Continued)

28. Reaction with triphenylmethyl chloride yields bis-(triphenylmethyl) ether, mp 142.
29. Tribenzoate, mp 71–2 (75–6); tri-p-toluenesulfonate, mp 103.
30. Oxidation yields 4-methoxyacetophenone, mp 38; odor of anise.
31. Also called oleyl alcohol; isomeric with elaidyl alcohol (trans), mp 36.7; β-naphthylurethane, mp 45.
32. Also named triethanolamine; bp 280/150 mm; see Table 10B.
33. Benzoyl derivative, mp 71–2; oxidizes to 3-nitrobenzoic acid, mp 140.
34. Mercury derivative with mercuric iodide, mp 175.
35. The d- and l-isomers melt at 37 and yield phenylurethanes, mp 110.
36. Diacetate, mp 65. Pinacol boils at 173.
37. Also named myristyl alcohol; p-bromobenzenesulfonyl derivative, mp 51.5; 4-methoxyphenyl urethane, mp 83; 3,4-dimethoxyphenylurethane, mp 79.5.
38. The d-α-isomer melts at 45 and boils at 201–2; this isomer yields a phenylurethane mp 82, and an acid phthalate, mp 145. The l-α-isomer melts at 47 and yields a phenylurethane, mp 82, an acid phthalate, mp 146 and a 4-nitrobenzoate, mp 109. The l-β-isomer melts at 3–4 and produces an acid phthalate, mp 153, and a 4-nitrobenzoate, mp 83.
39. Dimethyl ether, bp 180; diethyl ether, bp 208.
40. Exists in four allotropic forms, mp 44 (42), 35, 33, and 31. The acid phthalate exists in two forms, mp 110 (labile) and 122 (stable); benzoate, mp 55. The d-menthol melts at 38–40. The dl-form melts at 34, and yields a phenylurethane, mp 104, and an acid phthalate, mp 130.
41. Dibenzoyl derivative, mp 76–7.
42. Benzoate, mp 66.
43. Also named stearyl alcohol.
44. Also named styrene glycol; dibenzoyl derivative, mp 96–7.
45. Also named benzohydrol; easily oxidized to benzophenone by chromic acid; acetate, mp 41–2; forms adduct with 2-naphthol, mp 62.
46. Also named behenyl acohol; 4-methoxyphenylurethane, mp 98; 3,4-dimethoxyphenylurethane, mp 93.
47. Tetraacetate, mp 53.
48. See Table 21B.
49. Benzoate, mp 101–2; 2-nitrobenzoic acid, by oxidation, mp 146.
50. Oxidizes to the acid, mp 142.
51. Oxidizes to the acid, mp 242.
52. Acetyl derivative, mp 65.
53. Urethane, mp 86–7.
54. Also named saligenin; see Table 29B.
55. Benzoate, mp 118.5; 3-nitrobenzoate, mp 104.5; 2-nitrobenzoate, mp 124.5; see Table 21B.
56. Reacts with concentrated HCl to give a dichloride, mp 63–4; condenses with phenol in presence of $AlCl_3$ to give a product, mp 145.
57. Hexaacetyl derivative, mp 99; hexabenzoyl derivative, mp 216.
58. Acetate, mp 53; semicarbazone, mp 214.
59. Oxidizes to the acid, mp 241.
60. Dehydrates to cis-terpin, mp 105.
61. Tetraacetate, mp 89 (85); dibenzylidine derivative, mp 201–2.
62. Acetate, mp 124; phenylhydrazone, mp 178.
63. Oxidizes to 4-hydroxybenzoic acid, mp 215; monoacetate, mp 84; diacetate, mp 75.
64. Acetate, mp 83.
65. Acetate, mp 134; benzoate, mp 145. Also called cinchol.
66. Also named quinitol; diacetate, mp 102–3; dibenzoate, mp 151; di-p-toluenesulfonate, mp 159.
67. Benzoate, mp 151–2.
68. Acetate, mp 87–8.
69. Acetate, mp 176; benzoate, mp 168.
70. Hexaacetate, mp 119–20; hexabenzoate, mp 149–50. l-mannitol, mp 163–4. dl-α-mannitol (acritol), mp 168.
71. Hexaacetate, mp 171; hexabenzoate, mp 189–91.
72. Strong camphorlike odor; shaken with 50 per cent nitric acid for 3 hours, and then diluted, it yields d-camphor, mp 179. The dl-borneol, mp 210.5.
73. Hexaacetate, mp 212 (sublimes); hexabenzoate, mp 258.
74. Pentabenzoate, mp 155.
75. Tetraacetate, mp 84; tetrabenzoate, mp 99–101.
76. 1-Chloro-2-propanol yields a 2,4-dinitrobenzenesulfenate, mp 106–7. 2-Chloroethanol yields a 2,4-dinitrobenzenesulfenate, mp 121–2.
77. Dibenzoate, mp 53; di-4-chlorobenzoate, mp 110.
78. Tetrabromide, mp 70–1. Compound is 3,7-Dimethyl-2,7-octadien-1-ol.
79. 4-Aminobenzoate, mp 64.
80. 4-Aminobenzoate, mp 36.
81. Consult *Anal. Chem.*, **31**, 148 (1959) for the procedure to prepare the 2,4,6-trinitrobenzoates of the following alcohols: benzyl, 179–81; 1-butanol, 122–3; 2-butanol, 122–4; cyclohexanol, 151–2; 1-decanol, 122–3; ethanol, 157–8; 1-heptanol, 126–7; 1-hexanol, 128–30; methanol 160–1; 2-methyl-2-propanol, 123–4; 1-nonanol, 123; 1-octanol, 124; 1-pentanol, 123–4; 2-phenylethanol, 157–9; 1-propanol, 145–6; and 2-propanol, 155–8.

REFERENCES FOR TABLES 5 AND 6

Affsprung, H. E., and May, H. E., *Anal. Chem.*, **32**, 1164 (1960). Tetraphenylstilbonium Sulfate as a Reagent for Qualitative Analysis of Organic Acids. Gives the melting points of the derivatives of about 20 acids.

REFERENCES FOR TABLES 5 AND 6 (Continued)

Berger, J., *Acta Chem. Scand.*, **8**, 427 (1954). S-Benzylthiuronium Derivatives of Acids.

Blohm, H. W., and Becker, E. I., *Chem. Rev.*, **51**, 471 (1952). Allophanates of Alcohols. Lists the melting points of the allophanates of numerous alcohols.

Bohm, B. A., *J. Org. Chem.*, **24**, 728 (1959). Derivatives of Some Cycloalkylcarbinols.

Crosby, D. G., Boyd, J. B., and Johnson, H. E., *J. Org. Chem.*, **25**, 1826 (1960). Indole-3-alkanamides. Describes the preparation of indole-3-alkanamides and gives the melting points of five.

Dermer, O. C., and King, K., *J. Org. Chem.*, **8**, 168 (1943). N-Benzylamides as Derivatives for Identifying the Acyl Group in Esters. Lists 78 acids with the melting points of their N-benzylamides.

Eliel, E. L., and Hoover, T. E., *J. Org. Chem.*, **24**, 938 (1959). Birch Reduction of 2-Naphthoic Acid and of o-Methoxy-naphthoic Acids. Gives the method of reduction of various forms of 2-naphthoic acid and the melting points of the reduced products.

Fritz, J. S., and Schenck, G. H., *Anal. Chem.*, **31**, 1808 (1959). Acid-Catalyzed Acetylation of Organic Hydroxyl Groups. Describes the quantitative analysis of primary and secondary alcohols.

Fox, R. B., and Bailey, W. J., *J. Org. Chem.*, **25**, 1447 (1960). Organo Phosphorus Compounds. Reaction of Salts of Organo-Phosphorus Acids with Isocyanates. Describes the preparation and gives the melting points of nine derivatives.

Haslam, J., et al., *Analyst*, **86**, 256 (1961). Characterization of Organic Acids by Means of Their Benzylamine Salts. Lists melting points and I. R. spectra.

Hawkins, N. G., Webb, A. D., and Kepner, R. E., *Anal. Chem.*, **28**, 1975 (1956). Use of 2,4-Dinitrophenylhydrazone, of p-Phenylphenacyl Esters as Second Derivatives in Identification of Organic Acids. Gives the melting points of a number of acids as p-phenylphenacyl esters and their hydrazones.

Johnson, H. W., Jr., Kreyssler, H. A., and Needles, H. L., *J. Org. Chem.* **25**, 279 (1960). The Chemistry of β-Bromopropionyl Isocyanate. II. Use in Identification of Alcohols. Lists the melting points for a large number of alcohols derivatized by β-bromo-propionyl isocyanate. (β-Bromopropionyl carbamates or urethanes.)

Kirk, R. E., and Othmer, D. F., *Encyclopedia of Chemical Technology*, Interscience, New York, 1951. Vol. 6, p. 176 lists 87 fatty acids giving the melting points, boiling points, D_4^{20} n_D^{20}, neutralization values and theoretical iodine values.

Lawesson, S., *Acta. Chem. Scand.*, **11**, 1075 (1957). Plant Growth Regulators II. Methyl and Methoxy Substituted Naphthylboronic Acids. Gives the melting points of 4 substituted naphthylboronic acids with boron analysis.

Laskowski, D. E., and Adams, O. W., *Anal. Chem.*, **31**, 148–52 (1959). Identification of Alcohols by Microscopic Mixed Fusion Analysis.

Milligan, B., and Westhead, E. W., Jr., *J. Org. Chem.*, **20**, 1777 (1955). 3,5-Dinitrobenzoates and Their 1-Naphthylamine Addition Compounds. I. Preparation from Alcohols and Esters. Lists the melting points of the 1-naphthylamine adducts to 3,5-dinitrobenzoate esters of alcohols. Also gives the method of preparation of 3,5-dinitrobenzoates from the esters to identify the alcohol moiety.

Reid, E. E., *Organic Chemistry of Bivalent Sulfur*, Vol. V, Chemical Publishing Company, New York, 1963. For the identification of acids as S-substituted isothiouronium salts see pages 121–125.

Riebsomer, J. L., *J. Org. Chem.*, **2**, 182 (1946). Arylsulfonyl Esters of Nitro Alcohols. The melting points of arylsulfonyl esters of nitro alcohols are given.

Rodd, E. H. (ed.), *Chemistry of Carbon Compounds*, Elsevier, Houston, 1952. See Vol. IIIb, page 1332 for naphthalenedicarboxylic acid table. Gives the melting points of the acid, the dinitrile, and the dimethyl ester.

Schotte, L., and Veibel, S., *Acta. Chem. Scand.*, **7**, 1357 (1953). The Preparation of S-Alkylthiuronium Picrates and a New Method for the Estimation of Tertiary Alcohols. Lists the melting points of the S-alkylthiuronium picrates of twelve alcohols.

Stafford, R. W., Francel, R. J., and Shay, J. F., *Anal. Chem.*, **21**, 1454 (1949). Identification of Dicarboxylic Acids in Polymeric Esters. The melting points of the dibenzyl amides of several acids are given.

Thielcke, G. W., and Becker, E. I., *J. Org. Chem.*, **21**, 1003–1005 (1956). Some Reactions of Tetraphenylphthalic Anhydride and Esters of Tetraphenylphthalic Acid. Lists the melting points of the products obtained from the reaction of tetraphenylphthalic anhydride with sodium hydroxide, sodium alkoxides, alcohols, ammonia and aniline.

Woolfolk, E. O., Beach, E., and McPherson, S. P., *J. Org. Chem.*, **20**, 391 (1955). Use of p-Phenylazobenzoyl Chloride as a Reagent to Derivatize Alcohols. The melting points of the esters of 26 alcohols are listed. Chromatographic procedures are also discussed.

Wolff, M. E., and Owings, F. F., *J. Org. Chem.*, **25**, 1235 (1960). Esters and Ketones Related to Diphenyl Acetic Acid. Lists the melting points of 23 compounds related to diphenylacetic acid.

Zeiss, H. H., *Chem. Rev.*, **42**, 182 (1948). The Chemistry of Resin Acids. Gives the melting points and methyl esters for 13 acids.

TABLE 7A
ALDEHYDES—LIQUID

	Name of Compound	Note	BP	Melting Point of Derivatives Recommended	
				Semi-carbazone	2,4-Dinitro-phenyl-hydrazone
1	Methanal	1	−21	169	166
2	Ethanal	2	20.2	162	168.5N
3	Propanal	3	47.5–49	154N	148N
4	Glyoxal	5	50	279	328
5	Acrolein	4	52.4	171	165
6	Trifluoropropanal		56		151
			(745 mm)		
7	2-Methylpropanal		64	119	183
				(125–6)	(187)
8	2-Ketopropanal		72		299
					(di)
9	2-Methyl-2-propenal	4	73.5	198	206
10	Butanal		74.7	106	123
11	2,2-Dimethylpropanal		75	190	209
12	Chloroethanal		85	148	156–7
				(134–5)	
13	Dichloroethanal		89–90	155–6	
14	Methoxyethanal		92		124–5
15	3-Methylbutanal		92.5	107	123
16	2-Methylbutanal		92–3	103	120
17	Trichloroethanal		98	90d	131
18	Pentanal		103		107
					(98)
19	3,3-Dimethylbutanal		103		147
20	2-Butenal		104	199	190
21	2-Butynal		105–10		136
22	Ethoxyethanal		106		116–7
23	2-Isopropylacrolein		107–9		165
24	Bromoethanal		107–12	128	150
			(104–5)	(130)	
25	2,3-Dimethylbutanal		114		124
26	2-Ethylbutanal		117	99	94–5
					(130)
27	2-Methylpentanal		118.3	102	103
28	Propoxyethanal		120		86
			(750 mm)		
29	4-Methylpentanal		121	127	99
			(743 mm)		
30	Paraldehyde	6	128		
			(124)		
31	2-Methyl-3-methoxypropanal		129		102

TABLE 7A (Continued)
ALDEHYDES—LIQUID

Melting Point of Derivatives (Continued)

Others

	4-Nitro-phenyl-hydrazone	Phenyl-hydrazone	Oxime	Dimedone	Miscellaneous
1	181–2	145		191 (189)	
2	128.5	57(63) and 99	47	140	Thiosemicarbazone, 146
3	125	Oil	40	155	Picrate, 156–7
4		180	178	228 (186)	Phenylosazone, 169–70
5	150–1	52N		192	Benzoylhydrazone, 175–7
6					
7	131	Oil		154 (142)	
8	217(mono) 302–4(di)				
9		74N			
10	93–5 (87)			135	
11	119		41		
12			Oil		
13					
14	115				
15	110–11	Oil	48.5	154	Thiosemicarbazone, 52–3
16					
17			56 (40)		
18			52	104.5	Thiosemicarbazone, 65; Phenylsemicarbazone, 126–7
19					
20	185	56	119	186	
21					
22	113–4				
23					
24					
25					
26				102–3	
27					
28					
29					
30					
31					

(Continued)

TABLE 7A (Continued)
ALDEHYDES—LIQUID

	Name of Compound	Note	BP	Melting Point of Derivatives Recommended	
				Semi-carbazone	2,4-Dinitro-phenyl-hydrazone
32	Hexanal	7	131	106	104
33	2-Ethyl-3-methylbutanal		133.5		121
34	3,3-Dimethylpentanal		134		102
35	3-Methyl-2-butenal		135	223	182
36	Cyclopentylformaldehyde		136	124	
37	2-Methyl-2-penten-1-al		136.8	207	159
38	5-Methylhexanal		144 (750 mm)	117	117
39	3-Furaldehyde		144 (732 mm)	211	
40	Tetrahydrofurfural		144–5 (740 mm)	166	130
41	1-Cyclopentenylformaldehyde		146	208	
42	2-Hexenal		150	176	144
43	Heptanal		153	109	106
44	2-Propylpentanal		161	101	
45	Furfural	8	161.7	202	230N*
46	Hexahydrobenzaldehyde		162	173	
47	2-Ethylhexanal		163	254d	114–5 (120–1)
48	Butanedial		169–70		280
49	Octanal		171	102	106
50	2-Ethyl-2-hexenal		173	153	124–5
51	3-Fluorobenzaldehyde		173		
52	1,3-Hexadien-1-al		174	206	
53	Bromal		174		
54	2-Fluorobenzaldehyde		175		
55	4-Fluorobenzaldehyde		175		
56	Benzaldehyde	9	179	222N	239
57	Pyridine-2-aldehyde		180 (750 mm)	195–6	
58	2-Acetopropanal		186–8	180–2 (di)	235–6 (di)
59	5-Methylfurfural		187	211	212
60	Pentanedial		187–9d		
61	Nonanal	10	190 (185)	100 (84)	100* (96)
62	Phenylethanal (mp 34) (see Table 7B)		194		

TABLE 7A (Continued)
ALDEHYDES—LIQUID

| | Melting Point of Derivatives (Continued) | | | | |
| | | | Others | | |
	4-Nitro-phenyl-hydrazone	Phenyl-hydrazone	Oxime	Dimedone	Miscellaneous
32	80		51	108.5	Phenylsemicarbazone, 135–6
33					
34					
35					
36					
37		58–60	48–9		
38					
39		149.5			
40		67		123	
41	188				
42	139				
43	73		57 (55)	135 (103)	
44					
45	154 (127)	97	91–2N	160	p-Tolylhydrazone, 105–6 Phenylsemicarbazone, 180–1 p-Tolylsemicarbazone, 156–7
46			91		Oxime hydrochloride, 107–8d
47					
48			172		
49	80		60	90	
50					
51	202	114	63		Acid, 124
52		101	160d		
53					Bromal hydrate, 53–5
54	205	90	63		Acid, 127
55	212	147	syn: 116–7 anti: 86		Acid, 182.6
56	192N	158 and 154–5	35N	193	Phenylsemicarbazone, 180–1 Thiosemicarbazone, 160
57			113.5		Phenylhydrazone hydrochloride, 196
58	284–5 (di)		73–4 (di)		
59	130	147–8	syn: 112 anti: 51–2		
60	169		175(di)		Acid, 97
61			64 (69)	86	Thiosemicarbazone, 77
62					

(Continued)

TABLE 7A (Continued)
ALDEHYDES—LIQUID

	Name of Compound	Note	BP	Melting Point of Derivatives Recommended	
				Semi-carbazone	2,4-Dinitro-phenyl-hydrazone
63	Salicylaldehyde (see Table 29A)	11	197*	231	248N
64	2-Thiophenaldehyde		198		242
65	3-Methylbenzaldehyde (*m*-tolualdehyde)		199	204 (223–4)	195
66	2-Methylbenzaldehyde (*o*-tolualdehyde)		200 (197)	218 (212)	194
67	2-Phenylpropanal		202–5	153–4	136
68	4-Methylbenzaldehyde (*p*-tolualdehyde)		204	234 (215)	234*
69	*d*-Citronellal		207	84 (91–2)	78
70	Decanal		207–9	102	104
71	2-Chlorobenzaldehyde	12	210–11 (213–4)	229–30 and 146	209 (213.6)
72	Phenoxyethanal		215*d*	145	130
73	3-Chlorobenzaldehyde (mp 18)	13	216	228	256 (248)
74	3,5-Dimethylbenzaldehyde		220–2	201–2	
75	3-Phenylpropanal		224	127	155 (149)
76	Citral (geranial)	14	228	164N	110
77	Citral (neral)	14	228	171N	96
78	2-Nonenal		229–31	169	
79	2-Bromobenzaldehyde (mp 22)		230	214	
80	3-Methoxybenzaldehyde		230	233*d*	219
81	3-Bromobenzaldehyde		234	205	
82	4-Isopropylbenzaldehyde	15	236	211	244–5 (241)
83	2-Dimethylaminobenzaldehyde		244		
84	4-Methoxybenzaldehyde	16	248	210	253–4*d*
85	2-Phenylpropenal	17	252	215	255
86	4-Ethoxybenzaldehyde		255 (249)	202 (208)	
87	3,4-Diethoxybenzaldehyde		277–80		219
88	Diphenylethanal		315–6*d*	162	

TABLE 7A (Continued)
ALDEHYDES—LIQUID

Melting Point of Derivatives (Continued)

Others

	p-Nitro phenyl-hydrazone	Phenyl-hydrazone	Oxime	Dimedone	Miscellaneous
63	227	142	57 (63)	208N	Acid, 158
64		119 (139)			
65	157	91 (84)	60	172	Acid, 111–3
66	222	106 (101)	49	167	Acid, 104–5
67					
68	200.5*	113	80		Acid, 179–80
69				77–9	
70			69	91.7	p-Tolylsemicarbazone, 137–8
71	249	86	76 and 101–3	205d	Acid, 142
72		86	95		
73	216	134	70–1 and 118N		Acid, 158
74					Acid, 166
75	123		97*		Acid, 48
76			143–5		
77					
78					
79	240–1		102(α) and 126(β)		Acid, 128
80	171	76	40 and 112		Phenylthiosemicarbazone, 153
81	220	141	72d		Acid, 155
82	190	129	42 and 112	170–1	Acid, 116
83	191	72	87		Methiodide, 164
84	160–1	120–1	133N	145*	Acid, 146
85	195	168	64–5 and 138	213*	
86			157 and 118		Acid, 198
87			98		
88			α-120 β-106		Acid, 148

TABLE 7B
ALDEHYDES—SOLID

	Name of Compound	Note	MP	Melting Point of Derivatives Recommended	
				Semi-carbazone	2,4-Dinitro-phenyl-hydrazone
1	2,3,5,6-Tetramethylbenzaldehyde		20	270d	
2	2-Ethoxybenzaldehyde (bp 247–9)		20–2	219	
3	2-Bromobenzaldehyde (bp 230) (see Table 7A)		22		
4	Tetradecanal		23	106.5	108
5	Pentadecanal		24–5	108–9	107.5
6	Hexadecanal		34	108–9 (107)	105–7
7	1-Naphthaldehyde (bp 292)		34	221	
8	Phenylethanal (bp 194)		34	156 (153)	121 (110)
9	4-Methoxy-1-naphthaldehyde		34		
10	2-Chloro-3-phenylpropenal	18	34–6		
11	5-Hydroxymethyl-2-furaldehyde		35–6	195d (166–7)	184
12	Heptadecanal	19	36	108	
13	Piperonal	20	37	234	266d
14	2-Iodobenzaldehyde		37	206	
15	2-Methoxybenzaldehyde		38	215	253
16	Octadecanal	21	38N	108–9	101 (110)
17	Acetylsalicylaldehyde		38–9	167	
18	2-Aminobenzaldehyde (see Table 9B)		40	247	
19	4-Diethylaminobenzaldehyde		41	214d	
20	2-Nitrobenzaldehyde	22	41 (44)	256	250d (265)
21	3,4-Dichlorobenzaldehyde		42–4		300–1
22	3,3-Diphenylpropenal		44	214–5	196
23	2,4,5-Trimethylbenzaldehyde		44	243	
24	Dodecanal		44–5	106 (103)	106
25	4-Chlorobenzaldehyde	23	48	230 (233)	265
26	3-(2-Chlorophenyl)-propenal	24	50		
27	Pyrrole-2-aldehyde		50	183.5	
28	Quinoline-4-aldehyde		51–3		
29	Benzylglycollicaldehyde		52	137	

TABLE 7B (Continued)
ALDEHYDES—SOLID

Melting Point of Derivatives (Continued)

Others

	4-Nitrophenyl-hydrazone	Phenyl-hydrazone	Oxime	Dimedone	Miscellaneous
1			125		
2			57–9		
3					
4	95		82–3		Acid, 54
5	94–5		86		Thiosemicarbazone, 95–96.5
					Benzoylhydrazone, 87
6	96.5		88		Thiosemicarbazone, 109(106)
7	234	80	90		Acid, 162
8	151	63 (mono)	99	165	Dimer, 50
		101–2(di)			
9		113			
10		160	157–9		
11	185	140–1	77–8 and 108		
12			89.5		Acid, 61
13	200	102–3	110N	177–8 (193N)	
14		79	108		Acid, 162
15	205		92		Acid, 101
16	101		89		Thiosemicarbazone, 111
17	185–6	141–2			4-Bromophenylhydrazone, 137–8
18	220	221	135		
19		103	93		Anil, 108–9
20	263	156	102 and 154		
21	277		118–9		Acid 201–2 (208–9)
22		173			
23		127			
24	90		77–8		Thiosemicarbazone, 100
25	237 (220)	127	110 and 146		Acid, 240
26			96N		
27	182–3	139	164		
28	261–2		181–2		Monohydrate, 84–5
					Picrate, 179
29					Benzoate, 70

(Continued)

TABLE 7B (Continued)
ALDEHYDES—SOLID

Name of Compound	Note	MP	Melting Point of Derivatives Recommended	
			Semi-carbazone	2,4-Dinitro-phenyl-hydrazone
30 4-Chlorosalicylaldehyde		52.5	212	
31 Chloral hydrate (bp 96)		53		
32 2,3-Dimethoxybenzaldehyde		54	231	223–4
33 2,3-Diphenylpropanal		54	125	
34 3-Chlorosalicylaldehyde		54–5	240–3	
35 6-Nitrosalicylaldehyde		54–5		
36 Isoquinaldehyde		55.5	197	
37 Phthalaldehyde		56		
38 5-Methylsalicylaldehyde		56		
39 3-Iodobenzaldehyde		57	226	
40 3,4-Dimethoxybenzaldehyde		58 (44)	177	264–5
41 3-Nitrobenzaldehyde		58	246	289–90
42 2,5-Dichlorobenzaldehyde		58–9		
43 1-Hydroxy-2-naphthaldehyde		59–60		
44 3,5-Dinitrosalicylaldehyde		59–60		
45 2-Naphthaldehyde		60	245	270
46 4-Aldehydobiphenyl		60	243d	239d
47 3-Methoxy-1-naphthaldehyde		60	200	
48 3,5-Dichlorobenzaldehyde		65		
49 5-Methoxy-1-naphthaldehyde		66	246	
50 4-Bromobenzaldehyde	25	67	228	257–8
51 2,4-Dimethoxybenzaldehyde		69 (71)	258	
52 2,6-Dichlorobenzaldehyde	26	71		
53 Quinoline-2-aldehyde		71		
54 2,4-Dichlorobenzaldehyde		72	226	
55 4-Aminobenzaldehyde (see Table 9B)		72	153	
56 2-Chloro-4-nitrobenzaldehyde		74 (79)	234	247d
57 4-Dimethylaminobenzaldehyde		74	222	236
58 Quinoline-6-aldehyde		75–6	239	
59 2-Chloro-1-naphthaldehyde		76	215	
60 3-Ethoxy-4-hydroxybenzaldehyde		77–9	175	
61 4-Iodobenzaldehyde		78	224	
62 3,4,5-Trimethoxybenzaldehyde		78	219–20	
63 2-Chloro-5-nitrobenzaldehyde		78–9		276–7

Tables 737

TABLE 7B (Continued)
ALDEHYDES—SOLID

Melting Point of Derivatives (Continued)

Others

	4-Nitrophenyl-hydrazone	Phenyl-hydrazone	Oxime	Dimedone	Miscellaneous
30	257		155		
31					
32		138	99		
33					
34			167–8		5-Nitro, 129
35					Acid, 179–80(166–7)
36					
37		191(di)			Acid, 200–6
38		149	105		
39	212	155	62		Acid, 187
40		121	94–5	173	Acid, 181
41	247	120 (124)	120 (122)		Acid, 140
42		104–5	127–8		Acid, 153
43			145		
44					Acid, 174 (hyd.); (182 anh.)
45	230	206 (217–8)	156		Acid, 184
46		189–90	149–50		Acid, 228
47	197		102		
48		106.5	112		
49	246		104		
50	207–8	113	157 and 111		Acid, 251
51			106		5-Nitro, 188-9
52			150		4-Bromophenylhydrazone, 142
53	250s	204	188		Monohydrate, 51
54			136–7		Acid, 164
55		156	124		
56		154			
57	182	148	185		
58		185	191		Hydrate, 55 Methiodide, 218
59					Acid, 153
60		124–6			4-Acetyl, 48–9 4-Benzoyl, 57
61	201	121			Acid, 270
62	201–2		83–4		
63			176		

(Continued)

TABLE 7B (Continued)
ALDEHYDES—SOLID

	Name of Compound	Note	MP	Melting Point of Derivatives Recommended	
				Semi-carbazone	2,4-Dinitro-phenyl-hydrazone
64	4-Hydroxy-3-methoxybenzaldehyde (vanillin) (see Table 29B)		81	230	271d
65	2-Hydroxy-1-naphthaldehyde		82	240	
66	4-Chloro-1-naphthaldehyde		82		
67	6-Chloro-1-naphthaldehyde		84		
68	3,5-Dibromosalicylaldehyde		85		
69	4-Chloro-2-aminobenzaldehyde		86		
70	2-Naphthoxyethanal(hyd.)		87	182	
71	Acenaphthene-5-aldehyde		87 (108)	234	273
72	Isophthalaldehyde		89		
73	3,4,5-Trichlorobenzaldehyde		90–1	252–4	
74	Phenylglyoxal hydrate		91	217d	
75	Quinoline-8-aldehyde		94–5	238–9	
76	2,3-Diphenylpropenal		94–5	190 (195)	240
77	3,5-Dichlorosalicylaldehyde		95	227d	
78	Hydroxyethanal		96–7		
79	2-Aldehydobenzoic acid (hyd.; anh., m 240–50)		98–9	202	
80	2-Hydroxy-3-naphthaldehyde		99–100		
81	5-Chlorosalicylaldehyde		99.5	286–7	
82	Acenaphthene-3-aldehyde		100		
83	Phenanthrene-9-aldehyde		101	223	
84	3-Hydroxybenzaldehyde (see Table 29B)		104 (108)	198	257d
85	Anthracene-9-aldehyde		104–5	291	
86	5-Bromosalicylaldehyde		106	297d	
87	4-Nitrobenzaldehyde	27	106	221 (211)	322
88	2,3-Dihydroxybenzaldehyde		108	226d	
89	3-Nitrosalicylaldehyde		108–9		
90	6-Chloro-3-hydroxybenzaldehyde		110–1	236	
91	Phenanthrene-1-aldehyde		111.5		
92	2-Ethoxy-1-naphthaldehyde		112 (115)	214–5	258
93	4-Hydroxy-3,5-dimethoxy-benzaldehyde		113	188	
94	2,4,6-Trichloro-3-hydroxy-benzaldehyde		113–6		
95	4-Hydroxybenzaldehyde (see Table 29B)	28	116	224 (280N)	270–1
96	Terephthalaldehyde		116		
97	2,4,6-Trimethoxybenzaldehyde		118		

TABLE 7B (Continued)
ALDEHYDES—SOLID

Melting Point of Derivatives (Continued)
Others

	4-Nitrophenyl-hydrazone	Phenyl-hydrazone	Oxime	Dimedone	Miscellaneous
64	227 (223)	105	117 (122)	196–8*	
65			157		Picrate, 120
66					Acid, 210 (224)
67			126–7		Acid, 217
68			218–20		Acid, 228; 2-Acetyl, 90
69		230			
70		145	123.5		
71		140	126		Anil, 97
72		242	180		Acid, 348
73	342d	147			2-Nitro, 118–9; Acid, 210
74	309	152(di)	129(mono) 168(di)		
75		176	115		
76		125–6 (141)	165–6		
77		153	195–6		Acid, 220
78	177	162			
79		106	120(r.h.)		Anil, 174
80		246–8	207d		Anil, 158–9
81		150–2	128		Acid, 172
82					Acid, 256–7
83	265		157		
84	221–2	131 and 147	90		Acid, 200
85		207	186–7		
86			126		
87	249	159	133(129) and 182–4		Acid, 241
88		167			
89					Acid, 194–5(191–2)
90	250–1		146–7		Acid, 178(171)
91			189		
92		91			
93	216–7		91		Hydrazone, 208–9
94	272–3d		170–2		
95	266	184 (178)	72 112(anh.)	190* (184)	Acid, 215
96	294–5 (di)	278d (di)	212 (di)		Acid, 300
97			201–3		

(*Continued*)

TABLE 7B (Continued)
ALDEHYDES—SOLID

	Name of Compound	Note	MP	Melting Point of Derivatives Recommended	
				Semi-carbazone	2,4-Dinitro-phenyl-hydrazone
98	4-Chloro-3-hydroxybenzaldehyde		121	238–9	
99	3-Chloro-2-naphthaldehyde		121	268	
100	2-Amino-4-nitrobenzaldehyde		124	390	
101	5-Nitrosalicylaldehyde	29	126		
102	2-Amino-3-methoxy-4-hydroxy-benzaldehyde		128–9		
103	4-Nitrosalicylaldehyde		136		
104	2,4-Dihydroxybenzaldehyde		136	260d	302–3 (286)
105	3-Chloro-4-hydroxybenzaldehyde		139	210	
106	2-Chloro-3-hydroxybenzaldehyde		140	237	
107	2,6-Dichloro-3-hydroxybenzaldehyde		140–2		
108	2,4-Dichloro-3-hydroxybenzaldehyde		141		
109	dl-Glyceraldehyde (dimer)		142	160d	166–7*
110	2-Chloro-4-aminobenzaldehyde (see Table 9B)		147		
111	2-Chloro-4-hydroxybenzaldehyde		147–8	214	
112	3,5-Dibromo-4-aminobenzaldehyde		150	294	
113	3,4-Dihydroxybenzaldehyde		153–4	230d	275d
114	2,6-Dihydroxybenzaldehyde		155–6	245	288–91
115	3,5-Dihydroxybenzaldehyde		156–7	223–4	
116	3,5-Dichloro-4-hydroxybenzaldehyde		158–9	236–7	
117	3-Hydroxy-2-ketopropanal		160		
118	Diphenylglycollic aldehyde		163	242	
119	3-Aldehydobenzoic acid		175	265	
120	3-Aldehydo-2-hydroxybenzoic acid		179		
121	4-Hydroxy-1-naphthaldehyde		181	224	
122	4-Amino-3-nitrobenzaldehyde		191		
123	Pentachlorobenzaldehyde		202.5		
124	4,6-Diaminophthalaldehyde		208		
125	3,4,5-Trihydroxybenzaldehyde (hydrate)		212		
126	5-Aldehydo-2-hydroxybenzoic acid		248–9		
127	4-Aldehydobenzoic acid		256s		

NOTES FOR TABLES 7A AND 7B (ALDEHYDES)

1. Commercial "formalin" is a 37 to 40% solution of methanal (formaldehyde) in water. It generally contains 10 to 15% methanol to prevent polymerization.
2. Ethanal (acetaldehyde) yields iodoform by the usual technique, which fact distinguishes it from other aldehydes. The 2,4-dinitrophenylhydrazone exists as a stable isomer that melts at 168.5 (corrected) and as an unstable isomer that melts at 157. A mixture of these isomers melts at about 150.
3. The semicarbazone melts at 154 if

TABLE 7B (Continued)
ALDEHYDES—SOLID

Melting Point of Derivatives (Continued)
Others

	4-Nitrophenyl-hydrazone	Phenyl-hydrazone	Oxime	Dimedone	Miscellaneous
98	226–7		126(anh.)		Acid, 219–20
99	263		152		Acid, 217
100			193		Anil 147
101			218 (225)		Anil, 133
102		165	151–2		
103		168–9			
104		158	192		
105			145		Acid, 170
106	244–5		149		3-Acetyl, 62; 3-Benzoyl, 88; Acid, 157
107			174–5		
108	277–8		188		
109			117–8	197	
110					
111	288d		194		4-Acetyl, 52; 4-Benzoyl, 97; Acid, 159
112		147	164		Anil, 99; Acid, 330
113		175–6d	157	145d	Dibenzoate, 96–7
114			167		
115					
116			185		
117			135		
118			124		
119		164	188		Anil, 156; Amide, 190d
120		188	193		
121					
122	270–2	202	207		
123		152.5	201		
124		337(di)	220(di)		
125	226 (234–6)d		195–200		
126		219	179		
127		226	208–10		Anil, 222

NOTES FOR TABLES 7A AND 7B (ALDEHYDES) (Continued)

crystallized from water and at 89 if crystallized from benzene-ligroin. The 2,4-dinitrophenylhydrazone is also reported as red crystals melting at 150 (155).

4. The data given for the phenylhydrazone are for the pyrazoline.

5. Additional derivatives are: 4-bromophenylosazone, mp 215d; 4-chlorophenylosazone, mp 227d; and 4-nitrophenylosazone, mp 310d.

6. Ethanal is produced by warming paraldehyde with a few drops of concentrated sulfuric acid.

NOTES FOR TABLES 7A AND 7B (ALDEHYDES) (Continued)

7. Additional derivatives are: *p*-tolylsemicarbazone, mp 139–40; 1-naphthylsemicarbazone, mp 112–3; and 2-naphthylsemicarbazone, mp 126–8.
8. The 2,4-dinitrophenylhydrazone exists as red crystals, mp 230 (corrected) and as yellow crystals, mp 212–4. A mixture of the two isomers melts below 200. The α-oxime (from petroleum ether) melts at 75–6; β-oxime (from alcohol) melts at 91–2. Additional derivatives are: phenylsemicarbazone, mp 180–1, and *p*-tolylsemicarbazone, mp 156–7.
9. The semicarbazone melts at 233–5 when heated rapidly. The α-oxime (stable) melts at 35; a β-oxime (needles from ether) melts at 130. Other melting points reported for the 4-nitrophenylhydrazone are 234–6 and 262.
10. Additional derivatives are: phenylsemicarbazone, mp 131–2; *p*-tolylsemicarbazone, mp 155–6.
11. Salicylaldehyde dimedone crystallizes from 70% ethanol directly as the anhydride, mp 208 (corrected). The 2,4-dinitrophenylhydrazone melts at 248 (from absolute ethanol) and at 252 (from benzene).
12. α-Oxime, mp 76; β-oxime, mp 101–3. Semicarbazone, mp 229–30 (leaflets) and 146 (yellow prisms).
13. α-(*Anti*) oxime, mp 70–1; β-(*syn*) oxime, mp 118 (*r.h.*). The 2,4-dinitrophehylhydrazone crystallized from xylene melts at 248.
14. Ordinary citral reacts with semicarbazide, in the presence of sodium acetate, to yield a mixture of semicarbazones, mp 132, but in the absence of sodium acetate, only the geranial isomer precipitates, mp 164.
15. α-Oxime (from alcohol), mp 52 (61); β-oxime, mp 112.
16. β-(*syn*) oxime (needles from benzene), mp 133; α-(*anti*) oxime is more soluble than the β-form and exists as two forms: (1) leaflets, mp 45, and (2) needles (soluble in benzene), mp 65.

This aldehyde is also called anisaldehyde.
17. Also named cinnamaldehyde. The *syn*-oxime, mp 138.5, forms an acetyl derivative, mp 69–70. The *anti*-oxime, mp 64–5, forms an acetyl derivative, mp 35.
18. Oxidation produces 3-(2-chlorophenyl) propenoic acid which exists in two forms: *cis*- , mp 111, and *trans*- , mp 137–8. The *cis* isomer may be converted to the *trans* form by heating at 155.
19. Also called margaraldehyde. Crystallizes from absolute alcohol with 1 mole of alcohol. Mp 52.
20. Piperonaldimedone was reported to melt at 193 but a later report gives 177–8. The anhydride from the dimedone melts at 220 (corrected). The *anti*-oxime (from water), mp 110, yields an acetyl derivative, mp 86; the *syn*-oxime (from methanol), mp 146, yields an acetyl derivative, mp 99. Piperonal is also named 3,4-methylenedioxybenzaldehyde.
21. Also named stearaldehyde. A melting point of 55 has also been reported. It polymerizes readily to a white solid, mp 80.
22. *Anti*-oxime, mp 102 (needles); *syn*-oxime (from benzene), mp 154.
23. α-Oxime, mp 110; β-oxime, mp 146. The 2-nitrophenylhydrazone melts at 203.
24. The oxime sinters at 92 and melts at 96.
25. *Syn*-oxime, mp 157; *anti*-oxime, mp 111.
26. The hydrazone melts at 153.
27. *Syn*-oxime, mp 182–4; *anti*-oxime, mp 133 (129).
28. The 2,4-dinitrophenylhydrazone (from water, red crystals), mp 260, and (from acetic acid, purple crystals) mp 280*d*.
29. The methyl ether of this aldehyde melts at 89–90 and forms the following derivatives: oxime, mp 183; phenylhydrazone, mp 203–4; and semicarbazone, mp 234–5.

REFERENCES FOR TABLES 7A AND 7B

Allen, C. F. H., and Richmond, J. H., *J. Org. Chem.*, **2**, 222 (1937). Some Limitations of 2,4-Dinitrophenylhydrazine as a Reagent for Carbonyl Groups.

Braun, R. A., and Mosher, W. A., *J. Am. Chem. Soc.*, **80**, 3048 (1958). 2-Diphenylacetyl-1,3-indandione-1-hydrazone: A New Reagent for Carbonyl Compounds. Gives the melting points for the derivatives of about 65 aldehydes and ketones. The reagent is available from Nease Chemical Co., State College, Pa.

REFERENCES FOR TABLES 7A AND 7B (Continued)

Hinman, R. L., *J. Org. Chem.*, **25**, 1775 (1960). The Ultraviolet Absorption Spectra of Hydrazones of Aromatic Aldehydes. Lists six substituted benzaldehydes and the melting points of various hydrazones.

Kraft, W. M., and Herbst, R. M., *J. Org. Chem.*, **10**, 483 (1945). The Condensation of Aliphatic Aldehydes with Aliphatic Carbamates. Lists several long tables of data.

Smith, R. F., *J. Org. Chem.*, **25**, 453 (1960). Reaction of Diazoethane and 1-Diazopropane with Aliphatic Aldehydes. Lists five aldehydes and the melting points of their derivatives.

Tisler, M, *Z. Anal. Chem.*, **149**, 164 (1956). 4-Phenylthiosemicarbazid als Reagent zur Characterisierung von Aldehyden und Ketonen.

Tisler, M., *Z. Anal. Chem.*, **150**, 345 (1956). *p*-Tolylthiosemicarbazid als Reagenten zur Characterisierung von Aldehyden und Ketonen.

TABLE 8
AMIDES AND UREAS[a]

	Name of Compound	Note	BP		Name of Compound	Note	MP
	LIQUIDS			51	N,N-Diphenylformamide		73
				52	Oleamide		76
1	N,N-Dimethylformamide	19	153	53	dl-Lactamide		76
2	N,N-Dimethylacetamide	19	165	54	2-Methyl-2-butenamide		76
3	N,N-Dimethylpropionamide		177	55	Thioacetanilide		76
4	N,N-Diethylformamide		178	56	4-Methylhexananilide		77
5	N,N-Dimethylbutyramide		190–1	57	2-Methyl-2-butenanilide		77
6	Formamide	4, 19	195d	58	dl-2,3-Dimethylbutananilide		78
7	N-Isopropylacetamide		201–3	59	Pentadecananilide		78
8	N-Ethylacetamide		205	60	Dodecananilide		78
9	Formylpiperidine		222	61	Transbrassidanilide		79
10	Acetylpiperidine		226	62	N-Acetylacetamide		79
				63	dl-2-Methylpentanamide		80
	SOLIDS			64	dl-2-Chloropropionamide		80
		Note	MP	65	Tridecananilide		80
11	N-Methylacetamide	20	29–31	66	Propionamide	1,2,4	81
12	N,N-Diacetylaniline		38	67	N-Methylbenzamide		82
13	Oleanilide		41	68	Acetamide	1,2,3,4,19,25	82
14	N,N-Dimethylbenzamide		42	69	Ethoxyacetamide		82
15	Benzopiperidine		48	70	Aceto-N-methyl-p-toluidide		83
16	Ethyl urethane		49	71	α-Bromoisobutyranilide		83
17	Malonamide		50(mono)	72	γ-Phenylbutyramide		84
18	Formanilide		50	73	Tetradecananilide		84
19	N-Propylacetanilide		50	74	dl-2-Phenylbutyramide		84.5
20	Methyl urethane		52	75	Nicotinanilide		85(132)
21	Difluoracetamide		52	76	Propenamide		85
22	N-Phenylurethane		53	77	Allylurea		85
23	N-Benzylacetamide		53–4	78	Acetoacetanilide		85
24	N-Ethylacetanilide		54	79	α-Undecylenamide		87
25	Butyl urethane		54	80	Propynanilide		87
26	Acetoacetamide		54	81	dl-3-Methylpentananilide		87
27	Isobutyl urethane		55	82	2-Chloroacetanilide		88
28	N-Methyl-o-acetotoluidide		56	83	Butyl oxamate		88
29	Octananilide		56	84	d-Chaulmoogranilide		89
30	Nonananilide		57	85	Hexadecananilide		90
31	N-Methyl-2-phenylacetamide		57–9	86	Bromoacetamide		91
32	Methoxyacetanilide		58	87	Isopropyl urethane		92
33	dl-Lactanilide		59	88	Acetoacet-p-toluidide		92
34	Propyl urethane		60	89	2-Phenylpropionamide		92
35	Propynamide		61–2	90	Eicosananilide		92
36	Pentananilide		63	91	dl-2-Chloropropionanilide		92
37	3-Methylbutyl urethane		64	92	2,3-Dimethylbutananilide		92
38	Erucanilide		65	93	2-Nitroacetanilide		92
39	Aceto-m-toluidide		66	94	Elaidamide		93–4
40	Aceto-N-methyl-m-toluidide		66	95	4-Methyl-2-nitroacetanilide		94
41	Erucamide		66(84)	96	2-Ethylpentananilide		94
42	Ethyloxanilate		66–7	97	Transbrassidamide		94
43	Aminoacetamide			98	m-Toluamide	2	94(97)
	(see Table 9B)		66–7	99	N-Methyl-N-1-		
44	Undecylenanilide		67		naphthylacetamide		94
45	Decananilide		70	100	Ethylurea	4	94–6(92)
46	Undecananilide		71	101	Octadecananilide		95
47	Heptananilide		71	102	Iodoacetamide		95
48	2-Methylhexanamide		72	103	Azela-amide		95(mono)
49	N,N'-Diethylcarbanilide		72–3	104	dl-2-Methylpentananilide		95
50	3-Butenamide		73	105	Hexananilide		95(92)

[a] The melting points for hundreds of additional amides may be found by consulting the tables for acid halides, acid anhydrides, and acids where several types of amides are listed as derivatives. For the melting points of sulfonamides, see Tables 30 and 31.

(Continued)

TABLE 8 (Continued)
AMIDES AND UREAS

	Name of Compound	Note	MP		Name of Compound	Note	MP
106	Trichlorolactamide		96	164	2-Methylbutananilide		110
107	Butylurea		96	165	Propylurea	21	110
108	Heptamide	1	96	166	Octanamide		110(104)
109	Semicarbazide	5	96	167	3-Nitrophenylacetamide		110
110	Butyranilide		96	168	Dodecanamide		110(102)
111	Hydroxyacetanilide		97	169	N-Methyl-3-phenyl-propenamide		110–11
112	Trichloroacetanilide		97(94)	170	3-Bromopropionamide		111
113	Methoxyacetamide		97	171	3-Aminobenzamide		111
114	3-Phenylpropananilide		98(96)	172	2-Ethylbutanamide		112(107)
115	Dichloroacetamide		98s	173	4-Methylpentananilide		112(110)
116	2-Hydroxyisobutyramide		98	174	dl-2-Methylbutanamide		112
117	2-Methylhexananilide		98	175	Aceto-o-toluidide		112
118	4-Methylhexanamide		98	176	d-Hydnohexanamide		112–3
119	Nonanamide		99	177	Acetanilide		114*
120	2-Bromopropionanilide		99(110)	178	Ethyl oxamate		114
121	Phenoxyacetanilide		99	179	2-Chlorobenzanilide		114(118)
122	Pelargonamide	1	99	180	Acetoacet-4-methoxy-benzamide		115
123	Hexanamide	4	100	181	Dehydroacetanilide		115(mono)
124	Tridecanamide		100	182	Butyramide	1,2,4	115
125	Phenylpropynamide		100(109)	183	Methacrylamide		116
126	N,N-Diphenylacetamide		101	184	2-Bromo-3-methyl-butananilide		116
127	3-Iodopropionamide		101	185	Trimesanilide		118d
128	N-Benzylacetoacetamide		101	186	Phenylacetanilide		118
129	Phenoxyacetamide		101	187	α-Crotonanilide(trans)		118(115)
130	Methylurea	6	101	188	Dichloroacetanilide		118
131	N-Methylacetanilide		102	189	N-Ethyl-4-nitroacetanilide		118
132	Pentadecanamide		102	190	Chloroacetamide	1,22	118(121)
133	Levulinanilide		102	191	Cyanoacetamide	1	120
134	2-Ethylheptanamide		102	192	Hydroxyacetamide		120
135	Isocrotonanilide		102	193	2-Chlorophenoxyacetanilide		121
136	Isocrotonamide		102	194	4-Methylpentanamide	1	121
137	Undecanamide		103(99)	195	Tribromoacetamide		122
138	3,4-Dimethylbenzanilide		104	196	Sebacanilide		122(mono)
139	2-Ketopropionanilide		104s	197	3-Chlorobenzanilide		122
140	Acetoacet-o-toluidide		104–5	198	2-Bromopropionamide		123
141	Carbamylguanidine		105	199	Furanilide		123.5
142	2-Ethylpentanamide		105(103)	200	2-Ketopropionamide		124
143	Isobutyranilide		105	201	3-Methylpentanamide		125
144	Propenanilide		105	202	4-Chlorophenoxyacetanilide		125
145	3-Phenylpropionamide		105	203	Benzo-m-toluidide		125
146	Propionanilide		106	204	o-Toluanilide		[125]
147	Hexadecanamide	1	106	205	Butyl ethyl barbituric acid		125
148	d-Chaulmoogramide		106	206	Acetyl-β-phenylhydrazine		125–6
149	sym-Dimethylurea		106	207	Adipamide		125–30 (mono)
150	Pentanamide	4	106				
151	N-Cyclopropylacetamide		106	208	Ethyl hexylbarbituric acid		126
152	Tetradecanamide		107(103)	209	Succinimide	1,17	126
153	Heptadecanamide		108	210	m-Toluanilide		126
154	Decanamide		108(98)	211	Angelanilide		126
155	Thioacetamide		108	212	2,4-Dihydroxybenzanilide		126–7
156	Fluoroacetamide		108	213	2-Ethylbutananilide		127
157	Levulinamide		108d	214	Suberamide(octanedioic)		127(mono)
158	Azela-anilide		108(mono)	215	4-Methoxyacetanilide		127
159	Eicosanamide		108–9	216	Angelamide		127–8
160	Anthranilamide		109	217	Dibenzylacetamide		128
161	Pimelanilide		109(mono)				
162	Octadecanamide	1	109				
163	3-(2-Methylphenyl)-propanamide		109				

(Continued)

TABLE 8 (Continued)
AMIDES AND UREAS

#	Name of Compound	Note	MP	#	Name of Compound	Note	MP
218	Nicotinamide	13	128	272	α-Triphenylguanidine		145
219	Phenylpropynanilide		128(126)	273	3-Nitrosalicylamide (hyd.)		145
220	Suberanilide (Octanedioic)		128(mono)	274	dl-Phenylsuccinamide (β)		145
221	3-Methylbutan-4-bromoanilide		128		(α m 158)		
222	2,2-Dimethylpropionanilide		129	275	Diethylmalonamide		146(mono)
223	Piperine		129	276	Cyclohexancarboxylanilide		146
224	2-Methoxybenzamide		129	277	Hexahydrobenzanilide		146(131)
225	2-Ethoxybenzamide		129–30	278	Phenylurea		147
226	2-Methylpropionamide	1	130	279	3,4-Dimethoxyphenylacetamide		147
227	3,4-Dimethylbenzamide		130	280	Diphenylguanidine		147
228	2,3-Dibromopropionamide		130(133)	281	Succinanilide		148(mono)
229	Benzamide	1,2,3	130	282	3-Phenylpropenamide		148(142)
230	Anthranilanilide		131	283	2-Bromo-2-methylpropionamide		148
231	Bromoacetanilide		131				
232	2-Methoxybenzanilide		131	284	4-Methyl-3-nitroacetanilide		148
233	dl-2,3-Dimethylbutanamide		132	285	Benzylurea		149
234	2,2-Dimethylbutanamide		132(103)	286	Phthalamide (Phthalamic acid)		149(mono)
235	3,3-Dimethylbutananilide		132				
236	3,3-Dimethylbutanamide		132	287	N-Chlorosuccinimide		150
237	Malonanilide		132(mono)	288	Ethylmalonanilide		150
238	Nicotinanilide	12	132N(85)	289	3-Benzoylpropionanilide		150
239	dl-Mandelamide		132	290	2-Chlorophenoxyacetamide		150
240	Urea	3,7	132.8	291	3-Phenylpropenanilide		151(153)
241	3,5-Dimethylbenzamide		133	292	Adipanilide(mono)		151–3
242	4-Isopropylbenzamide		133	293	dl-Mandelanilide		152
243	2-Bromo-3-methylbutanamide		133	294	Aceto-p-toluidide		153
244	4-Chlorophenoxyacetamide		133	295	N-Methyl-4-nitroacetanilide		153
245	2,4-Dimethylacetanilide		133	296	Ethyl isoamylbarbituric acid		154
246	N-2-Naphthylacetamide		134	297	3,4-Dimethoxybenzanilide		154
247	Chloroacetamide		134	298	4-Fluorobenzamide		154
248	3-Chlorobenzamide	2	134	299	2,2-Dimethylpropionamide		154
249	Phenacetin		134	300	3-Nitrobenzanilide		154
250	3-Methylbutanamide	1	136	301	Isopropylurea	23	154
251	Acetylsalicylanilide		136	302	Benzilamide		155
252	3-Bromobenzanilide		136	303	2,5-Dichlorobenzamide		155
253	Salicylanilide		136	304	Pimelanilide		155(di)
254	2-Hydroxy-2-methylpropionanilide		136	305	2-Bromobenzamide	2	155
				306	Dibenzylacetamide		155
255	2-Ethoxyacetanilide	18	137–8	307	2-Nitrobenzanilide		155
256	3-Ethoxybenzamide		139	308	3-Bromobenzamide		155
257	3-Aminobenzanilide		140	309	1-Naphthylacetanilide		155
258	β-Phenylalanine amide		140	310	Phthalonamide (β)		155d
259	Trichloroacetamide		141		(α m 179d)		
260	2-Iodobenzanilide		141	311	3-Nitroacetanilide		155
261	2-Bromobenzanilide		141	312	4,4-Diphenylsemicarbazide	24	155–6
262	m-Tolylurea		142	313	N-Phenylsuccinimide		156
263	2-Chlorobenzamide	2	142	314	Dibromoacetamide		156
264	Salicylamide	2	142(139)	315	Phenylacetamide		156
265	o-Toluamide	1,2	265	316	l-Malamide		156.5–158.0
266	Furamide		143				(157)mono
267	3-Nitrobenzamide		143	317	Succinamide		157(155)
268	Iodoacetanilide		143–4	318	3-Hydroxybenzanilide		158
269	Benzo-o-toluidide		144	319	dl-Phenylsuccinamide (α)		158
270	p-Toluanilide		145(148)		(β m 145)		
271	2-Bromo-3-methylbutylurea		145	320	Benzo-p-toluidide		158
				321	p-Toluamide	1,2	158

(Continued)

TABLE 8 (Continued)
AMIDES AND UREAS

#	Name of Compound	Note	MP	#	Name of Compound	Note	MP
322	N-1-Naphthylacetamide		159	371	d-Camphoramide		177(mono)
323	Nitrourea		159(150)	372	Mesaconamide		177(di)
324	5-Bromo-3-isopropyl-6-methyluracil		159	373	4-Chloroacetanilide		179
325	N-1-Naphthylbenzamide		161	374	2,4-Dimethylbenzamide		179
326	Crotonamide		161(158)	375	Phthalonamide (α) (β m 155d)		179d
327	N-2-Naphthylbenzamide		162	376	4-Cyanobenzanilide		179
328	4-Aminoacetanilide		162	377	d-Tartaranilide		180d (mono)
329	4-Hydroxybenzamide (hyd.)		162	378	Diphenylacetanilide		180
330	1-Naphthanilide		163	379	1-Acetyl-2-methylurea		180
331	Benzanilide		163	380	1-Naphthylacetamide		380
332	Trichlorolactanilide		164	381	3,5-Dinitrosalicylamide		181
333	3,4-Dimethoxybenzamide		164	382	Maleamide (mono)		181
334	2-Benzoylbenzamide		165	383	Dimethylurea (*unsym.*)		182
335	l-Glutamamide		165	384	Thiourea		182
336	3,4-Dihydroxybenzanilide		166	385	3-Butyl-6-methyluracil		182–3
337	4-Bromoacetanilide		167	386	4-Aminobenzamide		183
338	Acetylanthranilanilide		167	387	4-Tolylurea		183
339	4-Methoxybenzamide		167(163)	388	Hippuramide		183
340	3-Furylpropenamide		168	389	3,5-Dinitrobenzamide		183
341	Diphenylacetamide		168	390	4-Iodoacetanilide		184
342	α-Benzoyl-β-phenylhydrazine		168	391	2-Iodobenzamide		184
343	Piperonylamide		169	392	3-(2-Nitrophenyl)-propenamide		185
344	dl-Tropamide		169				
345	Methyliminodiacetamide		169(mono) 169(di)	393	N,N'-Diacetyl-o-phenylenediamine		185
346	4-Hydroxyacetanilide		169	394	Cyclohexanecarboxylamide		185–6
347	4-Methoxybenzanilide		169–71	395	Citraconamide		185–7(di)
348	3-Hydroxybenzamide		170	396	3-Iodobenzamide		186
349	Alloxan	8	170d	397	Mesaconanilide		186(di)
350	Sebacamide		170(mono)	398	Hexahydrobenzamide		186
351	Aconitanilide (*cis*)		170d (mono)	399	Suberanilide		186(di)
352	Malonamide		170(di)	400	Hippuric acid		187
353	dl-Phenylsuccinanilide		β-170 (mono) (α-175 (mono))	401	Maleanilide		187(di)
				402	*meso*-Tartaramide		187(di)
				403	Diethylbarbituric acid		188
				404	*asym*-Diphenylurea		189
				405	Aconitanilide		189(di)
354	4-Ethoxybenzanilide		170	406	4-Bromobenzamide	2	189
355	2-Naphthanilide		171	407	Picramide (See Table 9B)		190
356	Azela-amide		172(di)	408	Itaconanilide		190(185)
357	Diallylbarbituric acid		173	409	N,N'-Diacetyl-m-phenylenediamine		191
358	p-Phenetylurea (Dulcin)		173				
359	N-Bromosuccinimide		173.5	410	Mucamide		192d (mono)
360	Maleanilide		173–5 (mono)	411	2-Naphthamide		192(195)
361	Ethyl phenylbarbituric acid		174	412	Itaconamide		192(di)
362	Pimelamide		175(di)	413	4-Nitrophthalanilide		192
363	Citraconanilide		175(di)	414	Biuret		192d
364	dl-Phenylsuccinanilide(α) (β m 170(mono))		175(mono)	415	o-Tolylurea		192
				416	d-Camphoramide		193(di)
365	4-Hydroxyphenylacetamide		175	417	4-Chlorobenzanilide		194
				418	4-Coumaramide		194
366	Glutaramide		175–6(di)	419	2-Benzoylbenzanilide		195
367	2-(p-Toluyl)benzamide		175–6	420	3-(3-Nitrophenyl)-propenamide		196
368	2-Nitrobenzamide	1	176				
369	Phthalonanilide		176(mono)	421	d-Tartaramide		196d
370	Acetylanthranilamide		177	422	2-Methyl-4-nitroacetanilide		196

(*Continued*)

TABLE 8 (Continued)
AMIDES AND UREAS

	Name of Compound	Note	MP		Name of Compound	Note	MP
423	4-Bromobenzanilide		197	474	dl-Phenylsuccinanilide		222(di)
424	l-Malanilide		197	475	2,4-Dihydroxybenzamide		222
425	4-Nitrophenylacetamide		198	476	4-Cyanobenzamide		223
426	4-Nitrophenylacetanilide		198(212)	477	sym-Di-m-tolylurea		223–5
427	4-Hydroxybenzanilide		198	478	Diethylmalonamide		224(di)
428	Cyanoacetanilide		198	479	Glutaranilide		224
429	Citranilide (monohyd.)		199(tri) (192)	480	5-Nitrosalicylanilide		224
				481	5-Nitrosalicylamide		225
430	4-Nitrophthalamide		200d	482	Pyrotartaramide		225(di)
431	Methylsuccinamide		200(di)	483	Benzylmalonamide		225(di)
432	Aconitanilide (cis)		200(di)	484	Malonanilide		225(di)
433	2-Naphthylacetamide		200	485	d,l-Tartaramide		226
434	Isatin		200–1	486	d-Camphoranilide		226(di)
435	3-Nitrophthalamide		210d(di)	487	Succinanilide		230(di)
436	Sebacanilide		201(di)	488	Diphenanilide		230
437	4-Nitrobenzamide	1	201	489	Benzo-2,4,6-tribromanilide		232
438	Ethyl isopropylbarbituric acid		201	490	Caffeine	9	234
				491	3,5-Dinitrobenzanilide		234
439	4-Ethoxybenzamide		202	492	3-Nitrophthalanilide		234(di)
440	1-Naphthamide		202	493	Terephthalanilide		234–7
441	2,6-Dichlorobenzamide		202	494	dl-Tartaranilide		236
442	2,4-Dinitrobenzamide		203	495	Nitroguanidine		237–9
443	d-Camphoranilide		204(mono)	496	Phthalimide		238
444	3-(4-Nitrophenyl)-propenamide		204	497	sym-Diphenylurea		240
				498	Muconamide		240(di)
445	N-Phenylphthalimide		205	499	4-Hydroxy-N-methyl-acetanilide		240
446	Gallanilide		207				
447	Tricarballylamide		207d(tri)	500	Adipanilide		241(di)
448	Dicyanodiamide		207	501	2-Hydroxy-3-naphthanilide		244(249)
449	Phthalonanilide		208(di)	502	Gallamide		245
450	Hippuranilide		208	503	Barbituric acid	15	245
451	2-Hydroxyacetanilide		209(201)	504	Tricarballylanilide		252(di)
452	2-Coumaramide		209d	505	Phthalanilide		253–5
453	Sebacamide		210(di)	506	Oxanilide		254(di)
454	4-Iodobenzanilide		210	507	sym-Di-o-tolylurea		255
455	Citramide		210–5d(tri)	508	2,4,6-Trinitrobenzamide		264d
456	dl-Phenylsuccinamide		211(di)	509	Theophylline	16	264
457	4-Nitrobenzanilide		211(204)	510	d-Tartaranilide		264d
458	3,4-DI-hydroxy-benzamide		212	511	Fumaramide		266d(di)
459	Diphenamide		212(di)	512	sym-Di-p-tolylurea		268
460	Ethylmalonamide		214(di)	513	Dimethylmalonamide		269(di)
461	4-Nitroacetanilide		215	514	Succinamide		273d
462	Suberamide		217(di)	515	Isophthalamide		280
463	Methylmalonamide		217(206)	516	Creatinine		292d
464	4-Iodobenzamide		217	517	5,5-Diphenylhydantoin		293–5
465	Benzylmalonanilide		217(di)	518	Mucanilide		310
466	4-Hydroxy-2-naphthamide		217–8	519	N,N′-Diaceto-p-phenylene-diamine		310
467	Acetourea		218	520	Fumaranilide		314
468	2-Hydroxy-3-naphthamide		218	521	Creatine		315d
469	Hydantoin	14	218	522	Theobromine	16	337
470	Phthalamide		220(di)	523	Xanthine		360
471	Mucamide		220(di)	524	Trimesamide		365d
472	Adipamide		220(di)	525	Uric acid	10	400d
473	Saccharin	1	220	526	Oxamide	11	419d(di)

NOTES FOR TABLE 8 (AMIDES)

1. *J. Am. Chem. Soc.*, **65**, 1355 (1943). Xanthyl (primary amides)

Amide	MP of Derivative
Acetamide	238–40
Benzamide	222.5–223.5
Butyramide	185–7
Chloracetamide	208–9
Cyanoacetamide	222–3
Heptamide	154–5
Isobutyramide	210–11
4-Methylpentanamide	159–60
3-Methylbutyramide	182–3
4-Nitrobenzamide	231–3
Palmitamide	140–2
Pelargonamide	147.5–148.5
Phenylacetamide	194–5
Phthalimide	176–7
Propionamide	210–11
Saccharin	199
Stearamide	139–41

2. *J. Am. Chem. Soc.*, **64**, 1738 (1942). Mercuric oxide derivatives.

Amide	MP of Mercury Derivative
Acetamide	196–7
2-Methoxybenzamide	241
Benzamide	222
3-Bromobenzamide	235
4-Bromobenzamide	266
Butyramide	222–4
3-Chlorobenzamide	245
4-Chlorobenzamide	258
Proprionamide	201
o-Toluamide	196
m-Toluamide	200
p-Toluamide	260

3. *J. Am. Chem. Soc.*, **51**, 3651 (1929). Phthalimide.

Amide	MP of Phthalimide
Acetamide	135–6
Benzamide	168
Benzene sulfonamide	205
p-Toluenesulfonamide	231
o-Toluenesulfonamide	182
p-Toluene-3-nitro-sulfonamide	247
Urea	188–90
Succinimide	245–7
o-Toluamide	199–200.5
p-Toluamide	224–5

4. *Ind. Eng. Chem., Anal Ed.*, **12**, 737 (1940). Oxalates.

Oxalate	MP of Oxalate
Acetamide	127.3
Butyramide	65.9–66.2
Caproamide	71.1–71.3
Ethylurea	55–60
Formamide	107.4–107.7
Propionamide	80.8–81.0
Pentanamide	61.1–61.4

5. Semicarbazide is easily derivatized by reaction with carbonyl compounds; forms a hydrochloride salt, mp 172–3.
6. It forms a picrate, mp 127*d*; acetyl derivative, mp 180–2.
7. Forms a picrate, mp 148 (142); oxalate, mp 171; nitrate, mp 163.
8. On oxidation, it yields alloxantin, mp of hydrate 201–3.
9. Its aqueous solution reacts with a saturated solution of $HgCl_2$ to yield a derivative, mp 246 (corrected).
10. Oxidation by dilute nitric acid produces alloxan, mp 170*d*; gives the murexide test.
11. This melting point was determined in a sealed tube.
12. Anilide crystallizes from benzene-ligroin, mp 132; from water as the dihydrate, mp 85.
13. Chloroaurate, mp 205; hydrolyzes to nicotinic acid, mp 237.
14. Diacetyl derivative, mp 143–4; 5- nitro-hydantoin, mp 170*d*.
15. Phenylhydrazone, mp 284; dibromo-derivative, mp 235*d*.
16. See *Chemical Abstracts*, **38**, 2289 (1944) for color tests to distinguish it from caffeine and theobromine.
17. Forms an adduct with 1,4-naphthalenediol, mp 133.5, and an adduct with 2-naphthol, mp 87.
18. Picrate, mp 192.5
19. Tetraphenylboron forms adducts with: acetamide, 180–2; dimethylacetamide, 119–20; formamide, 187–90; dimethylformamide, 117–21.
20. Hydrochloride salt, 87–9.
21. 3-Acetyl derivative, 115.
22. N-Acetyl derivative (from benzene) 105–6.
23. N-Acetyl derivative, 68–72.
24. Picrate, 164–7*d*; hydrochloride, 218–20.
25. Forms an adduct with 2-naphthol, mp 63, and an adduct with urea, mp 105.

REFERENCES FOR TABLE 8

Williams, J. W., Rainey, W. T. Jr., and Leopold, R. S., *J. Am. Chem. Soc.*, **64**, 1738 (1942). Identification of Amides through the Mercury Derivatives. Lists 15 amides; their melting points and the melting points of their mercury derivatives.

Zeif, M., and Woodside, R., *J. Org. Chem.*, **24**, 1338 (1959). Tetraphenylboron Derivatives of Amides. Lists the melting points of the tetraphenylboron derivatives of seven amides.

TABLE 9A
AMINES (PRIMARY AND SECONDARY)—LIQUID

	Name of Compound	Note	BP	Melting Point of Derivatives		
				Recommended		
				Benzene sulfonamide	Acetamide	Benzamide
1	Methylamine		−6	30	28	80
2	Dimethylamine	1	7	47	Oil	41
3	Ethylamine		16.6	58		71
4	Isopropylamine		32.4	26		100
5	Ethyl methylamine		36			
6	*tert*-Butylamine		44.4			134
7	Propylamine		47.8	36		84
8	Methyl isopropylamine	2	50			
9	Cyclopropylamine		50			99
10	Ethyleneimine	3	56			
11	Diethylamine		56	42		42
12	Allylamine		56(58)			
13	*sec*-Butylamine		63	70		76
14	*unsym*-Dimethylhydrazine	4	63			
15	Trimethyleneimine		63			
16	Isobutylamine		69	53		57
17	Butylamine		77			42
18	2-Amino-2-methylbutane		78			
19	2-(Methylamino)butane		78–9			
20	Ethyl propylamine		80–1			
21	*sym*-Dimethylhydrazine	5	81			
22	Cyclobutylamine		82			
23	Di-isopropylamine	6	84(86)			
24	Methylhydrazine	7	87 (745 mm)			
25	Pyrrolidine	8	89		Oil	Oil
26	2-Methoxyethylamine	9	90.5			
27	Methyl butylamine		91			
28	1-Amino-4-pentene		91–4			
29	*dl*-2-Aminopentane		92			
30	3-Methylbutylamine		95			
31	*d*-2-Methylbutylamine		96			
32	2-Methylpyrrolidine	10	97–8			
33	3-Methylpyrrolidine		103–5			
34	Pentylamine		104			
35	N-Methyl-2-aminopentane		105			
36	Piperidine		106.3	93–4	Oil	48
37	2,5-Dimethylpyrrolidine		106–8			
38	2-Aminoethyl ethyl ether	11	108			
39	Dipropylamine		109	51	Oil	
40	Diallylamine		111–2			
41	2,4-Dimethylpyrrolidine		115–7			
42	Ethylenediamine	12	116	168 (di)	51 (mono) 172(di)	244 (di) (249)
43	*d*- or *l*-2-Methylpiperidine		117			

TABLE 9A (Continued)
AMINES (PRIMARY AND SECONDARY)—LIQUID

	Melting Point of Derivatives							
					Others			
	p-Toluene-sulfon-amide	Phenyl-thiourea	α-Naph-thyl-thiourea	β-Naph-thyl-thiourea	α-Naph-thylurea	2-Nitro-1,3-in-dandione derivative	Picrate	Hydro-chloride	Chloro-platinate or Chloroaurate
1	75	113	192 (198)	127 (130)	197	203–5	207 (215)	229–30	
2	79	135	168	173	159	210	158		
3	63	106 (135)	121	142	200	202–3	165	109–10	
4	51	101	143		200				
5							196	126–30	207(Pt)
6		120					198		
7	52	63	103	114	196	184.5	135		
8		120					135		
9	120(di)						149		
10							142		
11	60	34	108	90	158	180–1	155	228–9	
12	64	98	145	106–8		180–1	140		
13	55	101	137	120			139–40		
14							146–7	81–2	
15							166–7		203(Pt) 192(Au)
16	78	82	151		178	178	150		
17	65	108–9	119	149		151			
18							183		
19							78		151(Pt)
20								225	198(Pt) 86(Au)
21							147–50	166–7	
22									210–5(Pt)
23				169			140		186–9(Pt)
24							169		
25	123						112N		
26									
27							112	170–1	205(Pt)
28									166(Pt) 195(Au)
29								168	82–3(Au)
30	65	102	97	116	132		138		
31								176	240(Pt)
32									172–3N
33							106		194(Pt)
34		69	103	114		182	139		
35							77–8		138(Pt)
36	96	101					152		
37							117–8	188–90	225(Pt)
38							122		
39		69	161	109	93	210	75		
40									
41							116–7		210(Pt)
42	123 (mono) 360(di)	102		223		204–5	233(di)		
43							116–7		194(Pt)

(Continued)

TABLE 9A (Continued)
AMINES (PRIMARY AND SECONDARY)—LIQUID

	Name of Compound	Note	BP	Melting Point of Derivatives Recommended		
				Benzene-sulfonamide	Acetamide	Benzamide
44	*dl*-2-Methylpiperidine		118.2			45
45	*dl*-1,2-Diaminopropane	13	119–20		139(di)	192(di)
46	*dl*-3-Methylpiperidine		124.8			
47	4-Methyl-1-aminopentane	14	125			
48	4-Aminoazobenzene		126		145	211
49	2,6-Dimethylpiperidine	15	127.9	50		111
50	Morpholine		128.3	118		75
51	3-Aminohexane		130			
52	Pyrrole		130–1			
53	Di-*sec*-butylamine	16	132			
54	Hexylamine		132.7	96		40
55	Cyclohexylamine		134	89	104	149
56	2,2,6-Trimethylpiperidine		138–9			
57	Di-isobutylamine		139	55	86	
58	1,3-Diaminopropane		139.7	96	101(di) (126)	147–8 (di)
59	4-Aminoheptane		139–40			
60	1,3-Diaminobutane		141–2			
61	2-Aminoheptane	17	142			
62	2-(Diethylamino)ethylamine		145			
63	2-Furanmethylamine		145–6			
64	N-Methylcyclohexylamine		145–7			85–6
65	2-Ethylpiperidine		146–7	64–5		
66	2,2,4-Trimethylpiperidine	18	148			
67	*sym*-Diethylenediamine		149–50			
68	3-Ethylpiperidine		153			
69	Heptylamine		156.9			
70	4-Ethylpiperidine		156–8			
71	Dibutylamine		159			
72	1,4-Diaminobutane (mp 27) (see Table 9B)		159			
73	1-Amino-2-propanol (see Table 6A)		159–60			
74	Hexahydrobenzylamine		163.5			98 (107)
75	N-Ethylcyclohexylamine		164			
76	2-(Ethylamino)ethanol		167–9			
77	1-Amino-2-hydroxybutane	19	168–70			
78	2-(Methylamino)-ethanol		169–70			
79	2-Ethylcyclohexylamine		170–1	121–2		
80	2-Aminoethanol	20	171			
81	3-Amino-2-pentanol	21	172			
82	4-Amino-2-butanol		172			
83	*dl*-2-Aminopropanol		173–6			
84	2-Amino-3-pentanol	22	174			

TABLE 9A (Continued)
AMINES (PRIMARY AND SECONDARY)—LIQUID

Melting Point of Derivatives

	p-Toluene-sulfon-amide	Phenyl-thiourea	α-Naph-thyl-thiourea	β-Naph-thyl-thiourea	Others α-Naph-thylurea	2-Nitro-1,3-in-dandione Derivative	Picrate	Hydro-chloride	Chloro-platinate or Chloroaurate
44	55				179		135	210	
45							137(di)		
46							138(di)	172	
47							123–5	220	200d(Pt)
48									
49							162–4	281	212(Pt)
50	147	136			198		146		
51								227	190–200 (Pt)
52		142–3			162		69d		
53									
54		77	79	126					
55		148	142	172		213			
56							195–6	236	128(Au)
57		113		136	119	231			
58							250 (di)		240(Pt)
59								246–7	235(Pt)
60							240–5	171–2	
61								133	63–4(Au)
62					104		115 (mono) 211d(di)		211(Pt)
63							150		
64							170		
65							133	181	202(Pt)
66									215(Pt)
67								260	223(Pt) 220(Au)
68							63	141	183(Pt)
69		75	68–9	115		149–50	121		
70									173(Pt) 105(Au)
71		86	123				59		
72									
73							142	73–4	195(Pt)
74							184–6		
75							133	184	
76							125–6		127(Au)
77									
78					125		148–50		125–30(Pt) 145–6(Au)
79							190		239(Pt)
80		138			186		160	100	
81									
82							122		206d(Pt)
83								87	198–9(Pt)
84									154(Pt)

(Continued)

TABLE 9A (Continued)
AMINES (PRIMARY AND SECONDARY)—LIQUID

	Name of Compound	Note	BP	Melting Point of Derivatives Recommended		
				Benzene-sulfon-amide	Acetamide	Benzamide
85	2-Fluoroaniline		176		80	
86	1,5-Diaminopentane		178–80	119		135(di)
87	Octylamine		180			
88	5-Methyl-2-pyrazoline	23	180			156
89	N-Methylbenzylamine		181			
90	Aniline		184.4	112	114	163
91	Benzylamine		184–5	88	60	105
92	Di-isopentylamine	24	187–8			
93	dl-α-Phenylethylamine	25	187		57	120
94	1,2-Diaminocyclohexane		187		160 (di)	
95	3-Fluoroaniline		187		88(83)	
96	4-Fluoroaniline	26	188		152	185
97	3-Aminopropanol		188			
98	1,2,3-Triaminopropane		190		200–2	217–8
99	4-Amino-2,6-dimethylpiperidine		195			
100	N-Methylaniline		196	79	102	63
101	1-Phenyl-2-aminopropane	27	196–7			159
102	β-Phenylethylamine		198	69	51	116
103	N-Ethylbenzylamine		199			
104	o-Toluidine		200	124	112	144
105	Nonylamine		201		34–5	49
106	m-Toluidine		203	95	65	125
107	Dipentylamine		203			
108	N-Ethylaniline		205	Oil	54	60
109	dl-1-Phenyl-2-aminopropane	28	205		64N	
110	cis-Decahydroquinoline		205–6			96
111	N-Methyl-m-toluidine		206		66	
112	N-Isopropylaniline		206–8		38	
113	l-Menthylamine		207 (212)		145	156
114	4-Methylpyrazole	29	207			
115	3-Methylbenzylamine		207		235–40	150
116	2-Methylbenzylamine		208		69	88
117	N-Methyl-o-toluidine		208		56	66
118	4-Methylbenzylamine		208		107–8	137
119	3-Methylpyrazole		208		29–30	
120	dl-1-Phenyl-1-aminopropane		208	81		115–6
121	Bis-(2-aminoethyl)amine		208		220 (tri)	166 (tri)
122	2-Chloroaniline		209	129	88	99
123	2-Ethylaniline		210		111	147
124	N-Methyl-p-toluidine	30	210	64	83	53
125	2-Phenyl-1-aminopropane		210			85
126	2,2′-Diaminodiethyl sulfide		213–4			

TABLE 9A (Continued)
AMINES (PRIMARY AND SECONDARY)—LIQUID

	Melting Point of Derivatives							
				Others				
	p-Toluene-sulfon-amide	Phenyl-thiourea	α-Naph-thyl-thiourea	β-Naph-thyl-thiourea	α-Naph-thylurea	2-Nitro-1,3-in-dandione Derivative	Picrate	Hydro-chloride	Chloro-platinate or Chloroaurate
85									
86		148					237		
87			72				112		
88							126		
89	95						118		197(Pt)
90	103	154	158	182 (166–7)			198		
91	116	156	172	173	203	179–80	199		
92		72	118		95	190	94.5	276	
93						207	189	158	
94							210–5 (di)	280	
95									
96									
97							222		199(Pt)
98									
99							220		
100	94	87	135–6	124–5	99	186	145		
101								236	
102	64	135				169	174(167)		
103	50						118		
104	108	136	167	193–4		197–8	213	218–20	
105							111		
106	114	104 (92)				193–4	200		
107			72	126					
108	87	89	129.5	128.5		183	138 (132)		
109							143	145–7	
110							142–5 (135–6)	226	157–8 (Au)
111									
112									
113		135					215		
114							142		
115							198 (156)	208	214(Pt)
116							215		220–3(Pt)
117	120						90		
118							204		
119							144		
120								190	
121							212 (tri)	233 (tri)	
122	105(193)	156				136	134		
123						183	194–5		
124	60						131		
125							182	123–4	
126							212	131	

(Continued)

TABLE 9A (Continued)
AMINES (PRIMARY AND SECONDARY)—LIQUID

| | Name of Compound | Note | BP | Melting Point of Derivatives | | |
| | | | | Recommended | | |
				Benzene-sulfon-amide	Acetamide	Benzamide
127	2,4-Dimethylaniline		214	130	133	192
128	1-Amino-1-phenyl-2-methyl-propane	31	214			
129	N-Ethyl-o-toluidine	32	214		N	
130	N-tert-butylaniline		214–6		55–6	
131	2,5-Dimethylaniline		215	138	139	140
132	4-Chlorobenzylamine		215			140
133	3-Ethylaniline		215		24–5	
134	2-Chloro-6-methylaniline		215		120	
135	2,6-Dimethylaniline		215		177	168
136	2,6-Diethylaniline	33	216			
137	N-Ethyl-p-toluidine		217			39
138	4-Ethylaniline		217.8		94	151
139	2-Chloro-N-methylaniline		218			
140	2,4-Dimethylbenzylamine		218–9			
141	2-Amino-N,N-dimethylaniline		220		72	51
142	3,5-Dimethylaniline		220		144	136
143	1-Phenyl-1-aminobutane		220–1 (223)			128
144	N-Ethyl-m-toluidine		221			72
145	3,5-Dimethylbenzylamine		221			
146	2-Amino-4-phenylbutane		221–2			108
147	2,3-Dimethylaniline		221–2		135	189
148	N-Propylaniline		222	54	47	
149	1-Aminoindane		222			142–3
150	3-Phenylpropylamine		222			57–8
151	2-Methyl-4,5,6,7-tetrahydroindole	34	222	86–91		
152	2-Propylaniline		222–4		104–5	119
153	2-Chloro-4-methylaniline		223	110	113	137
154	trans-9-Aminodecalin		223		183	148–9
155	2-Amino-4-phenylbutane		223			108
156	2-Methoxyaniline (v.s.) (o-anisidine)		225	89	88	60 (84)
157	4-Isopropylaniline (p-cumidine)		225		102	162
158	4-Propylaniline		225		96 (87)	115
159	4-Isopropylbenzylamine		226–8		65	93
160	α-Methyl-α-phenylhydrazine		227	132	92	153
161	N-Isobutylaniline		227			
162	4-tert-Butylaniline (mp 17)		228		169–70	134–6
163	9-Aminodecalin-(cis)		228		127	
164	Dihydroindole		228–30	133	105	118
165	2-Aminophenyl ethyl ether		229		79	
166	2,4,6-Trimethylaniline		229 (232)	137	216	204
167	2-Ethoxyaniline		229	102	79	104
168	3-Chloroaniline		230 (236)	121	72 (78)	120

TABLE 9A (Continued)
AMINES (PRIMARY AND SECONDARY)—LIQUID

| | | | | | Melting Point of Derivatives | | | |
| | | | | | Others | | | |
	p-Toluene-sulfon-amide	Phenyl-thiourea	α-Naph-thyl-thiourea	β-Naph-thyl-thiourea	α-Naph-thylurea	2-Nitro-1,3-in-dandione Derivative	Picrate	Hydro-chloride	Chloro-platinate or Chloroaurate
127	181	152					209		
128							166–8	275–7	
129									
130	82–3						191–2d		
131	232–3 (119)	148					171		
132							210		
133									
134									
135	212	204				181	180		
136									
137	71								
138	104						209		
139							133		
140							233	212	226(Pt)
141							138–40d		
142		153					200d		
143								288	184(Pt)
144						190–1		159	182(Pt)
145							225	245	204(Pt)
146								144	220d(Pt)
147							221	254	
148	56	104				190–1			
149							207		
150							152–3	218	
151							141		187(Pt)
152							151	173	
153									
154									
155								144	220(Pt)
156	127	136					200		
157									
158								203–4	
159									
160									
161	123								
162	179–80								
163			147						
164	99						174		
165									
166	167	193					193		
167	164	137							
168	138 (210)	124 (116)			251–2		177		

(Continued)

TABLE 9A (Continued)
AMINES (PRIMARY AND SECONDARY)—LIQUID

	Name of Compound	Note	BP	Melting Point of Derivatives Recommended		
				Benzene-sulfon-amide	Acetamide	Benzamide
169	2-Aminoindane		230		127	155
170	1-Aminoacridine		232–3		117	
171	Tetrahydroisoquinoline	35	233	154	46	129
172	2-*tert*-Butylaniline		233–5		159–61	
173	2-Dimethylamino-5-methylaniline		234			
174	4-Isobutylaniline		235		127	
175	2,3,6-Trimethylaniline		235		186	
176	2,6-Diethylaniline		235–6		135–6	
177	4-Aminoindane		236		126	136
178	2-Aminoundecane		237		58	
179	*unsym*-Ethylphenylhydrazine		237			
180	*sym*-Ethylphenylhydrazine	36	238–9			100
181	N-(2-Hydroxyethyl)ethylenediamine		239–41			
182	2-Bromo-4-methylaniline (mp 26)(see Table 9B)		240			
183	N-Methyl-4-chloroaniline		240		92	
184	1-Aminoundecane		240		48	60
185	4-Chloro-2-methylaniline (mp 29)(see Table 9B)		241			
186	2-Methyl-5-isopropylaniline		241		71 66(di)	102
187	N-Butylaniline		241.6		Oil	56
188	Phenylhydrazine (mp 19)		243	148	128 107(di)	168 177(di)
189	2-(2-Aminoethylamino)-ethanol		243–4			
190	1,3-Di(aminomethyl)benzene		245–8		118 (N,N'-di)	
191	2-Chloro-6-methoxyaniline		246		123 146(di)	135
192	3-Ethoxyaniline		248		97	103
193	4-Ethoxyaniline		250	143	135	173
194	Tetrahydroquinoline (mp 20)	37	250	67	Oil	74
195	2-Aminoacetophenone (mp 20)(see Table 21A)		250–2d		76–7	98
196	3-Methoxyaniline		251(245)		81	
197	3-Bromoaniline (mp 18)		251		87	120 (136)
198	3-Bromo-2-methylaniline		253–5		163	176–7
199	N-Isopentylaniline		254			
200	Dicyclohexylamine		254–5		103	153
201	2,3,4,6-Tetramethylaniline (mp 23–4)		255		215–7	
202	6-Methyl-1,2,3,4-tetrahydroisoquinoline	38	256			
203	4-Butylaniline		258–60		105	126

TABLE 9A (Continued)
AMINES (PRIMARY AND SECONDARY)—LIQUID

Melting Point of Derivatives
Others

	p-Toluene-sulfon-amide	Phenyl-thiourea	α-Naph-thyl-thiourea	β-Naph-thyl-thiourea	α-Naph-thyl-urea	2-Nitro-1,3-indan-dione De-rivative	Picrate	Hydro-chloride	Chloro-platinate or Chloro-aurate
169							239		
170	170						220 (206)		
171					145		200 (195)		
172									
173							151	192–3	
174	136–7								
175									
176									
177									
178							111	84	
179								137	
180								164	
181							199–201	196	
182									
183							153		
184								190	
185									
186									
187	56								
188	151	172						240–2	
189							198.5–200.5	195.7–196.4	
190							185–90		
191									
192	157	138					158		
193	106	136					69		
194					164				
195	148								
196	68						169		
197			143		259–60		180		
198									
199		81							
200							173		
201									
202							205	195–7	
203									200–2d(Pt)

(Continued)

TABLE 9A (Continued)
AMINES (PRIMARY AND SECONDARY)—LIQUID

	Name of Compound	Note	BP	Melting Point of Derivatives — Recommended		
				Benzene-sulfonamide	Acetamide	Benzamide
204	Methyl anthranilate (mp 24–5)		260d	107	101	100
205	4-Amino-N,N-diethylaniline	39	260–2		104	172
206	7-Methyl-1,2,3,4-tetrahydroquinoline		264			70–2
207	4-Methylindole		267			
208	4-Chloro-2-amino-N,N-dimethylaniline		267–8		90	
209	2,3-Dihydroxy-1-aminopropane	40	268d			113 (0,0,N-tri)
210	2,2′-Iminodiethanol (Diethanolamine)	41	270			
211	Diheptylamine	43	271			
212	3-Amino-N,N-dimethylaniline	42	272		87 69(di)	163–4
213	5,6,7,8-Tetrahydro-1-naphthylamine		275		158	
214	3-Amino-N,N-diethylaniline		276–8			
215	1,2-Bis(aminoethylamino)-ethane (triethylenetetramine)		277.5			238 (tetra)
216	Ethyl 3-aminobenzoate		294		110	148
217	N-Methyl-1-naphthylamine		294		95	121
218	Dibenzylamine		300	68		112
219	Diphenylmethylamine		303–4		146–7	172 (167
220	2-Amino-N,N-diethylaniline		312–3 (744 mm)			
221	1-Amino-1,2-diphenylethane		313			
222	N-Methyl-2-naphthylamine	44	315	107	51	84
223	N-Ethyl-2-naphthylamine		315		49	
224	1-Amino-2,3-diphenylpropane		315–7		85(di)	
225	N-Ethyl-1-naphthylamine		325		68	

NOTES FOR TABLE 9A: *p*- AND *s*-AMINES; LIQUIDS

1. N-Nitroso derivative, mp 66–7.
2. Phenylurea, mp 131.
3. Oxalate, mp 115.
4. Oxalate, mp 142; sulfate, mp 105.
5. Oxalate, mp 119; sulfate, mp 120.
6. N-Nitroso, mp 48.
7. Sulfate, mp 142.
8. Yellow picrate, mp 112; red picrate, mp 163–4.
9. Picrolonate, mp 235.
10. Chloroplatinate (anh.), mp 172–3, slow heat; 206–7, rapid heat; oxalate, mp 178–9. Picrate of N-methyl, mp 235.
11. Picrolonate, mp 204.
12. Hydrate with 1 mole water, mp 10.
13. The *d*- and *l*-forms boil at 120.5 and their picrates are reported as melting at 237.
14. Also called isohexylamine; oxalate, mp 166.
15. The stereoisomeride boils at 132–3; picrate, mp 124–8; N-benzoyl deriv, mp 84.
16. Oxalate, mp 104.
17. Oxalate, mp 204–5.
18. Methiodide, mp 266.
19. Picrolonate, mp 154; oxalate 193.
20. Reacts with phthalic anhydride to yield an imide, mp 127. Forms a salt with oxalic acid, mp 199–200.
21. Oxalate, (mono) mp 166, (di) mp 204.
22. Picrolonate, mp 215.
23. Phenylurea, mp 127.
24. Methiodide, mp 221.
25. Oxalate, mp 238.
26. N-*p*-Nitrobenzoyl derivative, mp 181.
27. Oxalate, mp 131.
28. The acetamide derivative melts at 93 after standing.
29. *o*-Nitrobenzoyl derivative, mp 107.

TABLE 9A (Continued)
AMINES (PRIMARY AND SECONDARY)—LIQUID

	Melting Point of Derivatives							
					Others			
	p-Toluene-sulfon-amide	Phenyl-thiourea	α-Naph-thyl-thiourea	β-Naph-thyl-thiourea	α-Naph-thyl-urea	2-Nitro-1,3-indan-dione Derivative	Picrate	Hydro-chloride	Chloro-platinate or Chloro-aurate
204							106		
205									
206							153-4	175	
207							194-5		
208							191		
209									185(Pt)
210					164		110		160(Pt)
211							117-20		
212								218	
213									
214							152		
215							240 (tetra)	266-70	
216									
217	164								
218	159		131				145		
219							205-6		
220	64-5						162		
221							212-3		188(Pt)
222	78						145		
223									
224								188-90	144-5(Au)
225									

NOTES FOR TABLE 9A: p- AND s-AMINES; LIQUIDS (Continued)

30. N-Nitroso derivative, mp 52.
31. Oxalate, mp 120-2.
32. N-Acetyl derivative, boils at 254-6.
33. Nitration yields the 4-nitro derivative, mp 78-9.
34. Methiodide, mp 195.
35. The 1,2,3,4-tetrahydro isomer boils at 223; the 5,6,7,8-tetrahydro isomer boils at 218.
36. Oxalate, mp 167.
37. The 1,2,3,4-tetrahydro isomer boils at 250; the 5,6,7,8-tetrahydro isomer boils at 222 and forms a picrate, mp 157.
38. Methiodide, mp 144-5.
39. N-Chloroacetyl derivative, mp 83-4.
40. Picrolonate, mp 220; O,N-di-p-benzoyl derivative, mp 139.
41. Nitrate, mp 69.
42. N-Chloroacetyl derivative, mp 102.
43. Forms a trihydrate, mp 32-3.
44. N-Nitroso derivative, mp 88.

TABLE 9B
AMINES (PRIMARY AND SECONDARY)—SOLIDS

	Name of Compound	Note	MP	Melting Point of Derivatives Recommended		
				Benzene-sulfonamide	Acetamide	Benzamide
1	Phenylhydrazine (bp 243) (see Table 9A)		19			
2	Tetrahydroquinoline (bp 250) (see Table 9A)		20			
3	2-Aminoacetophenone (bp 250–2d) (see Table 9A)		20			
4	3-Chloro-6-methylaniline		21–2		139–40(131)	
5	4-Aminostyrene		23.5		142	160–1
6	2,3,4,6-Tetramethylaniline (bp 255) (see Table 9A)		23–4			
7	2,3-Dichloroaniline		24		157	
8	Methyl anthranilate (bp 260d) (see Table 9A)		24–5			
9	2-Pyrrolidone	1	25			
10	3-Bromo-4-methylaniline		25–6	117	113	132
11	3-Chloro-4-methylaniline		26		105	122
12	2-Bromo-4-methylaniline		26		118	149
13	2-Amino-3-methylpyridine	2	26		64	220
14	2-Aminothiophenol	3	26		135 (N,S-di) 137(di)	154 (N,S-di) 177(di)
15	1,4-Diaminobutane		27			
16	1-Aminotridecane		27			71
17	Dodecylamine	4	27–8			
18	2,2′-Iminodiethanol		28	130		99
19	4-Chloro-2-methylaniline		29	125	140	142
20	3-Aminobiphenyl		30		148	
21	Diheptylamine		30			
22	dl-2,6-Dimethyl-1,2,3,4-tetrahydroquinoline		31–2			103–5
23	5-Bromo-2-methylaniline		32		165	
24	2-Bromoaniline		32		99	116
25	2-Methyl-1-naphthylamine		32		188	180
26	4,4′-Dimethyldibenzylamine		32.5			
27	3-Iodoaniline		33(27)		119	157
28	5,6-Dichloro-o-toluidine		33		145	
29	2-Aminobibenzyl		33		117	166
30	1,1-Diphenylhydrazine		34		184	192
31	4-Amino-N-methylacetanilide		35.5			
32	3-Bromo-5-methylaniline		36		171–2	
33	4-Amino-N-methylaniline		36		63	165
34	Pentadecylamine		36 (33)		72	
35	N-Benzylaniline		37	119	58	107
36	Tetradecylamine		37 (29)			
37	2-Nitro-N-methylaniline		37		70	
38	3-Iodo-4-methylaniline	5	37–8		130	
39	2,7-Dimethylindole		37–8			
40	5-Aminoindane		37–8		106	137
41	5,6,7,8-Tetrahydro-2-naphthylamine		38		107	167
42	5-Bromo-2-naphthylamine	6	38		165	109

TABLE 9B (Continued)
AMINES (PRIMARY AND SECONDARY)—SOLIDS

	Melting Point of Derivatives							
				Others				
	p-Toluene-sulfon-amide	Phenyl-thiourea	α-Naph-thyl-thiourea	β-Naph-thyl-thiourea	α-Naph-thyl-urea	2-Nitro-1,3-indan-dione Derivative	Picrate	Hydro-chloride	Chloro-platinate or Chloro-aurate
1									
2									
3									
4								265–7	
5									
6									
7									
8									
9								128–31N	82(Au)
10									
11									
12		154							
13							229		
14								217d	
15	224						250–5d		
16									
17	73							98	215(Pt)
18							110		
19									
20									
21							117 20		
22								180–3	
23									
24	90	146 (161)					129 (165)		
25									
26							153	272	
27	128								
28									
29							167–8		
30									
31							206		
32									
33									
34								199	
35	148–9	103					48	214–6	155(Pt)
36									
37									
38									
39							152		
40									
41							204		
42							216		

(Continued)

TABLE 9B (Continued)
AMINES (PRIMARY AND SECONDARY)—SOLIDS

	Name of Compound	Note	MP	Melting Point of Derivatives — Recommended		
				Benzene-sulfon-amide	Acetamide	Benzamide
43	2,2-Diphenylethylamine		38		88	143 (123)
44	2,3-Dimethyl-1,2,3,4-tetrahydro-quinoline		38–9			92
45	2,6-Dichloroaniline		39		44	
46	2-Piperidone	7	39–40			112
47	2-Aminobenzaldehyde (see Table 7B)		40		70–1	73–4
48	2-Amino-1-phenylethanol		40			148–9
49	*l*-Ephedrine		40(43)			
50	2-Chloro-4,6-dimethylaniline		40		205–6	148
51	2-Iodo-4-methylaniline	8	40		133	161
52	4-Amino-N,N-dimethylaniline	9	41		132–3	228
53	2-Amino-6-methylpyridine		41		90	90
54	1,6-Diaminohexane		42	154(di)		155(di)
55	3-Aminopropiophenone		42			
56	1,3-Diamino-2-propanol		42			
57	Thalline		42–3		46–7	
58	1,4-Diamino-2-butyne		42–6			
59	2,3-Dibromoaniline		43		164	
60	4-Amino-3-methylbiphenyl	10	43		165 (158)	189
61	*p*-Toluidine		43.7	120	147	158
62	2-Chloro-5-bromoaniline		45		141	
63	4-(N-Methylamino)-2-nitrotoluene		45 (57)		128	
64	2,4'-Diaminobiphenyl		45		202	278
65	Hexadecylamine		45–6		39(di)	
66	4-Aminobenzyl cyanide		46		97 153(di)	176–7
67	4-Aminothiophenol		46		154(N-)	180(N-)
68	2-Bromo-3,4-dimethylaniline		47		197	
69	1,2-Dibenzylhydrazine		47		78 117–8(di)	87
70	5-Hydroxy-2-methylaniline	11	47	183	178(N-)	
71	4-Amino-2-thiocresol	12	47		95	
72	2-Aminopropiophenone		47		71	130
73	3-Bromo-4-ethoxyaniline		47		114	
74	*trans*-Decahydroquinoline		47–9			56
75	3,4-Dimethylaniline		48	118	99	
76	4-Aminobibenzyl		48			170–1
77	2-Bromophenylhydrazine		48			
78	2-Chlorophenylhydrazine		48			
79	4-Fluoro-1-naphthylamine		48			197
80	2-Aminobiphenyl	13	49		121	102
81	Heptadecylamine		49		62	91
82	4-Amino-2-methyldiphenylamine		49–50		139–40	
83	2-Bromo-4,6-dimethylaniline		49–50		196–7	186
84	4-Bromo-2,6-dimethylaniline		49–50		197	

TABLE 9B (Continued)
AMINES (PRIMARY AND SECONDARY)—SOLIDS

Melting Point of Derivatives

	p-Toluene sulfonamide	Phenylthiourea	α-Naphthylthiourea	β-Naphthylthiourea	α-Naphthylurea	2-Nitro-1,3-indandione Derivative	Picrate	Hydrochloride	Chloroplatinate or Chloroaurate
43							212–3		
44							178		
45									
46									
47									
48							157–8		
49								218	186(Pt)
50				154					
51								188	
52							188		
53							202	155	218(Pt)
54							220	248	200d(Pt)
55	113								
56							230	185	240(Pt)
57							162		
58							230–2(di)		
59									
60									
61	118	141	168	163–4	234*	192–3	182d	243–5	
62								190(hyd.)	
63									
64									
65	64–5							178	
								(130–3)	
66							185		
67									
68									
69							130		
70									
71									
72	125d								
73							178–9		
74							158	276	228d(Pt)
									126(Au)
75									
76								210	286–9(Pt)
77									
78								194d	
79								280	
80									
81									
82								185–7	
83							122		
84									

(Continued)

TABLE 9B (Continued)
AMINES (PRIMARY AND SECONDARY)—SOLIDS

	Name of Compound	Note	MP	Melting Point of Derivatives Recommended		
				Benzene-sulfon-amide	Acetamide	Benzamide
85	2,4-Dibromo-6-methylaniline		50		205 88(di)	
86	1-Naphthylamine		50	167	159	160
87	2,5-Dichloroaniline		50		132	120
88	3,5-Dichloroaniline		50.5		186–7	147
89	3-Aminobibenzyl		51		128–9	
90	1-Amino-1,2,3-triazole		51			151
91	1-Methyl-2-naphthylamine		51		188–9	222
92	3-Methyl-1-naphthylamine		51–2		175–6	188–9
93	4-Methyl-1-naphthylamine		51–2		166–7	238–9
94	4-Chloro-2-methoxyaniline		52 (46)		150	
95	Indole	14	52	254	157–8	68
96	2,5-Dibromoaniline		52		172	
97	2-Aminodiphenylmethane		52		135	116
98	Di-(o-tolylamine)		52–3			114–5
99	4-Aminobiphenyl		53		171 120(di)	230
100	N,N′-Dimethyl-1,4-diaminobenzene	15	53			
101	Diphenylamine	16	53–4	124	101	180N
102	4-Chloro-3-aminodimethylaniline		54		97	
103	dl-α-aminobenzyl cyanide		55			159–60
104	2-Chloro-1-naphthylamine		56		191(mono) 88(di)	
105	3,5-Dimethyl-2-nitroaniline		56		114	
106	3,5-Dibromoaniline		56 (51)		231	169
107	1,2-Dimethylindole		56			
108	3-Nitro-4-methoxyaniline	17	57		153	
109	2-Aminopyridine		57(60)		71	165(di)
110	5-Bromo-2-ethoxyaniline		57 (53)		133	
111	5-Bromo-2-aminobiphenyl		57		130	162
112	Butyl 4-aminobenzoate (Butesin)		58			
113	4-Methoxyaniline		58	95	130 (127)	154 (157)
114	7-Methyl-1-naphthylamine		58–9		182–3	204
115	5-Methylindole		58–9			
116	3,5-Dichloro-o-toluidine		58–60			186
117	4-Bromo-2-methylaniline		59		156	115
118	1-Chloro-2-naphthylamine		59		147	98–9
119	l-Scopolamine		59			
120	2-Aminoazobenzene		59		126	122
121	2-Methylindole	18	59–61			
122	5-Methoxyindole		59–61		82	
123	2-Aminopyridine		60 (58)		71	165
124	4-Amino-3,5-dichlorotoluene		60		201	

TABLE 9B (Continued)
AMINES (PRIMARY AND SECONDARY)—SOLIDS

	Melting Point of Derivatives							
					Others			
	p-Toluene-sulfon-amide	Phenyl-thiourea	α-Naph-thyl-thiourea	β-Naph-thyl-thiourea	α-Naph-thyl-urea	2-Nitro-1,3-indan-dione Derivative	Picrate	Hydro-chloride	Chloro-platinate or Chloro-aurate
85									
86	157 (147)	165				209–10	163 (181–2d)		
87		166							
88									
89									
90							130	114	
91									
92									
93									
94							200		
95							187		
96							149		
97								137	
98									
99	255								
100							186		
101	141	152					182		
102									
103							160–1		
104									
105									
106									
107							125		
108									
109							216–7 (223)		
110							135–7		
111									
112							109–10	198	
113	114	157 (171)					170		
114									
115							151		
116									
117									
118									
119							190–1		
120									
121							139		
122							145		
123							216–7	86	231(Pt)
124									

(Continued)

TABLE 9B (Continued)
AMINES (PRIMARY AND SECONDARY)—SOLIDS

	Name of Compound	Note	MP	Melting Point of Derivatives Recommended		
				Benzene-sulfon-amide	Acetamide	Benzamide
125	3-Nitro-N-ethylaniline		60		89	
126	Procaine		61			
127	2-Iodoaniline		61		109	139
128	4-Iodoaniline		62		184	222
129	3-Chloro-1-naphthylamine		62		197	162
130	1,10-Decanediamine		62			
131	N-Phenyl-1-naphthylamine		62		115	152
132	2-Methylpiperazine	19	62			146–7(di)
133	3-Chloro-4-methoxyaniline		62		94	
134	2-Amino-4,4'-dimethylbiphenyl		62–3		118–9	95–6
135	8-Amino-6-methylquinoline	20	62–4		91–2	
136	1,3-Diaminobenzene	21	63	194	87–9N	125N
137	4-Amino-N-methylacetanilide		63			
138	2,4-Dichloroaniline		63	128	145	117
139	1-Bromo-2-naphthylamine	22	63		140(mono) 105(di)	
140	2,3-Diaminotoluene		63–4			
141	5-Methyl-2-naphthylamine		63–4		123–4	155–6
142	2-Mercaptopropylamine		63–5			
143	3-Aminopyridine		64		133	119
144	2,5-Diaminotoluene	23	64	147N	220(di)	307
145	3-Bromo-4-methoxyaniline		64		111	
146	9-Aminofluorene		64		262	260–1
147	4-Methylphenylhydrazine	24	65		121	146N
148	2-Methylpiperazine	25	65			146–7(di)
149	4-Aminobenzyl alcohol		65		188 (0, N-di)	150–1(N-)
150	2-Bromo-1-naphthylamine		65		198	179
151	2-Bromo-6-methoxyaniline		65			90
152	1,3-Diamino-2,6-dimethylbenzene		65–6			232(di) (227)
153	3,4-Dimethyl-2-nitroaniline		65–6		115–6	199–200
154	Aminoacetamide		65–7			
155	4-Bromoaniline		66	134	168	204
156	2-Aminocyclohexanol		66			
157	6-Amino-3-methylbenzophenone		66		159	118
158	1,8-Diaminonaphthalene		66.5			311–2(di)
159	5-Bromo-3-aminopyridine		66–7		76–8(hyd.) 127(anh.)	
160	3-Chloro-5-bromoaniline		67–8		151	
161	8-Methyl-1-naphthylamine		67–8		183–4	195–6
162	4-Iodoaniline	26	67–8		184	222
163	3-Nitro-N-methylaniline		68	83	95	105
164	2,4,5-Trimethylaniline		68	136	162	167
165	2,2'-Diaminobibenzyl		68		249(di)	255(di)
166	2,5-Diaminophenol		68		265(di)	
167	4-Methyl-2-naphthylamine		68		172–3	194–5
168	2-Bromo-4-chloroaniline	27	69		137 (134)	130–5
169	4-Hydroxyindole		69.5			

TABLE 9B (Continued)
AMINES (PRIMARY AND SECONDARY)—SOLIDS

	Melting Point of Derivatives							
				Others				
	p-Toluene sulfonamide	Phenyl-thiourea	α-Naphthyl-thiourea	β-Naphthyl-thiourea	α-Naphthyl-urea	2-Nitro-1,3-indandione Derivative	Picrate	Hydrochloride	Chloroplatinate or Chloroaurate
125									
126							146–7	153–6	
127							112		
128		153							
129									
130							134		
131									
132	174(di)						276–8	248–9	
133							186d		
134									
135									
136	172					200	184		
137							206		
138	126						106		
139									
140									
141									
142							143–4d	87–8	
143									
144	150N								
145								255	
146								255	
147									
148	174(di)						276–8	248–9	
149									
150									
151								225	
152									
153									
154								186–9	197–8(Au)
155	101	148	188 (166)	175–6	272–4		180		
156								175	
157							145		
158									
159							212–3		187(Au)
160							141		
161									
162		153							
163									
164									
165							225–30		
166									
167									
168									
169							159–60		

(Continued)

TABLE 9B (Continued)
AMINES (PRIMARY AND SECONDARY)—SOLIDS

	Name of Compound	Note	MP	Melting Point of Derivatives		
				Recommended		
				Benzene-sulfon-amide	Acetamide	Benzamide
170	3-Aminoazobenzene	28	69–70N		130–1N	
171	2,4-Dimethyl-6-nitroaniline		70		176 (173)	185
172	3-Bromo-1-naphthylamine		70		174	166
173	Pyrazole		70			46
174	1-Amino-5-methyl-1,2,3-triazole	29	70			158(mono) 138(di)
175	8-Aminoquinoline		70 (65)		103	98
176	2-Nitroaniline($v.s.$)		71	104	92	110
177	3-Chloro-4-aminobiphenyl		71		147	
178	3,4-Diaminotriphenylmethane		71–2		226(di)	243(di)
179	4-Chloroaniline		72	122	179	192
180	2-Aminobenzyl cyanide		72		120	
181	4-Aminobenzaldehyde (see Table 7B)	30	72		153N	
182	4-Bromo-2-naphthylamine		72		189	
183	3,4-Dichloroaniline		72		121	
184	6-Chloro-8-aminoquinoline	31	73s			
185	7-Chloro-8-aminoquinoline		73(75)		184–6	
186	2-Chloro-4-bromoaniline		73(69)		151	145
187	4-Aminophenylurethane		73–4		202 (181)	230
188	6-Bromo-2-nitroaniline		74		190	
189	3,4-Dimethyl-5-nitroaniline		74–5		209–10	223–4
190	2-Nitrodiphenylamine	32	74–6			
191	4-Nitromesidine	33	75	163	191	169
192	4-Aminodiphenylamine(anh.)		75		158	203
193	2-Amino-1,4-dimethylnaphthalene		75		219–20	
194	3,5-Dimethylindole		75			
195	2,3,5,6-Tetramethylaniline		75		207	
196	3,4,5-Trimethyl-aniline		75		163–4	
197	4-Chloro-1,2-diaminobenzene		76		108(di)	230(di)
198	2-Bromo-1,4-diaminobenzene		76		200(di)	235(di)
199	2-Nitro-6-methoxyaniline		76		158–9	
200	2-Nitro-4,6-dimethylaniline	34	76 (70)		176 (173)	185
201	4-Chloro-3-methoxyaniline		77		122	
202	5-Methyl-1-naphthylamine		77–8		194–5	173–4
203	4-Methyl-3-nitroaniline		78 (72)	160	148 (93)	172
204	2,4,6-Trichloroaniline	35	78		204 (206)	174
205	4-Chloro-3-bromoaniline		78		135	
206	2,6-Dibromo-4-methylaniline		79 (73)		200(mono) (183) 101.5(di)	
207	N-(4-Tolyl)-1-naphthylamine		79		124	140
208	Di(4-Tolyl)amine		79		85	125

TABLE 9B (Continued)
AMINES (PRIMARY AND SECONDARY)—SOLIDS

	Melting Point of Derivatives							
				Others				
	p-Toluene-sulfon-amide	Phenyl-thiourea	α-Naph-thyl-thiourea	β-Naph-thyl-thiourea	α-Naph-thyl-urea	2-Nitro-1,3-indan-dione Derivative	Picrate	Hydro-chloride-	Chloro-platinate or Chloro-aurate
170									
171									
172									
173							160	104	
174									
175	154–6								
176	142		145	176			73		
177									
178									
179	95 (119)	152			235		178		
180									
181									
182									
183									
184								208	212–3(Pt)
185									
186									
187									
188									
189									
190									
191									
192									
193									
194							180		
195								260	
196									
197									
198									
199									
200									
201									
202									
203	164	171		159					
204							83		
205									
206									
207									
208									

(Continued)

TABLE 9B (Continued)
AMINES (PRIMARY AND SECONDARY)—SOLIDS

	Name of Compound	Note	MP	Melting Point of Derivatives Recommended		
				Benzene-sulfon-amide	Acetamide	Benzamide
209	2,4-Dibromoaniline		79		146	134
210	2,4-Diaminophenol		79–80		220–2(di)	253(di)
211	2-Aminodiphenylamine		79–80		121	136
212	3-Amino-2-chloropyridine		80		90–1	
213	4-Aminopyrazole		80–2			173(di)
214	4-Bromo-3-methylaniline		81		103–4	
215	2,2′-Diaminodibenzyl sulfide		81		209(di)	
216	2,2′-Diaminobiphenyl		81		89 161(di)	159 190(di)
217	3-Aminoacenaphthene		81		192–3	209
218	2,4-Dimethyl-3-nitroaniline		81–2		149	236
219	2,6-Dimethyl-3-nitroaniline		81–2		170	
220	2-Aminobenzyl alcohol		82		114 (N-)	198–9 (O-)
221	1-Amino-1,3,4-triazole		82			
222	3-Amino-4′-chlorobiphenyl		82		184	
223	4-Chloro-3-methylaniline		83		91	119.5
224	2,6-Dibromoaniline		83–4		210 101(di)	
225	5-Chloro-2-methoxyaniline		84		104	78
226	4-Aminobutyrophenone (see Table 21B)		84		142	
227	4-Aminotriphenylmethane		84		168	198
228	3-Iodo-1-naphthylamine		84		207	174
229	5-Chloro-1-naphthylamine		85		128	
230	3-Hydroxyindole (indoxyl)		85		139	123 (133)
231	2-Aminophenanthrene		85		225–6	216.5
232	7-Methylindole		85			84
233	4-Bromo-2-aminobenzaldehyde (see Table 7B)		85			
234	2,2′-Diaminodiphenyl sulfide		85–6		160(di)	162–3(di)
235	3,4-Dimethoxyaniline		85–6		133	177
236	4-Aminobenzonitrile	36	86		205	170
237	4-Chloro-2-aminobenzaldehyde (see Table 7B)		86			
238	4-Aminoacenaphthene		87		175–6	196
239	5-Chloro-8-aminoquinoline		87		141	
240	4-Iodo-2-methylaniline	37	87(92)		170 (162)	184
241	3-Aminoacetanilide		87–9		191	
242	3-Aminophenanthrene		87.5		200–1	213
243	4-Amino-3-methyl-1-phenylpyrazole		88		94–5(hyd) 120(anh.)	181
244	2-Hydroxy-3-methylaniline		89		78–9	
245	3,4-Diaminotoluene	38	89–90	178–9(di)	210(di)	263–4(di)
246	Ethyl 4-aminobenzoate		90		110	148
247	4-Chlorophenylhydrazine		90			
248	2-Aminothiazole		90		203	

TABLE 9B (Continued)
AMINES (PRIMARY AND SECONDARY)—SOLIDS

	Melting Point of Derivatives							
				Others				
	p-Toluene-sulfon-amide	Phenyl-thiourea	α-Naph-thyl-thiourea	β-Naph-thyl-thiourea	α-Naph-thyl-urea	2-Nitro-1,3-indan-dione Derivative	Picrate	Hydro-chloride	Chloro-platinate or Chloro-aurate
209	134	171					124		
210							120d		
211									
212									
213							193–4		
214									
215							203–4		
216									
217							221		
218									
219									
220							110	108	
221							194–5	153	230d(Pt) 170(Au) (anh.)
222									
223				158					
224							123–4		
225							194d		
226								178	
227									
228									
229									
230									
231									
232							176		
233									
234									
235									227(Pt)
236							150		
237									
238							190–200d		
239									
240							189		
241	241								
242									
243							138		226(Pt)
244									
245	140								
246							131		
247								225–30d	
248									

(Continued)

TABLE 9B (Continued)
AMINES (PRIMARY AND SECONDARY)—SOLIDS

	Name of Compound	Note	MP	Melting Point of Derivatives Recommended		
				Benzene-sulfonamide	Acetamide	Benzamide
249	8-Bromo-1-naphthylamine		90		139	
250	2-Nitrophenylhydrazine		90		140–1	166
					57–8(di)	
251	2,2′-Diaminodibenzyl disulfide	39	90–1		202–5(di)	
252	1-Amino-2,6-dimethylnaphthalene		91		211	219–20
253	2,4-Diaminochlorobenzene		91		242(di)	178(di)
254	6-Hydroxyindole		91–2			
255	4-Amino-1-naphthalenethiol		91–3		173(N-)	
256	Ethyl 4-aminobenzoate	40	92			150
257	2-Methyl-3-nitroaniline		92		158	168
258	4-Fluoro-2-nitroaniline		92		71.5	
259	2,6-Dibromo-4-chloroaniline		93		226	194
260	4,4′-Diaminodiphenylmethane	41	93		236–7(di)	
261	2,2′-Diaminodiphenyl disulfide		93		156(di)	
262	Bis(2-aminophenyl) disulfide		93		168–9	141
263	3-Nitrophenylhydrazine		93		145	151
					(150)(di)	(153)(di)
264	7-Aminoquinoline(anh.) (hyd. mp, 73)		93–4		167	189
265	4-Amino-2,6-dimethylnaphthalene		93–4		207–8	
266	3-Aminoquinoline		94		167	
267	2-Bromo-4-methylphenylhydrazine		94		130–1	144
268	2,4-Dibromo-6-chloroaniline		95		227	192
269	3-Methylindole (skatole)		95		68	
270	1-Amino-4,5-dimethyl-1,2,3-triazole		95			
271	8-Amino-1-naphthol	42	95–7		181N	193N
272	Semicarbazide		96		165	225
273	4-Nitro-N-ethylaniline		96		119	98
274	2,4-Di-iodoaniline		96		141	181
					(171)	
275	5-Amino-2-methylpyridine		96		126	110–1
276	6,6′-Diamino-3,3′-dimethyldiphenyl-methane		96		226(di)	
					152(tetra)	
277	2-Methyl-6-nitroaniline		97		158	167
278	8-Nitro-1-naphthylamine		97	194	191	
279	3-Aminobenzyl alcohol	43	97		106–7(N-)	115(N-)
280	4-Nitro-2-naphthylamine		97		241	
281a	5-Bromo-2-methoxyaniline		97–8		160	108
281b	Benzoxazolone(hyd.) (anh. mp 141–2) (see Table 21B)		97–8		95	173–4
281c	1,2-Diaminonaphthalene	44	98	215d,N	234(di)	291(di)
282	4-Chloro-1-naphthylamine		98		186.5	
283	2-Amino-4-methylpyridine		98		103	114
284	2,5-Dimethyl-3-nitroaniline		98		180	
285	2-Aminophenacyl alcohol (see Table 21B)		98		141 (N-)	167 (O,N-di)
286	4-Fluoro-3-nitroaniline		98		138.5	
287	2,4-Diaminotoluene	45	99	192(di)	224(di)	224(di)
288	4-Amino-4′-methylbiphenyl		99		221	
289	3-Aminoacetophenone (see Table 21B)		99		128–9	
290	2-Aminoethanethiol		99–100			

TABLE 9B (Continued)
AMINES (PRIMARY AND SECONDARY)—SOLIDS

Melting Point of Derivatives

	p-Toluene sulfon- amide	Phenyl- thiourea	α-Naph- thyl- thiourea	β-Naph- thyl- thiourea	α-Naph- thyl- urea	2-Nitro- 1,3-indan- dione De- rivative	Picrate	Hydro- chloride	Chloro- platinate or Chloro- aurate
249									
250									
251									
252									
253	215								
254							137		
255									
256							131	207–8	
257									
258									
259									
260						248		281d	
261							141(di)		
262									
263									
264									225(Pt)
265									
266							210d		
267								190d	
268									
269							170–1	167–8	
270							124–5	131	215(Pt)
271	189(N-)						163–4		
272							166d		
273	107								
274									
275							201		
276							199		
277	122		171						
278									
279									
280	145								
281a									
281b									
281c									
282									
283							227d		
284									
285									
286									
287	192–3(di)								
288								280–3	
289	130								
290							126	72	

(Continued)

TABLE 9B (Continued)
AMINES (PRIMARY AND SECONDARY)—SOLIDS

	Name of Compound	Note	MP	Melting Point of Derivatives		
				Recommended		
				Benzene-sulfonamide	Acetamide	Benzamide
291	5-Chloro-2-aminobenzophenone		100		117	
292	2-Mercaptoethylamine		100			
293	4-Amino-3,2′-dimethylazobenzene	46	100		185	
294	Benzotriazole		100		51	112
295	6-Fluoro-3-nitroaniline		101.5		178.4	301(di)
296	1,2-Diaminobenzene	47	102	185	185	
297	4-Bromo-1-naphthylamine	48	102		193	
298	1,2,3-Triaminobenzene		103			
299	N-(4-Tolyl)-2-naphthylamine		103		85	139
300	3,4-Diaminobiphenyl	49	103		163(di)	248(di)
301	6,5′-Diamino-3,3′-dimethyldiphenyl sulfide		103–4		165(di)	185(di)
302	1-Aminotriphenylene		103–4		252	
303	Piperazine	50	104	285(di)	144(di)	196(di)
304	4-Bromo-o-tolylhydrazine	51	104		172	172–3
305	8-Nitro-2-naphthylamine		104		196	162
306	3-Amino-4,4′-dimethylbiphenyl		104–5		156–7	160–1
307	dl-2-Amino-1-phenyl-1-propanol (d or l, mp 52)	52	104–5		135(N-)	143(N-)
308	4,4′-Diaminodibenzyl sulfide		104–5		188(di)	224(di)
309	2,6-Diaminotoluene		105		202–3(di)	
310	4-Aminophenanthrene	53	105		190	224
311	2-Bromo-4-nitroaniline		105		129	160
312	Guanylurea	54	105			187–8
313	1,3-Diamino-4,6-dimethylbenzene		105		295(di)	258–9(di)
314	Triphenylmethylamine		105		207–8	160–2
315	2-Aminobenzophenone (see Table 21B)	55	105–6		72 (89)	80
316	3-Amino-6-phenylpyridine		105–6		148–9	201
317	5-Hydroxyindole (see Table 29B)		105–7			
318	4-Aminoacetophenone (see Table 21B)		106	128	167	205
319	4-Bromophenylhydrazine	56	106			
320	3-Amino-4-methylpyridine	57	106		84N	81N
321	2-Chloro-4-nitroaniline		107		139	161
322	2-Methyl-5-nitroaniline	58	107	172	151	186
323	2,4-Diaminopyridine		107			191–2(di)
324	9-Phenanthrylmethylamine	59	107		182–5	167
325	2,5-Diaminopyridine		107–10		290(di)	230(di)
326	N-Phenyl-2-naphthylamine		108		93	148(136)
327	4-Amino-isoquinoline		108		168	188–9
328	5-Aminoacenaphthene		108		238(mono) 122(di)	210 (199)
329	1-Aminoacridine		108		117	
330	4-Aminophenethyl alcohol	60	108	93	105	139–40 (N-)
331	4,4′-Diamino-2,2′-dimethylbiphenyl	61	108		281	
332	4-Aminoantipyrine		109		199	
333	2-Aminobenzamide	62	109		177	214–5
334	3-Amino-4-methylbenzophenone	63	109		108	

TABLE 9B (Continued)
AMINES (PRIMARY AND SECONDARY)—SOLIDS

	Melting Point of Derivatives							
				Others				
	p-Toluene-sulfon-amide	Phenyl-thiourea	α-Naph-thyl-thiourea	β-Naph-thyl-thiourea	α-Naph-thyl-urea	2-Nitro-1,3-indan-dione De-rivative	Picrate	Hydro-chloride-	Chloro-platinate or Chloro-aurate
291									
292							125–6	70–2	
293									
294									
295									
296	260(di)					172–4	208		
297									
298							183d		
299									
300									
301							179(di)		
302									
303	173(mono)						280		
304								183-4	
305									
306								260d	
307								194	
308									
309									
310							216d		
311									
312							265		
313	221(di)								
314									
315									
316									
317									
318	203								
319									
320							180		
321	164								
322									
323									224(Pt)
324							241		
325									
326									
327									
328							190–200d		
329	169–70						220 (206)		
330								171	
331							225		
332							144		
333									
334									

(Continued)

TABLE 9B (Continued)
AMINES (PRIMARY AND SECONDARY)—SOLIDS

	Name of Compound	Note	MP	Melting Point of Derivatives		
				Recommended		
				Benzene-sulfon-amide	Acetamide	Benzamide
335	5-Methyl-2-nitroaniline		109		86–7	
336	5-Aminoquinoline		110		178	
337	2-Iodo-1,4-diaminobenzene		110.5		211	254
338	2-Nitro-4-aminostilbene		110–1		192–3	
339	4-Bromo-2-nitroaniline		111		104	137–8
340	3-Amino-4-methylbenzo phenone	64	111		139	
341	Methyl 4-aminobenzoate (see Table 12A)		111–2			
342	2,3-Dimethyl-5-nitroaniline		111–2		230–1	227–8
343	2-Naphthylamine		112	102	132	162
344	4-Amino-3-methylbenzophenone	65	112		175	158
345	β-Aminopropiophenone (unstable)		[112–4]		90–1	104–5
346	3-Mercaptopropylamine		112–3			
347	4-Ethoxy-2-nitroaniline		113	72	104	
348	2-Hydroxybenzylaniline		113		93	
349	5-Bromo-3-methyl-2-hydroxyaniline	66	113		119(N-)	195(N-)
350	2,3-Diaminopyridine		113			
351	1-(2-Naphthyl)-2-aminoethanol		113.5		96	207.8
352	3-Nitroaniline		114	136	155	157
353	2,3-Dimethyl-4-nitroaniline		114		149–50	208–9
354	6-Aminoquinoline	67	114		138 75(di)	169
355	2,5-Dimethylpiperazine(cis)	68	114			152(di)
356	2,5-Dimethylindole		114–5			
357	4-Methyl-2-nitroaniline		115	102 (99)	99 (96)	148 143)
358	3-Amino-2-phenylquinoline	69	115–6		124 173(di)	
359	4-Chloro-2-nitroaniline		115–6		104	
360	2-Fluoreneamine		115–6		194–5	
361	4-Bromo-8-nitro-1-naphthylamine		116		202	
362	5-Amino-3-methyl-1-phenylpyrazole		116		110	
363	4,4'-Diethyldiaminobiphenyl		116		168(di)	
364	4-Aminotriphenylcarbinol		116		176	
365	5-Bromo-4-methyl-2-hydroxyaniline		116		199(N-) 188 (O,N-di)	223(N)
366	4-Chloro-2-nitroaniline		116–7		104	
367	3-Aminoindole		117d		162–3d	
368	Hydrastinine		117		105	
369	5-Amino-2-methylquinoline(anh.)		117–8		205	
370	3-Aminopyrene		117–8		260	
371	2,5-Dimethylpiperazine(trans)	70	118			228–9(d)
372	1-Aminofluorenone		118 (110)		138	149–50i
373	2,3,4,6-Tetrabromoaniline	71	118		228–9	
374	5-Nitro-2-methoxyaniline		118		175–6	160–1
375	4-Hydroxypyrazole		118			109(di)
376	2,3-Dimethyl-6-nitroaniline		118–9			160
377	5-Nitro-1-naphthylamine		119	183	220	

TABLE 9B (Continued)
AMINES (PRIMARY AND SECONDARY)—SOLIDS

	Melting Point of Derivatives							
					Others			
	p-Toluene sulfon- amide	Phenyl- thiourea	α-Naph- thyl- thiourea	β-Naph- thyl- thiourea	α-Naph- thyl- urea	2-Nitro- 1,3-indan- dione De- rivative	Picrate	Hydro- chloride	Chloro- platinate or Chloro- aurate
335									
336	203–4								
337									
338								223	
339									
340									
341									
342									
343	133	129		203		193	195		
344									
345							164–5	187	205(Pt)
346								69	
347	94			175					
348								131	184(Pt)
349									
350							262–4		
351							191–2	186	
352	138	160	161–2	167–8	238–9		143		
353									
354	193								
355	146–7(di)								
356							159		
357	146 (166)			212					
358									
359	110								
360									
361									
362							160–2		
363									
364									
365									
366	110								
367									
368									
369									
370									
371	225(di)								
372									
373									
374	128								
375								129	
376	177–8								
377									

(Continued)

TABLE 9B (Continued)
AMINES (PRIMARY AND SECONDARY)—SOLIDS

	Name of Compound	Note	MP	Melting Point of Derivatives Recommended		
				Benzene-sulfon-amide	Acetamide	Benzamide
378	2,4,6-Tribromo-5-hydroxyaniline	72	119		136	
379	1,4-Diaminonaphthalene	73	120		303(di)	280(di)
380	1,9-Diaminofluorene		120		293(di)	
381	2-(2-Aminoethyl)indole	74	120			173–4
382	2,2′-Diamino-4,4′-dimethylbiphenyl		120		189(di)	170(di)
383	2-Aminobenzhydrol		120		118	
384	6-Hydroxy-3-aminoacetophenone		121 (110)		165(N-) 174(di)	
385	4,4′-Diaminostilbene(cis)		121		172(di)	253(di)
386	4-Aminobenzhydrol		121			145
387	2-Aminotriphenylcarbinol		121		192	
388	2,6-Diaminopyridine	75	121.5		203(di)	176(di)
389	2,4,6-Tribromoaniline	76	122		232	198
390	3-Hydroxyaniline	77	122		148N	174N
391	2-Hydroxy-4-aminoacetophenone	78	122–3		91(N-)	
392	1-Aminoisoquinoline		122–3		148	
393	2,4-Dimethyl-5-nitroaniline		123	149	159	200
394	3,4,5-Tribromoaniline		123		255–6	210
395	4,4′-Diamino-2,2′-dimethyldiphenyl-methane		123		228(di)	
396	2,2′-Diaminostilbene(cis)		123 (107)		214–5(di)	
397	4-Methoxy-2-nitroaniline		123–5		117–8	
398	4-Aminobenzophenone	79	124		153	152
399	2-Amino-4-nitrobenzaldehyde (see Table 7B)		124			
400	7-Amino-8-hydroxyquinoline		124		175(N-)	
401	4-Aminoazobenzene	80	125 (127)		146	211
402	2-Amino-5-nitrobiphenyl		125		133	
403	3-Hydroxy-4-methoxyaniline		125–7		116–9 (N-)	162–4 (O,N-di)
404	1-Nitro-2-naphthylamine		126	156	123	168
405	Hydrazobenzene	81	126–7		159(mono) 105(di)	126N 162(di)
406	5-Chloro-2-nitroaniline		126–7		121	
407	Benzidine		127	232(di)	199(mono) 317(di)	203–5(mono) 352(di)
408	1-Aminoanthracene		127		212	
409	3-Methyl-1-phenyl-5-pyrazolone		127			
410	2-Aminopyrimidine	82	127–8			
411	6-Bromo-2-naphthylamine		128		192	218
412	5-Aminoisoquinoline	83	128		166	158–9
413	2-Amino-3-methoxy-4-hydroxy-benzaldehyde (see Table 7B)		128–9		97	
414	o-Tolidine	84	129		314(di)	265(di)
415	2-Nitro-4-methoxyaniline	85	129 (123)		117	140
416	3,7-Dimethyl-2-naphthylamine		129 (134)		231	
417	2-Aminotriphenylmethane		129		154–5	

TABLE 9B (Continued)
AMINES (PRIMARY AND SECONDARY)—SOLIDS

	Melting Point of Derivatives							
				Others				
	p-Toluene-sulfon-amide	Phenyl-thiourea	α-Naph-thyl-thiourea	β-Naph-thyl-thiourea	α-Naph-thyl-urea	2-Nitro-1,3-indan-dione Derivative	Picrate	Hydro-chloride-	Chloro-platinate or Chloro-aurate
378	146–7								
379	187–8*N*								
380							205		
381									
382									
383									
384									
385									
386								270–3	
387							122–3		
388							240		
389									
390	157	156							
391									
392							290–1	233	
393	192								
394									
395							216		
396							155–6	230	
397									
398									
399									
400							205		
401									
402	169								
403									
404	160								
405									
406									
407	243(di)								
408							190		
409								96	110(Pt)
410							237–8	196	216(Pt)
411									
412									
413									
414			167			216			
415									
416								275	
417									

(Continued)

TABLE 9B (Continued)
AMINES (PRIMARY AND SECONDARY)—SOLIDS

	Name of Compound	Note	MP	Melting Point of Derivatives		
					Recommended	
				Benzene-sulfon-amide	Acetamide	Benzamide
418	2-Aminoquinoline	86	129			
419	2-Aminofluorene	87	129		191	
420	3,3'-Dimethylbenzidine		129		103(mono) 306(di)	
421	2-Methyl-4-nitroaniline		130	158	202	
422	2,4-Diaminodiphenylamine		130		188(di)	213(2-N-)
423	2,2'-Diaminobenzoquinone		130			
424	3-Aminocoumarin		130		201–2	173
425	4-Chloro-2-aminopyridine		130–1		115–6	120–1 165–6(di)
426	4-Hydroxy-3-nitroaniline	88	131 (127)		157–8(N-)	
427	4-Bromo-3-nitroaniline		131–2		146	
428	4-Bromo-3-nitro-1-naphthylamine		132		223	
429	2-Aminoacetanilide		132		185–6	
430	2-Aminobenzthiazole		132 (129)		186	186
431	3,5-Dimethyl-4-nitroaniline		132		163	
432	2-Amino-4-methylquinoline		133		232	
433	2,2'-Diaminobenzophenone		133 (135)		168 (154)(di)	
434	2,2'-Diaminoazobenzene		134		271(di)	
435	3,5-Diaminoacetophenone		134		210(di)	
436	2-Hydroxy-3,5-dimethylaniline		134–5		96(N-)	154 (O,N-di)
437	3-(2-Aminophenyl)propenonitrile		134–5		172–4	
438	3-Methyl-4-nitroaniline		135		102	
439	2-Hydroxy-5-methylaniline	89	135		160(N-)	191(N-)
440	3-Methyl-2-naphthylamine		135		181–2	190
441	2-Aminoacenaphthene		135			
442	dl-6,6'-Diamino-2,2'-dimethylbiphenyl		136		205(di)	182(di)
443	1,4-Diamino-2-nitrobenzene	90	137		186(di)	236(4-N)
444	3,3'-Dimethoxybenzidine (dianisidine)		137–8		242(di)	236(di)
445	3-Nitro-1-naphthylamine		137		259	220
446	9-Aminophenanthrene		137–8		207–8	199
447	4-Hydroxy-3,5-dimethylaniline		137–8		160(di)	
448	2,6-Dinitroaniline		138		197	
449	2-(4-Aminophenyl)quinoline	91	138		189 154(di)	234
450	2-Amino-4-chlorophenol (see Table 29B)	92	138		185(N-)	
451	3,6-Dimethyl-2-naphthylamine		139		207	
452	2-Methoxy-4-nitroaniline		139–40	181	153–4	149–50
453	3,4-Dimethyl-6-nitroaniline		139–40		107	149–50
454	4-Aminopropiophenone		140		161 (172)	190
455	3-Nitro-4-aminobenzophenone		140 (135)			154–5
456	2-Amino-4-methyldiphenylamine		140			161

TABLE 9B (Continued)
AMINES (PRIMARY AND SECONDARY)—SOLIDS

	Melting Point of Derivatives							
					Others			
	p-Toluene-sulfon-amide	Phenyl-thiourea	α-Naph-thyl-thiourea	β-Naph-thyl-thiourea	α-Naph-thyl-urea	2-Nitro-1,3-indan-dione Derivative	Picrate	Hydro-chloride	Chloro-platinate or Chloro-aurate
418							255–6		
419									
420									
421	174		165	154				200	
422									
423							164		
424									
425							243–4		
426									
427									
428									
429									
430							256	236–7	
431									
432									230(Pt)
433							164		
434									
435									
436									
437							192–3		
438				159					
439									
440									
441							260	270d	
442	162–3(di)								
443									
444							225(di)		
445	200								
446							190		
447									
448									
449									
450									
451								283	
452	175 (169–70)								
453									
454									
455									
456								200	

(Continued)

TABLE 9B (Continued)
AMINES (PRIMARY AND SECONDARY)—SOLIDS

	Name of Compound	Note	MP	Melting Point of Derivatives Recommended		
				Benzene-sulfon-amide	Acetamide	Benzamide
457	4-Phenyl-3-thiosemicarbazide	93	140–1			195
458	1,4-Diaminobenzene	94	142	247(di)	304(di)	300(di)
459	2-Hydoxy-5-nitroaniline (anh.)	95	142–3		278–9	
460	2-Aminotriphenylene		142–3		260–1	
461	4-Nitro-2-aminostilbene	96	142–3		221(N-)	
462	8-Hydroxy-5-aminoquinoline		143		206 (O,N-di)	205 (O,N-di)
463	1-Bromo-8-nitro-2-naphthylamine		143		180	
464	5-Nitro-2-naphthylamine		144		186	182
465	2-Nitro-1-naphthylamine		144		199	175
466	2-Methyl-5-hydroxyaniline	97	144	183(N-)	178(N-)	
467	2,5-Dimethyl-4-nitroaniline		144–5	162	168–9	
468	4'-Bromo-4-aminobiphenyl		145		247	
469	1-Aminophenanthrene		146		220	
470	N,N'-Diphenyl-1,4-diaminobenzene		146		191.7(di)	
471	2-Hydroxy-5-methyl-1,3-diamino-benzene		146		225–7 (1,3-di)	
472	4-Chloro-2,6-dinitroaniline		146		215	
473	3,5-Dihydroxyaniline		146–52		119–21 (tri)	
474	Indazole		146.5		42	92–3
475	2-Chloro-4-aminobenzaldehyde		147		152	
476	4-Nitroaniline		147	139	215	199
477	Papaverine		147			
478	2-Aminobenzoic acid (anthranilic acid) (see Tables 4 and 5B)		147	214	185	182 (181)
479	5-Amino-3-methyl-1,2,4-triazole		148			285–90
480	2,2'-Dihydroxyhydrazobenzene		148			186(di)
481	7-Amino-2-methylquinoline (anh.)		148			172–3
482	4,4'-Diamino-2,2'-dimethylazoxybenzene		148		281(di)	290
483	4-Bromo-3-hydroxyaniline		150		210–2	
484	3-Amino-1-phenyl-1,2,4-triazole		150		168 118(di)	
485	9-Aminoanthracene		150		273–4 159(di)	
486	3,5-Dibromo-4-aminobenzaldehyde (see Table 7B)	98	150			
487	2-Methyl-6-hydroxyaniline		150			189(N-)
488	2-Hydroxy-1-naphthylamine	99	150dN		235(N-)	248–9(N-)
489	1-Hydroxy-2-naphthylamine	100	150		130(N-)	191(N-)
490	Hypoxanthine (6-hydroxypurine)		150d			
491	4-Aminoindazole		151		145–8(4-N) 202(di)	
492	4-Aminochalcone (ω-(4-aminobenzal)acetophenone)	101	151		179	
493	4-Aminopyrimidine		151–2		202	
494	Pentamethylaniline		151–2		213	
495	4-Nitro-N-methylaniline		152	121	153	112
496	5-Bromo-2-nitroaniline		152		139	
497	3-Fluoro-4-nitroaniline		153		138	

TABLE 9B (Continued)
AMINES (PRIMARY AND SECONDARY)—SOLIDS

| | Melting Point of Derivatives ||||||||
| | | | | | Others |||
	p-Toluene sulfon- amide	Phenyl- thiourea	α-Naph- thyl- thiourea	β-Naph- thyl- thiourea	α-Naph- thyl- urea	2-Nitro- 1,3-indan- dione De- rivative	Picrate	Hydro- chloride	Chloro- platinate or Chloro- aurate
457									
458	266(di)			178		261–3			
459									
460									
461								219	
462									
463									
464									
465									
466									
467	185								
468	174								
469							204		
470									
471									
472									
473									
474									
475									
476	191		187		236		100		
477							183		
478	217								
479							225		
480									
481							213–4		
482									
483		135–6(O-)							
484							220	187	
485									
486									
487									
488							109–10	250–2	
489									
490							250–4d		
491									
492									
493							226		
494									
495									
496									
497									

(Continued)

TABLE 9B (Continued)
AMINES (PRIMARY AND SECONDARY)—SOLIDS

	Name of Compound	Note	MP	Melting Point of Derivatives — Recommended		
				Benzene-sulfonamide	Acetamide	Benzamide
498	4-Hydroxy-3-chloroaniline		153		144(N-) 124(di)	
499	4-Chloro-2-hydroxyaniline		154		140(di)	
500	3-Aminobenzpyrazole		154		177–8(di)	182(di)
501	4-Nitro-2,5-dihydroxyaniline		154		126	
502	2-Nitro-4-hydroxyaniline	102	154		218(N-)	
503	3-Aminoindazole		154		177–8(di)	182(di)
504	4-Aminoquinoline	103	154		178	
505	8-Chloro-5-aminoquinoline		154–5		172–3(hyd) (N-)	
506	3-Aminotriphenylcarbinol		155		164	
507	1-Bromo-4-nitro-2-naphthylamine		155		176	
508	7-Aminoindazole		155–6		161(di)	
509	*l*-6,6'-Diamino-2,2'-dimethylbiphenyl		156		205(di)	172(di)
510	3,3'-Diaminoazobenzene		156 (140)		272(di)	286(di)
511	4-Chloro-3-nitro-1-naphthylamine		156		223	
512	1-Hydroxy-7-naphthylamine		156–8		215(N-)	
513	Carbohydrazide		156–9			
514	5-Amino-1-phenyl-1,2,4-triazole		157			
516	4-Nitrophenylhydrazine		157d		205	193
517	4-Aminopyridine		158		150(anh.)	202
518	4,4'-Diaminodiphenylamine		158		178(4-N) 239(4,4'-di)	
519	2,6-Dimethyl-4-nitroaniline	104	158		178	
520	4',4-Diamino-3,3'-dimethyldiphenylmethane		158–9		224(di)	215(di)
521	2-Aminophenylpropenoic acid		158–9			
522	4-Aminopyridine		158–9		150	202
523	3-Aminofluorenone		158–9		215(N-)	
524	Ajmaline		158–60			214–16d
525	8-Nitro-2-aminoquinoline		159		211	166
526	3-Amino-1,2,4-triazole		159		295–300d(N-) 190–1(di)	
527	4'-Nitro-2-aminobiphenyl		159		199	
528	3-Amino-2-methylquinoline		159–60		165	161
529	3-Nitro-4-hydroxy-1-naphthylamine		160		250 (238)	330
530	2-Amino-4-methylpyrimidine	105	160		248	
531	5-Nitro-2,4-dihydroxyaniline		160–1		261(N) 176(tri)	
532	4-Nitro-1,3-diaminobenzene	106	161 (157)		246(di)	222(di)
533	3-Hydroxy-4-methylaniline	107	161		225(N-)	
534	2-Hydroxy-4-methylaniline	108	162		171(N-)	169(N-)
535	4-Nitro-3-hydroxyaniline	109	162 (158)		221(N-)	

TABLE 9B (Continued)
AMINES (PRIMARY AND SECONDARY)—SOLIDS

	Melting Point of Derivatives							
				Others				
	p-Toluene-sulfon-amide	Phenyl-thiourea	α-Naph-thyl-thiourea	β-Naph-thyl-thiourea	α-Naph-thyl-urea	2-Nitro-1,3-indan-dione De-rivative	Picrate	Hydro-chloride	Chloro-platinate or Chloro-aurate
498	116–7(O-)								
499								226–7	
500									
501									
502									
503									
504							274		
505								275	
506									
507									
508									
509									
510									
511									
512									
513							191–2(di)		
514							175		197(Pt)
515									
516							119–20		
517							215–6		
518									
519									
520							192–3		
521									
522							215–6	240	251(Pt)
523									
524							126–7	133–4	
525							257		
526							231d	153	
527	163								
528							235		
529									
530									
531									
532	169(di)								
533									
534									
535									

(Continued)

TABLE 9B (Continued)
AMINES (PRIMARY AND SECONDARY)—SOLIDS

	Name of Compound	Note	MP	Melting Point of Derivatives Recommended		
				Benzene-sulfon-amide	Acetamide	Benzamide
536	4-Aminoacetanilide		162		304	
537	6-Hydroxy-2,4-dimethylaniline	110	163		186–7	211
538	2-Aminofluorenone (see Table 21B)		163		227–8	
539	4-(2-Aminoethyl)-phenol	111	164			162(N-)
540	4-Aminophenacyl alcohol	112	165		176–7	
541	3-Bromo-4-hydroxyaniline	113	165 (155)		157(N-)	184–5(N-)
542	2,7-Diaminonaphthalene		166		261(di)	267(di)
543	4'-Iodo-4-aminobiphenyl	114	166–7 (159)			
544	3,4-Diaminophenol	115	167–8		205–7(di)	203(di)
545	4,4'-Diaminotriphenylcarbinol		168(s.h.) 175(r.h.)		267(di)	
546	4-Amino-2-phenylquinoline	116	168		108 117(di)	182
547	4-Amino-2-methylquinoline		168			
548	3-Bromo-2-naphthylamine		168		172	176
549	2-Chloro-3,5-dinitroaniline		168		153	
550	4-Methyl-2,6-dinitroaniline		168 (172)		195	186
551	6-Aminocoumarin		168–70	159	216–7	173
552	Picramic acid	117	169		201	230
553	Benzimidazole		170		113–4	
554	Di-(2-naphthyl)amine	118	170.5			
555	4-Methyl-3,5-dinitroaniline		171		223	
556	4-Amino-3-nitrobiphenyl		172–3		132	143
557	1-Phenylsemicarbazide	119	172		196–7	
558	3,3'-Diaminobenzophenone	120	173		226–7(di)	
559	2-Hydroxy-4,5-dimethylaniline	121	173–5		191(N-)	195–6(N-)
560	2-Hydroxyaniline	122	174	141	201 (209)	167(N-)
561	3-Aminobenzoic acid (see Tables 4 and 5B)		174		250	
562	4,4'-Sulfonyldianiline		174–6		286(di)	
563	4-Hydroxy-3-methylaniline	123	175		179	194(O,N-di)
564	6-Amino-5,7-dimethylquinoline		175		212	
565	1,5-Diphenylcarbohydrazide		175–7		98	
566	2,2'-Diaminostilbene(trans)		176 (168)		304(di)	
567	5,5-Dimethylhydantoin		176–8		192(3-N) 186–7 (1,3-di)	
568	1,4-Diphenylsemicarbazide	124	177		192	
569	2-Methylbenzimidazole		177–8		85–6	
570	5-Nitro-6-aminoquinoline		178 (174)			168
571	6-Aminothymol	125	178–9		174	178–9
572	4-Hydroxy-2-methylaniline		179		130(N-)	92(O-)
573	N-Methylanthranilic acid (see Table 5B)		179			
574	2,4-Dinitroaniline		180		120	202 (220)

TABLE 9B (Continued)
AMINES (PRIMARY AND SECONDARY)—SOLIDS

	Melting Point of Derivatives							
					Others			
	p-Toluene-sulfon-amide	Phenyl-thiourea	α-Naph-thyl-thiourea	β-Naph-thyl-thiourea	α-Naph-thyl-urea	2-Nitro-1,3-indan-dione Derivative	Picrate	Hydro-chloride	Chloro-platinate or Chloro-aurate
536									
537									
538									
539							206		
540									
541									
542							210(di)		
543								295	
544									
545									
546									
547							197–9		223(Pt)
548								170	
549									
550									
551									
552	191								
553							225–6		
554							164		
555									
556									
557									
558									
559									
560	146	146							
561									
562									
563									
564							182		
565								125d	
566							209	267	
567									
568									
569									
570							270		
571									
572								215	
573									
574	219								

(Continued)

TABLE 9B (Continued)
AMINES (PRIMARY AND SECONDARY)—SOLIDS

	Name of Compound	Note	MP	Melting Point of Derivatives		
				Recommended		
				Benzene-sulfon-amide	Acetamide	Benzamide
575	4-Chloro-1-aminoanthraquinone		180		203–4	
576	4-Hydroxy-2,6-dimethylaniline		181		178–80	
577	5-Aminoindazole		181		165	
578	3-Acetoindazole (see Table 21B)		182		123	
579	1-Aminopyrene		182		276	
580	4-Aminobenzamide	126	183		274–5	
581	4-Amino-2,6-dimethylpyrimidine	127	183			
582	4-Hydroxyaniline	128	184s	125	168(N-)	216–7N
583	Phenothiazine		184–5		197	
584	3-Hydroxy-1-naphthylamine	129	185		179(N-)	309N
585	2-Nitro-5-hydroxyaniline		185–6		266(N-)	
586	6,6'-Diamino-3,3'-dimethyltriphenyl-methane		185–6		217(di)	196(di)
587	4-Amino-2,6-dimethylpyridine		186		113	
588	4'-Hydroxy-4-aminoazobenzene		186		236–7 203(N-)	
589	4'-Amino-4-methylbenzophenone	130	186–7		155	
590	6-Amino-2-methylquinoline		187–8		168–9	
591	4-Chloro-3,5-dinitroaniline		188		228d	
592	4-Aminobenzoic acid (see Tables 4 and 5B)		188	212	251d	278
593	4-Amino-5-methyl-2,4 diethyl-pyrimidine		189		59	
594	1-Amino-3-nitroguanidine		189		194	196.5
595	2,4,6-Trinitroaniline	131	190	211	230	196
596	1,5-Naphthalenediamine		190		360(di)	
597	6-Hydroxy-1-naphthylamine	132	190 (185)		218(N-)	152(N-)
598	4,4'-Diaminoazobenzene		190		275	
599	3-Acetoindole (see Table 21B)		190–1		151	
600	3-Nitro-4-aminobenzaldehyde		191		155	
601	1,5-Dinitro-2-naphthylamine		191		201	
602	5-Hydroxy-1-naphthylamine		192			276(O,N-di)
603	dl-2,2'-Diamino-1,1'-binaphthyl		193		235–6(di)	235(di)
604	1-Aminocarbazole		193		186(di)	242
605	4-Amino-2,6-dibromophenol		193–4 (190)		173–4	
606	6-Nitro-8-aminoquinoline	133	194		224	
607	2-Aminonicotinamide	134	195			
608	1-Amino-4-nitronaphthalene		195	158 (173)	190	224
609	6-Methyl-2-(4-aminophenyl)benzthiazole		195		227	245
610	2,6-Dichloro-4-nitroaniline		195–6		214–5 142.5(di)	
611	2-Amino-4,6-dimethylpyrimidine	135	197			
612	4-Nitro-1,2-diaminobenzene	136	198 (200–2)		205N	235(di)
613	4-Amino-2-naphthol	137	198d		179(N-)	188N
614	6-Amino-1-naphthol	138	199.5		130N	203N
615	2,4-Dinitrophenylhydrazine		199–200		197–8	206–7

Tables 791

TABLE 9B (Continued)
AMINES (PRIMARY AND SECONDARY)—SOLIDS

	Melting Point of Derivatives							
				Others				
	p-Toluene-sulfon-amide	Phenyl-thiourea	α-Naph-thyl-thiourea	β-Naph-thyl-thiourea	α-Naph-thyl-urea	2-Nitro-1,3-indan-dione De-rivative	Picrate	Hydro-chloride	Chloro-platinate or Chloro-aurate
575									
576									
577									
578									
579									
580									
581							214		
582	253	150							
583	155–6								
584									
585									
586									
587							194–5		250d(Pt)
588									
589									
590									
591									
592									
593									
594									
595									
596									
597							170		
598									
599							183		
600									
601	182								
602									
603							185		
604									
605									
606									180(Pt)
607									
608	185								
609									
610									
611							230	181	225(Pt)
612									
613									
614									
615									

(Continued)

TABLE 9B (Continued)
AMINES (PRIMARY AND SECONDARY)—SOLIDS

	Name of Compound	Note	MP	Melting Point of Derivatives		
				Recommended		
				Benzene-sulfon-amide	Acetamide	Benzamide
616	4-Bromo-2-nitro-1-naphthylamine		200		232	
617	2,5-Diaminobenzoic acid	139	200d		262 (2,5-di)	
618	2-Hydroxy-4-nitro-5-methylaniline		200		242(N-)	
619	3-Nitro-4-aminopyridine		200			
620	4-Amino-4'-nitrobiphenyl		200	174	264 (240)	
621	2-Amino-5-nitrobenzaldehyde	140	200		160–1	181–2
622	7-Amino-2-naphthol	141	201		232N	243–6N
623	2-Amino-5-nitrophenol (see Table 29B)	142	201–2		189N	
624	Isatin	143	201–2		141	
625	4-Chloro-2-nitro-1-naphthylamine		202		219	
626	4,4'-Diamino-1,1'-binaphthyl		202		363(di)	320(di)
627	1-Aminothioxanthone		202–3		233–4	
628	8-Chloro-4-nitro-1-naphthylamine		203		167	
629	4,4',4''-Triaminotriphenylmethane	144	203		201(tri)	
630	3-Amino-5-phenylacridine		204		256	246
631	2-Amino-1,4-naphthoquinone		204–5		202	
632	1-Amino-2-methylanthraquinone		205		176–7 203–6(di)	
633	4,4',4''-Triaminotriphenylcarbinol		205		192(tri)	
634	7-Hydroxy-1-naphthylamine (see Table 29B)	145	207		167N	211N
635	4-Aminopyrene		207		229	
636	1-Amino-4-hydroxyanthraquinone (see Table 29B)		207–8		194 (O,N-di)	
637	4,6-Diaminoisophthalaldehyde (see Table 7B)		208		270(mono) 280(di)	
638	3-Nitroanthranilic acid (see Table 5B)		208–9 (204)		180–1	
639	6-Aminobenzpyrazole		210		248(6-N) 184–5(di)	
640	4,4'-Diamino-2,5,2',5'-tetramethyl-triphenylmethane		210		217(di)	250(di)
641	6-Amino-2-naphthol	146	213		223N	234N
642	2-Amino-4,5-dimethylpyrimidine		214–5			
643	4-Amino-2-chlorobenzoic acid		214–5		206	
644	5-Bromo-4-hydroxy-2-methylaniline		215 (205)		171–2(di)	229(di)
645	2-Hydroxy-6-nitroaniline		216		172(N-)	
646	3,4-Diaminopyridine		218–9			222–3(di)
647	3,6-Dimethylcarbazole	147	219		129	
648	4-Aminophenoxyacetic acid (see Table 5B)		220d			
649	3-Methyl-2-pyrazolin-5-one		220–2		140	
650	10-Hydroxy-2-aminophenanthrene	148	221		182N	225N
651	3-Aminothioxanthone		221–2		236–7	
652	4-Nitro-5-aminoacenaphthene		222		252	233
653	5-Chloro-4-hydroxy-2-methylaniline		223–5 (204–5)		162(di)	220(di)

TABLE 9B (Continued)
AMINES (PRIMARY AND SECONDARY)—SOLIDS

| | | | | Melting Point of Derivatives | | | | |
| | | | | Others | | | | |
	p-Toluene-sulfon-amide	Phenyl-thiourea	α-Naph-thyl-thiourea	β-Naph-thyl-thiourea	α-Naph-thyl-urea	2-Nitro-1,3-indan-dione Derivative	Picrate	Hydro-chloride	Chloro-platinate or Chloro-aurate
616									
617									
618									
619							197–8	258–9	256(Pt)
620									
621									
622									
623									
624									
625									
626							147		
627									
628									
629									
630									
631									
632	218								
633									
634									
635									
636	197								
637									
638									
639								230	
640									
641									
642							250		227(Pt)
643								185	
644									
645	136(O-)								
646							235–7		231(Pt)
647							192		
648								223–5	
649									
650									
651								230	
652									
653									

(Continued)

TABLE 9B (Continued)
AMINES (PRIMARY AND SECONDARY)—SOLIDS

	Name of Compound	Note	MP	Melting Point of Derivatives — Recommended		
				Benzene-sulfonamide	Acetamide	Benzamide
654	2,6-Dimethylcarbazole	149	224			
655	1,8-Dinitro-2-naphthylamine		226		238	
656	N,N′-Di-2-naphthyl-1,4-diaminobenzene		228–30 (235)		210(di)	220(di)
657	4,4′-Diaminostilbene(*trans*)		231		353(di)	352(di)
658	3-Hydroxy-2-naphthylamine	150	234		244*N*	233–5*N*
659	3-Aminosalicyclic acid		235	194		189
660	2-Aminoanthracene	151	238		240	
661	2,4-Dinitro-1-naphthylamine		242		259	252
662	4-Hydroxy-2,5-dimethylaniline		242		177–9	
663	3-Bromo-1-aminoanthraquinone		243		214	
664	4,4′-Diaminobenzophenone	152	244		237(di)	
665	N,N′-Diphenylbenzidine		244–5			
666	2,2′,4,4′,6,6′-Hexanitrodiphenylamine (dipicrylamine)		244–6 (242)		240*d*	
667	Carbazole	153	246		69	98
668	1-Aminoanthraquinone		248–50 (252)		209–11 (215)	255
669	2-Chloro-4-nitro-1-naphthylamine		249		231	
670	4-Aminothioxanthone		249–50		273	
671	3-Aminocarbazole	154	254*N*		217(3-N)	250(3-NO)
672	3,4,5-Triamino-1,2,4-triazole	155	257*d*		240(tri)	
673	2,7-Diaminocarbazole	156	260		320(di)	
674	1,8-Diaminoanthraquinone		262		284(di)	324(di)
675	1,3-Dimethylxanthine		264 (270–4)		158(7-NO)	202(7-N)
676	1,4-Diaminoanthraquinone		268		271(di)	284(di)
677	3-Nitro-2-aminofluorenone		269		245–6	
678	1-Amino-2-naphthol (see Table 29B)	157	276*d*		235*N*	248–9*N*
679	1,1′-Diamino-2,2′-binaphthyl		281		230(di)	278(di)
680	1,7-Diaminoanthraquinone		290		283(di)	325(di)
681	2,7-Diaminofluorenone (see Table 21B)		290		222(di)	
682	1,6-Diaminoanthraquinone		292		295(di)	275(di)
683	5-Nitro-1-aminoanthraquinone		293		275	237
684	4-Nitro-1-aminoanthraquinone		296		256–8	
685	2-Aminoanthraquinone		305–8	271	262(mono) 257(di)	228
686	2-Amino-3-bromoanthraquinone		307		259 (217)	279
687	1,5-Diaminoanthraquinone		319		317(di)	315
688	2,3-Diaminoanthraquinone		353			315(di)

NOTES FOR TABLE 9B: *p*- AND *s*-SOLID AMINES

1. N-Acetyl derivative, has a boiling point at 231/737 mm; B, HCl mp 128–9 and B$_2$, HCl mp 86–8.
2. Picrate of the N-acetyl derivative, mp 180.
3. N-Acetyl derivative, mp 102–3; N-benzoyl derivative, mp 96.
4. Acid succinate derivative, mp 123. Salt with acetic acid, mp 65.
5. Oxalate, mp 103.

TABLE 9B (Continued)
AMINES (PRIMARY AND SECONDARY)—SOLIDS

| | | | | | Melting Point of Derivatives | | | |
| | | | | | Others | | | |
	p-Toluene-sulfon-amide	Phenyl-thiourea	α-Naph-thyl-thiourea	β-Naph-thyl-thiourea	α-Naph-thyl-urea	2-Nitro-1,3-indan-dione Derivative	Picrate	Hydro-chloride	Chloro-platinate or Chloro-aurate
654							162		
655	221								
656							217(di)		
657									
658									
659								150d	
660									
661	166								
662									
663	227								
664									
665									
666									
667	138						185		
668	228–9								
669									
670									
671									
672							282–4		
673									
674									
675									
676									
677									
678									
679									
680									
681							230(di)		
682									
683									
684									
685	304								
686									
687									
688									

NOTES FOR TABLE 9B: p- AND s-SOLID AMINES (Continued)

6. N-Benzal derivative, mp 63.
7. N-Acetyl derivative, boils at 238.
8. Oxalate, mp 120.
9. Oxalate, mp 194–6d.
10. N-Benzal derivative, mp 108.
11. O,N-Diacetyl derivative, mp 128–9.
12. N,S-Diacetyl derivative, mp 125; See Table 21B.
13. Propionyl derivative, mp 65.
14. N-Nitroso derivative, mp 171.

NOTES FOR TABLE 9B: *p*- AND *s*-SOLID AMINES (Continued)

15. N,N'-Dinitroso derivative, mp 148.
16. Yields the N-nitroso derivative readily, mp 67. The benzoyl derivative has one modification that melts at 107–9 and this form melts, resolidifies and transforms into the stable modification between 130 and 140, and then melts at 180.
17. N-Chloroacetyl derivative, mp 150.
18. Picryl chloride derivative, mp 115–6; trinitrobenzene adduct, mp 152; trinitroaniline derivative, mp 166.
19. Dinitroso derivative, mp 71.
20. Trinitrobenzene adduct, mp 139.
21. The di-substituted derivatives are difficult to prepare: diacetamide, mp 191; dibenzamide, mp 240; 1,6-dihydroxynaphthalene adduct, mp 125.
22. Propionyl derivative, mp 139; N-benzal derivative, mp 93–4; trinitrobenzene adduct, mp 192.
23. The mono-2-N- derivative.
24. The 1-N-benzamide melts at 68–70, and the 2-N-benzamide at 146.
25. Hydrolyzes in hot water to ammonia and glycine.
26. N-*p*-Nitrobenzoyl derivative, mp 269; N-benzal derivative, mp 86.
27. N-Propionyl derivative, mp 129.
28. The yellow-orange form melts at 69–70, and the red form at 90–1; The N-acetyl derivative of both forms melts at 130–1.
29. N-Benzal derivative, mp 67–8.
30. The N-acetyl derivative forms a phenylcarbazone, mp 209.
31. Methiodide, mp 178.
32. N-Nitroso derivative, mp 99–100.
33. Recrystallized from alcohol, the melting point is 66, and from ligroin, mp 75.
34. 4-Nitrobenzoyl derivative, mp 139–40.
35. The acetyl derivative is easily prepared.
36. Propionyl derivative, mp 169.
37. N-Benzal derivative, mp 55.
38. The 4-*p*-toluenesulfonamide melts at 140; the 3-N-acetamide at 95; the 4-N-acetamide at 131–2; the 3-N-benzamide at 158; the 4-N-benzamide at 193–4; the 3-N-benzenesulfonamide at 134–5; and the 4-N-benzenesulfonamide at 146–7.
39. Dipropionamide, mp 190–1.
40. N-Chloroacetyl derivative, mp 116.
41. See the derivatives with acids (methylenebisanilides); dibenzal derivative, mp 130.
42. O,N-Diacetyl derivative, mp 118; O,N-dibenzoyl derivative, mp 206.
43. O,N-Dibenzoyl derivative, mp 113–4.
44. 1-N-Benzenesulfonamide, mp 215*d*.
45. Both the 2-N-, and the 4-N-benzenesulfonamides melt at 138; 2-N-acetamide mp 140; 4-N-acetamide, mp 162; 4-N-benzamide, mp 142; 4-N-*p*-toluenesulfonamide, mp 160.
46. N-Chloroacetyl derivative, mp 171–2.

Forms two diacetyl derivatives, mp 75 and 65.
47. 1,6-Naphthalenediol adduct, mp 95.
48. Trinitrobenzene adduct, mp 196.
49. 3-N-Acetamide, mp 211; 4-N-acetamide, mp 163; 3-N-benzamide, mp 186; 4-N-benzamide, mp 221.
50. Boils at 146; reacts with acetaldehyde and sodium nitroprusside to give an intense blue color which changes to rose-red in acetic acid. Monoacetamide, mp 52; monobenzamide, mp 75.
51. β-N-*p*-Nitrobenzoyl derivative, mp 199*d*.
52. The N-*p*-nitrobenzoyl derivative of the *dl* form melts at 184. The *d*- and *l*-isomers melt at 52, and have N-*p*-nitrobenzoyl derivative, mp 175–6, and a hydrochloride, mp 171–2; O,N-dibenzoyl derivative of the *dl* form melts at 167–8.
53. Picrolonate, mp 195.
54. Picrolonate, mp 253; styphnate, mp 224.
55. The oxime of the N-acetyl derivative, melts at 180.
56. The acetophenone derivative, mp 112.
57. Picrate of the N-acetyl derivative, mp 194; picrate of the N-benzoyl derivative, mp 164.
58. *p*-Nitrobenzoyl derivative, mp 214.
59. Benzal derivative, mp 104.
60. O,N-Dibenzoyl derivative, mp 136.
61. N,N'-Dibenzal derivative, mp 172–3.
62. Hydrolyzes to anthranilic acid mp. 147
63. Hydrobromide, mp 130.
64. Oxime, mp 146.
65. Propionyl derivative, mp 128.
66. O,N-Diacetyl derivative, mp 200.
67. Methiodide, mp 199.
68. Boils at 162; N,N'-Dinitroso derivative, mp 95.
69. Methiodide, mp 238; ethiodide, mp 202.
70. Boils at 162; N,N'-dinitroso derivative, mp 174; di-l-naphthalenesulfonyl derivative, mp 269–70.
71. Trinitrobenzene adduct, mp 108.
72. This is the O,N,N-triacetyl derivative.
73. 1-N-Benzamide, mp 186; 1-N-*p*-toluenesulfonamide, mp 187–8.
74. Benzal derivative, mp 122.
75. Two forms of the HCl salt: mp 81–3 and 156–7.
76. The acetyl derivative is easily prepared.
77. O,N-Diacetyl derivative, mp 101; N-benzoyl derivative, mp 174; O-benzoyl derivative, mp 153.
78. O,N-Diacetyl derivative, mp 210.
79. Propionamide, mp 139.
80. Hydrobromide, mp 206; trinitrobenzene adduct, mp 156–7; propionyl derivative, mp 170.
81. Recrystallized from alcohol, the monoacetamide melts at 38–9 but from benzene it melts at 126.
82. N-*p*-Nitrobenzoyl derivative, mp 208;

NOTES FOR TABLE 9B: p- AND s-SOLID AMINES (Continued)

N-p-aminobenzoyl derivative, mp 241; N-p-aminobenzenesulfonyl derivative, mp 255.
83. Methiodide, mp 228; ethiodide, mp 210.
84. The hydrated monoacetamide, mp 103; monobenzamide, mp 198–200.
85. p-Nitrobenzoyl derivative, mp 204.
86. Methiodide, mp 247; ethiodide, mp 232; trinitrobenzene adduct, mp 186.
87. N-Chloroacetyl derivative, mp 189; N-p-aminobenzenesulfonyl derivative, mp 239.
88. Reduces to 2,4-diaminophenol, mp 78–80.
89. O,N-Diacetyl derivative, mp 145; O,N-dibenzoyl derivative, mp 190; N-propionamide, mp 95–6; O,N-dipropionyl derivative, mp 91–2.
90. 1-N-Acetamide, mp 162; 4-N-acetamide, mp 189.
91. Methiodide, mp 220.
92. O-Acetyl derivative, mp 73–4.
93. 1,2-Dibenzoyl derivative, mp 195; the 1,1,2,4-tetrabenzoyl derivative, mp 148.
94. Monoacetamide, mp 162–3; monobenzamide, mp 128; 3,5-dinitrobenzamide, mp 178. 1,6-Naphthalenediol adduct, mp 170.
95. The hydrate decomposes between 80 and 90 to yield the anhydrous compound, mp 142–3; The N-acetyl derivative when crystallized from acetic acid melts at 278–9; the N-benzoyl derivative decomposes above 200; O-p-toluenesulfonyl derivative, mp 122.
96. See Table 21-B.
97. O,N Diacetyl derivative, mp 128–9.
98. Anil, mp 99; Oxidizes to the acid, mp 330.
99. This compound exists in two forms, melting at 150d and 175d these forms yield the following derivatives: N-acetyl derivatives, mp 225 and 235; N-benzoyl derivatives, mp 232 and 248–9; O,N-diacetyl derivatives, mp 202 and 206; O,N-dibenzoyl derivatives, mp 226 and 235.
100. O,N-Diacetyl derivative, mp 117; O-N-dibenzoyl derivative, mp 180.
101. Oxime, mp 139.
102. O,N-Diacetyl derivative, mp 146; O-p-toluenesulfonyl derivative, mp 134; O-m-nitrobenzoyl derivative, mp 184.
103. Methiodide, mp 224; ethiodide, mp 232.
104. Tetra-acetyl derivative, mp 119.
105. N-p-Aminobenzenesulfonyl derivative, mp 248.
106. 1-N-Acetyl derivative, mp 200.
107. N-Diacetyl derivative, mp 132–3; N-chloroacetyl derivative, mp 154–5; O-p-toluenesulfonyl derivative, mp 111–2.
108. O,N-Dibenzoyl derivative, mp 162.
109. O,N-Diacetyl derivative, mp 149.
110. O,N-Diacetyl derivative, mp 87–8; O,N-dibenzoyl derivative, mp 148–9.
111. O,N-Dibenzoyl derivative, mp 172.
112. O,N-Diacetyl derivative, mp 162; O-acetyl derivative, mp 130; O-benzoyl derivative, mp 188; phenylhydrazone, mp 199.
113. O,N-Dibenzoyl derivative, mp 192.
114. N-Benzal derivative, mp 209.
115. N,N',O-Tribenzoyl derivative, mp 225.
116. Methiodide, mp 274; ethiodide, mp 244.
117. O-Acetyl derivative, mp 193; O-benzoyl derivative, mp 220; N-p-nitrobenzoyl derivative, mp 300.
118. Trinitrobenzene adduct, mp 147.
119. 1-Nitroso derivative, mp 126–7d.
120. Oxime, mp 177–8.
121. O,N-Diacetyl derivative, mp 157; O,N-dibenzoyl derivative, mp 152–3.
122. O-Benzoyl derivative, mp 185.
123. O-p-Toluenesulfonyl derivative, mp 109–10.
124. 1-Nitroso derivative, 174–5.
125. Triacetyl derivative, mp 91; O,N-dibenzoyl derivative, mp 166–7.
126. N-Chloroacetyl derivative, mp 241–3.
127. N-p-Nitrobenzenesulfonyl derivative, mp 188–90.
128. Also named, p-aminophenol. The monoacetyl derivative, melts at 168 and the diacetyl derivative, at 150d; the N-benzoyl derivative has been reported to melt at 216–7, 205, and 227; O-benzoyl derivative, mp 235; O,N-dibenzoyl derivative, mp 234.
129. O,N-Dibenzoyl derivative, mp 309; O-p-toluenesulfonyl derivative, mp 137.
130. Phenylhydrazone, mp 163.
131. Naphthalene adduct, mp 168.
132. O,N-Diacetyl derivative, mp 187; O,N-dibenzoyl derivative, mp 223.
133. Methiodide, mp 176.
134. N-4-Aminobenzenesulfonyl derivative, mp 210.
135. N-p-Nitrobenzenesulfonyl derivative, mp 220–1.
136. 1-N-Acetyl derivative, mp 205; 2-N-acetyl derivative, mp 195; N,N-1,2-diacetyl derivative, mp 255 (227); 2-N-benzoyl derivative, mp 217–8.
137. This is the O,N-dibenzoyl derivative.
138. O,N-Diacetyl derivative, mp 130; N-benzoyl derivative, mp 203; O,N-dibenzoyl derivative, mp 230.
139. 5-N-Acetyl derivative, mp 240d; ethyl ester, mp 51.
140. Oxime, mp 203.
141. N-Acetyl derivative, mp 232; O,N-diacetyl derivative, mp 156; N-benzoyl derivative, mp 243–6; O-benzoyl derivative, mp 177; O,N-dibenzoyl derivative, mp 181; N-phenacyl derivative, mp 160.
142. O,N-Diacetyl derivative, mp 187.
143. 2-Oxime, mp 200–1; 3-oxime, mp 225.
144. Trinitrobenzene adduct, mp 140.

NOTES FOR TABLE 9B: p- AND s-SOLID AMINES (Continued)

145. O,N-Diacetyl derivative, mp 178; O,N-dibenzoyl derivative, mp 208.
146. This is the O,N-dibenzoyl derivative; O,N-diacetyl derivative, mp 220.
147. N-Nitroso derivative, mp 106.
148. This is the O,N-di- derivative.
149. N-Nitroso derivative, mp 113.
150. O,N-Diacetyl derivative, mp 188; O,N-dibenzoyl derivative, mp 184.
151. Trinitrobenzene adduct, mp 169.
152. Phenylhydrazone, mp 240.
153. Trinitrofluorenone derivative, mp 174.
154. Sinters at 240–3 and then melts at 254; diacetyl derivative is reported to melt at 200, and a triacetyl derivative, at 175.
155. A hydrated salt with acetic acid melts at 175.
156. N,N′-Dibenzal derivative, mp 290.
157. O,N-Diacetyl derivative, mp 206; O,N-dibenzoyl derivative, mp 235; N-chloroacetyl derivative, mp 192–3.

REFERENCES FOR TABLES 9A AND 9B

The references for Tables 9 and 10 are grouped together under Table 10 on p. 807.

TABLE 10A
AMINES (TERTIARY)—LIQUID

	Name of Compound	Note	BP	Melting Point of Derivatives					
				Recommended		Others			
				Picrate	Quaternary Methyl Iodide	Methyl p-Toluene Sulfonate	Chloroplatinate	Chloroaurate	Hydrochloride Salt
1	Trimethylamine	2	3	216 (225)	230d				
2	Dimethylethylamine		37.5	193	300–1				221
3	Diethylmethylamine		66–7	185	298–9		231		179
4	N-Methylpyrrolidine		78–80	221			233		
5	Triethylamine	1,3	89	173	280–1				
6	1,2-Dimethylpyrrolidine		96	235			223		
7	1,3-Dimethylpyrrolidine	4	96–7				58–9		
8	N-Methylpyrrole	10	112						
9	Pyridine	1,5	116	167	117	139	241N		
10	1,2,5-Trimethylpyrrolidine		116	163	310				
11	N-Methylmorpholine		116	225	246		197–8		
12	N-Methylmorpholine		117	225	246		199		
13	2-Dimethylamino-diethyl ether		121	119–21	160–5				
14	Dimethylaminoacetone	6	123				176		
15	N-Methylpyrazole		127	148	190		196–8		
16	2-Methylpyridine	1,7	129	166	230	150	195		200
17	N-Ethylpiperidine		131	168			202		
18	N,N-Dimethylpiperazine		132	280					252–3 (hyd.)
19	2-Dimethylaminoethanol	8	135	96–7 (anh.)			195 (190)		
20	1,3-Dimethylpyrazole		136	138	256				
21	2-Methylpyrazine	9	136–7	145–6	129–30				
22	N-Ethylmorpholine		139	189			197–8		
23	4-Methylpyrimidine	11	141–2	131–4					
24	2,6-Dimethylpyridine	12	144	161	233		210		230–1
25	3-Methylpyridine	13	144	147	92 (36)		202	182–4	
26	N,N′-Tetramethyl-1,3-diaminopropane		144	205			246–7d		
27	4-Methylpyridine	1	145.4	167	149–50		231	205	
28	4-Chloropyridine		147–8	146			202		142
29	1,4-Dimethylpyrazole		148	165					
30	1,5-Dimethylpyrazole		148–55	172	251		253		
31	3-Chloropyridine		149	135			168		160
32	2-Ethylpyridine		149	107.8			167		
33	N-Propylpiperidine		152	121 (108)	181–2		179		212–3
34	2,5-Dimethylpyrazine		155	157					
35	Tri-n-propylamine	14	156.5	116	207–8				
36	2,4-Dimethylpyridine	1	157	181			220		
37	2,5-Dimethylpyridine		157	169			214		
38	2,3-Dimethylpyrazine		156	150			150d		
39	1,3,4-Trimethylpyrazole		160	164					
40	2,3-Dimethylpyridine	16	160.7	188			216		
41	2-Ethyl-6-methylpyridine		160–1	127–8			188–90d	128–9	
42	2-Diethylaminoethanol	15	163						
43	2-Tropene		163	285			217		
44	3,4-Dimethylpyridine	17	164	163			205		
45	3-Ethylpyridine		165	128			208–9		
46	4-Ethylpyridine		167.7	169			213		

(Continued)

TABLE 10A (Continued)
AMINES (TERTIARY)—LIQUID

	Name of Compound	Note	P	Recommended		Others			
							Melting Point of Derivatives		
				Picrate	Quaternary Methyl Iodide	Methyl p-Toluene Sulfonate	Chloroplatinate	Chloroaurate	Hydrochloride Salt
47	2-Chloropyridine	44	168 (170)			120			
48	d,l or l-Coniine		166.5				160		
49	1,4-Bis(dimethylamino)butane		167	199					
50	Tropane		167	281			230		
51	3-Bromopyridine	45	170 (174)	154	165	156	175		
52	3,5-Dimethylpyridine	18	170–1	238			255		
53	6-Ethyl-3-methylpyridine		171	144					
54	2,4,6-Trimethylpyridine		172	156			223	112	
55	2-Ethyl-3-methylpyridine		172–3	138–40					
56	2-Ethyl-4-methylpyridine		172–3	120–1			176d	80	
57	5-Ethyl-2-methylpyridine		174–6	164–5			180–1	87	
58	2,3,6-Trimethylpyridine		176–8	146			250–2 (137)		
59	4-Chloro-2,6-dimethylpyridine		178	167 (156)	233–4				
60	4,5-Dimethylpyrimidine		178				242–3	120d	
61	4-Ethyl-2-methylpyridine		179–80	141–2			203	90	
62	N,N-Dimethylbenzylamine	20	181	93	179		192		
63	Pyridine-2-aldehyde (see Table 7A)		181						
64	2,3,5-Trimethylpyridine (2,3,5-Collidine)		184	183 (179)			227–8 (147)		
65	N,N-Dimethyl-2-methylaniline	21	185	122	210				
66	4-Ethyl-2,6-dimethylpyridine		186	119–20			210		
67	2,4,5-Trimethylpyridine		188 (165)	159–60			189–90		
68	2,4-Diethylpyridine		187–8	98–100			170–1		
69	5-Ethyl-3-methylpyridine		190	195					
70	3-Diethylamino-1-propanol (see Table 6A)		190		175				
71	2-Butylpyridine		191–3	94			146–7		
72	Methyl 2-pyridyl ketone	22	192	131	161		220		
73	Tri-isobutylamine	23	192						
74	4-Ethyl-3-methylpyridine		192–3	144–5					
75	2,3,4-Trimethylpyridine		192–3	163–4			259 (180)		
76	N,N-Dimethylaniline	24	194	163	228 (220)	161	173d		
77	2-Bromopyridine	46	194	105–6		127			
78	3-Ethyl-4-methylpyridine	34	195–6	148–50			234–5d	140–1	
79	4-tert-Butylpyridine		196–7	121			212–3d	184	
80	2-Ethyl-3,5-dimethylpyridine		198	152			189		
81	2-Methylbenzoazole		200–1	117–8					
82	N-Ethyl-N-methylaniline		201 (209)	134–5 (129)	125				114
83	N-Methyl-2-pyrrolidone		201–3						79–81
84	N,N-Dimethyl-2,5-dimethylaniline		204	158			196		
85	3,4,5-Trimethylpyridine		205–7	174			197–8		

(Continued)

TABLE 10A (Continued)
AMINES (TERTIARY)—LIQUID

	Name of Compound	Note	BP	Melting Point of Derivatives					
				Recommended		Others			
				Picrate	Quaternary Methyl Iodide	Methyl p-Toluene Sulfonate	Chloroplatinate	Chloroaurate	Hydrochloride Salt
86	N,N-Dimethyl-2,4-dimethylaniline	25	205	123–4			219		
87	3-Butylpyridine		205–8	90			187–8	95	126
88	N,N-Diethyl-2-methylaniline		206 (210)	180	224				
89	2-Chloro-N,N-dimethylaniline		207	132	152				
90	Pyridazine		208	169					
91	N,N-Dimethyl-4-methylaniline	26	210	129	219	85			
92	Tributylamine	1,27	211 (216)	106	180 (177)				
93	3,4-Diethylpyridine		211	139			221		
94	N,N-Dimethyl-3-methylaniline		212	131	177				
95	Methyl 4-pyridyl ketone	28	212–4	130			205		
96	N,N-Diethylaniline	29	218	142	104–6				
97	Methyl 3-pyridyl ketone	30	220						
98	N,N-Diethyl-4-methylaniline		229	110	184				
99	2,3,4,5-Tetramethylpyridine		232–4	170–2			210		
100	3-Chloro-N,N-dimethylaniline		234 (252)	145	187				
101	Quinoline	1,31	237	203	133N	126	218		
102	2-Methylbenzothiazole	32	238	154					
103	Isoquinoline (see Table 10B; mp 26.5)	33	243						
104	dl-Nicotine		243	218	219				
105	2-(Dimethylamino)benzaldehyde (see Table 7A)		244	164			205–6		
106	Tri-isopentylamine		245	125					
107	N,N-Dipropylaniline		245	261	156				
108	2-Ethylquinoline		245–6	148	180		188		
109	2-Methylquinoxaline		245–7	215					
110	1-Methylindole		247 (239)	150					
111	2-Methylquinoline (Quinaldine)	1,34	247	191	195	161 (134)	228		224
112	8-Methylquinoline		248	200					
113	l-Nicotine		248	218			275		
114	1-Ethylisoquinoline		250	207–10			200		
115	2,4-Dimethyl-5,6,7,8-tetrahydroquinoline (mp, 20)		250	144	157				195
116	7-Methylquinoline		252	237			223–4		
117	Tripentylamine		257			80d			
118	3-Ethylisoquinoline		257	171–2			180		
119	6-Methylquinoline		258	229	219	154			
120	3-Chloroquinoline		258–60	182	276				
121	3-Methylquinoline	35	259	187	221		249	145	
122	3-Bromo-N,N-dimethylaniline		259	135					
123	4-Methylquinoline	36	263	210–11	173–4		226–30		
124	2,4-Dimethylquinoline	37	264	194	264		229		265–7
125	5-Methylquinoline		264	218	105				

(Continued)

TABLE 10A (Continued)
AMINES (TERTIARY)—LIQUID

	Name of Compound	Note	BP	Recommended			Others		
				Picrate	Quaternary Methyl Iodide	Methyl p-Toluene Sulfonate	Chloroplatinate	Chloroaurate	Hydrochloride Salt
126	N,N,N',N'-Tetramethyl-1,3-diaminobenzene		267		192				
127	2-Phenylpyridine		268–9	175			204		
128	6,8-Dimethylquinoline		269	230 (288–9)			235		246
129	N,N-Dibutylaniline		271	125		180			
130	4-Ethylquinoline	38	272–4	178–80	149		204		
131	N,N-Dimethyl-1-naphthylamine	39	273	145					
132	5,8-Dimethylquinoline	40	273–5	186 (198)			234d		
133	3-Bromoquinoline	41	274–6	190					
134	3,5-Dimethyl-1-phenylpyrazole		275	103	190		186		
135	2-Benzylpyridine (mp 11–14)		276 (742 mm)	140–1			183		
136	6-Bromoquinoline (mp 19)		278	217	273				
137	4,6-Dimethylquinoline	42	280	249	235–9			192	262–4
138	2,4,7-Trimethylquinoline		280–1	232	322		272		
139	4,7-Dimethylquinoline	43	283	224 (230)			227d		
140	3,4-Dimethyl-1-phenylpyrazole		285	283			180		
141	4-Benzylpyridine		287 (742 mm)	140–1			207		
142	8-Chloroquinoline		288		165		235d		
143	8-Bromoquinoline		302–4		281		230		
144	N-Benzyl-N-methylaniline		306	127	164				
145	2-Benzoylpyridine (see Table 21A)		317	130					126–8
146	2,2',2''-Trihydroxytriethylamine (triethanolamine) (see Table 6A) (mp 20–22)		360	73–4			118–9	77–8	177

NOTES FOR TABLE 10A (TERTIARY AMINES) (Continued)

1. β-Resorcyclic acid readily forms salts with many tertiary amines. The following are the melting points (corrected) of these salts for certain amines [Anal. Chem., 23, 1042 (1951)].
   triethyl, 120   4-methylpyridine   125
   tri-n-butyl, 121   2,4-dimethylpyridine 143
   tribenzyl, 141   quinoline,   128
   pyridine, 141   quinaldine,   145
   2-methylpyridine 141.
2. p-Toluenesulfonate salt, mp 162.
3. 2,4-Dinitrobenzoate salt, mp 81–3.
4. Mercuric chloride adduct, mp 200.
5. The chloroplatinate melting point has also been reported at 262–4; 3,5-dinitrobenzoate salt, mp 171 (162); 2,4-dinitrobenzoate salt, mp 141; p-toluenesulfonate salt, mp 160.
6. Oxime mp 99.
7. p-Toluenesulfonate salt, mp 161; ethiodide, 123.
8. O-Acetyl derivative boils at 152–4. See Table 6A.
9. Mercuric chloride adduct, 194–5d.
10. Mercuric chloride adduct, 120–30
11. Mercuric chloride adduct, mp 198.
12. Picrolonate, mp 206.
13. Styphnate, mp 154; oxidizes to nicotinic acid, mp 228.
14. Ethiodide, mp 238.
15. p-Nitrophenylurethan, 60. See Table 6A.
16. Picrolonate, mp 223–5.
17. Picrolonate, mp 227–8.
18. Picrolonate, mp 239–40.
19. Mercuric chloride adduct, 207

NOTES FOR TABLE 10A (TERTIARY AMINES) (Continued)

20. Picrolonate, mp 151.
21. Trinitrobenzene adduct, mp 113.
22. Ethiodide, mp 205. See Table 21A.
23. Methochloroplatinate, mp 174; ethochloroplatinate, mp 170.
24. 3,5-Dinitrobenzoate salt, mp 115; 2,4-dinitrobenzoate, salt, mp 104; *p*-toluenesulfonate salt, mp 133; *p*-nitroso derivative, mp 85.
25. Trinitrobenzene adduct, mp 114.
26. Trinitrobenzene adduct, mp 124.
27. Butiodide, mp 146.
28. Mercuric chloride double salt, mp 183–4. See Table 21A.
29. *p*-Nitroso derivative, mp 82–3.
30. Mercuric chloride double salt, mp 158. See Table 21A.
31. Ethiodide, mp 159; *p*-toluenesulfonate salt, mp 155; styphnate, mp 207–8.
32. Ethiodide, mp 190–2.
33. Ethiodide, mp 148.
34. Ethiodide, mp 233; styphnate, mp 213–4.
35. Ethiodide, mp 220.
36. Ethiodide, mp 141–3.
37. Ethiodide, mp 214; styphnate, mp 212.
38. Mercuric chloride double salt, mp 154.
39. Trinitrobenzene adduct, mp 105–7.
40. Styphnate, mp 184.
41. Oxalate, mp 107.
42. Styphnate, mp 221.
43. Styphnate, mp 272.
44. Mercuric chloride double salt, mp 178.
45. Mercuric chloride double salt, mp 203.
46. Mercuric chloride double salt, mp 185.

TABLE 10B
AMINES (TERTIARY)—SOLID

	Name of Compound	Note	MP	Melting Point of Derivatives				
				Recommended	Others			
				Picrate	Quaternary Methyl Iodide	Methyl p-Toluene Sulfonate	Chloroplatinate	Chloroaurate	Hydrochloride Salt
1	Pyrimidine (bp 124)		21	156				226	
2	5,7-Dimethylquinoline	1	22	249	206				243–4
3	4,6-Dimethylpyrimidine		25	142–3			103–4		
4	Isoquinoline	2	26.5	222	159	163	263d		227
5	2,8-Dimethylquinoline	3	27	183	221				
6	6-Methoxyquinoline		28 (20)		236				
7	4-Nitro-3-methylpyridine		28–9	128–9					
8	4-Bromoquinoline		29–30		265–70				
9	Benzpyrazine (quinoxaline)	4	30		176				
10	5-Methylpyrimidine	5	30	141				209	
11	4-Chloroquinoline		31	212			278		
12	7-Chloroquinoline		31–2		250		253		
13	N-Benzyl-N-ethylaniline		34	120 (112)					
14	4-Chloro-N,N-dimethylaniline	6	36				251		
15	8-Iodoquinoline		36		200		187–91		
16	1,3,5-Trimethylpyrazole		37	147					
17	3-Nitro-2,6-dimethylpyridine		37	143					
18	2-Chloroquinoline		38	122					
19	3-Nitro-2,4,6-trimethylpyridine		38	175	40–1		216	128	
20	2,3-Dimethyl-5,6,7,8-tetrahydroquinoline		38	169	117				
21	7-Methylquinoline		39	237					
22	4-Bromoisoquinoline		40		233				
23	6-Chloroquinoline	7	41		248	143			
24	N,N-Diethyl-4-aminobenzaldehyde (see Table 7B)		41						
25	3-Nitropyridine		41				254		154
26	2,4,8-Trimethylquinoline		42	193	229		273	191	238
27	4-Chloro-2-methylquinoline		42–3	178	212				
28	5-Chloroquinoline		45		231 (172)		255		
29	2,6,8-Trimethylquinoline		46	187–9			206–7		207
30	4-Pyridylmethanol		47	137–9					175
31	N,N-Dimethyl-2-naphthylamine		47	206					
32	2,6-Dimethylpyrazine		47	175–6					
33	3-Chloroisoquinoline		48	177					
34	2,4,5,8-Tetramethylquinoline		48	161					254
35	5-Bromoquinoline		48 (52)		205				225
36	2-Bromoquinoline		49		210				
37	8-Methoxyquinoline		50	143	160				
38	5,6-Dimethylquinoline	8	50	201					
39	N,N,N',N'-Tetramethyl-1,4-diaminobenzene		51		265				
40	Benzylideneaniline		51–2	183 (173)					
41	7-Bromoquinoline		52		240				213
42	2-Iodoquinoline	9	52–3		211–2				
43	3-Iodopyridine		53				211		

(Continued)

TABLE 10B (Continued)
AMINES (TERTIARY)—SOLID

	Name of Compound	Note	MP	Melting Point of Derivatives					
				Recommended		Others			
				Picrate	Quaternary Methyl Iodide	Methyl p-Toluene Sulfonate	Chloroplatinate	Chloroaurate	Hydrochloride Salt
44	8-Chloroisoquinoline		55	191					
45	2,3,8-Trimethylquinoline		55–6	242d					260d
46	4,8-Dimethylquinoline	10	58	229			226–7d	181	
47	3,6-Dimethylquinoline	11	58	253					
48	6,7-Dimethylquinoline	12	58	278					
49	2,3-Dibromopyridine	45	59						
50	2,5-Dichloropyridine	46	59–60						
51	2,6-Dimethylquinoline	13	60	191 (178)	237	175			
52	3-Nitro-N,N-dimethylaniline		60	119	205				
53	2,5-Dimethylquinoline	14	61	223					
54	2,7-Dimethylquinoline	15	61	196					
55	3,4′-Bipyridyl		62	215					
56	4-Acetoacetylquinoline (see Table 21B)		64–5				192–3		180–1
57	2,4,6-Trimethylquinoline (anh.) (hyd., mp 39.5)		65.5	200–1	245–7 (225)				268–72
58	3,5-Dichloropyridine	47	67–8						
59	3,3′-Bipyridyl		68	232					
60	2,3-Dimethylquinoline	16	68–9	230–1	218		230		
61	2,2′-Bipyridyl		69.5	158					
62	3-Bromo-4-chloroquinoline		70	185					
63	N,N-Dibenzylaniline		70	133d	135				
64	3-Chloro-2-methylquinoline		70–1	233d					
65	3,4-Dibromopyridine	48	71–2						
66	4-Benzoylpyridine (see Table 21B)		72	159–60					
67	5-Nitroquinoline		72	215d					214d
68	N-Methyl-2-quinolone	17	72–4						112
69	3,4-Dimethylquinoline	18	73–4	221	191			177	290
70	4-Dimethylaminobenzaldehyde (see Table 7B)		74						
71	8-Hydroxyquinoline	19	75	204	143d				
72	N,N-Dimethyl-4-aminophenol (see Table 29B)		76						
73	N,N-Diethyl-3-aminophenol (see Table 29B)		78		140–3				
74	4,5-Dimethylquinoline	20	78	233					
75	3,7-Dimethylquinoline	21	80	244					
76	8-Bromoisoquinoline	22	80.5		274				
77	3-Acetoacetylpyridine (see Table 21B)		85	155			173–5		92
78	4-Nitroso-N,N-dimethylaniline		85	140					177
79	N,N-Dimethyl-3-aminophenol	23	85						
80	2-Phenylquinoline	24	86	191–2	200				
81	2,3,6-Trimethylquinoline		86–7	212					
82	6,6′-Dimethyl-2,2′-bipyridyl	25	89–90	170–1					
83	Bis(4-dimethylaminophenyl)methane	26	91	185 (mono) 178(di)	214(di)				
84	Tribenzylamine	27	91	190	184				

(Continued)

TABLE 10B (Continued)
AMINES (TERTIARY)—SOLID

	Name of Compound	Note	MP	Melting Point of Derivatives					
				Recommended		Others			
				Picrate	Quaternary Methyl iodide	Methyl-p-Toluene Sulfonate	Chloroplatinate	Chloroaurate	Hydrochloride Salt
85	6-Iodoquinoline		91				265		210
86	2,3,4-Trimethylquinoline		92	216	260		215		274
87	4-Dimethylaminobenzophenone (see Table 21B)		92		188–90				
88	2,5-Dibromopyridine	49	93–4						186–8
89	1-Phenylisoquinoline		95–6	165	242				235
90	6-Bromo-2-methylquinoline	28	96–7		237				
91	4-Iodoquinoline		97		251		185		
92	4,4'-Bis(dimethylamino)benzohydrol	29	98 (102)N		195(di)				
93	5-Iodoquinoline		100		245		263		
94	4,4'-Bis(dimethylamino)triphenylmethane	30	102 (93)		231 220(di)				
95	2,3-Dimethylquinoxaline		106–7	189					
96	2-Hydroxypyridine (see Table 29B)	31	107						
97	5-Nitro-2-methylpyridine		107–8 (112)	132					
98	Acridine	32	111	208	224				
99	6-Phenylquinoline		111	205	194				
100	3,5-Dibromopyridine	50	112		274	219			
101	3-Bromo-6-chloroquinoline		112		287				168
102	5,7-Dibromoquinoline		112		287			240	158
103	Antipyrine	33	113	188					
104	4,4'-Bipyridyl (anh.) (hyd., mp 73)	34	114	257					
105	4-Dimethylaminoazobenzene		117		174				
106	Triphenylamine	35	127						214
107	3-Methyl-1-phenyl-5-pyrazolone	36	127				110 (hyd.)		96 (hyd.)
108	Methyleneaminoacetonitrile	37	129	127					
109	7-Nitroquinoline	38	132–3		231–2				
110	6-Nitroquinoline	39	154		245				
111	4-Nitro-N,N-Dimethylaniline		164						53
112	6-Hydroxypyrimidine	40	164–5N	190					
113	N,N'-Diphenylpiperazine	41	167–8						
114	Phenazine		171	181					
115	2,3,6-Trimethylpyridine		171–2	155–6			217	112–3	
116	4,4'-Bis(dimethylamino)benzophenone (see Table 21B)		174	156	105				
117	6-Hydroxyquinoline		193	236					
118	N,N'-Tetramethylbenzidine		193–4		263				
119	3-Hydroxyquinoline		198	240–5					
120	2-Hydroxyquinoline	42	199						
121	4-Hydroxyquinoline		201						187
122	6-Nitrobenzimidazole		203–5	215d					
123	Purine		215	208					
124	2-Hydroxy-4-methylquinoline		224	165–7			214d		
125	5-Hydroxyquinoline		224		224		230		240
126	4-Hydroxy-2-methylquinoline		232	200	201		215		
127	7-Hydroxyquinoline	43	235	244–5	251				
128	Hexamethylenetetramine	44	280s	179d (157)	190 (204)	205			

NOTES FOR TABLE 10B (TERTIARY AMINES—SOLIDS)

1. Styphnate, mp 247.
2. Ethiodide, mp 148.
3. Ethiodide, mp 229; styphnate, mp 194.
4. Ethiodide, mp 146.
5. Mercuric chloride double salt, mp 246.
6. Trinitrobenzene adduct, mp 124.
7. Ethiodide, mp 168–9.
8. Styphnate, mp 205.
9. Ethiodide, mp 220.
10. Styphnate, mp 231.
11. Ethiodide, mp 181; styphnate, mp 234.
12. Styphnate, mp 259.
13. Ethiodide, mp 225–7; styphnate, mp 200.
14. Styphnate, mp 207.
15. Styphnate, mp 222.
16. Styphnate, mp 243.
17. Mercuric chloride double salt, mp 189.
18. Styphnate, mp 232.
19. Trinitrobenzene adduct, mp 124; O-benzoyl derivative, mp 120.
20. Styphnate, mp 227.
21. Ethiodide, mp 250; styphnate, mp 214.
22. Salt with nitric acid, mp 193.
23. Benzoate ester, mp 95.
24. Ethiodide, mp 195.
25. Mercuric chloride double salt, mp 238.
26. Trinitrobenzene adduct, mp 114.
27. Ethiodide, mp 190; β-resorcylic acid salt, mp 141.
28. Ethiodide, mp 218.
29. Crystallized from benzene, the compound melts at 102 (mp of 93 is also reported from alcohol); trinitrobenzene adduct, mp 76.
30. Trinitrobenzene adduct, mp 89.
31. p-Toluenesulfonyl derivative, mp 53.
32. T.N.F., mp 164–5.
33. Salicylate, mp 92; adduct with 2-naphthol, mp 80.
34. Nitrate, mp 256.
35. Nitrated by fuming nitric acid in acetic acid, it yields a trinitro derivative, mp 280.
36. Trinitrobenzene adduct, mp 92; 1,2-diaminoethane derivative, mp 204.
37. Yields glycine on acid hydrolysis.
38. Ethiodide, mp 220.
39. Styphnate, mp 190; hydrobromide salt, mp 245.
40. After melting, it resolidifies and remelts at 215–20; O-acetyl derivative, mp 180.
41. Trinitrobenzene adduct, mp 171.
42. 2-Phenyl ether, mp 69.
43. O-Benzoyl derivative, mp 88–9.
44. Ethiodide, mp 150–1; Compound reacts with dilute acids to yield formaldehyde and the ammonium salt of the acid used.
45. Mercuric chloride double salt, mp 177.
46. Mercuric chloride double salt, mp 192–3.
47. Mercuric chloride double salt, mp 183.
48. Mercuric chloride double salt, mp 113–4.
49. Mercuric chloride double salt, mp 181–2.
50. Mercuric chloride double salt, mp 207–8.

REFERENCES FOR TABLES 9 AND 10

Bergman, E. D., and Bentov, M., *J. Org. Chem.*, **19**, 1594 (1954). An Improved Method for the Preparation of Fluorine Substituted Aromatic Amines. Describes a method of preparing fluorine-substituted aromatic amines and lists their melting points.

Bottini, A. T., and Olsen, K. E., *J. Org. Chem.* **27**, 452–55 (1962). The Oxidation of N-Alkyl-2,4-dinitroanilines with Chromic Acid. Tables with the melting points of the products and the per cent yields are given.

Buehler, C. A., and Calfee, J. D., *Anal. Chem.*, **6**, 351 (1934). Identification of Amines as 2,4-Dinitrobenzoates. Gives the melting points of many salts of 2,4-dinitrobenzoic acid with amines.

Crane, F. E., *Anal. Chem.*, **28**, 1794 (1956). Identification of Amines as Tetraphenylborates. Table II of this article lists 52 amines and the melting points of their salts.

Crane, F. E., *Anal. Chim. Acta*, **16**, 370 (1957). Tetraphenylboron Derivatives of amines.

Christensen, B. E., Wang, C. H., Davies, I. W., and Harris, D., *Anal. Chem.*, **21**, 1573 (1949). 2-Nitro-1,3-Indandione: Reagent for Identification of Organic Bases. Discusses the preparation and lists the melting points of 100 salts.

Christie, W. H., Christie, J. B., Wethington, J. A., Jr., and Pollard, C. B., *J. Org. Chem.*, **24**, 247 (1959). Derivatives of Piperazine. XXXI. Salts of Piperazine and N-Phenyl-piperazine for Utilization in Identification of Perfluoro-organic acids. Lists the melting points of the salts of 16 perhalo-organic acids.

Dahlbom, R., and Misiorny, A., *Acta Chem. Scand.*, **14**, 861 (1960). Substituted 1,3-Diamino-2-propanols. Gives the melting point of some 1,3-diamino-2-propanols.

Dummel, R. J., and Mosher, H. S., *J. Org. Chem.*, **24**, 1007 (1959). Some Nitro-pyridine Derivatives. Gives the method of preparation and the melting points of the derivatives of 10 compounds.

Dunbar, R. E., and Ferrin, F. J., *Microchem. J.* **4**, 167 (1960). Dibenzofuran-2-Sulfonic Acid Salts of Amines in Qualitative Organic Analysis. Lists the melting points and photomicrographs for a large number of primary aryl amines.

Ecke, G. G., Napolitano, J. P., Filby, A. H., and Kolka, A. J., *J. Org. Chem.*, **22**, 639 (1957). Ortho-Alkylation of Aromatic Amines. Gives the boiling points and the refractive indices of the alkylation products.

Epstein, P. F., *J. Org. Chem.*, **24**, 70 (1959). 5-Membered Heterocyclic Compounds Derived from Piperonal. I. A Study of the Reactions Between Piperonal and 1,2-Diamines. Lists the melting points of the diamine derivatives of 21 compounds.

Gautier, J. A., Renault, J., and Pellerin, F.,

REFERENCES FOR TABLES 9 AND 10 (Continued)

Ann. Pharm. Franc., **72a,** 726 (1955). Tetraphenylboron derivatives of Amines.

Gayer, J., Biochem. Z., **328,** 39 (1956). Tetraphenylboron Derivatives of Amines.

Hannig, E., Arch. Pharm. 290, 131 (1957). Tetraphenylboron Derivatives of Amines.

Hunsberger, I. M., Shaw, E. R., Fugger, J., Ketcham, R., and Lednicer, D., J. Org. Chem., **21,** 396 (1956). The Preparation of Substituted Hydrazines. IV. Arylhydrazines via Conventional Methods. Includes a long list of hydrochloride oxalates and salts with the melting points for many substituted phenylhydrazines.

Hurd, R. N., De La Mater, G., McElheny, G. C., Turner, R. J., and Wallingford, V. H., J. Org. Chem., **26,** 3980 (1961). Preparation of Dithiooxamide Derivatives. Lists the melting points and the I.R. spectra for many derivatives.

King, L. C., and Ozog, F. J., J. Org. Chem., **20,** 448 (1955). Reactions of Pyrylium and Pyridinium Salts with Amines. Gives the preparation data and the melting points of salts of amines.

Lapin, G. R., J. Chem. Educ., **28,** 126 (1951). A New Derivative for the Identification of Primary Aromatic Amines. Thirty amines are listed with the melting points of their derivatives.

Mariella, R. P., Gruber, M. J., and Elder, J. W., J. Org. Chem., **26,** 3217 (1961). Alkyl-Substituted Picrates of Hydrocarbons and Amines. Lists the melting points of the picrates of many amines.

Plien, E. M., and Dewey, B. T., Anal. Chem., **15,** 534 (1943); ibid., 18, 515 (1946). Organic Bases: Identification by Optical Properties (As Diliturates.)

Reid, E. E., *Organic Chemistry of Bivalent Sulfur*, vol. II, Chemical Publishing Co., New York, 1960. See pp. 340–350 for amine sulfides.

Richardson, A. G., Pierce, J. S., and Reid, E. E., J. Am. Chem. Soc., **74,** 4011 (1952). N-Aryl-N′-Alkyloxamides of Primary Aryl Amines. Gives extensive tables of the melting points of N-Aryl-N′-alkyloxamides of primary amines.

Sondri, G., Microchem. Acta, **1959,** 214. Tests for Alkaloids and Organic Bases Using Bromoplatinic Acid vs. Chloroplatinic Acid. The slide tests for organic bases are more sensitive using bromoplatinic acid than chloroplatinic acid.

Spialter, L., and Papalardo, J. A., J. Org. Chem., **22,** 840–843 (1957). Amines. III. Characterization of Some Aliphatic Tertiary Amines. The melting points of the derivatives of several tertiary amines are given.

Tansjo, L., Acta. Chem. Scand., **14,** 2097 (1960). N-Substituted Alkyltriamino-silanes. IV. On Intermolecular Condensation. The physical data are given for the derivatives of several compounds.

Tisler, M., Z. Anal. Chem., **165,** 272 (1959). p-Chlorobenzoylisothiocyanat als Reagenten zur Characterisierung von Primaren und Secondaren Aminen.

Wilson, K. O., Herbst, R. M., and Haak, W. J., J. Org. Chem., **24,** 1046 (1959). Alkylation Studies with Aminotetrazoles. Gives the method of preparation and lists the melting points of 27 such compounds.

Wilson, K. W., Anderson, F. E., and Donohoe, R. W., Anal. Chem., **23,** 1032 (1951). Identification and Separation of Amines Employing β-Resorcyclic Acid. Describes the method of preparation and gives the melting points of 29 esters.

Winstead, M. B., Anthony, K. V., Thomcas, L. L., Strachan, R. G., and Richwine, H. J., J. Chem. and Eng. Data, 7, 414 (1962). Identification of Amines: Succinimidomethyl Derivatives of Primary Amines. Gives the procedure and the melting points for about 35 aryl and alkyl amines.

Woolfolk, E. O., and Roberts, E. H., J. Org. Chem., **21,** 436 (1956): Use of p-Phenylazobenzoyl Chloride as a Reagent to Derivatize Amines. Lists 38 compounds with the melting points of their amides. Also discusses the chromatographic procedure of analysis.

Woolfolk, E. O., Reynolds, W. E., and Mason J. L., J. Org. Chem., **24,** 1445 (1959). p-Phenylazobenzenesulfonyl Chloride. A New Reagent for Identification and Separation of Amines. Lists the melting points of about 45 amides of p-phenylazobenzenesulfonic acid.

TABLE II
CARBOHYDRATES[1,2]

	Name of Compound	Note	MP	$[\alpha]_D^{20°}$	Derivatives[3] Azoates MP	$[\alpha]_{6438,CHCl_3}^{25°}$
1	Desosamine hydrochloride		82–84			
2	β-Melibiose dihydrate ($C_{12}H_{22}O_{11} \cdot 2H_2O$)		82–85	+129.5	280	+172
3	2-Deoxy-β-D-xylose		82–86	$-25 \to -2$		
4	L-Ribose		85–87	+23		
5	2-Deoxy-D-ribose		90	$+2.88 \to +2.13$		
6	D-Ribose		95	$-23.7(-21.5)$		
7	β-Maltose hydrate ($C_{12}H_{22}O_{11} \cdot H_2O$)	(a)	102.5	+136	275	+2
8	β-D-Fructose		102–4(A)	$-133 \to -92.0$	125	−440
9	α-L-Lyxose		105	$-6 \to +13.5$		
10	Glucosamine		105–10	+48(+44)		
11	α-D-Lyxose		106–7(101)	$+5.5 \to -14.0$		
12	α-L-Altrose		107–109.5	$-29 \to +32$		
13	Raffinose	(b)	119	+123.1	145	+146
14	β-L-Rhamnose	(c)	122–6(Ac)	+9.1		
15	D-Tagatose		124–25	$[\alpha]_D^{25} - 1°$		
16	β-D-Allose		128–28.5	$+0.6 \to +14$		
17	β-L-Allose		128–29	$-2 \to -14$		
18	L-Mannose		128–32	$[\alpha]_D^{25} - 14.5$		
19	α-D-Mannose	(d)	133	$+29 \to +14.5$		
20	α-D-Fucose		140–45	$+127 \to +76$		
21	2-Deoxy-D-glucose		142–44			
22	α-D-Xylose	(e)	143*(153)	$+96 \to +19.1$ (W4½)	157	+244
23	α-L-Fucose		145	−75.9		
24	α-D-Glucose (anh.)		146	+52.7	266	+223
25	6-Deoxy-α-D-glucose		146	$+73 \to +30$		
26	α-L-Glucose		146–7	$[\alpha]_D^{22} - 53$		
27	α-L-Xylose		146–50*	$-75 \to -19$		
28	β-D-Glucose		148–50	$+18.7 \to 52.7$		
29	Melezitose dihydrate		155	+88.8	(Sinters) 130	+188
30	β-D-Arabinose		158–9	$-175 \to -105$		
31	β-D-Galacturonic acid		160	+55.3		
32	β-L-Arabinose		160	+104.5	262	+755
33	α-Lactulose		160	$-12 \to -51$		
34	DL-Galactose		163(144)			
35	L-Sorbose		165(159–60)	−43.7		
36	β-D-Glucuronic Acid		165*	+36.3		
37	α-L-Galactose		165	$[\alpha]_D^{11} - 130 \to -81$		

(*Continued*)

TABLE II (Continued)
CARBOHYDRATES[1,2]

	Name of Compound	Note	MP	$[\alpha]_D^{20°}$	Derivatives[3] Azoates MP	$[\alpha]_{6438,CHCl_3}^{25°}$
38	α-D-Galactose		165.5*	+81.7	276	+436
39	β-Inulin			−40		
40	Sucrose	(f)	185	+66.5	125	+35
41	L-Ascorbic acid		190(187)	+49		
42	β-Gentiobiose		190*	$[\alpha]_D^{25}$ − 3.0 → +10.5		
43	α-Lactose monohydrate		202	+52.3		
44	α,α-Trehalose		203	+197.1	134.5	+210
45	α-Lactose (anh.)		223		288	+320
46	β-Lactose (anh.)		252	+53.6		
47	β-Cellobiose		225	+35.2	273	+105
48	Starch (sol.)		Dec.	+189		
49	Glycogen		240d			

[1] Special symbols used in these notes:
(A) = ethyl alcohol as solvent
(W) = water as solvent
(P) = pyridine as solvent
(P–A) = 50–50 pyridine–alcohol as solvent
(M) = methyl alcohol as solvent
(Aa) = acetic acid as solvent
(Ac) = acetone as solvent
(Chl) = chloroform

Where a letter followed by a number appears within parentheses, the letter stands for the solvent and the number signifies the number of hours required to attain equilibrium in mutarotation. Thus, for the phenylhydrazone of α-D-mannose, +33.8(P 56) means that pyridine is used as a solvent and that equilibrium is attained in 56 hours. The initial value, not given in the table, is $[\alpha]_D = +26.3$. This value changes to −6.3 in 9 hours. After 56 hours it again becomes positive and attains a state of equilibrium with a value of $[\alpha]_D = +33.8$, which is the recorded value.

Where the solvent (for rotation or crystallization) is water, the (W) is omitted, unless admixed with alcohol or where for some special reason it is deemed best to specify it.

[2] For a more complete listing of carbohydrates and their derivatives, see *Methods in Carbohydrate Chemistry*, R. L. Whistler and M. L. Wolfrom, ed., Academic Press, New York, 1962.

[3] For other derivatives, see *Notes* under the number of the compound.

NOTES ON CARBOHYDRATES

(a) According to the literature, maltose, when carefully heated to 160 (in vacuo), loses water and is resolved into the anhydride maltosan ($C_{12}H_{20}O_{10}$), a brown amorphous powder, which does not regenerate maltose when boiled with water and melts at 145–50. β-Maltose hydrate is stated to melt at 160–5 and at 102–3; the anhydrous form at 108. Beilstein (XXI, 386) gives the following melting points for hydrate:

NOTES ON CARBOHYDRATES (Continued)

110–25d, 106–112, 115, 115–25, 125–30; for the anhydrous form, it gives no melting point but only the temperature of dehydration.

(b) Raffinose crystallizes with 5 molecules of water and is freed therefrom by very slow heating.

(c) α-L-Rhamnose crystallizes with 1 molecule of water, and by cautious heating for several days is converted into anhydrous β-L-Rhamnose, which can be recrystallized from acetone.

(d) D-Mannose does not occur in nature as such, but only in the form of the polymerized anhydride mannan, from which it may be obtained by hydrolysis. It can be made by oxidation of D-Mannitol.

(e) D-Xylose, formerly called L-xylose, is best identified microscopically by conversion into "Cd xylonobromide," Cd $(C_5H_9O_6)_2 \cdot Cd\ Br_2 \cdot 2H_2O$, by means of Br_2 and $CdCO_3$ [Bertrand, *Bull. soc. chim.* (3), **5**, 556; **7**, 501; **15**, 592; cf. Widtsoe, *Ber.* **33**, 136]; or as the brucine salt of D-xylonic acid (Neuberg, *Ber.*, **35**, 1470–3).

(f) Sucrose, when crystallized from methyl alcohol, m 169–70; from ethyl alcohol, m 179–80; from ethyl alcohol and water, m 184–5. [Heldermann, *Z. physik. Chem.* **130**, 396 (1927); and Pictet and Vogel, *Helv. Chim. Acta*, **11**, 436 (1928)].

NOTES ON OTHER DERIVATIVES OF CARBOHYDRATES

Other derivatives with their melting points and specific rotation are given as follows. The number corresponds to the number of the compound on Table 11. Following the number is the name of the derivative, the melting point, and the specific rotation in parenthesis. Thus, (6): 4-Bromophenylhydrazone 170*(+10A) signifies the 4-Bromophenylhydrazone of D-ribose which has a melting point of 170 and a specific rotation $[\alpha]_D^{20} = -5.7$ in ethanol as a solvent.

1. Di-O-acetyldesosamine hydrochloride, 194–5d.
2. Phenylosazone, 176–8 (+43.2P); 4-bromophenylosazone, 181–2; octaacetyl, 177.5 (+102.5); oxime, 186; 2-phenylhydrazone, 145, also 160.
3. Benzylphenylhydrazone, 116–8 ($[\alpha]_D^{25}$ = +13.5P).
4. p-Bromophenylhydrazone, 170–2 ($[\alpha]_D^{25} = -11$A).
5. Benzylphenylhydrazone, 127–9.
6. Phenylosazone, 160 and 164; 4-bromophenylhydrazone, 170* (+10A); p-bromophenylosazone, 180–5.
7. Phenylosazone, 208; semicarbazone, 213d(+80W); 2-naphthylhydrazone 176 (+10.6M).
8. p-Nitrophenylhydrazone, 180–1 and 176 (+16P-A); 2-nitrophenylhydrazone, 156–7* (+31P-A); Phenylosazone, 210; Pentaacetyl, α-form, 70 (+34.7 chl.), β-form, 108–9 (−120.5 chl.), 1,3,4,6-tetrabenzoate, 124–5 (−6 → −14 chl.).
9. p-Bromophenylhydrazone, 157; 4-nitrophenylhydrazone, 172; 2,4-dinitrophenylhydrazone, 171–2 (−31P).
10. Phenylosazone, 210; N-Acetyl, 190; oxime, 127; phenylurea, 210; semicarbazone, 165.
11. Phenylosazone, 164; 4-nitrophenylhydrazone, 172; 2,4-dinitrophenylhydrazone, 171–2; 4-bromophenylhydrazone, 162 and 156–7; benzylphenylhydrazone, 116 and 128 (+26.4A).
12. Benzylphenylhydrazone, 147–8.
13. α-Methylphenylhydrazone, 190*; trityl, 130.
14. 4-Nitrophenylhydrazone, 190–1 (+21.4); 4-bromophenylosazone, 222; α-methylphenylhydrazone, 124; phenylosazone, 222 and 185 (+94P).

NOTES ON OTHER DERIVATIVES OF CARBOHYDRATES (Continued)

15. 4-Bromophenylosazone, 182–3.
16. 4-Bromophenylhydrazone, 145–7*. β-D-Allose is dimorphous and melts at 141 as well as at 128.
17. 4-Bromophenylhydrazone, 141–5 (+6A). See also, Note 16.
18. Phenylhydrazone, 199–200 ($[\alpha]_D^{25}$ −34P).
19. 4-Bromophenylhydrazone, 208; methylphenylhydrazone, 181 (+8.6M); 4-nitrophenylhydrazone, 195A and 203 (+56.0P–A); phenylhydrazone, 199 (+33.8P56); phenylosazone (glucosazone), 210.
20. 4-Bromophenylhydrazone, 181–3.
21. Dimethyl acetal, 58–9 (+42 chl.).
22. β-Naphthylhydrazone, 124*; 3-nitrophenylhydrazone, 163A; phenylosazone, 164 (A–W); phenylosotriazole, 89*; benzylphenylhydrazone, 95 (−33A); methylphenylhydrazone, 108.
23. 4-Bromophenylhydrazone, 181–4 (A–W); 4-nitrophenylhydrazone, 210–11; phenylhydrazone, 170; p-toluene sulfonylhydrazone, 174 (−17.0P); methylphenylhydrazone, 174($[\alpha]_D^{19}$ −17.0P); oxime, 188–9.
24. 2-Naphthylhydrazone, 178* (+11A); 4-nitrophenylhydrazone, 190–[196] (+21.5P–M); α-pentaacetate, 112; β-pentaacetate, 132; phenylosazone, 210[211W] (−1.5P–A); phenylosotriazole, 195–6*(−81.6P); 4-bromophenylhydrazone, 164–6 (−43.6 → +18.9P); 2,4-dinitrophenylosazone, 256–7; 4-nitrophenylosazone, 257.
25. Phenylosotriazole, 140 (−67.5A); phenylosotriazole tribenzoate, 100* (−33 chl.).
26. Benzylphenylhydrazone, 163–4 (−48P).
27. Di-O-benzylidene dimethylacetal, 210–2 (+7 chl.).
28. Phenylosazone, 210; 2,4-Dinitrophenylosazone, 256–7; 4-nitrophenylosazone, 257; β-pentaacetate, 132.
29. Hendeca methyl ether, $[\alpha]_D$ +114 (CH$_3$OH).
30. Phenylosazone, 162–3 and 160; benzylphenylhydrazone, 173; oxime, 136.
31. 4-Bromophenylhydrazone, 151d (+11.5M); brucine salt, 180; phenylhydrazone, 140.
32. Benzylphenylhydrazone, 174* (−11.5M); cinchonine salt, 178 (+139W); diphenylhydrazone, 204–5* (+14.9P); methylphenylhydrazone, 165*; 2-naphthylphenylhydrazone, 177(A); 4-nitrophenylhydrazone, 186(W); oxime, 139 (+13.3W); phenylhydrazone, 153 (+7.95P); phenylosazone, 166; semicarbazone, 190 (+23.8W); 4-bromophenylhydrazone, 155; tetraacetyl, 96–6 (α-form ether), 86 (β-form W).
33. Octaacetyl, 138 (−7 chl.)
34. Phenylosazone, 206; methylphenylhydrazone, 183; phenylhydrazone, 158–60.
35. 4-Bromophenylosazone, 181; phenylosotriazole, 159*; phenylosazone, 156 and 168; 2-nitrophenylosazone, 211–2; pentaacetyl, 97 (+2.9 chl.).
36. Cinchonine salt, 204* (+139.9W); 4-nitrophenylhydrazone, 225; semicarbazone, 188; thiosemicarbazone, 223.
37. Methylphenylhydrazone, 189 ($[\alpha]_D^{11}$ −23M).
38. α-Methylphenylhydrazone, 190–1* (inactive); mucic acid, 214d; 4-nitrophenylhydrazone, 197W* (+45.6); α-pentaacetate, 95; β-pentaacetate, 142; phenylosazone, 182–4 and 201 (+0.34P–A24); phenylosotriazole, 111* m-tolylhydrazone, 154*; o-tolylhydrazone, 176* (95%A); 4-bromophenylhydrazone, 168; benzylphenylhydrazone, 157; diphenylhydrazone, 157.
39. Triacetate, 150–60 (−45.5 Ac).
40. Octaacetate, 75 and 69 (+59.6 chl.); trityl, 127.9 (+44.3).
41. 4-Nitrophenylhydrazone, 262 (di); 4-bromophenylhydrazone, 170 (di);

NOTES ON OTHER DERIVATIVES OF CARBOHYDRATES (Continued)

     diphenylhydrazone, 187; di-2,4-dinitrophenylhydrazone, 282.
42. β-Octaacetate, 193 (-5 chl.).
43. Mucic acid, 214d; 4-nitrophenylosazone, 258; octaacetate, 100; phenylosazone, 210 and 200 (7.9M9); phenylosotriazole, 181*.
44. Octaacetate, 100–2* (+162 chl.); hexaacetate, 93–6 ($[\alpha]_D^{19}$ +158.3 chl.); octanitrate, 124.
45. Phenylosazone, 200 and 210–2.
46. 4-Nitrophenylosazone, 258d; 2-naphthylhydrazone, 203.
47. Octaacetate (α), 229.5 (+42.0 chl.); octaacetate (β), 192 and 202 (-14.5 chl.); oxime, 123–5; phenylhydrazone, 90; phenylosazone, 200d (-6.5P–A); phenylosotriazole, 165*; semicarbazone, 183–5 (-5.2W); thiosemicarbazone, 170d.
49. Dibenzoate (+179.8 chl.); triacetate (+170 chl.).

REFERENCES FOR TABLE 11

Rachinskij, V. V. and Knyazayatova, E. J., *Doklady Akad. Nauk. SSSR*, **85,** 1119 (1952). Identification of the carbohydrates by their R_f values. (Chromatography.)

Rao, P. S. and Beri, R. M., *Proc. Indian Acad. Sci.*, **33A,** 368 (1951); *ibid;* **34A,** 236 (1951). Identification of carbohydrates by their R_f values. (Chromatography.)

Stoll, A. and Ruegger, A., *Helv. Physiol. Acta*, **10,** 385 (1952). Identification of carbohydrates by their R_f values.

TABLE 12A
ESTERS—LIQUID

	Name of Compound	Note	MP	BP	D_4^{20}	n_D^{20}
1	Ethyl nitrite			17	0.900-D_4^{15}	
2	Methyl formate		−99	31.5	0.97421	1.344
3	Ethyl formate	1	−79.4	54.2	0.9225	1.3598
4	Trifluoroethyl trifluoroacetate			55	1.4725-D^{18}	1.2812
5	Methyl acetate		−98.7	57.3	0.9274	1.3617
6	Ethyl trifluoroacetate			60.5		1.3093^{15}
7	Methyl nitrate			65	1.217-D^{15}	
8	Isopropyl formate			71(68)	0.8728	1.368
9	Vinyl acetate			72.5	0.9342	1.3956
10	Butyl nitrite			75	0.911	
11	Methyl chloroformate			75	1.2231	1.3868
12	Ethyl acetate	2	−83.6	77.15	0.9006	1.3724
13	Trifluoroethyl acetate			77.8	1.2426	1.3202
14	Methyl propionate		−87.5	79.9	0.9151	1.3779
15	Methyl acrylate			80.3	0.961	1.3984
16	Propyl formate		−92.9	81	0.904	1.3779
17	Allyl formate			83	0.946	
18	Ethyl nitrate		−102	87	1.106	1.3853
19	Dimethyl carbonate			90.5	1.0702	1.3687
20	Isopropyl acetate		−73.4	91(88)	0.872	1.377
21	Methyl isobutyrate		−87.7	92.6	0.8906	1.3840
22	Ethyl chloroformate			93	1.1352	1.3974
23	sec-Butyl formate			97	0.884	1.384
24	Methylvinyl acetate			97	0.9308-D_{25}^{25}	1.4001
25	tert-Butyl acetate			97.8	0.867	1.386
26	Isobutyl formate			98	0.8854	1.3857
27	Isopentyl nitrite			99	0.880-D^{15}	1.3870
28	Methyl methacrylate		−50	99	0.936	1.413
29	Ethyl propionate	2	−73.9	99.1	0.8889	1.3853
30	Ethyl acrylate			99.3	0.909	1.406
31	Ethyl difluoroacetate			99.4	1.1800-D^{17}	1.3463
32	Methyl orthoformate			101	0.968	1.3793
33	Methyl trimethylacetate			101	0.891-D^0	1.4228
34	Propyl acetate		−95	101.5	0.8834	1.3847
35	Methyl butyrate		−84.8	102.7	0.8982	1.3879
36	Trimethyl orthoformate			102	0.9676	1.3793
37	Allyl acetate			104	0.9276	1.4049
38	2,2-Difluoroethyl acetate			106.4	1.1781-D^{15}	
39	Butyl formate		−91.9	106.6	0.8885	1.3894
40	Propyl nitrate			110	1.063	1.3979
41	Ethyl isobutyrate		−88.2	110	0.8693	1.3903
42	Isopropyl propionate			111	0.8931-D^0	
43	Chloromethyl acetate			111	1.094-D^{15}	
44	sec-Butyl acetate			112	0.872	1.3865
45	Propyl chloroformate			115	1.0901	1.4035
46	Methyl isopentanoate			116.7	0.8808	1.3900
47	Isobutyl acetate			117.2	0.8747	1.3901
48	Ethyl trimethylacetate			118.3	0.8547	1.3906
49	Methyl crotonate			119	0.946	1.425
50	2-Fluoroethyl acetate			119.5	1.0982	1.3780^{15}
51	Isopropyl isobutyrate			120.7	0.8471	

(Continued)

TABLE 12A (Continued)
ESTERS—LIQUID

	Name of Compound	Note	MP	BP	D_4^{20}	n_D^{20}
52	Ethyl butyrate		−100.8	121.6	0.8792	1.4000
53	Ethyl fluoroacetate			121.7	1.0926	1.3766
54	Propyl acrylate			122	0.92-D^0	
55	Propyl propionate			123	0.8809	1.3933
56	Allyl propionate			123	0.914	1.410
57	Isopentyl formate			123	0.8773	1.3976
58	tert-Pentyl acetate			124	0.873	1.392
59	Isobutyl nitrate			124	1.0152	1.4052
60	Allyl propionate			124		
61	Butyl acetate		−73.5	126.1	0.881	1.3961
62	Diethyl carbonate		−43	126.8	0.9751	1.3846
63	Methyl pentanoate		−91	127.7	0.885	1.397
64	Isopropyl butyrate			128	0.8787-D^0	
65	Isobutyl chloroformate			129.8	1.2337	1.4218
66	Methyl chloroacetate			130	1.238	1.4221
67	Pentyl formate			131	0.885	1.3992
68	Ethyl methoxyacetate			132	1.007	
69	Propyl isobutyrate			134	0.864	1.396
70	Ethyl 3-methylbutyrate		−99.3	134.7	0.8657	1.4009
71	Butyl nitrate			136	1.048-D^0	
72	Isobutyl propionate		−71.4	137	0.8876-D^0	1.3975
73	Cyclopentyl formate			138	1.000	1.432
74	Methyl pyruvate	9		138	1.154	
75	Ethyl crotonate			138	0.9175	1.4252
76	Allyl butyrate			142	0.902	1.416
77	Isopentyl acetate			142	0.8674	1.4003
78	Methyl dichloroacetate		−51.8	142.8	1.3774	1.4429
79	2-Methoxyethyl acetate			143	1.088	
80	Propyl butyrate			143	0.872	1.4005
81	Methyl bromoacetate			144d	1.657	
82	dl-Methyl lactate			144.8	1.0931	1.4144
83	Butylacrylate			145	0.898	1.4185
84	Butyl chloroformate			145(139)	1.079	1.4178,4
85	Ethyl chloroacetate		−26	145	1.158	1.4227
86	Ethyl orthoformate			145	0.897	1.3922
87	2-Chloroethyl acetate			145	1.178	1.4234
88	Triethyl orthoformate			145.5	0.8909	1.3922
89	Ethyl pentanoate		−91.2	145.5	0.8739	1.4009
90	Butyl acrylate			146	0.901-D_{20}^{20}	1.418
91	Ethyl 2-chloropropionate			146	1.087	
92	2-(4-Methylpentyl) acetate			146.3	0.8595-D_{20}^{20}	
93	Butyl propionate			146.8	0.895	1.401
94	Benzyl chloroacetate			147	1.2223-D_4^4	1.5246[18]
95	Isobutyl isobutyrate		−80.8	147(149)	0.8749-D_4^0	1.3999
96	Methyl ethoxyacetate			148	1.006	
97	Pentyl acetate		−70.8	148.8	0.8756	1.4031
98	Methyl glycollate			151	1.166	
99	Methyl hexanoate		−71	151.2	0.8846	1.405
100	Methyl trichloroacetate			152	1.488	1.457
101	Ethyl ethoxyacetate			152(158)	0.9701	1.4029
102	Cyclopentyl acetate			153	0.975	1.432

(Continued)

TABLE 12A (Continued)
ESTERS—LIQUID

	Name of Compound	Note	MP	BP	D_4^{20}	n_D^{20}
103	Hexyl formate			154	0.879	1.407
104	Isopentyl chloroformate			154	1.032-D^{15}	
105	Isopropyl pentanoate			154	0.858	1.401
106	Ethyl lactate		−25	154.5	1.030	1.410
107	Ethyl pyruvate	10		155	1.055	1.406
108	Pentyl isobutyrate			155	0.8592-D^{13}	1.4076
109	Propyl 3-methylbutyrate			156	0.862	1.403
110	2-Ethoxyethyl acetate			156	0.976	
111	Isobutyl butyrate			157	0.8620	1.4029
112	Propyl crotonate			157	0.908	1.428
113	Ethyl dichloroacetate			158	1.2821	1.4386
115	Ethyl glycollate			160	1.082	
116	Isopentyl propionate			160	0.870	1.4065
117	Cyclohexyl formate			162	1.010	1.443
118	Ethyl 2-bromopropionate			162	1.329	
119	2-Bromoethyl acetate			163	1.524	
120	Butyl butyrate			165	0.869	1.406
121	Propyl pentanoate			167(164)	0.870	1.4065
122	Ethyl hexanoate		−67.5	167.9	0.8710	1.4073
123	Ethyl trichloroacetate			168(164)	1.380	1.450
124	Dipropyl carbonate			168	0.943	1.4014
125	Isopropyl lactate			168	0.998-D_{20}^{20}	1.4082^{25}
126	Pentyl propionate			169	0.881	
127	Ethyl bromoacetate			169	1.506	1.451
128	Methyl acetoacetate			170	1.077	1.4196
129	Isobutyl 3-methylbutyrate			171	0.853	1.406
130	Hexyl acetate			171.5	0.8779-D^{15}	1.409
131	Methyl heptanoate		−55.8	173.8	0.8801	1.412
132	Methyl iodoacetate			170		
133	2-Ethoxyethyl acrylate			174	0.9834-D_{20}^{20}	
134	Acetonyl acetate	3		174–5		
135	Cyclohexyl acetate			175	0.970	1.442
136	Butyl chloroacetate			175	1.081-D_4^{15}	
137	Butyl 3-methylbutyrate			176	0.861	1.409
138	Furfuryl acetate			176	1.118	1.4627
139	Ethyleneglycol diformate			177	1.229	
140	Methyl methylacetoacetate			177	1.030	1.418
141	Isopentyl butyrate			178	0.864	1.411
142	Heptyl formate			178.1	0.8828-D^{15}	
143	Ethyl 3-bromopropionate			179	1.425	
144	Ethyl iodoacetate			180	1.818-D^{13}	1.508^{13}
145	Ethyl acetoacetate			181	1.025	1.4197
146	Ethyl methylacetoacetate			181	1.006	1.419
147	Methyl furoate			181.3	1.180	1.4860
148	Dimethyl malonate		−62	181.5	1.1539	1.4140
149	Butyl chloroacetate			181–2	1.0704	1.4301
150	Methyl hexahydrobenzoate			183	0.990	1.451
151	Butyl pentanoate			184	0.868	1.412
152	Butyl dichloroacetate			184		
153	Pentyl butyrate		−72.3	185	0.866	1.412

(*Continued*)

TABLE 12A (Continued)
ESTERS—LIQUID

	Name of Compound	Note	MP	BP	D_4^{20}	n_D^{20}
154	Diethyl oxalate	2	−41.5	185.4	1.0785	1.4104
155	Propyl hexanoate		−74	187	0.867	1.417
156	Butyl lactate		−43	188	0.984-D_{20}^{20}	1.4216
157	Dimethyl sulfate	11	−27	188	1.3348-D^{15}	1.3874
158	Ethyl heptanoate		−66.3	189	0.880	1.413
159	Di-isopropyl oxalate			189	1.0010	1.4100
160	Methyl ethylacetoacetate			190	0.989	
161	Di-isobutyl carbonate			190	0.9138	1.4072
162	Isopentyl 3-methylbutyrate			190	0.870	1.4130
163	Hexyl propionate			190	0.870	1.419
164	Ethylene glycol diacetate			190	1.1040	1.4150
165	1,3-Propylene glycol diacetate			191	1.059	1.417
166	Methyl levulinate			191(196)	1.0475	1.4233
167	2-Butoxyethoxy acetate			192.2	0.9424-D_{20}^{20}	
168	Heptyl acetate		−50.2	192.5	0.865	1.414
169	Tetrahydrofurfuryl acetate			194(740 mm)	1.061	1.438
170	Methyl octanoate		−41	194.6	0.878	1.417
171	Ethyl hexahydrobenzoate			196	0.962	1.448
172	Dimethyl succinate		18.2	196	1.1192	1.4197
173	Ethyl methylmalonate			196	1.019-D_4^{15}	
174	Phenyl acetate			196.7	1.078	1.503
175	Propylene glycol di-3-chlorobenzoate			198		
176	Ethyl ethylacetoacetate			198	0.972	1.422
177	Octyl formate			198.8	0.8786-D^{15}	
178	2-Ethyl 1-hexylacetate			199	0.8733-D_{20}^{20}	
179	Diethyl malonate	2	−51.5	199.3	1.0551	1.4162
180	Methyl benzoate		−12.5	199.4	1.0885	1.5168
181	Benzyl formate			203	1.080	1.5154
182	Vinyl benzoate			203	1.065	
183	γ-Butyrolactone			203–4	1.1441-D^0	
184	Methyl cyanoacetate			205.1	1.1277	1.4180
185	Ethyl levulinate	12		205.8	1.0111	1.4229
186	γ-Valerolactone			205–7	1.0465-D^{25}	1.4303^{25}
187	Dibutyl carbonate			207	0.941-D_4^0	
188	Pentyl pentanoate			207.4	0.8825-D_4^0	1.4181^{15}
189	Hexyl butyrate			207.9	0.866	1.420
190	o-Tolyl acetate			208	1.048	
191	Diethyl sulfate	13	−24.5	208	1.172-D^{25}	1.4010^{18}
192	Butyl hexanoate			208	0.865	1.421
193	Propyl heptanoate			208	0.866	1.421
194	Ethyl octanoate	2	−43.1	208.5	0.8667	1.4178
195	Octyl acetate			210		
196	Heptyl propionate			210	0.8594-D^{30}	
197	Trimethylene glycol diacetate			210	1.069	
198	Phenyl propionate		20	211		
199	Propyl furoate			211	1.0745-D^{26}	
200	Ethylene glycol dipropionate			212	1.045-D^{25}	
201	m-Tolyl acetate			212	1.049	1.4978
202	p-Tolyl acetate			212.5	1.051	1.500
203	Ethyl benzoate	2	−34.2	212.5	1.0465	1.506

(Continued)

TABLE 12A (Continued)
ESTERS—LIQUID

	Name of Compound	Note	MP	BP	D_4^{20}	n_D^{20}
204	Propyl oxalate			213	1.038	1.416
205	2-Ethylhexyl acrylate			213.5	0.886-D_{20}^{20}	1.435
206	Methyl nonanoate			213–4	0.892	
207	Methyl o-toluate			215	1.068	
208	Methyl m-toluate			215	1.061	
209	Benzyl acetate	2		217	1.055	1.5200
210	Diethyl succinate	2	−21	217.7	1.0398	1.4198
211	Ethoxyethoxyethyl acetate			218	1.013	
212	Isopropyl benzoate			218	1.023	
213	Diethyl fumarate		0.2	218.4	1.052	1.4410
214	Methyl phenylacetate			220(215)	1.068	1.507
215	Propyl levulinate			221	0.9895	
216	1-Phenylethyl acetate			222		
217	Diethyl maleate		−17	222.7	1.066	1.4416
218	Methyl salicylate	4	−9	223	1.184	1.5369
219	Heptyl butyrate			225.8	0.8555-D^{30}	
220	Butyl heptanoate			226	0.864	1.426
221	Pentyl hexanoate			226	0.863	1.426
222	Methyl decanoate			226	0.873	1.426
223	Ethyl o-toluate			227	1.034	1.508
224	Ethyl m-toluate			227	1.028	1.506
225	Ethyl nonanoate	2		227	0.8657	1.4220
226	Phenyl butyrate			227	1.023	
227	Ethyl phenylacetate	2		227.5	1.0333	1.500
228	Octyl propionate			227.9	0.8581-D^{30}	
229	Ethyl p-toluate			228	1.025	1.507
230	Di-isopentyl carbonate			228	0.912-D^{14}	
231	Isobutyl oxalate			229	1.002-D^{14}	
232	Propyl benzoate			230	1.023	1.500
233	Allyl benzoate			230	1.052	
234	Methyl 3-chlorobenzoate		20	231		1.492
235	2-Phenylethyl acetate			231–3		
236	Methyl dihydrocinnamate			232	1.043	1.503
237	Ethyl salicylate	4		234	1.131	1.5226
238	Methyl 2-chlorobenzoate			234		1.536
239	Diethyl glutarate	2	−24.1	234	1.0223	1.4240
240	Ethyl bromomalonate			235	1.426-D_{15}^{15}	
241	Methyl 3-methoxybenzoate			237	1.131	1.522
242	Isopropyl salicylate			237	1.095	
243	Butyl levulinate	3		238		
244	Ethyl 4-chlorobenzoate			238	1.181	1.524
245	Benzyl butyrate			238	1.033-D^{16}	
246	Propyl salicylate			239	1.098-D^{15}	1.516
247	Guaiacol acetate			240	1.133	1.512
248	Propyl phenylacetate			241	1.010	1.493
249	Isobutyl benzoate			241	0.999	
250	Dimethyl l-malate			242	1.2334	1.4425
251	Ethyl 3-chlorobenzoate			242	1.181	1.524
252	Geranyl acetate			242	0.9174-D^{15}	1.4660
253	Ethyl 2-chlorobenzoate			243	1.190	1.522
254	Butyl oxalate			243	1.010	1.423

(Continued)

TABLE 12A (Continued)
ESTERS—LIQUID

	Name of Compound	Note	MP	BP	D_4^{20}	n_D^{20}
255	Octyl butyrate			244	0.8550-D^{30}	
256	Methyl 2-bromobenzoate			244		
257	Methyl phenoxyacetate			245	1.147	
258	Thymyl acetate			245	1.009	
259	Diethyl adipate	2	−21	245	1.009	1.4277
260	Ethyl decanoate	2		245	0.8650	1.4258
261	Propyl succinate			246	1.006	1.425
262	Butoxyethoxyethyl acetate			246	0.983	
263	Isobutyl phenylacetate			247	0.999-D^{18}	
264	Ethyl dihydrocinnamate			248	1.016	1.495
265	Methyl 2-methoxybenzoate			248	1.156	1.534
266	Butyl benzoate			249	1.000	1.497
267	Diethylene glycol diacetate		19	250		
268	Ethyl 3-methoxybenzoate			251	1.100	1.515
269	Ethyl phenoxyacetate			251	1.101	
270	Diethyl l-malate		−10.2	253	1.1290	1.4362
271	Butyl phenylacetate			254	0.994	1.489
272	Ethyl 2-bromobenzoate			255		
273	Diethyl pimelate	2	−23.8	255	0.9929	1.4298
274	Ethyl benzoylformate			256–7	1.222-D^{25}	1.5190^{25}
275	Glyceryl triacetate			258	1.161-D^{15}	
276	Ethyl 3-bromobenzoate			259		
277	4-Methoxybenzyl acetate			260–5	1.1044	1.515^{15}
278	Ethyl 2-methoxybenzoate			261	1.104	1.525
279	Propyl dihydrocinnamate			262	0.998	1.491
280	Isopentyl oxalate			262	0.961	1.427
281	Isopentyl benzoate			262	1.004	1.495
282	Ethyl 4-bromobenzoate			263		
283	Isobutyl succinate			265	0.974	1.427
284	Ethyl benzoylacetate			265(270d)	1.117	
285	Ethyl anthranilate		13	267	1.117	1.565
286	Methyl laurate			268	0.870	1.432
287	Butyl salicylate			268(260)	1.073	1.512
288	Ethyl laurate	2	−1.7	269	0.862	1.4321
289	Ethyl anisate	2	7	269	1.1038	1.5254
290	Ethyl cinnamate	2	6.5	271	1.0490	1.5598
291	Methyl 2-nitrobenzoate			275	1.286	
292	Ethyl 2-iodobenzoate			275		
293	Isopentyl salicylate			277	1.065^{15}	
294	Resorcinol diacetate			278	1.179	
295	Methyl 2-iodobenzoate			278		
296	Diethyl d-tartrate		18.6	280	1.2028	1.4468
297	Diethyl suberate	2	5.9	282	0.9807	1.4324
298	Dimethyl phthalate			283.8	1.191	1.5138
299	Propyl cinnamate			284	1.028	1.551
300	Ethyl isophthalate		11	285	1.121	1.507
302	Glyceryl tripropionate			289	1.083	
303	Diethyl phthalate			289.5(298)	1.1175	1.5019
304	Diethyl azelate	2	−18.5	291	0.9729	1.4351
305	Ethyl m-aminobenzoate			294		

(*Continued*)

TABLE 12A (Continued)
ESTERS—LIQUID

	Name of Compound	Note	MP	BP	D_4^{20}	n_D^{20}
306	Triethyl citrate			294	1.1369	1.4455
307	Ethyl myristate	2	11.9	295	$0.861\text{-}D^{25}$	
308	Isopentyl succinate			297	0.958	1.434
309	Ethyl benzylmalonate			300	$1.077\text{-}D^{15}$	
310	Isopropyl phthalate			302	1.065	
311	o-Tolyl benzoate			307	1.114	
312	Diethyl sebacate	2	1.3	307	0.9631	1.4366
313	Ethyl 1-naphthoate			309	1.122	
314	Glyceryl tributyrate			318	$1.033\text{-}D^{17}$	
315	Benzyl salicylate			320		
316	Methyl myristate		18.5	323		1.428
317	Benzyl benzoate		21	324*	1.1224	1.5681
318	Dibutyl phthalate	2		335	$1.0484\text{-}D_{20}^{20}$	1.4900
319	Isopentyl phthalate			349	$1.024\text{-}D^{17}$	
320	o-Tricresyl phosphate		−30	400d	$1.197\text{-}D^{25}$	1.5568

TABLE 12B
ESTERS—SOLID[a]

	Name of Compound	Note	MP		Name of Compound	Note	MP
1	Phenyl propionate (bp 211)		20	53	dl-Methyl mandelate (bp 250)		53.3(58)
2	Methyl 3-chlorobenzoate (bp 231)		21	54	Dimethyl oxalate	14	54
3	d-Butyl tartrate		22	55	m-Cresyl benzoate		54
4	Ethyl palmitate (bp 185)	2	24	56	Methyl 3-iodobenzoate (bp 277)		54
5	Cetyl acetate		24.2(22)	57	Diglycol stearate		54–9
6	Methyl anthranilate (bp 299.8)	2	24.4	58	Methyl diphenylacetate		55–7
7	m-Tricresyl phosphate		26	59	Ethyl 4-nitrobenzoate	2	56
8	Bornyl acetate (bp 221)		27	60	1-Naphthyl benzoate		56
9	Butyl stearate		27.5	61	Ethyl 2-benzoylbenzoate		56–8
10	dl-Ethyl mandelate (bp 254)		28.1	62	2-Phenylethyl cinnamate		57–8
11	Bis(2-ethoxyethyl) phthalate		29–31	63	Ethyl diphenylacetate		58
12	Methyl palmitate		30	64	Butyl 4-aminobenzoate	6	58
13	Ethyl 2-nitrobenzoate		30	65	o-Cresyl carbonate		60
14	Eugenol acetate (bp 282)		30	66	Methyl diphenylacetate		60
15	Octadecyl acetate		30–33	67	Octyl 3,5-dinitrobenzoate		61
16	Methyl 3-bromobenzoate		32	68	4-tert-Butylphenyl salicylate	15	62–4
17	Ethyl 2-naphthoate (bp 304)		32	69	Catechol diacetate		63
18	Ethyl stearate	2	33	70	Ethyl 3-aminocinnamate		64
19	Methyl p-toluate		33	71	Tripalmitin		64.8
20	Thymyl benzoate		33	72	Methyl 2-aminocinnamate		65
21	Ethyl furoate (bp 197)		34	73	Ethyl trichloroacetate		66
22	Diethyl 4-nitrophthalate		34	74	Dimethyl 4-nitrophthalate		66
23	Methyl cinnamate (bp 261)		36	75	Piperonyl benzoate		66
24	Diglycol myristate		36–7	76	4-Chlorophenacyl acetate		67
25	Ethyl mandelate (bp 254)		37	77	Ethyl oxanilate		66–7
26	Methyl sebacate (288d)		38	78	Coumarin (bp 290)		67
27	Methyl nicotinate (bp 204)		38	79	1,3-Propanediol di-3-chlorobenzoate		67
28	Methyl 3-aminobenzoate		38	80	Methyl isophthalate		68
29	Methyl stearate		38.8	81	Phenyl benzoate		69
30	Benzyl cinnamate		39	82	Dimethyl 3-nitrophthalate		69
31	Ethyl 2,4-dinitrobenzoate		41	83	Ethyl 4-aminocinnamate		69
32	Benzyl phthalate		42	84	Butyl 4-hydroxybenzoate		69–71
33	Phenyl salicylate	5	42	85	Phenyl phthalate		70
34	Benzyl succinate		42	86	Methyl 3-hydroxybenzoate		70
35	Cyclohexyl oxalate		42	87	Methyl 2,4-dinitrobenzoate		70
36	Ethyl diphenate		42	88	2-Naphthyl acetate		71
37	Cinnamyl cinnamate		44	89	Glyceryl tristearate		71
38	Ethyl 2-nitrocinnamate		44	90	p-Cresyl benzoate		71
39	Methyl 4-chlorobenzoate		44	91	Glyceryl tribenzoate		72
40	Ethyl terephthalate (bp 302)		44	92	Phenyl cinnamate		72
41	Methyl anisate (bp 255)		45	93	Ethyl 3-hydroxybenzoate (bp 282)		72
42	Diethyl 3-nitrophthalate		46	94	Ethyleneglycol dibenzoate		73
43	Dibenzyl succinate		46–8	95	Phthalide (bp 290)		73
44	Diglycol palmitate		46.5	96	Methyl 2-nitrocinnamate		73
45	Ethyl 3-nitrobenzoate (bp 296)	2	47	97	Methyl diphenate		74
46	Dimethyl tartrate (bp 280)		48	98	p-Tricresyl phosphate		76–8
47	1-Naphthyl acetate		49	99	Methyl 2-naphthoate (bp 290)		77
48	Triphenyl phosphate		49	100	Ethyl 2-aminocinnamate		78
49	Methyl 1-naphthoate		49	101	Trimethyl citrate (bp 283–7d)		78–9
50	Methyl 2-benzoylbenzoate (352)		52	102	Methyl 3-nitrobenzoate		78
51	Phenyl stearate		52	103	Diphenyl carbonate (bp 306)		78
52	Dimethyleneglycol dibenzoate		53				

[a] Many esters not included in this Table may be found as derivatives in Tables 5, 6, and 29.

(*Continued*)

TABLE 12B (Continued)
ESTERS—SOLID[a]

	Name of Compound	Note	MP		Name of Compound	Note	MP
104	Ethyl 3-nitrocinnamate		79	127	Ethyl oxamate		114–5
105	Benzyl oxalate		80	128	p-Cresyl carbonate		115
106	Methyl 4-bromobenzoate		81	129	Ethyl 4-hydroxybenzoate		116
107	Catechol dibenzoate		84	130	tert-Butyl 4-nitrobenzoate		116
108	Methyl 3-aminocinnamate		84	131	Resorcinol dibenzoate		117
109	Diguaiacyl carbonate		87	132	Ethyl 3-nitrosalicylate		118
110	Ethyl 4-aminobenzoate	7	92	133	Methyl 5-nitrosalicylate		119
111	Dimethyl dl-tartrate (bp 282)*		90(87)	134	Diphenyl succinate (bp 330)		121
112	Pyrogallol tribenzoate		90	135	Hydroquinone diacetate		124
			92	136	Methyl 3-nitrocinnamate		124
113	Ethyl 3,5-dinitrobenzoate	2	93	137	Methyl 3,5-dinitro-salicylate		127
114	2-Naphthyl salicylate	5	95.5	138	Lactide (bp 255)		128
115	Methyl 4-nitrobenzoate		96	139	Methyl 4-aminocinnamate		129
116	Propyl 4-hydroxybenzoate		96	140	Methyl 4-hydroxybenzoate	8	131
117	Ethyl 3,5-dinitrosalicylate		99	141	Methyl 3-nitrosalicylate		132
118	Ethyl 5-nitrosalicylate		102	142	Ethyl 4-nitrocinnamate		137
119	Methyl fumarate (trans)		102				(142)
120	Benzoin acetate		104	143	Methyl terephthalate		141
121	Phloroglucinol triacetate		105	144	Methyl 4-nitrocinnamate		161
122	Diphenyl adipate		106	145	Pyrogallol triacetate		164–5
123	2-Naphthyl benzoate		107	146	Phloroglucinol tribenzoate		185
124	Methyl 3,5-dinitrobenzoate		108	147	Hydroquinone dibenzoate		199
125	Methyl 4-aminobenzoate		111–2				
126	Methyl 4-iodobenzoate		114				

NOTES FOR TABLES 12A AND 12B (ESTERS)

1. 2,4-Dinitrophenylhydrazine derivative, mp 126–7.
2. Many esters may be derivatized by reacting them with N-(2-aminoethyl)-morpholine [*J. Am. Chem. Soc.*, **73**, 1967 (1951)]. The melting point of the derivatives of the numbered (in parentheses) esters follow:

   Liquids: (12) 95.5 (289) 68.5
   (29) 85 (290) 130.6
   (154) 170 (291) 121.9
   (179) 120.5 (298) 157.2
   (194) 59 (305) 141.3
   (203) 123.4 (308) 76
   (209) 95.2 (313) 146
   (210) 174 (319) 124
   (225) 61.3 Solids: (4) 81
   (227) 88.9 (6) 126
   (239) 152.7 (18) 85
   (259) 165 (45) 131.9
   (260) 60.1 (59) 186.3
   (274) 137.9 (113) 189.5

3. See Table 21A.
4. See Table 29A.
5. See Table 29B.
6. See Table 9B; also named butesin.
7. See Table 9B; also named benzocaine.
8. Also named methylparaben.
9. 2,4-Dinitrophenylhydrazone, mp 187.
10. Phenylhydrazone, mp 118.
11. 2-Naphthyl ether, mp 72; tribromophenyl ester, mp 87.
12. 2,4-Dinitrophenylhydrazone, mp 101–2.
13. 2-Naphthyl ether, mp 37; tribromophenyl ester, mp 72.
14. Reacts with concentrated ammonium hydroxide to precipitate oxamide, mp 417.
15. Light absorber, 290–330 millimicrons.

REFERENCES FOR TABLES 12A AND 12B

Dermer, O. C., and King, J., *J. Org. Chem.*, **8**, 168 (1943). N-Benzylamides as Derivatives for Identifying Acyl groups in Esters.

Olah, G. A., Pavlath, A. E., Olah, J. A., and Herr, F., *J. Org. Chem.*, **22**, 880 (1957). Synthesis and Investigation of Organic Fluorine Compounds. XXII. Preparation of Aromatic Fluorinated Esters as Local Anesthetics. Lists the melting points of many fluorinated esters.

Rauscher, W. H., and MacPeek, D. I., *Anal. Chem.*, **22**, 923 (1950). Identification of Esters of Monobasic Acids by the Use of Ethanolamine.

TABLE 13A
ETHERS—LIQUID

	Name of Compound	Note	BP	D_4^{20}	n_D^{20}
1	Ethyl methyl ether		7.6	$0.7252\text{-}D_0^0$	
2	Ethylene oxide (epoxyethane)	1	10.7	$0.882\text{-}D_{10}^{10}$	$1.3614\text{-}n^4$
3	Vinyl ether		28.3		
4	Furan	2	31.27	0.9366	1.42157
5	Ethyl ether	3	34.60	0.71425	1.3526
6	1,2-Epoxypropane	4	35	0.830	1.466 (1.383)
7	Ethyl vinyl ether		36	0.7589	1.3768
8	Methyl propyl ether		39	$0.7356\text{-}D_4^{13}$	1.3579
9	Allyl methyl ether	5	46		
10	Ethyl isopropyl ether		53–4	0.7211	
11	*tert*-Butyl methyl ether	6	55.2	0.7405	1.3689
12	1,2-Epoxy-2-methylpropane		56		
13	Chloromethyl methyl ether	7	59	1.015 (1.070)	1.397
14	2,3-Epoxybutane	8	N		
15	1,2-Epoxybutane		63.2	0.837^{17}	1.385^{17}
16	Ethyl propyl ether	9	63.6	0.7386	1.3695
17	2-Methylfuran		64	0.913	1.434
18	Tetrahydrofuran		65	0.889	1.407
19	Allyl ethyl ether		65	0.7651	1.3881
20	Isopropyl ether	10	67.5	0.726	1.3688
21	Butyl methyl ether		70	0.774	1.374
22	1-Chloroethyl methyl ether		73	0.991	1.400
23	*tert*-Butyl ethyl ether		73.1	0.7404	1.3760
24	Tetrahydro-2-methylfuran		79	0.855	1.407
25	Ethyl isobutyl ether		81.1	$0.7323\text{-}D_4^{25}$	1.3739^{25}
26	*sec*-Butyl ethyl ether		81.2	0.7503	1.3802
27	Chloromethyl ethyl ether		83d (80)	1.026	1.404
28	Isopropyl propyl ether		83	0.7370	1.376
29	Isobutyl vinyl ether		83.3	$0.7706\text{-}D_{20}^{20}$	
30	1,2-Dimethoxyethane		85.2	0.867	1.3792
31	Dihydropyran		86(84.3)	0.923	1.440
32	*tert*-Pentyl methyl ether	11	86.3	0.7703	1.3885
33	Tetrahydropyran		88	0.881	1.421
34	1-Chlorovinyl ethyl ether		88–90		1.4208
35	Propyl ether	12	90.1	0.74698	1.3885
36	2-Chloroethyl methyl ether		91	1.035	1.411
37	Butyl ethyl ether		92.3	0.7506	1.3820
38	Butyl vinyl ether		93.3	0.7888	1.4026
39	2,5-Dimethylfuran	13	93–4	0.888	1.4363
40	1-Chloroethyl ethyl ether		98	0.966	1.405
41	Pentyl methyl ether		99	0.761	1.387
42	*tert*-Pentyl ethyl ether	14	101	0.7657	1.3912
43	1,4-Dioxane		101.4	1.0336	1.4232

(*Continued*)

TABLE 13A (Continued)
ETHERS—LIQUID

	Name of Compound	Note	BP	D_4^{20}	n_D^{20}
44	1-Ethoxy-2-methoxyethane		102	0.8529	1.3868
45	Bis(chloromethyl) ether		105	1.3152_{20}^{20}	1.435
46	1,3-Dioxane	15	105	1.0342	1.4165
47	Methyl cyclopentyl ether		105	0.862	1.420
48	1-Bromoethyl ethyl ether		105	1.0632^{12}	
49	2-Chloroethyl ethyl ether		107	0.989	1.411
50	Allyl chloromethyl ether		108		1.4340
51	2-Aminoethyl ethyl ether (see Table 9A)		108	0.8512	1.4101
52	Butyl isopropyl ether		108 (738 mm)	$0.7594\text{-}D_4^{15}$	
53	Chloromethyl propyl ether		109		1.4106
54	1-Chlorovinyl propyl ether		112–4		1.4230
55	1-Chloroethyl ether		114 (116)	1.111	1.423
56	1,2-Epoxy-3-chloropropane		118	1.181	1.438
57	Pentyl ethyl ether		118	0.762	1.393
58	1,2-Diethoxyethane		121	0.841	
59	sec-Butyl ether	16	121	0.760	1.396
60	Cyclopentyl ethyl ether		122	0.853	1.423
61	Isobutyl ether	17	123	0.762	
62	1-Propoxy-2-methoxyethane		124.5	0.8472	1.3947
63	Hexyl methyl ether		126	0.772	1.397
64	2-Bromoethyl ethyl ether		127	$1.370\text{-}D_4^0$	1.4447
65	2-Methoxy-1-propanol (see Table 6A)		130		
66	2-Chloro-1,3-epoxypropane	18	132–4		
67	Butyl chloromethyl ether		134		1.4208
68	Cyclohexyl methyl ether		134	0.875	1.435
69	1,2-Epoxy-3-bromopropane		134–6		
70	2-Ethoxyethanol (see Table 6A)		134.8	0.9297	1.4080
71	Ethyl hexyl ether		142	0.772	1.401
72	Butyl ether	19	142.4	0.7683	1.400
73	2-Chloroethyl chloromethyl ether		145–7		1.4578
74	Cyclohexyl ethyl ether		149	0.864	1.435
75	2-Bromoethyl ether		150	2.201	
76	Methoxybenzene (anisole)	20	153.8	0.9939	1.5170
77	Bis(2-methoxyethyl) ether		162	$0.9451\text{-}D_{20}^{20}$	1.4078
78	1-Butoxy-2-ethoxyethane		164.2		
79	Benzyl methyl ether	21	171*	0.9649	1.5008
80	2-Methoxytoluene	22	171	0.9853	1.505
81	Butoxyethanol	23	171 (743 mm)	0.9188	1.4177
82	Ethoxybenzene (phenetole)	24	172	0.9666	1.5080
83	Isopentyl ether	25	172.5	0.778	1.409

(Continued)

TABLE 13A (Continued)
ETHERS—LIQUID

	Name of Compound	Note	BP	D_4^{20}	n_D^{20}
84	4-Methoxytoluene	26	173	0.970	1.512
85	3-Methoxytoluene	27	173	0.972	1.513
86	1,2,2-Trichloroethyl ethyl ether		174		1.4630
87	1,8-Cineole	28	176	0.9267-D_{20}^{20}	1.4584
88	Bis(2-Chloroethyl) ether		178	1.220	1.457
89	Phenyl isopropyl ether		178	0.975	1.4992
90	2-Ethylhexyl vinyl ether		178	0.8102-D_{20}^{20}	
91	2-Ethoxytoluene	29	184	0.953	1.505
92	Benzyl ethyl ether		184–6*	0.9478	1.4958
93	Pentyl ether	30	187.5	0.7830	1.416
94	2,2′-Dichloropropyl ether		188	1.109	1.447
95	Bis(2-Ethoxyethyl) ether		188	0.906	1.411
96	Phenyl propyl ether	31	188	0.949	1.501
97	4-Ethoxytoluene	32	191	0.949	1.505
98	3-Ethoxytoluene	33	191	0.949	1.506
99	Allyl phenyl ether		192–5	0.9856-D_{15}^{15}	
100	2-(2-Methoxyethoxy)ethanol (see Table 6A)	34	194	1.035-D_{20}^{20}	1.4244
101	Styrene oxide		194	1.0523-D_4^{16}	1.5329²⁵
102	3-Chloro-1-methoxybenzene	35	194		
103	2-Chloro-1-methoxybenzene	36	195	1.191	1.545
104	4-Chloro-1-methoxybenzene	37	200	1.1851-D_4^{13}	
105	2-(2-Ethoxyethoxy)ethanol (see Table 6A)	38	202(196)	1.023-D_{20}^{20}	1.4298
106	1,2-Dibutoxyethane		204		
107	Butyl phenyl ether	39	206	1.5049	
108	1,2-Dimethoxybenzene (mp 22.5) (veratrole)	40	206.2	1.080	
109	2-Chloro-1-ethoxybenzene	41	208	1.134	1.530
110	2-Bromo-1-methoxybenzene	42	210		
111	Benzyl isobutyl ether		210–2	0.9233	1.4826
112	4-Chloro-1-ethoxybenzene (mp 21)	43	212	1.231-D_{20}^{20}	1.5227
113	3-Chloropropyl ether		215	1.140	
114	4-Bromo-1-methoxybenzene	44	215	1.494-D_4^9	
115	Methyl thymyl ether	45	216	0.954-D_4^0	
116	1,2-Di(2-methoxyethoxy)ethane		216	0.9862-D_{20}^{20}	1.4233
117	1,3-Dimethoxybenzene	46	217*	1.050	1.4233
118	2-Bromo-1-ethoxybenzene	47	218		
119	Benzyl butyl ether		219–21	0.9227	1.4833
120	2,2′-Dipropoxydiethyl ether		222	0.8877-D_{15}^{15}	
121	2-Butoxytoluene	48	223	0.9437-D_0^0	
122	Hexyl ether	49	229		
123	Ethyl 2-aminophenyl ether (see Table 9A)		229		
124	Di(2-chloroethoxy)ethane		230		

(*Continued*)

TABLE 13A (Continued)
ETHERS—LIQUID

	Name of Compound	Note	BP	D_4^{20}	n_D^{20}
125	Butoxyethoxyethanol		231	0.9954-D_{20}^{20}	
126	1-Allyl-3,4-dimethylenedioxy-benzene (safrole)	50	232	1.100	1.5383
127	4-Bromo-1-ethoxybenzene	51	233(229)		
128	1-(4-Methoxyphenyl)-1-propene (anethole) (mp 22)	52	235	0.989	1.558
129	1,3-Diethoxybenzene	53	235		
130	2-Iodo-1-methoxybenzene	54	240	1.800	
131	1,2-Dimethoxy-4-allylbenzene (eugenol methyl ether)	55	244	1.050	1.5360
132	2-Phenoxyethanol		245	1.1094-D_{20}^{20}	
133	2-Iodo-1-ethoxybenzene	56	246 (740 mm)	1.800	
134	1,2-Dimethylenedioxy-4-propenyl-benzene (isosafrole) (*trans*)	57	248	1.122	1.5782
135	2,2′-Dibutoxydiethyl ether		254	0.8853	
136	Phenyl ether (mp 28) (see Table 13B)		258		
137	Heptyl ether	58	263(260)	0.805	1.427
138	1,2-Dimethoxy-4-propenyl-benzene (isoeugenol methyl ether)	59	264	1.0528	1.5692
139	2-Nitro-1-ethoxybenzene	60	267		
140	1-Methoxynaphthalene	61	271*	1.0916	1.6940
141	2-Nitro-1-methoxybenzene	62	272	1.254	1.562
142	Bis(methoxyethoxyethyl) ether (tetraethyleneglycol dimethyl ether)		275.3	1.009	1.4322
143	1-Ethoxynaphthalene	63	280.5*	1.074	1.5973
144	Octyl ether		288	0.806	1.433
145	Benzyl ether	64	300*d*	1.0428	
146	Isopentyl 1-naphthyl ether	65	317.5	1.0069-D_4^{14}	1.5705[14]

REFERENCES FOR TABLE 13

Summers, L., *Chem. Rev.* **55**, 304 (1955).

TABLE 13B
ETHERS—SOLIDS

	Name of Compound	Note	MP	BP	D_4^{20}	n_D^{20}
1	Allyl 2-naphthyl ether		16	dec		1.600^{25}
2	4-Chloro-1-ethoxybenzene (see Table 13A)		21	212		
3	1-(4-Methoxyphenyl)-1-propene (see Table 13A)		22	235		
4	1,2-Dimethoxybenzene (see Table 13A)		22.5	206.2		
5	2-Pentoxynaphthalene	66	24.5 (30)	327.5*		
6	4-Iodo-1-ethoxybenzene	67	27	252		
7	Phenyl ether	68	28	259	1.073	1.5826^{24}
8	2-(3-Methylbutoxy)naphthalene	69	28	321d	$1.0156\text{-}D_4^{12}$	1.5768^{12}
9	2-Methoxybiphenyl	70	29	274		
10	2-Bromoethyl phenyl ether	71	32			
11	2-(2-Methylpropoxy)-naphthalene	72	33	304		
12	Dodecyl ether	73	33			
13	3-Nitro-1-ethoxybenzene	74	34	284		
14	2-(2-Butoxy)naphthalene	75	34	298.5*		
15	2-Butoxynaphthalene	76	35.5	309*		
16	2-Ethoxynaphthalene	77	36	282*	1.064	1.5932^{47}
17	3-Nitro-1-methoxybenzene	78	39	258		
18	2-(2-Propoxy)naphthalene	79	40	285		
19	2-Propoxynaphthalene	80	40	297		
20	1,2-Diethoxybenzene	81	43	217		
21	2,4,6-Trichloro-1-ethoxybenzene	82	44	246		
22	1,4-Dibutoxybenzene		45			
23	8-Methoxyquinoline	83	45			
24	1,2,3-Trimethoxybenzene	84	47	241		
25	4-Iodo-1-methoxybenzene		52	139		
26	4-Methoxyphenol		52.5			
27	1,3,5-Trimethoxybenzene	85	52–3	255		
28	1-Phenoxynaphthalene	86	54			
29	Hexadecyl ether	87	54	dec		
30	4-Nitro-1-methoxybenzene	88	54	259	1.233	1.5707^{60}
31	1,4-Dimethoxybenzene	89	56	213		
32	4-Nitro-1-ethoxybenzene	90	60	283		
33	2,4,6-Trichloro-1-methoxybenzene	91	60			
34	1,3-Diphenoxypropane		61	338–40		
35	Trioxane		61–3			
36	1,2-Dibenzoxybenzene	92	63–4			
37	Bis(2-phenoxyethyl)ether		66			
38	2,2′-Dimethoxybiphenyl	93	70			
39	1,4-Diethoxybenzene	94	72			
40	2-Phenethyl 1-naphthyl ether	95	72			
41	2,4,6-Tribromo-1-ethoxybenzene	96	72			

(*Continued*)

TABLE 13B (Continued)
ETHERS—SOLIDS

	Name of Compound	Note	MP	BP	D_4^{20}	n_D^{20}
42	2-Methoxynaphthalene	97	73*	273		
43	1-Benzyloxynaphthalene		77(61)			
44	1,2′-Dinaphthyl ether	98	81			
45	Biphenyleneoxide	99	86	288*		
46	2,4,6-Tribromo-1-methoxybenzene		87			
47	4-Methoxybiphenyl	100	90			
48	2,4-Dinitro-1-methoxybenzene		95			
49	1,2-Diphenoxyethane	101	98			
50	2-Benzyloxynaphthalene	102	101.5*	dec		
51	4,4′-Dimethoxydiphenyl ether		103			
52	2-Naphthyl ether	103	105			
53	3,5-Dinitro-1-methoxybenzene		105–6			
54	1-Naphthyl ether	104	110			
55	1,4-Dibenzyloxybenzene	105	128			
56	7-Methoxyquinoline	106	210	287		

NOTES FOR TABLES 13A AND 13B (ETHERS)

1. Forms a polymer in aqueous alkali, 56.
2. Furan may be detected by applying it to a pine splinter that has been moistened with hydrochloric acid. A green color develops.
3. 3,5-Dinitrobenzoate, 93.
4. Heated with dilute sulfuric acid yields 1,3-propanediol, bp 187.4.
5. Dibromo derivative, bp 185.
6. With 4% water, forms constant boiling mixture, bp 52.6.
7. Picrate, 163.
8. *Trans*-2,3-epoxybutane, bp 53–4/741 mm; D_4^{25} 0.8010; *cis*-2,3-epoxybutane, bp 58–9/745 mm; D_4^{25} 0.8226. The normal crude mixture is 65% *trans* and 35% *cis*.
9. With 25% ethanol, forms constant boiling mixture, bp 61.2.
10. 3,5-Dinitrobenzoate, 123.
11. With 9% water, forms constant boiling mixture, bp 73.8. With 50% methanol, forms constant boiling mixture, bp 62.3.
12. 3,5-Dinitrobenzoate, 74.
13. Pentabromo derivative, 180.
14. With 13% water, gives constant boiling mixture, bp 81.2. Bromine derivative, 65–6. Iodine derivative, 84–5. Constant boiling mixture with water, bp 82.8, contains 48 mole per cent of dioxane.
15. Picrate, 57.
16. 3,5-Dinitrobenzoate, 75.5.
17. 3,5-Dinitrobenzoate, 87.
18. Reacts with sodium amalgam to yield allyl alcohol.
19. 3,5-Dinitrobenzoate, 64.
20. Forms two 2,4-dinitro 87 and 95.5. 2,4-Dibromo melts at 61. Picrate melts at 79–81. 3,5-Dinitrobenzoate, 87. Sulfonamide, 113.
21. Picrate, 115–6.
22. 3,5-Dinitro derivative, 69. 5-Bromo derivative, 63–4. Picrate, 116. (119). Sulfonamide, 137.
23. Bromo, 172. 3-Nitro phthalate 120.
24. Picrate, 92. Sulfonamide, 150.
25. 3,5-Dinitrobenzoate, 61. Constant boiling mixture with water, bp 97.2.
26. Picrate, 88–9. Oxidizes to *p*-anisic acid, 184. Sulfonamide, 182.
27. Picrate, 113–4. 2-Nitro, 54–5; 2,4,6-Trinitro, 92. Oxidizes to *m*-hydroxybenzoic acid, 110. Sulfonamide, 130.
28. HBr addition product, 56–7. Adduct with 2-naphthol, 48.
29. Picrate, 118. Dinitro, 51. Sulfonamide, 149.
30. 3,5-Dinitrobenzoate, 42–3.
31. Sulfonamide, 117.

NOTES FOR TABLES 13A AND 13B (ETHERS) (Continued)

32. Picrate, 111. *p*-Ethoxybenzoic acid, 198. Sulfonamide, 138.
33. Picrate, 115. *m*-Ethoxybenzoic acid, 137. Sulfonamide, 111.
34. *p*-Nitrophenyl urethane, 73.5.
35. Sulfonamide, 131.
36. Nitro, 95.
37. 2-Nitro, 95. Sulfonamide, 151.
38. *p*-Nitrophenyl urethane, 66.
39. Picrate, 110–2. Sulfonamide, 103–4.
40. Picrate, 57. Dibromo 93. Nitro, 95. Sulfonamide, 136.
41. Nitro, 82. Sulfonamide, 133.
42. Nitro, 106. Sulfonamide, 140.
43. 2,6-Dinitro derivative, 54(61). Sulfonamide, 134.
44. Nitro, 88. Sulfonamide, 148.
45. Trinitro, 92
46. Picrate, 57. Dibromo, 140. 2,4,6-Trinitro, 125. Sulfonamide, 167.
47. Nitro, 98. Sulfonamide, 135.
48. Sulfonamide, 95–6.
49. 3,5-Dinitrobenzoate, 55.
50. Picrate, 105. Tribromo, 108. Pentabromo, 170. Trinitrobenzene adduct, 51.
51. Nitro, 47. Sulfonamide, 145.
52. Picrate, 70. Dibromide, 67. Tribromide, 108. Oxidizes to *p*-anisic acid, 184–6.
53. Picrate, 109. Tribromo, 69. Sulfonamide, 184.
54. Nitro derivative, mp 95.
55. Picrate, 114–5. Tribromide, 78. Oxidizes to veratric acid, 181.
56. Nitro, 96.
57. Picrate, 84. Tribromo, 110. Piperonylic acid, 229.
58. 3,5-Dinitrobenzoate, 47.
59. Picrate, 42–5. Dibromo, 101. Trinitrobenzene adduct, 69–70. Oxidizes to veratric acid, 181.
60. *o*-Phenetidine, bp 228.
61. Picrate, 130. Dibromo derivative, 54–5. Trinitrobenzene adduct, 139–40. Sulfonamide, 157.
62. *o*-Anisidine, b.p. 225.
63. Picrate, 119. 4-Bromo, 48. Trinitrobenzene adduct, 125.5. Sulfonamide, 165.
64. Picrate, 78. Dibromo derivative, 108.
65. Picrate, 96–7.
66. Picrate, 67(75).
67. Nitro, 96.
68. Picrate, 110. 4,4′-Dibromo, 55(58). 4,4′-Dinitro, 135. 2,2′, 4,4,′-Tetranitro, 195–7. Sulfonamide, 159.
69. Picrate, 94–6.
70. 5-Nitro, 95–6.
71. Bromo, 56.
72. Picrate, 84–5.
73. 3,5-Dinitrobenzoate, 60.
74. *m*-Phenetidine picrate, 158.
75. Picrate, 86.
76. Picrate, 67.
77. Picrate, 101. 1-Bromo, 66. 1,6-Dibromo, 94. Sulfonamide, 163.
78. *m*-Anisidine picrate, 169*d*.
79. Picrate, 95.
80. Picrate, 81.
81. Picrate, 71. Trinitro, 122. Sulfonamide, 162.
82. Dinitro derivative, 100.
83. Picrate, 162.
84. Picrate, 81. 5-Nitro, 106. 4,5,6-Tribromo, 73–4. Sulfonamide, 124.
85. 2-Bromo, 96–7. 2,4-Dibromo, 129–30. 2,4,6-Tribromo, 145.
86. Trinitrobenzene adduct, 112.5.
87. 3,5-Dinitrobenzoate, 166.
88. *p*-Anisidine, 57.
89. Sulfonamide, 148. Picrate, 48. Dibromo, 142. 2-Nitro 72. 2,3-Dinitro, 177. 2,5-Dinitro, 202. Trinitrobenzene adduct, 86.5.
90. Phenacetin, 138.
91. Dinitro, 95.
92. 4-Nitro, 98.
93. 5,5′-Dibromo, 155.
94. Nitro, 49. 2,3-Dinitro, 130. 2,5-dinitro, 176. Trinitrobenzene adduct, 87. Sulfonamide, 155.
95. Picrate, 118.
96. Nitro, 79.
97. Picrate, 117. Monobromo, 63. 1-Nitro, 128. 1,6,8-Trinitro, 215*d*. Trinitrobenzene adduct, 93.5. Sulfonamide, 151.
98. Picrate, 121–2.
99. Picrate, 94. 3-Nitro, 182. Dinitro, 245. Trinitrobenzene adduct, 96.
100. 4′-Nitro, 144. 3,4′-Dinitro, 134; 3,4′-disulfonamide, 171.
101. 2,4-Dinitro, 215. 4,4′-Dibromo, 134–5. 4,4′-Disulfonamide, 228–9.
102. Picrate, 123.
103. Picrate, 122. Trinitrobenzene adduct, 128.
104. Picrate, 115.
105. Nitro, 83.
106. Picrate, 229.

TABLE 14
ALKYL AND CYCLOALKYL HALIDES[a]

						Melting Point of Derivatives			
						Recommended		Others	
	Name of Compound	Note	BP	D_4^{20}	n_D^{20}	S-Alkyl-isothiourea Picrate	α-Naph-thalide	Anilide	Alkyl-mercuric Halide
	FLUORIDES								
1	Fluoromethane		−78.4	0.5786[c]	1.1727[c]				
2	Fluoroethane		−37.7	0.7182[c]	1.2656[c]				
3	2-Fluoropropane		−9.4	0.7238[c]	1.3020[c]				
4	1-Fluoropropane		−2.5	0.7956[c]	1.3115[c]				
5	2-Fluoro-2-methylpropane		12.1	0.7421[c]	1.3201[c]				
6	2-Fluorobutane		25.1	0.7621	1.3326				
7	1-Fluorobutane		32.5	0.7789	1.3396				
8	2-Fluoro-2-methylbutane		44.8	0.7780	1.3502				
9	2-Fluoropentane		50						
10	1-Fluoropentane		62.8	0.7903	1.3591				
11	3-Fluorohexane		82.9	0.7949	1.3714				
12	2-Fluorohexane		86.3	0.7916	1.3693				
13	1-Fluorohexane		91.5	0.7995	1.3738				
14	1-Fluoroheptane		117.9	0.8062	1.3854				
15	α-Fluorotoluene		140	1.02278^{25}					
16	1-Fluorooctane		142.3	0.8116	1.3946				
17	1-Fluorononane		165	0.8159	1.4022				
18	1-Fluorodecane		186.2	0.8194	1.4085				
19	1-Fluoroundecane		206	0.8181^{25}	1.4138				
	CHLORIDES								
20	Chloromethane		−24.2	0.9159[c]	1.3389[c]	224	160	114	167*
21	Chloroethene		−13.4	0.9106[c] (0.99343^{-20})	1.370[c]			104	
22	Chloroethane		12.3	0.8978[c] (0.9028^{10})	1.3676[c] (1.3738^{10})	188	126	102	192
23	2-Chloro-1-propene		22.7	0.9017	1.3973				
24	1-Chloro-1-propene(cis)		32.8	0.9347	1.4055				
25	2-Chloropropane		35.7	0.8617	1.3777	196		103	
26	1-Chloro-1-propene(trans)		37.4	0.935	1.4054			114	
27	3-Chloro-1-propene		45	0.9376	1.4157	155		114	

[a] See Table 16 for compounds that contain two or more halogen atoms in the nonaromatic part of the molecule. The monohalogen derivatives of certain alkenes are included.

[b] See *J. Am. Chem. Soc.*, **73**, 1968 (1951) for details of methods of preparing the 2,4-dinitrophenyl sulfide from primary and secondary alkyl halides and some related compounds. Tertiary alkyl halides and allyl halides do not react with sodium 2,4-dinitrothiophenolate to give derivatives. The sulfides may be oxidized to sulfones.

[c] For the liquid at the saturation pressure.

[d] For the supercooled liquid at 20°C.

TABLE 14 (Continued)
ALKYL AND CYCLOALKYL HALIDES[a]

Melting Point of Derivatives
Others (Continued)

	Alkyl β-Naph-thyl Ethers	Picrate of β-Naphthyl Ethers	Alkyl-3-nitro-phthal-imide	Alkyl-saccharin	2,4,6-Tri-iodophenyl Ethers	Alkyl 6-Nitro-benzo-thiazolyl-Sulfides	Alkyl 6-Nitro-benzo-thiazolyl-2-sulfones	2,4-Di-nitro-phenyl Sulfides[b]	2,4-Di-nitro-phenyl Sulfones[b]
					FLUORIDES				
1						134	186		
2						103–4	160		
3									
4									
5									
6									
7									
8									
9									
10									
11									
12									
13									
14									
15									
16									
17									
18									
19									
					CHLORIDES				
20	70	117	113	132		134	186	128	185(189)
21									
22	37	104	106	94	83–4	103–4	160	115	156(160)
23									
24									
25	41	92		134		75–6	154		
26									
27	16	99	100–1	58					

(Continued)

TABLE 14 (Continued)
ALKYL AND CYCLOALKYL HALIDES[a]

	Name of Compound	Note	BP	D_4^{20}	n_D^{20}	Recommended S-Alkyl-isothiourea Picrate	α-Naph-thalide	Others Anilide	Alkyl-mercuric Halide
28	1-Chloropropane		46.6	0.8909	1.3879	177	121	92	147*
29	2-Chloro-2-methyl-propane	30	50.7	0.8420	1.3857	151N	147	128	122–3
30	Chloromethyl methyl ether	1	58–60	1.063^{10}	1.39737				
31	2-Chloro-1,3-butadiene		59.4	0.9583^{20}_{20}	1.4583				
32	2-Chlorobutane		68.3	0.8732	1.3971	166	129	108	30.5
33	1-Chloro-2-methyl-propane		68.9	0.878	1.398	174(167)	125	109	
34	3-Chloro-2-methyl-propene	2	72	0.9475	1.4340				
35	1-Chlorobutane		78.4	0.88621	1.40211	177	112	63	128
36	1-Chloro-2,2-dimethylpropane		84.3	0.8660	1.4044			130–1	117–8
37	2-Chloro-2-methylbutane		85.6	0.8653	1.4055	127d	138	92	
38	2-Chloro-3-methylbutane		92.8	0.862	1.402				
39	3-Chloropentene	3	93–4	0.8978	1.4254				
40	dl-3-Chloro-2-methyl-1-butene	4	94	0.9088	1.4304				
41	3-Chloro-2-methyl-2-butene	4	94	0.925	1.4320				
42	2-Chloropentane		96.9	0.8698	1.4069		155	93	
43	3-Chloropentane		97.8	0.8731	1.4082	159		124	
44	1-Chloro-3-methylbutane		101	0.8927	1.4096	173	111	108	86
45	1-Chloropentane		107.8	0.8818	1.4120	154	112	96	110*
46	2-Chloro-diethylether		108	0.9895	1.4113				
47	1-Chloro-2-pentene	5	109–10	$0.908^{21.5}_4$	$1.4352^{21.5}$				
48	2-Chloro-2-methyl-pentane		110–3	0.863	1.4126		116–8	71.4	
49	2-Chloro-3,3-dimethyl-butane		111	0.8767	1.4182				89–9•
50	2-Chloro-2,3-dimethyl-butane		112	0.8780	1.4191				
51	Chlorocyclopentane		114–5	1.005	1.4510				108
52	1-Chloro-3,3-dimethyl-butane		115	0.8670	1.4160			138–9	133
53	3-Chloro-3-methylpentane		116	0.8900	1.421			87–8	
54	3-Chloro-1,2-epoxy-propane		116–7	1.181	1.438				
55	3-Chlorohexane		123	0.870^{20}_{20}	1.4163				

TABLE 14 (Continued)
ALKYL AND CYCLOALKYL HALIDES[a]

| | Melting Point of Derivatives | | | | | | | |
| | Others (Continued) | | | | | | | |
	Alkyl β-Naph-thyl Ethers	Picrate of β-Naphthyl Ethers	Alkyl-3-nitro-phthal-imide	Alkyl-saccharin	2,4,6-Tri-iodophenyl Ethers	Alkyl 6-Nitro-benzo-thiazolyl Sulfides	Alkyl 6-Nitro-benzo-thiazolyl-2-sulfones	2,4-Di-nitro-phenyl Sulfides[b]	2,4-Di-nitro-phenyl Sulfones[b]
28	39	75(81)	85	74	82	94–5	181	84	126
29									
30									
31									
32	298*	85		81		(Oil)	121	66	120
33	33	84		95		68–9	124	76	105
34									
35	33	67	72	38		64–5	152	66	92
36									
37									
38									
39									
40									
41									
42									
43									
44		94	94			56–7	115	80	124
45		67		58		48–9	125	80	83
46									
47									
48									
49									
50									
51									
52									
53									
54									
55									

(Continued)

TABLE 14 (Continued)
ALKYL AND CYCLOALKYL HALIDES[a]

	Name of Compound	Note	BP	D_4^{20}	n_D^{20}	Recommended		Others	
						S-Alkyl-isothiourea Picrate	α-Naph-thalide	Anilide	Alkyl-mercuric Halide
56	2-Chlorohexane		125	0.8694^{21}	1.4142^{21}			91–2	
57	1-Chloro-2-ethylbutane		125–7	0.8914	1.4230			83	
58	1-Chlorohexane		134.5	0.8785	1.4196	157	106	69	125*
59	Chlorocyclohexane		143	0.989	1.462		188	146	
60	1-Chloro-3,4-dimethylpentane		152	0.8825	1.4299			80–1	
61	1-Chloroheptane		159.1	0.8758	1.4256	142	95	57	119.5*
62	1-Chloro-2-ethylhexane		173	0.8833^{20}_{20}					
63	α-Chlorotoluene	6	179.4	1.100	1.439	188	166	117	104
64	1-Chlorooctane		180	0.8737	1.4305	134	91	57	151
65	α-Chloroethylbenzene		195d	1.0620	1.5276			133	
66	β-Chloroethylbenzene		197–8	1.069^{25}	1.5294			97	
67	β-Chlorovinylbenzene		198	1.1085	1.5710^{25}		217	115	
68	α-Chlorovinylbenzene		199	1.101	1.5600				
69	1-Chlorononane		203.4	0.8720	1.4345	131			
70	α,2-Dichlorotoluene	7	213–4						
71	α,3-Dichlorotoluene	8	216		1.2695^{15}				
72	1-Chlorodecane		223.4	0.8705	1.4379	137			
73	α-Chloro-4-isopropyltoluene		228	$1.020^{21.5}$	$1.523^{21.5}$				
74	α,3,4-Trichlorotoluene	9	241						
75	1-Chloroundecane		242.2	0.8693	1.4408	139			
76	α,2,4-Trichlorotoluene	10	248	1.4068	1.5761				
77	1-Chlorododecane		259.9	0.8682	1.4433	139			114
78	1-Chlorotridecane		277	0.8673	1.4454				
79	1-Chlorotetradecane		292	0.8665	1.4473				
80	1-Chloropentadecane		308	0.8658	1.4490				
81	1-Chlorohexadecane (12-14)		322	0.8652	1.4505	155(137)			102
			MP						
82	α-Chloro-3-bromotoluene	31	23						
83	2-Phenoxychloroethane	11	25						
84	1-Chloroheptadecane		26.2	0.8646^d	1.4519^d				
85	1-Chlorooctadecane		28.6	0.8641^d	1.4531^d				
86	α,4-Dichlorotoluene	12	30					166	
87	1-Chlorononadecane		35.7	0.8636^d	1.4542^d				
88	α-Chloro-4-bromotoluene	13	36(41)			219			
89	2,4,6-Trimethylbenzyl chloride	14	37						
90	1-Chloroeicosane		37.6	0.8632^d	1.4552^d				
91	α,2,6-Trichlorotoluene	15	39–40						
92	α-Chloro-3-nitrotoluene	16	45						
93	1-Chloro-2,2,3,3-tetramethylbutane		52–3						170–1
94	2,4-Dimethylphenacyl chloride	17	62						
95	α-Chloro-4-nitrotoluene	18	71			205			

TABLE 14 (Continued)
ALKYL AND CYCLOALKYL HALIDES[a]

	Melting Point of Derivatives							
	Others (Continued)							
	Alkyl β-Naph-thyl Ethers	Picrate of β-Naphthyl Ethers	Alkyl-3-nitro-phthal-imide	Alkyl-saccharin	2,4,6-Tri-iodophenyl Ethers	Alkyl 6-Nitro-benzo-thiazolyl Sulfides	Alkyl 6-Nitro-benzo-thiazolyl-2-sulfones	2,4-Di-nitro-phenyl Sulfides[b]	2,4-Di-nitro-phenyl Sulfones[b]
56									
57									
58						55.5	111.5	74	97
59						100–1	189		
60									
61						39.5	123.5	82	101
62									
63	99	123	143	110–1 (118)	122	114–5	194	130	178(182)
64								78	98
65									
66		84							
67									
68									
69								86	92
70									
71									
72								85	93
73									
74									
75									
76									
77									
78									
79									
80									
81			101	98				95	105

(Continued)

TABLE 14 (Continued)
ALKYL AND CYCLOALKYL HALIDES[a]

	Name of Compound	Note	BP	D_4^{20}	n_D^{20}	Recommended		Others	
						S-Alkyl-isothio-urea Picrate	α-Naph-thalide	Anilide	Alkyl-mercuric Halide
96	Triphenylchloromethane	19	113						
97	Bornyl chloride		132						

BROMIDES

		Note	BP	D_4^{20}	n_D^{20}	S-Alkyl-isothio-urea Picrate	α-Naph-thalide	Anilide	Alkyl-mercuric Halide
98	Bromomethane		3.56	1.6755[c]	1.4218[c]	224	160	114	172*(160)
99	Bromoethene		15.8	1.4933[c]	1.441[c]			104	
100	Bromoethane		38.4	1.4605	1.4239	188	126	104	193(198)
101	1-Bromo-1-propene(cis)		57.8	1.4291	1.4560				
102	2-Bromopropane		59.4	1.3140	1.4251	196		103	93
103	1-Bromo-1-propene(trans)		63.2	1.4155	1.456				
104	1-Bromopropane		71	1.3537	1.4343	177	121	92	138
105	3-Bromo-1-propene		71.3	1.398	1.4655	155		114	
106	2-Bromo-2-methylpropane	30	73.3	1.2209	1.4278	151N	147	128	
107	2-Bromo-1-butene		81	1.3136	1.4537				
108	2-Bromo-2-butene(cis)		85.8	1.3237	1.4580				
109	1-Bromo-1-butene		86.2	1.3192	1.456				
110	2-Bromobutane		91.2	1.2585	1.4367	166	129	108	39
111	1-Bromo-2-methylpropane		93.0	1.2645	1.4350	174(167)	125	109	55
112	1-Bromo-1-butene(trans)		94.7	1.3202	1.456				
113	2-Bromo-2-butene(trans)		94.8	1.3291	1.457				
114	1-Bromobutane		101.6	1.3758	1.4401	177	112	63	136
115	1-Bromo-2-butene		103–6	1.3371[25]	1.4822				
116	1-Bromo-2,2-dimethylpropane		106	1.1997	1.4370			126	
117	2-Bromo-2-methylbutane		109.2	1.2160	1.4420	127d	138	92	
118	2-Bromo-3-methylbutane		115.3	1.2209	1.4454				
119	2-Bromopentane		117.4	1.2075	1.4413	155		93	
120	3-Bromopentane		118.6	1.2124	1.4441	159		124	
121	1-Bromo-3-methylbutane		120.4	1.2071	1.4420	173	111	108	80
122	1-Bromo-2-methylbutane		120.5	1.2205	1.4452				
123	1-Bromopentane		129.6	1.2182	1.4443	154	112	96	127
124	2-Bromo-4-methylpentane		131	1.1568	1.4421	147			
125	2-Bromo-3,3-dimethylbutane		132	1.17[25]	1.45[25]				
126	Bromocyclopentane		137	1.387	1.489				
127	2-Bromohexane		146	1.1658	1.4832[25]				
128	1-Bromohexane		155.3	1.1744	1.4475	157	106	69	127.5
129	1-Bromo-1-heptane	20	163	1.1532	1.461				
130	Bromocyclohexane		165	1.336	1.495		188	146	153
131	1-Bromo-1-heptene		178.9	1.1400	1.4502	142	95	57	118.5*
132	α-Bromotoluene		198	1.438		188	166	117	119
133	1-Bromooctane		200.8	1.1122	1.4524	134	91	57	109
134	α-Bromoethylbenzene		205	1.3108[23]				133	
135	β-Bromoethylcyclohexane		212.1	1.2357	1.4899				
136	β-Bromoethylbenzene		218	1.359	1.556			97	169
137	β-Bromostyrene (I)	21	221	1.422	1.6094[20.5]		217	215	91
138	1-Bromononane		221.4	1.0893	1.4542	131			109
139	1-Bromodecane		240.6	1.0702	1.4557	137			

TABLE 14 (Continued)
ALKYL AND CYCLOALKYL HALIDES[a]

	Melting Point of Derivatives							
	Others (Continued)							
	Alkyl β-Naphthyl Ethers	Picrate of β-Naphthyl Ethers	Alkyl-3-nitro-phthal-imide	Alkyl-saccharin	2,4,6-Tri-iodophenyl Ethers	Alkyl 6-Nitro-benzo-thiazolyl Sulfides	Alkyl 6-Nitro-benzo-thiazolyl-2-sulfones	2,4-Di-nitro-phenyl Sulfides[b]	2,4-Di-nitro-phenyl Sulfones[b]
96									
97									
				BROMIDES					
98	70	117	113	132		134	186	128	185(189)
99									
100	37	104	106	94	83–4	103–4	160	115	156(160)
101									
102	41	92		134	43	75–6	154	95	140
103									
104	39	75(81)	85	74	82	94–5	181	84	126
105		99				72–3		71	
106									
107									
108									
109									
110	298*	85		81		(Oil)	121	66	120
111		84		95	48	68–9	124	76	105
112									
113									
114	33	67	72	38	66	64–5	152	66	92
115									
116									
117									
118									
119									
120									
121		94	94			56–7	115	80	124
122									
123		67		58	47	48–9	125	80	83
124									
125									
126									
127									
128					44–5	55.1	111.5	74	97
129									
130						100–1	189		
131						39.5	123.5	82	101
132	99	123	143			114–5	194	130	178(182)
133								78	98
134									
135									
136		84			88				
137									
138								86	92
139									

(Continued)

TABLE 14 (Continued)
ALKYL AND CYCLOALKYL HALIDES[a]

	Name of Compound	Note	MP	D_4^{20}	n_D^{20}	Melting Point of Derivatives			
						Recommended		Others	
						S-Alkyl-isothiourea Picrate	α-Naphthalide	Anilide	Alkylmercuric Halide
140	1-Bromoundecane		258.8	1.0539	1.4571				
141	1-Bromododecane		275.9	1.0399	1.4583	139			108
142	1-Bromotridecane		292	1.0277	1.4593				
143	1-Bromotetradecane (mp 5)		307	1.0170	1.4603				
144	1-Bromopentadecane (mp 19)		322	1.0075	1.4611				
145	1-Bromohexadecane (mp 18)		336	0.9991	1.4618	155(137)			101.5
146	β-Bromostyrene (II)	21	108/26 mm	1.427	1.5990[22]				
			MP						
147	α-Bromo-2-methyltoluene	22	21	1.3811[23]					
148	1-Bromooctadecane		28.2	0.9848[d]	1.4631[d]				
149	1-Bromoheptadecane		29.6	0.9916[d]	1.4625[d]				
150	α,2-Dibromotoluene	23	31			222			
151	α-Bromo-4-methyltoluene	24	35						
152	1-Bromoeicosane		36.9	0.9730[d]	1.4643[d]				
153	1-Bromononadecane		38.5	0.9786[d]	1.4637[d]				
154	α,3-Dibromotoluene	25	41			205			
155	α-Bromo-2-nitrotoluene	26	46–7						
156	α-Bromo-4-chlorotoluene	27	51 v.s.			194			
157	α-Bromo-3-nitrotoluene	16	58–9						
158	α,4-Dibromotoluene	13	62			219			
159	α-Bromo-4-nitrotoluene	18	99			205			
			IODIDES						
			BP						
160	Iodomethane		42.4	2.2790	1.5308	224	160	114	152
161	Iodoethene		56	2.037	1.53845			104	
162	Iodoethane		72.3	1.9358	1.5133	188	126	104	186
163	2-Iodopropane		89.5	1.7033	1.49969	196		103	
164	2-Iodo-2-methylpropane	28,30	100d	1.571[0]		151N	147	128	
165	3-Iodo-1-propene		102	1.8494	1.5530	155	121	114	112
166	1-Iodopropane		102.5	1.7489	1.5058	177	121	92	113*
167	2-Iodobutane		118	1.5920	1.49909	166	129	108	
168	1-Iodo-2-methylpropane		120.4	1.606	1.49597	174 (167)	125	109	72
169	2-Iodo-2-methylbutane		126	1.524[0]		127d	138	92	
170	1-Iodobutane		130.5	1.6154	1.5001	177	112(110)	63	117
171	1-Iodo-2-butene		132–3d	1.6823[0]					
172	dl-2-Iodopentane		141	1.5086	1.4961	155		93	
173	l-2-Iodopentane		143	1.5067[17]					
174	3-Iodopentane		144/738 mm	1.5176	1.4968	159		124	
175	1-Iodo-3-methylbutane		148	1.503	1.493	173	111	108	122
176	1-Iodo-2-methylbutane		148	1.524	1.4981				
177	1-Iodopentane		157	1.5161	1.4959	154	112	96	110*
178	Iodocyclopentane		166–7	1.7096	1.5447				
179	Iodocyclohexane		179 sl.d.	1.626[15]			188	146	
180	1-Iodohexane		181.3	1.4397	1.4928	157	106	69	110*

TABLE 14 (Continued)
ALKYL AND CYCLOALKYL HALIDES[a]

	Melting Point of Derivatives							
	Others (Continued)							
	Alkyl β-Naph-thyl Ethers	Picrate of β-Naphthyl Ethers	Alkyl-3-nitro-phthal-imide	Alkyl-saccharin	2,4,6-Tri-iodophenyl Ethers	Alkyl 6-Nitro-benzo-thiazolyl Sulfides	Alkyl 6-Nitro-benzo-thiazolyl-2-sulfones	2,4-Di-nitro-phenyl Sulfides[b]	2,4-Di-nitro-phenyl Sulfones[b]
140									
141									
142									
143								94	104
144									
145			101	98				95	105
146									
147									
148									
149									
150									
151									
152									
153									
154									
155									
156									
157									
158									
159									
				IODIDES					
160	70	117	113	132	98–9	134	186	128	185(189)
161									
162	37	102	106	94	83–4	103–4	160	115	156(160)
163	41	92		134		75–6	154	95	140
164									
165		99				72–3			
166	39	(81)75	85	74	82	94–5	181	84	126
167		85		81		(Oil)	121	66	120
168		84		95		68–9	124	76	105
169									
170		67	72	38		64–5	152	66	92
171									
172									
173									
174									
175		94	94			56–7	115	80	124
176									
177		67		58		48–9	125	80	83
178									
179									
180						55.5	111.5	74	97

(Continued)

TABLE 14 (Continued)
ALKYL AND CYCLOALKYL HALIDES[a]

	Name of Compound	Note	MP	D_4^{20}	n_D^2	Recommended S-Alkyl-isothiourea Picrate	α-Naph-thalide	Others Anilide	Alkyl-mercuric Halide
181	1-Iodoheptane		204	1.3791	1.4904	142	95	57	103*
182	1-Iodooctane		225.1	1.3298	1.4885	134			
183	1-Iodononane		245	1.2890	1.4870	131			
184	1-Iododecane		263.7	1.2546	1.4858	137			
185	1-Iodoundecane		281.5	1.2253	1.4848				
186	1-Iodododecane		298.2	1.1999	1.4840	139			
187	1-Iodotridecane		314	1.1778	1.4833				
188	1-Iodotetradecane		329	1.1584	1.4827				
			MP						
189	1-Iodopentadecane		24	1.366^{25}	1.4801^{25}				
190	α-Iodotoluene		24	1.7335^{25}		188	166	117	
191	1-Iodohexadecane		24.7	1.1213^{25}	1.4797^{25}	155(137)			82
192	1-Iodoheptadecane	29	33.7	1.1119^d	1.4814^d				
193	1-Iodooctadecane	29	34	1.0994^d	1.4810^d				
194	1-Iodoeicosane	29	41.9	1.0778^d	1.4805^d				
195	1-Iodononadecane	29	42	1.0881^d	1.4807^d				

NOTES FOR TABLE 14 (ALKYL AND CYCLOALKYL HALIDES)

1. Picrate, mp 163.
2. Phthalimide, mp 89–90.
3. Phthalimide, mp 78–9.
4. Dibromo derivative, mp 197–8.
5. Phthalimide, mp 69–70.
6. Forms a quaternary salt with N,N-dimethylaniline, mp 110. Compound is also named benzyl chloride.
7. 5-Nitro derivative, mp 66; o-chlorobenzoic acid, mp 142. Compound is also named o-chlorobenzyl chloride.
8. Oxidizes to 3-chlorobenzoic acid, mp 158. Compound is also named m-chlorobenzyl chloride.
9. Oxidizes to 3,4-dichlorobenzoic acid, mp 202 (208).
10. Oxidizes to 2,4-dichlorobenzoic acid, mp 164.
11. Reacts with potassium tetrachlorophthalimide, mp 155–6.
12. Oxidizes to 4-chlorobenzoic acid, mp 242.
13. Oxidizes to 4-bromobenzoic acid, mp 251.
14. Phthalimide, mp 209–10; 2,4,6-trimethylbenzyl alcohol, mp 88–9.
15. Carbonation of the Grignard reagent produces 2,6-dichlorophenylacetic acid; oxidizes to 2,6-dichlorobenzoic acid, mp 144 (139).
16. Oxidizes to 3-nitrobenzoic acid, mp 140.
17. Oxidation by sodium hypobromite yields 2,4-dimethylbenzoic acid, mp 126.
18. Oxidizes to 4-nitrobenzoic acid, mp 241; 4-nitrobenzyl alcohol, mp 93.
19. Compound boiled with water yields triphenylmethanol, mp 162; phthalimide, mp 172.
20. Data are for the cis and trans isomers.
21. β-Bromostyrene exists in two forms, one boils at 221 and the other only distills under reduced pressure (108/26 mm).
22. Also named α-bromo-o-xylene, and 2-methylbenzyl bromide.
23. Oxidizes to 2-bromobenzoic acid, mp 150.
24. Also named α-bromo-p-xylene and 4-methylbenzyl bromide; oxidizes to terephthalic acid, mp 300.
25. Oxidizes to 3-bromobenzoic acid, mp 155.
26. Oxidizes to 2-nitrobenzoic acid, mp 146–8.
27. Oxidizes to 4-chlorobenzoic acid, mp 242.
28. Readily converts to 2-methyl-2-propanol by even cold water, mp 25.5 (bp 82.5).
29. The data for density and refractive index are for the supercooled liquid at 20°.
30. Melting points of 161 and 188 have also been reported.
31. Oxidizes to 3-bromobenzoic acid, mp 155.

TABLE 14 (Continued)
ALKYL AND CYCLOALKYL HALIDES[a]

	Melting Point of Derivatives							
	Others (Continued)							
Alkyl β-Naphthyl Ethers	Picrate of β-Naphthyl Ethers	Alkyl-3-nitrophthal imide	Alkyl-saccharin	2,4,6-Triiodophenyl Ethers	Alkyl 6-Nitrobenzothiazolyl Sulfides	Alkyl 6-Nitrobenzothiazolyl 2-sulfones	2,4-Dinitrophenyl Sulfides[b]	2,4-Dinitrophenyl Sulfones[b]
181					39.5	123.5	82	101
182							78	98
183							86	92
184								
185								
186								
187								
188								
189								
190	99	123	143			114–5	194	130	178(182)
191			101	98				95	105
192									
193									
194									
195									

REFERENCES FOR TABLE 14

Gavin, P. M. G., *Anal Chem.*, **32**, 554 (1960). Characterization of Alkyl Halides. Lists the method of preparation along with many melting points.

Boyd, R. N., and Meadow, M., *Anal. Chem.*, **32**, 551 (1960). Characterization of Alkyl Halides by Use of Ethylenethiourea. The melting points of a large number of derivatives are given.

Chappelow, C. C., Elliott, R. L., and Goodwin, J. T., Jr., *J. Org. Chem.*, **25**, 435 (1960). The Phenylation and Methylation of Alkoxychlorosilanes. Gives the preparation, melting points, and refractive indices for 11 compounds.

Devereux, Sister M. A., and Donahoe, H. B., *J. Org. Chem.*, **25**, 457 (1960). N-Substituted Imides. II. Potassium Naphthylimide as a Reagent for the Identification of Alkyl Halides. Lists the melting points of 17 N-alkyl naphthylimides.

Dreisbach, R. R., *Physical Properties of Chemical Compounds II*. (Washington: Amer. Chem. Soc., 1959).

Dreisbach, R. R., *Physical Properties of Chemical Compounds III*. (Washington: Amer, Chem. Soc., 1961).

Howell, W. C., et al., *J. Org. Chem.* **22**, 255 (1957). Organometallic Reactions of ω-Fluoroalkyl Halides. II. Reactions of ω-Fluoroalkylmagnesium Chlorides. The physical constants of these compounds are listed.

Huntress, E. II., *Organic Chlorine Compounds*, New York, Wiley, 1948.

Olah, G. A., Pavlath, A. E., Olah, J. A., and Herr, F., *J. Org. Chem.*, **22**, 879 (1957). Synthesis and Investigation of Organic Fluorine Compounds. Lists the melting points of the ethyl ester hydrochloride for 21 compounds.

Reid, E. E., *Organic Chemistry of Bivalent Sulfur*, Vol. V, Chemical Publishing Co., New York, 1963. p. 120. Lists the common alkyl derivatives of S-alkylthiuronium salts. Gives the melting points for the (1) picrates, (2) 3,5-dinitrobenzoates, (3) styphanates, and (4) picrolonates. *Ibid.* pp. 110–121. Substituted thioureas as picrates, picrolonates, and styphanates.

Scheflan, L., and Jacobs, M. B., *The Handbook of Solvents*, Van Nostrand, New York, 1953.

Schotte, L., and Veibel, S., *Acta Chem. Scand.*, **7**, 1357 (1953). The preparation of S-Alkylthiuronium Picrates and a New Method for the Estimation of Tertiary Alcohols. Lists the melting points of 12 S-alkylthiuronium picrates.

TABLE 15A[a]
ARYL HALIDES—LIQUID

	Name of Compound	BP	D_4^{20}	n_D^{20}
1	1,3-Difluorobenzene	83	1.1571	1.4392^{16}
2	Fluorobenzene	85.1	1.0225	1.46837
3	1,4-Difluorobenzene	88.8	1.1701	1.4422^{19}
4	1,2-Difluorobenzene	92	1.1496^{25}	1.4451^{18}
5	2-Fluorotoluene	114	$1.0041^{13.2}$	1.4704
6	3-Fluorotoluene	116	$0.9972^{13.4}$	1.4691
7	4-Fluorotoluene	117	0.998	1.469
8	Chlorobenzene	131.7	1.10578	1.52406
9	1-Bromo-4-fluorobenzene	152	1.597	1.5310^{15}
10	Bromobenzene	156.1	1.49500	1.55972
11	2-Chlorotoluene	159.3	1.08245	1.52680
12	3-Chlorotoluene	162.3	1.072	1.521
13	4-Chlorotoluene (mp 7)	162.4	1.071	1.521
14	1,3-Dichlorobenzene	173.1	1.28844	1.54586
15	1-Chloro-2-ethylbenzene	178.4	1.05690	1.52175
16	1,2-Dichlorobenzene	180.5	1.30570	1.55154
17	2-Bromotoluene	181.7	1.42322	1.55650
18	3-Bromotoluene	183.7	1.410	1.551
19	1-Chloro-3-ethylbenzene	183.8	1.05294	1.51949
20	2-Chloro-1,4-dimethylbenzene	184–5	1.0589^{15}	
21	1-Chloro-4-ethylbenzene	184.4	1.04553	1.51751
22	Iodobenzene	188.3	1.8308	1.6200
23	2-Chlorovinylbenzene	188.7	1.1000	1.5649
24	3-Chloro-1,2-dimethylbenzene	189		
25	1-Chloro-3,4-dimethylbenzene	191	1.0691_{15}^{15}	
26	1-Chloro-2-isopropylbenzene	191.1	1.03414	1.51678
27	4-Chlorovinylbenzene	192	1.0868	1.5660
28	2-Bromochlorobenzene	195	1.646	1.580
29	2-Chloromethoxybenzene	195	1.248	1.5480
30	3-Bromochlorobenzene	196	1.630	1.577

[a] See Table 14 for compounds that have one halogen in a side chain, and see Table 16 for compounds that have two or more halogens in side chains even if halogens are substituted on the ring.

TABLE 15A (Continued)
ARYL HALIDES—LIQUID

	Recommended Derivative Nitro		Other Derivatives	
	Position	MP	Name	MP
1	1,3	74		
2			4,4'-Difluorodiphenylsulfone	98
			4-Fluorobenzenesulfonamide	125
3				
4				
5			4-Fluoro-3-methylbenzenesulfonamide	105
			2-Fluorobenzoic acid	127
6			3-Fluorobenzoic acid	124
			4-Fluoro-2-methylbenzenesulfonamide	174
7			4-Fluorobenzoic acid	182
			5-Fluoro-2-methylbenzenesulfonamide	141
8	2,4	52	4-Chlorobenzenesulfonamide	144
			2,4-Dinitrobenzenesulfenyl chloride adduct	123–4
9				
10	2,4	70	1-Naphthalide	161
		(75)	4-Bromobenzenesulfonamide	166(161)
			2,4-Dinitrobenzenesulfenyl chloride adduct	140–1
11	3,5	63	2-Chlorobenzoic acid	142(140)
			4-Chloro-3-methylbenzenesulfonamide	128
12	4,6	91	3-Chlorobenzoic acid	158(155)
			4-Chloro-2-methylbenzenesulfonamide	185
13	2	38	4-Chlorobenzoic acid	240
			5-Chloro-2-methylbenzenesulfonamide	143
14	4,6	103	2,4-Dichlorobenzenesulfonamide	182
15			2-Chlorobenzoic acid	142(140)
16	4,5	110	3,4-Dichlorobenzenesulfonamide	140(135)
17	3,5	82	2-Bromobenzoic acid	150
			4-Bromo-3-methylbenzenesulfonamide	146
18	4,6	103	3-Bromobenzoic acid	155
			4-Bromo-2-methylbenzenesulfonamide	168
19			3-Chlorobenzoic acid	158(155)
20				
21			4-Chlorobenzoic acid	240
22	4	171	4-Bromoiodobenzene	91
23				
24			Dil. $HNO_3 \rightarrow$ 3-Chloro-2-methylbenzoic acid	159
25	5	63	2-Chloro-4,5-dimethylbenzenesulfonamide	207
26			2-Chlorobenzoic acid	142(140)
27				
28				
29				
30				

(Continued)

TABLE 15A (Continued)
ARYL HALIDES—LIQUID

	Name of Compound	BP	D_4^{20}	n_D^{20}
31	4-Chloromethoxybenzene	198	$1.1851^{12.8}$	
32	1-Chloro-4-isopropylbenzene	198.3	1.02078	1.51174
33	2,6-Dichlorotoluene	199	1.2686	1.5510
34	1-Bromo-2-ethylbenzene	199.3	1.35483	1.54856
35	2,4-Dichlorotoluene	200*	1.249	1.549
36	3,5-Dichlorotoluene	201		
37	3-Iodotoluene	204	1.698	
38	2-Chloro-1,3,5-trimethylbenzene	204–6	1.0337^{30}	1.5212^{30}
39	1-Bromo-4-ethylbenzene	205.1	1.34226	1.54475
40	2-Bromo-1,4-dimethylbenzene (mp 9)	207	1.356	$1.5514^{18.5}$
41	2,3-Dichlorotoluene	207		1.5511
42	3,4-Dichlorotoluene	208.9	1.25256	1.54712
43	2-Bromovinylbenzene	209.8	1.4160	1.5927
44	1-Bromo-2-isopropylbenzene	210.3	1.30195	1.54084
45	2-Iodotoluene	211	1.698	1.6085
46	4-Bromovinylbenzene	212	1.3983	1.5947
47	1-Bromo-2,3-dimethylbenzene	212–3	1.365	
48	1,2,4-Trichlorobenzene (mp 17)	213	1.45	1.5671
49	1-Bromo-3,4-dimethylbenzene	214	$1.37^{18.4}$	$1.5571^{18.4}$
50	1-Fluoronaphthalene	214	1.134	1.594
51	2-Chloro-4-isopropyl-toluene	217	1.015^{17}	1.5178^{17}
52	3-Chloro-4-isopropyl-toluene	217	1.018^{18}	1.5179^{18}
53	1,3-Dibromobenzene	219	1.952	1.606
54	4-Bromoisopropylbenzene	219	1.2854	1.5361
55	1,2-Dibromobenzene (mp 7.1)	225.5	1.98429	1.61553
56	2-Iodo-1,4-dimethylbenzene	230/722 mm	1.613	1.598
57	4-Iodo-1,3-dimethylbenzene	232	1.623	1.599
58	2-Bromo-1-methyl-4-isopropylbenzene	234	1.267	
59	2,5-Dibromotoluene	236	1.811	
60	4-Iodo-1-isopropylbenzene	236–8		
61	3,4-Dibromotoluene	240	1.811	
62	2-Bromoiodobenzene	257	2.262	1.665
63	1-Chloronaphthalene	259.3	1.191	1.633
64	1-Bromonaphthalene (mp 6.2)	281.2	1.484	1.658
65	3-Chlorobiphenyl (mp 16)	284–5		
66	2-Bromobiphenyl	297		
67	3-Bromobiphenyl	299–301		
68	1-Iodonaphthalene	305		

TABLE 15A (Continued)
ARYL HALIDES—LIQUID

	Recommended Derivative Nitro		Other Derivatives	
	Position	MP	Name	MP
31	2	95		
32			4-Chlorobenzoic acid	240
33	3	50	2,4-Dichloro-3-methylbenzene-sulfonyl chloride	55(60)
			2,6-Dichlorobenzoic acid	139
34			2-Bromobenzoic acid	150
35	3,5	104	2,4-Dichlorobenzoic acid	164
36	2	61–2	3,5-Dichlorobenzoic acid	188
			2,4-Dichloro-6-methylbenzenesulfonamide	168–9
37	4,6	108	3-Iodobenzoic acid	187
38	4,6	178	3-Chloro-2,4,6-trimethylbenzene-sulfonamide	165–6
			2-Chlorobenzenetricarboxylic acid	285 anh.
39			4-Bromobenzoic acid	251
40			2-Bromoterephthalic acid	299
41	4	51	2,3-Dichlorobenzoic acid	163
42	6	63	3,4-Dichlorobenzoic acid	206
43			2-Bromobenzoic acid	150
44			2-Bromobenzoic acid	150
45	6	103	2-Iodobenzoic acid	162
46			4-Bromobenzoic acid	251
47			3-Bromophthalic acid	188
48	5	56		
49			4-Bromophthalic acid	(166)173–5
50			Picrate	113
51	5,6	109–10	3-Chloro-4-methylbenzoic acid	196
52	2,6	102–3		
53	4	61	2,4-Dibromobenzenesulfonamide	190
54			4-Bromobenzoic acid	251
55	4,5	114	3,4-Dibromobenzenesulfonamide	176
56			2-Iodoterephthalic acid	274–6
57			4-Iodoisophthalic acid	285–6
58			2-Bromoterephthalic acid	299
59			2,5-Dibromobenzoic acid	157
60			4-Iodobenzoic acid	269–70
61			3,4-Dibromobenzoic acid	235
62				
63	4,5	180	Picrate	137
64	4	85	Naphthalide	236
			4-Bromonaphthalenesulfonamide	191–3
			Picrate	134
			T.N.F.	171–3
65	4,4′	202–3	3-Chlorobenzoic acid	158(155)
66			2-Bromobenzoic acid	150
67			3-Bromobenzoic acid	155
68			Picrate	128

TABLE 15B
ARYL HALIDES—SOLID

	Name of Compound	MP
1	4-Bromo-1,2-dichlorobenzene (bp 237)	24.5
2	2,4-Dinitrofluorobenzene	25–7
3	3,5-Dichlorotoluene (bp 201)	26
4	1,2-Diiodobenzene (bp 286.5)	27
5	4-Bromotoluene (bp 184)	28.5
6	1,2-Dichloro-4-iodobenzene	30.4
7	2,4,6-Trichlorotoluene	32–4
8	2-Chlorobiphenyl (bp 273)	34
9	4-Iodotoluene (bp 211)	35
10	1,2-Dichloronaphthalene (bp 296)	35
11	2-Bromo-1-methylnaphthalene	36
12	5-Iodo-1,2,4-trimethylbenzene (bp 256–8)	37
13	1,3-Diiodobenzene (bp 285)	40(36)
14	2,3,4-Trichlorotoluene (bp 230)	41
15	1,2,4-Tribromobenzene (bp 275)	44–5
16	1,2,3,4-Tetrachlorobenzene (bp 254)	45–6
17	3,4,5-Trichlorotoluene (bp 246)	45
18	2,3,5-Trichlorotoluene (bp 232)	45–6
19	3-Bromo-1-methylnaphthalene	47
20	1,6-Dichloronaphthalene	49
21	1,2,3,5-Tetrachlorobenzene (bp 246*)	51
22	1,2,3-Trichlorobenzene (bp 218.5)	52
23	1,4-Dichlorobenzene (bp 173)	53
24	2-Iodonaphthalene (bp 309)	55$v.s.$
25	4,4'-Dichlorodiphenylmethane (bp 337)	55
26	2-Chloronaphthalene (bp 256)	56
27	2-Bromonaphthalene (bp 281)	59
28	2,2'-Dichlorobiphenyl	60
29	2-Fluoronaphthalene	60
30	1,3-Dichloronaphthalene (bp 291/775 mm)	62
31	1,3,5-Trichlorobenzene (bp 208.4)	63

TABLE 15B (Continued)
ARYL HALIDES—SOLID

	Recommended Derivative Nitro		Other Derivatives	
	Position	MP	Name	MP
1				
2			2,4-Dinitrophenyl ethers of phenols	
3			3,5-Dichlorobenzoic acid	188
4				
5			4-Bromobenzoic acid	251
6				
7	3	54		
	3,5	178–80	2,4,6-Trichlorobenzoic acid	160–1
8			2-Chlorobenzoic acid	142
9			4-Iodobenzoic acid	270
10	di	169	5,6-Dichloro-1,4-naphthoquinone	181
11			Picrate,	105–6
12			Dichloride	66
13				
14	5,6	141	2,3,4-Trichlorobenzoic acid	186–7
15				
16	5	64.5		
	5,6	151		
17	2	81–2	3,4,5-Trichlorobenzoic acid	203
	2,6	163–4		
18	4,6	149–50	2,3,5-Trichlorobenzoic acid	162
19			Picrate,	84
20	4	119	1,6-Dichloronaphthalenesulfonamide	216
21	4	41		
	4,6	162		
22	4	56		
	5	72	Sulfonyl chloride	65
	4,6	92–3	1,2,3-Trichlorobenzene-4-sulfonamide	226–30
23	2	54	2,5-Dichlorobenzenesulfonamide	180(186)
	2,6	106		
24			Picrate	95
25	3,3′	198–9	4,4′-Dichlorobenzophenone	145
26	1,8	75	7-Chloronaphthalenesulfonamide	126
			Picrate	81.5
27			T.N.F.	138–40
			Picrate	86(79)
			7-Bromonaphthalenesulfonamide	208
28	5,5′	203–4		
	3,5,3′,5′	307–8		
29			Picrate	101
30			1,3-Dichloronaphthalene-5-sulfonamide	272
			Phthalic acid	200–6
31	2	68	1,3,5-Trichlorobenzene-2,4-disulfonamide	248
	2,4	131	Sulfonyl chloride	35–40

(*Continued*)

TABLE 15B (Continued)
ARYL HALIDES—SOLID

	Name of Compound	MP
32	1-Bromo-5-methylnaphthalene	63–4
33	1,7-Dichloronaphthalene (bp 286)	64
34	1,3-Dibromonaphthalene	64
35	1,2-Dibromonaphthalene	67
36	1-Bromo-4-chlorobenzene (bp 197)	67
37	1,4-Dichloronaphthalene (bp 287/740 mm)	68
38	2,5-Dichloro-1,4-dimethylbenzene (bp 224.3)	68.2
39	1-Chloro-8-methylnaphthalene	68–9
40	2,5-Dibromo-1,4-dimethylbenzene (bp 261)	75
41	4-Chlorobiphenyl (bp 293)	77
42	1-Bromo-8-methylnaphthalene	80
43	2,2′-Dibromobiphenyl	81
44	1-Chloroanthracene	81–2
45	1,4-Dibromonaphthalene (bp 310)	82
46	2,4,5-Trichlorotoluene (bp 230/715 mm)	82
47	Pentachlorobenzene (bp 276)	86
48	1,2,3-Tribromobenzene	87.8
49	1,8-Dichloronaphthalene	89
50	4-Bromobiphenyl (bp 310)	89
51	1,4-Dibromobenzene (bp 219)	89
52	1,2,4-Triiodobenzene	91.5
53	1-Bromo-4-iodobenzene (bp 251)	92
54	1,5-Dichloronaphthalene	107
55	1,8-Dibromonaphthalene	109–10
56	4-Iodobiphenyl (bp 320*d*)	114
57	2,7-Dichloronaphthalene	114–5
58	1,2,3-Triiodobenzene	116
59	2,3-Dichloronaphthalene	120
60	1,3,5-Tribromobenzene (bp 271)	120
61	1,4-Diiodobenzene (bp 289)	129
62	2,6-Dichloronaphthalene (bp 285)	135–6

TABLE 15B (Continued)
ARYL HALIDES—SOLID

	Recommended Derivative Nitro		Other Derivatives	
	Position	MP	Name	MP
32			Picrate	110–1
33			Sulfonyl chloride	118
			1,7-Dichloronaphthalene-4-sulfonamide	226
34			Trinitrobenzene adduct	123
35			3,4-Dibromophthalic acid	196
36	2	72		
37	8	92	1,4-Dichloronaphthalene-6-sulfonamide	244
			Sulfonyl chloride	132
			5,8-Dichloro-1,4-naphthoquinone	173–4
38			2,5-Dichloroterephthalic acid	306
39			Picrate	139
40				
41			4-Chlorobenzoic acid	240
42			Picrate	153
43				
44			Picrate	101–2
45			3,6-Dibromophthalic acid	135
46	3	92	2,4,5-Trichlorobenzoic acid	168
	3,6	227		
47	1	146		
48				
49			1,8-Dichloronaphthalene-4-sulfonamide	228
			Sulfonyl chloride	114
50			4-Bromobenzoic acid	251
			Adduct with 2,4-dinitrophenylhydrazine	101
51	2,5	84	2,5-Dibromobenzenesulfonamide	195
52				
53				
54	8	142	1,5-Dichloronaphthalene-3-sulfonamide	204
			3-Chlorophthalic acid	185–7
			Picrate	87
55				
56			Dichloride	102d
57	mono	141–2	2,7-Dichloronaphthalene-3-sulfonamide	218
			Sulfonyl chloride	166
58				
59			2,3-Dichloronaphthalene-8-sulfonamide	268
			Sulfonyl chloride	142
60			2,4,6-Tribromobenzenesulfonamide	222d
61	2,5	171	4-Iodonitrobenzene	171.5*
62			2,6-Dichloronaphthalene-4-sulfonamide	269
			Sulfonyl chloride	136
			2,6-Dichloro-1,4-naphthoquinone	148–9

(*Continued*)

TABLE 15B (Continued)
ARYL HALIDES—SOLID

	Name of Compound	MP
63	1,2,4,5-Tetrachlorobenzene (bp 245)	140
64	4,4′-Dichlorobiphenyl (bp 315)	149*
65	Pentabromobenzene	160
66	4,4′-Dibromobiphenyl (bp 355–60)	164
67	5,6,7,8-Tetrachlorotetralin	174
68	1,2,4,5-Tetrabromobenzene	180
69	1,3,5-Triiodobenzene	184
70	Octachloronaphthalene (bp 442)	198
71	1,2,3,4-Tetrachloronaphthalene	198(182)
72	9,10-Dichloroanthracene	209–10

REFERENCE FOR TABLE 15

Fletcher, T, Namkung, M., Pan, H., and Wetzel, W., *J. Org. Chem.*, **25**, 996 (1960). Derivatives of Fluorine.

VIII. Lists various fluorofluorines and their derivatives and the melting points.

TABLE 15B (Continued)
ARYL HALIDES—SOLID

	Recommended Derivative Nitro		Other Derivatives	
	Position	MP	Name	MP
63	3	99–100		
	3,6	227–8		
64	2	102	4-Chlorobenzoic acid	240
	2,2′	138–9		
	2,3′,5′	166–7		
65				
66			4-Bromobenzoic acid	251
67			1,2-Dibromo-5,6,7,8-tetra-chloronaphthalene	142
68	3	168		
69				
70			Hexachloro-1,4-naphthoquinone	222
71			1,3-Dichloronaphthalene	61
72			9,10-Dichloroanthracene-2-sulfonamide	279
			Sulfonyl chloride	221–5
			Anthraquinone	286

TABLE 16A[a]
DIHALIDES AND POLYHALIDES (NONAROMATIC)—LIQUIDS

	Name of Compound	Note	BP	D_4^{20}	n_D^{20}	Melting Point of Derivatives
1	2,2-Difluoropropane		−0.4	0.9205	1.2904	
2	1,1-Difluoropropane		7.5	0.94	1.30	
3	Dichlorofluoromethane		8.9			
4	1,1,1-Trifluorobutane		16.7	1.0144	1.2921	
5	2,2,3,3-Tetrafluorobutane		16.9	1.1633	1.2915	
6	Dibromofluoromethane		23.2	2.288-D^{15}	1.3999^{12}	
7	Trichlorofluoromethane		23.7			
8	2,2-Difluorobutane		30.9	0.9016	1.3138	
9	1,1-Dichloroethene		31.6	1.218	1.427	
10	1,1,1,4,4,4-Hexafluoro-2-chloro-2-butene		35	1.456	1.2999	
11	1,2-Dibromofluoroethylene		36	1.693		
12	Dichloromethane	1	39.8	1.3255	1.4242	2-Naphthyl ether, 133 6-Nitro-2-mercaptobenzothiazole, 232–3
13	1,1-Difluorobutane		41	0.92	1.32	
14	Trichloroiodomethane		42	2.36-D^{17}		
15	1,1,2-Trichloro-1,2,2-trifluoroethane		47.5	1.5635-D^{26}	1.3557^{25}	
16	1,2-Dibromo-1,1,2,2-tetrafluoroethane		47.5			
17	1,2-Dichloroethylene (*trans*)		48	1.2569	1.452	Dibromo deriv., 190–5
18	1,1-Dichloroethane		57.3	1.1755	1.4164	Di-1-naphthyl ether, 117
19	2,2-Difluoropentane		59.7	0.8932	1.3351	
20	3,3-Difluoropentane		60.3	0.9023	1.3380	
21	1,2-Dichloroethylene (*cis*)		60.5	1.282	1.4428^{25}	Dibromo deriv., 190–5
22	Trichloromethane	2	61.7	1.4832	1.4459	
23	α,α-Dichloro-3-nitrotoluene		65			3-Nitrobenzoic acid, 140
24	1,1,1,4,4,4-Hexafluoro-2,3-dichloro-2-butene		67.7	1.6233	1.3459	
25	Bromochloromethane		68.1	1.9344	1.4838	
26	1,1-Difluoropentane		69	0.90	1.34	
27	2,2-Dichloropropane		69.3	1.112	1.4148	
28	1,1,2-Trichloro-2-fluoroethylene		71	1.530		
29	1,1,1-Trichloroethane		74	1.3390	1.4379	
30	Tetrachloromethane		76.7	1.5940	1.4601	
32	α,α-Dibromo-4-nitrotoluene		82			4-Nitrobenzoic acid, 241
33	1-Bromo-1-chloroethane		82.7	1.667-D^{16}		
34	1,2-Dichloroethane		83.5	1.2531	1.4448	2-Naphthyl ether, 217
35	1,1,2-Trichloroethylene	3	87.1	1.4642	1.4773	
36	1,1-Dichloropropane		88.1	1.1321	1.4289	
37	Bromodichloromethane		90.1	2.0055-D^{15}	1.502^{15}	

[a] See Table 14 for alkyl halides that contain only one halogen atom and Table 15 for compounds with halogens attached to aromatic structures only.

(*Continued*)

TABLE 16A (Continued)
DIHALIDES AND POLYHALIDES (NONAROMATIC)—LIQUIDS

	Name of Compound	Note	BP	D_4^{20}	n_D^{20}	Melting Point of Derivatives
38	1,1,2,2-Tetrachloro 1,2-difluoroethane		92.8	1.6447-D^{25}	1.4130^{25}	
39	2,3-Dichloro-1-propene		92–4			
40	1,1-Difluorohexane		95	0.90	1.36	
41	1,2-Dichloropropane		96.4	1.1590	1.4394	1,2-Diphenoxypropane, 32
						1,2-Di(2-naphthoxy)-propane, 152
42	Dibromomethane	1	97 (98.6)	2.4970	1.5420	6-Nitro-2-mercapto-benzothiazole, 232–3
43	1,1,2-Trichloro-2-fluoroethane		103	1.550		
44	α,α,α-Trifluorotoluene		103.5	1.1886	1.14146	
45	1,1,1,4,4,4-Hexafluoro-2,2,3-trichlorobutane		104	1.6968	1.3636	
46	Bromotrichloromethane		104.7	2.0122	1.5063	
47	Bis(chloromethyl)ether		105	1.315-D_{20}^{20}	1.435	
48	1,1-Dichloro-2-methylpropane		106d	1.011-D^{12}		
49	1-Bromo-2-chloroethane		106.7	1.689	1.491	6-Nitro-2-mercapto-benzothiazole, 101.5–102.5
50	1,2-Dichloro-2-methylpropane		108	1.093	1.4370	
51	1,2-Dibromoethylene (trans)		108			
52	1,1-Dibromoethane		112	2.0555	1.5128	6-Nitro-2-mercapto-benzothiazole, 145–6
53	Trichloronitromethane		112			
54	1,2-Dibromoethylene (cis)		112.5	1.5437		
55	1,1,2-Trichloroethane		113.8	1.4397	1.4714	
56	1,1-Dichlorobutane		113.8	1.086	1.4355	
57	2,3-Dichlorobutane		116	1.105		
58	Dibromochloromethane		118–20	2.440-D_{25}^{25}	1.545^{25}	
59	1,1-Difluoroheptane		119.7	0.8959	1.3710	
60	1,1-Difluoroheptane		120	0.8960	1.3710	
61	1,3-Dichloropropane		120.4	1.8778	1.4487	6-Nitro-2-mercapto-benzothiazole, 194–5
62	1,1,2,2-Tetrachloroethylene	4	121	1.6227	1.5053	
63	1,1,1-Trichloro-2,2,4,4,4-pentafluorobutane		121.5	1.6299	1.3762	
64	1,1,1-Trichloro-2,2,3,3-tetrafluorobutane		123.1	1.5686	1.3908	
65	1,2-Dichlorobutane		123.5			6-Nitro-2-mercapto-benzothiazole, 164–5
66	1,1,1,2-Tetrachloroethane		130.5	1.5406	1.4821	
67	1,2-Dibromoethane		131.4	2.1792	1.5387	6-Nitro-2-mercapto-benzothiazole, 202
68	1,2-Dibromo-1-propene		132	2.1785	1.5379	
69	1,1-Dibromopropane		133.5	1.982	1.5100	
70	1,3-Dichlorobutane		134	1.1158	1.445	
71	1,3-Dichloro-2-methylpropane		135	1.138		

(Continued)

TABLE 16A (Continued)
DIHALIDES AND POLYHALIDES (NONAROMATIC)—LIQUIDS

	Name of Compound	Note	BP	D_4^{20}	n_D^{20}	Melting Point of Derivatives
72	1,1-Dichloropentane		139.8	1.053	1.434	
73	1,1,2-Trichloropropane		140	1.372-D^{25}		
74	1,2-Dibromopropane		140	1.9324	1.5201	6-Nitro-2-mercapto-benzothiazole, 194–5
75	Perfluorodecalin (*trans*)		141		1.3148	
76	2,3-Dibromopropene		141.2			
77	Perfluorodecalin (*cis*)		142		1.3179	
78	1,1-Difluorooctane		142	0.89	1.38	
79	1,1,1,2-Tetrachloro-4,4,4-trifluoro-2-butene		143.2	1.6236	1.4360[25]	
80	1-Bromo-3-chloropropane		143.4	1.5970	1.4864	1,3-Diphenoxypropane, 60 1,3-Di(1-naphthoxy)-propane, 103–4
81	1,1,3-Trichloropropane		145.6	1.3557	1.4718	
82	1,1,2,2-Tetrachloroethane		146.2	1.5953	1.4990	
83	2,3-Dibromo-1-butene		146–7		1.5464	
84	1,2-Dibromo-2-methyl-propane		149	1.783	1.512	
85	Tribromomethane		149.2	2.890	1.598	
86	1,2-Dibromo-1-butene		150	1.887-D^0		
87	1,1,1,2-Tetrachloro-propane		150	1.473	1.4867	
88	1,4-Dichlorobutane		153.9	1.1408	1.4542	
89	α,α,α,3-Tetrabromotoluene		154–6			
90	1,3-Dibromo-1-propene		156	2.097-D^0	1.538[25]	
91	1,1-Dibromo-2-methyl-propene		156–7	1.866-D_{20}^{20}	1.530	
92	1,2,3-Trichloropropane		156.9	1.3888	1.4832	
93	2,3-Dibromobutane		157	1.792	1.515	
94	1,1-Dibromobutane		158	1.791	1.4988	
95	1,1-Dibromocyclobutane		159–61	1.933-D_{20}^{20}	1.5362	
96	Pentachloroethane		161.9	1.681	1.504	
97	1,1-Difluorononane		163	0.89	1.39	
98	1,1-Dichlorohexane		164	1.029	1.437	
99	1,2-Dibromobutane		166.3	1.795	1.5150	6-Nitro-2-mercapto-benzothiazole, 164–5
100	1,3-Dibromopropane		167.3	1.9822	1.5232	
101	1,3-Dibromo-2-butene		168–9	1.877	1.548	
102	1,2,3-Trichlorobutane		169	1.302		
103	1,1,1,3,3-Pentachloro-2,2,4,4,4-Pentafluoro-butane		170.4	1.8128	1.4227	
104	2,3-Dibromo-2-methyl-butane		171	1.6723	1.5102	
105	1,2-Dibromo-2-methyl-butane		173	1.6652	1.5088	
106	2,2-Di-iodopropane		173	2.5755	1.651	
107	1,3-Dibromobutane		174	1.820-D^0	1.507	
108	1,3-Dibromo-2-methyl-propane		174.6	1.821-D_0^{20}		
109	Bis(2-chloroethyl)ether		178	1.220	1.457	

(*Continued*)

TABLE 16A (Continued)
DIHALIDES AND POLYHALIDES (NONAROMATIC)—LIQUIDS

	Name of Compound	Note	BP	D_4^{20}	n_D^{20}	Melting Point of Derivatives
110	1,1-Dibromo-2,2-dimethylpropane		180	1.6695	1.5047	
111	1,1-Dibromopentane		180	1.660	1.501	
112	1,5-Dichloropentane		180	1.1006	1.4564	
113	Di-iodomethane	1	181d	3.3345	1.741	6-Nitro-2-mercapto-benzothiazole, 232–3
114	1,1-Difluorodecane		183	0.89	1.39	
115	1,1-Dichloroheptane		187	1.009	1.440	
116	1,2-Di-iodoethene (*cis*)		188			
117	1,1,2-Tribromoethane		188.9	2.6211	1.5934	
118	1,4-Dibromobutane		197	1.789	1.5190	
119	1,4-Dibromopentane		200	1.6861	1.5078	
120	1,1-Dibromohexane		202	1.565	1.495	
121	2,2,3-Tribromobutane		206	2.1724	1.560	
122	α,α-Dichlorotoluene	5	207	1.2557-D^{14}	1.5503	Benzoic acid, 122
123	1,1-Dichlorooctane		208	0.994	1.543	
124	1,1,1,4,4,4-Hexachloro-2,2,3,3-tetrafluorobutane		209	1.8087	1.4502	
125	1,2,2-Tribromo-1,1-dichloroethane		210	2.6315-D^{15}	1.6072^{15}	
126	Hexachloropropene		214.1	1.7638	1.5496	
127	Hexachloro-1,3-butadiene	6	215	1.6820	1.5542	
128	1,2,4-Tribromobutane		215	2.17	1.5608	
129	1,2,3-Tribromobutane		220	2.190	1.568	
130	α,α,α-Trichlorotoluene (mp 16)		220.8	1.3741	1.5579	Benzoic acid, 122
131	1,1-Dibromoheptane		222	1.491	1.490	
132	1,2,3-Tribromopropane		222.2	2.4209	1.5862	
133	1,5-Dibromopentane	1	222.3	1.7018	1.5126	6-Nitro-2-mercapto-benzothiazole, mp 132–3
134	1,3-Di-iodopropane		227(224)	2.5755	1.6423	Diphenyl ether, 60 Di-2-naphthyl ether, 148 Di-1-naphthyl ether, 103–4
135	1,2-Dibromoheptane		227–9	1.5086	1.4986	
136	1,1-Dichlorononane		228	0.982	1.445	
137	Hexachlorocyclo-pentadiene	7	236	1.715-D^{16}		
138	1,6-Dibromohexane		239–41	1.5948-D^{15}	1.511^{15}	
139	1,2-Dibromooctane		240–2	1.4580	1.4970	
140	1,1-Dibromooctane		242	1.432	1.488	
141	1,1,2,2-Tetrabromoethane		243.5	2.9656	1.6353	
142	1,1-Dichlorodecane		247	0.972	1.447	
143	1,2,2,3-Tetrabromopropane		250	2.703	1.6200	
144	1,1-Dibromononane		260	1.382	1.486	
145	1,8-Dibromooctane (mp 15–16)	1	270–2	1.468-D^{15}	1.501^{15}	
146	1,1-Dibromodecane		277	1.338	1.484	
147	1,9-Dibromononane	1	285–8	1.415-D^{15}		
148	1,3-Dibromo-2-(bromomethyl)propane		320	2.14	1.5512	

TABLE 16B
DIHALIDES AND POLYHALIDES (NONAROMATIC)—SOLIDS

	Name of Compound	Note	MP	D_4^{20}	n_D^{20}	Melting Point of Derivatives
1	1,5-Di-iodopentane		8–10	2.1903-D^{15}	1.6046^{15}	
2	1,1,1,2,3,3,3-Heptachloropropane		11–13			
3	Dibromodichloromethane		21–2			
4	α,α,α,3,4-Pentachlorotoluene (bp 283.4)		25	1.5913	1.5886	3,4-Dichlorobenzoic acid, 202
5	1,10-Dibromodecane		28	1.335-D^{30}	1.4905^{30}	
6	α,α,α,2-Tetrachlorotoluene		29.4			2-Chlorobenzoic acid, 142
7	1,1,1,2,2,3,3-Heptachloropropane (bp 248)		29.4	1.8048-D^{34}		
8	α,α'-Dichloro-m-xylene (bp 350–5)		34.2			Isophthalic acid, 348s.
9	1,7-Dibromoheptane	1	41.7	1.5225-D^{15}	1.514^{15}	
10	1,3-Dibromo-2-butene (trans)		54			
11	α,α'-Dichloro-o-xylene		54–5			Phthalic acid, 206–8
12	1,2-Dibromo-1,2-dichloroethane (bp 178.3)		66.8			
13	1,2-Di-iodoethane		81			
14	Carbon tetrabromide (bp 189.5)		90	2.9109-D^{99}		
15	α,α'-Dichloro-p-xylene		98–100			Terephthalic acid, 115
16	1,1,1-Trichloro-2,2-bis(p-chlorophenyl)ethane	8	108			
17	1,2,3,4-Tetrabromobutane (bp 282.7)		115.8			
18	Tri-iodomethane	9	119	4.008		
19	α,α'-Dibromo-p-xylene		143–5			Terephthalic acid, 115
20	Hexachloroethane		187(s.t.)			

NOTES FOR TABLES 16A AND 16B (DI- AND POLYHALIDES)

1. The following may be derivatized as S-alkyl bis(thiourea picrates): dihalomethane, mp 267d; 1,5-dibromopentane, mp 247d; 1,7-dibromoheptane, mp 208; 1,8-dibromo-octane, mp 214; 1,9-dibromononane, mp 193.
2. Gives the carbylamine test with aniline or most other primary amines.
3. Forms mercury bis(trichloroethylenide), mp (from ether) 83 by reaction with mercuric oxide, potassium cyanide, and sodium ethoxide.
4. Forms 2,2-dichloro-3-hydroxypropionic acid when reacted with paraformaldehyde in sulfuric acid.
5. Hydrolyzes to benzaldehyde, the 2,4-dinitrophenylhydrazone of which melts at 237.
6. The compound does not react with chlorine, nor does it polymerize.
7. Compound gives reactions characteristic of allylic halides as well as of alkenes.
8. The impure product melts as low as 90 due to the presence of isomers. For identification, see *J. Amer. Chem. Soc.*, **66**, 2129 (1944).
9. Is reduced by sodium arsenite to di-iodomethane; forms a compound with quinoline, mp 65d.

REFERENCES FOR TABLES 16A AND 16B

Buu-Hoi, N. G. Ph., Xuong, Ng. D., and Lavit, D., *J. Org. Chem.*, **19**, 1562 (1954). Fluorine-Containing Diaryl- and Triaryl-Ethylenes. Lists the melting points of 16 such compounds. Also gives the preparations.

Cutter, H. B., and Kreuchunas, A., *Anal. Chem.*, **25**, 198 (1953). 6-Nitro-2-Mercaptobenzothiazole. Reagent for the Identification of Alkyl Dihalides. Lists 14 dihalides and the melting points of their derivatives.

TABLE 17
PERFLUORO COMPOUNDS[a]

ACIDS

	Name of Compound	BP	n_D^{20}	D_4^{20}	Melting Point of Derivatives
1	Perfluoroacetic	72.4	1.2850	1.4890	Amide, 75s; anilide, 88; S-Benzylisothiouronium, 172d
2	Perfluoropropionic	96	1.2838	1.561	Amide, 95
3	Perfluoroisobutyric	118		1.649	
4	Perfluorobutyric	120	1.290(25)	1.641(25)	Amide, 105s; anilide, 93; S-Benzylisothiouronium, 293d
5	Perfluoro 3-methylbutanoic	137			
6	Perfluoropentanoic	139 (749 mm)		1.713	
7	Perfluorohexanoic	157 (742 mm)		1.762	Amide, 117
8	Perfluorocyclohexanoic	170 (740 mm)	1.325(29)	1.798	Amide, 112
9	Perfluoroheptanoic	175 (742 mm)		1.792	
10	Perfluorooctanoic	189 (736 mm)			Amide, 138
11	Perfluorononanoic	218 (740 mm)			
12	Perfluorodecanoic	218			Amide, 150
13	Perfluoroundecanoic	245			
14	Perfluorododecanoic	270			
15	Perfluorotertadecanoic	270 (740 mm)			
16	Perfluoropentadecanoic	294 MP			
17	Perfluoroacrylic	36			
18	Difluoromalonic	118			Diamide, 207
19	Perfluorobutanedioc (perfluorosuccinic)	120			Diamide, 260; imide, 67
20	Perfluorohexanedioic	134			Diamide, 237

ACID ANHYDRIDES

	Name of Compound	BP	n_D^{25}	D_4^{25}	Melting Point of Derivatives
1	Perfluoroacetic	40	1.269	1.490	Amide, 75s
2	Perfluoropropionic	72	1.273	1.571	Amide, 95
3	Perfluoro-n-butyric	108	1.285^{20}	1.665^{20}	Amide, 105s
4	Perfluoropentanoic	138			
5	Perfluorohexanoic	176	1.295	1.769	Amide, 117
6	Perfluorobutanedioic	55	1.3240^{20}	1.6209^{20}	Diamide, 260
7	Perfluoropentanedioic	72	1.3190	1.6541	Imide, 27 (bp 159/27 mm)

[a] Dr. Elliot Bergman prepared the table on perfluoro compounds for the second edition of this work. Except for the addition of a few compounds and the minor changes in data indicated by more recent work, this table remains the work on Dr. Bergman. The nomenclature used has been made to conform to that used in the other tables of this book. It should be noted that the term "perfluoro," as applied to such classes as aldehydes and acids, does not imply that fluorine has replaced hydrogen in the aldehyde or carboxyl groups. For data on additional perfluoro compounds, and particularly for compounds that have not been completely fluorinated, see the specialized books on fluorine chemistry, such as *Aliphatic Fluorine Compounds* by H. M. Lovelace, D. A. Rausch, and W. Postelnek, Reinhold, New York, 1958.

(*Continued*)

TABLE 17 (Continued)
PERFLUORO COMPOUNDS
ACID HALIDES

	Name of Compound	BP	n_D^{20}	D_4^{20}	Melting Point of Derivatives
1	Trifluoroacetyl fluoride	−59			Amide, 75s Anilide, 88
2	Trifluoroacetyl chloride	−20			
3	Trifluoroacetyl bromide	−5			
4	Trifluoroacetyl iodide	22			
5	Perfluoropropionyl fluoride	−28			
6	Perfluoropropionyl chloride	5			Amide, 95
7	Perfluorobutyryl fluoride	8			Amide, 105s Anilide, 93 Hydrazide, 76 4-Bromoanilide, 106
8	Perfluorobutyryl chloride	39	1.288	1.55	
9	Perfluorobutyryl bromide	54	1.3261	1.735	
10	Perfluorobutyryl iodide	76	1.3562	2.00	
11	Perfluoropentanoyl chloride	68			
12	Perfluorohexanoyl chloride	86			Amide, 117

ALCOHOLS

	Name of Compound	BP	n_D^{20}	D_4^{20}	Melting Point of Derivatives
1	2,2,2-Trifluoroethanol	74	1.2907[22]	1.3739[22]	Tosylate, 41
2	2,2,3,3,3-Pentafluoro-n-propanol	80	1.2900	1.505	
3	2,2,3,3,4,4,4-Heptafluoro-n-butanol	95	1.2944	1.601	α-Naphthylurethan, 78

ALDEHYDES AND KETONES

	Name of Compound	BP	Melting Point of Derivatives	
			2,4-Dinitrophenylhydrazone	Others
1	Perfluoroacetaldehyde	−19	149	Hydrate, 70
2	Perfluoropropionaldehyde	2		Hydrate, 53
3	Perfluoro-n-butyraldehyde	29	107	Hydrate, 61
4	Perfluoroacetone	−28		Semicarbazone, 153
5	Perfluorobutanone	0		Semicarbazone, 138d
6	1,1,1-Trifluoroacetone	22	140	Semicarbazone, 127
7	Perfluorocyclopentanone	24		Semicarbazone monohydrate, 186
8	Perfluoropentanone-2	30		
9	Perfluoropentanal	48 (740 mm)		
10	Perfluorohexanone-3	52		
11	3,3,4,4,5,5,5-Heptafluoropentanone-2	58	78	
12	Perfluoroheptanone-4	75		Ammonia adduct, 58
13	Perfluorooctanal	112		

(Continued)

TABLE 17 (Continued)
PERFLUORO COMPOUNDS
ARYLFLUOROALKYL KETONES

	Name of Compound	BP	n_D^{25}	D_4^{25}	Melting Point of 2,4-Dinitrophenylhydrazone
1	Phenylperfluoromethyl	152	1.4583^{20}	1.279^{20}	95
2	Phenylperfluoroethyl	161	1.4245	1.372	120
3	Phenylperfluoro-n-propyl	174	1.4130^{20}	1.473^{20}	Semicarbazone, 145
4	Phenylperfluoro-n-butyl	189	1.3990	1.517	136
5	Phenylperfluoro-n-amyl	204	1.3910	1.538	145
6	p-Tolylperfluoromethyl	179	1.4664	1.240	188
7	p-Tolylperfluoroethyl	181	1.4380	1.317	163
8	p-Tolylperfluoro-n-propyl	193	1.4230	1.384	142
9	p-Tolylperfluoro-n-butyl	211	1.4126	1.445	153
10	p-Tolylperfluoro-n-pentyl	217	1.4039	1.504	161

AMINES

	Name of Compound	BP	n_D^{25}	D_4^{25}	Melting Point of Derivatives
1	2,2,2-Trifluoroethylamine	38	1.295^{30}	1.245^{30}	
2	2,2,3,3,3-Pentafluoro-n-propylamine	50	1.297^{20}	1.400^{20}	
3	2,2,3,3,4,4,4-Heptafluoro-n-butylamine	68	1.298^{20}	1.493^{20}	HCl, 130–5
4	Perfluorotrimethylamine	−11			
5	Perfluorotriethylamine	69	1.262	1.75	
6	Perfluorotri-n-propylamine	129	1.278	1.83	
7	Perfluorotri-n-butylamine	177	1.291	1.87	
8	Perfluorotri-n-hexylamine	256	1.303	1.93	
9	Diperfluoroethylperfluoro-n-propylamine	91	1.269	1.76	
10	Diperfluoro-n-propylperfluoroethylamine	111	1.273	1.79	
11	Diperfluoroethylperfluorocyclohexylamine	163	1.304		

ESTERS

	Name of Compound	BP	n_D^{20}	D_4^{20}
1	Methyl perfluoroacetate	44		
2	Methyl perfluoropropionate	61	1.2884	1.393
3	Methyl perfluoro-n-butyrate	79	1.293	1.483
4	Methyl perfluoro-n-hexanoate	122	1.297^{29}	1.618^{29}
5	Methyl perfluoro-n-octanoate	158	1.304^{27}	1.684
6	Ethyl perfluoroacetate	62	1.3093^{15}	1.1952^{17}
7	Ethyl perfluoropropionate	76	1.2990^{25}	1.294
8	Ethyl perfluoro-n-butyrate	95	1.3032	1.394
9	Vinyl perfluoroacetate	39	1.3151	1.203
10	Vinyl perfluoropropionate	58	1.3095	1.319
11	Vinyl perfluoro-n-butyrate	79	1.3086	1.418
12	Vinyl perfluoro-n-pentanoate	99	1.3116	1.493
13	Vinyl perfluoro-n-hexanoate	66/100 mm	1.3115	1.546
14	Vinyl perfluoro-n-decanoate	53/0.5 mm	1.3176	1.707
15	Vinyl perfluorocyclohexanecarboxylate	59/45 mm	1.3362	1.628

(*Continued*)

TABLE 17 (Continued)
PERFLUORO COMPOUNDS
ESTERS (Continued)

	Name of Compound	BP	n_D^{20}	D_4^{20}
16	iso-Propyl perfluoropropionate	87	1.3090	1.224
17	iso-Propyl perfluoro-n-butyrate	106	1.310	1.324
18	n-Butyl perfluoroacetate	100	1.353^{22}	1.0268^{22}
19	tert-Butyl perfluoro-n-butyrate	116	1.318	1.278
20	sec-Butyl perfluoro-n-butyrate	126	1.3212^{25}	1.284
21	n-Butyl perfluoro-n-butyrate	132	1.3249^{25}	1.296
22	n-Octyl perfluoro-n-butyrate	108/27 mm	1.3582^{25}	1.185
23	n-Dodecyl perfluoro-n-butyrate	158/23 mm	1.3802^{25}	1.120
24	n-Hexadecyl perfluoro-n-butyrate	208/31 mm	1.3950^{25}	1.074
25	n-Octadecyl perfluoro-n-butyrate	185/4 mm	1.4020^{25}	
26	Diethyl perfluoro-succinate	89/15 mm	1.369	1.264
27	Diethyl perfluoroglutarate	76/3 mm	1.3546	1.3444^{29}
28	Diethyl perfluoroadipate	71/2.5 mm	1.3541	1.4026^{29}

ETHERS

	Name of Compound	BP	n_D^{25}	D_4^{25}
1	Perfluorodimethyl	−59		
2	Perfluorodiethyl	1		
3	Perfluoro-1,2-dimethoxyethane	13		
4	Perfluorotetrahydrofuran	1		
5	Perfluorotetrahydropyran	31–3	1.260(29)	1.68(15)
6	Perfluorobutyl-perfluoromethyl	36	1.260	1.581
7	Perfluorodipropyl	56		
8	Perfluorocyclohexyl-perfluoromethyl	80	1.273	1.74
9	Perfluoro-2,2′-diethoxydiethyl	97 (737 mm)	1.250(30)	1.617(30)
10	Perfluorodibutyl	101	1.2619	1.689
11	Perfluorodipentyl	138	1.268	1.758
12	Perfluorodihexyl	179(172)	1.278	1.803

FLUOROCARBONS

Perfluoro Alkanes

	Name of Compound	BP	n_D^t	D_4^t
1	Perfluoromethane	−128		
2	Perfluoroethane	−78		
3	Perfluoropropane	−37		
4	Perfluorobutane	−3(+1)		
5	Perfluoro-2-methylbutane	29.3		1.620(20)
6	Perfluoropentane	29.4		
7	Perfluorohexane	57	1.2515(22)	1.6995(25)
8	Perfluoro-2-methylpentane	57.7	1.2564(22)	1.7326(20)
9	Perfluoroheptane	82.4	1.2572(30)	1.704(30)
10	Perfluorooctane	104(107)	1.282(20)	1.73(20)
11	Perfluorononane	123(127)	1.283(20)	1.80(20)
12	Perfluorodecane (mp 36)	144(150)	1.271(45)	1.770(45)
13	Perfluoroundecane (mp 57.8)	161	1.268(70)	1.745(70)
14	Perfluorododecane (mp 75)	178		
15	Perfluorotridecane (mp 71)	196		
16	Perfluorohexadecane	238–40		

(Continued)

TABLE 17 (Continued)
PERFLUORO COMPOUNDS

Perfluoro Cycloalkanes

	Name of Compound	BP	n_D^t	D_4^t
1	Perfluorocyclopropane	−33(−28)		
2	Perfluorocyclobutane	−5		
3	Perfluorocyclopentane	23.6		1.654(−20)
4	Perfluorocyclohexane (mp 48–9)	52(50)	1.2685(30)	1.684(20)
5	Perfluoro(dimethylcyclopentane)	71	1.2765(20)	1.7660(20)
6	Perfluoro(ethylcyclopentane)	75	1.2772(20)	1.7707(20)
7	Perfluoro(methylcyclohexane)	76	1.2815(20)	1.7994(20)
8	Perfluoro(1,4-dimethylcyclohexane)	100.5	1.2897(20)	1.8503(20)
9	Perfluoro(ethylcyclohexane)	101.5	1.283(25)	1.826(25)
10	Perfluoro(1,3-dimethylcyclohexane)	101.8	1.2908(20)	1.8560(20)
11	Perfluoro(1,2-dimethylcyclohexane)	102.2	1.2923(20)	1.8672(20)
12	Perfluoroindane	117	1.3077(20)	1.8948(20)
13	Perfluoro(1,3,5-trimethylcyclohexane)	125	1.2995(20)	1.9025(20)
14	Perfluorodecalin	140	1.312(20)	1.946(20)

Perfluoro Alkenes and Alkynes

	Name of Compound	BP	n_D^t	D_4^t
1	Perfluoroethene	−78.4		
2	Perfluoro(1,2-propadiene)	−38		
3	Perfluoropropene	−29.4		
4	Perfluoro(2-butyne)	−24		
5	Perfluoro(cyclobutene)	0(5–6)	1.298(−20)	1.602(−20)
6	Perfluoro(2-butene)	1.2		1.601(−20)
7	Perfluoro(1-butene)	4.8		1.615(−20)
8	Perfluoro(1,3-butadiene)	6.0(7.5)	1.378(−20)	1.553(−20)
9	Perfluoro(2-methyl-1,3-butadiene)	30–1	1.300(0)	1.527(0)
10	Perfluorocyclohexene	52–3 (750 mm)	1.296(15)	
11	Perfluoro(2,3-dimethyl-2-butene)	55		
12	Perfluoro(1,3-cyclohexadiene)	56–7 (743 mm)	1.3149(20)	1.601(20)
13	Perfluoro(1-hexene)	57		
14	Perfluoro(1,4-cyclohexadiene)	57–8	1.318(18)	
15	Perfluoro(1,3,5-hexatriene)	65–7 (748 mm)		1.537(25)
16	Perfluoro(4-methyl-1-cyclohexene)	75	1.293(20)	
17	Perfluoro(1-heptene)	81		
18	Perfluoro(1,4,7-octatriene)	99–100		
19	Perfluoro(1-octene)	105		
20	Perfluoro(1-nonene)	123	1.2868(20)	

TABLE 18
ALKANES AND CYCLOALKANES

	Name of Compound	BP	D_4^{20}	n_D^{20}
1	2,2-Dimethylpropane	9.5	0.5910[a]	1.342[a]
2	Cyclobutane	12.5	0.703-D^0	
3	2-Methylbutane	27.8	0.61967	1.35373
4	Pentane	36	0.62624	1.35748
5	Cyclopentane	49.3	0.74538	1.40645
6	2,2-Dimethylbutane	49.7	0.64916	1.36872
7	2,3-Dimethylbutane	58	0.66164	1.37495
8	2-Methylpentane	60.3	0.65315	1.37145
9	3-Methylpentane	63.3	0.66431	1.37652
10	Hexane	68.7	0.65937	1.37486
11	Methylcyclopentane	71.8	0.74864	1.40970
12	2,2-Dimethylpentane	79.2	0.67385	1.38215
13	2,4-Dimethylpentane	80.5	0.67270	1.38145
14	Cyclohexane (mp 6.6)	80.7	0.77855	1.42623
15	2,2,3-Trimethylbutane	80.9	0.69011	1.38944
16	3,3-Dimethylpentane	86.1	0.69327	1.39092
17	1,1-Dimethylcyclopentane	87.8	0.75448	1.41356
18	2,3-Dimethylpentane	89.8	0.69508	1.39196
19	2-Methylhexane	90	0.67859	1.38485
20	1,3-Dimethylcyclopentane (trans)	90.8	0.74479	1.40894
21	1,3-Dimethylcyclopentane (cis)	91.7	0.74880	1.41074
22	1,2-Dimethylcyclopentane (trans)	91.9	0.75144	1.41200
23	3-Methylhexane	91.9	0.68713	1.38864
24	3-Ethylpentane	93.5	0.69816	1.39339
25	Heptane	98.4	0.68376	1.38764
26	2,2,4-Trimethylpentane	99.2	0.69192	1.39145
27	1,2-Dimethylcyclopentane (cis)	99.5	0.77262	1.42221
28	Methylcyclohexane	100.9	0.76939	1.42312
29	Ethylcyclopentane	103.5	0.76647	1.41981
30	1,1,3-Trimethylcyclopentane	104.9	0.74825	1.41119
31	2,2-Dimethylhexane	106.8	0.69528	1.39349
32	2,5-Dimethylhexane	109.1	0.69354	1.39246
33	1-trans-2-cis-4-Trimethylcyclopentane	109.3	0.74727	1.41060
34	2,4-Dimethylhexane	109.4	0.70036	1.39534
35	2,2,3-Trimethylpentane	109.8	0.71602	1.40295
36	1-trans-2-cis-3-Trimethylcyclopentane	110.2	0.7535	1.4138
37	3,3-Dimethylhexane	112	0.71000	1.40009
38	2,3,4-Trimethylpentane	113.5	0.71906	1.40422
39	1,1,2-Trimethylcyclopentane	113.7	0.77252	1.42298
40	2,3,3-Trimethylpentane	114.8	0.72619	1.40750
41	2,3-Dimethylhexane	115.6	0.71214	1.40113
42	3-Ethyl-2-methylpentane	115.7	0.71932	1.40401
43	1-cis-2-trans-4-Trimethylcyclopentane	116.7	0.76345	1.41855

[a] For the liquid at saturation pressure.

(Continued)

TABLE 18 (Continued)
ALKANES AND CYCLOALKANES

	Name of Compound	BP	D_4^{20}	n_D^{20}
44	1-*cis*-2-*trans*-3-Trimethylcyclopentane	117.5	0.7704	1.4218
45	2-Methylheptane	117.6	0.69792	1.39494
46	3,4-Dimethylhexane	117.7	0.71923	1.40406
47	4-Methylheptane	117.7	0.70463	1.39792
48	1-*cis*-2-*cis*-4-Trimethylcyclopentane	118	0.766	1.422
49	3-Ethyl-3-methylpentane	118.3	0.72742	1.40775
50	3-Ethylhexane	118.5	0.71358	1.40162
51	Cycloheptane	118–20	0.8099	1.4440
52	3-Methylheptane	118.9	0.70582	1.39848
53	1,4-Dimethylcyclohexane (*trans*)	119.4	0.76255	1.42090
54	1,1-Dimethylcyclohexane	119.5	0.78094	1.42900
55	1,3-Dimethylcyclohexane (*cis*)	120.1	0.76603	1.42294
56	3-Ethyl-*trans*-1-methylcyclopentane	120.8	0.7619	1.4186
57	2-Ethyl-*trans*-1-methylcyclopentane	121.2	0.7690	1.4219
58	3-Ethyl-*cis*-1-methylcyclopentane	121.4	0.7724	1.4203
59	1-Ethyl-1-methylcyclopentane	121.5	0.78093	1.42718
60	2,2,4,4-Tetramethylpentane	122.3	0.71947	1.40694
61	1-*cis*-2-*cis*-3-Trimethylcyclopentane	123	0.7792	1.4262
62	1,2-Dimethylcyclohexane (*trans*)	123.4	0.77601	1.42695
63	2,2,5-Trimethylhexane	124.1	0.70721	1.39972
64	1,4-Dimethylcyclohexane (*cis*)	124.3	0.78285	1.42966
65	1,3-Dimethylcyclohexane (*trans*)	124.5	0.78472	1.43085
66	Octane	125.7	0.70252	1.39743
67	Isopropylcyclopentane	126.4	0.77653	1.42582
68	2,2,4-Trimethylhexane	126.5	0.7156	1.4033
69	2-Ethyl-*cis*-1-methylcyclopentane	128.1	0.78522	1.42933
70	1,2-Dimethylcyclohexane (*cis*)	129.7	0.79627	1.43596
71	2,4,4-Trimethylhexane	130.6	0.72381	1.40745
72	*n*-Propylcyclopentane	130.9	0.77633	1.42626
73	2,3,5-Trimethylhexane	131.3	0.7219	1.4061
74	Ethylcyclohexane	131.8	0.788	1.4332
75	2,2-Dimethylheptane	132.7	0.7105	1.4016
76	2,2,3,4-Tetramethylpentane	133	0.73895	1.41472
77	2,4-Dimethylheptane	133.5	0.716	1.4033
78	2,2,3-Trimethylhexane	133.6	0.7292	1.4105
79	4-Ethyl-2-methylhexane	133.8	0.723	1.4068
80	2,2-Dimethyl-3-ethylpentane	133.8	0.7348	1.4123
81	2,6-Dimethylheptane	135.2	0.7089	1.4007
82	4,4-Dimethylheptane	135.2	0.725	1.4076
83	2,5-Dimethylheptane	136	0.715	1.4038
84	3,5-Dimethylheptane	136	0.723	1.4067
85	2,4-Dimethyl-3-ethylpentane	136.7	0.7379	1.4137
86	3,3-Dimethylheptane	137.3	0.725	1.4085
87	2,2,5,5-Tetramethylhexane	137.5	0.71875	1.40550
88	2,3,3-Trimethylhexane	137.7	0.738	1.4141
89	3-Ethyl-2-methylhexane	138	0.731	1.4120

(*Continued*)

TABLE 18 (Continued)
ALKANES AND CYCLOALKANES

	Name of Compound	MP	BP	D_4^{20}	n_D^{20}
90	2,3,4-Trimethylhexane		139	0.7392	1.4144
91	2,2,3,3-Tetramethylpentane		140.3	0.75666	1.42360
92	4-Ethyl-3-methylhexane		140.4	0.742	1.416
93	3,3,4-Trimethylhexane		140.5	0.7454	1.4178
94	2,3-Dimethylheptane		140.5	0.7260	1.4085
95	3,4-Dimethylheptane		140.6	0.7314	1.4111
96	3-Ethyl-3-methylhexane		140.6	0.741	1.4142
97	4-Ethylheptane		141.2	0.730	1.4096
98	2,3,3,4-Tetramethylpentane		141.6	0.75473	1.42222
99	3-Ethyl-2,3-dimethylpentane		142	0.754	1.419
100	4-Methyloctane		142.5	0.7199	1.4061
101	3-Ethylheptane		143	0.727	1.4093
102	2-Methyloctane		143.3	0.134	1.4031
103	3-Methyloctane		144.2	0.7207	1.4062
104	2,4,6-Trimethylheptane		144.8	0.7225	1.4071
105	3,3-Diethylpentane		146.2	0.75359	1.42051
106	4-Ethyl-2,2-dimethylhexane		147	0.733	1.4131
107	2,2,4-Trimethylheptane		147.7	0.7275	1.4092
108	2,2,4,5-Tetramethylhexane		147.8	0.73546	1.41318
109	2,2,5-Trimethylheptane		148	0.726	1.409
110	2,2,6-Trimethylheptane		148.2	0.7195	1.4059
111	2,2,3,5-Tetramethylhexane		148.4	0.7378	1.4142
112	Nonane		150.8	0.71763	1.40542
113	Cyclo-octane	14	150/750 mm	0.8349	1.4586
114	2,5,5-Trimethylheptane		152.8	0.7368	1.4136
115	2,4-Dimethyloctane		153	0.7264	1.4093
116	2,4,4-Trimethylheptane		153	0.733	1.412
117	2,3,3,5-Tetramethylhexane		153	0.746	1.4196
118	2,2,4,4-Tetramethylhexane		153.3	0.7470	1.4208
119	Isopropylcyclohexane		154.5	0.80232	1.44095
120	2,2,3,4-Tetramethylhexane		154.9	0.7548	1.4226
121	2,2-Dimethyloctane		155	0.7245	1.4082
122	3-Ethyl-2,2,4-Trimethylpentane		155.3	0.7571	1.4223
123	3,3,5-Trimethylheptane		155.6	0.7428	1.4170
124	2,3,6-Trimethylheptane		155.7	0.7345	1.4125
125	Butylcyclopentane		156.6	0.7846	1.4316
126	Propylcyclohexane		156.7	0.79360	1.43705
127	2,3,5-Trimethylheptane		157	0.741	1.416
128	3-Ethyl-2,5-dimethylhexane		157	0.741	1.416
129	2,4,5-Trimethylheptane		157	0.741	1.4160
130	3-Isopropyl-2,4-dimethylpentane		157	0.75830	1.42463
131	2,2,3-Trimethylheptane		158	0.742	1.417
132	4-Ethyl-2,4-Dimethylhexane		158	0.747	1.419
133	2,5-Dimethyloctane		158	0.736	1.414
134	2,6-Dimethyloctane		158.5	0.7285	1.4113

(Continued)

TABLE 18 (Continued)
ALKANES AND CYCLOALKANES

	Name of Compound	MP	BP	D_4^{20}	n_D^{20}
135	3-Ethyl-2,2-dimethylhexane		159	0.749	1.420
136	2,2,3,4,4-Pentamethylpentane		159.3	0.76703	1.43069
137	5-Ethyl-2-methylheptane		159.7	0.736	1.4134
138	2,7-Dimethyloctane		159.9	0.7242	1.4086
139	2,3,3-Trimethylheptane		160	0.7488	1.4202
140	4-Isopropylheptane		160	0.741	1.417
141	3,5-Dimethyloctane		160	0.736	1.413
142	3,6-Dimethyloctane		160	0.7363	1.4145
143	4-Ethyl-2-methylheptane		160	0.736	1.413
144	2,2,3,3-Tetramethylhexane		160.3	0.76446	1.42818
145	4,4-Dimethyloctane		161	0.737	1.414
146	5-Ethyl-3-methylheptane		161	0.743	1.416
147	2,3,4,5-Tetramethylhexane		161	0.757	1.424
148	1-Methyl-4-isopropylcyclohexane (p-Menthane)(*trans*)		161	0.792	1.4393
149	3,3-Dimethyloctane		161.2	0.7390	1.4165
150	4-Propylheptane		162	0.7364	1.4150
151	3,4-Diethylhexane		162	0.754	1.420
152	4,5-Dimethyloctane		162.1	0.7470	1.4190
153	2,3,4,4-Tetramethylhexane		162.2	0.7639	1.4270
154	2,3,4-Trimethylheptane		163	0.751	1.421
155	3-Isopropyl-2-methylhexane		163	0.751	1.421
156	2,3-Dimethyloctane		163.8	0.7376	1.4148
157	3-Ethyl-3-methylheptane		163.8	0.7501	1.4208
158	3,3,4-Trimethylheptane		164	0.757	1.424
159	4-Ethyl-2,3-dimethylhexane		164	0.759	1.424
160	3,4,4-Trimethylheptane		164	0.757	1.424
161	3,4,5-Trimethylheptane		164	0.759	1.424
162	3-Ethyl-2,4-dimethylhexane		164	0.759	1.424
163	2,3,3,4-Tetramethylhexane		164.6	0.7694	1.4297
164	4-Ethyl-3,3-dimethylhexane		165	0.764	1.427
165	5-Methylnonane		165.1	0.7326	1.4122
166	4-Methylnonane		165.7	0.7323	1.4123
167	3,4-Dimethyloctane		166	0.746	1.4182
168	3-Ethyl-2-methylheptane		166	0.746	1.418
169	2,2,3,3,4-Pentamethylpentane		166.1	0.78009	1.43606
170	3,3-Diethylhexane		166.3	0.767	1.428
171	2-Methylnonane		166.8	0.7281	1.4099
172	4-Ethyl-4-methylheptane		167	0.752	1.421
173	3-Ethyl-4-methylheptane		167	0.753	1.422
174	4-Ethyl-3-methylheptane		167	0.753	1.422
175	d-m-Menthane		167	0.8116	1.446
176	3-Methylnonane		167.8	0.7334	1.4125
177	4-Ethyloctane		168	0.740	1.416
178	3-Ethyloctane		168	0.740	1.416
179	3-Ethyl-2,2,3-trimethylpentane		168	0.781	1.436

(*Continued*)

TABLE 18 (Continued)
ALKANES AND CYCLOALKANES

	Name of Compound	MP	BP	D_4^{20}	n_D^{20}
180	*l-m*-Menthane		168	0.7938	1.4358
181	*p*-Menthane (*cis*) (*cis*-1-methyl-4-isopropylcyclohexane)		168.5	0.816	1.4515
182	3-Ethyl-2,3-dimethylhexane		169	0.765	1.427
183	3-Ethyl-2,3,4-trimethylpentane		169.4	0.7773	1.4333
184	3-Ethyl-3,4-dimethylhexane		170	0.772	1.431
185	3,3,4,4-Tetramethylhexane		170	0.7824	1.4368
186	Cyclononane (mp 9.7)		170–2	0.8534[15]	1.4328[16]
187	*o*-Menthane		171	0.8135	1.447
188	3,3-Diethyl-2-methylpentane		174	0.780	1.435
189	Decane		174.1	0.73005	1.41189
190	Pentylcyclopentane		180	0.7912	1.4358
191	Butylcyclohexane		180.9	0.79918	1.44075
192	Decahydronaphthalene (*trans*)		187.3	0.8699	1.4695
193	Isopentylcyclohexane		193	0.802	1.4423
194	Decahydronaphthalene (*cis*)(decalin)		195.7	0.8965	1.4810
195	Undecane		195.9	0.47017	1.41716
196	Cyclodecane	9.6	201	0.8577[20.4]	1.4692
197	Pentylcyclohexane		202.8	0.8037	1.4437
198	Hexylcyclopentane		203	0.7965	1.4392
199	9-Methyl-*trans*-decahydronaphthalene		205	0.8620	1.4631
200	1,10-Dimethyl-*trans*-decahydronaphthalene		213	0.8633	1.4659
201	9-Methyl-*cis*-decahydronaphthalene		215	0.8910	1.4804
202	Dodecane		216.3	0.74869	1.42160
203	1,10-Dimethyl-*cis*-decahydronaphthalene		220	0.8896	1.4812
204	Hexylcyclohexane		224	0.8076	1.4462
205	Heptylcyclopentane		224	0.8010	1.4421
206	9-Ethyl-*trans*-decahydronaphthalene		225	0.8610	1.466
207	9-Ethyl-*cis*-decahydronaphthalene		233	0.8860	1.480
208	1-Methyl-*trans*-decahydronaphthalene		235		1.4720
209	Tridecane		235.4	0.7564	1.4256
210	Octylcyclopentane		243	0.8048	1.4446
211	Heptylcyclohexane		244	0.8109	1.4484
212	Tetradecane	5.9	253.6	0.7628	1.4289
213	Nonylcyclopentane		262	0.8081	1.4467
214	Octylcyclohexane		264	0.8138	1.4503
215	Pentadecane	9.9	270.6	0.7685	1.4319
216	Decylcyclopentane		279.3	0.81097	1.44862
217	Nonylcyclohexane		282	0.8163	1.4519
218	Hexadecane (cetane)	18.2	286.8	0.77344	1.43453

(*Continued*)

TABLE 18 (Continued)
ALKANES AND CYCLOALKANES

	Name of Compound	MP	BP	D_4^{20}	n_D^{20}
219	Undecylcyclopentane		296	0.8135	1.4503
220	Decylcyclohexane		299	0.81858	1.45338
221	Dodecylcyclopentane		312	0.8158	1.4518
222	Undecylcyclohexane	5.8	316	0.8206	1.4547
223	Tridecylcyclopentane	5	327	0.8178	1.4531
224	Dodecylcyclohexane	12.5	331	0.8223	1.4559
225	Tetradecylcyclopentane	9	341	0.8196	1.4543
226	Tridecylcyclohexane	18.5	346	0.8239	1.4570
227	Pentadecylcyclopentane	17	355	0.8213	1.4554
228	Hexadecylcyclopentane	21		0.8194^{25}	1.4543^{25}
229	Heptadecane	22	301.8	0.7745^{25}	1.4348^{25}
230	Tetradecylcyclohexane	24		0.8221^{25}	1.4559^{25}
231	Heptadecylcylopentane	27		0.8173^{30}	1.4532^{30}
232	Octadecane	28.2	316.1	0.7751^{30}	1.4191^{70}
233	Pentadecylcyclohexane	29		0.8201^{30}	1.4545^{30}
234	Octadecylcyclopentane	30		0.8186^{30}	1.4541^{30}
235	Nonadecane	32.1	329.7	0.7776^b	1.4211^{70}
236	Hexadecylcyclohexane	33.6		0.8279^c	1.4596^c
237	Nonadecylcyclopentane	35		0.8266^c	1.4588^c
238	Eicosane	36.8	342.7	0.7550^{70}	1.4230^{70}
239	Heptadecylcyclohexane	37.8		0.8290^c	1.4603^c
240	Eicosylcyclopentane	38		0.8276^c	1.4595^c
241	Heneicosane	40.5		0.7583^{70}	1.4260^{70}
242	Octadecylcyclohexane	41.6		0.8300^c	1.4610^c
243	Heneicosylcyclopentane	42		0.8286^c	1.4602^c
244	Docosane	44.4		0.7631^{70}	1.4260^{70}
245	Docosylcyclopentane	45		0.8295^c	1.4608^c
246	Nonadecylcyclohexane	45.2		0.8310^c	1.4616^c
247	Tricosane	47.6		0.7641^{70}	1.4276^{70}
248	Eicosylcyclohexane	48.5		0.8318^c	1.4622^c
249	Tricosylcyclopentane	49		0.8304^c	1.4614^c
250	Tetracosane	50.9		0.7657^{70}	1.4286^{70}
251	Tetracosylcyclopentane	51		0.8312^c	1.4619^c
252	Heneicosylcyclohexane	51.5		0.8326^c	1.4627^c
253	Pentacosane	53.7		0.7693^{70}	1.4302^{70}
254	Pentacosylcyclopentane	54		0.8319^c	1.4624^c
255	Docosylcyclohexane	54.4		0.8334^c	1.4632^c
256	Hexacosylcyclopentane	56		0.8326^c	1.4628^c
257	Hexacosane	56.4		0.7704^{70}	1.4310^{70}
258	Tricosylcyclohexane	57		0.8341^c	1.4637^c
259	Heptacosane	59		0.7732^{70}	1.4321^{70}
260	Heptacosylcyclopentane	59		0.8333^c	1.4633^c
261	Tetracosylcyclohexane	59.5		0.8347^c	1.4641^c
262	Cyclopentadecane	60–1		0.8364^b	1.4592^{61}

[b] Densities for compounds melting above 28° were determined at the melting point.
[c] For the supercooled liquid at 20°.

(Continued)

TABLE 18 (Continued)
ALKANES AND CYCLOALKANES

	Name of Compound	MP	BP	D_4^{20}	n_D^{20}
263	Octacosylcyclopentane	61		0.8339^c	1.4637^c
264	Octacosane	61.4		0.7750^{70}	1.4330^{70}
265	Pentacosylcyclohexane	61.9		0.8353^c	1.4645^c
266	Nonacosylcyclopentane	63		0.3845^c	1.4640^c
267	Nonacosane	63.7		0.7797^b	1.4361^{65}
268	Hexacosylcyclohexane	64		0.8359^c	1.4649^c
269	Triacontylcyclopentane	65		0.8350^c	1.4644^c
270	Triacontane	65.8		0.7797^b	1.4348^{70}
271	Heptacosylcyclohexane	66.1		0.8365^c	1.4653^c
272	Hentriacontylcyclopentane	67		0.8356^c	1.4648^c
273	Hentriacontane	67.9		0.8111^c	1.4543^c
274	Octacosylcyclohexane	68		0.8370^c	1.4656^c
275	Dotriacontylcyclopentane	69		0.3860^c	1.4651^c
276	Dotriacontane	69.7		0.7791^{75}	1.4550^c
277	Nonacosylcyclohexane	69.9		0.8374^c	1.4659^c
278	Tritriacontylcyclopentane	70		0.8365^c	1.4654^c
279	Tritriacontane	71.4		0.8136^c	1.4557^c
280	Triacontylcyclohexane	71.6		0.8379^c	1.4662^c
281	Tetratriacontylcyclopentane	72		0.8370^c	1.4657^c
282	Tetratriacontane	73.1		0.8148^c	1.4563^c
283	Hentriacontylcyclohexane	73.3		0.8383^c	1.4665^c
284	Pentatriacontylcyclopentane	74		0.3874^c	1.4660^c
285	Pentatriacontane	74.7		0.8157^c	1.4568^c
286	Dotriacontylcyclohexane	74.8		0.8388^c	1.4668^c
287	Hexatriacontylcyclopentane	75		0.8378^c	1.4662^c
288	Hexatriacontane	76.2		0.8169^c	1.4573^c
289	Tritriacontylcyclohexane	76.3		0.8391^c	1.4670^c
290	Heptatriacontane	77.7		0.8179^c	1.4578^c
291	Tetratriacontylcyclohexane	77.7		0.8395^c	1.4673^c
292	Octatriacontane	79		0.8188^c	1.4583^c
293	Pentatriacontylcyclohexane	79.1		0.8399^c	1.4675^c
294	Nonatriacontane	80.3		0.8197^c	1.4588^c
295	Hexatriacontylcyclohexane	80.4		0.8402^c	1.4678^c
296	Tetracontane	81.5		0.8205^c	1.4593^c
297	2,2,3,3-Tetramethylbutane (bp 106.3)	100.7		0.8242^b	$1.4695^{106.3}$

REFERENCES FOR TABLE 18

Bell, M. F., *Anal. Chem.*, **22**, 1005 (1950). Analysis of East Texas Virgin Naptha Fractions Boiling up to 210 F.: Infrared Method. Many boiling points are given.

Braun, W. G., Spooner, D. F., and Fenske, M. R., *Anal. Chem.*, **22**, 1074 (1950). Raman Spectra: Hydrocarbons and Oxygenated Compounds. Includes 119 spectra.

Glasgow, A. R. Jr., Gordon, R. J., Willingham, C. B., Mair, B. J., and Rossini, F. D., *Anal. Chem.*, **29**, 357 (1957). Hydrocarbons in the 116 degree to 126°C. fraction of Petroleum. The boiling points may be found on p. 360.

Rossini, J. D., *Anal. Chem.*, **20**, 110 (1948). Pure Compounds from Petroleum; Purification and Purity of

REFERENCES TO TABLE 18 (Continued)

Hydrocarbons. Lists the boiling points of many hydrocarbons.

Streiff, A. J., Hulme, A. R., and Cowie, P. A., Krouskop, N. C., and Rossini, F. D., *Anal. Chem.*, **27,** 411 (1955). Purification, Purity, and Freezing Points of 64 American Petroleum Institute Standard and Research Hydrocarbons. F.P.s on p. 414.

Tadayon, J., Nissan, A. H., and Garner, F. H., *Anal. Chem.* **21,** 1532 (1949). Magneto-Optical Rotatory Power of Hydrocarbons and their Mixtures. Boiling points, refractive indices, and densities are given.

Ward, A. L. and Hurtz, S. S. Jr., *Anal. Chem.*, **10,** 559 (1938). Refraction, Dispersion and Related Properties of pure Hydrocarbons. Lists the properties of hundreds of hydrocarbons.

TABLE 19
ALKENES, ALKYNES, CYCLOALKENES, AND DIENES

	Name of Compound	Note	MP	BP	D_4^{20}	n_D^{20}
1	2-Methylpropene	1,3	−140.4	−6.9	$0.6266^{-6.6}$	1.3467
2	1-Butene	3		−6.3	$0.6255^{-6.5}$	1.3465
3	1,3-Butadiene	5	−108.9	−4.4	0.650^{-6}	1.4292^{-25}
4	2-Butene (*trans*)		−105.5	0.88	0.6042^a	
5	2-Butene (*cis*)		−138.9	3.72	0.6303^1	
6	1-Butyne	24	−125.7	8.1	0.6682^8	1.3962
7	1,2-Butadiene		−136.2	10.9	0.652	1.4208^3
8	3-Methyl-1-Butene (pol)		−168.5	20.1	0.6272	1.3643
9	1,4-Pentadiene		−148.3	26	0.66076	1.38876
10	3-Methyl-1-butyne		−89.7	26.4	0.666	1.3785
11	2-Butyne	3,5	−32.3	27	0.6910	1.3921
12	1-Pentene	1	−165.2	30	0.64050	1.37148
13	2-Methyl-1-butene		−137.6	31.2	0.6504	1.3778
14	3-Methyl-3-buten-1-yne			32	0.6801^{11}	1.4158
15	2-Methyl-1,3-butadiene (isoprene) (pol)	12	−146	34.1	0.68095	1.42194
16	2-Pentene (*trans*)		−140.2	36.4	0.6482	1.3793
17	2-Pentene (*cis*)		−151.4	36.9	0.6556	1.3830
18	2-Methyl-2-butene		−133.8	38.6	0.6623	1.3874
19	3-Methyl-1,2-butadiene			40	0.680	1.410
20	1-Pentyne	24	−105.7	40.2	0.6901	1.3852
21	1,3-Cyclopentadiene (pol)	13	−85	41	0.7983	1.4461
22	3,3-Dimethyl-1-butene		−115.2	41.2	0.6529	1.3760
23	1-*trans*-3-Pentadiene		−87.5	42	0.67603	1.43008
24	1,3-Pentadiene (piperylene)	5,14	−88.9	42.3	0.6803	1.4309
25	1-*cis*-3-Pentadiene		−140.8	44.1	0.69102	1.43634
26	Cyclopentene	1,7	−135.1	44.2	0.77199	1.42246
27	1,2-Pentadiene		−137.3	44.9	0.69257	1.42091
28	2,3-Pentadiene		−125.7	48.3	0.69502	1.42842
29	4-Methyl-1-pentene		−153.6	53.9	0.6642	1.3828
30	3-Methyl-1-pentene		−153	54.1	0.6675	1.3842
31	3-Methyl-1,4-pentadiene			55	0.695	1.405
32	2,3-Dimethyl-1-butene		−57.3	55.7	0.6779	1.3904
33	2-Methyl-1,4-pentadiene			56	0.694	1.405
34	2-Pentyne		−109.3	56.1	0.7107	1.4039
35	4-Methyl-*cis*-2-pentene		−134.4	56.3	0.6690	1.3880
36	4-Methyl-*trans*-2-pentene		−140.8	58.6	0.6686	1.3889
37	1,5-Hexadiene			59.5	0.6923	1.4042
38	2-Methyl-1-pentene		−135.7	60.7	0.6799	1.3920
39	1-Hexene	3	−139.8	63.5	0.67317	1.38788
40	2-Ethyl-1-butene			64.7	0.6894	1.3969
41	3-Methylcyclopentene			65	0.7622	1.4207
42	1,(*cis* and/or *trans*) 4-Hexadiene			65	0.700	1.415
43	3-Hexene (*cis*)		−137.8	66.4	0.6796	1.3947
44	3-Hexene (*trans*)		−113.4	67.1	0.6772	1.3943
45	2-Methyl-2-pentene		−135.1	67.3	0.6863	1.4004
46	3-Methyl-*trans*-2-pentene		−134.8	67.6	0.6942	1.4016
47	2-Hexene (*trans*)		−133	67.9	0.6784	1.3935
48	2,3-Hexadiene			68	0.680	1.395
49	2-Hexene (*cis*)		−141.1	68.8	0.6869	1.3977
50	2,3-Dimethyl-1,3-butadiene (pol)	5,8	−76	68.8	0.7267	1.4394
51	3-Methyl-1,2-pentadiene			70	0.715	1.425
52	4-Methyl-1,2-pentadiene			70	0.708	1.424
53	3-Methyl-*cis*-2-pentene		−138.4	70.5	0.6986	1.4045
54	1-Hexyne	2,24	−131.9	71.3	0.7155	1.3989
55	2-Methyl-2,3-pentadiene			72	0.711	1.425
56	4,4-Dimethyl-1-pentene		−136.6	72.5	0.6827	1.3918

(*Continued*)

TABLE 19 (Continued)
ALKENES, ALKYNES, CYCLOALKENES, AND DIENES

	Name of Compound	Note	MP	BP	D_4^{20}	n_D^{20}
57	1,(cis and/or trans) 3-Hexadiene			73	0.705	1.438
58	2,3-Dimethyl-2-butene	5	−74.3	73.2	0.7080	1.4122
59	2-Ethyl-1,3-butadiene			75	0.717	1.445
60	4-Methylcyclopentene			75.2	0.7796	1.4306
61	1-Methylcyclopentene		−127	75.8	0.7802	1.4330
62	1,2-Hexadiene			76	0.7149	1.4282
63	2-Methyl-1, (cis and/or trans) 3-pentadiene			76	0.719	1.446
64	4-Methyl-1,3-pentadiene			76.3	0.719	1.451
65	4,4-Dimethyl-trans-2-pentene		−115.2	76.8	0.6889	1.3982
66	3-Methyl-1, (cis and/or trans) 3-pentadiene			77	0.735	1.452
67	3,3-Dimethyl-1-pentene		−134.3	77.5	0.6974	1.3984
68	2,3,3-Trimethyl-1-butene			77.9	0.7050	1.4029
69	1,3,5-Hexatriene			78	0.718	1.4330
70	2,4-Hexadiene (all cis-trans isomers)	16	−79	80	0.720	1.450
71	1,3-Cyclohexadiene	15	−95	80.3	0.8413	1.4740
72	4,4-Dimethyl-cis-2-pentene		−135.4	80.4	0.6996	1.4024
73	3,4-Dimethyl-1-pentene			81	0.701	1.3995
74	3-Hexyne	3		81	0.724	1.4115
75	2,4-Dimethyl-1-pentene		−123.8	81.6	0.6943	1.3986
76	Cyclohexene	1,3	−103.5	83	0.81096	1.44654
77	2,4-Dimethyl-2-pentene			83.4	0.6955	1.4040
78	3-Methyl-1-hexene			84	0.695	1.395
79	2,3-Dimethyl-1-pentene		−134.8	84.3	0.7051	1.4033
80	3-Ethyl-1-pentene		−127.4	85.1	0.6962	1.3980
81	5-Methyl-1-hexene			85.3	0.6920	1.3966
82	5-Methyl-trans-2-hexene			86	0.700	1.400
83	2-Methyl-(cis and/or trans) 3-hexene			86	0.694	1.399
84	4-Methyl-1-hexene		−141.5	86.7	0.6985	1.4000
85	3,4-Dimethyl-(cis and/or trans) 2-pentene			87	0.713	1.407
86	4-Methyl-cis-2-hexene			87.4	0.6996	1.4024
87	4-Methyl-trans-2-hexene		−126.5	87.6	0.6975	1.4023
88	3,3-Dimethylcyclopentene			88	0.771	1.423
89	4,4-Dimethylcyclopentene			88	0.771	1.423
90	2-Ethyl-3-methyl-1-butene			89	0.715	1.410
91	5-Methyl-cis-2-hexene			91	0.700	1.400
92	2-Methyl-1-hexene		−102.8	92	0.7030	1.4034
93	1,3-Dimethylcyclopentene			92	0.766	1.428
94	1,4-Dimethylcyclopentene			93.2	0.779	1.4283
95	3-Methyl-trans-3-hexene			93.6	0.7099	1.4107
96	1-Heptene		−119.1	93.6	0.69698	1.39980
97	3-Methyl-(cis and/or trans) 2-hexene			94	0.7120	1.410
98	2-Ethyl-1-pentene			94	0.708	1.405
99	3-Methyl-cis-3-hexene			95.4	0.7132	1.4123
100	2-Methyl-2-hexene		−130.3	95.4	0.7082	1.4106
101	3-Heptene (trans)		−136.6	95.7	0.6981	1.4043
102	3-Heptene (cis)			95.8	0.7030	1.4059
103	3-Ethyl-2-pentene			96	0.7204	1.4148
104	2,3-Dimethyl-2-pentene		−118.3	97.5	0.7277	1.4208
105	2-Heptene (trans)		−109.5	98	0.7012	1.4045
106	3-Ethylcyclopentene			98.1	0.7830	1.4319
107	2-Heptene (cis)			98.5	0.708	1.406
108	2-Heptene (trans)			98.5	0.703	1.4041
109	1-Heptyne	2,24	−80.9	99.7	0.7328	1.4087
110	2,2-Dimethyl-trans-3-hexene			100.9	0.7039	1.4063

(Continued)

TABLE 19 (Continued)
ALKENES, ALKYNES, CYCLOALKENES, AND DIENES

	Name of Compound	Note	MP	BP	D_4^{20}	n_D^{20}
111	2,4,4-Trimethyl-1-pentene (di-isobutylene)	1	−93.5	101.4	0.7150	1.4086
112	2,5-Dimethyl-(*cis* and/or *trans*) 3-hexene			102	0.710	1.406
113	1,5-Dimethylcyclopentene			102	0.780	1.4331
114	5,5-Dimethyl-1-hexene			102.5	0.709	1.4049
115	4-Methylcyclohexene	1	−115.5	102.7	0.7991	1.4414
116	2-Isopropyl-3-methyl-1-butene			104	0.722	1.4085
117	3-Methylcyclohexene			104	0.8010	1.4444
118	3,4,4-Trimethyl-1-pentene			104	0.719	1.412
119	3,5-Dimethyl-1-hexene			104	0.708	1.404
120	3,3-Dimethyl-1-hexene			104	0.7140	1.4070
121	5,5-Dimethyl-*trans*-2-hexene			104.1	0.7066	1.4055
122	2,4,4-Trimethyl-2-pentene		−106.3	104.9	0.7218	1.4160
123	3,3,4-Trimethyl-1-pentene			105	0.729	1.4144
124	2,2-Dimethyl-*cis*-3-hexene		−137.4	105.4	0.7128	1.4099
125	1,2-Dimethylcyclopentene		−90.4	105.8	0.7976	1.4448
126	4-Ethylcyclopentene			106	0.798	1.440
127	4,4-Dimethyl-(*cis* and/or *trans*) 2-hexene			106	0.722	1.413
128	1-Ethylcyclopentene			106.3	0.7982	1.4410
129	5,5-Dimethyl-*cis*-2-hexene			106.9	0.7169	1.4113
130	4,4-Dimethyl-1-hexene			107.2	0.7198	1.4102
131	3-Ethyl-4-methyl-1-pentene			107.5	0.7200	1.4097
132	2,4-Dimethyl-*trans*-3-hexene			107.6	0.7145	1.4126
133	2,3,4-Trimethyl-1-pentene			108	0.729	1.415
134	2,3,3-Trimethyl-1-pentene		−69	108.3	0.7352	1.4174
135	2,4-Dimethyl-*cis*-3-hexene			109	0.7178	1.4140
136	4,5-Dimethyl-1-hexene			109	0.728	1.414
137	3-Ethyl-2-methyl-1-pentene			110	0.730	1.415
138	4,5-Dimethyl-(*cis* and/or *trans*) 2-hexene			110	0.725	1.413
139	1-Methylcyclohexene	3	−121	110	0.8102	1.4503
140	2-Ethyl-3,3-dimethyl-1-butene			110	0.728	1.4159
141	2-Ethyl-4-methyl-1-pentene			110.3	0.7195	1.4105
142	3-Ethyl-1-hexene			110.3	0.715	1.407
143	2,3-Dimethyl-1-hexene			110.5	0.7214	1.4113
144	2,4-Dimethyl-2-hexene			110.6	0.7213	1.4118
145	3-Methyl-1-heptene			111	0.711	1.406
146	2,4-Dimethyl-1-hexene			111.2	0.720	1.411
147	2,5-Dimethyl-1-hexene			111.6	0.7172	1.4105
148	3-Ethyl-3-methyl-1-pentene			112	0.3705	1.418
149	3,4-Dimethyl-1-hexene			112	0.724	1.413
150	3,5-Dimethyl-(*cis* and/or *trans*) 2-hexene			112	0.725	1.416
151	5-Methyl-(*cis* and/or *trans*) 3-heptene			112	0.713	1.410
152	3,4,4-Trimethyl-(*cis* and/or *trans*) 2-pentene			112	0.739	1.423
153	2-Methyl-(*cis* and/or *trans*) 3-heptene			112	0.706	1.407
154	2,5-Dimethyl-2-hexene			112.2	0.720	1.4140
155	2-Ethyl-3-methyl-1-pentene			112.5	0.729	1.4142
156	4-Methyl-1-heptene			112.8	0.717	1.410
157	4-Ethyl-(*cis* and/or *trans*) 2-hexene			113	0.725	1.412
158	4-Ethyl-1-hexene			113	0.726	1.412
159	2-Isopropyl-1-pentene			113	0.725	1.414
160	6-Methyl-1-heptene			113.2	0.7120	1.4070
161	5-Methyl-1-heptene			113.3	0.7164	1.4094

(*Continued*)

TABLE 19 (Continued)
ALKENES, ALKYNES, CYCLOALKENES, AND DIENES

	Name of Compound	Note	MP	BP	D_4^{20}	n_D^{20}
162	2,3-Dimethyl-(cis and/or trans) 3-hexene			114	0.728	1.416
163	2,5-Dimethyl-1,5-hexadiene			114		1.4290
164	4-Methyl-(cis and/or trans) 2-heptene			114	0.716	1.410
165	3-Ethyl-4-methyl-trans-2-pentene			114.3	0.7350	1.4210
166	6-Methyl-(cis and/or trans) 3-heptene			115	0.713	1.410
167	Cycloheptene (suberene)	17	−56	115*	0.8228	1.4580
168	3-Ethyl-4-methyl-cis-2-pentene			115	0.739	1.424
169	3-Ethyl-3-hexene			116	0.729	1.418
170	3,4-Dimethyl-(cis and/or trans) 2-hexene			116	0.737	1.418
171	2,3,4-Trimethyl-2-pentene		−113.3	116.3	0.7434	1.4275
172	3-Ethyl-2-methyl-2-pentene			117	0.739	1.4247
173	6-Methyl-(cis and/or trans) 2-heptene			117	0.718	1.412
174	4,4-Dimethylcyclohexene		−80.5	117	0.7996	1.4420
175	2-n-Propyl-1-pentene			117.7	0.7240	1.4136
176	5-Methyl-(cis and/or trans) 2-heptene			118	0.723	1.414
177	2,5-Dimethyl-1,5-hexadiene			118	0.7423	1.4293
178	3,3-Dimethylcyclohexene			119	0.804	1.445
179	2-Methyl-1-heptene		−90.0	119.3	0.7205	1.4123
180	2-Ethyl-1-hexene			120	0.7270	1.4157
181	3-Ethyl-(cis and/or trans) 2-hexene			121	0.737	1.424
182	3-Methyl-(cis and/or trans) 3-heptene			121	0.728	1.418
183	1-Octene		−104.7	121.3	0.71492	1.40870
184	2,3-Dimethyl-2-hexene		−115.1	121.8	0.7408	1.4268
185	3,4-Dimethyl-(cis and/or trans) 3-hexene			122	0.747	1.430
186	4-Methyl-(cis and/or trans) 3-heptene			122	0.725	1.417
187	3-Methyl-(cis and/or trans) 2-heptene			122	0.729	1.419
188	4-Octene-trans		−93.8	122.3	0.7141	1.4118
189	4-Octene-cis		−118.7	122.5	0.7212	1.4148
190	2-Methyl-2-heptene			122.6	0.7241	1.4170
191	3-Octene (cis)			122.9	0.721	1.4135
192	3-Octene (trans)		−110	123.3	0.7152	1.4126
193	2-Octene (trans)		−87.7	125	0.7199	1.4132
194	2-Octene (cis)		−100.2	125.6	0.7243	1.4150
195	1-Octyne	2,24	−79.3	126.2	0.7461	1.4159
196	4-Vinylcyclohexene			128	0.832	1.464
197	1,5-Dimethylcyclohexene			128	0.8051	1.448
198	1,4-Dimethylcyclohexene		−59	128	0.802	1.446
199	2,6-Dimethyl-2-heptene			128.9	$0.722^{15/15}$	1.412
200	4-Ethylcyclohexene			133	0.810	1.449
201	1,6-Dimethylcyclohexene			133	0.815	1.454
202	2,5-Dimethyl-2,4-hexadiene			133–8	0.7646_4^{18}	$1.4796_D^{19.5}$
203	3-Ethylcyclohexene			134	0.814	1.451
204	1-Ethylcyclohexene			136	0.823	1.4575
205	1,2-Dimethylcyclohexene			137	0.8250	1.4588
206	1,3-Dimethylcyclohexene			137	0.802	1.445
207	Ethynylbenzene (phenylacetylene)		−40	141.7	0.9281	1.5485
208	Cyclo-octene			144	0.8500	1.4704
209	Styrene (pol) (vinylbenzene)	3,5,9	−30.6	145.2	0.90600	1.54682
210	1-Nonene		−81.4	146.9	0.72922	1.41572
211	1,5-Cyclooctadiene			149.2	0.88325	1.49330
212	1-Nonyne	24	−50	150.8	0.7568	1.4217
213	α-Pinene, d- or l-	10	−50	155–6	0.8595	1.47299
214	α-Pinene, dl-		−50	156.2	0.8582	1.4658
215	Allylbenzene			157	0.8912	1.5042^{25}

(*Continued*)

TABLE 19 (Continued)
ALKENES, ALKYNES, CYCLOALKENES, AND DIENES

	Name of Compound	Note	MP	BP	D_4^{20}	n_D^{20}
216	β-Pinene			163–4	0.8694	1.4782
217	α-Methylstyrene (Isopropenylbenzene)	1	−23.2	165.4	0.9106	1.5386
218	2-Methyl-6-methylene-2,7-octadiene (myrcene)	11		166	0.7982	1.4722
219	2-Methylstyrene (o-methylvinylbenzene)		−68.6	169.8	0.9036	1.54654
220	β-Methylstyrene (propenylbenzene)		−53.2	170	0.911	1.549
221	1-Decene		−66.3	170.6	0.74081	1.42146
222	3-Methylstyrene (m-methylvinylbenzene)		−86.3	171.6	0.9113	1.54390
223	4-Methylstyrene (p-methylvinylbenzene)		−34.2	172.8	0.9106	1.54496
224	1-Decyne	2,24	−44	174	0.7655	1.4265
225	1-Phenyl-2-butene			175	$0.8881^{15/15}$	1.511
226	1-Methyl-3-isopropenylcyclohexene, (sylvestrene) (d or l)	5,18		176–8	0.8479	1.4760
227	1-Methyl-4-isopropenylcyclohexene, (l-limonene)	19		177.7	0.8422	1.47468
228	1-Methyl-3-isopropenylcyclohexene,	20		178		
229	1-Methyl-4-isopropenylcyclohexene, (dl-limonene)	5		178	0.8402	1.4727
230	1-Methyl-4-isopropenylcyclohexene, (d-limonene)	3		178	0.8411	1.47428^{21}
231	Indene (pol)	4,5	−2	182	0.9915	1.5764
232	3-Ethylstyrene		−101.3	190.1	0.89449	1.53512
233	1-Undecene		−49.2	192.7	0.75032	1.42609
234	4-Ethylstyrene		−49.7	192.8	0.89249	1.53763
235	1-Undecyne	2,24	−25	195	0.7728	1.4306
236	4-Isopropylstyrene		−44.7	204.2	0.88497	1.52891
237	1-Dodecene		−35.4	213.4	0.75836	1.43002
238	1-Dodecyne	2	−19	215	0.7788	1.4340
239	4-Isopropyl-α-methylstyrene		−30.6	220.8	0.89363	1.52381
240	1-Tridecene		−23.1	232.8	0.7653	1.4336
241	1-Tridecyne		−5	234	0.7842	1.4371
242	1-Tetradecene		12.9	251.1	0.7713	1.43631
243	1-Tetradecyne		0.0	252	0.7888	1.4396
244	1-Allylnaphthalene			265–7	1.0228	1.6140
245	1-Pentadecyne		10	268	0.7928	1.4419
246	1-Pentadecene		−3.7	268.2	0.77641	1.4389
247	1,1-Diphenylethene	1,3	8	277	1.038^{14}	1.610^{14}
248	1-Hexadecyne		15	284	0.7965	1.4440
249	1-Hexadecene		4.1	284.4	0.78112	1.44120
250	1-Heptadecene		11.2	299.7	0.7852	1.4432
251	1-Octadecene	5	17.6	314.2	0.7888	1.4449
252	1-Heptadecyne		22	299	0.7961^{25}	1.4437^{25}
253	1-Nonadecene		23.4	328	0.7886^{25}	1.445^{25}
254	1,4-Dihydronaphthalene	21	24.5	212	0.998	1.5740
255	1-Octadecyne		27	313	0.7955^{30}	1.4432^{30}
256	1-Eicosene		28.6	341.2	0.7882^{30}	1.4439^{30}
257	Dicyclo-4,7-pentadiene	22	32–3			
258	1-Nonadecyne		33	327	0.8050^a	1.4488^a
259	1-Heneicosene		33.3	355	0.7977^a	1.4494^a
260	1-Eicosyne		35	340	0.8073^a	1.4501^a
261	1-Docosene		37.8		0.8002^a	1.4505^a
262	1-Heneicosyne		41		0.8094^a	1.4513^a
263	1-Tricosene		41.6	379	0.8023^a	1.4516^a

(Continued)

TABLE 19 (Continued)
ALKENES, ALKYNES, CYCLOALKENES, AND DIENES

	Name of Compound	Note	MP	BP	D_4^{20}	n_D^{20}
264	1-Docosyne		45	363	0.8114[a]	1.4524[a]
265	1-Tetracosene		45.3	390	0.8045[a]	1.4527[a]
266	1-Pentacosene		48.7	401	0.8063[a]	1.4536[a]
267	1-Tricosyne		49	374	0.8131[a]	1.4534[a]
268	dl-Camphene	1	50	159	0.879	1.4402^{80}
269	d-Camphene		51	160–2		
270	l-Camphene	3,23	51.3*	159–60	0.8555	1.46207
271	1-Hexacosene		51.8	411	0.8082[a]	1.4545[a]
272	1-Tetracosyne		52	385	0.8148[a]	1.4544[a]
273	1-Heptacosene		54.7	421	0.8097[a]	1.4552[a]
274	1-Pentacosyne		55	395	0.8163[a]	1.4552[a]
275	1-Hexacosyne		57	405	0.8177[a]	1.456[a]
276	1-Octacosyne		57.5	430	0.8114[a]	1.4560[a]
277	1-Nonacosene		60	440	0.8127[a]	1.4567[a]
278	1-Heptacosyne		60	415	0.8190[a]	1.4568[a]
279	1-Octacosyne		62		0.8202[a]	1.4575[a]
280	1-Triacontene		62.4	448	0.8141[a]	1.4573[a]
281	1,2-Diphenylethyne	3,4	62.5	298		
282	1-Hentriacontene		64.6	457	0.8153[a]	1.4580[a]
283	1-Nonacosyne		65	432	0.8213[a]	1.4581[a]
284	1-Dotriacontene		66.7	465	0.8165[a]	1.4585[a]
285	1-Triacontyne		67	441	0.8224[a]	1.4587[a]
286	1-Tritriacontene		68.7	473	0.8176[a]	1.4591[a]
287	1-Hentriacontyne		69	449	0.8234[a]	1.4593[a]
288	1,4-Diphenyl-1,3-butadiene[a] (cis)		70			
289	1-Tetratriacontene		70.5	481	0.8186[a]	1.4596[a]
290	1-Dotriacontyne		71	457	0.8243[a]	1.4598[a]
291	1-Pentatriacontene		72.3	489	0.8196[a]	1.4601[a]
292	1-Tritriacontyne		73	464	0.8252[a]	1.4603[a]
293	1-Hexatriacontene		73.9	496	0.8205[a]	1.4605[a]
294	1-Tetratriacontyne		74	472	0.8260[a]	1.4608[a]
295	1-Heptatriacontene		75.5	503	0.8214[a]	1.4610[a]
296	1-Pentatriacontyne		76	479	0.8268[a]	1.4612[a]
297	1-Hexatriacontyne		77	486	0.8275[a]	1.4617[a]
298	1-Octatriacontene		77	510	0.8223[a]	1.4614[a]
299	1-Nonatriacontene		78.4	517	0.8231[a]	1.4618[a]
300	1-Heptatriacontyne		79	493	0.8282[a]	1.4621[a]
301	1-Tetracontene		79.8	523	0.8238[a]	1.4622[a]
302	1-Octatriacontyne		80	499	0.8289[a]	1.4625[a]
303	1-Nonatriacontyne		82	505	0.8295[a]	1.4628[a]
304	1-Tetracontyne		83	512	0.8301[a]	1.4632[a]
305	Acenaphthylene	4	92–3			
306	1,2-Diphenylethene	1,4	124	306s	0.970_{13}^{125}	
307	1,4-Diphenyl-1,3-butadiene (trans)	4	152.5	350		

[a] For the supercooled liquid at 20°C.

NOTES FOR TABLE 19

1. The following compounds have been derivatized by J. G. Sharefkin and T. Salzberg [Anal. Chem., **32**, 955 (1960)] by acylation of the alkene to a ketone which was then derivatized:
    Ethene: semicarbazone, mp 138.
    4-Methylcyclohexene: semicarbazone, mp 187.5
    dl-Camphene: semicarbazone, mp 195–6

    1,2-Diphenylethene (trans): oxime, mp 151
    Cyclopentene: semicarbazone, mp 210; 2,4-dinitrophenylhydrazone, mp 202
    Cyclohexene: semicarbazone, mp 218; 2,4-dinitrophenylhydrazone, mp 200
    2,2,4-Trimethylpentene: 2,4-dinitrophenylhydrazone, mp 162.
    1-Pentene: semicarbazone, mp 123; 2,4-dinitrophenylhydrazone, mp 124–5

NOTES FOR TABLE 19 (Continued)

2-Methylpropene: semicarbazone, mp 162; 2,4-dinitrophenylhydrazone, mp 198–9

Propene: semicarbazone, mp 140–2; 2,4-dinitrophenylhydrazone, mp 153

α-Methylstyrene; semicarbazone, mp 185; 2,4-dinitrophenylhydrazone, mp 173

1,1-Diphenylethene: 2,4-dinitrophenylhydrazone, mp 149

2. A number of alkynes have been characterized by hydration to carbonyl compounds for derivatization by J. G. Sharefkin and E. M. Bohosion [*Anal. Chem.*, **33**, 643 (1961)]:

    1-Decyne: semicarbazone, mp 124
    1-Octyne: semicarbazone, mp 124–5; 2,4-dinitrophenylhydrazone, mp 58
    1-Undecyne: semicarbazone, mp 122–3; 2,4-dinitrophenylhydrazone, mp 63
    1-Dodecyne: semicarbazone, mp 122–3; 2,4-dinitrophenylhydrazone, mp 81
    1-Heptyne: semicarbazone, mp 123; 2,4-dinitrophenylhydrazone, mp 89
    1-Hexyne: semicarbazone, mp 121; 2,4-dinitrophenylhydrazone, mp 106

3. The following compounds yield derivatives with 2,4-dinitrobenzenesulfenyl chloride with the melting points indicated: 2-Methylpropene, 86–7; 1-butene, 78; 2-butyne, 75–6; 1-hexene, 61–2; 3-hexyne, 65–6; cyclohexene, 117–8; 1-methylcyclohexene, 139–40; styrene, 143; *d*-limonene, 195–6; 1,1-diphenylethene, 136; *l*-camphene, 121–2; and 1,2-diphenylethyne, 206–7.

4. The following melting points of picrate adducts were available: indene, 96; 1-allylnaphthalene, 69; 1,2-diphenylethyne, 111; acenaphthylene, 202; 1,2-diphenylethene, 94–5 (T.N.F. derivative, 148–9); 1,4-diphenyl-1,3-butadiene (*trans*), 152–3.

5. The following compounds have bromo derivatives as shown: 1,3-butadiene, 119 (tetra); 2-butyne, 243 (tetra); 1,3-pentadiene, 114 (tetra); 2,3-dimethyl-1,3-butadiene, 138 (tetra); 2,3-dimethyl-2-butene, 121 (di); 2,4-hexadiene, 185 (tetra); styrene, 74 (di); *dl*-α-pinene, 169 (di); sylvestrene, 135 (tetra); *l*-limonene, 104 (tetra) *dl*-limonene, 124; indene, 32 (di); 1-octadecene, 24 (di); and acenaphthylene, 121–3 (di).

6. Certain alkynes form derivatives with K_2HgI_4: 1-butyne, 162–3; 1-pentyne, 118.5; and 1-heptyne, 61.
7. Pseudonitrosite from N_2O_3, 70.
8. Maleic anhydride adduct, 78–9.
9. Dithiocyanate derivative, 103.
10. Nitrosochloride derivative, 109; nitroso derivative, 132.
11. 1,4-Naphthoquinone adduct, 81; maleic anhydride adduct, 33–4.
12. Thiocyanate derivative, 77; reacts with maleic anhydride, 64.
13. Reacts with maleic anhydride, 164; reacts with benzoquinone, 76.
14. Reacts with maleic anhydride, 61–2.
15. Reacts with bromine in chloroform to yield a dibromo derivative which melts at 68 but isomerizes rapidly to 1,4-dibromo-2-cyclohexene which melts at 108 and this compound does not add bromine. However, if the dibromo compound (mp 68) is treated with bromine before it isomerizes, two tetrabromo derivatives are produced, melting at 87–9 and 155–6. The maleic anhydride derivative melts at 147.
16. Maleic anhydride derivative, 95.
17. Nitrosochloride, 118; oxidizes by nitric acid to pimelic acid, 105.
18. Dihydrochloride, 72; nitrosochloride, 107.
19. Dihydrochloride, 50; nitrosochloride, 103–4 (*cis*) and 105–6 (*trans*); nitroso, 72.
20. Dihydrochloride, 50.
21. Nitrosochloride, 143–4; nitrolamine, 146.
22. Phenylazide adduct, 128.
23. Hydrochloride, 125–7; H-phthalate, 163–4.
24. A procedure is given by A. P. Hobbs on page 9 of Section 4 in the *Handbook of Analytical Chemistry*, L. Meites, ed. McGraw-Hill, New York, 1963, for the preparation of the mercury acetylide of 1-alkynes: 1-butyne, 162–3; 1-pentyne, 118; 1-hexyne, 96; 1-heptyne, 61; 1-octyne, 80; 1-nonyne, 68; 1-decyne, 80; and 1-undecyne, 79.

REFERENCES FOR TABLE 19

Hendirckson, J. G., and Hatch, L. F., *J. Org. Chem.*, **25**, 1747 (1960). S-Alkylmercaptosuccinic Acids as Solid Derivatives of Olefins, Alkyl Bromides, and Mercaptans. Describes the method of preparation and lists the melting points of the derivatives of 46 compounds.

Mathieu, J. P., and Ourisson, G., *Optical Rotatory Power. II.*, Pergamon, New York, 1959. Triterpenoids.

Sharefkin, J. G., and Sulzberg, T., *Anal. Chem.*, **32**, 993 (1960). Detection and Characterization of Olefins as 2,4-Dinitrophenylhydrazones of the Methyl Ketones Formed by Friedel-Crafts Acetylation.

Sharefkin, J. G., and Sulzberg, T., *Anal. Chem.*, **32**, 995 (1960). Detection and Characterization of Alkenes by Friedel-Crafts Acetylation.

Sharefkin, J. G., and Bohosian, E. M., *Anal. Chem.*, **33**, 643 (1961). Detection of Alkynes by Hydration to Carbonyl Compounds.

TABLE 20A
AROMATIC HYDROCARBONS—LIQUID[a]

	Name of Compound	Note	BP	D_4^{20}	n_D^{20}	Nitro Position	Nitro MP	Aroyl Benzoic Acid	Picrate[b]	Acetamino (Ac) or Diacetamino (di)	2,4-Dinitrobenzene Sulfenyl Chloride
1	Benzene		80.1	0.87901	1.5012	1,3	89	127	84u		120
2	Toluene		100.6	0.86694	1.49693	2,4	70	137	88u		102–3
3	Ethylbenzene		136.2	0.86702	1.49588	2,4,6	37	122	96u	221(di)	97
4	p-Xylene		138.4	0.86105	1.49581	2,3,5	139	132	90u	223(di)	134–5
5	m-Xylene		139.1	0.86417	1.49722	2,4,6	183	126	91u		
6	o-Xylene		144.4	0.88020	1.50545	4,5	118	178	88u		
7	Isopropylbenzene (cumene)		152.4	0.86179	1.49146	2,4,6	109	133		106(Ac) 216(di)	
8	Propylbenzene		159.2	0.86204	1.49202			125	103u	96(Ac) 208(di)	
9	3-Ethyltoluene		161.3	0.86452	1.49661						
10	4-Ethyltoluene		162	0.86118	1.49500						
11	1,3,5-Trimethylbenzene (mesitylene)		164.7	0.86518	1.49937	2,4 2,4,6	86 235	211	97u		
12	2-Ethyltoluene		165.2	0.88069	1.50456						
13	tert-Butylbenzene		169.1	0.86650	1.49266	2,4	62			170(Ac) 210(di)	130–1
14	1,2,4-Trimethylbenzene (pseudocumene)		169.4	0.87582	1.50484	3,5,6	185		97u		

(Continued)

[a] For a tabular compilation of the properties of all of the alkyl substituted benzenes described in the literature through 1947, see Francis, *Properties of Alkylbenzenes*, Chem. Reviews, **42**, 107-162 (1948).

[b] The letter "u" appearing after the melting point of the picrate indicates that the molecular compound is unstable, in that it cannot be purified with ease.

TABLE 20A (Continued)
AROMATIC HYDROCARBONS—LIQUID[a]

	Name of Compound	Note	BP	D_4^{20}	n_D^{20}	Nitro Position	Nitro MP	Aroyl Benzoic Acid	Picrate[b]	Acetamino (Ac) or Diacetamino (di)	2,4-Dinitrobenzene Sulfenyl Chloride
15	Isobutylbenzene		172.8	0.85321	1.48646						99–100
16	sec-Butylbenzene		173.3	0.86207	1.49020						88–9
17	3-Isopropyltoluene (m-cymene)		175.1	0.8610	1.4930						
18	1,2,3-Trimethylbenzene		176.1	0.89438	1.51393				90.5		
19	Indane		177v.s.	0.9645	1.5381						
20	4-Isopropyltoluene (p-cymene)		177.1	0.8573	1.4909	2,6	54	123			
21	2-Isopropyltoluene (o-cymene)		178.2	0.8766	1.5006						
22	1,3-Diethylbenzene		181.1	0.86394	1.49552	2,4,6	62	114			
23	3-Propyltoluene		181.8	0.8610	1.4936						
24	Indene		182	0.857							
25	1-Methylindane		182–3	0.940	1.5260				98		
26	Butylbenzene		183.3	0.86013	1.48979					105(Ac) 214(di)	72–3
27	4-Propyltoluene		183.3	0.8584	1.4919						
28	1,2-Diethylbenzene		183.4	0.87996	1.50346						
29	1,4-Diethylbenzene		183.8	0.86196	1.49483						
30	1,3-Dimethyl-5-ethylbenzene		183.8	0.8648	1.4981						
31	2-Propyltoluene		184.8	0.8744	1.4998						
32	2,2-Dimethyl-1-phenylpropane		186	0.858	1.488						
33	1,4-Dimethyl-2-ethylbenzene		186.9	0.8772	1.5043					142(Ac) 181(di)	
34	2-Methylindane		187	0.9034	1.5070						
35	3-Methyl-2-phenylbutane		188	0.870	1.486						
36	1,3-Dimethyl-4-ethylbenzene		188.4	0.8763	1.5038						

37	3-tert-Butyltoluene	189.3	0.8657	1.4944		
38	1,2-Dimethyl-4-ethylbenzene	189.8	0.8745	1.5031		
39	1,3-Dimethyl-2-ethylbenzene	190	0.8904	1.5107		
40	3-Phenylpentane	191	0.8649	1.4877		
41	1,3-Dimethyl-5-isopropylbenzene	191	0.859	1.4955		
42	1-Ethyl-3-isopropylbenzene	192	0.859	1.492		
43	2-Methyl-2-phenylbutane	192.4	0.8748	1.4958		
44	4-tert-Butyltoluene	192.8	0.8612	1.4918		
45	2-Phenylpentane	193	0.8585	1.4876		
46	1-Ethyl-2-isopropylbenzene	193	0.888	1.508		
47	1,2-Dimethyl-3-ethylbenzene	193.9	0.8921	1.5117		
48	3-Isobutyltoluene	194	0.8536	1.4888		
49	1-Phenyl-2-methylbutane	194	0.8617	1.4880		
50	3-sec-Butyltoluene	194	0.858	1.490		
51	1,3-Dimethyl-5-isopropylbenzene	194.5	0.862	1.495		
52	1,3-Dimethyl-4-isopropylbenzene	195	0.869	1.5018		
53	2-sec-Butyltoluene	196	0.873	1.497		
54	2-Isobutyltoluene	196	0.8649	1.4935		
55	4-Isobutyltoluene	196	0.8517	1.4874		
56	1-Phenyl-3-methylbutane	196	0.8558	1.4847		
57	1,4-Dimethyl-2-isopropylbenzene	196.2	0.8738	1.5010		
58	1-Ethyl-4-isopropylbenzene	196.6	0.8585	1.4923		
59	4-sec-Butyltoluene	197	0.866	1.493		
60	2-Methyl-1-phenylbutane	197	0.859	1.486		
61	1,2,3,5-Tetramethylbenzene (isodurene)	198	0.8903	1.5130	4,6	181 (157) 213
62	3-Methyl-1-phenylbutane	198.9	0.856	1.484		
63	1,3-Dimethyl-2-isopropylbenzene	199	0.890	1.509		
64	1,3-Dimethyl-4-isopropylbenzene	199.1	0.873	1.500		
65	1-Methyl-4-sec-butylbenzene	200	0.8650	1.4932		
66	2-tert-Butyltoluene	200.5	0.8897	1.5076		
67	3,5-Diethyltoluene	200.7	0.8630	1.4969		

(Continued)

TABLE 20A (Continued)
AROMATIC HYDROCARBONS—LIQUID[a]

	Name of Compound	Note	BP	D_4^{20}	n_D^{20}	Melting Point of Derivatives					
						Nitro		Aroyl Benzoic Acid	Picrate[b]	Acetamino (Ac) or Diacet- amino (di)	2,4-Di- nitrobenzene Sulfenyl Chloride
						Posi- tion	MP				
68	1-Ethyl-3-propylbenzene		201	0.8607	1.4930						
69	1-Methyl-2-butylbenzene		201	0.8721	1.4958						
70	1,2-Dimethyl-4-isopropylbenzene		201.8	0.8699	1.4993						
71	Phenylbenzene (mp 69)		202	0.866	1.475						
72	1,3-Dimethyl-5-propylbenzene		202.2	0.8607	1.4952						
73	1,2-Dimethyl-3-isopropylbenzene		202.6	0.888	1.508						
74	1-Ethyl-2-propylbenzene		203	0.8744	1.4992						
75	1,3-Diisopropylbenzene		203.2	0.85593	1.4883						
76	3,4-Diethyltoluene		203.6	0.8762	1.5039						
77	1,2-Diisopropylbenzene		203.8	0.87707	1.49603						
78	1,4-Dimethyl-2-propylbenzene		204.3	0.8717	1.4999						
79	3-Butyltoluene		205	0.859	1.491						
80	2,4-Diethyltoluene		205	0.8748	1.5027						
81	1-Ethyl-4-propylbenzene		205	0.8594	1.4921						
82	1,2,3,4-Tetramethylbenzene (prehnitene)		205	0.9052	1.5203	5,6	176				
83	Pentylbenzene		205.4	0.8585	1.4878				92–5		
84	1,2-Dihydronaphthalene		206	0.9931[25]	1.5817					101(Ac)	
85	3-Methyl-3-phenylpentane		206	0.8755	1.4958						
86	1,3-Dimethyl-5-tert-butylbenzene		206	0.8645	1.4958						
87	1,3-Dimethyl-4-propylbenzene		206.6	0.8723	1.4998						
88	2,3-Diethyltoluene		206.6	0.8910	1.5105						
89	4-Butyltoluene		207	0.857	1.490						
90	2,5-Diethyltoluene		207.1	0.8758	1.5034						

Tables 881

#	Name	bp	d	n			
91	1,3-Diethyl-2-propylbenzene	207.6	0.8856	1.5063			
92	1,2,3,4-Tetrahydronaphthalene (tetralin)	207.6	0.9702	1.54135	5,7	95	153
93	2-Butyltoluene	208	0.871	1.496			
94	2,6-Diethyltoluene	208.8	0.8907	1.5106			
95	1,2-Dimethyl-4-propylbenzene	208.9	0.8715	1.5000			
96	2-Methyl-3-phenylpentane	209	0.8678	1.4912			
97	1,3-Dimethyl-5-propylbenzene	209	0.861	1.4933			
98	1,4-Diisopropylbenzene	210.4	0.85676	1.48983			
99	1,2-Dimethyl-3-propylbenzene	210.7	0.8864	1.5075			
100	1-Ethyl-4-*tert*-butylbenzene	211	0.8635	1.4950			
101	3-Phenylhexane	211	0.8609	1.4877			
102	2-Ethyl-1,3,5-Trimethylbenzene	212.4	0.883	1.5074			
103	3-Ethyl-1-methyl-4-isopropylbenzene	213	0.8722	1.5006			
104	5-Ethyl-1,2,4-trimethylbenzene	213	0.883	1.5075			
105	6-Ethyl-1,2,4-trimethylbenzene	213	0.8897	1.5118			
106	2-Ethyl-1-methyl-4-isopropylbenzene	214	0.8673	1.4969			
107	2-Phenylhexane	214	0.8600	1.4882			
108	1-Propyl-4-isopropylbenzene	215	0.8614	1.4972			
109	2-Methyl-1-phenylpentane	215	0.8624	1.4847			
110	5-Ethyl-1,2,3-trimethylbenzene	215.8	0.8863	1.5101			
111	3-Ethyl-1,2,4-trimethylbenzene	216.6	0.895	1.5133			
112	1,2,4-Triethylbenzene	217.7	0.8791	1.4982			
113	2-Methyl-1,2,3,4-tetrahydronaphthalene	218	0.952	1.531			
114	1,3,5-Triethylbenzene	218			2,4,6	108	129
115	1-Methyl-1,2,3,4-tetrahydronaphthalene	219	0.9580	1.5357			
116	4-Ethyl-1,2,3-trimethylbenzene	220.4	0.9019	1.5180			
117	3-Methyl-1-phenylpentane	221	0.8605	1.4876			
118	1,3,5-Trimethyl-2-propylbenzene	221	0.8782	1.5033			
119	1,4-Di-propylbenzene	221	0.8564	1.4914			

(*Continued*)

882 Tables

TABLE 20A (Continued)
AROMATIC HYDROCARBONS—LIQUID[a]

	Name of Compound	Note	BP	D_4^{20}	n_D^{20}	Nitro Position	Nitro MP	Aroyl Benzoic Acid	Picrate[b]	Acetamino (Ac) or Diacetamino (di)	2,4-Dinitrobenzene Sulfenyl Chloride
120	1,1-Dimethyl-1,2,3,4-tetrahydronaphthalene		221	0.950	1.5292						
121	1-Isopropyl-3-*tert*-butylbenzene		222	0.8512	1.4832						
122	3-Pentyltoluene		223	0.8593	1.4911						
123	5-Isopropyl-1,2,4-trimethylbenzene		223	0.8802	1.5069						
124	4-*tert*-Butyl-1-isopropylbenzene		224	0.8610	1.4928						
125	4-Phenylheptane		224	0.8665	1.4872						
126	2-Methyl-2-phenylhexane		225	0.8737	1.4943						
127	2,4-Diisopropyltoluene		225	0.8664	1.499						
128	4-Isopropyl-1-methyl-2-propylbenzene		225.2	0.8650	1.4937						
129	3-Methyl-3-phenylhexane		226	0.8776	1.4980						
130	Hexylbenzene		226.1	0.8575	1.4864						
131	3-Phenylheptane		227	0.8607	1.4862						
132	5-Propyl-1,2,4-trimethylbenzene		228	0.887	1.5095						
133	2,6-Diisopropyltoluene		228	0.8768	1.5032						
134	6-Methyl-1,2,3,4-tetrahydronaphthalene		229	0.9537	1.53572						
135	2,2-Dimethyl-1,2,3,4-tetrahydronaphthalene		230	0.935	1.5200						
136	2-Phenylheptane		231	0.8610	1.4863						
137	2,*cis* and/or *trans*-3-Dimethyl-1,2,3,4-tetrahydronaphthalene		232	0.940	1.523						

138	1-cis and/or trans-3-Dimethyl-1,2,3,4-tetrahydronaphthalene	234	0.940	1.525	
139	1-cis and/or trans-4-Dimethyl-1,2,3,4-tetrahydronaphthalene	234	0.940	1.525	
140	5-Methyl-1,2,3,4-tetrahydro-naphthalene	234.4	0.9720	1.54395	
141	1-cis and/or trans-2-Dimethyl-1,2,3,4-tetrahydronaphthalene	235	0.9470	1.5286	
142	2-Ethyl-1,2,3,4-tetrahydro-naphthalene	235	0.938	1.523	
143	Cyclohexylbenzene	235–6	0.9502	1.5239	
144	1-Ethyl-1,2,3,4-tetrahydro-naphthalene	236	0.9535	1.5321	
145	2,5-Dimethyl-1,2,3,4-tetrahydro-naphthalene	236	0.946	1.526	
146	2,8-Dimethyl-1,2,3,4-tetrahydro-naphthalene	236	0.941	1.526	
147	2,7-Dimethyl-1,2,3,4-tetrahydro-naphthalene	237	0.941	1.526	
148	2,6-Dimethyl-1,2,3,4-tetrahydro-naphthalene	238	0.941	1.526	
149	3-Phenyl-3-ethylhexane	239	0.875	1.4943	
150	1,4-Di-sec-butylbenzene	239	0.8590	1.4892	4
151	1,5-Dimethyl-1,2,3,4-tetrahydro-naphthalene	239	0.941	1.526	
152	6-Ethyl-1,2,3,4-tetrahydro-naphthalene	241	0.9568	1.5331	71
153	5-Ethyl-1,2,3,4-tetrahydro-naphthalene	242	0.973	1.540	
154	Heptylbenzene	244	0.8595	1.4875	68
155	1-Methylnaphthalene	244.6	1.02025	1.6174	142
156	5,6-Dimethyl-1,2,3,4-tetrahydro-naphthalene	252	0.975	1.552	

(*Continued*)

TABLE 20A (Continued)
AROMATIC HYDROCARBONS—LIQUID[a]

	Name of compound	Note	BP	D_4^{20}	n_D^{20}	Nitro Position	Nitro MP	Aroyl Benzoic Acid	Picrate[b]	Acetamino (Ac) or Diacet- amino (di)	2,4-Di- nitrobenzene Sulfenyl Chloride
157	6,7-Dimethyl-1,2,3,4-tetrahydro- naphthalene		252	0.954	1.538						
158	5,7-Dimethyl-1,2,3,4-tetrahydro- naphthalene		253.1	0.9583	1.5405						
159	5,8-Dimethyl-1,2,3,4-tetrahydro- naphthalene		254	0.967	1.547						
160	2-Ethylnaphthalene		257.9	0.9922	1.5999						
161	1-Ethylnaphthalene		258.7	1.00816	1.6062				71		
162	2-Isopropylnaphthalene		262	0.9795	1.5784				98		
163	1,7-Dimethylnaphthalene		263	1.003	1.607			121	93–5		
164	1,6-Dimethylnaphthalene		263	1.003	1.6073			114			
165	1,3-Dimethylnaphthalene		263	1.0063	1.6090			118			
166	1-Isopropylnaphthalene		263–4		1.5756			85–6			
167	1-Phenyloctane		264.5	0.8562	1.4845						
168	1,2-Dimethylnaphthalene		266	1.013	1.6164			131			
169	1,4-Dimethylnaphthalene		268	1.0166	1.6127			140			
170	1,1-Diphenylethane		268–70	1.003	1.5761						
171	1-Propylnaphthalene		272.5	0.9918	1.5952						
172	2-Propylnaphthalene		273.5	0.9770	1.5872						
173	1-Phenylnonane		282	0.8558	1.4838						
174	1-Butylnaphthalene		289.3	0.97673	1.5819						
175	2-Butylnaphthalene		292	0.9659	1.5776						
176	1-Phenyldecane		300	0.85553	1.48319						
177	1-Pentylnaphthalene		307	0.9656	1.5725						

178	2-Pentylnaphthalene	310	0.9561	1.5694
179	1-Phenylundecane	316	0.8553	1.4828
180	1-Hexylnaphthalene	322	0.9566	1.5647
181	2-Hexylnaphthalene	324	0.9479	1.620
182	1-Phenyldodecane	331	0.8551	1.4824
183	1-Heptylnaphthalene	340	0.9491	1.5582
184	2-Heptylnaphthalene	341	0.9410	1.5556
185	1-Phenyltridecane	346	0.8550	1.4821
186	1-Octylnaphthalene	356	0.9427	1.5526
187	2-Octylnaphthalene	357	0.9350	1.5501
188	1-Phenyltetradecane	359	0.8549	1.4818
189	1-Nonylnaphthalene	372	0.9371	1.5477
190	2-Nonylnaphthalene	372	0.9298	1.5454
191	1-Decylnaphthalene	387	0.9322	1.5435

TABLE 20B
AROMATIC HYDROCARBONS—SOLID

	Name of Compound	Note	MP	BP	T.N.F.	Nitro Position	Nitro MP	Aroyl Benzoic Acid	Picrate	Styphnate
1	Diphenylmethane	3	25.4			2,4,2',4'	172			
2	2,6-Dimethylphenanthrene		33–4						135–6	148–50
3	1-Propylphenanthrene		34–5						100–1	
4	2-Methylnaphthalene	4	34.4	241.1	125.6	1	81		116	
5	5-Ethyl-1-methylnaphthalene		40						97	
6	2-Isopropylphenanthrene		44–5						108	
7	6-Ethyl-2-methylnaphthalene		44–5						100–1 (109)	
8	1,5-Dimethylphenanthrene		46–7						127–9	132–3
9	1,4-Dimethylphenanthrene		50–1						143.5	136
10	Bibenzyl	4,5	53			4,4' 2,2',4,4'	180 169			
11	Methylenefluorene		53						152–3	
12	3,5-Dimethylphenanthrene		53–4						139	124–5
13	Pentamethylbenzene	4	54.3	231.8		6	154		131	
14	7-Methyl-3,4-benzphenanthrene		54.0–4.5						134–4.5	
15	2,9-Dimethylphenanthrene		56–7						138	
16	1,5-Dimethylphenanthrene		57–8						134–5	
17	2-Benzylnaphthalene		58		124.3–5.4				93–4	
18	1-Benzylnaphthalene		58–9						103–4	
19	1,2-Dimethylazulene	4	58–9						129–30	
20	1,7-Dimethyl-4-isopropylnaphthalene		60						92	120
21	3,4-Dimethylphenanthrene		62–3						129–30	142–3
22	1-Ethylphenanthrene		62.5						108–9	144
23	9-Ethylphenanthrene		62.5–3.0	198–200					123–4	

Tables 887

#	Compound										
24	1,8-Dimethylnaphthalene							65			
25	8-Methyl-3,4-benzophenanthrene							65–6	156		
26	2-Ethylphenanthrene							67–8 (64–5)	107–8 95.5–6.0		160
27	3,4-Benzophenanthrene		170.8–1.1					68	126–7		
28	4,8-Dimethylazulene	4						69–70	157–8		
29	Biphenyl (s, v.s.)	5		4,4′	202		237	69.2			
30	2-Methyl-3,4-benzphenanthrene							70.4–1.0	191.8–3.2		
31	3-Methylpyrene							71–2	211–2		
32	1,4-Dimethylanthracene							74	140		
33	9-Propylphenanthrene							74	134		
34	4,9-Dimethyl-1,2-benzanthracene	4	120.4–1.4					75	116		
35	4,5-Dimethylphenanthrene						(229)	75–6			
36	Benzalfluorene	8					224	76	115–6		
37	1,3-Dimethylphenanthrene							76–7	153–5		
38	1-Methyl-3,4-benzphenanthrene							77.8	112–3	165–6	
39	9-Methylanthracene							78–80	137		
40	1,2′-Binaphthyl		145–6.9					79–80	127–7.5		
41	2,3-Dimethylphenanthrene							79–80	146–7	147–8	
42	1,2,4,5-Tetramethylbenzene (durene)			3,6	196.8		263	79.2			
43	1,5-Dimethylnaphthalene							80–1	140		
44	6-Methyl-3,4-benzphenanthrene							80–1	118.0–8.5		
45	Naphthalene (s, v.s.)	4,5	143–4	1	218	61 (57)	172	80.25	149		
46	1,3-Dimethylanthracene							83	136		
47	1-Isopropylphenanthrene							85–6	125–6		
48	1,7-Dimethylphenanthrene							86	132	159	
49	1,6-Dimethylphenanthrene							87–8	134		
50	1,9-Dimethylphenanthrene							88	163.5	181	
51	1,3-Diphenylbenzene				363			89			
52	Triphenylmethane			4,4′,4″	206			92			

(*Continued*)

TABLE 20B (Continued)
AROMATIC HYDROCARBONS—SOLID

	Name of Compound	Note	MP	BP	T.N.F.	Nitro Position	Nitro MP	Aroyl Benzoic Acid	Picrate	Styphnate
53	5-Isopropylnaphthanthracene		92						157	
54	3,9-Dimethyl-1,2-benzanthracene	4	93						137–8	
55	12-Isopropylnaphthanthracene		94–5						157–8	
56	Retene	4	95.2s						124	
57	Acenaphthene	4,5	96.2s		175–6	5	101	198	161	
58	Phenanthrene	4,5	96.3		197				144 (133)	
59	1-Methyl-7-isopropylfluorene		96.5–7.0			di	245			
60	2,7-Dimethylnaphthalene		98						136	
61	Azulene	4	99						120d	
62	2,7-Dimethylphenanthrene		101–2						152–3	
63	2-Phenylnaphthalene		102–3		169.5–70.5					
64	1,2,3,4-Tetrahydroanthracene		103–5		182.4					
65	2,3-Dimethylnaphthalene		105						124	
66	9,10-Dihydroanthracene	9	108v.s.							
67	Di-1-naphthylmethane		109						142	
68	Fluoranthene		110		216				185–6	
69	2,4-Dimethylphenanthrene		111						142	
70	2,6-Dimethylnaphthalene		111		156				143	
71	Aceanthrene		113						120	
72	Fluorene	4	113.5 (116)	293–5	179	2, 2,7	156 199	227	87 (77)	
73	4,10-Dimethyl-1,2-benzanthracene		114						162	
74	5-Methylchrysene	4	117.2–7.8						142.6–3.0	
75	1-Methylchrysene		118						143	

#	Compound					
76	1′,10-Dimethyl-1,2-dibenzanthracene		122–3			147–8
77	9,10-Dimethyl-1,2-benzanthracene	12	122–3			112–3N
78	3,4-Benzfluorene		124–5		191.8	130–1
79	9-Isopropylnaphthanthracene		125			152
80	5,8-Dimethyl-1,2-benzanthracene		131			175
81	8-Isopropylnaphthanthracene		132–3			118
82	2-Methyl-1′,2′-benzpyrene		138–9			184–5
83	1,5-Dimethylanthracene		139–40			166–7
84	3,6-Dimethylphenanthrene		141			172–3
85	1,2-Dimethylphenanthrene		142–3			148 153
86	8,10-Dimethyl-1,2-benzanthracene		146			166
87	9-Methyl-1′,2′-benzpyrene	4	146.8–8.0			
88	9-Phenylfluorene	10	147–8			226–7
89	1-Methylpyrene		147.5–8.5			179.5–80.0
90	3-Methyl-1′,2′-benzpyrene	4	147.6–8.1			220(227)
91	Pyrene		148		242–3	135–8
92	4-Methylchrysene		151–1.5			135
93	6-Methylchrysene	4	151		251–2	133
94	1,2-Benzanthracene		159–60		160	145
95	1,1′-Binaphthyl		160			170
96	2-Methylchrysene		161			199–200
97	2′,6-Dimethyl-1,2-benzanthracene		164	Di	240d	170
98	Hexamethylbenzene	4	165			164–4.5
99	3-Methylchrysene		170.0–0.5			181.5–2.5
100	6-Methyl-1′,2′-benzpyrene	4	171.1–1.5			164
101	5-Methylchrysene		173		234.2	
102	Cholanthrene		173		245–6	
103	6,7-Dimethyl-1,2-benzanthracene		174			170
104	1,2-Benzpyrene		176.5–7.5			197–8
105	5,10-Dimethyl-1,2-benzanthracene		177			174
106	4,5-Benzpyrene		178–9			229–30
107	9,10-Dimethylanthracene		180–1			176–7d

(*Continued*)

TABLE 20B (Continued)
AROMATIC HYDROCARBONS—SOLID

	Name of Compound	Note	MP	BP	T.N.F.	Melting Point of Derivatives				
						Nitro		Aroyl Benzoic Acid	Picrate	Styphnate
						Position	MP			
108	5,6-Dimethyl-1,2,-benzanthracene		187–8						191–3	
109	2,2′-Binaphthyl		188		171				184	
110	1,2-Benzfluorene		189–90		213.5–5.5				127.5(di)	
111	1,8-Dimethylphenanthrene		191		193–4					
112	1,2,7,8-Dibenzanthracene		196						212	
113	1,2,3,4-Dibenzanthracene		200–2						207	
114	Di-2-Fluorenylmethane	11	201–2			Di-	256–7			
115	2,3-Benzfluorene		208–9		221.2–2.0					
116	sym-Tetraphenylethane		211	358–62						
117	5-Methyl-1′,2′-benzpyrene	4	215.7–6.2						207–8	
118	Anthracene	4,6	216.2s		194				138u	
119	4-Methyl-1′,2′-benzpyrene		217.5–8.0						203–4	
120	2-Methylchrysene		224.5–5.5						143–6	
121	4-Methylchrysene		229						143	
122	Chrysene	4,7	251		248–9			214	273	
123	1-Methylchrysene	4	254–5*							
124	Anthanthrene		257		268–9				184	
125	2,3,6,7-Dibenzphenanthrene		257						213(di)	
126	2,3,5,6-Dibenzphenanthrene		261						214(di)	
127	1,2,5,6-Dibenzanthracene		262(267)							
128	Perylene		273–4		270–1					
129	Picene		364		257–8					

NOTES FOR TABLE 20A AND 20B (AROMATIC HYCROCARBONS)

1. 1,2,3-Tribromoindane, mp 134; treated with bromine in the presence of a trace of solid iodine at room temperature, 1,2,3,4-tetrabromoindane is produced, mp 200.
2. Compound treated with chlorine in the presence of a trace of solid iodine yields the 5,6,7,8-tetrachloro derivative, mp 172.
3. Oxidizes by nitric acid to benzophenone.
4. The following compounds (numbers in parentheses) form adducts with syn-trinitrobenzene, the melting points for which are given: (4) 123; (10) 102; (13) 121; (19) 166–7; (28) 179–80; (34) 124–5; (45) 153; (54) 145; (56) 169; (57) 168; (58) 145; (61) 176; (72) 105; (74) 173; (87) 219; (90) 211; (93) 190; (98) 174; (100) 209–10; (117) 230–1; (118) 164; (122) 186; (123) 174–6.
5. The following compounds (numbers in parentheses) form derivatives with 2,4-dinitrophenylsulfenyl chloride: (10) 132–3; (29) 142–3; (45) 173–4; (57) 187–9d; (58) 250–1.
6. Oxidizes to anthraquinone; dibromo derivative, mp 122.
7. Dibromo derivative, mp 275.
8. Dibromo derivative, mp 116d.
9. Adduct with maleic anhydride, mp 258–61.
10. Dibromo derivative, mp 181–2; tribromo derivative, mp 167–71.
11. Oxidizes to di-2-fluorenyl ketone, mp 297–8.
12. Forms a picrate (black), mp 112–3, and a dipicrate (red), mp 102–6.

REFERENCES FOR TABLE 20

Buckles, R. E., Johnson, R. C., and Probst, W. J., *J. Org. Chem.*, **22**, 55 (1957). Comparison of N-Bromoacetamide and N-Bromosuccinimide as Brominating Agents. Gives the melting points of the bromination products of alkanes, alkenes, and some aromatic compounds.

Buess, C. M., and Lawson, D., *Chem. Rev.*, **60**, 313 (1960). Preparations, Reactions, and Properties of Triphenylenes. Lists the melting points of the compounds and their derivatives.

Campbell, T. W., and McDonald, R. N., *J. Org. Chem.*, **24**, 1246 (1959). Synthesis of Hydrocarbon Derivatives by the Wittig Synthesis. I. Distyrylbenzenes. Gives the method of preparation and the melting points of 11 distyrylbenzenes and 5 bis(phenylethyl)benzenes.

Gray, F. W., Gerecht, J. F., and Krems, I. J., *J. Org. Chem.*, **20**, 511 (1955). The Preparation of Model Long Chain Alkyl-benzenes and a Study of Their Isomeric Sulfonation Products. Gives the melting points of 7 isomeric alkylbenzenesulfonates. Considerable discussion of spectra data.

Pan, H., and Fletcher, L., *J. Org. Chem.*, **25**, 1106 (1960). Derivatives of Fluorene. IX. 4-Hydroxy-2-Fluorenamine.

Rodd, E. H., *Chemistry of Carbon Compounds*, Elsevier, Houston, 1952. See Vol. IIIb, p. 1286. Table of monoalkylnaphthalenes gives BP, MP, and picrates of 24 compounds. See also p. 1287 for table of dimethyl to octamethyl naphthalenes giving BP, MP, picrates, and styphnates for about 40 compounds.

Williams, R. B., Hastings, S. H., and Anderson, J. A., Jr., *Anal. Chem.*, **24**, 1191 (1952). Determination of Individual Alkyl Aromatic Hydrocarbons from Benzene Through the C_{10} Aromatics by Infrared Spectrometry. Gives the boiling point data on many hydrocarbons.

TABLE 21A
KETONES—LIQUID

				Melting Point of Derivatives				
				Recommended		Others		
	Name of Compound	Note	BP	Semi-carbazone	2,4-Di-nitro-phenyl-hydrazone	4-Nitro-phenyl-hydra-zone-	Phenyl-hydrazone	Oxime
1	Trifluoroacetone		21		139			
2	Acetone	29	56.11	190 (187)	126	148–9	42	59
3	1,1,1,5,5,5-Hexafluoro-2,4-pentanedione	30	63–5					
4	2-Butanone	31	80	146	118–9	128–9	Oil	bp 152
5	3-Buten-2-one		81	141				
6	3-Butyn-2-one		86		181	143		
7	Butane-2,3-dione	32	88	235 (mono) 278–9(di)	314–5* (di)	230 (mono)	134 (mono) 243(di)	76(mono) 245–6(di)
8	3-Methyl-2-butanone		94.3	113–4	124–5	108–9	Oil	Oil
9	3-Methyl-3-buten-1-one		97 (734 mm)	173	181			
10	Cyclobutanone		100	204	147			
11	3-Pentanone		102	138–9	156	144	Oil	69
12	2-Pentanone		102.3	112 (106)	145	117	Oil	58
13	1-Penten-3-one		102 (740 mm)		129			
14	3,3-Dimethyl-2-butanone (pinacolone)		106*	157–8	127		Oil	75(79)
15	1,1,1-Trifluoro-2,4-pentane-dione	5	107					
16	Methyl cyclopropyl ketone	6	114	118–20 (110–2)	150			50–1
17	2-Methyl-3-pentanone		114–5	95 (80)	111–2			
18	1-Methoxy-2-propanone		115		163 (159)	111 (109)		
19	4-Methoxy-2-butanone		116 (739 mm)	141				
20	4-Methyl-2-pentanone		116.8	132 (134)	81 (95)			58
21	3-Methyl-2-pentanone		118	94–5	71			
22	1-Chloro-2-propanone	7	119	150N	125	83		
23	2-Methyl-1-penten-3-one		119 (751 mm)	161				
24	1,1-Dichloro-2-propanone		120	163				
25	Dimethylaminoacetone	4	121					99
26	3-Penten-2-one		122	142	155			
27	3-Hexanone		123–4	117 (112)	130			
28	2,4-Dimethyl-3-pentanone		124	160*	88			34
29	1,1,1-Trifluoro-2,4-hex-anedione	8	124					
30	4,4-Dimethyl-2-pentanone		125 (122)		100			
31	2,2-Dimethyl-3-pentanone		125–6	144	175			79–80
32	2,3-Hexanedione		128				138(di)	175(di)
33	2-Hexanone	9	128	125* (122)	110	88	Oil	49
34	1-Ethoxy-2-propanone		128	96				bp 188

(Continued)

TABLE 21A (Continued)
KETONES—LIQUID

	Name of Compound	Note	BP	Melting Point of Derivatives				
				Recommended		Others		
				Semi-carbazone	2,4-Dinitro-phenyl-hydrazone	4-Nitro-phenyl-hydrazone	Phenyl-hydrazone	Oxime
35	5-Hexen-2-one		128–9	100–2	104			bp 190
36	4-Methyl-3-penten-2-one	10	130	164N (133)	205–6	134	142	48–9
37	3,4-Hexanedione		130	270d	145		161(di)	185
38	3,3-Dimethyl-2-pentanone		130 (733 mm)		112			
39	Cyclopentanone	11	130.7v.s.	210 (203)	146*N	154	55	56.5
40	1-Methoxy-2-butanone		133		198			
41	2,2,4-Trimethyl-3-pentanone		135	132				144
42	5-Methyl-3-hexanone		135 (735 mm)	152				
43	1-Bromo-2-propanone		136	135d				36
44	3-Methyl-2-hexanone		136	119				
45	4-Methyl-3-hexanone		136	137	78			
46	2-Methoxy-3-pentanone		136	120				
47	Cyclobutyl methyl ketone		137–9	149				61
48	4-Methyl-5-hexen-2-one		138	112				
49	1-Chloro-2-butanone		138	151				
50	5-Methyl-2,3-hexanedione		138				116–7 (di)	171–2 (di) 75–6
51	3-Methyl-4-penten-2-one		138	201				
52	3,4-Dimethyl-2-pentanone		138	113				
53	3-Ethyl-2-pentanone		138–40	99				
54	4-Hexen-3-one		139	157				
55	2-Methylcyclopentanone		139	187 (171)			60d	
56	2,4-Pentanedione	12	139	112N	209N		99–100N	149(di)
57	3-Hydroxy-3-methyl-2-butanone		140	165				87
58	5-Methyl-2-cyclopentenone		140	175–6				
59	4-Methyl-2-hexanone		142 (139)	120 (128)				
60	1-Propoxy-2-propanone		144		142			
61	3,4-Dimethyl-4-penten-2-one		144	114				
62	5-Methyl-2-hexanone		144	142–3	95			Oil
63	4-Heptanone		144v.s.	132	75			bp 193
64	3-Hydroxy-2-butanone	13	145 (148)	185 (202)	318(bis)			
65	2,4-Dimethyl-3-hexanone		145		71			
66	3-Methylcyclopentanone		145	185				
67	1-Methoxy-3-methyl-2-butanone		145 (750 mm)		163			
68	2,2-Dimethyl-3-hexanone		145 (740 mm)		124 (116)			
69	1-Hydroxy-2-propanone		146v.s.	196	128.5*	173	103	71
70	4-Chloro-3-methyl-2-butanone		146	116				
71	1-Hepten-4-one		146–7	110				
72	3,4-Dimethyl-3-penten-2-one		147	200				
73	3,4-Heptanedione		147 (732 mm)					167–8(di)
74	2,5-Dimethyl-3,4-hexanedione		148					125(mono) 172(di)

(Continued)

TABLE 21A (Continued)
KETONES—LIQUID

	Name of Compound	Note	BP	Melting Point of Derivatives				
				Recommended		Others		
				Semi-carbazone	2,4-Di-nitro-phenyl-hydrazone	4-Nitro-phenyl-hydrazone	Phenyl-hydrazone	Oxime
75	3-Heptanone		148	101 (111)				
76	5-Methyl-4-hexen-3-one		148	163				
77	5-Methyl-5-hexen-2-one		149	137				
78	2-Methyl-4-heptanone		150	124				
79	4,4-Dimethyl-3-hexanone		150–2	98				
80	2-Heptanone		151.2	123	89		207	
81	3-Ethyl-5-hexen-2-one		152		53			
82	5-Hepten-2-one		153	105				
83	3,3-Dimethylcyclopentanone		153 (748 mm)	178				
84	Cyclopentyl methyl ketone		155 (158)	143 (145)				
85	1-Methoxy-3-hexanone		155 (746 mm)	170				
86	Cyclohexanone		156	166–7	162 (160)	146–7	81 (77)	91
87	4-Methyl-6-hepten-3-one		156		80			
88	2-Hepten-4-one		156–7	147				
89	2-Methylcyclopenten-3-one		157					128
90	2,3-Hexanedione		158					175(di)
91	3,4-Dimethyl-2-hexanone		158 (155)	120 (126)				
92	3,4-Dimethyl-3-hexen-2-one		158	142				
93	2,2,4-Trimethyl-3-hexanone		158	145				
94	2-Ethyl-1-hexen-3-one		158 (742 mm)	119				
95	1-Isopropoxy-3-methyl-2-butanone		160		88			
96	1-Methylcyclopenten-3-one		161	220				127
97	2-Ethylcyclopentanone		161	189				
98	3-Methyl-2-heptanone		162	82				
99	4,5-Dimethyl-5-hexen-3-one		162	110				
100	3,5-Dimethyl-4-heptanone		162 (170–3)	83–4				
101	Ethyl 2-ketobutyrate		162			151	86	62–3
102	6-Methyl-3-heptanone		163	132				
103	2-Methylcyclohexanone		165	197d	137*	132		43
104	2-Ketopropionic acid (pyruvic acid)	1	165		218		192	
105	3-Octanone		165–6	117(r.h.)				
106	4,5-Dimethyl-4-hexen-3-one		166 (750 mm)	209				
107	2,2-Dimethyl-3-heptanone		166 (745 mm)	145				
108	2,6-Dimethyl-4-heptanone		168	122 (126)	66			205–10
109	2-Methyl-4-octanone		168	132				
110	2,5-Dimethyl-4-heptanone		169	133				
111	3-Methylcyclohexanone		169	180	155	119	94	43
112	4-Hydroxy-4-methyl-2-pentanone	2	169.2		202–3	209		58

(Continued)

TABLE 21A (Continued)
KETONES—LIQUID

	Name of Compound	Note	BP	Melting Point of Derivatives				
				Recommended		Others		
				Semi-carbazone	2,4-Di-nitro-phenyl-hydrazone	4-Nitro-phenyl-hydra-zone	Phenyl-hydrazone	Oxime
113	Methyl acetoacetate		170	152	120			
114	4-Octanone		170	96	41			
115	2,2-Dimethylcyclohexanone		170–1	201 (193)	140–2			
116	6-Methyl-2-heptanone		171	154	77			
117	2,4-Dimethylcyclohexanone (trans)		171	136				
118	4-Methylcyclohexanone		171	203	134	128	110	39
119	dl-2,5-Dimethylcyclohexanone	14	171–3	122 and 173				111
120	2,3-Octanedione	15	172–3 (733 mm)				117–8(di)	173(di)
121	2-Octanone		173	124–5	58	92–3		
122	5-Ethyl-3-heptanone		173	134				
123	2-Methyl-3-ethylcyclopentanone		174	170				
124	2-Isopropylcyclopentanone		174	202				
125	Acetoxyacetone	16	174–5	145		144	60	
126	3-Methyl-3-hepten-2-one		175	164				
127	3-Thiophanone		175	192				36
128	2,4-Dimethylcyclohexanone (cis)		176	200 (190)				98–9
129	Methyl methylacetoacetate		177	138				
130	2-Hydroxypentan-4-one		177				102–3	
131	4-Ethyl-4-hydroxy-3-hexanone		178 (742 mm)	177				
132	2,3-Dimethylcyclohexanone		178–9	203–4				
133	2-Ketopentanoic acid	1	179	220(di)	205	201–2	145	
134	2,2,6-Trimethylcyclohexanone		179	209	141			
135	3,3-Dimethylcyclohexanone		179 (748 mm)	219				
136	5-Ethyl-4-hepten-3-one		179 (740 mm)	105				
137	Cyclohexyl methyl ketone		180	177	140	154		60
138	4-Ethyl-3-methylcyclopentanone		180	208–9				
139	dl-3,5-Dimethylcyclohexanone(trans)	17	180–1	193–4N				
140	Cycloheptanone		181	163	148	137		23
141	Ethyl acetoacetate		181	133 (129d)	93			
142	Ethyl methylacetoacetate		181	86				
143	3,5-Dimethylcyclohexanone(cis)		182–3	202–3				74
144	2-Propylcyclopentanone		183	214d				
145	5-Nonanone		186–7	90	70			
146	4-Nonanone		187–8	73–4	57–8	84–5		
147	3-Nonanone		188	119 (112)				
148	2,5-Dimethyl-2-cyclohexenone		189–90	165				93(169)
149	2-Acetylpyridine	4	190				155	121
150	3-Propylcyclopentanone		190–1	178–9				
151	3-Ethylcyclohexanone		192	182 (175)				
152	4,6-Dimethyl-3-cyclohexenone		194					102

(Continued)

TABLE 21A (Continued)
KETONES—LIQUID

	Name of Compound	Note	BP	Recommended		Others		
				Semi-carbazone	2,4-Dinitro-phenyl-hydrazone	4-Nitro-phenyl-hydrazone	Phenyl-hydrazone	Oxime
153	2,5-Hexanedione		194	185d (mono) 224(di)	257 (di)	210–2 (di)	120 (di)	137(di)
154	d-Fenchone	18	195–6v.s.	184 (172)	140			165N and 123N
155	2-Nonanone		195.3	118–9	56			
156	4-Methyl-1-acetylcyclohexane		195–7	159 and 175				57–9
157	Methyl levulinate		196	143	142		96	
158	3-Methylhexane-2,5-dione		196 (745 mm)	219–20 (di)		112–3 (di)		
159	4-Fluoroacetophenone		196	219				
160	dl-2-Ethyl-5-methylcyclohexanone		197	178–81				
161	Phorone (mp 28) (see Table 21B)		198					80
162	2-Propylcyclohexanone		198–9 (748 mm)	133				67–8
163	Ethyl ethylacetoacetate		198	154				
164	α-l-Thujone	19	199–201	186–8	116–7			
165	1-Acetyl-1-cyclohexene		201–2	221d				99
166	Acetophenone (mp 20)	20	202v.s.	198–9* (203)	238–40N	184–5	105–6	60
167	β-d-Thujone		202	174 and 170–2	114			55
168	4-Furfuryl-2-butanone		203	143				
169	3-Methyl-2,6-heptanedione	21	203–4	192N				
170	Ethyl levulinate		206	148	102		104	
171	4-Decanone		206–7	51–2				
172	3,5-Dimethyl-2-cyclohexenone	22	208–9	179–80 (168–71)			76–8	
173	l-Menthone		209 (207)	189 (187)	146		53	59
174	2-Decanone		210	124				
175	3,4-Dimethyl-2,5-hexanedione		210					195(di) (202)
176	3-Decanone		211	101				
177	4-Acetyl pyridine	4	212				150	142
178	2-Acetylthiophene		213–4	190–1	243–4		96	81
179	2-Methylacetophenone		214	203	159			61
180	1,5,5-Trimethylcyclohexen-3-one		215	199.5d (191)	130		68	79.5 (76)
181	o-Hydroxyacetophenone (mp 28) (see Table 21B)		215					
182	2-Pyridyl propyl ketone	23	217–8				82	48
183	Propiophenone (mp 20)		218	182 (174)	191		147	54
184	Phenyl propyl ketone		218–21	188				50
185	3-Methylacetophenone		220	198	207			55
186	3-Acetylpyridine		220				137	133
187	Isopropyl phenyl ketone		222	181 (167)	163		73	94 (61)
188	Methylphenylglyoxal	24	222	229–32N (di)		256–7 (di)	104–5N (di)	238–40N (di)
189	2,5-Dimethylacetophenone		223–5	169				58

(Continued)

TABLE 21A (Continued)
KETONES—LIQUID

	Name of Compound	Note	BP	Melting Point of Derivatives				
				Recommended		Others		
				Semi-carbazone	2,4-Di-nitro-phenyl-hydrazone	4-Nitro-phenyl-hydra-zone	Phenyl-hydrazone	Oxime
190	Pulegone		224 (221–2)	174 (175–6)	142			119
191	*tert*-Butyl phenyl ketone		224 (750 mm)	150 (168)	194–5			167
192	Isobutyl phenyl ketone		225	210	124			72 (64.5)
193	1-Phenyl-2-butanone		226	135 (146)				
194	3-Undecanone		227	90				
195	2-Undecanone		228	122–3	63	90–1		44–5
196	3-Chloroacetophenone		228	232		176		88
197	2-Chloroacetophenone		229	160	206	215		113
198	1-Phenoxy-2-propanone		229–30	176				
199	*d*-Carvone	25	230	162–3*N* and 142–3	191	174–5	109–10	72–3*N* and 56–7
200	Phenyl propyl ketone		230 (218–21)	188 (191)	190			50
201	4-Chloroacetophenone		232	204 (160)(146)	236	239	114	95
202	4-Phenyl-2-butanone		235	142	127–8			87
203	3,5-Dimethylacetophenone		236–7			180		114
204	Butyl levulinate		238	102–3	66		78–81	
205	2-Methoxyacetophenone		239 (245)	183			114	83
206	3-Methoxyacetophenone		240 (252)	196				
207	1-Phenyl-3-pentanone		244	80				
208	5-Isopropyl-2-methylaceto-phenone		245	147				91–3
209	2,4,5-Trimethylacetophenone		246–7	204				85–6
210	3,4-Dimethylacetophenone		246–7	233–4	251–2			85
211	Butyl phenyl ketone		248	166	166 (124)	162 (122)		52
212	Ethyl acetonedicarboxylate (see Table 12A)		250*d*	95	86–7			
213	2,5-Dichloroacetophenone		251					130
214	2-Aminoacetophenone (mp 20)	3	250–2	290*d*			108	109
215	4-Isopropylacetophenone		252–4		183		70–1	
216	Ethyl benzoylacetate		265	125				
217	4-Acetobutyric acid	1	275*d*	175*d*				104–5
218	Hexyl phenyl ketone		283	119		127–8		75
219	3-Phenylcyclohexanone		287–8 (736 mm)	167				128–9
220	1-Acetylnaphthalene	26	302	288–9 (232)			146	140 (138)
221	Ethyl 1-naphthyl ketone	27	305–7					58
222	Phenyl 2-pyridyl ketone (see Table 10A)	4, 28	317		199		136	150–2 and 165–7
223	2,4-Dimethylbenzophenone		319					126 and 152
224	2,4,4'-Trimethylbenzophenone		340					132
225	Dypnone		340–5	151				134(*syn*) and 78(*anti*)

NOTES FOR TABLE 21A

1. For additional data, see Table 5.
2. For additional data, see Table 6.
3. For additional data, see Table 9.
4. For additional data, see Table 10.
5. Copper chelate, mp 189.
6. Odor like camphor.
7. The melting point of the semicarbazone has been recorded from a low of 136–7 to a high of 165. Compound reacts with potassium 3-nitrophthalimide to yield a derivative with mp 152–3.
8. Copper chelate, mp 155.
9. Thiosemicarbazone, mp 110.
10. Mesityl oxide semicarbazone exists as two isomers: α, mp 164; β (from benzene), mp 133–4.
11. The 2,4-dinitrophenylhydrazone melts at 140 when crystallized from acetic acid and at 142 from ethanol.
12. The compound consists of a keto-enol mixture. Compound reacts with 4-nitrophenylhydrazine hydrochloride to yield the pyrazole, mp 99–100; with 2,4-dinitrophenylhydrazine, the phenylhydrazone forms, mp 209, and this converts to the pyrazole, mp 122; with semicarbazide hydrochloride, 3,5-dimethylpyrazole-1-carbonamide, mp 112, which converts on warming with hydrochloric acid to 3,5-dimethylpyrazole, mp 107.
13. Phenylosazone, mp 243.
14. d-Semicarbazone melts at 176–7, and the oxime at 97–8.
15. The 2-oxime, mp 39; 3-oxime, mp 59; and 2-phenylhydrazone, mp 103–4.
16. 4-Bromophenylhydrazone, mp 137–8.
17. dl-Semicarbazone, mp 193–4; d-, mp 193–4; l-, mp 189.
18. d- or l-Fenchone oxime exists as two isomers; α, mp 165, and β, mp 123. dl-Fenchone oxime has two isomers; α, mp 159, and β, mp 129.
19. 3-Nitrobenzoylhydrazone, mp 156.
20. The derivative is reported to melt at 238–40 when crystallized from alcohol, and at 249–50 when crystallized from acetic acid or chloroform.
21. The disemicarbazone first melts at 192; on solidification, it remelts at 225.
22. Thiosemicarbazone, mp 195d.
23. Picrate, mp 75.
24. α-Oxime, mp 166–7; β-oxime, mp 114–5; α-phenylhydrazone, mp 143–5; β-semicarbazone, mp 213.
25. d-Carvone forms two semicarbazones, mp 162–3 and 141–2. The semicarbazone of the l isomer melts at 162–3 and that of the dl isomer at 154–6. d-Carvone forms two oximes; α, 72–3 and β, 56–7. l-Carvone has two oximes: α-(−), mp 72 and β-(+), mp 57–8, dl-Carvoxime melts at 93–4.
26. Picrate, mp 119–20. T.N.F., mp 125–6.
27. Picrate, mp 78.
28. Picrate, mp 130. Hydrochloride salt, mp 126–8.
29. Thiosemicarbazone, mp 179.
30. Copper chelate, mp 113–5.
31. 2-Phenylsemicarbazone, mp 168.
32. Anil, mp 139 (di).

TABLE 21B
KETONES—SOLID

	Name of Compound	Note	MP	Recommended		Others		
				Semi-carbazone	2,4-Di-nitro-phenyl-hydrazone	4-Nitro-phenyl-hydra-zone	Phenyl-hydrazone	Oxime
1	3-Dodecanone		19	100–1(89)				
2	Acetophenone (bp 202) (see Table 21A)		20					
3	2-Chloroacetophenone (bp 232) (see Table 21A)		20					
4	Propiophenone (bp 218) (see Table 21A)		20					
5	2-Aminoacetophenone (bp 250–2d) (see Table 21A)		20					
6	2-Dodecanone (bp 247)		21	122	81			
7	Butyl p-tolyl ketone (bp 266–7)		22	212				
8	Pentyl phenyl ketone (bp 265)		24.5	132	168			
9	2-Pyrrolidone (bp 245)	3	24.6					
10	3,5-Dichloroacetophenone		26					138
11	4-Phenyl-2-butanone (bp 216)		27	198	156	145	87	68–70
12	4-Methylacetophenone (bp 226)		28	205	258	198	96	88
13	2-Tridecanone		28	123	70–1	101–2		
14	2-Hydroxyacetophenone (bp 215)		28	210	214		110	118
15	Phorone (bp 198)	95	28	221	112			48
16	1-Furfuryl-2-butanone (bp 183)		30	189				
17	3-Methyl-2-ketobutanoic acid (bp 171)	1	31		194		143	163–5
18	2-Acetobutyric acid (see Table 5B)	1	31.5			168–9	125	
19	2-Ketobutyric acid		32	210	191	232	210	
20	2-Acetylfuran		33	150	220	185–6	86.5	104
21	7-Tridecanone		33			97		
22	3-Acetylpropionic acid		33		206	174–5	108	45–6
23	2,4-Dichloroacetophenone		33–4					148
24	2-Tetradecanone		33–4	115–6				
25	3-Tetradecanone		34	89	45–7			40
26	Methyl 1-naphthyl ketone	96	34	235 (229)			146	136 (139)
27	1,3-Diphenyl-2-propanone		34	145–6 (125–6)	100		121 (128–9)	125
28	2,2,6,6-Tetramethyl-4-piperidone	97	35	219–20				153
29	4-Chloropropiophenone		36	175–6	223			62–3
30	Acetoacetic acid	98	36–7N	151–2	124			
31	2-Phenylcyclopentanone		37	214				
32	3-Methoxybenzophenone		37		234			
33	Furfurylideneacetone		38		241		131–2	
34	4-Carbomethoxy-3-thiophanone		38	190				
35	4-Methoxyacetophenone		38	198	228	195	142	87
36	1,2-Cyclohexanedione (bp 193–5)		38–40				124 (mono) 152–3(di)	187–8(di)

(Continued)

TABLE 21B (Continued)
KETONES—SOLID

	Name of Compound	Note	MP	Melting Point of Derivatives				
				Recommended		Others		
				Semi-carbazone	2,4-Di-nitro-phenyl-hydrazone	4-Nitro-phenyl-hydra-zone	Phenyl-hydrazone	Oxime
37	1-Phenyl-1-hepten-3-one (styryl butyl ketone)	99	38–9				98	
38	4,4-Dimethylcyclohexanone		38–40	204				
39	3,4-Methylenedioxypropio-phenone		39	187–8			97	104
40	2-Hydroxybenzophenone		39	250–1			155	141 and 143
41	2-Methoxybenzophenone		39		251			145–8
42	4-Ketohexanoic acid		40	176d				
43	3-Bromopropiophenone		40	183				
44	4-Phenyl-3-buten-2-one (benzalacetone)	6	42v.s.	187–8	227 (223)	165–7	159 (157)	117
45	3-Benzoylpyridine	7	42				144	141–3 and 162–3
46	1-Indanone	8	42v.s.	233	258	234–5	130–1N	146
47	3-Aminopropiophenone	3	42					112–3
48	2-Bromobenzophenone		42					133
49	8-Pentadecanone		42					120
50	1-Chloro-4-phenyl-2-butanone		42		147			
51	Bis(thioanisyl) ketone		43	117				
52	8-Acetylguanine		43.5		253			
53	Phenyl 2-pyridyl ketone	9	43–5				136–7	150–2 and 165–70
54	2,7-Octanedione		44	224–5(di) and 260(di)				158(di)
55	2-Benzoylfuran		44					122
56	1,3-Dichloro-2-propanone		45	120	132–3			
57	2-Methyl-5-isopropyl-1,4-benzoquinone (thymoquinone)	10	45	201–2d (mono) 237(di)	179–80N		93N	160–2
58	2-Tetradecyl-3-thiophanone		45.5	158				
59	2-Acetylquinoline		46				45	
60	3-Chloropropiophenone		46	180				
61	4-Bromopropiophenone		46	171				
62	2,2'-Dichlorobenzophenone		46–7		206–8			
63	Phenyl undecyl ketone		47		93–4			63
64	2-Aminopropiophenone	3	47	190				88–9
65	Benzophenone	12	48	167	238–9	154–5 and 144N	137–8	144
66	4-Fluorobenzophenone		48				105	135
67	4-Phenyl-2-ketobutyric acid		48–50	175d			144–5	165
68	2,4,6-Heptanetrione		49	203			142(di)	68.5 2,6(di)
69	2,8-Nonanedione		49	198(di)	156(bis)			85(di)
70	2-Bromoacetophenone		49–50	177				
71	2-Hydroxy-5-methylaceto-phenone		50	212	273–4			145
72	2-Acetoacetylpyridine		50					78(mono) 146–7(di)
73	Phenacyl bromide (α-bromoacetophenone)		50	146				89.5 and 97

(Continued)

TABLE 21B (Continued)
KETONES—SOLID

	Name of Compound	Note	MP	Melting Point of Derivatives				
				Recommended		Others		
				Semi-carbazone	2,4-Di-nitro-phenyl-hydrazone	4-Nitro-phenyl-hydrazone	Phenyl-hydrazone	Oxime
74	5-Phenoxy-2-pentanone		50		110			
75	4-Bromoacetophenone		51	208	237* (230)		126	128
76	3-Bromophenacyl bromide		51	163–4d				
77	Phenyl ethynyl ketone		51		214			
78	3,4-Dimethoxyacetophenone		51	218		227	131	140
79	Diphenylacetoin	13	52	169				
80	Propyl 2-naphthyl ketone	14	52	185–6				89
81	9-Heptadecanone		53					112
82	Phenyl p-tolyl ketone		55	121	199–200			153–4
83	2,3-Dimethyl-1,4-benzo-quinone		55					166
84	5-Phenyl-3-thiophanone		55	206				
85	2-Acetylnaphthalene	15	56	235–7	262d		176–7	145
86	4,4'-Dimethoxybenzophenone		56		200–1			
87	2-Nonadecanone		56					77
88	2-Chlorobenzoquinone	16	57	185d				148 and 158N
89	2-Indanone		57					153
90	Benzylideneacetophenone	17	58N	168N	245*		120	140 and 68 (116 and 75)
91	Phenacyl chloride (α-chloroacetophenone)		59	156 (149)	212			89
92	2-Chloroacetophenone		59	159–60				112–3
93	4-Methylbenzophenone	18	60 (55)	121 2	202.4*		109	154 and 115N
94	Desoxybenzoin		60	148	204*	163	116	98
95	Hexyl 2-naphthyl ketone		60	132				74
96	Ethyl 2-naphthyl ketone		60	202				133
97	2,4,6-Trimethylphenyl methyl ketone		60	205 (197)				
98	Difurfurylideneacetone		61				122	
99	1-Phenyl-1,3-butanedione		61		151	100	150–3	
100	1,1-Diphenylacetone		61	170			131	165
101	4-Methoxybenzophenone	11	62		228–9N (180)	198–9	132 and 90	146–7 and 115–6
102	4-Acetoacetylpyridine	19	62					165(mono)
103	2,4'-Dibromobenzophenone		62					141–2
104	2-Phenylcyclohexanone		63	190	139			169
105	4-Methoxybenzil		63					124
106	Butyl 4-hydroxyphenyl ketone	20	63	164			78	
107	3-Benzoylpropenoic acid	1,21	64	190	229–30		197	168d
108	4-Acetoacetylquinoline	22	64–5					171(mono)
109	Benzoylformic acid	1,23	66		196–7			127 and 145d
110	4-Ketopentamethylene sulfide		66	151				85
111	2,4'Dichlorobenzophenone		66–7		230–1			
112	Benzyl 1-naphthyl ketone	24	67				101	148–52
113	2,2'-Dimethylbenzophenone		67					105
114	3,5-Dibromoacetophenone		68	268		109–10		
115	6-Phenyl-5-hexen-2-one (cinnamalacetone)		68	186	222–3		180	153

(Continued)

TABLE 21B (Continued)
KETONES—SOLID

	Name of Compound	Note	MP	Melting Point of Derivatives				
				Recommended		Others		
				Semi-carbazone	2,4-Di-nitro-phenyl-hydrazone	4-Nitro-phenyl-hydra-zone	Phenyl-hydrazone	Oxime
116	2-Methyl-1,4-benzoquinone	25	69v.s.	178–9N	128N		130N	134–5dN
117	12-Tricosanone		69.5	179				
118	2-Chlorobenzyl phenyl ketone		71					86
119	3-Acetylphenanthrene	26	72	230			193–4	143–4
120	Dihydroxyacetone	27	72N		277–8	160		84
121	4-Benzoylpyridine	28	72				181–2	176–7 and 152.5
122	2,6-Dimethyl-1,4-benzo-quinone		72–3s					175(1) and 170–1(4)
123	Phenyl phenylethyl ketone		73	144				87
124	4-Methoxybenzylidene-acetone		73		229			
125	Phenoxymethyl phenyl ketone		74	187				
126	9-Acetylphenanthrene	29	74.5	201				154–5
127	1-Naphthyl phenyl ketone		75					161
128	9-Acetylfluorene		75.5				139	
129	2-Benzofuryl methyl ketone		76 (72)	207			154	
130	Chloralacetone		76					104–6
131	3-Hexene-2,5-dione		76		291–2(di)			
132	Bis(phenacyl) sulfide		77				147(di)	151
133	1,4-Cyclohexanedione		77–9	221–2	240			188
134	3,4,5-Trimethoxybenzo-phenone		78		237–9			
135	2-Naphthoxyacetone		78	203			154	123
136	Dibenzoylmethane		78 (81)	205			105 (mono)	
137	4-Chlorobenzophenone		78		185		106	163
138	3-Nitroacetophenone	30	81	257	228		128 (135)	132
139	4-Nitroacetophenone		81		257–8		132	174
140	4-Bromobenzophenone		82		230		126	169
141	2-Benzoylpropionic acid		82–3				100–4	
142	9-Fluorenone	31	83v.s.		283–4	269	151–2	195–6*
143	4-Aminobutyrophenone	3	84					
144	2,2′-Diacetylbiphenyl		84				178(di)	212(di)
145	3-Acetoacetylpyridine	32	85					79(di)
146	Phenacyl alcohol	2	86	146			112	70
147	18-Pentatricontanone		88.5					67 (62–3)
148	4-Acetylphenanthrene	33	90					
149	4-Dimethylaminobenzo-phenone	4, 34	92				105	
150	4,4′-Dimethylbenzophenone	35	95	140	229 (229.4*)		100	163
151	Benzil	36	95	243–4N (di)	189(di)	290N (di)	225N (di)	237N (di) 180
152	4,5-Dimethyltrithione		95–7					
153	3-Hydroxyacetophenone		96	195	257			

(Continued)

TABLE 21B (Continued)
KETONES—SOLID

	Name of Compound	Note	MP	Melting Point of Derivatives				
				Recommended		Others		
				Semi-carbazone	2,4-Di-nitro-phenyl-hydrazone	4-Nitro-phenyl-hydrazone	Phenyl-hydrazone	Oxime
154	3-Nitrophenacyl bromide	37	96					127
155	Benzoxazolone (hydrate)	3, 38	97–8N				280	
156	2,3-Dihydroxyacetophenone	5	98					
157	4-Nitrophenacyl bromide	39	98					
158	2-Aminophenacyl alcohol	3	98				198	
159	3-Aminoacetophenone	40	99	196d	266			192–4
160	2-Benzoylacrylic acid (hydrate, mp 64)	41	99 (anh.)	190			197	168d
161	Di-1-naphthyl ketone		100					200
162	2-Amino-5-chlorobenzo-phenone	3	100					
163	Benzyl 2-naphthyl ketone	42	100				165–7	129
164	2,4-Diketopentanoic acid	1	101					
165	1,3-Dibenzoylbenzene		101–2					201(mono) 70–3(di)
166	Cumaranone		102	231				159
167	2-Acetyl-1-naphthol	43	102	245–50			136–7	168–9
168	1,5-Diphenyl-2,4-pentadien-1-one	44	102N		222	135		135N
169	1,3-Cyclohexanedione		104					156
170	2-Aminobenzophenone	3, 45	106					156N and 127N
171	4-Aminoacetophenone	3	106	250	267			148
172	4-Chlorophenyl benzyl ketone		108					123
173	4-Bromophenacyl bromide	46	109–10					115
174	4-Hydroxyacetophenone	47	109	199	261		151	145
175	1,2-Dibenzoylethylene(trans)		110					211
176	Piperonalacetone		110–1	217 and 168			163	186
177	5-(4-Methoxyphenyl)-trithione		109–11					170
178	1,5-Diphenyl-3-pentadienone (dibenzalacetone)	48	112	187–90	180	173	153	142–4
179	2,4-Diacetyl-3-methylphenol		112					191(di)
180	4,4′-Dimethoxybenzoin	49	113	185				240
181	1,4-Diacetylbenzene		114					159
182	2-Acetylbenzoic acid	1	114–5		186			
183	4-Hydroxy-3-methoxyaceto-phenone	5, 50	115	166			125	95
184	3,4-Dihydroxyacetophenone	5, 51	116					184d
185	3-Hydroxybenzophenone		116					76 and 126
186	1,4-Benzoquinone	52	116v.s.	243N 181d	231*(di)		152	240N
187	Phenyl-4-ketobutanoic acid		116					
188	2-Ketopentanedioic acid		117–8	220	213d		152–3	152d
189	4-Acetylbiphenyl		120–1		241–2			184–6
190	7-Acenaphthenone	53	121				90	175(mono) 222(di)
191	3-(2-Nitrophenyl)-2-ketopropionic acid	54	121				103.5 (mono) 107(di)	121.2

(Continued)

TABLE 21B (Continued)
KETONES—SOLID

	Name of Compound	Note	MP	Melting Point of Derivatives				
				Recommended		Others		
				Semi-carbazone	2,4-Di-nitro-phenyl-hydrazone	4-Nitro-phenyl-hydrazone	Phenyl-hydrazone	Oxime
192	4-Nitro-3-methyl-1-*p*-nitrophenylpyrazolone-5		121					
193	2,6-Dichloro-1,4-benzoquinone		121	218*d*(4)				140*d*(4)
194	4-Phenyltrithione		123					175
195	4-Aminobenzophenone	3, 55	124		189–91			168 and 127
196	4-Phenylphenacyl bromide	56	124–5					
197	2,5-Dimethyl-1,4-benzoquinone		124–5					173(mono) 272(di) (254)
198	1,4-Naphthoquinone		125*v.s.*	247(mono)	278(mono)	277–9 (mono)	205–6*d* (mono)	198(mono)
199	2-Acetopropenoic acid	57	125				169 (158)	206*N* and 189*N*
200	2,5-Dihydroxybenzophenone	58	125				144	
201	4-Phenylphenacyl chloride	59	126–7					
202	5-Phenyltrithione		126–7					139
203	Succinimide		126–7					197*d* (mono)
205	Vanillalacetone		129		230		127–8	
206	1,5-Bis(4-methoxyphenyl)-1,4-pentadien-3-one		129–31					148–51
207	4,4′-Dimethoxybenzil	60	133	254–5 (di)				195(di)
208	1,2-Benzfluoren-9-one	61	133					202*d*
209	1,2-Dibenzoylethylene(*cis*)		134					210(di)
210	2,2′-Diaminobenzophenone	3	135					
211	3,4,5-Tribromoacetophenone		135	265*d*			129–34*d*	
212	Furoin		135		217		81	161
213	4-Hydroxybenzophenone	5, 62	135	194	242.4*		144	152 and 81*N*
214	Benzoin	2, 63	137 (133)	205–6*d*	236		158–9*N* and 106	99*N* and 151–2
215	Phenyl 2,4,6-trimethylphenyl ketone		137		232			
216	4-Aminopropiophenone	3	140					153
217	3,3′-Dibromobenzophenone		141					181–2*d*
218	2,5-Dihydroxy-3-dodecyl-1,4-benzoquinone		143				189(di)	75
219	1,9-Diphenyl-1,3,6,8-nonatetraen-5-one (dicinnamalacetone)	64	144		195.7		166	
220	2-Acetylphenanthrene	65	144	260*N*			187–8	
221	2,4-Dihydroxybenzophenone	5	144					
222	3,4-Dihydroxybenzophenone	5	145					
223	1,2-Naphthoquinone	66	145–7*d* (120)	184*d*		236*N*	138*N*	169(di)
224	2,4-Dihydroxyacetophenone	67	147 (144)	218			159	198–200

(Continue

TABLE 21B (Continued)
KETONES—SOLID

	Name of Compound	Note	MP	Recommended		Others		
				Semi-carbazone	2,4-Di-nitro-phenyl-hydrazone	4-Nitro-phenyl-hydra-zone	Phenyl-hydrazone	Oxime
225	1,2-Dibenzoylethane		147					204
226	4,4′-Dichlorobenzophenone	68	147–8		240–1N			135
227	3,5-Dihydroxyacetophenone	5, 69	148	205–6		236–7		
228	4-Hydroxypropiophenone	70	148		240–1 (229)N			
229	1,2-Dibenzoylbenzene		148					150(mono)
230	5,5-Dimethyl-1,3-cyclo-hexanedione		148					176
231	Methone (5,5-dimethyldihydroresorcinol)		148–9					115(mono)
232	2,4′-Dihydroxybenzo-phenone	5, 71	151					
233	2-Acetyltriphenylene		152–3					202
234	5-Hydroxy-1,4-naphtho-quinone	72	154					N
235	3-Phenyl-2-ketopropanoic acid	1, 73	157	N	N	187–8		57–8
236	Anthrone	74	157–9					
237	3-Aminofluorenone	3	158–9					
238	2,5-Dichloro-1,4-benzo-quinone	75	161–2					155–6d
239	1,4-Dibenzoylbenzene		161					212–3 (mono) 235(di)
240	2-Aminofluorenone	3, 76	163					
241	2,4,6-Trihydroxybenzo-phenone	5	165					
242	Furil	77	165		215		184(di)	100(di)N
243	Benzanthrone	78	170					
244	3,3′-Dihydroxybenzophenone	5	170					
245	Quinhydrone	79	171					
246	2,3,4-Trihydroxyaceto-phenone	5, 80	172	225r.h.				162–3
247	Xanthone	81	174				152	161
248	4,4′-Bis(dimethylamino)benzophenone (Michler's ketone)	82	174 (179)		273–4		174–5	233
249	4,4′-Dibromobenzophenone	83	177					150–2
250	dl-Camphor	84	178N	238N	177	217	233	118–9
251	4-Nitrophenanthrenequinone	85	179–80	210N				170N
252	3-Acetoindazole	86	182					222
253	3,4,5-Trihydroxyaceto-phenone	87	187–8	216–7		260d		
254	2-Acetoindole	3	190–1					144–7
255	1,8-Dihydroxyanthraquinone	5	193					
256	1-Hydroxyanthraquinone	5	193					
257	Camphorquinone	88	198v.s.	236d and 147N	190(di)	239	170	N
258	Retenequinone		198	200				130–1
259	dl-4-Chlorocamphor		198–9	260–4d				158–60

(Continued)

TABLE 21B (Continued)
KETONES—SOLID

	Name of Compound	Note	MP	Melting Point of Derivatives				
				Recommended		Others		
				Semi-carbazone	2,4-Di-nitro-phenyl-hydrazone	4-Nitro-phenyl-hydra-zone	Phenyl-hydrazone	Oxime
260	1,4-Dihydroxyanthraquinone	5	202					
261	2,5-Dihydroxyacetophenone	5	202			215–6		149–50
262	Isatin	3, 89	203.5					198–200N and 214
263	2,3,5-Trihydroxyaceto-phenone	5, 90	206–7			241–2d		
264	9,10-Phenanthraquinone		207s	220d	312d	245	165	158(mono) 202(di)(r.h)
265	1-Amino-4-hydroxyanthra-quinone	5	207–8					
266	Acetophenone-4-carboxylic acid		208	269			234	
267	4,4'-Dihydroxybenzophenone	5	210		190–2			
268	2,5-Dihydroxybenzoquinone	5	211					
269	3,4,3',4'-Tetrahydroxy-benzophenone		227–8					145
270	2,5-Dinitrophenanthraquinone		228					190(mono)
271	Ninhydrin		243				207–8(di)	201
272	4,4'-Diaminobenzophenone	3	244				240	
273	2-Chlorophenanthraquinone		252–3	220				180–5 (mono)
274	2-Benzoyl-7-hydroxy-naphthalene		253		123			
275	2-Nitrophenanthraquinone	91	258–60					213(mono)
276	Acenaphthenequinone	92	261	271N(di)		247 (mono)	219N (di)	223d(di)
277	3-Chloroanthraquinone		265					204d (mono)
278	1,9-Oxalylanthracene		270				203(mono)	251(mono)
279	3-Nitrophenanthraquinone	93	279–80	254(mono)				240N
280	9,10-Anthraquinone		286N			238–40	183	224
281	Alizarin	94	290N					
282	Chloranil		290(s.t.)				220	
283	1,7-Dihydroxyanthraquinone	5	292					
284	2,7-Diaminofluorenone	3	290			280	230	255
285	Dianthrone	5	300					
286	2,7-Dinitrophenanthra-quinone		301–3					246–8 (mono)
287	2,3-Diaminoanthraquinone	3	353					
288	1,5-Dinitroanthraquinone		384–5					253(mono)

NOTES FOR TABLE 21B

1. For additional data, see Table 5.
2. For additional data, see Table 6.
3. For additional data, see Table 9.
4. For additional data, see Table 10.
5. For additional data, see Table 29.
6. Thiosemicarbazone, mp 148.
7. Picrate, mp 161. Hydrochloride, mp 160–2.
8. The phenylhydrazone melts at 134–5 if extracted with hot methanol and recrystallized.
9. Picrate, mp 130.
10. The 1-phenylhydrazone melts at 93, and the 1-(2,4-dinitrophenyl)hydrazone melts at 179–80. These are adducts and not true hydrazones. The diphenylsemicarbazone, mp 24
11. The melting point given is that obtained when the derivative is crystallized from chloroform
12. The 4-nitrophenylhydrazone has two forms, yellow, mp 154–5, and red, mp 144.
13. 4-Nitrobenzoate, mp 84.
14. Picrate, mp 68–9.

NOTES FOR TABLE 21B (Continued)

15. Picrate, mp 82; T.N.F., mp 132–3.
16. The 1-oxime, mp 158; 4-oxime, mp 148; 4-semicarbazone, mp 185d.
17. Benzylideneacetophenone (benzalacetophenone) dimerizes to produce four dimers, mp 124, 178, 195, and 225–6. There are three forms of the semicarbazone: α-(colorless), mp 168, β-(yellow), mp 170, and γ-(colorless) mp 178–9. The picrate, mp 93–7.
18. The ketoxime precipitates as a mixture of isomers that can be separated by fractional crystallization by adding water to an acetic acid solution. The less soluble isomer melts at 154 and the other at 115.
19. Anil, mp 103–4; chloroplatinate, mp 228d.
20. The ethyl ether melts at 31 and forms a semicarbazone that melts at 192.
21. Dibromide, mp 148.
22. Mono-anil, mp 129; chloroplatinate, mp 192–3d; hydrochloride, mp 180–1.
23. Thiosemicarbazone, mp 188–9.
24. Picrate, mp 101–2.
25. The data given are for the mono derivatives. Disemicarbazone, mp 240; di-2,4-dinitrophenylhydrazone, mp 269; dioxime, mp 220d.
26. Picrate, mp 125–6; T.N.F., mp 185.
27. This compound dimerizes, mp 78–81; diacetate, mp 48; dibenzoate, mp 120.5.
28. Picrate, mp 159–60; hydrochloride, mp 195–7.
29. Picrate, mp 107–8.
30. o-Tolylhydrazone, mp 135–6; m-tolysemicarbazone, mp 232–3.
31. Hydrazone, mp 149–50.
32. Picrate, mp 155; chloroplatinate, mp 173–5; hydrochloride, mp 92.
33. Picrate, mp 129–30.
34. Anil, mp 151.
35. Hydrazone, mp 108–10.
36. The monosemicarbazone, mp 174–5 (182); monophenylhydrazone, mp 134; the monooxime exists in two forms: mp 138 and 114; the dioxime exists in three forms: *amphi*-, mp 164–5, *syn*-, mp 207 (soluble in alcohol), and the *anti*-, mp 237 (practically insoluble in alcohol); the mono- p-nitrophenylhydrazone, mp 192–3.
37. Oxidizes to 3-nitrobenzoic acid.
38. Crystallizes from hot diluted hydrochloric acid, mp 141–2 (anh.).
39. Oxidizes to 4-nitrobenzoic acid.
40. Hydrazone, mp 98.
41. Hydrazone, mp 185–6.
42. Picrate, mp 143.
43. Acetate, mp 107.5; benzoate, mp 128.
44. The compound, crystallized from alcohol, mp 102, but another form is sparingly soluble in alcohol and melts at 235. Two oximes, mp 135 and 140.
45. Two forms of the oxime: (1) alkali stable, mp 156, and (2) acid stable, mp 127.
46. Acetate, mp 85; chloroacetate, mp 104.
47. Acetate, mp 54; benzoate, mp 134–5.
48. Picrate, mp 114.
49. Acetate, mp 94–5.
50. Acetate, mp 58; benzoate, mp 106.
51. Diacetate, mp 91; benzoate, mp 118.
52. Picrate, mp 79; reacts with equimolecular quantity of hydroquinone to produce quinhydrone, mp 171; monosemicarbazone, mp 178 (166); mono-2,4-dinitrophenylhydrazone, mp 185; the mono-oxime is tautomeric with 4-nitrosophenol; the hydrazones are adducts rather than true hydrazones.
53. Picrate, mp 113.
54. *unsym*-Diphenylhydrazone, mp 125; anil, mp 92.
55. Oxime: (1) prisms, mp 168, and (2) needles, mp 126; hydrazone, mp 139–40.
56. Acetate, mp 110; benzoate, mp 167.
57. Acetyl derivatives of the oxime, mp 206, melts at 155, whereas the acetyl derivative of the oxime, mp 189, melts at 143; the compound forms a methyl ester, mp 60.5, which yields a phenylhydrazone, mp 156, and an oxime, mp 100.
58. 5-Methyl ether, mp 84; dimethyl ether, mp 51.
59. Acetate, mp 110; benzoate, mp 167.
60. The mono-oxime has two forms: mp 133 and 217; diphenylhydrazone, mp 118.
61. T.N.F., mp 218–20.
62. The low melting oxime, mp 81, changes to the higher melting form, mp 152, by continued heating at its melting point; acetate, mp 81; benzoate, mp 94–5.
63. Phenylhydrazone, two forms: (1) needles, mp 158–9, and (2) prisms, mp 106; oxime: (*syn*) mp 99, and (*anti*) 151–2; hydrazone, mp 75.
64. Decomposes in the light.
65. The semicarbazone melts at 260, then solidifies and remelts at 297–8; T.N.F., mp 155.
66. 1-*p*-Nitrophenylhydrazone, mp 250–1 (246); 2-*p*-nitrophenylhydrazone, mp 236; The two isomeric mono-oximes are tautomeric with the corresponding nitrosophenols: (1) 1-oxime, mp 110, and the 2-oxime, mp 162–4.
67. Diacetate, mp 38; dibenzoate, mp 81.
68. The 2,4-dinitrophenylhydrazone was crystallized from chloroform; hydrazone, mp 91–3.
69. Diacetate, mp 91–2.
70. The data for the 2,4-dinitrophenylhydrazone, mp 240–1, was given when crystallized from chloroform.
71. Anil, mp 214.
72. 1-Oxime, mp 187; 4-oxime, mp 203; the dioxime explodes at 225; acetyl derivative, mp 154–5.
73. The free acid does not form a semicarbazone but its salts do form the semicarbazone, mp 180d; *p*-tolylhydrazone, mp 158. The 2,4-dinitrophenylhydrazone, from alcohol-water, mp 192–4; from petroleum ether, mp 162–4.
74. Oxidizes to anthraquinone by chromic oxide or nitric acid.
75. The oxime yields an acetyl derivative, mp 149.
76. Hydrazone, mp 209.
77. Monophenylhydrazone, mp 82–3. Five oximes are of record: α-mono, mp 106; β-mono, mp 97–8; α-di, mp 100 and 166; β-di, mp 188–90.

NOTES FOR TABLE 21B (Continued)

78. May be oxidized to anthraquinone-1-carboxylic acid, mp 291–2.
79. Easily oxidized to quinone, mp 116, and may be reduced to hydroquinone, mp 171.
80. Picrate, mp 133; triacetate, mp 85.
81. Anil, mp 134.5.
82. Picrate, mp 156–7; hydrazone, mp 150; di-methyl iodide, mp 105.
83. Hydrazone, mp 92–4.
84. d-Camphor has a melting point of 179.9, and its semicarbazone melts at 247–8. l-Camphor melts at 178.6 and forms a semicarbazone, mp 238 and an oxime, mp 115. d-Camphor forms a hydrazone, mp 55, and a p-bromophenylhydrazone, mp 101.
85. Monosemicarbazone, mp 210; mono-oxime, mp 169–70; dioxime, mp 210.
86. N-Acetyl derivative, mp 123.
87. Triacetate, mp 111–2.
88. The 3-semicarbazone has two forms: α-(from ethanol), mp 236, and β-(from benzene-petroleum ether), mp 147; α-mono-oxime, mp 153; β-mono-oxime, mp 114–5; the dioxime has four forms, mp 136, 194, 201, and 248; 3-p-bromophenylhydrazone (from acetic acid), mp 215–6.
89. α-oxime, mp 198–200; β-oxime, mp 214; the diacetyl derivative of the β-oxime melts at 175.
90. Triacetate, mp 106–7.
91. Monothiosemicarbazone, mp 234–5.
92. Monosemicarbazone, mp 192–3; monophenylhydrazine, mp 179; mono-p-tolylhydrazone, mp 163; the mono-oxime has two forms: (1) mp 230, which yields an acetyl derivative, mp 247, and (2), mp 207d; monohydrazone, mp 241; 4-bromophenylhydrazone, mp 193.
93. Mono-oxime, mp 240; dioxime, mp 200.
94. Compound sublimes above 110 (which differentiates it from flavopurpurin and antrapurpurin which sublime above 160 and 170); the diacetate may be prepared by reaction with acetic anhydride (in presence of sulfuric acid), mp 182.
95. Tetrabromide, mp 88–9.
96. Picrate, mp 116 (119).
97. Monohydrate, mp 58.
98. This melting point for the acid is given by one source—others list it as unstable. The 2-chlorophenylhydrazone melts at 74–7.
99. The dimer melts at 175–6.

REFERENCES FOR TABLES 21A AND 21B

Aldous, D. L., Reibsomer, J. L., and Castle, R. N., *J. Org. Chem.*, **25**, 1151 (1960). Synthesis of Diaryloxazoles. Lists the melting points of the derivatives of several ketones.

Allen, C. F. H., and Gates, J. W. Jr., *J. Org. Chem.*, **6**, 596 (1941). The Identification of Carbonyl Compounds by the Use of N-Methyl-beta-carbohydrazidopyridinium-p-toluene Sulfonate. A listing of the compounds is given with the melting points of the derivatives.

Buu-Hoi, N. G. Ph., Lavit, D., and Xuong, N. G. D., *J. Org. Chem.*, **19**, 1617 (1954). Some Syntheses From p-Fluoroanisole. Gives the method of preparation and lists the melting points of 10 fluoroketones prepared from 4-fluoroanisole.

Campbell, R. D., and Schultz, F. J., *J. Org. Chem.*, **25**, 1877 (1960). Nitroolefins. II. Derivatives of α-Nitroacetophenone. Gives the method of preparation and the melting points of the derivatives obtained.

Hurd, C. D., and Nilson, M. E., *J. Org. Chem.*, **20**, 927 (1955). Aliphatic Nitro Ketones. Gives the method of preparation and lists the melting points of the derivatives of nitro alcohols and ketones.

Magnusson, R., *Acta Chem. Scand.*, **14**, 1643 (1960). Reactions Between Quinones and Carbonyl Compounds Catalysed by Aluminum Oxide. Describes the addition of acetone to various o-benzoquinones and gives the melting points of 11 derivatives.

Massie, S. P., et al., *J. Org Chem.*, **21**, 1006 (1956). Ring Derivatives of Phenothiazine. II. 2-Phenothiazinyl Ketones and Their Derivatives. A method for preparing 2-phenothiazinyl ketones is described. The melting points are listed along with infrared spectra data.

Reid, E. E., *Organic Chemistry of Bivalent Sulfur*, Vol. II. Chemical Publishing Co., New York, 1960. See pp. 360–368 for keto sulfides.

Reid, E. E., *Organic Chemistry of Bivalent Sulfur*, Vol. V. Chemical Publishing Co., New York, 1963. For the identification of ketones as thiosemicarbazones and their melting points see pp. 233–245.

Truitt, P., Wood, F. M. Jr., and Hall, R. L., *J. Org. Chem.*, **25**, 1460 (1960). Amebicides. II. Acyl Derivatives of 2-Amino-1,4-Naphthoquinone. Lists the melting points of 31 derivatives.

Veibel, S., *Acta Chem. Scand.*, **1**, 54 (1947). On the Application of p-Carboxyphenylhydrazones in the Identification of Carbonyl Compounds. The melting points of the phenylhydrazones of about 75 common carbonyl compounds are given.

Woolfrom, M. L., and Arsenault, G. P., *J. Org. Chem.*, **25**, 205 (1960). Preparation of 2,4-Dinitrophenylhydrazine Derivatives of Highly Oxygenated Carbonyl Compounds.

TABLE 22
AZO COMPOUNDS

	Name of Compound	MP	Melting Point of Derivative
1	2-Methylazobenzene	18	
2	3,3'-Dimethylazobenzene	54	Hydrazo, 165, Azoxy, *cis*, 38; *trans*, 83
3	2,2'-Dimethylazobenzene	55	Hydrazo, 38 Azoxy, *cis*, 53; *trans*, 82
4	2-Aminoazobenzene	59	N-Acetyl, 126, N-Benzoyl, 122
5	3-Aminoazobenzene	67	N-Acetyl, 131
6	3-Chloroazobenzene	67.5	
7	Azobenzene(*trans*); *cis*, 71	68	Hydrazo, 130
8	3-Bromoazobenzene	69	
9	1-Benzeneazonaphthalene	70	
10	2-Nitroazobenzene	70	
11	4-Methylazobenzene	72	
13	3-Azopyridine (*cis*)	82	
14	2-Azopyridine (*trans*); *cis*, 87 (*cis-trans* eutectic, 56)	83	Picrate, 180*d*
15	2-Bromoazobenzene	87	
16	4-Bromoazobenzene	88	
17	4-Chloroazobenzene	89	
18	3-Nitroazobenzene	95	
19	3,3'-Dichloroazobenzene	101	
20	4-Azopyridine	108	
21	2,4-Diaminoazobenzene	117	
22	4,4'-Bis-(dimethylamino)azobenzene	117	
23	2-(3-Methylbenzeneazo)-1-naphthol	118	Methyl ether, 50
24	4-Aminoazobenzene	127	N-acetyl, 147
25	4-Hydroxy-3-methylazobenzene	128	Acetate, 82
26	1-(2-Methylbenzeneazo)-2-naphthol	131	Methyl ether, 58
27	2-Benzeneazonaphthalene	131	
28	2,2'-Diethoxyazobenzene	131	
29	1-Benzeneazo-2-naphthol	134	Acetate, 117
30	2,2'-Diaminoazobenzene	134	
31	4-Nitroazobenzene	135	
32	2,2'-Dichloroazobenzene	136	Azoxy: *cis*, 93; *trans*, 57
33	1,2'-Azonaphthalene	136	
34	1-(4-Methylbenzeneazo)-2-naphthol	137	Acetate, 99
35	2-Benzeneazo-1-naphthol	138	Acetate, 121
36	3-Azopyridine(*trans*)	140	
37	2,4,6-Trinitroazobenzene	142	
38	1-(3-Methylbenzeneazo)-2-naphthol	142	Methyl ether, 81
39	4,4'-Dimethylazobenzene	144	Hydrazo, 134; Azoxy: *cis*, 76; *trans*, 83–5
40	2,2'-Azobiphenyl	145	Hydrazo, 182
41	2-(4-Methylbenzeneazo)-1-naphthol	145	Dibromo, 236

(*Continued*)

TABLE 22 (Continued)
AZO COMPOUNDS

	Name of Compound	MP	Melting Point of Derivative
42	4-(2-Methylbenzeneazo)-1-naphthol	147	Methyl ether, 93
43	4-Hydroxyazobenzene	152	Acetate, 89
44	3,3'-Dinitroazobenzene	153	Azoxy: *cis*, 146; *trans*, 83
45	3,3'-Diaminoazobenzene	154–6	
46	2-(2-Methylbenzeneazo)-1-naphthol	156	Dibromo, 254
47	1-(4-Aminobenzeneazo)-2-naphthol	160	N-Benzenesulfonamide, 239
48	4,4'-Diethoxyazobenzene	160	
49	1-(3-Chlorobenzeneazo)-2-naphthol	161	Acetate, 81
50	1-(4-Chlorobenzeneazo)-2-naphthol	162	Acetate, 134
51	1-(2-Bromobenzeneazo)-2-naphthol	165	Acetate, 157
52	Benzeneazoresorcinol	170	Diacetate, 104
53	2,2'-Dihydroxyazobenzene	172	Diacetate, 150
54	1-(3-Bromobenzeneazo)-2-naphthol	172	Acetate, 88
55	1-(4-Bromobenzeneazo)-2-naphthol	175	Acetate, 136
56	1-(2-Methoxybenzeneazo)-2-naphthol	178	Methyl ether, 94
57	1-(3-Aminobenzeneazo)-2-naphthol	179	N-benzenesulfonamide, 218
58	2-Hydroxy-1,2'-azonaphthalene	179	Acetate, 117
59	4-(2-Bromobenzeneazo)-1-naphthol	183	Acetate, 123
60	4-(2-Chlorobenzeneazo)-1-naphthol	185	
61	2,2'-Dibromoazobenzene	185	Azoxy, 114
62	4,4'-Dichloroazobenzene	187	
63	2-(4-Chlorobenzeneazo)-1-naphthol	187	Methyl ether, 111
64	1,1'-Azonaphthalene	190	
65	2,2'-Dinitroazobenzene	194	Azoxy: *cis*, 175; *trans*, 81
66	1-(3-Nitrobenzeneazo)-2-naphthol	194	Acetate, 162
67	4-(3-Methylbenzeneazo)-1-naphthol	199d	
68	4,4'-Dibromoazobenzene	205	Azoxy, 176
69	3,3'-Dihydroxyazobenzene	205	Azoxy, 183
70	4-Benzeneazo-1-naphthol	206d	Acetate, 128
71	2,2'-Azonaphthalene	208	
72	1-(2-Nitrobenzeneazo)-2-naphthol	210	Methyl ether, 137
73	4-(3-Bromobenzeneazo)-1-naphthol	211	Acetate, 112
74	2-(2-Nitrobenzeneazo)-1-naphthol	218	
75	4,4'-Dinitroazobenzene	220	Azoxy: *cis*, 192; *trans*, 85–6
76	4-(3-Chlorobenzeneazo)-1-naphthol	222d	
77	2-Hydroxy-1,1'-azonaphthalene	230	Methyl ether, 67
78	4-(4-Chlorobenzeneazo)-1-naphthol	230d	
79	2-(4-Nitrobenzeneazo)-1-naphthol	235	Acetate, 180
80	4-(4-Bromobenzeneazo)-1-naphthol	238	Acetate, 141
81	2-(3,3'-Dibromo)azopyridine	240–1	
82	4,4'-Diaminoazobenzene	240–1 (246)	
83	2-(3,3'-Dichloro)azopyridine	248d	
84	4,4'-Azobiphenyl	250	Hydrazo, 170
85	1-(4-Nitrobenzeneazo)-2-naphthol	250	Acetate, 193
86	4-(4-Nitrobenzeneazo)-1-naphthol	283	Acetate, 166

REFERENCES FOR TABLE 22

Badger, G. M. et al, *J. Chem. Soc.*, **1953**, 2143. Oxidation of Azo Compounds; includes melting points of azo and azoxy compounds.

Koch, L., Milligan, R. F., and Zuckerman, S., *Anal. Chem.*, **16**, 755 (1944). Azo-2-Naphthol Dyes. Lists the melting points of unsulfonated azo-2-naphthol dyes and the melting points of the benzoyl derivatives from the catalytic reduction process.

Rodd, E. H., *Chemistry of Carbon Compounds* Elsevier, Houston, 1952. See Vol. III, pp. 334–344 for azo compounds.

TABLE 23

AZOXY COMPOUNDS[a]

	Name of Compound	MP		Name of Compound	MP
1	Azoxybenzene(*trans*)	36	16	2,2'-Dichloroazoxybenzene(*cis*)	93
2	3,3'-Dimethylazoxybenzene(*cis*)	38	17	2,2'-Diethoxyazoxybenzene	102
3	3,3'-Diethoxyazoxybenzene	50	18	3,3'-Dibromoazoxybenzene	113
4	3,3'-Dimethoxyazoxybenzene	52	19	2,2'-Dibromoazoxybenzene	114
5	2,2'-Dimethylazoxybenzene(*cis*)	53	20	4,4'-Dimethoxyazoxybenzene	119
6	2,2'-Dichloroazoxybenzene(*trans*)	57	21	1,1'-Azoxynaphthalene	127
7	4,4'-Dimethylazoxybenzene(*cis*)	76	22	4,4'-Diethoxyazoxybenzene	138
8	2,2'-Dinitroazoxybenzene(*trans*)	81	23	3,3'-Dinitroazoxybenzene(*cis*)	146
9	2,2'-Dimethoxyazoxybenzene	81	24	4,4'-Dichloroazoxybenzene	154
10	2,2'-Dimethylazoxybenzene(*trans*)	82	25	2,2'-Azoxybiphenyl	158
11	3,3'-Dimethylazoxybenzene(*trans*)	83	26	2,2'-Azoxynaphthalene	168
12	3,3'-Dinitroazoxybenzene(*trans*)	83	27	2,2'-Dinitroazoxybenzene(*cis*)	175
13	4,4'-Dimethylazoxybenzene(*trans*)	83–5	28	4,4'-Dibromoazoxybenzene	176
14	4,4'-Dinitroazoxybenzene(*trans*)	85–6	29	4,4'-Dinitroazoxybenzene(*cis*)	192
15	Azoxybenzene(*cis*)	87	30	4,4'-Azoxybiphenyl	212

[a] Unless specified, azoxy compounds are assumed to be the *trans* isomer. Azoxy compounds may be derivatized by reduction to the corresponding azo, hydrazo, or amino compounds.

TABLE 24

HYDRAZO COMPOUNDS[a]

	Name of Compound	MP		Name of Compound	MP
1	3,3'-Dimethylhydrazobenzene	38	12	3,3'-Diethoxyhydrazobenzene	119
2	2,2'-Dibromohydrazobenzene	82	13	4,4'-Dichlorohydrazobenzene	122
3	4,4'-Diethoxyhydrazobenzene	86	14	Hydrazobenzene	130
4	3-Nitrohydrazobenzene	86	15	4,4'-Dibromohydrazobenzene	130
5	2,2'-Dichlorohydrazobenzene	87	16	4,4'-Dimethylhydrazobenzene	134
6	2,2'-Diethoxyhydrazobenzene	89	17	2,2'-Hydrazonaphthalene	140
7	4-Chlorohydrazobenzene	90	18	2,2'-Dihydroxyhydrazobenzene	148
8	3,3'-Dichlorohydrazobenzene	94	19	1,1'-Hydrazonaphthalene	153
9	2,2'-Dimethoxyhydrazobenzene	102	20	2,2'-Dimethylhydrazobenzene	165
10	3,3'-Dibromohydrazobenzene	107–9	21	4,4'-Hydrazobiphenyl	170
11	4-Bromohydrazobenzene	115	22	2,2'-Hydrazobiphenyl	182

[a] Hydrazo compounds may be derivatized by oxidation to the corresponding azo compounds, or by reduction to the amino compounds.

TABLE 25
ISOCYANATES AND ISOCYANIDES

	Name of Compound	Note	BP		Name of Compound	Note	BP
1	Methyl isocyanate	2	44	13	Phenyl isocyanide	8	165–6
2	Methyl isocyanide	8	59.6	14	o-Tolyl isocyanide	8	183–4
3	Ethyl isocyanate	3	60	15	o-Tolyl isocyanate	5	186
4	Isopropyl isocyanate		67	16	p-Tolyl isocyanate	6	187
			(70–75)	17	m-Tolyl isocyanate	7	195–6
5	Ethyl isocyanide	8	78–9				(183)
6	tert-Butyl isocyanate	9	85	18	1-Naphthyl isocyanate	1	269–70
7	Isopropyl isocyanide	8	87				MP
8	Propyl isocyanate		87	19	4-Chlorophenyl isocyanate	1	30
9	Propyl isocyanide	8	100	20	2-Nitrophenyl isocyanate	1	41
10	Isobutyl isocyanate	10	102	21	4-Bromophenyl isocyanate	1	43
11	Butyl isocyanate	11	115	22	3-Nitrophenyl isocyanate	1	51
12	Phenyl isocyanate	1,4	162–3	23	2-Naphthyl isocyanate	1	55–6
			(166)	24	4-Nitrophenyl isocyanate	1	57

NOTES FOR TABLE 25

1. The numbered compounds yield urethane derivatives with 1-naphthol and 2-naphthol which have the melting points indicated in parentheses (the 1-naphthol derivative is listed first in each case): 12 (178.5; 158); 18 (152; 157); 19 (163; 192); 20 (sinters at 125, melts at 130; sinters at 135, melts at 143); 21 (149; 190–2); 22 (144; 153); 23 (174–5; 203); 24 (189; 182).
2. N-Methyl-N'-phenylurea, mp 152.
3. N-Ethyl-N'-phenylurea, mp 104 (99).
4. N,N'-Diphenylurea, mp 238–9.
5. N-Phenyl-N'-o-tolylurea, mp 212 (196).
6. N-Phenyl-N'-p-tolylurea, mp 226.
7. N-Phenyl-N'-m-tolylurea, mp 173–4.
8. Isocyanides may be derivatized by reducing them to the corresponding secondary amines which may then be acylated.
9. N-tert-Butyl-N'-phenylurea, mp 167–8.
10. N-Isobutyl-N'-phenylurea, mp 151–2.
11. N-Butyl-N'-phenylurea, mp 129–30.

TABLE 26A
NITRILES—LIQUID

Name of Compound	BP	Carboxylic Acid Formed by Hydrolysis		Name of Compound	BP	Carboxylic Acid Formed by Hydrolysis
1 Acrylonitrile	77.2	bp 140	31	3-Cyanothiophene	179	mp 138
2 Fluoroacetonitrile	80	mp 33	32	dl-Lactonitrile	182–4	mp 25–6(18)
3 Acetonitrile	81.6	bp 118	33	Glycollonitrile	183d	mp 80
4 Trichloroacetonitrile	86	mp 58	34	Heptanonitrile	184.5	bp 223
5 2-Methylacrylonitrile	90–1	bp 160	35	Benzonitrile	190.2	mp 122
6 2-Ketopropionitrile	93	bp 165	36	2-Cyanothiophene	196	mp 192
7 Propionitrile	97.4	bp 141	37	4-Chlorobutyronitrile	196–7	mp 16
8 Isobutyronitrile	104	bp 155	38	Methyl cyanoacetate	200	
9 Trimethylacetonitrile	106	mp 35	39	o-Tolunitrile	205	mp 104
10 2-Ethylacrylonitrile	111	bp 180	40	Octanonitrile	205.2	bp 239
11 2-Chloroisobutyronitrile	116	mp 31				(mp 16)
12 Butyronitrile	117.9	bp 162.5	41	Ethyl cyanoacetate	207	
13 Crotononitrile	119	mp 72	42	m-Tolunitrile	212	mp 113(110)
14 2-Butenonitrile	119	bp 163(169)	43	3-Hydroxypropionitrile	220	
15 Methoxyacetonitrile	120	bp 203–4	44	2-Cyanopyridine	222	mp 137
16 2-Hydroxyisobutyronitrile	120d	mp 79	45	Nonanonitrile	224	bp 255
17 2-Methylbutyronitrile	125	bp 176				(mp 15)
18 Chloroacetonitrile	127	mp 63	46	Ethylene cyanohydrin	229.7	mp 80
19 Isopentanonitrile	130	bp 176	47	2-Phenylpropionitrile	232	
20 2-Bromoisobutyronitrile	139–40	mp 48–9	48	Phenylacetonitrile	234	mp 76.5
21 3-Pentenonitrile	140	bp 188–9	49	4-Hydroxybutyronitrile	240	bp 204
22 2,2-Dimethylacrylonitrile	140–2	bp 70	50	Decanonitrile	243	mp 31
23 Pentanonitrile	141.3	bp 186	51	Undecanonitrile	260.8	mp 30
24 Diethylacetonitrile	145	bp 190	52	3-Phenylpropionitrile	261	
25 2-Cyanofuran	147	mp 133–4	53	Dodecanonitrile	277	mp 44
26 2-Methylpentanonitrile	147	bp 196	54	Glutaronitrile	286	mp 97
27 4-Methylpentanonitrile	155	bp 199	55	Tridecanonitrile	293	mp 44
28 Hexanonitrile	163.6	bp 205			(mp 8–9)	
29 2-Methylhexanonitrile	165	bp 210	56	Adiponitrile	295	mp 153
30 3-Chloropropionitrile	178	mp 41				

TABLE 26B
NITRILES—SOLIDS

No.	Name of Compound	MP	Carboxylic Acid Formed by Hydrolysis	No.	Name of Compound	MP	Carboxylic Acid Formed by Hydrolysis
1	Cyanodiphenymethane (bp 313–4)	19	mp 117	28	4-Aminobenzyl cyanide (bp 312) (See Table 9B)	46	
2	Tetradecanonitrile (bp 306)	19.5	mp 54	29	3-Bromo-p-tolunitrile	47	mp 140
3	3-Phenylpropenonitrile (bp 255–6)	20	mp 133	30	4-Bromophenylacetonitrile	47	mp 114
4	dl-Mandelonitrile	22	mp 118–9	31	Eicosanonitrile	49.5	mp 77
5	Pentadecanonitrile (bp 322)	23	mp 52	32	3-Cyanopyridine (bp 240–5) (nicotinonitrile)	50 (52)	mp 232
6	2-Chlorophenylacetonitrile (bp 251)	24	mp 95	33	4-Iodoacetophenylacetonitrile	50–1	mp 135
7	2-Methoxybenzonitrile (255–6)	24–5	mp 100–1	34	4-Cyanodiphenylmethane	51	mp 157–8
8	2-Cyanopyridine	26	mp 136–7	35	5-Cyanothiazole	53	mp 218
9	Methylmalononitrile	26	mp 135	36	2-Bromobenzonitrile (bp 253)	53	mp 150
10	p-Tolunitrile (bp 217)	27	mp 179–80	37	2-Iodobenzonitrile	55	mp 162
11	d-Mandelonitrile (bp 170d)	28–9	mp 133	38	Succinonitrile (bp 265–7)	56.6	mp 186–8
12	4-Chlorophenylacetonitrile (bp 265–7)	30	mp 105	39	2-Iodo-p-tolunitrile	57–8	mp 205–6
13	Malononitrile (bp 218–9)	31	mp 135	40	1,2-Benzenediacetonitrile	60	mp 150
14	Hexadecanonitrile	31.5	mp 63	41	3-Chloro-p-tolunitrile	61–2	mp 155–6
15	Maleonitrile	31	mp 130	42	4-Methoxybenzonitrile (bp 256–7)	61–2	mp 184–6
16	3,3-Dimethylbutanonitrile (bp 138)	32–3	bp 186–8	43	4-Methoxyphenylpropenonitrile	64	mp 170
17	1-Naphthylacetonitrile	33	mp 131	44	1,4-Dicyanocyclohexane(cis)	65	mp 168–9
18	Heptadecanonitrile	34	mp 61	45	Bromomalononitrile	65–6	mp 113
19	4-Fluorobenzonitrile (bp 189–90)	35	mp 183	46	2-Naphthonitrile	66(62)	mp 184
20	1-Naphthonitrile (bp 299)	36	mp 162	47	Cyanoacetic acid	66	mp 135–6
21	Indole-3-acetonitrile	36	mp 164–5	48	5-Chloro-o-tolunitrile	67	mp 172
22	3-Bromobenzonitrile (bp 255)	38	mp 155	49	4-Biphenylcarbonitrile	85–7	mp 226
23	Stearonitrile	41(43)	mp 70–1	50	Phenylmalononitrile	69	mp 152–3
24	3-Chlorobenzonitrile	41	mp 158(155)	51	6-Nitro-o-tolunitrile	69–70	mp 184
25	2-Chlorobenzonitrile (bp 232)	43	mp 142(140)	52	5-Bromo-o-tolunitrile	70	mp 187
26	Nonadecanonitrile	43	mp 69				
27	2-Bromo-p-tolunitrile	44	mp 204				

NITRILES—SOLIDS

	Name of Compound	MP	Carboxylic Acid Formed by Hydrolysis		Name of Compound	MP	Carboxylic Acid Formed by Hydrolysis
53	2-Aminobenzyl cyanide (see Table 9B)	72		82	3-Nitro-*o*-tolunitrile	109–10	mp 151–2
54	Diphenylacetonitrile	75	mp 148	83	2-Nitrobenzonitrile	110	mp 146
55	6-Nitro-*m*-tolunitrile	80	mp 219	84	4-Chloro-1-naphthonitrile	110	mp 210
56	Benzoylacetonitrile	81	mp 103–4	85	4-Bromobenzonitrile	112	mp 251
57	4-Cyanopyridine (isonicotinonitrile)	83(79)	mp 315(306)	86	4-Nitrophenylacetonitrile	116	mp 153
58	2-Nitro-*m*-tolunitrile	84	mp 225	87	6-Bromo-3-nitrobenzonitrile	117	mp 180
59	4-Cyanobiphenyl	85–6	mp 228	88	3-Nitrobenzonitrile	118	mp 140
60	3-Aminobenzonitrile (see Table 9B)	86		89	4-Bromo-3-nitrobenzonitrile	120	mp 203–4
61	4-Aminobenzonitrile (see Table 9B)	86		90	1-Cyanophenanthrene	128	mp 232–3
62	2-Naphthylacetonitrile	86(81)	mp 142	91	N-Methylaminoacetonitrile	129	mp 232*d*
63	2,6-Dimethylbenzonitrile	90–1	mp 116	92	5-Bromo-3-nitro-*p*-tolunitrile	130	mp 206
64	5-Chloro-2-nitro-*p*-tolunitrile	93	mp 180–1	93	5-Bromo-2-nitro-*p*-tolunitrile	132	mp 203
65	4-Nitro-*m*-tolunitrile	93–4	mp 134	94	4-Nitro-1-naphthonitrile	133	mp 220–1
66	2-Cyanoquinoline	94	mp 157	95	1-Nitro-2-naphthonitrile	138	mp 239
67	4-Chlorobenzonitrile	96(92)	mp 240	96	1,4-Dicyanocyclohexane(*trans*)	140	mp 300
68	4-Chloro-2-nitrobenzonitrile	98	mp 142	97	Phthalonitrile	141	mp 200–6
69	4-Chloro-3-nitrobenzonitrile	100–1	mp 181–2	98	8-Nitro-2-naphthonitrile	143	mp 295(288)
70	3-Nitro-*p*-tolunitrile	101	mp 164–5	99	5-Chloro-2-naphthonitrile	144	mp 270
71	3-Cyanophenanthrene	102	mp 269	100	5-Chloro-1-naphthonitrile	145	mp 245
72	4-Cyanoquinoline	102	mp 253–4	101	4-Nitrobenzonitrile	147	mp 241
73	4-Bromo-1-naphthonitrile	102–3	mp 212(220)	102	5-Bromo-1-naphthonitrile	147	mp 261(256)
74	5-Nitro-*m*-tolunitrile	104–5	mp 174	103	5-Bromosalicylnitrile	158–9	mp 165
75	4-Nitro-*o*-tolunitrile	105	mp 179	104	Isophthalonitrile	162*s*	mp 345–7
76	6-Chloro-3-nitrobenzonitrile	105–6	mp 165	105	4,4'-Dicyanodiphenylmethane	169(165)	mp 334(290)
77	5-Bromo-3-nitro-*o*-tolunitrile	106–7	mp 226	106	5-Nitro-2-naphthonitrile	172–3	mp 295
78	2-Nitro-*p*-tolunitrile	107	mp 190	107	9-Cyanoanthracene	175	mp 207
79	9-Cyanophenanthrene	107	mp 252	108	Indole-3-cyanide	178	mp 210–8
80	3-Cyanoquinoline	108	mp 275	109	5-Nitro-1-naphthonitrile	205	mp 241–2
81	2-Cyanophenanthrene	109	mp 259–60	110	Terephthalonitrile	222*s*	*s*

TABLE 27A
NITRO COMPOUNDS[a]—LIQUID

	Name of Compound	BP	Data for Derivatives of the Amine Formed by the Reduction of All Nitro Groups			Others
			Benzene-sulfonamide	Benzamide	Picrate	
1	Nitroethylene	98–9	58	71	165	
2	Nitromethane	101	30	80	207 (215)	Acetamide, mp 28
3	Chloropicrin	113				
4	Nitroethane	114	58	71	165	
5	2-Nitropropane	120.3		26		
6	3-Nitropropene	125–30	39		140	
7	Tetranitromethane (mp 13)	126				
8	1-Nitropropane	131.6	36	84	135	
9	2-Nitrobutane	140	70	76	140	
10	2-Methyl-2-nitrobutane	150			183	
11	2-Nitropentane	152–4				
12	1-Nitrobutane	153			151	
13	2-Methyl-1-nitropropane	154–8	53	57	150	
14	3-Methyl-1-nitrobutane	164			138	Acetamide, bp 230
15	1-Nitropentane	173			139	
16	2-Nitrohexane	176				
17	1-Nitrohexane	193	96	40	126	
18	1-Nitroheptane	193–5			121	
19	2-Nitroheptane	194–8				
20	1-Fluoro-3-nitrobenzene	200				3-Fluoroaniline, bp 186
21	Nitrocyclohexane	205–6		147		Acetamide, mp 104
22	1-Nitro-octane	206–10 (sl d)			112	
23	Nitrobenzene	210.85*	112	160		1,3-Dinitrobenzene, mp 90 Acetamide, mp 114
24	1-Fluoro-2-nitrobenzene	215				2-Fluoroaniline, bp 175
25	2-Nitrotoluene	221.7	124	143	213	2,4-Dinitrotoluene, 70 Acetamide, 110–1
26	2-Nitroethylbenzene	224		147	195	Acetamide, 111–2
27	1,3-Dimethyl-2-nitrobenzene	226		168	180	2,4,6-Trinitro-deriv., 182 Acetamide, mp 177
28	Phenylnitromethane	226d	88	105	194	Acetamide, 60
29	3-Nitrotoluene	233	95	125	200	3-Nitrobenzoic acid, 140 Acetamide, 65
30	4-Nitroethylbenzene	241		151		2,4,6-Trinitro-deriv., 37 Acetamide, 94
31	2,5-Dimethylnitrobenzene	241–2	138	140	171	Acetamide, 139
32	2,4-Dimethylnitrobenzene	246	129	192	209	Acetamide, 133(130)
33	2,3-Dimethylnitrobenzene	250		189	221	Acetamide, 135
34	2-Nitro-4-isopropyltoluene	264		102		2,6-Dinitrocymene, 54 Acetamide, 71
35	2-Methoxynitrobenzene (mp 10)	265	89	60(84)	200	2,4,6-Trinitro-deriv., 68 Acetamide, 85(88)
36	4-Nitro-tert-butylbenzene	267		134–6		Acetamide, 169–70
37	2-Ethoxynitrobenzene (mp 5–6)	268	102	104		Acetamide, 179 2,4-Dinitroethoxybenzene, 86

[a] Additional nitro compounds may be found in other tables; for example, in the table of nitriles (p. 913).

TABLE 27B
NITRO COMPOUNDS—SOLIDS

	Name of Compound	MP	Benzene-sulfon-amide	Benz-amide	Pic-rate	Others
			Data for Derivatives of the Amine Formed by the Reduction of All Nitro Groups			
1	Trinitromethane (BP 45/22 mm)	15				
2	2-Methyl-2-nitropropane (bp 127)	25.5		134	198	
3	2,4-Dinitrofluorobenzene (see Table 15B)	25–7				
4	4-Nitrofluorobenzene	27	185			Acetamide, 152
5	3-Nitrobenzyl alcohol	27				3-Nitrobenzoic acid, 140
6	3,4-Dimethylnitrobenzene	30				1,2-Dinitro-deriv., 82 Acetamide, 99
7	3-Nitro-1,2,4-trimethyl-benzene	30				Acetamide, 186
8	2-Chloronitrobenzene	32	129	99	134	1,4-Dinitro-deriv., 52(50) Acetamide, 87
9	2,4-Dichloronitrobenzene (bp 258)	33	128	115	106	Acetamide, 143–6
10	4-Bromo-3-nitrotoluene	33				Acetamide, 121(114)
11	3-Ethoxynitrobenzene	34		103	158	Acetamide, 97
12	2-Nitrobiphenyl	37(33)		102		Acetamide, 121
13	6-Chloro-2-nitrotoluene (bp 238)	37		173		6-Chloro-2-nitrobenzoic acid, 161 Acetamide, 157–9(136)
14	3-Iodonitrobenzene	38		157		3-Iodoaniline, 33 Acetamide, 119
15	3-Methoxynitrobenzene (bp 258)	38			169	1,5-Dinitro-deriv., 106 Acetamide, 81
16	4-Chloro-2-nitrotoluene (bp 240)	38				4-Chloro-2-nitrobenzoic acid, 142 Acetamide, 139–40
17	5-Nitroindane	40		137		Acetamide, 106
18	3,4-Dichloronitrobenzene (bp 255–6)	43				3,4-Dichloroaniline, 72
19	2-Bromonitrobenzene	43		116	129	2,4-Dinitrobromobenzene, 72 Acetamide, 99
20	4-Nitrobenzal chloride	43				4-Nitrobenzoic acid, 241
21	4-Nitroindane	44		136		Acetamide, 126
22	2,4,6-Trimethylnitrobenzene	44		204		Dinitromesitylene, 86 Acetamide, 216–7
23	3-Chloronitrobenzene	45	121	118	177	Acetamide, 72(78)
24	3-Nitrobenzyl chloride	45				3-Nitrobenzoic acid, 140–1
25	Ethyl 3-nitrobenzoate	47				
26	2-Nitroazoxybenzene	49				Acetamide, 156
27	2-Iodonitrobenzene	49		139		2-Iodoaniline, 56 Acetamide, 109
28	2-Nitrobenzyl chloride	49				2-Nitrobenzoic acid, 146 2,4-Dinitrobenzyl chloride, 34
29	4-Nitrotoluene (bp 238.34)	51.65*	120	158	182	2,4-Dinitrotoluene, 70 Acetamide, 147
30	2,4-Dinitrochlorobenzene	52		178		2,4-Dinitrophenol, 114 Acetamide, 242 2,4,6-Trinitro-deriv., 183

(Continued)

TABLE 27B (Continued)
NITRO COMPOUNDS—SOLIDS

	Name of Compound	MP	Data for Derivatives of the Amine Formed by the Reduction of All Nitro Groups			Others
			Benzene-sulfon-amide	Benz-amide	Pic-rate	
31	4-Methoxynitrobenzene	52.5	95	154		2,4-Dinitro-deriv., 89 Acetamide, 130(127)
32	2,5-Dichloronitrobenzene	54		120		1,3-Dinitro-deriv., 104 Acetamide, 132
33	4-Iodo-3-nitrotoluene	55				4-Iodo-3-aminotoluene, 38 3-Nitro-p-cresol, 36.5
34	3-Bromonitrobenzene	56		136(120)	180	3,4-Dinitrobromobenzene, 59 Acetamide, 87
35	Ethyl 4-nitrobenzoate	56				Ethyl 4-aminobenzoate, 90
36	1,2,4-Trichloro-5-nitro-benzene (bp 288)	57				4,5-Dichloro-2-nitroaniline, 175
37	5-Nitro-3,4-dimethoxytoluene	58				
38	3-Nitrobenzyl bromide	58				
39	β-Nitrostyrene	58				
40	2-Nitro-1-methylnaphthalene	58–9		222		Acetamide, 188–9
41	4-Ethoxynitrobenzene	60(57–8)	143	173	69	2,4-Dinitroethoxybenzene, 86 Acetamide, 137
42	1-Nitronaphthalene	61	167	160	163 (181)	Acetamide, 159
43	1,2,4-Trinitrobenzene	61				Naphthalene adduct, 52
44	3-Nitrobiphenyl	61(59)				3-Aminobiphenyl, 30 Acetamide, 148
45	3,4-Dinitrotoluene	61	178–9	263–4		Acetamide, 210 3,4-Dinitrobenzoic acid, 165(161)
46	4-Iodo-2-nitrotoluene	61				4-Iodo-2-aminotoluene, 48
47	2,3-Dinitrotoluene	63				2,3-Dinitrobenzoic acid, 201
48	2-Chloro-4-nitrotoluene	63				2-Chloro-4-nitrobenzoic acid 140–1
49	8-Nitro-1-methylnaphthalene	63–4		195–6		Acetamide, 183–4
50	3,5-Dichloro-1-nitrobenzene	65				3,5-Dichloroaniline, 51
51	3-Nitrobenzal chloride	65				3-Nitrobenzaldehyde, 58
52	2,6-Dinitrotoluene	66				2,4,6-Trinitrotoluene, 80(82) Acetamide, 202–3 2,6-Dinitrobenzoic acid, 202–3
53	2,4,6-Trinitromethoxybenzene	68				Picric acid, 122.5
54	2,4-Dinitrotoluene	70				2,4,6-Trinitrotoluene, 80(82) Naphthalene adduct, 60 Diacetamide, 224
55	2-Nitroazobenzene	71		122		Acetamide, 126
56	4-Nitro-1-methylnaphthalene	71–2		238–9		Acetamide, 166–7
57	4-Nitrobenzyl chloride	72				4-Nitrobenzoic acid, 241
58	2-Nitro-4-chlorobromobenzene	72				
59	2,4-Dinitrobromobenzene	72				2,4-Dinitrophenol, 114
60	3,4,6-Trinitro-1,2-dimethyl-benzene	72				
61	2-Nitrobenzyl alcohol	74				2-Nitrobenzoic acid, 146
62	3,5-Dimethylnitrobenzene (bp 273)	75				Acetamide, 144
63	3,5-Dinitro-1,2-dimethyl-benzene	76				Reduces by ammonium sulfide to the 5-amino derivative, mp 75; Acetamide, 209–10

(Continued)

TABLE 27B (Continued)
NITRO COMPOUNDS—SOLIDS

	Name of Compound	MP	Data for Derivatives of the Amine Formed by the Reduction of All Nitro Groups			Others
			Benzene-sulfonamide	Benzamide	Picrate	
64	2-Nitro-5-methylnaphthalene	76–7		155–6		
65	2-Nitronaphthalene	78	102	162		Acetamide, 132
66	2,4,6-Trinitro-1-ethoxybenzene	78				Naphthalene adduct, 39; Picric acid, 122.5
67	2,4,6-Trinitrotoluene	80(82)				Naphthalene adduct, 97; 2,4,6-Trinitrobenzoic acid, 220; Adduct with 2-Naphthol, 110
68	1-Nitro-2-methylnaphthalene	81		180		1,8-Dinitro deriv., 209; Acetamide, 188
69	4-Nitrophenanthrene	81		224		Acetamide, 190
70	3,4-Dinitro-1,2-dimethylbenzene	82				4-Amino deriv., 66; Acetamide, 115–6
71	4-Nitrobenzal bromide	82				4-Nitrobenzoic acid, 241
72	1-Nitro-5-methylnaphthalene	82–3		173–4		Acetamide, 194–5
73	Picryl chloride	83	211			Picric acid, 122.5
74	4-Chloronitrobenzene	84	121	192		4-Nitrophenol, 114
75	2,5-Dinitro-1,4-dibromobenzene	84				3,5-Dibromo-4-nitroaniline, 175
76	2,4-Dinitro-1,3-dimethylbenzene	84(82)		232(di)		4-Amino deriv., 84
77	2-Nitro-1,3-dihydroxybenzene	85				Dimethyl ether, 131
78	2,4-Dinitro-1,3,5-trimethylbenzene	85				2,4,6-Trinitro deriv., 232
79	1,4-Dibromo-2-nitrobenzene	85				Acetamide, 171–2
80	4-Bromo-1-nitronaphthalene	85				4-Bromo-1-naphthylamine, 102; Acetamide, 193
81	2,4-Dinitro-1-ethoxybenzene	86				2,4,6-Trinitro deriv., 78; 4-Amino deriv., 40
82	4-Chloro-1-nitronaphthalene	87(85)				4-Chloro-1-naphthylamine, 98; Acetamide, 186
83	8-Nitroquinoline	88–9		98		8-Aminoquinoline, 70
84	1,3-Dinitrobenzene (v.s.)	90	194	240	184	3-Nitroaniline, 114; Naphthalene adduct, 52
85	2′,4-Dibromo-3-nitrobiphenyl	90				3-Amino deriv., mp 88; Acetamide, 118
86	2,4,5-Trinitro-1,3-dimethylbenzene	90				
87	1,4-Dinitro-2,3-dimethylbenzene	90				3-Amino deriv., 114
88	3,5-Dinitrotoluene	92				Naphthalene adduct, 63; 3,5-Dinitrobenzoic acid, 204–5
89	2,3-Dinitro-1,4-dimethylbenzene	93				2,3-Dinitro-p-toluic acid, 249
90	4,6-Dinitro-1,3-dimethylbenzene	93		258–9(di)		2,4,6-Trinitro deriv., mp 125
91	2,4′-Dinitrobiphenyl	93		276–8(di)		Acetamide, 202(di)
92	4-Nitrobenzyl alcohol (see Table 6B)	93				

(Continued)

TABLE 27B (Continued)
NITRO COMPOUNDS—SOLIDS

	Name of Compound	MP	Data for Derivatives of the Amine Formed by the Reduction of All Nitro Groups			Others
			Benzene-sulfon-amide	Benz-amide	Pic-rate	
93	8-Chloro-1-nitronaphthalene	94				8-Chloro-1-naphthylamine, 96(88); Acetamide, 137
94	2,4-Dinitro-1-methoxy-benzene	95				2,4-Dinitrophenol, 114 Naphthalene adduct, 50
95	3-Nitroazobenzene	96				3-Aminoazobenzene, mp 56–7; Acetamide, 130–1
96	4-Chloro-2-nitro-1-methoxybenzene	98		78	194	Amino deriv., 84; Acetamide, 104
97	4-Nitrobenzyl bromide (see Table 14)	99				Esters of acids
98	2-Nitrophenanthrene	99		216		Amino deriv., 85; Acetamide, 225
99	8-Bromo-1-nitronaphthalene	99–100				Amino deriv., 90; Acetamide, 138–9
100	2,5-Dinitro-1,3-dimethyl-benzene	101				2-Amino deriv., 158; Diamino deriv., 104
101	1,2-Dinitronaphthalene	102–3	215(1–N)	291(di)		Diamino deriv., 98; Acetamide, 234
102	2,4,5-Trinitrotoluene	104				Naphthalene adduct, 89
103	5-Nitroacenaphthene	106(101)		210(199)		Amino deriv., 108; Acetamide, 238
104	4,5-Dibromo-2-nitrobiphenyl	108				2-Amino deriv., 86; Acetamide, 215*d*
105	2,3′-Dinitrobiphenyl	110				
106	2,4-Dinitrobiphenyl	110				
107	5-Chloro-1-nitronaphthalene	111				Amino deriv., 85 Acetamide, 128
108	2,3,4-Trinitrotoluene	112				Naphthalene adduct, 100
109	3-Nitro-1,2,4,5-tetra-methylbenzene	112–3				Amino deriv., 75; Acetamide, 207
110	4-Nitrobiphenyl	114		230		Amino deriv., 53; Acetamide, 171
111	3,4,5-Trinitro-1,2-dimethyl-benzene	115				
112	9-Nitrophenanthrene	116–7		199	79	Acetamide, 207–8
113	3-Nitrobenzonitrile	117–8				3-Nitrobenzoic acid, 140
114	1,2-Dinitrobenzene	118	186	301		2-Nitroaniline, 71 1,2-Diaminobenzene, 102
115	2,4′-Dinitrodiphenylmethane	118				2,4′-Dinitrobenzophenone, 197 Diamino deriv., 88–9
116	4,5-Dinitro-1,2-dimethyl-benzene	118(115)				4-Amino deriv., 140 Diamino deriv., 126 Diacetamide, 227–8
118	1,3,5-Trinitronaphthalene	122				Diacetamide, 250 2-Naphthol adduct 146–8
119	1,3,5-Trinitrobenzene	122				Anthracene adduct, 164 2-Naphthol adduct, 155(159) Naphthalene adduct 152.5

(Continued)

TABLE 27B (Continued)
NITRO COMPOUNDS—SOLIDS

	Name of Compound	MP	Data for Derivatives of the Amine Formed by the Reduction of All Nitro Groups			Others
			Benzene-sulfonamide	Benzamide	Picrate	
120	5-Bromo-1-nitronaphthalene	122.5				Amino deriv., 69
						Acetamide, 215
121	2,4,6-Trinitrophenol	122.5				Naphthalene adduct, 149
						Anthracene adduct, 138
						Acetate, 76
122	2,6-Dinitro-1,4-dimethyl-benzene	123				2-Amino deriv., 98
123	4,4′-Dibromo-2-nitro-biphenyl	124				2-Amino deriv., 132
						Acetamide, 192
124	4,5,6-Trinitro-1,3-dimethyl-benzene	125				
125	2,4,6-Tribromo-1-nitrobenzene	125		198		Amino deriv., 122
						Acetamide, 232
126	4-Bromo-1-nitrobenzene	126	136	204	180	4-Bromoaniline, mp 66;
						Acetamide, 168
127	2,2′-Dinitrobiphenyl	128(124)		190–1(di)		Diamino deriv., 81
						Diacetamide, 161
128	1,4-Dinitronaphthalene	131–2		280(di)		Diamino deriv., 120
						Diacetamide, 303–4
129	1,2-Dinitro-3,5-dimethyl-benzene	132				Diamino deriv., 78
130	4-Nitroazobenzene	135			211 (205)	Amino deriv., 126 Acetamide, 144–6
131	2,4′-Dibromo-4-nitro-biphenyl	137				4-Amino deriv., 105
						Acetamide, 195
132	4-Chloro-1,5-dinitro-naphthalene	138				4,8-Dinitro-1-naphthol, 235d
133	2,3,5-Trinitro-1,4-dimethyl-benzene	139				
134	4-Bromo-1,5-dinitro-naphthalene	143				4,8-Dinitro-1-naphthol, 235d
135	2,4-Dinitrostilbene	143–5				Diamino deriv., 119–20
136	1,3-Dinitronaphthalene	144				Diamino deriv., 96
						Diacetamide, 263–5
137	4-Nitrobenzonitrile	145–6				4-Nitrobenzoic acid, 241
138	9-Nitroanthracene	146				Amino deriv., 145–50
						Acetamide, 273–4
139	4-Chloro-1,3-dinitro-naphthalene	146.5				2,4-Dinitro-1-naphthol, 140
140	2,5-Dinitro-1,4-dimethyl-benzene	147(142)				2-Amino deriv., 144–5
						Diamino deriv., 150
141	1,4,5-Trinitronaphthalene	148–9				
142	6-Nitroquinoline	149–50		169		Amino deriv., 114
143	3-Nitroacenaphthene	151–2		209–10	221	Amino deriv., 81–2
						Acetamide, 192–3
144	4-Nitroazoxybenzene	153				Amino deriv., 138
145	3,8-Dinitroacenaphthene	155–6				Diamino deriv., 167–8
146	2-Nitrofluorene	156				Amino deriv., 129
						Acetamide, 191
147	2,2′-Dinitrodiphenylmethane	159				Diamino deriv., 160
						2,2′-Dinitrobenzophenone. 188–9
148	1,6-Dinitronaphthalene	167(162)		265(di)		Diamino deriv., 85–6(77)

(Continued)

TABLE 27B (Continued)
NITRO COMPOUNDS—SOLIDS

	Name of Compound	MP	Data for Derivatives of the Amine Formed by the Reduction of All Nitro Groups			Others
			Benzene-sulfonamide	Benzamide	Picrate	
149	1,8-Dinitronaphthalene	170		311–2(di)		1,3,8-Trinitro deriv., 218 Diamino deriv., 66
150	4-Bromo-1,8-dinitronaphthalene	170				4,5-Dinitro-1-naphthol, 235d(230d)(r.h.)
151	3-Nitrophenanthrene	170–1		213		Amino deriv., 87–8 Acetamide, 200–1
152	4-Iodo-1-nitrobenzene	173		222		Amino deriv., 62 Acetamide, 184
153	1,4-Dinitrobenzene	173	247	300		Naphthalene adduct, 119 Diamino deriv., 140(147) Diacetamide, 304
154	3,3′-Dinitrodiphenylmethane	175				Diamino deriv., 53–4 Diacetamide, 193; 3,3′-Dinitrobenzophenone, 155
155	4-Chloro-1,8-dinitronaphthalene	180				4,5-Dinitro-1-naphthol, 235(230d)r.h.
156	4,4′-Dinitrobibenzyl	180.5				4-Nitrobenzoic acid, 241
157	9-Nitrofluorene	181–2		260–1		Amino deriv., 64(47) Acetamide, 262
158	2,4,6-Trinitro-1,3-dimethylbenzene	183				
159	4,4′-Dinitrodiphenylmethane	183				4,4′-Dinitrobenzophenone, 189 Diamino deriv., 93 Diacetamide, 236–7
160	2-Nitroanthraquinone	185		227–8		Amino deriv., 306 Acetamide, 262
161	2,2′-Dinitrostilbene	196			209	Diamino deriv., 176 Diacetamide, 304
162	3,3′-Dinitrobiphenyl	200				Diamino deriv., 94 Diactamide, 257–8
163	2,5-Dinitrofluorene	207				Diamino deriv., 175 Diacetamide, 279
164	2,2′-Dinitroazobenzene	209–10 (194–5)				Diamino deriv., 134 Diacetamide, 271
165	1,5-Dinitronaphthalene	214				1,4,5-Trinitro deriv., 154 Diamino deriv., 190 Diacetamide, 360
166	1,3,8-Trinitronaphthalene	218				
167	4,4′-Dinitroazobenzene	222–3 (216)				Diamino deriv., 250–1 Mono acetyl, 212
168	1-Nitroanthraquinone	230		255		Amino deriv., 252(243) Acetamide, 218
169	2,7-Dinitronaphthalene	234		267	210	Diamino deriv., 166(159) Diacetamide, 261
170	4,4′-Dinitrobiphenyl	237(240)		203–5 (mono) 352(di)		Diamino deriv., 128 Mono acetamide, 199 Diacetamide, 317
171	1,3-Dinitroanthraquinone	240				Diamino deriv., 290
172	1,6-Dinitroanthraquinone	255–7		275		Diamino deriv., 262 Diacetamide, 295
173	2,7-Dinitroanthraquinone	280(262)		300		
174	4,4′-Dinitrostilbene(*trans*)	288		352		Diamino deriv., 231 Diacetamide, 353

(*Continued*)

TABLE 27B (Continued)
NITRO COMPOUNDS—SOLIDS

	Name of Compound	MP	Data for Derivatives of the Amine Formed by the Reduction of All Nitro Groups			Others
			Benzene-sulfon-amide	Benz-amide	Pic-rate	
175	1,7-Dinitroanthraquinone	295		225		Diamino deriv., 290 Diacetamide, 283
176	1,8-Dinitroanthraquinone	311–2		324		Diamino deriv., 262 Diacetamide, 284
177	1,5-Dinitroanthraquinone	384–5				Diamino deriv., 319 Diacetamide, 317

REFERENCES FOR TABLE 27

Davis, R. B., and Pizzini, L. B., *J. Org. Chem.*, **25**, 1844 (1960). Condensation of Aromatic Nitro Compounds with Acrylacetonitriles. II. Some *p*-Substituted Nitrobenzenes. Lists 29 condensation products with nitro compounds.

Jones, R. L., and Riddick, J. A., *Anal. Chem.*, **28**, 1137 (1956). Volumetric Determination of Primary and Secondary Nitroparaffins.

Kapil, R. S., *J. Org. Chem.*, **25**, 1036 (1960). The Nitration of 4-Nitro-2-iodotoluene. Gives the method of preparation and the melting points of the products.

TABLE 28A
C-NITROSO COMPOUNDS[a]

	Name of Compound	MP		Name of Compound	MP
1	1-Nitroso-4-ethoxybenzene	34	19	1-Nitroso-2-methoxybenzene	103
2	1-Nitroso-4-methoxybenzene	34	20	5-Nitroso-*o*-anisidine	107
3	1-Nitroso-2,4-dimethylbenzene	41	21	1-Nitroso-2-naphthol	110
4	1-Nitroso-3,4-dimethylbenzene	45	22	4-Nitroso-N-methylaniline	118
5	4-Nitrosotoluene	48	23	4-Nitrosophenol	
6	1-Nitroso-3-methoxybenzene	48		(from acetone-benzene, mp 133)	125
7	3-Nitrosotoluene	53	24	5-Nitroso-*o*-cresol	135
8	4-Nitroso N-benzyl-N-ethylaniline	62	25	4-Nitrosodiphenylamine	144
9	4-Nitroso N-ethyl-N-methylaniline	67	26	1-Nitroso-2,6-dimethylbenzene	145
10	Nitrosobenzene	68	27	6-Nitroso-*o*-cresol	155
11	2-Nitrosotoluene	72	28	2-Nitroso-1-naphthol	162*d*
12	4-Nitroso-N-ethylaniline	78	29	Nitrosothymol	162
13	4-Nitroso-N,N-diethylaniline	84(87)	30	5-Bromo-2-nitrosobenzoic acid	173
14	1-Nitroso-2,3-dimethylbenzene	91	31	4-Nitrosoaniline	174
15	4-Nitroso-N,N-dimethylaniline	92.5	32	5-Chloro-2-nitrosobenzoic acid	179
	(forms a HCl salt, mp 177)	(87)	33	4-Nitroso-1-naphthol	194
16	3-Nitrosopyridine	94	34	2-Nitrosobenzoic acid	210
17	1-Nitrosonaphthalene	98	35	1-Nitrosoanthraquinone	224
18	1-Nitroso-2,5-dimethylbenzene	101			

[a] Nitroso compounds may be derivatized by reduction to the corresponding amine.

TABLE 28B
N-NITROSO COMPOUNDS[a]

	Name of Compound	MP		Name of Compound	MP
1	N-Nitrosoformanilide	45–6	6	N-Nitrosodiphenylamine	67
2	N-Nitrosoacetanilide	51*d*	7	N-Nitroso-4-nitrophenylacetamide	72
3	N-Nitrosopropionanilide	52	8	N-Nitroso-4-chlorophenylacetamide	83–4
4	N-Nitrosobromoacetanilide	54–5	9	N-Nitroso-4-bromophenylacetamide	88
5	N-Nitroso-2-chlorophenylacetamide	59			

[a] For others, see *Chemistry of Carbon Compounds*, E. H. Rodd, ed., Vol. III, p. 252, Elsevier, Amsterdam, 1954.

TABLE 29A
PHENOLS—LIQUID

	Name of Compound	Note	MP	BP	Melting Point of Derivatives		
					Recommended		Others
					α-Naph-thyl-urethan	3,5-Dinitro-benzoate	N-Phenyl-urethan
1	2-Chlorophenol	10	7	175.6	120	143	121
2	Phenol	1	41.8	182			
3	o-Cresol	1,10	31	191			
4	2-Bromophenol	10	5	195	129		
5	2-Chloro-4-methylphenol			196			
6	Salicylaldehyde	7		197			133
7	p-Cresol	1	34.8	202.1			
8	m-Cresol		11.5	202.2	128	165.4*	125
9	2-Ethylphenol			207		108	143.5
10	2,4-Dimethylphenol	1	27	211.5*			
11	2-Isopropylphenol	11	16	212			
12	2-Bromo-4-methylphenol			213–4			
13	2-Hydroxyacetophenone	2	28	215			
14	3-Ethylphenol			217			137
15	2-Ethoxyphenol		28	217			
16	Methyl salicylate	12		224			
17	4-Propylphenol		22	232			129
18	Ethyl salicylate	13		234			98–100
19	2-Butylphenol			234–7			
20	4-Isobutylphenol			236			
21	2-Methyl-5-isopropylphenol (carvacrol)			237.8	116	83 (77)	135 (138)
22	Isopropyl salicylate	14		240–2			
23	3-Methoxyphenol			243	129		
24	3-Ethoxyphenol	15		246–7			
25	4-Butylphenol		22	248			115
26	4-Pentylphenol		23	248–53			
27	4-Allyl-2-methoxyphenol (Eugenol)			254.8	122	130.8*	95.5
28	Isobutyl salicylate	16		260–2			
29	2-Methoxy-4-propenylphenol (Isoeugenol)			267.5	150	158.4*	118(cis) 152(trans)
30	Butylsalicylate	17		270–2 (260)			
31	Isopentyl salicylate	18		276–8			

TABLE 29A (Continued)
PHENOLS—LIQUID

Melting Point of Derivatives
Others (Continued)

	p-Xenyl-urethan	N,N-Di-phenyl-urethan	4-Nitro-benzoate	(Mono, di, and tri) Benzoate	(Mono, di, and tri) Acetate	(Mono, di, tri) Bromo Derivative	Aryloxy-acetic Acid	p-Toluene-Sulfonate	2,4-Dinitro-phenyl Ether
1			115			48–9 (mono) 76(di)	145	74	99
2									
3									
4				86		95	143	78	89
5				71–2			108		
6			128		39		132	63–4	
7									
8	164	101	90	55	12	84(tri)	103	51(56)	74
9			57	39			141		
10									
11									
12								121	
13									
14			68	72			77		
15				31					
16			128	92	52				
17				38					
18			107–8	79–80					
19					106				
20							125		
21	166		51			46	151		
22									
23						104(tri)		111–13	87–8
24									114–5
25			68	27			81		
26				51.5			90		
27		108	81	70	30	118(tetra)	81 and 100	85	115
28									
29			109	106	80		94 and 116		128
30									
31									

TABLE 29B
PHENOLS—SOLID

	Name of Compound	Note	MP	BP	Melting Point of Derivatives — Recommended — α-Naphthylurethan	Melting Point of Derivatives — Recommended — 3,5-Dinitrobenzoate	Others — N-Phenylurethan
1	4-Butylphenol	3	22	248			
2	4-Propylphenol	3	22	232			
3	4-Pentylphenol	3	23	248–53			
4	2,4-Dimethylphenol	10	27	211.5*	135	164.4	103
5	2-Ethoxyphenol	19	28	217			
6	2-Hydroxyacetophenone	2	28	215			
7	o-Cresol	10	31	191–2	142	138.4*	142
8	2-Methoxyphenol (Guaiacol)		32	205v.s.	118	141.2*	136
9	3-Bromophenol	10	32	236	108		
10	3-Chlorophenol	10	33	214	158	156	
11	2-Bromo-4-chlorophenol		33–4				
12	2-Nitro-4-methylphenol		36				
13	p-Cresol	10	36	202	146	188.6*	113
14	2,4-Dibromophenol		36	238–9			
15	2,6-Diethylphenol		37–8	219			170–1
16	3-Iodophenol		40			183	138
17	2-Hydroxybenzophenone	2	41				
18	2-Nitro-3-methylphenol		41				
19	Phenol (bp 182)	10	41.8	182	133	145.8*	126
20	Phenyl salicylate	20	42N				112N (242)
21	4-Ethyl-1-naphthol	21	42				
22	4-Chlorophenol	10,22	43	217	166	186	148.5
23	2,4-Dichlorophenol	23	43	209			
24	2-Iodophenol	10	43				122
25	2-Nitrophenol(v.s.)	10	45		113	155	
26	2-Chloro-5-methylphenol(v.s.)	24	46	196			
27	5-Chloro-2-hydroxybiphenyl		46				
28	4-Ethylphenol	10	47	219	128	133	120
29	2,4,6-Trichloro-3-methylphenol		47	265			
30	5-Hydroxy-2-methylaniline	6	47				
31	4-Chloro-2-methylphenol		49	222–5			
32	2,6-Dimethylphenol		49	203	176.5	158.8*	133
33	2,6-Dibromo-4-methylphenol		49(54)				
34	5-Methyl-2-isopropylphenol (thymol)	10	49.7	233	160	103.2	107
35	2-Chloro-3-methylphenol	25	49–50 (55–6)	194 (199)			
36	2-Hydroxy-5-methylacetophenone	2	50				
37	3-Ethyl-1-naphthol	26	51				
38	2-Benzylphenol		54	312			118
39	3-Chloro-4-methylphenol		55	228			
40	4-Hexanoylresorcinol		56	343d			
41	4-Methoxyphenol		56	244			
42	3-Bromo-4-methylphenol		56	245			
43	5-Methyl-2-nitrophenol		56				
44	2-Cyclohexylphenol		56–7				111.5
45	3-Bromo-5-methylphenol		56–7				
46	2,3-Dichlorophenol		56–7	206			

TABLE 29B (Continued)
PHENOLS—SOLID

Melting Point of Derivatives
Others (Continued)

	p-Xenyl-urethan	N,N-Di-phenyl-urethan	p-Nitro-benzoate	(Mono, di, and tri) Benzoate	(Mono, di, and tri) Acetate	(Mono, di, tri) Bromo Derivative	Aryloxy-acetic Acid	p-Toluene-Sulfonate	2,4-Dinitro-phenyl Ether
1									
2									
3									
4	184		102	38			141		102–3
5				31					
6									
7	151	73	94			56(di)	152	55	90
8		118	93	57–8		116(tri)	116	85	97
9				86			108	52.4	
10			99	71			110		75
11				99–100			139–40*		
12			192	102					
13	198	94	98	70		108(tetra)	135	70	93
						49(di)			
14			183	97.5	36	95(tri)	153	120	135
15							67–8		
16			133	72–3	38		115	60–1	
17			124						
18				79	59				
19	173	105	127	69		95(tri)	99	96	70–1
20		144	111	81	99.5				
21									
22		97	171	88	7–8		156	71	126
23				97		68	135	125	119
24				34			135		95
25		114	141	59	41	117(di)	158	83	142
		(119)							
26				31(40)				96	
27				88					
28			81	60			97		
29					35			92–3	
30									
31							115–7		
32	198					79	139.5		
33			141–2	94–5	67				
34	194		70	33		55	149	71	67
35				55–6				96	
36									
37									
38									
39									
40			89–91						
41				87	32		110–12		
42				75					
43				77	48				
44									76–7
45					83				
46						90(di)			

(Continued)

TABLE 29B (Continued)
PHENOLS—SOLID

	Name of Compound	Note	MP	BP	Melting Point of Derivatives		
					Recommended		Others
					α-Naph-thyl-urethan	3,5-Dinitro-benzoate	N-Phenyl-urethan
47	3,5-Dihydroxytoluene	28	56–8				
48	2,4-Dibromo-6-methylphenol		57				
49	2-Hydroxybiphenyl	10	57.5	275			
50	2,3,6-Trichlorophenol		58				
51	2,5-Dichlorophenol(v.s.)		58–9	212			
52	1,2-Naphthalenediol (anh. mp 103–4)		60(hyd.)				
53	4-Chloro-5-methyl-2-isopropylphenol	29	62.4				
54	4-Isopropylphenol		61	223–5			
55	2,3,5-Trichlorophenol		62				
56	3,4,6-Trichloro-2-methylphenol		62				
57	3,4-Dimethylphenol	10,30	62.5 (65)	225	141–2	181.6*	120
58	2,6-Dinitrophenol	31	63–4				
59	1-Acetyl-2-naphthol		64				
60	4-Bromo-3-methylphenol		64				
61	4-Bromo-2-methylphenol		64	235			
62	4-Bromophenol	10,32	64		169	191	140
63	3,5-Dimethylphenol(v.s.)	10	64(68)	220		195.4*	149
64	2-Methyl-1-naphthol	33	64–5				
65	3,4-Dihydroxytoluene		65	252			166(bis)
66	4-Chloro-3-methylphenol(v.s.)	27	66	235	153–4		
67	2,3,6-Trichloro-4-methylphenol		66–7				
68	2,6-Dichlorophenol		67				
69	4-Hexylresorcinol	34	67.5 (69)	335d			
70	3,5-Dichlorophenol(v.s.)		68	233			
71	3,4-Dichlorophenol		68	252			
72	2,4,5-Trichlorophenol		68				
73	2,4,6-Trichlorophenol		68	245			
74	2-Bromo-4,6-Dichlorophenol		68–9	268d			
75	2,3,4,6-Tetrachlorophenol	35	69				
76	2,4,6-Trimethylphenol		69	220			143
77	2-Methyl-6-nitrophenol	36	70				
78	Methyl 3-hydroxybenzoate		70	280			115–6
79	2-Ethyl-1-naphthol	37	70				
80	2,4,5-Trimethylphenol		71	232			110
81	2,4-Di-iodophenol		72				
82	1-Chloro-2-naphthol	38	72				
83	5-Chloro-2-methylphenol		73–4				
84	2,5-Dibromophenol		73–4				
85	1,2-Dihydroxy-3,5-dimethylbenzene		73–4				
86	2,5-Dimethylphenol	10	74.5	212	173	137.2*	166
87	2,3-Dimethylphenol	10	75				173.5
88	1,3-Dibromo-2-naphthol		75				
89	8-Hydroxyquinoline	8	75–6				
90	4-Dimethylaminophenol		76				
91	3-Chloro-4-hydroxybiphenyl		76–7 (80)				
92	2,3,4-Trichloro-6-methylphenol		77				
93	3-Hydroxybiphenyl		78 (75)				

TABLE 29B (Continued)
PHENOLS—SOLID

Melting Point of Derivatives

Others (Continued)

	p-Xenyl-urethan	N,N-Di-phenyl-urethan	p-Nitro-benzoate	(Mono, di, and tri) Benzoate	(Mono, di, and tri) Acetate	(Mono, di, tri) Bromo Derivative	Aryloxy-acetic Acid	p-Toluene Sulfonate	2,4-Dinitro-phenyl Ether
47									
48			136–7	62					
49				76	62–3			65	113–4
50									
51				69	100.5(di)				
					206(tri)				
52				106(di)					
53									
54				71–2					
55				103					
56				110					
57	183			58.5		171(tri)	162.5		105–6
58								135	
59				85–6					
60				84				85	
61				68					
62		99	180	102	21.5	95(tri)	157	94	141
63	150		109	24		166(tri)	111(81)	83	100–1
64				94–5	81–2				
65	193			58(di)			58(di)		
66				86				98	112
67				89	37–8				
68				75					
69									
70				55	38	189(tri)		116*	
71									
72				93			157		
73		143	105–6	75.5			182–6		136
74								82–3	140–1
75				108	65–6				
76				62		158(di)	142		
77				42				66	
78									
79									
80	196			63	34.5	35	132		
81				98	71			165–7*	
82					42–3				
83				53–4					
84								110	
85					161(di)				
86	162		87	61		178(tri)	118		
87							187		
88					102				
89			174–5		118–20			115	165d
90					78			130	
91				110–1					109–11
				(95–7)					
92					45				
93				60–1					
				(57–8)					

(Continued)

TABLE 29B (Continued)
PHENOLS—SOLID

	Name of Compound	Note	MP	BP	Melting Point of Derivatives		
					Recommended		Others
					α-Naph-thyl-urethan	3,5-Dinitro-benzoate	N-Phenyl-urethan
94	4-Chloro-2-iodophenol		78				128
95	3-(Diethylamino)-phenol	8,39	78	276–80			
96	4-Bromo-2,6-dinitrophenol		78				
97	4-Methyl-3-nitrophenol		79				
98	2,3,4,6-Tetramethylphenol		79–81				178–9
99	5-Bromo-2-methylphenol		80				
100	3,5-Dibromophenol	40	81				
101	4-Chloro-2,6-dinitrophenol		81				
102	4-Hydroxy-3-methoxybenzaldehyde (vanillin)	7	81	285			
103	4-Methyl-2-naphthol		81–2				
104	2,4,6-Tribromo-3-methylphenol	10	81–2(84)				
105	3,4-Di-iodophenol		83				
106	2,3,4-Trichlorophenol		83.5				
107	4-Benzylphenol		84				
108	4-Methyl-1-naphthol		84–5				
109	2,4-Dimethyl-1-naphthol	41	84–5				
110	1-Bromo-2-naphthol	42	85				
111	3-Bromo-2-naphthol	43	85				
112	2-Hydroxybenzyl alcohol		85–6				
113	3-Chloro-2-methylphenol		86	225			
114	6-Methyl-2,4-Dinitrophenol		86.5				
115	4-Chloro-2-nitrophenol	44	86–7				
116	2,4,5-Tribromophenol		87				
117	2-Amino-6-methylphenol	6	89				
118	4-Bromo-2-nitrophenol	45	89				
119	3,4,5-Tribromo-2-methylphenol		89				
120	2,4,6-Tribromo-3-nitrophenol		89–90				
121	2,4,5-Tribromo-6-methylphenol		91				
122	4-Chloro-2,6-dibromophenol		92				
123	4-*tert*-Pentylphenol		92–3 (96)	260–5			108
124	4-Iodophenol		93–4				148
125	1-Naphthol	10	94	280	152	217.4*	178
126	6-Chloro-1-naphthol		94				
127	3-Bromo-4-hydroxybiphenyl		94–5				
128	1,3,6-Naphthalenetriol	46	95				
129	5-Chloro-2-hydroxybenzophenone	47	95				
130	2,4,6-Tribromophenol	10	95		153	174	
131	2-Naphthyl salicylate	48	95.5				268
132	2,3,5-Trimethylphenol		96	233			
133	3-Hydroxyacetophenone	2	96				
134	2-Methyl-4-nitrophenol		96				
135	2,4,5-Trinitrophenol	49	96				
136	3-Nitrophenol	10	97		167	159	129
137	5-Methyl-1-naphthol		97–8				
138	6-Ethyl-2-naphthol	50	98				
139	2,3-Dihydroxyacetophenone	2	98				

TABLE 29B (Continued)
PHENOLS—SOLID

Melting Point of Derivatives

Others (Continued)

	p-Xenyl-urethan	N,N-Di-phenyl-urethan	p-Nitro-benzoate	(Mono, di, and tri) Benzoate	(Mono, di, and tri) Acetate	(Mono, di, tri) Bromo Derivative	Aryloxy-acetic Acid	p-Toluene Sulfonate	2,4-Dinitro-phenyl Ether
94				88 (84)	57				
95				22–3					
96				154	111			136	
97								91	
98				71–2					
99				41					
100				77	53				
101									
102				78	102		187	115	131
103				117–8					
104				84–5	68			113–4	
105				123					
106				141					
107				87					75–6
108				81					
109									
110					56				
111					94				
112				51(di)			120		
113									
114				135 (132)	95–6				
115					47–8				
116				99					
117									
118					75				
119					106–7				
120								146–7	
121				133	76–7				
122								107–8	145–6
123				61					
124		127		119	32		156	99	156
125	190		143 (140)	56	48–9	105 (2,4-di)	193.5	89	128
126					47				
127				93–4	74–5				
128					112–3(tri)				
129									
130		153	153	81	82(87) 136	120(tetra)	200	113	137–8
131									
132				50					
133									
134				128				107	
135									
136			174	95	55–6	91(di)	156	112–3	136
137				77–8					
138									
139					109(di)				

(Continued)

TABLE 29B (Continued)
PHENOLS—SOLID

	Name of Compound	Note	MP	BP	Melting Point of Derivatives Recommended		Others
					α-Naph-thyl-urethan	3,5-Dinitro-benzoate	N-Phenyl-urethan
140	3,5-Dibromo-2-methylphenol		98–101				
141	2,5-Di-iodophenol		99				
142	2-Hydroxy-3-naphthaldehyde	7	99–100				
143	5-Chlorosalicylaldehyde	7	99.5				
144	4-*tert*-Butylphenol	51	100	237	110		148.5
145	3,4,5-Trichlorophenol	52	101(91)				
146	2-Acetyl-1-naphthol		102	325			
147	2,3,6-Tribromo-4-methylphenol		102				
148	1-Nitro-2-naphthol		103				
149	3,3'-Dihydroxydiphenylmethane		103				
150	3,5-Di-iodophenol		104				
151	4-Chloro-2-naphthol		104				
152	3-Hydroxybenzaldehyde	7	104				
153	5-Methyl-2-naphthol	53	104–5				
154	1,2-Dihydroxybenzene		105	245.6	175	152	169
155	5-Hydroxyindole	6,54	105				
156	3-Chloro-2,4,6-tribromophenol		105–6				
157	5-Bromosalicylaldehyde	7	106				
158	Chlorohydroquinone		106				
159	4-Chlororesorcinol		106–8				
160	2-Hydroxypyridine	4,55	107				
161	2,4-Dichloro-1-naphthol	56	107–8				
162	2,5-Dinitrophenol	57	108				
163	1,2-Naphthalenediol		108				
164	3,5-Dihydroxytoluene	28	108		160	190	154
165	4-Chloro-2,6-di-iodophenol		109				
166	4-Hydroxyacetophenone	2	109				
167	1-Nitroso-2-naphthol	58	109.5				
168	2,4,6-Trinitro-3-methylphenol	59	109–10				
169	2,2'-Dihydroxybiphenyl		110	326			145
170	Resorcinol (1,3-Dihydroxybenzene)		110	275.9		201	164
171	Bromohydroquinone		110				
172	7-Methyl-1-naphthol	60	110–1				
173	6-Chloro-3-hydroxybenzaldehyde	7	111				
174	2-Chloro-4-nitrophenol	61	111				
175	1-Methyl-2-naphthol	62	111				
176	2,4-Dibromo-1-naphthol		111				
177	2,4,6-Tribromoresorcinol		112				
178	2,4-Diacetyl-3-methylphenol	2	112				
179	4-Hydroxy-3,5-dimethoxybenz-aldehyde	7	112–3				
180	2,2'-Dihydroxy-3,3'-dimethyl-biphenyl		113				
181	4,4'-Dihydroxy-2,2'-dimethyl-biphenyl		114				

TABLE 29B (Continued)
PHENOLS—SOLID

Melting Point of Derivatives
Others (Continued)

	p-Xenyl-urethan	N,N-Di-phenyl-urethan	p-Nitro-benzoate	(Mono, di, and tri) Benzoate	(Mono, di, and tri) Acetate	(Mono, di, tri) Bromo Derivative	Aryloxy-acetic Acid	p-Toluene-Sulfonate	2,4-Dinitro-phenyl Ether
140				91–3					
141					70				
142									
143									
144				81–2		50	86	109–10	108–10
145				120					
146				128	107.5				
147					77				
148					61				
149					57–8(di)				
150				93	79				
151					56				
152				38			148		
153				107–8					
154			159 (mono) 169(di)	131 (mono) 84(di)	57–8 (mono) 65(di)	193(tetra)			136–8
155									
156									
157									
158					62(mono) 72(di)				
159				66(di)	46–7(di)				
160				42				53	
161					74–6				
162									
163					109(di)				
164	196		214	88(di)	25(di)	104(tri)	217		153–4
165									
166									
167									
168				140	135				
169				101	95(di)	188(di)		190	
170		130	182(di) (175)	135–6 (mono) 117(di)		112(tri)	195	80–1(di)	194
171					72(di)	186(di)			
172					39–41				
173									
174									
175				116–7	66				
176					92–3				
177				120(mono)	114(mono) 108(di)				
178									
179									
180				147(di)					
181				127(di)	75(di)				

(Continued)

TABLE 29B (Continued)
PHENOLS—SOLID

	Name of Compound	Note	MP	BP	Melting Point of Derivatives		
					Recommended		Others
					α-Naphthylurethan	3,5-Dinitrobenzoate	N-Phenylurethan
182	2-Bromo-4-nitrophenol	63	114				
183	4-Nitrophenol	10	114		150–1	186	156
184	2,4-Dinitrophenol		114				
185	2,3,5-Tri-iodophenol		114				
186	Ethyl 4-hydroxybenzoate		115	298			
187	4-Hydroxy-3-methoxyacetophenone	2	115				
188	4-Chloro-3,5-dimethylphenol		115–6				
189	4-Bromo-1,5-naphthalenediol	65	116				
190	3,4-Dihydroxyacetophenone	2	116				
191	4-Hydroxybenzaldehyde	7	116–7				
192	2,6-Dihydroxytoluene		117	271			
193	1,3,5-Trihydroxybenzene (phloroglucinol)(anh.)		117				
194	2,4′-Dihydroxydiphenylmethane		117				
195	3-Bromo-1,2-naphthalenediol		117				
196	4,6-Dibromo-2-nitrophenol		117.5				
197	2-Methyl-5-nitrophenol		118				
198	2,3,5,6-Tetramethylphenol		118	249			
199	6-Bromo-2,4-dinitrophenol		118–9				
200	2,3,6-Trinitrophenol	66	119				
201	2,6-Dihydroxytoluene		119–20	264			
202	3,4,5-Trihydroxytoluene		120s				
203	2,2′-Dihydroxy-4,4′-dimethylbiphenyl		120				
204	4-Nitro-2-naphthol		120				
205	9-Hydroxyanthracene	67	120(r.h.)				
206	4-Chloro-1-naphthol		120–1				
207	4-Chloro-3-hydroxybenzaldehyde	7	121				
208	4-Chloro-3,5-dibromophenol		121				
209	3-Aminophenol	6	122			179	
210	2,4,6-Trinitrophenol	68	122				
211	4-Nitroresorcinol		122				
212	4-Bromo-2-naphthol	69	122				
213	4-Benzyloxyphenol		122				
214	4,6-Dichloro-2-nitrophenol	70	122–3				
215	2-Naphthol	10	123	286	157	210.2*	156
216	1,4-Dichloro-2-naphthol		123–4				
217	3,3′-Dihydroxybiphenyl		123–4				
218	1,3-Naphthalenediol	71	124				
219	2,5-Dihydroxytoluene		124				
220	1,3-Dihydroxy-4,6-dimethylbenzene		125	276–9			
221	2,6-Dichloro-4-nitrophenol		125d				
222	2,5-Dihydroxybenzophenone	2,72	125				
223	4-Hydroxybenzyl alcohol		125				
224	4-Benzyl-1-naphthol		125–6				
225	5-Nitrosalicylaldehyde	7	126				
226	3,5-Dinitrophenol		126(122)				
227	7-Chloro-2-naphthol		126.5				
228	3,4,6-Tribromo-2-nitrophenol	73	127				

TABLE 29B (Continued)
PHENOLS—SOLID

	Melting Point of Derivatives							
	Others (Continued)							
	p-Xenyl-urethan	N,N-Di-phenyl-urethan	p-Nitro-benzoate	(Mono, di, and tri) Benzoate	(Mono, di, and tri) Acetate	(Mono, di, tri) Bromo Derivative	Aryloxy-acetic Acid	p-Toluene Sulfonate	2,4-Dinitro-phenyl Ether
182				131–2	62				
183		112	159	142	81–2	142 (2,6-di)	187	97	120
184			139	132	72	118(6)		121	248
185					123				
186				94					
187				106	58				
188					49				
189					138(di)				
190				118(di)	91(di)				
191				90			198		
192				105–6(di)					
193			283	185	104	151(tri)			
194				108(di)	70(di)				
195					160(di)				
196					89			140	
197					74			123–4	
198						118(4)			
199				94	105			157	
200									
201				105–6(di)					
202					99(tri)				
203				148(di)					
204								122	
205					134				
206					44				
207									
208				132					
209			143						
210					76				
211				110(di)	90–1(di)				
212					61				
213									129–30
214					77				
215		141	169	107	72	84	154	125	95
216					90–1				
217				92(di)	82.5(di)				
218					56				
219				119–20 (di)	92(mono) 49(di)		153		
220					45(di)				
221									
222									
223					84(mono) 75(di)				
224				103	87–8				
225									
226					127				
227					104.5				
228					118				

(Continued)

TABLE 29B (Continued)
PHENOLS—SOLID

	Name of Compound	Note	MP	BP	Melting Point of Derivatives		
					Recommended		Others
					α-Naph-thyl-urethan	3,5-Dinitro-benzoate	N-Phenyl-urethan
229	2,4-Dibenzoylresorcinol	74	127–8				
230	2-Nitro-1-naphthol	75	127–8				
231	4-Iodo-2-naphthol	76	128.5				
232	4-Bromo-1-naphthol		129				
233	3-Methyl-4-nitrophenol		129				
234	6-Bromo-2-naphthol	77	129–30				
235	8-Nitro-1-naphthol		130				
236	Methyl 4-hydroxybenzoate		131				
237	4-Amino-2-nitrophenol	6	131				
238	2,4-Dibromo-1,3-naphthalenediol		131				
239	5-Chloro-1-naphthol		131.5				
240	2,4,5-Trihydroxytoluene		131–2				
241	4-Cyclohexylphenol		132			168*	145.5
242	1,2,3-Trihydroxybenzene (pyrogallol)		133	309		205(tri)	173(tri)
243	o-Cumaraldehyde	7	133				
244	3,4-Dinitrophenol	78	134				
245	2-Amino-4-methylphenol	6	135				
246	4-Hydroxybenzophenone	2	135				
247	1,4-Dimethyl-2-naphthol		135–6				
248	Trichlorohydroquinone	79	136				
249	Trichlorophloroglucinol	80	136				
250	2,4-Dinitro-1-naphthol		138				
251	1,6-Naphthalenediol	81	138				
252	2-Amino-4-chlorophenol	6	138				
253	1,4,6-Naphthalenetriol		138–40				
254	3-Chloro-4-hydroxybenzaldehyde	7	139 (132–4)				
255	4-Phenyl-1-naphthol		140				
256	2-Methyl-1,3-naphthalenediol		140				
257	2-Chloro-3-hydroxybenzaldehyde	7	140				
258	1,2,4-Trihydroxybenzene		140.4				
259	1,8-Naphthalenediol	82	142				
260	2,5-Dihydroxy-3-dodecyl-1,4-benzoquinone	2	143	290	147–8		232–3
261	2-Methyl-5-isopropylhydroquinone	83	143				
262	3,4-Phenanthrenediol	84	143				
263	2,6-Dibromo-4-nitrophenol		144				
264	3-Amino-4-methylphenol	6	144				
265	2,3-Dinitrophenol	85	144				
266	2,4-Dihydroxybenzophenone	2	144(146)				
267	8-Nitro-2-naphthol	86	144–5				
268	3,4-Dihydroxybiphenyl		145				
269	3,4-Dihydroxybenzophenone	87	145(134)				
270	2,4,6-Tri-iodoresorcinol		145				
271	2,4-Dihydroxyacetophenone	2	147				
272	2-Methyl-3-nitrophenol		147				
273	2,4-Dinitroresorcinol		147.8				
274	2-Chloro-4-hydroxybenzaldehyde	7	147–8				
275	4-Hydroxypropiophenone	2	148				

TABLE 29B (Continued)
PHENOLS—SOLID

Melting Point of Derivatives

Others (Continued)

	p-Xenyl-urethan	N,N-Diphenyl-urethan	p-Nitro-benzoate	(Mono, di, and tri) Benzoate	(Mono, di, and tri) Acetate	(Mono, di, tri) Bromo Derivative	Aryloxy-acetic Acid	p-Toluene Sulfonate	2,4-Dinitro-phenyl Ether
229									
230					118				
231					59				
232					51				
233				74	34				
234					103				
235									
236				135	85				
237									
238					125(di)				
239					53				
240					114–5(tri)				
241			137	118.5	35				
242		212 (tri)	230 (tri)	140(mono) 108(di)	110(di) 165(tri)	158(di)	198		
243									
244									
245									
246				115 (94–5)	81				
247				124–5	77–8				
248					153(di)				
249					167–8(tri)				
250				174					
251				103–4	73				
252									
253					94–5				
254									
255				73–4					
256					118				
257									
258				120(tri)	97(tri)				
259				175(di)	155(di)				
260				97–8	54				
261					141–2(di)				
262					159(di)				
263					181			128–9	
264									
265									
266					78(di)				
267					101–2				
268					78(di)				
269				95(di)					
270					170(di)				
271									
272								94	
273							155	126–7	
274				97	52				
275									

(*Continued*)

TABLE 29B (Continued)
PHENOLS—SOLID

	Name of Compound	Note	MP	BP	Melting Point of Derivatives		
					Recommended		Others
					α-Naphthylurethan	3,5-Dinitrobenzoate	N-Phenylurethan
276	9,10-Phenanthrenediol		148				
277	3,5-Dihydroxyacetophenone	2	148				
278	3,4,5-Phenanthrenetriol		148				
279	2-Amino-1-naphthol	6	150				
280	2-Amino-3-methylphenol	6	150				
281	2,4'-Dihydroxybenzophenone	2	150				
282	9-Phenanthrol	88	153				
283	2,2'-Dihydroxy-5,5'-dimethylbiphenyl		153–4				
284	1,2,4-Naphthalenetriol		154				
285	5-Hydroxy-1,4-naphthoquinone	2	154				
286	4-Amino-3-nitrophenol	6	154				
287	3,5-Stilbenediol		155–6				
288	2,6-Dihydroxybenzaldehyde	7	155–6				
289	1,4,5-Trichloro-2-naphthol		157–8				
290	7-Amino-1-naphthol	6	158				
291	4,4'-Dihydroxydiphenylmethane		158				
292	Salicylic acid	9	158.3				
293	2,4,6-Tri-iodophenol		159			181	
294	4,4'-Dihydroxy-3,3'-dimethylbiphenyl		160–1				
295	2,3-Naphthalenediol	89	161(164)				
296	5-Amino-2-methylphenol	6	161				
297	4,4'-Dihydroxytriphenylmethane		161				
298	3-Hydroxyphthalic acid	9	161d				
299	1,2-Anthracenediol		161–2				
300	2-Amino-5-methylphenol	6	162				
301	1,3,4-Trichloro-2-naphthol		162				
302	2,4'-Dihydroxybiphenyl		162	342			
303	2,6-Dihydroxy-1,4-dimethylbenzene		163				
304	4-Nitro-1-naphthol	90	164				
305	2,2'-Dihydroxy-6,6'-dimethylbiphenyl		164				
306	2,6-Dibromohydroquinone		164				
308	4-Hydroxybiphenyl	10	165	305–8			167.5
309	1,2,3,5-Tetrahydroxybenzene	91	165				
310	2,4,6-Trihydroxybenzophenone	92	165				
311	3,4-Diaminophenol	6	167–8				
312	2,4-Dinitrosoresorcinol		168				
313	2,3,4-Trichloro-1-naphthol		168				
314	4-Hydroxyacetanilide		169				
315	2-Amino-4,6-dinitrophenol (Picramic acid)	6	169				
316	2-Phenanthrol	93	169				
317	5-Amino-1-naphthol	6	170d				
318	3,3'-Dihydroxybenzophenone	2	170				
319	2-Methyl-1,4-naphthalenediol (β, mp 60)	94	170N				
320	2,4-Dihydroxyazobenzene		170				
321	5-Nitro-1-naphthol		171				
322	Hydroquinone (1,4-Hydroxybenzene)		172	286		317	224(di)

TABLE 29B (Continued)
PHENOLS—SOLID

Melting Point of Derivatives

Others (Continued)

	p-Xenyl-urethan	N,N-Di-phenyl-urethan	p-Nitro-benzoate	(Mono, di, and tri) Benzoate	(Mono, di, and tri) Acetate	(Mono, di, tri) Bromo Derivative	Aryloxy-acetic Acid	p-Toluene Sulfonate	2,4-Dinitro-phenyl Ether
276				216–7(di)	202(di)				
277					91–2(di)				
278					138(tri)				
279									
280									
281					84–5				
282					77				
283					88(di)				
284					134–5(tri)				
285					154–5				
286									
287				150–1(di)	100–1(di)				
288									
289					129				
290									
291				156	69–70				
292			205		135		191		
293				137	156				
294				185	131(di) (136)				
295				235(di)					
296									
297					109–10(di)				
298				118	114–6				
299					157(di)				
300									
301					134				
302					94(di)				
303					69(di)				
304									
305									
				136(di)	87(di)				
306					117(di)				
308				150(121)	88–9			177	118
309									
310				125–6(tri)					
311									
312				182–4	119–20				
313					123–4				
314					150–1				
315									
316									
317									
318				101–2	89–90				
319				181	113(di)				
320					104(di)				
321				109	114				
322	230		258	199(di)	123(di)	186(di)	250	159(di)	243–6d

(Continued)

TABLE 29B (Continued)
PHENOLS—SOLID

	Name of Compound	Note	MP	BP	Melting Point of Derivatives		
					Recommended		Others
					α-Naphthylurethan	3,5-Dinitrobenzoate	N-Phenylurethan
323	2,2′-Dihydroxyazobenzene		172				
324	2,3,4-Trihydroxyacetophenone	2,95	173				
325	3,5-Dinitrosalicylic acid	9	173–4				
326	2-Aminophenol	6	174				
327	1,6,7-Naphthalenetriol		175				
328	1,8-Dihydroxy-2-methylanthraquinone		175				
329	4-Amino-2-methylphenol	6	175				
330	1,2-Dihydroxy-4-nitrobenzene		176				
331	1,8,9-Anthracenetriol		176–7 (178–80)				
332	1,7-Naphthalenediol		178(181)				203–4
333	1,2-Phenanthrenediol		178				
334	4-Amino-3-methylphenol	6	178–9				
335	2,4,6-Trinitroresorcinol (styphnic acid)		179–80				
336	2,5-Phenanthrenediol		180				
337	9,10-Anthracenediol		180				
338	1,2,5,8-Naphthalenetetraol		180d				
339	4-Hydroxynaphthaldehyde	7	181				
340	6-Nitro-1-naphthol		181–2				
341	4-Benzylideneaminophenol		183 (185–6)				
342	4-Aminophenol	6,96	184s				
343	4-Amino-2-naphthol	6	185				
344	6-Bromo-1,4-dihydroxy-9,10-anthraquinone		185.5				
345	1,5-Dinitro-2-naphthol	97	187d				
346	1,2,6-Naphthalenetriol		188				
347	Pentachlorophenol		190				
348	2,7-Naphthalenediol	98	190				
349	1,4,5,8-Naphthalenetetraol		190				
350	5-Amino-2-naphthol	6	191				
351	1,4-Naphthalenediol		192				
352	5-Amino-1-naphthol	6	192				
353	1-Hydroxy-9,10-anthraquinone	2	193(200)				
354	1,8-Dihydroxyanthraquinone	2,99	193				
355	4-Amino-2,6-dibromophenol	6	193–4				
356	2,4,5,6-Tetrabromo-3-methylphenol		194				
357	3,4-Dihydroxyphenylpropenoic acid	9,100	195d				
358	1,6-Dinitro-2-naphthol		195d				
359	1,2,7-Naphthalenetriol		197				
360	1,8-Dinitro-2-naphthol	101	198d				
361	2,3,5,6-Tetrabromo-4-methylphenol		199				
362	2-Hydroxyquinoline (anh.)	8	199				
363	2,5-Dihydroxybenzoic acid	102	199–200				
364	6-Amino-1-naphthol	6	199.5				
365	Hexahydroxybenzene		200d				
366	3-Hydroxybenzoic acid	9	200				

TABLE 29B (Continued)
PHENOLS—SOLID

| | Melting Point of Derivatives | | | | | | | |
| | Others (Continued) | | | | | | | |
	p-Xenyl-urethan	N,N-Di-phenyl-urethan	p-Nitro-benzoate	(Mono, di, and tri) Benzoate	(Mono, di, and tri) Acetate	(Mono, di, tri) Bromo Derivative	Aryloxy-acetic Acid	p-Toluene Sulfonate	2,4-Dinitro-phenyl Ether
323					150(di)				
324					107–8(di)				
					85(tri)				
325				163					
326									
327					143–4(tri)				
328					205				
329									
330				156(di)	98(di)				
331					209–10(tri)				
332			182–3	101.5(di) (113–5)	108(di)				
333					147(di)				
334									
335									
336					144(di)				
337				292(di)	260(di)				
338					202(tetra)				
339									
340					121				
341				144	92				
342									
343									
344									
345					220.5(di)				
346					262(tri)				
347				164–5 (159)	149–50		196	145	
348		176		139(di)	136(di)		149	150	
349					277–9(tetra)				
350									
351				169(di)	128–30(di)				
352									
353					172				
354					231–2				
355									
356				153–4	165–6				
357									
358									181
359					181–2				
360									
361					156				
362									
363					118–9N				
364									
365				313(hexa)	203(hexa)				
366					131		206		

(Continued)

TABLE 29B (Continued)
PHENOLS—SOLID

	Name of Compound	Note	MP	BP	Melting Point of Derivatives		
					Recommended		Others
					α-Naphthylurethan	3,5-Dinitrobenzoate	N-Phenylurethan
367	Methyl 3,4,5-trihydroxybenzoate		200				
368	7-Amino-2-naphthol	6	201(208)				
369	1,4-Dihydroxyanthraquinone	103	201–2				
370	2-Amino-5-nitrophenol	6	201–2				
371	2-Hydroxyacetanilide		201–3				
372	2,8-Phenanthrenediol		202				
373	2,5-Dihydroxyacetophenone	2	202				
374	4-Hydroxyphthalic acid	9	204–5				
375	4-Hydroxy-3,5-dimethoxybenzoic acid		205				
376	4-Hydroxycoumarin		206(232)				
377	2,3,5-Trihydroxyacetophenone	2	206–7				
378	6-Chloro-3-hydroxy-p-toluic acid	9	206–7				
379	3,3'-Dihydroxyazobenzene		207				
380	8-Amino-2-naphthol	6	207d				
381	2,3,4,5-Tetrabromo-6-methylphenol		208				
382	4,4'-Dihydroxybenzophenone	2,104	210				
383	2,5-Dihydroxybenzoquinone		211 (215d)				
384	6-Amino-2-naphthol	6	213d				
385	(2-Hydroxyphenyl)propenoic acid	9	214				
386	2,4,6-Trihydroxytoluene		214–6				
387	4-Hydroxybenzoic acid	9	215				
388	4-Chloro-2,3,5,6-tetrabromophenol		215				
389	2,6-Naphthalenediol		215(222)				
390	4,4'-Dihydroxyazobenzene		216				
391	1,2,4,5-Tetrahydroxybenzene		215–20				
392	2,4-Dihydroxybenzoic acid	9	216d				
393	2,2'-Dihydroxy-1,1'-binaphthyl	105	218				
394	1,3,5-Trihydroxybenzene (dihydrate)		218r.h.		162		191(tri)
395	1,1'-Dihydroxy-2,2'-binaphthyl		220				
396	3,5-Dihydroxypyrene		220d				
397	2-Propylphenol		220(226)				111
398	3,6-Phenanthrenediol		221				
399	1,4-Dihydroxy-2,3-dimethylbenzene		221				
400	3-Hydroxy-2-naphthoic acid	9	222				
401	2,6-Naphthalenediol		222				
402	2,4,6-Trihydroxyacetophenone		222				
403	1,8-Anthracenediol		225				
404	1,2,3,4-Naphthalenetetraol		225				
405	Pentabromophenol		225–6				
406	5-Nitrosalicylic acid	9	229–30				
407	4-Chloro-1,2,3-trihydroxyanthraquinone		233				
408	2,6-Phenanthrenediol		234				
409	3-Amino-2-naphthol	6	235				

TABLE 29B (Continued)
PHENOLS—SOLID

Melting Point of Derivatives — Others (Continued)

	p-Xenyl-urethan	N,N-Di-phenyl-urethan	p-Nitro-benzoate	(Mono, di, and tri) Benzoate	(Mono, di, and tri) Acetate	(Mono, di, tri) Bromo Derivative	Aryloxy-acetic Acid	p-Toluene Sulfonate	2,4-Dinitro-phenyl Ether
367				139(tri)	120(tri)				
368									
369					200(di) and 207–8(di)				
370									
371				140					
372					125(di)				
373				113(di)	68(di)				
374									
375				229–32	191				
376					103				
377					106–7(tri)				
378					146				
379				188(di)	144(di)				
380									
381					154				
382					156(152)				
383					150–2				
384									
385					154–5				
386					76(tri)				
387					187		278		
388				203					
389				215(di)	175(di)				
390				210–2(di) and 249–51(di)	198–9(di)				
391					226–7(tetra)				
392									
393				204(mono) 160(di)	109(di)				
394		283		174(tri)	105(tri)	151(tri)			
395					169(di)				
396					155(di)				
397									
398					125(di)				
399				182(di)					
400					184–6				
401				215(di)	175(di)				
402				117–8(tri)	103(tri)				
403					184(di)				
404					220(tetra)				
405					197				
406									
407									
408					187(tri)				
409					122–3(di)				

(Continued)

TABLE 29B (Continued)
PHENOLS—SOLID

	Name of Compound	Note	MP	BP	Melting Point of Derivatives		
					Recommended		Others
					α-Naphthylurethan	3,5-Dinitrobenzoate	N-Phenylurethan
410	5-Hydroxy-1-naphthoic acid		235				
411	4,8-Dinitro-1-naphthol	106	235d				
412	2,3,5,6-Tetrachlorohydroquinone		236–7				
413	1,2,8-Trihydroxy-9,10-anthraquinone		239–40				
414	3,8-Phenanthrenediol		247				
415	4,4'-Dihydroxydiphenylsulfone		247–9				
416	3-Hydroxy-1-naphthoic acid	9	248				
417	1,5-Naphthalenediol	107	258(265)				
418	1,2,4-Trihydroxy-9,10-anthraquinone		259				
419	Phenolphthalein		265*				135
420	1,5-Dihydroxyanthracene		265d				
421	1,3-Dihydroxy-4-methylanthraquinone		265–6				
422	2-Bromo-1,4-dihydroxy-9,10-anthraquinone		265–8				
423	1,2,5-Trihydroxy-9,10-anthraquinone		273–4				
424	4,4'-Dihydroxybiphenyl		274–5				
425	1,6-Dihydroxyanthraquinone	108	276				
426	1-Amino-2-naphthol	6	276d				
427	1,5-Dihydroxyanthraquinone		280				
428	Chloranilic acid	109	283–4				
429	4,4'-Stilbenediol		284				
430	Bis-hydroxycoumarin	110	288–9				
431	Alizarin	111	290				
432	1,7-Dihydroxyanthraquinone		293				
433	4,4'-Dihydroxy-1,1'-binaphthyl		300 (266–8)				
434	2-Hydroxy-9,10-anthraquinone		305				
435	1,2,3-Trihydroxy-9,10-anthraquinone		313–4d				
436	3,8-Dihydroxypyrene		330				
437	1,2,7-Trihydroxy-9,10-anthraquinone	112	369				

NOTES FOR TABLES 29A AND 29B

1. See Table 29B.
2. See Table 21B.
3. See Table 29A.
5. See Table 6B.
6. See Table 9B.
7. See Table 7B.
8. See Table 10B.
9. See Table 5B.
10. See *J. Org. Chem.*, **22**, 828 (1957) for the preparation of *p*-phenylazobenzoates as derivatives of phenols. The melting point of the numbered (in parentheses) phenols follows: *liquids*: (1) 120–1; (3) 110–1; (4) 126–7: *solids*: (4) 110–13; (7) 110–1; (9) 125–6; (10) 127–8; (13) 134–5; (19) 148–50; (22) 153–4; (24) 126–7; (28) 117–8; (35) 85–88; (48) 141–44; (56) 104–7; (61) 167–8; (62) 105–6; (85) 96–7; (103) 130–2; (123) 118–9; (128) 116–9; (134) 160–2; (181) 203–6; (214) 190–3; (307) 213–4.
11. 4,6-Dinitro derivative, mp 53.
12. Acetyl derivative, mp 52; 3,5-dinitro derivative, mp 126–7.
13. 3,5-Dinitro derivative, mp 92–3.
14. 3,5-Dinitro derivative, mp 101–2.
15. Picrate, mp 105–6.
16. 3,5-Dinitro derivative, mp 72–3.

TABLE 29B (Continued)
PHENOLS—SOLID

Melting Point of Derivatives
Others (Continued)

	p-Xenyl-urethan	N,N-Di-phenyl-urethan	p-Nitro-benzoate	(Mono, di, and tri) Benzoate	(Mono, di, and tri) Acetate	(Mono, di, tri) Bromo Derivative	Aryloxy-acetic Acid	p-Toluene-Sulfonate	2,4-Dinitro-phenyl Ether
410				241	202				
411									
412				233(di)	245(di)				
413					224				
414					184				
415					163–5(di)				
416				222–3	169–70				
417				235(242) (di)	161(di)				
418					193(tri)				
419				169(di)	143				
420					198(di)				
421									
422					181–2				
423					226–9(di)				
					228–9(tri)				
424				241(di)	161(di) (164)		274	189–90 (di)	
425					206(di)				
426									
427					244–5(di)				
428					182.5(di)				
429					213(di)				
430					250–2d				
431				187(di)	184(di)				
432					199(di)				
433					217(di)				
434				202–4	159–60				
435				213–5(tri)	181–2(tri)			196–8(di)	
436					224(di)				
437					223(tri)N				

NOTES FOR TABLES 29A AND 29B (Continued)

17. 3,5-Dinitro derivative, mp 60–1; 4-nitrobenzyl ether, mp 92.
18. 3,5-Dinitro derivative, mp 61–2.
19. Allophanate, mp 212.
20. Also called salol; crystallizes in three modifications: 42, 38.8, and 28.5; for the phenylurethan, See C.A., **26**, 5556. 1932.
21. Picrate, mp 152.5.
22. 2,6-Dibromo derivative, mp 90
23. 2-Naphthylurethan, mp 166.
24. Benzenesulfonyl derivative, mp 99.
25. Benzenesulfonyl derivative, mp 58.
26. Picrate, mp 145.
27. Benzenesulfonyl derivative, mp 66.
28. Also named orcinol; the hydrated compound melts at 56–8 and the anhydrous compound melts at 108.
29. Also named 6-chlorothymol; benzyl ether, mp 55; 2-nitrobenzyl ether, mp 117.
30. Picrate, mp 83.8.
31. Picrate acid, mp 122; naphthalene adduct, mp 59; methyl ether, mp 118.
32. Dinitro derivative, mp 76.
33. Picrate, mp 133–4.
34. Attempts to prepare the three benzoate esters yielded only tarry products.

NOTES FOR TABLES 29A AND 29B (Continued)

35. Methyl ether, mp 64–5; ethyl ether, mp 55.
36. Methyl ether, mp 30; ethyl ether, bp 249–50.
37. Picrate, mp 123.
38. Methyl ether, mp 70–1;
39. Methiodide, mp 140–3; methyl urethan, mp 85–6.
40. Methyl ether, mp 140.
41. Picrate, mp 143–4.
42. Methyl ether, mp 85.
43. Methyl ether, mp 77–8.
44. Methyl ether, mp 98; ethyl ether, mp 61–2.
45. Benzenesulfonyl derivative, mp 83–4; methyl ether, mp 47(43).
46. Benzenesulfonyl derivative, mp 102–3.
47. Methyl ether, mp 103–4; a light absorber at 320–380 millimicrons.
48. Acetyl derivative, mp 136.
49. Naphthalene adduct, mp 72–3.
50. Picrate, mp 167.
51. 2,6-Dibromo derivative, mp 64–7; benzenesulfonyl derivative, mp 70–1.
52. Methyl ether, mp 130; m-nitrobenzenesulfonate, mp 176.
53. Picrate, mp 156–7.
54. The methyl ether, mp 55, yields an N-acetyl derivative, mp 82, that forms a picrate, mp 145.
55. Benzyl ether, mp 42.
56. Methyl ether, mp 58.
57. Methyl ether, mp 97; ethyl ether, mp 85.
58. Tautomeric with 1,2-naphthoquinone-1-oxime; the oxime yields: methyl ether, mp 75, and a benzenesulfonyl derivative, mp 124–5.
59. Methyl ether, mp 94.
60. Picrate, mp 164–5.
61. Reduces to 2-chloro-4-aminophenol, mp 153.
62. Picrate, mp 163–4.
63. Methyl ether, mp 106.
64. Picrate, mp 163–4.
65. Dimethyl ether, mp 115.
66. Naphthalene adduct, mp 100.
67. Reduces to anthracene and oxidizes to anthraquinone.
68. Commonly called picric acid; naphthalene adduct, mp 151.
69. Methyl ether, mp 64.
70. 3-Nitrobenzoyl derivative, mp 149–50.
71. 1,3,5-Trinitrobenzene adduct, mp 174.5.
72. 5-Methyl ether, mp 84; dimethyl ether, mp 51.
73. Methyl ether, mp 72.
74. Light absorber at 280–370 millimicrons.
75. Methyl ether, mp 80; ethyl ether, mp 84.
76. Methyl ether, mp 67.
77. Methyl ether, mp 108.
78. Methyl ether, mp 70; ethyl ether, mp 87.
79. Diethyl ether, mp 68.5.
80. Methyl ether (from benzene), mp 93–5; trimethyl ether (from alcohol), mp 130–1.
81. 2-Naphthylamine adduct, mp 110.5.
82. Picrate, mp 135–7; 2-naphthylamine adduct, mp 124.
83. Oxidizes to thymoquinone, mp 45.
84. Methyl ether, mp 62–3; dimethyl ether, mp 45.
85. Methyl ether, mp 119; ethyl ether, mp 101.
86. 3-Nitrobenzenesulfonate, mp 144–6.
87. Methyl ether, mp 131–2; dimethyl ether, mp 103–4.
88. Picrate, mp 185.
89. Adduct with 2-naphthylamine, mp 168; dimethyl ether, 116.5.
90. 3-Nitrobenzenesulfonate, mp 135.
91. 1,3-Dimethyl ether, mp 83; 1,2,3-triethyl ether, mp 105.
92. 2,4-Dimethyl ether, mp 83; 2,6-dimethyl ether, mp 178–9.
93. Picrate, mp 156.
94. Exists in two forms: α-, 170, and β-, mp 60.
95. Picrate, mp 133.
96. 2,4-Dinitrobenzoate, mp 204.5.
97. m-Nitrobenzenesulfonate, mp 153.
98. Methyl ether, mp 117.
99. Methyl ether, mp 197–8; dimethyl ether, mp 219.
100. Dimethyl ether, mp 180.
101. 1,2,3-Trinitrobenzene adduct, mp 162; 2-naphthylamine adduct, mp 163.
102. 2-Acetyl derivative, mp 172; 5-acetyl derivative, mp 131–2.
103. Dimethyl ether, mp 143; diethyl ether, mp 177.
104. Methyl ether, mp 151–2; dimethyl ether, mp 146.
105. Di-picrate, mp 175–6.
106. 3-Nitrobenzene sulfonate, mp 165; ethyl ether, mp 115.
107. 2-Naphthylamine adduct, mp 229.5.
108. Dimethyl ether, mp 185.
109. Systematic name: 2,5-dichloro-3,6-dihydroxy-p-benzoquinone; dimethyl ether, mp 141–2; reagent for many cations.
110. Also named dicumarol.
111. Systematic name: 1,2-dihydroxy-9,10-anthraquinone; the dibenzoyl derivative has two forms, mp 160 (unstable) and 187 (stable).
112. 2-Acetyl derivative, mp 296–8; 2,7-diacetyl derivative, mp 192–3.

REFERENCES FOR TABLES 29A AND 29B

Blicke, F. F., and McCarty, F. J., *J. Org. Chem.*, **24,** 1061 (1959). The Use of Substituted Phenols in the Mannich Reaction and the Dehalogenation of Aminomethylhalophenols. Gives the experimental data and the melting points of 79 derivatives.

Laskowski, D. E., *Anal. Chem.*, **32,** 1171 (1960). Gives the melting points for a number of eutectic mixtures and adducts of quinones with hydrocarbons.

McKinley, J. B., Nickels, J. E., and Sidhu, S. S., *Anal. Chem.*, **16,** 304 (1944). Phenyl Isocyanate Derivatives of Certain Alkylated Phenols. Gives the melting points and X-ray powder diffraction data.

Smith, B., *Acta Chem. Scand.*, **10,** 1589 (1956). Quantitative Bromination of Phenols. Discusses the method for the quantitative bromination of 47 phenols.

Woolfolk, E. O., and Taylor, J. M., *J. Org. Chem.*, **22,** 827 (1957). Lists the melting points for about 30 phenols and their 4-phenylazobenzoyl chloride derivatives.

TABLE 30
SULFONAMIDES

	Name of Compound[a]	Note	Melting Point of Amide	Melting Point of Derivatives			
				Acid	Chloride	Anilide	1-Naph-thylamide
1	3-Methylbutane-1-		3		bp 98/13 mm	42	90–1
2	2-Methylpropane-1-		14–6		bp 80/13 mm	38	107
3	Butane-1-		45	−15	bp 75/10 mm	10–15	60.5
4	Propane-1-		52	7.5	(bp 78/15 mm)	10	84
5	Ethane-	19	58	−17	bp 178	58	66
6	Propane-2-		60		bp 61/9 mm	84	134
7	Heptane-1-		75		mp 16		
8	2,4,5-Trimethoxybenzene-1-		76		mp 130	170	
9	Methane-		90	20	bp 60/21 mm	100.5	125.5
10	3,3-Dimethylbutane-	1	96–7		mp 43–4		
11	Hexadecane-1-		97	54	mp 54		
12	2-Ethylbenzene-1-		100		11.7		
13	Toluene-α-; (benzyl-)		105		92	102	166(146)
14	Toluene-3-		108		12	96	
15	2,4-Dimethyl-6-nitrobenzene-1-		108		97		
16	4-Ethylbenzene-1-	2	110		12		
17	3-Hydroxynaphthalene-2-	19	110		112		
18	Pyridine-3-		110–1			145	
19	3-Methoxynaphthalene-2-		113		138	174	
20	2,6-Dimethylbenzene-1-		113(96)	98	39		
21	8-Methylnaphthalene-2-		116		88		
22	2-Phenylethane-1-		122	91	33	77	
23	2-Methylnaphthalene-1-		124		83–5		
24	4-Fluorobenzene-1-		125		36(30)		
25	4-Chloro-3-methylbenzene-1-		128		63	92	
26	3-Chloro-4-methoxybenzene-1-		131		82		
27	4-Aminonaphthalene-2-	3	131(hyd.)				
28	D-Camphor-10-		132	193	67	121(88)	
29	6-Fluoronaphthalene-2-		133	105(hyd.)	97	129	
30	DL-Camphor-8-		133–5	56–8	106		
31	3-Chloro-4-methylbenzene-1-		134		38	96	
32	Tetralin-6-		135		58	155–6	
33	3,5-Dimethylbenzene-1-		135		94(90)	119	
34	D-Camphor-8-		137		138		
35	Toluene-4-	4, 19	137	92	69	103	157
36	2,4-Dimethylbenzene-1-	5, 19	138	62(hyd.)	34	110	
37	2-Bromonaphthalene-1-		140		97		
38	3,4-Dichlorobenzene-1-		140(135)		22(19)		
39	2,4,6-Trimethylbenzene-1-	6	142	78	56	109	
40	7-Ethoxynaphthalene-2-		142		103	153	
41	3-Aminobenzene-1-	19	142				
42	D-Camphor-3-		143	77	88	124	
43	5-Chloro-2-methylbenzene-1-		143	21(24)			
44	4-Methylnaphthalene-2-		144		125		
45	3,4-Dimethylbenzene-1-	19	144	64(55)	52		
46	4-Chlorobenzene-1-	19	144	69(93)	53	104	190
47	4-Bromo-3-methylbenzene-1-		146		50		
48	3-Chlorobenzene-1-	19	148				
49	2,5-Dimethylbenzene-1-	7, 19	148	86(hyd.) 48(anh.)	24–6		
50	6-Methoxynaphthalene-1-		149.5		80.5	177.5	
51	Naphthalene-1-	19	150	90	68(66)	112(152)	
52	4,6-Dichloro-2,5-dimethylbenzene-1-		150		81	175	
53	3-Bromo-4-methylbenzene-1-		151		60		
54	2-Chloronaphthalene-1-		153		75		

(Continued)

TABLE 30 (Continued)
SULFONAMIDES

	Name of Compound[a]	Note	Melting Point of Amide	Melting Point of Derivatives			
				Acid	Chloride	Anilide	1-Naphthylamide
55	8-Chloro-7-methoxynaphthalene-1-		153		137	196	
56	2-Aminobenzene-1-	19	153				
57	Benzene-	8, 19	153	66(anh.)	14.5	112	170–1
58	2-Iodonaphthalene-1-		154		110		
59	6-Ethoxynaphthalene-1-		154		118	194.5	
60	2,6-Dichloro-4-methylbenzene-1-		154–5		56		
61	4-Amino-2-hydroxybenzene-1-		155		169		
62	Toluene-2-	9	156	57	68	136	
63	4-Methoxynaphthalene-2-		157		75.5	145	
64	2,4-Dinitrobenzene-1-		157(154)	130(anh.) 106–8(hyd.)	102		
65	2-Ethoxynaphthalene-1-		158		116	187	
66	2-Methoxynaphthalene-1-		159		121	196.5	
67	3-Nitrotoluene-α-		159d	74(hyd.)	100		
68	5-Chloro-2-nitrobenzene-1-		159		93		
69	7-Methylnaphthalene-2-		163–4		63–4		
70	4-Aminobenzene-1-		165			200	196
71	3,6-Dichloro-2,5-dimethylbenzene-1-		165		71	171	
72	4-Bromobenzene-1-	19	166(161)	88–90 (102–3)	76	119	183.5
73	5-Bromo-2-methylbenzene-1-		166–7		33–5		
74	3-Nitrobenzene-1-	19	167	48	64	126	166.5
75	2,4-Dimethoxybenzene-1-		167		70		
76	5-Chloro-2-methyl-3-nitrobenzene-1-		167		60		
77	4-Hydroxynaphthalene-1-		167			199–200	
78	2,3-Dimethylbenzene-1-		167		47		
79	4,6-Dichloro-2-methylbenzene-1-		168		43		
80	4,5-Dibenzylnaphthalene-1-		168		151		
81	4-Bromo-2-methylbenzene-1-		168		50		
82	4-Ethoxynaphthalene-1-		170		103	180	
83	3-Carboxybenzene-	19	170(di)	148(anh.) 98(hyd.)	20(di)		
84	2,4-Dimethyl-3-nitrobenzene-1-		172		96		
85	2,5-Dimethyl-3-nitrobenzene-1-	10	173	128(200)	61	143–4	
86	7-Ethoxy-8-nitronaphthalene-1-		173.4		155		
87	4-Methylnaphthalene-1-		174(177)		81	158	
88	3,4-Dibromobenzene-1-		175		34		
89	4-Chloro-3-nitrobenzene-1-		175–6		40–1(60–2)		
90	7-Chloronaphthalene-2-		176		87		
91	4-Bromo-3-nitrobenzene-1-		176–7		55–7		
92	4-Hydroxybenzene-1-		176–7			141	
93	4-Methylnaphthalene-1-		177		81	158	
94	2-Chloro-5-methyl-6-nitrobenzene-1-		177		122		
95	4-Nitrobenzene-1-		180	109–11(95)	80	171(136)	
96	3,4-Dimethyl-5-nitrobenzene-1-		180		70		
97	8-Aminonaphthalene-2-		181(hyd.)			147	
98	2,5-Dichlorobenzene-1-	19	181	93–7	38	160	
99	2,4,5-Trimethylbenzene-1-		181	112	61		
100	5-Chloro-4-methyl-2-nitrobenzene-1-		181	128	99		
101	5-Ethoxynaphthalene-1-		182.5		121	130	
102	2,4-Dichlorobenzene-1-		182		54		
103	6-Ethoxynaphthalene-2-		183		107.5	153	
104	4-Iodobenzene-1-		183		85	143	
105	4-Ethoxynaphthalene-2-		183		85	143.5	
106	4,5-Dichloro-3-methylbenzene-1-		183–5		85–8		

(Continued)

TABLE 30 (Continued)
SULFONAMIDES

	Name of Compound[a]	Note	Melting Point of Amide	Melting Point of Derivatives			
				Acid	Chloride	Anilide	1-Naphthylamide
107	6-Chloronaphthalene-2-		184		110		
108	5-Nitronaphthalene-2-	19	184	118–9	125		
109	8-Chloronaphthalene-2-		185		94		
110	4-Chloro-2,5-Dimethylbenzene-1-		185	100	50	155	
111	2-Chloro-5-nitrobenzene-1-		185–6	168–9	90		
112	2-Bromobenzene-1-		186		51		
113	3,5-Dichloro-2-methylbenzene-1-		186		54		
114	2-Chloro-4-methylbenzene-1-		186		46(52)		
115	4-Methylbenzene-1,3-di-		186–7(191)		54(56)	189	
116	4-Chloronaphthalene-1-		187	130–3	94–5	145–6	162
117	8-Iodonaphthalene-1-		187		115	140	
118	2,4-Dimethyl-5-nitrobenzene-1-		187(179)	132(122)	98		
119	2,4-Diaminobenzene-1,5-di-		187		275	236	
120	2-Methyl-5-nitrobenzene-1-		187	137–40	44(47)	148	
121	8-Bromonaphthalene-2-		187–8		121		
122	2-Chlorobenzene-1-		188		28.5		
123	4-Nitronaphthalene-1-		188		99		
125	2,6-Dichloro-3-methylbenzene-1-		188		19.5		
126	5-Methylnaphthalene-2-		188–9		120–2	133–4	
127	6-Methoxynaphthalene-2-		189		93	120	
128	Sulfanilguanidine (anh.) (hyd., 143)	18	189–90				
129	2,4-Dibromobenzene-1-		190	110(anh.)	79		
130	Phenanthrene-3-	19	190	175–6(anh.)	110–1		
131	8-Nitronaphthalene-1-		191		167.5		
132	Sulfapyridine	17	191–2				
133	3,5-Dichloro-4-methylbenzene-1-		191		69		
134	5,6-Dichloronaphthalene-2-		192		167		
135	Anthraquinone-2.7-di-		192		186		
136	2,5-Dimethyl-6-nitrobenzene-1-	11	192	145(anh.)	110	182	
137	2-Nitrobenzene-1-		193	70(85)	69	115	
138	Phenanthrene-9-	19	193–4	174(anh.)	127		
139	5-Methoxynaphthalene-1-		194.5		119.5	157	
140	4-Bromonaphthalene-1-		195		87		
141	2,5-Dibromobenzene-1-		195	128(anh.)	71		
142	4,7-Dichloronaphthalene-2-		196		156		
143	5-Fluoronaphthalene-1-		196–7	105(hyd.)	122–3		
144	8-Chloronaphthalene-1-		197		101		
145	4,5-Dichloronaphthalene-2-		197		158		
146	7-Methylnaphthalene-1-		197		96	162–4	
147	2,5-Dimethyl-4-nitrobenzene-1-		197–8	140	75	131	
148	Acenaphthene-3-		199	87–9	113–4		
149	4-Chloro-3-methyl-5-nitrobenzene-1-		201		52		
150	Sulfathiazole	16	202.5				
151	Anthracene-1-		205		90		
152	4,8-Dichloronapthalene-2-		205		141		
153	4-Iodonaphthalene-2-		206(204)		124(121)	136	
154	4-Fluoronaphthalene-1-		206	100(hyd.)	86	144	
155	6-Methylnaphthalene-2-		206		98		
156	4-Aminonaphthalene-1-	19	206				
157	4-Hydroxy,2,6-dimethylbenzene-1,3-di-		206–8		117–8	205–7	
158	3,7-Diethylnaphthalene-1-		207		105–7		
159	6-Bromonaphthalene-2-		207		124		

(Continued)

TABLE 30 (Continued)
SULFONAMIDES

	Name of Compound[a]	Note	Melting Point of Amide	Melting Point of Derivatives			
				Acid	Chloride	Anilide	1-Naphthylamide
160	7-Bromonaphthalene-1-		209		147		
161	7-Iodonaphthalene-2-		210		100		
162	2,4,6-Trichlorobenzene-1-		210–2d		35–40		
163	4-Aminonaphthalene-1-	12	212			190	
164	6-Iodonaphthalene-1-		213		92.5		
165	Fluorene-2-		213d	155(hyd.)	164		
166	1-Nitronaphthalene-2-		214	105	121	202	
167	6-Chloronaphthalene-1-		214		70		
168	2,5-Dichlorobenzene-1,3-di-		215–7		114		
169	5-Methylbenzene-1,3-di-		216		94	153	
170	5-Chloronaphthalene-2-		216		115		
171	Naphthalene-2-	19	217(213)	91(122)	76(79)	132	
172	4,7-Dichloronaphthalene-1-		217		151		
173	6-Bromonaphthalene-1-		217		77		
174	4,6-Dichloronaphthalene-2-		218		130		
175	7-Bromonaphthalene-2-		218		100		
176	3,6-Dichloronaphthalene-2-		218		166		
177	6-Ethoxy-1-nitronaphthalene-2-		218		146		
178	5-Aminonaphthalene-2-	13, 19	219			127–8	
179	4-Acetamidobenzene-1-		219		149	214	215
180	5-Bromonaphthalene-2-		220		96		
181	7-Methoxynaphthalene-2-		220		83	121	
182	7,8-Dichloronaphthalene-1-		221		138		
183	8-Hydroxynaphthalene-1-		222d	107(hyd.)			
184	6-Iodonaphthalene-2-		222		140		
185	6-Nitronaphthalene-1-		223		127		
186	5,6-Dichloronaphthalene-1-		223		106		
187	Acenaphthene-5-		223		111	178	
188	2-Methylbenzene-1,4-di-		224		98	178(di)	
189	5-Chlorobenzene-1,3-di-		224		106		
190	4-Nitronaphthalene-2-		225		139.5		
191	4,6-Dichloronaphthalene-1-		226		119		
192	5-Chloronaphthalene-1-		226		95		
193	4-Methoxynaphthalene-1-		226		98.5	147.5	
194	7,8-Dichloronaphthalene-2-		227		124		
195	2,3,4-Trichlorobenzene-1-		227–30		64–5		
196	3,4-Dichloro-2-methylbenzene-1-		228		51–2		
197	2,4,6-Tribromobenzene-1-		228	64		220–2d	
198	8-Nitronaphthalene-2-		228		169		
199	6,8-Dichloronaphthalene-2-		228		121		
200	Benzene-1,3-di-	14, 19	229		63	148–50	245
201	4,5-Dichloronaphthalene-1-		229		117		
202	Biphenyl-4-		230		115	125	
203	5-Bromonaphthalene-1-		232–3		95		
204	3,5-Dinitrobenzene-1-		235		99		
205	7-Chloronaphthalene-1-		235		129		
206	5,6,8-Trichloronaphthalene-2-		235		158		
207	4,6,7,8-Tetrachloronaphthalene-2-		235		176		
208	5-Nitronaphthalene-1-		236		113	123	
209	4-Carboxybenzene-1-		236(di)	260(anh.) 94(hyd.)	57(di)	252(di)	
210	2,3-Dichloro-4-methylbenzene-1-		237		41		
211	4-Chloro-2-nitrobenzene-1-		237	82	75	138	
212	6-Hydroxynaphthalene-2-	19	238	167(anh.) 129(hyd.)		161	
213	4-Methylbenzene-1,2-di-		237–9		109–11	190	

(Continued)

TABLE 30 (Continued)
SULFONAMIDES

	Name of Compound[a]	Note	Melting Point of Amide	Melting Point of Derivatives			
				Acid	Chloride	Anilide	1-Naphthylamide
214	5-Iodonaphthalene-1-		239		114		
215	4,5-Dimethylbenzene-1,3-di-		239		79	200	
216	7-Iodonaphthalene-1-		240		165		
217	4-Methoxybenzene-1,3-di-		240		86	209	
218	4-Acetamidonaphthalene-1-		241			170	
219	Naphthalene-2,7-di-	19	242		158(162)		
220	5,8-Dichloronaphthalene-2-		244		134		
221	6,7,8-Trichloronaphthalene-2-		245		157		
222	Anthraquinone-1,5-di-		246(350)	310d	265–70	270d	
223	1-Iodonaphthalene-2-		247		94		
224	4,6-Dimethylbenzene-1,3-di-		249		130	196	
225	5,6,7-Trichloronaphthalene-1-		249		131		
226	Phenanthrene-2-	19	253–4	150	156	157–8	
227	Benzene-1,2-di-	19	254		143	241	
228	2-Amino-5-methylbenzene-1,3-di-		257		156	192	
229	5-Aminonaphthalene-1-	15, 19	260			171	
230	2-Methylbenzene-1,3-di-		260		88	162	
231	Anthracene-2-		261		122	201	
232	7-Nitronaphthalene-1-		261		170		
233	Anthraquinone-2-		261		197	193	
234	6,7-Dichloronaphthalene-1-		268		142		
235	3,7-Dichloronaphthalene-1-		269		136		
236	1-Bromonaphthalene-2-		271		93		
237	5,7-Dichloronaphthalene-1-		272		149		
238	Naphthalene-1,4-di-		273		160(166)	179	
239	4,6-Dichlorobenzene-1,3-di-		276		123		
240	9,10-Dichloroanthracene-2-		279		221	248	
241	1,5-Dichloronaphthalene-2-		282		125		
242	Benzene-1,4-di-		288		131(139)	249	
243	2,5-Dimethylbenzene-1,3-di-		295		81	174	
244	Naphthalene-1,6-di-	19	298	125(anh.)	129		
245	Biphenyl-4,4'-di-	19	300	72	203		
246	Naphthalene-2,6-di-	19	305		225		
247	2,5-Dimethylbenzene-1,4-di-		310		164	223	
248	Naphthalene-1,5-di-	19	310(340)	245(anh.)	183	249	
249	Benzene-1,3,5-tri-		310–5(tri)		187(tri)	237(tri)	
250	Anthraquinone-2,6-di-		321		250		
251	Anthracene-1,8-di-		333		225	224	
252	8-Cyanonaphthalene-1-		333–4		139		

[a] The substituted naphthalenesulfonamides are numbered in such a way that the sulfonamide group will have as small a number as possible.

NOTES FOR TABLE 30

1. Benzamide, mp 121–2.
2. N-Xanthylsulfonamide, mp 196.
3. Acetyl derivative of the amide, mp 220.
4. N-Xanthylsulfonamide, mp 197.
5. N-Xanthylsulfonamide, mp 188.
6. N-Xanthylsulfonamide, mp 203.
7. N-Xanthylsulfonamide, mp 176.
8. N-Xanthylsulfonamide, mp 200.
9. N-Xanthylsulfonamide, mp 183.
10. p-Toluidide, mp 136.
11. p-Toluidide, mp 159.
12. Acetyl derivative, of the anilide, mp 23 ; acetyl derivative of the amide, 247.
13. Acetyl derivative of the amide, mp 238–9.
14. N-Xanthylsulfonamide, mp 170.
15. Acetyl derivative of the amide, mp 231–2.
16. Acetyl derivative, mp 266–7.
17. Acetyl derivative, mp 226–7.
18. Acetyl derivative, mp 262–6; hydrochloride salt, 205–6.
19. See Table 32.

TABLE 31
SULFONYL CHLORIDES

	Name of Compound[a]	MP	Melting Point of Derivatives	
			Amide	Anilide
1	2-Ethylbenzene-	11.7	100	
2	4-Ethylbenzene-	12	110	
3	Toluene-3-	12	108	96
4	Benzene-	14.5	153	112
5	Heptane-1-	16	75	
6	2,6-Dichloro-3-methylbenzene-	19.5	188	
7	3-Carboxybenzene-	20(di)	170(di)	
8	5-Chloro-2-methylbenzene-	21	143	
9	3,4-Dichlorobenzene-	22(19)	140(135)	
10	2,5-Dimethylbenzene-	24–6	148	
11	2-Chlorobenzene-	28.5	188	
12	2-Phenylethane-1-	33	122	77
13	2,4-Dimethylbenzene-	34	139(137)	110
14	3,4-Dibromobenzene-	34	175	
15	5-Bromo-2-methylbenzene-	34–5	167	
16	4-Fluorobenzene-	36	125	
17	2,5-Dichlorobenzene-	38	180(186)	160
18	3-Chloro-4-methylbenzene-	38	134	96
19	2,6-Dimethylbenzene-	39	113(96)	
20	4-Chloro-3-nitrobenzene-	40–1	175–6	
21	2,3-Dichloro-4-methylbenzene-	41	237	
22	4,6-Dichloro-2-methylbenzene-	43	168	
23	3,3-Dimethylbutane-1- (N-Benzylamide, mp 121–2)	43–4	96–7	
24	2-Methyl-5-nitrobenzene-	44	187	148
25	2-Chloro-4-methylbenzene-	46(52)	186	
26	2,3-Dimethylbenzene-	47	167	
27	2,3-Dichloro-6-methylbenzene-	49	186	
28	4-Bromo-2-methylbenzene-	50	168	
29	4-Chloro-2,5-dimethylbenzene-	50	185	155
30	4-Bromo-3-methylbenzene-	50	146	
31	2-Bromobenzene-	51	186	
32	3,4-Dimethylbenzene-	52	144	
33	4-Chloro-3-methyl-5-nitrobenzene-	52	201	
34	3,4-Dichloro-2-methylbenzene-	52	228	
35	4-Chlorobenzene-	53	144	104
36	2,4-Dichlorobenzene-	54	182	
37	Hexadecane-	54	97	
38	3,5-Dichloro-2-methylbenzene-	54	186	
39	4-Methylbenzene-1,3-di-	54(56)	191	189
40	2,6-Dichloro-4-methylbenzene-	56	154–5	
41	2,4,6-Trimethylbenzene-	56	142	109
42	4-Bromo-3-nitrobenzene-	55–7	176–7	
43	4-Carboxybenzene-	57(di)	236(di)	252(di)

(*Continued*)

TABLE 31 (Continued)
SULFONYL CHLORIDES

	Name of Compound[a]	MP	Melting Point of Derivatives	
			Amide	Anilide
44	Tetralin-6-	58	135	155–6
45	5-Chloro-2-methyl-3-nitrobenzene-	60	167	
46	3-Bromo-4-methylbenzene-	60	151	
47	2,5-Dimethyl-3-nitrobenzene-	61	173	143–4
48	2,4,5-Trimethylbenzene-	61–2	181	
49	Benzene-1,3-di-	63	229	148–50
50	4-Chloro-3-methylbenzene-	63	128	
51	7-Methylnaphthalene-2-	63–4	163–4	
52	3-Nitrobenzene-	64	167	126
53	2,4,6-Tribromobenzene-	64	228d	220–2d
54	2,3,4-Trichlorobenzene-	64–5	227–30	
55	D-Camphor-10-	67	132	121
56	Naphthalene-1-	68(66)	150	112(152)
57	Toluene-2-	68	156	136
58	Nitrobenzene-2-	69	193	115
59	Toluene-4-	69	137	103
60	3,5-Dichloro-4-methylbenzene-	69	191	
61	2,4-Dimethoxybenzene-	70	167	
62	6-Chloronaphthalene-1-	70	214	
63	3,4-Dimethyl-5-nitrobenzene-	70	180	
64	2,5-Dibromobenzene-	71	195	
65	3,6-Dichloro-2,5-dimethylbenzene-	71	165	171
66	2,5-Dimethyl-4-nitrobenzene-	75	197–8	131
67	4-Chloro-2-nitrobenzene-	75	237	138
68	2-Chloronaphthalene-1-	75	153	
69	4-Methoxynaphthalene-2-	75.5	157	145
70	Bromobenzene-4-	76	166(161)	119
71	Naphthalene-2-	76(79)	217(213)	132
72	6-Bromonaphthalene-1-	77	217	
73	2,4-Dibromobenzene-	79	190	
74	4,5-Dimethylbenzene-1,3-di-	79	239	200
75	4-Nitrobenzene-	80	180	171(136)
76	6-Methoxynaphthalene-1-	80.5	149.5	177.5
77	2,5-Dimethylbenzene-1,3-di-	81	295	174
78	4,6-Dichloro-2,5-dimethylbenzene-	81	150	175
79	4-Methylnaphthalene-1-	81	174(177)	158
80	3-Chloro-4-methoxybenzene-	82	131	
81	7-Methoxynaphthalene-2-	83	220	
82	2-Methylnaphthalene-1-	83–5	124	
83	4-Ethoxynaphthalene-2-	85	183	143.5
84	4-Iodobenzene-	85	183	143
85	4-Methoxybenzene-1,3-di-	86	240	209
86	4-Fluoronaphthalene-1-	86	206	144
87	7-Chloronaphthalene-2-	87	176	

(*Continued*)

TABLE 31 (Continued)
SULFONYL CHLORIDES

	Name of Compound[a]	MP	Melting Point of Derivatives	
			Amide	Anilide
88	4-Bromonaphthalene-1-	87	195	
89	8-Methylnaphthalene-2-	88	116	
90	D-Camphor-3-	88	143	124
91	2-Methylbenzene-1,3-di-	88	260d	162
92	6-Chloro-3-nitrobenzene-	90	186	
93	Anthracene-1-	90	205	
94	Toluene-α-	92	105	102
95	6-Iodonaphthalene-1-	92.5	213	
96	1-Bromonaphthalene-2-	93	271	
97	5-Chloro-2-nitrobenzene-	93	159	
98	3,5-Dimethylbenzene-	94(90)	135	119
99	5-Methylbenzene-1,3-di-	94	216	153
100	6-Methoxynaphthalene-2-	94	189	120
101	8-Chloronaphthalene-2-	94	185	
102	1-Iodonaphthalene-2-	94	247	
103	Ethane-1,2-di-	95		69
104	4-Chloronaphthalene-1-	95	187	145–6
105	5-Chloronaphthalene-1-	95	226	
106	5-Bromonaphthalene-1-	95	232–3	
107	7-Methylnaphthalene-1-	96	197	162–4
108	2,4-Dimethyl-3-nitrobenzene-	96	172	
109	5-Bromonaphthalene-2-	96	220	
110	3-Fluoronaphthalene-1-	97	133	129
111	2-Bromonaphthalene-1-	97	140	
112	2,4-Dimethyl-6-nitrobenzene-	97	108	
113	2-Methylbenzene-1,4-di-	98	224	178
114	6-Methylnaphthalene-2-	98	206	
115	4-Methoxynaphthalene-1	98.5	226	147.5
116	3,5-Dinitrobenzene-	99	235	
117	5-Chloro-4-methyl-2-nitrobenzene-	99	181	
118	4-Nitronaphthalene-1-	99	188	
119	3-Nitrotoluene-α-	100	159d	
120	7-Bromonaphthalene-2-	100	218	
121	6-Fluoronaphthalene-2-	100	133	
122	7-Iodonaphthalene-2-	100	210	
123	8-Chloronaphthalene-1-	101	197	
124	2,4-Dinitrobenzene-	102	157(154)	
125	4-Ethoxynaphthalene-1-	103	170	180
126	7-Ethoxynaphthalene-2-	103	142	153
127	3,7-Diethylnaphthalene-1-	105–7	207	
128	DL-Camphor-8-	106	133–5	
129	5-Chlorobenzene-1,3-di-	106	224	
130	5,6-Dichloronaphthalene-1-	106	223	
131	6-Ethoxynaphthalene-2-	107.5	183	153

(Continued)

TABLE 31 (Continued)
SULFONYL CHLORIDES

	Name of Compound[a]	MP	Melting Point of Derivatives	
			Amide	Anilide
132	4-Methylbenzene-1,2-di-	109–11	237–9	190
133	6-Chloronaphthalene-2-	110	184	
134	2,5-Dimethyl-6-nitrobenzene-	110	192	192
135	2-Iodonaphthalene-1-	110	154	
136	Phenanthrene-3-	110–11	190	
137	Acenaphthene-5-	111	223	178
138	3-Hydroxynaphthalene-2-	112	110	
139	5-Nitronaphthalene-1-	113	236	
140	Acenaphthene-3-	113–4	199	
141	5-Iodonaphthalene-1-	114	239	
142	2,5-Dichlorobenzene-1,3-di-	114	215–7	
143	Biphenyl-4-	115	230	125
144	8-Iodonaphthalene-1-	115	187	140
145	5-Chloronaphthalene-2-	115	216	
146	2-Ethoxynaphthalene-1-	116	158	187
147	4,5-Dichloronaphthalene-1-	117	229	
148	6-Ethoxynaphthalene-1-	118	154	194.5
149	4-Hydroxy-2,6-dimethylbenzene-1,3-di-	119	206–8	205–7
150	4,6-Dichloronaphthalene-1-	119	226	
151	5-Methoxynaphthalene-1-	119.5	194	157
152	5-Methylnaphthalene-2-	120–2	188–9	133–4
153	5-Ethoxynaphthalene-1-	121	182.5	130
154	1-Nitronaphthalene-2-	121	214	202
155	8-Bromonaphthalene-2-	121	187–8	
156	2-Methoxynaphthalene-1-	121	159	196.5
157	6,8-Dichloronaphthalene-2-	121	228	
158	2-Chloro-5-methyl-6-nitrobenzene-	122	177	
159	Anthracene-2	122	261	201
160	5-Fluoronaphthalene-1-	122–3	196–7	
161	4,6-Dichlorobenzene-1,3-di-	123	276	
162	4-Iodonaphthalene-1-	124(121)	206(204)	136
163	6-Bromonaphthalene-2-	124	207	
164	7,8-Dichloronaphthalene-2-	124	227	
165	4-Methylnaphthalene-2-	125	144	
166	1,5-Dichloronaphthalene-2-	125	282	
167	5-Nitronaphthalene-2-	125	184	
169	Phenanthrene-9-	127	193–4	
170	6-Nitronaphthalene-1	127	223	
171	Naphthalene-1,6-di-	129	298	
172	7-Chloronaphthalene-1-	129	235	
173	4,6-Dimethylbenzene-1,3-di-	130	249	196
174	2,4,5-Trimethoxybenzene-	130	76	170
175	4,6-Dichloronaphthalene-2-	130	218	

(*Continued*)

TABLE 31 (Continued)
SULFONYL CHLORIDES

	Name of Compound[a]	MP	Melting Point of Derivatives	
			Amide	Anilide
176	Benzene-1,4-di-	131(139)	288	249
177	5,6,7-Trichloronaphthalene-1-	131	249	
178	5,8-Dichloronaphthalene-2-	134	244	
179	3,7-Dichloronaphthalene-1-	136	269	
180	8-Chloro-7-methoxynaphthalene-1-	137	153	196
181	D-Camphor-8-	138	137	
182	7,8-Dichloronaphthalene-1-	138	221	
183	3-Methoxynaphthalene-2-	138	113	
184	8-Cyanonaphthalene-1-	139	333–4	
185	4-Nitronaphthalene-2-	139.5	225	
186	6-Iodonaphthalene-2-	140	222	
187	4,8-Dichloronaphthalene-2-	141	205	
188	6,7-Dichloronaphthalene-1-	142	268	
189	Benzene-1,2-di-	143	254	241
190	6-Ethoxy-5-nitronaphthalene-2-	146	218	
191	7-Bromonaphthalene-1-	147	209	
192	5,7-Dichloronaphthalene-1-	149	272	
193	4-Acetamidobenzene-1-	149	219	214
194	4,5-Dibenzylnaphthalene-1-	151	168	
195	4,7-Dichloronaphthalene-1-	151	217	
196	7-Ethoxy-8-nitronaphthalene-1-	155	173.4	
197	4,7-Dichloronaphthalene-2-	156	196	
198	2-Amino-5-methylbenzene-1,3-di-	156	257	192
199	Phenanthrene-2-	156	253–4	157–8
200	6,7,8-Trichloronaphthalene-2-	157	245	
201	4,5-Dichloronaphthalene-2-	158	197	
202	Naphthalene-2,7-di-	158(162)	242	
203	5,6,8-Trichloronaphthalene-2-	158	235	
204	Naphthalene-1,4-di-	160(166)	273	179
205	2,5-Dimethylbenzene-1,4-di-	164	310	223
206	Fluorene-2-	164	213d	
207	7-Iodonaphthalene-1-	165	240	
208	3,6-Dichloronaphthalene-2-	166	218	
209	8-Nitronaphthalene-1-	167.5	191	
210	5,6-Dichloronaphthalene-2-	167	192	
211	4-Amino-2-hydroxybenzene-1-	169	155	
212	8-Nitronaphthalene-2-	169	228	
213	7-Nitronaphthalene-1-	170	261	
214	4,6,7,8-Tetrachloronaphthalene-2-	176	235	
215	Naphthalene-1,5-di-	183	310(340)	249
216	Anthraquinone-2,7-di-	186	192	
217	Benzene-1,3,5-tri-	187	310–5	237
218	Anthraquinone-2-	197	261	193
219	Anthraquinone-1,6-di-	197–8		227–8

(*Continued*)

TABLE 31 (Continued)
SULFONYL CHLORIDES

	Name of Compound[a]	MP	Melting Point of Derivatives	
			Amide	Anilide
220	Biphenyl-4,4′-di-	203	300	
221	9,10-Dichloroanthracene-2-	221	279	248
222	Anthraquinone-1,8-di-	223	>340	238
223	Anthracene-1,8-di-	225	333	224
224	Naphthalene-2,6-di-	225	305	
225	Anthraquinone-1,7-di-	232		238
226	Anthracene-1,5-di-	249	>300	293
227	Anthraquinone-2,6-di-	250	321	
228	Anthraquinone-1,5-di-	265–70	246	270d
229	2,4-Diaminobenzene-1,5-di-	275	187	

[a] The substituted naphthalenesulfonyl chlorides are numbered to give the sulfonyl chloride group as small a number as possible.

TABLE 32
SULFONIC ACIDS

	Name of Compound	Note	MP	Melting Point of Derivatives				
				Amide	S-Benzyl-thiuronium Salt	p-Toluidine Salt	Aniline Salt	o-Toluidine Salt
1	4-Acetamidobenzene-1-	1,21						
2	3-Aminobenzene-1-	1		142	148			
3	4-Aminobenzene-1-	1,21		164	185			
4	2-Aminobenzene-1-			153	132			
5	4-Hydroxy-5-amino-naphthalene-2,7-di-				312d	335d	340d	320d
6	Anthracene-1,5-di-	2						
7	Anthraquinone-1-	3	218		191		284	
8	Anthraquinone-2-			261	211	308	309	
9	Anthraquinone-1,5-di-	4						
10	Anthraquinone-1,6-di-	5	207(hyd.)					
11	Anthraquinone-1,7-di-	6	120(hyd.)					
12	Anthraquinone-1,8-di-	7						
13	Benzene-	1,21	66	153	148	205	240	176
14	Benzene-1,2-di-	1		254	206			
15	Benzene-1,3-di-	1,21		229	214			
16	Benzothiazole-2-				171			
17	Toluene-α-	21						
18	Biphenyl-4,4'-di-	1		300	171	330d		
19	4-Bromobenzene-1-	1,21	103 (88–9)	166	170	215–6	237–8	182–3
20	Butane-1-	21						
21	dl-Camphane-10-	16	200–2					
22	Camphor-3-	17		143	210			
23	3-Chlorobenzene-1-	1		148		199	206–7	
24	4-Chlorobenzene-1-	1,20	69(93)	144	175	208–10	222–3	163–4
25	2,5-Dichlorobenzene-1-	1	97	181	170			
26	3-Diethylaminobenzene-1-				183			
27	2,4-Dimethylbenzene-1-	1	62(hyd.)	138	146			
28	2,5-Dimethylbenzene-1-	1	86(hyd.)	148	184			
29	3,4-Dimethylbenzene-1-	1	64(hyd.)	144	208			
30	5,7-Dinitro-8-hydroxy-naphthalene-2-	15	151					
31	Diphenylamine-4-						206	
32	Diphenylamine-4,4'-di						239	
33	Ethane-	1,21		58	115			
34	Methane-	21						
35	Ethane-1,2-di-	8	104					
36	2-Methylbenzene-1,3-di-	9						
37	Naphthalene-1-	1	90	150	137	181	183	237
38	Naphthalene-1,5-di-	1	240(anh.)	310	257(251)	332		
39	Naphthalene-1,6-di-	1		298	235	314–5	298–9	323–4
40	Naphthalene-2-	1	91	217	191	221	269	213
41	Naphthalene-2,6-di-	1		305	256		360	
42	Naphthalene-2,7-di-	1		243	212	299	251–2	238
43	1-Hydroxynaphthalene-2-				169			
44	4-Hydroxynaphthalene-1-	10			103	196	186–7	203–4
45	5-Hydroxynaphthalene-1-	11						
46	4-Hydroxynaphthalene-2,6-di-				217			
47	4-Hydroxynaphthalene-1,5-di-				205			
48	2-Hydroxynaphthalene-1-					162	182	179
49	3-Hydroxynaphthalene-2-	1		110			241–2	
50	6-Hydroxynaphthalene-2-	1	129(hyd.)	238	217(207)	248	264	208

(Continued)

TABLE 32 (Continued)
SULFONIC ACIDS

	Name of Compound	Note	MP	Melting Point of Derivatives				
				Amide	S-Benzyl-thiuronium Salt	p-Toluidine Salt	Aniline Salt	o-Toluidine Salt
51	7-Hydroxynaphthalene-2-		109(hyd.)			237	249	
52	7-Hydroxynaphthalene-1-	19		195	218	232	240	242
53	2-Hydroxynaphthalene-1,7-di-					219		
54	3-Hydroxynaphthalene-1,7-di-				233	250	254	257
55	7-Hydroxynaphthalene-1,3-di-	12			228	294		271
56	4-Aminonaphthalene-1-	1		206	195			
57	5-Aminonaphthalene-1-	1		260	179			
58	5-Aminonaphthalene-2-	1		219	191			
59	8-Aminonaphthalene-1-	13			300d			
60	8-Aminonaphthalene-1,3,5-tri-					292d	312d	304d
61	2-Aminonaphthalene-1-				139			
62	6-Aminonaphthalene-2-				330d			
63	7-Aminonaphthalene-1,5-di-				209–11			
64	7-Aminonaphthalene-1,4-di				276			
65	7-Aminonaphthalene-1,3-di				276d			
66	3-Nitrobenzene-1-	1,21	48	167	146	222	222	193
67	1-Nitronaphthalene-2-			105	214		202	
68	5-Nitronaphthalene-1-						265d	
69	5-Nitronaphthalene-2-	1	118–9	184			260d	
70	4-Hydroxybenzene-	1,18		177	168	202	170	192
71	Propane-1-	21						
72	Propane-2-	21						
73	2-Carboxybenzene-1-	14			206	197	165	127
74	3-Carboxybenzene-1-	1	98(hyd.)	170	163	224–6		
75	4-Carboxybenzene-1-	1		236	213			
76	Thymol-		15–16		212			
77	Toluene-2-	1	57	156	170	203–4	218	
78	Toluene-4-	1,21	92	137	181–2	198	238	190
79	Phenanthrene-2-	1		253		291		
80	Phenanthrene-3-	1		190		222		
81	Phenanthrene-9-	1		193–4		235		

NOTES FOR TABLE 32 ON SULFONIC ACIDS

1. See Table 30 for additional derivatives.
2. Sulfonyl chloride, mp 249(di); anilide, mp 293(di); amide melts above 330.
3. Anilide, mp 214.
4. Sulfonyl chloride, mp 265–70; anilide mp 269–70; amide melts above 350.
5. Sulfonyl chloride, mp 197–8; anilide, mp 69.
6. Sulfonyl chloride, mp 232; anilide, mp 238.
7. Sulfonyl chloride, mp 223; anilide, mp 238, amide melts above 340.
8. Sulfonyl chloride, mp 95; anilide, mp 69.
9. Sulfonyl chloride, mp 88; anilide, mp 162.
10. Anilide, mp 199; 2-naphthylamide, mp 204.
11. Anilide, mp 201.
12. Sulfonyl chloride, mp 162; anilide, mp 195.
13. Anilide, mp 139–40; aniline salt of the N-acetyl derivative, mp 273.
14. Sulfonyl chloride, mp 79; anilide, mp 195.
15. Methylamine salt, mp 265–8; trimethylamine salt, mp 217–23.
16. Sulfonyl chloride, mp 82–4; sulfonyl morpholide, mp 101–3; sulfonyl benzyl amide, mp 114–115.
17. Anilide, mp 143; chloride, mp 88; p-toluidide, mp 196–7.
18. Anilide, mp 141.
19. 1-Naphthylamine salt, mp 241.
20. 4-Bromoanilide, mp 163; p-toluidide, mp 88; phenyl ester, mp 90–91.
21. The melting points of the 1-naphthylamides of the numbered acids are given in parentheses: 1 (215); 3 (196); 13 (171); 15 (245); 17 (146); 19 (183.5); 20 (60.5); 33 (66); 34 (125.5); 66 (166.5); 71 (84); 72 (134); 78 (157).

REFERENCES FOR TABLE 32

Allen, C. F. H., and Allan, J. V., *J. Org. Chem.*, **10**, 1 (1945). Addendum to The Identification of Aromatic Sulfonic Acids Containing an Amino Group. Lists the melting points of 11 substituted chlorobenzenesulfonamides.

Hazlet, S. E., *J. Am. Chem. Soc.*, **59**, 287 (1937). Some New Sulfonic Acid Esters. Lists the melting points of the esters of several phenols with benzenesulfonic acid and with *p*-toluenesulfonic acid.

Kahrasch, N., Potempa, S. J., and Wehrmeister, L., *Chem. Rev.*, **39**, 269 (1946). The Sulfenic Acids and Their Derivatives. Lists 32 sulfenic acids and the melting points of their halogen, amide, and sulfenate derivatives.

Latimer, P. H., and Bost, R. W., *J. Am. Chem. Soc.*, **59**, 2500 (1937). Identification of Some Aliphatic Sulfonic Acids. Lists the melting points of the phenylhydrazine salts of the alkylsulfonic acids up through octanesulfonic acid.

TABLE 33A
THIOCYANATES[a]

	Name of Compound	BP	MP		Name of Compound	BP	MP
1	Methyl-	131		7	Phenyl-	231	
2	Ethyl-	147		8	Benzyl-	256	43(38)
3	Isopropyl-	150		9	1,4-Butanedi-		61
4	Allyl-	161		10	1,2-Ethanedi-		90
5	Propyl-	165		11	*o*-Nitrophenyl-		136
6	Butyl-	186(182)					

[a] For the boiling points under reduced pressure of higher molecular weight thiocyanates, see *J. Am. Chem. Soc.*, **57**, 198 (1935).

TABLE 33B
ISOTHIOCYANATES

	Name of Compound	BP	MP	Melting Points of Derivatives with:	
				Aniline	Benzylamine
1	Methyl-	118			78
2	Ethyl-	131–2			103
3	*tert*-Butyl-	140	10.5		
4	Isopropyl-	150			126
5	Allyl-	152		98	94.5
6	Propyl-	153(743 mm)			88
7	*sec*-Butyl-	159			78
8	Isobutyl-	162			112
9	Butyl-	167			50
10	Phenyl-	220		153	
11	*o*-Tolyl-	239		139	
12	*p*-Tolyl-	239	26	141	
13	Benzyl-	243			148
14	4-Chlorophenyl-		45	154	
15	1-Naphthyl-		56	158	173
16	3-Nitrophenyl-		60.5	155	
17	2-Naphthyl-		63	182	
18	2-Nitrophenyl-		72		
19	4-Nitrophenyl-		113		

TABLE 34A
THIOETHERS (SULFIDES)—NONCYCLIC

	Name of Compound[a]	Note	Sulfide (BP)	d_4^{20}	n_D^{20}	Sulfone (MP)	Disulfide (BP)
1	Methyl		37.3	0.8458	1.4356	109	109.7
2	Ethyl, methyl	1	66.6	0.8422	1.4404	36	130
3	Methyl, vinyl		67.3	0.9026	1.4845		
4	Vinyl		84	0.9174			
5	Methyl, isopropyl		84.7	0.8291	1.4390		
6	Allyl, methyl		92	0.8767	1.4712		
7	Ethyl	1	92	0.8363	1.4425	71(74)	154
8	Ethyl, vinyl		92	0.873	1.4756		
9	Methyl, propyl		95.5	0.8438	1.4442		
10	tert-Butyl, methyl		99	0.8257	1.4402		
11	Chloromethyl, methyl		106				
12	Ethyl, isopropyl		107.4	0.8246	1.4407		165.5
13	Isobutyl, methyl		112.5	0.8335	1.4433		
14	2-Methylallyl, methyl		113				
15	Allyl, ethyl		116	0.8676			
16	Ethyl, propyl		118.5	0.8246	1.4407	25	173.7
17	tert-Butyl		120.4		1.4505		175.7
18	Isopropyl		120.7	0.8146	1.4388	36	176
19	Butyl, methyl	1	122.5	0.8427	1.4477		
20	Chloromethyl, ethyl		128			33	
21	Propyl, isopropyl		132	0.8269	1.4440		
22	sec-Butyl, ethyl		133.6	0.8353	1.4477		
23	Isobutyl, ethyl	1	134.2	0.8306	1.4452		
24	Allyl		138.6	$0.8876^{27/4}$	1.4895		
25	Chloromethyl, isopropyl		138–9				
26	2-Chloroethyl, methyl		140				
27	Propyl	1	142	0.8444	1.4787	30	194
28	Butyl, ethyl		144.3	0.8376	1.4491	50	193
29	Methyl, pentyl	1	145	0.843	1.448		
30	Bis(methylthio)methane		148			142(di)	
31	tert-Butyl		149		1.4505		201
32	Chloromethyl, propyl		150				
33	Chloromethyl		156			71–2	
34	2-Chloroethyl, ethyl		157				
35	Ethyl, 3-methylbutyl	1	160	0.8349	1.4495	13.5	
36	Isobutyl, chloromethyl		160–1				
37	4-Chlorophenyl, methyl		170			57–8	
38	Isobutyl	1	171	0.8285	1.4471	17	215
39	2-Methylallyl		173	0.8836	1.4862		
40	Bis(ethylthio)methane		184			104(di)	
41	Methyl, 2-thienyl		186				
42	Crotyl		186–7	$0.9032^{0/4}$	$1.495^{25/D}$		
43	Butyl	1,2	188	0.8386	1.4525	44	231
44	Methyl, phenyl		188	1.053	1.5869	88	
45	Dichloromethyl		189				
46	Benzyl, methyl		195–8		$1.5550^{25/D}$		
47	Ethyl, phenyl		204	$1.024^{15/4}$	$1.5701^{15/D}$	42	
48	Isopropyl, phenyl		208	0.9855	1.5468		

(Continued)

TABLE 34A (Continued)
THIOETHERS (SULFIDES)—NONCYCLIC

	Name of Compound[a]	Note	Sulfide (BP)	d_4^{20}	n_D^{20}	Related Compounds Sulfone (MP)	Disulfide (BP)
49	1,2-Bis(ethylthio)ethane		210			136.5(di)	
50	Methyl, p-tolyl	2	211–2	1.026	1.573	89	
51	3-Methylbutyl		215	0.8285	1.4471	31	250
52	2-Chloroethyl (mp 14.5)	2	215–7			56	
53	Allyl, phenyl		215–8	1.0275	1.5760	Oil	
54	Phenyl, propyl		220	0.9995	1.5571	44	
55	Ethyl, p-tolyl		220	1.0016[17.5/4]	1.5568	55–6	
56	Benzyl, ethyl	1	220			84	
57	1,3-Bis(ethylthio)propane		228–31			181–2(di)	
58	3-Methylbutyl, phenyl		240–2	0.9681	1.5380	36	
59	2-Chloroethyl, phenyl		245			45	
60	Bis(butylthio)methane		250			96(di)	
61	2-Chloroethyl, p-tolyl		255–7			78	
62	2-Hydroxyethyl, p-tolyl		282–3			55	
63	m-Tolyl		290			94	
64	Phenyl	2,3	296	1.116	1.6312	128	mp 60
65	Heptyl		298			80	
66	Phenyl, m-tolyl		309.5	1.0937[15/4]			
67	Phenyl, o-tolyl		309.9	1.1012[15/4]			
68	Octyl		310	0.8412[25/4]		76	
69	Phenyl, p-tolyl (mp 16)		311.5	1.090[16/4]		124.5	

[a] Another system of nomenclature uses "thia" to indicate sulfur in a chain and numbers both carbon and sulfur atoms in that chain. For example, ethyl sulfide would be 3-thiapentane; bis(methylthio)methane would be 1,3-dithiapentane; ethyl phenyl sulfide would be (1-thiapropyl)benzene; and isopropyl phenyl sulfide would be (2-methyl-1-thiapropyl)benzene.

NOTES FOR TABLE 34A

1. The adducts with mercuric chloride have the melting points given in the parentheses for the compounds as numbered in the table: 2 (128); 7 (119.5); 19 (116.5); 23 (108); 27 (122 (127)); 29 (127); 35 (87); 38 (131); 43 (113 (111)); 56 (142).

2. The sulfoxides of the numbered compounds have melting points as shown in parentheses: 43 (33); 50 (50–54); 52 (109.5); 64 (70.4).

3. The disulfone of the disulfide melts at 193–4.

REFERENCES FOR TABLE 34A

Decker, Q. W., and Post, H. W., *J. Org. Chem.*, **25**, 249 (1960). Studies in Organosilicon Chemistry. XXXVIII. Further Studies in Sila-Organic Polysulfides. Method of preparation is given along with the boiling points, refractive indices, and infrared spectra data for a number of compounds.

Krug, R. C., and Rigney, J., *J. Org. Chem.*, **23**, 1697 (1958). The Chemistry of Some Unsaturated Sulfones.

Reid, E. E., *Organic Chemistry of Bivalent Sulfur*, Vol. II, Chemical Publishing Co., New York, 1960. For data on hydroxy sulfides see pp. 305–313; alkoxysulfides, pp. 314–319; halogenated sulfides, pp. 322–335; nitrophenyl sulfides, pp. 354–359; ketosulfides, pp. 360–368; aminosulfides, pp. 340–350, sulfides acids, pp. 203–227. Considerable data on cyclic sulfides is given on pp. 89–109.

TABLE 34B
THIOETHERS (SULFIDES)—CYCLIC

	Name of Compound[a]	Note	BP	$D^{20/4}$	$n^{20/D}$	Sulfone (MP)
1	Ethylene sulfide	3	55–6	1.0046	1.4914	
2	α-Methylethylene sulfide	4	76	$0.946^{18/4}$	1.473	
3	Thiophene		84.2	1.06485	1.52890	
4	α,α-Dimethylethylene sulfide		87		1.4641	
5	Trimethylene sulfide	5	95	1.0200	$1.506^{23/D}$	76
6	α-Ethylethylene sulfide		104	$0.930^{18/4}$	$1.475^{19/D}$	
7	α-Methyltrimethylene sulfide		106	0.9571	1.4830	bp 251–3
8	2-Methylthiophene		112.4	1.02183	1.52042	
9	α,α′-Dimethyltrimethylene sulfide		113–4	$0.8710^{18/4}$	$1.4502^{18/D}$	bp 255–6
10	3-Methylthiophene		115.4	1.02183	1.52042	
11	Tetramethylene sulfide	6	119–21		1.5047	28.4
12	β,β-Dimethyltrimethylene sulfide		120		$1.4739^{18/D}$	55
13	α-Methyltetramethylene sulfide		133.2	0.9555	1.4909	bp 280
14	3-Ethylthiophene		136	0.9980	1.5146	
15	2,5-Dimethylthiophene		136.7	0.9850	1.5129	
16	β-Methyltetramethylene sulfide		138.7	0.9634	1.4924	
17	2,4-Dimethylthiophene		140.7	0.9956	1.5104	
18	2,3-Dimethylthiophene		141.6	1.0021	1.5192	
19	Pentamethylene sulfide		141.8	0.9856	1.5067	97
20	α,α′-Dimethyltetramethylene sulfide (*trans*)	1	142	0.9188	1.4776	bp 278
21	α,α′-Dimethyltetramethylene sulfide (*cis*)	2	142.3	0.9222	1.4799	bp 278
22	3,5-Dimethylthiophene		145	1.008	1.5212	
23	1,4-Thioxane		148	1.1177	1.5070	130
24	α-Methylpentamethylene sulfide		152	0.9428	1.4905	68.5
25	2-Isopropylthiophene		153	0.9678	1.5038	
26	3-Isopropylthiophene		157	0.9733	1.5052	
27	β-Methylpentamethylene sulfide		158	0.9473	1.4922	83
28	2-Propylthiophene		158.5	0.9687	1.5049	
29	γ-Methylpentamethylene sulfide		158.6	0.9471	1.4923	
30	5-Ethyl-2-methylthiophene		160.1	0.9661	1.5073	
31	2,6-Dimethyl-1,4-thioxane		160–1		1.4733	105.5
32	3-Propylthiophene		161	0.9716	1.5057	
33	2-Ethyl-3-methylthiophene		161	0.9815	1.5105	
34	3,5-Dimethyl-1,4-thioxane		162			102
35	2-Methyl-4-ethylthiophene		163	0.9742	1.5098	
36	2,3,5-Trimethylthiophene		164.5	0.9753	1.5112	
37	Hexamethylene sulfide		170		1.5125	71
38	2,2-Dimethyldithiolane		171			232(di)
39	2,3,4-Trimethylthiophene		172.7	0.995	1.5208	
40	2-Methyldithiolane		172.3			198(di)
41	Dithiolane		175			205(di)
42	α,γ′-Dimethyltetramethylene sulfide		178	0.9265	1.4818	
43	2-Ethyldithiolane		191–2			124

[a] See notes 3 to 6 for additional current names for these and related compounds.

NOTES FOR TABLE 34B

1. The *trans* isomer forms an adduct with one formula weight of mercuric chloride, mp 111.
2. The *cis* isomer forms an adduct containing two formula weights of mercuric chloride, mp 180.
3. Other names: thiacyclopropane and thiirane.
4. Other names: 2-methylthiacyclopropane and 2-methylthiirane.
5. Other names: thiacyclobutane and thietane.
6. Other names: thiacyclopentane, thilane, and tetrahydrothiophene.

TABLE 34C
THIOETHERS (SULFIDES)—SOLIDS

	Name of Compound	Note	Sulfide (MP)	Sulfoxide	Sulfone	Disulfide
1	Ethyl, 2-naphthyl		16		43–5	
2	Ethyl, hexadecyl		19		88	
3	2-Methyl-1,4-dithiane		20		304(di)	
4	Decyl		27		206–7	
5	Undecyl		35			
6	2-Bromoethyl		35		111–2	
7	4-Iodophenyl, phenyl		35		141	
8	4-Bromophenyl, methyl		37.5		56–7	
9	Dodecyl		40.5			
10	Benzyl, phenyl		41	123	146	
11	4-Chlorobenzyl		41			
12	1-Naphthyl, phenyl		41.5		100	
13	Ethyl, 4-nitrophenyl		44		138.5	
14	Isopropyl, 4-nitrophenyl		44.5		115.3	
15	4-Methoxyphenyl	1	46	96	130	N
16	Benzyl		50	134.8	151.7	73
17	2,5-Dichlorophenyl, methyl		51		88	
18	2-Naphthyl, phenyl	2	51.8		115–6	67.5N
19	Bis(phenylthio)methane		52		120–1(di)	
20	Tetradecyl		53.8			
21	1,3-Dithiane		54		330(308)(di)	
22	Bis(benzylthio)methane		55		216(di)	
23	4-Nitrophenyl, phenyl		55		142	
24	p-Tolyl		57	93	158	48
25	Octyl		57		76	
26	Hexadecyl		61.3	99.8	103.4	
27	2-Chloroethyl, 4-nitrophenyl		62		128	
28	Methyl, 2-naphthyl		64			
29	o-Tolyl		64		134–5	
30	Methyl, 2-nitrophenyl		65		106	
31	Benzyl, 4-bromophenyl		65		159	
32	1,2-Bis(phenylthio)ethane		69–70		180(di)	
33	Octadecyl		71			
34	2-Phenyl-1,3-dithiane		71–2		265(di)	
35	Methyl, 4-nitrophenyl		72		141	
36	3-Phenylpropyl		73		117	
37	4-Methylbenzyl		76		197	
38	2-Nitrophenyl, phenyl		77(80)		145	
39	Cinnamyl, phenyl		78	90–1	111–2	
40	4-Nitrobenzyl, phenyl		79		209.5	
41	1,2-Bis(p-tolylthio)ethane		80–1		200–1(di)	
42	4-Chlorophenyl		89(98)		148	
43	2-Phenylethyl		92.5(90)	69	100.6	
44	1-Naphthyl		110		187	
45	1,4-Dithiane	3	111	N	200(mono)N	
46	4-Bromophenyl		112	153	172	
47	Benzyl, 4-nitrophenyl		123		172	
48	1,3-Bis(2-nitrophenylthio)propane		140		156–7(di)	
49	2-Naphthyl		151		177	
50	4-Nitrobenzyl		158–9	212	260.5	
51	2,4-Dinitrophenyl		193–7		240–1	
52	2,4,6-Trinitrophenyl		232		307	

NOTES FOR TABLE 34C (THIOETHERS)

1. The disulfone is reported to melt at 221, rapid heat, and to decompose at 210–2 when heated slowly.
2. The disulfone melts at 166.
3. The sulfoxide-sulfone melts at 279; the disulfone does not melt at 330.

TABLE 35A
THIOLS (MERCAPTANS AND THIOPHENOLS) (LIQUIDS)

	Name of Compound	Note	BP	Melting Point of Derivatives					
				Recommended		Others			
				2,4-Dinitrophenyl Thioether	2,4-Dinitrophenyl Sulfone	3,5-Dinitrothiobenzoate	3-Nitrothiophthalate	1-Anthraquinone Thioether	Mercury Salt
1	Methanethiol		6	128*	189.5			221	176
2	Ethanethiol		36	115	160	62	149	184	85
3	2-Propanethiol		56	94	140.5	84	145	134	63
4	Propanethiol		67	84	127.5	52	137	151	72
5	2-Butanethiol		84.5	66	120				189
6	2-Methyl-1-propanethiol		88	76	105.5	64	136	144	95
7	2-Propene-1-thiol		90	72	105				
8	Butanethiol		97	66	92	49	144	112.5	86
9	2-Methyl-1-butanethiol		97						60
10	2-Methoxyethanethiol		110–12	90					
11	1-Phenylethane-1-thiol		119–20		161				
12	2-Chloropropanethiol		125	76–7					
13	2-Chloroethanethiol		125–6	95–7					
14	2-Ethoxyethanethiol		125–6	66					
15	Pentanethiol		126	80	83	40	132	114	75
16	1,2-Ethanedithiol		147	248					
17	Hexanethiol		151	74	97			129	58
18	2-Hydroxyethanethiol	1	158	100–2					123
19	Cyclohexanethiol		159	148	172				78
20	2-Thiophenethiol		166(171)	119	143				
21	1,3-Propanedithiol		169(173)	194					
22	Thiophenol	11	169	121	161	149	131		
23	Heptanethiol		176	82	101	53	96	132	77
24	o-Thiocresol		194	101	155				
25	Phenylmethanethiol	2	194	130	182.5	120	137		
26	m-Thiocresol		195	91	145				
27	1,4-Butanedithiol	3	195.6						
28	Octanethiol		199	78	98			95	71
29	2-Phenylethanethiol		199	89	133				
30	2-Chlorothiophenol		205	138					
31	Bis(2-mercaptoethyl) ether	4	217						
32	1,5-Pentanedithiol	5	217						
33	Nonanethiol		220	86	92			117.5	
34	1,6-Hexanedithiol	6	237						

TABLE 35B
THIOLS (MERCAPTANS AND THIOPHENOLS) (SOLIDS)

	Name of Compound	Note	MP	Recommended		Other
				2,4-Dinitrophenyl Thioether	2,4-Dinitrophenyl Sulfone	Derivative (MP)
1	Hexadecanethiol		19	91(96)	105	
2	1,10-Decanedithiol		20			Dibenzoyl, 57
3	2-Aminothiophenol		26	152		Diacetyl, 135
4	m-Phenylenedithiol	7	26–7			T.N.B adduct, 76–7
5	4-Dimethylaminothiophenol		28.5	176		
6	o-Phenylenedithiol	8	29			Diacetyl, 88.5
						Dibenzoyl, 74–5
7	p-Thiocresol	9	43(30)	103	189.5	Diacetyl, 66
						Dibenzoyl, 161
8	4-Aminothiophenol	10	45			
9	4-Chlorothiophenol		54	123		
10	2-Nitrothiophenol		58–61	131–3		
11	4-Bromothiophenol		75	142	190	
12	3-Nitrothiophenol		75			
13	4-Nitrothiophenol		77	160		
14	4-Nitro-1-naphthalenethiol		77–9	193		
15	2-Naphthalenethiol		81	145	228	
16	4-Iodothiophenol		85–6	140.5		
17	4-Amino-1-naphthalenethiol		91–3			N-acetyl, 173
18	Bis(4-mercaptophenyl) ether		98			Diacetyl, 68
19	p-Phenylenedithiol		98			Diacetyl, 126
20	2-Aminoethanethiol		98–100	94.5		
21	Triphenylmethanethiol		107	190		
22	2-Nitro-4-bromothiophenol		110	142		
23	4-Biphenylthiol		111	146	170	
24	4-Mercapto-1-naphthol		114			Diacetyl, 77
25	Bis(4-mercaptophenyl) sulfide		114			Diacetyl, 65
26	2,4,6-Trinitrothiophenol		114	217		
27	1,5-Naphthalenedithiol		119			Diacetyl, 187–9
						Dibenzoyl, 232
28	2-Nitro-4-chlorothiophenol		120–2	141		
29	2,4-Dinitrothiophenol		131–2	193–7	240–1	
30	6-Mercapto-2-naphthol		137			Diacetyl, 107
31	2,7-Naphthalenedithiol		181(174)			Diacetyl, 110
						Dibenzoyl, 152–3

NOTES FOR TABLES 35A AND 35B

1. S-4-Nitrophenyl derivative (from ether), mp 62; O,S-dibenzoyl derivative, mp 39; S-Phenylurethane, mp 59–60 (from benzene).
2. Phenyl ether, mp 42; compound oxidizes to dibenzyl sulfide.
3. Dibenzoyl derivative, mp 49.
4. 4-Nitrobenzoate derivative, mp 106.5.
5. Dibenzoyl derivative, mp 45.
6. Dibenzoyl derivative, mp 57.
7. Compound boils at 245.
8. Compound boils at 238–9.
9. S-p-Tolyl derivative, mp 57; the O-ethyl ether, mp 41, forms a S-benzoyl derivative, mp 106. N-Phenylurethan, mp 132.
10. N-Acetyl derivative, mp 154 (may be white or yellow); N-benzoyl derivative, mp 182 (from ethanol).
11. Phenyl urethan, mp 128S.

REFERENCES FOR TABLES 35A AND 35B

Hendrickson, J. G., and Hatch, L. F., *J. Org. Chem.*, **25**, 1747 (1960). S-Alkylmercaptosuccinic Acids as Solid Derivatives of Olefins, Alkyl Bromides, and Mercaptans. Gives the method of preparation and lists the melting points of the derivatives of 46 compounds.

Reid, E. E., *Organic Chemistry of Bivalent Sulfur*, Vol. I., Chemical Publishing Co., New York, 1958. For data on thiols, see pp. 401–410. Data on mercapto acids may be found on pp. 469–473.

Appendix

Apparatus

Most of the procedures outlined in this book may be carried out by using the glassware and laboratory equipment normally found in all organic laboratories and stockrooms. The glassware should be of borosilicate quality, and only the smaller volume items are generally used. Specialized equipment for specific procedures was described in several of the chapters and reference was made to a source of supply for these items. However, in most cases, these procedures may be adequately carried out by using other equipment that may already be available in the laboratory.

List of Reagents and Chemicals

It is assumed that all laboratories are supplied with concentrated hydrochloric acid, nitric acid, sulfuric acid, and ammonium hydroxide, and with the usual $6N$ dilutions of these chemicals and sodium hydroxide; hence, they are not included in the listings. In many instances the same chemical is used in several different concentrations in different parts of the text and where this is the case only the most concentrated solution is listed; for example, sodium hydroxide in aqueous solution is used as $6N$, 10 per cent, 5 per cent, and 2.5 per cent solutions, but it is expected that the lower concentrations will be prepared as needed by diluting the 10 per cent stock solution. All of the chemicals that are recommended for any use in the text have been listed. A large percentage of the experiments can be made with a much less extensive list of chemicals, since many of the listed chemicals are for single tests or are for the preparation of derivatives in cases where several different reagents can be effectively used.

Special Reagents

Barfoed's reagent. Dissolve 16.6 g of crystallized copper acetate in 245 ml of water and add 2.4 ml of glacial acetic acid.

Benedict's Reagent. Dissolve 4.3 g of finely pulverized copper sulfate in 25 ml of hot water, cool and dilute to 40 ml with water. Dissolve separately 43 g of sodium citrate and 25 g of anhydrous sodium carbonate (or an equivalent amount of the hydrate form) in 150 ml of water. Heat to effect solution; cool; then add the copper sulfate solution and dilute to 250 ml. Keep the solution in a cork-stoppered bottle.

Ceric Nitrate Reagent. Dissolve 90 g of ceric ammonium nitrate, $Ce(NH_4)_2(NO_3)_6$, in 225 ml of warm $2N$ nitric acid.

"Doctor" Solution (Sodium Plumbite Reagent). Dissolve 45 g of sodium hydroxide in 240 ml of water and then dissolve 7.5 g of litharge (PbO) in the hot caustic solution.

Schiff's Fuchsin Aldehyde Reagent. Prepare 100 ml of a 0.1 per cent aqueous solution of *p*-rosaniline hydrochloride. Add 4 ml of a saturated aqueous solution of sodium bisulfite. Allow the mixture to stand for 1 hour and then add 2 ml of concentrated hydrochloric acid.

Another commonly used method of preparing this reagent is as follows. Dissolve 250 mg of *p*-rosaniline hydrochloride (fuchsin) in 50 ml of warm water. Cool and saturate the solution with sulfur dioxide until the pink color has disappeared. Add 250 mg of decolorizing carbon, shake, filter and dilute to 250 ml. If the stock solution develops a pink color, repeat the saturation with sulfur dioxide before using it.

Seliwanoff's Reagent. Dissolve 125 mg of pure resorcinol in 250 ml of dilute hydrochloric acid (83 ml of concentrated acid and 167 ml of water).

Tollens' Reagent. This reagent is unstable and should be made at the time it is to be used by following the directions given in Test P.6F.

General Reagents

All solutions are aqueous unless some other solvent is specified. In cases where the salt is usually obtained in the hydrated form the actual weights of chemicals are given to make the desired concentration of the anhydrous compound. *It is assumed that the less concentrated solutions will be made as needed from the stock solutions listed here.*

Ammonium molybdate, 5 per cent.

Barium chloride, 5 per cent.

Bromine, 2 per cent in carbon tetrachloride.

Chloranil (tetrachloroquinone), a saturated solution in dioxane.

Ferric chloride, 10 per cent (16 g of the hexahydrate per 100 ml of solution).

Formaldehyde, 40 per cent (formalin).

Hydrochloric acid, 10 per cent; also $1.2N$ and $2N$.

Hydrogen peroxide, 6 per cent.

Hydroxylammonium chloride, $1N$ in methanol (7 g in 100 ml).

Iodine, 10 per cent iodine in a 20 per cent solution of potassium iodide.

Lead acetate, 5 per cent; also, a saturated solution in ethanol.

Methone (5,5-dimethylcyclohexane-1,3-dione), 5 per cent in ethanol.

Nickelous chloride, 10 per cent.

Phenol, 4 per cent in water.

Potassium dichromate, 5 per cent.

Potassium hydroxide, $5N$ in 80 per cent methanol (28 g dissolved in 20 ml of water, then diluted to 100 ml with methanol); also, $2N$ in 80 per cent methanol (10 ml of the $5N$ diluted to 25 ml with 80 per cent methanol); also, $6N$ in water (33.6 g in 100 ml).

Potassium iodate, 4 per cent.

Potassium iodide, 20 per cent.

Potassium periodate, 5 per cent.

Potassium permanganate, 1 per cent.

Silver nitrate, 5 per cent; also, a saturated solution in ethanol.

Sodium acetate, 10 per cent (16.5 g of the trihydrate per 100 ml of solution).

Sodium bicarbonate, $1.1N$.

Sodium bisulfide (or potassium bisulfide), 10 per cent. (This reagent may be prepared by bubbling hydrogen sulfide into a 10 per cent solution of sodium or potassium hydroxide until the pH reaches about 8.)

Sodium carbonate, 10 per cent.

Sodium hydroxide, 10 per cent; also, $6N$ (24 g per 100 ml of solution) and $2.5N$.

Sodium hypochlorite, 5 per cent ("Clorox").

Sodium nitroprusside, 10 per cent.

Pure Chemicals

The following pure chemicals are used in one or more of the procedures, aside from the formation of derivatives. In many cases, alternative methods are provided and, thus, a much-restricted list of chemicals may be adequate.

Acetaldehyde
Acetic acid, glacial
Acetic anhydride
Acetone
Acetyl chloride

cis-Aconitic anhydride
Aluminum chloride, anhydrous
Aluminum ethoxide
4-Aminoantipyrine
4-Aminodimethylaniline

Aminoguanidine
Ammonium vanadate
Anthrone
Aniline
Barium hydroxide
Benzene
Benzenesulfonyl chloride
Benzidine
Benzoyl chloride
Benzoyl peroxide
Benzylamine
Bromine
N-Bromosuccinimide
Brucine sulfate
Butanol
tert-Butyl alcohol
Calcium chloride
Calcium hydroxide
Carbon disulfide
Carbon tetrachloride
Charcoal, decolorizing
Chloranil
Chloroform
Collidine
Chromic acid
Citric acid
Copper acetate
Copper sulfate
o-Dianisidine
Diethylene glycol (2,2'-oxydiethanol)
Diglyme (bis(2-methoxyethyl) ether)
N,N-Dimethylaniline
4-Dimethylaminobenzaldehyde
5,5-Dimethyl-1,3-cyclohexanedione
N,N-Dimethyl-1-naphthylamine
Dimethyl oxalate
2-Diphenylacetyl-1,3-indandione-1-hydrazone
N,N'-Diphenylbenzidine
3,5-Dinitrobenzoic acid
2,4-Dinitrofluorobenzene
2,4-Dinitrophenylhydrazine
Dioxane
Diphenylamine
Ethyl acetate
Ethanol
Ethyl ether
Ethylene glycol (1,2-ethanediol)
Ferric ammonium sulfate
Ferric chloride
Ferrous ammonium sulfate
Ferrous sulfate
Fluorescein chloride
Glycerol
Hexane
2-Hydrazinobenzothiazole
Hydroxylammonium chloride
N-Hydroxylbenzenesulfonamide
8-Hydroxyquinoline
Indicators:
   Alizarin yellow R
   Bromothymol blue
   Bromocresol purple
   Phenolphthalien
   Thymol blue
   Benzeneazodiphenylamine
   Methylene blue
N-Iodosuccinimide
Isatin
Isopropyl alcohol (2-propanol)
Lanthanum chloranilate (2,5-dichloro-3,6-dihydroxy-*p*-benzoquinone, lanthanum salt)
Lead oxide
Magnesium
Magnesium sulfate, anhydrous
Methanol
Molybdic acid
Morpholine
1-Naphthol
2-Naphthol
1-Naphthylamine
Nickel chloride
Nickel sulfate
Ninhydrin
2-Nitrobenzaldehyde
4-Nitrobenzenediazonium fluoroborate
Periodic acid
Peroxyacetic acid
Phenanthracene
Phenol
Phenylhydrazine hydrochloride
2-Phenylindole
Phloroglucinol
Phosphoric acid, 85 per cent
Phosphorous pentachloride
Phosphorous pentoxide
Phthalic anhydride
Picric acid
Piperonal
Potassium carbonate, anhydrous
Potassium dichromate
Potassium fluoride
Potassium hydroxide

Potassium iodate
Potassium iodide
Potassium thiocyanate
Propionic acid
Propylene glycol
Pyridine
Quinhydrone
Resorcinol
Rhodamine B
Sodium
Sodium acetate
Sodium azide
Sodium bisulfite
Sodium carbonate, anhydrous
Sodium hydroxide
Sodium nitrite
Sodium sulfate, anhydrous

Stannous chloride
Sulfanilic acid
Tetraisopropyl titanate
Tetraphenyl borate
3,3′,5,5′-Tetrabromophenolphthalein, ethyl ester
Thiobarbituric acid
Thionyl chloride
Thiourea
Tin
Toluene-3,4-dithiol, zinc complex
p-Toluenesulfonyl chloride
p-Toluidine
Uranyl acetate
Zinc, dust
Zinc chloride, anhydrous

Selected Derivatizing Reagents

The following reagents, not listed under the heading *Pure Chemicals*, are available for the formation of derivatives. The needed reagents may be determined after a decision has been reached as to the derivatives desired.

Benzyl chloride (α-chlorotoluene)
S-Benzylisothiuronium chloride
Benzylphenylhydrazine
4-Bromoaniline
4-Bromobenzenesulfonyl chloride
4-Bromophenacyl bromide
4-Bromophenylhydrazine hydrochloride
4-Bromophenyl isocyanate
Chloroacetic acid
Chloroplatinic acid
Chlorosulfonic acid
4,4′-Diaminodiphenylmethane
N,N′-Dicyclohexylcarbodiimide
3,5-Dinitrobenzoyl chloride
2,4,7-Trinitrofluorenone
2,4-Dinitrobenzenesulfenyl chloride
2-Diphenylacetyl-1,3-indandione-1-hydrazone
Diphenylcarbamyl chloride
Diphenylhydrazine
2-Mercaptobenzothiazole
Mercuric iodide
Mercuric sulfate
Methylphenylhydrazine
Methyl iodide
Methyl p-toluenesulfonate

2-Naphthylamine
1-Naphthylhydrazine
1-Naphthyl isocyanate
2-Naphthyl isocyanate
Nitric acid, fuming
3- and 4-Nitrobenzenesulfonyl chloride
4-Nitrobenzoyl chloride
4-Nitrobenzyl chloride
4-Nitrophenylcarbamyl chloride
6-Nitro-2-mercaptobenzothiazole
4-Nitrophenylhydrazine
2-, 3-, and 4-Nitrophenyl isocyanate
3-Nitrophthalic anhydride
Phenacyl chloride
p-Phenetidine
o-Phenylenediamine
Phenylhydrazine hydrochloride
Phenyl isocyanate
Phenyl isothiocyanate
Picrolonic acid
Piperazine
Potassium phthalimide
Resorcylic acid
Semicarbazide hydrochloride
Silver 3,5-dinitrobenzoate
Sodium anthraquinone-1-sulfonate

Sodium saccharin
Tetraphenylstibonium sulfate
Thioglycolic acid
Thiosemicarbazide

o-Toluidine
p-Tolyl isocyanate
Triiodophenol
Xanthydrol

COMMON DRYING AGENTS FOR ORGANIC COMPOUNDS

Anhydrous Substance	Applicable to	Not Applicable to	Drying Power	Relative Efficiency
Calcium chloride	Hydrocarbons, halides, ethers, esters	Hydroxy and amino compounds	High below 30°	Medium
Calcium sulfate ("Drierite")	All compounds	None	Low	Good
Magnesium sulfate	All compounds	None	High	Good
Potassium carbonate	Amines, alcohols, ketones	Acids	Medium	Medium
Potassium hydroxide	Amines, hydrazines, saturated hydrocarbons	Most compounds	High	Good
Phosphorus pentoxide	Halides, hydrocarbons, nitriles	Most compounds	High	Excellent
Sodium hydroxide	Amines, hydrazines, saturated hydrocarbons	Most compounds	High	Excellent
Sodium metal	Ethers, saturated hydrocarbons	Most compounds	High	Excellent

Cleaning Solutions

Chromic acid cleaning mixture. Dissolve 10 g of sodium dichromate in 10 ml of water in a 400 ml beaker. Add slowly with careful stirring 200 ml of concentrated sulfuric acid. The temperature will rise to nearly 80°. Allow the mixture to cool to about 40°, and place in a dry glass-stoppered bottle. Care should be exercised in making and handling this solution.

Trisodium phosphate solution. Glassware that does not contain tars may be cleaned with a 15 per cent solution of trisodium phosphate. A warm solution with the aid of an abrasive powder, such as pumice, is safer to handle and cleans as well as or better than chromic acid solutions.

The Neutralization Equivalent

The neutralization equivalent of an acid or acidic compound is actually its equivalent weight and hence is of value in identifying the compound. If the substance is a liquid, weigh it in a dropping bottle and then transfer a few drops of the liquid to the titration flask and reweigh the dropping bottle. The weight of the sample is the difference in the two weights of the dropping bottle. If the substance is even reasonably volatile, it is best to have the solvent (water, alcohol, or a mixture of water and alcohol depending on the solubility of the substance) in the titration flask before adding the substance so as to retard its evaporation.

Transfer an accurately weighed sample (100 to 300 mg) of the acidic substance to a 125 to 250 ml flask and add 25 to 35 ml of water, alcohol, or a mixture of these solvents to dissolve the substance. Titrate the solution with a standardized solution of sodium hydroxide (approximately $0.1N$) using phenolphthalein as the indicator. Calculate the neutralization equivalent of the compound as follows:

$$N.E. = \frac{\text{wt. of sample in grams} \times 1000}{\text{ml of NaOH} \times N \text{ of the NaOH}}$$

This formula may be written as:

$$N.E. = \frac{\text{milligrams of sample}}{\text{milliequivalents of NaOH}}$$

The Saponification Equivalent of an Ester

The objective is to hydrolyze the ester and determine the amount of alkali that reacts with the acid produced. Saponification of an ester by aqueous sodium hydroxide is inhibited because most esters are sparingly soluble in the aqueous solution. Solutions of potassium hydroxide in diethylene glycol, potassium hydroxide in ethylene glycol, and potassium hydroxide in a mixture of glycol monoethyl ether have been used. It has been found that a solution of potassium hydroxide in 80 to 90 per cent methanol, ethanol, propanol, or 2-propanol makes a satisfactory saponifying agent for esters except for those with very high weights or that for some reason are difficult to hydrolyze. In such cases, one of the glycols should be substituted for the lower alcohols. The alcoholic solution of potassium hydroxide tends to attack glass and to absorb carbon dioxide from the air during the refluxing of the mixture, hence it is wise to standardize the alkaline solution under conditions which are similar to those used for the saponification of the ester. Since many esters are at least moderately volatile it has been found advisable to have the alcoholic solution of potassium hydroxide in the flask before adding the weighed sample of the ester so as to dissolve the ester and decrease its loss by evaporation. Liquid esters should be *weighed by difference* from a dropping bottle.

Dissolve approximately 3 g of potassium hydroxide in 60 ml of alcohol and allow any sediment to settle. Add exactly 25 ml of the alcoholic solution of potassium hydroxide to each of two 125 to 250 ml flasks. To one of the flasks, add an accurately weighed sample (300 to 400 mg) of the ester and use the other flask as a "blank." Attach reflux condensers to both flasks and gently boil the solution for one hour. Cool the solutions and rinse each condenser with 10 ml of water, catching the washings in the solutions. Titrate each of the solutions with standardized hydrochloric acid (approximately 0.5N) using phenolphthalein as an indicator.

The difference in the volumes of hydrochloric acid required for the solution which contained the ester and that required for the blank represents the amount of potassium hydroxide which reacted with the ester. The number of milliliters of a solution multiplied by the normality of the solution equals the number of milliequivalents, hence the volume of the hydrochloric acid (the difference in the volumes required for the two titrations) multiplied by the normality of the hydrochloric acid equals the milliequivalents of potassium hydroxide required for the ester. Calculate the saponification equivalent of the ester as follows:

$$S.E. = \frac{\text{milligrams of ester}}{\text{milliequivalents of KOH}}$$

Methods have been developed that eliminate the use of the blank in the above method by using double-indicator titrations instead. Details for this method may be found in the references. These are semimicro and micro methods.

Saponification number. Industrially, the saponification number is more generally used than the saponification equivalent. The saponification number is defined as the number of milligrams of potassium hydroxide required to saponify one gram of the oil or fat. It may be calculated from the data obtained from the determination of the saponification equivalent by the the use of the following formula:

$$S.N. = \frac{\text{milliequivalents of KOH} \times 56.1}{\text{grams of ester}}$$

The Iodine Number of a Fat or Oil

The chemical literature records several fundamental methods and many modifications of those methods for the determination of the iodine number (centigrams of iodine absorbed per gram of the sample). Two procedures have been selected, one for a macro quantity, and one for a micro quantity. The iodine numbers, as determined by various methods, do not always agree.

Macro Method

The method used here is that of Wijs. More details of this method and also other methods are given in the *Official and Tentative Methods of the American Oil Chemists Society* and the *Official Methods of the Association of Official Agricultural Chemists* publications.

Add 100 to 400 mg of the sample to a 250 to 500 ml glass-stoppered flask containing 20 ml of carbon tetrachloride (or chloroform). Add 25 ml of the iodine solution (Wijs) which has been prepared by dissolving 13 g of pure iodine in a liter of pure acetic acid and then passing chlorine gas into the solution until the quantity of sodium thiosulfate required for titration is not quite doubled. Moisten the stopper with 5 drops of a 15 per cent solution of potassium iodide. Let the flask stand in the dark for 30 minutes and then add 20 ml of 15 per cent potassium iodide and 100 ml of water. Titrate the solution with sodium thiosulfate (about $0.1N$) until the yellow color almost disappears and then add 3 drops of a 1 per cent starch suspension and continue the titration until the blue color entirely disappears. Near the end of the titration, shake the solution vigorously. Run a blank determination (without the fat or oil). The iodine number is the number of centigrams of iodine that reacts with one gram of fat or oil.

$$I.N. = [(V \times N \text{ of } Na_2S_2O_3 \text{ for blank}) - (V \times N \text{ of } Na_2S_2O_3 \text{ for sample})] \times 0.127 \times \frac{100}{\text{wt. of sample}}$$

Micro Method

The Kaufman method[1] is given here because it has proved satisfactory in our laboratory. The reagents required are: (a) $0.1N$ bromine in methanol saturated with dry sodium bromide; (b) 10 per cent potassium iodide; (c) $0.05N$ sodium thiosulfate; and (d) alcohol-free chloroform.

The fat or oil (10 to 30 mg) is dissolved in 2 ml of chloroform in a 150 ml iodine flask. Place 2 ml of chloroform in another flask and use it as a blank to be treated by the same procedures as the sample. Exactly 5 ml of the $0.1N$ bromine solution are added from a microburet. The mixture

---

[1] *Ber.* **70B**, 2554 (1937); [c.f. *C.A.* **32**, 1961 (1938)].

is shaken. The reaction is complete in 1 to 5 minutes. Add 3 ml of the potassium iodide solution and shake the mixture. The excess bromine reacts with the potassium iodide to liberate free iodine which is then titrated with the sodium thiosulfate solution. Since the number of milliequivalents of sodium thiosulfate is equal to the number of milliequivalents of excess bromine and since the milliequivalent weight of iodine may be considered as 0.127, the following equation may be used to calculate the iodine number.

$$I.N. = [V \times N \text{ of } Na_2S_2O_3 \text{ for blank)} \\ - (V \times N \text{ of } Na_2S_2O_3 \text{ for sample})] \times 0.127 \times \frac{100}{\text{wt. of sample}}$$

REFERENCES

Neutralization Equivalents

E. Ellenbogen and E. Brand, *Anal. Chem.*, **27**, 2007 (1955). N. E. of amino acids, peptides, and peptones.

N. Esenberg and M. A. Settle, *Chemist-Analyst*, **50**, 111 (1961). N. E. of 5-bromovaleric acid.

J. Mitchell, I. M. Kolthoff, E. S. Proskauer, and A. W. Weissberger, *Organic Analysis*, Vol. II, pp. 10–14, Interscience, New York, 1954.

W. T. Robinson, Jr., et al., *Talanta*, **3**, 307 (1960). N. E. of alcohols.

Saponification Numbers and Equivalents

P. A. G. Amo and D. Martin, *C.A.*, **54**, 925 (1960). Potentiometric method.

B. Baekler et al., *Pharm. Acta Helv.*, **36**, 338 (1961). Determination of strongly colored substances.

P. Budowski and A. Bondi, *C.A.*, **54**, 21790 (1960). Use of 0.5N ethanolic KOH compared with higher concentrations.

D. T. Englis and J. E. Reinschreiber, *Anal. Chem.*, **21**, 602 (1949).

E. A. Fehnel and E. D. Amstutz, *Ind Eng. Chem.*, *Anal. Ed.*, **16**, 53 (1944). New indicator for colored solutions.

G. A. Gallagher et al., *J. Am. Chem. Soc.*, **79**, 4324 (1957). Low-weight esters.

W. Hessler and H. Marsen, *C.A.*, **56**, 8866 (1962). Use of fluorescence indicators with waxes.

D. Ketchum, *Ind. Eng. Chem.*, *Anal. Ed.*, **18**, 273 (1946). Micro method.

R. N. Lal and J. B. Lal, *C.A.*, **55**, 18019 (1961); **54**, 23203 (1960).

F. A. Lee, *J. Assoc. Offic. Ag. Chemists*, **41**, 899 (1958). Microdetermination of saponification numbers.

K. Mercali and W. Reimann III, *Ind. Eng. Chem.*, *Anal. Ed.*, **18**, 144, 460 (1946). Semimicro method.

J. Mitchell, Jr., et al., *Ind. Eng. Chem.*, *Anal. Ed.*, **16**, 410 (1944).

W. Reimann III, *Ind. Eng. Chem.*, *Anal. Ed.*, **15**, 325 (1943).

W. E. Shaefer and W. J. Balling, *Anal. Chem.*, **23**, 1126 (1951). Difficultly saponifiable esters.
W. B. Swann et al., *Anal. Chem.*, **30**, 1830 (1958). Ion exchange method.
C. H. Van Etten, *Anal. Chem.*, **23**, 1697 (1951). Micro determinations.
C. O. Willits, *Anal. Chem.*, **21**, 136 (1949).

Iodine Number

R. Armitrano, *C.A.*, **53**, 16515 (1959). Use of mercuric acetate to accelerate Hanus method on naphthas.
W. Arve and B. Grote, *C.A.*, **52**, 19411 (1958). Use of mercuric acetate.
V. I. Babaev and N. M. Danshina, *C.A.*, **56**, 3581 (1962). Acids and alcohols by Hanus method.
I. Bellucci and R. De Gori, *C.A.*, **51**, 11735 (1957). Impurities in ICl_3.
S. Galanos and E. K. Voudouris, *C.A.*, **54**, 8111 (1960). Modified Winkler method.
A. Jovtscheff, *C.A.*, **54**, 4000 (1960).
S. Kohn and G. Taguet, *C.A.*, **53**, 760 (1959).
A. Romeo and M. De Leo, *C.A.*, **53**, 5707 (1959).
R. P. A. Sims and B. Stone, *J. Am. Oil Chemist's Soc.*, **33**, 287 (1956).
W. T. Smith and R. L. Shriner, *The Examination of New Organic Compounds*, Wiley, New York, 1956, pp. 120–125. Several methods with references.
P. Smits, *Rec. trav. chim.*, **78**, 713 (1959). Spectrophotometric method.
H. Sulzer, *C.A.*, **51**, 11925 (1957). Study of Stahli's modification of Wijs method (ICl_3).
H. Sulzer and O. Högl, *C.A.*, **52**, 8589 (1958).
F. Takei et al., *C.A.*, **53**, 13889 (1959).
E. A. Timofeeva et al., *C.A.*, **52**, 13540 (1958).

Bromine Number

V. Hamann and A. Hermann, *Mikrochim. Acta*, **1961**, 105; *C.A.*, **56**, 10647 (1962). Bromate-bromide reagent with mercuric acetate catalyst.
R. S. Hartley and B. C. Hobson, *C.A.*, **56**, 1541 (1962). Bromine in CCl_4.
W. Heidbrink, *C.A.*, **53**, 16561 (1959). Bromine vapor with air-sensitive compounds.
I. Napoli, *C.A.*, **54**, 16873 (1960). Use of a platinum double electrode with a transistor indicator.
J. Mitchell, I. M. Kolthoff, E. S. Proskauer, and A. W. Weissberger, *Organic Analysis*, Vol. III, pp. 210–218, Interscience, New York, 1956.
J. Morganer, *C.A.*, **51**, 14464 (1957). Bromine vapor (Rossman technique).

Index of Text

acetals: derivs. of, 514–515
  hydrolysis of, 514–515
  tables, 667
acetamides: from amines, 589–593; xanthyl derivs. of, 610
acetamido derivs. of aromatic hydrocarbons, 581–582
acetates: of carbohydrates, 525–526; of phenols, 486–487
acetylation: of amines, 589–593; of amino acids, 454; of β-naphthol, 487; of sulfanilamide, 634
acetylenes, derivs. of, 547–577
acids (*see also* carboxylic acids): dibasic, solubility and m.p., 311; effect of substituents on strength, 331–333; indicator method of classifying, 335–342; inorganic, esters of, 540–541; solubility classification, 315–317
acid-base solutes, effect of solvents on, 330–331
acid anhydrides: derivs. of, 459
  tables, 668–669
  test for, 366
acid chlorides, prepn. of, 444
acid functions, derivs. of, 437–464
acid halides: apparatus for prepn. of, 472
  derivs. of, 458–459
  tables, 670–671
  tests for, 366–367
acidic substances, tests for, 359–360
aconitic anhydride test for tertiary amines, 387–388

N-acyl and N-aroyl derivs. of amino acids, 452–455
N-acylphthalimides, from amides, 611
adsorbents: for column chromatography, 136–138; for thin-layer chromatography, 147
adsorption chromatography, *see* column chromatography
alcohols: chromatographic procedures for, 125, 141, 482–484
  derivs. of, 465–484; in aqueous solution, 472; perfluorinated, 568; small amounts of, 477–479, 483
  prepn. from: acetals, and identn., 515; aldehydes, 509; esters, and identn. of, 537–538
  solubilities in water, 310
  tables, 714–725
  tests for: acid chloride test, 368; chromic acid test, 371; oxidations, 355; specific class tests, 367–369; subclass tests, 369–371; vanadium oxine test, 368; xanthate test, 368–369
aldehydes (*see also* carbonyl compounds): derivs. of, 496–510
  chromatographic identn. of small amounts of, 505
  perfluorinated, reactivity of, 567
  tables, 728–741
  tests for, 394–396
alkanes, characterization of, 572–573
  tables, 862–868
alkenes, derivs. of, 574–577

alkenes, tables, 870–875
alkyl and aryl amines, see amines
alkyl and aryl halides, distinguishing tests for, 372–373
alkyl halides: derivs. of, 548–558; tables, 830–856
alkyl nitrates, nitrites, and sulfates, 540–541
2-alkylbenzimidazoles, from acids, 450
9-alkylfluorene-9-carboxylates, from alkyl halides, 557–558
2,2'-(alkylenedithio)bis(6-nitrobenzothiazoles), from alkyl dihalides, 557
S-alkyl-2-mercapto-4,5-dihydroglyoxalinium picrates, from alkyl halides, 558
alkylmercuric salts, from alkyl halides, 551, 554–555
alkyl β-naphthyl ethers, from alkyl halides, 551
N-alkylphthalimides, from alkyl halides, 555–556
S-alkylthiuronium picrates, from alkyl halides, 549–551
alkynes, 574–577
aluminum alkoxide test for water, 277
aluminum chloride test for benzenoid structure, 350–351
amides: derivs. of, 607–611
  as derivs. of: acid anhydrides and acid halides, 458–459; amino acids, 453; carboxylic acids, 439–441; esters, 534; nitriles, 618–621
  hydrolysis of, 608–609
  tables, 744–748
  tests for, 373–378
amines: chromatographic procedures for, 125, 141, 150, 606–607
  derivs. of, 589–606; of primary and secondary, 589–601; of tertiary, 601–606
  prepn. from: azo compounds, 616; nitriles, 621–622; nitro compounds, 625–626; nitroso compounds, 628–629
  salts of: hydrolysis test for, 349; with sulfonic acids, 636–637
  separation of, 596–597
  tables, 750–806
  tests for: chloranil, 356; copper ion, 379; ferricyanide, 357; fluorescein chloride, 379; specific class, 379–388; tetraphenylborate, 379
amino acids: chromatographic procedures for, 125–131, 150, 457–458
  derivs. of, 450–457; N-acyl and N-aroyl derivs., 452–455; N-ureido, 455–456; salts, 456–457
  ninhydrin test for, 365
  tables, 672–681
4-aminoantipyrine: test for phenols, 409; derivs. of carbonyl compounds, 507
p-aminodimethylaniline, prepn. and use, 529–530
aminoguanidine test for carbohydrates, 391
amino-nitrogen functions, derivs. of, 589–614
ammonia, hypochlorite-phenol test, 349
ammonium salts, hydrolysis test for, 349
ammonolysis of sulfonyl chlorides, 639
ampholytes, indicator methods, 338
amphoteric compounds, 314–315
anhydrides, see acid anhydrides
anilides: prepn. from: acid anhydrides, 459; acid halides, 458; alkyl halides, 551–553; carboxylic acids, 439–446; esters, 536–537
  tests for, 377
  tables, 744–748, 948–952
anilines, substituted, basicity of, 335
aniline acetate test for carbohydrates, 390
aniline test for acid halides, 366
antazoline reagent, 401
anthraquinonyl thioethers, from thiols, 642
anthrone test for carbohydrates, 388–389
apparatus for small-scale experimentation (see microapparatus)
applicators for thin-layer chromatography, 147–148
aromatic compounds: chromatography of, 140
  tests for, 350–352
aromatic ethers: bromination of, 545–546; from alkyl halides, 556
aromatic hydrocarbons, derivs. of, 577–585
  tables, 877–890
aroylbenzoic acids, from aromatic hydrocarbons, 583–584
aryl amines, see amines
aryl and alkyl halides, distinguishing test, 372–373

aryl halides, derivs. of, 558–564
  tables, 842–851
aryloxyacetic acids, from phenols, 489–490
arsenic, detection of, 298, 299
association, and solubility, 58–60, 303–306
atomic refractions, 235
autoprotolysis constants of some solvents, 335
azides, use in derivn. of: alcohols, 475, 479; thiols, 643
azoates, of sugars, prepn. of, 526
azo compounds: derivs. of, 615–616; prepn. from nitroso compounds, 628–629; tables, 909–910
azoxy compounds: derivs. of, 617; tables, 911

Baeyer test, *see* permanganate test
balance, for small-scale work, 9–12
Barfoed reagent, 390
bases: effect of substituents on strength, 333–335; indicator method of classifying, 335–343; solubility classification, 313–315
basicity constant, 330
baths, heating: for melting point, 181–185; for small-scale work, 22–24
Beilstein test for halogens, 293–294
Benedict reagent, 395
benzamides: from amines, 589–594; from benzonitrile, 620–621; mercuric derivative of, 611
benzenesulfonyl chloride test for amines, 382–383
benzenesulfenamides, from amines, 595
benzenesulfonamides: from amines, 595–598; from aromatic ethers, 545, 546; from aryl halides, 564
benzenesulfonyl derivs. of amines, 595–598
benzenoid structure, test for, 350–352
benzoates from: carbohydrates, 525–526; phenols, 486–487
benzoylation of: amines, 589–594; amino acids, 453
N-benzylamides, from esters, 534–535
benzylamine salts of carboxylic acids, 449
benzylamine test for acid halides, 366
S-benzylthiuronium chloride, prepn. of, 635–636

S-benzylthiuronium derivs. of: carboxylic acids, 449; sulfonic acids, 635–636
boiling points, 202–213: determination of, 211–213; of solutions, 207–211; and structure, 203–205; variation with pressure, 205–207
bond types and solubility, 308–309, 312, 318–319
bromide ions, in fusion extract, 293–295
bromination of: aromatic ethers, 544–545; phenols, 490
bromine in carbon tetrachloride test, 353–354; comparison with permanganate test, 353
bromine derivs. of alkenes, 575–576
p-bromoanilides, from carboxylic acids, 440
p-bromobenzenesulfonamides, from amines, 597–598
bromo derivs. of: ethers, 545–546; phenols, 490
p-bromophenacyl esters, from carboxylic acids, 447
N-bromosuccinimide: alcohol subclass test, 369–370; amine subclass test, 380–381; for derivs. of alcohols, 480
burners, micro, 19–20
brucine test for nitrates, 402

calcium oxide fusion, 290
calibration of thermometers, 195–197
camphor, use in cryoscopic methods, 230–232
carbamates, *see* urethans
carbohydrates: chromatographic procedures for, 125, 151, 526, 528–529: derivs. of, 519–529; osazone formation, 522–525; oxidation of, 357–358, 527; periodic acid test, 357–358; sepn. by ion-exchange resins, 156; specific rotations of, 227, 528; subclass tests for, 390–392; tables, 809–810; tests for, 388–392
carbon, detection of, 282–286
carbonyl compounds (*see also* aldehydes and ketones): chromatographic sepn., 125, 141; derivs. of, 496–518; 3,5-dinitrobenzoic acid test for, 393–394; sepn. by ion-exchange resins, 156; specific class tests, 392–397

carboxylic acids (*see also* acids): chromatographic sepn. of, 125, 141, 150, 155, 439
  derivs. of, 437–450
  prepn. of: by hydrolysis of esters, 537–540; from nitriles, 618–620; by oxidation of a cycloalkane, 573; by oxidation of sugars, 527–528; by oxidation of unsaturated hydrocarbons, 573–574; by oxidation of side chains, 589
  tables, 684–711
  tests for, 363–365
carbylamine test for primary amines, 386
carotenoid pigments, chromatographic sepn., 143–144
catalytic reduction, *see* reduction
centrifuge in sepn. of crystals, 71–74
ceric nitrate test, 356–357
chelation, and solubility, 305, 317
chemicals needed, 969, 971–974
chloranil test, 356, 401
chloride ions, in fusion extract, 293–295
chlorosulfonylation: of ethers, 546; of aryl halides, 564
chromatographic procedures: column, 131–145
  gas, 157–159
  ion-exchange, 152–157
  paper, 120–131
  principles of, 117–120
  thin-layer, 145–152
  use in separation or identification of: alcohols, 482–484; aldehydes, 509–510; amino acids, 457–458; carboxylic acids, 439; nitro compounds, 624; phenols, 491–492; sugars, 125, 526, 528–529
citric acid-acetic anhydride test for tertiary amines, 388
classification by: chemical tests, 343–423; indicator method, 328–342; solubility tests, 303–327
cleaning solutions, 974
color, in general tests, 347–348
column chromatography, 131–145; adsorbents, 136–138; detection of zones, 138–143; examples, 143–145; solvents, 138; techniques, 131–136
copper ion reducing test, 390, 395
copper ion test for amines, 379
copper sulfate test for water, 277

cresols, derivs. of, 490
cryoscopic methods for molecular weights, 228–233
crystallization, 54–82; drying of crystals, 75–82; sepn. of crystals by centrifugation, 71–74; sepn. of crystals by filtration, 63–74; solution prepn. 61–63; solvent selection, 55–60
crystallization of derivatives, 57
crystals, behavior on heating and cooling, 277–282
cyanide ion, tests for, 292–293
cyclization of methone derivs. of aldehydes, 504–506
cycloalkanes, characterization of, 572–573
cycloalkenes: derivs. of, 574–577; tables, 870–875
cycloalkyl halides: derivs. of, 548–558; tables, 830–841
cyclohexanol, use in cryoscopic work, 232–233

Davidson's indicators, 336
decomposition: and m.p., 193; and rate of heating, 450–451
density: determination of, 221–225; and structure, 221
derivatives, preparation from:
  acetals, 514–515
  acid anhydrides, 459
  acid halides, carboxylic, 458–459
  acid halides, sulfonic, 634
  acids, amino, 450–457
  acids, carboxylic, 437–450
  acids, sulfonic, 635–640
  alcohols, 465–484
  aldehydes, 496–510
  amides, carboxylic, 607–611
  amides, sulfonic, 631–634
  amines, primary and secondary, 589–607
  amines, tertiary, 601–604
  amino acids, 450–457
  azo compounds, 615–616
  azoxy compounds, 617
  carbohydrates, 519–529
  carboxylic acids, 437–450
  esters, 532–541
  ethers, 541–547
  halogen compounds, aryl, 558–564
  halogen compounds, alkyl and cycloalkyl, 548–558

derivatives, preparation from: halogen compounds, perfluoro, 564–569
hydrazines, 607
hydrazo compounds, 617
hydrocarbons, aromatic, 577–585
hydrocarbons, saturated, 571–573
hydrocarbons, unsaturated, 574–577
isocyanates, 617
isocyanides, 617
isothiocyanates, 640–641
ketals, 514–515
ketones, 510–514
lactones, 540
mercaptans, 641–644
nitrates, 540–541
nitriles, 617–623
nitrites, 540–541
nitro compounds, 623–628
nitroso compounds, 628–629
phenols, 484–492
sulfates, 540–541
sulfides, 641
sulfonamides, 631–634
sulfonic acids, 635–640
sulfonyl chlorides, 634
thiocyanates, 640–641
thioethers, 641
thiols, 641–644
thiophenols, 641–644
derivatization, problems in, 429–436; evaluation of m.p., 436; prepn. of small amounts, 434–435; selection of derivs., 430–434
desiccants and desiccators, 77–79
dialysis, 160
diazotization test, 384
dicarboxylic acids; from cycloalkanes, 573; from cycloalkenes, 576–577; derivs. of, 443–444; tables, 684–711
dienes, 574–577
dimethone derivs. of aldehydes, 504–506
p-dimethylaminoanils, derivs. of aldehydes, 506
N,N-dimethyl-α-naphthylamine, in sulfonamide test, 275
dinitro and trinitro hydrocarbons, test for, 404
2,4-dinitrobenzenesulfenamides, from amines, 598

2,4-dinitrobenzenesulfenyl chloride, derivn. of: aromatic hydrocarbons, 585; olefins, 577
3,5-dinitrobenzoates: chromatography of, 125
prepn. from: acetals, 515; alcohols, 467–473, 482–484; alkyl halides, 555; amines, 602–603; amino acids, 453; esters, 538–539; ethers, 543–544; phenols, 485–486
3,5-dinitrobenzoic acid test for carbonyl compounds, 393–394
3,5-dinitrobenzoyl chloride: derivn. of mercaptans, 643; prepn. of 472–473
2,4-dinitrochlorobenzene: reaction with: mercaptans, 642; phenols, 490–491; amines, 606
2,4-dinitrofluorobenzene, test for amines, 385
2,4-dinitrophenyl derivs. of amino acids, 454–455; of phenols, 490–491
2,4-dinitrophenylhydrazine test for carbonyl compounds, 392–393
2,4-dinitrophenylhydrazones, as derivs. of: carbonyl compounds, 499–502; fluorine compounds, 568
2,4-dinitrophenyl sulfides and sulfones as derivs. of: alkyl halides, 556–557; aryl halides, 585; of olefins, 577
2,4-dinitrophenyl thioethers: oxidation of, 641; prepn. from thiols, 642
3,5-dinitrothiobenzoates, from thiols, 643–644
2-diphenylacetyl-1,3-indandione 1-hydrazone, use for derivs. of carbonyl compounds, 506–507
diphenylamine test: for nitrates and nitrites, 401; for nitro compounds, 403–404
diphenylcarbamyl chloride, use in derivn. of phenols, 488
disposable micropipets, 17
distillation: accessory equipment, 92–93; apparatus and procedures for, 82–101; at atmospheric pressure, 83–90; fractional, 87–90; at reduced pressure, 90–92; simple, 83–87; small-scale, 96–99; steam, 99–101; with molecular stills, 94–96
disulfides, test for, 410

986 Index of Text

dithiocyanates, from unsaturated hydrocarbons, 575
divisions, solubility, 322–325
Doctor test: in column chromatography, 142; for thiols, 412
droppers and pipets for small-scale, 12–15
dry box, 25–26
drying agents, 974
drying of crystals, 75–82

electronegativity of radicals, order, 315
electrophoresis, 159–160
elements: analysis for, 282–299; decomposition methods, 286–291; detection of ions, 291–296
Emich microdetection test for carbon, 285
enolization, and solubility, 316–317
enols, ferric chloride test for, 406–407
epoxides, test for, 399
equipment for small-scale work (*see also* microapparatus), 8–53
essential oils, chromatographic separation of, 149–150
esters: chromatographic separation of, 150; derivs. of, 532–541; hydrolysis of, 537–538; prepn. from acid anhydrides and acid halides, 458–459; prepn. from carboxylic acids, 447–448; tables, 814–822; tests for, 397–398
ethanol (*see also* alcohols): chromatographic procedure for small amounts, 483; derivn. of, 470
ethers: from alkyl halides, 551; derivs. of, 541–547; 2,4-dinitrophenyl, 490–491; phenolic, picrates of, 547; pseudosaccharin, from alcohols, 482; tables, 823–828; tests for, 398–399; thioethers, 641; trityl, 481
ethyl ether, in solubility classification, 312–313
eutectic melting points, *see* melting points
evaporation, small-scale, 31–34
extraction, 109–111; small-scale, 29–30

ferric chloride test, 367, 406–407
ferric hydroxamate test, 366–367, 368, 373–374, 397–398
ferricyanide test, 357
ferrous hydroxide test, 403
ferrox test, 296–297

filtration, 30, 63–71
fluoride ion, in fusion extract, 295–296
fluorine compounds: derivs. of, 564–569; tables, 857–861
formaldehyde-sulfuric acid test, 351–352; 372–373
fractionation procedures, 54–116; chromatography, 117–158; crystallization, 54–82; dialysis, 160; distillation, 82–101; electrophoresis, 159–160; extraction, 109–111; foams, 161–162; inclusion compounds, 161; molecular sieves, 160–161; sublimation, 101–108; zone refining, 162
Friedel-Crafts reaction: for derivs. 583–584; as test, 350–351
functional groups: detcn. by infrared spectroscopy, 647; general tests, 362–414
fusion: with sodium hydroxide, 373; in tests for elements, 287–291
fusion techniques, fusion and cooling observations, 277–282

gas chromatography, 157–158; examples, 158
glove box, 25–26
general reagents, 970–971
glycols, *see* alcohols
glycosides, chromatographic separation of, 150
gravitometer, *see* density, determination of
Griess reagent, 360
Grignard reagent: addition of nitriles to, 622–623; apparatus and procedure for prepn., 45–48; in derivn. of esters, 536–537; prepn. from alkyl halides, 551–554
grinding, 29

halides: derivs. of, 548–569; distinguishing alkyl and aryl, 372–373
haloform reaction, with methyl ketones, 513
halogen compounds, derivs. of, 548–569
halogens, detcn. of, 293–296
N-halosuccinimide test, for amines, 380–381
handling in an inert atmosphere, 25–26
heating apparatus, small-scale, 19–24; blocks and stages, 20–21; baths, 22–24, 181–184; burners, micro, 19–20;

hot plates, 20; immersion heaters, 22–23; mantles, 23
high vacuum line, 39–40
Hinsberg method for amines, 382–383, 596–597
Hoesch synthesis, 623
hydantoins: from amino acids, 455–456; from ketones, 513
hydrazides, prepn. from esters, 534–536
hydrazines: derivn. of, 607; in derivn. of isothiocyanates, 640; from N-nitrosoamines, 629; test for, 399
2-hydrazinobenzothiazole, for detection of aldehydes, 396
hydrazo compounds, derivn. of, 617
hydrazones: hydrolysis of, 406; substituted, from ketones, 510–511; substituted, from sugars, 523–525; tests for, 406
hydrocarbons: aromatic, derivs. of, 577–585; tables, 877–890
saturated, derivs. of, 571–573; tables, 862–868
selected, physical constants of, 236
solubility in sulfuric acid, 320
tests for, 400–401
unsaturated, derivs. of, 574–577; column chromatography of, 142; tables, 870–875
hydrochloric acid in solubility classification, 313–315
hydrogen, detcn. of, 282–286
hydrogenations, catalytic, small-scale, 40–45
hydrogen bonds, and solubility, 58–59, 305–306, 308–313, 317
hydrolysis: of acetals, 514–515; of amides and substituted amides, 373, 608–609; of amine salts, 349; of ammonium salts, 349; of esters, 537–539; of nitriles, 618–621; of sulfonamides, 633–634
hydrolysis test, distinguishing alkyl and aryl halides, 372
hydroxamate derivs. of acids, chromatography of, 125
hydroxamate test: for acid halides, 366–367; for acid anhydrides, 366; for alcohols, 368; for amides, 373–374
hydroxamic acids, ferric chloride test for, 367

hydroxy acids, chromatography of, 125
N-hydroxybenzenesulfonamide test for aldehydes, 394
hydroxy compounds (*see also* alcohols; carbohydrates): ceric nitrate test for, 356–357; ferric chloride test for, 406–407
hydroxylamines: from nitroso compounds, 629; reacn. with carbonyl compounds, 512–513
hypochlorite test for methyl ketones, 513
hypochlorite-phenol test for ammonia, 349

identification of organic substances: correlation and interpretation of data, 197–202; general directions, 3–6; tentative, 275–423
ignition test, 276, 348–349
imides: aromatic cyclic, in derivn. of alkyl halides, 555–556; derivs. of, 607–611
α-iminoalkylmercaptoacetic acid hydrochlorides, from nitriles, 622
immiscible liquids, separation of, 29–30
inclusion compounds, 161
indicators, classification by, 328–342; formation in phenol test, 407–408
indophenols, formation of, 408
inert atmosphere, apparatus for working in an, 25–26
infrared spectroscopy, 647, 648–649, 650
instrumental methods, 646–656; advantages, 646–647; information obtained, 647–648; micro samples, 649–651; sample handling, 648–649
iodate-iodide test for acids, 359
iodic acid oxidation test, 357–358
iodide ions, in fusion extract, 293–295
iodine-azide test for thiols, 411
iodine numbers, 976–978
iodoform test, 361
N-iodosuccinimide, in amine subclass test, 380–381
ion-exchange chromatography, 152–157; applications and limitations, 154; examples, 155–157; techniques, 153–154
isatin test for mercaptans, 412
isocyanates: derivs. of, 617
use in derivatization of: alcohols, 474–480; alkyl halides, 551–554; amines, 598–601; phenols, 487–489
tables, 912

isocyanides, derivs. of, 617
isocyanide test for amines, 386
isomerism, and solubility, 309–311, 317
isothiocyanates: derivs. of, 640
  reactions with amines, 598–601
  tables, 961

keto acids, chromatography of, 125
ketones (*see also* aldehydes and carbonyl compounds): derivs. of, 510–514
  prepn. from: acetylenes, 576; nitriles, 622
  perfluorinated, activity of, 567
  tables, 892–906
  tests for, 396–397
ketoses, Seliwanoff test for, 391

lactic acid, nitrochromic acid test for, 358
lactones, derivs. of, 540
lead acetate test for mercaptans, 412
lignin test for amines, 384–385
Lucas reagent for alcohol subclasses, 370

magnesium-sodium carbonate fusion, 289
magnetic stirrers, 26–27
manometers, 92–93
manostats, 93–94
mass spectroscopy, 648, 649
melting, microscopic study, 277–282
melting points: apparatus and procedures for, 178–195; bars or blocks, 185–187; calibration of thermometers, 195–197; capillary tubes, 179–181; and decomposition, 193; eutectic, 178, 279, 281; evaluation of data, 197–202; freezing points, 192; liquid heating baths for, 181–185; microscope hot stage, 187–192; mixed, 177–178, 193–195; and solubility, 311; and structure, 172–177
mercaptans: addition to olefins, 525; chromatography of, 142, 150; derivs. of, 641–644; tables, 966–967; tests for, 411–413
2-mercaptobenzothiazole, reaction with alkyl halides, 557
mercury, detcn. of, 298
mercuric salts, as derivs. of: acetylenes, 575; alkyl halides, 554–555; amides, 610–611
metallic elements, detcn. of, 298–299

methanol (*see also* alcohols), chromatographic procedure for small amount, 483–489
methiodides of tertiary amines, 604
methone reagent and test, 395, 504
methylcycloalkanes, characterization, 572
methyl ketones, tests for, 361, 397
methylphenylhydrazine, reacn. with sugars, 522
methyl *p*-toluenesulfonates of tertiary amines, 604
microapparatus, 8–53; burets, 19; condensers, 34, 35; desiccators, 77–79; evaporators, 31–34; extraction, 109–111; filtration, 63–71; flasks, 34, 35; Grignard reaction, 45–48; heaters, 19–24; hydrogenation, 40–45; microcones, prepn. and use, 477; mortars, 29; pipets, 15–19; pycnometers, 221–224; separatory funnels, 29–30; spatulas, 24–25; stirrers, 26–29; sublimators, 101–108
microcrystallization, 71–74
microhydrogenation, apparatus and procedure for, 40–45
micro quantities: definition of, 6; preparation of, 434–435
microscope: in fusion study, 277–282
microscope hot stage, 187–192
Millon test, for phenols, 408–409
"mixed" fusion, 279–282
"mixed" melting points, 193–195
mixtures: fractionation procedures, 54–170; sepn. of, 244–271; examples, 248–250
molal depression constants, 229
molar refraction, 234–237
molecular sieves, 160–161
molecular weights, 228–234
Molisch test, 390
molybdenum oxide, in carbon test, 285
morpholides, prepn. from esters, 536
morpholine, derivn. of aryl halides by, 559
mortars, 29
mustard oils, derivs. of, 640

α-naphthalides, from alkyl halides, 551–553
β-naphthyl ethers, picrates of, 551
α-naphthyl isocyanate, in derivn. of: alcohols, 474–479; phenols, 487–488

α-naphthylureido derivs. of amino acids, 455–457
α-naphthylurethans, 474–479
neutralization equivalent, 974–975
nickel-dithiocarbamate test for secondary amines, 387
ninhydrin test, 365
nitrates and nitrites, 401–402, 540–541
nitration: of aromatic hydrocarbons, 578–580; of aryl halides, 559–564; of mononitro compounds, 627–628
nitric acid, 100%, prepn. of, 561–562
nitriles: derivs. of, 617–623; tables, 913–915; test for, 402
nitrites, alkyl, 402
nitroanilines: chromatographic sepn. of, 144–145; acetylation of, 593
m-nitrobenzaldehyde, oxidation of, 508
o-nitrobenzaldehyde test for methyl ketones, 397
m-nitrobenzazide, reacn. with thiols, 643
nitrobenzene, catalytic reduction, 626
p-nitrobenzenediazonium fluoroborate test, 384, 407
p-nitrobenzoates, prepn.: from alcohols, 467–470, 471–472; from phenols, 485
p-nitrobenzoyl derivs. of amines, 591
p-nitrobenzyl esters, from carboxylic acids, 447–448
nitrochromic acid test, 358
nitro compounds: derivs. of, 623–628; tables, 916–923; tests for, 402–405
nitro derivs. of aryl halides, 559–564
nitrogen, detcn. of, 292–293
nitrogen compounds, basic properties, 313–315
nitrogen functions (*see also* amino acids), derivs. of, 589–630
nitrogen-oxygen functions, 360
2-nitro-1,3-indanedione salts of amines, 605
nitroparaffins, tests for, 404–405
nitrophenols: data on acidity of, 334; test for, 405; catalytic hydrogenation of, 491; sepn. of, 248–249
p-nitrophenylhydrazones: from aldehydes, 499–501; from sugars, 523
m-nitrophenylurethans, from thiols, 643
3-nitrophthalates, from alcohols, 473–474
3-nitrophthalic anhydride, reacn. with: mercaptans, 532; amines, 605

3-nitrophthalimides, substituted, from amines, 605
nitroprusside test: for mercaptans, 411–412; for secondary alkyl amines, 387; for sulfur, 291
N-nitrosoamines, deriv. of, 629
nitroso compounds: derivs. of, 628–629; tables, 923; test for, 405–406
nitrotoluenes: oxidation of, 627; nitration of, 627–628
nitrous acid test: for acids, 360; for nitroparaffins, 404–405; for phenols, 408
nomenclature of compounds, 659–661
nuclear magnetic resonance spectroscopy, 647–648, 649, 651

"oiling out," 64
olefins: chromatographic procedures for, 142, 150
  derivs. of, 574–577
  tables, 870–875
optical rotation, 225–228
optical rotatory dispersion, 226
osazones: formation and conversion to osotriazoles, 520–525; rates of formation, 523–524
osotriazoles, formation of, 525
oxalates of amides, 611
oxidation: of alcohols, 356–358, 371; of aldehydes, 358, 395–396; of azo compounds, 615; of carbohydrates, 356, 358, 390, 527–528; of cyclohexane, 573; of cyclohexene to adipic acid, 576–577; of polyhydroxy compounds, 356, 358, 390; of side chains of hydrocarbons, 585; of side chains of nitro compounds, 626–627; of thioethers to sulfones, 641; of unsaturated hydrocarbons, 574–577; wet, 285–286
oxidizable compounds, tests for, 355–359
oxidizing agents: ferrous hydroxide test for, 403
  selective action of: ceric nitrate, 356–357; chloranil, 356; ferricyanide, 357; iodic acid, 357–358; nitrochromic acid, 358; Tollens reagent, 358–359
oximes: from ketones, 512–513
  tests for: ferric chloride, 406; specific class test, 406
oxygen, detcn. of, 296

oxygen compounds, reacn. with sulfuric acid, 319

paper chromatography, 120–131: examples, 130–131; solvent systems, 124–126; techniques, 120–124, 126–129
paraffins and cycloparaffins, 571–573, 862–868
pentanols, solubility in water, 310
pentoses, Tollens test for, 392
perfluorination, effect on reactivity, 565–569
perfluoro compounds: derivs. of, 567–569; prepn. of, 565; tables, 857–861
periodic acid oxidation test, 357
permanganate: oxidation of nitro compounds by, 626–627; test for active unsaturation (Baeyer test), 354–355; test compared with bromine test, 353
peroxides, test for, 398
pH test, 328–342, 359
phenacyl esters, from carboxylic acids, 447
phenols: chromatographic procedures for, 125, 142, 150, 491–492; derivs. of, 484–492; tables, 924–945; tests for, 406–410
phenolic ethers: picrates of, 547; substituted, prepn. of, 556
phenylazobenzenesulfonates, chromatography of, 125
p-phenylazobenzoyl chloride: for derivs. of alcohols, 481; for derivs. of phenols, 492; for derivs. of amines, 591; preparation of, 526
o-phenylenediamine, reacn. with carboxylic acids, 450
phenylhydantoins, derivs. of amino acids, 456
phenylhydrazine: derivn. of, 607; salts of carboxylic acids, 449
phenylhydrazones, prepn. of: general, 497–498; from aryl ketones, 510–511; substituted, from carbohydrates, 520–525
2-phenylindole, in nitrite test, 402
phenyl isocyanate, derivn. of phenols by, 487–489
phenylthioureas, substituted, 598–601
phenylurea derivs. of amino acids, 455–456
phloroglucinol, reacn. with nitriles, 623
phosphorus, detcn. of, 297–298

phthalic anhydride, condensation with hydrocarbons, 583–584
phthalimides: from amides, 611; derivn. of alkyl halides by, 555–556
physical methods, 4–5
physical properties, 171–243; boiling point, 202–213; density, 219–225; melting point, 172–202; molar refraction and dispersion, 234–237; molecular weight, 228–234; optical rotation, 225–228; refractive index, 213–219
picrates: of S-alkylisothioureas, 549–551; of aromatic ethers, 544, 547; of aromatic hydrocarbons, 582–583; of β-naphthyl ethers, 551; of tertiary amines, 603
picrolonates of amino acids, 457
pipets, capillary, in sepn. of solvents, 30–31
pipets and droppers, 12–19
pipettors, 15–16
plant pigments, separation by chromatography, 143–144, 151–152
polyhydroxy compounds, see alcohols
preliminary examination of unknown, 275–302
pressure, effect on boiling points, 205–207
proteins, ninhydrin test for hydrolysis products, 365
pulverizing, 29
pseudosaccharin ethers, derivs. of alcohols, 481–482
pycnometers, see density determination

quaternary salts of tertiary amines, 601–604
quinhydrone test for amines, 380

Rast molecular weight determinations: with camphor, 230–232; with cyclohexane, 232–233
reactions, small-scale apparatus for, 34–48
reagents for detecting zones in chromatography: in column chromatography, 138–143; in paper chromatography, 125; in thin-layer chromatography, 148–149, 150
reagents for indicator method of classification, 336
reducing compounds, tests for: copper ion, 390; ferricyanide, 357; silver ion, 358

reduction: of azo compounds, 616; of nitriles, 621–622; of nitro compounds, 625–626; of nitrophenols, 491; of nitroso compounds, 629; of aldehydes, 508–509; of small quantities, by hydrogen, 40–45
reduction, catalytic, 40–45, 491, 508–509, 626, 629; by metals, metal hydrides, and metal ions, 615–616, 621–622, 625–626, 628–629
reflux, small-scale apparatus for, 34–37
reference works in organic qualitative analysis, 661–663
refractive index: of melt, 218; and refractometers, 215–218; and temperature, 214; and wavelength, 214–215
R_f value, 118–119, 121
resorcinol test for carbohydrates, 390
β-resorcylates of amines, 603
rhodamine B-uranyl acetate test for acids, 359–360
rings, weighing, 12
p-rosaniline hydrochloride test for aldehydes, 394–395

saccharides, nitrochromic acid test for, 358
salts: as derivs. of amino acids, 456–457; carboxylic acids, 449–450; sulfonic acids, 636–637
  tests for, 349–350
sample size, 6
saponification equivalent, 975–976
Schiff test for aldehydes, 394–395
Schotten-Baumann reaction, 591–594
Seliwanoff test for ketoses, 391
semicarbazones: chromatographic separation of, 150
  prepn. from: acetals, 514–515; carbonyl compounds, 502–504, 511–512; fluorine compounds, 568; ketones, 511–512
  test for, 406
semimicro apparatus (see also microapparatus), 8–53
separation of immiscible solvents, small-scale apparatus for, 29–30
separation of mixtures, 244–271
side chains, oxidation of, 585, 626–628
silicone fluids for heating baths, 23–24, 184
silver ion reducing test, 395–396

silver nitrate: alcoholic, in halide test, 372; aqueous, 395–396
Simon test, for amines, 387
small-scale experimentation, apparatus and equipment for (see also microapparatus), 8–53
sodium anthraquinone-α-sulfonate, reacn. with thiols, 642
sodium bicarbonate, in solubility classification, 317–318
sodium fusion, 287–289
sodium hydroxide, in classification by solubility, 315–317
sodium hypoiodite test (iodoform test), 361
sodium hypochlorite-phenol test, 349
sodium saccharin, in derivn. of alkyl halides, 555–556
solubility: classification by, 303–327; determination of, 60–61, 320–322; divisions, 322–325; of fluorine compounds, 566
solutions, boiling points of, 207–211
solvents: for catalytic hydrogenations, 43; for column chromatography, 138; for cryoscopic methods, 229; for derivatives, 57; effects on acids and bases, 330–331; for paper chromatography, 124–125; for Rast molecular weight determination, 229; selection for crystallization, 55–59
solvent pairs, 56, 57
solvent systems, for paper chromatography, 125
special reagents, 969–970
specific rotation, 227–228; of carbohydrates and derivs., 528
steam distillation, 99–101
stirrers, 26–29
sublimation and sublimators, 101–108; fractional, 105–108
substituents, effect on acidity and basicity, 313–318
sugars, see carbohydrates
sulfates, alkyl: derivs. of, 540–541
sulfenamides, from amines, 595–598
sulfides: column chromatography of, 142; tables, 962–965; tests for, 410; sodium hydroxide fusion test, 410
sulfinic acids, ferric chloride test, 407

sulfonamides: chromatographic separation of, 150; derivs. of, 631–634
  hydrolysis of, 633–634
  prepn. from: amines, 595–598; aryl halides, 564; ethers, 546; sulfonic acids, 639–640; tables, 948–952
  tests for, 375
sulfo-α-naphthylamides, from sulfonic acids, 637–640
sulfonyl chlorides: ammonolysis of, 639; derivs. of, 634; prepn. of, 638–639; tables, 953–958
sulfones: from oxidation of thioethers, 641; sodium hydroxide fusion test, 410
sulfonic acids, derivs. of, 635–640; tests for, 365; tables, 959–960
sulfur functions, derivs. of, 631–644
sulfur, detcn. of, 291
sulfuric acid, in solubility classification, 318–320
systematic approach, 3–4

tables of compounds and their derivatives, 667–967
  list of tables and page numbers, 665–666
  nomenclature used in the tables, 659–661
tartaric acid, nitrochromic acid test for, 358
terpenes: derivs. of, 574; chromatographic sepn., 150
tetraphenylborate test for amines, 379
textbooks on qualitative organic analysis, 6–7
thermometers, calibration of, 195–197
thin-layer chromatography, 145–152; adsorbents, 147; advantages, 145–146; applicators, 147–148; examples, 149–152; technique, 146–147, 148–149
thioacids, addition to olefins, 575
thioalcoholysis of nitriles, 622
thiobenzhydrazide, in derivn. of sugars, 527
thiocarbamic esters, substituted, 643
thioesters, prepn. from thiols, 643–644
thioethers: derivs. of, 641; prepn. of, 556–557, 642
  tables, 962–965

thiols (mercaptans and thiophenols): derivs. of, 641–644; test for, 411–413; tables, 966–967
thiophenols: addition to olefins, 575; derivs. of, 641–644; tests for, 411–413; tables, 966–967
thiosemicarbazides, from isothiocyanates, 640
thiosemicarbazones, from carbonyl compounds, 502–503, 511–512
thioureas, in derivn. of alkyl halides, 549–551; prepn. from amines, 598–601; prepn. from isothiocyanates, 640; substituted, from nitriles, 621–622; test for, 375–376
thymol-α-naphthylurethan, prepn. of, 488
tin: reduction by, 625–628; stannous ion as reducing agent, 616
Tollens test for pentoses, 392
Tollens reagent: selective action of, 358–359; test for aldehydes, 395–396
toluene-3,4-dithiol-zinc complex, test for ketoses, 391
p-toluenesulfonamides, prepn. of, 595–596
p-toluenesulfonyl derivs. of amino acids, 453–454
p-toluidides, prepn. from: acid anhydrides, 459; acid halides, 458; alkyl halides, 551–553; carboxylic acids, 439–446; esters, by Grignard reaction, 536–537
o-toluidine, benzoylation of, 594
o- and p-toluidine, sulfonates from, 636–637
p-toluidine acetate test for carbohydrates, 389
tosyl derivs. of carbohydrates, 527
2,4,7-trinitrofluorenone (T.N.F.) adducts of aromatic hydrocarbons, 352, 582–583
trityl ethers: from alcohols, 481; from carbohydrates, 435
tubes, weighing, 12

ultramicro quantities, definition, 6
ultraviolet and visible spectroscopy, 647, 648, 649–650
unknown material: fractionation procedures, 54–170; general directions, 3–6; preliminary examinations, 275–302; separation procedures, 244–271

unsaturated hydrocarbons: derivs. of, 574–577; tables, 870–875
unsaturation, active: bromine test, 353–354; permanganate ion test, 354–355
ureas, derivs. of, 607–611; substituted, as derivs. of amines, 598–601; test for substitutes, 375–376
N-ureido derivs. of amino acids, 455–456
ureides, derivs. of, 609–610
urethans: from phenols, 487–489; from perfluoro alcohols, 568–569; substituted, from alcohols, 474–480, 487–489

vacuum drying apparatus, 77–82
vacuum line, 39–40
vanadium-oxine test, 368

visible and ultraviolet spectroscopy, 647, 648, 649–650
volume measurement, small-scale, 12–19

water, in solubility classification, 308–311; test for, 277
weighing apparatus, small-scale, 9–12
wet oxidation method for carbon, 285–286

xanthate test for alcohols, 368–369
xanthyl derivs. of amides, 609
N-xanthylsulfonamides, from sulfonamides, 634

zinc, as reducing agent, 616–617, 629
zinc chloride: in alcohol subclass test, 370; anhydrous, prepn. of, 543
zinc-sodium carbonate fusion, 289–290
zone refining, 162

Index of Compounds*

aceanthrene, 888(71)
acenaphthene, 888(57)
acenaphthene-3-carboxaldehyde, 738(82)
acenaphthene-5-carboxaldehyde, 738(71)
acenaphthenequinone, 906(276)
acenaphthene-3-sulfonamide, 950(148)
acenaphthene-5-sulfonamide, 951(187)
acenaphthene-3-sulfonyl chloride, 956(140)
acenaphthene-5-sulfonyl chloride, 956(137)
7-acenaphthenone, 903(190)
acenaphthylene 875(305)
acetal, 667(6)
acetaldehyde (ethanal), 728(2)
acetamide, 744(68)
4-acetamidobenzenesulfonamide, 951(179)
4-acetamidobenzenesulfonic acid, 959(1)
4-acetamidobenzenesulfonyl chloride, 957(193)
4-acetamidonaphthalene-1-sulfonamide, 952(218)
acetanilide, 745(177)
acetic acid, 684(5)
acetic anhydride, 668(2)
acetic propionic anhydride, 668(3)
acetoacetamide, 744(26)
acetoacetanilide, 744(78)
acetoacetic acid, 688(22), 899(30)
acetoacet-4-methoxybenzamide, 745(180)
acetoacet-o-toluidide, 744(88)
2-acetoacetylpyridine, 900(72)
3-acetoacetylpyridine, 805(77), 902(145)
4-acetoacetylpyridine, 901(102)
4-acetoacetylquinoline, 805(56), 901(108)
2-acetobenzoic acid, 694(138), 903(182)
4-acetobenzoic acid, 706(355), 906(266)
2-acetobutyric acid, 688(14), 899(18)
4-acetobutyric acid, 897(217)
acetoin (3-hydroxy-2-butanone), 716(60), 893(64)
2-acetoindazole, 905(254)
3-acetoindazole, 790(578), 905(252)
3-acetoindole, 790(599)
acetol (1-hydroxy-2-propanone), 716(61), 893(69)
aceto-N-methyl-m-toluidide, 744(40)
aceto-N-methyl-p-toluidide (N-methylaceto-p-toluidide), 744(70)
2-acetonaphthone, 901(85)
acetone, 892(2)
acetone-1,3-dicarboxylic acid, 696(191)
acetonitrile, 913(3)
acetonylacetone (2,5-hexanedione), 896(153)
acetophenone, 896(166), 899(2)
acetophenone-4-carboxylic acid, 706(352), 906(266)
2-acetopropanal, 730(58)
2-acetopropenoic acid, 696(166), 904(199)
aceto-m-toluidide, 744(39)
aceto-o-toluidide, 745(175)
aceto-p-toluidide, 746(294)
acetourea, 748(467)
acetoxyacetone, 895(125)
N-acetylacetamide, 744(62)
acetylacetone (2,4-pentanedione), 663(21), 893(56)
acetylanthranilamide, 747(370)
acetylanthranilanilide, 747(338)
acetylanthranilic acid, 702(305)
2-acetylbenzoic acid, 694(138), 903(182)
4-acetylbenzoic acid, 706(355), 906(266)
4-acetylbiphenyl, 903(189)
acetyl bromide, 671(73)
acetyl chloride, 670(7)
1-acetyl-1-cyclohexene, 896(165)
acetylenedicarboxylic acid, 702(289)
9-acetylfluorene, 902(128)
acetyl fluoride, 670(1)
acetylglycine, 706(345)
8-acetylguanine, 900(52)
acetyl iodide, 671(85)
1-acetyl-2-methylurea, 747(379)
1-acetylnaphthalene, 897(220)
2-acetylnaphthalene, 901(85)
1-acetyl-2-naphthol, 928(59)
2-acetyl-1-naphthol, 903(167), 932(146)

* The numbers without parentheses are page numbers of the tables. The numbers in parentheses are the compound numbers within the particular table where a compound is listed.

996 Index of Compounds

2-acetylphenanthrene, 904(220)
3-acetylphenanthrene, 902(119)
4-acetylphenanthrene, 902(148)
9-acetylphenanthrene, 902(126)
acetyl-β-phenylhydrazine, 745(206)
acetylpiperidine, 744(10)
3-acetylpropionic acid, 899(22)
2-acetylpyridine, 895(149)
3-acetylpyridine, 895(186)
4-acetylpyridine, 896(177)
2-acetylquinoline, 900(59)
acetylsalicylaldehyde, 734(17)
acetylsalicylanilide, 746(251)
acetylsalicylic acid, 696(190)
2-acetylthiophene, 896(178)
2-acetyltriphenylene, 905(233)
aconitanilide, 747(405)
aconitanilide(*cis*), 747(351), 748(432)
aconitic acid(*cis*), 696(169)
aconitic acid(*trans*), 704(322)
aconitic anhydride(*cis*), 668(39)
aconitic anhydride(*trans*), 669(55)
acridine, 806(98)
acrolein, 728(5)
acrylamide, 744(76)
acrylanilide, 745(144)
acrylic acid, 684(7)
acrylonitrile, 913(1)
acryloyl chloride, 670(11)
adipamide, 745(207), 748(472)
adipanilide, 746(292), 748(500)
adipic acid (hexanedioic acid), 700(236)
adiponitrile, 913(56)
ajmaline, 786(524)
β-alanine, 674(36)
DL-alanine, 678(112)
D- or L-alanine, 678(114)
2-aldehydobenzoic acid, 692(107), 738(79)
3-aldehydobenzoic acid, 702(285), 740(119)
4-aldehydobenzoic acid, 708(421), 740(127)
4-aldehydobiphenyl, 736(46)
3-aldehydo-2-hydroxybenzoic acid, 740(120)
5-aldehydo-2-hydroxybenzoic acid, 740(126)
alizarin, 906(281), 944(431)
allo-2-chlorocinnamic acid(*cis*), 694(128)
L-(+)-alloisoleucine, 678(95)
β-D-allose, 809(16)
β-L-allose, 809(17)
DL-allothreonine, 676(73)
alloxan, 747(349)
allyl acetate, 814(37)
allyl alcohol, 714(7)
allylamine, 750(12)
allylbenzene, 873(215)
allyl benzoate, 818(233)

allyl bromide, 836(105)
allyl butyrate, 815(76)
allyl chloride, 830(27)
allyl chloromethyl ether, 824(50)
allyl cyanide (3-butenonitrile), 913(14)
1-allyl-3,4-dimethylenedioxybenzene, 826(126)
allyl ethyl ether, 823(19)
allyl ethyl sulfide, 962(15)
allyl formate, 814(17)
allyl iodide, 838(165)
allyl isothiocyanate, 961(5)
allyl mercaptan, 966(7)
4-allyl-2-methoxyphenol, 924(27)
allyl methyl ether, 823(9)
allyl methyl sulfide, 962(6)
1-allylnaphthalene, 874(244)
allyl 2-naphthyl ether, 827(1)
allyl phenyl ether, 825(99)
allyl phenyl sulfide, 963(53)
allyl propionate, 815(56), 815(60)
allyl sulfide, 962(24)
allyl thiocyanate, 961(4)
allylurea, 744(77)
α-L-altrose, 809(12)
2-aminoacenaphthene, 782(441)
3-aminoacenaphthene, 772(217)
4-aminoacenaphthene, 772(238)
5-aminoacenaphthene, 776(328)
aminoacetamide, 744(43), 768(154)
2-aminoacetanilide, 782(429)
3-aminoacetanilide, 772(241)
4-aminoacetanilide, 747(328), 788(536)
2-aminoacetophenone, 758(195), 762(3), 897(214), 899(5)
3-aminoacetophenone, 774(289), 903(159)
4-aminoacetophenone, 776(318), 903(171)
1-aminoacridine, 758(170), 776(329)
1-aminoanthracene, 780(408)
2-aminoanthracene, 794(660)
9-aminoanthracene, 784(485)
1-aminoanthraquinone, 794(668)
2-aminoanthraquinone, 794(685)
2-aminoazobenzene, 766(120), 909(4)
3-aminoazobenzene, 770(170), 909(5)
4-aminoazobenzene, 752(48), 780(401), 909(24)
2-aminobenzaldehyde, 734(18), 764(47)
4-aminobenzaldehyde, 736(55), 770(181)

2-aminobenzamide, 776(333)
3-aminobenzamide, 745(171)
4-aminobenzamide, 747(386), 790(580)
3-aminobenzanilide, 746(257)
1-(3-aminobenzeneazo)-2-naphthol, 910(57)
1-(4-aminobenzeneazo)-2-naphthol, 910(47)
2-aminobenzenesulfonamide, 949(56)
3-aminobenzenesulfonamide, 948(41)
4-aminobenzenesulfonamide (sulfanilamide), 949(70)
2-aminobenzenesulfonic acid, 959(4)
3-aminobenzenesulfonic acid, 959(2)
4-aminobenzenesulfonic acid (sulfanilic acid), 959(3)
2-aminobenzhydrol, 780(383)
4-aminobenzhydrol, 780(386)
2-aminobenzoic acid (anthranilic acid), 672(11), 698(220), 784(478)
3-aminobenzoic acid, 672(19), 788(561), 702(282)
4-aminobenzoic acid, 674(30), 704(317), 790(592)
3-aminobenzonitrile, 915(60)
4-aminobenzonitrile, 772(236), 915(61)
2-aminobenzophenone, 776(315), 903(170)
4-aminobenzophenone, 780(398), 904(195)
3-aminobenzpyrazole, 786(500)
6-aminobenzpyrazole, 792(639)
2-aminobenzthiazole, 782(430)
2-aminobenzyl alcohol, 772(220)
4-aminobenzyl alcohol, 768(149)
dl-α-aminobenzyl cyanide, 766(103)
2-aminobenzyl cyanide, 770(180), 915(53)
4-aminobenzyl cyanide, 764(66), 914(28)
2-aminobibenzyl, 762(29)
3-aminobibenzyl, 766(89)
4-aminobibenzyl, 764(76)
2-aminobiphenyl, 764(80)
3-aminobiphenyl, 762(3)
4-aminobiphenyl, 766(99)
2-amino-3-bromoanthraquinone, 794(686)
2-amino-4-bromo-5-methylphenol, 778(365)
2-amino-4-bromo-6-methylphenol, 778(349)
4-amino-2-bromo-5-methylphenol, 792(644)

Index of Compounds 997

3-amino-6-bromophenol, 784(483)
4-amino-2-bromophenol, 788(541)
1-amino-2-butanol, 718(97)
4-amino-2-butanol, 718(103), 752(82)
L-(+)-2-aminobutyric acid, 678(108)
DL-2-aminobutyric acid, 678(120)
DL-3-aminobutyric acid, 674(32)
4-aminobutyric acid, 674(39)
4-aminobutyrophenone, 772(226), 902(143)
1-aminocarbazole, 790(604)
3-aminocarbazole, 794(671)
4-aminochalcone (ω-(4-aminobenzal)acetophenone), 784(492)
2-amino-5-chlorobenzoic acid, 706(361)
3-amino-2-chlorobenzoic acid, 672(16)
3-amino-4-chlorobenzoic acid, 674(56)
3-amino-5-chlorobenzoic acid, 674(55)
4-amino-2-chlorobenzoic acid, 674(54), 792(643)
2-amino-5-chlorobenzophenone, 903(162), 776(291)
3-amino-4'-chlorobiphenyl, 772(222)
4-amino-2-chloro-5-methylphenol, 792(653)
2-amino-4-chlorophenol, 782(450), 936(252)
2-amino-5-chlorophenol, 786(499)
4-amino-2-chlorophenol, 786(498)
3-amino-2-chloropyridine, 772(212)
5-amino-8-chloroquinoline, 786(505)
5-amino-3-chlorosalicylic acid, 676(83)
2-aminocinnamic acid, 672(15)
3-aminocinnamic acid(trans), 672(24)
3-aminocinnamic acid(cis), 674(33)
4-aminocinnamic acid, 672(20)
3-aminocoumarin, 782(424)
6-aminocoumarin, 788(551)
3-amino-p-cresol, 764(70)
2-aminocyclohexanol, 768(156)
9-aminodecalin(cis), 756(163)
9-aminodecalin(trans), 756(154)
2-aminodecanoic acid, 676(87)
4-amino-3,5-dibromobenzaldehyde, 784(486)
4-amino-2,6-dibromophenol, 790(605), 940(355)

4-amino-3,5-dichlorotoluene, 766(124)
2-amino-N,N-diethylaniline, 760(220)
3-amino-N,N-diethylaniline, 760(214)
4-amino-N,N-diethylaniline, 760(205)
2-amino-N,N-dimethylaniline, 756(141)
3-amino-N,N-dimethylaniline, 760(212)
4-amino-N,N-dimethylaniline, 764(52)
4-amino-3,2'-dimethylazobenzene, 776(293)
2-amino-1,3-dimethylbenzene (2,6-dimethylaniline), 756(135)
2-amino-1,4-dimethylbenzene (2,5-dimethylaniline), 756(131)
4-amino-1,3-dimethylbenzene (2,4-dimethylaniline), 754(127)
5-amino-1,3-dimethylbenzene (3,5-dimethylaniline), 756(142)
2-amino-4,4'-dimethylbiphenyl, 768(134)
3-amino-4,4'-dimethylbiphenyl, 776(306)
1-amino-3,7-dimethylnaphthalene, 774(265)
1-amino-2,6-dimethylnaphthalene, 774(252)
2-amino-1,4-dimethylnaphthalene, 770(193)
2-amino-3,5-dimethylphenol, 788(537)
2-amino-4,5-dimethylphenol, 788(559)
2-amino-4,6-dimethylphenol, 782(436)
4-amino-3,5-dimethylphenol, 790(576)
4-amino-2,5-dimethylphenol, 794(662)
4-amino-2,6-dimethylphenol, 782(447)
4-amino-2,6-dimethylpiperidine, 754(99)
4-amino-2,6-dimethylpyridine, 790(587)
2-amino-4,6-dimethylpyrimidine 790(611)
4-amino-4,5-dimethylpyrimidine, 792(642)
4-amino-2,6-dimethylpyrimidine, 790(581)
6-amino-5,7-dimethylquinoline, 788(563)
1-amino-4,5-dimethyl-1,2,3-triazole, 774(270)
2-amino-4,6-dinitrophenol, 938(315)

2-aminodiphenylamine, 772(211)
4-aminodiphenylamine, 770(192)
1-amino-1,2-diphenylethane, 760(221)
2-aminodiphenylmethane, 766(97)
1-amino-2,3-diphenylpropane, 760(224)
2-aminoethanethiol, 774(290), 967(20)
2-aminoethanol, 718(99), 752(80)
2-aminoethyl alcohol (ethanolamine), 718(99), 752(80)
2-(2-aminoethylamino)ethanol, 758(189)
dl-α-aminoethylbenzene (α-phenylethylamine), 754(93)
β-aminoethylbenzene (β-phenylethylamine), 754(102)
2-aminoethyl ethyl ether, 750(38), 824(51)
2-(2-aminoethyl)indole, 780(381)
4-(2-aminoethyl)phenol, 788(539)
2-aminofluorene, 782(419)
9-aminofluorene, 768(146)
1-aminofluorenone, 778(372)
2-aminofluorenone, 788(538), 905(240)
3-aminofluorenone, 786(523), 905(237)
2-aminoheptane, 752(61)
4-aminoheptane, 752(59)
dl-2-aminoheptanoic acid, 678(97)
3-aminohexane, 752(51)
4-aminohexanoic acid, 674(42)
6-aminohexanoic acid, 674(38)
3-aminohydrocinnamic acid, 672(2)
4-aminohydrocinnamic acid, 672(9)
3-amino-6-hydroxyacetophenone, 780(384)
4-amino-2-hydroxyacetophenone, 780(391)
1-amino-4-hydroxyanthraquinone, 792(636), 906(265)
4-amino-2-hydroxybenzenesulfonamide, 949(61)
4-amino-2-hydroxybenzenesulfonyl chloride, 957(211)
1-amino-2-hydroxybutane, 752(77)
α-amino-β-hydroxybutyric acid (DL-threonine), 676(65)
2-amino-10-hydroxyphenanthrene, 792(650)
α-amino-β-hydroxypropionic acid (DL-serine), 676(77)

5-amino-8-hydroxyquinoline, 784(462)
7-amino-8-hydroxyquinoline, 780(400)
1-aminoindane, 756(149)
2-aminoindane, 758(169)
4-aminoindane, 758(177)
5-aminoindane, 762(40)
3-aminoindazole, 786(503)
4-aminoindazole, 784(491)
5-aminoindazole, 790(577)
7-aminoindazole, 786(508)
3-aminoindole, 778(367)
2-aminoisobutyric acid, 678(96)
1-aminoisoquinoline, 780(392)
4-aminoisoquinoline, 776(327)
5-aminoisoquinoline, 780(412)
β-aminolactic acid (DL-isoserine), 674(37), 676(78)
aminomalonic acid, 672(5)
2-amino-3-methoxy-4-hydroxybenzaldehyde, 740(102), 780(413)
3-amino-6-methoxyphenol, 780(403)
4-amino-N-methylacetanilide, 762(31), 768(137)
4-amino-N-methylaniline, 762(33)
1-amino-2-methylanthraquinone, 792(632)
2-amino-5-methylbenzene-1,3-disulfonamide, 952(228)
2-amino-5-methylbenzene-1,3-disulfonyl chloride, 957(198)
2-amino-4-methylbenzoic acid, 672(18)
3-amino-4-methylbenzophenone, 776(334), 778(340)
4-amino-3-methylbenzophenone, 778(344)
4'-amino-4-methylbenzophenone, 790(589)
6-amino-3-methylbenzophenone, 768(157)
4-amino-3-methylbiphenyl, 764(60)
4-amino-4'-methylbiphenyl, 774(288)
2-amino-2-methylbutane, 750(18)
2-amino-2-methylbutyric acid, 674(57)
4-amino-5-methyl-2,4-diethylpyrimidine, 790(593)
2-amino-4-methyldiphenylamine, 782(456)
4-amino-2-methyldiphenylamine, 764(82)
2-amino-4-methyl-5-nitrophenol, 792(618)
dl-2-amino-2-methylpentanoic acid, 678(111)
2-amino-3-methylphenol, 784(487), 938(280)

2-amino-4-methylphenol, 782(439), 936(245)
2-amino-5-methylphenol, 786(534), 938(300)
2-amino-6-methylphenol, 772(244), 930(117)
3-amino-4-methylphenol, 784(466), 936(264)
3-amino-6-methylphenol, 786(533), 938(296)
4-amino-2-methylphenol, 788(563), 940(329)
4-amino-3-methylphenol, 788(572), 940(334)
4-amino-3-methyl-1-phenylpyrazole, 772(243)
5-amino-3-methyl-1-phenylpyrazole, 778(362)
2-amino-3-methylpyridine, 762(13)
2-amino-4-methylpyridine, 774(283)
2-amino-6-methylpyridine, 764(53)
3-amino-4-methylpyridine, 776(320)
3-amino-6-methylpyridine, 774(275)
2-amino-4-methylpyrimidine, 786(530)
2-amino-4-methylquinoline, 783(432)
3-amino-2-methylquinoline, 786(528)
4-amino-2-methylquinoline, 788(547)
5-amino-2-methylquinoline, 778(369)
6-amino-2-methylquinoline, 790(590)
7-amino-2-methylquinoline, 784(481)
8-amino-6-methylquinoline, 768(135)
1-amino-5-methyl-1,2,3-triazole, 770(174)
7-aminonaphthalene-1,3-disulfonic acid, 960(65)
7-aminonaphthalene-1,4-disulfonic acid, 960(64)
7-aminonaphthalene-1,5-disulfonic acid, 960(63)
4-aminonaphthalene-1-sulfonamide, 950(156), 951(163)
4-aminonaphthalene-2-sulfonamide, 948(27)
5-aminonaphthalene-1-sulfonamide, 952(229)
5-aminonaphthalene-2-sulfonamide, 951(178)
8-aminonaphthalene-2-sulfonamide, 949(97)
2-aminonaphthalene-1-sulfonic acid, 960(61)

4-aminonaphthalene-1-sulfonic acid, 960(56)
5-aminonaphthalene-1-sulfonic acid, 960(57)
5-aminonaphthalene-2-sulfonic acid, 960(58)
6-aminonaphthalene-2-sulfonic acid, 960(62)
8-aminonaphthalene-1-sulfonic acid, 960(59)
4-amino-1-naphthalenethiol, 774(255), 967(17)
8-aminonaphthalene-1,3,5-trisulfonic acid, 960(60)
1-amino-2-naphthoic acid, 674(44)
2-amino-1-naphthoic acid, 672(7)
3-amino-1-naphthoic acid, 672(25)
3-amino-2-naphthoic acid, 674(53)
4-amino-1-naphthoic acid, 672(22)
5-amino-1-naphthoic acid, 674(47)
6-amino-1-naphthoic acid, 674(43)
6-amino-2-naphthoic acid, 676(63)
7-amino-1-naphthoic acid, 676(61)
1-amino-2-naphthol, 794(678), 944(426)
2-amino-1-naphthol, 938(279)
3-amino-2-naphthol, 942(409)
4-amino-2-naphthol, 790(613), 940(343)
5-amino-1-naphthol, 938(317), 940(352)
5-amino-2-naphthol, 940(360)
6-amino-1-naphthol, 790(614), 940(364)
6-amino-2-naphthol, 792(641), 942(384)
7-amino-1-naphthol, 938(290)
7-amino-2-naphthol, 792(622), 942(368)
8-amino-1-naphthol, 774(271)
8-amino-2-naphthol, 942(380)
2-aminonicotinamide, 790(607)
5-amino-4-nitroacenaphthene, 792(652)
2-amino-4-nitrobenzaldehyde, 740(100), 780(399)
2-amino-5-nitrobenzaldehyde, 792(621)
4-amino-3-nitrobenzaldehyde, 740(122)
4-amino-3-nitrobenzoic acid, 710(43)
2-amino-5-nitrobiphenyl, 780(402)
4-amino-3-nitrobiphenyl, 788(556)

Index of Compounds

4-amino-4'-nitrobiphenyl, 792(620)
1-amino-3-nitroguanidine, 790(594)
1-amino-4-nitronaphthalene, 790(607)
2-amino-3-nitrophenol, 792(645)
2-amino-4-nitrophenol, 784(459), 792(623)
2-amino-5-nitrophenol, 942(370)
3-amino-6-nitrophenol, 786(535)
4-amino-2-nitrophenol, 782(426), 936(237)
4-amino-3-nitrophenol, 786(502), 938(286)
2-amino-4-nitrostilbene, 784(461)
4-amino-2-nitrostilbene, 778(338)
dl-2-aminononanoic acid, 678(92)
2-aminooctanoic acid, 678(91)
3-aminooctanoic acid, 674(41)
dl-2-aminopentane, 750(29)
5-amino-1-pentene, 750(28)
dl-2-aminopentanoic acid, 672(17)
dl-3-aminopentanoic acid, 672(51)
5-aminopentanoic acid, 672(14)
2-amino-3-pentanol, 752(84)
3-amino-2-pentanol, 752(81)
2-aminophenacyl alcohol, 903(158)
4-aminophenacyl alcohol, 788(540)
1-aminophenanthrene, 784(469)
2-aminophenanthrene, 772(231)
3-aminophenanthrene, 772(242)
4-aminophenanthrene, 776(310)
9-aminophenanthrene, 782(446)
4-aminophenethyl alcohol, 776(330)
2-aminophenol (2-hydroxyaniline), 788(560), 940(326)
3-aminophenol (3-hydroxyaniline), 780(390), 934(209)
4-aminophenol (4-hydroxyaniline), 790(582), 940(342)
4-aminophenoxyacetic acid, 706(380), 792(648)
dl-α-aminophenylacetic acid, 676(82)
l-α-aminophenylacetic acid, 680(124)
2-aminophenylacetic acid, 672(6)
3-aminophenylacetic acid, 672(12)
4-aminophenylacetic acid, 674(35), 704(329)
2-aminophenylacetonitrile, 914(53)

4-aminophenylacetonitrile, 914(28)
3-amino-5-phenylacridine, 792(630)
2-amino-4-phenylbutane, 756(146) (155)
2-amino-1-phenylethanol, 764(48)
2-aminophenyl ethyl ether, 756(165)
3-amino-3-phenylisobutyric acid, 676(76)
1-amino-1-phenyl-2-methylpropane, 756(128)
dl-2-amino-1-phenyl-1-propanol, 776(307)
2-aminophenylpropenoic acid, 786(521)
3-(4-aminophenyl)propenoic acid, 672(20)
3-(2-aminophenyl)propenonitrile, 782(437)
dl-2-amino-2-phenylpropionic acid, 676(85)
d or l-2-amino-2-phenylpropionic acid, 678(113)
dl-3-amino-3-phenylpropionic acid, 676(70)
d or l-3-amino-3-phenylpropionic acid, 676(71)
3-amino-6-phenylpyridine, 776(316)
2-(4-aminophenyl)quinoline, 782(449)
3-amino-2-phenylquinoline, 778(358)
4-amino-2-phenylquinoline, 788(546)
3-amino-1-phenyl-1,2,4-triazole, 784(484)
5-amino-1-phenyl-1,2,4-triazole, 786(514)
4-aminophenylurethane, 770(187)
1-amino-2-propanol, 716(82), 752(73)
dl-2-aminopropanol, 752(83), 718(107)
3-aminopropanol, 718(128), 754(97)
β-aminopropiophenone, 778(345)
2-aminopropiophenone, 900(64)
3-aminopropiophenone, 764(55), 900(47)
4-aminopropiophenone, 764(72), 782(454), 904(216)
4-aminopyrazole, 772(213)
1-aminopyrene, 790(579)
3-aminopyrene, 778(370)
4-aminopyrene, 792(635)
2-aminopyridine, 766(109) (123)
3-aminopyridine, 768(143)
4-aminopyridine, 786(517) (522)

2-aminopyrimidine, 780(410)
4-aminopyrimidine, 784(493)
2-aminoquinoline, 782(418)
3-aminoquinoline, 774(266)
4-aminoquinoline, 786(509)
5-aminoquinoline, 778(336)
6-aminoquinoline, 778(354)
7-aminoquinoline, 774(264)
8-aminoquinoline, 770(175)
5-aminoresorcinol, 784(473)
3-aminosalicylic acid, 676(72), 708(398), 794(659)
5-aminosalicylic acid, 678(101), 710(429)
2-aminothiazole, 772(248)
2-aminothiophenol, 762(14), 967(3)
4-aminothiophenol, 967(8)
1-aminothioxanthone, 792(627)
3-aminothioxanthone, 792(651)
4-aminothioxanthone, 794(670)
6-aminothymol, 788(571)
2-amino-p-toluic acid, 672(18)
3-amino-p-toluic acid, 672(23)
1-amino-1,2,3-triazole, 766(90)
1-amino-1,3,4-triazole, 772(221)
3-amino-1,2,4-triazole, 786(526)
3-amino-2,4,6-tribromophenol, 780(378)
1-aminotridecane, 762(16)
ω-aminotridecylic acid, 672(21)
2-aminotriphenylcarbinol, 780(387)
3-aminotriphenylcarbinol, 786(506)
4-aminotriphenylcarbinol, 778(364)
1-aminotriphenylene, 776(302)
2-aminotriphenylene, 784(460)
2-aminotriphenylmethane, 780(417)
4-aminotriphenylmethane, 772(227)
1-aminoundecane, 758(184)
2-aminoundecane, 758(178)
amylacetal, 667(38)
amyl acetate, 815(97)
amylal, 667(36)
amyl alcohol(active) (2-methylbutanol), 714(31)
n-amyl alcohol (1-pentanol), 716(49)
sec-amyl alcohol (dl-2-pentanol), 714(23)
tert-amyl alcohol, 714(10)
amylamine, 750(34)
n-amylbenzene (n-pentylbenzene), 880(83)
sec-amylbenzene, 879(40)
n-amyl bromide, 836(123)
tert-amyl bromide, 836(117)
n-amyl butyrate, 816(153)
n-amyl chloride, 832(45)
tert-amyl chloride, 832(37)
amylcyclohexane, 866(197)

Index of Compounds

amylene (1-pentene), 870(12)
amyl ether, 825(93)
amyl ethyl ether, 824(57)
amyl formate, 815(67)
n-amyl iodide, 838(177)
tert-amyl iodide, 838(169)
amyl mercaptan, 966(15)
amyl methyl ether, 823(41)
amyl methyl ketone (2-heptanone), 894(80)
amyl 2-naphthyl ether, 827(5)
4-n-amylphenol, 924(26), 926(3)
4-tert-amylphenol, 930(123)
amyl valerate, 817(188)
anethole, 826(128)
angelamide, 745(216)
angelanilide, 745(211)
angelic acid, 688(32)
aniline, 754(90)
anilinomalonic acid, 694(155)
anisaldehyde (4-methoxybenzaldehyde), 732(84)
2-anisamide, 746(224)
4-anisamide, 747(339)
4-anisanilide, 747(347)
4-anisic acid, 702(304)
anisic anhydride, 668(46)
2-anisidine (2-methoxyaniline), 756(156)
3-anisidine, 758(196)
4-anisidine, 766(113)
anisole (methoxybenzene), 824(76)
2-anisoyl chloride, 671(50)
3-anisoyl chloride, 671(49)
4-anisoyl chloride, 671(67)
2-anisyl alcohol, 720(164)
4-anisyl alcohol, 720(166), 722(6)
anthanthrene, 890(124)
anthracene, 890(118)
anthracene-9-carboxaldehyde, 738(85)
anthracene-1-carboxylic acid, 708(414)
anthracene-9-carboxylic acid, 706(353)
1,2-anthracenediol, 938(299)
1,8-anthracenediol, 942(403)
9,10-anthracenediol, 940(337)
anthracene-1,8-disulfonamide, 952(251)
anthracene-1,5-disulfonic acid, 959(6)
anthracene-1,5-disulfonyl chloride, 958(226)
anthracene-1,8-disulfonyl chloride, 958(223)
anthracene-1-sulfonamide, 950(151)
anthracene-2-sulfonamide, 952(231)
anthracene-1-sulfonyl chloride, 955(93)
anthracene-2-sulfonyl chloride, 956(159)

1,8,9-anthracenetriol, 940(331)
anthranilamide, 745(160)
anthranilanilide, 746(230)
anthranilic acid (2-aminobenzoic acid), 672(11), 698(220), 784(478)
9,10-anthraquinone, 906(280), 944(429)
anthraquinone-1-carboxylic acid, 710(434)
anthraquinone-2-carboxylic acid, 710(433)
anthraquinone-1,5-disulfonamide, 952(222)
anthraquinone-2,6-disulfonamide, 952(250)
anthraquinone-2,7-disulfonamide, 950(135)
anthraquinone-1,5-disulfonic acid, 959(9)
anthraquinone-1,6-disulfonic acid, 959(10)
anthraquinone-1,7-disulfonic acid, 959(11)
anthraquinone-1,8-disulfonic acid, 959(12)
anthraquinone-1,5-disulfonyl chloride, 958(228)
anthraquinone-1,6-disulfonyl chloride, 957(219)
anthraquinone-1,7-disulfonyl chloride, 958(225)
anthraquinone-1,8-disulfonyl chloride, 958(222)
anthraquinone-2,6-disulfonyl chloride, 958(227)
anthraquinone-2,7-disulfonyl chloride, 957(216)
anthraquinone-2-sulfonamide, 952(233)
anthraquinone-1-sulfonic acid, 959(7)
anthraquinone-2-sulfonic acid, 959(8)
anthraquinone-2-sulfonyl chloride, 957(218)
anthrone, 905(236)
apocamphoric acid, 704(341)
β-D-arabinose, 809(30)
β-L-arabinose, 809(32)
arachidamide, 745(159)
arachidanilide, 744(90)
arachidic acid, 690(73)
arachidic anhydride, 668(41)
DL-arginine, 676(74)
L-arginine, 674(45)
L-ascorbic acid, 810(41)
L-β-asparagine (β-aspartamide), 676(64)
L-α-asparagine, 674(50)
β-aspartamide (L-α-asparagine), 676(64)
DL-aspartic acid, 676(89)
D- or L-aspartic acid, 676(90)
azelamide, 744(103), 747(356)

azelaanilide, 745(158)
azelaic acid (nonanedioic acid), 694(123)
azobenzene, 909(7)
2,2'-azobiphenyl, 909(40)
4,4'-azobiphenyl, 910(84)
1,1'-azonaphthalene, 910(64)
1,2'-azonaphthalene, 909(33)
2,2'-azonaphthalene, 910(71)
2-azophenetole, 909(28)
4-azophenetole, 910(48)
2-azophenol, 910(53)
2-azopyridine, 909(14)
3-azopyridine(cis), 909(13)
3-azopyridine(trans), 909(36)
4-azopyridine, 909(20)
2-azotoluene, 909(3)
3-azotoluene, 909(2)
4-azotoluene, 909(39)
2-azoxyanisole, 911(9)
3-azoxyanisole, 911(4)
4-azoxyanisole, 911(20)
azoxybenzene(cis), 911(15)
azoxybenzene(trans), 911(1)
2,2'-azoxybiphenyl, 911(25)
4,4'-azoxybiphenyl, 911(30)
1,1'-azoxynaphthalene, 911(21)
2,2'-azoxynaphthalene, 911(26)
2-azoxyphenetole, 911(17)
3-azoxyphenetole, 911(3)
4-azoxyphenetole, 911(22)
2-azoxytoluene, 911(5)
3-azoxytoluene, 911(2)
4-azoxytoluene, 911(7)
azulene, 888(61)

barbituric acid, 708(415), 748(503)
behenic anhydride, 668(43)
benzalacetone (4-phenyl-3-buten-2-one), 900(44)
benzalacetophenone (chalcone), 901(90)
benzal chloride, 855(122)
benzaldehyde, 730(56)
benzaldehyde dibutyl acetal, 667(41)
benzaldehyde diethyl acetal, 667(39)
benzaldehyde dimethyl acetal, 667(32)
benzalfluorene, 887(36)
benzamide, 746(229)
benzanilide, 747(331)
1,2-benzanthracene, 889(94)
benzanthrone, 905(243)
benzene, 877(1)
benzeneazo-o-cresol (4-hydroxy-3-methylazobenzene), 909(25)
1-benzeneazonaphthalene, 909(9)
2-benzeneazonaphthalene, 909(27)
1-benzeneazo-2-naphthol, 909(29)

Index of Compounds 1001

2-benzeneazo-1-naphthol, 909(35)
4-benzeneazo-1-naphthol, 910(70)
benzeneazoresorcinol, 910(52)
1,2-benzenediacetonitrile, 914(40)
1,2-benzenediol, 932(154)
1,3-benzenediol, 932(170)
1,4-benzenediol, 938(322)
benzene-1,2-disulfonamide, 952(227)
benzene-1,3-disulfonamide, 951(200)
benzene-1,4-disulfonamide, 952(242)
benzene-1,2-disulfonic acid, 959(14)
benzene-1,3-disulfonic acid, 959(15)
benzene-1,2-disulfonyl chloride, 957(189)
benzene-1,3-disulfonyl chloride, 954(49)
benzene-1,4-disulfonyl chloride, 957(176)
benzenesulfonamide, 949(57)
benzenesulfonic acid, 959(13)
benzenesulfonyl chloride, 953(4)
benzene-1,3,5-trisulfonyl chloride, 957(217)
1,2-benzfluoren-9-one, 904(208)
1,2-benzfluorene, 890(110)
2,3-benzfluorene, 890(115)
3,4-benzfluorene, 889(78)
benzhydrol (diphenylcarbinol), 722(45)
benzidine, 780(407)
benzil, 902(151)
benzilamide, 746(302)
benzilic acid, 700(229)
benzimidazole, 788(553)
2-benzofuryl methyl ketone, 902(129)
benzoic acid, 696(162)
benzoic anhydride, 668(24)
benzoin, 724(66), 904(214)
benzoin acetate, 822(120)
benzonitrile, 913(35)
3,4-benzophenanthrene, 887(27)
benzophenone, 900(65)
benzopiperidine, 744(15)
1,4-benzoquinone (quinone), 903(186)
benzothiazole, 776(294)
benzothiazole-2-sulfonic acid, 959(16)
benzo-2,4,6-tribromoanilide, 748(489)
benzotrichloride, 855(130)
benzotrifluoride, 853(44)
benzoylacetic acid, 692(119)
benzoylacetone, 901(99)
benzoylacetonitrile, 915(56)

2-benzoylacrylic acid, 903(160)
N-benzoylanthranilic acid, 702(294)
2-benzoylbenzamide, 747(334)
2-benzoylbenzanilide, 747(419)
2-benzoylbenzoic acid, 692(96), 696(174)
benzoyl bromide, 671(81)
2-benzoylbutyric acid, 692(87)
3-benzoylbutyric acid, 696(175)
benzoyl chloride, 670(36)
benzoylformic acid, 690(59), 901(109)
2-benzoylfuran, 900(55)
2-benzoyl-7-hydroxynaphthalene, 906(274)
2-benzoylphenol, 926(17)
α-benzoyl-β-phenylhydrazine, 747(342)
3-benzoylpropenoic acid, 690(58), 901(107)
3-benzoylpropionanilide, 746(289)
2-benzoylpropionic acid, 690(81), 902(141)
3-benzoylpropionic acid, 694(141)
2-benzoylpyridine, 802(145)
3-benzoylpyridine, 900(45)
4-benzoylpyridine, 805(66), 902(121)
benzoyl-m-toluidine, 745(203)
benzoyl-o-toluidine, 746(269)
benzoyl-p-toluidine, 746(320)
bcnzpyrazine (quinoxaline), 804(9)
1,2-benzpyrene, 889(104)
4,5-benzpyrene, 889(106)
benzyl acetate, 818(209)
N-benzylacetoacetamide, 745(128)
N-benzylacetamide, 744(23)
benzyl alcohol, 720(141)
benzylamine, 754(91)
N-benzylaniline, 762(35)
benzyl benzoate, 820(317)
2-benzylbenzoic acid, 694(137)
benzyl bromide, 836(132)
benzyl 4-bromophenyl sulfide, 965(31)
benzyl butyl ether, 825(119)
benzyl butyrate, 818(245)
benzyl chloride, 834(63)
benzyl chloroacetate, 815(94)
benzyl cinnamate, 821(30)
benzyl ether, 826(145)
benzyl ethyl ether, 825(92)
N-benzyl-N-ethylaniline, 804(13)
benzyl ethyl ketone (1-phenyl-2-butanone), 897(193)
benzyl ethyl sulfide, 963(56)
benzyl fluoride, 830(15)
benzyl formate, 817(181)
benzylglycolicaldehyde, 734(29)

benzylideneacetophenone, 901(90)
4-benzylideneaminophenol, 940(341)
benzylideneaniline, 804(40)
2-benzylidenepropionic acid, 690(80)
3-benzylidenepropionic acid, 692(91)
benzyl iodide, 840(190)
benzyl isobutyl ether, 825(111)
benzyl isothiocyanate, 961(13)
benzylmalonamide, 748(483)
benzylmalonanilide, 748(465)
benzylmalonic acid, 694(146)
benzyl mercaptan, 966(25)
N-benzyl-N-methylaniline, 802(144)
benzyl methyl ether, 824(79)
benzyl methyl ketone, 899(27)
benzyl methyl sulfide, 962(46)
1-benzylnaphthalene, 886(18)
2-benzylnaphthalene, 886(17)
4-benzyl-1-naphthol, 934(224)
benzyl 1-naphthyl ether, 828(43)
benzyl 2-naphthyl ether, 828(50)
benzyl 2-naphthyl ketone, 903(163)
benzyl 4-nitrophenylsulfide, 965(47)
benzyl oxalate, 822(105)
1-benzyloxynaphthalene, 828(43)
2-benzyloxynaphthalene, 828(50)
4-benzyloxyphenol, 934(213)
2-benzylphenol, 926(38)
4-benzylphenol, 930(107)
benzyl phenyl sulfide, 965(10)
benzyl phthalate, 821(32)
2-benzylpyridine, 802(135)
4-benzylpyridine, 802(141)
benzyl salicylate, 820(315)
benzyl succinate, 821(34)
benzyl sulfide, 965(16)
benzylsulfonamide, 948(13)
benzylsulfonic acid (toluene-α-sulfonic acid), 959(17)
benzylsulfonyl chloride, 955(94)
benzyl thiocyanate, 961(8)
benzylurea, 746(285)
betaine (trimethylglycine), 680(123)
biacetyl, 892(7)
bibenzyl, 886(10)
1,1'-binaphthyl, 889(95)
1,2'-binaphthyl, 887(40)
2,2'-binaphthyl, 890(109)
o,o'-biphenol (2,2'-dihydroxybiphenyl), 932(169)
p,p'-biphenol (4,4'-dihydroxybiphenyl), 944(424)
biphenyl, 887(29), 881(71)

Index of Compounds

biphenyl-4-carbonitrile, 914(49)
biphenyl-4-carboxylic acid, 708(389)
biphenyl-4,4′-disulfonamide, 952(245)
biphenyl-4,4′-disulfonic acid, 959(18)
biphenyl-4,4′-disulfonyl chloride, 958(220)
biphenylene oxide (dibenzofuran), 828(45)
biphenyl-4-sulfonamide, 951(202)
biphenyl-4-sulfonyl chloride, 956(143)
2,2′-bipyridyl, 805(61)
3,3′-bipyridyl, 805(59)
3,4′-bipyridyl, 805(55)
4,4′-bipyridyl, 806(104)
bis(2-aminoethyl)amine, 754(121)
1,2-bis(aminoethylamino)ethane (triethylenetetramine), 760(215)
1,3-bis(aminomethyl)benzene, 758(190)
bis(2-aminophenyl) disulfide, 774(262)
bis(benzylthio)methane, 965(22)
bis(butylthio)methane, 963(60)
bis(2-chloroethoxy)ethane, 825(124)
bis(2-chloroethoxy)methane, 667(35)
bis(2-chloroethyl) ether, 825(88), 854(109)
1,2-bis(chloromethyl)benzene, 856(11)
1,3-bis(chloromethyl)benzene, 856(8)
1,4-bis(chloromethyl)benzene, 856(15)
bis(chloromethyl) ether, 824(45), 853(47)
4,4′-bis(dimethylamino)azobenzene, 909(22)
4,4′-bis(dimethylamino)benzhydrol, 806(92)
4,4′-bis(dimethylamino)benzophenone, 905(248), 806(116)
1,4-bis(dimethylamino)butane, 799(49)
bis(4-dimethylaminophenyl)methane, 805(83)
4,4′-bis(dimethylamino)triphenylmethane, 806(94)
bis(2-ethoxyethyl) ether, 825(95)
bis(2-ethoxyethyl) phthalate, 821(11)
1,2-bis(ethylthio)ethane, 963(49)
bis(ethylthio)methane, 962(40)
1,3-bis(ethylthio)propane, 963(57)

1,2-bis(2-hydroxyethoxy)ethane, 720(169)
bis(2-mercaptoethyl) ether, 966(31)
bis(4-mercaptophenyl) ether, 967(18)
bis(4-mercaptophenyl) sulfide, 967(25)
bis(methoxyethoxyethyl) ether, 826(142)
bis(2-methoxyethyl) ether, 824(77)
1,5-bis(4-methoxyphenyl)-1,4-pentadien-3-one, 904(206)
1,1-bis(2-methylpropoxy)ethane, 667(27)
bis(2-methylpropoxy)methane, 667(24)
bis(methylthio)methane, 962(30)
1,3-bis(2-nitrophenylthio)propane, 965(48)
bis(phenacyl) sulfide, 902(132)
bis(2-phenoxyethyl) ether, 827(37)
1,2-bis(phenylthio)ethane, 965(32)
bis(phenylthio)methane, 965(19)
bis(thioanisyl) ketone, 900(51)
1,2-bis(*p*-tolylthio)ethane, 965(41)
biuret, 747(414)
d-borneol, 724(74)
bornyl acetate, 821(8)
bornyl chloride, 836(97)
brassidamide, 744(97)
brassidanilide, 744(61)
brassidic acid (*trans*-13-docosenoic acid), 690(49)
bromal, 730(53)
bromoacetamide, 744(86)
bromoacetanilide, 746(231)
4-bromoacetanilide, 747(337)
bromoacetic acid, 688(38)
bromoacetic anhydride, 668(22)
ω-bromoacetophenone (phenacyl bromide), 900(73)
2-bromoacetophenone, 900(70)
4-bromoacetophenone, 901(75)
bromoacetyl bromide, 671(79)
bromoacetyl chloride, 670(27)
3-bromo-1-aminoanthraquinone, 794(663)
4-bromo-2-aminobenzaldehyde, 772(233)
4-bromo-4′-aminobiphenyl, 784(468)
5-bromo-2-aminobiphenyl, 766(111)
5-bromo-3-aminopyridine, 768(159)
2-bromoaniline, 762(24)
3-bromoaniline, 758(197)
4-bromoaniline, 768(155)
2-bromoanisole, 825(110)

4-bromoanisole, 825(114)
2-bromoazobenzene, 909(15)
3-bromoazobenzene, 909(8)
4-bromoazobenzene, 909(16)
2-bromobenzaldehyde, 732(79), 734(3)
3-bromobenzaldehyde, 732(81)
4-bromobenzaldehyde, 736(50)
2-bromobenzamide, 746(305)
3-bromobenzamide, 746(308)
4-bromobenzamide, 747(406)
2-bromobenzanilide, 746(261)
3-bromobenzanilide, 746(252)
4-bromobenzanilide, 748(423)
bromobenzene, 842(10)
1-(2-bromobenzeneazo)-2-naphthol, 910(51)
1-(3-bromobenzeneazo)-2-naphthol, 910(54)
1-(4-bromobenzeneazo)-2-naphthol, 910(55)
4-(2-bromobenzeneazo)-1-naphthol, 910(59)
4-(3-bromobenzeneazo)-1-naphthol, 910(73)
4-(4-bromobenzeneazo)-1-naphthol, 910(80)
2-bromobenzenesulfonamide, 950(112)
4-bromobenzenesulfonamide, 949(72)
4-bromobenzenesulfonic acid, 959(19)
2-bromobenzenesulfonyl chloride, 953(31)
4-bromobenzenesulfonyl chloride, 954(70)
2-bromobenzoic acid, 700(228)
3-bromobenzoic acid, 700(238)
4-bromobenzoic acid, 710(422)
4-bromobenzoic anhydride, 669(67)
2-bromobenzonitrile, 914(36)
3-bromobenzonitrile, 914(22)
4-bromobenzonitrile, 915(85)
2-bromobenzophenone, 900(48)
4-bromobenzophenone, 902(140)
3-bromobenzotribromide, 854(89)
3-bromobenzoyl chloride, 671(48)
4-bromobenzoyl chloride, 671(62)
2-bromobenzyl bromide, 838(150)
3-bromobenzyl bromide, 838(154)
4-bromobenzyl bromide, 838(158)
3-bromobenzyl chloride, 834(82)
4-bromobenzyl chloride, 834(88)
2-bromobiphenyl, 844(66)

Index of Compounds

3-bromobiphenyl, 844(67)
4-bromobiphenyl, 848(50)
1-bromobutane, 836(114)
2-bromobutane, 836(110)
1-bromo-1-butene, 836(109)
1-bromo-1-butene(*trans*), 836(112)
1-bromo-2-butene, 836(115)
2-bromo-1-butene, 836(107)
2-bromo-2-butene(*cis*), 836(108)
2-bromo-2-butene(*trans*), 836(113)
4-bromo-2-butenoic acid, 690(72)
2-bromobutyric acid, 686(42)
3-bromobutyric acid, 688(2)
2-bromo-4-chloroaniline, 768(168)
1,2-bromochlorobenzene, 842(28)
1,3-bromochlorobenzene, 842(30)
1,4-bromochlorobenzene, 848(36)
1-bromo-1-chloroethane, 852(33)
1-bromo-2-chloroethane, 853(49)
bromochloromethane, 852(25)
2-bromo-4-chlorophenol, 926(11)
1-bromo-3-chloropropane, 854(80)
1-bromo-3-chloro-2-propanol, 720(135)
3-bromo-4-chloroquinoline, 805(62)
3-bromo-6-chloroquinoline, 806(101)
α-bromo-4-chlorotoluene, 838(156)
4-bromocinnamic acid, 708(418)
2-bromo-1,4-diaminobenzene, 770(198)
bromocyclohexane, 836(130)
bromocyclopentane, 836(126)
2-bromocymene, 844(58)
1-bromodecane, 836(139)
4-bromo-1,2-dichlorobenzene, 846(1)
bromodichloromethane, 852(37)
2-bromo-4,6-dichlorophenol, 928(74)
2-bromo-1,4-dihydroxy-9,10-anthraquinone, 944(422)
6-bromo-1,4-dihydroxy-9,10-anthraquinone, 940(344)
2-bromo-3,4-dimethylaniline, 764(68)
2-bromo-4,6-dimethylaniline, 764(83)
3-bromo-N,N-dimethylaniline, 801(122)
4-bromo-2,6-dimethylaniline, 764(84)

1-bromo-2,3-dimethylbenzene, 844(47)
1-bromo-3,4-dimethylbenzene, 844(49)
2-bromo-1,4-dimethylbenzene, 844(40)
4-bromo-2,5-dimethylbenzoic acid, 702(278)
5-bromo-2,4-dimethylbenzoic acid, 702(292)
3-bromo-2,2-dimethylbutane, 836(125)
1-bromo-2,2-dimethylpropane, 836(116)
2-bromo-3,5-dinitrobenzoic acid, 706(364)
4-bromo-1,5-dinitronaphthalene, 921(134)
4-bromo-1,8-dinitronaphthalene, 922(150)
4-bromo-2,6-dinitrophenol, 930(96)
6-bromo-2,4-dinitrophenol, 934(199)
1-bromododecane, 838(141)
1-bromoeicosane, 838(152)
bromoethanal, 728(24)
bromoethane, 836(100)
2-bromoethanol (ethylene bromohydrin), 716(65)
bromoethene, 836(99)
3-bromo-4-ethoxyaniline, 764(73)
5-bromo-2-ethoxyaniline, 766(110)
2-bromo-1-ethoxybenzene, 825(118)
4-bromo-1-ethoxybenzene, 826(127)
2-bromoethyl acetate, 816(119)
α-bromoethylbenzene (α-phenylethyl bromide), 836(134)
β-bromoethylbenzene, 836(136)
1-bromo-2-ethylbenzene, 844(34)
1-bromo-4-ethylbenzene, 849(39)
β-bromoethylcyclohexane, 836(135)
2-bromoethyl ether, 824(75)
1-bromoethyl ethyl ether, 824(48)
2-bromoethyl ethyl ether, 824(64)
2-bromoethyl phenyl ether, 827(10)
2-bromoethyl sulfide, 965(6)
1,4-bromofluorobenzene, 842(9)
bromoform, 854(85)
1-bromoheptadecane, 838(149)
1-bromoheptane, 836(131)
1-bromo-1-heptene, 836(129)
1-bromohexadecane, 838(145)
1-bromohexane, 836(128)

2-bromohexane, 836(127)
4-bromohydrazobenzene, 911(11)
bromohydroquinone, 932(171)
3-bromo-4-hydroxyaniline, 788(54)
4-bromo-3-hydroxyaniline, 784(483)
5-bromo-2-hydroxybenzoic acid, 700(264)
3-bromo-4-hydroxybiphenyl, 930(127)
5-bromo-4-hydroxy-2-methylaniline, 792(644)
5-bromo-2-hydroxy-3-methylbenzoic acid, 708(401)
4-bromo-3-hydroxy-2-naphthoic acid, 708(396)
5-bromo-2-hydroxy-*m*-toluic acid, 708(401)
1,2-bromoiodobenzene, 844(62)
1,4-bromoiodobenzene, 848(53)
2-bromoisobutyramide, 746(283)
2-bromoisobutyranilide, 744(71)
2-bromoisobutyric acid, 688(36)
2-bromoisobutyronitrile, 913(20)
1-bromo-2-isopropylbenzene, 844(44)
1-bromo-4-isopropylbenzene, 844(54)
5-bromo-3-isopropyl-6-methyluracil, 747(324)
2-bromo-4-isopropyltoluene, 844(58)
4-bromoisoquinoline, 804(22)
8-bromoisoquinoline, 805(76)
bromomalonic acid, 694(135)
bromomalononitrile, 914(45)
bromomethane, 836(98)
2-bromo-6-methoxyaniline, 768(151)
3-bromo-4-methoxyaniline, 768(145)
5-bromo-2-methoxyaniline, 774(281a)
2-bromo-1-methoxybenzene, 825(110)
4-bromo-1-methoxybenzene, 825(114)
2-bromo-4-methylaniline, 758(182), 762(12)
3-bromo-2-methylaniline, 758(198), 762(10)
3-bromo-4-methylaniline, 762(10)
3-bromo-5-methylaniline, 762(32)
4-bromo-2-methylaniline, 766(117)
4-bromo-3-methylaniline, 772(214)

Index of Compounds

5-bromo-2-methylaniline, 762(23)
3-bromo-4-methylbenzene-sulfonamide, 948(53)
4-bromo-2-methylbenzene-sulfonamide, 949(81)
4-bromo-3-methylbenzene-sulfonamide, 948(47)
5-bromo-2-methylbenzene-sulfonamide, 949(73)
3-bromo-4-methylbenzene-sulfonyl chloride, 954(46)
4-bromo-2-methylbenzene-sulfonyl chloride, 953(28)
4-bromo-3-methylbenzene-sulfonyl chloride, 953(30)
5-bromo-2-methylbenzene-sulfonyl chloride, 953(15)
2-bromo-4-methylbenzoic acid, 698(203)
4-bromo-2-methylbenzoic acid, 704(313)
2-bromo-3-methylbutanamide, 746(243)
2-bromo-3-methylbutananilide, 745(184)
1-bromo-2-methylbutane, 836(122)
1-bromo-3-methylbutane, 836(121)
2-bromo-2-methylbutane, 836(117)
2-bromo-3-methylbutane, 836(118)
2-bromo-3-methylbutylurea, 746(271)
2-bromo-3-methylbutyric acid, 688(29)
5-bromo-3-methyl-2-hydroxyaniline, 778(349)
5-bromo-4-methyl-2-hydroxyaniline, 778(365)
1-bromo-5-methylnaphthalene, 848(32)
1-bromo-8-methylnaphthalene, 848(42)
2-bromo-1-methylnaphthalene, 846(11)
3-bromo-1-methylnaphthalene, 846(19)
6-bromo-4-methyl-3-nitrobenzoic acid, 704(336)
4-bromo-2-methyl-6-nitrobenzoic acid, 708(386)
2-bromo-4-methylpentane, 836(124)
2-bromo-4-methylphenol, 924(12)
3-bromo-4-methylphenol, 926(42)
3-bromo-5-methylphenol, 926(45)
4-bromo-2-methylphenol, 928(61)
4-bromo-3-methylphenol, 928(60)
5-bromo-2-methylphenol, 930(99)
2-bromo-4-methylphenyl-hydrazine, 774(267)
1-bromo-2-methylpropane, 836(111)
2-bromo-2-methylpropane, 836(106)
2-bromo-2-methylpropionamide, 746(283)
6-bromo-2-methylquinoline, 806(90)
5-bromo-3-methylsalicylic acid, 708(401)
α-bromo-2-methyltoluene, 838(147)
α-bromo-4-methyltoluene, 838(151)
7-bromo-1-naphthoic acid, 708(402)
8-bromo-1-naphthoic acid, 702(287)
4-bromo-1-naphthonitrile, 915(73)
1-bromonaphthalene, 844(64)
2-bromonaphthalene, 846(27)
3-bromo-1,2-naphthalenediol, 934(195)
4-bromo-1,5-naphthalenediol, 934(189)
1-bromonaphthalene-2-sulfonamide, 952(236)
2-bromonaphthalene-1-sulfonamide, 948(37)
4-bromonaphthalene-1-sulfonamide, 950(140)
5-bromonaphthalene-1-sulfonamide, 951(203)
5-bromonaphthalene-2-sulfonamide, 951(180)
6-bromonaphthalene-1-sulfonamide, 951(173)
6-bromonaphthalene-2-sulfonamide, 950(159)
7-bromonaphthalene-1-sulfonamide, 951(160)
7-bromonaphthalene-2-sulfonamide, 951(175)
8-bromonaphthalene-2-sulfonamide, 950(121)
1-bromonaphthalene-2-sulfonyl chloride, 955(96)
2-bromonaphthalene-1-sulfonyl chloride, 955(111)
4-bromonaphthalene-1-sulfonyl chloride, 955(88)
5-bromonaphthalene-1-sulfonyl chloride, 955(106)
5-bromonaphthalene-2-sulfonyl chloride, 955(109)
6-bromonaphthalene-1-sulfonyl chloride, 954(72)
6-bromonaphthalene-2-sulfonyl chloride, 956(163)
7-bromonaphthalene-1-sulfonyl chloride, 957(191)
7-bromonaphthalene-2-sulfonyl chloride, 955(120)
8-bromonaphthalene-2-sulfonyl chloride, 956(155)
1-bromo-2-naphthol, 930(110)
3-bromo-2-naphthol, 930(111)
4-bromo-1-naphthol, 936(232)
4-bromo-2-naphthol, 934(212)
6-bromo-2-naphthol, 936(234)
5-bromo-1-naphthonitrile, 915(102)
1-bromo-2-naphthylamine, 768(139)
2-bromo-1-naphthylamine, 768(150)
3-bromo-1-naphthylamine, 770(172)
3-bromo-2-naphthylamine, 788(548)
4-bromo-1-naphthylamine, 776(297)
4-bromo-2-naphthylamine, 770(182)
5-bromo-2-naphthylamine, 762(42)
6-bromo-2-naphthylamine, 780(411)
8-bromo-1-naphthylamine, 774(249)
2-bromo-4-nitroaniline, 776(311)
4-bromo-3-nitroaniline, 782(427)
5-bromo-2-nitroaniline, 784(496)
6-bromo-2-nitroaniline, 770(188)
1,2-bromonitrobenzene, 917(19)
1,3-bromonitrobenzene, 918(34)
1,4-bromonitrobenzene, 921(126)
4-bromo-3-nitrobenzene-sulfonamide, 949(91)
4-bromo-3-nitrobenzene-sulfonyl chloride, 953(42)
2-bromo-5-nitrobenzoic acid, 702(290)
4-bromo-3-nitrobenzoic acid, 704(338)
4-bromo-3-nitrobenzonitrile, 915(89)
6-bromo-3-nitrobenzonitrile, 915(87)
6-bromo-3-nitro-4-methylbenzoic acid, 704(336)
4-bromo-1-nitronaphthalene, 919(80)
5-bromo-1-nitronaphthalene, 921(120)
8-bromo-1-nitronaphthalene, 920(99)

Index of Compounds

1-bromo-4-nitro-2-naphthylamine, 786(507)
1-bromo-8-nitro-2-naphthylamine, 784(463)
4-bromo-2-nitro-1-naphthylamine, 792(616)
4-bromo-3-nitro-1-naphthylamine, 782(428)
4-bromo-8-nitro-1-naphthylamine, 778(361)
2-bromo-4-nitrophenol, 934(182)
4-bromo-2-nitrophenol, 930(118)
5-bromo-2-nitrosobenzoic acid, 923(30)
α-bromo-2-nitrotoluene, 838(155)
α-bromo-3-nitrotoluene, 838(157)
α-bromo-4-nitrotoluene, 838(159)
4-bromo-3-nitrotoluene, 917(10)
4-bromo-6-nitro-o-toluic acid, 708(386)
5-bromo-2-nitro-p-tolunitrile, 915(93)
5-bromo-3-nitro-o-tolunitrile, 915(77)
5-bromo-3-nitro-p-tolunitrile, 915(92)
1-bromononadecane, 838(153)
1-bromononane, 836(138)
1-bromooctadecane, 838(148)
1-bromooctane, 836(133)
1-bromopentadecane, 838(144)
1-bromopentane, 836(123)
2-bromopentane (2-pentyl bromide), 836(119)
3-bromopentane (3-pentyl bromide), 836(120)
3-bromophenacyl bromide, 901(76)
4-bromophenacyl bromide, 903(173)
2-bromophenetole, 825(118)
4-bromophenetole, 826(127)
2-bromophenol, 924(4)
3-bromophenol, 926(9)
4-bromophenol, 928(62)
2-bromophenoxyacetic acid, 698(211)
dl-2-bromo-2-phenylacetic acid, 692(88)
2-bromophenylacetic acid, 692(118)
4-bromophenylacetic acid, 694(136)
4-bromophenylacetonitrile, 914(30)
2-bromophenylenediamine, 770(198)
2-bromophenylhydrazine, 764(77)

4-bromophenylhydrazine, 776(319)
4-bromophenyl isocyanate, 912(21)
4-bromophenyl methyl sulfide, 965(8)
2-bromo-3-phenylpropenoic acid(cis), 696(160)
4-bromophenyl sulfide, 965(46)
1-bromopropane, 836(104)
2-bromopropane, 836(102)
3-bromo-1-propanol, 718(112)
3-(4-bromophenyl)propenoic acid, 708(418)
1-bromo-2-propanone, 893(43)
1-bromo-1-propene(cis), 836(101)
1-bromo-1-propene(trans), 836(103)
3-bromopropene, 836(105)
2-bromopropionamide, 745(198)
3-bromopropionamide, 745(170)
2-bromopropionanilide, 745(120)
dl-2-bromopropionic acid, 686(37), 688(6)
3-bromopropionic acid, 690(53)
2-bromopropionyl bromide, 671(80)
3-bromopropiophenone, 900(43)
4-bromopropiophenone, 900(61)
3-bromopropylene oxide, 824(69)
2-bromopyridine, 800(77)
3-bromopyridine, 799(51)
2-bromoquinoline, 804(36)
3-bromoquinoline, 802(133)
4-bromoquinoline, 804(8)
5-bromoquinoline, 804(35)
6-bromoquinoline, 802(136)
7-bromoquinoline, 804(41)
8-bromoquinoline, 802(143)
5-bromosalicylaldehyde, 738(86), 932(157)
3-bromosalicylic acid, 702(303)
5-bromosalicylic acid, 700(264)
5-bromosalicylnitrile, 915(103)
β-bromostyrene, 836(137), 838(146)
2-bromostyrene, 844(43)
4-bromostyrene, 844(46)
N-bromosuccinimide, 747(359)
1-bromotetradecane, 838(143)
bromoterephthalic acid, 710(435)
4-bromothiophenol, 967(11)
α-bromotoluene, 836(132)
2-bromotoluene, 842(17)
3-bromotoluene, 842(18)
4-bromotoluene, 846(5)
2-bromo-p-toluic acid, 698(203)

5-bromo-o-toluidine (4-bromo-2-methylaniline), 766(117)
2-bromo-p-tolunitrile, 914(27)
3-bromo-p-tolunitrile, 914(29)
5-bromo-o-tolunitrile, 914(52)
4-bromo-o-tolylhydrazine, 776(304)
bromotrichloromethane, 853(46)
1-bromotridecane, 838(142)
1-bromoundecane, 838(140)
2-bromo-vinylbenzene, 844(43)
4-bromo-vinylbenzene, 844(46)
1,2-butadiene, 870(7)
1,3-butadiene, 870(3)
butanal (butyraldehyde), 728(10)
butanedial, 730(48)
butanedioic acid (succinic acid), 704(314)
butanedioic anhydride, 669(50)
1,4-butanediol (tetramethylene glycol), 720(157)
1,3-butanediol, 720(142)
2,3-butanediol, 718(121), 722(15)
butane-2,3-dione, 892(7)
butanedioyl chloride, 670(34)
1,4-butane dithiocyanate, 961(9)
1,4-butanedithiol, 966(27)
butane-1-sulfonamide, 948(3)
butane-1-sulfonic acid, 959(20)
1-butanethiol, 966(8)
2-butanethiol, 966(5)
1-butanol, 714(21)
2-butanol, 714(9)
2-butanone (ethyl methyl ketone), 892(4)
2-butenal, 728(20)
3-butenamide, 744(50)
1-butene, 870(2)
2-butene(cis), 870(5)
2-butene(trans), 870(4)
2-butenedioic acid(cis), 696(179)
2-butenedioic acid(trans), 704(331)
2-butenoic acid(cis), 684(17)
2-butenoic acid(trans), 690(70)
3-butenoic acid (vinylacetic acid), 684(15)
2-butenoic anhydride, 668(14)
3-buten-1-ol, 714(16)
dl-3-buten-2-ol (methylvinylcarbinol), 714(5)
3-buten-2-one (methyl vinyl ketone), 892(5)
3-butenonitrile (allyl cyanide), 913(14)
2-butenoyl chloride, 670(25)
butesin (butyl 4-aminobenzoate), 766(112), 821(64)
2-butoxyethanol ("butylcellosolve") (ethylene glycol monobutyl ether), 718(100), 824(81)

1006 Index of Compounds

2-butoxyethyl acetate, 817(167)
1-butoxy-2-ethoxyethane, 824(78)
2-(2-butoxyethoxy)ethanol, 720(155), 826(125)
butoxyethoxyethyl acetate, 819(262)
2-(2-butoxy)naphthalene, 827(14)
2-butoxynaphthalene, 827(15)
2-butoxytoluene, 825(121)
n-butylacetal (1,1-dibutoxy-ethane), 667(31)
sec-butylacetal (1,1-di(2-butoxy)ethane), 667(26)
n-butyl acetate, 815(61)
sec-butyl acetate, 814(44)
tert-butyl acetate, 814(25)
butyl acrylate, 815(83), 815(90)
butylal (dibutoxymethane), 667(28)
n-butyl alcohol, 714(21)
sec-butyl alcohol, 714(9)
tert-butyl alcohol, 714(4), 722(8)
n-butylamine, 750(17)
sec-butylamine, 750(13)
tert-butylamine, 750(6)
butyl 4-aminobenzoate (butesin), 766(112), 821(64)
N-butylaniline, 758(187)
N-*tert*-butylaniline, 756(130)
2-*tert*-butylaniline, 758(172)
4-*n*-butylaniline, 758(203)
4-*tert*-butylaniline, 756(162)
n-butylbenzene, 878(26)
sec-butylbenzene, 878(16)
tert-butylbenzene, 877(14)
butyl benzoate, 819(266)
n-butyl bromide, 836(114)
sec-butyl bromide, 836(110)
tert-butyl bromide, 836(106)
butyl butyrate, 816(120)
butyl caproate, 817(192)
butyl carbamate (butylurethane), 744(25)
sec-butylcarbinol (2-methylbutanol), 714(31)
"butylcarbitol," 720(155)
butyl carbonate, 817(187)
"butylcellosolve," (2-butoxyethanol), 718(100), 824(81)
n-butyl chloride, 832(35)
sec-butyl chloride, 832(32)
tert-butyl chloride, 832(29)
butyl chloroacetate, 816(136), 816(149)
butyl chloroformate, 815(84)
butyl chloromethyl ether, 824(67)
butylcyclohexane, 866(191)
butylcyclopentane, 864(125)
butyl dichloroacetate, 816(152)
α-butylene oxide (1,2-epoxybutane), 823(15)

n-butyl ethyl sulfide, 962(28)
sec-butyl ethyl sulfide, 962(22)
tert-butyl ethyl sulfide, 962(31)
n-butyl fluoride, 830(7)
sec-butyl fluoride, 830(6)
tert-butyl fluoride, 830(5)
n-butyl formate, 814(39)
sec-butyl formate, 814(23)
butyl heptanoate, 818(220)
butyl hexanoate, 817(192)
butyl 4-hydroxybenzoate, 821(84)
butyl 4-hydroxyphenyl ketone, 901(106)
n-butyl iodide, 838(170)
sec-butyl iodide, 838(167)
tert-butyl iodide, 838(164)
n-butyl isocyanate, 912(11)
tert-butyl isocyanate, 912(6)
4-*tert*-butyl-1-isopropyl-benzene, 882(124)
butyl isopropyl ether, 824(52)
n-butyl isothiocyanate, 961(9)
sec-butyl isothiocyanate, 961(7)
tert-butyl isothiocyanate, 961(3)
butyl lactate, 817(156)
butyl levulinate, 818(243), 897(204)
butylmalonic acid, 692(115)
n-butyl mercaptan, 966(8)
sec-butyl mercaptan, 966(5)
butyl 3-methylbutyrate, 816(137)
n-butyl methyl ether, 823(21)
tert-butyl methyl ether, 823(11)
butyl methyl ketone (2-hexanone), 892(33), 904(204)
n-butyl methyl sulfide, 962(19)
tert-butyl methyl sulfide, 962(10)
n-butyl 2-naphthyl ether, 827(15)
sec-butyl 2-naphthyl ether, 827(14)
3-butyl-6-methyluracil, 747(385)
1-butylnaphthalene, 884(174)
2-butylnaphthalene, 884(175)
butyl nitrate, 815(71)
tert-butyl 4-nitrobenzoate, 822(130)
butyl nitrite, 814(10)
butyl oxalate, 818(254)
butyl oxamate, 744(83)
butyl pentanoate, 816(151)
butyl perfluoroacetate, 860(18)
n-butyl perfluorobutyrate, 860(21)
sec-butyl perfluorobutyrate, 860(20)
tert-butyl perfluorobutyrate, 860(19)
2-butylphenol, 924(19)
4-*n*-butylphenol, 924(25), 926(1)
4-*tert*-butylphenol, 932(144)

butyl phenylacetate, 819(271)
butyl phenyl ether, 825(107)
n-butyl phenyl ketone, 897(211)
tert-butyl phenyl ketone, 897(191)
4-*tert*-butylphenyl salicylate, 821(68)
butyl propionate, 815(93)
2-butylpyridine, 800(71)
3-butylpyridine, 801(87)
4-*tert*-butylpyridine, 800(79)
butyl salicylate, 819(287), 924(30)
butyl stearate, 821(9)
n-butyl sulfide, 962(43)
tert-butyl sulfide, 962(17)
butyl *d*-tartrate, 821(3)
butyl thiocyanate, 961(6)
2-*n*-butyltoluene, 881(93)
2-*sec*-butyltoluene, 879(53)
2-*tert*-butyltoluene, 879(66)
3-*n*-butyltoluene, 880(79)
3-*sec*-butyltoluene, 879(50)
3-*tert*-butyltoluene, 879(37)
4-*n*-butyltoluene, 880(89)
4-*sec*-butyltoluene, 879(59)
4-*tert*-butyltoluene, 879(44)
butyl *p*-tolyl ketone, 899(7)
butylurethane (butyl carbamate), 744(25)
butylurea, 745(107)
butyl valerate, 816(151)
butyl vinyl ether, 823(38)
2-butynal, 728(21)
1-butyne, 870(6)
2-butyne, 870(11)
2-butyne-1,4-diol, 722(35)
3-butyn-2-one, 892(6)
butyraldehyde (butanal), 728(10)
butyramide, 745(182)
butyranilide, 745(110)
butyric acid, 684(14)
butyric anhydride, 668(7)
γ-butyrolactone, 668(8), 817(183)
butyronitrile, 913(12)
butyrophenone, 897(200)
butyryl bromide, 671(77)
butyryl chloride, 670(15)
butyryl fluoride, 670(5)
butyryl iodide, 671(87)

caffeine, 748(490)
dl-camphane-10-sulfonic acid, 959(21)
d-camphene, 875(269)
l-camphene, 875(270)
dl-camphene, 875(268)
dl-camphor, 905(250)
d-camphoramide, 747(371), 747(416)
d-camphoranilide, 748(443), 748(486)
d-camphoric acid, 704(316)

Index of Compounds

d-camphoric anhydride, 669(68)
camphorquinone, 905(257)
d-camphor-1-sulfonamide, 948(28)
d-camphor-3-sulfonamide, 948(42)
d-camphor-8-sulfonamide, 948(34)
dl-camphor-8-sulfonamide, 948(30)
camphor-3-sulfonic acid, 959(22)
d-camphor-3-sulfonyl chloride, 955(90)
d-camphor-8-sulfonyl chloride, 957(181)
dl-camphor-8-sulfonyl chloride, 955(128)
d-camphor-10-sulfonyl chloride, 954(55)
l-canaline, 674(52)
dl-canaline, 674(31)
l-canavanine, 674(27)
capraldehyde (decanal), 732(70)
capramide, 745(154)
capranilide, 744(45),
capric acid (decanoic acid), 688(12)
capric anhydride, 668(19)
caproaldehyde (hexanal), 730(32)
caproamide, 745(123)
caproanilide, 744(105)
caproic acid (hexanoic acid), 686(44)
caproic anhydride, 668(15)
capronitrile, 913(28)
caproyl chloride, 670(30)
caproylresorcinol, 926(40)
caprylaldehyde (octanal), 730(49)
caprylamide, 745(119), 745(166)
caprylanilide, 744(29)
capryl chloride, 670(45)
caprylic acid (octanoic acid), 686(54)
caprylic anhydride, 668(17)
caprylonitrile, 913(40)
caprylyl chloride, 670(35)
carbamide (urea), 746(240)
carbamylguanidine, 745(141)
carbanilide (sym-diphenylurea), 748(497)
carbazole, 794(667)
"carbitol" (2-(2-ethoxyethoxy)-ethanol), 720(140), 825(105)
carbohydrazide, 786(513)
4-carbomethoxy-3-thiophanone, 899(34)
carbon tetrabromide, 856(14)
carbon tetrachloride, 852(30)
carbon tetrafluoride, 860(1)
3-carboxybenzenesulfonamide, 949(83)

4-carboxybenzenesulfonamide, 951(209)
2-carboxybenzenesulfonic acid, 960 (73)
3-carboxybenzenesulfonic acid, 960(74)
4-carboxybenzenesulfonic acid, 960(75)
3-carboxybenzenesulfonyl chloride, 953(7)
4-carboxybenzenesulfonyl chloride, 953(43)
4-carboxy-1,3-dithiolane, 690(44)
4-carboxypentamethylene sulfide, 694(132)
2-carboxytetramethylene sulfide, 690(45)
carvacrol, 924(21)
d-carvone, 897(199)
catechol, 932(154)
catechol diacetate, 821(69)
catechol dibenzoate, 822(107)
catechol diethyl ether, 827(20)
β-cellobiose, 810(47)
"cellosolve acetate" (ethylene glycol monoethyl ether acetate), 816(110)
cetane (hexadecane), 860(218)
cetyl acetate, 821(5)
cetyl alcohol (hexadecanol) (hexadecyl alcohol), 722(29)
cetylamine, 764(81)
cetyl bromide, 838(145)
cetyl chloride, 834(81)
cetyl ether, 827(29)
cetyl iodide, 840(191)
cetyl mercaptan, 967(1)
cetyl sulfide, 965(26)
chalcone (benzalacetophenone) 901(90)
d-chaulmoogramide, 745(148)
d-chaulmoogranilide, 744(84)
d-chaulmoogric acid, 690(66)
chelidonic acid, 710(423)
chloral, 728(17)
chloral hydrate, 736(31)
chloranil, 906(282)
chloranilic acid, 944(428)
chloroacetaldehyde, 728(12)
chloroacetamide, 745(190)
chloroacetanilide, 746(247)
2-chloroacetanilide, 744(82)
4-chloroacetanilide, 747(373)
chloroacetic acid, 690(55)
chloroacetic anhydride, 668(25)
chloroacetone, 892(22)
chloroacetonitrile, 913(18)
ω-chloroacetophenone (phenacyl chloride), 901(91)
2-chloroacetophenone, 897(197), 899(3), 901(92)
3-chloroacetophenone, 897(196)
4-chloroacetophenone, 897(201)
chloroacetyl bromide, 671(76)

chloroacetyl chloride, 670(17)
chloroacetyl fluoride, 670(6)
4-chloro-1-aminoanthraquinone, 790(575)
2-chloro-4-aminobenzaldehyde, 740(110), 784(475)
4-chloro-2-aminobenzaldehyde, 738(69), 772(237)
2-chloro-3-aminobenzoic acid, 672(16)
2-chloro-4-aminobenzoic acid, 674(54)
4-chloro-3-aminobenzoic acid 674(56)
5-chloro-3-aminobenzoic acid, 674(55)
5-chloro-2-aminobenzophenone, 776(291)
3-chloro-4-aminobiphenyl, 770(177)
4-chloro-2-amino-N,N-dimethylaniline, 760(208)
4-chloro-3-amino-N,N-dimethyl-aniline, 766(102)
4-chloro-2-aminopyridine, 782(425)
5-chloro-8-aminoquinoline, 772(239)
6-chloro-8-aminoquinoline, 770(184)
7-chloro-8-aminoquinoline, 770(185)
8-chloro-5-aminoquinoline, 786(505)
3-chloro-5-aminosalicylic acid, 676(83)
2-chloroaniline, 754(122)
3-chloroaniline, 756(168)
4-chloroaniline, 770(179)
2-chloroanisole, 825(103), 842(29)
3-chloroanisole, 825(102)
4-chloroanisole, 825(104), 844(31)
1-chloroanthracene, 848(44)
5-chloroanthranilic acid, 706(361)
3-chloroanthraquinone, 906(277)
3-chloroazobenzene, 909(6)
4-chloroazobenzene, 909(17)
2-chlorobenzaldehyde, 732(71)
3-chlorobenzaldehyde, 732(73)
4-chlorobenzaldehyde, 734(25)
2-chlorobenzamide, 746(263)
3-chlorobenzamide, 746(248)
2-chlorobenzanilide, 745(179)
3-chlorobenzanilide, 745(197)
4-chlorobenzanilide, 747(417)
chlorobenzene, 842(8)
1-(3-chlorobenzeneazo)-2-naphthol, 910(49)
1-(4-chlorobenzeneazo)-2-naphthol, 910(50)
2-(4-chlorobenzeneazo)-1-naphthol, 910(63)

4-(2-chlorobenzeneazo)-1-naphthol, 910(60)
4-(3-chlorobenzeneazo)-1-naphthol, 910(76)
4-(4-chlorobenzeneazo)-1-naphthol, 910(78)
5-chlorobenzene-1,3-disulfonamide, 951(189)
5-chlorobenzene-1,3-disulfonyl chloride, 955(129)
2-chlorobenzenesulfonamide, 950(122)
3-chlorobenzenesulfonamide, 948(48)
4-chlorobenzenesulfonamide, 948(46)
3-chlorobenzenesulfonic acid, 959(23)
4-chlorobenzenesulfonic acid, 959(24)
2-chlorobenzenesulfonyl chloride, 953(11)
4-chlorobenzenesulfonyl chloride, 953(35)
2-chlorobenzoic acid, 698(209)
3-chlorobenzoic acid, 700(249)
4-chlorobenzoic acid, 708(409)
2-chlorobenzoic anhydride, 668(42)
3-chlorobenzoic anhydride, 668(44)
4-chlorobenzoic anhydride, 669(65)
2-chlorobenzonitrile, 914(25)
3-chlorobenzonitrile, 914(24)
4-chlorobenzonitrile, 915(67)
4-chlorobenzophenone, 902(137)
2-chlorobenzotrichloride, 856(6)
4'-chlorobenzophenone-2-carboxylic acid, 698(221)
2-chlorobenzoquinone, 901(88)
2-(4-chlorobenzoyl)benzoic acid, 698(221)
2-chlorobenzoyl chloride, 670(46)
3-chlorobenzoyl chloride, 670(44)
4-chlorobenzoyl chloride, 670(43)
2-chlorobenzyl alcohol, 724(50)
4-chlorobenzyl alcohol, 724(51)
4-chlorobenzylamine, 756(132)
4-chlorobenzyl bromide, 838(156)
2-chlorobenzyl chloride, 834(70)
4-chlorobenzyl chloride, 834(86)
2-chlorobenzyl phenyl ketone, 902(118)
4-chlorobenzyl sulfide, 965(11)
2-chlorobiphenyl, 846(8)

3-chlorobiphenyl, 844(65)
4-chlorobiphenyl, 848(41)
2-chloro-4-bromoaniline, 770(186)
2-chloro-5-bromoaniline, 764(62)
3-chloro-5-bromoaniline, 768(160)
4-chloro-3-bromoaniline, 770(205)
α-chloro-3-bromotoluene, 834(820)
α-chloro-4-bromotoluene, 834(88)
1-chlorobutane, 832(35)
2-chlorobutane, 832(32)
2-chloro-1,3-butadiene, 832(31)
1-chloro-2-butanone, 893(49)
2-chloro-2-butenoic acid(cis), 690(64)
4-chloro-2-butenoic acid, 692(84)
4-chlorobutyronitrile, 913(37)
dl-4-chlorocamphor, 905(259)
2-chlorocinnamic acid, 706(362)
4-chlorocumene (1-chloro-4-isopropylbenzene), 844(32)
chlorocyclohexane, 834(59)
chlorocyclopentane, 832(51)
1-chlorodecane, 834(72)
4-chloro-1,2-diaminobenzene, 770(197)
4-chloro-2,6-dibromophenol, 930(122)
4-chloro-3,5-dibromophenol, 934(208)
2-chlorodiethyl ether, 832(46)
4-chloro-2,6-diiodophenol, 932(165)
2-chloro-N,N-dimethylaniline, 801(89)
2-chloro-4,6-dimethylaniline, 764(50)
3-chloro-N,N-dimethylaniline, 801(100)
4-chloro-N,N-dimethylaniline, 804(14)
1-chloro-3,4-dimethylbenzene, 842(25)
2-chloro-1,4-dimethylbenzene, 842(20)
3-chloro-1,2-dimethylbenzene, 842(24)
4-chloro-2,5-dimethylbenzenesulfonamide, 950(110)
4-chloro-2,5-dimethylbenzenesulfonyl chloride, 953(29)
2-chloro-2,3-dimethylbutane, 832(50)
3-chloro-2,2-dimethylbutane, 832(49)
4-chloro-2,2-dimethylbutane, 832(52)
1-chloro-3,4-dimethylpentane, 834(60)

4-chloro-3,5-dimethylphenol, 934(188)
1-chloro-2,2-dimethylpropane, 832(36)
4-chloro-2,6-dimethylpyridine, 799(59)
2-chloro-3,5-dinitroaniline, 788(549)
4-chloro-2,6-dinitroaniline, 784(472)
4-chloro-3,5-dinitroaniline, 790(591)
4-chloro-1,3-dinitronaphthalene, 921(139)
4-chloro-1,5-dinitronaphthalene, 921(132)
4-chloro-1,8-dinitronaphthalene, 922(155)
4-chloro-2,6-dinitrophenol, 930(101)
2-chloro-2,2-diphenylacetic acid, 694(151)
1-chlorododecane, 834(77)
1-chloroeicosane, 834(90)
2-chloro-1,3-epoxypropane, 824(66)
3-chloro-1,2-epoxypropane, 832(54)
2-chloroethanal, 728(12)
chloroethane, 830(22)
2-chloroethanethiol, 966(13)
2-chloroethanol (ethylene chlorohydrin), 714(33)
chloroethene, 830(21)
2-chloro-1-ethoxybenzene, 825(109)
4-chloro-1-ethoxybenzene, 825(112), 827(2)
2-chloroethyl acetate, 815(87)
α-chloroethylbenzene, 834(65)
β-chloroethylbenzene, 834(66)
1-chloro-2-ethylbenzene, 842(15)
1-chloro-3-ethylbenzene, 842(19)
1-chloro-4-ethylbenzene, 842(21)
1-chloro-2-ethylbutane, 834(57)
2-chloroethyl chloromethyl ether, 824(73)
1-chloroethyl ether, 824(55)
2-chloroethyl ether, 832(46)
1-chloroethyl ethyl ether, 823(40)
2-chloroethyl ethyl ether, 824(49)
2-chloroethyl ethyl sulfide, 962(34)
1-chloro-2-ethylhexane, 834(62)
1-chloroethyl methyl ether, 823(22)
2-chloroethyl methyl ether, 823(36)
2-chloroethyl methyl sulfide, 962(26)

Index of Compounds 1009

2-chloroethyl 4-nitrophenyl sulfide, 965(27)
2-chloroethyl phenyl sulfide, 963(59)
2-chloroethyl sulfide, 963(52)
2-chloroethyl p-tolyl sulfide, 963(61)
chloroform, 852(22)
1-chloroheptadecane, 834(84)
1-chloroheptane, 834(61)
1-chlorohexadecane, 834(81)
1-chlorohexane, 834(58)
2-chlorohexane, 834(56)
3-chlorohexane, 832(55)
4-chlorohydrazobenzene, 911(7)
chlorohydroquinone, 932(158)
4-chloro-2-hydroxyaniline, 786(499)
2-chloro-3-hydroxybenzaldehyde, 740(106), 936(257)
2-chloro-4-hydroxybenzaldehyde, 740(111), 936(274)
3-chloro-4-hydroxybenzaldehyde, 740(105), 936(254)
4-chloro-3-hydroxybenzaldehyde, 740(98), 934(207)
6-chloro-3-hydroxybenzaldehyde, 738(90), 932(173)
5-chloro-2-hydroxybenzoic acid, 702(279)
5-chloro-2-hydroxybenzophenone, 930(129)
3-chloro-4-hydroxybiphenyl, 928(91)
5-chloro-2-hydroxybiphenyl, 926(27)
5-chloro-4-hydroxy-2-methylaniline, 792(653)
5-chloro-2-hydroxy-4-methylbenzoic acid, 706(354)
4-chloro-1-hydroxy-2-naphthoic acid, 708(397)
4-chloro-3-hydroxy-2-naphthoic acid, 708(395)
5-chloro-2-hydroxy-3-nitrobenzoic acid, 700(255)
5-chloro-2-hydroxy-p-toluic acid, 706(354)
6-chloro-3-hydroxy-p-toluic acid, 942(378)
4-chloro-2-iodophenol, 930(94)
2-chloroisocrotonic acid, 690(51)
1-chloro-2-isopropylbenzene, 842(26)
1-chloro-4-isopropylbenzene (p-chlorocumene), 844(32)
α-chloro-4-isopropyltoluene, 834(73)
2-chloro-4-isopropyltoluene, 844(51)
3-chloro-4-isopropyltoluene, 844(52)
3-chloroisoquinoline, 804(33)

8-chloroisoquinoline, 805(44)
4-chloromandelic acid, 694(153)
2-chloro-6-methoxyaniline, 758(191)
3-chloro-4-methoxyaniline, 768(133)
4-chloro-2-methoxyaniline, 766(94)
4-chloro-3-methoxyaniline, 770(201)
5-chloro-2-methoxyaniline, 772(225)
2-chloro-1-methoxybenzene, 825(103), 842(29)
4-chloro-1-methoxybenzene, 825(104), 844(31)
3-chloro-4-methoxybenzenesulfonamide, 948(26)
3-chloro-4-methoxybenzenesulfonyl chloride, 954(80)
8-chloro-7-methoxynaphthalene-1-sulfonamide, 949(55)
8-chloro-7-methoxynaphthalene-1-sulfonyl chloride, 957(180)
chloromethyl acetate, 814(43)
2-chloro-4-methylaniline, 756(123), 776(321)
2-chloro-6-methylaniline, 756(134)
3-chloro-4-methylaniline, 762(11)
4-chloro-2-methylaniline, 758(185), 762(19)
4-chloro-3-methylaniline, 772(223)
2-chloro-N-methylaniline, 756(139), 756(153)
2-chloro-4-methylbenzenesulfonamide, 950(114)
3-chloro-4-methylbenzenesulfonamide, 948(31)
4-chloro-3-methylbenzenesulfonamide, 948(25)
5-chloro-2-methylbenzenesulfonamide, 948(43)
2-chloro-4-methylbenzenesulfonyl chloride, 953(25)
3-chloro-4-methylbenzenesulfonyl chloride, 953(18)
4-chloro-3-methylbenzenesulfonyl chloride, 954(50)
5-chloro-2-methylbenzenesulfonyl chloride, 953(8)
6-chloro-2-methylbenzoic acid, 692(117)
2-chloro-4-methylbenzoic acid, 700(240)
4-chloro-2-methylbenzoic acid, 702(280)
1-chloro-3-methylbutane, 832(44)
2-chloro-2-methylbutane, 832(37)

2-chloro-3-methylbutane, 832(38)
4-chloro-3-methyl-2-butanone, 893(70)
3-chloro-2-methyl-2-butene, 832(41)
dl-3-chloro-2-methyl-1-butene, 832(40)
chloromethyl ethyl ether, 823(27)
chloromethyl ethyl sulfide, 962(20)
4-chloro-5-methyl-2-isopropylphenol, 928(53)
chloromethyl isopropyl sulfide, 962(25)
chloromethyl methyl ether, 823(13), 832(30)
1-chloro-8-methylnaphthalene, 848(39)
2-chloro-5-methyl-6-nitrobenzenesulfonamide, 949(94)
4-chloro-3-methyl-5-nitrobenzenesulfonamide, 950(149)
chloromethyl methyl sulfide, 962(20)
5-chloro-2-methyl-3-nitrobenzenesulfonamide, 949(76)
5-chloro-4-methyl-2-nitrobenzenesulfonamide, 949(100)
2-chloro-5-methyl-6-nitrobenzenesulfonyl chloride, 956(158)
4-chloro-3-methyl-5-nitrobenzenesulfonyl chloride, 953(33)
5-chloro-2-methyl-3-nitrobenzenesulfonyl chloride, 954(45)
5-chloro-4-methyl-2-nitrobenzenesulfonyl chloride, 955(117)
2-chloro-2-methylpentane, 832(48)
3-chloro-3-methylpentane, 832(53)
2-chloro-3-methylphenol, 926(35)
2-chloro-4-methylphenol, 924(5)
3-chloro-2-methylphenol, 930(113)
3-chloro-4-methylphenol, 926(39)
4-chloro-2-methylphenol, 926(31)
4-chloro-3-methylphenol, 928(66)
5-chloro-2-methylphenol, 928(83)
(4-chloro-2-methylphenoxy)-acetic acid, 694(156)
1-chloro-2-methylpropane, 832(33)
2-chloro-2-methylpropane, 832(29)

Index of Compounds

3-chloro-2-methylpropene, 832(34)
2-chloro-2-methylpropanoic acid, 688(8)
chloromethyl propyl ether, 824(53)
chloromethyl propyl sulfide, 962(32)
3-chloro-2-methylquinoline, 805(64)
4-chloro-2-methylquinoline, 804(27)
5-chloro-4-methylsalicylic acid, 706(354)
chloromethyl sulfide, 962(33)
2-chloro-1-naphthaldehyde, 736(59)
3-chloro-2-naphthaldehyde, 740(99)
4-chloro-1-naphthaldehyde, 738(66)
6-chloro-1-naphthaldehyde, 738(67)
1-chloronaphthalene, 844(63)
2-chloronaphthalene, 846(26)
2-chloronaphthalene-1-sulfonamide, 948(54)
4-chloronaphthalene-1-sulfonamide, 950(116)
5-chloronaphthalene-1-sulfonamide, 951(192)
5-chloronaphthalene-2-sulfonamide, 951(170)
6-chloronaphthalene-1-sulfonamide, 951(167)
6-chloronaphthalene-2-sulfonamide, 950(107)
7-chloronaphthalene-1-sulfonamide, 951(205)
7-chloronaphthalene-2-sulfonamide, 949(90)
8-chloronaphthalene-1-sulfonamide, 950(144)
8-chloronaphthalene-2-sulfonamide, 950(109)
2-chloronaphthalene-1-sulfonyl chloride, 954(68)
4-chloronaphthalene-1-sulfonyl chloride, 955(104)
5-chloronaphthalene-1-sulfonyl chloride, 955(105)
5-chloronaphthalene-2-sulfonyl chloride, 956(145)
6-chloronaphthalene-1-sulfonyl chloride, 954(62)
6-chloronaphthalene-2-sulfonyl chloride, 956(133)
7-chloronaphthalene-1-sulfonyl chloride, 956(172)
7-chloronaphthalene-2-sulfonyl chloride, 954(87)
8-chloronaphthalene-1-sulfonyl chloride, 955(123)
8-chloronaphthalene-2-sulfonyl chloride, 955(101)

3-chloro-2-naphthoic acid, 706(372)
4-chloro-1-naphthoic acid, 706(359)
5-chloro-1-naphthoic acid, 708(411)
5-chloro-2-naphthoic acid, 710(426)
6-chloro-1-naphthoic acid, 706(373)
7-chloro-1-naphthoic acid, 708(410)
8-chloro-1-naphthoic acid, 702(277)
1-chloro-2-naphthol, 928(82)
4-chloro-1-naphthol, 934(206)
4-chloro-2-naphthol, 932(151)
5-chloro-1-naphthol, 936(239)
6-chloro-1-naphthol, 930(126)
7-chloro-2-naphthol, 934(227)
4-chloro-1-naphthonitrile, 975(84)
5-chloro-1-naphthonitrile, 915(100)
5-chloro-2-naphthonitrile, 915(99)
1-chloro-2-naphthylamine, 766(118)
2-chloro-1-naphthylamine, 766(104)
3-chloro-1-naphthylamine, 768(129)
4-chloro-1-naphthylamine, 774(282)
5-chloro-1-naphthylamine, 772(229)
4-chloro-2-nitroaniline, 778(359), 778(366)
5-chloro-2-nitroaniline, 780(406)
4-chloro-2-nitroanisole, 920(96)
2-chloro-4-nitrobenzaldehyde, 736(56)
2-chloro-5-nitrobenzaldehyde, 736(63)
1,2-chloronitrobenzene, 917(8)
1,3-chloronitrobenzene, 917(23)
1,4-chloronitrobenzene, 919(74)
2-chloro-5-nitrobenzenesulfonamide, 950(111)
4-chloro-2-nitrobenzenesulfonamide, 951(211)
4-chloro-3-nitrobenzenesulfonamide, 949(89)
5-chloro-2-nitrobenzenesulfonamide, 949(68)
2-chloro-5-nitrobenzenesulfonyl chloride, 955(92)
4-chloro-2-nitrobenzenesulfonyl chloride, 954(67)
4-chloro-3-nitrobenzenesulfonyl chloride, 953(20)
5-chloro-2-nitrobenzenesulfonyl chloride, 955(97)

2-chloro-4-nitrobenzoic acid, 698(201)
2-chloro-5-nitrobenzoic acid, 700(265)
3-chloro-2-nitrobenzoic acid, 708(400)
4-chloro-2-nitrobenzoic acid, 698(207)
4-chloro-3-nitrobenzoic acid, 702(296)
5-chloro-2-nitrobenzoic acid, 698(199)
4-chloro-2-nitrobenzonitrile, 915(68)
4-chloro-3-nitrobenzonitrile, 915(69)
6-chloro-3-nitrobenzonitrile, 915(76)
4-chloro-2-nitro-1-methoxybenzene, 920(96)
4-chloro-1-nitronaphthalene, 919(82)
5-chloro-1-nitronaphthalene, 920(107)
8-chloro-1-nitronaphthalene, 920(93)
2-chloro-4-nitro-1-naphthylamine, 794(669)
4-chloro-2-nitro-1-naphthylamine, 792(625)
4-chloro-3-nitro-1-naphthylamine, 786(511)
8-chloro-4-nitro-1-naphthylamine, 792(628)
2-chloro-4-nitrophenol, 932(174)
4-chloro-2-nitrophenol, 930(115)
5-chloro-2-nitrosobenzoic acid, 923(32)
α-chloro-3-nitrotoluene, 834(92)
α-chloro-4-nitrotoluene, 834(95)
2-chloro-4-nitrotoluene, 918(48)
4-chloro-2-nitrotoluene, 917(16)
6-chloro-2-nitrotoluene, 917(13)
5-chloro-2-nitro-p-tolunitrile, 915(64)
1-chlorononadecane, 834(87)
1-chlorononane, 834(69)
1-chlorooctadecane, 834(85)
1-chlorooctane, 834(64)
1-chloropentadecane, 834(80)
1-chloropentane, 832(45)
2-chloropentane, 832(42)
3-chloropentane, 832(43)
1-chloropentene, 832(47)
3-chloropentene, 832(39)
4-chlorophenacyl acetate, 821(76)
2-chlorophenanthraquinone, 906(273)
2-chlorophenetole, 825(109)
4-chlorophenetole, 825(112), 827(2)

Index of Compounds 1011

2-chlorophenol, 924(1)
3-chlorophenol, 926(10)
4-chlorophenol, 926(22)
2-chlorophenoxyacetamide, 746(290)
4-chlorophenoxyacetamide, 746(244)
2-chlorophenoxyacetanilide, 745(193)
4-chlorophenoxyacetanilide, 745(202)
2-chlorophenoxyacetic acid, 698(214)
4-chlorophenoxyacetic acid, 700(250)
2-chlorophenylacetic acid, 692(100)
2-chlorophenylacetonitrile, 914(6)
4-chlorophenylacetonitrile, 914(12)
4-chlorophenyl benzyl ketone, 903(172)
1-chloro-4-phenyl-2-butanone, 900(50)
2-chloro-p-phenylenediamine, 774(253)
4-chloro-o-phenylenediamine, 770(197)
2-chlorophenylhydrazine, 764(78)
4-chlorophenylhydrazine, 772(247)
4-chlorophenyl isocyanate, 912(19)
4-chlorophenyl isothiocyanate, 961(14)
4-chlorophenyl methyl sulfide, 962(37)
2-chloro-3-phenylpropanol, 734(10)
3-(2-chlorophenyl)propenol, 734(26)
2-chloro-3-phenylpropenoic acid, 698(196)
3-(2-chlorophenyl)propenoic acid(cis), 696(173)
3-(2-chlorophenyl)propenoic acid($trans$), 706(362)
3-(3-chlorophenyl)propenoic acid, 696(183)
3-(3-chlorophenyl)propionic acid, 690(75)
4-chlorophenyl sulfide, 965(42)
3-chlorophthalic acid, 704(308)
4-chlorophthalic acid, 700(246)
chloropicrin, 916(3)
chloroprene, 832(31)
1-chloropropane, 832(28)
2-chloropropane, 830(25)
2-chloropropanethiol, 966(12)
1-chloro-2-propanol, 714(5)
dl-2-chloro-1-propanol, 714(38)
3-chloro-1-propanol, 716(85)

1-chloro-2-propanone, 892(22)
1-chloro-1-propene(cis), 830(24)
1-chloro-1-propene($trans$), 830(26)
2-chloro-1-propene, 830(23)
3-chloro-1-propene, 830(27)
dl-2-chloropropionamide, 744(64)
dl-2-chloropropionanilide, 744(91)
dl-2-chloropropionic acid, 684(22)
3-chloropropionic acid, 688(24)
3-chloropropionitrile, 913(30)
2-chloropropionyl chloride, 670(19)
3-chloropropiophenone, 900(60)
4-chloropropiophenone, 899(29)
3-chloropropyl alcohol, 716(85)
2-chloropropylene oxide (β-epichlorohydrin), 824(66)
3-chloropropyl ether, 825(113)
2-chloropyridine, 800(47)
3-chloropyridine, 799(31)
4-chloropyridine, 799(28)
2-chloroquinoline, 804(18)
3-chloroquinoline, 801(120)
4-chloroquinoline, 804(11)
5-chloroquinoline, 804(28)
6-chloroquinoline, 804(23)
7-chloroquinoline, 804(12)
8-chloroquinoline, 802(142)
4-chlororesorcinol, 932(159)
3-chlorosalicylaldehyde, 736(34)
4-chlorosalicylaldehyde, 736(30)
5-chlorosalicylaldehyde, 738(81), 932(143)
5-chlorosalicylic acid, 702(279)
α-chlorostyrene, 834(68)
β-chlorostyrene, 834(67)
2-chlorostyrene, 842(23)
4-chlorostyrene, 842(27)
N-chlorosuccinimide, 746(287)
4-chloro-2,3,5,6-tetrabromophenol, 942(388)
1-chlorotetradecane, 834(79)
1-chloro-2,2,3,3-tetramethylbutane, 834(93)
2-chlorothiophenol, 966(30)
4-chlorothiophenol, 967(9)
α-chlorotoluene, 834(63)
2-chlorotoluene, 842(11)
3-chlorotoluene, 842(12)
4-chlorotoluene, 842(13)
2-chloro-p-toluic acid, 700(240)
4-chloro-o-toluic acid, 702(280)
6-chloro-m-toluidine (4-chloro-3-methylaniline), 772(223)

3-chloro-2,4,6-tribromophenol, 932(156)
1-chlorotridecane, 834(78)
4-chloro-1,2,3-trihydroxyanthraquinone, 942(407)
2-chloro-1,3,5-trimethylbenzene, 844(38)
3-chloro-p-tolunitrile, 914(41)
5-chloro-o-tolunitrile, 914(48)
1-chloroundecane, 834(75)
α-chlorovinylbenzene, 834(68)
β-chlorovinylbenzene, 834(67)
2-chlorovinylbenzene, 842(23)
4-chlorovinylbenzene, 842(27)
1-chlorovinyl ethyl ether, 823(34)
1-chlorovinyl propyl ether, 824(54)
cholanic acid, 700(256)
cholanthrene, 889(102)
l-cholesterol, 724(69)
chrysene, 890(122)
cinchol (β-sitosterol), 724(67)
1,8-cineole ("eucalyptol"), 825(87)
cinnamaldehyde, 732(85)
cinnamalacetone, 901(115)
cinnamamide, 746(282)
cinnamanilide, 746(291)
cinnamic acid($trans$), 696(186)
cinnamic anhydride, 669(57)
cinnamonitrile, 914(3)
cinnamoyl chloride, 671(61)
cinnamyl alcohol, 722(14)
cinnamyl cinnamate, 821(37)
cinnamyl phenyl sulfide, 965(39)
citraconamide, 747(395)
citraconanilide, 747(363)
citraconic acid, 693(97)
citraconic anhydride, 668(10)
citral (geranial), 732(76)
citral (neral), 732(77)
citramide, 748(455)
citranilide (monohyd.), 748(429)
citric acid, 692(110), 700(232)
d-citronellal, 732(69)
d-citronellic acid, 686(50)
citronellol, 720(151)
L-citrulline, 676(60)
2,3,5-collidine, 800(64)
2,4,6-collidine (2,4,6-trimethylpyridine), 800(54)
coniine, 800(48)
2-coumaramide, 748(452)
4-coumaramide, 747(418)
2-coumaric acid($trans$), 706(356)
4-coumaric acid, 706(346)
coumarin, 821(37)
creatine, 678(118), 748(521)
creatinine, 678(122), 748(516)
m-cresol, 924(8)
o-cresol, 924(3), 926(7)

1012 Index of Compounds

p-cresol, 924(7), 926(13)
m-cresyl acetate (m-tolyl acetate), 817(201)
o-cresyl acetate (o-tolyl acetate), 817(190)
p-cresyl acetate (p-tolyl acetate), 817(202)
m-cresyl benzoate, 821(55)
o-cresyl benzoate (o-tolyl benzoate), 820(311)
p-cresyl benzoate, 821(90)
o-cresyl carbonate, 821(65)
p-cresyl carbonate, 822(128)
m-cresyl ethyl ether, 825(98)
o-cresyl ethyl ether, 825(91)
p-cresyl ethyl ether, 825(97)
m-cresyl methyl ether, 825(85)
o-cresyl methyl ether (methylanisole), 824(80)
p-cresyl methyl ether, 825(84)
crotonaldehyde, 728(20)
crotonamide, 747(326)
α-crotonanilide(trans), 745(187)
crotonic acid(cis), 684(22)
α-crotonic acid(trans), 690(70)
crotonic anhydride, 668(14)
crotononitrile, 913(13)
crotonoyl chloride, 670(25)
crotyl sulfide, 962(42)
cumaldehyde, 732(82)
o-cumaraldehyde, 936(243)
cumaranone, 903(166)
cumene (isopropylbenzene), 877(7)
cumidine (4-isopropylaniline, 756(157))
cyanoacetamide, 745(191)
cyanoacetanilide, 748(428)
cyanoacetic acid, 690(60), 914(47)
cyanoanthracene, 915(107)
4-cyanobenzamide, 748(476)
4-cyanobenzanilide, 747(376)
4-cyanobenzoic acid, 706(376)
4-cyanobiphenyl, 915(59)
4-cyanobutyric acid, 688(31)
cyanodiphenylmethane, 914(1)
4-cyanodiphenylmethane, 914(34)
2-cyanofuran, 913(25)
8-cyanonaphthalene-1-sulfonamide, 952(252)
8-cyanonaphthalene-1-sulfonyl chloride, 957(184)
1-cyanophenanthrene, 915(90)
2-cyanophenanthrene, 915(81)
3-cyanophenanthrene, 915(71)
9-cyanophenanthrene, 915(79)
2-cyano-3-phenylpropionic acid, 692(116)
2-cyanopropionic acid, 684(9)
3-cyanopropionic acid, 688(37)
2-cyanopyridine, 914(8), 913(44)

3-cyanopyridine (nicotinonitrile), 914(32)
4-cyanopyridine, 915(57)
2-cyanoquinoline, 915(66)
3-cyanoquinoline, 915(80)
4-cyanoquinoline, 915(72)
5-cyanothiazole, 914(35)
2-cyanothiophene, 913(36)
3-cyanothiophene, 913(31)
cyclobutane, 862(2)
cyclobutanecarboxylic acid, 686(31)
cyclobutane-1,1-dicarboxylic acid, 700(248)
cyclobutanone, 892(10)
cyclobutylamine, 750(22)
cyclobutylmethanol, 716(57)
cyclobutyl methyl ketone, 893(47)
cyclodecane, 866(196)
cycloheptane (suberane), 863(51)
cycloheptanone, 895(140)
cycloheptene (suberene), 873(167)
1,3-cyclohexadiene (1,2-dihydrobenzene), 871(71)
1,3-cyclohexadiene-1-carboxylic acid, 692(99)
cyclohexane, 862(14)
cyclohexanecarboxaldehyde (hexahydrobenzaldehyde), 730(46)
cyclohexanecarboxamide, 747(394)
cyclohexanecarboxanilide, 746(272)
cyclohexanecarboxylic acid, 688(8)(11)
1,2-cyclohexanedicarboxylic anhydride(trans), 669(58)
1,4-cyclohexanediol(trans), 724(68)
1,2-cyclohexanedione, 899(36)
1,3-cyclohexanedione, 903(169)
1,4-cyclohexanedione, 902(133)
cyclohexanethiol, 966(19)
cyclohexanol, 716(84), 722(7)
cyclohexanone, 894(86)
cyclohexene, 871(76)
cyclohexene-1,2-dicarboxylic acid, 696(170)
cyclohexene-1,2-dicarboxylic anhydride, 668(38)
cyclohexyl acetate, 816(135)
cyclohexylacetic acid, 686(47)
cyclohexylamine, 752(55)
cyclohexylbenzene, 883(143)
cyclohexyl bromide, 836(130)
cyclohexylcarbinol (hexahydrobenzyl alcohol, 718(122)
cyclohexyl chloride, 834(59)
cyclohexyl ethyl ether, 824(74)
cyclohexyl formate, 816(117)

cyclohexyl iodide, 838(179)
cyclohexyl mercaptan, 966(19)
cyclohexylmethanol, 718(122)
cyclohexyl methyl ether, 824(68)
cyclohexyl methyl ketone, 895(137)
cyclohexyl oxalate, 821(35)
2-cyclohexylphenol, 926(44)
4-cyclohexylphenol, 936(241)
cyclononane, 866(186)
1,5-cyclooctadiene, 873(211)
cyclooctane, 864(113)
cyclooctene, 873(208)
cyclopentadecane, 867(262)
1,3-cyclopentadiene, 870(21)
cyclopentanecarboxaldehyde, 730(36)
cyclopentane, 862(5)
cyclopentane-1,3-dicarboxylic acid(cis), 694(152)
cyclopentanol, 716(55)
cyclopentanone, 893(39)
cyclopentene, 870(26)
1-cyclopentenecarboxaldehyde, 730(41)
1-cyclopentenecarboxylic acid, 696(159)
cyclopentyl acetate, 815(102)
cyclopentyl bromide, 836(126)
cyclopentyl chloride, 832(51)
cyclopentyl ethyl ether, 824(60)
cyclopentyl formate, 815(73)
cyclopentyl methyl ether, 824(47)
cyclopentyl iodide, 838(178)
cyclopentylmethanol, 716(86)
cyclopentyl methyl ketone, 894(84)
cyclopropanecarboxylic acid, 684(23)
N-cyclopropylacetamide, 745(151)
cyclopropylacetic acid, 686(27)
cyclopropylamine, 750(9)
m-cymene (m-isopropyltoluene), 878(17)
o-cymene, (o-isopropyltoluene), 878(21)
p-cymene (p-isopropyltoluene), 878(20)
L-cysteine, 680(134)
L-cystine, 676(84)

2,4-D, 698(204)
D.D.T. (1,1,1-trichloro-2,2-bis(p-chlorophenyl)ethane), 856(16)
decahydronaphthalene(cis) (decalin), 866(194)
decahydronaphthalene(trans), 866(192)
cis-decahydroquinoline, 854(110)

Index of Compounds 1013

trans-decahydroquinoline, 764(74)
decalin (decahydronaphthalene-(*cis*)), 866(194)
decamethylenediamine, 768(130)
decanal, 732(70)
decanamide, 745(154)
decananilide, 744(45)
decane, 866(189)
1,10-decanediamine, 768(130)
decanedioic acid (sebacic acid), 696(187)
1,10-decanedithiol, 967(2)
decanoic acid (capric acid), 688(12)
decanoic anhydride, 668(19)
1-decanol (decyl alcohol), 720(158)
2-decanol, 720(143)
decanonitrile, 913(50)
2-decanone, 896(174)
3-decanone, 896(176)
4-decanone, 896(171)
decanoyl chloride, 670(45)
1-decene, 874(221)
decyl alcohol (1-decanol), 720(158)
decyl bromide, 836(139)
decyl chloride, 834(72)
decylcyclohexane, 867(220)
decylcyclopentane, 866(216)
decyl fluoride, 830(18)
decyl iodide, 840(184)
decylnaphthalene, 885(191)
decyl sulfide, 965(4)
1-decyne, 874(224)
2-dehydracetic acid, 694(126)
dehydroacetanilide, 745(181)
2-deoxy-D-glucose, 809(21)
6-deoxy-α-D-glucose, 809(25)
deoxy-D-ribose, 809(5)
deoxy-β-D-xylose, 809(3)
desosamine hydrochloride, 809(1)
desoxybenzoin, 901(94)
diacetone alcohol (4-hydroxy-4-methyl-2-pentanone), 716(93), 894(112)
N,N-diacetylaniline, 744(12)
1,4-diacetylbenzene, 903(181)
2,2′-diacetylbiphenyl, 902(144)
2,4-diacetyl-3-methylphenol, 903(179), 932(178)
N,N′-diacetyl-*m*-phenylenediamine, 747(409)
N,N′-diacetyl-*o*-phenylenediamine, 747(393)
N,N′-diacetyl-*p*-phenylenediamine, 748(519)
diallylamine, 750(40)
diallylbarbituric acid, 747(357)
3,5-diaminoacetophenone, 782(435)
1,4-diaminoanthraquinone, 794(676)

1,5-diaminoanthraquinone, 794(687)
1,6-diaminoanthraquinone, 794(682)
1,7-diaminoanthraquinone, 794(680)
1,8-diaminoanthraquinone, 794(674)
2,3-diaminoanthraquinone, 906(287), 794(688)
2,2′-diaminoazobenzene, 782(434), 909(30)
2,4-diaminoazobenzene, 909(21)
3,3′-diaminoazobenzene, 910(45), 786(510)
4,4′-diaminoazobenzene, 790(598), 910(82)
1,2-diaminobenzene, 776(296)
1,3-diaminobenzene, 768(136)
1,4-diaminobenzene, 784(458)
4,6-diaminobenzene-1,3-disulfonamide, 950(119)
4,6-diaminobenzene-1,3-disulfonyl chloride, 958(229)
2,5-diaminobenzoic acid, 792(617)
2,2′ diaminobenzophenone, 782(433), 904(210)
3,3′-diaminobenzophenone, 788(558)
4,4′-diaminobenzophenone, 906(272), 794(664)
2,2′-diaminobibenzyl, 768(165)
2,2′-diaminobiphenyl, 772(216)
2,4′-diaminobiphenyl, 764(64)
3,4-diaminobiphenyl, 776(300)
1,1′-diamino-2,2′-binaphthyl, 794(679)
dl-2,2′-diamino-1,1′-binaphthyl, 790(603)
4,4′-diamino-1,1′-binaphthyl, 792(626)
1,3-diaminobutane, 752(60)
1,4-diaminobutane (putrescine), 752(72), 762(15)
1,4-diamino-2-butyne, 764(58)
2,7-diaminocarbazole, 794(673)
2,4-diaminochlorobenzene, 774(253)
1,2-diaminocyclohexane, 754(94)
2,2′-diaminodibenzyl disulfide, 774(251)
2,2′-diaminodibenzyl sulfide, 772(215)
4,4′-diaminodibenzyl sulfide, 776(308)
2,2′-diaminodiethyl sulfide, 754(126)
4,4′-diamino-2,2′-dimethylazoxybenzene, 784(482)
l-6,6′-diamino-2,2′-dimethyldiphenyl, 786(509)
4,4′-diamino-3,3′-dimethyldiphenylmethane, 786(520)

1,3-diamino-2,6-dimethylbenzene, 768(152)
1,3-diamino-4,6-dimethylbenzene, 776(313)
2,2′-diamino-4,4′-dimethylbiphenyl, 780(382)
4,4′-diamino-2,2′-dimethylbiphenyl, 776(331)
dl-6,6′-diamino-2,2′-dimethylbiphenyl, 782(442)
4,4′-diamino-2,2′-dimethyldiphenylmethane, 780(395)
6,6′-diamino-3,3′-dimethyldiphenylmethane, 774(276)
6,5′-diamino-3,3′-dimethyldiphenyl sulfide, 776(301)
6,6′-diamino-3,3′-dimethyltriphenylmethane, 790(586)
2,4-diaminodiphenylamine, 782(422)
4,4′-diaminodiphenylamine, 786(518)
2,2′-diaminodiphenyl disulfide, 774(261)
4,4′-diaminodiphenylmethane, 774(260)
2,2′-diaminodiphenyl sulfide, 772(234)
1,9-diaminofluorene, 780(380)
2,7-diaminofluorenone, 906(284), 794(681)
1,6-diaminohexane, 764(54)
4,6-diaminoisophthalaldehyde, 792(637)
1,2-diaminonaphthalene, 774(281c)
1,4-diaminonaphthalene, 780(379)
1,8-diaminonaphthalene, 768(158)
2,7-diaminonaphthalene, 782(542)
1,4-diamino-2-nitrobenzene, 782(443)
1,5-diaminopentane, 754(86)
2,4-diaminophenol, 772(210)
2,5-diaminophenol, 768(166)
3,4-diaminophenol, 938(311), 788(544)
4,6-diaminophthalaldehyde, 740(124)
dl-1,2-diaminopropane, 750(45)
1,3-diaminopropane, 752(58)
1,3-diamino-2-propanol, 764(56)
2,3-diaminopyridine, 778(350)
2,4-diaminopyridine, 776(323)
2,5-diaminopyridine, 776(325)
2,6-diaminopyridine, 780(388)
3,4-diaminopyridine, 792(646)
2,2′-diaminostilbene(*cis*), 780(396)
2,2′-diaminostilbene(*trans*), 788(566)
4,4′-diaminostilbene(*cis*), 780(385)

4,4'-diaminostilbene(*trans*), 794(657)
diaminosuccinic acid(*meso*), 680(125)
4,4'-diamino-2,5,2',5'-tetramethyltriphenylmethane, 792(640)
2,3-diaminotoluene, 768(140)
2,4-diaminotoluene, 774(287)
2,5-diaminotoluene, 768(144)
2,6-diaminotoluene, 776(309)
3,4-diaminotoluene, 772(245)
4,4'-diaminotriphenylcarbinol, 788(545)
3,4-diaminotriphenylmethane, 770(178)
diamylamine, 754(107)
dianisidine, 782(444)
dianthrone, 906(285)
dibenzalacetone (1,5-diphenyl-3-pentadienone), 903(178)
1,2,3,4-dibenzanthracene, 890(113)
1,2,5,6-dibenzanthracene, 890(127)
1,2,7,8-dibenzanthracene, 890(112)
dibenzofuran (biphenylene oxide), 828(45)
1,2-dibenzoylbenzene, 905(229)
1,3-dibenzoylbenzene, 903(165)
1,4-dibenzoylbenzene, 905(239)
1,2-dibenzoylethane, 905(225)
1,2-dibenzoylethylene(*cis*), 904(209)
1,2-dibenzoylethylene(*trans*), 903(175)
dibenzoylmethane, 902(136)
2,4-dibenzoylresorcinol, 936(229)
2,3,5,6-dibenzphenanthrene, 890(126)
2,3,6,7-dibenzphenanthrene, 890(125)
dibenzylacetamide, 745(217)
dibenzylacetanilide, 746(306)
dibenzylacetic acid, 692(92)
dibenzylamine, 760(218)
N,N-dibenzylaniline, 805(63)
1,2-dibenzylhydrazine, 764(69)
dibenzyl ketone (1,3-diphenyl-2-propanone), 899(27)
4,5-dibenzylnaphthalene-1-sulfonamide
4,5-dibenzylnaphthalene-1-sulfonyl chloride, 957(194)
1,2-dibenzyloxybenzene, 827(36)
1,4-dibenzyloxybenzene, 828(55)
dibenzyl succinate, 821(43)
dibromoacetamide, 746(314)
dibromoacetic acid, 688(34)
3,5-dibromoacetophenone, 901(114)
3,5-dibromo-4-aminobenzaldehyde, 740(112), 784(486)

2,3-dibromoaniline, 764(59)
2,4-dibromoaniline, 772(209)
2,5-dibromoaniline, 766(96)
2,6-dibromoaniline, 772(224)
3,5-dibromoaniline, 766(106)
2,2'-dibromoazobenzene, 910(61)
4,4'-dibromoazobenzene, 910(68)
2-(3,3'-dibromo)azopyridine, 910(81)
2,2'-dibromoazoxybenzene, 911(19)
3,3'-dibromoazoxybenzene, 911(18)
4,4'-dibromoazoxybenzene, 911(28)
1,2-dibromobenzene, 844(55)
1,3-dibromobenzene, 844(53)
1,4-dibromobenzene, 848(51)
2,4-dibromobenzenesulfonamide, 950(129)
2,5-dibromobenzenesulfonamide, 950(141)
3,4-dibromobenzenesulfonamide, 949(88)
2,4-dibromobenzenesulfonyl chloride, 954(73)
2,5-dibromobenzenesulfonyl chloride, 954(64)
3,4-dibromobenzenesulfonyl chloride, 953(14)
2,4-dibromobenzoic acid, 702(281)
2,5-dibromobenzoic acid, 700(244)
2,6-dibromobenzoic acid, 700(230)
3,5-dibromobenzoic acid, 706(377)
2,4'-dibromobenzophenone, 901(103)
3,3'-dibromobenzophenone, 904(217)
4,4'-dibromobenzophenone, 905(249)
2,2'-dibromobiphenyl, 848(43)
4,4'-dibromobiphenyl, 850(66)
1,3-dibromo-2-(bromomethyl)-propane, 855(148)
1,1-dibromobutane, 854(94)
1,2-dibromobutane, 854(99)
1,3-dibromobutane, 854(107)
1,4-dibromobutane, 855(118)
2,3-dibromobutane, 854(93)
1,2-dibromo-1-butene, 854(86)
1,3-dibromo-2-butene, 854(101), 856(10)
2,3-dibromo-1-butene, 854(83)
3,4-dibromobutyric acid, 688(39)
2,4-dibromo-6-chloroaniline, 774(268)
2,6-dibromo-4-chloroaniline, 774(259)

dibromochloromethane, 853(58)
1,1-dibromocyclobutane, 854(95)
1,1-dibromodecane, 855(146)
1,10-dibromodecane, 856(5)
1,2-dibromo-1,2-dichloroethane, 856(12)
dibromodichloromethane, 856(3)
2,5-dibromo-1,4-dimethylbenzene, 848(40)
1,1-dibromo-2,2-dimethylpropane, 855(110)
1,1-dibromoethane (ethylidene bromide), 853(52)
1,2-dibromoethane (ethylene bromide), 853(67)
2,2-dibromoethanol, 718(118)
1,2-dibromoethylene(*cis*), 853(54)
1,2-dibromoethylene(*trans*), 853(51)
dibromoethyl ether, 824(75)
1,2-dibromofluoroethylene, 852(11)
dibromofluoromethane, 852(6)
1,1-dibromoheptane, 855(131)
1,2-dibromoheptane, 855(135)
1,7-dibromoheptane, 856(9)
1,1-dibromohexane, 855(120)
1,6-dibromohexane, 855(138)
2,2'-dibromohydrazobenzene, 911(2)
3,3'-dibromohydrazobenzene, 911(10)
4,4'-dibromohydrazobenzene, 911(15)
2,6-dibromohydroquinone, 938(306)
dibromomethane, 853(42)
2,4-dibromo-6-methylaniline, 766(85)
2,6-dibromo-4-methylaniline (3,5-dibromo-*p*-toluidine), 770(206)
1,2-dibromo-2-methylbutane, 854(105)
2,3-dibromo-2-methylbutane, 854(104)
2,4-dibromo-6-methylphenol, 928(48)
2,6-dibromo-4-methylphenol, 926(33)
3,5-dibromo-2-methylphenol, 932(140)
1,2-dibromo-2-methylpropane, 854(84)
1,3-dibromo-2-methylpropane, 854(108)
1,1-dibromo-2-methylpropene, 854(91)
1,2-dibromonaphthalene, 848(35)
1,3-dibromonaphthalene, 848(34)

Index of Compounds 1015

1,4-dibromonaphthalene, 848(45)
1,8-dibromonaphthalene, 848(55)
2,4-dibromo-1,3-naphthalenediol, 936(238)
1,3-dibromo-2-naphthol, 928(88)
2,4-dibromo-1-naphthol, 932(176)
1,4-dibromo-2-nitrobenzene, 919(79)
2′,4-dibromo-3-nitrobiphenyl, 919(85)
4,4′-dibromo-2-nitrobiphenyl, 921(123)
4,4′-dibromo-2-nitrobiphenyl, 921(123)
4,5-dibromo-2-nitrobiphenyl, 920(104)
2,6-dibromo-4-nitrophenol, 936(263)
4,6-dibromo-2-nitrophenol, 934(196)
α,α-dibromo-4-nitrotoluene, 852(32)
1,1-dibromononane, 855(144)
1,9-dibromononane, 855(147)
1,1-dibromooctane, 855(140)
1,2-dibromooctane, 855(139)
1,8-dibromooctane, 855(145)
1,1-dibromopentane, 855(111)
1,4-dibromopentane, 855(119)
1,5-dibromopentane, 855(133)
2,4-dibromophenol, 926(14)
2,5-dibromophenol, 928(84)
3,5-dibromophenol, 930(100)
1,1-dibromopropane, 853(69)
1,2-dibromopropane, 854(74)
1,3-dibromopropane, 854(100)
2,3-dibromopropanol, 720(148)
1,2-dibromopropene, 853(68)
1,3-dibromopropene, 854(90)
2,3-dibromopropene, 854(76)
2,3-dibromopropionamide, 746(228)
2,3-dibromopropionic acid, 690(61)
2,3-dibromopyridine, 805(49)
2,5-dibromopyridine, 806(88)
3,4-dibromopyridine, 805(65)
3,5-dibromopyridine, 806(100)
5,7-dibromoquinoline, 806(102)
3,5-dibromosalicylaldehyde, 738(68)
3,5-dibromosalicylic acid, 708(390)
1,2-dibromo-1,1,2,2-tetrafluoroethane, 852(16)
α,2-dibromotoluene, 838(150)
α,3-dibromotoluene, 838(154)
α,4-dibromotoluene, 838(158)
2,5-dibromotoluene, 844(59)
3,4-dibromotoluene, 844(61)

3,5-dibromo-*p*-toluidine (2,6-dibromo-4-methylaniline), 770(206)
α,α′-dibromo-*p*-xylene, 856(19)
1,4-dibutoxybenzene, 827(22)
1,1-dibutoxybutane, 667(34)
2,2′-dibutoxydiethyl ether, 826(135)
1,1-dibutoxyethane, 667(31)
1,1-di(2-butoxy)ethane (*sec*-butylacetal), 667(26)
1,2-dibutoxyethane, 825(106)
dibutoxymethane, 667(28)
α,α-dibutoxytoluene, 667(41)
di-*n*-butylamine, 752(71)
di-*sec*-butylamine, 752(53)
N,N-dibutylaniline, 802(129)
1,4-di-*sec*-butylbenzene, 883(150)
dibutyl carbonate, 817(187)
di-*sec*-butyl ether, 824(59)
dibutyl ketone (nonanone), 895(145)
di-*sec*-butyl ketone (3,5-dimethyl-4-heptanone), 894(100)
dibutyl phthalate, 820(318)
dicetyl ether (dihexadecyl ether), 827(29)
dichloroacetamide, 745(115)
dichloroacetanilide, 745(188)
dichloroacetic acid, 686(29)
dichloroacetic anhydride, 668(12)
1,1-dichloroacetone, 892(24)
2,4-dichloroacetophenone, 899(23)
2,5-dichloroacetophenone, 897(213)
3,5-dichloroacetophenone, 899(10)
dichloroacetyl chloride, 670(18)
2,3-dichloroaniline, 762(7)
2,4-dichloroaniline, 768(138)
2,5-dichloroaniline, 766(87)
2,6-chloroaniline, 764(45)
3,4-dichloroaniline, 770(183)
3,5-dichloroaniline, 766(88)
9,10-dichloroanthracene, 850(72)
9,10-dichloroanthracene-2-sulfonamide, 952(240)
9,10-dichloroanthracene-2-sulfonyl chloride, 958(221)
2,2′-dichloroazobenzene, 909(32)
3,3′-dichloroazobenzene, 909(19)
4,4′-dichloroazobenzene, 910(62)
2-(3,3′-dichloro)azopyridine, 910(83)
2,2′-dichloroazoxybenzene(*cis*), 911(16)

2,2′-dichloroazoxybenzene (*trans*), 911(6)
4,4′-dichloroazoxybenzene, 911(24)
2,4-dichlorobenzaldehyde, 736(54)
2,5-dichlorobenzaldehyde, 736(42)
2,6-dichlorobenzaldehyde, 736(52)
3,4-dichlorobenzaldehyde, 734(21)
3,5-dichlorobenzaldehyde, 736(48)
2,5-dichlorobenzamide, 746(303)
2,6-dichlorobenzamide, 748(441)
1,2-dichlorobenzene, 842(16)
1,3-dichlorobenzene, 842(14)
1,4-dichlorobenzene, 846(23)
2,5-dichlorobenzene-1,3-disulfonamide, 951(168)
4,6-dichlorobenzene-1,3-disulfonamide, 952(239)
2,5-dichlorobenzene-1,3-disulfonyl chloride, 956(142)
4,6-dichlorobenzene-1,3-disulfonyl chloride, 956(161)
2,4-dichlorobenzenesulfonamide, 949(102)
2,5-dichlorobenzenesulfonamide, 949(98)
3,4-dichlorobenzenesulfonamide, 948(38)
2,5-dichlorobenzenesulfonic acid, 959(25)
2,4-dichlorobenzenesulfonyl chloride, 953(36)
2,5-dichlorobenzenesulfonyl chloride, 953(17)
3,4-dichlorobenzenesulfonyl chloride, 953(9)
2,3-dichlorobenzoic acid, 700(260)
2,4-dichlorobenzoic acid, 700(259)
2,5-dichlorobenzoic acid, 700(235)
2,6-dichlorobenzoic acid, 698(213)
3,4-dichlorobenzoic acid, 706(350)
2,2′-dichlorobenzophenone, 900(62)
2,4′-dichlorobenzophenone, 901(111)
4,4′-dichlorobenzophenone, 905(226)
2,5-dichloro-1,4-benzoquinone, 905(238)
2,6-dichloro-1,4-benzoquinone, 904(193)
3,4-dichlorobenzotrichloride, 856(4)

Index of Compounds

3,4-dichlorobenzoyl chloride, 670(47)
2,4-dichlorobenzyl chloride, 834(76)
2,6-dichlorobenzyl chloride, 834(91)
3,4-dichlorobenzyl chloride, 834(74)
2,2'-dichlorobiphenyl, 846(28)
4,4'-dichlorobiphenyl, 850(64)
1,1-dichlorobutane, 853(56)
1,2-dichlorobutane, 853(65)
1,3-dichlorobutane, 853(70)
1,4-dichlorobutane, 854(88)
2,3-dichlorobutane, 853(57)
3,6-dichloro-2,5-dimethylbenzenesulfonamide, 949(71)
4,6-dichloro-2,5-dimethylbenzenesulfonamide, 948(52)
3,6-dichloro-2,5-dimethylbenzenesulfonyl chloride, 954(65)
4,6-dichloro-2,5-dimethylbenzenesulfonyl chloride, 954(78)
4,4'-dichlorodiphenylmethane, 846(25)
dichloroethanal, 728(13)
1,1-dichloroethane (ethylidene chloride), 852(18)
1,2-dichloroethane (ethylene chloride), 852(34)
1,1-dichloroethene, 852(9)
di(2-chloroethoxy)ethane, 825(124)
di(2-chloroethoxy)methane, 667(35)
1,2-dichloroethylene(*cis*), 852(21)
1,2-dichloroethylene(*trans*), 852(17)
α,α'-dichloroethyl ether, 824(55)
β,β'-dichloroethyl ether, 825(88)
dichlorofluoromethane, 852(3)
1,1-dichloroheptane, 855(115)
1,1-dichlorohexane, 854(98)
2,2'-dichlorohydrazobenzene, 911(5)
3,3'-dichlorohydrazobenzene, 911(8)
4,4'-dichlorohydrazobenzene, 911(13)
2,4-dichloro-3-hydroxybenzaldehyde, 740(108)
2,6-dichloro-3-hydroxybenzaldehyde, 740(107)
3,5-dichloro-4-hydroxybenzaldehyde, 740(116)
1,2-dichloro-4-iodobenzene, 846(6)
dichloromaleic acid, 694(142)
dichloromaleic anhydride, 669(51)

3,5-dichloro-4-methoxybenzoic acid, 704(334)
dichloromethane (methylene chloride), 852(12)
1,2-dichloro-4-methylbenzene (3,4-dichlorotoluene), 844(42)
2,3-dichloro-4-methylbenzenesulfonamide, 951(210)
2,6-dichloro-3-methylbenzenesulfonamide, 950(125)
2,6-dichloro-4-methylbenzenesulfonamide, 949(60)
3,4-dichloro-2,-methylbenzenesulfonamide, 951(196)
3,5-dichloro-2-methylbenzenesulfonamide, 950(113)
3,5-dichloro-4-methylbenzenesulfonamide, 950(133)
4,5-dichloro-3-methylbenzenesulfonamide, 949(106)
4,6-dichloro-2-methylbenzenesulfonamide, 949(79)
2,3-dichloro-6-methylbenzenesulfonyl chloride, 953(27)
2,3-dichloro-4-methylbenzenesulfonyl chloride, 953(21)
2,6-dichloro-3-methylbenzenesulfonyl chloride, 953(6)
2,6-dichloro-4-methylbenzenesulfonyl chloride, 953(40)
3,4-dichloro-2-methylbenzenesulfonyl chloride, 953(34)
3,5-dichloro-2-methylbenzenesulfonyl chloride, 953(38)
3,5-dichloro-4-methylbenzenesulfonyl chloride, 954(60)
4,6-dichloro-2-methylbenzenesulfonyl chloride, 953(22)
α,α'-dichloromethyl ether, 824(45)
1,1-dichloro-2-methylpropane, 853(48)
1,2-dichloro-2-methylpropane, 853(50)
1,3-dichloro-2-methylpropane, 853(71)
dichloromethyl sulfide, 962(45)
1,2-dichloronaphthalene, 846(10)
1,3-dichloronaphthalene, 846(30)
1,4-dichloronaphthalene 848(37)
1,5-dichloronaphthalene, 848(54)
1,6-dichloronaphthalene, 846(20)
1,7-dichloronaphthalene, 848(33)
1,8-dichloronaphthalene, 848(49)
2,3-dichloronaphthalene, 848(59)
2,6-dichloronaphthalene, 848(62)

2,7-dichloronaphthalene, 848(57)
1,5-dichloronaphthalene-2-sulfonamide, 952(241)
3,6-dichloronaphthalene-2-sulfonamide, 951(176)
3,7-dichloronaphthalene-1-sulfonamide, 952(235)
4,5-dichloronaphthalene-1-sulfonamide, 951(201)
4,5-dichloronaphthalene-2-sulfonamide, 950(145)
4,6-dichloronaphthalene-1-sulfonamide, 951(191)
4,6-dichloronaphthalene-2-sulfonamide, 951(174)
4,7-dichloronaphthalene-1-sulfonamide, 951(172)
4,7-dichloronaphthalene-2-sulfonamide, 950(142)
4,8-dichloronaphthalene-2-sulfonamide, 950(152)
5,6-dichloronaphthalene-1-sulfonamide, 951(186)
5,6-dichloronaphthalene-2-sulfonamide, 950(134)
5,7-dichloronaphthalene-1-sulfonamide, 952(237)
5,8-dichloronaphthalene-2-sulfonamide, 952(220)
6,7-dichloronaphthalene-1-sulfonamide, 952(234)
6,8-dichloronaphthalene-2-sulfonamide, 951(199)
7,8-dichloronaphthalene-1-sulfonamide, 951(182)
7,8-dichloronaphthalene-2-sulfonamide, 951(194)
1,5-dichloronaphthalene-2-sulfonyl chloride, 956(166)
3,6-dichloronaphthalene-2-sulfonyl chloride, 957(208)
3,7-dichloronaphthalene-1-sulfonyl chloride, 957(179)
4,5-dichloronaphthalene-1-sulfonyl chloride, 956(147)
4,5-dichloronaphthalene-2-sulfonyl chloride, 957(201)
4,6-dichloronaphthalene-1-sulfonyl chloride, 956(150)
4,6-dichloronaphthalene-2-sulfonyl chloride, 956(175)
4,7-dichloronaphthalene-1-sulfonyl chloride, 957(195)
4,7-dichloronaphthalene-2-sulfonyl chloride, 957(197)
4,8-dichloronaphthalene-2-sulfonyl chloride, 957(187)
5,6-dichloronaphthalene-1-sulfonyl chloride, 955(130)
5,6-dichloronaphthalene-2-sulfonyl chloride, 957(210)
5,7-dichloronaphthalene-1-sulfonyl chloride, 957(192)

Index of Compounds 1017

5,8-dichloronaphthalene-2-sulfonyl chloride, 957(178)
6,7-dichloronaphthalene-1-sulfonyl chloride, 957(188)
6,8-dichloronaphthalene-2-sulfonyl chloride, 956(157)
7,8-dichloronaphthalene-1-sulfonyl chloride, 957(182)
7,8-dichloronaphthalene-2-sulfonyl chloride, 956(164)
1,4-dichloro-2-naphthol, 934(216)
2,4-dichloro-1-naphthol, 932(161)
2,6-dichloro-4-nitroaniline, 790(610)
2,4-dichloronitrobenzene, 917(9)
2,5-dichloronitrobenzene, 918(32)
3,4-dichloronitrobenzene, 917(18)
3,5-dichloronitrobenzene, 918(50)
2,6-dichloro-4-nitrophenol, 934(221)
4,6-dichloro-2-nitrophenol, 934(214)
α,α-dichoro-3-nitrotoluene, 852(23)
1,1-dichlorononane, 855(136)
1,1-dichlorooctane, 855(123)
1,1-dichloropentane, 854(72)
1,5-dichloropentane, 855(112)
2,3-dichlorophenol, 926(46)
2,4-dichlorophenol, 926(23)
2,5-dichlorophenol, 928(51)
2,6-dichlorophenol, 928(68)
3,4-dichlorophenol, 928(71)
3,5-dichlorophenol, 928(70)
2,4-dichlorophenoxyacetic acid, 698(204)
2,5-dichlorophenylacetic acid, 694(124)
2,5-dichlorophenyl sulfide, 965(17)
1,1-dichloropropane, 852(36)
1,2-dichloropropane, 853(41)
1,3-dichloropropane, 853(61)
2,2-dichloropropane, 852(27)
1,3-dichloro-2-propanol, 718(111)
2,3-dichloropropanol, 718(120)
1,1-dichloro-2-propanone, 892(24)
1,3-dichloro-2-propanone, 900(56)
2,3-dichloropropene, 853(39)
2,2′-dichloropropyl ether, 825(94)
2,5-dichloropyridine, 805(50)
3,5-dichloropyridine, 805(58)
3,5-dichlorosalicylaldehyde, 738(77)
3,5-dichlorosalicylic acid, 706(381)

α,α-dichlorotoluene, 855(122)
α,2-dichlorotoluene, 834(70)
α,3-dichlorotoluene, 834(71)
α,4-dichlorotoluene, 834(86)
2,3-dichlorotoluene, 844(41)
2,4-dichlorotoluene, 844(35)
2,6-dichlorotoluene, 844(33)
3,4-dichlorotoluene, 844(42)
3,5-dichlorotoluene, 844(36), 844(36)
3,5-dichloro-o-toluidine, 766(116)
5,6-dichloro-o-toluidine, 762(28)
α,α′-dichloro-m-xylene, 856(8)
α,α′-dichloro-o-xylene, 856(11)
α,α′dichloro-p-xylene, 856(15)
2,5-dichloro-p-xylene, 848(38)
dicinnamalacetone, 904(219)
1,4-dicyanocyclohexane(cis), 914(44)
1,4-dicyanocyclohexane(trans), 915(96)
dicyanodiamide, 748(448)
4,4′-dicyanodiphenylmethane, 915(105)
dicyclohexylamine, 758(200)
dicyclo-4,7-pentadiene, 874(257)
diethanolamine (2,2′-iminodiethanol), 720(167), 722(10)
2,2′-diethoxyazobenzene, 909(28)
3,3′-diethoxyazobenzene, 911(3)
4,4′-diethoxyazobenzene, 910(48)
2,2′-diethoxyazoxybenzne, 911(17)
4,4′-diethoxyazoxybenzene, 911(22)
3,4-diethoxybenzaldehyde, 732(87)
1,2-diethoxybenzene, 827(20)
1,3-diethoxybenzene, 826(129)
1,4-diethoxybenzene, 827(39)
1,1-diethoxy-2-bromoethane, 667(25)
1,1-diethoxybutane (ethylbutyral), 667(18)
1,1-diethoxy-2-chloroethane, 667(23)
1,1-diethoxy-2,2-dichloroethane, 667(30)
1,1-diethoxyethane, 667(6)
1,2-diethoxyethane, 824(58)
1,1-diethoxyheptane, 667(33)
2,2′-diethoxyhydrazobenzene, 911(6)
3,3′-diethoxyhydrazobenzene, 911(12)
4,4′-diethoxyhydrazobenzene, 911(3)
diethoxymethane, 667(5)

1,1-diethoxy-2-methylpropane, 667(15)
3,3-diethoxypentane, 667(22)
1,1-diethoxypropane, 667(13)
3,3-diethoxypropene, 667(14)
α,α-diethoxytoluene, 667(39)
diethylacetal, 667(6)
diethylacetaldehyde, 728(26)
diethylacetic acid (2-ethylbutanoic acid), 686(36)
diethylacetyl chloride, 670(28)
diethylacetonitrile, 913(24)
diethyl adipate, 819(259)
diethylamine, 750(11)
3-diethylaminobenzenesulfonic acid, 959(26)
4-diethylaminobenzaldehyde, 734(19), 804(24)
4-diethylaminobenzoic acid, 688(26)
2-diethylaminoethanol, 716(87), 799(42)
2-diethylaminoethyl alcohol, 716(87), 799(42)
2-diethylaminoethylamine, 752(62)
3-diethylaminophenol, 930(95), 805(73)
3-diethylaminopropanol, 718(130), 800(70)
N,N-diethylaniline, 801(96)
2,6-diethylaniline, 756(136), 758(176)
diethyl azelate, 819(304)
diethylbarbituric acid, 747(403)
1,2-diethylbenzene 878(28)
1,3-diethylbenzene, 878(22)
1,4-diethylbenzene 878(29)
N,N′-diethylcarbanilide, 744(49)
diethylcarbinol (3-pentanol), 714(20)
diethyl carbonate, 815(62)
4,4′-diethyldiaminobiphenyl, 778(363)
sym-diethylenediamine, 752(67)
diethylene glycol diacetate, 819(267)
diethylene glycol diethyl ether, 825(95)
diethylene glycol (2,2′-dihydroxydiethyl ether), 720(162)
diethylene glycol dimethyl ether, 824(77)
diethylene glycol monobutyl ether, 720(155), 826(125)
diethylene glycol monobutyl ether acetate, 621(195)
diethylene glycol monoethyl ether (2-(2-ethoxyethoxy)-ethanol), 718(140), 825(105)
diethylene glycol monoethyl ether acetate, 819(262)

1018 Index of Compounds

diethylene glycol monomethyl ether ("Methylcarbitol"), 720(132), 825(100)
N,N-diethylformamide, 744(4)
diethyl fumarate, 818(213),
diethyl glutarate, 821(239)
3,3-diethylhexane, 865(170)
3,4-diethylhexane, 865(151)
diethyl ketone (3-pentanone), 892(11)
diethyl *l*-malate, 819(270)
diethyl maleate, 818(217)
diethylmalonamide, 746(275)
diethyl malonate, 817(179)
diethylmalonic acid, 696(168)
diethylmalonyl dichloride, 670(37)
diethylmethylacetic acid, 686(42)
diethylmethylamine, 799(3)
N,N-diethyl-2-methylaniline, 801(88)
N,N-diethyl-4-methylaniline, 801(98)
3,3-diethyl-2-methylpentane, 866(188)
3,7-diethylnaphthalene-1-sulfonamide, 950(158)
3,7-diethylnaphthalene-1-sulfonyl chloride, 955(127)
diethyl 3-nitrophthalate, 821(42)
diethyl 4-nitrophthalate, 821(22)
diethyl oxalate, 817(154)
3,3-diethylpentane, 864(105)
2,6-diethylphenol, 926(15)
diethyl perfluoroadipate, 860(28)
diethyl perfluoroglutarate, 860(7)
diethyl perfluorosuccinate, 860(26)
diethyl phthalate, 819(303)
diethyl pimelate, 819(273)
diethylpropylal, 667(13)
1,3-diethyl-2-propylbenzene, 881(91)
2,4-diethylpyridine, 800(68)
3,4-diethylpyridine, 801(93)
diethyl sebacate, 820(312)
diethyl suberate, 819(297)
diethyl succinate, 818(210)
diethyl sulfate, 817(191)
diethyl *d*-tartrate, 819(296)
2,3-diethyltoluene, 880(88)
2,4-diethyltoluene, 880(80)
2,5-diethyltoluene, 880(90)
2,6-diethyltoluene, 881(94)
3,4-diethyltoluene, 880(76)
3,5-diethyltoluene, 879(67)
N,N-diethyl-*o*-toluidine, 801(88)
N,N-diethyl-*p*-toluidine, 801(98)

di-2-fluorenylmethane, 890(114)
difluoroacetamide, 744(21)
difluoroacetic acid, 684(6)
1,2-difluorobenzene, 842(4)
1,3-difluorobenzene, 842(1)
1,4-difluorobenzene, 842(3)
1,1-difluorobutane, 852(13)
2,2-difluorobutane, 852(8)
1,1-difluorodecane, 855(114)
difluoroethanol, 714(6)
2,2-difluoroethyl acetate, 814(38)
1,1-difluoroheptane, 853(59) (60)
1,1-difluorohexane, 853(40)
difluoromalonic acid, 857(18)
1,1-difluorooctane, 854(78)
1,1-difluoropentane, 852(26)
2,2-difluoropentane, 852(19)
3,3-difluoropentane, 852(20)
1,1-difluoropropane, 852(2)
2,2-difluoropropane, 852(1)
difurfurylideneacetone, 901(98)
diglycine (glycylglycine), 676(86)
diglycolic acid, 698(222)
diglycol myristate, 821(24)
diglycol palmitate, 821(44)
diglycol stearate, 821(57)
diguaiacyl carbonate (guaiacol carbonate), 822(109)
diheptylamine 760(211), 762(21)
dihexadecyl ether (dicetyl ether), 827(29)
dihexoxymethane, 667(40)
dihexyl ketone, 899(21)
9,10-dihydroanthracene, 888(66)
1,2-dihydrobenzene (1,3-cyclohexadiene), 871(71)
dihydroindole, 756(164)
1,2-dihydronaphthalene, 880(84)
1,4-dihydronaphthalene, 874(254)
dihydropyran, 823(31)
dihydroxyacetone, 724(48), 902(120)
2,3-dihydroxyacetophenone, 930(139), 903(156)
2,4-dihydroxyacetophenone (resacetophenone), 905(224), 936(271)
2,5-dihydroxyacetophenone, 906(261), 942(373)
3,4-dihydroxyacetophenone, 903(184), 934(190)
3,5-dihydroxyacetophenone, 905(227), 938(277)
2,3-dihydroxy-1-aminopropane, 760(209)
3,5-dihydroxyaniline, 784(473)
1,5-dihydroxyanthracene, 944(420)
1,4-dihydroxyanthraquinone, 906(260), 942(369)

1,5-dihydroxyanthraquinone, 944(427)
1,6-dihydroxyanthraquinone, 944(425)
1,7-dihydroxyanthraquinone, 906(283), 944(432)
1,8-dihydroxyanthraquinone, 905(255), 940(354)
2,2'-dihydroxyazobenzene, 940(323), 910(53)
2,4-dihydroxyazobenzene, 938(320)
3,3'-dihydroxyazobenzene, 910(69), 942(379)
4,4'-dihydroxyazobenzene, 942(390)
2,3-dihydroxybenzaldehyde, 738(88)
2,4-dihydroxybenzaldehyde (β-resorcylic aldehyde), 740(104)
2,6-dihydroxybenzaldehyde, 740(114), 938(288)
3,4-dihydroxybenzaldehyde (protocatachualdehyde), 740(113)
3,5-dihydroxybenzaldehyde (α-resorcylic aldehyde), 740(115)
2,4-dihydroxybenzamide, 748(475)
2,4-dihydroxybenzanilide, 745(212)
3,4-dihydroxybenzanilide, 747(336)
1,2-dihydroxybenzene, 932(154)
1,3-dihydroxybenzene, 932(170)
1,4-dihydroxybenzene, 938(322)
2,3-dihydroxybenzoic acid, 704(339)
2,4-dihydroxybenzoic acid (β-resorcylic acid), 706(371), 942(392)
2,5-dihydroxybenzoic acid, 704(333), 940(363)
3,4-dihydroxybenzoic acid, 704(330)
2,4-dihydroxybenzophenone, 936(266)
2,4'-dihydroxybenzophenone, 904(221), 905(232), 938(281)
2,5-dihydroxybenzophenone, 904(200), 934(222)
3,3'-dihydroxybenzophenone, 905(244), 938(318)
3,4-dihydroxybenzophenone, 904(222), 936(269)
4,4'-dihydroxybenzophenone, 906(267), 942(382)
2,5-dihydroxybenzoquinone, 906(268), 942(383)
1,1'-dihydroxy-2,2'-binaphthyl, 942(395)
2,2'-dihydroxy-1,1'-binaphthyl, 942(393)

Index of Compounds 1019

4,4'-dihydroxy-1,1'-binaphthyl, 944(433)
2,2'-dihydroxybiphenyl (o,o'-biphenol), 937(169)
2,4'-dihydroxybiphenyl, 938(302)
3,3'-dihydroxybiphenyl, 934(217)
3,4-dihydroxybiphenyl, 936(268)
4,4'-dihydroxybiphenyl (p,p'-biphenol), 944(424)
3,4-dihydroxycinnamic acid, 704(325), 940(357)
2,2'-dihydroxydiethyl ether (diethylene glycol), 720(162)
1,2-dihydroxy-3,5-dimethylbenzene, 928(85)
1,3-dihydroxy-2,5-dimethylbenzene, 938(303)
1,3-dihydroxy-4,6-dimethylbenzene, 934(220)
1,4-dihydroxy-2,3-dimethylbenzene, 942(399)
2,2'-dihydroxy-3,3'-dimethylbiphenyl, 932(180)
2,2'-dihydroxy-4,4'-dimethylbiphenyl, 934(203)
2,2'-dihydroxy-5,5'-dimethylbiphenyl, 938(283)
2,2'-dihydroxy-6,6'-dimethylbiphenyl, 938(305)
4,4'-dihydroxy-2,2'-dimethylbiphenyl, 932(181)
4,4'-dihydroxy-3,3'-dimethylbiphenyl, 938(294)
2,4'-dihydroxydiphenylmethane, 934(194)
3,3'-dihydroxydiphenylmethane, 932(149)
4,4'-dihydroxydiphenylmethane, 938(291)
4,4'-dihydroxydiphenylsulfone, 944(415)
2,5-dihydroxy-3-dodecyl-1,4-benzoquinone, 904(218), 936(260)
di(2-hydroxyethyl)amine (2,2'-iminodiethanol), 720(167), 722(10), 760(210)
2,2'-dihydroxyhydrazobenzene, 784(480), 911(18)
1,3-dihydroxy-4-methylanthraquinone, 944(421)
1,8-dihydroxy-2-methylanthraquinone, 940(328)
1,2-dihydroxynaphthalene, 928(52)
1,3-dihydroxynaphthalene, 934(218)
1,5-dihydroxynaphthalene, 944(417)
1,8-dihydroxynaphthalene, 936(259)
2,6-dihydroxynaphthalene, 942(401)

1,2-dihydroxy-4-nitrobenzene, 940(330)
3,6-dihydroxyphthalic acid, 706(379)
3-(3,4-dihydroxyphenyl)-propenoic acid, 704(325), 940(357)
3,5-dihydroxypyrene, 942(396)
3,8-dihydroxypyrene, 944(436)
2,5-dihydroxytoluene, 934(219)
2,6-dihydroxytoluene, 934(192), 934(201)
3,4-dihydroxytoluene, 928(65)
3,5-dihydroxytoluene, 928(47), 932(164)
4,4'-dihydroxytriphenylmethane, 938(297)
2,4-diiodoaniline, 774(274)
1,2-diiodobenzene, 846(4)
1,4-diiodobenzene, 848(61)
1,3-diiodobenzene, 846(13)
1,2-diiodoethane (ethylene iodide), 856(13)
1,2-diiodoethane(cis), 855(116)
diiodomethane, 855(113)
1,5-diiodopentane, 856(1)
2,4-diiodophenol, 928(81)
2,5-diiodophenol, 932(141)
3,4-diiodophenol, 930(105)
3,5-diiodophenol, 932(150)
1,3-diiodopropane, 855(134)
2,2-diiodopropane, 854(106)
L-3,5-diiodotyrosine, 674(49)
1,1-diisobutoxyethane (isobutylal), 667(27)
diisobutylamine, 752(57)
diisobutylcarbinol, 718(109)
diisobutyl carbonate, 817(161)
diisobutylene (2,4,4-trimethyl-1-pentene), 872(111)
diisobutyl ketone (isovalerone), 894(108)
diisopentylamine, 754(92)
diisopentyl carbonate, 818(230)
diisopropoxymethane, 667(12)
1,1-diisopropoxypropane, 667(21)
diisopropylacetal, 667(9)
diisopropylamine, 750(23)
1,2-diisopropylbenzene, 880(77)
1,3-diisopropylbenzene, 880(75)
1,4-diisopropylbenzene, 881(98)
diisopropyl ketone (2,4-dimethyl-3-pentanone), 892(28)
diisopropyl oxalate, 817(159)
2,4-diisopropyltoluene, 882(127)
2,6-diisopropyltoluene, 882(133)
2,4-diketopentanoic acid, 692(114), 903(164)
dilauryl ether (dodecyl ether), 827(12)
3,4-dimethoxyacetophenone, 901(78)

1,2-dimethoxy-4-allylbenzene, 826(131)
3,4-dimethoxyaniline, 772(235)
2,2'-dimethoxyazoxybenzene, 911(9)
3,3'-dimethoxyazoxybenzene, 911(4)
4,4'-dimethoxyazoxybenzene, 911(20)
2,3-dimethoxybenzaldehyde, 736(32)
2,4-dimethoxybenzaldehyde, 736(51)
3,4-dimethoxybenzaldehyde (veratraldehyde), 736(40)
3,4-dimethoxybenzamide, 747(333)
3,4-dimethoxybenzanilide, 746(297)
1,2-dimethoxybenzene, 825(108), 827(4)
1,3-dimethoxybenzene, 825(117)
1,4-dimethoxybenzene, 827(31)
2,4-dimethoxybenzenesulfonamide, 949(75)
2,4-dimethoxybenzenesulfonyl chloride, 954(61)
3,3'-dimethoxybenzidine (dianisidine), 782(444)
4,4'-dimethoxybenzil, 904(207)
4,4'-dimethoxybenzophenone, 901(86)
3,4-dimethoxybenzoic acid, 702(295)
2,2'-dimethoxybiphenyl, 827(38)
1,1-dimethoxy-2-bromoethane, 667(19)
1,1-dimethoxybutane, 667(11)
1,1-dimethoxy-2-chloroethane, 667(17)
4,4'-dimethoxydiphenyl ether, 828(51)
1,1-dimethoxyethane (dimethylacetal), 667(2)
1,2-dimethoxyethane, 823(30)
1,2-di(2-methoxyethoxy)ethane 825(116)
2,2'-dimethoxyhydrazobenzene, 911(9)
dimethoxymethane, 667(1)
3,4-dimethoxyphenylacetamide, 746(279)
3,4-dimethoxyphenylacetic acid, 692(103)
1,1-dimethoxy-2-phenylethane, 667(37)
2,2-dimethoxypropane, 667(4)
1,2-dimethoxy-4-propenylbenzene, 826(138)
α,α-dimethoxytoluene, 667(32)
dimethylacetal (1,1-dimethoxyethane), 667(2)
N,N-dimethylacetamide, 744(2)
2,4-dimethylacetanilide, 746(245)

1020 Index of Compounds

2,5-dimethylacetophenone, 896(189)
3,4-dimethylacetophenone, 897(210)
3,5-dimethylacetophenone, 897(203)
dimethylamine, 750(2)
dimethylaminoacetone, 799(14) 892(25)
4-dimethylaminoazobenzene, 806(105)
2-dimethylaminobenzaldehyde, 732(83), 801(105)
4-dimethylaminobenzaldehyde, 736(57), 805(70)
2-dimethylaminobenzoic acid, 690(71)
4-dimethylaminobenzoic acid, 676(75), 708(412)
4-dimethylaminobenzophenone, 902(149), 806(87)
2-dimethylaminoethanol, 799(19), 714(39)
2-dimethylaminodiethyl ether, 799(13)
2-dimethylaminoethyl alcohol, 714(39), 719(19)
2-dimethylamino-5-methylaniline, 758(173)
3-dimethylaminophenol, 805(79)
4-dimethylaminophenol, 805(72), 928(90)
4-dimethylaminothiophenol, 967(5)
N,N-dimethylaniline, 800(76)
2,3-dimethylaniline, 756(147)
2,4-dimethylaniline, 756(127)
2,5-dimethylaniline, 756(131)
2,6-dimethylaniline, 756(135)
3,4-dimethylaniline, 764(75)
3,5-dimethylaniline, 756(142)
1,3-dimethylanthracene, 887(46)
1,4-dimethylanthracene, 887(32)
1,5-dimethylanthracene, 889(83)
9,10-dimethylanthracene, 889(107)
2,2'-dimethylazobenzene, 909(3)
3,3'-dimethylazobenzene, 909(2)
4,4'-dimethylazobenzene, 909(39)
2,2'-dimethylazoxybenzene(*cis*), 911(5)
2,2'-dimethylazoxybenzene-(*trans*), 911(10)
3,3'-dimethylazoxybenzene(*cis*), 911(2)
3,3'-dimethylazoxybenzene-(*trans*), 911(11)
4,4'-dimethylazoxybenzene(*cis*), 911(7)
4,4'-dimethylazoxybenzene (*trans*), 911(3)
1,2-dimethylazulene, 886(19)

4,8-dimethylazulene, 887(28)
3,5-dimethylbenzaldehyde, 732(74)
N,N-dimethylbenzamide, 744(14)
2,4-dimethylbenzamide, 747(374)
3,4-dimethylbenzamide, 746(227)
3,5-dimethylbenzamide, 746(241)
3,4-dimethylbenzanilide, 745(138)
2,6-dimethyl-1,2-benzanthracene, 889(97)
3,9-dimethyl-1,2-benzanthracene, 888(54)
4,9-dimethyl-1,2-benzanthracene, 887(34)
4,10-dimethyl-1,2-benzanthracene, 888(73)
5,6-dimethyl-1,2-benzanthracene, 890(108)
5,8-dimethyl-1,2-benzanthracene, 889(80)
5,10-dimethyl-1,2-benzanthracene, 889(105)
6,7-dimethyl-1,2-benzanthracene, 889(103)
8,10-dimethyl-1,2-benzanthracene, 889(86)
9,10-dimethyl-1,2-benzanthracene, 889(77)
2,5-dimethylbenzene-1,3-disulfonamide, 952(243)
2,5-dimethylbenzene-1,4-disulfonamide, 952(247)
4,5-dimethylbenzene-1,3-disulfonamide, 952(215)
4,6-dimethylbenzene-1,3-disulfonamide, 952(224)
2,5-dimethylbenzene-1,3-disulfonyl chloride, 954(77)
2,5-dimethylbenzene-1,4-disulfonyl chloride, 957(205)
4,5-dimethylbenzene-1,3-disulfonyl chloride, 954(74)
4,6-dimethylbenzene-1,3-disulfonyl chloride, 956(173)
2,3-dimethylbenzenesulfonamide, 949(78)
2,4-dimethylbenzenesulfonamide, 948(36)
2,5-dimethylbenzenesulfonamide, 948(49)
2,6-dimethylbenzenesulfonamide, 948(20)
3,4-dimethylbenzenesulfonamide, 948(45)
3,5-dimethylbenzenesulfonamide, 948(33)
2,4-dimethylbenzenesulfonic acid, 959(27)
2,5-dimethylbenzenesulfonic acid, 959(28)

3,4-dimethylbenzenesulfonic acid, 959(29)
2,3-dimethylbenzenesulfonyl chloride, 953(26)
2,4-dimethylbenzenesulfonyl chloride, 953(13)
2,5-dimethylbenzenesulfonyl chloride, 953(10)
2,6-dimethylbenzenesulfonyl chloride, 953(19)
3,4-dimethylbenzenesulfonyl chloride, 953(32)
3,5-dimethylbenzenesulfonyl chloride, 955(98)
3,3'-dimethylbenzidine, 782(420)
2,4-dimethylbenzoic acid, 696(172)
2,5-dimethylbenzoic acid, 696(182)
2,6-dimethylbenzoic acid, 694(143)
3,4-dimethylbenzoic acid 702(269)
3,5-dimethylbenzoic acid (mesitylenic acid), 702(271)
2,6-dimethylbenzonitrile, 915(63)
4,4'-dimethylbenzophenone (di-*p*-tolyl ketone), 902(150)
2,6-dimethylbenzonitrile, 915(63)
2,2'-dimethylbenzophenone, 901(113)
2,4-dimethylbenzophenone, 897(223)
2,3-dimethyl-1,4-benzoquinone, 901(83)
2,5-dimethyl-1,4-benzoquinone, 904(197)
2,6-dimethyl-1,4-benzoquinone, 902(122)
2,6-dimethylbenzoyl chloride, 670(40)
α,4-dimethylbenzyl alcohol (methyl-*p*-tolylcarbinol), 720(147)
N,N-dimethylbenzylamine, 800(62)
2,4-dimethylbenzylamine, 756(140)
3,5-dimethylbenzylamine, 756(145)
6,6'-dimethyl-2,2'-bipyridyl, 805(82)
2,3-dimethyl-1,3-butadiene, 870(50)
2,3-dimethylbutanal, 728(25)
3,3-dimethylbutanal, 728(19)
2,2-dimethylbutanamide, 746(234)
dl-2,3-dimethylbutanamide, 746(233)
3,3-dimethylbutanamide, 746(236)

Index of Compounds

2,3-dimethylbutananilide, 744(58) (92)
3,3-dimethylbutananilide, 746(235)
2,2-dimethylbutane, 862(6)
2,3-dimethylbutane, 862(7)
2,3-dimethyl-2,3-butanediol, 722(17)
3,3-dimethylbutane-1-sulfonamide, 948(10)
3,3-dimethylbutane-1-sulfonyl chloride, 953(23)
2,2-dimethylbutanoic acid, 684(25)
dl-2,3-dimethylbutanoic acid, 686(28)
3,3-dimethylbutanoic acid, 684(21)
2,2-dimethylbutanol, 716(46)
2,3-dimethylbutanol, 716(59)
2,3-dimethyl-2-butanol, 714(22)
3,3-dimethyl-2-butanol (dl-pinacolyl alcohol), 714(24)
3,3-dimethyl-2-butanone, 892(14)
3,3-dimethylbutanonitrile, 914(16)
2,3-dimethyl-1-butene, 870(32)
2,3-dimethyl-2-butene (tetramethylethylene), 871(58)
3,3-dimethyl-1-butene, 870(22)
1,3-dimethyl-5-*tert*-butylbenzene, 880(86)
N,N-dimethylbutyramide, 744(5)
2,6-dimethylcarbazole, 794(654)
3,6-dimethylcarbazole, 792(647)
dimethyl carbonate, 814(19)
N,N-dimethyl-2-chloroaniline, 801(89)
1,1-dimethylcyclohexane, 863(54)
1,2-dimethylcyclohexane(*cis*), 863(70)
1,2-dimethylcyclohexane(*trans*), 863(62)
1,3-dimethylcyclohexane(*cis*), 863(55)
1,3-dimethylcyclohexane(*trans*), 863(65)
1,4-dimethylcyclohexane(*cis*), 863(64)
1,4-dimethylcyclohexane(*trans*), 863(53)
5,5-dimethyl-1,3-cyclohexanedione, 905(230)
2,4-dimethylcyclohexanol-(*trans*), 718(110)
3,3-dimethylcyclohexanol, 718(125)
3,4-dimethylcyclohexanol, 718(129)
2,2-dimethylcyclohexanone, 895(115)
2,3-dimethylcyclohexanone, 895(132)

2,4-dimethylcyclohexanone-(*cis*), 895(128)
2,4-dimethylcyclohexanone-(*trans*), 895(117)
dl-2,5-dimethylcyclohexanone, 895(119)
3,3-dimethylcyclohexanone, 895(135)
3,5-dimethylcyclohexanone-(*cis*), 895(143)
dl-3,5-dimethylcyclohexanone-(*trans*), 895(139)
4,4-dimethylcyclohexanone, 900(38)
1,2-dimethylcyclohexene, 873(205)
1,3-dimethylcyclohexene, 873(206)
1,4-dimethylcyclohexene, 873(198)
1,5-dimethylcyclohexene, 873(197)
1,6-dimethylcyclohexene, 873(201)
3,3-dimethylcyclohexene, 873(178)
4,4-dimethylcyclohexene, 873(174)
2,5-dimethyl-2-cyclohexenone, 895(148)
3,5-dimethyl-2-cyclohexenone, 896(172)
4,6-dimethyl-3-cyclohexenone, 895(152)
1,1-dimethylcyclopentane, 862(17)
1,2-dimethylcyclopentane(*cis*), 862(27)
1,2-dimethylcyclopentane(*trans*), 862(22)
1,3-dimethylcyclopentane(*cis*), 862(21)
1,3-dimethylcyclopentane-(*trans*), 862(20)
3,3-dimethylcyclopentanone, 894(83)
1,2-dimethylcyclopentene, 872(125)
1,3-dimethylcyclopentene, 871(93)
1,4-dimethylcyclopentene, 871(94)
1,5-dimethylcyclopentene, 872(113)
3,3-dimethylcyclopentene, 871(88)
4,4-dimethylcyclopentene, 871(89)
1,10-dimethyl-*cis*-decahydronaphthalene, 866(203)
1,10-dimethyl-*trans*-decahydronaphthalene, 866(200)
N,N′-dimethyl-1,4-diaminobenzene, 766(100)

1′,10-dimethyl-1,2-dibenzanthracene, 889(76)
4,4′-dimethyldibenzylamine, 762(26)
5,5-dimethyldihydroresorcinol (methone), 905(231)
N,N-dimethyl-2,4-dimethylaniline, 801(86)
N,N-dimethyl-2,5-dimethylaniline, 800(84)
2,2-dimethyldithiolane, 964(38)
1,2-dimethylenedioxy-4-propenylbenzene, 826(134)
dimethylethylamine, 799(2)
1,2-dimethyl-3-ethylbenzene, 879(47)
1,2-dimethyl-4-ethylbenzene, 879(38)
1,3-dimethyl-2-ethylbenzene, 879(39)
1,3-dimethyl-4-ethylbenzene, 878(36)
1,3-dimethyl-5-ethylbenzene, 878(30)
1,4-dimethyl-2-ethylbenzene, 878(33)
α,α-dimethylethylene sulfide, 964(4)
2,2-dimethyl-3-ethylpentane, 863(80)
2,4-dimethyl-3-ethylpentane, 863(85)
N,N-dimethylformamide, 744(1)
2,5-dimethylfuran, 823(39)
2,2-dimethylheptane, 863(75)
2,3-dimethylheptane, 864(94)
2,4-dimethylheptane, 863(77)
2,5-dimethylheptane, 863(83)
2,6-dimethylheptane, 863(81)
3,3-dimethylheptane, 863(86)
3,4-dimethylheptane, 864(95)
3,5-dimethylheptane, 863(84)
4,4-dimethylheptane, 863(82)
2,6-dimethyl-4-heptanol, 718(109)
3,3-dimethyl-4-heptanol, 718(117)
3,5-dimethyl-4-heptanol, 718(101)
2,2-dimethyl-3-heptanone, 894(107)
2,5-dimethyl-4-heptanone, 894(110)
2,6-dimethyl-4-heptanone (isovalerone), 894(108)
3,5-dimethyl-4-heptanone (di-*sec*-butyl ketone), 894(100)
2,6-dimethyl-2-heptene, 873(199)
2,5-dimethyl-1,5-hexadiene, 873(163) (177)
2,5-dimethyl-2,4-hexadiene, 873(202)

2,2-dimethylhexane, 862(31)
2,3-dimethylhexane, 862(41)
2,4-dimethylhexane, 862(34)
2,5-dimethylhexane, 862(32)
3,3-dimethylhexane, 862(37)
3,4-dimethylhexane, 863(46)
2,5-dimethyl-2,5-hexanediol, 724(56)
2,5-dimethyl-3,4-hexanedione, 893(74)
3,4-dimethyl-2,5-hexanedione, 896(175)
2,2-dimethylhexanoic acid, 686(44)
2,2-dimethyl-3-hexanol, 716(73)
2,3-dimethyl-2-hexanol, 716(80)
2,3-dimethyl-3-hexanol, 716(78)
2,2-dimethyl-3-hexanone, 893(68)
2,4-dimethyl-3-hexanone, 893(65)
3,4-dimethyl-2-hexanone, 894(91)
4,4-dimethyl-3-hexanone, 894(79)
2,2-dimethyl-cis-3-hexene, 872(124)
2,2-dimethyl-trans-3-hexene, 871(110)
2,3-dimethyl-1-hexene, 872(143)
2,3-dimethyl-2-hexene, 873(184)
2,3-dimethyl-3-hexene, 873(162)
2,4-dimethyl-1-hexene, 872(146)
2,4-dimethyl-2-hexene, 872(144)
2,4-dimethyl-cis-3-hexene, 872(135)
2,4-dimethyl-trans-3-hexene, 872(132)
2,5-dimethyl-1-hexene, 872(147)
2,5-dimethyl-2-hexene, 872(154)
2,5-dimethyl-3-hexene, 872(112)
3,3-dimethyl-1-hexene, 872(120)
3,4-dimethyl-1-hexene, 872(149)
3,4-dimethyl-2-hexene, 873(170)
3,4-dimethyl-3-hexene, 873(185)
3,5-dimethyl-1-hexene, 872(119)
3,5-dimethyl-2-hexene, 872(150)
4,4-dimethyl-1-hexene, 872(130)
4,4-dimethyl-2-hexene, 872(127)
4,5-dimethyl-1-hexene, 872(136)
4,5-dimethyl-2-hexene, 872(138)
5,5-dimethyl-1-hexene, 872(114)
5,5-dimethyl-cis-2-hexene, 872(129)
5,5-dimethyl-trans-2-hexene, 872(121)
3,4-dimethyl-3-hexen-2-one, 894(92)
4,5-dimethyl-4-hexen-3-one, 894(106)
4,5-dimethyl-5-hexen-3-one, 894(99)

5,5-dimethylhydantoin, 788(567)
sym-dimethylhydrazine, 750(21)
unsym-dimethylhydrazine, 750(14)
2,2′-dimethylhydrazobenzene, 911(20)
3,3′-dimethylhydrazobenzene, 911(1)
4,4′-dimethylhydrazobenzene, 911(16)
1,2-dimethylindole, 766(107)
2,5-dimethylindole, 778(356)
2,7-dimethylindole, 762(39)
3,5-dimethylindole, 770(194)
1,2-dimethyl-3-isopropylbenzene, 880(73)
1,2-dimethyl-4-isopropylbenzene, 880(70)
1,3-dimethyl-2-isopropylbenzene, 879(63)
1,3-dimethyl-4-isopropylbenzene, 879(52), 879(64)
1,3-dimethyl-5-isopropylbenzene, 879(41), 879(51)
1,4-dimethyl-2-isopropylbenzene, 879(57)
1,7-dimethyl-4-isopropylnaphthalene, 886(20)
dimethyl l-malate, 818(250)
dimethylmalonamide, 748(513)
dimethyl malonate, 816(148)
dimethylmalonic acid, 704(320)
N,N-dimethyl-2-methylaniline, 800(65)
N,N-dimethyl-3-methylaniline, 901(94)
N,N-dimethyl-4-methylaniline, 901(91)
1,2-dimethylnaphthalene, 884(168)
1,3-dimethylnaphthalene, 884(165)
1,4-dimethylnaphthalene, 884(169)
1,5-dimethylnaphthalene, 887(43)
1,6-dimethylnaphthalene, 884(164)
1,7-dimethylnaphthalene 884(163)
1,8-dimethylnaphthalene 887(24)
2,3-dimethylnaphthalene, 888(65)
2,6-dimethylnaphthalene 888(70)
2,7-dimethylnaphthalene, 888(60)
1,4-dimethyl-2-naphthol, 936(247)
2,4-dimethyl-1-naphthol, 930(109)
N,N-dimethyl-1-naphthylamine, 802(131)

N,N-dimethyl-2-naphthylamine, 804(31)
1,4-dimethyl-2-naphthylamine, 770(193)
3,6-dimethyl-2-naphthylamine, 782(451)
3,7-dimethyl-1-naphthylamine, 774(265)
3,7-dimethyl-2-naphthylamine, 780(416)
2,3-dimethyl-4-nitroaniline, 778(353)
2,3-dimethyl-5-nitroaniline, 778(342)
2,3-dimethyl-6-nitroaniline, 778(376)
2,4-dimethyl-3-nitroaniline, 772(218)
2,4-dimethyl-5-nitroaniline, 780(393)
2,4-dimethyl-6-nitroaniline, 770(171)
2,5-dimethyl-3-nitroaniline, 774(284)
2,5-dimethyl-4-nitroaniline (5-nitro-p-2-xylidine), 784(467)
2,6-dimethyl-3-nitroaniline, 772(219)
2,6-dimethyl-4-nitroaniline, 786(519)
3,4-dimethyl-2-nitroaniline, 768(153)
3,4-dimethyl-5-nitroaniline, 770(189)
3,4-dimethyl-6-nitroaniline, 782(453)
3,5-dimethyl-2-nitroaniline, 766(105)
3,5-dimethyl-4-nitroaniline, 782(431)
4,6-dimethyl-2-nitroaniline, 770(200)
1,3-dimethyl-2-nitrobenzene, 916(27)
2,3-dimethylnitrobenzene (3-nitro-o-xylene), 916(33)
2,4-dimethylnitrobenzene, 916(32)
2,5-dimethylnitrobenzene, 916(31)
3,4-dimethylnitrobenzene, 917(6)
3,5-dimethylnitrobenzene, 918(62)
2,4-dimethyl-3-nitrobenzenesulfonamide, 949(84)
2,4-dimethyl-5-nitrobenzenesulfonamide, 950(118)
2,4-dimethyl-6-nitrobenzenesulfonamide, 948(15)
2,5-dimethyl-3-nitrobenzenesulfonamide, 949(85)
2,5-dimethyl-4-nitrobenzenesulfonamide, 950(147)

Index of Compounds

2,5-dimethyl-6-nitrobenzene-sulfonamide, 950(136)
3,4-dimethyl-5-nitrobenzene-sulfonamide, 949(96)
2,4-dimethyl-3-nitrobenzene-sulfonyl chloride, 955(108)
2,4-dimethyl-6-nitrobenzene-sulfonyl chloride, 955(112)
2,5-dimethyl-3-nitrobenzene-sulfonyl chloride, 954(47)
2,5-dimethyl-4-nitrobenzene-sulfonyl chloride, 954(66)
2,5-dimethyl-6-nitrobenzene-sulfonyl chloride, 955(134)
3,4-dimethyl-5-nitrobenzene-sulfonyl chloride, 954(63)
dimethyl 3-nitrophthalate, 821(82)
dimethyl 4-nitrophthalate, 821(74)
2,2-dimethyloctane, 864(121)
2,3-dimethyloctane, 865(156)
2,4-dimethyloctane, 864(115)
2,5-dimethyloctane, 864(133)
2,6-dimethyloctane, 864(134)
2,7-dimethyloctane, 865(138)
3,3-dimethyloctane, 865(149)
3,4-dimethyloctane, 865(167)
3,5-dimethyloctane, 865(141)
3,6-dimethyloctane, 865(142)
4,4-dimethyloctane, 865(145)
4,5-dimethyloctane, 865(152)
dimethyl oxalate, 821(54)
3,3-dimethylpentanal, 730(34)
2,2-dimethylpentane, 862(12)
2,3-dimethylpentane, 862(18)
2,4-dimethylpentane, 862(13)
3,3-dimethylpentane, 862(16)
2,2-dimethyl-3-pentanol, 716(54)
2,4-dimethylpentanol, 716(48)
2,4-dimethyl-2-pentanol, 714(36)
2,4-dimethyl-3-pentanol, 716(48)
2,2-dimethyl-3-pentanone, 892(31)
2,4-dimethyl-3-pentanone (diisopropyl ketone), 892(28)
3,3-dimethyl-2-pentanone, 893(38)
3,4-dimethyl-2-pentanone, 893(52)
4,4-dimethyl-2-pentanone, 892(30)
2,3-dimethyl-1-pentene, 871(79)
2,3-dimethyl-2-pentene, 871(104)
2,4-dimethyl-1-pentene, 871(75)
2,4-dimethyl-2-pentene, 871(77)
3,3-dimethyl-1-pentene, 871(67)
3,4-dimethyl-1-pentene, 871(73)
3,4-dimethyl-2-pentene, 871(85)

4,4-dimethyl-1-pentene, 870(56)
4,4-dimethyl-*cis*-2-pentene, 871(72)
4,4-dimethyl-*trans*-2-pentene, 871(65)
3,4-dimethyl-3-penten-2-one, 893(72)
3,4-dimethyl-4-penten-2-one, 893(61)
3,4-dimethyl-1-pentyn-3-ol, 714(37)
2,4-dimethylphenacyl chloride, 834(94)
1,2-dimethylphenanthrene, 889(85)
1,3-dimethylphenanthrene, 887(37)
1,4-dimethylphenanthrene, 886(9)
1,5-dimethylphenanthrene, 886(8), 886(16)
1,6-dimethylphenanthrene, 887(49)
1,7-dimethylphenanthrene, 887(48)
1,8-dimethylphenanthrene, 890(111)
1,9-dimethylphenanthrene, 887(50)
2,3-dimethylphenanthrene, 887(41)
2,4-dimethylphenanthrene, 888(69)
2,6-dimethylphenanthrene, 886(2)
2,7-dimethylphenanthrene, 888(62)
2,9-dimethylphenanthrene, 886(15)
3,4-dimethylphenanthrene, 886(21)
3,5-dimethylphenanthrene, 886(12)
3,6-dimethylphenanthrene, 889(84)
4,5-dimethylphenanthrene, 887(35)
2,3-dimethylphenol, 928(87)
2,4-dimethylphenol, 924(10), 926(4)
2,5-dimethylphenol, 928(86)
2,6-dimethylphenol, 926(32)
3,4-dimethylphenol, 928(57)
3,5-dimethylphenol, 928(63)
N,N-dimethyl-*p*-phenylene-diamine, 764(52)
2,6-dimethyl-*m*-phenylene-diamine, 768(313)
4,6-dimethyl-*m*-phenylene-diamine, 776(313)
2,2-dimethyl-1-phenylpropane, 878(32)
3,4-dimethyl-1-phenylpyrazole, 802(140)

3,5-dimethyl-1-phenylpyrazole, 802(134)
dimethyl phthalate, 819(298)
N,N-dimethylpiperazine, 799(18)
2,5-dimethylpiperazine(*cis*), 778(355)
2,5-dimethylpiperazine(*trans*), 778(371)
2,6-dimethylpiperidine, 752(49)
2,2-dimethylpropanal, 728(11)
2,2-dimethylpropane (neopentane), 862(1)
2,2-dimethylpropanol (neopentyl alcohol), 722(31)
N,N-dimethylpropionamide, 744(3)
2,2-dimethylpropionamide, 746(299)
2,2-dimethylpropionanilide, 746(222)
1,2-dimethyl-3-propylbenzene, 881(99)
1,2-dimethyl-4-propylbenzene, 881(95)
1,3-dimethyl-4-propylbenzene, 880(87)
1,3-dimethyl-5-propylbenzene, 881(97), 880(72)
1,4-dimethyl-2-propylbenzene, 880(78)
2,3-dimethylpyrazine, 799(38)
2,5-dimethylpyrazine, 799(34)
2,6-dimethylpyrazine, 804(32)
1,3-dimethylpyrazole, 799(20)
1,4-dimethylpyrazole, 799(29)
1,5-dimethylpyrazole, 799(30)
2,3-dimethylpyridine, 790(40)
2,4-dimethylpyridine (lutidine), 799(36)
2,5-dimethylpyridine, 799(37)
2,6-dimethylpyridine, 799(24)
3,4-dimethylpyridine, 799(44)
3,5-dimethylpyridine, 800(52)
4,5-dimethylpyrimidine, 800(60)
4,6-dimethylpyrimidine, 804(3)
1,2-dimethylpyrrolidine, 799(6)
1,3-dimethylpyrrolidine, 799(7)
2,4-dimethylpyrrolidine, 750(41)
2,5-dimethylpyrrolidine, 750(37)
2,3-dimethylquinoline, 805(60)
2,4-dimethylquinoline, 801(124)
2,5-dimethylquinoline, 805(53)
2,6-dimethylquinoline, 805(51)
2,7-dimethylquinoline, 805(54)
2,8-dimethylquinoline, 804(5)
3,4-dimethylquinoline, 805(69)
3,6-dimethylquinoline, 805(47)
3,7-dimethylquinoline, 805(75)
4,5-dimethylquinoline, 805(74)

1024 Index of Compounds

4,6-dimethylquinoline, 802(137)
4,7-dimethylquinoline, 802(139)
4,8-dimethylquinoline, 805(46)
5,6-dimethylquinoline, 804(38)
5,7-dimethylquinoline, 804(2)
5,8-dimethylquinoline, 802(132)
6,7-dimethylquinoline, 805(48)
6,8-dimethylquinoline, 802(128)
2,3-dimethylquinoxaline, 806(95)
dimethyl succinate, 817(172)
dimethyl sulfate, 817(157)
dimethyl tartrate, 821(46)
dimethyl dl-tartrate, 822(111)
1,1-dimethyl-1,2,3,4-tetrahydronaphthalene, 882(120)
1,5-dimethyl-1,2,3,4-tetrahydronaphthalene, 883(151)
1,2-dimethyl-1,2,3,4-tetrahydronaphthalene, 883(141)
1,3-dimethyl-1,2,3,4-tetrahydronaphthalene, 883(138)
1,4-dimethyl-1,2,3,4-tetrahydronaphthalene, 883(139)
2,2-dimethyl-1,2,3,4-tetrahydronaphthalene, 882(135)
2,3-dimethyl-1,2,3,4-tetrahydronaphthalene, 882(137)
2,5-dimethyl-1,2,3,4-tetrahydronaphthalene, 883(145)
2,6-dimethyl-1,2,3,4-tetrahydronaphthalene, 883(148)
2,7-dimethyl-1,2,3,4-tetrahydronaphthalene, 883(147)
2,8-dimethyl-1,2,3,4-tetrahydronaphthalene, 883(146)
5,6-dimethyl-1,2,3,4-tetrahydronaphthalene, 883(156)
5,7-dimethyl-1,2,3,4-tetrahydronaphthalene, 884(158)
5,8-dimethyl-1,2,3,4-tetrahydronaphthalene, 884(159)
6,7-dimethyl-1,2,3,4-tetrahydronaphthalene, 884(157)
2,3-dimethyl-1,2,3,4-tetrahydroquinoline, 764(44)
2,3-dimethyl-5,6,7,8-tetrahydroquinoline, 804(20)
2,4-dimethyl-5,6,7,8-tetrahydroquinoline, 801(115)
dl-2,6-dimethyl-1,2,3,4-tetrahydroquinoline, 762(22)
α,α'-dimethyltetramethylene sulfide(cis), 964(21)
α,α'-dimethyltetramethylene sulfide(trans), 964(20)
α,γ-dimethyltetramethylene sulfide, 964(42)
cis-2,5-dimethylthiacyclopentane, 964(21)
trans-2,5-dimethylthiacyclopentane, 964(20)
2,3-dimethylthiophene, 964(18)
2,4-dimethylthiophene, 964(17)

2,5-dimethylthiophene, 964(15)
3,5-dimethylthiophene, 964(22)
2,6-dimethyl-1,4-thioxane, 964(31)
3,5-dimethyl-1,4-thioxane, 964(34)
N,N-dimethyl-m-toluidine, 800(94)
N,N-dimethyl-o-toluidine, 800(65)
N,N-dimethyl-p-toluidine, 801(91)
α,α'-dimethyltrimethylene sulfide, 964(9)
β,β-dimethyltrimethylene sulfide, 964(12)
4,5-dimethyltrithione, 902(152)
1,3-dimethylxanthine, 794(675)
di-1-naphthyl ketone, 903(161)
sym-dimethylurea, 745(149)
unsym-dimethylurea, 747(383)
di-α-naphthol, 944(433)
di-β-naphthol, 942(393)
di(2-naphthyl)amine, 788(554)
N,N'-di-2-naphthyl-1,4-diaminobenzene, 794(656)
1,2'-dinaphthyl ether, 828(44)
di-1-naphthylmethane, 888(67)
N,N'-di-2-naphthyl-p-phenylenediamine, 794(656)
3,8-dinitroacenaphthene, 921(145)
2,4-dinitroaniline, 788(574)
2,6-dinitroaniline, 782(448)
2,4-dinitroanisole, 828(48), 920(94)
3,5-dinitroanisole, 828(53)
1,3-dinitroanthraquinone, 922(171)
1,5-dinitroanthraquinone, 906(288), 923(177)
1,6-dinitroanthraquinone, 922(172)
1,7-dinitroanthraquinone, 923(175)
1,8-dinitroanthraquinone, 923(176)
2,7-dinitroanthraquinone, 922(173)
2,2'-dinitroazobenzene, 910(65), 922(164)
3,3'-dinitroazobenzene, 910(44)
4,4'-dinitroazobenzene, 910(75), 922(167)
2,2'-dinitroazoxybenzene(cis), 911(27)
2,2'-dinitroazoxybenzene-(trans), 911(8)
3,3'-dinitroazoxybenzene(cis), 911(23)
3,3'-dinitroazoxybenzene-(trans), 911(12)
4,4'-dinitroazoxybenzene(cis), 911(29)

4,4'-dinitroazoxybenzene-(trans), 911(14)
2,4-dinitrobenzamide, 748(442)
3,5-dinitrobenzamide, 747(389)
3,5-dinitrobenzanilide, 748(491)
1,2-dinitrobenzene, 920(114),
1,3-dinitrobenzene, 919(84)
1,4-dinitrobenzene, 922(153)
2,4-dinitrobenzenesulfonamide, 949(64)
3,5-dinitrobenzenesulfonamide, 951(204)
2,4-dinitrobenzenesulfonyl chloride, 955(124)
3,5-dinitrobenzenesulfonyl chloride, 956(116)
2,4-dinitrobenzoic acid, 702(298)
2,6-dinitrobenzoic acid, 704(335)
3,4-dinitrobenzoic acid, 700(261)
3,5-dinitrobenzoic acid, 706(348)
2,4-dinitrobenzoic anhydride, 669(60)
3,5-dinitrobenzoic anhydride, 668(48)
3,5-dinitrobenzoyl bromide, 671(83)
2,4-dinitrobenzoyl chloride, 671(64)
3,5-dinitrobenzoyl chloride, 671(66)
4,4'-dinitrobibenzyl, 922(156)
2,2'-dinitrobiphenyl, 921(127)
2,3'-dinitrobiphenyl, 920(105)
2,4-dinitrobiphenyl, 919(91), 920(106)
3,3'-dinitrobiphenyl, 922(162)
4,4'-dinitrobiphenyl, 922(170)
2,4-dinitro-1-bromobenzene, 918(59)
2,4-dinitro-1-chlorobenzene, 917(30)
2,5-dinitro-1,4-dibromobenzene, 919(75)
1,2-dinitro-3,5-dimethylbenzene, 921(129)
1,4-dinitro-2,3-dimethylbenzene, 919(87)
2,3-dinitro-1,4-dimethylbenzene, 919(89)
2,4-dinitro-1,3-dimethylbenzene, 919(76)
2,5-dinitro-1,3-dimethylbenzene, 920(100)
2,5-dinitro-1,4-dimethylbenzene, 921(140)
2,6-dinitro-1,4-dimethylbenzene, 921(122)
3,4-dinitro-1,2-dimethylbenzene, 919(70)
4,5-dinitro-1,2-dimethylbenzene, 920(116)

Index of Compounds 1025

4,6-dinitro-1,3-dimethylbenzene, 919(90)
2,2'-dinitrodiphenylmethane, 921(147)
2,4'-dinitrodiphenylmethane, 920(115)
3,3'-dinitrodiphenylmethane, 922(154)
4,4'-dinitrodiphenylmethane, 922(159)
2,4-dinitro-1-ethoxybenzene, 919(81)
2,5-dinitrofluorene, 922(163)
2,4-dinitrofluorobenzene, 846(2), 917(3)
5,7-dinitro-8-hydroxynaphthalene-2-sulfonic acid, 959(30)
2,4-dinitromesitylene, 919(78)
2,4-dinitro-1-methoxybenzene, 828(48), 920(94)
3,5-dinitro-1-methoxybenzene, 828(53)
1,2-dinitronaphthalene, 920(101)
1,3-dinitronaphthalene, 921(136)
1,4-dinitronaphthalene, 921(128)
1,5-dinitronaphthalene, 922(165)
1,6-dinitronaphthalene, 921(148)
1,8-dinitronaphthalene, 922(149)
2,7-dinitronaphthalene, 922(169)
1,5-dinitro-2-naphthol, 940(345)
1,6-dinitro-2-naphthol, 940(358)
1,8-dinitro-2-naphthol, 940(360)
2,4-dinitro-1-naphthol, 936(250)
4,8-dinitro-1-naphthol, 944(411)
1,5-dinitro-2-naphthylamine, 790(601)
1,8-dinitro-2-naphthylamine, 794(655)
2,4-dinitro-1-naphthylamine, 794(661)
2,5-dinitrophenanthraquinone, 906(27)
2,7-dinitrophenanthraquinone, 906(286)
2,4-dinitrophenetole, 919(81)
2,3-dinitrophenol, 936(265)
2,4-dinitrophenol, 934(184)
2,5-dinitrophenol, 932(162)
2,6-dinitrophenol, 928(58)
3,4-dinitrophenol, 936(244)
3,5-dinitrophenol, 934(226)
2,4-dinitrophenylacetic acid, 702(306)
2,4-dinitrophenylhydrazine, 790(615)
2,4-dinitrophenyl sulfide, 965(51)
2,4-dinitroresorcinol, 936(273)

3,5-dinitrosalicylaldehyde, 736(44)
3,5-dinitrosalicylamide, 747(381)
3,5-dinitrosalicylic acid, 702(283)
4-Nitroso-N,N-dimethylaniline, 805(78)
2,4-dinitrosoresorcinol, 938(312)
2,2'-dinitrostilbene, 922(161)
2,4-dinitrostilbene, 921(135)
4,4'-dinitrostilbene(trans), 922(174)
2,4-dinitrothiophenol, 967(29)
2,3-dinitrotoluene, 918(47)
2,4-dinitrotoluene, 918(54)
2,6-dinitrotoluene, 918(52)
3,4-dinitrotoluene, 918(45)
3,5-dinitrotoluene, 919(88)
2,6-dinitro-p-toluic acid, 708(385)
2,4-dinitro-1,3,5-trimethylbenzene, 919(78)
1,3-dioxane, 667(7), 824(46)
1,4-dioxane, 823(43)
dipentene (dl-limonene), 874(229)
1,1-dipentoxyethane, 667(38)
dipentoxymethane, 667(36)
dipentylamine, 754(107)
diperfluoroethylperfluorocyclohexylamine, 859(11)
diperfluoropropylperfluoroethylamine, 859(10)
diperfluoroethylperfluoropropylamine, 859(9)
diphenamide, 748(459)
diphenanilide, 748(488)
2,2'-diphenic acid, 708(391)
diphenic anhydride, 669(66)
1,2-diphenoxyethane, 828(49)
1,3-diphenoxypropane, 827(34)
diphenylacetamide, 747(341)
N,N-diphenylacetamide, 745(126)
diphenylacetanilide, 747(378)
diphenylacetic acid, 698(223)
diphenylacetoin, 901(79)
1,1-diphenylacetone, 901(100)
diphenylacetonitrile, 915(54)
diphenylacetylene, 875(281)
diphenyl adipate, 822(122)
diphenylamine, 766(101)
diphenylamine-4,4'-disulfonic acid, 959(32)
diphenylamine-4-sulfonic acid, 959(31)
1,3-diphenylbenzene, 887(51)
N,N'-diphenylbenzidine, 794(665)
1,4-diphenyl-1,3-butadiene(cis), 875(288)
1,4-diphenyl-1,3-butadiene-(trans), 875(307)

diphenylcarbamyl chloride, 671(70)
diphenylcarbinol (benzhydrol), 722(45)
diphenyl carbonate, 821(103)
1,5-diphenylcarbohydrazide, 788(565)
N,N'-diphenyl-1,4-diaminobenzene, 784(470)
diphenyl-4,4'-disulfonic acid, 959(18)
diphenylethanal, 732(88)
1,1-diphenylethane, 884(170)
1,2-diphenyl-1,2-ethanediol, 724(65)
1,2-diphenylethanol, 722(43)
1,1-diphenylethene, 874(247)
1,2-diphenylethene, 875(306)
diphenylethylamine, 764(43)
1,1-diphenylethylene, 874(247)
1,2-diphenylethyne, 875(281)
N,N-diphenylformamide, 744(51)
diphenylglycolaldehyde, 740(118)
diphenylguanidine, 746(280)
5,5-diphenylhydantoin, 748(517)
1,1-diphenylhydrazine, 762(30)
diphenylmethane, 886(1)
diphenylmethylamine, 760(219)
1,9-diphenyl-1,3,6,8-nonatetraen-5-one, 904(219)
1,5-diphenyl-2,4-pentadien-1-one, 903(168)
1,5-diphenyl-1,4-pentadien-3-one (dibenzalacetone), 903(178)
N,N'-diphenylpiperazine, 806(113)
2,3-diphenylpropanal, 736(33)
1,3-diphenyl-2-propanone (dibenzyl ketone), 899(27)
2,3-diphenylpropenal, 738(76)
3,3-diphenylpropenal, 734(22)
1,4-diphenylsemicarbazide, 788(568)
4,4-diphenylsemicarbazide, 746(312)
diphenyl succinate, 822(134)
diphenyl-4-sulfonamide, 951(202)
diphenyl-4-sulfonyl chloride, 956(143)
1,4-diphenyltriphenylene-2,3-dicarboxylic acid, 710(440)
asym-diphenylurea, 747(404)
sym-diphenylurea (carbanilide), 748(497)
dipropenyl (2,4-hexadiene), 871(70)
1,1-dipropoxybutane, 667(29)
2,2'-dipropoxydiethyl ether, 825(120)
1,1-dipropoxyethane, 667(20)

1026 Index of Compounds

1,1-di(2-propoxy)ethane (diisopropylacetal), 667(9)
dipropoxymethane (propylal), 667(16)
dipropylamine, 750(39)
N,N-dipropylaniline, 801(107)
1,4-dipropylbenzene, 881(119)
dipropylcarbinol (4-heptanol), 716(76)
dipropyl carbonate, 816(124)
dipropyl ketone (4-heptanone), 893(63)
2,2'-dipyridyl, 805(61)
1,3-dithiane. 965(21)
1,4-dithiane, 965(45)
dithiolane, 964(41)
1,3-dithiolane-4-carboxylic acid, 690(44)
di(2-tolyl)amine, 766(98)
di(4-tolyl)amine, 770(208)
di-4-tolyl ketone (4,4'-dimethylbenzophenone), 902(150)
sym-di-2-tolylurea, 738(507)
sym-di-3-tolylurea, 748(477)
sym-di-4-tolylurea, 748(512)
djenkolic acid, 680(130)
docosane, 867(244)
docosanoic anhydride, 668(43)
docosanoic acid, 690(78)
docosanol, 722(46)
docosene, 874(261)
10-docosenoic acid, 688(17)
13-docosenoic acid(*cis*), 688(19)
docosylcyclohexane, 867(255)
docosylcyclopentane, 867(245)
1-docosyne, 875(264)
dodecanal, 734(24)
dodecanamide, 745(168)
dodecananilide, 744(60)
dodecane, 866(202)
dodecanedioic acid, 696(177)
dodecanedioic anhydride, 668(40)
dodecanoic acid, 688(27)
dodecanoic anhydride, 668(23)
dodecanol, 720(165), 722(4)
2-dodecanone, 899(6)
3-dodecanone, 899(1)
dodecanonitrile, 913(53)
dodecanoyl chloride, 671(53)
1-dodecene, 874(237)
dodecyl alcohol (lauryl alcohol), 720(165), 722(4)
dodecylamine, 762(17)
dodecyl bromide, 838(141)
dodecylcyclohexane, 867(224)
dodecylcyclopentane, 867(221)
dodecyl ether (dilauryl ether), 827(12)
dodecyl iodide, 840(186)
dodecyl sulfide, 965(9)
dodecyl perfluorobutyrate, 860(23)
dodecyne, 874(238)
1-dotriacontene, 875(284)

1-dotriacontyne, 875(290)
dotriacontylcyclohexane, 868(286)
dotriacontylcyclopentane, 868(275)
dotricontane, 868(276)
dulcin (*p*-phenetylurea), 747(358)
dulcitol, 724(73)
durene (1,2,4,5-tetramethylbenzene), 887(42)
dypnone, 897(225)

eicosanamide, 745(159)
eicosananilide, 744(90)
eicosane, 867(238)
eicosanoic acid, 690(73)
eicosanoic anhydride, 668(41)
eicosanol, 722(41)
eicosanonitrile, 914(31)
eicosene, 874(256)
eicosylcyclohexane, 867(248)
eicosylcyclopentane, 867(240)
eicosyne, 874(260)
elaidamide, 744(94)
elaidic acid, 688(30)
elaidyl alcohol (*trans*-9-octadecenol), 722(19)
enanthaldehyde (heptanal), 730(43)
enanthoanilide, 744(47)
enanthonitrile, 913(34)
enanthoyl chloride, 670(32)
l-ephedrine, 764(49)
α-epibromohydrin (3-bromopropylene oxide), 824(69)
α-epichlorohydrin, 824(56)
β-epichlorohydrin (2-chloropropylene oxide), 824(66)
1,2-epoxy-3-bromopropane, 824(69)
1,2-epoxybutane (α-butylene oxide), 823(15)
2,3-epoxybutane, 823(14)
1,2-epoxy-3-chloropropane, 824(56)
epoxyethane (ethylene oxide), 823(2)
1,2-epoxy-2-methylpropane, 823(12)
1,2-epoxypropane (propylene oxide), 823(6)
2,3-epoxy-1-propanol, 716(94)
ergosterol, 724(71)
erucamide, 744(41)
erucanilide, 744(38)
erucic acid, 688(19)
erythritol (*dl*-1,2,3,4-tetrahydroxybutane), 724(47)
meso-erythritol, 724(62)
ethanal (acetaldehyde), 728(2)
1,2-ethanediol, 720(136)
ethane-1,2-disulfonic acid, 959(35)
ethane-1,2-disulfonyl chloride, 955(103)

1,2-ethanedithiocyanate, 961(10)
1,2-ethanedithiol, 966(16)
ethanesulfonamide, 948(5)
ethanesulfonic acid, 959(33)
ethanethiol, 966(2)
ethanol, 714(2)
ethanolamine (2-aminoethyl alcohol), 718(99)
ethoxyacetamide, 744(69)
ethoxyacetic acid, 686(39)
2-ethoxyaniline, 756(167)
3-ethoxyaniline, 758(192)
4-ethoxyaniline, 758(193)
2-ethoxybenzaldehyde, 734(2)
4-ethoxybenzaldehyde, 732(86)
2-ethoxybenzamide, 746(225)
3-ethoxybenzamide, 746(256)
4-ethoxybenzamide, 748(439)
4-ethoxybenzanilide, 747(354)
ethoxybenzene, 824(82)
2-ethoxybenzoic acid, 688(4)
3-ethoxybenzoic acid, 698(194)
4-ethoxybenzoic acid, 704(327)
4-ethoxybenzoic anhydride, 668(47)
ethoxyethanal, 728(22)
2-ethoxyethanethiol, 966(14)
2-ethoxyethanol ("ethylcellosolve")(ethylene glycol monoethyl ether), 714(42), 824(70)
2-(2-ethoxyethoxy)ethanol ("carbitol")(diethylene glycol monoethyl ether), 720(140), 825(105)
ethoxyethoxyethyl acetate, 818(211)
2-ethoxyethyl acetate, 816(110)
2-ethoxyethyl acrylate, 816(133)
3-ethoxy-4-hydroxybenzaldehyde, 736(60)
1-ethoxy-2-methoxyethane, 824(44)
2-ethoxy-1-naphthaldehyde, 738(92)
1-ethoxynaphthalene, 826(143)
2-ethoxynaphthalene, 827(16)
2-ethoxynaphthalene-1-sulfonamide, 949(65)
4-ethoxynaphthalene-1-sulfonamide, 949(82)
4-ethoxynaphthalene-2-sulfonamide, 949(105)
5-ethoxynaphthalene-1-sulfonamide, 949(101)
6-ethoxynaphthalene-1-sulfonamide, 949(59)
6-ethoxynaphthalene-2-sulfonamide, 949(103)
7-ethoxynaphthalene-2-sulfonamide, 948(40)
2-ethoxynaphthalene-1-sulfonyl chloride, 956(146)

Index of Compounds

4-ethoxynaphthalene-1-sulfonyl chloride, 955(125)
4-ethoxynaphthalene-2-sulfonyl chloride, 954(83)
5-ethoxynaphthalene-1-sulfonyl chloride, 956(153)
6-ethoxynaphthalene-1-sulfonyl chloride, 956(148)
6-ethoxynaphthalene-2-sulfonyl chloride, 955(131)
7-ethoxynaphthalene-2-sulfonyl chloride, 955(126)
4-ethoxy-2-nitroaniline, 778(347)
2-ethoxynitrobenzene, 916(37)
3-ethoxynitrobenzene, 917(11)
4-ethoxynitrobenzene, 918(41)
6-ethoxy-1-nitronaphthalene-2-sulfonamide, 951(177)
7-ethoxy-8-nitronaphthalene-1-sulfonamide, 949(86)
6-ethoxy-5-nitronaphthalene-2-sulfonyl chloride, 957(190)
7-ethoxy-8-nitronaphthalene-1-sulfonyl chloride, 957(196)
3-ethoxyphenol, 924(24)
1-ethoxy-2-propanone, 892(34)
2-ethoxytoluene, 825(91)
3-ethoxytoluene, 825(98)
4-ethoxytoluene, 825(97)
2-ethoxyphenol, 924(15), 926(5)
N-ethylacetamide, 744(8)
N-ethylacetanilide, 744(24)
ethyl acetate, 814(12)
2-ethylacrylonitrile, 913(10)
ethyl acetoacetate, 816(145), 895(141)
ethyl acetonedicarboxylate, 897(212)
ethylacral (1,1-diethoxy-2-propene), 667(14)
ethyl acrylate, 814(30)
ethylal, 667(5)
ethyl alcohol, 714(2)
ethylamine, 750(3)
ethyl 3-aminobenzoate, 760(216), 819(305)
ethyl 4-aminobenzoate, 772(246), 774(256), 822(110)
ethyl 2-aminocinnamate, 821(100)
ethyl 3-aminocinnamate, 821(70)
ethyl 4-aminocinnamate, 821(83)
2-(ethylamino)ethanol, 718(96), 752(76)
ethyl 2-aminophenyl ether, 825(123)
N-ethylaniline, 754(108)
2-ethylaniline, 754(123)
3-ethylaniline, 756(133)
4-ethylaniline, 756(138)

ethyl anisate, 819(285)
ethyl anthranilate, 819(285)
ethylbenzene, 877(3)
2-ethylbenzenesulfonamide, 948(12)
4-ethylbenzenesulfonamide, 948(16)
2-ethylbenzenesulfonyl chloride, 953(1)
4-ethylbenzenesulfonyl chloride, 953(2)
ethyl benzoate, 817(203)
2-ethylbenzoic acid, 690(65)
ethyl benzoylacetate, 819(284), 897(216)
ethyl 2-benzoylbenzoate, 821(61)
2-ethylbenzoyl chloride, 670(42)
ethyl benzoylformate, 819(274)
α-ethylbenzyl alcohol (α-phenylpropyl alcohol) (ethylphenylcarbinol), 720(149)
N-ethylbenzylamine, 754(103)
N-ethyl-N-benzylaniline, 804(13)
ethyl benzylmalonate, 820(309)
ethyl bromide, 836(100)
ethyl bromoacetate, 816(114), 819(127)
ethyl 2-bromobenzoate, 189(272)
ethyl 3-bromobenzoate, 819(276)
ethyl 4-bromobenzoate, 819(282)
ethyl bromomalonate, 818(240)
ethyl 2-bromopropionate, 816(118)
ethyl 3-bromopropionate, 816(143)
2-ethyl-1,3-butadiene, 871(59)
2-ethylbutanal, 728(26)
2-ethylbutanamide, 745(172)
2-ethylbutananilide, 745(213)
2-ethylbutane-1,1-dicarboxylic acid, 690(48)
2-ethylbutanoic acid (diethylacetic acid), 686(30)
2-ethylbutanol, 716(64)
2-ethylbutanoyl chloride, 670(28)
2-ethyl-1-butene, 870(40)
ethylbutylal (1,1-diethoxybutane), 667(18)
1-ethyl-4-tert-butylbenzene, 881(100)
α-ethylbutyraldehyde (diethylacetaldehyde), 728(26)
ethyl butyrate, 815(52)
ethyl caprate, 819(260)
ethyl caproate, 816(122)
α-ethylcaproic acid (2-ethylhexanoic acid), 686(52)

ethyl caprylate, 817(194)
ethyl carbamate (ethylurethane), 744(16)
ethyl carbonate, 815(62)
"ethylcellosolve" (2-ethoxyethanol), 714(42), 824(70)
ethyl chloride, 830(22)
ethyl chloroacetate, 815(85)
ethyl 2-chlorobenzoate, 818(253)
ethyl 3-chlorobenzoate, 818(251)
ethyl 4-chlorobenzoate, 818(244)
ethyl chloroformate, 670(14), 814(22)
ethyl 2-chloropropionate, 815(91)
ethyl cinnamate, 819(290)
ethyl crotonate, 815(75)
ethyl cyanoacetate, 913(41)
ethylcyclohexane, 863(74)
ethyl cyclohexanecarboxylate, 817(171)
3-ethylcyclohexanone, 895(151)
1-ethylcyclohexene, 873(204)
3-ethylcyclohexene, 873(203)
4-ethylcyclohexene, 873(200)
N-ethylcyclohexylamine, 752(75)
2-ethylcyclohexylamine, 752(79)
ethylcyclopentane, 862(29)
2-ethylcyclopentanone, 894(97)
1-ethylcyclopentene, 872(128)
3-ethylcyclopentene, 871(106)
4-ethylcyclopentene, 872(126)
9-ethyl-cis-decahydronaphthalene, 866(207)
9-ethyl-trans-decahydronaphthalene, 866(206)
ethyl decanoate, 819(260)
ethyl dichloroacetate, 816(113)
ethyl difluoroacetate, 814(31)
ethyl dihydrocinnamate, 819(264)
2-ethyl-3,3-dimethyl-1-butene, 872(140)
3-ethyl-2,2-dimethylhexane, 865(135)
3-ethyl-2,3-dimethylhexane, 866(182)
3-ethyl-2,4-dimethylhexane, 865(162)
3-ethyl-2,5-dimethylhexane, 864(128)
3-ethyl-3,4-dimethylhexane, 866(184)
4-ethyl-2,2-dimethylhexane, 864(106)
4-ethyl-2,3-dimethylhexane, 865(159)
4-ethyl-2,4-dimethylhexane, 864(132)
4-ethyl-3,3-dimethylhexane, 865(164)

1028 Index of Compounds

3-ethyl-2,3-dimethylpentane, 864(99)
2-ethyl-3,5-dimethylpyridine, 800(80)
4-ethyl-2,6-dimethylpyridine, 800(66)
ethyl 2,4-dinitrobenzoate, 821(31)
ethyl 3,5-dinitrobenzoate, 822(113)
ethyl 3,5-dinitrosalicylate, 822(117)
ethyl diphenate, 821(36)
ethyl diphenylacetate, 821(63)
2-ethyldithiolane, 964(43)
ethylene bromide, 853(67)
ethylene bromohydrin (2-bromoethanol), 716(65)
ethylene chloride (1,2-dichloroethane), 714(33)
ethylene chlorohydrin (2-chloroethanol), 714(33)
ethylene cyanohydrin, 913(46)
ethylenediamine, 750(42)
ethylene glycol (glycol), 720(136)
ethylene glycol bis-2-chloroethyl ether, 825(124)
ethylene glycol diacetate, 817(164)
ethylene glycol dibenzoate, 821(52) (94)
ethylene glycol diethyl ether, 824(58)
ethylene glycol diformate, 816(139)
ethylene glycol dimethyl ether, 824(30)
ethylene glycol dipropionate, 817(200)
ethylene glycol monobutyl ether (2-butoxyethanol), 824(81), 718(100)
ethylene glycol monoethyl ether (2-ethoxyethanol), 714(42), 824(70)
ethylene glycol monoethyl ether acetate ("cellosolve acetate"), 816(110)
ethylene glycol monomethyl ether (2-methoxyethanol), 714(28)
ethylene glycol monomethyl ether acetate ("methylcellosolve acetate"), 815(79)
ethylene glycol monophenyl ether, (2-phenoxyethanol), 720(163), 826(132)
ethylene iodide (1,2-diiodoethane), 852(31), 856(13)
ethylene oxide (epoxyethane), 823(2)
ethylene sulfide, 964(1)
ethylenimine, 750(10)
ethyl ether, 823(5)
ethyl ethoxyacetate, 815(101)

ethyl ethylacetoacetate, 817(176)
α-ethylethylene sulfide, 896(163), 964(6)
ethyl fluoride, 830(2)
ethyl fluoroacetate, 815(53)
ethyl formate, 814(3)
ethyl furoate (pyromucate), 821(21)
N-ethylglycine, 672(26)
ethyl glycolate, 816(115)
2-ethylheptanamide, 745(134)
3-ethylheptane, 864(101)
4-ethylheptane, 864(97)
ethyl heptanoate, 817(158)
5-ethyl-3-heptanone, 895(122)
5-ethyl-4-hepten-3-one, 895(136)
ethyl hexadecyl sulfide, 965(2)
ethyl hexahydrobenzoate, 817(171)
2-ethylhexanal, 730(47)
3-ethylhexane, 863(50)
ethyl hexanoate, 816(122)
2-ethylhexanoic acid (α-ethylcaproic acid), 686(46)
2-ethylhexanol, 718(124)
3-ethyl-3-hexanol, 716(83)
2-ethyl-2-hexenal, 730(50)
2-ethyl-1-hexene, 873(180)
3-ethyl-1-hexene, 872(142)
3-ethyl-2-hexene, 873(181)
3-ethyl-3-hexene, 873(169)
4-ethyl-1-hexene, 872(158)
4-ethyl-2-hexene, 872(157)
2-ethyl-1-hexen-3-one, 894(94)
3-ethyl-5-hexen-2-one, 894(81)
2-ethyl-1-hexyl acetate, 817(178)
2-ethylhexyl acrylate, 818(205)
ethylhexylbarbituric acid, 745(208)
ethyl hexyl ether, 824(71)
2-ethylhexyl vinyl ether, 825(90)
ethyl 3-hydroxybenzoate, 821(93)
ethyl 4-hydroxybenzoate, 822(129), 934(187)
4-ethyl-4-hydroxy-3-hexanone, 895(131)
ethylidene bromide (1,1-dibromoethane), 853(52)
ethylidene chloride (1,1-dichloroethane), 852(18)
ethyl iodide, 838(162)
ethyl iodoacetate, 816(144)
ethyl 2-iodobenzoate, 819(292)
ethylisoamylbarbituric acid, 746(296)
ethyl isobutyl ether, 823(25)
ethyl isobutyrate, 814(41)
ethyl isocyanate, 912(3)
ethyl isocyanide, 912(5)
ethyl isophthalate, 819(300)
ethylisopropylbarbituric acid, 748(438)

1-ethyl-2-isopropylbenzene, 879(46)
1-ethyl-3-isopropylbenzene, 879(42)
1-ethyl-4-isopropylbenzene, 879(58)
ethyl isopropyl ether, 823(10)
ethyl isopropyl sulfide, 962(12)
1-ethylisoquinoline, 801(114)
3-ethylisoquinoline, 801(118)
ethyl isothiocyanate, 961(2)
ethyl 2-ketobutyrate, 894(101)
ethyl lactate, 816(106)
ethyl laurate, 819(288)
ethyl levulinate, 817(185), 896(170)
ethylmalonamide, 748(460)
ethylmalonanilide, 746(288)
ethylmalonic acid, 694(130)
ethyl mandelate, 821(25)
ethyl dl-mandelate, 821(10)
ethyl mercaptan, 966(2)
ethyl methoxyacetate, 815(68)
ethyl 2-methoxybenzoate, 819(278)
ethyl 3-methoxybenzoate, 819(268)
ethylmethylacetic acid, (dl-2-methylbutanoic acid), 684(24)
ethyl methylacetoacetate, 816(146), 895(142)
3-ethyl-2-methylacrylic acid-(trans), 688(3)
ethylmethylamine, 750(5)
N-ethyl-N-methylaniline, 800(82)
1-ethyl-2-methylbenzene (2-ethyltoluene), 877(12)
1-ethyl-3-methylbenzene (3-ethyltoluene), 877(9)
1-ethyl-4-methylbenzene (4-ethyltoluene), 877(10)
2-ethyl-3-methylbutanal, 730(33)
2-ethyl-2-methyl-1-butanol, 716(72)
2-ethyl-2-methylbutanoic acid, 686(36)
2-ethyl-3-methyl-1-butene, 871(90)
ethyl 3-methylbutyl sulfide, 962(35)
ethyl 3-methylbutyrate, 815(70)
dl-2-ethyl-5-methylcyclohexanone, 896(160)
1-ethyl-1-methylcyclopentane, 863(59)
2-ethyl-cis-methylcyclopentane, 863(69)
2-ethyl-trans-1-methylcyclopentane, 863(57)
3-ethyl-cis-1-methylcyclopentane, 863(58)
3-ethyl-trans-1-methylcyclopentane, 863(56)

Index of Compounds 1029

4-ethyl-3-methylcyclopenta-
  none, 895(138)
ethyl methyl ether, 823(1)
3-ethyl-2-methylheptane,
  865(168)
3-ethyl-3-methylheptane,
  865(157)
3-ethyl-4-methylheptane,
  865(173)
4-ethyl-2-methylheptane,
  865(143)
4-ethyl-3-methylheptane,
  865(174)
4-ethyl-4-methylheptane,
  865(172)
5-ethyl-2-methylheptane,
  865(137)
5-ethyl-3-methylheptane,
  865(146)
3-ethyl-2-methylhexane,
  863(89)
3-ethyl-3-methylhexane, 864(96)
4-ethyl-2-methylhexane, 863(79)
4-ethyl-3-methylhexane, 864(92)
2-ethyl-1-methyl-4-isopropyl-
  benzene, 881(106)
3-methyl-1-methyl-4-isopropyl-
  benzene, 881(103)
ethyl methyl ketone (2-
  butanone), 892(4)
ethyl methylmalonate, 817(173)
5-ethyl-1-methylnaphthalene,
  886(5)
6-ethyl-2-methylnaphthalene,
  886(7)
3-ethyl-2-methylpentane,
  862(42)
3-ethyl-3-methylpentane,
  863(49)
2-ethyl-3-methyl-1-pentene,
  872(155)
2-ethyl-4-methyl-1-pentene,
  872(141)
3-ethyl-2-methyl-1-pentene,
  872(137)
3-ethyl-2-methyl-2-pentene,
  873(172)
3-ethyl-3-methyl-1-pentene,
  872(148)
3-ethyl-4-methyl-1-pentene,
  872(131)
3-ethyl-4-methyl-*cis*-2-pentene,
  873(168)
3-ethyl-4-methyl-*trans*-2-
  pentene, 873(165)
2-ethyl-3-methylpyridine,
  800(55)
2-ethyl-4-methylpyridine,
  800(56)
2-ethyl-6-methylpyridine,
  799(41)
3-ethyl-4-methylpyridine,
  800(78)
4-ethyl-2-methylpyridine,
  800(61)

4-ethyl-3-methylpyridine,
  800(74)
5-ethyl-2-methylpyridine,
  800(57)
5-ethyl-3-methylpyridine,
  800(69)
6-ethyl-3-methylpyridine,
  800(53)
ethyl methyl sulfide, 962(2)
2-ethyl-3-methylthiophene,
  964(33)
5-ethyl-2-methylthiophene,
  964(30)
N-ethylmorpholine, 799(11),
  799(22)
ethyl myristate, 820(307)
1-ethylnaphthalene, 884(161)
2-ethylnaphthalene, 884(160)
ethyl 1-naphthoate, 820(313)
ethyl 2-naphthoate, 821(17)
2-ethyl-1-naphthol, 928(79)
3-ethyl-1-naphthol, 926(37)
4-ethyl-1-naphthol, 926(21)
6-ethyl-2-naphthol, 930(138)
N-ethyl-1-naphthylamine,
  760(225)
N-ethyl-2-naphthylamine,
  760(223)
ethyl 1-naphthyl ether, 826(140)
ethyl 2-naphthyl ether, 827(16)
ethyl 1-naphthyl ketone,
  897 221)
ethyl 2-naphthyl ketone,
  901(96)
ethyl 2-naphthyl sulfide, 965(1)
ethyl nitrate, 814(18)
ethyl nitrite, 814(1)
N-ethyl-4-nitroacetanilide,
  745(189)
ethyl 2-nitrobenzoate, 821(13)
ethyl 3-nitrobenzoate, 821(45),
  917(25)
ethyl 4-nitrobenzoate, 821(59),
  918(35)
ethyl 2-nitrocinnamate, 821(38)
ethyl 3-nitrocinnamate,
  822(104)
ethyl 4-nitrocinnamate,
  822(142)
ethyl 4-nitrophenyl sulfide,
  965(13)
ethyl 3-nitrosalicylate, 822(132)
ethyl 5-nitrosalicylate, 822(118)
ethyl nonanoate, 818(225)
3-ethyloctane, 865(178)
4-ethyloctane, 865(177)
ethyl octanoate, 817(194)
ethyl orthoformate, 815(86)
ethyloxalyl chloride, 670(26)
ethyl oxamate, 745(178),
  822(127)
ethyl oxanilate, 744(42),
  821(77)
ethyl palmitate, 821(4)

ethyl pelargonate, 818(225)
2-ethylpentanamide, 745(142)
2-ethylpentananilide, 744(96)
3-ethylpentane (triethyl-
  methane), 862(24)
ethyl pentanoate, 815(89)
2-ethylpentanoic acid, 686(40)
2-ethylpentanol, 716(92)
3-ethyl-3-pentanol (triethylcarbinol), 716(58)
3-ethyl-2-pentanone, 893(53)
2-ethyl-1-pentene, 871(98)
3-ethyl-1-pentene, 871(80)
3-ethyl-2-pentene, 871(103)
3-ethyl-1-pentyn-3-ol, 714(45)
ethyl perfluoroacetate, 859(6)
ethyl perfluorobutyrate, 859(8)
ethyl perfluoropropionate,
  859(7)
1-ethylphenanthrene, 886(22)
2-ethylphenanthrene, 887(26)
9-ethylphenanthrene, 886(23)
2-ethylphenol, 924(9)
3-ethylphenol, 924(14)
4-ethylphenol, 926(28)
ethyl phenoxyacetate, 819(269)
dl-α-ethylphenylacetamide
  (α-phenylbutyramide),
  744(74)
ethyl phenylacetate, 818(227)
dl-α-ethylphenylacetic acid
  (α-phenylbutyric acid),
  688(25)
ethylphenylbarbituric acid,
  747(361)
ethylphenylcarbinol (α-ethylbenzyl alcohol), 720(149)
sym-ethylphenylhydrazine,
  758(180)
unsym-ethylphenylhydrazine,
  758(179)
ethyl phenylpropionate,
  819(264)
ethyl phenyl sulfide, 962(47)
N-ethylpiperidine, 799(17)
2-ethylpiperidine, 752(65)
3-ethylpiperidine, 752(68)
4-ethylpiperidine, 752(70)
ethyl propionate, 814(29)
ethylpropylamine, 750(20)
1-ethyl-2-propylbenzene,
  880(74)
1-ethyl-3-propylbenzene,
  880(68)
1-ethyl-4-propylbenzene,
  880(81)
ethyl propyl ether, 823(16)
ethyl propyl sulfide, 962(16)
2-ethylpyridine, 799(32)
3-ethylpyridine, 799(45)
4-ethylpyridine, 799(46)
ethyl pyromucate (ethyl
  furoate), 821(21)
ethyl pyruvate, 816(107)

1030 Index of Compounds

2-ethylquinoline, 801(108)
4-ethylquinoline, 802(130)
ethyl salicylate, 818(237), 924(18)
ethyl stearate, 821(18)
3-ethylstyrene, 874(232)
4-ethylstyrene, 874(234)
ethyl sulfide, 962(7)
ethyl terephthalate. 821(40)
1-ethyl-1,2,3,4-tetrahydro-
  naphthalene, 883(144)
2-ethyl-1,2,3,4-tetrahydro-
  naphthalene, 883(142)
5-ethyl-1,2,3,4-tetrahydro-
  naphthalene, 883(153)
6-ethyl-1,2,3,4-tetrahydro-
  naphthalene, 883(152)
ethyl thiocyanate, 961(2)
3-ethylthiophene, 964(14)
ethyl m-toluate, 818(224)
ethyl o-toluate, 818(223)
ethyl p-toluate, 818(229)
2-ethyltoluene (1-ethyl-2-
  methylbenzene), 877(12)
3-ethyltoluene (1-ethyl-3-
  methylbenzene), 877(9)
4-ethyltoluene (1-ethyl-4-
  methylbenzene). 877(10)
N-ethyl-m-toluidine, 756(144)
N-ethyl-o-toluidine, 756(129)
N-ethyl-p-toluidine, 756(137)
ethyl p-tolyl sulfide, 963(55)
ethyl trichloroacetate, 816(123),
  821(73)
ethyl trifluoroacetate, 814(48)
ethyl trimethylacetate, 814(48)
2-ethyl-1,3,5-trimethylbenzene,
  881(102)
3-ethyl-1,2,4-trimethylbenzene,
  881(111)
4-ethyl-1,2,3-trimethylbenzene,
  881(116)
5-ethyl-1,2,3-trimethylbenzene,
  881(110)
5-ethyl-1,2,4-trimethylbenzene,
  881(104)
6-ethyl-1,2,4-trimethylbenzene,
  881(105)
3-ethyl-2,2,3-trimethylpentane,
  865(179)
3-ethyl-2,2,4-trimethylpentane,
  864(122)
3-ethyl-2,3,4-trimethylpentane,
  866(183)
ethylurea 744(100)
ethylurethane (ethyl carbamate),
  744(16)
ethyl valerate, 815(89)
3-ethylvinylbenzene, 874(232)
ethyl vinyl ether, 823(7)
ethyl vinyl sulfide, 962(8)
ethynylbenzene (phenylacetylene), 873(207)
ethynyl-1-cyclohexanol, 722(11)
"eucalyptol" (1,8-cineole),
  825(87)

eugenol, 924(27)
eugenol acetate, 821(14)
eugenol methyl ether, 826(131)

d-fenchone, 896(154)
dl-fenchyl alcohol. 722(22)
fluoranthene, 888(68)
fluorene, 888(72)
2-fluoreneamine, 778(360)
fluorene-2-sulfonamide,
  951(165)
fluorene-2-sulfonyl chloride,
  957(206)
9-fluorenone, 902(142)
fluoroacetamide, 745(108)
fluoroacetic acid, 684(13),
  688(10)
fluoroacetonitrile, 913(2)
4-fluoroacetophenone, 896(159)
fluoroacetyl chloride, 670(10)
fluoroacetyl fluoride, 670(3)
2-fluoroaniline, 754(85)
3-fluoroaniline, 754(95)
4-fluoroaniline, 754(96)
2-fluorobenzaldehyde, 730(54)
3-fluorobenzaldehyde, 730(51)
4-fluorobenzaldehyde, 730(55)
4-fluorobenzamide, 746(298)
fluorobenzene, 842(2)
4-fluorobenzenesulfonamide,
  948(24)
4-fluorobenzenesulfonyl
  chloride, 953(16)
2-fluorobenzoic acid, 696(176)
3-fluorobenzoic acid, 696(164)
4-fluorobenzoic acid, 702(299)
4-fluorobenzonitrile, 914(19)
4-fluorobenzophenone, 900(66)
1-fluorobutane, 830(7)
2-fluorobutane, 830(6)
1-fluorodecane. 830(18)
fluoroethane, 830(2)
2-fluoroethanol, 714(12)
2-fluoroethyl acetate, 814(50)
1-fluoroheptane, 830(14)
1-fluorohexane, 830(13)
2-fluorohexane, 830(12)
3-fluorohexane, 830(11)
fluoromethane, 830(1)
2-fluoro-2-methylbutane,
  830(8)
2-fluoro-2-methylpropane,
  830(5)
1-fluoronaphthalene, 844(50)
2-fluoronaphthalene, 846(29)
4-fluoronaphthalene-1-
  sulfonamide, 950(154)
5-fluoronaphthalene-1-
  sulfonamide, 950(143)
6-fluoronaphthalene-2-
  sulfonamide, 948(29)
3-fluoronaphthalene-1-sulfonyl
  chloride, 955(110)
4-fluoronaphthalene-1-sulfonyl
  chloride, 954(86)

5-fluoronaphthalene-1-sulfonyl
  chloride, 956(160)
6-fluoronaphthalene-2-sulfonyl
  chloride, 955(121)
4-fluoro-1-naphthylamine,
  764(79)
3-fluoro-4-nitroaniline, 784(497)
4-fluoro-2-nitroaniline, 774(258)
4-fluoro-3-nitroaniline, 774(286)
6-fluoro-3-nitroaniline, 776(295)
1-fluoro-2-nitrobenzene, 916(24)
1-fluoro-3-nitrobenzene, 916(20)
1-fluoro-4-nitrobenzene, 917(4)
1-fluorononane, 830(17)
1-fluorooctane, 830(16)
1-fluoropentane, 830(10)
2-fluoropentane, 830(9)
1-fluoropropane, 830(4)
2-fluoropropane, 830(3)
α-fluorotoluene, 830(15)
2-fluorotoluene, 842(5)
3-fluorotoluene, 842(6)
4-fluorotoluene, 842(7)
1-fluoroundecane, 830(19)
formaldehyde (methanal),
  728(1)
formamide, 744(6)
formanilide, 744(18)
formic acid, 684(4)
formylpiperidine, 744(9)
β-D-fructose, 809(8)
α-D-furose, 809(20)
α-L-fucose, 809(23)
fumaramide, 748(511)
fumaranide, 748(520)
fumaric acid, 704(331)
fumaryl chloride, 670(31)
3-furaldehyde, 730(39)
furamide, 746(266)
furan, 823(4)
furan-3-carboxylic acid,
  696(161)
furanacrylamide, 747(340)
furanilide, 745(199)
furfural, 730(45)
furfuryl acetate, 816(138)
furfuryl alcohol, 718(102)
1-furfuryl-2-butanone, 899(16)
4-furfuryl-2-butanone, 896(168)
furfurylideneacetone, 899(33)
furil, 905(242)
2-furoic acid (pyromucic acid),
  696(185)
furoin, 904(212)
furonitrile, 913(25)
2-furylacetic acid, 690(63)
2-furylmethylamine, 752(63)
3-furylpropenamide, 747(340)

α-L-galactose, 809(34), 809(37)
α-D-galactose, 810(38)
β-D-galacturonic acid, 809(31)
gallamide, 748(502)
gallanilide, 748(446)

Index of Compounds 1031

gallic acid, 708(420)
β-gentiobiose, 810(42)
geranial (citral), 732(76)
geraniol, 720(156)
geranyl acetate, 818(252)
glucosamine, 809(10)
α-D-glucose, 809(24)
α-L-glucose, 809(26)
β-D-glucose, 809(28)
β-D-glucuronic acid, 809(36)
glutaconic (cis) and (trans) acid, 698(193)
L-glutamamide, 747(335)
DL-glutamic acid, 674(34), 704(328)
L-glutamic acid, 674(46), 706(363)
L-glutamine, 674(28)
glutaramide, 747(366)
glutaranilide, 748(479)
glutaric acid, 692(105)
glutaric anhydride, 668(29)
glutaronitrile, 913(54)
glutaryl chloride, 670(41)
DL-glyceraldehyde, 740(109)
glycerol, 720(170)
glyceryl triacetate, 819(275)
glyceryl tribenzoate, 821(91)
glyceryl tributyrate, 820(314)
glyceryl tripropionate, 819(302)
glyceryl tristearate, 821(89)
glycine, 676(67)
glycogen, 810(49)
glycol (ethylene glycol), 720(136)
glycolic acid, 690(79)
glycolamide, 745(192)
glycolanilide, 745(111)
glycol monophenyl ether (2-phenoxyethanol), 720(163), 826(132)
glycolonitrile, 913(33)
glycylglycine (diglycine), 676(86)
glyoxal, 728(4)
guaiacol (2-methoxyphenol), 926(8)
guaiacol acetate, 818(247)
guaiacol benzoate, 821(58)
guaiacol carbonate (diguaiacyl carbonate), 822(109)
guanylurea, 776(312)

heliotropin (piperonal), 734(13)
hemimellitene (1,2,3-trimethylbenzene), 878(18)
hendecyl alcohol (undecanol), 720(161)
heneicosane, 867(241)
1-heneicosene, 874(259)
heneicosylcyclohexane, 867(252)
heneicosylcyclopentane, 867(243)
1-heneicosyne, 874(262)
hentriacontane, 868(273)

hentriacontene, 875(282)
hentriacontylcyclohexane, 868(283)
hentriacontylcyclopentane, 868(272)
hentriacontyne, 875(287)
1,1,1,2,3,3,3-heptachloropropane, 856(2)
1,1,1,2,2,3,3-heptachloropropane, 856(7)
heptacosane, 867(259)
1-heptacosyne, 865(278)
heptacosylcyclohexane, 868(271)
heptacosylcyclopentane, 867(260)
heptadecanamide, 745(153)
heptadecanal, 734(12)
heptadecane, 867(229)
heptadecanoic acid (margaric acid), 690(52)
heptadecanoic anhydride, 668(32)
heptadecanol, 722(33)
9-heptadecanone, 901(81)
heptadecanonitrile, 914(18)
1-heptadecene, 874(250)
heptadecylamine, 764(81)
heptadecylcyclohexane, 867(239)
heptadecylcyclopentane, 867(231)
1-heptadecyne, 874(252)
2,2,3,3,4,4,4-heptafluorobutanol, 858(3)
2,2,3.3,4,4,4-heptafluorobutylamine, 589(3)
3,3,4,4,5,5,5-heptafluoropentanone-2, 858(11)
heptaldehyde (heptanal), 730(43)
heptanal (enanthaldehyde) (heptaldehyde), 730(43)
heptanamide, 745(108)
heptananilide, 744(47)
heptane, 862(25)
heptanedioic acid, 694(121)
2,4-heptanediol, 720(153)
heptane-3,4-dione, 893(73)
heptane-1-sulfonamide, 948(7)
heptane-1-sulfonyl chloride, 953(5)
heptanethiol, 966(23)
heptane-2,4,6-trione, 900(68)
heptanoic acid, 686(45)
heptanoic anhydride, 668(16)
heptanol (heptyl alcohol), 718(113)
dl-2-heptanol, 716(79)
3-heptanol, 716(75)
4-heptanol, 716(76)
3-heptanone, 894(75)
2-heptanone (amyl methyl ketone), 894(80)
4-heptanone (dipropyl ketone), 893(63)

heptanonitrile (enanthronitrile), 913(34)
heptanoyl chloride (enanthoyl chloride), 670(32)
heptatriacontane, 868(290)
1-heptatriacontene, 875(295)
1-heptatriacontyne, 875(300)
1-heptene, 871(96)
2-heptene(trans), 871(105), 871(108)
2-heptene(cis), 871(107)
3-heptene(trans), 871(101)
3-heptene(cis), 871(102)
dl-1-hepten-3-ol, 716(70)
1-hepten-4-one, 893(71)
2-hepten-4-one, 894(88)
5-hepten-2-one. 894(85)
heptyl acetate, 817(168)
heptyl alcohol (heptanol), 718(113)
heptylamine, 752(69)
heptylbenzene, 883(154)
heptyl bromide, 836(131)
heptyl butyrate, 818(219)
heptyl chloride, 834(61)
heptylcyclohexane, 866(211)
heptylcyclopentane, 866(205)
heptyl ether, 826(137)
heptyl fluoride, 830(14)
heptyl formate, 816(142)
heptyl iodide, 840(181)
1-heptylnaphthalene, 885(183)
2-heptylnaphthalene, 885(184)
heptyl propionate, 817(196)
heptyl sulfide, 963(65)
1-heptyne, 871(109)
hexachloro-1,3-butadiene, 855(127)
hexachlorocyclopentadiene, 855(137)
hexachloroethane, 856(20)
hexachloropropane, 855(126)
1,1,1,4,4,4-hexachloro-2,2,3,3-tetrafluorobutane, 855(124)
hexacosane, 867(257)
hexacosanoic acid, 692(90)
hexacosanol, 724(52)
1-hexacosene, 875(271)
hexacosylcyclohexane, 868(268)
hexacosylcyclopentane, 867(256)
1-hexacosyne, 875(275)
hexadecanal, 734(6)
hexadecanamide, 745(147)
hexadecananilide (palmitanilide), 744(85)
hexadecane (cetane), 860(218)
hexadecane-1-sulfonamide, 948(11)
hexadecane-1-sulfonyl chloride, 953(37)
hexadecanethiol, 967(1)
hexadecanoic acid (palmitic acid), 690(54)
hexadecanoic anhydride, 668(31)

hexadecanol (cetyl alcohol), 722(29)
hexadecanonitrile, 914(14)
hexadecanoyl chloride, 671(55)
1-hexadecene, 874(249)
1-hexadecoxyglycerol, 722(40)
hexadecyl alcohol (cetyl alcohol), 722(29)
hexadecylamine, 764(65)
hexadecyl bromide, 838(145)
hexadecyl chloride, 834(81)
hexadecylcyclohexane, 867(236)
hexadecylcyclopentane, 867(228)
hexadecyl ether, 827(29)
hexadecyl iodide, 840(191)
hexadecyl mercaptan, 967(1)
hexadecyl perfluorobutyrate, 860(24)
hexadecyl sulfide, 965(26)
1-hexadecyne, 874(248)
1,3-hexadienal, 730(52)
1,2-hexadiene, 871(62)
1,3-hexadiene, 871(57)
1,4-hexadiene, 870(42)
1,5-hexadiene, 870(37)
2,3-hexadiene, 870(48)
2,4-hexadiene (dipropenyl), 871(70)
2,4-hexadien-1-ol, 722(13)
1,1,1,4,4,4-hexafluoro-2-chloro-2-butene, 852(10)
1,1,1,5,5,5-hexafluoro-2,4-pentanedione, 892(3)
1,1,1,4,4,4-hexafluoro-2,2,3-trichlorobutane, 853(45)
hexahydrobenzaldehyde (cyclohexanecarboxaldehyde), 730(46)
hexahydrobenzamide, 747(398)
hexahydrobenzanilide, 746(277)
hexahydrobenzoic acid, 688(11)
hexahydrobenzyl alcohol (cyclohexylmethanol), 718(122)
hexahydrobenzylamine, 752(74)
hexahydro-o-cresol (2-methylcyclohexanol, cis or β), 716(91)
hexahydroxybenzene, 940(365)
hexamethylbenzene, 889(98)
hexamethylenediamine, 764(54)
hexamethylene sulfide, 964(37)
hexamethylenetetramine, 806(128)
hexanal (caproaldehyde), 730(32)
hexanamide, 745(123)
hexananilide (capranilide), 744(105)
hexane, 862(10)
hexanedioic acid (adipic acid), 700(236)
1,6-hexanediol, 722(23)
2,3-hexanedione, 892(32), 894(90)

2,5-hexanedione (acetonylacetone), 896(153)
3,4-hexanedione, 893(37)
1,6-hexanedithiol, 966(34)
hexanethiol, 966(17)
2,2′,4,4′,6,6′-hexanitrodiphenylamine, 794(666)
hexanoic acid (caproic acid), 686(38)
hexanoic anhydride, 668(15)
1-hexanol (hexyl alcohol), 716(77)
2-hexanol, 716(50)
3-hexanol, 714(44)
2-hexanone (butyl methyl ketone), 892(33)
3-hexanone, 892(27)
hexanonitrile, 913(28)
hexanoyl chloride, 670(30)
4-hexanoylresorcinol, 926(40)
hexatriacontane, 868(288)
1-hexatriacontene, 875(293)
hexatriacontylcyclohexane, 868(295)
hexatriacontylcyclopentane, 868(287)
1-hexatriacontyne, 875(297)
1,3,5-hexatriene, 871(69)
1-hexene, 870(39)
2-hexene($trans$), 870(47)
2-hexenal, 730(42)
2-hexene(cis), 870(49)
3-hexene($trans$), 870(44)
3-hexene(cis), 870(43)
3-hexene-2,5-dione, 902(131)
2-hexenoic acid, 688(13)
1-hexen-3-ol, 714(41)
4-hexen-1-ol, 714(74)
4-hexen-3-one, 893(54)
5-hexen-2-one, 893(35)
hexyl acetate, 816(130)
hexylal, 667(40)
hexyl alcohol (1-hexanol), 716(77)
hexylamine, 752(54)
hexylbenzene, 882(130)
hexyl bromide, 836(128)
hexyl chloride, 834(58)
hexylcyclohexane, 866(204)
hexylcyclopentane, 866(198)
hexyl ether, 825(122)
hexyl ethyl ether, 824(71)
hexyl fluoride, 830(13)
hexyl formate, 816(103)
hexyl iodide, 838(180)
hexyl mercaptan, 966(17)
hexyl methyl ether, 824(63)
hexyl methyl ketone (2-octanone), 895(121)
1-hexylnaphthalene, 885(180)
2-hexylnaphthalene, 885(181)
hexyl 2-naphthyl ketone, 901(95)
hexyl phenyl ketone, 897(218)
hexyl propionate, 817(163)

4-hexylresorcinol, 928(69)
1-hexyne, 870(54)
3-hexyne, 871(74)
hippuramide, 747(388)
hippuranilide, 748(450)
hippuric acid, 674(29), 704(309), 747(400)
histamine, 672(1)
L-histidine, 678(104)
4-homopyrocatechol, 928(65)
hydantoin, 748(469)
d-hydnohexanamide, 745(176)
hydrastinine, 778(368)
2-hydrazoanisole, 911(9)
hydrazobenzene, 780(405), 911(14)
2,2′-hydrazobiphenyl, 911(22)
4,4′-hydrazobiphenyl, 911(21)
1,1′-hydrazonaphthalene 911(19)
2,2′-hydrazonaphthalene, 911(17)
2-hydrazophenetole, 911(6)
3-hydrazophenetole, 911(12)
4-hydrazophenetole, 911(3)
3-hydrazophenol, 911(18)
2-hydrazotoluene, 911(20)
4-hydrazotoluene, 911(16)
α-hydrinidone (1-indanone), 900(46), 901(89)
hydrocinnamaldehyde, 732(75)
hydrocinnamamide, 745(145)
hydrocinnamanilide, 745(114)
hydrocinnamic acid (3-phenylpropionic acid), 688(33)
hydroquinone, 938(322)
hydroquinone diacetate, 922(135)
hydroquinone dibenzoate, 822(147)
hydroquinone dibenzyl ether, 828(55)
hydroquinone diethyl ether, 827(39)
hydroquinone dimethyl ether, 827(31)
hydroquinone monomethyl ether (4-methoxyphenol), 827(26), 926(41)
hydroxyacetamide, 745(192)
2-hydroxyacetanilide, 748(451), 942(371)
4-hydroxyacetanilide, 747(346), 938(314)
hydroxyacetic acid, 690(79)
2-hydroxyacetophenone, 926(6)
3-hydroxyacetophenone, 902(153), 930(133)
4-hydroxyacetophenone, 903(174), 932(166)
2-hydroxy-4-aminoacetophenone, 780(391)
6-hydroxy-3-aminoacetophenone, 780(384)

Index of Compounds

4-hydroxy-4-aminoazobenzene, 790(588)
4-hydroxy-5-aminonaphthalene-2,7-disulfonic acid, 959(5)
10-hydroxy-2-aminophenanthrene, 790(650)
2-hydroxyaniline (2-aminophenol), 788(560), 950(326)
3-hydroxyaniline (3-aminophenol), 934(209), 780(390)
4-hydroxyaniline (4-aminophenol), 790(582), 940(342)
9-hydroxyanthracene, 934(205)
1-hydroxyanthraquinone, 905(256)
1-hydroxy-9,10-anthraquinone, 940(353)
2-hydroxy-9,10-anthraquinone, 944(434)
2-hydroxyazobenzene, 909(12)
4-hydroxyazobenzene, 910(43)
2-hydroxy-1,1'-azonaphthalene, 910(77)
2-hydroxy-1,2'-azonaphthalene, 910(58)
2-hydroxybenzaldehyde (salicylaldehyde), 732(63), 924(6)
3-hydroxybenzaldehyde, 738(84), 932(152)
4-hydroxybenzaldehyde, 738(95), 934(191)
3-hydroxybenzamide, 747(348)
4-hydroxybenzamide, 747(329)
3,4-dihydroxybenzamide, 748(458)
3-hydroxybenzanilide, 746(318)
4-hydroxybenzanilide, 748(427)
4-hydroxybenzenesulfonamide, 949(92)
4-hydroxybenzenesulfonic acid, 960(70)
2-hydroxybenzoic acid (salicylic acid), 700(251), 938(292)
3-hydroxybenzoic acid, 704(332), 940(366)
4-hydroxybenzoic acid, 706(368), 942(387)
2-hydroxybenzophenone, 900(40), 926(17)
3-hydroxybenzophenone, 903(185)
4-hydroxybenzophenone, 936(246), 904(213)
2-hydroxybenzoyl chloride, 671(56)
2-hydroxybenzyl alcohol (saligenin), 724(54), 930(112)
4-hydroxybenzyl alcohol, 724(64), 934(223)
2-hydroxybenzylaniline, 778(348)
2-hydroxybiphenyl, 928(49)
3-hydroxybiphenyl, 928(93)

4-hydroxybiphenyl (4-phenylphenol), 938(308)
2-hydroxybutanedioic acid (malic acid), 692(112)
3-hydroxy-2-butanone (acetoin), 716(60), 893(64)
4-hydroxybutyronitrile, 913(49)
4-hydroxy-3-chloroaniline, 786(498)
2-hydroxy-2-(4-chlorophenyl)-acetic acid, 694(153)
2-hydroxycinnamic acid, 706(366), 942(385)
4-hydroxycoumarin, 942(376)
2-hydroxycyclohexanecarboxylic acid, 694(129)
1-hydroxydecalin(cis), 722(34), 724(59)
1-hydroxydecalin(trans), 722(28), 722(34)
2-hydroxydecanoic acid, 690(69)
4-hydroxy-3,5-dimethoxybenzaldehyde, 738(93), 932(179)
4-hydroxy-3,5-dimethoxybenzoic acid, 704(344), 942(375)
2-hydroxy-3,5-dimethylaniline, 782(436)
4-hydroxy-2,5-dimethylaniline, 794(662)
2-hydroxy-4,5-dimethylaniline, 788(559)
4-hydroxy-2,6-dimethylaniline, 790(576)
4-hydroxy-3,5-dimethylaniline, 782(447)
6-hydroxy-2,4-dimethylaniline, 788(537)
4-hydroxy-2,6-dimethylbenzene-1,3-disulfonamide, 950(157)
4-hydroxy-2,6-dimethylbenzene-1,3-disulfonyl chloride, 956(149)
hydroxyethanal, 738(78)
2-hydroxyethanethiol, 966(18)
N-(2-hydroxyethyl)ethylenediamine, 758(181)
2-hydroxyethyl p-tolyl sulfide, 963(62)
3-hydroxyindole (indoxyl), 772(230)
4-hydroxyindole, 768(169)
5-hydroxyindole, 776(317), 932(155)
6-hydroxyindole, 774(254)
2-hydroxyisobutyramide, 745(116)
2-hydroxyisobutyranilide, 746(254)
2-hydroxyisobutyric acid, 690(76)
2-hydroxyisobutyronitrile, 913(16)

2-hydroxyisophthalic acid, 708(404)
4-hydroxyisophthalic acid, 710(439)
3-hydroxy-2-ketopropanal, 740(117)
2-hydroxymalonic acid, 692(109)
4-hydroxy-3-methoxyacetophenone, 903(183), 934(187)
3-hydroxy-4-methoxyaniline, 780(403)
4-hydroxy-3-methoxybenzaldehyde (vanillin), 738(64), 930(102)
4-hydroxy-N-methylacetanilide, 748(499)
2-hydroxy-5-methylacetophenone, 900(71), 926(36)
2-hydroxy-3-methylaniline, 772(244)
2-hydroxy-4-methylaniline, 786(534)
2-hydroxy-5-methylaniline, 782(439)
3-hydroxy-4-methylaniline, 786(533)
4-hydroxy-2-methylaniline, 788(572)
4-hydroxy-3-methylaniline, 788(563)
5-hydroxy-2-methylaniline, 764(70), 926(30)
4-hydroxy-3-methylazobenzene, 909(25)
2-hydroxy-3-methylbenzoic acid, 702(270)
2-hydroxy-5-methylbenzoic acid, 700(231)
2-hydroxy-6-methylbenzoic acid, 702(274)
3-hydroxy-2-methylbenzoic acid, 698(217)
4-hydroxy-3-methylbenzoic acid, 702(284)
5-hydroxy-2-methylbenzoic acid, 702(302)
3-hydroxy-3-methyl-2-butanone, 893(57)
2-hydroxy-5-methyl-1,3-diaminobenzene, 784(471)
5-hydroxymethyl-2-furaldehyde, 734(11)
4-hydroxy-4-methyl-2-pentanone (diacetone alcohol), 716(93), 894(112)
2-hydroxy-2-methylpropionanilide, 746(254)
2-hydroxy-2-methylpropionic acid, 690(76)
2-hydroxy-4-methylquinoline, 806(124)
4-hydroxy-2-methylquinoline, 806(126)
1-hydroxy-2-naphthaldehyde, 736(43)

1034 Index of Compounds

2-hydroxy-1-naphthaldehyde, 736(62)
2-hydroxy-3-naphthaldehyde, 738(80), 932(142)
4-hydroxy-1-naphthaldehyde, 740(121), 940(339)
2-hydroxynaphthalene-1,7-disulfonic acid, 960(53)
3-hydroxynaphthalene-1,7-disulfonic acid, 960(54)
4-hydroxynaphthalene-1,5-disulfonic acid, 959(47)
4-hydroxynaphthalene-2,6-disulfonic acid, 959(46)
7-hydroxynaphthalene-1,3-disulfonic acid, 960(55)
3-hydroxynaphthalene-2-sulfonamide, 948(17)
4-hydroxynaphthalene-1 sulfonamide, 949(77)
6-hydroxynaphthalene-2-sulfonamide, 951(212)
8-hydroxynaphthalene-1-sulfonamide, 951(183)
1-hydroxynaphthalene-2-sulfonic acid, 959(43)
2-hydroxynaphthalene-1-sulfonic acid, 959(48)
3-hydroxynaphthalene-2-sulfonic acid, 959(49)
4-hydroxynaphthalene-1-sulfonic acid, 959(44)
5-hydroxynaphthalene-1-sulfonic acid, 959(45)
6-hydroxynaphthalene-2-sulfonic acid, 959(50)
7-hydroxynaphthalene-1-sulfonic acid, 960(52)
7-hydroxynaphthalene-2-sulfonic acid, 960(51)
2-hydroxy-3-naphthamide, 748(468)
4-hydroxy-2-naphthamide, 748(466)
2-hydroxy-3-naphthanilide, 748(501)
1-hydroxy-2-naphthoic acid, 704(323)
2-hydroxy-1-naphthoic acid, 700(243)
3-hydroxy-1-naphthoic acid, 944(416)
3-hydroxy-2-naphthoic acid, 708(383), 942(400)
4-hydroxy-1-naphthoic acid, 702(301)
4-hydroxy-2-naphthoic acid, 702(300), 708(384)
5-hydroxy-1-naphthoic acid, 708(399), 944(410)
5-hydroxy-2-naphthoic acid, 706(360)
6-hydroxy-1-naphthoic acid, 706(357)

6-hydroxy-2-naphthoic acid, 708(413)
7-hydroxy-2-naphthoic acid, 710(427)
8-hydroxy-1-naphthoic acid, 702(275)
8-hydroxy-2-naphthoic acid, 708(393)
5-hydroxy-1,4-naphthoquinone, 905(234), 938(285)
1-hydroxy-2-naphthylamine, 784(489)
1-hydroxy-7-naphthylamine, 786(512)
2-hydroxy-1-naphthylamine, 784(488)
3-hydroxy-1-naphthhylamine, 790(584)
3-hydroxy-2-naphthylamine, 794(658)
6-hydroxy-1-naphthylamine, 790(597)
7-hydroxy-1-naphthylamine, 792(634)
2-hydroxy-5-nitroaniline, 784(459)
2-hydroxy-6-nitroaniline, 792(645)
4-hydroxy-3-nitroaniline, 782(426)
2-hydroxy-4-nitro-5-methyl-aniline, 792(618)
β-hydroxynorvaline, 676(68)
2-hydroxy-4-pentanone, 718(114)
4-hydroxy-2-pentanone, 895(130)
4-hydroxyphenylacetamide, 747(365)
2-hydroxyphenylacetic acid, 698(219)
dl-2-hydroxy-2-phenylacetic acid, 694(149)
4-hydroxyphenylacetic acid, 698(226)
2-hydroxy-p-phenylenediamine, 768(166)
4-hydroxyphenylglycine, 676(79)
3-hydroxyphthalic acid, 700(252), 702(273), 938(298)
4-hydroxyphthalic acid, 704(342), 942(374)
L-hydroxyproline, 678(93)
3-(2-hydroxyphenyl)propenoic acid, 706(366), 942(385)
2-hydroxy-3-phenylpropionic acid, 692(104)
dl-3-hydroxy-2-phenylpropionic acid, 694(148)
1-hydroxy-2-propanone (acetol), 716(61), 893(69)
3-hydroxypropionitrile, 913(43)

4-hydroxypropiophenone, 905(228), 936(275)
4-hydroxypyrazole, 778(375)
2-hydroxypyridine, 806(96), 932(160)
6-hydroxypyrimidine, 806(112)
2-hydroxyquinoline, 806(120), 940(362)
3-hydroxyquinoline, 806(119)
4-hydroxyquinoline, 806(121)
5-hydroxyquinoline, 806(125)
6-hydroxyquinoline, 806(117)
7-hydroxyquinoline, 806(127)
8-hydroxyquinoline, 805(71), 928(89)
hydroxysuccinic acid (l-malic acid), 692(112)
3-hydroxytriphenylene-2-carboxylic acid, 710(438)
β-hydroxyvaline, 674(58)
hypoxanthine (6-hydroxypurine), 784(490)

imidazole-4-carboxylic acid, 710(428)
2,2'-iminodiethanol (bis(2-hydroxyethyl)amine) (diethanolamine), 720(167), 722(10), 760(210), 762(18)
indane, 878(19)
1-indanone (α-hydrindone), 900(46), 901(89)
indazole, 784(474)
indene, 874(231). 878(24)
indole, 766(95)
indole-3-acetonitrile, 914(21)
indole-3-cyanide, 915(108)
indole-3-propionic acid, 696(184)
meso-inositol, 724(76)
β-inulin, 810(39)
1-iodooctadecane, 840(193)
iodoacetamide, 744(102)
4-iodoacetanilide, 747(390)
iodoacetanilide, 746(268)
iodoacetic acid, 690(83)
4-iodophenylacetonitrile, 914(33)
4'-iodo-4-aminobiphenyl, 788(543)
2-iodoaniline, 768(127)
3-iodoaniline, 762(27)
4-iodoaniline, 768(128), 768(162)
2-iodoanisole, 826(130)
4-iodoanisole, 827(25)
2-iodobenzaldehyde, 734(14)
3-iodobenzaldehyde, 736(39)
4-iodobenzaldehyde, 736(61)
2-iodobenzamide, 747(391)
3-iodobenzamide, 747(396)
4-iodobenzamide, 748(464)
2-iodobenzanilide, 746(260)
4-iodobenzanilide, 748(456)

Index of Compounds 1035

iodobenzene, 842(22)
4-iodobenzenesulfonamide, 949(104)
4-iodobenzenesulfonyl chloride, 954(84)
2-iodobenzoic acid, 700(254)
3-iodobenzoic acid, 704(312)
4-iodobenzoic acid, 710(425)
2-iodobenzonitrile, 914(37)
4-iodobenzoyl bromide, 671(82)
4-iodobiphenyl, 848(56)
1-iodobutane, 838(170)
2-iodobutane, 838(167)
1-iodo-2-butene, 838(171)
iodocyclohexane, 838(179)
iodocyclopentane, 838(178)
1-iododecane, 840(184)
1-iodododecane, 840(186)
2-iodo-1,4-diaminobenzene, 844(56)
4-iodo-1,3-dimethylbenzene, 844(57)
1-iodoeicosane, 840(194)
iodoethene, 834(161), 838(162)
2-iodo-1-ethoxybenzene, 826(133)
4-iodo-1-ethoxybenzene, 827(6)
iodoform, 856(18)
1-iodoheptadecane, 840(192)
1-iodoheptane, 840(181)
1-iodohexadecane, 840(191)
1-iodohexane, 838(180)
4-iodo-1-isopropylbenzene, 844(60)
iodomethane, 838(160)
2-iodo-1-methoxybenzene, 826(130)
4-iodo-1-methoxybenzene, 827(25)
2-iodo-4-methylaniline, 764(51)
3-iodo-4-methylaniline, 762(38)
4-iodo-2-methylaniline, 772(240)
1-iodo-2-methylbutane, 838(176)
1-iodo-3-methylbutane, 838(175)
2-iodo-2-methylbutane, 838(169)
1-iodo-2-methylpropane, 838(168)
2-iodo-2-methylpropane, 838(164)
1-iodonaphthalene, 844(68)
2-iodonaphthalene, 846(24)
1-iodononadecane, 840(195)
1-iodonaphthalene-2-sulfonamide, 952(223)
2-iodonaphthalene-1-sulfonamide, 949(58)
4-iodonaphthalene-1-sulfonamide, 950(153)
5-iodonaphthalene-1-sulfonamide, 952(214)
6-iodonaphthalene-1-sulfonamide, 951(164)
6-iodonaphthalene-2-sulfonamide, 951(184)
7-iodonaphthalene-1-sulfonamide, 952(216)
7-iodonaphthalene-2-sulfonamide, 951(161)
8-iodonaphthalene-1-sulfonamide, 950(117)
1-iodonaphthalene-2-sulfonyl chloride, 955(102)
2-iodonaphthalene-1-sulfonyl chloride, 956(135)
4-iodonaphthalene-1-sulfonyl chloride, 956(162)
5-iodonaphthalene-1-sulfonyl chloride, 956(141)
6-iodonaphthalene-1-sulfonyl chloride, 955(95)
6-iodonaphthalene-2-sulfonyl chloride, 957(186)
7-iodonaphthalene-1-sulfonyl chloride, 957(207)
7-iodonaphthalene-2-sulfonyl chloride, 955(122)
8-iodonaphthalene-1-sulfonyl chloride, 956(144)
4-iodo-2-naphthol, 936(131)
3-iodo-1-naphthylamine, 772(228)
2-iodonitrobenzene, 917(27)
3-iodonitrobenzene, 917(14)
4-iodonitrobenzene, 922(152)
4-iodo-2-nitrotoluene, 918(46)
4-iodo-3-nitrotoluene, 918(33)
1-iodononane, 840(183)
1-iodooctane, 840(182)
1-iodopentadecane, 840(189)
1-iodopentane, 838(177)
dl-2-iodopentane, 838(172)
l-2-iodopentane, 838(173)
3-iodopentane, 838(174)
2-iodophenetole, 826(133)
4-iodophenetole, 827(6)
2-iodophenol, 926(24)
3-iodophenol, 926(16)
4-iodophenol, 930(124)
2-iodo-*p*-phenylenediamine, 778(337)
4-iodophenyl phenyl sulfide, 965(7)
1-iodopropane, 838(166)
2-iodopropane, 838(163)
2-iodo-1-propene, 838(165)
3-iodopropionamide, 745(127)
2-iodopropionic acid, 690(82)
3-iodopyridine, 804(43)
2-iodoquinoline, 804(42)
4-iodoquinoline, 806(91)
5-iodoquinoline, 806(93)
6-iodoquinoline, 806(85)
8-iodoquinoline, 804(15)
1-iodotetradecane, 840(188)
4-iodothiophenol, 967(16)

α-iodotoluene, 840(190)
2-iodotoluene, 844(45)
3-iodotoluene, 844(37)
4-iodotoluene, 846(9)
2-iodo-*p*-tolunitrile, 914(39)
1-iodotridecane, 840(187)
5-iodo-1,2,4-trimethylbenzene, 846(12)
1-iodoundecane, 840(185)
isatin, 748(434), 792(624), 906(262)
isoamyl acetate, 815(77)
isoamyl alcohol (3-methylbutanol) (isobutylcarbinol), 714(35)
sec-isoamyl alcohol (*dl*-3-methyl-2-butanol), 714(17)
isoamylamine, 750(30)
N-isoamylaniline, 758(199)
isoamyl benzoate, 819(281)
isoamyl bromide, 836(121)
isoamyl butyrate, 816(141)
isoamyl carbonate, 818(230)
isoamyl chloride 832(44)
isoamyl chloroformate, 816(104)
isoamylcyclohexane, 866(193)
isoamyl ether, 824(83)
isoamyl formate, 815(57)
isoamyl iodide, 838(175)
isoamyl isovalerate, 817(162)
isoamyl 2-naphthyl ether, 827(8)
isoamyl nitrite, 814(27)
isoamyl oxalate, 819(280)
isoamyl phthalate, 820(319)
isoamyl propionate, 816(116)
isoamyl salicylate, 819(293)
isoamyl succinate, 820(308)
isoamylurethane, 744(37)
dl-isoborneol, 724(75)
isobutylacetal, 667(27)
isobutyl acetate, 814(47)
isobutylal, 667(27)
isobutyl alcohol, 714(13)
isobutylamine, 750(16)
N-isobutylaniline, 756(161)
4-isobutylaniline, 758(174)
isobutylbenzene, 878(15)
isobutyl benzoate, 818(249)
isobutyl bromide, 836(111)
isobutyl butyrate, 816(111)
isobutylcarbinol (isoamyl alcohol), 714(34)
isobutyl carbonate, 817(161)
isobutyl chloride, 832(33)
isobutyl chloroformate, 815(65)
isobutyl chloromethyl sulfide, 962(36)
isobutylene bromide, 854(84)
isobutylene glycol (2-methyl-1,2-propanediol), 718(116)
isobutyl ether, 824(61)
isobutyl ethyl sulfide, 962(23)
isobutyl formate, 814(26)

Index of Compounds

isobutyl iodide, 825(168)
isobutyl isobutyrate, 815(95)
isobutyl isocyanate, 912(10)
isobutyl isothiocyanate, 961(8)
isobutyl mercaptan, 966(9)
isobutyl 3-methylbutyrate, 816(129)
isobutyl methyl ketone (4-methyl-2-pentanone), 892(20)
isobutyl methyl sulfide, 962(13)
isobutyl 2-naphthyl ether, 827(11)
isobutyl nitrate, 815(59)
isobutyl oxalate, 818(231)
4-isobutylphenol, 924(20)
isobutyl phenylacetate, 819(263)
isobutyl phenyl ketone, 897(192)
isobutyl propionate, 815(72)
isobutyl salicylate, 924(28)
isobutyl succinate, 819(283)
isobutyl sulfide, 962(38)
2-isobutyltoluene, 879(54)
3-isobutyltoluene, 879(48)
4-isobutyltoluene, 879(55)
isobutylurethane, 744(27)
isobutyl vinyl ether, 823(29)
isobutyraldehyde, 728(7)
isobutyramide, 746(226)
isobutyranilide, 745(143)
isobutyric acid, 684(11)
isobutyric anhydride, 668(5)
isobutyronitrile, 913(8)
isobutyrophenone (isopropyl phenyl ketone), 896(187)
isobutyryl bromide, 671(75)
isobutyryl chloride, 670(13)
isocaproamide, 745(194)
isocaproanilide (4-methylpentananilide), 745(173)
isocaproic acid (4-methylpentanoic acid), 686(40)
isocapronitrile, 913(26)
isocaproyl chloride, 670(29)
isocrotonamide, 745(136)
isocrotonanilide, 745(135)
isocrotonic acid (*cis*-crotonic acid), 684(22)
isodurene (1,2,3,5-tetramethylbenzene), 879(61)
isodurenol (2,3,4,6-tetramethylphenol), 930(98)
isoduridine, 758(201), 762(6)
isoeugenol, 924(29)
isoeugenol methyl ether, 826(138)
DL-isoleucine, 678(106)
L-(+)-isoleucine, 678(103)
isonicotinonitrile (4-cyanopyridine), 915(57)
isopentane (2-methylbutane), 862(4)
isopentanonitrile, 913(9)
isopentyl acetate, 815(77)
N-isopentylaniline, 758(199)

isopentyl benzoate, 819(281)
isopentyl butyrate, 816(141)
isopentyl chloroformate, 816(104)
isopentylcyclohexane, 866(193)
isopentyl ether, 824(83)
isopentyl formate, 815(57)
isopentyl 3-methylbutyrate, 817(162)
isopentyl 1-naphthyl ether, 826(146)
isopentyl nitrite, 814(27)
isopentyl oxalate, 819(280)
isopentyl propionate, 816(116)
isopentyl phthalate, 820(319)
isopentyl salicylate, 819(293), 924(31)
isopentyl succinate, 820(308)
isophorone, 896(180)
isophthalaldehyde, 738(72)
isophthalamide, 748(515)
isophthalic acid, 710(442)
isophthalonitrile, 915(104)
isoprene (2-methyl-1,3-butadiene), 870(15)
isopropenylbenzene (α-methylstyrene), 874(217)
1-isopropoxy-3-methyl-2-butanone, 894(95)
isopropylacetal, 667(12)
isopropyl acetate, 814(20)
4-isopropylacetophenone, 897(215)
N-isopropylacetamide, 744(7)
2-isopropylacrolein, 728(23)
isopropyl alcohol, 714(3)
isopropylamine, 750(4)
N-isopropylaniline, 754(112)
4-isopropylaniline (cumidine), 756(157)
4-isopropylbenzaldehyde, 732(82)
4-isopropylbenzamide, 746(242)
isopropylbenzene (cumene), 877(7)
isopropyl benzoate, 818(21)
2-isopropylbenzoic acid, 690(56)
3-isopropylbenzoic acid, 688(41)
4-isopropylbenzoic acid, 694(145)
α-isopropylbenzyl alcohol (isopropylphenylcarbinol), 720(152)
4-isopropylbenzylamine, 756(159)
4-isopropylbenzyl chloride, 834(73)
isopropyl bromide, 836(102)
1-isopropyl-3-*tert*-butylbenzene, 882(121)
isopropyl chloride, 830(25)
isopropylcyclohexane, 864(119)
isopropylcyclopentane, 863(67)

2-isopropylcyclopentanone, 895(124)
3-isopropyl-2,4-dimethylpentane, 864(130)
isopropyl ether, 823(20)
isopropyl fluoride, 830(3)
isopropyl formate, 814(8)
4-isopropylheptane, 865(140)
isopropyl iodide, 838(163)
isopropyl isobutyrate, 814(51)
isopropyl isocyanate, 812(4)
isopropyl isocyanide, 912(7)
isopropyl isothiocyanate, 961(4)
isopropyl lactate, 816(125)
isopropyl mercaptan, 966(3)
5-isopropyl-2-methylacetophenone, 897(208)
2-isopropyl-3-methyl-1-butene, 872(116)
3-isopropyl-2-methylhexane, 865(155)
isopropyl methyl ketone (3-methyl-2-butanone), 892(8)
4-isopropyl-1-methyl-2-propylbenzene, 882(128)
4-isopropyl-α-methylstyrene, 874(239)
1-isopropylnaphthalene, 884(166)
2-isopropylnaphthalene, 884(162)
5-isopropylnaphthanthracene, 888(53)
8-isopropylnaphthanthracene, 889(81)
9-isopropylnaphthanthracene, 889(79)
12-isopropylnaphthanthracene, 888(55)
isopropyl 2-naphthyl ether, 827(18)
isopropyl 4-nitrophenyl sulfide, 965(14)
isopropyl pentanoate, 816(105)
2-isopropyl-1-pentene, 872(159)
isopropyl perfluorobutyrate, 860(17)
isopropyl perfluoropropionate, 860(16)
1-isopropylphenanthrene, 887(47)
2-isopropylphenanthrene, 886(6)
2-isopropylphenol, 924(11)
4-isopropylphenol, 928(54)
isopropylphenylcarbinol (α-isopropylbenzyl alcohol), 720(152)
isopropyl phenyl ketone (isobutyrophenone), 896(187)
isopropyl phenyl sulfide, 962(48)
3-(4-isopropylphenyl)propenoic, acid, 700(267)
isopropyl phthalate, 820(310)

Index of Compounds

isopropyl propionate, 814(42)
isopropyl propyl ether, 823(28)
isopropyl salicylate, 818(242), 924(22)
4-isopropylstyrene, 874(236)
isopropyl sulfide, 962(18)
isopropyl thiocyanate, 961(3)
2-isopropylthiophene, 964(25)
3-isopropylthiophene, 964(26)
2-isopropyltoluene (2-cymene), 878(21)
3-isopropyltoluene (3-cymene), 878(17)
4-isopropyltoluene (4-cymene), 878(20)
5-isopropyl-1,2,4-trimethyl-benzene, 882(123)
isopropylurea, 746(301)
isopropylurethane, 744(87)
isopropyl valerate, 816(105)
isoquinaldehyde, 736(36)
isoquinoline, 801(103), 804(4)
isosafrole, 826(134)
D- or L-isoserine, 674(37)
DL-isoserine (β-aminolactic acid), 676(78)
isovaleraldehyde, 728(15)
isovaleramide, 746(250)
isovaleranilide, 745(164)
isovaler-p-bromanilide, 746(221)
isovaleric acid, 684(23)
isovaleric anhydride, 668(11)
isovalerone (2,6-dimethyl-4-heptanone) (diisobutyl ketone), 894(108)
isovaleronitrile, 913(19)
isovaleryl bromide, 671(78)
isovaleryl chloride, 670(21)
itaconamide, 747(412)
itaconanilide, 747(408)
itaconic acid, 700(263)
itaconic anhydride, 668(33)

2-ketobutyric acid, 688(15), 899(19)
3-ketobutyric acid, 688(22)
4-ketohexanoic acid, 688(23), 900(42)
5-ketohexanoic acid, 686(53)
2-keto-3-methylbutanoic acid, 688(9)
2-keto-3-(2-nitrophenyl) propionic acid, 696(158)
4-ketopentamethylene sulfide, 901(110)
2-ketopentanedioic acid, 694(147), 903(188)
2-ketopentanoic acid, 684(20), 895(133)
2-ketopropanol, 728(8)
2-ketopropananilide, 745(139)
2-ketopropionamide, 745(200)
2-ketopropionic acid, 684(16), 894(104)
2-ketopropionitrile, 913(6)

dl-lactamide, 744(53)
dl-lactanilide, 744(33)
dl-lactic acid, 686(56)
lactide, 822(138)
dl-lactonitrile, 813(32)
α-lactose (anh.), 810(45)
α-lactose (monohydrate), 810(43)
β-lactose (anh.), 810(46)
lactulose, 809(33)
DL-lanthionine, 678(100)
lanthionine(meso), 678(121)
lauraldehyde, 734(24)
lauramide, 745(168)
lauranilide, 744(60)
lauric acid, 688(12)
lauric anhydride, 668(23)
laurone (12-tricosanone), 902(117)
lauronitrile, 913(53)
lauroyl chloride, 671(53)
lauryl alcohol (dodecyl alcohol), 720(165), 722(4)
lauryl bromide, 838(141)
lauryl chloride, 834(77)
lauryl iodide, 840(186)
lepidine (4-methylquinoline), 801(123)
DL-leucine, 678(109)
D- or L-leucine, 678(110)
levulinamide, 745(157)
levulinanilide, 745(133)
levulinic acid, 688(18)
lignoceric acid (tetracosanoic acid), 692(86)
l-limonene, 874(227)
dl-limonene (dipentene), 874(229)
d-limonene, 874(230)
l-linaloöl, 720(138)
2,4-lutidine (2,4-dimethyl-pyridine), 799(36)
DL-lysine, 680(132)
D- or L-lysine, 676(62)
lyxose, 809(9)
α-D-lyxose, 809(11)

l-malamide, 746(316)
l-malanilide, 748(424)
maleamide, 747(382)
maleanilide, 747(360), 747(401)
maleic acid, 696(179)
maleic anhydride, 668(27)
maleonitrile, 914(15)
l-malic acid (hydroxysuccinic acid), 692(112)
malonamide, 744(17), 747(352)
malonanilide, 746(237), 748(484)
malonic acid (propanedioic acid), 696(189)
malononitrile, 914(13)
β-maltose hydrate, 809(7)
dl-mandelamide, 746(239)
dl-mandelanilide, 746(293)

d-mandelonitrile, 914(11)
dl-mandelic acid, 694(149)
dl-mandelonitrile, 914(4)
d-mannitol, 724(72)
L-mannose, 809(18)
α-D-mannose, 809(19)
margaramide, 745(153)
margaric acid (heptadecanoic acid), 690(52)
margaric anhydride, 668(32)
melezitose dihydrate, 809(29)
β-melibiose dihydrate, 809(2)
4-menthane, 865(148)
d-m-menthane, 865(175)
l-m-menthane, 866(180)
p-menthane(cis), 866(181)
o-menthane, 866(187)
l-menthol, 722(26)
l-menthone, 896(173)
l-menthylamine, 754(113)
4-mercapto-1-naphthol, 967(24)
6-mercapto-2-naphthol, 967(30)
2-mercaptobenzoic acid, 700(266)
4-mercaptobenzoic acid, 706(378)
2-mercaptoethylamine, 776(292)
2-mercaptopropylamine, 768(142)
3-mercaptopropylamine, 778(346)
mesaconamide, 747(372)
mesaconanilide, 747(397)
mesaconic acid, 708(406)
mesaconyl dichloride, 670(9)
mesidine (2,4,6-trimethyl-aniline), 756(166)
mesitol, 928(76)
mesitylenamide, 746(241)
mesitylene (1,3,5-trimethyl-benzene), 877(11)
mesitylenic acid (3,5-dimethyl-benzoic acid), 702(271)
mesityl oxide, 893(36)
methacrylamide, 745(183)
methacrylic acid, 684(17), 688(12)
methanal (formaldehyde), 728(1)
methanesulfonamide, 948(9)
methanesulfonic acid, 959(34)
methanethiol, 966(1)
methanol, 714(1)
DL-methionine, 678(98)
L-methionine, 678(99)
methone (5,5-dimethyldihydro-resorcinol), 905(231)
methoxyacetamide, 745(113)
methoxyacetanilide, 744(32)
4-methoxyacetanilide, 745(215)
methoxyacetic acid, 868(35)
methoxyacetonitrile, 913(15)

1038 Index of Compounds

2-methoxyacetophenone, 897(205)
3-methoxyacetophenone, 897(206)
4-methoxyacetophenone, (899)35
methoxyacetyl chloride, 670(20)
2-methoxyaniline (o-anisidine), 756(156)
3-methoxyaniline, 758(196)
4-methoxyaniline, 766(113)
2-methoxybenzaldehyde, 734(15)
3-methoxybenzaldehyde, 732(80)
4-methoxybenzaldehyde (anisaldehyde), 732(84)
2-methoxybenzamide, 746(224)
4-methoxybenzamide, 747(339)
2-methoxybenzanilide, 746(232)
4-methoxybenzanilide, 747(347)
methoxybenzene, 824(76)
1-(2-methoxybenzeneazo)-2-naphthol, 910(56)
4-methoxybenzene-1,3-disulfonamide, 952(217)
4-methoxybenzene-1,3-disulfonyl chloride, 954(85)
4-methoxybenzil, 901(105)
2-methoxybenzoic acid, 692(111)
3-methoxybenzoic acid, 694(127)
4-methxoybenzoic acid, 702(304)
4-methoxybenzoic anhydride, 668(46)
2-methoxybenzonitrile, 914(7)
4-methoxybenzonitrile, 914(42)
2-methoxybenzophenone, 900(41)
3-methoxybenzophenone, 899(32)
4-methoxybenzophenone, 901(101)
2-methoxybenzoyl chloride, 671(50)
3-methoxybenzoyl chloride, 671(49)
4-methoxybenzoyl chloride, 671(59)
4-methoxybenzyl acetate, 819(277)
2-methoxybenzyl alcohol, 720(164)
4-methoxybenzyl alcohol, 720(166), 722(6)
4-methoxybenzylideneacetone, 902(124)
2-methoxybiphenyl, 827(9)
4-methoxybiphenyl, 828(47)
1-methoxy-2-butanone, 893(40)
4-methoxy-2-butanone, 892(19)
2-methoxycinnamic acid, 704(311)
2-methoxyethanethiol, 966(10)
methoxyethanal, 728(14)
2-methoxyethanol ("methyl-cellosolve") (ethylene glycol monomethyl ether), 714(28)
2-(2-methoxyethoxy)ethanol, 720(132), 825(100)
2-methoxyethyl acetate, 815(79)
2-methoxyethylamine, 750(26)
1-methoxy-3-hexanone, 894(85)
5-methoxyindole, 766(122)
1-methoxy-3-methyl-2-butanone, 893(67)
3-methoxy-2-methyl-1-propanol, 716(71)
3-methoxy-1-naphthaldehyde, 736(47)
4-methoxy-1-naphthaldehyde, 734(9)
5-methoxy-1-naphthaldehyde, 736(49)
1-methoxynaphthalene, 826(140)
2-methoxynaphthalene, 828(42)
2-methoxynaphthalene-1-sulfonamide, 949(66)
3-methoxynaphthalene-2-sulfonamide, 948(19)
4-methoxynaphthalene-2-sulfonamide, 949(63)
5-methoxynaphthalene-1-sulfonamide, 950(139)
6-methoxynaphthalene-1-sulfonamide, 948(50)
6-methoxynaphthalene-2-sulfonamide, 950(127)
7-methoxynaphthalene-2-sulfonamide, 951(181)
2-methoxynaphthalene-1-sulfonyl chloride, 956(156)
3-methoxynaphthalene-2-sulfonyl chloride, 957(183),
4-methoxynaphthalene-1-sulfonyl chloride, 955(115)
4-methoxynaphthalene-2-sulfonyl chloride, 954(69)
5-methoxynaphthalene-1-sulfonyl chloride, 956(151)
6-methoxynaphthalene-1-sulfonyl chloride, 954(76)
6-methoxynaphthalene-2-sulfonyl chloride, 955(100)
7-methoxynaphthalene-2-sulfonyl chloride, 954(81)
2-methoxy-4-nitroaniline, 782(452)
4-methoxy-2-nitroaniline, 780(397)
2-methoxynitrobenzene, 916(35)
3-methoxynitrobenzene, 917(15)
4-methoxynitrobenzene, 918(31)
4-methoxy-3-nitrobenzoic acid, 704(310)
2-methoxy-3-pentanone, 893(46)
2-methoxyphenol (guaiacol), 926(8)
3-methoxyphenol (resorcinol monomethyl ether), 924(23)
4-methoxyphenol (hydroquinone monomethyl ether), 827(26), 926(41)
4-methoxyphenylacetic acid, 692(89)
1-(4-methoxyphenyl)-1-ethanol, 720(171)
1-(4-methoxyphenyl)-1-propene, 826(128), 827(3)
3-(2-methoxyphenyl)propenoic acid(*trans*), 704(311)
4-methoxyphenyl sulfide, 965(15)
5-(4-methoxyphenyl)trithione, 903(177)
2-methoxy-1-propanol, 714(32), 824(65)
1-methoxy-2-propanone, 892(18)
2-methoxy-4-propenylphenol, 924(29)
6-methoxyquinoline, 612(2), 804(6)
7-methoxyquinoline, 828(56)
8-methoxyquinoline, 804(37), 827(23)
2-methoxytoluene, 824(80)
3-methoxytoluene, 825(85)
4-methoxytoluene, 825(84)
N-methylactamide, 744(11)
N-methylacetanilide, 745(131)
methyl acetate, 814(5)
methyl acetoacetate, 816(128), 895(113)
2-methylacetophenone, 896(179)
3-methylacetophenone, 896(185)
4-methylacetophenone, 899(12)
N-methyl-o-acetotoluidide, 744(28)
N-methyl-p-acetotoluidide (aceto-N-methyl-p-toluidide), 744(70)
N-methylacetylanthranilic acid, 704(319)
4-methyl-1-acetylcyclohexane, 896(156)
methyl acrylate, 814(15)
methylacrylonitrile, 913(5)
methylal, 667(1)
dl-N-methyl-α-alanine, 678(107)
N-methyl-β-alanine, 672(4)
methyl alcohol, 714(1)
2-methylallyl methyl sulfide, 962(14)
2-methylallyl sulfide, 962(39)
methylamine, 750(1)

Index of Compounds

methylaminoacetonitrile, 915(91)
4-methylamino-2-nitrotoluene, 764(63)
4-methyl-1-aminopentane, 752(47)
2-methylaminopentane, 750(35)
6-methyl-2-(4-aminophenyl)-benzthiazole, 790(609)
N-methylaniline, 754(100)
2-methylaniline, 754(104)
3-methylaniline, 754(106)
4-methylaniline, 764(61)
methyl anisate, 821(41)
2-methylanisole (o-cresyl methyl ether), 824(80)
9-methylanthracene, 887(39)
methyl anthranilate, 760(204), 762(8), 821(6)
N-methylanthranilic acid, 702(297), 788(573)
2-methylazobenzene, 909(1)
4-methylazobenzene, 909(11)
2-methylbenzaldehyde (o-tolualdehyde), 732(66)
3-methylbenzaldehyde (m-tolualdehyde), 732(65)
4-methylbenzaldehyde (p-tolualdehyde), 732(68)
N-methylbenzamide, 744(67)
1-(2-methylbenzeneazo)-2-naphthol, 909(26)
1-(3-methylbenzeneazo)-2-naphthol, 909(38)
1-(4-methylbenzeneazo)-2-naphthol, 909(34)
2-(2-methylbenzeneazo)-1-naphthol, 910(46)
2-(3-methylbenzeneazo)-1-naphthol, 909(23)
2-(4-methylbenzeneazo)-1-naphthol, 909(41)
4-(2-methylbenzeneazo)-1-naphthol, 910(42)
4-(3-methylbenzeneazo)-1-naphthol, 910(67)
2-methylbenzene-1,3-disulfonamide, 952(230)
2-methylbenzene-1,4-disulfonamide, 951(188)
4-methylbenzene-1,2-disulfonamide, 951(213)
4-methylbenzene-1,3-disulfonamide, 950(115)
5-methylbenzene-1,3-disulfonamide, 951(169)
2-methylbenzene-1,3-disulfonic acid, 959(36)
2-methylbenzene-1,3-disulfonyl chloride, 955(91)
2-methylbenzene-1,4-disulfonyl chloride, 955(113)
4-methylbenzene-1,2-disulfonyl chloride, 956(132)

4-methylbenzene-1,3-disulfonyl chloride, 953(39)
5-methylbenzene-1,3-disulfonyl chloride, 955(99)
2-methylbenzimidazole, 788(569)
methyl benzoate, 817(180)
2-methylbenzoazole, 800(81)
4-methylbenzophenone, 901(93)
4'-methylbenzophenone-2-acrboxylic acid, 698(216)
8-methyl-3,4-benzophenanthrene, 887(25)
2-methyl-1,4-benzoquinone, 902(116)
2-methylbenzothiazole, 801(102)
methyl 2-benzoylbenzoate, 821(50)
2-(4 methylbenzoyl)benzoic acid, 698(216)
1-methyl-3,4-benzphenanthrene, 887(38)
2-methyl-3,4-benzphenanthrene, 887(30)
6-methyl-3,4-benzphenanthrene, 887(44)
7-methyl-3,4-benzphenanthrene, 886(14)
2-methyl-1',2'-benzpyrene, 889(82)
3-methyl-1',2'-benzpyrene, 889(90)
4-methyl-1',2'-benzpyrene, 890(119)
5-methyl-1',2'-benzpyrene, 890(117)
6-methyl-1',2'-benzpyrene, 889(100)
9-methyl-1',2'-benzpyrene, 889(87)
α-methylbenzyl alcohol (methylphenylcarbinol) (1-phenyl ethyl alcohol), 720(139)
2-methylbenzyl alcohol (o-tolylcarbinol), 722(18)
3-methylbenzyl alcohol (m-tolylcarbinol), 720(146)
4-methylbenzyl alcohol (p-tolylcarbinol), 722(37)
N-methylbenzylamine, 754(89)
2-methylbenzylamine, 754(116)
3-methylbenzylamine, 754(115)
4-methylbenzylamine, 754(118)
2-methylbenzyl bromide, 838(147)
4-methylbenzyl bromide, 838(151)
4-methylbenzyl sulfide, 965(37)
methyl bromide, 836(98)
methyl bromoacetate, 815(81)
2-methyl-5-bromobenzenesulfonamide, 949(73)
4-methyl-3-bromobenzenesulfonamide, 948(53)

methyl 2-bromobenzoate, 818(256)
methyl 4-bromobenzoate, 822(106)
methyl 5-bromobenzoate, 821(16)
2-methyl-1,3-butadiene (isoprene), 870(15)
3-methyl-1,2-butadiene, 870(19)
2-methylbutanal (α-methylbutyraldehyde), 728(16)
3-methylbutanal, 728(15)
dl-2-methylbutanamide, 745(174)
3-methylbutanamide, 746(250)
2-methylbutananilide, 745(164) (174)
3-methylbutan-4-bromoanilide, 746(221)
2-methylbutane (isopentane), 862(3)
3-methylbutane-1-sulfonamide, 948(1)
2-methylbutane-1-thiol, 966(9)
dl-2-methylbutanoic acid (ethylmethylacetic acid), 684(19)
3-methylbutanoic acid, 684(18)
3-methylbutanoic anhydride, 668(11)
2-methylbutanol (active amyl alcohol) (sec-butylcarbinol), 714(31)
2-methyl-2-butanol, 714(10)
3-methylbutanol (isoamyl alcohol), 714(35)
dl-3-methyl-2-butanol (sec-isoamyl alcohol) (methylisopropylcarbinol), 714(17)
3-methyl-2-butanone (isopropyl methyl ketone), 892(8)
3-methylbutanoyl bromide, 671(78)
dl-2-methylbutanoyl chloride, 670(22)
3-methylbutanoyl chloride, 670(21)
3-methyl-2-butenal, 730(35)
2-methyl-2-butenamide, 744(54)
2-methyl-2-butenanilide, 744(57)
2-methyl-1-butene, 870(13)
2-methyl-2-butene, 870(18)
3-methyl-1-butene, 870(8)
2-methyl-2-butenoic acid(cis), 690(57)
2-methyl-2-butenoic acid(trans), 688(32)
3-methyl-2-butenoic acid, 690(62)
3-methyl-3-buten-2-ol, 714(15)
3-methyl-3-buten-2-one, 892(9)
3-methyl-3-buten-1-yne, 870(14)

1040 Index of Compounds

2-(3-methylbutoxy)naphthalene, 827(8)
methylbutylamine, 750(27)
d-2-methylbutylamine, 750(31)
3-methylbutylamine, 750(30)
1-methyl-2-butylbenzene, 880(69)
1-methyl-4-sec-butylbenzene, 879(65)
3-methylbutyl phenyl sulfide, 963(58)
3-methylbutyl sulfide, 963(51)
3-methylbutylurethane, 744(37)
2-methyl-3-butyn-2-ol, 714(11)
3-methyl-1-butyne, 870(10)
α-methylbutyraldehyde (2-methylbutanal), 728(16)
methyl butyrate, 814(35)
2-methylbutyronitrile, 913(17)
methyl caprate, 818(222)
methyl caproate, 815(99)
methyl caprylate, 817(170)
methyl carbamate (methylurethane), 744(20)
"methylcarbitol" (diethylene glycol monomethyl ether), 720(132)
methyl carbonate, 814(19)
"methylcellosolve" (2-methoxyethanol), 714(28)
"methylcellosolve acetate" (ethylene glycol monoethyl ether acetate), 815(79)
methyl chloride, 830(20)
methyl chloroacetate, 815(66)
N-methyl-4-chloroaniline, 758(183)
methyl 2-chlorobenzoate, 818(238)
methyl 3-chlorobenzoate, 818(234), 821(2)
methyl 4-chlorobenzoate, 821(39)
methylchloroform (1,1,1-trichloroethane), 852(29)
methyl chloroformate, 814(11)
1-methylchrysene, 888(75), 890(123)
2-methylchrysene, 889(96), 890(120)
3-methylchrysene, 889(99)
4-methylchrysene, 889(92), 890(121)
5-methylchrysene, 888(74), 889(101)
6-methylchrysene, 889(93)
methyl cinnamate, 821(23)
methyl crotonate, 814(49)
methyl cyanoacetate, 817(184), 913(38)
methylcyclohexane, 862(28)
methyl cyclohexanecarboxylate, 816(150)
2-methylcyclohexanol (cis or β) (hexahydro-o-cresol), 716(91)
2-methylcyclohexanol (trans or α), 718(95), 722(3)
3-methylcyclohexanol (cis or β), 718(104)
3-methylcyclohexanol (trans or α), 718(108)
4-methylcyclohexanol (cis or β), 718(105)
4-methylcyclohexanol (trans or α), 718(106)
2-methylcyclohexanone, 894(103)
3-methylcyclohexanone, 894(111)
4-methylcyclohexanone, 895(118)
1-methylcyclohexene, 872(139)
3-methylcyclohexene, 872(117)
4-methylcyclohexene, 872(115)
N-methylcyclohexylamine, 752(64)
methyl cyclohexyl ketone, 895(137)
methylcyclopentane, 862(11)
1-methylcyclopentanol, 714(43)
2-methylcyclopentanone, 893(55)
3-methylcyclopentanone, 893(66)
1-methylcyclopentene, 871(61)
3-methylcyclopentene, 870(41)
4-methylcyclopentene, 871(60)
1-methylcyclopenten-3-one, 894(96)
2-methylcyclopenten-3-one, 894(89)
5-methylcyclopenten-2-one, 893(58)
methyl cyclopentyl ether, 824(47)
methyl cyclopropyl ketone, 892(16)
1-methyl-trans-decahydronaphthalene, 866(208)
9-methyl-cis-decahydronaphthalene, 866(201)
9-methyl-trans-decahydronaphthalene, 866(199)
methyl decanoate, 818(222)
methyl dichloroacetate, 815(78)
methyl dihydrocinnamate, 818(236)
4-methyl-2,6-dinitroaniline, 788(550)
4-methyl-3,5-dinitroaniline, 788(555)
methyl 2,4-dinitrobenzoate, 821(87)
methyl 3,5-dinitrobenzoate, 822(124)
4-methyl-2,6-dinitrobenzoic acid, 708(385)
6-methyl-2,4-dinitrophenol, 930(114)
methyl 3,5-dinitrosalicylate, 822(137)
2-methyl-1,3-dioxane, 667(8)
2-methyl-1,3-dioxolane, 667(3)
methyl diphenate, 821(97)
methyl diphenylacetate, 821(66)
2-methyl-1,4-dithiane, 965(3)
2-methyldithiolane, 964(40)
methyl enanthate (methyl heptanoate), 816(131)
methyleneaminoacetonitrile, 806(108)
methylene bromide, 853(42)
methylene chloride, (dichloromethane), 852(12)
methylenefluorene, 886(11)
methylene iodide, 855(113)
methyl ethoxyacetate, 815(96)
1-methyl-2-vinylbenzene (2-methylstyrene), 874(219)
1-methyl-3-vinylbenzene (3-methylstyrene), 874(222)
1-methyl-4-vinylbenzene (4-methylstyrene), 874(223)
3,4-methylenedioxypropiophenone, 900(39)
methyl ethylacetoacetate, 817(160)
methylethylacetyl chloride, 670(22)
2-methyl-3-ethylcyclopentanone, 895(123)
α-methylethylene sulfide, 964(2)
methyl ethyl ether, 823(1)
2-methyl-4-ethylthiophene, 964(35)
methyl fluoride, 830(1)
methyl formate, 814(2)
methyl fumarate, 822(119)
2-methylfuran (sylvan), 823(17)
5-methylfurfural, 730(59)
methyl furoate (pyromucate), 816(147)
methyl gallate (methyl 3,4,5-trihydroxybenzoate), 942(367)
N-methylglycine (sarosine), 674(48)
methyl glycolate, 815(98)
2-methylheptane, 863(45)
3-methylheptane, 863(52)
4-methylheptane, 863(47)
3-methyl-2,5-heptanedione
methyl heptanoate, 816(131)
4-methylheptanol, 718(123)
3-methyl-3-heptanol, 716(88)
6-methylheptanol, 718(127)
2-methyl-4-heptanone, 894(78)
3-methyl-2-heptanone, 894(98)
6-methyl-2-heptanone, 895(116)
6-methyl-3-heptanone, 894(102)
2-methyl-1-heptene, 873(179)
2-methyl-2-heptene, 873(190)
2-methyl-3-heptene, 872(153)
3-methyl-1-heptene, 872(145)
3-methyl-2-heptene, 873(187)

Index of Compounds 1041

3-methyl-3-heptene, 873(182)
4-methyl-1-heptene, 872(156)
4-methyl-2-heptene, 873(164)
4-methyl-3-heptene, 873(186)
5-methyl-1-heptene, 872(161)
5-methyl-2-heptene, 873(176)
5-methyl-3-heptene, 872(151)
6-methyl-1-heptene, 872(160)
6-methyl-2-heptene, 873(173)
6-methyl-3-heptene, 873(166)
3-methyl-3-hepten-2-one, 895(126)
4-methyl-6-hepten-3-one, 894(87)
methyl hexahydrobenzoate, 816(150)
5-methylhexanal, 730(38)
2-methylhexanamide, 744(48)
4-methylhexanamide, 745(118)
2-methylhexananilide, 745(117)
4-methylhexananilide, 744(56)
2-methylhexane, 862(19)
3-methylhexane, 862(23)
3-methylhexane-2,5-dione, 896(158)
5-methylhexane-2,3-dione, 893(50)
methyl hexanoate, 815(99)
2-methylhexanoic acid, 686(41)
4-methylhexanoic acid, 686(43)
2-methyl-1-hexanol, 716(89)
3-methyl-3-hexanol, 716(53)
dl-4-methyl-1-hexanol, 716(90)
4-methyl-3-hexanol, 716(66)
5-methyl-3-hexanol, 716(62)
3-methyl-2-hexanone, 893(44)
4-methyl-2-hexanone, 893(59)
4-methyl-3-hexanone, 893(45)
5-methyl-2-hexanone, 893(62)
2-methylhexanonitrile, 913(29)
2-methyl-1-hexene, 871(92)
2-methyl-2-hexene, 871(100)
2-methyl-3-hexene, 871(83)
3-methyl-1-hexene, 871(78)
3-methyl-2-hexene, 871(97)
3-methyl-cis-3-hexene, 871(99)
3-methyl-trans-3-hexene, 871(95)
4-methyl-1-hexene, 871(84)
4-methyl-cis-2-hexene, 871(86)
4-methyl-trans-2-hexene, 871(87)
5-methyl-1-hexene, 871(81)
5-methyl-cis-2-hexene, 871(91)
5-methyl-trans-2-hexene, 871(82)
4-methyl-5-hexen-2-one, 893(48)
5-methyl-4-hexen-3-one, 894(76)
5-methyl-5-hexen-2-one, 894(77)
α-methylhydrocinnamic acid, 688(21)
2-methyl-5-hydroxyaniline, 784(466)

2-methyl-6-hydroxyaniline, 784(487)
methyl 3-hydroxybenzoate, 821(86), 928(78)
methyl 4-hydroxybenzoate, 822(140), 936(236)
methyliminodiacetamide, 747(345)
methyliminodiacetic acid, 708(387)
1-methylindane, 878(25)
2-methylindane, 878(34)
1-methylindole, 801(110)
2-methylindole, 766(121)
3-methylindole, 774(269)
4-methylindole, 760(207)
5-methylindole, 766(115)
7-methylindole, 772(232)
methyl iodide, 838(160)
methyl iodoacetate, 816(132)
methyl 2-iodobenzoate, 819(295)
methyl 3-iodobenzoate, 821(56)
methyl 4-iodobenzoate, 822(126)
methylisobutylcarbinol acetate, 815(92)
methyl isobutyrate, 814(21)
methyl isocyanate, 812(1)
methyl isocyanide, 912(2)
methyl isopentanoate, 814(46)
methyl isophthalate, 821(80)
1-methyl-3-isopropenylcyclohexene, 874(226), 874(228)
1-methyl-4-isopropenylcyclohexene, 874(227)(229)(230)
methylisopropylamine, 750(8)
2-methyl-5-isopropylaniline, 758(186)
2-methyl-5-isopropyl-1,4-benzoquinone, 900(57)
trans-1-methyl-4-isopropylcyclohexane, 865(148)
cis-1-methyl-4-isopropylcyclohexane, 866(181)
methylisopropylcarbinol (dl-3-methyl-2-butanol), 714(17)
2-methyl-5-isopropylhydroquinone, 936(261)
2-methyl-5-isopropylphenol, 924(21)
5-methyl-2-isopropylphenol, 926(34)
methyl isopropyl sulfide, 962(5)
methyl isothiocyanate, 961(1)
methyl isovalerate, 814(46)
3-methyl-2-ketobutanoic acid, 899(17)
methyl dl-lactate, 815(82)
methyl laurate, 819(286)
methyl levulinate, 817(166), 896(157)
methylmalonamide, 748(463)
methylmalonic acid, 698(198)

methylmalononitrile, 914(9)
methyl dl-mandelate, 821(53)
methyl mercaptan, 966(1)
methyl 2-methoxybenzoate, 819(265)
methyl 3-methoxybenzoate, 818(241)
2-methyl-3-methoxypropanal, 728(31)
methyl methylacetoacetate, 816(140), 895(129)
methyl methacrylate, 814(28)
2-methyl-6-methylene-2,7-octadiene, 874(218)
N-methylmorpholine, 799(12)
methyl myristate, 820(316)
1-methylnaphthalene, 883(1)
2-methylnaphthalene, 886(4)
2-methyl-1,3-naphthalenediol, 936(256)
2-methyl-1,4-naphthalenediol, 938(319)
2-methylnaphthalene-1-sulfonamide, 948(23)
4-methylnaphthalene-1-sulfonamide, 949(87) (93)
4-methylnaphthalene-2-sulfonamide, 948(44)
5-methylnaphthalene-2-sulfonamide, 950(126)
6-methylnaphthalene, 2-sulfonamide, 950(155)
7-methylnaphthalene-1-sulfonamide, 950(146)
7-methylnaphthalene-2-sulfonamide, 949(69)
8-methylnaphthalene-2-sulfonamide, 948(21)
2-methylnaphthalene-1-sulfonyl chloride, 954(82)
4-methylnaphthalene-1-sulfonyl chloride, 954(79)
4-methylnaphthalene-2-sulfonyl chloride, 956(165)
5-methylnaphthalene-2-sulfonyl chloride, 956(152)
6-methylnaphthalene-2-sulfonyl chloride, 955(114)
7-methylnaphthalene-1-sulfonyl chloride, 955(107)
7-methylnaphthalene-2-sulfonyl chloride, 954(50)
8-methylnaphthalene-2-sulfonyl chloride, 955(89)
methyl 1-naphthoate, 821(49)
methyl 2-naphthoate, 821(99)
1-methyl-2-naphthol, 937(175)
2-methyl-1-naphthol, 928(64)
4-methyl-1-naphthol, 930(108)
4-methyl-2-naphthol, 930(103)
5-methyl-1-naphthol, 930(137)
5-methyl-2-naphthol, 932(153)
7-methyl-1-naphthol, 932(172)
N-methyl-N-naphthylacetamide, 744(99)

N-methyl-1-naphthylamine, 760(217)
N-methyl-2-naphthylamine, 760(222)
1-methyl-2-naphthylamine, 766(91)
2-methyl-1-naphthylamine, 762(25)
3-methyl-1-naphthylamine, 766(92)
3-methyl-2-naphthylamine, 782(440)
4-methyl-1-naphthylamine, 766(93)
4-methyl-2-naphthylamine, 768(167)
5-methyl-1-naphthylamine, 770(202)
5-methyl-2-naphthylamine, 768(141)
7-methyl-1-naphthylamine, 766(114)
8-methyl-1-naphthylamine, 768(161)
methyl 1-naphthyl ether, 826(140)
methyl 2-naphthyl ether, 828(42)
methyl 1-naphthyl ketone, 899(26)
methyl 2-naphthyl sulfide, 965(28)
methyl nicotinate, 821(27)
methyl nitrate, 814(7)
N-methyl-4-nitroacetanilide, 746(295)
2-methyl-4-nitroacetanilide, 747(422)
4-methyl-2-nitroacetanilide, 744(95)
4-methyl-3-nitroacetanilide, 746(284)
2-methyl-3-nitroaniline (6-nitro-o-toluidine), 774(257)
2-methyl-4-nitroaniline (5-nitro-o-toluidine), 782(421)
2-methyl-5-nitroaniline (4-nitro-o-toluidine), 776(322)
2-methyl-6-nitroaniline (3-nitro-o-toluidine), 774(277)
3-methyl-4-nitroaniline (6-nitro-m-toluidine), 782(438)
4-methyl-2-nitroaniline (3-nitro-p-toluidine), 778(357)
4-methyl-3-nitroaniline (2-nitro-p-toluidine), 770(203)
2-methyl-5-nitrobenzenesulfonamide, 950(120)
2-methyl-5-nitrobenzenesulfonyl chloride, 953(24)
methyl 2-nitrobenzoate, 819(291)
methyl 3-nitrobenzoate, 821(102)
methyl 4-nitrobenzoate, 822(115)

2-methyl-2-nitrobutane, 916(10)
3-methyl-1-nitrobutane, 916(14)
methyl 2-nitrocinnamate, 821(96)
methyl 3-nitrocinnamate, 821(136)
methyl 4-nitrocinnamate, 822(144)
2-methyl-3-nitrophenol, 936(272)
2-methyl-4-nitrophenol, 930(134)
2-methyl-5-nitrophenol, 934(197)
2-methyl-6-nitrophenol, 928(77)
3-methyl-4-nitrophenol, 936(233)
4-methyl-3-nitrophenol, 930(97)
5-methyl-2-nitrophenol, 926(43)
methyl 2-nitrophenyl sulfide, 965(30)
methyl 4-nitrophenyl sulfide, 965(35)
2-methyl-1-nitropropane, 916(13)
2-methyl-2-nitropropane, 917(2)
methyl 3-nitrosalicylate, 822(141)
methyl 5-nitrosalicylate, 822(133)
2-methylnonane, 865(171)
3-methylnonane, 865(176)
4-methylnonane, 865(166)
5-methylnonane, 865(165)
methyl nonanoate, 818(206)
2-methyloctane, 864(102)
3-methyloctane, 864(103)
4-methyloctane, 864(100)
methyl octanoate, 817(170)
2-methyl-4-octanone, 894(109)
methyl orthoformate, 814(32)
methyl palmitate, 821(12)
methyl pelargonate, 818(206)
methyl pivalate (trimethylacetic acid), 688(20)
methyl pyromucate (methyl furoate), 816(147)
2-methyl-1,3-pentadiene, 871(63)
2-methyl-1,4-pentadiene, 870(33)
2-methyl-2,3-pentadiene, 870(55)
3-methyl-1,2-pentadiene, 870(51)
3-methyl-1,3-pentadiene, 871(66)
3-methyl-1,4-pentadiene, 870(31)
4-methyl-1,2-pentadiene, 870(52)
4-methyl-1,3-pentadiene, 871(64)
α-methylpentamethylene sulfide, 964(24)

β-methylpentamethylene sulfide, 964(27)
γ-methylpentamethylene sulfide, 964(29)
2-methylpentanal, 728(27)
4-methylpentanal, 728(29)
3-methylpentanamide, 745(201)
4-methylpentanamide, 745(194)
dl-2-methylpentananilide, 744(104)
dl-3-methylpentananilide, 744(81)
4-methylpentananilide, 745(173)
2-methylpentane, 862(8)
3-methylpentane, 862(9)
methyl pentanoate, 815(63)
dl-2-methylpentanoic acid, 686(32)
dl-3-methylpentanoic acid, 686(33)
4-methylpentanoic acid (isocaproic acid), 686(34)
2-methyl-1-pentanol, 716(63)
2-methyl-2-pentanol, 714(25)
2-methyl-3-pentanol, 714(30)
3-methyl-1-pentanol, 716(69)
3-methyl-2-pentanol, 714(40)
3-methyl-3-pentanol, 714(27)
4-methyl-1-pentanol, 716(68)
4-methyl-2-pentanol, 714(34)
2-methyl-3-pentanone, 892(17)
3-methyl-2-pentanone, 892(21)
4-methyl-2-pentanone (isobutyl methyl ketone), 892(20)
2-methylpentanonitrile, 913(26)
4-methylpentanonitrile, 913(27)
4-methylpentanoyl chloride, 670(29)
2-methyl-2-pentenal, 730(37)
2-methyl-1-pentene, 870(38)
2-methyl-2-pentene, 870(29)
3-methyl-1-pentene, 870(30)
3-methyl-cis-2-pentene, 870(53)
3-methyl-trans-2-pentene, 870(46)
4-methyl-1-pentene, 870(29)
4-methyl-cis-2-pentene, 870(35)
4-methyl-trans-2-pentene, 870(36)
2-methyl-1-penten-3-one, 892(23)
3-methyl-4-penten-2-one, 893(51)
4-methyl-3-penten-2-one, 893(36)
2-(4-methylpentyl) acetate, 815(92)
methyl pentyl sulfide, 962(29)
3-methyl-1-pentyn-3-ol, 714(26)
methyl perfluorobutyrate, 859(3)
methyl perfluorocaproate, 859(4)

Index of Compounds

methyl perfluorocaprylate, 859(5)
methyl perfluorohexanoate, 859(4)
methyl perfluorooctanoate, 859(5)
methyl perfluoropropionate, 859(2)
methyl phenoxyacetate, 819(257)
N-methyl-2-phenylacetamide, 744(31)
methyl phenylacetate, 818(214)
2-methyl-1-phenylbutane, 879(60)
2-methyl-2-phenylbutane, 879(43)
3-methyl-1-phenylbutane, 879(62)
3-methyl-2-phenylbutane, 878(35)
methylphenylcarbinol (dl-α-methylbenzyl alcohol), 720(139)
2-methyl-m-phenylenediamine, 776(309)
2-methyl-p-phenylenediamine, 768(144)
3-methyl-o-phenylenediamine, 768(140)
methyl 2-phenylethyl ketone, 897(202)
3-methyl-3-phenylhexane, 882(129)
α-methyl-α-phenylhydrazine, 756(160)
4-methylphenylhydrazine, 768(147)
3-methyl-1-phenyl-5-pyrazolone, 806(107)
3-(2-methylphenyl)propanamide, 745(163)
2-methyl-1-phenyl-1-propanol, 720(152)
N-methyl-3-phenylpropenamide, 745(169)
2-methyl-3-phenylpropionic acid, 688(21)
methylphenylglyoxal, 896(188)
2-methyl-2-phenylhexane, 882(126)
2-methyl-1-phenylpentane, 881(109)
2-methyl-3-phenylpentane, 881(96)
3-methyl-1-phenylpentane, 881(117)
3-methyl-3-phenylpentane, 880(85)
methyl 3-phenylpropionate, 818(236)
3-methyl-1-phenyl-5-pyrazolone, 780(409)
methyl phenyl sulfide, 962(44)
2-methylpiperazine, 768(132), 768(148)

d- or l-2-methylpiperidine, 750(43)
dl-2-methylpiperidine, 752(44)
dl-3-methylpiperidine, 752(46)
2-methylpropanal, 728(7)
2-methyl-1,2-propanediol (isobutylene glycol), 718(116)
2-methylpropane-1-sulfonamide, 948(2)
2-methyl-1-propanol, 714(13)
2-methyl-2-propanol, 714(4), 722(8)
2-methyl-1-propanethiol, 966(6)
2-methyl-2-propenal, 728(9)
2-methylpropene, 870(1)
2-methylpropenoic acid, 684(12), 688(1)
2-methylpropionamide, 746(226)
methyl propionate, 814(14)
2-methylpropionyl bromide, 671(75)
2-(2-methylpropoxy)naphthalene, 827(11)
1-methyl-3-propylbenzene (3-propyltoluene), 878(23)
1-methyl-4-propylbenzene (4-propyltoluene), 878(27)
methyl propyl ether, 823(8)
methyl propyl ketone (2-pentanone), 892(12)
methyl propyl sulfide, 962(9)
N-methylpyrazole, 799(15)
2-methylpyrazine, 799(21)
3-methylpyrazole, 754(119)
4-methylpyrazole, 754(114)
3-methyl-2-pyrazolin-5-one, 792(649)
5-methyl-2-pyrazoline, 754(88)
1-methylpyrene, 889(89)
3-methylpyrene, 887(31)
2-methylpyridine (α-picoline), 799(16)
3-methylpyridine (β-picoline), 799(25)
4-methylpyridine (γ-picoline), 799(27)
methyl 2-pyridyl ketone, 800(72)
methyl 3-pyridyl ketone, 801(97)
methyl 4-pyridyl ketone, 801(95)
4-methylpyrimidine, 799(23)
5-methylpyrimidine, 804(10)
N-methylpyrrole, 799(8)
N-methylpyrrolidine, 799(4)
2-methylpyrrolidine, 750(32)
3-methylpyrrolidine, 750(33)
N-methyl-2-pyrrolidone, 800(83)
methyl pyruvate, 815(74)
2-methylquinoline, 801(111)
3-methylquinoline, 801(121)

4-methylquinoline (lepidine), 801(123)
5-methylquinoline, 801(125)
6-methylquinoline, 801(119)
7-methylquinoline, 801(116), 804(21)
8-methylquinoline, 801(112)
N-methyl-2-quinolone, 805(68)
2-methylquinoxaline, 801(109)
5-methylsalicylaldehyde, 736(38)
methyl salicylate, 818(218), 924(16)
3-methylsalicylic acid, 702(270)
5-methylsalicylic acid, 700(231)
6-methylsalicylic acid, 702(274)
methyl sebacate, 821(26)
methyl stearate, 821(29)
α-methylstyrene (isopropenylbenzene), 874(217)
2-methylstyrene, 874(219)
3-methylstyrene, 874(222)
4-methylstyrene, 874(223)
methylsuccinamide, 748(431)
methylsuccinic acid (pyrotartaric acid), 694(139)
methyl sulfide, 962(1)
methyl terephthalate, 822(143)
2-methyl-4,5,6,7-tetrahydroindole, 756(151)
6-methyl-1,2,3,4-tetrahydroisoquinoline, 758(202)
1-methyl-1,2,3,4-tetrahydronaphthalene, 881(115)
2-methyl-1,2,3,4-tetrahydronaphthalene, 881(113)
5-methyl-1,2,3,4-tetrahydronaphthalene, 883(140)
6-methyl-1,2,3,4-tetrahydronaphthalene, 882(134)
7-methyl-1,2,3,4-tetrahydroquionoline, 760(206)
α-methyltetramethylene sulfide, 964(13)
β-methyltetramethylene sulfide, 964(16)
2-methylthiacyclohexane, 964(24)
3-methylthiacyclohexane, 964(27)
4-methylthiacyclohexane, 964(29)
2-methylthiacyclopentane, 964(13)
3-methylthiacyclopentane, 964(16)
methyl 2-thienyl ketone, 896(178)
methyl 2-thienyl sulfide, 962(41)
methyl thiocyanate, 961(1)
2-methylthiophene, 964(8)
3-methylthiophene, 964(10)
methyl thymyl ether, 825(115)
methyl m-toluate, 818(208)
methyl o-toluate, 818(207)

1044 Index of Compounds

methyl p-toluate, 821(19)
N-methyl-m-toluidine, 754(111)
N-methyl-o-toluidine, 754(117)
N-methyl-p-toluidine, 754(124)
methyl-p-tolylcarbinol (p,α-dimethylbenzyl alcohol), 720(147)
methyl m-tolyl ketone, 896(185)
methyl o-tolyl ketone, 896(179)
methyl p-tolyl sulfide, 963(50)
methyl trichloroacetate, 815(100)
methyl 3,4,5-trihydroxybenzoate (methyl gallate), 942(367)
methyl trimethylacetate (methyl pivalate), 814(33)
α-methyltrimethylene sulfide, 964(7)
methylurea, 745(130)
methylurethane (methyl carbamate), 744(20)
methyl valerate, 815(63)
methylvinyl acetate, 814(24)
m-methylvinylbenzene, 874(222)
o-methylvinylbenzene, 874(219)
p-methylvinylbenzene, 874(223)
methylvinylcarbinol (dl-3-buten-2-ol), 714(5)
methyl vinyl ketone (3-buten-2-one), 892(5)
methyl vinyl sulfide, 962(3)
Michler's ketone, 905(248)
morpholine, 752(50)
mucamide, 747(410), 748(471)
mucanilide, 748(518)
mucic acid, 706(365)
muconamide, 748(498)
muconic acid, 710(432)
myrcene, 874(218)
myristamide, 745(152)
myristanilide, 744(73)
myristic acid (tetradecanoic acid), 690(46)
myristic anhydride, 668(28)
myristonitrile, 914(2)
myristoyl chloride, 671(54)
myristyl alcohol (tetradecanol), 722(21)
myristyl bromide, 838(143)
myristyl chloride, 834(79)
myristyl iodide, 840(188)

1-naphthaldehyde, 734(7)
2-naphthaldehyde, 736(45)
naphthalene, 887(45)
1-naphthaleneacetic acid, 696(181)
2-naphthaleneacetic acid, 698(208)
1,5-naphthalenediamine, 790(596)
naphthalene-1,2-dicarboxylic acid, 702(286)
naphthalene-2,3-dicarboxylic acid, 708(408)

1,2-naphthalenediol, 928(52), 932(163)
1,3-naphthalenediol, 934(218)
1,4-naphthalenediol, 940(351)
1,5-naphthalenediol, 944(417)
1,6-naphthalenediol, 936(251)
1,7-naphthalenediol, 940(332)
1,8-naphthalenediol, 936(259), 938(307)
2,3-naphthalenediol, 938(295)
2,6-naphthalenediol, 942(389), 942(401)
2,7-naphthalenediol, 940(348)
naphthalene-1,4-disulfonamide, 952(238)
naphthalene-1,5-disulfonamide, 952(248)
naphthalene-1,6-disulfonamide, 952(244)
naphthalene-2,6-disulfonamide, 952(246)
naphthalene-2,7-disulfonamide, 952(219)
naphthalene-1,5-disulfonic acid, 959(38)
naphthalene-1,6-disulfonic acid, 959(39)
naphthalene-2,6-disulfonic acid, 959(41)
naphthalene-2,7-disulfonic acid, 959(42)
naphthalene-1,4-disulfonyl chloride, 957(204)
naphthalene-1,5-disulfonyl chloride, 957(215)
naphthalene-1,6-disulfonyl chloride, 956(171)
naphthalene-2,6-disulfonyl chloride, 958(224)
naphthalene-2,7-disulfonyl chloride, 957(202)
1,5-naphthalenedithiol, 967(27)
2,7-naphthalenedithiol, 967(31)
naphthalene-1-sulfonamide, 948(51)
naphthalene-2-sulfonamide, 951(171)
naphthalene-1-sulfonic acid, 959(37)
naphthalene-2-sulfonic acid, 959(40)
naphthalene-1-sulfonyl chloride, 954(56)
naphthalene-2-sulfonyl chloride, 954(71)
1,2,3,4-naphthalenetetraol, 942(404)
1,2,5,8-naphthalenetetraol, 940(338)
1,4,5,8-naphthalenetetraol, 940(349)
2-naphthalenethiol, 967(15)
1,2,4-naphthalenetriol, 938(284)
1,2,6-naphthalenetriol, 940(346)
1,2,7-naphthalenetriol, 940(359)

1,3,6-naphthalenetriol, 930(128)
1,4,6-naphthalenetriol, 936(253)
1,6,7-naphthalenetriol, 940(327)
1,2-naphthalic anhydride, 669(63)
1,8-naphthalic anhydride, 669(71)
2,3-naphthalic anhydride, 669(69)
1-naphthamide, 748(440)
2-naphthamide, 747(411)
1-naphthanilide, 747(330)
2-naphthanilide, 747(355)
1-naphthoic acid, 700(253)
2-naphthoic acid, 702(307)
1-naphthoic anhydride, 669(59)
1-naphthol (α-naphthol), 930(125)
2-naphthol (β-naphthol), 934(215)
1-naphthonitrile, 914(20)
2-naphthonitrile, 914(46)
1,2-naphthoquinone (β-naphthoquinone), 904(223)
1,4-naphthoquinone, 904(198)
1-naphthoxyacetic acid, 704(321)
2-naphthoxyacetic acid, 700(241)
2-naphthoxyacetone, 902(135)
2-naphthoxyethanal, 738(70)
d- or l-2-(1-naphthoxy)propionic acid, 696(171)
d- or l-2-(2-naphthoxy)propionic acid, 694(144)
dl-2-(2-naphthoxy)propionic acid, 694(125)
N-1-naphthylacetamide, 747(322)
N-2-naphthylacetamide, 746(246)
1-naphthylacetamide, 747(380)
2-naphthylacetamide, 748(433)
1-naphthylacetanilide, 746(309)
1-naphthyl acetate, 821(47)
2-naphthyl acetate, 821(88)
α-naphthylacetic acid, 696(181)
β-naphthylacetic acid, 698(208)
1-naphthylacetonitrile, 914(17)
2-naphthylacetonitrile, 915(62)
1-naphthylamine, 766(86)
2-naphthylamine, 778(343)
1-(2-naphthyl)-2-aminoethanol, 778(351)
N-1-naphthylbenzamide, 747(325)
N-2-naphthylbenzamide, 747(327)
1-naphthyl benzoate, 821(60)
2-naphthyl benzoate, 822(123)
2-(2-naphthyl)benzoic acid, 704(318)
1-(1-naphthyl)-1-ethanol,

Index of Compounds 1045

722(42)
1-naphthyl ether, 828(54)
2-naphthyl ether, 828(52)
1-naphthyl isocyanate, 912(18)
2-naphthyl isocyanate, 912(23)
1-naphthyl isothiocyanate, 961(15)
2-naphthyl isothiocyanate, 961(17)
1-naphthyl methyl ketone, 899(26)
2-naphthyl methyl ketone, 901(85)
1-naphthyl phenyl ketone, 902(127)
1-naphthyl phenyl sulfide, 965(12)
2-naphthyl phenyl sulfide, 965(18)
2-naphthyl salicylate, 822(114), 930(131)
1-naphthyl sulfide, 965(44)
2-naphthyl sulfide, 965(49)
neopentane (2,2-dimethylpropane), 862(1)
neopentyl alcohol (2,2-dimethylpropanol), 722(31)
neopentyl bromide, 836(116)
neopentyl chloride, 832(36)
neral (citral), 732(77)
nicotinamide, 746(218)
nicotinanilide, 744(75), 746(238)
l-nicotine, 801(113)
dl-nicotine, 801(104)
nicotinic acid, 708(403)
nicotinic anhydride, 668(52)
nicotinonitrile, 914(33)
ninhydrin, 906(271)
2,2′,2″-nitrilotriethanol, 720(173), 722(1)
3-nitroacenaphthene, 921(143)
5-nitroacenaphthene, 920(103)
2-nitroacetanilide, 744(93)
3-nitroacetanilide, 746(23)
4-nitroacetanilide, 748(461)
3-nitroacetophenone, 902(138)
4-nitroacetophenone, 902(139)
4-nitro-5-aminoacenaphthene, 792(652)
4-nitro-1-aminoanthraquinone, 794(684)
5-nitro-1-aminoanthraquinone, 794(683)
3-nitro-4-aminobenzaldehyde, 790(600)
3-nitro-4-aminobenzophenone, 782(455)
4-nitro-2-aminobiphenyl, 786(527)
3-nitro-2-aminofluorenone, 794(677)
3-nitro-4-aminopyridine, 792(619)

5-nitro-6-aminoquinoline, 788(570)
6-nitro-8-aminoquinoline, 790(606)
8-nitro-2-aminoquinoline, 786(525)
2-nitro-4-aminostilbene, 778(338)
4-nitro-2-aminostilbene, 784(461)
2-nitroaniline, 770(176)
3-nitroaniline, 778(352)
4-nitroaniline, 784(476)
3-nitroanisic acid, 704(310)
2-nitroanisole, 826(141)
3-nitroanisole, 827(17), 917(15)
4-nitroanisole, 827(30), 918(31)
4-nitroanthracene, 921(138)
3-nitroanthranilic acid, 706(351), 792(638)
1-nitroanthraquinone, 922(168)
2-nitroanthraquinone, 922(160)
2-nitroazobenzene, 909(10), 918(55)
3-nitroazobenzene, 909(18), 920(95)
4-nitroazobenzene, 909(31), 921(130)
2-nitroazoxybenzene, 917(26)
4-nitroazoxybenzene, 921(144)
4-nitrobenzal bromide, 852(32), 919(71)
3-nitrobenzal chloride, 852(23), 918(51)
4-nitrobenzal chloride, 917(20)
2-nitrobenzaldehyde, 734(20)
3-nitrobenzaldehyde, 736(41)
4-nitrobenzaldehyde, 738(87)
2-nitrobenzamide, 747(368)
3-nitrobenzamide, 746(267)
4-nitrobenzamide, 748(437)
2-nitrobenzanilide, 746(307)
3-nitrobenzanilide, 746(300)
4-nitrobenzanilide, 748(457)
nitrobenzene, 916(23)
1-(2-nitrobenzeneazo)-2-naphthol, 910(72)
1-(3-nitrobenzeneazo)-2-naphthol, 910(66)
1-(4-nitrobenzeneazo)-2-naphthol, 910(85)
2-(2-nitrobenzeneazo)-1-naphthol, 910(74)
2-(4-nitrobenzeneazo)-1-naphthol, 910(79)
4-(4-nitrobenzeneazo)-1-naphthol, 910(86)
2-nitrobenzenesulfonamide, 950(137)
3-nitrobenzenesulfonamide, 949(74)
4-nitrobenzenesulfonamide, 949(95)
3-nitrobenzenesulfonic acid,

960(66)
2-nitrobenzenesulfonyl chloride, 954(58)
3-nitrobenzenesulfonyl chloride, 954(52)
4-nitrobenzenesulfonyl chloride, 954(75)
2-nitrobenzoic acid, 698(215)
3-nitrobenzoic acid, 698(200)
4-nitrobenzoic acid, 708(407)
2-nitrobenzoic anhydride, 669(56)
3-nitrobenzoic anhydride, 669(61)
4-nitrobenzoic anhydride, 669(64)
2-nitrobenzonitrile, 915(83)
3-nitrobenzonitrile, 920(113) 915(88)
4-nitrobenzonitrile, 921(137), 915(101)
4-nitrobenzoyl bromide, 671(84)
2-nitrobenzoyl chloride, 671(57)
3-nitrobenzoyl chloride, 671(60)
4-nitrobenzoyl chloride, 671(67)
2-nitrobenzyl alcohol, 724(49), 918(61)
3-nitrobenzyl alcohol, 722(9), 917(5)
4-nitrobenzyl alcohol, 724(60) 919(92)
2-nitrobenzyl bromide, 838(155)
3-nitrobenzyl bromide, 838(157), 918(38)
4-nitrobenzyl bromide, 838(159), 920(97)
2-nitrobenzyl chloride, 917(28)
3-nitrobenzyl chloride, 834(92), 917(23)
4-nitrobenzyl chloride, 834(95), 918(57)
4-nitrobenzyl phenyl sulfide, 965(40)
4-nitrobenzyl sulfide, 965(50)
2-nitrobiphenyl, 917(12)
3-nitrobiphenyl, 918(44)
4-nitrobiphenyl, 920(110)
2-nitro-4-bromothiophenol, 967(22)
1-nitrobutane, 916(12)
2-nitrobutane, 916(9)
4-nitro-*tert*-butylbenzene, 916(36)
4-nitrocatechol, 940(330)
2-nitro-4-chloro-1-bromobenzene, 918(58)
2-nitro-4-chlorothiophenol, 967(28)
2-nitrocinnamic acid(*trans*), 708(405)
4-nitrocinnamic acid(*trans*), 710(431)

1046 Index of Compounds

2-nitrocinnamamide, 747(392)
3-nitrocinnamamide, 747(420)
4-nitrocinnamamide, 748(444)
nitrocyclohexane, 916(21)
4-nitro-1,2-diaminobenzene, 790(612)
4-nitro-1,3-diaminobenzene, 786(532)
4-nitro-2,5-dihydroxyaniline, 786(501)
5-nitro-2,4-dihydroxyaniline, 786(531)
2-nitro-1,3-dihydroxybenzene, 919(77)
5-nitro-3,4-dimethoxytoluene, 918(37)
2-nitro-4,6-dimethylaniline, 770(200)
3-nitro-N,N-dimethylaniline, 805(52)
4-nitro-N,N-dimethylaniline, 806(11)
3-nitro-2,6-dimethylpyridine, 804(17)
2-nitrodiphenylamine, 770(190)
nitroethane, 916(4)
2-nitroethanol, 718(131)
2-nitro-1-ethoxybenzene, 826(139)
3-nitro-1-ethoxybenzene, 827(13)
4-nitro-1-ethoxybenzene, 827(32)
3-nitro-N-ethylaniline, 768(125)
4-nitro-N-ethylaniline, 774(273)
2-nitro-1-ethylbenzene, 916(26)
4-nitro-1-ethylbenzene, 916(30)
nitroethylene, 916(1)
2-nitrofluorene, 921(146)
9-nitrofluorene, 922(157)
1,4-nitrofluorobenzene, 917(4)
nitroguanidine, 748(495)
1-nitroheptane, 916(18)
2-nitroheptane, 916(19)
1-nitrohexane, 916(17)
2-nitrohexane, 916(16)
3-nitrohydrazobenzene, 911(4)
2-nitro-4-hydroxyaniline, 786(502)
2-nitro-5-hydroxyaniline, 790(585)
4-nitro-3-hydroxyaniline, 786(535)
3-nitro-4-hydroxy-1-naphthylamine, 786(529)
4-nitroindane, 917(21)
5-nitroindane, 917(17)
3-nitro-4-iodotoluene, 918(33)
2-nitro-4-isopropyltoluene, 916(34)
4-nitromesidine, 770(191)
nitromesitylene, 917(22)
nitromethane, 916(2)
2-nitro-4-methoxyaniline, 780(415)

2-nitro-6-methoxyaniline, 770(199)
3-nitro-4-methoxyaniline, 766(108)
5-nitro-2-methoxyaniline, 778(374)
2-nitro-1-methoxybenzene, 826(141)
3-nitro-1-methoxybenzene, 827(17)
4-nitro-1-methoxybenzene, 827(30)
2-nitro-N-methylaniline, 762(37)
3-nitro-N-methylaniline, 768(163)
4-nitro-N-methylaniline, 784(495)
1-nitro-2-methylnaphthalene, 919(68)
1-nitro-5-methylnaphthalene, 919(72)
2-nitro-1-methylnaphthalene, 918(40)
2-nitro-5-methylnaphthalene, 919(64)
4-nitro-1-methylnaphthalene, 918(56)
8-nitro-1-methylnaphthalene, 918(49)
4-nitro-3-methyl-1-(p-nitrophenyl)pyrazolone-5, 904(192)
2-nitro-3-methylphenol, 926(18)
2-nitro-4-methylphenol, 926(12)
4-nitro-3-methylpyridine, 804(7)
5-nitro-2-methylpyridine, 806(97)
1-nitronaphthalene, 918(42)
2-nitronaphthalene, 919(65)
1-nitronaphthalene-2-sulfonamide, 951(166)
4-nitronaphthalene-1-sulfonamide, 950(123)
4-nitronaphthalene-2-sulfonamide, 951(190)
5-nitronaphthalene-1-sulfonamide, 951(208)
5-nitronaphthalene-2-sulfonamide, 950(124), 950(108)
6-nitronaphthalene-1-sulfonamide, 951(185)
7-nitronaphthalene-1-sulfonamide, 952(232)
8-nitronaphthalene-1-sulfonamide, 950(131)
8-nitronaphthalene-2-sulfonamide, 951(198)
1-nitronaphthalene-2-sulfonic acid, 960(67)
5-nitronaphthalene-1-sulfonic acid, 960(68)
5-nitronaphthalene-2-sulfonic acid, 960(69)

1-nitronaphthalene-2-sulfonyl chloride, 956(154)
4-nitronaphthalene-1-sulfonyl chloride, 955(118)
4-nitronaphthalene-2-sulfonyl chloride, 957(185)
5-nitronaphthalene-1-sulfonyl chloride, 956(139)
5-nitronaphthalene-2-sulfonyl chloride, 956(168), 956(167)
6-nitronaphthalene-1-sulfonyl chloride, 956(170)
7-nitronaphthalene-1-sulfonyl chloride, 957(213)
8-nitronaphthalene-1-sulfonyl chloride, 957(209)
8-nitronaphthalene-2-sulfonyl chloride, 957(212)
4-nitro-1-naphthalenethiol, 967(14)
1-nitro-3-naphthol, 932(148)
2-nitro-1-naphthol, 936(230)
4-nitro-1-naphthol, 938(304)
4-nitro-2-naphthol, 934(204)
5-nitro-1-naphthol, 938(321)
6-nitro-1-naphthol, 940(340)
8-nitro-1-naphthol, 936(235)
8-nitro-2-naphthol, 936(267)
1-nitro-2-naphthonitrile, 915(95)
4-nitro-1-naphthonitrile, 915(94)
5-nitro-1-naphthonitrile, 915(109)
5-nitro-2-naphthonitrile, 915(106)
8-nitro-2-naphthonitrile, 915(98)
1-nitro-2-naphthylamine, 780(404)
2-nitro-1-naphthylamine, 784(465)
3-nitro-1-naphthylamine, 782(445)
4-nitro-2-naphthylamine, 774(280)
5-nitro-1-naphthylamine, 778(377)
5-nitro-2-naphthylamine, 784(464)
8-nitro-1-naphthylamine, 774(278)
8-nitro-2-naphthylamine, 776(305)
1-nitrooctane, 916(22)
1-nitropentane, 916(15)
2-nitropentane, 916(11)
3-nitrophenacyl alcohol, 724(58)
4-nitrophenacyl alcohol, 724(63)
3-nitrophenacyl bromide, 903(154)
4-nitrophenacyl bromide, 903(157)

Index of Compounds 1047

3-nitrophenanthrene, 922(151)
4-nitrophenanthrene, 919(69)
9-nitrophenanthrene, 920(119)
2-nitrophenanthraquinone, 906(275)
3-nitrophenanthraquinone, 906(279)
4-nitrophenanthrenequinone, 905(251)
2-nitrophenetole, 826(139)
3-nitrophenetole, 827(13), 917(11)
4-nitrophenetole, 827(32)
2-nitrophenol, 926(25)
3-nitrophenol, 930(136)
4-nitrophenol, 934(183)
4-nitrophenoxyacetic acid, 704(315)
3-nitrophenylacetamide, 745(167)
4-nitrophenylacetamide, 748(425)
4-nitrophenylacetanilide, 748(426)
2-nitrophenylacetic acid, 698(206)
3-nitrophenylacetic acid, 694(154)
4-nitrophenylacetic acid, 700(233)
4-nitrophenylacetonitrile, 915(86)
4-nitrophenylacetyl chloride, 671(65)
2-nitro-p-phenylenediamine, 782(443)
4-nitro-m-phenylenediamine, 786(532)
4-nitro-o-phenylenediamine, 790(612)
2-nitrophenylhydrazine, 774(250)
3-nitrophenylhydrazine, 774(263)
4-nitrophenylhydrazine, 786(516)
2-nitrophenyl isocyanate, 912(20)
3-nitrophenyl isocyanate, 912(22)
4-nitrophenyl isocyanate, 912(24)
2-nitrophenyl isothiocyanate, 961(18)
3-nitrophenyl isothiocyanate, 961(16)
4-nitrophenyl isothiocyanate, 961(19)
3-(2-nitrophenyl)-2-ketopropionic acid, 696(158), 903(191)
2-nitrophenyl phenyl sulfide, 965(38)
4-nitrophenyl phenyl sulfide, 965(23)
3-(2-nitrophenyl)propenamide, 747(392)
3-(3-nitrophenyl)propenamide, 747(420)
3-(4-nitrophenyl)propenamide, 748(444)
3-(2-nitrophenyl)propenoic acid, 708(405)
3-(3-nitrophenyl)propenoic acid, 704(343)
3-(4-nitrophenyl)propenoic acid, 710(431)
4-nitrophenylpropynoic acid, 702(293)
2-nitrophenyl thiocyanate, 961(11)
3-nitrophthalamide, 748(435)
4-nitrophthalamide, 748(430)
3-nitrophthalanilide, 748(492)
4-nitrophthalanilide, 747(413)
3-nitrophthalic acid, 706(374)
4-nitrophthalic acid, 700(165)
3-nitrophthalic anhydride, 669(62)
4-nitrophthalic anhydride, 669(49)
3-nitrophthaloyl chloride, 671(68)
1-nitropropane, 916(8)
2-nitropropane, 916(5)
3-nitropropene, 916(6)
3-nitropyridine, 804(25)
5-nitroquinoline, 805(67)
6-nitroquinoline, 806(110), 921(142)
7-nitroquinoline, 806(109)
8-nitroquinoline, 919(83)
2-nitroresorcinol, 919(77)
4-nitroresorcinol, 934(211)
4-nitrosalicylaldehyde, 740(103)
5-nitrosalicylaldehyde, 740(101), 934(225)
6-nitrosalicylaldehyde, 736(35)
3-nitrosalicylamide, 746(273)
5-nitrosalicylamide, 748(481)
5-nitrosalicylanilide, 748(480)
3-nitrosalicylic acid, 696(167), 698(224)
5-nitrosalicylic acid, 708(394), 942(406)
N-nitrosoacetanilide, 923(2)
4-nitrosoaniline, 923(31)
5-nitroso-o-anisidine, 923(20)
2-nitrosoanisole, 923(19)
3-nitrosoanisole, 923(6)
1-nitrosoanthraquinone, 923(35)
nitrosobenzene, 923(10)
2-nitrosobenzoic acid, 923(34)
4-nitroso-N-benzyl-N-ethylaniline, 923(8)
N-nitrosobromoacetanilide, 923(4)
N-nitroso-4-bromophenylacetamide, 923(9)
N-nitroso-2-chlorophenylacetamide, 923(5)
N-nitroso-4-chlorophenylacetamide, 923(8)
5-nitroso-o-cresol, 923(24)
6-nitroso-o-cresol, 923(27)
4-nitroso-N,N-diethylaniline, 923(13)
4-nitroso-N,N-dimethylaniline, 805(78), 923(15)
1-nitroso-2,3-dimethylbenzene, 923(14)
1-nitroso-2,4-dimethylbenzene, 923(3)
2-nitroso-1,4-dimethylbenzene, 923(18)
2-nitroso-1,3-dimethylbenzene, 923(26)
4-nitroso-1,2-dimethylbenzene, 923(4)
4-nitrosodiphenylamine, 923(25)
N-nitrosodiphenylamine, 923(6)
1-nitroso-4-ethoxybenzene, 923(1)
4-nitroso-N-ethylaniline, 923(12)
4-nitroso-N-ethyl-N-methylaniline, 923(9)
N-nitrosoformanilide, 923(1)
1-nitroso-2-methoxybenzene, 923(19)
1-nitroso-3-methoxybenzene, 923(6)
1-nitroso-4-methoxybenzene, 923(2)
4-nitroso-N-methylaniline, 923(22)
1-nitrosonaphthalene, 923(17)
1-nitroso-2-naphthol, 923(21), 932(167)
2-nitroso-1-naphthol, 923(28)
4-nitroso-1-naphthol, 923(33)
N-nitroso-4-nitrophenylacetamide, 923(7)
4-nitrosophenol, 923(23)
N-nitrosopropionanilide, 923(3)
3-nitrosopyridine, 923(16)
nitrosothymol, 923(29)
2-nitrosotoluene, 923(11)
3-nitrosotoluene, 923(7)
4-nitrosotoluene, 923(5)
β-nitrostyrene, 918(39)
3-nitro-1,2,4,5-tetramethylbenzene, 920(109)
2-nitrothiophenol, 967(10)
3-nitrothiophenol, 967(12)
4-nitrothiophenol, 967(13)
2-nitrotoluene, 916(25)
3-nitrotoluene, 916(29)
4-nitrotoluene, 917(29)
3-nitrotoluene-α-sulfonamide, 949(67)

1048 Index of Compounds

3-nitrotoluene-α-sulfonyl chloride, 955(119)
3-nitro-*p*-toluidine (4-methyl-2-nitroaniline), 778(357)
6-nitro-*o*-toluidine (2-methyl-3-nitroaniline), 774(257)
2-nitro-*p*-toluidine (4-methyl-3-nitroaniline), 770(203)
4-nitro-*o*-toluidine (2-methyl-5-nitroaniline), 776(322)
6-nitro-*m*-toluidine (3-methyl-4-nitroaniline), 782(438)
5-nitro-*o*-toluidine (2-methyl-4-nitroaniline), 782(421)
4-nitro-*o*-toluidine, 776(322)
3-nitro-*o*-toluidine (2-methyl-6-nitroaniline), 774(277)
2-nitro-*m*-tolunitrile, 915(58)
2-nitro-*p*-tolunitrile, 915(78)
3-nitro-*o*-tolunitrile, 915(82)
3-nitro-*p*-tolunitrile, 915(70)
4-nitro-*m*-tolunitrile, 915(65)
4-nitro-*o*-tolunitrile, 915(75)
5-nitro-*m*-tolunitrile, 915(74)
6-nitro-*m*-tolunitrile, 915(55)
6-nitro-*o*-tolunitrile, 914(51)
3-nitro-1,2,4-trimethylbenzene, 917(7)
3-nitro-2,4,6-trimethylpyridine, 804(19)
nitrourea, 747(323)
2-nitro-*m*-xylene (2,6-dimethylnitrobenzene), 916(27)
2-nitro-*p*-xylene (2,5-dimethylnitrobenzene), 916(31)
3-nitro-*o*-xylene (2,3-dimethylnitrobenzene), 916(33)
4-nitro-*m*-xylene, 916(32)
4-nitro-*o*-xylene (3,4-dimethylnitrobenzene), 917(6)
nonacosane, 868(267)
nonacosanoic acid, 692(95)
1-nonacosene, 875(277)
nonacosylcyclohexane, 868(277)
nonacosylcyclopentane, 868(266)
1-nonacosyne, 875(283)
nonadecane, 867(235)
nonadecanoic acid, 690(67)
nonadecanol, 722(38)
2-nonadecanone, 901(87)
nonadecanonitrile, 914(26)
1-nonadecene, 874(253)
nonadecylcyclohexane, 867(246)
nonadecylcyclopentane, 867(237)
1-nonadecyne, 874(258)
nonanal (pelargonaldehyde), 730(61)
nonanamide, 745(119)
nonananilide, 744(30)
nonane, 864(112)
nonanedioic acid, 694(123)
2,8-nonanedione, 900(69)
nonanethiol, 966(33)

nonanoic acid (pelargonic acid), 686(49)
1-nonanol (nonyl alcohol), 720(144)
2-nonanol, 720(133)
2-nonanone, 895(155)
3-nonanone, 895(147)
4-nonanone, 895(146)
5-nonanone (dibutyl ketone), 895(145)
nonanonitrile, 913(45)
nonanoyl chloride (pelargonyl chloride), 670(39)
nonatriacontane, 868(294)
nonatriacontene, 875(299)
nonatriacontyne, 875(303)
2-nonenal, 732(78)
1-nonene, 873(210)
nonyl alcohol (1-nonanol), 720(144)
nonylamine, 754(105)
nonyl bromide, 836(138)
nonyl chloride, 834(69)
nonylcyclohexane, 866(217)
nonylcyclopentane, 866(213)
nonyl fluoride, 830(17)
nonyl iodide, 840(183)
nonyl mercaptan, 966(33)
1-nonylnaphthalene, 885(189)
2-nonylnaphthalene, 885(190)
1-nonyne, 873(212)
DL-norleucine, 678(115)
L-(+)- or D-(—)-norleucine, 678(117)
DL-norvaline, 678(119), 680(133)
L-(+) or D-(—)-norvaline, 680(126)

octachloronaphthalene, 850(70)
octacosane, 868(264)
octacosanoic acid, 692(94)
octacosylcyclohexane, 868(274)
octacosylcyclopentane, 868(263)
1-octacosyne, 875(276), 875(279)
octadecanal, 734(16)
octadecanamide, 745(162)
octadecananilide, 744(101)
octadecane, 867(232)
octadecanoic acid (stearic acid), 690(68)
octadecanoic anhydride, 668(35)
octadecanol (stearyl alcohol), 722(36)
octadecanonitrile, 914(23)
octadecanoyl chloride, 671(58)
1-octadecene, 874(251)
cis-9-octadecenol (oleyl alcohol), 720(172)
trans-9-octadecenol (elaidyl alcohol), 722(19)

octadecyl acetate, 821(15)
octadecylamine, 764(65)
octadecyl bromide, 838(148)
octadecyl chloride, 834(85)
octadecylcyclohexane, 867(242)
octadecylcyclopentane, 867(234)
octadecyl iodide, 840(191)
octadecyl perfluorobutyrate, 860(25)
octadecyl sulfide, 965(33)
1-octadecyne, 874(255)
octafluoroadipic acid, 857(20)
octanal (caprylaldehyde), 730(49)
octanamide, 745(166)
octane, 863(66)
octanedioic acid (suberic acid), 698(210)
2,3-octanedione, 895(120)
2,7-octanedione, 900(54)
octanethiol, 966(28)
octanoic acid (caprylic acid), 686(48)
octanoic anhydride, 668(17)
1-octanol (octyl alcohol), 720(134)
2-octanol, 718(119)
2-octanone (hexyl methyl ketone), 895(121)
3-octanone, 894(105)
4-octanone, 895(114)
octanonitrile, 913(40)
octanoyl chloride, 670(35)
octatriacontane, 868(292)
1-octatriacontene, 875(298)
1-octatriacontyne, 875(302)
1-octene, 873(183)
2-octene(*cis*), 873(194)
2-octene(*trans*), 873(193)
3-octene(*cis*), 873(191)
3-octene(*trans*), 873(192)
4-octene(*cis*), 873(189)
4-octene(*trans*), 873(188)
octyl acetate, 817(195)
octyl alcohol (octanol), 720(134)
octylamine, 754(87)
octyl bromide, 836(133)
octyl butyrate, 819(255)
octyl chloride, 834(64)
octylcyclohexane, 866(214)
octylcyclopentane, 866(210)
octyl 3,5-dinitrobenzoate, 821(67)
octyl ether, 826(144)
octyl fluoride, 830(16)
octyl formate, 817(177)
octyl iodide, 840(182)
octyl mercaptan, 966(28)
1-octylnaphthalene, 885(186)
2-octylnaphthalene, 885(187)
octyl perfluorobutyrate, 860(22)
octyl propionate, 821(228)
octyl sulfide, 963(68), 965(25)

Index of Compounds 1049

1-octyne, 873(195)
oleamide, 744(52)
oleanilide, 744(13)
oleic acid, 686(55)
oleic anhydride, 668(18)
oleyl alcohol (cis-9-octadecenol), 720(172)
orcinol (3,5-dihydroxytoluene), 928(47), 932(164)
DL-ornithine, 680(131)
L-ornithine, 672(10)
oxalic acid, 692(113)
1,9-oxalylanthracene, 906(278)
oxalyl chloride, 670(8)
oxamic acid, 706(358)
oxamide, 748(526)
oxanilic acid, 698(225)
oxanilide, 748(506)
oxydiethanoic acid, 698(222)

palmitaldehyde, 734(6)
palmitamide, 745(147)
palmitanilide, 744(85)
palmitic acid (hexadecanoic acid), 690(54)
palmitic anhydride, 668(31)
palmitonitrile, 914(14)
palmitoyl chloride, 671(55)
papaverine, 784(477)
paraldehyde, 728(30)
pelargonaldehyde (nonanal), 730(61)
pelargonamide, 745(122)
pelargonanilide, 744(30)
pelargonic acid (nonanoic acid), 686(55)
pelargonyl chloride, 670(39)
pentabromobenzene, 850(65)
pentabromophenol, 942(405)
pentachlorobenzaldehyde, 740(123)
pentachlorobenzene, 848(47)
pentachloroethane, 854(96)
1,1,1,3,3-pentachloro-2,2,4,4,4-pentafluorobutane, 854(103)
pentachlorophenol, 940(347)
α,α,α,3,4-pentachlorotoluene, 856(4)
pentacosane, 867(253)
pentacosanoic acid, 692(85)
1-pentacosene, 875(266)
1-pentacosyne, 875(274)
pentacosylcyclohexane, 868(265)
pentacosylcyclopentane, 867(254)
pentadecanal, 734(5)
pentadecanamide, 745(132)
pentadecananilide, 744(59)
pentadecane, 866(215)
pentadecanoic acid, 688(43)
pentadecanol, 722(25)
8-pentadecanone, 900(49)
pentadecanonitrile, 914(5)
1-pentadecene, 874(246)

pentadecylamine, 762(34)
pentadecylcyclohexane, 867(233)
pentadecylcyclopentane, 867(227)
1-pentadecyne, 874(245)
1,2-pentadiene, 870(27)
1,3-pentadiene (piperylene), 870(24)
1-cis-3-pentadiene, 870(25)
1-trans-3-pentadiene, 870(23)
1,4-pentadiene, 870(9)
2,3-pentadiene, 870(28)
1,3-pentadiene-1-carboxylic acid (cis, trans), 688(16)
1,3-pentadiene-1-carboxylic acid (trans, trans), 696(188)
pentaerythritol, 724(78)
2,2,3,3,3-pentafluoropropanol, 858(2)
2,2,3,3,3-pentafluoropropylamine, 859(2)
pentahydroxycyclohexane (d-quercitol), 724(77)
pentamethylaniline, 784(494)
pentamethylbenzene, 886(13)
pentamethylene glycol, 720(160)
pentamethylene sulfide, 964(19)
2,2,3,3,4-pentamethylpentane, 865(169)
2,2,3,4,4-pentamethylpentane, 865(136)
pentanal (valeraldehyde), 728(18)
pentanamide, 745(150)
pentananilide, 744(36)
pentane, 862(4)
pentanedial, 730(60)
pentanedioic acid, 692(105)
1,5-pentanediol, 720(160)
2,4-pentanedione (acetylacetone), 893(56)
pentanedioyl chloride, 670(41)
1,5-pentanedithiol, 966(32)
pentanethiol, 966(15)
pentanoic acid (valeric acid), 684(24)
pentanoic anhydride, 668(13)
1-pentanol (amyl alcohol), 716(49)
dl-2-pentanol (sec-amyl alcohol), 714(23)
3-pentanol (diethylcarbinol), 714(20)
2-pentanone (methyl propyl ketone), 892(12)
3-pentanone (diethyl ketone), 892(11)
pentanonitrile, 913(23)
pentanoyl chloride, 670(24)
pentatriacontane, 868(285)
pentatriacontanone, 902(147)
1-pentatriacontene, 875(291)
pentatriacontylcyclohexane, 868(293)

pentatriacontylcyclopentane, 868(284)
1-pentatriacontyne, 875(296)
1-pentene (amylene), 870(12)
cis-2-pentene, 870(17)
trans-2-pentene, 870(16)
3-pentenoic acid, 684(26)
1-penten-3-ol, 714(19)
2-penten-1-ol, 716(56)
3-penten-2-ol, 714(38)
4-penten-1-ol, 716(51)
1-penten-3-one, 890(13)
3-penten-2-one, 890(26)
3-pentenonitrile, 913(21)
2-pentoxynaphthalene, 827(5)
n-pentyl acetate, 815(97)
tert-pentyl acetate, 815(58)
pentylamine, 750(34)
pentylbenzene (amylbenzene), 880(83)
1-pentyl bromide, 836(123)
2-pentyl bromide (2-bromopentane), 836(119)
3-pentyl bromide (3-bromopentane), 836(120)
pentyl butyrate, 816(153)
pentyl caproate, 818(221)
pentyl chloride, 832(45)
pentylcyclohexane, 866(197)
pentylcyclopentane, 866(190)
pentyl ether, 825(93)
n-pentyl ethyl ether, 824(57)
tert-pentyl ethyl ether, 823(42)
pentyl fluoride, 830(10)
pentyl formate, 815(67)
pentyl hexanoate, 818(221)
pentyl iodide, 838(177)
pentyl isobutyrate, 816(108)
2-(3-pentyl)malonic acid, 690(48)
pentyl mercaptan, 966(15)
n-pentyl methyl ether, 823(41)
tert-pentyl methyl ether, 823(32)
1-pentylnaphthalene, 884(177)
2-pentylnaphthalene, 885(178)
pentyl pentanoate, 817(188)
4-pentylphenol, 924(26), 926(3)
pentyl phenyl ketone, 899(8)
4-tert-pentylphenol, 930(123)
pentyl propionate, 816(126)
3-pentyltoluene, 882(122)
pentyl valerate, 817(188)
1-pentyne, 870(20)
2-pentyne, 870(34)
2-pentynoic acid, 688(40)
perchloroethane, 856(20)
perchloroethylene, 853(62)
perchloropropylene, 855(126)
perfluoroacetaldehyde, 858(1)
perfluoroacetic acid, 857(1)
perfluoroacetic anhydride, 857(1)
perfluoroacetone, 858(4)

1050   Index of Compounds

perfluoroacrylic acid, 857(17)
perfluoroadipic acid, 857(20)
perfluoro-1,3-butadiene, 861(8)
perfluorobutane, 860(4)
perfluorobutanedioic acid, 857(19)
perfluorobutanedioic anhydride, 857(6)
perfluorobutanone, 858(5)
perfluoro-1-butene, 861(6)
perfluoro-2-butene, 861(6)
perfluorobutyl perfluoromethyl ether, 860(6)
perfluoro-2-butyne, 861(4)
perfluorobutyraldehyde, 858(3)
perfluorobutyric acid, 857(4)
perfluorobutyric anhydride, 857(3)
perfluorobutyryl bromide, 858(9)
perfluorobutyryl chloride, 858(8)
perfluorobutyryl fluoride, 858(7)
perfluorobutyryl iodide, 858(10)
perfluorocapric acid, 857(12)
perfluorocaproic acid, 857(7)
perfluorocaproic anhydride, 857(5)
perfluorocaproyl chloride, 858(12)
perfluorocaprylic acid, 857(10)
perfluorocyclobutane, 861(2)
perfluorocyclobutene, 861(5)
perfluoro-1,3-cyclohexadiene, 861(12)
perfluoro-1,4-cyclohexadiene, 861(14)
perfluorocyclohexane, 861(4)
perfluorocyclohexanecarboxylic acid, 857(8)
perfluorocyclohexyl perfluoromethyl ether, 860(8)
perfluorocyclopentane, 861(3)
perfluorocyclopentanone, 858(7)
perfluorocyclopropane, 861(1)
perfluorodecalin, 861(14)
perfluorodecalin(*cis*), 854(77)
perfluorodecalin(*trans*), 854(75)
perfluorodecane, 860(12)
perfluorodecanoic acid, 857(12)
perfluoro-2,2'-diethoxydiethyl ether, 860(9)
perfluorodiethyl ether, 860(2)
perfluorodihexyl ether, 860(12)
perfluoro-1,2-dimethoxyethane, 860(3)
perfluoro-2,3-dimethyl-2-butene, 861(11)
perfluoro-1,2-dimethylcyclohexane, 861(11)
perfluoro-1,3-dimethylcyclohexane, 861(10)
perfluoro-1,4-dimethylcyclohexane, 861(8)
perfluorodimethylcyclopentane, 861(5)
perfluorodimethyl ether, 860(1)
perfluorodipentyl ether, 860(11)
perfluorodipropyl ether, 860(7)
perfluorododecane, 860(14)
perfluorododecanoic acid, 857(14)
perfluoroethane, 860(2)
perfluoroethene, 861(1)
perfluoroethylcyclohexane, 861(9)
perfluoroethylcyclopentane, 861(6)
perfluoroethylene, 861(1)
perfluoroglutaric anhydride, 857(7)
perfluoroheptane, 860(9)
perfluoroheptanoic acid, 857(9)
perfluoroheptanone-4, 858(12)
perfluoro-1-heptene, 861(17)
perfluorohexadecane, 860(16)
perfluorohexane, 860(7)
perfluorohexanedioic acid, 857(20)
perfluorohexanoic acid, 857(7)
perfluorohexanoic anhydride, 857(5)
perfluorohexanone-3, 858(10)
perfluorohexanoyl chloride, 858(12)
perfluoro-1,3,5-hexatriene, 861(15)
perfluoro-1-hexene, 861(13)
perfluoroindane, 861(12)
perfluoroisobutyric acid, 857(3)
perfluoroisovaleric acid, 857(5)
perfluorolauric acid, 857(14)
perfluoro-2-methyl-1,3-butadiene, 861(9)
perfluoromethane, 860(1)
perfluoro-2-methylbutane, 860(5)
perfluoro-3-methylbutanoic acid, 857(5)
perfluoromethylcyclohexane, 861(7)
perfluoro-4-methyl-1-cyclohexene, 861(16)
perfluoro-2-methylpentane, 860(8)
perfluoromethyl perfluorobutyl ether, 860(6)
perfluoromethyl perfluorocyclohexyl ether, 860(8)
perfluorononane, 860(11)
perfluorononanoic acid, 857(11)
perfluoro-1-nonene, 861(20)
perfluorooctane, 860(10)
perfluorooctanoic acid, 857(10)
perfluorooctanol, 858(13)
perfluoro-1,4,7-octatriene, 861(18)
perfluoro-1-octene, 861(19)
perfluoropentadecanoic acid, 857(16)
perfluoropentanal, 858(9)
perfluoropentane, 860(6)
perfluoropentanedioic anhydride, 857(7)
perfluoropentanoic acid, 857(6)
perfluoropentanoic anhydride, 857(4)
perfluoropentanone-2, 858(8)
perfluoropentanoyl chloride, 858(11)
perfluoro-1,2-propadiene, 861(2)
perfluoropropane, 860(3)
perfluoropropene, 861(3)
perfluoropropionaldehyde, 858(2)
perfluoropropionic acid, 857(2)
perfluoropropionic anhydride, 857(2)
perfluoropropionyl chloride, 858(6)
perfluoropropionyl fluoride, 858(5)
perfluoropropylene, 861(3)
perfluorosuccinic acid, 857(19)
perfluorosuccinic anhydride, 857(6)
perfluorotetradecanoic acid, 857(15)
perfluorotetrahydrofuran, 860(4)
perfluorotetrahydropyran, 860(5)
perfluorotributylamine, 859(7)
perfluorotridecane, 860(15)
perfluorotriethylamine, 859(4)
perfluorotrihexylamine, 859(8)
perfluorotrimethylamine, 859(4)
perfluoro-1,3,5-trimethylcyclohexane, 861(13)
perfluorotripropylamine 859(5)
perfluoroundecane, 860(13)
perfluoroundecanoic acid, 857(13)
perfluorovaleric acid, 857(6)
perfluorovaleric anhydride, 857(4)
perfluorovaleryl chloride, 858(11)
perylene, 890(128)
phenacetin, 746(249)
phenacyl alcohol, 724(55), 902(146)

Index of Compounds 1051

phenacyl bromide (ω-bromo-acetophenone), 900(73)
phenacyl chloride (ω-chloroacetophenone), 901(91)
9,10-phenanthraquinone, 906(264)
phenanthrene, 888(58)
phenanthrene-1-carboxaldehyde, 738(91)
phenanthrene-9-carboxaldehyde, 738(83)
1,2-phenanthrenediol, 940(333)
2,5-phenanthrenediol, 940(336)
2,6-phenanthrenediol, 942(408)
2,8-phenanthrenediol, 942(372)
3,4-phenanthrenediol, 936(262)
3,6-phenanthrenediol, 942(398)
3,8-phenanthrenediol, 944(414)
9,10-phenanthrenediol, 938(276)
phenanthrene-2-sulfonamide, 952(226)
phenanthrene-3-sulfonamide, 950(130)
phenanthrene-9-sulfonamide, 950(138)
phenanthrene-2-sulfonic acid, 960(79)
phenanthrene-3-sulfonic acid, 960(80)
phenanthrene-9-sulfonic acid, 960(81)
phenanthrene-2-sulfonyl chloride, 956(199)
phenanthrene-3-sulfonyl chloride, 956(136)
phenanthrene-9-sulfonyl chloride, 956(169)
3,4,5-phenanthrenetriol, 938(278)
2-phenanthrol, 938(316)
9-phenanthrol, 938(282)
9-phenanthrylmethylamine, 776(324)
phenazine, 806(114)
phenethyl alcohol (2-phenylethyl), 720(150)
phenethyl bromide (β-bromoethylbenzene), 836(136)
phenethyl chloride (β-chloroethylbenzene), 834(66)
2-phenethyl 1-naphthyl ether, 827(40)
m-phenetidine, 758(192)
o-phenetidine, 756(167)
p-phenetidine, 758(193)
phenetole, 824(82)
p-phenetylurea (dulcin), 747(358)
phenol, 924(2), 926(19)
phenolphthalein, 944(419)
phenol-4-sulfonamide, 949(92)
phenol-4-sulfonic acid, 960(70)
phenothiazine, 790(583)
phenoxyacetaldehyde, 732(72)
phenoxyacetamide, 745(129)

phenoxyacetanilide, 745(121)
phenoxyacetic acid, 692(108)
2-phenoxybenzoic acid, 694(134)
2-phenoxychloroethane, 834(83)
phenoxyethanal, 732(72)
2-phenoxyethanol (glycol monophenyl ether) (ethylene glycol monophenyl ether), 720(163), 826(132)
phenoxymethyl phenyl ketone, 902(125)
1-phenoxynaphthalene, 827(28)
5-phenoxy-2-pentanone, 901(74)
1-phenoxy-2-propanone, 897(198)
2-phenoxypropionic acid, 694(140)
3-phenoxypropionic acid, 692(106)
phenylacetaldehyde, 734(8)
phenylacetamide, 746(315)
phenylacetanilide, 745(186)
phenyl acetate, 817(174)
phenylacetic acid, 690(74)
phenylacetic anhydride, 668(37)
phenylacetonitrile, 913(48)
phenylacetyl chloride, 670(38)
phenylacetylene (ethynylbenzene), 873(207)
DL-phenylalanine, 676(88), 710(424)
D- or L-phenylalanine, 678(102)
β-phenylalanineamide, 746(258)
1-phenyl-1-aminobutane, 756(143)
dl-1-phenyl-1-aminopropane, 754(120)
dl-1-phenyl-2-aminopropane, 754(109), 754(101)
2-phenyl-1-aminopropane, 754(125)
4-phenylazobenzoyl chloride, 671(71)
phenylbenzene, 880(71), 887(29)
phenyl benzoate, 821(81)
2-phenylbenzoic acid, 694(133)
4-phenylbenzoyl chloride, 671(72)
2-phenylbutane (sec-butylbenzene), 878(16)
2-phenylbutanedioic acid, 702(272)
1-phenyl-1,3-butanedione, 901(99)
1-phenyl-1-butanol, 722(27)
1-phenyl-2-butanone (benzyl ethyl ketone), 897(193)
4-phenyl-2-butanone, 897(202), 899(11)
1-phenyl-2-butene, 874(225)
4-phenyl-3-buten-2-one (benzalacetone), 900(44)

dl-2-phenylbutyramide (dl-α-ethylphenylacetamide), 744(74)
4-phenylbutyramide, 744(72)
phenyl butyrate, 717(226)
2-phenylbutyric acid (dl-α-ethylphenylacetic) 688(25)
4-phenylbutyric acid, 688(42)
phenyl chloroformate, 670(33)
phenyl cinnamate, 821(92)
2-phenylcyclohexanone, 901(104)
3-phenylcyclohexanone, 897(219)
2-phenylcyclopentanone, 899(31)
1-phenyldecane, 884(176)
2-phenyl-1,3-dithiane, 865(34)
1-phenyldodecane, 885(182)
m-phenylenediamine, 768(136)
o-phenylenediamine, 776(296)
p-phenylenediamine, 784(458)
m-phenylenedithiol, 967(4)
o-phenylenedithiol, 967(6)
p-phenylenedithiol, 967(19)
phenylethanal, 730(62), 734(8)
1-phenyl-1,2-ethanediol, 722(44)
2-phenylethane-1-sulfonamide, 948(22)
2-phenylethane-1-sulfonyl chloride, 953(12)
1-phenylethane-1-thiol, 966(11)
2-phenylethane-1-thiol, 966(29)
1-phenyl-1-ethanol, 720(139), 722(2)
2-phenyl-1-ethanol, 720(150)
phenyl ether, 826(136), 827(7)
1-phenylethyl acetate, 818(216)
2-phenylethyl acetate, 818(235)
1-phenylethyl alcohol (α-methylbenzyl alcohol), 720(139), 722(2)
2-phenylethyl alcohol (phenethyl alcohol), 720(150)
dl-α-phenylethylamine (α-aminoethylbenzene), 754(93)
β-phenylethylamine (β-aminoethylbenzene), 754(102)
α-phenylethyl bromide (α-bromoethylbenzene), 836(134)
β-phenylethyl bromide (phenethyl bromide), 836(136)
α-phenylethyl chloride (α-chloroethylbenzene), 834(65)
β-phenylethyl chloride (phenethyl chloride), 834(66)
2-phenylethyl cinnamate, 821(62)
3-phenyl-3-ethylhexane, 883(149)

1052 Index of Compounds

2-phenylethyl sulfide, 965(43)
phenyl ethynyl ketone, 901(77)
9-phenylfluorene, 990(88)
phenylglycine, 672(8)
phenylglyoxal, 738(74)
2-phenylheptane, 882(136)
3-phenylheptane, 882(131)
4-phenylheptane, 882(125)
1-phenyl-1-heptanol, 720(168)
1-phenyl-1-hepten-3-one, 900(37)
2-phenylhexane, 881(107)
3-phenylhexane, 881(101)
6-phenyl-5-hexen-2-one, 901(115)
phenylhydrazine, 758(188), 762(1)
phenyl isocyanate, 912(12)
phenyl isocyanide, 912(13)
phenyl isopropyl ether, 825(89)
1-phenylisoquinoline, 806(89)
phenyl isothiocyanate, 961(10)
4-phenyl-4-ketobutanoic acid, 903(187)
4-phenyl-2-ketobutanoic acid, 688(35), 900(67)
3-phenyl-2-ketopropanoic acid, 700(242), 905(235)
phenylmalonic acid, 700(237)
phenylmalononitrile, 914(50)
phenyl mercaptan, 966(22)
phenylmethanethiol, 966(25)
1-phenyl-2-methylbutane, 879(49)
1-phenyl-3-methylbutane, 879(56)
1-phenyl-2-methyl-2-propanol, 772(5)
2-phenylnaphthalene, 888(63)
4-phenyl-1-naphthol, 936(255)
N-phenyl-1-naphthylamine, 768(131)
N-phenyl-2-naphthylamine, 776(326)
phenylnitromethane, 916(28)
1-phenylnonane, 884(173)
1-phenyloctane, 884(167)
2-phenylpentane, 879(45)
3-phenylpentane, 879(40)
5-phenylpentanoic acid, 690(50)
1-phenyl-3-pentanone, 897(207)
phenyl phenylethyl ketone, 902(123)
phenyl perfluoroamyl ketone, 859(50)
phenyl perfluorobutyl ketone, 859(4)
phenyl perfluoroethyl ketone, 859(2)
phenyl perfluoromethyl ketone, 859(1)
phenyl perfluoropropyl ketone, 859(3)
4-phenylphenacyl bromide, 904(196)

4-phenylphenacyl chloride, 904(201)
2-phenylphenol (2-hydroxybiphenyl), 928(49)
3-phenylphenol (3-hydroxybiphenyl), 928(93)
4-phenylphenol (4-hydroxybiphenyl), 938(308)
phenyl phthalate, 921(85)
N-phenylphthalimide, 748(445)
2-phenylpropanal, 732(67)
3-phenylpropanal, 732(75)
3-phenylpropananilide, 745(114)
1-phenyl-1-propanol, 720(145)
2-phenyl-2-propanol, 722(20)
3-phenylpropanol, 720(159)
2-phenylpropenal, 732(85)
3-phenylpropenamide, 746(282)
3-phenylpropenanilide, 746(291)
2-phenylpropenoic acid, 694(122)
3-phenylpropenoic acid(*trans*), 696(186)
3-phenyl-2-propen-1-ol, 722(14)
3-phenylpropenonitrile, 914(3)
phenylpropiolamide, 745(125)
phenylpropiolanilide, 746(219)
phenylpropiolic acid, 698(195)
2-phenylpropionamide, 744(89)
3-phenylpropionamide, 745(145)
phenyl propionate, 817(198), 821(1)
2-phenylpropionic acid, 686(51)
3-phenylpropionic acid (hydrocinnamic acid), 688(33)
2-phenylpropionitrile, 913(47)
3-phenylpropionitrile, 913(52)
α-phenylpropyl alcohol (α-ethylbenzyl alcohol), 720(149)
3-phenylpropylamine, 756(150)
phenyl propyl ether, 825(96)
phenyl propyl ketone, 896(184), 897(200)
phenyl propyl sulfide, 963(54)
3-phenylpropyl sulfide, 965(36)
phenylpropynamide, 745(125)
phenylpropynanilide, 746(219)
phenylpropynoic acid, 698(195)
2-phenylpyridine, 802(127)
phenyl 2-pyridyl ketone, 897(222), 900(53)
2-phenylquinoline, 805(80)
6-phenylquinoline, 806(99)
phenyl salicylate, 821(33), 926(20)
1-phenylsemicarbazide, 788(557)
phenyl stearate, 821(151)
dl-phenylsuccinamide, 748(456)
dl-phenylsuccinamide(α), 746(319)
dl-phenylsuccinamide(β), 746(274)
dl-phenylsuccinanilide, 747(353), 748(474)

dl-phenylsuccinanilide(α), 747(364)
dl-phenylsuccinic acid, 702(272)
N-phenylsuccinimide, 746(313)
phenyl sulfide, 963(64)
1-phenyltetradecane, 885(188)
phenyl thiocyanate, 961(7)
5-phenyl-3-thiophanone, 901(84)
4-phenyl-3-thiosemicarbazide, 784(457)
phenyl *p*-tolyl ketone, 901(82)
phenyl *m*-tolyl sulfide, 963(66)
phenyl *o*-tolyl sulfide, 963(67)
phenyl *p*-tolyl sulfide, 963(69)
1-phenyltridecane, 885(185)
phenyl 2,4,6-trimethylphenyl ketone, 904(215)
4-phenyltrithione, 904(194)
5-phenyltrithione, 904(202)
1-phenylundecane, 885(179)
phenyl undecyl ketone, 900(63)
phenylurea, 746(278)
N-phenylurethane, 744(22)
phloroglucinol (anh.), 934(193)
phloroglucinol (dihydrate), 942(394)
phloroglucinol triacetate, 822(121)
phloroglucinol tribenzoate, 822(146)
phorone, 896(161), 899(15)
phthalaldehyde, 736(37)
phthalamic acid (monophthalamide), 746(286)
phthalamide (di), 748(470)
phthalamide (mono) (phthalamic acid), 746(286)
phthalanilide, 748(505)
phthalic acid, 706(347)
phthalic anhydride, 669(54)
phthalide, 821(95)
phthalimide, 748(490)
phthalonamide (α), 747(375)
phthalonamide (β), 746(310)
phthalonanilide, 747(369), 748(449)
phthalonic acid, 698(218)
phthalonitrile, 915(97)
phthaloyl chloride, 671(52)
picene, 890(129)
α-picoline (2-methylpyridine), 799(16)
β-picoline (3-methylpyridine), 799(25)
γ-picoline (4-methylpyridine), 799(27)
picolinic acid, 698(192)
picolinyl chloride, 671(63)
picramic acid, 788(552), 938(315)
picramide (2,4,6-trinitroaniline), 747(407), 790(595)
picric acid (trinitrophenol), 696(163)

Index of Compounds

picrolonic acid, 694(157)
picryl chloride, 671(69), 919(73)
pimelamide, 747(362)
pimelanilide, 745(161), 746(304)
pimelic acid, 694(121)
pinacol, 722(17)
pinacolone, 892(14)
pinacolyl alcohol (3,3-dimethyl-2-butanol), 714(24)
dl-α-pinene, 873(214)
β-pinene, 874(216)
d- or l-α-pinene, 873(213)
piperazine, 776(303)
piperic acid, 706(370)
piperidine, 750(36)
2-piperidone, 764(46)
piperine, 746(223)
piperonal (heliotropin), 734(13)
piperonalacetone, 903(176)
piperonyl alcohol, 722(32)
piperonylamide, 747(343)
piperonyl benzoate, 821(75)
piperonylic acid, 708(392)
piperylene (1,3-pentadiene), 870(24)
pivaldehyde (trimethylacetaldehyde), 728(11)
pivalamide, 746(299)
pivalanilide, 746(222)
pivalate (methyl trimethylacetate), 814(33)
pivalic anhydride, 668(6)
pivalyl chloride, 670(16)
prehnitene, 880(82)
procaine, 768(126)
DL-proline, 674(40)
L-(—)-proline, 674(59)
propanal (propionaldehyde), 728(3)
propanedioic acid (malonic acid), 696(189)
1,2-propanediol (α-propylene glycol), 718(126)
1,3-propanediol (trimethylene glycol), 720(145)
1,3-propanediol di-3-chlorobenzoate, 821(79)
1,3-propanedithiol, 966(21)
propane-1-sulfonamide, 948(4)
propane-2-sulfonamide, 948(6)
propane-1-sulfonic acid, 960(71)
propane-2-sulfonic acid, 960(72)
1-propanethiol, 966(4)
2-propanethiol, 966(3)
1-propanol, 714(8)
2-propanol, 714(3)
propane-1,2,3-tricarboxylic acid, 700(268)
propenamide, 744(76)
propenanilide, 745(144)
propene-2,3-dicarboxylic acid, 700(263)

2-propene-1-thiol, 966(7)
propenoic acid, 684(7)
2-propen-1-ol, 714(7)
propenoyl chloride, 670(11)
propenylbenzene (β-methylstyrene), 874(22)
propenyl bromide (1-bromopropene), 836(101)(103)
propiolamide, 744(35)
propiolanilide, 744(80)
propiolic acid, 684(13)
propionaldehyde (propanal), 728(3)
propionamide, 744(66)
propionanilide, 745(146)
propionic acid, 684(8)
propionic anhydride, 668(4)
propionitrile, 913(7)
propiophenone, 896(183), 899(4)
propionyl bromide, 671(74)
propionyl chloride, 670(12)
propionyl fluoride, 670(2)
propionyl iodide, 671(86)
propoxyethanal, 728(28)
1-propoxy-2-methoxyethane, 824(62)
2-propoxynaphthalene, 827(19)
2-(2-propoxy)naphthalene, 827(18)
1-propoxy-2-propanone, 893(60)
propylacetal (1,1-dipropoxyethane), 667(20)
N-propylacetanilide, 744(19)
propyl acetate, 814(34)
propyl acrylate, 815(54)
propylal (dipropoxymethane), 667(16)
propyl alcohol, 714(8)
propylamine, 750(7)
N-propylaniline, 756(148)
2-propylaniline, 756(152)
4-propylaniline, 756(158)
propylbenzene, 877(8)
propyl benzoate, 818(232)
propyl bromide, 836(104)
propyl butyrate, 815(80)
propyl caproate, 817(155)
propyl carbonate, 816(124)
propyl chloride, 832(28)
propyl chloroformate, 814(45)
propyl cinnamate, 819(299)
propyl crotonate, 816(112)
propylcyclohexane, 864(126)
propylcyclohexanone, 896(162)
propylcyclopentane, 863(72)
2-propylcyclopentanone, 895(144)
3-propylcyclopentanone, 895(150)
propyl dihydrocinnamate, 819(279)

1,3-propylene glycol diacetate, 817(165)
propylene glycol di-3-chlorobenzoate, 817(175)
α-propylene glycol (1,2-propanediol), 718(126)
propylene oxide (1,2-epoxypropane), 823(6)
propyl ether, 823(35)
propyl fluoride, 830(4)
propyl formate, 814(16)
propyl furoate, 817(199)
4-propylheptane, 865(150)
propyl heptanoate, 817(193)
propyl hexanoate, 817(155)
propyl 4-hydroxybenzoate, 822(116)
propyl iodide, 838(166)
propyl isobutyrate, 815(69)
propyl isocyanate, 912(8)
propyl isocyanide, 912(9)
1-propyl-4-isopropylbenzene, 881(108)
propyl isopropyl sulfide, 962(21)
propyl isothiocyanate, 961(6)
propyl levulinate, 818(215)
propyl mercaptan, 966(4)
propyl 3-methylbutyrate, 816(109)
1-propylnaphthalene, 884(171)
2-propylnaphthalene, 884(172)
propyl 2-naphthyl ether, 827(19)
propyl 2-naphthyl ketone, 901(80)
propyl nitrate, 814(40)
propyl oxalate, 818(204)
2-propylpentanal, 730(44)
propyl pentanoate, 816(121)
2-propyl-1-pentene, 873(175)
isopropyl perfluorobutyrate, 860(17)
isopropyl perfluoropropionate, 860(16)
1-propylphenanthrene, 886(3)
9-propylphenanthrene, 887(33)
4-propylphenol, 924(17), 926(2)
propyl phenylacetate, 818(248)
2-propylphenol, 942(397)
propyl 3-phenylpropionate, 819(279)
N-propylpiperidine, 799(33)
propyl propionate, 815(55)
propyl salicylate, 818(246)
propyl succinate, 819(261)
propyl sulfide, 962(27)
propyl thiocyanate, 961(5)
2-propylthiophene, 964(28)
3-propylthiophene, 964(32)
2-propyltoluene, 878(31)
3-propyltoluene (1-methyl-3-propylbenzene), 878(23)
4-propyltoluene (1-methyl-4-propylbenzene), 878(27)

1054   Index of Compounds

5-propyl-1,2,4-trimethylbenzene, 882(132)
propylurea, 745(165)
propylurethane, 844(34)
propyl valerate, 816(121)
propynamide, 744(35)
propynanilide, 740(80)
propynoic acid, 684(10)
2-propyn-1-ol, 714(18)
protocatechualdehyde (3,4-dihydroxybenzaldehyde), 740(113)
protocatechuamide, 748(458)
protocatechuanilide, 747(336)
protocatechuic acid, 704(330)
pseudocumene (1,2,4-trimethylbenzene), 877(14)
pseudocumenol (2,4,5-trimethylphenol), 928(80)
pseudocumidine (2,4,5-trimethylaniline), 768(164)
pulegone, 897(190)
purine, 806(123)
putrescine (1,4-diaminobutane), 752(72), 762(15)
pyrazole, 770(173)
pyrene, 889(91)
pyridazine, 801(90)
pyridine, 799(9)
pyridine-2-carboxaldehyde, 730(57), 800(63)
pyridine-2-carboxylic acid, 698(192)
pyridine-4-carboxylic acid, 710(437)
pyridine-2,3-dicarboxylic acid, 704(324)
pyridine-3-sulfonamide, 948(18)
pyridylmethanol, 804(30)
2-pyridyl propyl ketone, 896(182)
pyrimidine, 804(1)
pyrocatechol, 932(154)
pyrogallol, 936(242)
pyrogallol triacetate, 822(145)
pyrogallol tribenzoate, 822(112)
pyrogallol trimethyl ether, 827(24)
pyromucic acid (furoic acid), 696(185)
pyrotartaramide, 748(482)
pyrotartaric acid (methylsuccinic acid), 694(139)
pyrrole, 752(52)
pyrrole-2-carboxaldehyde, 734(27)
pyrrole-1-carboxylic acid, 692(102)
pyrrolidine, 750(25)
pyrrolidone, 762(9), 899(9)
pyruvamide, 745(200)
pyruvanilide, 745(139)

pyruvic acid, 684(21), 894(104)
pyruvonitrile, 913(6)

d-quercitol (pentahydroxycyclohexane), 724(77)
quinaldine, 801(111)
quinhydrone, 905(245)
quinoline, 801(101)
quinoline-2-carboxaldehyde, 736(53)
quinoline-4-carboxaldehyde, 734(28)
quinoline-6-carboxaldehyde, 736(58)
quinoline-8-carboxaldehyde, 738(75)
quinoline-2-carboxylic acid, 700(247)
quinone (benzoquinone), 903(186)
quinoxaline, 804(9)

raffinose, 809(13)
resacetophenone (2,4-dihydroxyacetophenone), 905(227), 936(271)
resorcinol, 932(170)
resorcinol diacetate, 819(294)
resorcinol dibenzoate, 822(131)
resorcinol diethyl ether, 826(129)
resorcinol dimethyl ether, 825(117)
resorcinol monomethyl ether, (3-methoxyphenol), 924(23)
β-resorcylamide, 748(475)
β-resorcylanilide, 745(212)
β-resorcylic acid (2,4-dihydroxybenzoic), 706(371), 942(392)
α-resorcylaldehyde (3,5-dihydroxybenzaldehyde), 740(115)
β-resorcylaldehyde (2,4-dihydroxybenzaldehyde) 740(104)
retene, 888(56)
retenequinone, 905(258)
β-L-rhamnose, 809(14)
D-ribose, 809(6)
L-ribose, 809(4)

saccharin (o-sulfobenzimide) 748(473)
safrole, 826(126)
salicylaldehyde, 732(63), 924(6)
salicylamide, 746(264)
salicylanilide, 746(253)
salicylic acid, 700(251) 938(292)
salicylyl chloride, 671(56)
saligenin (o-hydroxybenzyl alcohol), 724(54), 930(112)
sarcosine (N-methylglycine), 674(48)

L-scopolamine, 766(119)
sebacamide, 747(350), 748(453)
sebacanilide, 745(196), 748(436)
sebacic acid (decanedioic acid), 696(187)
sebacic anhydride, 668(34)
semicarbazide, 745(109), 774(272)
DL-serine (α-amino-β-hydroxypropionic acid), 676(77)
L-serine, 676(66)
β-sitosterol (cinchol), 724(67)
sorbic acid, 688(16), 696(188)
D-sorbitol, 724(57)
L-sorbose, 809(35)
starch, 810(48)
stearaldehyde, 734(16)
stearamide, 745(162)
stearanilide, 744(101)
stearic acid (octadecanoic acid), 690(68)
stearic anhydride, 668(35)
stearone, 902(147)
stearonitrile, 914(23)
stearoyl chloride, 671(58)
stearyl alcohol (octadecanol), 722(36)
stilbene, 875(306)
3,5-stilbenediol, 938(287)
4,4'-stilbenediol, 944(430)
styphnic acid (2,4,6-trinitroresorcinol), 940(335)
styrene (vinylbenzene), 873(209)
styrene oxide, 825(101)
styryl butyl ketone, 900(37)
suberamide, 745(214), 748(462)
suberane (cycloheptane), 863(51)
suberanilide, 746(220), 747(399)
suberene (cycloheptene), 873(167)
suberic acid (octanedioic acid), 698(210)
suberic anhydride, 668(30)
succinamide, 746(317), 748(514)
succinanilide, 746(281), 748(487)
succinic acid (butanedioic acid), 704(314)
succinic anhydride, 669(50)
succinimide, 745(209), 904(203)
succinonitrile, 914(38)
succinyl chloride, 670(34)
sucrose, 810(40)
sulfanilguanidine, 950(128)
sulfanilic acid (4-aminobenzenesulfonic acid), 949(70), 959(3)
sulfapyridine, 950(132)
sulfathiazole, 950(150)
2-sulfobenzimide (saccharin), 748(473)
2-sulfobenzoic acid, 960(73)
3-sulfobenzoic acid, 698(205), 960(74)

Index of Compounds

4-sulfobenzoic acid, 960(74)
o-sulfobenzoic anhydride, 669(53)
4,4′-sulfonyldianiline, 788(562)
sylvan (2-methylfuran), 823(17)
sylvestrene, 874(226)

D-tagatose, 809(15)
d-tartaramide, 747(421)
dl-tartaramide, 748(485)
meso-tartaramide, 747(402)
d-tartaranilide, 747(377), 748(510)
dl-tartaranilide, 748(494)
d-tartaric acid, 702(276)
dl-tartaric acid, 704(340)
meso-tartaric acid, 698(202)
terephthalaldehyde, 738(96)
terephthalanilide, 748(493)
terephthalic acid, 710(436)
terephthalonitrile, 915(110)
dl-α-terpineol, 722(16)
terpin hydrate, 724(61)
2,3,4,6-tetrabromoaniline, 778(373)
1,2,4,5-tetrabromobenzene, 850(68)
1,2,3,4-tetrabromobutane, 856(17)
tetrabromo-o-cresol, 942(381)
sym-tetrabromoethane, 855(141)
1,1,2,2-tetrabromoethane, 855(141)
tetrabromomethane, 856(14)
2,4,5,6-tetrabromo-6-methylphenol, 942(381)
2,3,4,6-tetrabromo-3-methylphenol, 940(356)
2,3,5,6-tetrabromo-4-methylphenol, 940(361)
tetrabromophthalic anhydride, 669(72)
1,2,2,3-tetrabromopropane, 855(143)
1,2,3,4-tetrachlorobenzene, 846(16)
1,2,3,5-tetrachlorobenzene, 846(21)
1,2,4,5-tetrachlorobenzene, 850(63)
1,1,2,2-tetrachloro-1,2-difluoroethane, 853(38)
1,1,1,2-tetrachloroethane, 853(66)
1,1,2,2-tetrachloroethane, 854(82)
sym-tetrachloroethane, 854(82)
1,1,2,2-tetrachloroethylene, 853(62)
2,3,5,6-tetrachlorohydroquinone, 944(412)
tetrachloromethane, 852(30)
1,2,3,4-tetrachloronaphthalene, 850(71)

4,6,7,8-tetrachloronaphthalene-2-sulfonamide, 951(207)
4,6,7,8-tetrachloronaphthalene-2-sulfonyl chloride, 957(214)
2,3,4,6-tetrachlorophenol, 928(75)
tetrachlorophthalic acid, 708(417)
tetrachlorophthalic anhydride, 669(70)
1,1,1,2-tetrachloropropane, 854(87)
5,6,7,8-tetrachlorotetralin, 850(67)
$\alpha,\alpha,\alpha,$2-tetrachlorotoluene, 856(6)
1,1,1,2-tetrachloro-4,4,4-trifluoro-2-butene, 854(79)
tetracontane, 868(296)
1-tetracontene, 875(301)
1-tetracontyne, 875(304)
tetracosane, 867(250)
tetracosanoic acid (lignoceric acid), 692(86)
tetracosene, 875(265)
tetracosylcyclohexane, 867(261)
tetracosylcyclopentane, 867(251)
tetracosyne, 875(272)
tetradecanal, 734(4)
tetradecanamide, 745(152)
tetradecananilide, 744(73)
tetradecane, 866(212)
tetradecanoic acid (myristic acid), 690(46)
tetradecanoic anhydride, 668(28)
tetradecanol (myristyl alcohol), 722(21)
2-tetradecanone, 899(24)
3-tetradecanone, 899(25)
tetradecanonitrile, 914(2)
tetradecanoyl chloride, 671(54)
1-tetradecene, 874(242)
tetradecyl bromide, 838(143)
tetradecylamine, 762(36)
tetradecylcyclohexane, 867(230)
tetradecylcyclopentane, 867(225)
tetradecyl sulfide, 965(20)
2-tetradecyl-3-thiophanone, 900(58)
1-tetradecyne, 874(243)
tetraethylene glycol dimethyl ether, 826(142)
2,2,3,3-tetrafluorobutane, 852(5)
tetrafluorosuccinic acid, 857(19)
1,2,3,4-tetrahydroanthracene, 888(64)
tetrahydrofuran, 823(18)
tetrahydrofurfural, 730(40)
tetrahydrofurfuryl acetate, 817(169)

tetrahydrofurfuryl alcohol, 718(115)
tetrahydroisoquinoline, 758(171)
tetrahydro-2-methylfuran, 823(24)
tetrahydronaphthalene (tetralin), 881(92)
5,6,7,8-tetrahydro-1-naphthylamine, 760(213)
5,6,7,8-tetrahydro-2-naphthylamine, 762(41)
tetrahydrophthalic acid, 706(369)
tetrahydropyran, 823(33)
tetrahydroquinoline, 758(194), 762(2)
tetrahydrosilvan, 823(24)
tetrahydrothiophene-2-carboxylic acid, 690(45)
1,2,3,5-tetrahydroxybenzene, 938(309)
1,2,4,5-tetrahydroxybenzene, 942(391)
3,4,3′,4′-tetrahydroxybenzophenone, 906(269)
dl-1,2,3,4-tetrahydroxybutane (erythritol), 724(47)
l-2,3,4,5-tetrahydroxypentanoic acid, 694(150)
tetraiodophthalic anhydride, 669(73)
tetralin (tetrahydronaphthalene), 881(92)
tetralin-6-sulfonamide, 948(32)
tetralin-6-sulfonyl chloride, 954(44)
N,N,2,4-tetramethylaniline, 801(86)
2,3,4,6-tetramethylaniline, 758(201), 762(6)
2,3,5,6-tetramethylaniline, 770(195)
1,2,3,4-tetramethylbenzene, 880(820)
1,2,3,5-tetramethylbenzene (isodurene), 879(61)
1,2,4,5-tetramethylbenzene (durene), 887(42)
N,N,N′,N′-tetramethylbenzidine, 806(118)
2,3,5,6-tetramethylbenzaldehyde, 734(1)
2,2,3,3-tetramethylbutane, 868(297)
N,N,N′,N′-tetramethyl-1,3-diaminobenzene, 802(126)
N,N,N′,N′-tetramethyl-1,4-diaminobenzene, 804(39)
N,N,N′,N′-tetramethyl-1,3-diaminopropane, 799(26)
tetramethylene glycol (1,4-butanediol), 720(157)
tetramethylene sulfide, 964(11)
tetramethylethylene (2,3-dimethyl-2-butene), 870(58)

Index of Compounds

tetramethylethylene glycol (pinacol), 722(17)
2,2,3,3-tetramethylhexane, 865(144)
2,2,3,4-tetramethylhexane, 864(120)
2,2,3,5-tetramethylhexane, 864(111)
2,2,4,4-tetramethylhexane, 864(118)
2,2,4,5-tetramethylhexane, 864(108)
2,2,5,5-tetramethylhexane, 863(87)
2,3,3,4-tetramethylhexane, 865(163)
2,3,3,5-tetramethylhexane, 864(117)
2,3,4,4-tetramethylhexane, 865(153)
2,3,4,5-tetramethylhexane, 865(147)
3,3,4,4-tetramethylhexane, 865(185)
2,2,3,3-tetramethylpentane, 864(91)
2,2,3,4-tetramethylpentane, 863(76)
2,2,4,4-tetramethylpentane, 863(60)
2,3,3,4-tetramethylpentane, 864(98)
2,3,4,6-tetramethylphenol (isodurenol), 930(98)
2,3,5,6-tetramethylphenol, 934(198)
N,N,N′,N′-tetramethyl-m-phenylenediamine, 802(126)
N,N,N′,N′-tetramethyl-p-phenylenediamine, 804(39)
2,2,6,6-tetramethyl-4-piperidone, 899(28)
2,3,4,5-tetramethylpyridine, 801(99)
2,4,5,8-tetramethylquinoline, 804(34)
tetranitromethane, 916(7)
sym-tetraphenylethane, 890(116)
tetratriacontane, 868(282)
1-tetratriacontene, 875(289)
tetratriacontylcyclohexane, 868(291)
tetratriacontylcyclopentane, 868(281)
1-tetratriacontyne, 875(294)
thalline, 764(57)
theobromine, 758(522)
theophylline, 748(509)
thiacyclohexane, 964(19)
thiazole-4-carboxylic acid, 704(326)
thiazole-5-carboxylic acid, 706(375)
2-thienyl mercaptan, 966(20)
thioacetamide, 745(155)

thioacetanilide, 744(55)
thioacetic acid, 684(2)
m-thiocresol, 966(26)
o-thiocresol, 966(24)
p-thiocresol, 967(7)
dl-thiomalic acid, 698(227)
3-thiophanone, 895(127)
thiophene, 963(3)
thiophene-2-carboxaldehyde, 732(64)
thiophene-2-carboxylic acid, 696(178)
thiophene-3-carboxylic acid, 698(197)
2-thiophenethiol, 966(20)
thiophenol (benzenethiol), 966(22)
thiopropionic acid, 684(3)
thiosalicylic acid, 700(258)
thiourea, 747(384)
1,4-thioxane, 964(23)
DL-threonine (α-amino-β-hydroxybutyric acid), 676(65)
2,4,6-tribromoresorcinol, 932(177)
trichloroacetonitrile, 913(4)
2,4,5-trimethylaniline, 768(164)
triphenyl phosphate, 821(48)
β-d-thujone, 896(167)
α-l-thujone, 896(164)
thymol, 926(34)
thymolsulfonic acid, 960(76)
thymoquinone, 900(57)
thymyl acetate, 819(258)
thymyl benzoate, 821(20)
dl-thyroxine, 676(69)
tiglamide, 744(54)
tiglanilide, 744(57)
tiglic acid, 690(57)
o-tolidine, 780(414)
2-tolualdehyde, 732(66)
3-tolualdehyde, 732(65)
4-tolualdehyde, 732(68)
2-toluamide, 746(265)
3-toluamide, 744(98)
4-toluamide, 746(321)
2-toluanilide, 745(204)
3-toluanilide, 745(210)
4-toluanilide, 746(270)
toluene, 877(2)
1-m-tolueneazo-2-naphthol, 909(38)
1-o-tolueneazo-2-naphthol, 909(26)
1-p-tolueneazo-2-naphthol, 909(34)
2-m-tolueneazo-1-naphthol, 909(23)
2-o-tolueneazo-1-naphthol, 910(46)
2-p-tolueneazo-1-naphthol, 909(41)
4-m-tolueneazo-1-naphthol, 910(67)

4-o-tolueneazo-1-naphthol, 910(42)
toluene-α-sulfonamide, 948(13)
toluene-2-sulfonamide, 949(62)
toluene-3-sulfonamide, 948(14)
toluene-4-sulfonamide, 948(35)
toluene-α-sulfonic acid, 959(17)
toluene-2-sulfonic acid, 960(77)
toluene-4-sulfonic acid, 960(78)
toluene-α-sulfonyl chloride, 955(94)
toluene-2-sulfonyl chloride, 954(57)
toluene-3-sulfonyl chloride, 953(3)
toluene-4-sulfonyl chloride, 954(59)
toluhydroquinone, 934(219)
2-toluic acid, 692(120)
3-toluic acid, 694(131)
4-toluic acid, 702(288)
2-toluic anhydride, 668(21)
3-toluic anhydride, 668(36)
4-toluic anhydride, 668(45)
2-toluidine, 754(104)
3-toluidine, 754(106)
4-toluidine, 764(61)
2-tolunitrile, 913(39)
3-tolunitrile, 913(42)
4-tolunitrile, 914(10)
4-toluquinone, 902(116)
2-(4-toluoyl)benzamide, 747(367)
2-(4-toluoyl)benzoic acid, 698(216)
3-tolyl acetate (m-cresyl acetate), 817(201)
2-tolyl acetate (o-cresyl acetate), 817(190)
4-tolyl acetate (p-cresyl acetate), 817(202)
2-tolylacetic acid, 692(93)
4-tolylacetic acid, 692(101)
4-tolyl benzoate (o-cresyl benzoate), 820(311)
2-tolylcarbinol (o-methylbenzyl alcohol), 722(18)
3-tolylcarbinol (3-methylbenzyl alcohol), 720(146)
4-tolylcarbinol (4-methylbenzyl alcohol), 722(37)
1-(4-tolyl)-1-ethanol, 720(147)
4-tolylhydrazine, 602(58)
2-tolyl isocyanate, 912(15)
3-tolyl isocyanate, 912(17)
4-tolyl isocyanate, 912(16)
2-tolyl isocyanide, 912(14)
2-tolyl isothiocyanate, 961(11)
4-tolyl isothiocyanate, 961(12)
N-(4-tolyl)-1-naphthylamine, 770(207)
N-(4-tolyl)-2-naphthylamine, 776(299)
4-tolyl perfluoroamyl ketone, 859(10)

Index of Compounds 1057

4-tolyl perfluorobutyl ketone, 859(9)
4-tolyl perfluoroethyl ketone, 859(7)
4-tolyl perfluoromethyl ketone, 859(6)
4-tolyl perfluoropentyl ketone, 859(10)
4-tolyl perfluoropropyl ketone, 895(8)
2-tolyl sulfide, 965(29)
3-tolyl sulfide, 963(63)
4-tolyl sulfide, 965(24)
2-tolylurea, 747(415)
3-tolylurea, 746(262)
4-tolylurea, 747(387)
α,α-trehalose, 810(44)
triacontane, 868(270)
triacontanoic acid, 692(98)
1-triacontene, 875(280)
triacontylcyclohexane, 868(280)
triacontylcyclopentane, 868(269)
1-triacontyne, 875(285)
1,2,3-triaminobenzene, 776(298)
1,2,3-triaminopropane, 754(98)
3,4,5-triamino-1,2,4-triazole, 794(672)
4,4′,4″-triaminotriphenylcarbinol, 792(633)
4,4′,4″-triaminotriphenylmethane, 792(629)
triamylamine, 801(117)
tribenzylamine, 805(84)
tribromoacetamide, 745(195)
tribromoacetic acid, 696(180)
3,4,5-tribromoacetophenone, 409(211)
2,4,6-tribromoaniline, 780(389)
3,4,5-tribromoaniline, 780(394)
2,4,6-tribromoanisole, 838(46)
1,2,3-tribromobenzene, 848(48)
1,2,4-tribromobenzene, 846(15)
1,3,5-tribromobenzene, 848(60)
2,4,6-tribromobenzenesulfonamide, 951(197)
2,3,4-tribromobenzenesulfonyl chloride, 954(54)
2,4,6-tribromobenzenesulfonyl chloride, 954(53)
1,2,3-tribromobutane, 855(129)
1,2,4-tribromobutane, 855(128)
2,2,3-tribromobutane, 855(121)
1,2,2-tribromo-1,1-dichloroethane, 855(125)
2,2,2-tribromoethanal, 730(53)
1,1,2-tribromoethane, 855(117)
2,2,2-tribromoethanol, 724(53)
2,4,6-tribromo-1-ethoxybenzene, 827(41)
2,2,2-tribromoethyl alcohol, 724(53)
2,4,6-tribromo-5-hydroxyaniline, 780(378)
tribromomethane, 854(85)

2,4,6-tribromo-1-methoxybenzene, 828(46)
2,3,6-tribromo-4-methylphenol, 932(147)
2,4,5-tribromo-6-methylphenol, 930(121)
2,4,6-tribromo-3-methylphenol, 930(104)
3,4,5-tribromo-2-methylphenol, 930(119)
2,4,6-tribromonitrobenzene, 931(125)
2,4,6-tribromo-3-nitrophenol, 930(120)
3,4,6-tribromo-2-nitrophenol, 934(228)
2,4,6-tribromophenetole, 827(41)
2,4,5-tribromophenol, 930(116)
2,4,6-tribromophenol, 930(130)
1,2,3-tribromopropane, 855(132)
tributylamine, 801(92)
tricarballylamide, 748(447)
tricarballylanilide, 748(504)
tricarballylic acid, 560(127)
trichloroacetamide, 746(259)
trichloroacetanilide, 745(112)
trichloroacetic acid, 690(47)
trichloroacetyl chloride, 670(23)
trichloroacetyl fluoride, 670(4)
2,4,6-trichloroaniline, 770(204)
2,4,6-trichloroanisole, 827(33)
1,2,3-trichlorobenzene, 846(22)
1,2,4-trichlorobenzene, 844(48)
1,3,5-trichlorobenzene, 846(31)
2,3,4-trichlorobenzenesulfonamide, 951(195)
2,4,6-trichlorobenzenesulfonamide, 951(162)
2,4,6-trichlorobenzoic acid, 700(257)
3,4,5-trichlorobenzoic acid, 704(337)
1,2,3-trichlorobutane, 854(102)
trichloroethanal, 728(17)
1,1,1-trichloroethane (methylchloroform), 852(29)
1,1,1-trichloro-2,2-bis(p-chlorophenyl)ethane (D.D.T.), 856(16)
1,1,2-trichloroethane, 853(55)
trichloroethanol, 716(67)
2,4,6-trichloro-1-ethoxybenzene, 827(21)
1,1,2-trichloroethylene, 852(35)
1,2,2-trichloroethyl ethyl ether, 825(86)
1,1,2-trichloro-2-fluoroethane, 853(43)
1,1,2-trichloro-2-fluoroethylene, 852(38)
trichlorofluoromethane, 852(7)
trichlorohydroquinone, 936(248)

2,4,6-trichloro-3-hydroxybenzaldehyde, 738(94)
trichloroiodomethane, 852(14)
trichlorolactamide, 745(106)
trichlorolactanilide, 747(332)
trichlorolactic acid, 696(165)
trichloromethane, 852(22)
2,4,6-trichloro-1-methoxybenzene, 827(33)
2,3,4-trichloro-6-methylphenol, 928(92)
2,3,6-trichloro-4-methylphenol, 928(67)
2,4,6-trichloro-3-methylphenol, 926(29)
3,4,6-trichloro-2-methylphenol, 928(56)
5,6,7-trichloronaphthalene-1-sulfonamide, 952(225)
5,6,8-trichloronaphthalene-2-sulfonamide, 951(206)
6,7,8-trichloronaphthalene-2-sulfonamide, 952(221)
5,6,7-trichloronaphthalene-1-sulfonyl chloride, 957(177)
5,6,8-trichloronaphthalene-2-sulfonyl chloride, 957(203)
6,7,8-trichloronaphthalene-2-sulfonyl chloride, 957(200)
1,3,4-trichloro-2-naphthol, 938(301)
1,4,5-trichloro-2-naphthol, 938(289)
2,3,4-trichloro-1-naphthol, 938(313)
1,2,4-trichloro-5-nitrobenzene, 918(36)
trichloronitromethane, 853(53), 916(3)
1,1,1-trichloro-2,2,4,4,4-pentafluorobutane, 853(63)
2,4,6-trichlorophenetole, 827(21)
2,3,4-trichlorophenol, 930(106)
2,3,5-trichlorophenol, 928(55)
2,3,6-trichlorophenol, 928(50)
2,4,5-trichlorophenol, 928(72)
2,4,6-trichlorophenol, 928(73)
3,4,5-trichlorophenol, 932(145)
trichlorophloroglucinol, 936(249)
1,1,2-trichloropropane, 854(73)
1,1,3-trichloropropane, 854(81)
1,2,3-trichloropropane, 854(92)
1,1,1-trichloro-2,2,3,3-tetrafluorobutane, 853(64)
α,α,α-trichlorotoluene, 855(130)
α,2,4-trichlorotoluene, 834(76)
α,2,6-trichlorotoluene, 834(91)
α,3,4-trichlorotoluene, 834(74)
2,3,4-trichlorotoluene, 846(14)
2,3,5-trichlorotoluene, 846(18)
2,4,5-trichlorotoluene, 848(46)
2,4,6-trichlorotoluene, 846(7)
3,4,5-trichlorotoluene, 846(17)

1058   Index of Compounds

1,1,2-trichloro-1,2,2-trifluoro-
  ethane, 852(15)
tricosane, 867(247)
12-tricosanone (laurone),
  902(117)
1-tricosene, 874(263)
tricosylcyclohexane, 867(258)
tricosylcyclopentane, 867(249)
1-tricosyne, 875(267)
tri-*m*-cresyl phosphate, 821(7)
tri-*o*-cresyl phosphate, 820(320)
tri-*p*-cresyl phosphate, 821(98)
tridecanamide, 745(124)
tridecananilide, 744(65)
tridecane, 866(209)
tridecanoic acid, 688(28)
tridecanoic anhydride, 668(26)
tridecanol, 722(12)
2-tridecanone, 899(13)
7-tridecanone, 899(21)
tridecanonitrile, 913(55)
1-tridecene, 874(240)
1-tridecyne, 874(241)
tridecylcyclohexane, 867(226)
tridecylcyclopentane, 867(223)
triethylamine, 799(5)
1,2,4-triethylbenzene, 881(112)
1,3,5-triethylbenzene, 881(114)
triethylcarbinol (3-ethyl-3-
  pentanol), 761(58)
triethyl citrate, 820(306)
triethylene glycol, 720(169)
triethylene glycol dimethyl
  ether, 825(116)
triethylmethane (3-ethyl-
  pentane), 862(24)
triethyl orthoformate, 815(88)
trifluoroacetic acid, 684(1)
trifluoroacetic anhydride, 668(1)
1,1,1-trifluoroacetone, 858(6),
  892(1)
trifluoroacetyl bromide, 858(3)
trifluoroacetyl chloride, 858(2)
trifluoroacetyl fluoride, 858(1)
trifluoroacetyl iodide, 858(4)
2,2,2-trifluoroethanol, 858(1)
trifluoroethyl acetate, 814(13)
2,2,2-trifluoroethylamine,
  859(1)
trifluoroethyl trifluoroacetate,
  814(4)
1,1,1-trifluoro-2,4-hexanedione,
  892(29)
1,1,1-trifluoro-2,4-
  pentanedione, 892(15)
trifluoropropanal, 728(6)
1,1,1-trifluoro-2-propanol,
  722(30)
α,α,α-trifluorotoluene, 853(44)
2,3,4-trihydroxyacetophenone,
  905(246), 940(324)
2,3,5-trihydroxyacetophenone,
  905(263), 942(377)
2,4,6-trihydroxyacetophenone,
  942(402)

3,4,5-trihydroxyacetophenone,
  905(253)
1,2,3-trihydroxy-9,10-
  anthraquinone, 944(435)
1,2,4-trihydroxy-9,10-anthra-
  quinone, 944(418)
1,2,5-trihydroxy-9,10-anthra-
  quinone, 944(423)
1,2,7-trihydroxy-9,10-anthra-
  quinone, 944(437)
1,2,8-trihydroxy-9,10-anthra-
  quinone, 944(413)
3,4,6-trihydroxybenzaldehyde,
  740(125)
1,2,3-trihydroxybenzene,
  936(242)
1,2,4-trihydroxybenzene,
  936(258)
1,3,5-trihydroxybenzene,
  934(193), 942(394)
2,4,6-trihydroxybenzophenone,
  905(241), 938(310)
2,4,5-trihydroxytoluene,
  936(240)
2,4,6-trihydroxytoluene,
  942(386)
3,4,5-trihydroxytoluene,
  934(202)
2,2′,2″-trihydroxytriethylamine,
  802(146)
1,2,3-triiodobenzene, 848(58)
1,2,4-triiodobenzene, 848(52)
1,3,5-triiodobenzene, 850(69)
triiodomethane, 856(18)
2,3,5-triiodophenol, 934(185)
2,4,6-triiodophenol, 938(293)
2,4,6-triiodoresorcinal, 936(270)
triisoamylamine, 801(106)
triisobutylamine, 800(73)
triisopentylamine, 801(106)
trimesamide, 748(524)
trimesanilide, 745(185)
trimesic acid, 710(443)
2,4,6-trimethoxybenzaldehyde,
  738(97)
3,4,5-trimethoxybenzaldehyde,
  736(62)
1,2,3-trimethoxybenzene,
  827(24)
1,3,5-trimethoxybenzene,
  827(27)
2,4,5-trimethoxybenzene-
  sulfonyl chloride, 956(174)
2,4,5-trimethoxybenzoic acid,
  698(212)
3,4,5-trimethoxybenzophenone,
  902(134)
trimethylacetaldehyde, 728(11)
trimethylacetic acid (pivalic
  acid), 688(20)
trimethylacetic anhydride,
  668(6)
trimethylacetonitrile, 813(9)
2,4,5-trimethylacetophenone,
  897(209)

trimethylacetyl chloride, 670(16)
trimethylamine, 799(1)
2,3,6-trimethylaniline, 758(175)
2,4,5-trimethylaniline, 768(164)
2,4,6-trimethylaniline
  (mesidine), 756(166)
3,4,5-trimethylaniline, 770(196)
2,4,5-trimethylbenzaldehyde,
  734(23)
1,2,3-trimethylbenzene
  (hemimellitene), 878(18)
1,2,4-trimethylbenzene
  (pseudocumene), 877(14)
1,3,5-trimethylbenzene
  (mesitylene), 877(11)
2,4,5-trimethylbenzene-
  sulfonamide, 949(99)
2,4,6-trimethylbenzenesulfon-
  amide, 948(39)
2,4,5-trimethylbenzenesulfonyl
  chloride, 954(48)
2,4,6-trimethylbenzenesulfonyl
  chloride, 953(41)
2,4,6-trimethylbenzoic acid,
  700(239)
2,4,4-trimethylbenzophenone,
  897(224)
2,4,6-trimethylbenzyl chloride,
  834(89)
2,2,3-trimethylbutane, 862(15)
2,3,3-trimethyl-1-butene,
  871(68)
trimethyl citrate, 821(101)
2,2,6-trimethylcyclohexanone,
  895(134)
3,5,5-trimethylcyclo-3-hexen-
  one, 896(180)
1,1,2-trimethylcyclopentane,
  862(39)
1,1,3-trimethylcyclopentane,
  862(30)
1-*cis*-2-*cis*-3-trimethylcyclo-
  pentane, 863(61)
1-*cis*-2-*trans*-3-trimethylcyclo-
  pentane, 863(44)
1-*trans*-2-*cis*-3-trimethylcyclo-
  pentane, 862(36)
1-*cis*-2-*cis*-4-trimethylcyclo-
  pentane, 863(48)
1-*cis*-2-*trans*-4-trimethylcyclo-
  pentane, 862(43)
1-*trans*-2-*cis*-4-trimethylcyclo-
  pentane, 862(33)
trimethylene bromohydrin,
  718(112)
trimethylene chlorohydrin,
  716(85)
trimethylene glycol (1,3-
  propanediol), 720(145)
trimethylene glycol diacetate,
  817(197)
trimethylenimine, 750(15)
trimethylene sulfide, 964(5)
trimethylglycine (betaine),
  680 (123)

Index of Compounds 1059

2,2,3-trimethylheptane, 864(131)
2,2,4-trimethylheptane, 864(107)
2,2,5-trimethylheptane, 864(109)
2,2,6-trimethylheptane, 864(110)
2,3,3-trimethylheptane, 865(139)
2,3,4-trimethylheptane, 865(154)
2,3,5-trimethylheptane, 864(127)
2,3,6-trimethylheptane, 864(124)
2,4,4-trimethylheptane, 864(116)
2,4,5-trimethylheptane, 864(129)
2,4,6-trimethylheptane, 864(104)
2,5,5-trimethylheptane, 864(114)
3,3,4-trimethylheptane, 865(158)
3,3,5-trimethylheptane, 864(123)
3,4,4-trimethylheptane, 865(160)
3,4,5-trimethylheptane, 865(161)
2,2,3-trimethylhexane, 863(78)
2,2,4-trimethylhexane, 863(68)
2,2,5-trimethylhexane, 863(63)
2,3,3-trimethylhexane, 863(88)
2,3,4-trimethylhexane, 864(90)
2,3,5-trimethylhexane, 863(73)
2,4,4-trimethylhexane, 863(71)
3,3,4-trimethylhexane, 864(93)
2,2,4-trimethyl-3-hexanone, 894(93)
trimethyl orthoformate, 814(36)
2,2,3-trimethylpentane, 862(35)
2,2,4-trimethylpentane, 862(26)
2,3,3-trimethylpentane, 862(40)
2,3,4-trimethylpentane, 862(38)
2,2,4-trimethyl-3-pentanone, 893(41)
2,3,3-trimethyl-1-pentene, 872(134)
2,3,4-trimethyl-1-pentene, 872(133)
2,3,4-trimethyl-2-pentene, 873(171)
2,4,4-trimethyl-1-pentene, 872(111)
2,4,4-trimethyl-2-pentene, 872(122)
3,3,4-trimethyl-1-pentene, 872(123)
3,4,4-trimethyl-1-pentene, 872(118)
3,4,4-trimethyl-2-pentene, 872(152)
2,4,5-trimethylphenol (pseudocumenol), 928(80)
2,3,5-trimethylphenol, 930(132)
2,4,6-trimethylphenol, 928(76)
2,4,6-trimethylphenyl methyl ketone, 901(97)
2,2,4-trimethylpiperidine, 752(66)
2,2,6-trimethylpiperidine, 752(56)
1,3,5-trimethyl-2-propylbenzene, 881(118)
1,3,4-trimethylpyrazole, 799(39)
1,3,5-trimethylpyrazole, 804(16)
2,3,4-trimethylpyridine, 800(75)

2,3,5-trimethylpyridine, 800(64)
2,3,6-trimethylpyridine, 806(115)
2,4,5-trimethylpyridine, 800(67)
2,4,6-trimethylpyridine (2,4,6-collidine), 800(54)
3,4,5-trimethylpyridine, 800(85)
1,2,5-trimethylpyrrolidine, 799(10)
2,3,4-trimethylquinoline, 806(86)
2,3,6-trimethylquinoline, 805(91)
2,3,8-trimethylquinoline, 805(45)
2,4,6-trimethylquinoline, 805(57)
2,4,7-trimethylquinoline, 802(138)
2,4,8-trimethylquinoline, 804(26)
2,6,8-trimethylquinoline, 804(29)
2,3,4-trimethylthiophene, 964(39)
2,3,5-trimethylthiophene, 964(36)
2,4,6-trinitroaniline (picramide), 747(407)
2,4,6-trinitroanisole, 918(53)
2,4,6-trinitroazobenzene, 909(37)
2,4,6-trinitrobenzamide, 748(508)
1,2,4-trinitrobenzene, 918(43)
1,3,5-trinitrobenzene, 920(119)
2,4,6-trinitrobenzoic acid, 708(388)
2,4,6-trinitrochlorobenzene, 919(73)
2,4,6-trinitro-1-ethoxybenzene, 919(66)
2,3,5-trinitro-1,4-dimethylbenzene, 921(133)
2,4,5-trinitro-1,3-dimethylbenzene, 919(86)
2,4,6-trinitro-1,3-dimethylbenzene, 922(158)
3,4,5-trinitro-1,2-dimethylbenzene, 920(111)
3,4,6-trinitro-1,2-dimethylbenzene, 918(60)
4,5,6-trinitro-1,3-dimethylbenzene, 921(124)
trinitromethane, 917(1)
2,4,6-trinitro-1-methoxybenzene, 918(53)
2,4,6-trinitro-3-methylphenol, 932(168)
1,3,5-trinitronaphthalene, 920(118)
1,3,8-trinitronaphthalene, 922(166)
1,4,5-trinitronaphthalene, 921(141)

2,4,6-trinitrophenetole, 919(66)
2,3,6-trinitrophenol, 934(200)
2,4,6-trinitrophenol (picric acid), 696(163), 921(121), 934(210)
2,4,6-trinitrophenyl sulfide, 965(52)
2,4,6-trinitroresorcinol (styphnic acid), 940(335)
2,3,4-trinitrotoluene, 920(108)
2,4,5-trinitrotoluene, 920(102)
2,4,6-trinitrotoluene, 919(67)
2,4,6-trinitrothiophenol, 967(26)
trioxane, 827(35)
tripalmitin, 821(71)
tripentylamine, 806(117)
triphenylamine, 806(106)
triphenylchloromethane, 836(96)
triphenylene-2-carboxylic acid, 710(441)
α-triphenylguanidine, 746(272)
triphenylmethane, 887(52)
triphenylmethanethiol, 967(21)
triphenylmethanol, 724(70)
triphenylmethylamine, 776(314)
triphenylmethyl chloride, 836(96)
triphenyl phosphate, 821(48)
tripropylamine, 799(35)
tritriacontane, 868(279)
tritriacontene, 875(286)
tritriacontylcyclohexane, 868(289)
tritriacontylcyclopentane, 868(778)
tritriacontyne, 875(292)
tritylamine, 776(314)
trityl chloride, 836(96)
dl-tropamide, 747(344)
tropane, 800(50)
2-tropene, 799(43)
dl-tropic acid, 694(148)
DL-tryptophane, 678(94)
L-tryptophane, 678(105)
DL-tyrosine, 680(129)
L-(—)-tyrosine, 680(127)

undecanamide, 745(137)
undecananilide, 744(46)
undecane, 866(195)
undecanoic acid (undecylic acid), 686(54), 688(7)
undecanoic anhydride, 668(20)
undecanol (hendecyl alcohol) (undecyl alcohol), 720(161)
2-undecanone, 897(195)
3-undecanone, 897(194)
undecanonitrile, 913(51)
1-undecene, 874(233)
undecyl alcohol (undecanol), 720(161)
undecyl bromide, 838(140)
undecyl chloride, 834(75)

undecylcyclohexane, 867(222)
undecylcyclopentane, 867(219)
α-undecylenamide, 744(79)
undecylenanilide, 744(44)
undecylenic acid, 686(52), 688(5)
undecylic acid (undecanoic acid), 686(60)
dl-2-undecanol, 720(154)
undecyl fluoride, 830(19)
undecyl iodide, 840(185)
undecyl sulfide, 965(5)
1-undecyne, 874(235)
urea (carbamide), 746(240)
ureidoacetic acid, 702(291)
uric acid, 758(525)

valeraldehyde (pentanal), 728(18)
valeramide, 745(150)
valeranilide, 744(36)
valeric acid (pentanoic acid), 684(29)
valeric anhydride, 668(13)
γ-valerolactone, 668(9), 817(186)
valeronitrile, 913(23)
valerophenone, 897(211)
valeryl chloride, 670(24)
D-(—)-valine, 672(13)
L-(+)-valine, 672(3), 680(128)
DL-valine, 678(116)
vanillalacetone, 904(205)
vanillic acid, 706(349)
vanillin, 930(102), 738(64)
veratraldehyde, 736(40)
veratramide, 747(333)
veratranilide, 746(297)
veratric acid (3,4-dimethoxybenzoic acid), 702(295)
veratrole, 827(4), 825(108)
vinyl acetate, 814(9)
vinylacetic acid (3-butenoic acid), 684(20)
vinylbenzene (styrene) 873(209)
vinyl benzoate, 817(182)
vinyl bromide, 836(99)
vinyl chloride, 830(21)
4-vinylcyclohexene, 873(196)
vinyl ether, 823(3)
vinylidene chloride, 852(9)
vinyl iodide, 838(161)
vinyl perfluoroacetate, 859(9)
vinyl perfluorobutyrate, 859(11)
vinyl perfluorocaprate, 859(14)
vinyl perfluorocaproate, 859(13)
vinyl perfluorocyclohexanecarboxylate, 859(15)
vinyl perfluorodecanoate, 859(14)
vinyl perfluorohexanoate, 859(13)
vinyl perfluoropentanoate, 859(12)
vinyl perfluoropropionate, 859(10)
vinyl perfluorovalerate, 859(12)
vinyl sulfide, 962(4)

xanthine, 748(523)
xanthone, 905(247)
m-xylene, 877(5)
o-xylene, 877(6)
p-xylene, 877(4)
α-D-xylose, 809(22)